GRAM NEGATIVE COCCI

Anaerobic

Veillonella, Others
Table 38.3 (p. 549)

Aerobic

Neisseria and
Moraxella
Table 27.1 (p. 358)

Aerobic

Growth on sheep blood agar

(−)

Haemophilus
Tables 30.3 (p. 414),
30.4 (p. 418)

Campylobacter
Table 31.6 (p. 441)

Legionella

Others

(+)

Ferments glucose

(+)

Aeromonas

Plesiomonas
Table 31.4
(p. 436)

Vibrio
Table 31.2
(p. 434)

Kingella, others
Table 30.5
(p. 421)

Pasteurella
Table 30.7
(p. 423)

(−)

Alcaligenes
Table 29.3 (p. 400)

Brucella
Table 30.1 (p. 410)

Pseudomonas
Table 29.2 (p. 397)

Bordetella
Table 30.2 (p. 413)

Campylobacter
Table 31.6 (p. 441)

Moraxella
Table 29.4 (p. 401)

Others
Table 29.1 (p. 393)

See inside back cover for **Gram positive bacilli** and **Gram positive cocci** flowcharts.

DIAGNOSTIC
MICROBIOLOGY

BAILEY & SCOTT'S

DIAGNOSTIC MICROBIOLOGY

ELLEN JO BARON, PHD
Diplomate, American Board of Medical Microbiology
Associate Professor of Medicine
University of California School of Medicine
Los Angeles, California

LANCE R. PETERSON, MD
Diplomate, Special Competency in Medical Microbiology
American Board of Pathology
Director, Clinical Microbiology
Northwestern Memorial Hospital;
Professor of Pathology and of Medicine
Northwestern University Medical School
Chicago, Illinois

SYDNEY M. FINEGOLD, MD, SM (AAM)
Diplomate, American Board of Medical Microbiology
Wadsworth V.A. Medical Center;
Professor of Medicine and of Microbiology and Immunology
University of California School of Medicine
Los Angeles, California

NINTH EDITION
with 510 illustrations, 412 in color

 Mosby

St. Louis Baltimore Boston Chicago London Madrid Philadelphia Sydney Toronto

Mosby

Dedicated to Publishing Excellence

Editor: James F. Shanahan
Developmental Editor: Lisa M. Potts
Project Manager: John Rogers
Production Editor: Chris Murphy
Design: William Seabright and Associates
Design Coordinators: Julie Taugner, Renée Duenow
Manufacturing Supervisor: Betty Richmond
Cover photo: Institute Pasteur-CNRI/Phototake

NINTH EDITION

Printed in the United States of America
Composition by the Clarinda Company
Printing/binding by Von Hoffmann Press, Inc.

Mosby–Year Book, Inc.
11830 Westline Industrial Drive
St. Louis, Missouri 63146

Library of Congress Cataloging in Publication Data
Baron, Ellen Jo.
    Bailey and Scott's diagnostic microbiology. –– 9th ed. / Ellen Jo
  Baron, Lance R. Peterson, Sydney M. Finegold.
      p.   cm.
    Includes bibliographical references and index.
    ISBN 0-8016-6987-1
    1. Diagnostic microbiology. I. Peterson, Lance R. II. Finegold,
 Sydney M., 1921- . III. Title. IV. Title: Bailey and Scott's
 diagnostic microbiology. V. Title: Diagnostic microbiology.
    [DNLM: 1. Microbiological Techniques. QW 25 B265b 1994]
 QR67.B37   1994
 616'.01'028 –– dc20
 DNLM/DLC
 for Library of Congress   93-40119
                                                  CIP

94 95 96 97 98 / 9 8 7 6 5 4 3 2 1

CONTRIBUTORS

CHAPTER AUTHORS

DIANE M. CITRON, BS

Microbiologist/Associate Director
R.M. Alden Research Laboratory
Santa Monical Hospital
Santa Monica, California

MARTHA A.C. EDELSTEIN, BA, MT (ASCP)

Clinical Microbiology Laboratory
Hospital of the University of Pennsylvania
Philadelphia, Pennsylvania

LYNNE S. GARCIA, MS

Manager, Clinical Microbiology
University of California Los Angeles Medical
    Center
Los Angeles, California

GLENN D. ROBERTS, PHD

Professor of Microbiology and Laboratory
    Medicine
Mayo Medical School;
Director, Clinical Mycology and
    Mycobacteriology Laboratories
Mayo Clinic and Mayo Foundation
Rochester, Minnesota

RICHARD B. THOMSON JR, PHD

Director, Microbiology and Virology
Associate Professor of Clinical Pathology
Northwestern University Medical School
Department of Pathology and Laboratory
    Medicine
Evanston Hospital
Evanston, Illinois

JOHN A. WASHINGTON II, MD

Professor, Department of Pathology
The Ohio State University
Columbus, Ohio;
Vice Chairman, Division of Pathology and
    Laboratory Medicine
Chairman, Department of Clinical Pathology
Head, Section of Microbiology
Medical Director, Reference Laboratory
The Cleveland Clinic Foundation
Cleveland, Ohio

CHAPTER CONSULTANTS

JILL E. CLARRIDGE III, PHD

Assistant Professor
Baylor College of Medicine;
Chief, Microbiology Section
Veterans Affairs Medical Center
Houston, Texas

MARTIN D. COHEN, MT (ASCP)

Clinical Laboratory Technologist
University of California Los Angeles Medical Center
Los Angeles, California

ISABELL COOPER, BS, MT (ASCP)

Senior Technologist, Clinical Microbiology
Supervisor, Molecular Epidemiology
Northwestern Memorial Hospital
Chicago, Illinois

DONNA HACEK, BS, MT (ASCP)

Technical Coordinator
Clinical Microbiology
Northwestern Memorial Hospital
Chicago, Illinois

JANET A. HINDLER, MCLS, MT (ASCP)

Senior Specialist, Clinical Microbiology
University of California Los Angeles Medical Center
Los Angeles, California

CLAUDIA J. HINNEBUSCH, BS, MT (ASCP)

Clinical Laboratory Technologist
University of California Los Angeles Medical Center
Los Angeles, California

J. MICHAEL JANDA, PHD

Research Scientist
Microbial Diseases Laboratory
California Department of Health Services
Berkeley, California

PAMELA JEAN KELLY, BS, MT (ASCP)

Senior Technologist
Supervisor, Anaerobe Laboratory
Clinical Microbiology
Northwestern Memorial Hospital
Chicago, Illinois

LINDA M. MANN, PHD

Assistant Professor
Associate Director
Clinical Microbiology and Immunology
University of Texas Medical Branch at Galveston
Galveston, Texas

PAMELA JACOB MIZEN, BS, MT (ASCP)

Senior Technologist, Parasitology
Clinical Microbiology
Northwestern Memorial Hospital
Chicago, Illinois

PATRICK R. MURRAY, PHD

Professor, Departments of Medicine and Pathology
Washington University;
Director, Clinical Microbiology
Barnes Hospital
St. Louis, Missouri

MARIE T. PEZZLO, MA, MT (ASCP)

Medical Microbiology Division
Department of Pathology
University of California Irvine Medical Center
Orange, California

M. JOHN PICKETT, PHD

Professor of Microbiology, Emeritus
Department of Microbiology and Molecular Genetics
University of California
Los Angeles, California

KATHRYN L. RUOFF, PHD

Assistant Professor
Department of Microbiology and Molecular Genetics
Harvard Medical School;
Assistant Microbiologist
Massachusetts General Hospital
Boston, Massuchusetts

DAVID F. WELCH, PHD

Associate Professor of Pediatrics
University of Oklahoma Health Sciences Center;
Director of Clinical Microbiology
The University Hospitals
Oklahoma City, Oklahoma

*Our work is dedicated to those
who faithfully support and care about us*

To Jim, LoAnn, and Mary

FOREWORD

Forty-five years ago many knowledgeable scientists and physicians were convinced that the advent of antimicrobial agents spelled the death knell for clinical microbiology. They expected that the ensuing decade would provide the means to successfully eradicate all infectious agents. They did not count on the fantastic metabolic ingenuity of the microbial world, a behavior that permitted prokaryotic and eukaryotic microbes to triumph over the adversities presented by the constantly changing environments of the last 3 billion years. Surely, many if not most of the wonder drugs we call antimicrobial agents are but human modifications of molecules elaborated by microorganisms seeking an advantage in the perennial struggle for sustenance.

Indeed bacteria, fungi, protozoa, and viruses responded vigorously to the anthropocentric convictions of the optimists. Enzymes that modify the "killer" drugs became common. The inexhaustible microbial pool in nature permitted the intrusion of hitherto unknown organisms and those considered commensal saprophytes into the human biosphere, where they complicated the recovery of the very patients helped most effectively by the advance in science and medicine. The more antimicrobial agents were produced, the greater the variety of emerging resistance mechanisms and the appearance of heretofore unidentified pathogens.

Advances in our understanding of immunity offered an additional aspect in our comprehension of infectious disease and underlined the important role played by the individual host in overt expression of infectious disease. Thus the need for diagnostic clinical microbiology grew logarithmically, whereas competent practitioners of the dis-

cipline had diminished almost to the point of extinction. One of the few available guides for the neophyte clinical microbiologist was *Bailey & Scott's Diagnostic Microbiology*. It provided a logical approach to the often confusing world of clinically significant microorganisms obfuscated further by the dynamics of nomenclature and refinements in taxonomy.

This ninth edition continues in bolstering the accomplishments of previous contributors. It enhances its leadership by providing significant information on the clinical manifestations of infectious diseases that demand laboratory guidance for diagnosis and therapy. This edition, of course, has maintained its tradition of updating diagnostic procedures and provides in a singular fashion explanations of the many emerging technologies for clinical microbiology.

The authors have continued the tradition, started in earlier volumes, of highlighting procedures in simple terms explaining principles, methods, quality control, describing expected results, and providing performance schedules. Singling out procedures in the text where needed is an exceptional practice that deserves the applause of all users, be they student, technologist, microbiologist, pathologist, or infectious disease practitioner. The many texts that address the problems facing workers in diagnostic clinical microbiology do not provide such timely and ready access to procedural information. In addition, the organization of the entire text follows the logical work flow in the clinical microbiology laboratory and reflects not only the authors' superb command of the entire subject matter but also the fact that they know the problems confronting the worker at the bench. The major tasks of most clinical microbiology laboratories concern all aspects of bacteriology. This vast subject is addressed precisely and completely. The authors bring the same thorough approach to virology, mycology, and parasitology with due respect for the limitation in those fields encountered in the average clinical microbiology setting.

I congratulate the authors for continuing to improve the pertinence of this respected and indispensable tool for all individuals who would venture into the fascinating world of clinical microbiology.

HENRY D. ISENBERG, PHD
Chief, Division of Microbiology
Professor, Laboratory Medicine
Long Island Jewish Medical Center
The Long Island Campus for the
Albert Einstein College of Medicine
New Hyde Park, New York

PREFACE

The problem of controlling infectious disease and the important role of diagnostic microbiology in this effort are steadily increasing. During 1991, 4.3 million children died worldwide from acute respiratory diseases. In that year a total of 3.5 million people died from diarrhea, 1 million from malaria, and 900,000 from measles. By the end of 1991 there were a cumulative 1.5 million AIDS deaths. As a single cause of death among the infectious diseases, tuberculosis is the leader with nearly 3 million deaths annually. We as clinical microbiologists clearly need to remain current in our diagnostic acumen and active in developing new tests for management of infectious diseases.

The ninth edition of *Bailey & Scott's Diagnostic Microbiology* continues to present a comprehensive view of medical microbiology from the standpoint of the organization and function of a clinical microbiology laboratory; likely agents associated with infectious disease syndromes; and procedures for the detection, identification, and susceptibility testing of etiological agents. While the book has changed and grown over its life span of more than 50 years, its audience has developed to include students of medical technology, medical microbiology, pathology, infectious diseases, infection control, and other health-related disciplines. In addition, the book serves as a laboratory reference and procedure manual for practicing microbiologists. We have striven to continue serving the needs of all of these groups.

This edition has been modified dramatically from previous editions. With the addition of a new principal author, Lance Peterson, we have again redesigned the book to keep current with the rapidly changing times. The new format and greatly expanded number of full color illustrations (60% new or modified drawings and photographs) are testimony to our commitment to af-

firm *Bailey & Scott's Diagnostic Microbiology* as representative of the state of the art. We have tried to use color more specifically and effectively in this edition than has been done in previous editions. Color is used to highlight key tables, lists, charts, and procedures, and to aid the student in retaining the material visualized.

We have accommodated the need for more pages of scientific content by deleting the appendices that contained media and stain formulations and addresses of laboratory suppliers. That information is readily available from other sources. The additional space has been used to broaden the scope of the book. The glossary remains, and numerous new terms have been added. As in the previous editions, the glossary terms are printed in boldface in the text when they are first used. We have expanded identification tables and augmented our coverage of new technology, new etiological agents of infectious diseases, and the evolving interest in public health and preventive medicine.

An innovative comprehensive flow chart for bacterial identification with reference to the appropriate tables in the text has been added to the inside front and back covers. We hope that students new to the field will use this chart as a quick source of information and that experienced microbiologists who use the book as a bench reference will find the chart handy as a guide for quickly locating the information pertinent to their needs.

The ninth edition continues our tradition of providing the most up-to-date information possible concerning important agents of infectious diseases, such as the resurgent *Mycobacterium tuberculosis*, the newly discovered *Rochalimaea henselae*, and the recently newsworthy *Escherichia coli* O157:H7. Readers of the ninth edition will find state-of-the-art coverage of these and similarly infamous other microbes, including the latest strategies for laboratory diagnosis.

We are particularly excited about our new Chapter 5, prepared by the distinguished John A. Washington II, which addresses microbiologists' expanded responsibilities as advisors to clinicians and shapers of laboratory policy to benefit patient care. We continue to profit from the expertise of Lynne Shore Garcia (Parasitology) and Glenn D. Roberts (Mycology), whose authoritative chapters are essential components of this book. In addition, Richard B. Thomson, Jr. has built on the excellent foundation provided by W. Lawrence Drew, who created the original virology section in the seventh edition, by providing diagnostic virology studies for the first time. This edition also benefits from the expertise of Diane M. Citron, who updated the anaerobic chapters conceived originally by Martha A.C. Edelstein for the seventh edition.

As always, the authors are indebted to the many experts who reviewed our chapters. Their suggestions and modifications greatly enhance the usefulness and accuracy of the book. In an undertaking of this scope, errors and omissions seem to be inescapable, and we welcome your corrections, comments and suggestions. We have tried to respond to all of our readers who took the time to give us their input regarding the previous editions, and we urge you to continue to do so.

As reflected in the foreword by the eminent Henry D. Isenberg, the role of microbiologists continues to expand as they struggle to keep pace with the ever-evolving microbial world. And as we stated in the preface to the eighth edition, there will always be a need for conventional microbiology and competent microbiologists. *Bailey & Scott's Diagnostic Microbiology* endeavors to support these needs.

ELLEN JO BARON
LANCE R. PETERSON
SYDNEY M. FINEGOLD

ACKNOWLEDGMENTS

We want to acknowledge the support and commitment of our new editor, Jim Shanahan, and the energetic efforts of our new developmental editor, Lisa M. Potts. Jim is responsible for the new look of the book, and his attention to our needs is appreciated greatly. Chris Murphy, our production editor, has also been especially diligent.

We also thank all those students, scientists, and practitioners of diagnostic microbiology who share their work and observations by writing and speaking. We all learn through the sharing of insights in our discipline, and the field of diagnostic microbiology moves forward.

CONTENTS

PART FOUR
METHODS FOR IDENTIFICATION OF ETIOLOGICAL AGENTS OF INFECTIOUS DISEASE

Glossary

DIAGNOSTIC MICROBIOLOGY

PART ONE

ORGANIZATION AND FUNCTION

OF THE CLINICAL

MICROBIOLOGY LABORATORY

Despite the encroachments of new technology, the Petri dish is still an essential part of the modern microbiology laboratory. (Courtesy Phototake.)

DIAGNOSTIC MICROBIOLOGY: PURPOSE AND PHILOSOPHY

Clinical microbiologists are part of the health care team and serve an important role in the diagnosis, management, and prevention of infections in patients. In 1978 Dr. Harold Neu published his impression of the services required by clinicians from a clinical microbiology laboratory. They were (1) relevant information needed to make medical decisions; (2) establishment of guidelines for proper specimen collection; (3) ensuring timely specimen transport; (4) identification and (5) susceptibility testing of microbes; (6) rapid reporting of test results; and (7) close consultation between the clinician and clinical microbiologist. In today's world of rapidly changing health care needs, properly trained, experienced clinical microbiologists should have an even greater role in the diagnostic approach to infectious diseases, the control of nosocomial pathogens, and the allocation of health care resources. The diagnostic microbiologist should take pride in the important role of clinical microbiology and feel responsibility commensurate with it.

ROLE OF MODERN DIAGNOSTIC MICROBIOLOGY

A principal role of a clinical microbiologist is to consult with other health care professionals, to provide accurate infectious disease diagnosis, and to devise laboratory testing to assist in optimum management of infectious diseases in patients. This role often includes expanded responsibilities relating to preventing the spread of infection to other individuals, and the efficient use of medical resources. In general, it is important to document the presence of infection, to determine its specific nature, and to provide appropriate therapeutic guidelines early in the course of the illness. Early diagnoses for treatment are usually more urgent with infectious diseases than with any other type of disease process. In many infections mortality and morbidity may be reduced significantly if proper treatment is provided early. In others, hospitalization may be avoided or shortened, and surgery may be avoided if specific, early antimicrobial therapy is given. Although the responsibility is large, it is very gratifying to see speedy improvement in a critically ill, infected patient; all members of the health team share in this gratification.

RESPONSIBILITY TO THE CLINICIAN AND TO THE PATIENT

The appropriate level of microbiology laboratory service must be available at all times. This means that a microbiologist should be on call if it is not reasonable to have the laboratory staffed at all times that expert microbiology service is required. Other laboratory staff may be able to inoculate certain types of specimens, but the expertise of the microbiologist usually is required to properly interpret a Gram stain or colonial morphology late at night. In assessing early findings, one should provide as much detail as possible and should not hesitate to indicate the various possibilities as to the microbiological diagnosis. The clinician determines his or her own differential diagnosis based on the clinical picture, roentgenograms, blood studies, and rapid laboratory test information. The diagnostic possibilities, as seen by a competent microbiologist, are of paramount importance in narrowing down or substantiating a diagnosis. The microbiologist must be alert to unusual findings. As suggested previously, speed is critical in serious illness caused by infectious agents. Experienced microbiologists can usually find one or more ways to expedite presumptive identification of microorganisms in clinical specimens. If the health team is to function with maximum efficiency, the members of that team need to communicate freely and frequently, particularly when dealing with seriously ill patients. The physician should lead in this respect, but if he or she does not, the others on the team can initiate communication when it may be helpful. The primary concern always must be the patient. The clinician should supply the microbiologist with pertinent historical and other data regarding the patient's illness; the microbiologist should be able to utilize this information.

The microbiologist can also serve the clinician by instructing him or her about the importance of proper specimen collection and transport, and providing directions for specific techniques to be used. Microbiologists also have a responsibility to notify infection control personnel or epidemiologists and public health authorities when easily transmissible infectious agents with high pathogenic potential are isolated. Urgent matters should be communicated by telephone.

The microbiologist also has an obligation to the patient aside from considerations of rapid, accurate diagnosis. Professionalism demands that microbiologists, as well as other members of the health team, not hesitate to deal with certain specimens or patients and that they respect the patient's rights and feelings with regard to sensitive information. At times the microbiologist may be pressured to release information on a patient to the patient, his or her family, or others; such pressure must be resisted. Demands of this nature should be referred to the physician who is in charge of the patient's care.

SPECIMEN COLLECTION AND TRANSPORT

Specimen collection and transportation are critical considerations because the quality of a laboratory's work may be limited by the nature of a specimen and its condition on arrival in the laboratory. Chapter 7 includes a detailed discussion of correct collection and transportation methods.

Specimens should be obtained to preclude or minimize the possibility of introducing extraneous microorganisms that are not involved in the infectious process. The problem is greatest when the specimen may become contaminated with elements of the endogenous (or colonizing) flora that may serve as pathogens in the clinical disease entity being evaluated (e.g., *Klebsiella* as an oral cavity colonizer in a patient with pneumonia). Even normally sterile body fluids may be contaminated during specimen collection in a way that causes significant difficulty in interpretation. For example, if blood cultures have become contaminated with coagulase-negative staphylococci during venipuncture, a patient suspected of having bacterial endocarditis may be treated for the infection, although he or she does not really have it. Since this organism can cause endocarditis in numerous clinical settings, this patient's contaminated cultures may lead to additional testing, administration of unnecessary antibiotics, and a prolonged hospital stay. Careful skin preparation before procedures such as blood cultures and spinal taps and using techniques for bypassing areas of normal flora when this is important and feasible (e.g., covered brush bronchoscopy in critically ill patients with pneumonia) will prevent many problems associated with "false-positive" results. Personnel must be instructed to label carefully all specimens submitted to the laboratory, including data such as the patient's name, hospital number, date and time of collection, antibiotic treatment, and the exact nature and source of the specimen, as discussed in Chapter 3.

NONCULTURAL METHODS OF DIAGNOSIS

As noted previously, the physician depends on data from many sources in arriving at tentative and definitive diagnoses. In addition to the usual medical history and physical examination, the thorough clinician explores epidemiological information (travel history; exposure to animals, ticks, or other vectors; illness in the family or neighborhood; previous illness in the patient that might have relapsed). He or she also obtains a number of laboratory tests (e.g., complete blood count, blood chemistry tests, liver function tests, spinal fluid examination) and special tests such as a chest roentgenogram, computed tomography or magnetic resonance imaging, radionuclide scans, and ultrasound studies. In the case of suspected infection, examination of direct preparations such as wet mounts, Gram stains, and dark-field examinations may provide information quickly. The use of monoclonal antibodies or nucleotide probes may provide remarkable specificity in direct examination of clinical specimens (see Chapter 11). Certain skin tests also may be useful diagnostically. Rapid examination for products of microorganisms (e.g., gas-liquid chromatography for demonstration of volatile metabolic endproducts) may be helpful on occasion. Finally, the physician looks for an immunological response to a given microorganism, usually in the patient's serum (see Chapter 13). This, of course, requires an infection to be present sufficiently long for an antibody response to take place.

REJECTION OF SPECIMENS

Criteria for specimen rejection, which are indicated in appropriate places in this book, must be set up. However, it is an important rule always to talk to the requesting physician or a member of the health care team before rejecting specimens, since they are primarily responsible for the patient's welfare. They may need supportive information and advice. They may believe that an unorthodox specimen may provide useful information. If this is not the case, the microbiologist must explain why it does not and must work with physicians to resolve any disagreement.

Frequently it is necessary to do the best possible job on a less than optimal specimen. It may often be necessary to educate the clinician regarding the disadvantages of such an approach, and the best way to do this is to review individual specimens in a friendly and cooperative manner.

EXTENT OF IDENTIFICATION REQUIRED

Microbiologists lean heavily toward definitive identification. For example, identification of species within the *Bacteroides* genus may be useful. Merely identifying an organism as "*Bacteroides ureolyticus* group" is inadequate. *B. gracilis* is more

pathogenic and more resistant to antimicrobial agents than *B. ureolyticus* and other members of the group.

An example of this was an outbreak of infantile diarrhea of serious proportion that persisted for many months in a large general hospital in the southwestern United States because stool cultures did not reveal any pathogens. Until the *Escherichia coli* isolates were tested for enterotoxin production and found to be positive, all therapeutic and control measures attempted proved futile. Another example of the benefit of definitive identification was that of the unusual blood culture isolates from a number of outbreaks—and these were not really recognized initially as outbreaks—which finally permitted determination that commercial intravenous fluid bottles were contaminated; this led to definitive control measures.[3] Many cases of bacteremia and many outbreaks across the country were attributable to this source.

Identification of a clostridial blood culture isolate such as *Clostridium septicum* will alert the clinician to the possibility of malignancy or other disease, usually in the colon, because of the strong association between these two events.[1] Identification of an organism thought to be *Bacteroides fragilis* in a patient with bacteremia of unknown source as actually being *Bacteroides splanchnicus* could indicate that the source is probably in the gastrointestinal tract, whereas *B. fragilis* might also be in the genital tract or even another site.

Precise identification of organisms from a patient with two distinct episodes of infection at an interval of some months might permit establishing that the second infection represents a recurrence rather than a new infection, or it might permit excluding that possibility. Recurrent infection may suggest a foreign body or an abscess from prior surgery; in the absence of prior surgery, it suggests the possibility of malignancy or other abnormal underlying process. Definitive identification of an atypical mycobacterium may permit more accurate assessment as to whether the organism is truly involved in the disease, what the source might be, its drug susceptibility pattern, and the likelihood of its responding to the therapeutic regimen chosen. In the case of certain organisms, especially the anaerobes, incomplete identification may lead to very serious errors. Things that we learn in our first course of microbiology—that certain organisms produce spores, that some are gram-positive and some are gram-negative, and so on—do not necessarily help with the anaerobes. It may be very difficult to demonstrate spores in sporulating anaerobes. Gram-positive anaerobes are very commonly decolorized and may appear gram-negative, even early in the course of growth. Cocci may be mistaken for bacilli, or, more commonly, bacilli may be mistaken for cocci. It may even be difficult to define what an anaerobe actually is, considering that certain organisms such as some clostridia (aerotolerant) are capable of relatively good growth under aerobic conditions. There are several outstanding examples in the literature of serious identification errors of this type. For example, *B. melaninogenicus* (now called *Prevotella melaninogenica*) has been incorrectly reported as being an anaerobic gram-positive streptococcus, and even such a great anaerobist as Prévot mistakenly classified as a *Fusobacterium (F. biacutum)* an organism subsequently shown to be gram-positive and a spore-former *(Clostridium symbiosum)*. Determination of the specific species of a group D streptococcus (e.g., *S. bovis* as opposed to *Enterococcus faecalis)*[2] may permit the use of less toxic and less expensive therapeutic agents. It may also have implications with regard to serious underlying disease. Organisms as diverse as *Listeria monocytogenes* and *Actinomyces viscosus* have been incorrectly identified as diphtheroids (even in reports in the literature) when shortcuts were taken. Careful bacteriological studies permit us to define the role of various organisms in different infectious processes, the prognosis associated with these, and the likely epidemiology of specific infections. Finally, definitive identification is important in educating clinicians as to the role of various organisms in infectious processes and, also important, prevents deterioration of the skills and interest of highly trained microbiologists.

At the same time, some shortcuts and use of limited identification procedures in certain cases are necessary in most clinical laboratories. Careful application of knowledge of the significance of various organisms in specific situations and thoughtful use of limited approaches will keep microbiology testing cost effective and the laboratory's workload manageable while providing for optimum patient care.

QUANTITATION OF RESULTS

The concept of quantitation has not been adequately evaluated in microbiology except in the case of urine cultures. Quantitation can be useful in other situations as well. It may help to distinguish between an organism present as a "contaminant" and an organism actively involved in infec-tion. It is often useful in determining the relative importance of different organisms recovered from mixed infections. Ordinarily, formal quantitation by dilution procedure or even by means of a quantitative loop is not necessary or desirable. Numbers of organisms present can be graded as "many" ("heavy growth"), "moderate numbers," or "few" ("light growth") on the basis of Gram stain appearance and the amount of growth on an isolation plate. For example, does a given colony type extend to the secondary streak, tertiary streak, and so on? The Gram stain is important because differences in amount of growth or rate of growth of different organisms may be related to the fastidiousness of the organism.

EXPEDITING RESULTS

The need for speed in identification and susceptibility testing is another crucial area that has been neglected, as noted previously. Few diseases are as dynamic and rapid acting as the infectious diseases. In certain serious infections, such as bacteremia, endocarditis, meningitis, and certain pneumonias, delay of a few hours in providing proper therapy may lead to death. It is not always feasible or safe to try to "cover" all possible types of infecting organisms with antimicrobial therapy. Clearly, the clinician must use this type of empirical approach initially, but the microbiologist must be prepared to assist in choosing a rational approach through knowledge of the role of various organisms in different disease processes and through interpretation of direct smears. Clinicians must communicate with the laboratory about problem cases. When consulted by the clinician in such cases, the microbiologist should offer something more than routine daily inspection and subculture so that any necessary modification of therapy can be made as soon as possible. For example, cultures can be examined at 6- to 12-hour intervals, certain rapid diagnostic tests can be applied, and the processing can be speeded up considerably for selected cases (see Chapter 4). In addition to lowering mortality and morbidity, rapid provision of data may shorten hospitalization and thus save money for the patient and the hospital. It may also help avoid a surgical procedure (such as in the case of enteric fever).

A limited number of organisms are responsible for the majority of infections. Simple procedures (e.g., Gram stain, colony morphology, catalase, coagulase, bile solubility, oxidase, spot indole, and rapid direct fluorescent antibody [DFA] stains) often may provide accurate presumptive identification quickly.

INTERACTION WITH THE CLINICIAN

In communicating with the physician, the microbiologist can avoid confusion and misunderstanding by not using jargon or abbreviations and by providing reports with clear-cut conclusions. Do not assume that the clinician is fully familiar with laboratory procedures or the latest taxonomic schemes. When feasible, provide interpretation on the written report, along with the specific results. However, never provide only your interpretation without the actual bacteriological data (e.g., "Only normal flora isolated").

Monthly newsletters may be used to supplement laboratory manuals in order to provide physicians with material such as details of your procedures, new nomenclature, and changes in usual susceptibility patterns of given organisms (very useful to clinicians when selecting empirical therapy) in your hospital's setting.

REFERENCES

1. Alpern, R.J., and Dowell, V.R. Jr. 1969. *Clostridium septicum* infections and malignancy. J.A.M.A. 209:385.

2. Facklam, R.R., and Carey, R.B. 1985. Streptococci and aerococci. p. 154-175. In Lennette, E.H., Balows, A., Hausler, W.J. Jr., and Shadomy, H.J., editors. Manual of clinical microbiology, ed 4. American Society for Microbiology, Washington, D.C.

3. Maki, D.G., Rhame, F.S., Goldmann, D.A., and Mandell, G.L. 1973. The infection hazard posed by contaminated intravenous infusion fluid. In Sonnenwirth, A.C., editor. Bacteremia: laboratory and clinical aspects. Charles C Thomas, Publisher, Springfield, Ill.

BIBLIOGRAPHY

Neu, H.C. 1978. What should the clinician expect from the microbiology laboratory? Ann. Intern. Med. 89(Part 2):781.

Sturm, A.W. 1988. Rational use of antimicrobial agents and diagnostic microbiology facilities. J. Antimicrob. Chemother. 22:257-260.

Washington, II, J.A. 1988. Effective use of the clinical microbiology laboratory. J. Antimicrob. Chemother. 22(Suppl. A):101-112.

Ellner, P.D. 1987. Diagnostic laboratory procedures in infectious diseases: what to expect from the microbiology laboratory. Med. Clin. North Am. 71:1065-1078.

LABORATORY SAFETY

Microbiology laboratories are unique environments in relation to the safety of those who work within them. Clinical specimens received from patients pose a hazard to personnel because of infectious agents they may contain. Spurred on by the spread of **acquired immunodeficiency syndrome (AIDS),** the United States Occupational Safety and Health Administration (OSHA) published a rule in the Federal Register that defines risks to workers of contracting a bloodborne pathogen, primarily human immunodeficiency virus (HIV) and hepatitis B virus (HBV), and requires practices to follow to minimize those risks.[8,9] Employers are required to provide education and materials necessary to reduce the exposure of employees to infectious agents. Every facility must have an Exposure Control Plan that describes the extent of risk for all job classifications and tasks and that details methods for implementing and monitoring risk reduction activities. Centers for Disease Control (CDC) also recommends safety precautions concerning the handling of patient materials by health care workers.[5] These recommendations, known as Universal Precautions (or universal blood and body-fluid precautions), stress that, since not all patients carrying bloodborne pathogens can be reliably identified, all persons whose activities involve contact with patients or with blood or other body fluids from patients in a health care setting should exercise the same consistent precautions.[4] These precautions are detailed later.

Cultures of infectious agents also pose a threat. The most important measures available for preventing laboratory-acquired infection are (1) education of employees who routinely handle the potentially infectious agents and (2) education (by means of biohazard notices as well as written and verbal instruction) of those people who have only limited exposure to the environment, such as custodial personnel, delivery personnel, and visitors.

Risks from a microbiology laboratory may extend to adjacent laboratories and to families of those who work in the microbiology laboratory. For example, Blaser and Feldman[1] noted that 5 of 31 individuals who contracted typhoid fever from proficiency testing specimens did not work in a microbiology laboratory. Two patients were family members of a microbiologist who had worked with *Salmonella typhi*, two were students whose afternoon class was in a laboratory where the organism had been cultured that morning, and one worked in an adjacent chemistry laboratory. The last several documented cases of smallpox in the world were acquired by employees working in a different area in the building where a smallpox research laboratory was located. The principal investigator was so dismayed that his possibly deficient safety practices led to these smallpox cases that he committed suicide. It is the legal responsibility of the laboratory director and supervisor to ensure that an Exposure Control Plan has been implemented and that the mandated safety guidelines are followed.

GENERAL SAFETY CONSIDERATIONS

In addition to the threat of infection, microbiology laboratories also contain the safety risks associated with any laboratory or institutional environment, those of fire, electrical hazards, chemical hazards, hazardous environmental situations (slippery floors, faulty air-handling systems), radioactive materials, equipment malfunction, and dangers imposed by natural disasters. Each section of the microbiology laboratory should have a safety manual containing information about procedures to follow in case of fire, flood, earthquake, or other natural disaster; location of all safety equipment; and practices to follow in the event of a situation posing a safety hazard. These considerations are universal to all laboratories and are addressed in references on general laboratory safety. Guidelines for the storage of flammable materials and caustic chemicals have been delineated by the College of American Pathologists and should be strictly enforced. A good way to ensure that all laboratory personnel are familiar with current safety practices is to hold periodic unannounced safety drills. By varying the hypothetical hazardous situation, the safety director allows all personnel the opportunity to become aware of their knowledge of proper action in an emergency and to improve their performance before a real disaster strikes. Documentation of these activities is an important component of the Exposure Control Plan.

BIOHAZARDS AND PRACTICES SPECIFIC TO MICROBIOLOGY IN GENERAL

The biohazards present in a clinical microbiology laboratory are a more specific concern of this text. Certain aspects of the organization, design, and functioning of such a laboratory should be ac-cepted generally as good safety practices throughout all specialty areas of the laboratory.

LABORATORY ENVIRONMENT

The microbiology laboratory poses many hazards to unsuspecting and untrained people; therefore access should be limited to those who have been informed of which actions they can perform safely. Visitors, especially small children, should be discouraged. Certain areas of high risk, such as the mycobacteriology and virology laboratories, should be closed to visitors. Custodial personnel should be trained to discriminate among the waste containers, dealing only with those that contain noninfectious material.

Care should be taken to prevent insects from infesting any laboratory areas. Mites, for example, have been known to crawl over the surface of media, carrying bacteria from colonies on a plate to other areas. Houseplants can serve as a source of such potential hazards and should be carefully observed for infestation if they are not excluded altogether from the laboratory environment.

OSHA regulations require that health care facilities provide employees with all devices and mechanisms, called **engineering controls,** necessary to protect them from the hazards encountered during the course of work. This usually includes plastic shields to protect workers from droplets, sharps disposal containers, holders for glass bottles, trays in which to carry smaller hazardous items (such as blood culture bottles), hand-held pipetting devices, and other devices.

The air-handling system of a microbiology laboratory should move air from low to higher risk areas, never the reverse. Ideally, the microbiology laboratory should be under negative pressure, and air should not be recirculated after it passes through microbiology. The selected use of biological safety cabinets for those procedures that generate infectious aerosols is critical to laboratory safety. Many infectious diseases, such as plague, tularemia, brucellosis, tuberculosis, and Legionnaires' disease, may be contracted by inhalation of the infectious particles, often present in a small droplet of liquid. Because blood is a primary specimen that may contain infectious virus particles, subculturing blood cultures by puncturing the septum with a needle should be performed behind a barrier to protect the worker from droplets. Several other common procedures used to process specimens for culture, notably mincing, grinding, vortexing, and preparing direct smears, are known to produce aerosol droplets.

These procedures must be performed in a biological safety cabinet.

BIOLOGICAL SAFETY CABINET

A biological safety cabinet (BSC) is a device that encloses a workspace in such a way as to protect workers from aerosol exposure to infectious disease agents. Air that contains the infectious material is sterilized, either by heat, ultraviolet light, or, most commonly, by passage through a **high-efficiency particulate air (HEPA)** filter that removes most particles larger than 0.3 μm in diameter. These cabinets are designated by class according to the degree of biological containment they afford. Class I cabinets allow room (unsterilized) air to pass into the cabinet and around the material within, sterilizing only the air to be exhausted (Figure 2.1). Class II cabinets sterilize air that flows over the infectious material, as well as air to be exhausted. The air flows in "sheets," which serve as barriers to particles from outside the cabinet, and that direct the flow of contaminated air into the filters (Figure 2.2). Such cabinets are called **laminar flow BSCs.** Type IIA cabinets have a fixed opening, and Type IIB cabinets have a variable sash opening through which the operator gains access to the work surface. Class III cabinets afford the most protection to the worker since they are completely enclosed, with negative pressure. Air coming into and going out of the cabinet is filter sterilized, and the infectious material within is handled with rubber gloves that are attached and sealed to the cabinet (Figure 2.3).

Most hospital clinical microbiology laboratory technologists use Class II cabinets routinely. The use of BSCs for certain procedures will be discussed further. The routine inspection and documentation of adequate function of these cabinets is a critical factor in an ongoing quality assurance program. Important to proper operation of laminar flow cabinets is maintenance of an open area for 3 feet (90 cm) from the cabinet during operation of the air-circulating system to ensure that infectious material is directed through the HEPA filter. An excellent quality control program for BSCs is found in the *Clinical Microbiology Procedures Handbook.*[7]

PROTECTIVE CLOTHING

Microbiologists should wear laboratory coats over their street clothes, and these coats should be removed before leaving the laboratory. Most exposures to blood-containing fluids occur on the hands or forearms, so gowns with closed wrists or forearm covers and gloves that cover all poten-

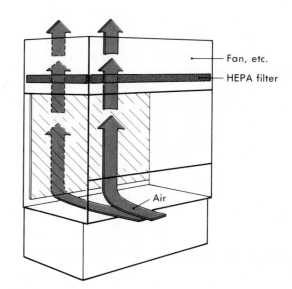

FIGURE 2.1

Class I biological safety cabinet.

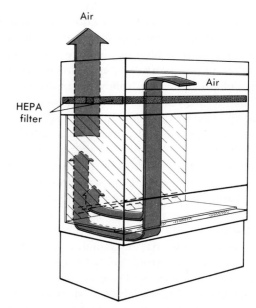

FIGURE 2.2

Class II biological safety cabinet.

tially exposed skin on the arms would be most beneficial. If the laboratory protective clothing becomes contaminated with body fluids or a potential pathogen, it should be sterilized in an autoclave immediately and cleaned before reusing. Obviously, laboratory workers who plan to enter an area of the hospital in which patients at special risk of acquiring infection are present, such as intensive care units, the nursery, operating rooms, or areas in which immunosuppressive therapy is being administered, should take every precaution to cover their street clothes with clean or sterile

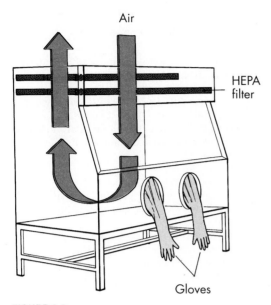

FIGURE 2.3

Class III biological safety cabinet.

protective clothing appropriate to the area being visited. Special impervious protective clothing is advisable for certain activities, such as working with radioactive substances or caustic chemicals. Solid-front gowns are indicated for those working with specimens being cultured for mycobacteria. Unless large volume spills of potentially infectious material are anticipated, impervious laboratory gowns are not necessary in microbiology laboratories.

DECONTAMINATION

All materials contaminated with potentially infectious agents must be decontaminated before disposal. These include unused portions of patient specimens, as well as media that have been inoculated, whether pathogens have grown or not. Infectious material to be disposed of should be placed into containers that are labeled as to their biohazard risk. Certain instruments, such as scissors, forceps, and scalpel blade holders, should be placed into a closed metal container until they can be sterilized in an autoclave. Sharp objects, needles, and scalpel blades should be inserted into a hard-sided puncture-proof container *without resheathing.* Such containers are commercially available. The container should be clearly labeled as to contents, sterilized in an autoclave, and discarded. Only needles that lock onto syringes or one-piece units should be used. Specially designed holders for resheathing needles without posing a hazard to the worker are now available commercially; two-handed recapping of needles with their original protective sheaths after use has

led to numerous accidental needle-puncture wounds in hospital and laboratory personnel. Attempts should be made to modify procedures so as to avoid the use of needles. A constant educational effort has been shown to be the most effective activity for prevention of exposure incidents. OSHA mandates that employers provide initial training for employees determined to have occupational exposure risk and update safety training at least annually.

Decontamination of potentially infectious material to be discarded from the microbiology laboratory is most often accomplished by steam sterilization in an **autoclave,** an instrument that uses moist heat at greater than atmospheric pressure to kill infectious agents. The pressure allows the steam to be heated above the normal boiling point, thus decreasing the time needed to sterilize. Material to be sterilized should be packed loosely so that the steam can circulate freely around it. Material of potentially greater hazard, such as cultures and specimens from the mycobacteriology laboratory, is often sterilized for 1 hour instead of the usual 30 to 45 minutes (at 121° C and 15 pounds per square inch [psi] pressure). It is best to locate the autoclave within the laboratory; if infectious material must be transported for sterilization, however, it should be double-bagged and carried in leakproof, sterilizable containers.

Autoclaves should be monitored regularly for adequate sterilization performance as part of regular quality control practice (discussed in Chapter 3). The proper loading and placement of material in the autoclave does, however, help to determine whether material has been adequately sterilized. An important safety precaution for those operating an autoclave is to open the door slowly after completion of a cycle. Several workers have been severely burned by the rush of steam that escapes through the crack in the autoclave door as it is first being opened. The use of a protective face mask for this procedure will prevent such burn incidents.

Workbenches and other horizontal surfaces should be decontaminated at least after every shift and immediately after every spill by washing with a liquid antiseptic agent such as a phenolic compound, 70% ethanol, or a 0.5% solution of sodium hypochlorite (10% solution of household bleach in water). Diluted bleach is the most effective agent against viral contaminants such as HBV and HIV. For those laboratories in which viral contamination is a significant hazard, diluted bleach is recommended as the antiseptic of choice.

Bleach can also inactivate the infectious proteins called **prions** (discussed in Chapter 41) that may be the etiological agents of Creutzfeldt-Jakob disease. The longer the surface is allowed to remain wet with the agent, the more effective the antiseptic action will be. A minimum of 10 minutes is recommended. Germicides are not active unless they are in solution. Most disinfectants do not destroy bacterial spores, but these forms of pathogens, with the exception of anthrax spores, do not usually pose a hazard to personnel. It is good practice to wipe down the outside surfaces of other laboratory equipment, such as centrifuges, vortex mixers, telephones, and even fluorescent lamps above work surfaces, with disinfectant on a regular basis.

PERSONNEL PRACTICES

Modifying personnel practices to minimize biohazard risk is the most important safety activity available. The use of engineering or personal protective controls is helpful, but careful handling of infectious materials is a more effective measure. Potentially infectious materials should be handled in a BSC if at all possible. If open flame burners are used, care must be taken to prevent spattering of material from inoculating needles and loops. The burner should be in a protective container to prevent accidental burns. Incinerator burners offer a less hazardous alternative to an open flame, although as they age and the insulation wears thin, they can pose an electrical hazard. Incinerator burners should be inspected every 6 months for signs of wear in the ceramic core, which must be replaced as necessary. CDC's Universal Precautions for personnel handling blood and body fluids (which in microbiology would include all secretions and excretions containing visible blood in addition to blood, serum, semen, all sterile body fluids, saliva from dental procedures, and vaginal secretions[6]) are summarized in the box at right.

Universal Precautions do not apply to feces, nasal secretions, saliva (except in dental procedures), sputum, sweat, tears, urine, or vomitus unless they are grossly bloody.[6] If the risk of aerosol infection during specimen handling is great, such as for tuberculosis, the use of a molded surgical mask and solid-front gown is recommended, even though work is performed in a BSC. Additional safety recommendations for laboratory workers are shown in the box on p. 13. Personnel practices also include the use of impervious gowns in areas where the work entails a significant risk of splatters of blood, solid front gowns in areas where droplets on the clothing

---

Essentials of Universal Precautions

1. Use of appropriate barrier precautions to prevent skin and mucous membrane exposure, including wearing gloves at all times and masks, goggles, gowns, or aprons if there is a risk of splashes or droplet formation
2. Thorough washing of hands and other skin surfaces after gloves are removed and immediately after any contamination
3. Taking special care to avoid injuries with sharp objects such as needles and scalpels
4. Use of mouthpieces or other ventilation devices with one-way valves for performing any resuscitation measures
5. Refraining from handling patient care equipment or patients if health care workers have exudative lesions or weeping dermatitis
6. Very stringent adherence to precautions by pregnant workers

---

may be hazardous (such as in mycobacteriology), full face masks or goggles for certain benchtop activities that generate splatters, and plenty of disposable gloves.

Mouth pipetting is strictly prohibited; mechanical devices (Figure 2.4) are used for drawing all liquids into pipettes. Eating, drinking, smoking, and applying cosmetics are strictly forbidden in work areas. Food and drink are stored in refrigerators in areas separate from the work area. All personnel should wash their hands with soap and water after removing gloves, handling infectious material, and before leaving the laboratory area. The risk of laboratory-acquired infection with HIV is extremely low (0.9% of 351 health care workers [laboratory workers would typically have less risk] with percutaneous exposures seroconverted[4]), although the potential consequences are grave. The consistent practice of Universal Precautions by health care workers handling all patient material will lessen the risks associated with such specimens. However, it is good practice to store sera collected periodically from all health care workers so that a seroconversion can be documented. Serum should be drawn immediately in the event of a percutaneous contamination accident, whether it is examined at the time of the accident or later. More detailed measures are described in the CDC guidelines.[5,6]

An alert and responsible physician will warn the microbiology laboratory personnel about specimens that may contain the agents of highly contagious, dangerous disease such as tularemia or Lassa fever. Again, however, microbiologists should exert caution in working with *all* speci-

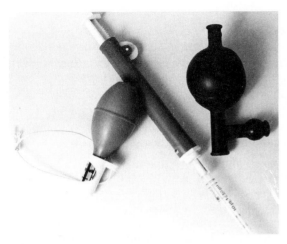

FIGURE 2.4
Several devices used for drawing liquids into pipettes as alternatives to mouth pipetting.

mens because the physician may not suspect a special problem.

CLASSIFICATION OF BIOLOGICAL AGENTS ON THE BASIS OF HAZARD

The Centers for Disease Control published a booklet entitled *Classification of Etiologic Agents on the Basis of Hazard,*[2] which has served as a reference for assessing the relative risks of working with biological agents since 1969. An updated list is found in the 1988 CDC/NIH booklet *Biosafety in Microbiological and Biomedical Laboratories.*[3] Although many unusual agents will not be mentioned here, we will list a few of the more common agents likely to be encountered in clinical microbiology practice in the United States. In general, patient specimens pose a greater risk to laboratory workers than do microorganisms in culture because the nature of etiological agents in patient specimens is unknown.

SPECIFIC AGENTS
Biosafety Level 1 agents include those that have no known pathogenic potential for immunocompetent persons. These agents are useful in laboratory teaching exercises for beginning level students of microbiology. Level 1 agents include *Bacillus subtilis* and *Mycobacterium gordonae.* Precautions for working with Level 1 agents include standard good laboratory technique as described previously.

Biosafety Level 2 agents are those most commonly being sought in clinical specimens. They include all of the common agents of infectious disease, as well as HIV and several more unusual

pathogens.[5] The most common causes of laboratory-acquired infection are hepatitis B virus, *Coxiella burnetii, Brucella* species, *Francisella tularensis, Mycobacterium tuberculosis, Salmonella* and *Shigella* species, and arboviruses. Many other agents have been documented to cause laboratory-acquired infections. For handling clinical specimens suspected of harboring any of these pathogens, Biosafety Level 2 precautions are sufficient. This level of safety includes the principles outlined previously and the added measures of limited access to the laboratory during working procedures, training in the handling of pathogenic agents by laboratory personnel, direction by competent supervisors, and the performance of aerosol-generating procedures in a biological safety cabinet. Employers must offer hepatitis B vaccine to all employees determined to be at risk of exposure.

Biosafety Level 3 procedures have been recommended for the handling of material suspected of harboring certain viruses, including certain arboviruses and arenaviruses, unlikely to be encountered in a routine clinical laboratory, and for cultures growing *M. tuberculosis,* the mold stages of systemic fungi, and some other organisms present in quantities greater than those found in primary specimens. These precautions, in addition to those undertaken for Level 2, include design and engineering features that increase the containment of potentially dangerous material by careful control of air movement and the requirement that personnel wear protective clothing and devices. Persons working with Biosafety Level 3 agents should have baseline sera stored for comparison with acute sera that should be drawn in the event of an unexplained illness, when appropriate.

Biosafety Level 4 agents, which include only certain viruses of the arbovirus, arenavirus, or filovirus groups, none of which are commonly found in the United States, require the use of maximum containment facilities; personnel and all materials are decontaminated before leaving the facility, and all procedures are performed under maximum containment (special protective clothing, cabinets). Most of these facilities are research laboratories.

SPECIAL PRECAUTIONS FOR SPECIFIC AREAS OF CLINICAL MICROBIOLOGY

BACTERIOLOGY AND GENERAL ACCESSIONING

Guidelines for the laboratory should require that all respiratory tract specimens must be manipulated in a BSC. All tissues, whether ground, minced, or touch-cultured, are handled in the cabinet. Inoculating material from Isolator tubes (described in Chapter 15) is also performed under laminar flow air-handling conditions in the cabinet. As recommended by CDC's Universal Precautions, laboratory workers should wear gloves when handling patient specimens. Specimen containers should be opened inside the BSC. Safety caps should be used when centrifuging materials such as blood (Isolator tubes), sputum, and feces. Always keep in mind that the outer surface of specimen containers may have been contaminated inadvertently with specimen. Work with patient material should be performed slowly and carefully to prevent accidental inoculation or production of aerosols.

Certain bacteria, when present on laboratory media after incubation, should be processed only within a Class II or higher biological safety cabinet. These include situations or colonies suspicious for *Yersinia pestis*, *Brucella* species, *Francisella tularensis*, *Rickettsia* and *Coxiella* species, *Bacillus anthracis*, and *Pseudomonas pseudomallei*. If production of aerosols is possible, *Neisseria gonorrhoeae*, *N. meningitidis*, and *Legionella* species should be manipulated in a cabinet.

MYCOBACTERIOLOGY

Processing of all specimens for mycobacteria should take place within a biological safety cabinet. Personnel should wear solid-front gowns and they may choose to wear face-molding masks as an added protective measure. All centrifugations should take place in sealed centrifuge safety cups. The placement of centrifuges under an exhaust system or within a biological safety cabinet will further decrease the hazards associated with centrifugation. Further information can be found in Chapter 42.

MYCOLOGY

After inoculation, all plated media should be *taped shut* or sealed with a cellulose band (Shrink Seal) to prevent accidental opening. Masking tape, certain cellophane tapes, and colored labeling tape have been found satisfactory (the tape must not lose its stickiness in a moist incubator). Complete sealing of edges is not necessary.

All moldlike fungi (those with aerial mycelia, as described in Chapter 44) must be manipulated within a biological safety cabinet. Many laboratories have been plagued by contaminating molds spread throughout all areas by spores from a single culture. When a mold appears on any medium within the laboratory, even if fungi are not being sought, the plate should be sealed immediately and opened thereafter only in the cabinet.

Any white, fuzzy fungus that appears on culture plates should be regarded with extreme suspicion as a possible dimorphic pathogen. Manipulations should be kept to a minimum and performed only by trained technologists with impeccable technique. Slide cultures should not be attempted for identification of fungi that grossly resemble the systemic pathogens *Coccidioides*, *Blastomyces*, and *Histoplasma*. Further discussion of handling such cultures can be found in Chapter 44.

PARASITOLOGY

The eggs of certain cestodes, if ingested during the infective stage, can cause disease. Microbiologists should exercise caution when handling suspected infected fecal material, and gloves should be worn if contact with infective material is possible. The use of a biological safety cabinet for handling most parasitic specimens is not necessary. However, for respiratory secretions that may harbor pathogens other than *Pneumocystis carinii* or flukes, manipulation within a safety cabinet is recommended. Cultures of the protozoan agents of primary meningoencephalitis (e.g., *Naegleria fowleri*) should be handled in a safety cabinet. Chapter 45 includes more discussion of handling of these agents.

VIROLOGY

All virology procedures performed in a routine clinical virology laboratory should be carried out within a biological safety cabinet. All specimens should be handled with gloves. Technologists should hold a 4 × 4 inch square gauze pad soaked in 10% bleach solution so as to completely cover the rubber stopper of serum tubes or other containers as the stopper is pulled off to help to absorb droplets formed during removal of the tops of such containers. Barrier devices, including plastic decapping devices and antiseptic-impregnated material squares, are available for this purpose. Special virology procedures, such as dissection of the heads of animals suspected of rabies infection, require stringent safety precautions. The precautions necessary for laboratories performing routine virology will be discussed further in Chapter 43.

SEROLOGY

Sera can contain agents that potentially may lead to laboratory-acquired infection. Gloves should be worn during manipulations that might allow skin contact with sera. The production of aerosols during the removal of tube tops or during other manipulations should be avoided, as discussed previously for handling of viral specimens. Procedures that generate aerosols, such as vortexing, pouring, shaking incubations, and washing steps for microtiter assays, should be performed in a biological safety cabinet. The careful technique of laboratory workers is the best single safety measure available.

The availability of a proved effective and safe vaccine against hepatitis B should serve to reduce significantly the rate of laboratory-acquired hepatitis. We recommend that all individuals who are required by their duties in the laboratory to have even moderate contact with sera receive the vaccine. Employers are mandated to offer HBV vaccine to all employees at risk. *As described previously, Universal Precautions require that all sera and other body fluids should be handled as if they were infectious.*

MAILING BIOLOGICALS

Potentially infectious material must often be sent to distant laboratories for further studies. Etiological agents, specimens, and biological materials may be shipped following the requirements of the Interstate Shipment of Etiologic Agents code. Cultures should be grown on solid media in a tube, preferably plastic but most often glass. The

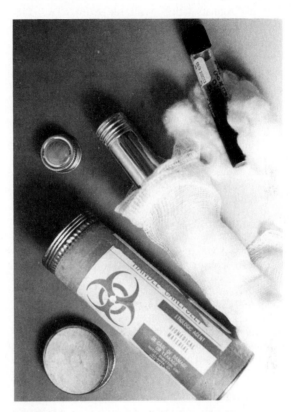

FIGURE 2.5
Proper containers for mailing etiological agents.

FIGURE 2.6
Label for etiological agents and biomedical material.

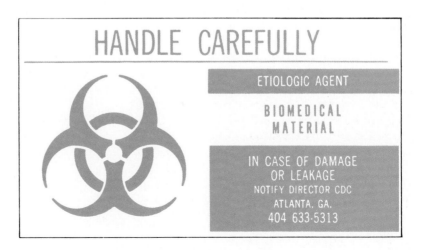

cap must be sealed with waterproof tape. This primary container is packed in enough absorbent material to absorb the entire volume of the culture if it should leak or break; it is inserted into a second container, often a metal tube. The second container is capped and is inserted into a shipping container, such as a cardboard mailing tube. Any labeling should be affixed to the outside of the secondary container. The mailing label is affixed to the shipping container along with an official Etiologic Agents label, available from CDC (Figures 2.5 and 2.6). Shipments are limited to 50 ml volume each.

REFERENCES

1. Blaser, M.J., and Feldman, R.A. 1980. Acquisition of typhoid fever from proficiency-testing specimens. N. Engl. J. Med. 303:1481.
2. Centers for Disease Control and National Institutes of Health. 1974. Classification of etiologic agents on the basis of hazard, ed 4. U.S. Department of Health, Education, and Welfare, Public Health Service, Washington, D.C.
3. Centers for Disease Control and National Institutes of Health. 1988. Biosafety in microbiological and biomedical laboratories, ed 2. H.H.S. Publication No. (NIH) 88-8395. U.S. Department of Health and Human Services, Public Health Service, Washington, D.C.
4. Centers for Disease Control. 1987. Recommendations for prevention of HIV transmission in health-care settings. M.M.W.R. 36(suppl):3S.
5. Centers for Disease Control. 1988. 1988 Agent summary statement for human immunodeficiency virus and report on laboratory-acquired infection with human immunodeficiency virus. M.M.W.R. 37(suppl):1.
6. Centers for Disease Control. 1988. Update: Universal precautions for prevention of transmission of human immunodeficiency virus, hepatitis B virus, and other bloodborne pathogens in health-care settings. M.M.W.R. 37:377.
7. Mann, L.M. 1992. Biological safety cabinet. Sect. 12.19. In Isenberg, H.D., editor. Clinical microbiology procedures handbook. American Society for Microbiology, Washington, D.C.
8. OSHA. 1991. Occupational exposure to bloodborne pathogens: final rule. Federal Register 56:64175-64182.
9. OSHA. 1991. Occupational exposure to bloodborne pathogens: correction July 1, 1992. 29CFR Part 1910. Federal Register 57:127:29206.

BIBLIOGRAPHY

College of American Pathologists. 1977. Safety inspection checklist. The College, Skokie, Ill.

Gilchrist, M.J.R., Hindler, J., and Fleming, D.O. 1992. Laboratory safety management. p. xxix-xxxvii. In Isenberg, H.D., editor. Clinical microbiology procedures handbook. American Society for Microbiology, Washington, D.C.

Inderlied, C.B. 1985. The laboratory environment. In Wagner, K.P., and Kelley, S.G., editors: Clinical microbiology: listen, look, and learn series. Health and Education Resources, Inc., American Society for Microbiology, Washington, D.C.

Kubica, G.P. 1990. Your tuberculosis laboratory: are you really safe from infection? Clin. Microbiol. Newsletter 12:85.

Miller, B.M., editor. 1986. Laboratory safety: principles and practices. American Society for Microbiology, Washington, D.C.

Gröschel, D.H.M. and Strain, B.A. 1991. Laboratory safety in clinical microbiology. p. 49-58. In Balows, A., Hausler, W.J. Jr., Herrmann, K.L., et al., editors. Manual of clinical microbiology, ed 5. American Society for Microbiology, Washington, D.C.

Salkin, I.F. and Gershon, R. 1992. Biohazards and safety. Sect. 14. In Isenberg, H.D., editor. Clinical microbiology procedures handbook. American Society for Microbiology, Washington, D.C.

3

LABORATORY ORGANIZATION AND CONTINUOUS QUALITY ASSESSMENT

With the ongoing oversight of health care delivery by the medical, allied health, and governmental communities (see Chapter 4), the microbiologist must ensure that the laboratory contributes to the highest quality of patient care by its expertise, its accuracy, and its relevance in the diagnosis of infectious diseases. As stated in the revised *Accreditation Manual for Hospitals* published by the Joint Commission on Accreditation of Healthcare Organizations (JCAHO, 1992), the standard that all hospitals must meet, a hospital can improve the quality of patient care "by assessing and improving those governance, managerial, clinical, and support services that most affect patient outcomes." Specifically addressed toward the laboratory is the requirement that there exist "quality control systems and measures of the pathology and clinical laboratory services (that) are designed to assure the medical reliability of laboratory data." It is only by constant self-evaluation of the laboratory's performance through a comprehensive quality assessment program that such a level of excellence can be developed and maintained. Quality control, the process-oriented monitoring of activities within the laboratory, is only one facet of such a program.[1] The ease with which technologists can optimally perform their work and the extent to which they can function as a team to accomplish the goal of high quality patient care are determined, in part, by the organization of both the physical environment and the workflow within the laboratory. This chapter

discusses general considerations for specimen handling and laboratory operations that contribute to the overall quality testing that technologists, physicians, and administrators should expect of the laboratory, and addresses some actual components of an active quality control/quality assessment program.

SPECIMEN PROCUREMENT AND LABORATORY ACCESSIONING

The quality of results from the microbiology laboratory and their contribution to good patient care are directly dependent on the freshness, quality, and appropriateness of the submitted specimens. Education of nurses and physicians in the proper collection and transport of specimens, and in the appropriate utilization of the diagnostic capabilities of the laboratory are the key steps in cost-effective medical care. An easily accessible handbook containing all information needed for obtaining appropriate specimens, ordering tests correctly, and transporting specimens to the laboratory should be available at every patient care unit. A periodic bulletin or newsletter can inform nurses and physicians of changes in procedures and other important information when these occur. Laboratories should provide a central receiving area to which all specimens are brought. This simplifies the delivery process and allows the laboratory to properly process all incoming specimens in a timely manner. The ability to determine whether a specimen has been received, and at what point in the processing system a given specimen may be found, is an important part of the overall organization of the laboratory. This may be called an "audit trail." It is also at this central point that specimens with requests for testing in more than one area of the laboratory can be divided and distributed. Monitoring to be certain that specimens are being transported to the laboratory within an acceptable period of time is a worthwhile technical measure of laboratory organization. The laboratory also needs to monitor other technical performance functions such as quality control and participation in proficiency testing programs. However, to be truly effective to the hospital, the laboratory must have broad based involvement in the institution's delivery of cost-efficient, quality patient care, which entails input into the use of laboratory-provided services (laboratory utilization).

SPECIMEN IDENTIFICATION AND ACCEPTANCE

Proper identification of each specimen requires the patient's name and unique identification number. The specimen should be tightly capped and in the proper container. The practice of Universal Precautions (Chapter 2) mandates that all incoming specimens are treated as though they may contain a hazardous agent. At the accessioning point, specific criteria for rejection of specimens should be enforced. Table 3.1 contains some suggested reasons for refusal to process a specimen. Other reasons for specimen rejection may be appropriate for individual laboratories. Under certain circumstances, such as receipt of an unlabeled specimen from a site from which it would be difficult or impossible to collect a new sample (e.g., cerebrospinal fluid or tissue biopsies), there should be a policy that would permit overriding the original rejection. One often used is the acceptance of an unlabeled specimen after the attending physician (or other person who actually collected the specimen has personally come to the laboratory to identify and label the specimen and has signed a statement assuming responsibility for the proper identification. Caution must be exercised to ensure this occurs only in those instances where re-collection can cause physical harm to the patient, and where the number of a specific specimen type received by the laboratory is so low that proper identification of the unlabeled specimen can be done with a high degree of certainty.

LABORATORY REQUISITION FORM

The request slip accompanying the specimen must contain the patient's identification parameters, the patient's location, the date and time that the specimen was collected, the exact source and type of specimen, the requested studies, and other pertinent information (such as the antimicrobial agents the patient is receiving), as well as the name of the responsible physician. It is very useful to include the time at which the specimen was received in the laboratory; many laboratories use a time-stamp machine for this purpose. A clean and legible request form is essential to optimal processing. The format of the request form can often influence the way in which tests are ordered. As mentioned in Chapter 4, listing a limited number of possible tests on the request form will contribute to limiting the test requests to those listed. The use of screening procedures and test panels dealing with a series of tests often aids in identification of a particular infectious agent or type of infection (such as "strep throat screen" or "respiratory virus panel"). This can simplify the ordering process for physicians and clarify the request for the technologists. If the patient's clinical diagnosis (or suspected diagnoses) is (are) included on the request, the laboratory can more easily determine which tests to perform to detect the suspected pathogens.

REPORTING RESULTS

Part of the organization of the microbiology laboratory must include a system for delivery of test results to the physician (or patient-care unit). Assuring minimal turnaround time from test receipt to reporting results and ascertaining that accurately relayed results actually reach the health care provider are important aspects of quality medical care. Certain information is important enough to warrant a telephone call to notify the physician of a positive (or significant negative) result. Care must be taken when giving telephone reports since verbal reporting is easily misinterpreted. When verbal reports are provided, it is worthwhile telling the physician when a computer (or written) report can be expected so he or she can double-check the telephone message. Suggested results to report immediately include positive Gram stain results, other positive smears of any sort, positive antigen detection results from sterile body fluids, growth from any sterile body fluid or tissue (such as blood and cerebrospinal fluid cultures), positive acid-fast smears, presence of organisms that would require the patient to be

TABLE 3.1

CRITERIA FOR REJECTION OF REQUESTS FOR MICROBIOLOGICAL TESTS*

CATEGORY	CRITERION FOR REJECTION	ACTION
Identification	Container not identified	Write "container not identified" and call ordering unit to send someone to verify the ID if the request form is attached to the specimen. Otherwise do not process. (See text.)
	Container and request form have different IDs	Document on report form. Call ordering unit to send correct request forms or resolve the problem.
Request form	Insufficient information marked on form	Call ordering unit for additional necessary information.
Specimen	Specimen grossly contaminated	Call ordering unit to send repeat specimen. If they are unable to re-collect, ask them to come down to clean it in the laboratory. Final resort: clean it with germicide, wear gloves, work in biohazard chamber.
	Specimen submitted in improper container	Notify ordering unit of problem. If unable to re-collect, perform test if possible and make note on report form that improper container was used. If unable to perform test, document reason on report form and discard specimen.
	Excessive delay between specimen collection and arrival in laboratory	If specimen was not stored properly, notify ordering unit and request repeat specimen. Do not process unless patient care is likely to be compromised by delay. Note on report that delay may result in less accurate results.
	Inadequate specimen for number of tests requested	Call physician for priority of requests and to request additional material. Note on report that specimen had inadequate volume as reason for lack of performance of certain tests.
	Some factor renders specimen inadequate for request. Examples: sputum on a swab or in tissue paper, stool for ova and parasites with gross barium visible, specimen placed into formalin, container broke or leaked during transport	Call ordering unit or physician to inform them of problem. Note problem on report form and do not process specimen.

Continued.

isolated, such as difficult to treat nosocomial pathogens (MRSA), and growth of mycobacteria. It is equally important for the technologist to document what information was relayed by telephone, when this was done, and to whom. A log book near the telephone to record a summary of any significant verbal contact is worthwhile. All results also must be delivered in written form, either directly from the laboratory or via a computerized laboratory information system (LIS). The work of the laboratory will be worthwhile only if it is useful in providing care for the patient. One aspect of the more careful scrutiny of hospital practices by government and other third-party payers (Chapter 4) has been a close examination of result reporting from laboratories to the patient care areas. Most administrative policies now require an interim or preliminary written report within 24 or at the most 48 hours after specimen receipt. Those laboratories that have hospital-wide on-line computer capability have a system that facilitates such reporting. Other laboratories must rely on multiple-copy report forms and mail or facsimile (FAX) delivery to accomplish this purpose.

Certain results are important to people other than physicians directly responsible for patients. Infection control practitioners monitor the incidence and spread of various infections throughout the hospital and must be given access to relevant information, both on a daily basis and in cumulative fashion. Policies that govern which culture results must be immediately communicated to infection control practitioners should be clearly delineated. The laboratory also has an obligation to report certain infections, such as preventable venereal diseases, tuberculosis, agents of transmissible gastroenteritis, and diseases of grave concern, such as botulism and plague, to the local public health authorities. All technologists should be informed as to which results require additional reporting, and standardized procedures should be established for this function.

TABLE 3.1—cont'd

CRITERIA FOR REJECTION OF REQUESTS FOR MICROBIOLOGICAL TESTS*

CATEGORY	CRITERION FOR REJECTION	ACTION
	Pooled 24-hour urine or sputum received for culture	Call ordering unit to inform them that 24-hour specimens are unacceptable for microbiology. Do not process and send report out with note indicating reason for not performing test.
	Dry swab received for culture	Notify ordering unit that dry swabs are not suitable for culture. Do not process. Send report out with note indicating reason for not performing test.
	Improper specimen for test. Examples: blood in serum tube with clot for malaria smear, slide with material smeared on for culture, pleural fluid for serologic test	Call ordering unit to verify test request. Do not process, and send documented reason for not performing test on report.
	Foley catheter tip	Call ordering unit to inform them that clinically relevant results cannot be obtained from this specimen. Request a urine for culture. Send out report noting the reason for not processing the specimen.
	Duplicate specimen received (other than blood)	Send notice that only one of the duplicate specimens will be processed unless the laboratory is notified and reasons are given for processing duplicate(s).
	Anaerobic culture requested on improper specimen. Examples: expectorated sputum, voided urine (not suprapubic tap), vaginal discharge, prostatic secretions, gastrointestinal tract (except in blind loop syndrome), oral material, respiratory secretions not collected by needle aspiration or by special plugged double-lumen catheter device, lochia, environmental material, any specimen exposed to air for extended time	Notify ordering unit that the specimen is unacceptable for such a request. Process aerobically only.

* All specimens not processed should be refrigerated for 3 days before they are discarded as a safeguard against erroneous discarding of an irretrievable specimen. Always discuss with clinician before discarding.

PROCEDURE MANUAL

Every laboratory must provide procedure manuals detailing all aspects of operations within that laboratory. Such manuals should be easily accessible at all times to workers who may wish to consult them. Manuals should discuss safety aspects of working within the laboratory, administrative and technical policies that must be followed, specimen collection information, specimen acceptability criteria, and criteria for performance of all tests and reporting of results. The manual should detail the steps to take in the event of a failure to meet quality control standards established for each procedure. The procedure manual should be the working reference standard for the entire staff. The manuals are reviewed regularly (at least annually) by supervisory staff and the procedures are updated as they are modified or replaced. Documenting laboratory procedures, as is done by means of a manual, is as important as documenting the results of quality control procedures outlined later. The National Committee for Clinical Laboratory Standards has established guidelines for writing laboratory procedure manuals.[2] An excellent resource is the *Clinical Microbiology Procedures Handbook* published by the American Society for Microbiology.

CONTINUOUS QUALITY ASSESSMENT

Quality personnel performance determines the quality of laboratory results; thus establishing and maintaining high levels of performance are crucial to quality laboratory results. The laboratory's physical layout and environment should be adequate for the number of employees and conducive to good performance. New employees should be given thorough orientation to policies and procedures. Maintenance and upgrading of the technical skills of all personnel should be pursued through laboratory rounds, in-service education, opportunities to attend continuing education programs, and by encouraging self-directed learning. Individual personnel performance also should be monitored. Periodic comparisons of specimen processing to work cards and results reports, daily checking by a supervisor of all reports produced during all shifts for errors and omissions, and an occasional evaluation of each technologist's/technician's performance by complete investigation of their current culture workups may point out hidden sources of error. Both internally and externally generated proficiency test specimens can help identify when an individual needs additional training to achieve satisfactory work performance. Quality control activities are both internal and external.[4] Internal activities involve monitoring the performance of media, reagents, instruments, products, and equipment. Documentation of performance of quality control measures is as important as the performance itself. The written record for each laboratory procedure or function that is monitored should also contain a section that details any deviation from expected results, problems, or failures and the corrective actions taken in response to such

problems. A special logbook for corrective actions is very helpful in documenting an ongoing quality control program's effectiveness. Incidents occurring in the laboratory that are not directly related to performance of tests, such as spills of infectious materials, personal injuries, abusive interactions between technologists and physicians, or any unusual occurrence, should also be documented. Quality control records must be held for a minimum of 2 years; instrument records must be held for the life of the instrument.

Instrument quality control monitoring should include preventive maintenance, temperature charts if applicable, and other records of performance. An example of a quality control log for an incubator is shown in Figure 3.1. Volumetric instruments must be calibrated at least annually. All perishable products and all solutions must be labeled with the date of receipt or preparation, strength, safety warnings or additional information, when they were first used, and the expiration date. All products prepared in the laboratory should be checked for proper performance. In the case of microbiological media, tests should include sterility and growth of control microorganisms with both positive and negative outcomes. Selective media should be monitored for their ability to inhibit the expected organisms as well as for the growth of the sought-for organisms. The National Committee for Clinical Laboratory Standards has recently established guidelines for the testing of microbiological media, including the names of organisms to be tested, expected results, and fre-

FIGURE 3.1

Example of incubator quality control performance record.

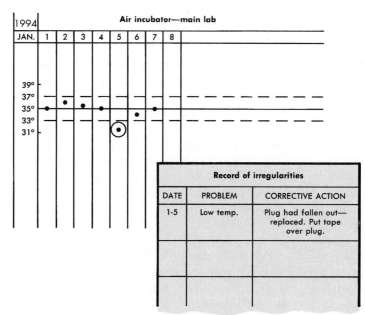

quency of testing.[3] These guidelines are further discussed in Chapter 9.

Organisms for use in monitoring media and reagent performance are available commercially in lyophilized form from several suppliers. Test organisms not available commercially should be obtained from a reference laboratory. Organisms can be kept frozen indefinitely at $-70°$ C in skim milk (diluted 1:1 in distilled water and sterilized in an autoclave at 121° C for 10 minutes) in plastic or glass freezer vials. If small (1 to 2 mm) beads are included in the freezing solution, these can be "chipped out" individually and directly placed onto sterile agar to revive frozen organisms without thawing the entire frozen aliquot. Alternatively, organisms may be maintained for quite some time on nutrient or blood agar slants under a layer of sterile mineral oil. Viability is prolonged when the medium is prepared without added sodium chloride or carbohydrate. Yeasts and mycobacteria retain viability for several months stored in suspension in sterile distilled water. Freezing and lyophilization are generally preferred as storage methods because they more reliably preserve antibiotic susceptibility/resistance characteristics in the stored strains.

Test organisms appropriate for performance monitoring will be mentioned in relationship to tests as they are described throughout this book. The Bibliography at the end of this chapter contains more detailed lists of those media and reagents that should be evaluated and suggestions for organization of quality control programs. Susceptibility testing procedures require particularly stringent quality control, as detailed in Chapter 14. The minimum amount of quality control performed by a laboratory is dictated by the agency under which the laboratory is accredited, such as the Joint Commission on Accreditation of Healthcare Organizations, Health Care Financing Administration, or the state in which the laboratory operates. A laboratory must adjust its practices to satisfy the most stringent regulating agency. Many workers believe that more extensive media quality control is required than those described as minimum standards.[5] The documented history of results within established limits can help to determine how often quality control procedures must be performed by any particular laboratory.

Laboratories can also monitor their overall performance by using a system of internal proficiency testing of processing outcomes. This is accomplished by occasionally submitting the same sample, divided into two specimens, as a second specimen, unknown to the technologists. Lapses in consistency are often spotted in this way.

External quality control activities include periodic inspection by regulatory agencies responsible for accreditation of the laboratory and performance of routine procedures on check samples and unknowns provided by an outside resource (proficiency testing). These programs are valuable for comparing one's results with those of many other laboratories, spotting errors in technique, and learning to recognize unusual agents or rare results. It can be argued that outcome measurements as determined by proficiency testing may assess only a laboratory's best performance. Nevertheless, they are a valuable and necessary component of an ongoing quality assessment program and should be used to improve delivery of patient care when deficiencies are noted.

The laboratory is often responsible for helping maintain quality control for other areas of the hospital. Monitoring the performance of autoclaves in surgery, laundry, and other sections may be a function of the microbiology laboratory. Often the laboratory provides sterility checks for the pharmacy, blood bank, milk bank, dialysis unit, and other areas. Unless a documented nosocomial infection outbreak has occurred, cultures of environmental surfaces are not recommended. Before environmental cultures are undertaken the problem being evaluated, the laboratory methods to be used, and how the results will be interpreted must be carefully defined.

STATISTICS

Maintaining adequate records of specimens received, procedures performed, and test results is extremely helpful for determining workload, staffing needs, and organization of the laboratory. Use of some method for assessing work accomplished, such as the College of American Pathologists laboratory management index program, will enable a laboratory to determine productivity, trends in test requests, and proper pricing structures. Other statistics that are helpful include cumulative susceptibility testing results and a trend analysis indicating if more than the "usual" number of a given isolate is being found (pointing to a possible nosocomial outbreak caused by that pathogen). An updated compilation of these data, provided to physicians on a regular basis, will aid them in selecting the most efficacious antimicrobial agents, as well as alert them to new trends in susceptibility patterns or outbreaks among microbial isolates at their institution. In addition, by comparing a particular organism's susceptibility pattern to the established norm, a technologist may find an error in identification or an unusual isolate that requires further study.

INFORMATION PROCESSING

The number and diversity of the records and statistics that must be generated, evaluated, and stored by the microbiology laboratory have increased just as technological advances have blossomed. Even with the development of information-processing systems suitable for use by the laboratory, continued improvement is needed. It is hoped that microbiology laboratories can make use of some of the new technology to aid in the giant task of recordkeeping. The use of computers for many tasks within the clinical microbiology laboratory is discussed in Chapter 4.

REFERENCES

1. Barros, A. 1987. Is quality control quality assurance? Med. Laboratory Observer 19:21.
2. NCCLS. 1992. Clinical laboratory procedure manuals; approved guideline, GP2-A2. National Committee for Clinical Laboratory Standards, Villanova, Penn.
3. NCCLS. 1990. Quality assurance for commercially prepared microbiological culture media; approved standard, M22-A. National Committee for Clinical Laboratory Standards, Villanova, Penn.
4. Sommers, H.M. 1985. Towards more effective quality control. In Smith, J.W., editor. The role of clinical microbiology in cost-effective health care. College of American Pathologists, Skokie, Ill.
5. Shanholtzer, C.J., and Peterson, L.R. 1987. Laboratory quality assurance testing of microbiologic media from commercial sources. Am. J. Clin. Pathol. 88:210-215.

BIBLIOGRAPHY

August, M.J., Hindler, J.A., Huber, T.W., and Sewell, D.L. 1990. Cumitech 3A, Quality control and quality assurance practices in clinical microbiology. Coordinating editor, Weissfeld, A.S. American Society for Microbiology, Washington, D.C.

Bartlett, R.C. 1974. Medical microbiology: quality, cost, and clinical relevance. John Wiley & Sons, New York.

Bartlett, R.C. 1985. Quality control in clinical microbiology. In Lennette, E.H., Balows, A., Hausler, W.J. Jr., and Shadomy, H.J., editors. Manual of clinical microbiology, ed 4. American Society for Microbiology, Washington, D.C.

Blazevic, D.J., Hall, C.T., and Wilson, M.E. 1976. Cumitech 3, Practical quality control procedures for the clinical microbiology laboratory. In Balows, A., coordinating editor. American Society for Microbiology, Washington, D.C.

Environmental Protection Agency. 1986. EPA guide for infectious waste management. (Publication No. PB86.199130.) National Technical Information Service, Port Royal Road, Springfield, VA 22161.

Isenberg, H.D., editor. 1992. Clinical microbiology procedures handbook. American Society for Microbiology, Washington, D.C.

Miller, J.M. 1983. Quality control in the microbiology laboratory. Centers for Disease Control, U.S. Department of Health and Human Services, Public Health Service, Atlanta.

Miller, J.M., and Wentworth, B.B., editors. 1985. Methods for quality control in diagnostic microbiology. American Public Health Association, Washington, D.C.

Sewell, D.L. 1987. Quality control in the new environment: microbiology. Med. Lab. Observer 19:45.

Weissfeld, A.S., and Bartlett, R.C. 1987. Quality control. In Howard, B., Weissfeld, A.S., and Tilton, R.C. Clinical and pathogenic microbiology. Mosby, St. Louis.

4

MANAGING THE CLINICAL MICROBIOLOGY LABORATORY: EFFECTIVE PATIENT CARE IN A COST-CONSCIOUS ENVIRONMENT

The federally mandated system in which hospitals receive reimbursement for costs largely based on the admission and discharge diagnosis of a patient, rather than on the cost of medical care generated by that patient (diagnosis related groups [DRGs]), has been in existence since 1983. Although numerous factors influence the amount of reimbursement (Medicare, Medicaid, or private insurance) paid for each episode of care, the financial advantage of keeping costs as low as possible is obvious. The challenge that faces microbiologists relying on reimbursement from health care providers is to continue to deliver quality results under increasing budgetary constraints. Many laboratories have responded to this as an opportunity, finding new (and sometimes better) ways to deliver quality results rapidly in a cost-efficient manner. Industry has also responded by becoming more involved in developing rapid testing methods. The ultimate beneficiary of laboratory efforts is the patient; physicians and health care workers try to ensure that he or she receives timely and appropriate treatment, with the overall goal being quality patient care.

Strategies that can be employed to achieve rapid, accurate, and efficient diagnoses are presented in this chapter. It is intended as a sampler of ideas as a basis upon which microbiologists can build. It is not an exhaustive catalogue of the available choices for alternate test procedures to achieve maximum cost containment. Microbiologists, in choosing strategies that will work in their unique situations, are helping to shape the future of health care.

GENERAL APPROACHES TO LIMITING COST

Knowledge about the true costs of microbiology laboratory procedures and information about the budget allocated to operations are essential for administrators of the laboratory to monitor its costs and performance. Laboratory administration must receive timely, accurate reports of the appropriate budget, broken down into divisions representing personnel costs, supplies, services, and overhead (if charged to the laboratory). Almost all direct costs in microbiology laboratories are included within salaries, benefits, and supplies; it is these areas in which cost containment should be pursued most aggressively. It is also essential to know the actual costs incurred in performance of each procedure. Bartlett and others[2] and Garcia and Bruckner[10] have discussed procedures for gathering this information. An ongoing record of work-load units (testing time, usually recorded in minutes) performed by technologists in each area of the microbiology laboratory is important for assessing trends, rearranging staffing patterns to fit changing needs, and forecasting future developments, as well as for determining the technologist time expended on each procedure performed. The College of American Pathologists' workload recording units (CAP units) are widely used for this purpose.[4] The newest concept, laboratory management index program, is designed to provide similar data.

Savings in supplies can be realized by substituting clean, nonsterile containers for certain specimens such as stool, urine, and sputum, the microbiological results from which will not be compromised.[14] Throat swabs for isolation of group A β-streptococci can be transported in glassine envelopes, rather than transport media, without losing viability. Forming consortia with other laboratories to buy common supplies in bulk can be an effective measure; sharing or splitting the performance of tests between neighboring laboratories to maximize utilization of resources can also pay off. A careful evaluation of disposable costs coupled with creative thinking allows microbiologists to modify methodology to save resources.

EDUCATION

The physician has the ultimate responsibility for patient care, and cost-generating procedures originate with a physician's order. Microbiologists need to actively educate physicians in proper use of the microbiology laboratory. Periodic orientation sessions detailing the functions and capabilities of the microbiology laboratory are helpful. Physicians should be aware of the extent of diagnostic services provided and the expected turnaround time for test results. Clinicians should meet microbiologists with whom they may interact so that communication is ongoing. Guidelines for reporting microbiological data to facilitate physician understanding have been suggested (see box).

Proper laboratory utilization must be discussed with the nursing staff. One strategy that pays off in terms of helping nurses and physicians to submit specimens properly for microbiological analysis is the availability of a small, printed manual containing detailed instructions on proper collection methods, including containers, request forms, special handling requirements, timing and numbers of specimens to submit, and other needed information. Updated annually with procedural changes, such a book rapidly becomes a useful reference manual, and the laboratory benefits both in decreased numbers of phone calls and in better speci-

---

Characteristics of Good Microbiological Result Reports

1. Inclusion of interpretive statements
2. Notification of improper specimen collection or handling
3. Reporting of relevant susceptibility results only
4. Clear and unambiguous terminology
5. Explanation of unusual organism names
6. Inclusion of normal values with numerical data
7. Reporting negative results from special tests (e.g., "no *Campylobacter* isolated")
8. Reporting normal flora from contaminated sites (e.g., "normal skin flora")
9. Reporting of results in semiquantitative form
10. Reporting of all isolates from normally sterile sites
11. Naming a source or reference for further information
12. Reports must appear uncluttered so significant findings are easily seen

Modified from Lee, A., and Gooden, H. 1982. Improving communication of microbiology test results. Am. J. Clin. Pathol. 77:443.

---

men quality. Some information to include in such a manual is in Chapter 4. A quarterly or semiannual publication originating in the microbiology laboratory that has a positive impact on patient care is a cumulative summary of antimicrobial susceptibility of selected microorganisms isolated from patients at the hospital. Information to include in this publication, which is best reduced to a pocket-sized folded card, can include maximum serum, urine, and cerebrospinal fluid levels achievable for selected antimicrobial agents and dosage regimens, combined with representative costs. By knowing susceptibility patterns within the institution, clinicians can prescribe antimicrobial agents that are likely to be the most effective at the least cost. The antimicrobial susceptibility results for third generation cephalosporins, broad-spectrum penicillins, and more expensive aminoglycosides should be reported only for those bacteria that exhibit resistance to a predetermined group of first-choice antimicrobial agents, thus shaping the treatment patterns of physicians who consider in vitro susceptibility results in therapeutic decisions. In addition, microbiologists must participate on those committees, usually consisting of representatives from the infectious disease service, medical and surgical services, and pharmacy, that choose which antimicrobial agents are routinely used in the hospital and tested in the microbiology laboratory. By educating practitioners to appropriate use of antimicrobial agents in their institution, often in such subtle ways as reporting patterns, microbiologists are contributing to cost savings.

LIMITATIONS ON TESTING

Laboratories can limit tests in several ways without reducing quality patient care. As suggested by Ellner,[6,7] limits on the numbers of specimens in several categories that are accepted by the laboratory will contribute to more rational patient management. By offering test panels that automatically offer increasingly specialized tests based on the results of screening tests, laboratories can decrease the number of nonproductive tests run. When choosing such a strategy, the laboratory must select screening tests that have a very high predictive value for negative test results (e.g., those with high sensitivity). Screening test requests for appropriateness is crucial. A number of specimens are inappropriate for culture and may even yield misleading information. For example, Foley catheter tip cultures, surface cultures of decubiti, anaerobic cultures from the vagina or cervix, bowel contents, or oral sources, and cultures of periodontal material (except for specialized laboratories) do not contribute to good patient management and should be avoided. The laboratory should limit the testing of routine throat cultures to detection of β-hemolytic streptococci and respiratory viruses. Laboratory handling of specimens also affects resource utilization. Pooling the three formalin-preserved stool specimens received from outpatients for detection of ova and parasites has been shown to yield the same results as examination of the specimens separately,[20] saving both technologist time and staining materials. Cultures of samples should be avoided if Gram stains or other direct visual examinations will yield sufficient diagnostic information. For example, a 10% potassium hydroxide (KOH) or calcofluor white wet preparation of scrapings from lesions on the tongue of a child suspected of having thrush or on the vaginal wall of a woman suspected of candidal vaginitis, if positive, provides the diagnosis of yeast infection without culture. The presence of clue cells, lack of lactobacilli, increased pH, and disagreeable odor of vaginal discharge are indicative of bacterial vaginosis (nonspecific vaginitis), and cultures are unnecessary. The presence of multinucleated giant cells in a direct fluorescent antibody stain from a suspected herpetic vesicle precludes the necessity for culture. However, when using screening methods such as those described previously, the microbiologist must remember that without isolation of a microorganism there is currently no method for determining the susceptibility or resistance of the pathogenic microbe(s) to the action of antimicrobial agents.

Limitations that may be practiced within the laboratory should be undertaken only after careful study and discussions with clinicians to determine that no negative effect on patient care will result. The number of primary plates that are routinely inoculated with clinical material may be decreased. Stool cultures can be plated to two selective and differential media instead of three or four, in addition to blood and *Campylobacter* and MacConkey agars (Chapter 17). It may be possible to eliminate inoculation of stool into enrichment broth for subsequent subculture if specimens are from patients with acute diarrhea, since it has been shown that no significant additional isolates are recovered by enrichment in this setting. The number of media set up for cultivation of mycobacteria and fungi can be limited to support the growth of all etiological agents in a more general sense. For example, instead of incubating separate sets of fungal media at 35° to 37° and 25° C, incubating one set of plates at 30° C, a temperature at which all fungi can grow, is sufficient. Biplates, such as anaerobic *Bacteroides* bile esculin and laked sheep blood with kanamycin and vancomycin, can be inoculated instead of two separate plates.

Limitations on all anaerobic bacteriology procedures should be implemented based on patient-care needs.[8] Blood cultures and stool cultures from hospitalized patients should be limited based on clinical usefulness.[1,11,13] Other cost-saving measures can be taken depending on individualized situations present in the patient population served by the laboratory. The physicians always must be kept well informed of laboratory policy so that special needs will be accommodated and patient care will not suffer when tests are limited to save money.[27] In some circumstances, what appears to be a costly procedure may, in fact, prove cost-effective in the overall achievement of better patient care. Virus culture of cerebrospinal fluids, a seemingly low-yield procedure, was found to shorten hospitalization and prevent unnecessary antibiotic use in a significant number of patients from whom viruses were isolated. The overall cost savings and patient-care improvements more than offset the initial cost of the laboratory test.[3] In other circumstances, such as direct antigen detection in cerebrospinal fluid, the rapid results cannot compensate for lack of sensitivity in serious infections such as meningitis; and antigen test results are unlikely to alter patient management.[12] Usefulness of such tests in each laboratory should be evaluated very carefully, with input from clinicians, before they are implemented.

SCREENING PROCEDURES

Tremendous cost savings can be achieved if specimens unlikely to yield diagnostic information are not processed beyond initial evaluation. The screening of expectorated sputum for the numbers of squamous epithelial cells (with or without polymorphonuclear neutrophils) per low-power field is an effective way to improve the quality of these specimens and to prevent unnecessary work. Sputum can be screened unstained for even greater time and cost savings.[26] Screening should be extended to wound cultures received on swabs. Even when only one swab is received in the laboratory, both cultures and a Gram-stained smear may be prepared using the following method. Typically medium is inoculated first followed by preparation of a Gram-stained smear. Alternatively, the microbiologist vortexes the swab for at least 30 seconds in 0.5 ml of the appropriate diluent (tryptic soy broth or brucella broth) to suspend the material on the swab, and, with a sterile capillary pipette, inoculates several drops of the suspension to appropriate media and onto the surface of a slide. The presence of polymorphonuclear neutrophils indicates an infectious process; the presence of squamous epithelial cells indicates surface skin contamination. The appearance of the Gram stain can further be used by technologists to determine the extent of definitive identification and susceptibility testing to be performed on the organisms recovered. The immediate information conveyed to practitioners by the results of the Gram stain combined with the subsequent decrease in unnecessary studies performed on specimens of dubious value can outweigh the inconvenience and time invested in the staining procedure.

A number of recent technical advances, including the use of bioluminescent analysis, colorimetric methods, and cytospin Gram stains have been directed toward screening urine (Chapter 19). Since the largest numbers of specimens received by many laboratories are urines, of which more than half are culture negative or not diagnostically significant, a urine screening system can reduce unnecessary work. Physicians will benefit by receiving negative results at least 1 day sooner than by conventional culture, which aids in preventing antibiotic overusage and allows earlier evaluation for additional causes of urinary tract related symptoms. A urine screen that has been evaluated for its applicability to the specific clinical milieu in which the laboratory functions should be considered.[21]

New technologies are being developed for screening procedures for many other specimens from infection sites. Tests are available for rapid testing of stool filtrates for the presence of *Clostridium difficile* antigen and toxins, screening of enrichment broths from stool specimens for presence of *Salmonella* or *Shigella*, and presence of antigens of several infectious agents in urine and other body fluids. Some tests on clinical specimens will have the greatest value as accurate predictors of negative results, thus limiting the number of specimens that are assessed by more cumbersome methods. Gas chromatography has long been studied as a means to quickly evaluate body fluids for the presence of bacterial metabolic products, although it has not gained widespread usage.

STRATEGIES FOR CHOOSING METHODS

Microbiology laboratories should make available to physicians tests necessary for diagnosis and management of the patient population served. Not all tests can be performed reliably or economically by the laboratory. A laboratory must decide which tests to perform in-house and which methods to use, as well as which tests to send to an outside reference laboratory and which reference laboratory to use.

Based on historical and predicted numbers of test requests, laboratories can remove low-volume tests from their menu to realize cost savings. Cost accounting information can help determine which tests should be sent out based on direct costs alone. Additional cost factors to be considered include the cost of quality control testing that must be performed with each in-house patient test, the supplies that must be discarded because of outdating if test volume does not use all purchased materials, cost of proficiency testing surveys to cover the test, clerical time for record keeping and preparing additional forms, and mailing or other transportation costs to the reference laboratory. Non-cost considerations include the ability of an outside laboratory to provide results in time to be useful clinically, the loss of expertise among in-house staff who no longer perform a test, the quality of available reference laboratories, and the enhanced prestige of the in-house laboratory that offers special tests among clinician-users and the hospital administration (which uses such information in public relations campaigns to attract more patients).

Once it has been decided to perform a test in-house, a method must be chosen. Patient population demographics influences this decision, because tests vary in their efficiency of detection of a positive result (e.g., particular organism or disease state) in relation to the prevalence of that disease in the population.

Commonly used measurements of a test's performance include sensitivity, specificity, predictive value of a positive test result (PVP), and predictive value of a negative test (PVN) result.[9] Inherent in determining the sensitivity and specificity of a test is the concept of a "gold standard" against which the test's performance is measured. This standard may not always exist. For example, new tests for detection of respiratory syncytial virus antigen by fluorescent antibody stain or enzyme-linked immunosorbent assay (ELISA) may be more sensitive than, and as specific as, conventional culture. In such a case, the performance of one of these new test systems should not be judged against standard methods to determine sensitivity. Given a suitable standard of comparison, the sensitivity and specificity of a test method can be judged independently of the patient population tested. Therefore laboratories need not always conduct extensive studies to evaluate new tests but should make use of evaluations carried out by respected, competent microbiologists and published in scientific literature. The predictive value of a positive test decreases as the prevalence of the disease in the population decreases. Prevalence can thus be used to make initial decisions on types of tests to perform. Only hands-on experience, however, can ultimately determine the appropriateness of a given test for each laboratory setting. An algorithm for choosing a diagnostic test method has been published.[22]

A similar approach should be applied to the decision whether to automate a given procedure, which instrument to use, and whether to purchase or lease the instrument. Published reviews of automated systems are very helpful.[25] Initial equipment costs, ongoing reagent costs, equipment maintenance charges, and labor-saving costs (less technologist hands-on time or fewer skilled workers required to perform the test) must be taken into account when comparing overall economics of instrumentation. Many automated testing methods require considerably more hands-on time than is expected, and this needs to be carefully considered when a purchase of such equipment is anticipated.[24] The skills and numbers of the available workforce, turnaround time, specimen preparation requirements, record-keeping and report-generating capabilities, and space requirements are noneconomic considerations that affect any decision for or against automation. Microbiology, perhaps the least automated of the

clinical laboratory disciplines, has adopted such methods slowly. Automation in microbiology has found a place in today's cost-conscious environment, particularly in smaller laboratories where specimen volume is too low to justify sufficient numbers of specialized technologists performing only microbiology related testing.

RAPID DETECTION OF INFECTIOUS AGENTS

VISUAL TESTS

As discussed throughout this book, there are numerous methods for rapidly detecting the presence of infectious agents or their products. The easiest to perform are direct (microscopic) wet preparations and simple stains. The Gram stain is probably the single most cost-effective test used in clinical microbiology. Other visual methods employ immunological reagents coupled to marker substances. Examples include the rapid detection of *Chlamydia* elementary bodies by a monoclonal antibody conjugated with a fluorescein marker. Horseradish peroxidase, alkaline phosphatase, and other enzymatic markers have been used, particularly for detection of viral subunits, either specific nucleic acid sequences or surface proteins. Fluorescent reagents are available for specific staining of microbes such as *Legionella* species, *Neisseria gonorrhoeae*, *Bordetella pertussis*, *Yersinia pestis*, *Francisella tularensis*, *Treponema pallidum*, *Bacteroides fragilis*, and a number of fungal and parasitic agents of infectious disease. For specimens that are suspected of harboring unique infectious agents, the use of rapid detection methods can save days and directly influence patient care.

PARTICLE AGGLUTINATION METHODS

Latex agglutination and staphylococcal coagglutination reagents coated with antibodies are available for rapid detection of many antigens. Group A β-hemolytic streptococci can be detected in throat swabs within 15 minutes of specimen collection. Commercial kits are available for rapid detection of the antigens for many common causes of meningitis: *Haemophilus influenzae* type b, *Neisseria meningitidis*, *Streptococcus pneumoniae*, and *Streptococcus agalactiae*. These tests can detect antigens in urine or serum as well. Cryptococci can be detected with more sensitivity using latex reagents for cryptococcal antigen than by India ink for direct microscopy. Latex tests for detection of rotavirus in stool filtrates have compared favorably with electron microscopy and ELISA methods. Latex agglutination procedures for detection of herpesvirus in lesion scrapings, *Legionella* antigen in urine, *C. difficile* protein in stool, and so forth are available. Because of limitations of sensitivity, such tests are usually better predictors of positive results than of negative results, although false-positive results also occur with some of the tests (e.g., *C. difficile*).

AUTOMATION

The plethora of automated microbiology instruments allows rapid detection of the presence or absence of infectious agents, and some provide an identification (Chapter 12). The appropriate use of such instruments may increase laboratory productivity and improve accuracy. The Vitek AMS instrument was developed to rapidly detect the presence of certain etiological agents, particularly in urine. Urine with numbers of microorganisms greater than 10^5/ml can be identified within hours with greater than 95% sensitivity. Other instruments such as the Microbial ID Incorporated (Newark, Delaware) software for the Hewlett-Packard silica-capillary column gas-liquid chromatography instrument can rapidly and accurately identify many different microbial isolates. A modification of this methodology using high-performance liquid chromatography has also been applied to the rapid and accurate identification of mycobacteria. Detection of microbial metabolism, as used by the Becton-Dickinson BACTEC® and other similar systems, can significantly decrease the time to detection of mycobacteria in sterile body fluids, including blood, and from concentrated decontaminated specimens such as sputum. Other automated blood screening devices are now available, utilizing colorimetric, electrical impedance, gas utilization or production, and other technological tools. The Limulus lysate test (Chapter 11), although somewhat labor intensive, may be automated for detection of endotoxin. The Abbott Quantum makes use of enzyme-linked immunosorbent assay (ELISA) technology to detect antigens of gonococci and chlamydia in genital swab specimens. An ELISA test is commonly used to detect the presence of rotavirus antigen in feces. Radioimmunoassay and ELISA methods can be used for rapid diagnosis of hepatitis A and B.

OTHER STRATEGIES

Modest additional labor when the specimen is first received may save time and effort later. Two such approaches used for blood cultures are lysis centrifugation (DuPont Isolator) and antimicro-

bial removal (Becton Dickinson Antimicrobial Removal Device and Bactec resin bottles). These systems entail specialized initial processing of blood cultures (Bactec simply requires a specialized culture collection bottle), but both systems decrease the time to detection of positive results from patients receiving antimicrobial agents. Rapid initiation of appropriate treatment based on blood culture results should save hospitals money and more importantly lead to more successful therapy of patients. Blood cultures not handled by an automated system or by direct plating may reveal positive results more rapidly if an early detection procedure is employed. These include internal subculture to agar that is enclosed within the system (Gibco biphasic bottle; Roche Septi-Check) and early external subculture or acridine orange stain. Almost 80% of all positive blood cultures will be detected within 24 hours of collection if the vented or aerobic bottle is agitated during initial incubation and examined microscopically by acridine orange, or subcultured within 6 to 12 hours after receipt.

DECREASING ANALYSIS TIME FOR IDENTIFICATION RESULTS

NONCOMMERCIAL METHODS

Once microbial growth has occurred, several methods can be used to decrease the time required to ascertain identification and susceptibility results. Spot biochemical tests can presumptively identify several bacteria. The most cost-effective measure available to microbiologists is to presumptively identify bacteria based on colony morphology and simple biochemical tests. Several organisms lend themselves to such methods. The indole and oxidase tests are two of the most useful and rapid tests available. Indole-positive, oxidase-negative, lactose-positive, nonmucoid, flat colonies on MacConkey agar can be reported presumptively as *Escherichia coli*. These organisms comprise a substantial percentage of the gram-negative bacilli isolated in most laboratories. Swarming *Proteus* species seen on blood agar may also be identified presumptively with spot indole. Indole-positive *Proteus* is likely to be *P. vulgaris*. Indole-negative *Proteus* is likely to be *P. mirabilis*. Colonies that morphologically resemble *Pseudomonas aeruginosa* may be identified presumptively based on β-hemolysis, blue-green pigment, characteristic odor, and positive oxidase spot test results. Technicolor (Tech) agar, which is essentially Medium A of King et al. who formulated the medium to enhance the production of pyocyanin, a water-soluble blue pigment, is useful for detection of blue-green pigment. The only gram-negative rod known to produce pyocyanin is *P. aeruginosa*. Reyes et al. found that 94% of *P. aeruginosa* strains produced pyocyanin on Tech agar after overnight incubation, whereas 98% were positive after 48 hours' incubation.[23] Most of their *P. aeruginosa* isolates showing delayed or no pyocyanin production were mucoid strains.

When assessing the results of cultures of genital specimens taken from sexually active adults, technologists may presumptively identify colonies of gram-negative diplococci growing on Thayer-Martin media that are oxidase-positive as *Neisseria gonorrhoeae* or *N. meningitidis*. The PYR test can be used for rapid identification of *Streptococcus pyogenes* and *Enterococcus* species from morphologically characteristic colonies. From cerebrospinal fluid and blood cultures, slightly β-hemolytic colonies of catalase-positive gram-positive short bacilli or coccobacilli can be presumptively identified as *Listeria monocytogenes* if they exhibit characteristic tumbling motility on wet preparation. A positive slide coagulase result on a colony of catalase-positive gram-positive cocci in clusters identifies the isolate as *Staphylococcus aureus*. Yeast cells from certain sources can be presumptively identified as *Candida albicans* if they form germ tubes within 3 hours at 35° C in fetal bovine serum. The Quellung rapid serological slide test can also be used to identify *S. pneumoniae* and *H. influenzae* type b.

Suspicious colonies on primary stool media can be subcultured to blood agar plates or to Kliger's iron or triple sugar iron agar slants immediately upon discovery. Rapid urease or phenylalanine tests can be set up at the same time to rule out certain nonpathogens. After 4 to 5 hours of incubation, enough of a film of growth will have occurred on the blood agar or the slant surface to allow the technologist to perform slide agglutination serological typing tests. Carbohydrate fermentation reactions of a number of non-Enterobacteriaceae can be determined within 4 hours using a broth containing 20% carbohydrate, peptone, meat extract, and buffers, described in Chapter 10 (Procedure 10.11). Fastidious bacteria, including *Kingella* species, *Actinobacillus*, *Cardiobacterium*, DF-2, and *Neisseria*, can be rapidly identified biochemically using this medium. Other approaches have also been suggested.[16]

All microbiologists need to consider ideas about the implementation of rapid methods for identification that will fit into the workflow, capabilities, and needs of their individual situations.

Many settings exist in which rapid, presumptive information is more important than delayed, definitive information. It is important that the microbiologist ascertain from infectious disease specialists and other physicians who utilize laboratory results which definitive results can be delayed or even deleted in exchange for early, presumptive information. Sources cited in the Bibliography and References contain additional ideas.

COMMERCIAL METHODS

More commercial systems for rapid identification of microorganisms exist than can be mentioned here.[25] Many of them are discussed in Chapter 10. By using enzymatic substrates that yield a colored end product, enzyme activity can be used to identify bacteria, even in the absence of growth. Rapid enzymatic or biochemical tests systems, made by Key Scientific, bioMérieux Vitek, Innovative Diagnostics, Becton-Dickinson Microbiology Systems, Remel Laboratories, Austin Biologicals, and many others, are available for preliminary screening and identification of gram-negative bacilli, gram-negative cocci, anaerobic bacteria, and some gram-positive cocci. Immunological reagents for identification of organisms by particle agglutination are also available. Such systems (as discussed in Chapter 26) for identification of β-hemolytic streptococci yield results within minutes or a few hours. An important source of inocula for direct identification and susceptibility testing is growth from samples of normally sterile body sites, such as growth from blood cultures. Broth from a positive blood culture medium may be inoculated directly to identification systems, such as rapid enzymatic test kits, or to coagulase plasma, depending on the Gram stain morphology. Even the particle agglutination identification systems have been shown to accurately identify organisms found in blood culture media when testing is done on properly prepared material. One technique for preparing a sample suitable for biochemical inoculation is to remove 10 ml of medium from a blood culture bottle containing only one morphological type of organism. The culture medium is centrifuged at $1500\times g$ for 10 minutes to concentrate the bacteria, the supernatant is discarded, and the pellet is resuspended in distilled water (which serves to lyse blood cells). If desired, a second centrifugation can be performed to wash the bacteria free of blood cell components, but this has not usually been necessary. This suspension is then inoculated to the rapid identification system of choice. Another method is occasionally applied to suspicious colonies from stool cultures; blood culture broth may be plated to a blood agar plate and incubated for 3 to 4 hours to allow a sufficient film of growth to develop. This film is removed with a cotton swab, emulsified in water or saline, and used to inoculate the rapid systems.

The strategies and products that address cost containment should be evaluated for their utility in each laboratory, and their use should be considered where appropriate. It is possible that microbiology laboratories may have to respond to the DRG challenge by rearranging workflow, expanding the hours of operation, hiring more laboratory assistants, utilizing more automated methods, and otherwise changing to conform to the needs of the patients served.

IMPROVING TIMELINESS OF SUSCEPTIBILITY RESULTS

The susceptibility of a microorganism to antimicrobial agents is often more important than definitive microbial identification. Several strategies have been developed to yield faster results. Exhaustively reviewed by Isenberg and D'Amato,[15] rapid susceptibility testing methods range from reading disk diffusion agar plates after only 5 or 6 hours' incubation to direct inoculation of susceptibility testing media with specimen material. Not all are reliable or appropriate, but this review discusses uses of these potential tests. The most effective method for rapid susceptibility testing of isolated bacteria appears to be that of Lorian,[19] in which isolated colonies are suspended in trypticase soy broth to a density equal to a McFarland 1.0 turbidity standard and the suspension then inoculated onto Mueller-Hinton agar without preincubation. Standard zone sizes were used for interpretation of results, but Mueller-Hinton with 5% sheep blood was used for all gram-positive cocci. Qualitative results obtained after 5 hours of incubation agreed with conventional Bauer-Kirby disk diffusion determinations over 97% of the time for organisms evaluated. Rapid methods have not been so standardized for quantitative susceptibility results. New technologies exploiting the ability of growing organisms to produce fluorescent marker metabolic end products or the technological advances that allow laser imaging of tiny amounts of bacterial mass are being actively pursued as possible candidates for rapid susceptibility testing use. However, it may be very difficult to shorten incubation times to less than 6 hours and still obtain reliable results since some antimi-

crobial agent resistance, such as that caused by inducible β-lactamases, is not reliably detected with shorter incubation times. Direct susceptibility testing of urine has not yielded results that correlate well with conventional methods. Positive blood cultures, however, since they are almost universally monomicrobic, can be adapted for direct testing. Methods cited previously for preparation of inocula for identification systems can also be used to prepare inocula for direct susceptibility testing. Greater than 95% agreement with interpretative results of conventional testing procedures typically can be achieved by dropping 2 to 3 drops of well-agitated blood culture broth (Gram stained to ascertain the presence of only one morphological type) directly to the surface of Mueller-Hinton agar. This inoculum is spread and handled in the same manner as standard disk diffusion.[5] Results are thus available 1 day sooner using this method. However, routine testing should always be done to confirm the rapid results.

Even quantitative susceptibility testing can be speeded up, as reported by Kiehn et al.[17] A 1-ml suspension made of equal amounts of turbid blood culture broth and brain-heart infusion broth, incubated for 3 to 6 hours and diluted 1:500, can be used as the inoculum for a microbroth dilution susceptibility test with results that have greater than 98% agreement with conventional methods (within two dilutions). Of course, all of these strategies work only with rapidly growing organisms, and accepted practice is to subsequently confirm the results with conventional methods. The equivalent of direct susceptibility testing of at least two selected bacterial species can be accomplished extremely rapidly by using chromogenic substrates, as discussed in Chapter 14. The chromogenic cephalosporin nitrocefin can be used with pure culture material as well as with pelleted material from cerebrospinal fluid (e.g., for detection of β-lactamases). Such information should be generated for isolates of gonococci and *H. influenzae*. Detection of β-lactamases produced by *S. aureus* or *Bacteroides* spp. may require increased reaction time, but the chromogenic substrate is valuable for rapid detection of antimicrobial agent (penicillin or ampicillin) resistance. β-lactamases can be detected using other systems, such as acidometric or iodometric, but these methods are slightly more cumbersome than the chromogenic substrate tests. A rapid test for detection of the acetyltransferase enzyme that renders an organism resistant to chloramphenicol can be used to test isolates of *H. influenzae* from cerebrospinal fluid in locations where chloramphenicol resistance has been reported, and where that agent is used in the initial treatment of meningitis. This test is available as a filter paper–impregnated disk (Remel Laboratories). Additional new detection systems for resistance to other agents need development.

Automation has made an impact on susceptibility testing in microbiology. A number of systems are available, as detailed in Chapter 12, that yield results within 4 to 6 hours after the test is inoculated. Although certain types of resistance will not be detected with these systems as they are currently produced (such as inducible cephalosporinases and small numbers of methicillin-resistant staphylococci in a population of bacteria, which need at least a 6-hour incubation for detection of the resistant bacterial population), the information generated by rapid systems is potentially valuable. Caution is required here, too, since any test requiring 4 or more hours' incubation time often cannot be reported to physicians before they leave for the day. Therefore the information from such tests may not be used clinically until the next day, even if available to laboratory personnel much earlier.

Use of the radiometric system (Bactec) for detection of metabolic activity has revolutionized susceptibility testing of *M. tuberculosis*. Results that required up to 3 weeks are now available in several days. When using this system it is critical for the laboratory to report out an identification for *M. tuberculosis* or MOTT (mycobacterium other than tuberculosis) with the initial report of a positive culture for acid-fast bacteria since clinicians will often treat after receiving the report of a positive culture containing acid-fast bacteria. Reporting numerous positive cultures that eventually are identified as containing atypical mycobacteria will add to medical care costs by causing unnecessary hospitalization and treatment. The in vitro effect of antimycobacterial agents alone and in combinations also can be assessed in this system.

COMPUTERIZATION

The final product of a microbiologist's work is information, and the thrust of this discussion has been toward strategies that decrease the time until information is available. The speed with which this data is generated will have no impact on patient care if results are not quickly relayed to the physician with responsibility for patient care. It is in the rapid dissemination of information that the modern laboratory differs most from those of the past. A well-designed computer system is essential for information storage, data analysis, and retrieval.

Cost-effectiveness measures that can be aided by computerization include logging in of new specimens. A computer can flag a second specimen from the same patient on a single day to alert the laboratory to (probably) unnecessary test requests.[27] Without an automated system, duplicate specimens in a large laboratory are rarely recognized until after they have been inoculated, if at all. A microbiology computer can be used to maintain inventory and avoid costly delays caused by lack of supplies, as well as loss of operating reagents because of outdating or overstocking. The information critical to patient care will reach the intended physician or hospital ward directly, without the misunderstandings that occur when messages are relayed by telephone.

The computer can also be used to assess workload units of procedures, to effect direct billing, and to maintain quality control records. Of course, computers have potential for preparing cumulative susceptibility data reports, and for epidemiological evaluation. Direct communication between microbiology instruments and laboratory computers can allow data to be transmitted easily after verification. Ultimately, computers will help to effect change in microbiology practices mandated by the need to control rising costs without adversely affecting patient care. The change from methods that require time for bacterial growth to more rapid methods for detection of pathogens and prediction of susceptibility patterns has begun late for microbiology in comparison to other laboratory specialities, such as chemistry and hematology. It is still uncertain how quickly this type of rapid testing can develop in view of the new pathogens that continue to be recognized and the multiple mechanisms by which a given microorganism can become resistant to any therapeutic agent. Clinicians, however, realize the value of rapid availability of results and will utilize the information for better patient care whenever it is available; they will support the development of more and better systems. The revolution in microbiology, boosted by DRGs, has only begun.

REFERENCES

1. Aronson, M.D., and Bor, D.H. 1987. Blood cultures. Ann. Intern. Med. 106:246.
2. Bartlett, R.C., Kohan, T.S., and Rutz, C. 1979. Comparative costs of microbial identification employing conventional and prepackaged commercial systems. Am. J. Clin. Pathol. 71:194.

3. Chonmaitree, T., Menegus, M.A., and Powell, K.R. 1982. The clinical relevance of `CSF viral culture.' J.A.M.A. 247:1843.

4. College of American Pathologists. 1988. Manual for laboratory workload recording method. The College, Skokie, Ill.

5. Coyle, M.B., McGonagle, L.A., Plorde, J.J., et al. 1984. Rapid antimicrobial susceptibility testing of isolates from blood cultures by direct inoculation and early reading of disk diffusion tests. J. Clin. Microbiol. 20:473.

6. Ellner, P.D. 1987. Cost-containment strategies for the diagnostic microbiology laboratory. Clin. Microbiol. Newsletter 9:117.

7. Ellner, P.D. 1985. A dozen ways to achieve more cost-effective microbiology. Med. Lab. Observer 17:40.

8. Finegold, S.M., and Edelstein, M.A.C. 1988. Coping with anaerobes in the 80s. pp. 1-10. In Hardie, J.M., and Borriello, S.P., editors. Anaerobes today. John Wiley & Sons, New York.

9. Galen, R.S., and Gambino, S.R. 1975. Beyond normality: the predictive value and efficiency of medical diagnosis. John Wiley & Sons, New York.

10. Garcia, L.S., and Bruckner, D.A. 1985. Microbiology's economics: dissecting procedures and costs. Med. Lab. Observer 17:51.

11. Gilligan, P.H. 1986. Diarrheal disease in the hospitalized patient. Infect. Control 7:607.

12. Granoff, D.M., Murphy, T.V., Ingram, D.L., and Cates, K.L. 1986. Use of rapidly generated results in patient management. Diagn. Microbiol. Infect. Dis. 4(Suppl.):157S.

13. Gross, P.A., VanAntwerpen, C.L., Hess, W.A., and Reilly, K.A. 1988. Use and abuse of blood cultures: program to limit use. Am. J. Infect. Control 16:114.

14. Harris, P.C., and Sealey, L.B. 1986. Practical cost savings in microbiology. Med. Lab. Observer 18:32.

15. Isenberg, H.D., and D'Amato, R.F. 1984. Rapid methods for antimicrobic susceptibility testing. In Cunha, B.A., and Ristuccia, A.M., editors. Antimicrobial therapy. Raven Press, New York.

16. Kelly, M.T. 1986. Making organism identification cost-effective. Med. Lab. Observer 18:55.

17. Kiehn, T.E., Capitolo, C., and Armstrong, D. 1982. Comparison of direct and standard microtiter broth dilution susceptibility testing of blood culture isolates. J. Clin. Microbiol. 16:96.

18. Lee, A., and Gooden, H. 1982. Improving communication of microbiology test results. Am J. Clin. Pathol. 77:443.

19. Lorian, V. 1977. A five hour disc antibiotic susceptibility test. In Lorian, V., editor. Significance of medical microbiology in the care of patients. Williams & Wilkins, Baltimore.

20. Peters, C.S., Hernandez, L., Sheffield, et al. 1988. Cost containment of formalin-preserved stool specimens for ova and parasites from outpatients. J. Clin. Microbiol. 26:1584.

21. Pezzlo, M. 1988. Detection of urinary tract infections by rapid methods. Clin. Microbiol. Rev. 1:268.

22. Radetsky, M., and Todd, J.K. 1984. Criteria for the evaluation of new diagnostic tests. Ped. Infect. Dis. 3:461.

23. Reyes, E.A.P., Bale, M.J., Cannon, W.H. and Matsen, J.M. 1981. Identification of *Pseudomonas aeruginosa* by pyocyanin producion on tech agar. J. Clin. Microbiol. 13:456

24. Spencer, R.C. 1990. Sixth international congress on rapid methods and automation in microbiology and immunology: Meeting commentary. J. Antimicrob. Chemother. 26:739.

25. Stager, C.E., and Davis J.R. 1992. Automated systems for identification of microorganisms. Clin. Micro. Rev. 5:302.

26. Wetterau, L.M., Zeimis, R.T., and Hollick, G.E. 1986. Direct examination of unstained smears for the evaluation of sputum specimens. J. Clin. Microbiol. 24:143.

27. Winkel, P., and Statland, B.E. 1984. Assessing cost savings when unnecessary utilization of laboratory tests can be abolished. Am. J. Clin. Pathol. 82:418.

BIBLIOGRAPHY

Anargyros, P., Astill, D.S.J., and Lim, I.S.L. 1990. Comparison of improved BACTEC and Lowenstein-Jensen media for culture of *Mycobacteria* from clinical specimens. J. Clin. Microbiol. 28:1288.

Bartlett, R.C. 1974. Medical microbiology: quality, cost and clinical relevance. John Wiley & Sons, New York.

Becker, B.L. 1985. Test menus and profiles: signs of change under DRG's. Med. Lab. Observer 17:26.

Butler, W.R., Jost, Jr., K.C., and Kilburn, J.O. 1991. Identification of mycobacteria by high-performance liquid chromatography. J. Clin. Microbiol. 29:2468.

Doern, G.V., Scott, D.R., and Rashad, A.L. 1982. Clinical impact of rapid antimicrobial susceptibility testing of blood culture isolates. Antimicrob. Agents Chemother. 21:1023.

Everett, G.D., de Blois, S., Chang, P.F., and Holets, T. 1983. Effect of cost education, cost audits, and faculty chart review on the use of laboratory services. Arch. Intern. Med. 143:942.

Garcia, L.S. 1985. A cost containment checklist. Med. Lab. Observer 17:67.

Hugh, R, and Gilardi, G.L. 1980. *Pseudomonas*. In Lennette, E.H., Balows, A., Hausler, W.J., Jr., and Truant, J.P., editors. pp. 291-293. Manual of clinical microbiology, ed 3, American Society for Microbiology, Washington, D.C.

Jorgenson, J.H., editor. 1987. Automation in clinical microbiology. CRC Press, Boca Raton, Fla.

King, E.O., Ward, M.K., and Raney, D.E. 1954. Two simple media for the demonstration of pyocyanin and fluorescein. J. Lab. Clin. Med. 44:301.

Laboratory Medicine. 1988. Symposium: the practice of real-time diagnostic microbiology. Lab. Med. 19:336 (Entire issue devoted to rapid testing in microbiology.)

Matsen, J.M. 1982. How rapid should rapid methods be? Clin. Microbiol. Newsletter 4:164.

Matsen, J.M., and Kelly, M.T. 1985. Practical clinical microbiology—laboratory perspectives for the era of prospective payment and DRG's. American Society for Clinical Pathology Commission on Continuing Education, Council on Microbiology, Chicago. (Workshop manual.)

McPherson, K.A., and Needham, C.A. 1987. Method evaluation and test selection. pp. 27-33. In

Wentworth, B.B., Baselski, V.S., Doern, G.V., et al., editors. Diagnostic procedures for bacterial infections, ed 3, American Public Health Association, Washington, D.C.

Miller, J.M. 1984. The impact of DRG's on hospital laboratories. Clin. Microbiol. Newsletter 6:57.

Sanders, C.B., and Aldridge, K.E. 1983. The microbiology laboratory: garbage in, garbage out. Clin. Microbiol. Newsletter 5:123.

Tilton, R.C., editor. 1982. Rapid methods and automation in microbiology. American Society for Microbiology, Washington, D.C.

5

THE ROLE OF THE MICROBIOLOGIST IN MEDICAL PRACTICE

In recent years clinical laboratories have evolved from being profit centers to being cost centers, and laboratory charges are often included in the amount paid by the government or other third-party payer according to diagnosis. Moreover, physicians' use of laboratory tests are coming under increasing scrutiny. One result of all of these changes is the increasing need for medical laboratory professionals to develop and establish information that is required for making clinical decisions. In this setting it is crucial to reassess the consultative role of the clinical microbiologist as a part of high quality, cost-effective care.

The elimination of misuse or overuse of laboratory tests must be based on sound clinical judgment and a thorough knowledge of what constitutes clinically relevant microbiology. The clinical microbiologist must be available for consultation by phone and, if necessary, on site.

Of increasing importance and value is the use of quality assurance and utilization review programs in which the appropriateness or adequacy of requests made of the laboratory are regularly evaluated. Such programs are of utility, however, only when carefully planned to address a problem that lends itself to a measurable audit before and after some form of intervention takes place. The microbiologist is an important consultant in these global (as opposed to single patient) consultation opportunities through knowledge of appropriate laboratory medicine in clinical practice. In some instances it may be possible to study ordering patterns by individual physicians for a particular test and identify outliers for audit. Current efforts by

government and medical organizations to develop practice guidelines for a variety of medical conditions will result in further scrutiny of laboratory utilization patterns.

STRATEGIC APPROACHES

SPECIMEN COLLECTION

One of the major responsibilities of the clinical microbiologist is to develop and promulgate guidelines for specimen collection. Proper specimen collection is critical for two major reasons. First, properly collected and transported specimens have the best likelihood of yielding a positive result when they contain pathogenic microbes. Second, only the correct specimen (including type, volume of material, and transport conditions) will provide a high probability that a negative result is truly indicative that a pathogen is not present. Clinicians often use both positive and negative results to rule in or rule out a diagnosis and for directing treatment; therefore high quality specimens are critical. Such guidelines must be made widely available to the medical staff, residents, and nursing personnel. In-service education at medical staff and resident conferences are important points of communicaton of such information and also of feedback from clinicians regarding laboratory services. Because nursing personnel are often involved in specimen collection, they must participate in nursing education programs. Operating room personnel usually provide surgeons with specimen collection and transport systems or devices, so careful instruc-

35

tion of these personnel as to the rationale for certain specimen collection procedures is of vital importance and frequently more productive than any written memoranda to surgical staff and residents. Special protocols need to be developed for the collection and handling of certain specimens to ensure that a complete examination of irreplaceable material is performed. Examples of specimens requiring special protocols are bronchoalveolar lavage fluids and open lung biopsies from immunocompromised patients. A sample of such a protocol is in Figure 5.1. This type of approach suggests which tests should be considered routine for this type of specimen. Such a protocol should be in place for specimens originating from invasive procedures that cannot be repeated without undue expense and risk to the patient. The clinical microbiologist also must work with the surgical pathologist in developing procedures for examination of tissue.

Availability of the medical laboratory professional for consultation by phone or on site is also important. Clinical training and experience usually facilitate consultations as does a knowledge of problems specific to individual subspecialties of medicine.

The laboratory needs to ensure that all necessary materials for proper specimen collection are available on nursing units and in operating rooms. Examples include sterile anaerobic transport vials for use in operating rooms for transport of abscess fluid and normally sterile body fluids. Pneumatic tube specimen transport systems are becoming increasingly popular in hospitals, so the clinical microbiologist needs to work with the purchasing department to ensure that the specimen containers included in lumbar puncture trays, for example, are leakproof under the conditions present in a pneumatic tube.

Design of an appropriate test requisition form requires close attention. The form should allow the ordering physician to clearly define the tests being requested. Although a menu-driven requisition form risks abuse in terms of over-ordering, there are certainly instances in which, as already discussed, it is preferable to provide special requisitions to meet the needs of a special protocol. For example, with a menu type of requisition it may be all too easy for a medical student, physician, or nurse to order bacteriological examination only on a bronchoalveolar lavage fluid from an immunocompromised patient. Although bacteriological examination of such a specimen is not inappropriate, it is clearly insufficient and risks misdiagnosis of infection caused by mycobacteria, fungi, viruses, or *Pneumocystis carinii*. Recommended guidelines for specimen procurement, transport, and processing have been published by the American College of Chest Physicians.[1,2,3] In the years ahead more and more hospitals can be expected to install order entry systems for laboratory tests and thereby allow the clinical microbiologist to define specimen and test requirements in a far more systematic and controlled manner than is currently the case with paper requisition forms.

A major role of the clinical microbiologist is to define rejection criteria for specimens. Establishing such criteria requires a clear understanding of disease processes, appropriate transport conditions, and the availability of alternative specimens

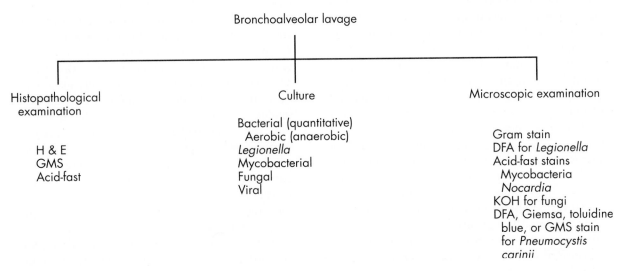

FIGURE 5.1
Protocol for processing bronchoalveolar lavage specimens from immunocompromised patients.

that would more clearly define the disease process. An example of this is the unsuitability of using catheter tips as material for the diagnosis of urinary tract infections. In this situation, a properly collected sample of urine is the appropriate clinical material to process. Rejection criteria must be clearly explained to the clinical and nursing staffs. It is also important to realize that no rejection criterion will apply in 100% of cases, and one needs to understand that physicians and nurses can be enormously resourceful in circumventing the rejection criteria by describing a specimen for which rejection criteria exist as being something else (e.g., tracheal aspirate instead of sputum). The availability of a pediatric blood culture bottle or device may suddenly gain popularity among housestaff (e.g., for use in adults), so keeping track of the use of such systems according to patient demographic information is an important quality assurance step. Speciation of coagulase-negative staphylococci isolated from urine to detect *Staphylococcus saprophyticus* is, for example, unnecessary except in women of child-bearing age. In many institutions much time, energy, and money is expended on cultures for *Salmonella, Shigella,* and *Campylobacter* in stool samples from patients with diarrhea occurring more than 3 days following hospitalization. Thus monitoring patients' demographic data at the time of specimen accession to obtain information regarding age, sex, and date of hospitalization is important.

Other examples of the application of rejection criteria are the use of cytology to assess the quality (i.e., extent of contamination) of sputum specimens, based on the numbers of squamous epithelial cells observed per low-power field of a Gram-stained smear, and institution of a program to encourage submission of aspirates of abscess material in anaerobic transport vials rather than on swabs.

Communication of rejection criteria to physicians and nurses is important. In some instances, direct communication is essential to ensure timely collection of a second, and it is hoped acceptable, specimen.

To minimize the frequency of unacceptable specimens, some hospitals have adopted, as a matter of policy, phlebotomy services for blood cultures and have expanded the responsibilities of respiratory therapists to include sputum collection, urinary catheter care teams to include urine collection, and infectious disease nurses or pharmacists to include blood collection for antibiotic monitoring. Such measures have been shown to be cost effective in terms of reducing the number

of repeat specimens obtained to take the guesswork out of confusing laboratory test results.

PROCESS CONTROL

One of the most important functions that a clinical microbiologist performs is to decide what is clinically relevant as regards specimen processing, culturing, identification, and susceptibility testing. Considerable judgment is required to decide what organisms to look for and report. It is essential to recognize what constitutes indigenous flora and not to clutter reports with the identification of organisms that have little or nothing to do with the disease process. There is a tendency to ascribe an etiological role to what is on a printed report and to treat the organism(s) rather than the disease process. Indiscriminate reporting of indigenous flora by the laboratory contributes to indiscriminate use of antibiotics, unnecessary costs, potential morbidity caused by superinfection or other side effects, and emergence of resistant flora. Moreover, timely reports of the isolation of certain groups of organisms such as anaerobic bacteria usually provide more than enough information for a clinical decision to be made. For example, a report that anaerobic gram-negative bacilli plus other mixed anaerobic flora have been isolated from an intraabdominal abscess usually suffices and is certainly of greater clinical value than a report issued 2 to 3 weeks later with the identification of each of the species isolated. What is clinically relevant to identify and report varies by site. Thus throat cultures should be processed to optimize the recovery of group A streptococci. Non–group A β-hemolytic streptococci are often present in throat cultures; however, they are frequently present in asymptomatic children and adults and may cause self-limited pharyngitis but are not associated with the nonsuppurative sequelae characteristic of group A streptococci. *Haemophilus influenzae* is also commonly found in the upper respiratory tract, is not a documented cause of pharyngitis, and therefore should not be part of a throat culture report. *Staphylococcus aureus* is a common inhabitant of the upper respiratory tract, but identifying and reporting it in a throat culture can be misconstrued and lead to inappropriate and unnecessary therapy. Similarly misleading reports can be generated from vaginal and cervical cultures in which the list of documented pathogens is limited to a few sexually transmitted diseases, but the diversity of indigenous flora is great. Once again, group B streptococci, except in the perinatal period, have no documented pathogenic role in this site. Nor do the

Enterobacteriaceae. Yet, reports of group B streptococci or *Escherichia coli* from vaginal or cervical cultures of women with vaginitis or cervicitis often lead to unnecessary therapy.

ANTIMICROBIAL SUSCEPTIBILITY TESTING

Coordination between the laboratory and the hospital's formulary committee is important to ensure that antimicrobial agents selected for testing correspond with those in the hospital's formulary. A commercially prepared system may not contain the same antimicrobial agents as those present in the hospital's formulary, resulting in confusion and neglect on the part of clinicians as regards susceptibility test results. Selection of bacteria to be tested should be based on the predictability (or lack thereof) of susceptibility to antimicrobial agents of choice. Thus there is no point in testing a group A β-hemolytic *Streptococcus* against penicillin since resistance has not spread to a significant level. Anaerobic bacteria are virtually always susceptible to ampicillin/sulbactam, chloramphenicol, imipenem, ticarcillin/clavulanate, and, with the exception of some anaerobic gram-positive bacilli, metronidazole. Broad-spectrum cephalosporins are the currently recommended therapy for invasive *Haemophilus influenzae* infections. Since resistance to these agents has not yet been reported, there is little need to test isolates on a routine basis, except perhaps for resistance to ampicillin.

Selection of antimicrobial agents to be tested should be made in consultation with the hospital's formulary committee so that what is tested and what is available correspond. This process can be challenging since commercially available kits and devices often do not contain the same drugs as those in the formulary, forcing one to use more than one kit or device or a custom-prepared device, both at greater expense. For this reason, many hospitals have reverted to using the disk diffusion test, which offers greater flexibility at less expense. Laboratories therefore need to assess carefully the cost-effectiveness of semiautomated and automated systems for susceptibility testing. In some instances, choice of a system is made because it provides a quantitative result or MIC (minimum inhibitory concentration) rather than the qualitative result (susceptible, intermediate, or resistant) provided by the disk diffusion test. The quantitative result is incorrectly perceived as being more accurate and what clinicians want. The reality is, first, that interpretation of the disk

diffusion test is based on the direct correlation between MICs and zone diameters of inhibition; second, MICs are correctly interpretable by only a minority of physicians; and, third, MICs are seldom required for the management of antimicrobial therapy. Therefore, dilution tests are not more accurate than diffusion tests, and MICs must be interpreted on the laboratory report.

Selection of antimicrobial agents to be tested is based not only on the hospital's formulary composition but also on the spectrum class representative (e.g., oxacillin for penicillinase-resistant penicillins: methicillin, oxacillin, nafcillin, cloxa-cillin, dicloxacillin to be tested against staphylococci). Cephalothin or cefazolin serves as the spectrum class representative for first generation oral and parenteral cephalosporins. Any one of the third generation cephalosporins may be tested against the Enterobacteriaceae. Guidelines for selection of antimicrobial agents to be tested against various categories of bacteria and interpretative criteria for zone diameters of inhibition obtained from the disk diffusion test of MICs derived from dilution tests are provided by the National Committee for Clinical Laboratory Standards.[4,5]

Although interpretative guidelines of susceptibility tests are helpful to physicians, it must be recognized that a number of other factors guide antimicrobial therapy. These include the concentrations achieved at various tissue (infection) sites by the antibiotic and its route of excretion, the relative frequency and potential severity of toxicity that may result from administration of the antibiotic, possible interaction of antibiotic with other drugs being administered concurrently, preexisting or developing renal or hepatic impairment, history of hypersensitivity to the antibiotic, and prior experience with the antibiotic. It must also be recognized that antimicrobial therapy is often started empirically based on the site of infection and the most likely etiological agent(s) involved, so that the susceptibility test report may serve only to confirm the activity of the antimicrobials being administered or allow modifications in the therapy to be made when a less expensive, equally active agent is available. In some instances, test results will demonstrate inactivity of the antimicrobial agent being administered. Unfortunately, the reality is that because of the time lag between specimen submission and reporting of susceptibility test results, reports are often not seen or are ignored by the ordering physician. The role of the clinical microbiologist in this setting is important. Identifying significant positive cultures (i.e., from blood and cerebrospinal, pericardial, pleural, synovial fluids) and informing the patient's physician, while inquiring about the antimicrobial therapy being administered, is a simple first step and establishes rapport between two health care professionals. Many variations on this theme can be developed, depending on one's resources. In our institution pathology residents and microbiology fellows review the histories and discuss laboratory results of such patients with their counterparts on clinical services and then present the cases at daily laboratory rounds with laboratory staff. Management problems are discussed and recommendations made to the attending residents and staff. The recommendations can be made either verbally or as part of a written consultation. In other institutions the clinical microbiologists will make rounds to review the charts and discuss patients with significant isolates. As an alternative, if an infectious diseases practitioner is available, a list of significant isolates may be provided to her or him for follow-up. Finally, an increasing number of hospitals are hiring infectious disease pharmacists who can advise the attending staff on antibiotic therapy according to susceptibility testing information. Whatever the mechanism, the point is that the clinical microbiologist bears a responsibility beyond establishing the most appropriate laboratory procedures and ensuring appropriateness of antimicrobial therapy in cases with significant positive cultures.

The other major area in which the clinical laboratory can provide support for antimicrobial therapy is antibiotic assay. Assays are usually indicated for antibiotics with narrow toxic to therapeutic ratios, such as vancomycin and the aminoglycosides. Proper interpretation of such assays depends on proper timing of specimen collection in relation to dosing and proper specimen storage. Although phlebotomists are frequently responsible for collection of the blood sample, it is not always possible for them to coordinate specimen collection with a specific time interval following administration of the antibiotic. Actual antibiotic administration times may differ from specified dosage schedules so that it is not unusual for putative peak serum levels to be lower than putative trough serum levels. For these reasons, some hospitals have implemented therapeutic drug monitoring services in which a nurse or pharmacist assumes responsibility for both antibiotic administration and blood collection. Such approaches have actually been shown to save money by taking the guesswork out of interpretation of spurious results, which in turn lead to repetitive assays

in an effort to figure out the real answer. Since assays for vancomycin and aminoglycosides are usually performed today by immunoassay in the clinical chemistry laboratory where there is often 24-hour coverage and equipment back-up, it is important for the clinical microbiologist to work with the clinical chemist, pharmacist, and/or nurse involved in therapeutic drug monitoring to ensure that proper collection of samples takes place.

In the area of antimicrobial therapy the clinical microbiologist is often consulted regarding selection of drugs to be tested, interpretation of susceptibility test results, and, when an infectious disease specialist is unavailable, therapeutic guidelines. Examples of instances in which the clinical microbiologist might provide limited consultation include information regarding the unreliability of trimethoprim-sulfamethoxazole susceptibility test results against enterococci because of the intrinsic in vivo resistance of these organisms to this combination of antimicrobial agents and recommendations regarding the necessity of combination therapy with a cell wall active antimicrobial agent and an aminoglycoside for treatment of serious enterococcal infections. Examples of more comprehensive consultations might include indications for and interpretation of antimicrobial assays in therapeutic drug monitoring programs, particularly concerning issues relating to dosage of antimicrobial agents with narrow therapeutic to toxic ratios (e.g., aminoglycosides), and discussions with clinical staff regarding bactericidal or synergistic testing, with specific reference to methodological problems of testing, interpretation of results, and applicability of results to patient care. Responsibilities for more specific guidelines regarding antimicrobial therapy and specific tests for directing or evaluating therapy require both clinical experience and review of the patient's clinical history and current status. As a result, the clinical microbiologist must exercise special caution in dispensing advice based strictly on laboratory data. Detailed knowledge of the pharmacokinetic properties of the antimicrobial agents or agents under consideration for administration and the patient's underlying conditions that may affect either the drugs' pharmacokinetics or potentially toxic side effects need to be fully understood. In such instances the clinical consultation services of the medical director of the laboratory can be reimbursed under the appropriate CPT code (80500-26 for Clinical Consultation: Limited, 80502-26 for Clinical Consultation: Comprehensive). Other examples related to antimicrobial susceptibility testing in which there may be a chargeable professional component include special tests, such as the serum bactericidal test or synergy tests.

ROLE FOR THE CLINICAL MICROBIOLOGIST

The role of the clinical microbiologist varies to some degree with the size and type of hospital in which he or she works. In all instances, the clinical microbiologist must define specimen requirements, specimen transport, process control, and report transmission. In most instances the clinical microbiologist participates in discussions regarding the selection of antimicrobial agents to be included in the hospital's formulary and in the resultant selection of antimicrobial agents to be tested in vitro. In most instances the clinical microbiologist is a member of the hospital's infection control committee. It is crucial for the clinical microbiologist to be active in institution-wide committees such as these, since microbiology has a broad impact on patient care and the microbiology professional needs to be visible as a knowledgeable and helpful resource for the delivery of medical care.

Opportunities for reimbursement of professional components are relatively limited in microbiology. As stated previously, billing for limited and comprehensive clinical consultations is available to laboratory physicians for the interpretation of antimicrobial susceptibility data and guidance of antimicrobial chemotherapy. Other reimbursable activities for the professional component of microbiological services include the review of special stains (CPT code 83112-26), review of cytology smears (CPT code 88104-26), and clinical consultation (limited or comprehensive) regarding evaluation and interpretation of a patient's culture results, provided these consultations meet strictly defined guidelines. The evolution of clinical microbiology and the clinical microbiologist will continue into the future. We have an important role to play in the delivery of high quality patient care. Assuming this responsibility and defining the future role for our discipline are significant tasks to be undertaken seriously.

REFERENCES

1. Pingleton S.K., Fagon J-Y, Leeper K.V. Jr. 1992. Patient selection for clinical investigation of ventilator-associated pneumonia: Criteria for evaluating diagnostic techniques. Chest. 102(suppl):553S-556S.
2. Meduri U., Chastre J. 1992. The standardization of bronchoscopic techniques for ventilator-associated pneumonia. Chest. 102(suppl):557S-564S.
3. Baselski V.S., Lel-Torkey M., Coalson J.J., Griffin J.P. 1992. The standardization of criterial for processing and interpreting laboratory specimens in patients with suspected ventilator-associated pneumonnia. Chest 102(suppl):571S-579S.
4. National Committee for Clinical Laboratory Standards. 1990. Performance Standards for Antimicrobial Disk Susceptibility Tests—Fourth Edition; Approved Standard. NCCLS Document M2-A4. Villanova, Penn: NCCLS.
5. National Committee for Clinical Laboratory Standards. 1990. Methods for Dilution Antimicrobial Susceptibility Tests for Bacteria That Grow Aerobically—Second Edition; Approved Standard. NCCLS Document M7-A2. Villanova, Penn: NCCLS.

6

HOSPITAL EPIDEMIOLOGY

In addition to its important role in assisting in the diagnosis and treatment of patients, the microbiology laboratory has the opportunity and responsibility to recognize infections of public health importance. The laboratory identifies infections that are spread from one individual to another, including tuberculosis, venereal disease, and certain diarrheal diseases. More broadly, data from clinical microbiology laboratories may suggest or document an epidemic outbreak in the community.

The balance of this chapter is devoted to the identification and investigation of nosocomial infections, which is of particular importance to the hospital-based microbiology laboratory.

NOSOCOMIAL INFECTIONS

Nosocomial infections are those that are acquired in an institutional (usually hospital or nursing home) setting (Chapter 24). These may, of course, first appear after a patient has been discharged from the hospital, depending on the incubation period of the infection in question. Thus distinguishing between community-acquired and nosocomial infections occasionally is difficult. Identification by the laboratory of an unusual number of isolates of an uncommon pathogen, of a number

of strains of the same organism with an unusual antimicrobial susceptibility pattern or unique biochemical feature (e.g., H_2S-positive *Escherichia coli*), or clustering of isolates of a given type from an intensive care unit or other discrete area within the hospital suggests the likelihood of an outbreak. The laboratory works closely with the infection control team to investigate the epidemiology of such outbreaks in order to find corrective measures. The laboratory assists in preventing infection of other patients and of hospital personnel by identifying infections that are likely to require isolation and by providing guidance in the use of disinfectants and sterilizing procedures.

INCIDENCE
The National Nosocomial Infections Study (NNIS) carried out by the Centers for Disease Control indicates that 5% to 6% of hospitalized patients develop nosocomial infection. Each nosocomial infection adds 5 to 10 days to the affected patient's time in the hospital. The mortality of patients with either lung or bloodstream nosocomial infections can be as high as 75%. These infections add direct charges of at least $1800. Thus, for the United States as a whole, nosocomial infections cost far in excess of $4 billion per year in direct

costs alone. This does not take into account charges of physicians, loss of productivity, and unanticipated death. Approximately 1% of nosocomially infected patients die from their infection, and these infections contribute to the deaths of an additional 2% to 3% of afflicted patients. Attack rates vary according to the type of hospital. Large referral hospitals often have higher rates of nosocomial infection than small community hospitals, and large teaching hospitals have higher infection rates than small teaching hospitals. The difference in risk of infection is probably related to several factors—severity of the illness, frequency of invasive diagnostic and therapeutic procedures, and variation in the effectiveness of infection control programs. Within hospitals, surgical and medical services have the highest rates of infection, and pediatric and nursery units the lowest. Convalescent care facilities and nursing homes also contribute to the number of nosocomial infections and are required to have active infection control programs.

EPIDEMIOLOGY

Three principal factors determine the likelihood that a given patient will acquire a nosocomial infection: (1) susceptibility of the patient to the infection, (2) the inoculum and virulence of the infecting organism, and (3) the nature of the patient's exposure to the infecting organism. In general, of course, hospitalized persons have an increased susceptibility to infection, and it is not possible to immunize patients against nosocomial infections. Corticosteroids, cancer chemotherapeutic agents, and antimicrobial agents all contribute to the likelihood of nosocomial infection by suppressing immunity or altering the host's normal flora to that of resistant (hospital) microbes but are nonetheless important drugs that are generally much more beneficial than harmful. Exerting influence over the virulence of the infecting organisms is not possible. Patients with serious community-acquired infections are admitted to hospitals frequently. In the case of an influenza epidemic, it is important to avoid hospitalizing patients without serious complications. Nonetheless, it is necessary to admit a number of patients, and the disease may spread nosocomially in this manner. Our best hope in attempting to minimize nosocomial infections is to eliminate sources of exposure and to interrupt means of spread of these organisms in the hospital environment through the use of Universal Precautions. The most important means of transmission of nosocomial infections is by direct contact. In-

fected patients are the primary source of organisms in nosocomial infections, although asymptomatic carriers may transmit such infection. Handwashing by personnel before and after contact with each patient is still the most important way to prevent spreading nosocomial infections. Appropriate isolation of infected patients with highly transmissible diseases is also important. The second most common means of spread of hospital-acquired infection is by contaminated vehicles as in so-called *common source outbreaks.* Included here are contaminated food, water, medications, or medical devices. Other types of materials (fomites) in the hospital may serve as sources of nosocomial infection. A recently described example was wet mattresses, which served as environmental reservoirs of *Acinetobacter* that was involved in an outbreak of infection in a burn center. For the most part airborne spread of infection is less significant in the hospital setting, although it can be implicated in situations such as influenza, pulmonary tuberculosis, and even methicillin-resistant *Staphylococcus aureus.*

TYPES OF NOSOCOMIAL INFECTION

Urinary tract infection, the most common type of hospital-acquired infection, accounts for some 40% of nosocomial infections. Surgical wound infections account for an additional 20%, lower respiratory tract infections (primarily pneumonia) account for 15%, and nosocomial bacteremia accounts for an additional 5%. Thus these four categories together account for about 80% of nosocomial infections. *E. coli* is the most common pathogen isolated from urinary tract infections and is usually, but not always, of endogenous origin. However, it is less important in nosocomial infection than *Klebsiella, Enterobacter, Pseudomonas,* and other antimicrobial-resistant organisms. Important factors in nosocomial urinary tract infection are indwelling urinary catheters and other types of urological manipulations carried out for diagnostic or therapeutic purposes. Surgical wound infections can involve indigenous flora, but *S. aureus,* enterococci, coagulase-negative staphylococci, and a number of different types of gram-negative aerobic or facultative bacilli are important pathogens in nosocomial infection. Lower respiratory tract infection acquired in the hospital setting often is related to aspiration and thus involves normal oral flora; however, patients in the hospital setting commonly acquire nosocomial pathogens such as *S. aureus, Pseudomonas* species, and various gram-negative bacilli in their oropharyngeal flora. If such patients should then aspirate, these organisms, in addition to anaerobes and streptococci from the indigenous flora, play a role in a subsequent pneumonia. Respiratory therapy procedures, such as endotracheal suctioning or inhalation therapy, may introduce nosocomial pathogens into the patient's respiratory tract. Contaminated vehicles such as medications or water used for inhalation therapy may also be responsible. Nosocomial bacteremia is most often related to the use of intravascular devices. These catheters or cannulas may be contaminated by direct contact spread via health care personnel or from the patient's own skin flora. Coagulase-negative staphylococci are now the most common agents of nosocomial bacteremia. Contaminated vehicles such as intravenous fluids or arterial pressure transducers have been important sources of nosocomial infections.

Approximately two thirds of nosocomial infections involve a single pathogen, and about 20% involve multiple pathogens. Pathogens are identified in 85% of nosocomial infections. Of these, about 85% are aerobic or facultative bacteria, and 7% fungi. The most frequently reported pathogens have been *E. coli, S. aureus,* enterococci, and *P. aeruginosa.* Additional pathogens, in order of frequency of isolation from nosocomial infections, are *Klebsiella* spp., coagulase-negative staphylococci, *Enterobacter, Proteus, Candida,* and *Serratia* spp. The frequency of isolation of different pathogens varies, of course, according to the service on which the patient is hospitalized and the site of infection. *Legionella* spp. also have been implicated in outbreaks in hospitals, with the source of the organisms being the water system, air-conditioning system, or cooling tower. Enteric pathogens such as *Salmonella* and *Shigella* have also caused hospital outbreaks.

Many or most of the nosocomial bacterial pathogens are resistant to a number of antimicrobial agents. Methicillin-resistant *S. aureus* infections, which initially were encountered at large teaching hospitals, are increasing in frequency in all hospitals. Various gram-negative bacilli, such as *K. pneumoniae, S. marcescens,* and *P. aeruginosa,* are manifesting increased resistance to aminoglycosides, β-lactam agents, and the new fluoroquinolones. The incidence of nosocomial viral infections, including rotavirus, respiratory syncytial virus, and influenza, is now appreciated with the increase of virus identification capabilities. More than 75,000 nosocomial viral infections likely occur in the United States annually.[7]

CONTROL PROGRAMS AND ROLE OF MICROBIOLOGY LABORATORY

Hospital infection control programs are designed to detect and monitor hospital-acquired infections and to prevent or control their spread. Hospitals have multidisciplinary infection control committees, which should include one or more representatives from the microbiology laboratory. Most hospitals have one or more infection control nurses who collect and analyze surveillance data, monitor patient care practices, and participate in epidemiological investigations. Some hospitals have physicians or sanitarians as hospital epidemiologists. Infection control personnel are responsible for systematic surveillance of patient disease to recognize problems of hospital-acquired infection and to detect epidemics as early as possible. The microbiology laboratory provides important data for this surveillance effort. Obviously, there must be good communication between laboratory personnel and the infection control team, and it is desirable for infection control personnel to review the results of cultures in the laboratory on a daily basis. This provides a good opportunity for such people and laboratory microbiologists to exchange information. In the past it was common for microbiology laboratories to culture the inanimate environment of the hospital routinely, as well as certain fomites. Personnel now recognize that such activities done on a routine basis provide little or no useful information and are not cost effective. However, investigation of a specific nosocomial infection problem may be facilitated by cultures of the environment and/or personnel.

The accurate performance of routine microbiology procedures may be helpful in infection control work. The importance of full identification of isolates is underlined by a recent report of an outbreak of bacteremia and meningitis caused by *Streptococcus faecium* in a neonatal intensive care unit.[2] Such outbreaks may not be recognized without full characterization of organisms. In addition, the microbiology laboratory must be able to develop methods for investigating unexpected infection control problems. The laboratory may be asked to characterize organisms in detail, to perform additional susceptibility tests, and to screen a large number of specimens of various types for a particular pathogen. Larger laboratories now routinely use molecular techniques to determine the relatedness, or clonality, of microbes involved in an apparent nosocomial outbreak.[3] The laboratory director and senior personnel must assist the infection control team in planning strategies that are consistent with the nature of the outbreak and the resources of the laboratory. A laboratory should be prepared to ask for help from county or state health departments or other resources, when necessary. Obviously, laboratory budgets should be adequate to cover routine infection control testing, as well as special tests and emergency procedures. More extensive coverage of microbiology procedures as they relate to a comprehensive infection control program can be found in works by McGowan and co-workers.[5,6]

REPORTING OF RESULTS

As noted previously, infection control personnel often review laboratory results daily, and this should include direct discussions between the two groups. Certain types of results (similar to panic values in other laboratories) should initiate a phone call to the infection control team or infectious disease physicians, in addition to the attending physician. For example, all positive blood cultures, all positive spinal fluid cultures, all positive acid-fast smears, isolations of enteric pathogens such as *Shigella,* and isolation of resistant organisms such as methicillin-resistant *S. aureus* should be reported by phone. Isolation of a new or unusual pathogen should result in prompt reporting to the infection control team. An example of this is emerging highly drug-resistant nosocomial pathogens such as vancomycin-ampicillin-aminoglycoside–resistant *Enterococcus faecium,* multidrug resistant *Mycobacterium tuberculosis,* and highly penicillin-resistant pneumococci.

Needed laboratory results should be kept in a readily accessible format such as a logbook if the laboratory is not computerized, and laboratory records, including microbiology work cards, should be kept for at least 2 years. If the laboratory changes any of its technical practices that might influence the surveillance data, such as culturing urines with a 0.01-ml calibrated loop instead of the previously used 0.001-ml loop, all affected parties should be notified of the change and a record of the date of institution of the new practice should be kept with the infection control data files. Isolations of unusual organisms or organisms with unique susceptibility patterns are spotted readily. However, it may be much more difficult to recognize a cluster of cases of infection involving a frequently isolated organism. Computers are useful for this purpose. Infection control personnel should be supplied with periodic summaries of selective results. Tables or graphs indicating the frequency with which particular organisms have been isolated from various wards or units and on specific services are helpful. The antimicrobial agent susceptibility of frequent or difficult to treat isolates should also be tabulated and published on a regular basis.

CHARACTERIZING EPIDEMIC STRAINS

A good system for typing microbial strains involved in outbreaks is standardized, reproducible, sensitive, stable, readily available, inexpensive, applicable to a wide range of microorganisms, and field tested in conjunction with epidemiological investigation. No such ideal typing system is available, but it is important to appreciate the strengths and weaknesses of the systems that we use. Typing methods used include biological or biochemical typing (biotype), antimicrobial susceptibility patterns (antibiogram), and serological typing (serotype). Although these are typically readily available, they are no longer considered as reliable as newer molecular methods. Bacteriocin typing (including both susceptibility to bacteriocin and production of bacteriocin) and bacteriophage typing (phage type) can be useful but are restricted to only a few laboratories. Plasmid analysis and restriction enzyme analysis of plasmid or genomic (chromosomal plus plasmid) nucleic acids are now being widely applied. Genomic restriction enzyme analysis can be undertaken by many laboratories, with only modest capital equipment outlay, and can be very useful for epidemiological analysis of nosocomial infections.[3]

Routine identification procedures may be useful in permitting recognition of hospital-acquired infection outbreaks. Failure to fully identify *Pseudomonas* species might lead to overlooking less commonly encountered species such as *P. cepacia* or *X. maltophilia* as causes of an outbreak. Identification of an unusual Enterobacteriaceae strain such as *Klebsiella oxytoca* in sputum from several patients with pneumonia should alert the laboratory and infection control to a potential outbreak. Accurate reproducible susceptibility testing is important to the epidemiologist just as it is to the clinician. Recognition of outbreaks of infection caused by gentamicin-resistant Enterobacteriaceae, methicillin-resistant *S. aureus,* and amikacin-resistant *P. aeruginosa,* for example, is very important.

SPECIAL PROCEDURES FOR DETERMINING MICROBIAL RELATEDNESS

As indicated previously, a number of other procedures aside from biotyping, antibiograms, and serotyping can be used for characterization of epidemic strains. Although some are primarily useful in reference laboratory or research settings, others can be performed in larger clinical laboratories. Bacteriophage typing, by definition, is limited to bacteria. It has been used with *S. aureus, P. aeruginosa* and *Salmonella* species. Phage typing consists of spotting a series of bacteriophages on a "lawn" of the organism to be typed on an agar plate culture. Bacteriophages have a very specific host range, each one attacking only a particular type of a given organism such as *S. aureus.* Thus a number of phages are active against *S. aureus,* but each one lyses only particular types. If only phages numbered 52A and 80 among the whole set of phages used produce zones of lysis with a given strain of *S. aureus,* that strain is designated phage type 52A, 80.

Bacteriocin typing is also limited to bacteria. Bacteriocins are a group of heterologous substances produced by bacteria that inhibit the growth of closely related species; they are usually proteins. There are two systems of bacteriocin typing: bacteriocin production, which involves determination of the inhibitory spectrum of bacteriocins produced by a strain to be typed, and bacteriocin susceptibility, in which the susceptibility of the strain to be typed to bacteriocins produced by a set of standard producer strains is determined. These two methods are being less widely used since only labs with a collection of phages or bacteriocins can do this testing. Also, this typing may not be reproducible as desired since response to phages and bacteriocins may be plasmid mediated, and loss of plasmid DNA can alter a typing result done by these methods.

Plasmids are extrachromosomal bits of genetic material that self-replicate. Plasmids may be transferred from one bacterial cell to another by conjugation or transduction. Plasmid analysis has often been used to explain the occurrence of unusual or linked antibiotic resistance patterns. It has been shown that plasmid or R factor (resistance genes carried on plasmids) epidemics occur when a specific plasmid is transmitted from one genus of bacteria to another. Plasmid profiles, patterns created when plasmids are separated on agarose gel by electrophoresis, can also be used to characterize the similarity of bacterial strains that carry the plasmids. Relatedness of strains is based on the number and size of plasmids, with strains from identical sources showing identical plasmid profiles. Plasmids themselves or total genomic DNA (plasmid plus chromosomal DNA) may be typed by means of restriction endonuclease digestion patterns (Figure 6.1). Restriction enzymes recognize specific nucleotide sequences in DNA and produce double-stranded cleavages that

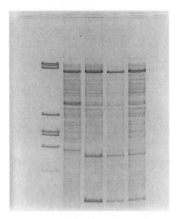

FIGURE 6.1

Example of restriction endonuclease analysis patterns for four *S. aureus* isolates. The three right side isolates have identical restriction endonuclease gel patterns. The strain on the far left has a single band change in the bottom half of the band pattern, indicating a genetic difference between it and the other strains. The seven bands in the farthest left lane represent molecular weight standard.

break the DNA into smaller fragments. There are a great many of these enzymes for which the specific recognition sequence and cleavage site have been defined. Figure 6.2 briefly outlines two methods used epidemiologically for nucleic acid analysis of bacterial strains. This method has the advantage of being highly reproducible, very accurate in determining relatedness of microbial strains, and well within the technical capabilities of experienced microbiology technologists.[3] Modifications of this basic technique have been developed to reduce the number of bands generated to less than 20 in an attempt to make the gels easier to interpret. These include pulsed-field electrophoresis and probing with short DNA fragments of ribosomal RNA. However, these modifications increase the difficulty of specimen preparation and testing and may not be as sensitive in detecting differences between microbial isolates as is the simpler method of genomic DNA restriction analysis. Restriction enzyme analysis may also be used for characterizing parasites, viruses, fungi, and mycobacteria (Table 6.1).

SCREENING POPULATIONS

When a hospital infection outbreak occurs, typically discovered from noting an increase in the number of isolates of a single microbial species, determining the extent to which the infecting strain has spread among patients may be necessary. As part of this effort, the laboratory may need to screen a large number of patients in a short period of time. It is important to choose culture sites and culture media to avoid unnecessary effort. Sites known to be colonized by a particular pathogen should be selected for culture. For ex-

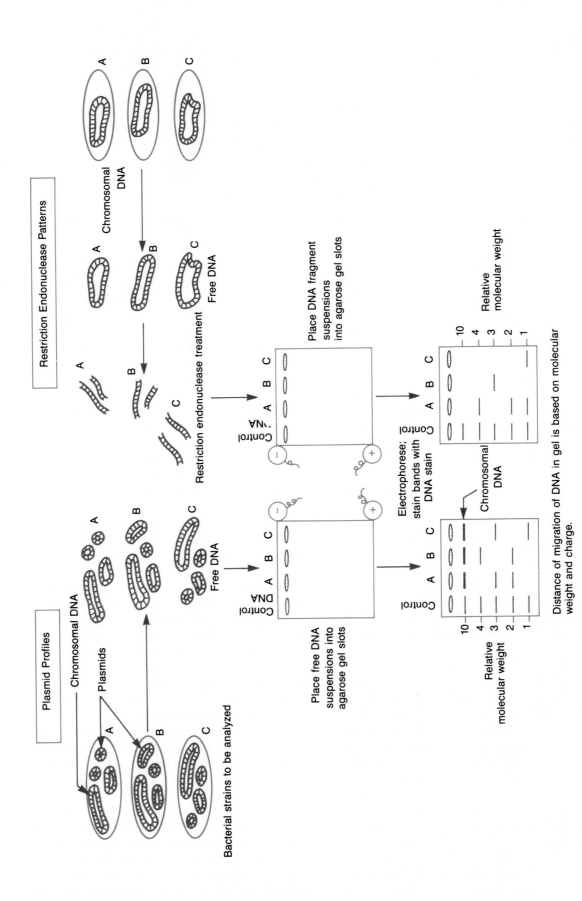

FIGURE 6.2

Two methods used to type bacteria based on genetic patterns.

TABLE 6.1

EPIDEMIOLOGICAL TYPING SYSTEMS MOST USEFUL FOR NOSOCOMIAL PATHOGENS

TYPING SYSTEM	MICROORGANISMS
Biotype	*Salmonella, Shigella,* other species of Enterobacteriaceae, *Pseudomonas,* fungi, *Staphylococcus*
Antibiogram	Enterobacteriaceae, *Pseudomonas* and other nonfermentative bacilli, *Staphylococcus*
Serotype	Viruses, Enterobacteriaeae, *Pseudomonas,* streptococci, *Legionella, Chlamydia*
Bacteriocin type	*Shigella, Pseudomonas, Serratia, Clostridium*
Bacteriophage type	*Staphylococcus, Pseudomonas, Salmonella, Mycobacterium, Clostridium*
Plasmid profiles	All bacteria, resistant strains
Restriction enzyme analysis	Herpes simplex virus, adenovirus, bacterial plasmids, bacterial DNA, HIV

Modified from Aber, R.C., and Mackel, D.C. 1981. Epidemiologic typing of nosocomial microorganisms. Am. J. Med. 70:899.

ample, in the case of a group B streptococcal outbreak, good sources of cultures would be the umbilicus, throat, and perhaps stool. In the case of a staphylococcal outbreak in a nursery setting, cultures of the umbilicus and nose would be appropriate. For enteric gram-negative bacilli, throat and stool are the major reservoirs. Enterococci are usually carried in the stool. Selective or differential media, or both, should be used to facilitate recovery of the organism being sought. In the case of an outbreak with an antibiotic-resistant organism, the microbiologist may incorporate an appropriate concentration of the drug or drugs in question in appropriate media. With this approach, testing each batch is important to be certain that it supports the growth of the epidemic strain being sought.

Occasionally culturing hospital personnel who may be involved in spreading an epidemic strain is desirable. It is usually necessary to culture only those individuals who are epidemiologically incriminated. Culturing large numbers of the medical staff indiscriminately is expensive, may yield misleading information, and causes anxiety or hostility.

Demonstration of bacteria on the hands of personnel may not only confirm the mechanism of cross-infection but also may serve as an important educational device to impress personnel with the importance of handwashing. A simple hand-culturing procedure is to have personnel press their fingertips, including nail tips, into the surface of nutrient agar media. Unfortunately, this technique is not very sensitive. Detection of hand colonization is much more likely if personnel vigorously wash their hands with 10 to 20 ml of a nutrient broth in an ethylene oxide sterilized plastic bag or plastic glove containing the broth. The broth is then inoculated onto agar by a semiquantitative method.

STORAGE OF STRAINS

Microbiology laboratories must save unique isolates from patients so that they may be retrieved for further study if an epidemic is recognized later. Such organisms should be retained for at least a month to give the infection control team adequate time to review the situation to determine whether further analysis is indicated. Such storage, of course, is also valuable to clinicians who may wish to have additional identification procedures or susceptibility testing performed. Organisms that are documented as being involved in nosocomial infections and organisms with unique susceptibility patterns should be saved for extended periods of time. Long-term storage of this type is best done by freezing at −70° C to −85° C since many antibiotic resistance genes are located on plasmids, which can be lost using other types of storage.

MONITORING

Bacteriological monitoring of personnel and of certain medical products or solutions is indicated on occasion, as noted previously.

PERSONNEL MONITORING

Some institutions require preemployment stool cultures on food handlers, which is of limited value. Stressing that food handlers submit cultures if they develop diarrhea is much more important. There should be no routine screening of personnel for nasal carriage of *S. aureus.* A significant percentage of the general population, as well as hospital personnel, carry this organism, but most individuals rarely shed enough organisms to pose a hazard and no simple way exists to predict which nasal carriers disseminate staphylococci. Culture of skin or other infections in members of the hospital staff is clearly an important function of the microbiology laboratory. The employee health service should be notified promptly of recovery of significant pathogens from such infections.

MONITORING OF HOSPITAL SUPPLIES AND ENVIRONMENT

Because there have been a few outbreaks of nosocomial infection related to contaminated intravenous fluids and other commercial products, some hospitals have considered routine sampling of products before use. This is not practical and is not recommended since this type of contamination typically involves only a small fraction of production lots, the level of contamination is quite low, and "routine sampling" would not be sufficient to detect such a small amount of contamination. The quality control provided by the manufacturer is usually much more extensive than would be practical in the hospital microbiology laboratory. The better approach is for the infection control team to monitor patients for the development of nosocomial infections that might be related to the use of contaminated commercial products. In the event of an outbreak or an incident related to suspected contamination, an appropriate microbiological study is indicated. Most often, such infections are actually caused by in-use contamination, rather than contamination during the manufacturing process. Suspect lots of fluid and catheter trays should be saved, and the Food and Drug Adminstration notified if contamination is suspected.

Some hospitals prepare their own parenteral solutions, especially hyperalimentation fluids. Microbiological testing is indicated as part of the quality control procedure. Since contamination of such fluids is generally rare and sporadic, it is not practical for most institutions to perform rigorous quality control to guarantee sterility of each lot that is produced. When culture is indicated, approximately 20 ml of fluid should be collected aseptically. A calibrated loop of 0.01 ml capacity may be used to inoculate the surface of a nutrient agar plate. The remaining fluid should be inoculated into a suitable nutrient broth in dilutions as low as 1:2. Vacuum-assisted filtration of the fluid through sterile nitrocellulose filters and subsequent incubation of the filter on agar is an alternative technique for detecting contamination in large volumes of fluids (>100 ml). Hospitals that prepare infant formula and those that bank human milk should perform microbiological studies to detect possible contamination. Several bottles of formula from each lot produced should be cultured by preparing a nutrient agar pour plate with 1 ml of milk. It has been suggested that no more than 25 colonies should be present after 48 hours of incubation, and none of these should be potential pathogens such as *S. aureus* or group A *Streptococcus*. In the event of a *Legionella* outbreak, environmental sampling should be done. In particular, the potable water supply, the air-conditioning cooling towers, and other elements of the air-conditioning system should be cultured. Details of suitable procedures may be found in the laboratory manual by Edelstein.[4]

REFERENCES

1. Aber, R.C., and Mackel, D.C. 1981. Epidemiologic typing of nosocomial microorganisms. Am. J. Med. 70:899.
2. Coudron, P.E., Mayhall, C.G., Facklam, R.R., Spadora, A.C., Lamb, V.A., Lybrand, M.R., and Dalton, H.P. 1984. *Streptococcus faecium* outbreak in a neonatal intensive care unit. J. Clin. Microbiol. 20:1044.
3. Peterson, L.R., Petzel, R.A., Clabots, C.R., Fasching, C.E., and Gerding, D.N. 1993. Medical technologists using molecular epidemiology as part of the infection control team. Diag. Microbiol. Infect. Dis. 16:303.
4. Edelstein, P.H. 1984. Legionnaires' disease laboratory manual. Document PB 84 156 827. National Technical Information Service, Springfield, Va.
5. McGowan, J.E. 1985. Role of the microbiology laboratory in prevention and control of nosocomial infections. In Lennette, E.H., Balows, A., Hausler, W.J. Jr., and Shadomy, H.J., editors. Manual of clinical microbiology, ed. 4. American Society for Microbiology, Washington, D.C.
6. McGowan, J.E. Jr., Weinstein, R.A., and Mallison, G.F. 1986. The role of the laboratory in control of nosocomial infections. In Bennett, J.V., and Brachman, P.S., editors. Hospital infections, ed. 2. Little, Brown & Co., Boston.
7. Valenti, W.M., et al. 1980. Nosocomial viral infection. I. Epidemiology and significance. Infect. Control 1:33.

BIBLIOGRAPHY

Archer, G.L., Dietrick, D.R., and Johnston, J.L. 1985. Molecular epidemiology of transmissible gentamicin resistance among coagulase-negative staphylococci in a cardiac surgery unit. J. Infect. Dis. 151:243.

Centers for Disease Control. 1986 (published annually). National nosocomial infection study report. U.S. Department of Health and Human Services, Public Health Service, Atlanta.

Coyle, M.B., and Schoenknecht, F.D. 1986. The clinical laboratory. In Bennett, J.V., and Brachman, P.S., editors. Hospital infections, ed. 2. Little, Brown & Co., Boston.

Favero, M.S., and Bond, W.W. 1991. Sterilization, disinfection, and antisepsis in the hospital. pp. 183-200. In Balows, A., Hausler, W.J. Jr., Herrmann, K.L., Isenberg, H.D., and Shadomy, H.J., editors. Manual of clinical microbiology, ed. 5. American Society for Microbiology, Washington, D.C.

Gilchrist, M.J.R., and Brooks, L.H. 1989. Nosocomial infections. In Davis, B.G., Bishop, M.L., Mass, D. editors. Clinical laboratory science: strategies for practice. J.B. Lippincott, Philadelphia.

Hayes, J.S., Soule, B.M., and LaRocco, M.T. 1987. Nosocomial infections: an overview. In Howard, B.J., Klaas, J. II, Rubin, S.J., Weissfeld, A.S., and Tilton, R.C. Clinical and pathogenic microbiology. Mosby, St. Louis.

Jarvis, W.R., White, J.W., Munn, V.P., Mosser, J.L., Emori, T.G., Culver, D.H., Thornsberry, C., and Hughes, J.M. 1983. Nosocomial infection surveillance, 1983. M.M.W.R. 33:9SS.

Mayer, L.W. 1988. Use of plasmid profiles in epidemiologic surveillance of disease outbreaks and in tracing the transmission of antibiotic resistance. Clin. Microbiol. Rev. 1:228.

Schaberg, D.R., and Culver, D.H. 1991. Major trends in the microbial etiology of nosocomial infection. Am. J. Med. 91(suppl 3B):72S. (Entire issue devoted to nosocomial infections.)

Shands, K.N., Ho, J.L., Meyer, R.D., Gorman, G.W., Edelstein, P.H., Millison, G.F., Finegold, S.M., and Fraser, D.W. 1985. Potable water as a source of Legionnaires' disease. J.A.M.A. 253:1412.

Smith, P.B. 1983. Biotyping—its value as an epidemiologic tool. Clin. Microbiol. Newsletter 5:165.

Wenzel, R.P. 1991. Epidemiology of hospital-acquired infection. pp. 147-150. In Balows, A., Hausler, W.J. Jr., Herrmann, K.L., Isenberg, H.D., and Shadomy, H.J., editors. Manual of clinical microbiology, ed. 5. American Society for Microbiology, Washington, D.C.

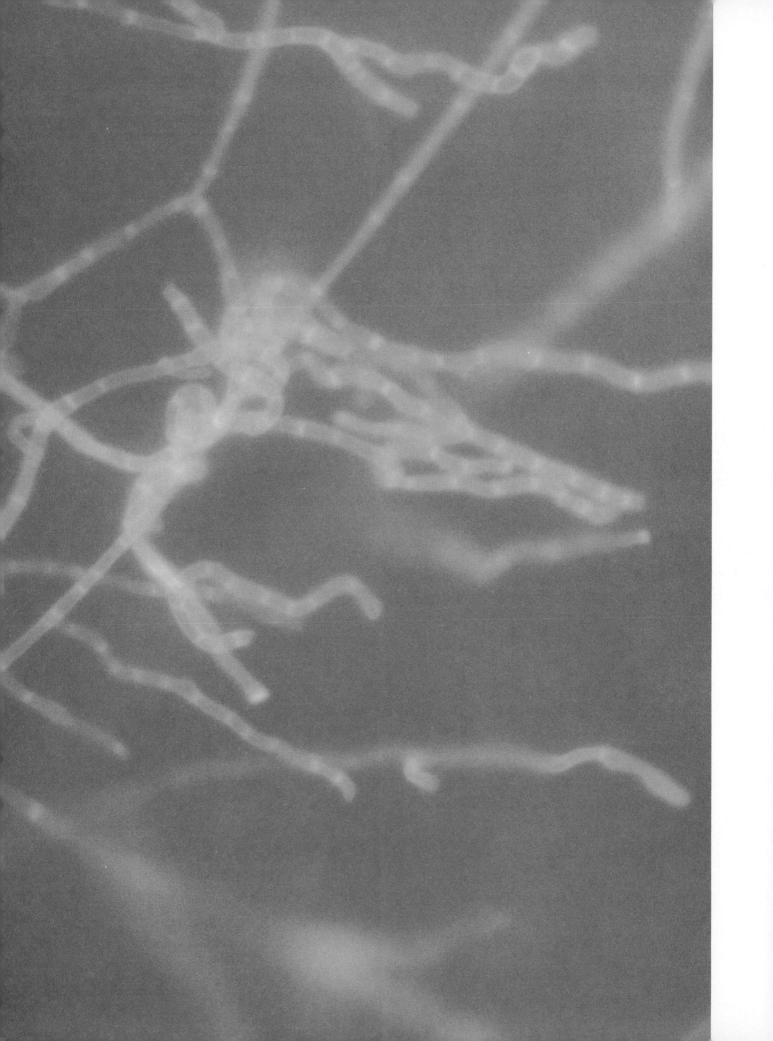

PART TWO

HANDLING CLINICAL

SPECIMENS FOR

MICROBIOLOGICAL STUDIES

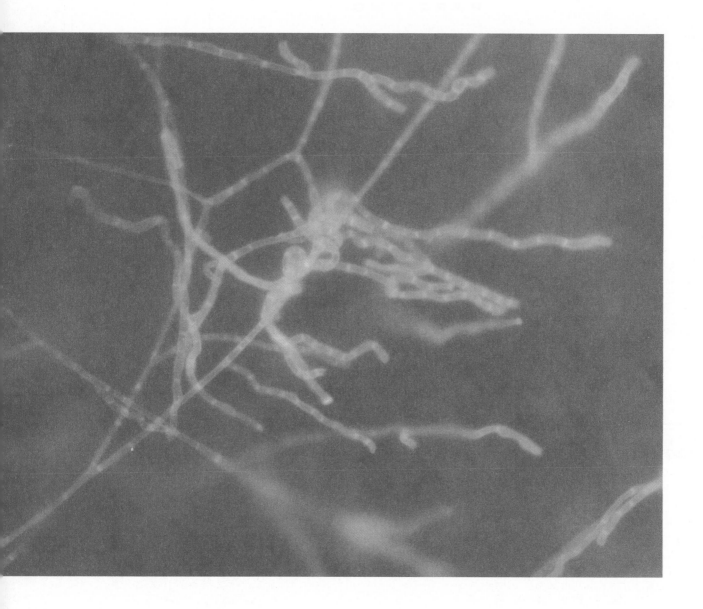

Rapid visual detection of
pathogenic microbes, such
as these septate hyphae
stained with calcofluor white
and viewed under UV light,
in clinical specimens can
provide guidance to
physicians during the critical
early stages of infectious
disease.

7

SELECTION, COLLECTION, AND TRANSPORT OF SPECIMENS FOR MICROBIOLOGICAL EXAMINATION

CHAPTER OUTLINE

SPECIMEN SELECTION

COLLECTION PROCEDURES
  Anaerobic collection procedures

SPECIMEN TRANSPORT
  Anaerobic specimen transport

SHIPPING SPECIMENS

HANDLING OF SPECIMENS IN THE LABORATORY

INOCULATION OF PRIMARY CULTURE MEDIA

SPECIMEN SELECTION

Specimens obtained for microbiological studies should be representative of the disease process, and sufficient material must be collected to ensure a complete and accurate examination. For example, serous drainage from the surface of a diabetic foot ulcer with underlying osteomyelitis usually does not contribute to determining the etiology of the bone infection. In the example cited, the ideal specimen would be a bone biopsy (to be studied histologically as well as bacteriologically). Although often it may not be feasible to obtain infected tissue, tissue is clearly the ideal specimen. Frankly purulent drainage is next in terms of desirability and is entirely satisfactory. In the case of a spreading lesion of the skin and subcutaneous tissue (such as progressive synergistic bacterial gangrene), material from the active margin of the lesion, rather than from the central portion of the lesion, is most likely to accurately reflect the true bacteriology of the process. Material obtained on a swab from a sinus tract opening often does not yield the true infecting organisms. A deep biopsy of the sinus tract would be much more reliable. Expectorated sputum, particularly if it is not a good purulent specimen (*Legionella pneumophila* is one exception) with minimal salivary contamination, presents major problems, particularly since pneumonia is often a serious infection. Ways around this problem include obtaining blood cultures, examining pleural fluid when present, screening the sputum specimen under 100× magnification to determine its quality (Chapters 4 and 8), obtaining bronchoalveolar lavage, and ultimately performing a transbronchial biopsy or other invasive procedure. Material from normally sterile sites in the body always provides an excellent specimen if care is taken to avoid contamination with skin flora.

Since anaerobic bacteria may be involved in infections of any type anywhere in the body, one should always culture anaerobically specimens of any variety that are free of contamination with normal flora. Certain specimens are essentially always contaminated with normal flora and therefore should not ordinarily be cultured anaerobically: throat swabs, nasopharyngeal swabs, gingival swabs, expectorated sputum, sputum obtained by nasotracheal or orotracheal suction, specimens obtained via a bronchoscope, gastric contents, small bowel contents (the latter two types of specimens may yield valuable information on culture in the case of "blind loop" and similar syndromes and gastric biopsy in the case of *Helicobacter pylori* infection), large bowel contents or feces (except for *Clostridium difficile* and *Clostridium botulinum*), ileostomy and colostomy effluents, voided or catheterized urine, and vaginal or cervical swabs (except as discussed under specimen collection from patients with endometritis).

COLLECTION PROCEDURES

Failure to isolate the causative organism in an infectious process is not necessarily the fault of inadequate cultural methods; it frequently results from faulty collecting or transport technique. The microbiologist must provide adequate supplies and proper instructions to ensure correctly collected samples. In-service programs to nurses and physicians and newsletter updates should be used to address this critical aspect of laboratory studies. The following are general considerations regarding the collection of material for culture. Specific instructions for the handling of a variety of specimens are given in subsequent chapters.

Whenever possible, specimens should be obtained **before antimicrobial agents have been administered.** Often cerebrospinal fluid (CSF) from a patient with bacterial meningitis reveals no bacterial pathogens on smear or culture when an antibiotic has been given within the previous 24 hours. A patient with salmonellosis may have a negative stool culture if the specimen has been collected while he or she was receiving antibacterial therapy that was only suppressive, only to reveal a positive culture several days after therapy has been terminated. If the culture has been taken after initiation of antibacterial therapy, the laboratory should be informed so that specific counteractive measures such as adding penicillinase or merely diluting the specimen may be carried out.

It is axiomatic that material should be collected where the suspected organism is **most likely to be found, with as little external contamination as possible.** The skin and all mucosal surfaces are populated with an indigenous flora and may often also acquire a transient flora or even become colonized for extended periods with potential pathogens from the hospital environment. The latter is particularly true of individuals who are quite ill, especially if they are receiving antimicrobial therapy (resistant organisms colonize as normal flora is suppressed). Accordingly, special procedures must be employed to help distinguish between organisms involved in an infectious process and those representing normal flora or "abnormal" colonizers that are not actually causing infection. Four major approaches are used to resolve this problem:

1. Cleanse skin surface with germicide using enough friction for mechanical cleansing as well. Start centrally and go out in ever enlarging circles. Repeat this several times, using a new swab each time. Alcohol (70%) is satisfactory for skin, but a full 2 minutes of wet contact time is needed. Iodine (2% tincture of iodine) and povidone-iodine work more quickly (1 minute) and are effective against spore-forming organisms. Collection of normally sterile body fluids (such as blood, joint, pleural, or cerebrospinal fluid) by percutaneous needle aspiration should always be preceded by thorough skin decontamination as described here.
2. Bypass areas of normal flora entirely (e.g., percutaneous transtracheal aspiration rather than coughed sputum).
3. Culture only for a specific pathogen (e.g., group A streptococci in the throat).
4. Quantitate culture results as a means of determining the likelihood of organisms being involved in infection (e.g., quantitative urine culture). Less formal quantitation may also be satisfactory and should **routinely** be used; this may involve grading on a scale of 1+ to 4+, or simply heavy growth, moderate growth, light growth, four colonies, and so forth.

Another factor contributing to the successful isolation of the causative agent is the **stage of the disease** at which the specimen is collected for culture. Enteric pathogens are present in much greater numbers during the **acute,** or diarrheal, stage of intestinal infections and are more likely to be isolated at that time. Viruses responsible for causing meningoencephalitis are isolated from CSF with greater frequency when the fluid is obtained soon after the **onset** of the disease rather than when the symptoms of acute illness have subsided. Subsequent chapters discuss the type of specimen most likely to contain the etiological agent during different stages of disease.

There are occasions when patients must participate actively in the collection of a specimen, such as a sputum sample. They should be given full instructions (Chapter 17), and cooperation should be encouraged by the caregiver. Too often a container is placed on the patient's bedside table, with the only instructions being to "spit in this cup."

Another specimen for which proper collection procedures are essential for reliable culture results is the clean-catch, midstream urine specimen. These specimens, usually collected by the patient without the assistance of a trained medical care worker, often comprise a large portion of the specimens received by a clinical microbiology laboratory. Microbiologists may be asked to help prepare guidelines for proper specimen collection. The use of a printed card (bilingual, if necessary) with the procedure clearly described and preferably illustrated can help to ensure patient compliance. Separate cards should be given to males and females. Careful patient education should improve the quality of such urine specimens received by the laboratory. Specimens should be of a **sufficient quantity** to permit complete examination and should be placed in sterile containers that avoid hazard to the patient, nurse, or ward messenger. A serious danger to the laboratory worker, as well as to all others involved, is the soiled outer surface of a sputum container, a leaking stool sample, or possible contact with blood-containing exudates. The hazard of spreading disease by inadequately trained nonprofessional workers is frequently overlooked. Its control requires continued education and constant vigilance by those in responsible and supervisory positions.

Provision must be made for the **prompt delivery** of specimens to the laboratory if the results of analysis are to be valid. For example, isolating *Shigella* from a fecal specimen is difficult when it has remained on the hospital ward too long, permitting overgrowth by commensal organisms and an increasing death rate of the shigellae. In some instances it may be necessary to take culture (room temperature) media and other equipment to the patient's bedside to ensure prompt inoculation of the specimen. This is an unusual circumstance and requires prior arrangements with the laboratory.

Although most pathogenic microorganisms are not greatly affected by small changes in temperature, they are generally susceptible to drying out, particularly when on cotton applicator sticks. However, some bacteria, such as the meningococcus in CSF, are quite sensitive to low temperatures and require immediate culturing.

Specimens for gonococci can be inoculated directly onto selective media such as modified Thayer-Martin medium with carbon dioxide provided in a bottle (Transgrow) or by placing a generating tablet in a chamber (Jembec; Chapters 9 and 20). Alternate methods include placing inoculated Thayer-Martin plates immediately into a candle jar.

Clinical material likely to contain abundant microbial flora may be held in most instances at 5° C in a refrigerator for several hours before culturing if it cannot be processed right away. This is particularly true with specimens such as urine, feces, and material on swabs taken from a variety of sources, with the exception of wound cultures, which may contain oxygen-sensitive anaerobes.

These should be inoculated promptly. Not only will refrigeration preserve the viability of most pathogens, but it will also minimize overgrowth of commensal organisms, increased numbers of which could make the isolation of a significant microbe more difficult. However, the sooner an organism leaving the sheltered environment of its host is transferred to an appropriate artificial culture medium, the better are the chances of its survival and subsequent multiplication.

Although not a function of specimen collection, a prerequisite that the **laboratory be given sufficient clinical information** to guide the microbiologist in selection of suitable media and appropriate techniques is essential. Likewise, the clinician must appreciate the **limitations and potentials** of the bacteriology laboratory and realize that a negative report does not necessarily invalidate the diagnosis. Close cooperation and frequent consultation must occur among the clinician, nurse, and microbiologist.

Laboratory personnel should reject specimens not obtained in a proper manner and should be supported in this position by infectious disease clinicians or pathologists. In rejecting specimens, of course, the reasons should be explained to the requesting physician. Specimens should never be discarded before discussion with the requesting clinician. Some specimens, such as those taken at the time of surgery, are difficult or impossible to replace. Those specimens that cannot be replaced should be Gram stained and interpreted as carefully as possible. Guidelines for specimen rejection are discussed in Chapter 3 (Table 3.1).

ANAEROBIC COLLECTION PROCEDURES

Proper collection, or taking care to avoid inclusion of normal flora, cannot be overemphasized because indigenous anaerobes are often present in such large numbers that even minimal contamination of a specimen with normal flora can give very misleading results and cause much wasted effort.

Coughed sputum is unsuitable because it becomes contaminated with normal flora anaerobes on its passage through the mouth and pharynx. For the same reason, bronchial washings or those obtained by nasotracheal tube suctioning also should not be cultured anaerobically. The sampling tube always contacts normal flora on its downward path. The efficacy of a double-lumen plugged catheter in preventing such contamination is such that it is possible to obtain reliable results for diagnosis of bacterial pneumonia by careful attention to detail in carrying out these procedures and by doing quantitative cultures (including anaerobes). Adequate pleuropulmonary specimens for anaerobic culture can be obtained by transtracheal aspiration (TTA), thoracentesis, or direct percutaneous needle puncture and aspiration of lung. Tracheostomy tube specimens may provide useful material when the tube is first placed; when it has been in place for a while there is inevitable contamination with oropharyngeal secretions, whether or not there is an inflated cuff.

Voided urine specimens are unsuitable for anaerobic culture because the distal portion of the urethra and the meatus are colonized with a normal flora containing anaerobes, which will contaminate urine passing through these areas. Percutaneous aspiration of urine from a full bladder provides a reliable specimen.

Endometritis presents a very difficult problem. Anaerobes are clearly very important in this infection. However, most cases of endometritis follow childbirth, and it has been demonstrated that in the postpartum period, **whether or not there is endometrial infection,** significant numbers of anaerobes and other organisms from the cervical and vaginal flora may be found in the uterine cavity. There may not even be quantitative differences between infected and uninfected patients. This situation may apply also in postabortal endometritis. Thus one should obtain blood cultures and culture any better sources of material that may be available. Culture for the *Bacteroides fragilis* group may be useful. Since the presence of the *B. fragilis* group in this type of specimen indicates a poorer prognosis and since this organism is more resistant to antimicrobials than other anaerobes, it makes sense to determine whether it is present. Biopsy of endometrial tissue obtained with an endometrial suction curette (Pipelle; Unimar, Inc.) or collection of uterine contents using a protected swab will provide a satisfactory specimen.[5,7]

Infections of decubitus ulcers commonly involve anaerobic organisms, particularly when the decubitus is in the vicinity of the anus (sacral decubiti and decubiti of the hips and buttocks). Since these areas are subject to fecal contamination (this is how they become infected), the area must be thoroughly cleansed with an antiseptic agent before cultures are taken. Cultures from the base of the lesion may be obtained after thorough debridement of surface debris. Whenever possible, one should aspirate collections of pus from under skin flaps or from deep pockets, using a syringe and a needle. The same considerations apply to other lesions in these areas (e.g., perirectal abscess).

Specimens that are normally sterile (such as spinal fluid and blood and joint fluid) may be collected in the usual fashion after thorough skin decontamination.

In general, material for anaerobic culture is best obtained by tissue biopsy or by using a needle and syringe. All air must be expelled from the syringe and needle. Use of swabs is a poor alternative because of excessive exposure of the specimen to the deleterious effects of oxygen and drying.

SPECIMEN TRANSPORT

The container bearing a specimen must not contribute its own microbial flora. Furthermore, the original flora should neither multiply nor decrease because of prolonged standing on the ward or prolonged refrigeration in the laboratory. In other words, **a sterile container should be used and the specimen should be plated as soon as possible.** Although these are not hard-and-fast rules, any deviation should be the responsibility of the microbiologist. A variety of containers have been devised for collecting bacteriological specimens. Many of these can be used repeatedly after proper sterilization and cleaning, whereas others must be incinerated after use. Apart from the sterile disposable plastic cup, the most used (but not necessarily the most desirable) piece of collecting equipment is the cotton-, calcium alginate-, or Dacron (polyester)-tipped swab, usually with a plastic shaft. Wooden shafts are best avoided for specimen collection because they may contain toxic products. Commercial systems are available (Figure 7.1) consisting of a plastic tube containing a sterile polyester-tipped swab and a small glass ampule or pouch of modified Stuart's holding medium. The unit is removed from its sterile envelope, and the swab is used to collect the specimen. It is then returned to the tube, and the holding medium is brought into contact with the swab. Such systems will provide sufficient moisture for storage up to 72 hours at room temperature.

Some cotton used for applicators may contain fatty acids that may be detrimental to microbial growth.[8] An excellent substitute is calcium alginate wool, derived from alginic acid, a natural plant product. Calcium alginate–tipped wooden applicators or flexible aluminum nasopharyngeal swabs (Figure 7.2) are available commercially. Calcium alginate should not be used where her-

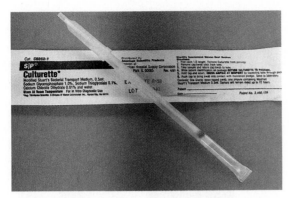

FIGURE 7.1

Sterile specimen collection unit consisting of a polyester swab with a plastic shaft inserted into a tight-fitting cap within a sterile tube (Culturette, Becton-Dickinson Microbiology Systems). After the specimen is collected, the glass ampule containing transport medium is broken to moisten and protect the specimen on the swab.

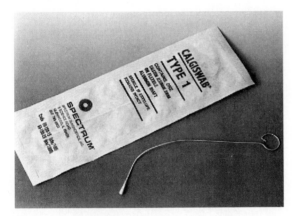

FIGURE 7.2

Calcium alginate nasopharyngeal swabs on a flexible wire that can be bent to conform to the shape of the nasal passage (Spectrum Diagnostics).

pesvirus is anticipated, because it may inhibit replication of this virus, or when swabs will be used for subsequent direct antigen detection, because the calcium alginate will interfere with the extraction reagents. Follow manufacturer's instructions when swabs are used in conjunction with commercial systems for antigen detection or other extraction procedures.

A modification of the applicator uses 28-gauge Nichrome or thin aluminum wire in place of the plastic stick. Because of its thinness and flexibility, the wire applicator is recommended for collecting specimens from the nasopharynx or the urethra.

A variety of transport media have been devised for prolonging the survival of microorganisms when a significant delay occurs between collection and culturing. Stuart and others[1,10] advo-

cated a medium that has proved effective in preserving the viability of pathogenic agents in clinical material. The medium consists of buffered semisolid agar devoid of nutrients and containing sodium thioglycolate as a reducing agent and is used in conjunction with swabs. Stuart's medium (available commercially) maintains a favorable pH and prevents both dehydration of secretions during transport and oxidation and enzymatic self-destruction of the pathogen present. However, the glycerophosphate present permits multiplication of certain organisms.

Cary-Blair transport medium[4] (Figure 7.3) allows recovery of salmonellae and shigellae, *Vibrio cholerae*, and *Campylobacter* from fecal specimens for adequate time periods; *Yersinia pestis* survives for at least 75 days. The medium itself remains good after storage at room temperature for longer than 1½ years.

Polyester-tipped swabs, even when dried, may be used for delayed recovery of group A streptococci from throat cultures. A report by Hosty and co-workers[6] indicates that the incorporation of a small amount of silica gel in the glass tube containing the polyester swab will further maintain the viability of group A streptococci in dry throat swabs for as long as 3 days before plating. Other transport media including Amies charcoal medium (good for *Neisseria gonorrhoeae* and other fastidious organisms) have been used successfully.[1,9]

ANAEROBIC SPECIMEN TRANSPORT

The collection of specimens for anaerobic culturing poses a special problem in that the conventional methods previously described will not lead

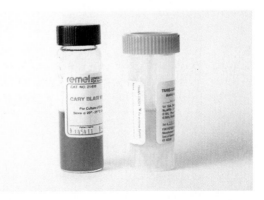

FIGURE 7.3

Small vial containing Cary-Blair medium for transport and maintenance of fecal specimens for culture and second vial containing PVA for transport of feces for parasitic studies (Remel Laboratories).

to optimal recovery of these air-intolerant microorganisms. A crucial factor in the final success of anaerobic culturing is the transport of clinical specimens; the lethal effect of atmospheric oxygen must be nullified until the specimen has been processed anaerobically in the laboratory. We recommend using a rubber-stoppered collection tube or vial, gassed out with oxygen-free CO_2 or nitrogen and containing an agar indicator system (Figure 7.4). One injects the specimen (pus, body fluid, or other liquid material) through the rubber stopper after first expelling all air from the syringe and the needle. In the laboratory, the material is aspirated from the transport container by needle and syringe, inoculated to media, and incubated under anaerobic conditions. If only a swab specimen can be obtained, it should be placed immediately into a rubber-stoppered tube with a deep column of prereduced and anaerobically sterilized transport medium (Figure 7.5).

A plastic collection device with its own anaerobic transport medium (Anaerobic AccuCul-Shure, Medical Laboratory Automation, Inc., Pleasantville, N.Y.) has been found to maintain anaerobic specimens adequately.[3] Tissue placed into loosely sealed plastic tubes or a Petri dish in a Bio-Bag (Becton-Dickinson Microbiology Systems), Anaerobic Pouch (EM Diagnostic Systems, Gibbstown, N.J.), or GasPak Pouch (BBL Microbiology Systems) (Figure 7.6) is an excellent transport method for anaerobic culture.

The materials recommended for anaerobic transport actually constitute ideal **universal transport setups,** since **all** types of microorganisms should survive well in them. It should be stressed that, although the swab is the most widely used transport vehicle, it is preferable to

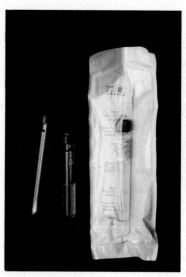

FIGURE 7.5

Examples of additional transport methods for anaerobic specimens. Tube on left contains swab in oxygen-free atmosphere (prepared in-house). Sterile pack on right contains screwcap tube with agar deep suitable for inserting swab (Becton-Dickinson Microbiology Systems). Tube in middle incorporates both agar deep and screwcap with a rubber septum (Anaerobe Systems) suitable for either swab or injected liquid specimen. (Photograph by Pete Rose.)

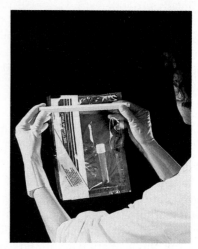

FIGURE 7.6

Use of anaerobic self-contained atmosphere generating bag system (Becton-Dickinson Microbiology Systems) to transport tissue. Cap of tube is left loose until indicator indicates anaerobiosis in the bag.

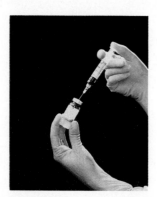

FIGURE 7.4

Anaerobic transport vial containing agar with oxygen tension indicator (Anaerobe Systems, San Jose, Calif.). Liquid specimen is injected into tube through the rubber septum.

submit a larger specimen whenever possible (e.g., aspirated pus or tissue). When organisms may be scarce (as in some forms of tuberculosis), the larger the specimen the better. Throughout this text, specimen collection and handling methods are discussed in relation to the disease process or the etiological agent. Procedures should be chosen based on the needs of the patients served by the laboratory.

SHIPPING SPECIMENS

On occasion it is necessary to submit specimens to a **reference laboratory** in a distant city, thus requiring transportation by mail or express. If, for example, there is no viral diagnostic service available in the immediate area, shipping specimens at a low temperature will preserve their viability. This is especially true of virus-containing material, such as CSF, throat and rectal swabs, stools, and tissue, which should be refrigerated immediately and shipped in a Styrofoam box with commercial refrigerant packs (e.g., 3M Cryogel). Specimens should not be frozen. Whole blood is **not** shipped. Rather, the serum is separated and sent in a sterile tube. If a culture of an isolated organism is to be sent to a reference laboratory (state or public health laboratory), it is not necessary to send it in the frozen state.

In most instances microbiological specimens can be satisfactorily shipped through the mail, provided that **special precautions** are taken against breakage and subsequent contamination of the mailing container. According to the Code of Federal Regulations,* a viable organism or its toxin or a diagnostic specimen of a volume less than 50 ml shall be placed in a securely closed, watertight specimen container, which is enclosed in a second durable watertight container, further described in Chapter 2.

Double mailing containers and standardized biohazard labels (see Figures 2.5 and 2.6) must be used when the microbiologist considers the specimen to be of a hazardous nature or thinks it may constitute a definite danger to the person handling the container. This includes clinical specimens, cultures for identification, and any potential biohazardous materials.

For the shipment of fecal specimens containing salmonellae or shigellae over long distances, the filter paper method may also be employed.[2] In collecting the specimen by this technique, fresh fecal material must be spread fairly thinly over a strip of filter paper or clean blotting paper and allowed to dry at room temperature. Using forceps, the smeared strip is then folded inward from the ends in such a way that the fecal material is covered. The folded specimen may then be inserted in a plastic envelope (polyethylene is recommended) and placed in a container to conform with postal regulations. It is possible to ship a large number of such specimens in one container through the mail. The pathogens are unaffected by this treatment, whereas the normal intestinal flora dies. On receipt at the laboratory the paper specimen may be cut into three pieces. One piece is placed in physiological saline for suspension and direct plating, and the remaining pieces are placed in selenite and tetrathionate broths, respectively, for enrichment and subsequent plating. When it is necessary to ship fecal specimens or when such specimens must be held for some time before culturing, they should be placed in a preservative solution. Cary-Blair medium has proved satisfactory. It should be discarded if it becomes acid. Approximately 1 g of feces is emulsified in not more than 10 ml of preservative, and the preserved specimen is shipped in a heavy glass container (universal type) with a screw cap in the regular double mailing container just described.

HANDLING OF SPECIMENS IN THE LABORATORY

In previous sections of this book the importance of a properly collected specimen was stressed and the responsibility of personnel in its collection was indicated. In the following section the subsequent handling of the specimen in the laboratory is considered. In addition, the responsibility of the laboratory worker is pointed out.

Since it is not always practical for many specimens to be inoculated as soon as they arrive in the laboratory, refrigeration at 4° C to 6° C offers a safe and dependable method of storing many clinical samples until they can be conveniently handled. However, some may require immediate plating (e.g., specimens that might contain gonococci or *Bordetella pertussis*), whereas others must be immediately frozen (e.g., serum for subsequent antimicrobial agent assay). The length of time of refrigeration varies with the type of specimen: swabs from wounds (except for anaerobic cultures), the urogenital tract (although *N. gonorrhoeae* is susceptible to cold), throat, and rectum, and samples of feces or sputum can be refrigerated for 2 to 3 hours without appreciable loss of pathogens. Urine specimens for culture may be refrigerated at least 24 hours without affecting the bacterial flora (except the tubercle bacillus, which may be adversely affected by the urine); on the other hand, CSF from a patient with suspected meningitis should be examined **at once.**

Specimens submitted for the isolation of virus should be refrigerated immediately; even for storage up to 5 days, they should be refrigerated and **never frozen.** Specimens of clotted blood for virus serology may also be refrigerated but never frozen, as discussed in Chapter 43.

---

* Section 72.25 of Part 72, Title 42, amended.

TABLE 7.1

SELECTION OF PRIMARY PLATING MEDIA FOR BACTERIOLOGY SPECIMENS

SPECIMEN	DIRECT SMEAR	ROUTINE MEDIA	SUGGESTED ADDITIONAL MEDIA	SPECIAL SITUATIONS	COMMENTS
• Abscess/pus* (closed wound)	+	B, M, An, Thio	As smear indicates; CNA		Wash any granules and "emulsify" in saline
• Autopsy Blood, tissue			Follow procedure for "living" material		
Blood Peripheral blood		Choc, An	As smear indicates	Brucellosis, tularemia, cell wall–deficient bacteria, leptospirosis, mycobacteria	Smear when evidence of growth
• Bone marrow		B, M, Choc			
Body fluids (except blood, CSF, urine)		Centrifuge fluids for 15 min at 2500 rpm except when grossly purulent. Stain and culture sediment. Incubate supernatant for 24 h at 37° C. Filtration is also useful if the filter can be cultured directly.			
Bile	+	B, M, An, Thio	CNA, HE, XLD, GN		
Hematoma	+	Treat as abscess	CNA	PRAS if anaerobes suspected	Do not use An if collected perorally
Joint	+	B, Choc, Thio		Brucellosis	
Pericardial, peritoneal	+	B, M, An, Choc, Thio	CNA		
Pleural empyema	+	B, M, An, Choc, Thio	CNA	Legionella, Chlamydia in infants, Nocardia, Actinomyces, mycobacteria	
Catheters Foley		Do not culture			
• Central venous pressure lines, umbilical or intravenous catheters		B	Thio		Roll segment back and forth across agar with sterile forceps 4 times; ≥15 colonies are associated with clinical significance(see Chapter 15)
• Central nervous system Brain tissue	+	Treat as abscess			

B, blood agar; M, MacConkey agar; An, anaerobic blood agar; Choc, chocolate agar; Thio, thioglycolate broth; TM, Thayer-Martin agar; CNA, Columbia agar with colistin and nalidixic acid; HE, Hektoen enteric agar; XLD, xylose lysine deoxycholate agar; GN, gram-negative broth; HBT, human blood–Tween bilayer agar; Campy, Campylobacter agar; PRAS, prereduced anaerobically sterilized media.

* Bullets. (•) before entry indicate specimens that require immediate attention.

Modified from Isenberg, H.D., Schoenknecht, F.D., von Graevenitz, A., and Rubin, S.J. 1979. Collection and processing of bacteriological specimens. Cumitech 9. American Society for Microbiology. Washington, D.C.

Specimen		Media (routine)	Media (additional)	Also culture for	Comments
CSF, shunt, meningomyelocele, and ventricular fluid	+	If large enough volume, centrifuge for 15 min at 2500 rpm at 37° C. B, Choc, Thio	As smear indicates; M, CNA, An	Leptospirosis, mycobacteria	If large enough volume, centrifuge for 15 min at 2500 rpm and smear and culture sediment. Incubate supernatant for 48 h. *Haemophilus influenzae* may not sediment at 2500 rpm and may require 10,000 × g for 10 min. Alternatively, culture entire specimen
Ear					
Internal	+	B, M, Choc, Thio	CNA	An if aspiration or biopsy material	
External	+	B, M. Choc	CNA		
Eye				*Acanthamoeba, Chlamydia*	Direct inoculation of media by physician is recommended
Conjunctiva	+	B, M, Choc, Thio	TM, CNA		
Other	+	B, M, Choc, An, Thio	CNA		
Genital tract—female					
• Amniotic fluid	+	B, M, TM, Choc, An, Thio	CNA		Should be collected without contamination by normal flora
Cervix	±	TM	Choc with 1% Iso-VitaleX	*Chlamydia*	A small number of *Neisseria gonorrhoeae* are susceptible to vancomycin. Choc with 1% IsoVitaleX, especially when there is clinical evidence of gonorrhea and negative cultures (see Chapter 20)
• Cul de sac	+	B, M, Choc, TM, An, Thio	CNA, HBT		Anaerobic culture is done on material collected without contamination by vaginal flora
• Uterine material (endometrium, products of conception: fetus, placenta, lochia), tubes, ovaries	+	B, M, Choc, TM, An, Thio	CNA, HBT		
Intrauterine device		Thio			
Urethra	±	TM	Choc with 1% Iso-VitaleX	*Mycoplasma, Chlamydia* (nongonococcal urethritis)	Smears of urethral exudate that are positive for intracellular gram-negative diplococci from females are presumptively diagnostic and must be confirmed by culture
Vagina (except cuff)	±	TM	Choc with 1% Iso-VitaleX, B	Selective group B streptococcal media in pregnancy	Vaginal cuff infections are treated as abscess/pus. Gram stain for bacterial vaginosis
• Vulva (Bartholin abscess)	+	B, M, TM, An, Thio	CNA, Choc with 1% Iso-VitaleX, HBT		Aspirated specimen
Genital tract—male					
Prostate fluid	+	B, M, TM, Thio	CNA		
Testes and epididymis (aspirate)	+	B, CNA, TM, M			
Urethra	+	TM		*Chlamydia* (nongonococcal urethritis), *Mycoplasma*	

Continued.

TABLE 7.1

SELECTION OF PRIMARY PLATING MEDIA FOR BACTERIOLOGY SPECIMENS—cont'd

SPECIMEN	DIRECT SMEAR	ROUTINE MEDIA	SUGGESTED ADDITIONAL MEDIA	SPECIAL SITUATIONS	COMMENTS
Intestinal tract					
Colostomy, ileostomy, feces, rectal swab	Presence of cells only (polymorphonuclear leukocytes)	B, M, HE, XLD, GN, Campy (42° C)		*Yersinia enterocolitica, Vibrio cholerae, V. parahaemolyticus, Neisseria gonorrhoeae* (rectal swab), *Mycobacterium avium* complex	
Gastric aspirate	+	B, M, Choc	CNA	Mycobacteria	In young child, may be submitted in place of sputum
Plastic prostheses	–	Thio			
Pus—closed wound		See Abscess/pus			
Respiratory tract					
Throat/pharynx	–	Streptococcal selective agar		*Legionella* Vincent's angina, *N. gonorrhoeae, N. meningitidis, Corynebacterium diphtheriae, Arcanobacterium haemolyticum*	Routine culture for group A streptococci only. May be submitted instead of sputum in cystic fibrosis
Epiglottis	–	B, Choc	CNA		
Nasal sinus	+	B, M, Choc, An, Thio	CNA		
Nasopharynx	–	B, Choc	CNA	*Bordetella pertussis*	
Nose	–	B, CNA	Mannitol salt		
Pleural fluid		See Body fluids			
Sputum	+	B, M, Choc	CNA	*Mycoplasma, Nocardia,* mycobacteria	See Chapter 17
Bronchial secretions	+	B, M, Choc	CNA	Mycobacteria	
Tracheal aspirate	+	B, M, Choc	CNA		
• Transtracheal aspirate	+	B, M, Choc, An, Thio	CNA	*Mycoplasma,* mycobacteria	
Oral cavity	+				
Dental abscess	+	B, M, An, Thio	As smear indicates CNA, Mannitol salt		
Skin					
• Deep wound (open)	+	B, M	CNA, Choc	Petechiae/pustules (*Neisseria*)	

Specimen		Media		Comments
Superficial wound	+	B, M		Quantitative culture of burn tissue may be useful (see procedure in Chapter 22)
Traumatized areas (burns, bites, decubitus ulcers)	+	B, M, Thio	As smear indicates: CNA, An	Foul-smelling or gas-containing lesions; culture anaerobically
Rash, nonpurulent lesions	+	B, M, Choc, Thio	CNA	Smear may be diagnostic, especially in meningococcemia
Tissue		Grind with sterile Alundum (aluminum oxide) with a sterile mortar and pestle or sterile tissue grinder or homogenize (Chapter 22). Use enough broth to give a 10%-20% suspension.		
Surgical/biopsy	+	B, M, An, Thio	CNA	Other media on the basis of history, body site, and smear (i.e., tularemia or brucellosis)
Urine				
Clean-voided, catheterized or ileal loop urine	±	B, M (see Chapter 19)	CNA	Leptospirosis, mycobacteria. Gram stain on unspun urine may be useful in selected patients.
Suprapubic bladder tap; cystoscopy or ureterostomy urine	+	B, M, An, Thio (0.1 ml per plate)	CNA	

Gastric washings and resected lung tissue submitted for culture of *Mycobacterium tuberculosis* should be processed soon after delivery, since tubercle bacilli may die rapidly in either type of specimen. Alternatively, gastric specimens and urine may be adjusted to neutral pH if some storage is necessary before processing. Further discussion of handling material for mycobacterial culture is found in Chapter 42.

Pieces of hair or scrapings from the skin and nails submitted for the isolation of fungi may be kept at room temperature (protected from dust) for several days before inoculation. On the other hand, sputum, bronchial secretions, bone marrow, and purulent material from patients suspected of having systemic fungal infection should be inoculated to appropriate media as soon as possible, especially when the diagnosis of histoplasmosis is considered. Further guidelines for collection of specimens for fungal isolation are given in Chapter 44.

INOCULATION OF PRIMARY CULTURE MEDIA

Depending on the etiological agents suspected, specimens are plated and inoculated to several growth media. Special media for unusual agents (e.g., Bordet-Gengou or Jones-Kendrick charcoal agar for *B. pertussis*) are inoculated by request. Laboratories in some geographical areas may choose to inoculate routinely media that other laboratories could not justify economically (e.g., selective agar for *E. coli* H7:0157 or *Yersinia enterocolitica*). Table 7.1 suggests one approach to the inoculation of specimens for routine bacteriological studies. As discussed in Chapters 15 through 24, etiological agents other than bacteria (fungi, viruses, parasites) must often be considered and sought in clinical specimens.

REFERENCES

1. Amies, C.R. 1967. A modified formula for the preparation of Stuart's transport medium. Can. J. Public Health 58:296.
2. Bailey, W.R., and Bynoe, E.T. 1953. The "filter paper" method for collecting and transporting stools to the laboratory for enteric bacteriological examination. Can. J. Public Health 44:468.
3. Baron, E.J., Väisänen, M.-L., McTeague, M., et al. 1993. Comparison of Accu-CulShure system and swab placed in B-D Port-a-Cul tube for culture collection and transport. Clin. Infect. Dis. 16(suppl. 4): S235.
4. Cary, S.G., and Blair, E.B. 1964. New transport medium for shipment of clinical specimens. J. Bacteriol. 88:96.
5. Eschenbach, D.A., Rosene, K., Tompkins, L.S., et al. 1986. Endometrial cultures obtained by a triple-lumen method from afebrile and febrile postpartum women. J. Infect. Dis. 153:1038.
6. Hosty, T.S., Johnson, M.B., Freear, M.A., et al. 1964. Evaluation of the efficiency of four different types of swabs in the recovery of group A streptococci. Health Lab. Sci. 1:163.
7. Martens, M.G., Faro, S., Hammill, H.A., et al. 1989. Transcervical uterine cultures with a new endometrial suction curette: a comparison of three sampling methods in postpartum endometritis. Obstet. Gynecol. 74:273.
8. Pollock, M.R. 1948. Unsaturated fatty acids in cotton plugs. Nature 161:853.
9. Sautter, R.L. and Wilson, M.T. 1988. Specimen transport containers are not created equal. Clin. Microbiol. Newsletter 10:181.
10. Stuart, R.D., Tosach, S.R., and Patsula, T.M. 1954. The problem of transport of specimens for culture of gonococci. Can. J. Public Health 45:73.

BIBLIOGRAPHY

Forney, J.E., editor. 1968. Collection, handling, and shipment of microbiological specimens. Public Health Serv. Pub. No. 976, Nov. 1968. U.S. Government Printing Office, Washington, D.C.

Higgins, M. 1950. A comparison of the recovery rate of organisms from cotton-wool and calcium alginate wool swabs. Ministry Health Public Lab. Serv. Bull. 43.

Isenberg, H.D., Washington, J.A. II, Doern, G.V., and Amsterdam, D. 1991. Specimen collection and handling. pp. 15-28. In Balows, A., Hausler, W.J., Jr., Herrmann, K.L., Isenberg, H.D., and Shadomy, H.J., editors. Manual of clinical microbiology, ed 5. American Society for Microbiology, Washington, D.C.

Rubin, S.J. 1987. Specimen collection and processing. In Howard, B.J., Klass, J. II, Rubin, S.J., et al. Clinical and pathogenic microbiology. Mosby, St. Louis, Mo.

Shea, Y.R. 1992. Specimen collection and transport. Sect. 1.1. In Isenberg, H.D., editor. Clinical microbiology procedures handbook. American Society for Microbiology, Washington, D.C.

Summanen, P., Baron, E.J., Citron, D.M., et al. 1992. Wadsworth anaerobic bacteriology manual, ed 5. Star Publishing, Belmont, Calif.

Washington, J.A. II. 1985. Bacteria, fungi and parasites. In Mandell, G.L., Douglas, R.G. Jr., and Bennett, J.E., editors. Principles and practice of infectious diseases, ed 2. John Wiley & Sons, New York.

8

OPTICAL METHODS FOR LABORATORY
DIAGNOSIS OF INFECTIOUS DISEASES

Since the observations of van Leeuwenhoek during the late seventeenth century, in which the presence of infectious agents smaller than the human eye could see was first described, the light microscope has become a most important tool for the diagnosis of infection. Even today, with the emphasis on rapid methods for diagnosis, many of which require complicated instruments or immunological reagents, direct visual inspection of clinical specimens obtained from patients is still the fastest and most specific way to immediately augment a physician's clinical diagnosis.[2] Many infectious agents can be reliably identified with only a few simple stains and a basic microscope. The Gram stain is still the single most efficient and cost-effective test for rapid early diagnosis of bacterial infection. Methods discussed in this chapter include many common visual techniques that have been employed by microbiologists for decades, as well as a few relatively recent innovations in microscopy.

EXAMINATION OF FRESH MATERIAL

Most microscopic examinations of material are carried out with **brightfield microscopy,** in which the object to be viewed is illuminated with light from below the field of focus, usually pro-

vided by a coiled filament tungsten lamp. The glowing filaments are prevented from causing glare by focusing their light on the substage condenser, rather than on the object. This is known as **Köhler illumination,** named for the inventor of the method. Since most pathogens visible with the microscope are of a refractive index similar to that of aqueous suspending liquids, they are invisible unless they or the technique are modified. Closing the aperture of the condenser slightly may aid in detection of certain organisms, particularly protozoans and fungi, because the light will be hitting the edges of the object at a sharper angle, increasing contrast. For Köhler illumination, which ensures the best brightfield viewing of unstained clinical material, the condenser should be properly focused. This is achieved by placing a stained slide on the stage and viewing it through the 10× objective. After the radiant field diaphragm (the control in the base of the microscope where the light source is) is stopped down, the condenser is moved up or down until the leaves around the edge of the diaphragm are in sharp focus and the condenser is centered so that the circle of light is in the center of the field of view. At this point, the leaves of the radiant field diaphragm are opened until they just disappear from the field of view. The condenser aperture diaphragm must now be adjusted. For best viewing, the aperture should be closed slowly until the sharpest image is obtained. Since the human eye is most sensitive to green light, a green filter over the light source may enhance visualization.

For direct observation of unstained material, **phase-contrast microscopy** may be helpful. In this lighting method, beams of light pass through the object and are partially deflected by different densities or thicknesses (refractive indices) of the object. These light beams are deflected again when they impinge on a special objective lens, increasing in light wave amplitude (and brightness) when aligning in phase, as when passing through material of uniform refractive index, and decreasing in amplitude (visualized as darkness) when out of phase, as when passing through areas of differing refractive index. Resolution for phase-contrast microscopy is also heightened with the use of a green filter. Some laboratories routinely perform phase microscopic examination of bacterial colonies emulsified in water or saline or the supernatant of broth cultures (such as blood cultures) and covered by a coverslip and examined under an oil immersion lens (1000×). With practice, microbiologists can learn to recognize morphologies, motility, and various other characteristics helpful in aiding early presumptive identification. In **oil immersion microscopy** the oil, which fills the space between the objective and the coverglass, helps to keep light rays from dispersing and, since the oil is of the same refractive index as glass, prevents changes in wavelength caused by changes in refractivity of the medium through which the light passes. Immersion oil is necessary to achieve enough resolution for the visualization of most bacteria; the 100× objective is usually used for this purpose together with 10× oculars (total magnification = 1000×).

DIRECT EXAMINATION OF CLINICAL SPECIMENS

Many clinical specimens may be examined in their native state, under brightfield or phase-contrast microscopy, preferably as soon as they are collected from the patient. Specimens that can be applied directly to the surface of a slide for this purpose include sputum, exudate from lesions, aspirated fluid, stool, vaginal discharge, and urine sediment. Material too thick for easy differentiation of suspended material can be diluted with equal parts of physiological sterile saline. It is best to prepare all slides made from fresh clinical material within a biohazard hood, since emulsifying material on the surface of a slide often creates aerosols. A coverslip is then gently laid over the surface of the material, and excess liquid is blotted

from around the edges with a paper towel or tissue, which is discarded as contaminated material. The technologist should wear gloves during this procedure. This sort of preparation, known as a **direct wet mount,** can be preserved for longer viewing time by ringing the edge of the coverslip with nail polish or histological mounting medium. Wet preparations are often used to detect motile trophozoites of fecal parasites such as *Giardia lamblia, Entamoeba histolytica,* and *Dientamoeba fragilis.* The eggs and cysts of other parasites, larvae, and adult worms are also often seen in wet mounts made from freshly passed stool. *Trichomonas vaginalis* can be seen moving in wet mounts prepared from vaginal discharge material or spun sediment from fresh urine specimens. Parasites can also be found in direct wet mounts made from material aspirated from the duodenum, lung, or abscess contents.

Stool may be examined directly by phase-contrast microscopy for evidence of certain bacterial diseases and parasites. Experienced microbiologists can recognize the characteristic darting motility and short, curved morphology of *Campylobacter jejuni.* In areas of endemic cholera the presence of *Vibrio cholerae* in stool can be established presumptively by microscopy. Examination of blood, often diluted in saline, can establish the diagnosis of relapsing fever or leptospirosis (Chapter 32). Microfilariae, including trypanosomes and hemoflagellates, may also be seen in direct preparations of blood.

SLIGHTLY MODIFIED DIRECT PREPARATIONS OF CLINICAL MATERIAL

The use of **10% potassium hydroxide (KOH preparation)** will help to distinguish fungal elements in a direct wet preparation of clinical material. Proteinaceous components, such as host cells, are partially digested by the alkali, leaving intact the polysaccharide-containing fungal cell walls. The material to be examined, whether fluid or skin or nail scrapings, is added to a drop of 10% aqueous KOH on a glass slide. The KOH may be preserved with 0.1% thimerosal (Sigma Chemical Co.), or its digestive capabilities may be enhanced by 40% dimethyl sulfoxide (DMSO), as suggested by McGinnis.[7] A coverslip is laid over the preparation, and excess fluid is removed from the edges by gently pressing the slide, coverslip side down, onto several thicknesses of paper towels. If this procedure is followed, the slide may be examined several minutes later for the presence of fungal elements. Gentle heating may speed the activity of the KOH, but it is unnecessary and may be harmful to the specimen if overdone. A small amount of lactophenol cotton blue or Quink black ink (available commercially) can be added to the 10% KOH for enhanced visibility of fungal elements, or lactophenol cotton blue may be used alone for wet mount preparations of fungi. One of the most useful stains for visualizing fungal elements and *Pneumocystis carinii* is calcofluor white or cellu-fluor, described later in the chapter.

The presence of encapsulated yeast, suspicious for *Cryptococcus neoformans*, particularly in cerebrospinal fluid specimens, can be determined by adding equal parts of India ink (Pelikan brand) or nigrosin stain to the spun sediment of the spinal fluid. The polysaccharide capsules will exclude the particles of ink, and the capsule will appear as a clear halo around the organisms. Direct visualization of fungi in clinical specimens will be discussed further in Chapter 44. Capsules of bacteria may also be demonstrated with India ink or nigrosin stains, although these preparations are rarely used in laboratories today.

Lugol's iodine is often added to direct wet mounts of fecal material to aid the microscopist in differentiating parasitic cysts from host white blood cells. Many cysts will take up the iodine, appearing light brown in color. Other objects remain clear. The glycogen vacuole of cysts of *Iodamoeba bütschlii* is particularly visible after it absorbs the iodine. The use of iodine is delineated in Chapter 45.

Another simple stain, Loeffler's methylene blue, may be added to wet mounts of feces (equal amounts of stain and feces) for determination of the presence of leukocytes (Figure 8.1). The presence of many polymorphonuclear leukocytes is indicative of invasive disease such as bacterial dysentery or campylobacteriosis, as opposed to the noninflammatory nature of the diarrhea of most parasitic diseases or certain food poisonings.

By adding specific antiserum to a wet preparation of selected clinical material, certain organisms may be identified by a visible antigen-antibody reaction, the **Quellung reaction.** Organisms with capsules, such as *Haemophilus influenzae* type b and *Streptococcus pneumoniae,* exhibit apparent capsular swelling in the presence of homologous antibody. Pathogens in cerebrospinal fluid and sputum can be identified rapidly by this method, although it is less commonly used today than in the past.

DARKFIELD MICROSCOPY

Certain bacteria are so thin that they cannot be resolved in direct preparations, even with phase-contrast microscopy. However, their characteristic motility is an important feature of presumptive

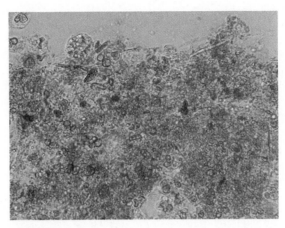

FIGURE 8.1

Loeffler's methylene blue–stained preparation of feces from a patient with invasive bowel disease, characterized by the presence of many leukocytes.

identification. These bacteria, primarily spirochetes such as *Borrelia* and treponemes, are best visualized by **darkfield microscopy,** a method of allowing light to be reflected or refracted off the surface of the object, which appears brightly lit against a black background (Figure 8.2). Light from below the object is blocked in a central circle so that only light from the outer ring reaches the object at a sharp angle. The object reflects and scatters this light around the object's edges, and the scattered light is viewed through the objective. Large, flat objects such as host cells and clear liquid will not refract much light and will appear very dark. To control the path of light, a drop of immersion oil is placed on the top lens of the darkfield condenser, which is then slowly raised until the oil comes into contact with the bottom of the slide containing the specimen. The condenser height is adjusted until the brightest light is visible reflecting off the numerous small particles in the specimen. The light coming into the condenser from below should be the brightest possible. The microbiologist should set the darkfield lighting with a sample slide of a scraping made from the inside of one's cheek emulsified in saline before the actual sample is obtained. In that way, the delicate clinical material can be examined without delay. This method is used most often for the demonstration of motile treponemes in exudate expressed from a primary chancre of syphilis. To best identify these treponemes in darkfield preparations, oil is used on the top of the coverslip to allow viewing with the oil immersion lens at 1000×. Specimen collection and preparation are discussed in Chapter 20. Motile campylobacters in stool may also be seen under darkfield microscopy.

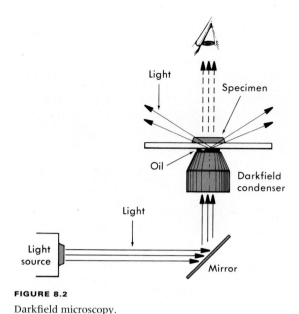

FIGURE 8.2
Darkfield microscopy.

EXAMINATION OF FIXED, STAINED MATERIAL

Examination of stained material, either direct clinical specimens or samples of growth from cultures, is the most useful method for presumptive identification of bacteria and the presence of certain viruses and for definitive identification of most parasites and many fungi. Individual stains used for special purposes will be mentioned in this chapter, but specific procedures will usually be found in the sections of the book that detail techniques for handling particular specimens or identifying specific pathogens.

Stains are either (1) **simple,** consisting of the addition of one dye that serves to delineate morphology but renders all structures the same hue, or (2) **differential.** Differential stains consist of more than one dye added in several steps, and the stained structures are differentiated by color, as well as by shape. Certain classical stains are commonly used in clinical microbiology, and selected procedures for these stains will be discussed later. Formulas for reagents not shown here can be found in the previous edition of this book[1] and other standard texts, and procedures not listed here are found in the sections of the book describing specific uses for certain staining methods.

PREPARATION OF A "SMEAR"

Material to be stained is dropped (if liquid) or rolled (if present on a swab) onto the surface of a clean, dry glass slide. Once a swab has touched the surface of a nonsterile slide, it cannot be used for inoculating culture media. For staining

colonies, a sterile needle may be used to transfer a small amount of the bacterial growth from a solid medium to the surface of the slide. This material is emulsified in a drop of sterile water or saline on the slide. For very tiny colonies that might become lost in even a small drop of saline, a sterile wooden applicator stick can be used to touch the colony and obtain a bit of growth. This material is then rubbed directly onto the slide, where it can be easily seen. Bacterial morphology is well preserved with this technique. If more than one specimen is to be stained on the same slide, a wax pencil may be used to indicate divisions. It is helpful to draw a "map" of the slide so that different Gram stain results can be recorded in an organized fashion (Figure 8.3). In the case of certain critical specimens, such as cerebrospinal fluid, the use of alcohol-cleaned, sterilized slides is recommended. The material placed on the slide to be stained is allowed to dry or may be heated on a slide warmer to 60° C for at least 10 minutes to kill any pathogens that may be present. If the material is not heated, formalinized, or sterilized in an autoclave, organisms may survive the staining procedure. Therefore all stained slides should be treated as though they were potentially infectious and should be discarded with other contaminated material after use. Flaming a slide by passing the slide through the blue flame of a Bunsen burner several times, until the slide is too hot to touch comfortably, will affix the material to the glass but may not be bactericidal. In fact, cellular morphology may be distorted by the heat. Slides fixed in this way must be allowed to cool before they can be stained.

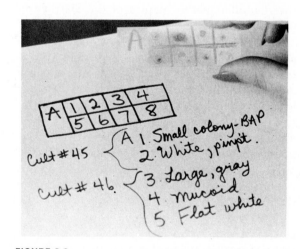

FIGURE 8.3

Preparation of a slide map for staining several colonies on one slide.

PROCEDURE 8.1

CONVENTIONAL GRAM STAIN

PRINCIPLE

Prokaryotes will differentially retain crystal violet depending on cell wall characteristics. Bacteria can be grouped initially based on their Gram stain reactions.

METHOD

1. Prepare reagents as follows:
   a. Crystal violet
Crystal violet, 90% dye content	10g
Absolute methyl alcohol	500 ml
   b. Iodine
Iodine crystals	6 g
Potassium iodide	12 g
Distilled water	800 ml
   c. Decolorizer
Acetone	400 ml
Ethyl alcohol (95%)	200 ml
   d. Counterstain
Safranin, 99% dye content	10 g
Distilled water	1000 ml

2. Fix material on slide with methanol or heat fix as detailed in text. If slide is heat fixed, allow it to cool to the touch before applying stain.

3. Flood slide with crystal violet and allow it to remain on the surface without drying for 10 to 30 seconds.

4. Rinse the slide with tap water, shaking off all excess.

5. Flood the slide with iodine and allow it to remain on the surface without drying for twice as long as the crystal violet was in contact with the slide surface (20 seconds of iodine for 10 seconds of crystal violet, for example).

6. Rinse with tap water, shaking off all excess.

7. Flood the slide with decolorizer for 10 seconds and rinse it off immediately with tap water. Repeat this procedure until the blue dye no longer runs off the slide with the decolorizer. Thicker smears require more prolonged decolorizing. Rinse with tap water and shake off excess.

8. Flood the slide with counterstain and allow it to remain on the surface without drying for 30 seconds. Rinse with tap water and gently blot the slide dry with paper towels or bibulous paper or air dry. For delicate smears, such as thin body fluids, air drying is the best method.

9. Examine microscopically under an oil immersion lens at 1000× for white cells, bacteria, and other structures.

QUALITY CONTROL

Prepare a suspension of mixed *Streptococcus pyogenes* ATCC 19615 and *Escherichia coli* ATCC 25922 in saline. Place a thin drop on the surface of a slide and allow it to air dry. (These can be prepared in advance and stored indefinitely at room temperature in a covered box.) Fix and stain the smear in the same manner as the test slides. Examine it microscopically under oil immersion.

EXPECTED RESULTS

One should see purple cocci in chains (streptococci) and pink rods *(E. coli)*.

PERFORMANCE SCHEDULE

Test each new batch of stains used and test at weekly intervals.

GRAM STAIN

First devised by Hans Christian Gram during the late nineteenth century, the Gram stain can be used to divide most bacterial species effectively into two large groups: those that take up the basic dye, crystal violet (gram-positive), and those that allow the crystal violet dye to wash out easily with the decolorizer alcohol or acetone (gram-negative). The classic Gram stain procedure (Procedure 8.1) entails fixing the material to be stained, either by flaming as described previously or by fixing in alcohol. Methanol fixation preserves the morphology of red blood cells, as well as bacteria, and is especially useful for examining bloody specimen material and blood culture su-pernatant fluid.[5] Slides are overlayed with 95% methanol for 1 minute, the methanol is allowed to run off, and the slides are air dried before staining.

After fixation, the first step in the Gram stain is the application of the crystal violet. A mordant, Gram's iodine, is applied after the crystal violet to chemically bond the alkaline dye to the cell wall. The decolorization step distinguishes gram-positive from gram-negative cells. It has recently been shown that the difference in composition between gram-positive cell walls, which contain thick peptidoglycan with numerous teichoic acid cross-linkages, and gram-negative cell walls, which consist of a thin layer of peptidoglycan and a thick external coat of lipopolysaccharides and protein islands, accounts for the differing Gram stain characteristics of these two major groups of organisms.[2,3] Presumably their extensive teichoic acid cross-links contribute to gram-positive organisms' ability to resist alcohol decolorization. Gram-positive organisms that have lost cell wall integrity because of antibiotic treatment, old age, or action of autolytic enzymes will allow the crystal violet to wash out with the decolorizer step. Animal cells such as red and white blood cells allow the stain to wash out with the decolorizer as well. At this stage, those organisms that stain gram positive still retain the crystal violet and those that stain gram negative are clear. Addition of the counterstain safranin (or 0.1% aqueous basic fuchsin) will stain these clear organisms and cells pink or red. Although the counterstain may be taken up by the gram-positive organisms as well, their purple appearance will not be altered. Yeast cells may also stain gram positive, although fungal mycelia take up the Gram stain variably. Modifications of the classic Gram stain include changes in reagents and timing.

In addition to determining the Gram reaction and morphology of isolated colonies of bacteria, the Gram stain should be used to examine clinical material directly. Sputum can be assessed for suitability for culture by determining the numbers of squamous epithelial cells and polymorphonuclear leukocytes present in the specimen (Chapter 17). If more than 10 epithelial cells are found in an average low-power field (100×), a sputum sample can be assumed to be contaminated with normal oral flora and unsuitable for culture. The presence of few or no epithelial cells and more than 25 polymorphonuclear leukocytes represents a very good specimen. Obviously, patients with profound granulocytopenia will show few or no polymorphonuclear leukocytes; in sputum specimens from such patients the number of epithelial cells is used to judge the quality of the specimen. The numbers and morphology of bacteria seen in direct smears of clinical material are very valuable as early clues to the cause of disease, as well as for comparison to the growth resulting after incubation. The presence of anaerobic organisms may be signaled, for example, by a culture showing many bacteria on Gram stain that yields light or no growth after aerobic incubation on standard media. Comparing Gram stain results to culture results is an excellent internal method for monitoring quality assurance. Several grading systems for reporting Gram stain results have been published.

Unspun urine can also be Gram stained to determine the presence of significant bacteriuria. One µl drop of well-mixed unspun urine is allowed to dry on the surface of a slide without being spread out. The presence of one or more bacteria in most oil immersion fields (1000×) is indicative of more than 100,000 organisms per milliliter of urine. White blood cells may also be noted. Screening procedures for urine are discussed further in Chapter 19.

ACID-FAST STAINS

The cell walls of certain parasites and bacteria contain long-chain (50 to 90 carbon atoms) fatty acids (mycolic acids), lending them the property of resistance to destaining of basic dyes by acid alcohol. Thus, they are called **"acid-fast."** Mycobacteria such as *Mycobacterium tuberculosis* and *Mycobacterium marinum* and coccidian parasites such as *Cryptosporidium* species (Figure 8.4) are characterized by their acid-fast staining properties. The classic acid-fast stain, Ziehl-Neelsen (Procedure 8.2), requires the addition of heat to allow the stain to enter the wax-containing cell wall. The cold Kinyoun acid-fast stain modification is detailed in Chapter 42. A modified acid-fast stain has been developed for differentiating *Nocardia* species, filamentous, branching bacteria that contain fatty acid chains in their cell walls of approximately 50 carbon atoms, from similar non–acid-fast actinomycetes, *Actinomyces* and *Propionibacterium* sp. *Nocardia* are decolorized by the standard acid-alcohol step but not by a milder decolorizer, 0.5% to 1% sulfuric acid. These organisms are known as partially or weakly acid-fast bacteria (Figure 8.5). A procedure for performing this weak acid-fast stain is given in Chapter 34. Another modification of the acid-fast stain for *Cryptosporidium* sp. is described in Chapter 45.

METHYLENE BLUE STAIN

A simple methylene blue stain of growth from Loeffler's agar slant cultures of possible diphtheria specimens may reveal bacilli with characteristic metachromatic granules of *Corynebacterium diphtheriae*. This same stain will reveal the morphology of fusiform bacteria and of spirochetes in oral specimens.

DIFFERENTIAL STAINS FOR PARASITES

Several differential stains are used to highlight the visibility and internal structures of cysts, trophozoites, or other forms of parasites, particularly those found in stool specimens. These stains, further described in Chapter 45, include Wheatley-

PROCEDURE 8.2

ZIEHL-NEELSEN ACID-FAST STAIN

PRINCIPLE

Certain bacteria, parasitic cysts, and rare fungal forms, because of mycolic acids in their cell walls, retain the basic dye carbolfuchsin despite acid-alcohol rinsing. This characteristic differentiates them from other bacteria and is an initial step in their identification.
Note: Smears should be prepared in a biohazard hood.

METHOD

1. Prepare reagents as follows:
    a. Carbolfuchsin
       Basic fuchsin   0.3 g
       Ethanol (95%)   10 ml

    All ingredients are available from Sigma Chemical Co. or other chemical supply houses. Add together and mix. Add the fuchsin solution to 100 ml of 5% phenol solution in distilled water. This solution should then be placed on a stirring platform with a stir-bar in a 37° C incubator overnight; let the stain stand for several days to allow all components to go into solution before using the stain.

    b. Decolorizer
       Ethanol (95%)   97 ml
       Concentrated HCl   3 ml

    Add the hydrochloric acid to the ethanol slowly, working in a chemical fume hood.

    c. Counterstain
       Methylene blue   0.3g
       Distilled water   100 ml

2. Use only new slides. Fix smears on heated surface (60° C for at least 10 minutes). Flood smears with carbolfuchsin and heat to almost boiling by performing the procedure on an electrically heated platform or by passing the flame of a Bunsen burner underneath the slides on a metal rack. The stain on the slides should steam.

3. Allow slides to sit for 5 minutes after heating; do not allow them to dry out.

4. Wash the slides in distilled water (tap water may contain acid-fast bacilli). Shake off excess liquid.

5. Flood slides with decolorizer for approximately 1 minute. Check to see that no more red color runs off the surface when the slide is tipped. Add a bit more decolorizer for very thick slides or those that continue to "bleed" red dye.

6. Wash thoroughly with water and remove the excess.

7. Flood slides with counterstain and allow to remain on surface of slides for 1 minute.

8. Wash with distilled water and stand slides upright on paper towels to air dry. Do not blot dry.

9. Examine microscopically, screening at high power (400×) and confirming all suspicious organisms at 1000× with an oil-immersion lens.

QUALITY CONTROL

Prepare separate suspensions of *M. tuberculosis H37Ra* (suggest ATCC 25177) and *Nocardia asteroides* (suggest ATCC 3308) in Dubos 7H9 broth (use glass beads to disperse the organisms) and place a separate drop of each suspension onto opposite ends of a slide. Allow to dry on a heated slide warmer. (These can be prepared in advance in a biohazard hood and stored indefinitely at room temperature in a covered box.) Fix and stain the slide along with the test slides.

EXPECTED RESULTS

The mycobacteria will stain dark red, and the nocardia will stain blue.

PERFORMANCE SCHEDULE

Perform each time new stain reagents are used and each time any acid-fast stains are performed.

---

trichrome and iron hematoxylin. Toluidine O stain, used for rapid examination of respiratory tract material for the presence of *Pneumocystis carinii*, is described in Chapter 45. Silver stains, which stain bacterial and fungal cells, are also used to visualize parasites. Although such stains are usually performed by pathologists, microbiologists may be required to add these procedures to their protocols.

DIFFERENTIAL STAINS FOR BLOOD SMEARS AND TISSUE SECTIONS

Parasites that circulate in the bloodstream may be found within erythrocytes or free in the plasma. Several stains have been developed to help differentiate these parasites from human host cell components. The two stains most commonly used are Wright's and Giemsa. Thin films of blood are fixed with methanol to preserve the red cell morphol-

FIGURE 8.4
Acid-fast stain of *Cryptosporidium* in unconcentrated stool.

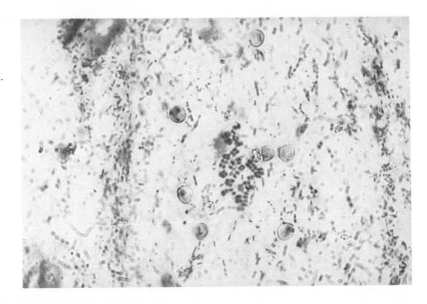

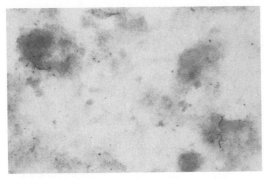

FIGURE 8.5
Partially acid-fast *Nocardia* species isolated from respiratory secretions. See organism at lower right. (Photograph by Jay Packer.)

ogy so that the relationship of the parasites to the red cells can be seen clearly. The slides are then stained, revealing the nuclei and cytoplasmic features of the parasites. To search a larger quantity of blood for the presence of parasites, a thick film may be prepared by placing a large drop of blood (the size of a nickel) in one area of a slide and defibrinating it by making circles of ever-increasing size from the center outward, using the edge of another glass slide. To visualize the parasites in such a thick smear, the red cells must be lysed with water. Preparation of stains of blood films is discussed in Chapter 45. The Giemsa stain is also used to visualize inclusions in virally or other infected cells, either directly from clinical material, such as the base of a suspected herpetic vesicle or a corneal scraping from a suspected case of *Chlamydia trachomatis* conjunctivitis, or for staining a monolayer of infected cell cultures, such as

the McCoy cells of an in vitro *Chlamydia* culture (discussed further in Chapter 39). The presence of other parasitic infections, such as toxoplasmosis in brain tissue, may be detected using the Giemsa stain. Neither the Wright's nor the Giemsa stain will reliably stain fungal or bacterial elements, so they must be used in conjunction with other stains if the etiology of a suspected infection is totally unknown.

Other stains are used for special purposes, such as the iodine stain for the inclusions of chlamydia-infected monolayer cells, Seller's stain for the Negri bodies in rabies-infected tissues, and the Giménez stain for *Chlamydia* and *Legionella*. These stains are discussed in the chapter in which their use is described.

FUNGAL STAINS

Fungal elements can be seen in fixed material with several stains. The periodic acid-Schiff (PAS) is probably one of the best general stains since most fungal elements in clinical material (such as sputum and tissue) will take up the stain. Methenamine silver, in addition to staining certain parasites, will also stain fungal cell walls a dark blackish brown. This stain, however, usually takes increased time to perform. The polysaccharide capsular material of certain fungi, notably *Cryptococcus neoformans*, stains bright pink with mucicarmine stain, which may prove useful for differentiating this fungus in tissue.

FLUORESCENCE MICROSCOPY

Certain dyes, called fluors or **fluorochromes,** have the property of becoming excited (raised to

FIGURE 8.6

Fluorescence microscope.

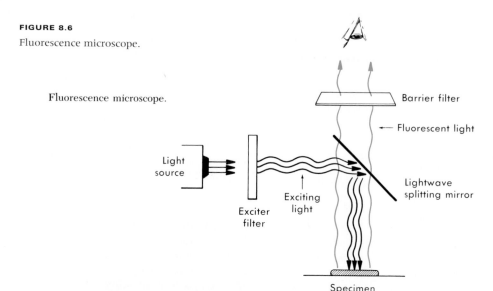

Fluorescence microscope.

FIGURE 8.7

Acridine orange–stained staphylococci from blood.

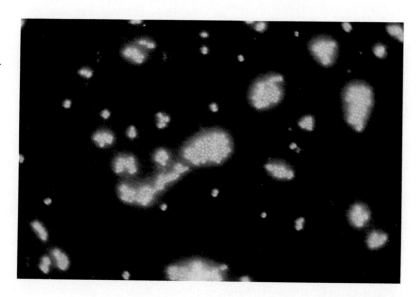

a higher energy level) after absorbing ultraviolet (UV) light (light of short wavelength). As the excited molecules return to their normal state, they release the excess energy in the form of visible light of longer wavelength than that which first excited them. This property of becoming self-luminous is called **fluorescence.** Modern microscopic methods have been developed to exploit the enhanced detection possible with this system. Figure 8.6 diagrams a modern fluorescence microscope, in which the light is emitted from above (epifluorescence). An excitation filter passes light of the desired wavelength to excite the fluorochrome, and a barrier filter in the objective prevents the exciting wavelengths from damaging the eyes of the observer. Fluorescing objects appear brightly lit against a dark back-

ground, with the color dependent on the dye being used. Many agents will absorb fluorescent dyes directly.

ACRIDINE ORANGE

The fluorochrome acridine orange binds to nucleic acid, either in the native or the denatured state. In some formulations of acridine orange, depending on the pH and concentration, the color of the fluorescence will vary (Figure 8.7). The use of acridine orange for highlighting bacteria in blood culture media (Procedure 8.3) has become widely accepted. Studies have shown the staining of blood cultures with acridine orange to be as sensitive as blind subculture for initial detection of positive cultures.[6] This stain has also been used for detection of cell wall–deficient bacteria such as

---

PROCEDURE 8.3

ACRIDINE ORANGE STAIN

PRINCIPLE

Acridine orange, a vital stain, will intercalate with nucleic acid, changing the dye's optical reflecting characteristics so that it will fluoresce bright orange under UV light. Any nucleic acid–containing object will fluoresce.

METHOD

1. Fix slide, either in methanol or with heat, as described previously.

2. Flood slide with acridine orange stain (available from BBL Microbiology Products, Difco Laboratories, Remel Laboratories, and other suppliers). Allow stain to remain on surface of slide for 2 minutes without drying.

3. Rinse with tap water and allow the slide to air dry by leaning it upright to drain on paper towels.

4. Examine the slide microscopically under UV light with the same light source as that used for fluorescein. Bacteria will fluoresce bright orange against a green-fluorescing or dark background. The nuclei of blood cells may also fluoresce.

QUALITY CONTROL

Save a culture-positive blood culture broth and prepare a smear at the same time as the unknown sample smear is prepared. Fix and stain both smears simultaneously and examine microscopically under oil immersion.

EXPECTED RESULTS

The known positive should show bright orange–fluorescing organisms of the morphology of the organism isolated previously.

PERFORMANCE SCHEDULE

Perform with new reagents and each time the stain is used.

---

mycoplasmas in broth cultures and from suspected colonies on the surface of agar media.

RHODAMINE-AURAMINE

The mycolic acid in the cell walls of mycobacteria has an affinity for the fluorochromes auramine and rhodamine. These dyes will bind to mycobacteria, which appear bright yellow or orange against a greenish background. The counterstain, potassium permanganate, helps to prevent nonspecific fluorescence. Most acid-fast objects, including many of the sporozoan parasites, will stain with auramine or rhodamine as well. One widely used staining procedure, that of Truant and others,[9] is shown in Procedure 8.4. Some species of mycobacteria resist this stain, so specimens from peripheral lesions should also be stained with a carbol fuchsin stain.

An important aspect of the rhodamine-auramine stain is that slides may be restained with Ziehl-Neelsen or Kinyoun stain directly over the fluorochrome stain, as long as the oil has been removed. In this way, positive slides can be confirmed with the traditional stain, which will also aid in differentiating morphology.

CALCOFLUOR WHITE

The cell walls of fungi will bind the stain calcofluor white, greatly enhancing their visibility in tissue and other specimens. As described by Hageage and Harrington,[4] this stain is used in place of 10% KOH for initial examination of clinical material. It is also used to enhance visualization of morphological elements of pure cultures of fungi. For many applications, it has supplanted the lactophenol cotton blue stain in some laboratories. Calcofluor white or cellufluor also stains *Pneumocystis carinii*, highlighting the parentheses-like internal structures[8] (Figure 8.8). Organisms fluoresce blue-white or apple green depending on the light source. The modification of the stain that we suggest for use as a fungal stain is outlined in Procedure 8.5. The fluorescent stain and the KOH are added separately, since they tend to precipitate if combined and stored as a single solution. The stain is also available commercially (Fungi-Fluor; Polysciences, Inc., Warrington, Pa. and others).

ANTIBODY-CONJUGATED STAINS

The most specific detectors are antibodies that bind tightly to antigens against which they are directed (Chapter 13). Antibodies can be produced against entire classes of agents that share common antigens or against specific determinants, such as the carbohydrate that is common to the cell walls of *Streptococcus pyogenes*. If antibodies are conju-

PROCEDURE 8.4

RHODAMINE-AURAMINE ACID-FAST STAIN

PRINCIPLE

The nonspecific chromophores rhodamine and auramine bind to mycolic acids in the cell walls of acid-fast organisms and are refractory to rinsing by acid-alcohol. These stains thus exhibit the same characteristics as fuchsin-based, acid-fast stains although they are visualized under UV light. Organisms fluorescing orange-yellow or red are more easily detected than traditionally stained organisms.

METHOD

1. Prepare reagents as follows:
   a. Rhodamine-auramine

Auramine O (C.I. 41000)	1.5 g
Rhodamine B (C.I. 45170)	0.75 g
Glycerol	75 ml
Melted phenol crystals	10 ml
Distilled water	50 ml

   Reagents are available from Sigma Chemical Co. and other chemical supply companies. Add the phenol and water together; then add the other ingredients. Place on a magnetic stirrer with a stir bar in the incubator. Allow to stir at 37° C overnight, or until all particulates are dissolved. Filter through glass wool and store in a brown glass bottle with a glass stopper. The stain is stable for at least 6 months in the refrigerator.

   b. Decolorizer

Hydrochloric acid (concentrated)	0.5 ml
Ethanol (70%)	100 ml

   Add acid slowly to alcohol; work in a fume hood. Store in a glass bottle; cap tightly. Solution is stable indefinitely.

   c. Counterstain

Potassium permanganate	0.5 g
Distilled water	100 ml

   Mix together, filter through coarse filter paper or glass wool, and store in a brown glass bottle. The solution is stable at room temperature for 6 months.

2. Heat-fix slides as described previously.

3. Flood slides with rhodamine-auramine. Allow the stain to remain on the slide for 15 minutes. Do not allow the surface to dry.

4. Rinse the slides with distilled water and shake off excess liquid.

5. Flood slides with decolorizer for 2 to 3 minutes. Slide will still appear pink.

6. Rinse thoroughly with distilled water, shake off excess.

7. Flood slides with counterstain for 3 to 4 minutes. Do not allow slides to dry.

8. Rinse thoroughly with distilled water and allow to air dry as described previously for other acid-fast stains.

9. Examine microscopically under the same UV light source as used for fluorescein or other light source dictated by the instrument. Acid-fast bacilli will be visible as bright yellow-orange organisms against a green background. Slides can be screened with high power (400×) and verified under oil immersion.

QUALITY CONTROL

Same procedure as for Ziehl-Neelsen stain discussed previously.

EXPECTED RESULTS

Acid-fast organisms will exhibit bright orange-yellow to red fluorescence, depending on the filter system used. Non–acid-fast organisms will not be visible.

PERFORMANCE SCHEDULE

As for Ziehl-Neelsen stain.

Modified from Truant, J.P., Brett, W.A., and Thomas, W. Jr. 1962. Henry Ford Hosp. Med Bull. 10:287.

gated to a dye or chromogenic substrate that allows their reactive sites to interact with the homologous antigens, they serve as visible flags for the presence of that antigen.

FLUORESCEIN-CONJUGATED STAINS

Antibodies bound to the fluorochrome fluorescein isothiocyanate (FITC) are used to visualize many bacteria in direct specimens. Fluorescein fluoresces an intense apple green when excited. Examples include *Bordetella pertussis* in nasopharyngeal smears from children suspected of having whooping cough and *Legionella* species in respiratory specimens or tissue of patients with Legionnaires' disease. Monoclonal antibodies (Chapter 11) have been successfully conjugated to fluores-

PROCEDURE 8.5

EXAMINING CLINICAL MATERIAL FOR THE PRESENCE OF FUNGAL ELEMENTS

PRINCIPLE

Cellulose in the cell walls of certain prokaryotes and eukaryotic fungi will bind the nonspecific fluorochrome calcofluor white, allowing detection of fungal elements in wet preparations more easily and rapidly than with conventional KOH preparations and more rapid screening with lower magnification. The dye concentrates in fungal cell walls and fluoresces blue-white or green under UV light, depending on the filter system used.

METHOD

1. Prepare calcofluor white stain as follows:
   Calcofluor white
   M2R (Polysciences or
   Sigma Chemical Co.)   0.1 g
   Evans blue (Sigma)   0.05 g
   Distilled water   100 ml

   Mix thoroughly and store in a brown bottle at room temperature. The solution is stable for 1 year.

2. Add one drop of the calcofluor white solution and one drop of 10% KOH (described previously) to the specimen to be examined on a clean glass slide. Place a coverslip over the material and turn the slide face down on several thicknesses of paper toweling, pushing gently to flatten the coverslip down and express excess fluid out from the edges of the coverslip and into the paper towels.

3. Examine the slide under UV light, using a K530 excitation filter and a BG 12 barrier filter, as for fluorescein. Other filter combinations that produce light of blue-white wavelength, such as a G-365 excitation filter and an LP 420 barrier filter, can be used.

4. Fungal elements will appear apple green (or blue-white, depending on the filter

combination used), with a much dimmer reddish tinted fluorescing background.

QUALITY CONTROL

Prepare a dilute saline suspension of *Candida albicans* ATCC 10231 and *E. coli* ATCC 25922. Add the calcofluor reagents as described and examine this material on the other end of the test slide or on a separate slide.

EXPECTED RESULTS

The large yeast cells will fluoresce brightly, and the small rod-shaped *E. coli* will be barely visible as pale outlines.

PERFORMANCE SCHEDULE

Test a quality control suspension when new reagents are prepared and each time the stain is performed.

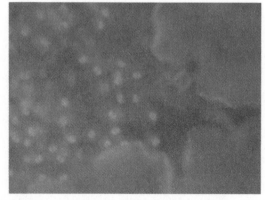

FIGURE 8.8

Pneumocystis carinii cysts stained with the Fungi-Fluor Pneumocystis kit from Polysciences, Inc. (cellufluor stain). Note the internal "double-parentheses"-like structures within the cyst wall. (Courtesy Barbara Waggett, Polysciences, Inc., Warrington, Pa.)

cein for detection of chlamydiae; herpes, respiratory syncytial (Figure 8.9), rabies, and other viruses; and treponemes and other pathogens in directly stained clinical material. Other fluorescein-conjugated antibodies are used to identify pure cultures of organisms, such as certain *Actinomyces* species, *Legionella* sp., *S. pyogenes*, and *Neisseria gonorrhoeae*.

ENZYME-CONJUGATED STAINS

For those laboratories that do not have access to a fluorescent microscope, enzymes that catalyze the production of a colored precipitin product are an excellent alternative conjugate for specific antibody detector reagents. Horseradish peroxidase is a small enzyme that produces an orange-brown precipitate as its easily visible endpoint. Conjugated to antibodies, it is known as **immunoperoxidase stain,** and it is used to detect cytomegalovirus and other virus proteins or nucleic

FIGURE 8.9

Direct fluorescent antibody stain of respiratory syncytial virus from a nasopharyngeal aspirate. (Courtesy Dr. Paul D. Swenson, Seattle-King County Department of Public Health.)

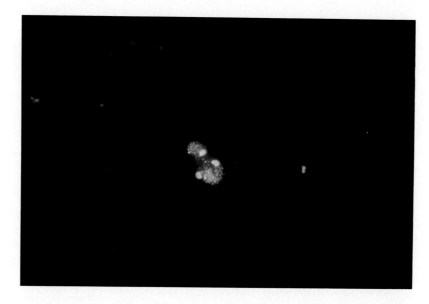

FIGURE 8.10

Immunoperoxidase-stained cell culture monolayer infected with herpes simplex virus. (From Kaplan, M.H., and Swenson, P.D. 1986. Herpes virus infections. In Sun, T. Sexually related diseases: clinical and laboratory aspects. Mosby, St. Louis.)

FIGURE 8.11

Rotavirus particles stained by immunoelectron microscopy.

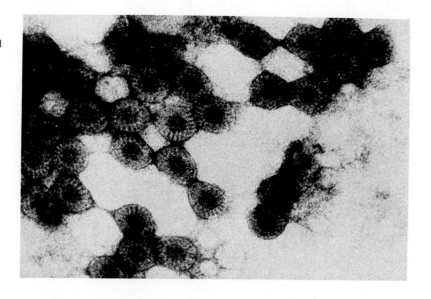

acids in cells (Figure 8.10). Other enzymes have been conjugated to antibodies. Alkaline phosphatase produces a blue precipitate as its end product, and it, too, has been used as a detector for viral antigens, as well as inclusions of chlamydiae.

BIOTIN-AVIDIN-ENZYME-CONJUGATED STAINS

Single-stranded nucleic acid probes (Chapter 11), antimicrobial antibodies, or antibiotin antibodies can be bound to the small molecule, biotin. This molecule has a strong affinity for the protein avidin, which has four binding sites. Thus biotin bound to avidin or antibody can be complexed to fluorescent dyes or to color-producing enzymes to form specific detector systems. Products are available for the detection of nucleic acids of cytomegalovirus, herpes I and II viruses, hepatitis B virus, adenovirus 2, Epstein-Barr virus, *Chlamydia,* and other etiological agents.

ELECTRON MICROSCOPY

The electron microscope uses electrons instead of light to visualize small objects; thus the resolution is increased relative to the shortness of the wavelength of the electron beams. Instead of lenses, the electrons are focused by electromagnetic fields and form an image on a fluorescent screen, like a television screen. Many new morphological features of bacteria, bacterial components, fungi, and parasites have been discovered using electron microscopy. For routine microbiological diagnosis, however, the greatest use of electron microscopy has been for detection of viral causes of gastroenteritis. All enteric viruses can be identified in either direct electron microscopic preparations of fecal material or by immunoelectron microscopy, in which fecal samples are mixed with specific antiviral antibody before staining (Figure 8.11). Since an electron microscope is a major capital investment, few laboratories have the capability to use these techniques on a routine basis.

REFERENCES

1. Baron, E.J., and Finegold, S.M. 1990. Appendix B: Formulas for commonly used stains. p. A-35-A-49. Bailey & Scott's diagnostic microbiology, ed 8. Mosby, St. Louis, Mo.
2. Bottone, E.J. 1988. The Gram stain: the century-old quintessential rapid diagnostic test. Lab. Med. 19:288.
3. Davies, J.A., Anderson, G.K., Beveridge, T.J., et al. 1983. Chemical mechanism of the Gram stain and synthesis of a new electron-opaque marker for electron microscopy which replaces the iodine mordant of the stain. J. Bacteriol. 156:837.
4. Hageage, G.J. Jr., and Harrington, B.J. 1984. Use of calcofluor white in clinical mycology. Lab. Med. 15:109.
5. Mangels, J.I., Cox, M.E., and Lindberg, L.H. 1984. Methanol fixation: an alternative to heat fixation of smears before staining. Diagn. Microbiol. Infect. Dis. 2:129.
6. McCarthy, L.R., and Senne, J.E. 1980. Evaluation of acridine orange stain for detection of microorganisms in blood cultures. J. Clin. Microbiol. 11:281.
7. McGinnis, M.R. 1980. Laboratory handbook of medical mycology. Academic Press, New York.
8. Stratton, N., Hryniewicki, J., Aarnaes, S.L., et al. 1991. Comparison of monoclonal antibody and calcofluor white stains for the detection of *Pneumocystis carinii* from respiratory specimens. J. Clin. Microbiol. 29:645.
9. Truant, J.P., Brett, W.A., and Thomas, W. Jr. 1962. Fluorescence microscopy of tubercle bacillus stained with auramine and rhodamine. Henry Ford Hosp. Med. Bull. 10:287.

BIBLIOGRAPHY

Clancy, M.N., Cohen, M., and Garcia, L.S. 1992. Brightfield microscopy. Sect. 12.14. In Isenberg, H.D., editor. Clinical microbiology procedures handbook. American Society for Microbiology, Washington, D.C.

Clarridge, J.E., and Mullins, J.M. 1987. Microscopy and staining. pp. 87-104. In Howard, B.J., Klass, J. II, Rubin, S.J., et al. Clinical and pathogenic microbiology. Mosby, St. Louis.

Goodman, N.L. 1985. Direct microscopy in diagnosing fungal disease. Diagn. Med. 8:14.

Smith, R.F. 1990. Microscopy and photomicrography: a working manual. CRC Press, Boca Raton, Fla.

Yong, D.C.T., and Peter, J.B. 1984. Using DEM to detect pediatric viral gastroenteritis. Diagn. Med. 7:45.

9

CULTIVATION AND ISOLATION OF VIABLE PATHOGENS

Although future trends in clinical microbiology continue to point in the direction of rapid, non–growth-dependent methods for detecting the presence of infectious agents, the isolation and identification of viable pathogens is still the "gold standard" for diagnosis of infectious diseases today and will always remain important. A pure culture of a clone of identical cells has been necessary for performing biochemical differentiation tests and susceptibility studies, since single cells are impossible to work with easily, and mixed cultures yield no useful or even misleading information. In the late nineteenth century, the development of solid agar in Robert Koch's laboratory by Walther and Lina Hesse aided the expansion of the newly burgeoning science of clinical microbi-

ology. Before solid media plates were available, microbiologists had to rely on the method developed by Lister, employing the principle of limiting dilutions from cultures in broth media for obtaining pure isolates with which to work. General concepts applicable to the in vitro cultivation of pathogens are outlined in the following section.

Methods used to create optimal conditions for the cultivation of pathogens have been developed over years of experimentation. At this time, commercially produced environmental systems, artificial media, cell culture lines, and all other items necessary for the practice of clinical microbiology are readily available to microbiologists in the industrialized world. Laboratorians in less well-developed parts of the world may still need to bleed their own sheep (or other animals) and perform other basic tasks necessary to practice diagnostic microbiology, but most microbiologists now purchase most materials needed for cultivation, identification, and susceptibility testing of pathogenic microorganisms.

Clinical microbiologists used to rely heavily on inoculation of experimental animals for the isolation and identification of infectious agents. Today, such activities are usually confined to research facilities or public health laboratories. For this reason, animal inoculation techniques are not mentioned extensively in the rest of this book, unless their application is particularly important for clinical diagnosis.

ARTIFICIAL MEDIA

GENERAL CONCEPTS OF ARTIFICIAL MEDIA

Over the years, varied strategies have been developed for the cultivation of pathogens. Ingredients necessary for the growth of pathogens can be supplied by a living system, as in the human or animal host or in cell culture, or by mixing together the required nutrients in an artificial system. During the nineteenth century, media were prepared in glass containers, from which the phrase *in vitro* (which means "in glass") originated. To encourage the growth of particular organisms from a milieu containing only a few of the desired organisms among large numbers of indigenous flora, various types of laboratory-prepared nutrient-containing solutions were concocted and called **enrichment media.** An example of such a medium is selenite broth, which encourages the growth of small numbers of stool pathogens and suppresses the growth of the much larger numbers of normal stool organisms. A second class of artificial media is called **supportive.** These media contain nutrients that allow most nonfastidious organisms to grow at their natural rates, without affording any particular organism a growth advantage. Examples of supportive media are nutrient agar and brain-heart infusion agar. Media containing one or more agents inhibitory to all organisms except the organism being sought were developed, first using dyes that exhibited antibacterial characteristics, later using antibiotics, and still later incorporating components that take into consideration certain metabolic activities of the organisms sought. Such media are known as **selective media,** since they select for certain organisms to the disadvantage of others. An example of a selective medium is phenylethyl alcohol agar, which inhibits the growth of aerobic and faculta-

tively anaerobic gram-negative rods and allows gram-positive cocci to grow. The fourth type of medium, **differential,** employs some factor (or factors) that allows colonies of organisms that possess certain metabolic or cultural characteristics to be morphologically distinguished from those organisms with different characteristics. The most supportive differential medium is sheep blood agar, which allows many organisms to grow and additionally allows different organisms to be distinguished on the basis of their hemolytic reactions against the sheep red blood cells, production of pigment, and so forth. Table 9.1 lists various media frequently used in clinical microbiology, along with the ingredients that allow for differential or selective ability.

PREPARATION OF DEHYDRATED ARTIFICIAL MEDIA THAT ARE TO BE STERILIZED IN AN AUTOCLAVE

Dehydrated media powders should be kept in their original bottles with the caps tightly closed. The date received and the date first opened for use should be recorded on the bottle itself. Media makers should always *read the label* before preparing any medium. Media should be prepared in clean glassware that has been rinsed in distilled or deionized water. To avoid boiling over during heating, a vessel holding a liquid solution to be sterilized in an autoclave should never be filled more than two-thirds full. The proper amount of powder is weighed onto nonabsorbent paper and poured into the container in which it is to be prepared. Distilled water is added next with vigorous swirling to achieve an even suspension. If other liquid ingredients are to be added, they should be incorporated at this point. If the solution is clear, as most broths usually are, it requires no further manipulation before autoclaving; however, most agar solutions require heating almost to the boiling point, with constant agitation, to achieve an even solution. The use of a stirrer–hot plate and a magnetic stir-bar will greatly increase the efficiency of this stage of media making. The hot solution must be watched extremely carefully as soon as tiny bubbles begin to appear, since these media tend to boil over very easily. Some media are ready to be dispensed at this point. If the medium is to be sterilized in an autoclave, it is capped with either a plastic screw cap or a plug. A good plug can be made from a large wad of nonabsorbable cotton wrapped in a square of gauze one layer thick. For sterilization of flasks of agar that are to be poured into plates by hand, we have found a large square of aluminum foil molded to

TABLE 9.1

PRIMARY PLATING MEDIA

MEDIUM	COMPONENTS/COMMENTS	PRIMARY PURPOSE
Bacteroides bile esculin agar (BBE)	Trypticase soy agar base with ferric ammonium citrate, enriched with hemin (5 mg/ml). Bile salts and gentamicin act as inhibitors.	Selective and differential for *Bacteroides fragilis* group; good for presumptive identification
Bile esculin agar (BEA)	Nutrient agar base with ferric citrate. Hydrolysis of esculin by group D streptococci imparts a brown color to medium; sodium desoxycholate inhibits many bacteria.	Differential isolation and presumptive identification of group D streptococci
Bismuth sulfite agar (BS)	Peptone agar with dextrose and ferrous sulfate. Gram-positive organisms and other Enterobacteriaceae inhibited by bismuth sulfite and brilliant green.	Selective for isolation of *Salmonella* from stool
Blood agar	Trypticase soy agar, *Brucella* agar, or beef heart infusion base with 5% sheep blood.	Cultivation of fastidious microorganisms, determination of hemolytic reactions
Bordet-Gengou agar	Potato-glycerol-based medium enriched with 15%-20% defibrinated blood. Contaminants inhibited by methicillin (final concentration of 2.5 μg/ml).	Isolation of *Bordetella pertussis*
Buffered charcoal yeast extract agar (BCYE)	Yeast extract, agar, charcoal and salts supplemented with L-cysteine HCl, ferric pyrophosphate, ACES buffer, and α-ketoglutarate.	Selective for *Legionella* sp.
Campy-blood agar	Contains vancomycin (10 mg/L), trimethoprim (5 mg/L), polymixin B (2500 U/L), amphotericin B (2 mg/L), and cephalothin (15 mg/L) in a *Brucella* agar base with sheep blood.	Selective for *Campylobacter* sp.
CDC anaerobic blood agar	Trypticase soy agar with 5% sheep blood enriched with hemin, L-cystine, and vitamin K_1.	Isolation of anaerobic and other organisms; enhanced growth of peptostreptococci
Cefsulodin-irgasan-novobiocin (CIN) agar	Peptone base with yeast extract, mannitol, and bile salts. Supplemented with cefsulodin, irgasan, and novobiocin; neutral red and crystal violet indicators.	Selective for *Yersinia* sp.; may be useful for isolation of *Aeromonas sp.*
Chocolate agar	Peptone base, enriched with solution of 2% hemoglobin or IsoVitaleX (BBL).	Cultivation of *Haemophilus* and *Neisseria* sp.
Columbia colistin-nalidixic acid (CNA) agar	Columbia agar base with 10 mg colistin per liter, 15 mg nalidixic acid per liter, and 5% sheep blood.	Selective isolation of gram-positive cocci
Cooked meat (CM; also called chopped meat)	Solid meat particles initiate growth of bacteria, reducing substances lower oxidation-reduction potential (Eh).	Cultivation of anaerobic organisms
Cycloserine-cefoxitin fructose agar (CCFA)	Egg yolk base with fructose, cycloserine (500 mg/L), and cefoxitin (16 mg/L) added to inhibit stool flora. Neutral red indicator.	Selective for *Clostridium difficile*
Cystine-lactose-electrolyte-deficient (CLED) agar	Peptone base agar with lactose and L-cystine; bromthymol blue indicator inhibits swarming of *Proteus* sp.	Isolation and enumeration of bacteria in urine
Cystine-tellurite blood agar	Infusion agar base with 5% sheep blood. Reduction of potassium tellurite by *Corynebacterium diphtheriae* produces black colonies.	Isolation of *C. diphtheriae*
Dermatophyte test medium (DTM) agar	Nutrient base with glucose and phenol red indicator. Contaminants inhibited by cycloheximide, gentamicin, and tetracycline.	Isolation and identification of dermatophytes
Eosin methylene blue (EMB) agar (Levine)	Peptone base with lactose and sucrose. Eosin and methylene blue as indicators.	Isolation and differentation of lactose-fermenting and non–lactose-fermenting enteric bacilli
Gram-negative broth (GN)	Peptone base broth with glucose and mannitol. Sodium citrate and sodium desoxycholate act as inhibitory agents.	Selective (enrichment) liquid medium for enteric pathogens

Continued.

TABLE 9.1

PRIMARY PLATING MEDIA—cont'd

MEDIUM	COMPONENTS/COMMENTS	PRIMARY PURPOSE
Hektoen enteric (HE) agar	Peptone base agar with bile salts, lactose, sucrose, salicin, and ferric ammonium citrate. Indicators include bromthymol blue and acid fuchsin.	Differential, selective medium for the isolation and differentiation of *Salmonella* and *Shigella* from other gram-negative enteric bacilli
Kanamycin-vancomycin laked blood agar (KVLB)	*Brucella* agar base with kanamycin (75 µg/ml), vancomycin (7.5 µg/ml), vitamin K_1 (10 µg/ml), and 5% laked blood.	Selective isolation of *Bacteroides* sp.
Lombard-Dowell agar	Casein digest agar enriched wtih hemin (10 mg/L), vitamin K_1 (10 mg/L), L-cystine (0.4 g/L), and yeast extract.	Isolation and initial testing of anaerobic organisms
Löwenstein-Jensen (L-J) agar	Egg-based medium; contaminants inhibited by malachite green.	Isolation of mycobacteria
MacConkey agar	Peptone base with lactose. Gram-positive organisms inhibited by crystal violet and bile salts. Neutral red as indicator.	Isolation and differentiation of lactose fermenting and non–lactose-fermenting enteric bacilli
Mannitol salt agar	Peptone base, mannitol, and phenol red indicator. Salt concentration of 7.5% inhibits most bacteria.	Selective isolation of coagulase-positive staphylococci
Middlebrook 7H10 agar	Complex base with albumin, salts, enzymatic digest of casein enrichment, and malachite green inhibitor.	Isolation of and antimicrobial susceptibility testing of mycobacteria
Mycosel or mycobiotic agar	Peptone base with glucose; contaminants inhibited by chloramphenicol and cycloheximide.	Isolation of dermatophytes
New York City (NYC) agar	Peptone agar base with cornstarch, supplemented with yeast dialysate, 3% hemoglobin, and horse plasma. Antibiotic supplement includes vancomycin (2 µg/ml), colistin (5.5 µg/ml), amphotericin B (1.2 µg/ml), and trimethoprim (3µg/ml).	Selective for *Neisseria gonorrhoeae*
Petragnani agar	Coagulated-egg medium with malachite green to inhibit commensals.	Isolation of mycobacteria
Phenylethyl alcohol (PEA) agar	Nutrient agar base. Phenylethanol inhibits growth of gram-negative organisms.	Selective isolation of gram-positive cocci and anaerobic gram-negative bacilli
Sabouraud dextrose agar	Peptone base agar. Final pH of medium (5.6) favors growth of fungi over bacteria.	Isolation of dermatophytes
Salmonella-Shigella (SS) agar	Peptone base with lactose, ferric citrate, and sodium citrate. Neutral red as indicator; inhibition of coliforms by brillant green and bile salts.	Selective for *Salmonella* and *Shigella* sp.
Schaedler agar	Peptone and soy protein base agar with yeast extract, dextrose, and buffers. Addition of hemin, L-cystine, and 5% blood enriches for anaerobes.	Nonselective medium for the recovery of anaerobes and aerobes
Selenite broth	Peptone base broth. Sodium selenite toxic for most Enterobacteriaceae.	Enrichment of isolation of *Salmonella*
Skirrow agar	Peptone and soy protein base agar with lysed horse blood. Vancomycin inhibits gram-positive organisms; polymyxin B and trimethoprim inhibit most gram-negative organisms.	Selective for *Campylobacter*, particularly recommended for *Helicobacter pylori*
Streptococcal selective agar (SSA)	Contains crystal violet, colistin, and trimethoprim-sulfamethoxazole in 5% sheep blood agar base.	Selective for *Streptococcus pyogenes* and *Streptococcus agalactiae*
Tetrathionate broth	Peptone base broth. Bile salts and sodium thiosulfate inhibit gram-positive organisms and Enterobacteriaceae.	Selective for *Salmonella* and *Shigella*

TABLE 9.1

PRIMARY PLATING MEDIA—cont'd

MEDIUM	COMPONENTS/COMMENTS	PRIMARY PURPOSE
Thayer-Martin agar	Blood agar base enriched with hemoglobin and supplement B; contaminating organisms inhibited by colistin, nystatin, vancomycin, and trimethoprim.	Selective for *N. gonorrhoeae* and *N. meningitidis*
Thioglycolate broth	Pancreatic digest of casein, soy broth, and glucose enrich growth of most microorganisms. Thioglycolate and agar reduce Eh.	Supports growth of anaerobes, aerobes, microaerophilic, and fastidious microorganisms
Thiosulfate citrate–bile salts (TCBS) agar	Peptone base agar with yeast extract, bile salts, citrate, sucrose, ferric citrate, and sodium thiosulfate. Bromthymol blue acts as indicator.	Selective and differential for vibrios
Vaginalis (V) agar	Columbia agar base supplemented with 5% human blood.	Selective and differential for *Gardnerella vaginalis*
Xylose lysine desoxycholate (XLD) agar	Yeast extract agar with lysine, xylose, lactose, sucrose, and ferric ammonium citrate. Sodium desoxycholate inhibits gram-positive organisms; phenol red as indicator.	Isolation and differentiation of *Salmonella* and *Shigella* from other gram-negative enteric bacilli

the shape of the flask to be an excellent cover that can be removed and replaced many times and that allows the neck of the flask to remain sterile until the foil is lifted off (Figure 9.1). The solutions are placed in the autoclave and sterilized; the timing of the sterilization should start from the moment the temperature reaches 121° C.[3] Very large quantities of media may require a longer sterilization time than is recommended on the package label. Once the sterilization cycle is completed, the autoclave chamber is slowly returned to atmospheric pressure to prevent the liquid from bubbling over. Sterilized media should not be kept in the autoclave once the pressure has equalized, since prolonged heat may alter some of the ingredients.

Sterilization in an autoclave can be dangerous; all the safety precautions outlined in Chapter 2 and the quality control practices outlined in Chapter 3 should be carefully adhered to by operators. Autoclave performance should be monitored for microbial killing ability regularly and for achievement of proper temperature during every use (e.g., with autoclave tape). Media removed from the autoclave should be placed into a 55° C water bath to cool before any supplements are added or before plates are poured. Liquid media or media that are not to be dispensed may be allowed to cool on the bench top. Agar will tend to settle to the bottom of a flask during sterilization, so all flasks should be swirled in a large circle on the bench top (to avoid making bubbles) before

pouring. If a few bubbles appear on the surface of a freshly poured plate, they can be removed by quickly passing the flame from a Bunsen burner over the surface. We have found that leaving the covers of Petri dishes slightly ajar allows contaminants to reach the agar surfaces. Our recommendation is to leave tops closed and incubate plates overnight after they have set to remove surface moisture and detect contamination.

FIGURE 9.1

Gauze-covered nonabsorbable cotton plug and aluminum foil cover for flasks used in media preparation.

METHODS OF STERILIZATION OTHER THAN AUTOCLAVE

Media that contain serum or certain proteins are often sterilized by **inspissation.** In this intermittent sterilization method, the media are placed in a chamber through which steam continuously flows for approximately 30 minutes each day for several successive days. Loeffler agar and Löwenstein-Jensen agar are prepared by this method.

Delicate media may also be sterilized by allowing the flasks or tubes to remain in a chamber through which steam actively flows. Such a chamber, the Arnold steam sterilizer, is also often used to remelt media that have been prepared in advance and allowed to harden. This method is particularly useful for pouring fresh agar plates made from previously prepared agar deep tubes, to which a sterile additive not suitable for autoclave sterilization (e.g., blood) is added immediately before pouring. Plates seldom used but that require fresh enrichments (egg yolk agar, Bordet-Gengou, etc.) can be prepared as needed from agar deep tubes, which have a much longer shelf life than poured plates. Carbohydrate solutions and other liquids that may be denatured by heat can be **filter sterilized** by injecting the liquid through a syringe attached to a membrane filter with pores no larger than 0.2 or 0.45 μm in diameter (Figure 9.2). Disposable closed membrane filtration systems, often used to sterilize cell culture media, are available commercially. Most hospitals use **gas sterilization** for instruments and equipment. This method uses ethylene oxide to destroy bacteria and spores. Microbiologists may make use of the gas sterilizer to process reusable plastic ware and instruments. Gas sterilization takes much longer than the other methods described, usually requiring an overnight cycle and aeration time for the ethylene oxide to diffuse away.

CONDITIONS NECESSARY FOR GROWTH OF PATHOGENS

In order for bacteria or fungi to multiply on or in artificial media, they must have available the required nutrients, a permissive temperature, enough moisture in the medium and in the atmosphere, the proper gaseous atmosphere, proper salt concentration, and an appropriate pH, and there must be no growth-inhibiting factors (e.g., other bacteria or fungi or artificial compounds that antagonize growth).

ESTABLISHING ATMOSPHERIC CONDITIONS REQUIRED FOR GROWTH OF PATHOGENS OTHER THAN ANAEROBES

Pathogenic organisms are either **aerobic,** utilizing oxygen as a terminal electron acceptor and showing good growth in an atmosphere of room air; **anaerobic,** relatively intolerant to the presence of oxygen; or **microaerobic** (formerly called microaerophilic), growing best in atmospheres of reduced oxygen tension. Aerobic organisms can be incubated in room air without much difficulty. Most clinically significant "aerobic" organisms are actually **facultatively anaerobic;** they grow under either aerobic or anaerobic conditions. True aerobic organisms include *Pseudomonas* sp., members of the Neisseriaceae family, *Brucella* sp., *Bordetella* sp., and *Francisella* sp., as well as mycobacteria, filamentous fungi, and others. To achieve an atmosphere of incubation other than room air, several strategies have been developed. Organisms that grow best with greater CO_2 concentrations than are found in room air, called **capnophilic,** may be incubated in an atmosphere of 5% to 10% CO_2 in a special incubator with sealed doors. Gas of the proper mixture is fed into the incubator automatically from nearby cylinders. The CO_2 concentration should be checked on a routine basis.

By placing inoculated cultures into a sealable container, evacuating the room air down to a negative pressure of 25 pounds of mercury, and replacing it with a commercially produced artificial mixture of gases placed under pressure in a gas cylinder, microbiologists can produce any desired

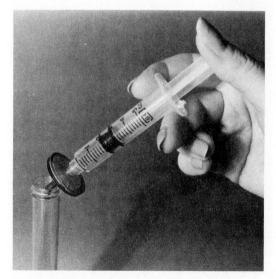

FIGURE 9.2

Use of a membrane filter system to sterilize solutions nonsterilizable in autoclave.

atmosphere. Three evacuation-replacement cycles are usually necessary to remove all residual normal atmosphere. Sealable plastic bags into which fit the components of a CO_2 and-hydrogen-generating system or oxygen-binding components and an indicator of the proper atmosphere are also available commercially for the production of specialized atmospheres of incubation. Bags have been designed for creating either anaerobic or capnophilic atmospheres (Figure 9.3). These bags are especially convenient for incubating primary plates from important cultures or in other circumstances where few plates are used. The bags accommodate only one or a few culture plates each, but all plates can be examined through the plastic without opening the bag. Some organisms that appear to be microaerobic or capnophilic are actually **humidophilic;** they require increased moisture in the atmosphere.

Most pathogenic *Campylobacter* sp. require a high CO_2 content (5% to 10%) and no more than 6% oxygen (microaerobic). This atmosphere can be achieved safely by using a premixed gas for evacuation-replacement or by creating the atmosphere in a sealed jar or plastic bag with a commercially available self-contained generator system. These generator systems, which are similar to those available for creating an anaerobic atmosphere in a closed container, consist of an envelope that contains the components of a small hydrogen-generating (and sometimes a CO_2-generating) system. Chemical components required to catalyze the reaction between hydrogen and oxygen, forming water, removing oxygen, and thus creating the anaerobic atmosphere in such a system (alumina-coated palladium pellets), may be incorporated into the generator or supplied separately. Other methods for creating the atmo-

sphere required by campylobacters are discussed in Chapter 31.

The proper atmosphere for microaerobic organisms requires an oxygen tension lower than that of room air. A CO_2 concentration of approximately 3% can be achieved in a **candle jar.** A small, white wax candle is lit in a jar with a sealable lid, such as a commercial mayonnaise jar. The candle uses up just enough oxygen before it goes out (from lack of oxygen) to lower the oxygen tension. The products of combustion are CO_2 and water, both growth factors for the organisms. The candle jar is often used to cultivate *Neisseria gonorrhoeae.* A self-contained culture medium and increased CO_2-producing system have been developed for culturing gonococci. A tablet of sodium bicarbonate (e.g., Alka-Seltzer) dissolves in the moisture created by sealing the medium in a Ziploc plastic bag and produces enough CO_2 to allow growth of the pathogen (Figure 9.4). Microaerobic conditions can also be created by adding a small concentration of agar to a liquid medium. By preventing oxygen at the surface from being dispersed throughout the liquid by inhibiting circulating convection currents, the agar serves to create a minimicroaerobic environment about 1 to 2 cm below the surface of the medium. *Leptospira* sp. are cultivated in this way.

METHODS FOR ESTABLISHING ANAEROBIC ATMOSPHERIC CONDITIONS FOR INCUBATION OF PRIMARY CULTURE PLATES

CONVENTIONAL METHOD. The most frequently used system for creating specialized anaerobic and capnophilic atmospheres is the anaerobic jar. Available anaerobic jars include the GasPak (BBL Microbiology Systems), those made by EM Diag-

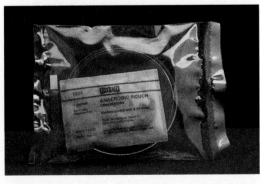

FIGURE 9.3
Anaerobic atmosphere-generating pouch uses iron filings to remove oxygen. This product is now available from EM Diagnostic Systems, Gibbstown, NJ.

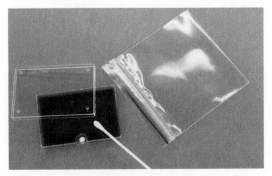

FIGURE 9.4
Carbon dioxide–generating Thayer-Martin medium in plastic tray (Jembec-type plate) for early growth and transport of cultures for *Neisseria gonorrhoeae.*

nostic Systems, and Oxoid U.S.A. These systems use a clear, heavy plastic jar with a lid that is clamped down to make it airtight (Figure 9.5).

Anaerobic jars are used primarily with plated media. The introduction of a gas mixture containing hydrogen into a jar is followed by catalytic conversion of the oxygen in the jar with hydrogen to water, thus establishing anaerobiosis. A catalyst composed of palladium-coated alumina pellets either integrated into the gas-generating envelope or held in a wire mesh is preferred, since there is no explosion hazard with this "cold" catalyst. However, the pellets can be inactivated by excess moisture and H_2S. Therefore, they should be reactivated *after each use* by heating the basket or sachet of pellets to 160° C in a drying oven for 1½ to 2 hours. It is convenient to have a few extra baskets or sachets of catalysts for this purpose. Reactivated catalysts should be stored in a dry area until used. Integrated generator/catalyst systems are disposable, and the catalysts are not reused. We do not recommend the use of jars without catalysts.

Anaerobic jars can be set up by two different methods. The easiest method uses a commercially available hydrogen and CO_2 generator envelope,

which is activated by simply adding 10 ml of water. The open envelope is placed in the jar with the inoculated plates, water is added, and the jar is sealed. Production of heat within a few minutes (detected by touching the top of the jar) and subsequent development of moisture on the walls of the jar are indications that the catalyst and generator envelope are functioning properly. Reduced conditions are achieved in 1 to 2 hours, although the methylene blue or resazurin indicators take longer to decolorize. Alternatively, the "evacuation-replacement" system may be used. Air is removed from the sealed jar by drawing a vacuum of 25 inches (62.5 cm) of mercury. This process is repeated two times, filling the jar with an oxygen-free gas such as nitrogen between evacuations. The final fill of the jar is made with a gas mixture containing 80% to 90% nitrogen, 5% to 10% hydrogen, and 5% to 10% CO_2. Many anaerobes require CO_2 for maximal growth. The atmosphere in the jars should be monitored by including an indicator to check anaerobiosis. Anaerobiosis is achieved more quickly by the evacuation-replacement method. However, both methods give comparable yields of anaerobes from clinical specimens if the specimen is properly transported and set up in jars immediately after plates are streaked.

Stringent anaerobic conditions required by some organisms are attainable in an enclosed system, called a **glove box** or an anaerobic chamber. Made of molded or flexible clear plastic, these chambers allow materials to enter through an air lock. The operator uses gloves or sleeves that form airtight seals around his or her arms to handle items inside the chamber (Figure 9.6). Media

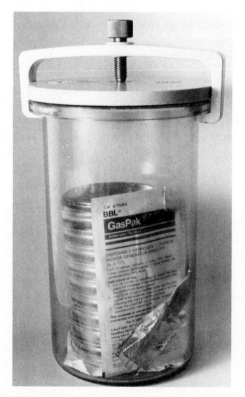

FIGURE 9.5

GasPak anaerobic jar (BBL Microbiology Systems) containing inoculated plates, gas-generating envelope, catalysts, and indicator strip.

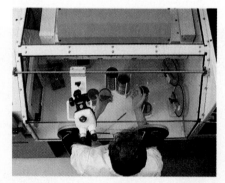

FIGURE 9.6

View from above of microbiologist working in an anaerobic chamber (Sheldon Manufacturing, Cornelius, Ore.). The rubber sleeves form airtight seals around the operator's arms, allowing the operator to work without gloves. (Photo courtesy of Anaerobe Systems, San Jose, Calif.)

stored in the chamber are kept oxygen free and thus are able to support the growth of even oxygen-sensitive anaerobic organisms. For practical purposes, most pathogenic anaerobes can tolerate a minimal exposure to oxygen. Methods that make use of fresh media and that allow inoculated cultures to be brought under anaerobic conditions quickly (oxygen-reduction potential less than −10 mV), such as the anaerobic jars or plastic pouches discussed previously, should be adequate for isolation of most clinically significant anaerobes.

PREREDUCED, ANAEROBICALLY STERILIZED (PRAS) TECHNIQUES. Media produced under anaerobic conditions are called **prereduced, anaerobically sterilized (PRAS)** media. PRAS tubes are made by combining the constituents of the medium, boiling the liquid to remove dissolved air, and then gassing out with an oxygen-free gas. Air is prevented from gaining entrance into the containers by a gassing procedure or by keeping the container stoppered or sealed. To lower the oxidation-reduction potential (Eh) of the medium, a reducing agent may be added before sterilization. Details of PRAS media preparation and inoculation are given in the VPI *Anaerobe Laboratory Manual.*[1] PRAS media, both tubed and plates, are available from commercial sources. Cultivation of anaerobes is discussed further in Chapter 35.

TEMPERATURES OF INCUBATION

Human pathogens generally multiply best at temperatures similar to those in the host. Isolation of most pathogens therefore can be carried out using incubators of only two temperatures: 35° C, close to the normal internal human body temperature, and 30° C, the temperature of the body's surface. With the few exceptions noted later, all bacteria and viruses of pathogenic importance may be isolated from cultures incubated at 35° C. Material from lesions suspected of being infected with *Mycobacterium marinum,* blood and urine being tested for leptospires, and all cultures from which fungi are being sought should be incubated at 30° C. Recovery of certain organisms can be enhanced by incubation at unusual temperatures; *Campylobacter jejuni* and *E. coli* grow at 42° C, although most other fecal pathogens cannot. Incubation at this temperature, therefore, acts as an enrichment procedure. Cold enrichment for *Listeria* and *Yersinia enterocolitica* uses the same principle. Certain viruses, such as respiratory syncytial virus, multiply best in tubes incubated in roller drums at 33° to 36° C.

The temperature of incubators should be monitored by checking each area of the incubator where cultures will be placed, since temperature variations do occur within incubators. The humidity can be controlled automatically by feeding water from an external source into the system as needed or manually by placing a large pan filled with water on the bottom shelf of the incubator. A little detergent in the water will discourage contaminants.

PH OF ARTIFICIAL MEDIA

Although commercially produced dehydrated powders are so consistent that even the ultimate pH is usually correct, the hydrogen ion concentration should be checked using a pH meter. Especially after altering the pH of a medium with additives or for special uses, the pH should be reestablished. It is important to remember that pH electrodes are calibrated according to temperature; thus the pH reading of a hot solution will be different than that taken at room temperature. If the calibration temperature for the electrode cannot be adjusted, the pH must be measured at approximately 28° C, or room temperature. Solid media pose an additional problem, since the pH is best adjusted while the media are still in liquid form and thus hotter than 50° C. A surface electrode may be used to measure the pH of agar, but these electrodes are very expensive and the agar must be set, which makes it difficult to alter the pH later if needed. A workable strategy is to allow a small quantity of the molten agar to solidify in a tiny beaker or plastic cup. The pH of this material can be easily determined with a regular pH electrode thrust into the agar after it has been vigorously broken up with a tongue depressor. The moisture in the medium will be enough to allow proper operation of the pH meter.

CHARACTERISTICS OF CERTAIN FREQUENTLY USED ARTIFICIAL MEDIA

Only a few of the hundreds of available media are mentioned here as a sample of the types of media available. The choices of which media should be used to culture clinical specimens will be discussed in the chapters in Part Three that detail laboratory handling of specimens and in the chapters in Part Four that deal with individual isolates. Complete descriptions of the composition and use of these and other media are found in the *Difco Manual* (Difco Laboratories), the *Oxoid Manual* (Oxoid U.S.A.), and the *BBL Manual* (Becton-Dickinson).

BACTEROIDES BILE ESCULIN AGAR (BBE)

BBE agar is useful for the rapid isolation and presumptive identification of the *Bacteroides fragilis* group. It contains 100 μg/ml of gentamicin, which inhibits most aerobic organisms; 20% bile, which inhibits most anaerobes except for the *B. fragilis* group and a few other species; and esculin, which aids in detecting the *B. fragilis* group that are usually esculin positive. Among other non–*B. fragilis* group organisms that may grow on this medium are *Fusobacterium mortiferum, Klebsiella pneumoniae, Enterococcus,* and yeast. However, unlike the *B. fragilis* group, their colony size is less than 1 mm in diameter.

BLOOD AGAR

Most specimens received in a clinical microbiology laboratory are plated onto blood agar, since it supports all but the most fastidious, clinically significant isolates and since most microbiologists have become adept at making decisions about the identification of bacteria from their colonial morphologies on blood agar. These media consist of a base containing a protein source (e.g., tryptones), soybean protein digest (containing a slight amount of natural carbohydrate), sodium chloride, agar, and 5% blood. In the United States the blood source is usually sheep, whereas horse blood is often used in Europe. Certain bacteria produce extracellular enzymes that act on the red cells to lyse them completely (beta hemolysis) or to produce a greenish discoloration around the colony (alpha or incomplete hemolysis), whereas others have no effect (sometimes called gamma hemolysis); this is determined on blood agar (Figure 9.7). Production of hemolysins by bacteria depends on many environmental factors, such as pH and atmosphere of incubation. Microbiologists often use colony morphology and hemolysin production as initial screening tests to assist in the decision as to what further steps may be necessary for identification of an isolate. To read the hemolytic reaction on a blood agar plate accurately, the technologist must hold the plate up to the light and observe the plate with the light coming from behind. If the loop used to streak the culture on the blood agar has been stabbed into the medium to cause organisms to grow below the surface, production of oxygen-sensitive beta hemolysin may be enhanced with certain organisms. Alternatively, plates may be incubated anaerobically to demonstrate oxygen-sensitive hemolysis.

BRAIN-HEART INFUSION MEDIA

Another nutritionally rich formula, brain-heart infusion (BHI), can be used to grow a variety of microorganisms, either as a broth or hardened with agar, with or without added blood. Key ingredients include infusion from several animal tissue sources, added peptone, phosphate buffer, and a small concentration of dextrose. The carbohydrate provides a readily accessible source of energy for the organisms. BHI broths are often used as blood culture media and as basal media for many metabolic tests, particularly for identification of streptococci. The usual base for blood agar plates is heart infusion agar, but BHI, trypticase soy agar, or *Brucella* agar is preferred by many workers.

BHI agar with 5% to 10% sheep blood and the antimicrobial agents chloramphenicol (16

FIGURE 9.7

Sheep blood agar plate exhibiting all three types of hemolysis.

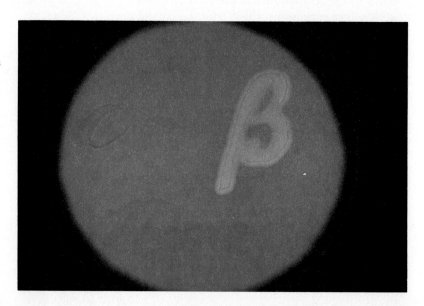

µg/ml) and gentamicin (5 µg/ml) will inhibit the growth of bacteria while allowing the growth of even the most fastidious dimorphic fungi. This agar should be used as a primary plating medium for the growth of fungi, since it has been shown to yield better recovery than the previously recommended Sabouraud dextrose.

CHOCOLATE AGAR

This medium uses the same base as blood agar. Originally, red blood was added to the molten base and the temperature raised enough to lyse partially the red blood cells (about 85° C), causing the medium to turn a chocolate-brown color. Now, hemoglobin and the other nutrients present in the lysed red cells, hemin (also known as "X" factor), and the coenzyme nicotine adenine dinucleotide (called "V" factor) are added as supplements to a nutritionally rich agar base. *Neisseria gonorrhoeae* and *Haemophilus* sp., among other fastidious organisms, will grow best in the presence of the nutrients supplied by chocolate agar.

CHOPPED MEAT BROTH

With or without added glucose, this medium is used to enrich and preserve the growth of anaerobic organisms. The pieces of meat provide substrates for proteolytic enzymes, serve as reducing substances to maintain the low Eh, and somehow prevent rapidly growing bacteria from overgrowing slower forms. A mineral oil or vaspar overlay on a chopped meat culture of an anaerobic bacterium will allow many anaerobes to remain viable at room temperature for several months. Chopped meat need not be incubated anaerobically to support the growth of anaerobes.

COLUMBIA CNA AGAR WITH BLOOD

Columbia agar base is a nutritionally rich formula containing three peptone sources. Five percent defibrinated blood provides more nutrients and the capability of displaying hemolytic reactions. The antibacterial agents colistin (10 µg/ml) and nalidixic acid (15 µg/ml) completely suppress the growth of Enterobacteriaceae and *Pseudomonas* sp. while allowing yeast, staphylococci, streptococci, and enterococci to grow. Certain gram-negative organisms, such as *Gardnerella vaginalis* and some *Bacteroides* sp., can grow very well on Columbia CNA agar with blood.

GN BROTH

Used as a selective broth for the cultivation of *Salmonella* and *Shigella* from stool specimens and rectal swabs, GN (gram-negative) broth contains several active ingredients. Sodium citrate and sodium desoxycholate (a bile salt) destroy gram-positive organisms and inhibit the early multiplication of coliforms. The addition of more mannitol than dextrose serves to encourage the growth of mannitol-fermenting pathogens and discourage the growth of *Proteus* sp. The medium is buffered to remain at neutral pH, even after production of acid metabolites by bacterial growth. To obtain the optimal benefit of the selective nature of GN broth, it should be subcultured 6 to 8 hours after initial inoculation and incubation. After this time, the coliforms begin to overgrow the pathogens.

HEKTOEN ENTERIC AGAR

This medium is included as an example of a selective, differential agar that is not autoclave sterilized. The concentrations of bile salts and the dyes bromthymol blue and acid fuchsin are high enough to inhibit the growth of most normal fecal flora while inhibiting the growth of *Salmonella* and *Shigella* sp. only slightly. Because so few organisms can grow on the medium, it is unnecessary to sterilize it before dispensing plates. Lactose-fermenting organisms, by lowering the pH of the medium in the area of colonies, turn the colonies yellow. The addition of ferric ammonium citrate, a source of iron common to many media formulas, allows the production of H_2S from sodium thiosulfate to be visualized by formation of a black precipitate around colonies.

KANAMYCIN-VANCOMYCIN LAKED RABBIT BLOOD AGAR (KVLB)

KVLB agar is useful for the selective isolation of *Bacteroides* and *Prevotella* sp. The medium contains 75 µg/ml kanamycin, which inhibits most aerobic, facultative, and anaerobic gram-negative rods except for *Bacteroides,* and 7.5 µg/ml vancomycin, which inhibits most gram-positive organisms. The laked blood allows earlier pigmentation of the pigmented anaerobic gram-negative rods. However, many strains of *Porphyromonas asaccharolytica* and *P. gingivalis* will not grow on this medium; a modification containing only 2 µg/ml vancomycin is used for selection of these species. Yeast and other kanamycin-resistant organisms sometimes grow on this medium; therefore, one should Gram stain and check the aerotolerance of all isolates.

LÖWENSTEIN-JENSEN AGAR

This medium was developed as a selective enrichment agar for mycobacteria. Malachite green dye inhibits the growth of contaminants that are able to survive the initial specimen processing. Löwenstein-based media are the only frequently used formulas that require the addition of homogenized eggs, which necessitates sterilization by inspissation. The utilization of egg protein by some mycobacteria results in production of niacin, an important differentiating characteristic. Many modifications of the basic medium are used, including those of Gruft (added antibiotics and ribonucleic acid growth factor) and Wallenstein (added glycerol).

MACCONKEY AGAR

The most frequently used primary selective and differential agar, MacConkey agar, contains crystal violet dye to inhibit the growth of gram-positive cocci and the pH indicator neutral red to impart differential characteristics. Gram-negative bacilli grow readily; lactose fermenters produce acid products of metabolism that cause the pH of the medium close to the colony to fall. The neutral red then turns red at the acid pH (Figure 9.8). Nonlactose fermenters remain colorless and translucent. MacConkey is the most supportive of the selective and differential media used for isolation of *Shigella* sp.

MIDDLEBROOK AGARS AND BROTHS

These media, made of more defined components than egg base media, are also used for cultivation of mycobacteria. The malachite green and low pH serve to inhibit contaminants, and the clear nature of the agar allows technologists to observe colony morphology easily. Middlebrook formulas are also used for antimicrobial susceptibility testing of mycobacteria, since the antimicrobial

agents would be damaged by the repeated heating used to prepare inspissated egg base media. The enrichment supplements of oleic acid, albumin, dextrose, and catalase must be added to support the growth of most mycobacteria.

PHENYLETHYL ALCOHOL AGAR

By adding phenylethyl alcohol to a peptone and beef extract base, an agar is created that inhibits the growth of gram-negative bacteria. Five percent sheep blood provides nutrients for streptococci and staphylococci. This medium is also used to cultivate anaerobic gram-negative organisms, which are not inhibited by the alcohol. After preparation, phenylethyl alcohol plates smell like roses.

PPLO AGAR

An enrichment medium for mycoplasma (formerly called **pleuropneumonia-like organisms, PPLO**), this formula contains slightly less agar than most bacteriological media and is prepared at a slightly higher pH. Beef heart infusion, peptones, and the added nutrients of serum or ascitic fluid allow the mycoplasma to grow as small colonies with dense centers. Agents such as penicillin and crystal violet can be added to inhibit contaminants.

SABOURAUD DEXTROSE AGAR

Developed for the cultivation of pathogenic fungi, particularly the agents of superficial mycoses, this medium contains peptones, dextrose, and agar. It is recommended only for primary isolation of dermatophytes at this time, with the addition of the

FIGURE 9.8

Appearance of lactose-fermenting and non–lactose-fermenting colonies on MacConkey agar.

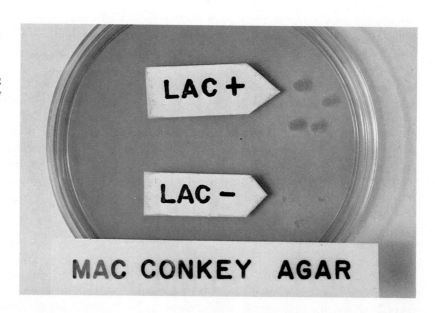

antimicrobial agents cycloheximide (0.5 µg/ml) and chloramphenicol (16 µg/ml). Subcultures of fungi originally isolated on BHI may exhibit more standard morphology on Sabouraud dextrose; thus, it is still useful for identification of molds once they have been isolated. The final pH, about 5.6, is much lower than that of most media and tends to inhibit the growth of bacteria.

STREPTOCOCCAL SELECTIVE AGAR
Recently improved, streptococcal selective agar (SSA) is available commercially. A modification of sheep blood agar, this medium contains crystal violet, trimethoprim-sulfamethoxazole, and colistin in concentrations adequate to inhibit most streptococci except for *Streptococcus pyogenes* and *S. agalactiae*. Beta hemolysis is readily observed. The medium is effective for primary plating of throat swabs for detection of group A streptococci.

THAYER-MARTIN AGAR
Chocolate agar has been modified to be selective for pathogenic *Neisseria* by the addition of antibiotics, including colistin (to inhibit other gram-negative bacteria), vancomycin (to inhibit gram-positive bacteria), and nystatin or anisomycin (to inhibit yeast). Thayer-Martin agar incorporates nystatin to inhibit yeasts. Modified Thayer-Martin (MTM) includes trimethoprim to inhibit *Proteus*. Although several modifications are available, the most common is Martin-Lewis agar, which substitutes anisomycin (characterized by a longer shelf life and greater activity against yeast) for nystatin.

THIOGLYCOLATE BROTH
The most frequently used enrichment broth in clinical microbiology, thioglycolate, uses 0.075% agar to prevent convection currents from carrying atmospheric oxygen throughout the broth. Thioglycolic acid also acts as a reducing agent, lowering the Eh of the medium. With the addition of many nutrient factors, such as casein, yeast and beef extracts, vitamins, and others, this medium enhances the growth of most pathogenic bacteria. Other nutrient supplements, an oxidation-reduction indicator (resazurin), dextrose, vitamin K_1, and hemin, have been added in various modified formulas. Technologists can visualize the difference between the diffuse, even growth of gram-negative, facultative bacilli and the discrete, puff-ball-type growth of gram-positive cocci. Strict aerobes, such as pseudomonads and yeast, tend to grow in a thin layer on the surface of the broth (Figure 9.9). For cultivation of anaerobes, thioglycolate supplemented with hemin (5 µg/ml), vita-

min K_1 (0.1 µg/ml), and a marble chip of sodium bicarbonate (1 mg/ml) is best.

TRICHOMONAS (DIAMOND'S) MEDIUM
Developed for isolation of human protozoa, this medium contains antibiotics such as chloramphenicol to inhibit growth of contaminating bacteria. The pH, about 6.0, favors the growth of trichomonads. Nutrients are provided by peptones, maltose, and cysteine.

XYLOSE-LYSINE-DESOXYCHOLATE AGAR
As with Hektoen agar, xylose-lysine-desoxycholate (XLD) agar is selective for *Shigella* and *Salmonella* and is not autoclave sterilized. The salts inhibit many Enterobacteriaceae and gram-positive organisms. The phenol red indicator accounts for the differentiation of non–lactose fermenters *(Shigella, Salmonella)* as colorless (pale pink) colonies. Ferric ammonium citrate allows the visualization of H_2S-producing organisms as colonies with black centers. Organisms that ferment the carbohydrates in the medium (xylose, lactose, sucrose) produce yellow colonies.

• • •

Many other media are available that can be used to cultivate pathogens. We have mentioned examples of supportive, enrichment, selective, and differential media. Our basic recommendations for handling specimens are covered in Part Three. However, for an individual laboratory, the ultimate choice of which primary and secondary media to use for inoculating clinical specimens depends on the patient population served by the

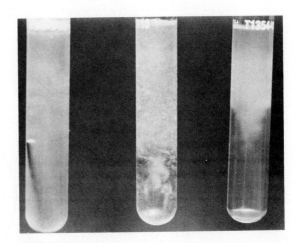

FIGURE 9.9

Growth of gram-negative bacilli *(left tube)*, gram-positive cocci *(center tube)*, and yeast *(right tube)* in thioglycolate broth.

laboratory, the extent of services offered by the laboratory, and the cost-benefit ratio of each additional medium. The laboratory director often needs to make difficult decisions concerning limiting the numbers of media used to obtain the most generally beneficial results. Some of these issues are addressed in Chapter 4.

USING STREAK PLATES TO ISOLATE AND ENUMERATE GROWTH OF PATHOGENS

Inocula are usually spread over the surface of agar plates in a standard pattern so that the quantity of bacterial growth can be determined, either semiquantitatively or relatively. A useful streaking pattern is illustrated in Figure 9.10. The relative numbers of organisms in the original specimen can be estimated based on the extent of growth of colonies past the original area of inoculation. For some viscous specimens, such as sputum, and for some highly selective media, such as stool agars, better isolation of all colony types can be achieved if the technologist returns the loop to the original inoculum area on the plate several times during the initial streaking pattern. We have found that flaming the loop between streaking areas is not

necessary for most specimens, although turning the loop to access a previously unused edge will enhance isolation. It may be beneficial, however, to flame the inoculating loop between streaking areas when the original inoculum is material from an already-growing bacterial colony.

Streaking plates with a measured amount of inoculum, such as that found in a standard calibrated loop (used to quantify colony-forming units [CFUs] in fluid specimens such as urine), should be done to facilitate counting colonies. For this purpose, the inoculum should be spread out more evenly over the entire plate (Figure 9.11).

QUALITY CONTROL OF MICROBIOLOGICAL PRIMARY CULTURE MEDIA

Until recently, laboratories were required by regulatory agencies to perform in-house quality control tests on all media prepared or purchased by the laboratory. The time and expense of in-house testing placed a strain on already overworked microbiologists, and its value, particularly for commercial media, was questioned by many critics of the system. The National Committee for Clinical Laboratory Standards (NCCLS) studied the issue

FIGURE 9.10
Streaking pattern for primary inoculation of plates to achieve isolated colonies. Growth in the initial half of the plate only is semiquantitated as 1+ or 2+ (sparse); growth into the third quadrant is reported as 3+ (moderate); and growth into the fourth quadrant is reported as 4+ (heavy). If numbers of colonies can be counted, this number should be reported.

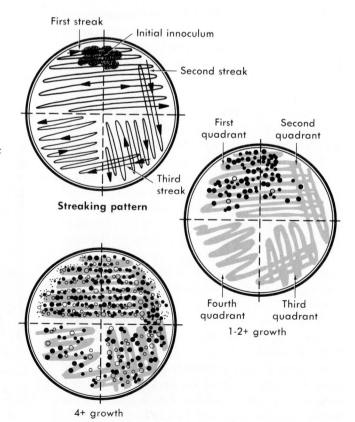

FIGURE 9.11
Inoculation of plates
streaked with a calibrated
loop.

and subsequently developed recommendations for abbreviated quality control testing procedures for commercially prepared media that would be more relevant.[2] When manufacturers follow the testing methods recommended, the performance of their media has been shown to be adequate and consistent.

In addition to listing those purchased media that microbiologists need not test in-house, the NCCLS guidelines include methods for testing the performance of media that are applicable to all primary growth media that require testing, either by the manufacturer or within the laboratory (Procedure 9.1). For quality control organisms and expected results used for commercially prepared media, the reader is referred to the NCCLS publication M22-A. Organisms and results expected for several of the major purchased media that must still be tested within each laboratory are listed in Table 9.2. Guidelines did not include commercial identification kit media, susceptibility testing media, or media used for isolation of parasites, viruses, chlamydiae, or mycoplasmas. Throughout this text, quality control suggestions accompany the procedures presented. Of course, all media prepared in the laboratory must be fully tested before placing into use. In the absence of standardized guidelines, testing should be performed with clones of stock strains of organisms for which the medium is intended to show expected characteristics, both positive and negative. Quality control organisms are available commercially.

CELL CULTURES FOR CULTIVATION OF PATHOGENS

Although fungi and most bacteria grow readily on artificial media, certain pathogens require factors provided only by living cells. Often these organisms are obligate intracellular parasites, but occa-

sionally they are organisms for which the critical growth factors needed to support growth on wholly artificial media are not known. An example of a pathogen of the latter type is *Legionella pneumophila,* which was isolated only in chicken embryo culture before the necessary growth factors were discovered and incorporated into artificial media. All viruses and chlamydiae are obligate intracellular parasites, growing outside of the host only in tissue cultures or cell cultures. These media are made up of layers of living cells growing on the surface of a solid matrix such as the inside of a glass tube or the bottom of a plastic flask (Figure 9.12). Pathogenic viruses are cultivated in clinical microbiology laboratories on cell cultures of two types, **primary cell lines** and **continuous cell lines.** Primary cell lines are produced by cutting up fresh tissue, often kidney, into tiny pieces. When treated with the proteolytic agent trypsin, the tissue pieces break up into individual cells, which are seeded into a flask or tube containing growth-supportive media. The cells attach to the inside bottom of the container and multiply until they reach a single layer, a **monolayer,** of confluent growth. Normal cells are inhibited from growing on top of each other. Fibroblast cells, such as human foreskin primary cell line, are spindle

FIGURE 9.12

Tissue culture flask used for maintaining cell lines.

PROCEDURE 9.1

QUALITY CONTROL TESTING FOR PRIMARY CULTURE MEDIA

PRINCIPLE

Certain media, particularly those with blood product additives, must be monitored for sterility, support of growth of desired microorganisms, and performance characteristics within the expected parameters of the media. Organisms of known growth characteristics can be used to verify media performance standards.

METHOD

1. User must inspect plate media for cracks in the agar or in the plastic Petri dish, unequal filling of plates, hemolysis of blood additives, the mushy appearance that signifies that the plates had been frozen during transport, excessive bubbles, and contamination.

2. User should incubate selected plates before use to ensure that they are sterile.

3. User should test the growth-supportive characteristics by inoculating the appropriate quality control organism(s) as follows:
   a. Prepare a fresh subculture of the test organism(s). Organisms can be stored frozen at −70° C in soybean casein digest broth with a final concentration of 10% glycerol, frozen in skim milk or purchased lyophilized. After incubation sufficient to produce isolated colonies, the organisms can be used for testing.
   b. Prepare a suspension of several colonies in sterile saline to match a McFarland 0.5 turbidity standard (Procedure 14.1), approximately 2×10^8 colony-forming units (CFUs) per milliliter. This is the stock suspension.
   c. Dilute this suspension 1:100 in sterile saline, and inoculate each plate to be tested with 0.01 ml (use a 10 μL calibrated loop) of the diluted suspension. Spread the inoculum evenly on the plate so that colonies can be counted.

4. User should test the inhibitory characteristics of a selective medium as follows:
   a. Dilute the stock suspension 1:10 in sterile saline, and inoculate each plate to be tested with 0.01 ml of the diluted suspension as above.

5. User should test the tubed media by inoculating 0.01 ml of the stock suspension into the tube.

6. Incubate all media under the conditions required for normal use.

EXPECTED RESULTS

Media are within acceptable limits if they are sterile, if test organisms grow well and display typical colony morphology, and if appropriate test organisms are inhibited. Tubed media should display proper positive and negative reactions with appropriate test microorganisms.

PERFORMANCE SCHEDULE

Media prepared in-house should be tested with each batch. Purchased media should be tested with each lot number and with each shipment received at a separate time.

---

shaped (Figure 9.13). Cells from other tissue sources, such as kidney cells, may be shaped more irregularly, such as polygons (Figure 9.14). Primary cell lines usually carry the same number of chromosomes as the natural cells from which they were derived, the diploid number of normal somatic cells, and they will multiply for approximately only 50 generations before they begin to die off. If cells are obtained from a malignant tissue source, such as a human epithelial cell cancer, they will multiply indefinitely in cell culture and they often contain aberrations in the number of chromosomes they possess (aneuploid). Such cell cultures are called continuous cell lines.

In addition to their use for cultivation of viruses, tissue cultures may be used for other pur-

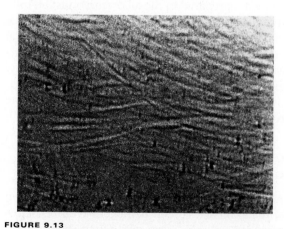

FIGURE 9.13

Monolayer of uninoculated human fibroblasts, visualized at 400×.

TABLE 9.2

ABBREVIATED LIST OF QUALITY CONTROL PROCEDURES FOR MEDIA THAT REQUIRE IN-HOUSE TESTING

MEDIUM	CONTROL ORGANISM (SUGGESTED ATCC NO.)*	EXPECTED RESULTS
Campylobacter agar	Campylobacter jejuni (33290)	Growth
	Escherichia coli (25922)	Inhibition (partial)
Agar for isolation of pathogenic Neisseria (not modified Thayer-Martin)	Neisseria gonorrhoeae (43070)	Growth
	Neisseria meningitidis (13090)	Growth
	Neisseria sicca (9913)	Inhibition (complete)
	Candida albicans (60193)	Inhibition (partial)
	Proteus mirabilis (43071)	Inhibition (partial)
	Staphylococcus epidermidis (12228)	Inhibition (partial)
Rabbit or horse blood agar	Haemophilus influenzae (10211)	Growth, no hemolysis
	Haemophilus haemolyticus (33390)	Growth, beta hemolysis
Bordet-Gengou agar	Bordetella pertussis (9340)	Growth
Charcoal yeast and buffered charcoal yeast extract agar	Legionella pneumophila (33152)	Growth
Thiosulfate-citrate–bile salts–sucrose agar	Vibrio parahaemolyticus (17802)	Blue-green colonies
	Vibrio alginolyticus (17749)	Yellow colonies
	E. coli (25922)	Inhibition
Cycloserine-cefoxitin-fructose agar	Clostridium difficile (9689)	Yellow colonies
	E. coli (25922)	Inhibition
KVLB agar (anaerobic)	Prevotella intermedia (25261)	Brown colonies
	Bacteroides fragilis (25285)	Growth
	E. coli (25922)	Inhibition

* American Type Culture Collection available from ATCC, Rockville, Md.

FIGURE 9.14

Monolayer of African green monkey kidney cells, visualized at 400×.

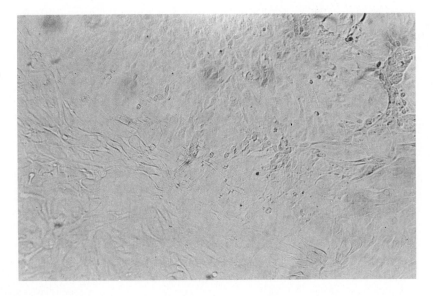

poses in clinical microbiology. Several cell lines, including Hep-2, HeLa, and Vero, have been used for detection of the cytotoxin of *Clostridium difficile*. Chlamydiae are often cultivated in McCoy cells, the monolayer of which is treated with cycloheximide to prevent multiplication before the specimen is inoculated. Maintenance and inoculation of the tissue culture lines used for routine clinical virology are reviewed in Chapter 43.

REFERENCES

1. Holdeman, L.V., Cato, E.P., and Moore, W.E.C. 1977. Anaerobe laboratory manual, ed. 4. Anaerobe Laboratory, Virginia Polytechnic Institute and State University, Blacksburg, Va.
2. National Committee for Clinical Laboratory Standards. 1990. Quality assurance for commercially prepared microbiological culture media, Approved Standard M22-A. Order No. M22-A, available from NCCLS, 771 East Lancaster Ave., Villanova, PA 19085.

3. Nelson, E.A., and Molitoris, E. 1992. Autoclave (steam sterilizer). Sect. 12.3. In Isenberg, H.D., editor. Clinical microbiology procedures handbook. American Society for Microbiology, Washington, D.C.

BIBLIOGRAPHY

Atlas, R.M. and Parks, L.C. (editor). 1993. Handbook of microbiological media. CRC Press, Boca Raton, Fla.

Difco manual, ed. 10. 1984. Difco Laboratories, Detroit.

Emmons, C.W., Binford, C.H., Utz, J.P., and Kwon-Chung, K.J. 1977. Medical mycology, ed. 3. Lea & Febiger, Philadelphia.

Estevez, E.G. 1984. Bacteriologic plate media: review of mechanisms of action. Lab. Med. 15:258.

Hesse, W. 1992. Walther and Angelina Hesse—early contributors to bacteriology. ASM News 58:425.

Isenberg, H.D., Washington, J.A. II, Doern, G.V., and Amsterdam, D. 1991. Specimen collection and handling. p. 15. In Balows, A., Hausler, W.J., Jr., Herrmann, K.L., et al., editors. Manual of clinical microbiology, ed 5. American Society for Microbiology, Washington, D.C.

Joklik, W.K. 1984. The nature, isolation, and measurement of animal viruses. p. 801. In Joklik, W.K., Willett, H.P., and Amos, D.B., editors. Zinsser's microbiology, ed. 18. Appleton-Century-Crofts, Norwalk, Conn.

MacFaddin, J.F. 1985. Media for isolation-cultivation-identification-maintenance of medical bacteria, vol. 1. Williams & Wilkins, Baltimore.

Nash, P., and Krenz, M.M. 1991. Culture media. p. 1226. In Balows, A., Hausler, W.J., Jr., Herrmann, K.L., et al., editors. Manual of clinical microbiology, ed. 5. American Society for Microbiology, Washington, D.C.

Oxoid manual, ed. 5. 1982. Oxoid Ltd., Basingstoke, U.K.

Summanen, P., Baron, E.J., Citron, D.M., et al. 1992. Wadsworth anaerobic bacteriology manual, ed. 5. Star Publishing Co., Belmont, Calif.

Washington, J.A., II. 1985. Laboratory procedures in clinical microbiology, ed. 2. Springer-Verlag, New York.

10

CONVENTIONAL AND RAPID MICROBIOLOGICAL METHODS FOR IDENTIFICATION OF BACTERIA AND FUNGI

Cultivation and identification of specific pathogens from material collected from patients suspected of having infection is often still the most reliable diagnostic tool, even though it is not the fastest. In some cases (e.g., with *Rickettsia* sp. and human immunodeficiency virus [HIV]), recovery of the infecting organisms is difficult or impossible (as with *Treponema pallidum*). In those cases, reliance must be placed on serologic, molecular, or other methods for diagnosis. However, until molecular biological techniques are readily available and cost-effective for routine diagnostic work, definitive diagnosis of most infections continues to require isolation of an etiological agent. Chapter 9 discusses various growth media and strategies for encouraging the growth of cultivatable microorganisms. This chapter discusses basic morphological clues and enzymatic and biochemical tests that are used to identify such pathogens once they have been isolated. In addition, we mention some rapid methods and basic concepts of the commercial systems currently available for performing biochemical and enzymatic tests. Certain biochemical tests used to identify bacteria are also used for identification of some fungi.

Identification of viruses does not depend primarily on biochemical or enzymatic means. Virus identification relies more heavily on the visual detection of specific damage inflicted on tissue culture cells by invasion and proliferation of viruses. Immunological assays for viral antigens in tissue or tissue culture are also used. Fungi and eukaryotic parasites are identified primarily by visualizing characteristic morphological features of the parasites themselves. Techniques for identification of these agents are discussed in Chapters 43 to 45. In addition, many new methods based on antigenic detection and molecular manipulations are not covered in this chapter; several are discussed in Chapter 11.

BASIC APPROACHES TO IDENTIFICATION OF PATHOGENS

This section discusses general strategies for determining the category of pathogen isolated and for deciding which further tests are needed for identification.

PRELIMINARY IDENTIFICATION OF COLONIES GROWING ON SOLID MEDIA

Since most clinical specimens are inoculated onto and into several media, including some selective or differential agars (Chapter 9), the first clue to identification of an isolated colony is the nature of the medium on which the organism is growing. For example, with rare exceptions such as enterococci, only gram-negative bacteria grow well on MacConkey agar or other selective or differential agars for gram-negative bacteria. Plates that contain substances inhibitory to gram-negative bacteria, such as Columbia agar with colistin and nalidixic acid, support growth of gram-positive organisms. Most bacteria and fungi proliferate on nutrient or supplemented media, such as 5% sheep blood agar, chocolate agar, and brain-heart infusion agar.

Moldlike fungi produce fuzzy or fluffy growth because of their aerial hyphae. If a fuzzy colony is noted on agar plates (Figure 10.1), the plate should be placed into a biological safety cabinet before the plate is opened for further examination. All bacterial colonies should be subcultured to other media. The fungus can then be studied by methods outlined in Chapter 44. Before the plate is removed from the cabinet, it should be sealed

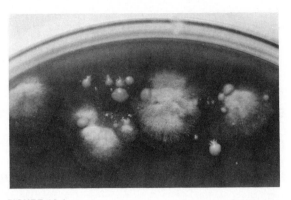

FIGURE 10.1

A moldlike fungus is growing among the bacterial colonies on this primary blood agar plate inoculated with sputum. The plate must be examined in a biohazard hood to avoid dissemination of the fungus.

on two sides with tape to prevent dissemination of spores. Such precautions decrease the chances of laboratory acquisition of systemic fungal infection and prevent contamination of other media during subsequent plate handling or incubation.

It is unwise to place total confidence on colonial morphology for preliminary identification of isolates on primary media, since a microorganism may produce a colony that is not different from colonies of many other species. For example, yeast colonies often resemble those of staphylococci. Unless colony morphology is distinctive (e.g., that of typical *Pseudomonas aeruginosa*) or growth appears on selective media, the cellular morphology of the microorganism must be determined. Even colonies on selective media usually require some additional verification (e.g., some *Escherichia coli* colonies are indistinguishable from those of *Shigella* sp. on Hektoen agar). Examination of a wet preparation of bacterial colonies under oil immersion (1000× magnification), with or without phase microscopy, can rapidly provide many clues as to possible identity. For example, a wet preparation prepared from a translucent, alpha-hemolytic colony on blood agar may yield cocci in chains, a strong indication that the bacteria are probably streptococci; or long, thin, uniform, chaining rods, suspicious for lactobacilli. Motility can often be detected by this initial examination; the presence of yeast cells, mimicking bacterial colonies, can also be discovered.

GRAM STAIN MORPHOLOGY FOR INITIAL IDENTIFICATION OF BACTERIA: CLASSICAL AND NONTRADITIONAL METHODS

The microbiologist should perform a Gram stain of material from isolated colonies to gain the most valuable cellular morphological information. Most bacteria can be divided into four distinct

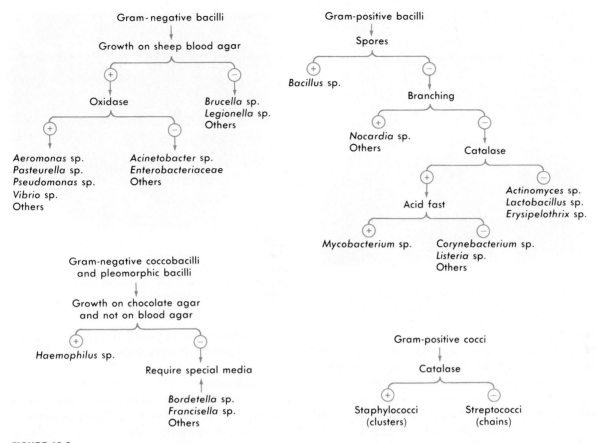

FIGURE 10.2

Flow chart illustrating basic identification parameters for bacteria.

groups: gram-positive cocci, gram-negative cocci, gram-positive bacilli, and gram-negative bacilli. Some species are morphologically indistinct and are described by combining the preceding terms, such as "gram-negative coccobacilli" or "gram-variable bacilli." Other morphological shapes encountered include curved rods and spirals. Identification procedures are based on the cellular morphology of bacteria. Once this basic information is known, definitive identification can proceed (Figure 10.2). Gram stain results are not always indicative of the organism's cell wall structure. For example, certain gram-positive bacteria (and fungi) lose some of their cell wall integrity with age or under adverse conditions (e.g., exposure to antimicrobial agents) and appear to stain gram negative. Many *Bacillus* and *Clostridium* sp. appear gram negative on stains, resulting in incorrect additional tests for identification (and much confusion). To try to capture the proper stain reaction of an organism suspected of being gram positive, it may be useful to stain a very young broth subculture. If this fails, at least two nonstain systems are available to aid in determination of true Gram stain reaction. Carlone et al.[2] have shown the reagent L-alanine-4-nitroanilide (LANA) to differ-

entiate Gram reactivity of aerobic and facultatively anaerobic organisms. This reagent is commercially available impregnated in cotton swabs or on filter paper disks, which turn yellow when touched to the colony of a gram-negative bacterium. The potassium hydroxide (KOH) test has also been used successfully for problem organisms.[4] A loop of growth from a colony of the organism is emulsified on the surface of a glass slide in a suspension of 3% KOH. The suspension is stirred continuously for 60 seconds, after which the loop is gently pulled from the suspension. Gram-negative cell walls are broken down, releasing viscid chromosomal material, which causes the suspension to become thick and stringy (Figure 10.3). Although useful, neither of these tests is foolproof, as reviewed by von Graevenitz and Bucher.[7] Many gram-positive bacteria (with a few exceptions, such as certain lactobacilli, *Leuconostoc,* and *Pediococcus* sp.) are susceptible to vancomycin, an antimicrobial agent that acts on the gram-positive cell wall. Inhibition of growth by vancomycin at concentrations as low as 3 µg/ml can often presumptively identify a gram-negative-staining organism as gram positive. Certain gram-negative organisms, notably

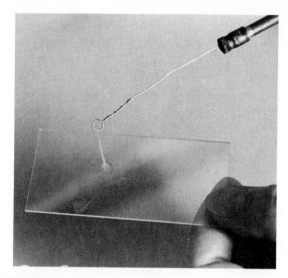

FIGURE 10.3

The KOH test. Addition of 3% KOH to a gram-negative bacillus causes release of viscous nuclear material.

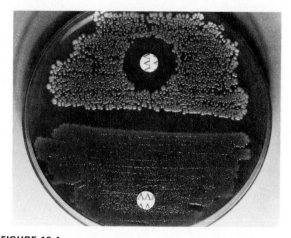

FIGURE 10.4

Susceptibility to 5 µg vancomycin (impregnated in a filter paper disk) can differentiate gram-positive bacilli *(upper half of plate)* from gram-negative bacilli (no zone of inhibition, *lower half of plate*).

Moraxella and *Acinetobacter* sp., may be vancomycin susceptible. The vancomycin screening test for gram-negative organisms can be easily performed by heavily inoculating the organism onto a portion of the surface of a sheep blood agar plate and placing a 5 µg vancomycin disk (available from disk manufacturers, typically used for assisting in Gram reaction determination with anaerobic bacteria) on the inoculated agar surface. Any zone of inhibition after overnight incubation is usually indicative of a gram-positive bacterium (Figure 10.4). Conversely, with a few exceptions, most truly gram-negative organisms are resistant to vancomycin and susceptible to colistin or polymyxin at 10 µg/ml. Differentiation of the vancomycin-resistant streptococci-like organisms is discussed in Chapter 26. A 30 µg vancomycin disk is used for that purpose.

EXTREMELY RAPID BIOCHEMICAL OR ENZYMATIC TESTS THAT CAN BE PERFORMED ON SINGLE COLONIES GROWING ON PRIMARY MEDIA

All the tests described here can be performed directly using colonies seen on primary isolation plates. They all can produce positive reactions in less than an hour, and most take only a minute or two, although incubation for up to 4 hours may be required. Such tests can help the technologist place an isolate of known morphology into a further subdivision, thus directing the additional procedures needed for identification (Figure 10.2) and providing valuable tentative or preliminary identification in cases of serious infection. In addition to those tests mentioned here, others are

used in laboratories around the world; still others are constantly being developed.

CATALASE TEST

The enzyme catalase catalyzes the liberation of water and oxygen from hydrogen peroxide, a metabolic end product toxic to bacteria. All members of the staphylococci are catalase positive, whereas members of the genus *Streptococcus* are negative. Catalase reactions can also differentiate *Listeria monocytogenes* (catalase positive) from beta-hemolytic streptococci. Most *Neisseria* sp. are catalase positive. Catalase can also help distinguish *Bacillus* sp. (catalase positive) from *Clostridium* sp. (mostly catalase negative). The test can be performed with a very small amount of growth removed from an agar surface (Procedure 10.1).

CLUMPING FACTOR TEST ("SLIDE COAGULASE TEST")

Gram-positive cocci that are catalase positive belong to the family Micrococcaceae, which includes the staphylococci. The clumping factor test (Procedure 10.2) is used to screen quickly for isolates of *S. aureus*, which are almost always coagulase positive. Although other species of staphylococci may be coagulase positive (e.g., *S. intermedius*, *S. hyicus* ss. *hyicus*), they are not important agents of human disease. Clumping factor is a cell-associated substance that binds plasma fibrinogen, causing agglutination of the organisms by binding them together with aggregated fibrinogen. With the exception of *Staphylococcus lugdunensis*, which is clumping factor positive but not coagulase positive, those organisms that produce clumping factor also elaborate the coagulase enzyme and can be identified presumptively as *S. au-*

PROCEDURE 10.1

CATALASE TEST

PRINCIPLE

The breakdown of hydrogen peroxide into oxygen and water is mediated by the enzyme catalase. When a small amount of an organism that produces catalase is introduced into hydrogen peroxide, rapid elaboration of bubbles of oxygen, the gaseous product of the enzyme's activity, is produced.

METHOD

1. With a loop or sterile wooden stick, transfer a small amount of pure growth from the agar onto the surface of a clean, dry glass slide.

2. Immediately place a drop of 3% hydrogen peroxide (H_2O_2) onto a portion of a colony on the slide.

3. Observe for the evolution of bubbles of gas, indicating a positive test (Figure 10.5).

QUALITY CONTROL

Colonies of *Staphylococcus aureus* ATCC 25923 and *Streptococcus pyogenes* ATCC 19615 are tested.

EXPECTED RESULTS

The staphylococci are catalase positive and produce copious bubbles; streptococci are catalase negative and do not yield any visible bubbling of the hydrogen peroxide.

PERFORMANCE SCHEDULE

Test quality control organisms with each new batch of reagents placed into use and each day that tests are performed.

COMMENTS

The test should be performed only on isolates grown on non-blood-containing media. Although red blood cells contain some catalase, a technologist can distinguish the very weak reaction of contaminating red blood cells by performing a control slide catalase test with a small loopful of the blood-containing agar on the same slide with the organism. If the catalase reaction from the colony is much stronger than that from the agar alone, the test can be considered positive. Some bacteria produce a peroxidase that catalyzes the breakdown of hydrogen peroxide; the catalase test may thus appear to be weakly positive (a few bubbles slowly elaborated). This reaction should not be confused with a truly positive catalase test.

---

reus. Not all *S. aureus* strains produce clumping factor. The reaction must be read within 10 seconds, and the inclusion of an autoagglutination control decreases the number of false-positive results. Rabbit coagulase plasma with ethylenediaminetetraacetic acid (EDTA) or citrate is preferable to human plasma, which may contain substances inhibitory to the reaction. The clumping factor test for presumptive identification of *S. aureus* has been supplanted in many laboratories by specific immunological reagent particle agglutination tests that detect cell surface antigens such as protein A and clumping factor, providing a rapid and definitive identification without the need to perform further tests on clumping factor–negative isolates (Chapter 11).

OXIDASE TEST

Performed to identify *Neisseria* sp. presumptively and to characterize gram-negative bacilli initially, the oxidase test indicates the presence of the enzyme cytochrome oxidase. This iron-containing porphyrin enzyme participates in the electron transport mechanism and in the nitrate metabolic pathways of some bacteria. Although the test can be performed by flooding the agar surface of an inoculated plate with the reagent after incubation, it is most easily interpreted when carried out by the Kovac's method (Procedure 10.3). If an iron-containing wire is used to transfer growth, a

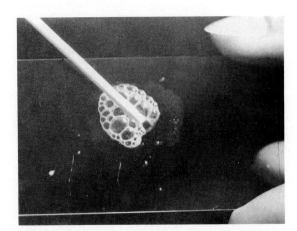

FIGURE 10.5

Catalase test. Bubbles of gas are released when a catalase-positive organism is emulsified in a drop of hydrogen peroxide on a slide.

PROCEDURE 10.2

CLUMPING FACTOR ("SLIDE COAGULASE") TEST

PRINCIPLE
The presence of a cell surface–associated substance that binds fibrinogen and thus allows aggregation of organisms in plasma-containing fibrinogen is detected by observation of clumping of cells.

METHOD
1. Place a drop of coagulase plasma (rabbit plasma with EDTA or citrate, available commercially from BBL Microbiology Systems, Difco Laboratories, and others) on a clean, dry glass slide.

2. Place a drop of distilled water or saline next to the drop of plasma as a control.

3. With a loop, straight wire, or wooden stick, emulsify an amount of the isolated colony being tested in each drop, inoculating the water or saline first. Try to create a smooth suspension.

4. Observe for clumping in the coagulase plasma drop and a smooth, homogeneous suspension in the control. Clumping in both drops indicates that the organism autoagglutinates and is unsuitable for the slide coagulase test.

QUALITY CONTROL
Colonies of *S. aureus* ATCC 25923 and *S. epidermidis* ATCC 12228 should be tested.

EXPECTED RESULTS
Positive organisms such as *S. aureus* exhibit immediate aggregation visible to the naked eye, usually resulting in complete clearing of the background of the suspension. Negative organisms such as *S. epidermidis* retain the smooth, milky appearance of the original suspension.

PERFORMANCE SCHEDULE
Quality control organisms should be tested with each new batch of coagulase plasma prepared and each day that tests are performed.

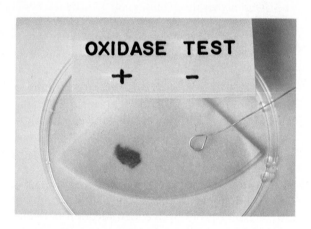

FIGURE 10.6
Kovac's oxidase test. A purple color results when a small portion from a colony of an oxidase-positive organism is rubbed onto filter paper saturated wtih the oxidase reagent. (From Baron, E.J. 1985. Clinical microbiology, vol. 3: Listen, look, and learn. Health and Educaiton Resources, Bethesda, Md.)

false-positive reaction may result; therefore, platinum wire or wooden sticks are recommended. Certain organisms may show slight positive reactions after the initial 10 seconds have passed; such results are not considered definitive. Passing a culture several times on artificial media and testing a very fresh subculture may sometimes induce an organism that shows a questionable or negative reaction to display its true oxidase positivity. Several manufacturers, including Becton Dickinson Microbiology Systems, Difco Laboratories, Austin Biological Laboratories, E-Y Laboratories, and others, produce oxidase test systems consist-

ing of reagent-impregnated filter paper strips or disposable glass ampules containing small amounts of reagent. Such systems yield results comparable to the conventional test and save preparation time.

SPOT INDOLE TEST

Organisms that produce the enzyme tryptophanase are able to degrade the amino acid tryptophan into pyruvic acid, ammonia, and the product indole. Indole is detected by its combination with the indicator aldehyde (available commercially) to form a colored end product (Procedure

PROCEDURE 10.3

SPOT OXIDASE TEST (KOVAC'S METHOD)

PRINCIPLE

The cytochrome oxidase enzyme is able to oxidize the substrate tetramethyl-*p*-phenylenediamine dihydrochloride, forming a colored end product, indophenol. The dark-purple end product will be visible if a small amount of growth from a strain that produces the enzyme is rubbed on substrate-impregnated filter paper.

METHOD

1. Prepare a solution of 1% tetramethyl-*p*-phenylenediamine dihydrochloride (available from Kodak Chemicals, Sigma Chemical Co., and other suppliers) in sterile distilled water each day. To eliminate some work, 50 mg of reagent powder can be weighed into each of many plastic, snap-top tubes at one sitting. Each day of use, 5 ml of water is added to one tube (e.g., to a predrawn line), which is discarded after that day's use. The reagent is also commercially produced in individually sealed glass ampules for daily use.

2. Place a filter paper circle into a sterile, plastic, disposable Petri dish, and moisten the filter paper with several drops of the fresh reagent.

3. Remove a small portion of the colony to be tested (preferably not more than 24 hours old) with a platinum wire or wooden stick, and rub the growth on the moistened filter paper.

4. Observe for a color change to blue or purple (Figure 10.6) within 10 seconds (timing is critical).

QUALITY CONTROL

Test *Escherichia coli* ATCC 25922 and *Neisseria gonorrhoeae* ATCC 43069.

EXPECTED RESULTS

Positive organisms such as *Neisseria* sp. turn the filter paper dark purple within 10 seconds; negative organism material such as that from *E. coli* remain colorless or the color of the colony within 10 seconds.

PERFORMANCE SCHEDULE

Perform when a new lot number of reagent is received and each day that tests are performed.

10.4 and Figure 10.7). This test can be used to differentiate swarming *Proteus* sp. from one another and to begin to characterize *E. coli* presumptively. It has also been found useful for examining anaerobic organisms.

BILE SOLUBILITY TEST

Streptococcus pneumoniae possesses an active autocatalytic enzyme that lyses the organism's own cell wall during cell division. Under the influence of a bile salt (sodium deoxycholate), the organisms rapidly autolyze. Procedure 10.5 outlines the performance of this test. Other alpha-hemolytic streptococci do not possess such an active enzyme and will not dissolve in bile. The bile solubility test may not always work, since old colonies may have lost their active enzyme. Therefore, non-bile-soluble streptococci that resemble pneumococci should be further identified by another method.

PYR (L-PYRROLIDONYL-β-NAPHTHYLAMIDE HYDROLYSIS) TEST

S. pyogenes, Enterococcus sp., some staphylococci, and miscellaneous other gram-positive strains are able to hydrolyze the substrate PYR via the enzyme L-pyroglutamyl aminopeptidase. This test is as specific as the more labor-intensive and slower bile esculin agar and salt broth tests used classically to identify enterococci and more specific

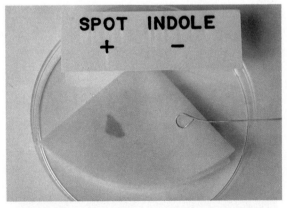

FIGURE 10.7

Spot indole test. A blue-green color results when a small portion from a colony of a indole-positive organism is rubbed onto filter paper saturated with the indole reagent. (From Baron, E.J. 1985. Clinical microbiology, vol. 3: Listen, look, and learn. Health and Education Resources, Bethesda, Md.)

PROCEDURE 10.4

SPOT INDOLE TEST

PRINCIPLE

The indole end product of the action of tryptophanase on tryptophan can be detected by its ability to combine with certain aldehydes to form a colored compound. The blue-green compound formed by indole and cinnamaldehyde is visualized by rubbing bacteria that produce tryptophanase on filter paper impregnated with the substrate.

METHOD

1. Prepare indole reagent (1% paradimethylaminocinnamaldehyde, available from Sigma Chemical Co. and other chemical supply houses, dissolved in 10% [vol/vol] concentrated hydrochloric acid). Store in a dark bottle in the refrigerator.

2. Saturate a qualitative filter paper (Whatman No. 1 is fine) in the bottom of a Petri dish with the reagent.

3. Using a wooden stick or loop, rub a portion of the colony on the filter paper. Rapid development of a blue color indicates a positive test. Most indole positive organisms turn blue within 30 s.

QUALITY CONTROL

Test a fresh subculture of *E. coli* ATCC 25922 and *Enterobacter cloacae* ATCC 23355.

EXPECTED RESULTS

Positive organisms such as *E. coli* display a blue-green color on the filter paper; negative organisms such as *E. cloacae* remain colorless or turn slightly pink.

PERFORMANCE SCHEDULE

Perform when a new lot number of reagent is received and each day that tests are performed.

---

than the overnight bacitracin test used classically to identify group A streptococci presumptively. Several configurations of the test are available commercially (Figure 10.9). One method is given in Procedure 10.6. Colonies should be catalase negative and morphologically consistent with these two groups of organisms, since certain Micrococcaceae and occasional viridans streptococci may also produce the enzyme. *Lactococcus garviae*, a recently recognized streptococcus-like species, can hydrolyze the substrate slowly, resulting in

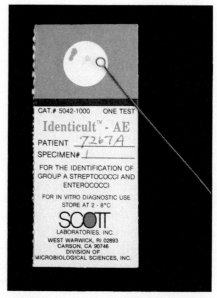

FIGURE 10.9

PYR test. The presence of the aminopeptidase enzyme results in a pink color change when the indicator is added (organism smeared on left side of paper circle). PYR-negative organisms show no color change (organisms on right side of filter paper circle). The system pictured (Scott Laboratories, now sold by Adams Scientific) is one of several available commercially. (Photo by Pete Rose.)

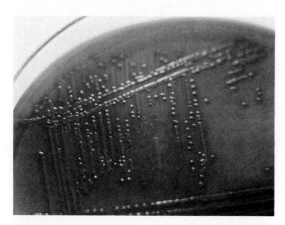

FIGURE 10.8

Bile solubility test. Within 15 minutes after placing a drop of bile reagent on a colony of *S. pneumoniae*, the colony dissolves.

PROCEDURE 10.5

BILE SOLUBILITY TEST

PRINCIPLE

S. pneumoniae will rapidly autocatalyze in the presence of the surfactant sodium deoxycholate (one of the major components of bile). The "melting" or dissolution of a colony within 30 minutes after exposure to the detergent is indicative of this species.

METHOD

1. Make a solution of bile salt (10% sodium deoxycholate, available from Difco Laboratories, Sigma Chemical Co., and other sources, in distilled water), and store it at room temperature. The reagent is good for 6 months as long as it remains sterile.

2. Select a well-isolated suspicious colony from a blood agar or chocolate agar plate, and place 1 drop of the reagent directly on the colony.

3. Keeping the plate very level to prevent the reagent from running and washing a nonpneumococcal colony away, producing a false-positive result, allow the reagent to dry. Placing the plate in an aerobic incubator will speed up this process.

4. When the reagent has dried (in approximately 15 minutes), examine the area for the absence or flattening of the original colony (Figure 10.8), indicators of bile solubility. Ascertain that an alpha-hemolytic streptococcal colony has not merely been washed to another portion of the plate by making certain that the bile drop has remained where it was placed (the blood under the drop becomes slightly hemolyzed).

QUALITY CONTROL

Test a fresh subculture of *S. pneumoniae* ATCC 27336 and *S. mitis* ATCC 15909.

EXPECTED RESULTS

Species of *S. pneumoniae* disappear under the drop, and the area appears flat. Other streptococci are not affected, and the colony elevation is obvious once the reagent has dried.

PERFORMANCE SCHEDULE

Perform when a new lot number of reagent is received and monthly.

PROCEDURE 10.6

PYR TEST

PRINCIPLE

S. pyogenes and *Enterococcus* sp. possess the enzyme L-pyroglutamyl aminopeptidase, which hydrolyzes an amide substrate with formation of the free β-naphthylamine, which combines with a cinnamaldehyde reagent to form a bright-red end product.

METHOD

1. Rub a small amount of a colony to be tested on the surface of a filter paper impregnated with PYR (available commercially).

2. Add a drop of freshly reconstituted detector reagent, *N,N*-dimethylaminocinnamalde-hyde, with detergent (available commercially), and observe for a red color within 5 minutes.

QUALITY CONTROL

Test a fresh subculture of *S. pyogenes* ATCC 19615 and *S. agalactiae* ATCC 13813.

EXPECTED RESULTS

Positive organisms such as *S. pyogenes* yield a bright-red color change within 5 min; negative organisms, including *S. agalactiae,* yield either an orange color or no change.

PERFORMANCE SCHEDULE

Test each new batch of reagents received and each day that the test is performed.

PROCEDURE 10.7

RAPID UREASE TEST

PRINCIPLE

Hydrolysis of urea by the enzyme urease releases the end product ammonia, the alkalinity of which causes the indicator phenol red to change from yellow to red. The broth method employs buffers that control the pH change and speed the reaction.

METHOD

1. Prepare urea broth as follows:

Yeast extract	0.1 g
Monopotassium phosphate	0.091 g
Disodium phosphate	0.095 g
Urea	20 g
Phenol red	0.01 g
Distilled water	1000 ml

Mix ingredients together and store in refrigerator in small, plastic snap-top or borosilicate glass screw-capped tubes in aliquots of 0.5 ml. The reagent should be stable for 6 months. Do not use if the color is other than pale straw. The rapid urea broth is available commercially.

2. Inoculate a tube of broth with a heavy suspension of the organism to be tested.

3. Incubate the tube at 35° C and observe at 15, 30, and 60 minutes and up to 4 hours for a change in color to pink or red.

QUALITY CONTROL

Test a fresh subculture of *E. coli* ATCC 25922 and *Proteus mirabilis* ATCC 12453.

EXPECTED RESULTS

Urease positive organisms such as *Proteus* sp. yield a bright-pink or bright-red color to the broth; negative organisms such as *E. coli* do not cause a change in the color of the broth.

PERFORMANCE SCHEDULE

Test each new batch of broth and monthly.

---

late development of color (10 minutes), as can a few other coccal genera.

RAPID UREASE TEST

Proteus sp., *Klebsiella* sp., some *Citrobacter* sp., some *Haemophilus* sp., *Bilophila wadsworthia,* the yeast *Cryptococcus neoformans,* and several other bacteria and fungi produce the enzyme urease, which hydrolyzes urea into ammonia, water, and carbon dioxide. The alkaline end products cause the indicator phenol red to change from yellow to pink or red. This test can be used as a component of screening tests for lactose-negative colonies on differential media plated with material from stool specimens, helping to differentiate *Salmonella* and *Shigella* sp., which are urease negative, from the urease-positive non-pathogens. (Note that *Yersinia enterocolitica,* a stool pathogen, is also urease positive.) Commercial reagents are available for performance of rapid urease tests, including reagent tablets and swabs. A broth method is outlined in Procedure 10.7. Other methods specific for *Cryptococcus* species are discussed in the mycology section of the text.

RAPID THERMONUCLEASE TEST

Another way to distinguish between *S. aureus* and coagulase-negative staphylococci (particularly useful for supernatants of blood culture broths growing gram-positive cocci in clusters, for which rapid presumptive identification is essential) is the demonstration of the production of a thermostable deoxyribonuclease by *S. aureus* (Procedure 10.8). Only plates designated for thermonuclease testing are adequate for this particular procedure.[3]

RAPID HIPPURATE HYDROLYSIS TEST

Several species, including group B beta-hemolytic streptococci, *Gardnerella vaginalis, Listeria* sp., and others, are able to hydrolyze hippurate. The presence of the constitutive enzyme, hippuricase, is detected by seeing a colored end product that is formed when ninhydrin oxidizes the amino acids produced during hippurate hydrolysis (Procedure 10.9)

MISCELLANEOUS RAPIDLY DETERMINED INFORMATION

Colonial morphology, fluorescence under ultraviolet light (pigmenting *Prevotella* or *Porphyromonas*), pigment production (e.g., *Pseudomonas, Chromobacterium*), spreading (e.g., *Capnocytophaga, Proteus*), pitting of agar (e.g., *Eikenella corrodens*), hemolytic reaction on blood agar, odor, and many other characteristics mentioned throughout the text are very

PROCEDURE 10.8

RAPID THERMONUCLEASE TEST

PRINCIPLE

S. aureus can be reliably identified by production of a heat-stable deoxyribonuclease enzyme. Organisms are heated to destroy non-heat-stable nucleases and are allowed to interact with media containing DNA intercalated with a dye, toluidine blue, that looks blue when bound to DNA and changes color when its conformation is altered. Breakdown of the DNA by the nuclease changes the dye's structure and causes it to reflect a different wavelength of light, appearing red or pink instead of blue.

METHOD

1. For isolated colonies, inoculate several colonies into 1 ml of brain-heart infusion broth and incubate for 2 h at 35° C; then follow the procedure outlined next.

2. Boil approximately 2 ml of blood culture broth (can be mixed or unmixed) or brain-heart infusion broth

suspension of organisms (as above) in a water bath or microwave oven for 15 minutes. The red cells will clot in the bottom of the tube. Allow the tube to cool to room temperature.

3. Punch a well in DNAse test agar (Thermal agar, Edge Biologicals, Inc. or Thermonuclease agar, Remel) with the large end of a Pasteur pipette or a plastic drinking straw, remove the agar plug, and fill the well with the heated and cooled supernatant above the red cell clot (approximately 2 drops) or broth suspension.

4. Incubate the plate at 37° C in an upright position, and inspect the plate after 1 hour and again after 2 hours.

5. The plate may be refrigerated and reused several times.

QUALITY CONTROL

A portion of a positive control, a negative blood culture seeded

with *S. aureus* and passed to a new negative blood culture weekly, should be boiled and tested as just outlined.

EXPECTED RESULTS

Positive results, as for *S. aureus*, are indicated by a pink or red halo around the well. No change is observed around wells filled with supernatants from negative organisms. A small, clear zone around a well is not indicative of a positive test, as some coagulase-negative staphylococci can degrade the dye without denaturing the DNA.

PERFORMANCE SCHEDULE

The test should be performed with each new batch or lot number of agar media and each time the test is performed.

Modified from Madison, B.M., and Baselski, V.S. 1983. J. Clin. Microbiol. 18:722; and Ratner, H.B., and Stratton, C.W. 1985. J. Clin. Microbiol. 21:995.

helpful in narrowing down identification possibilities.

CONVENTIONAL METABOLIC BIOCHEMICAL TESTS

A relatively small proportion of the total genetic makeup of bacteria is involved in production of the enzymes that metabolize various substrates. These enzymes have traditionally been used as markers for separating species groups. If an organism possessed a given enzyme and was able to utilize a substrate, an end product would be formed that changed the pH of the medium, causing a visual color change in a pH indicator substance. In some cases the organism's ability to grow in a medium could be detected by increased turbidity or the presence of colonies on the surface. This approach has been modified in many ways in re-

cent years, but it is still used for identification of unusual pathogens, difficult-to-identify pathogens, and as the standard against which all other metabolic test methods are compared. There are many conventional tests other than those described in this chapter; some of them are mentioned in later sections of the book. Detailed explanations of the principles of metabolic reactions can be found in other references, such as *Manual of Clinical Microbiology*, edited by Balows et al., *Zinsser's Microbiology*, edited by Joklik et al., and *Microbiology, Including Immunology and Molecular Genetics*, edited by Davis et al.

OXIDATION AND FERMENTATION TESTS

The metabolic pathways used by a microorganism during utilization of a substrate (usually with the concomitant production of acid byproducts) can be either **oxidative** or **fermentative.** Oxidation

PROCEDURE 10.9

RAPID HIPPURATE HYDROLYSIS

PRINCIPLE

The end products of hydrolysis of hippuric acid include glycine and benzoic acid. Glycine is deaminated by the oxidizing agent, ninhydrin, which becomes reduced during the process. The end products of the ninhydrin oxidation react to form a purple colored dye. The test medium must contain only hippurate, since ninhydrin might react with any free amino acids present in growth media or other broths.

METHOD

1. Prepare 1% sodium hippurate substrate as follows:

Sodium hippurate
(Sigma Chemical Co.)   1 g
Distilled water   100 ml

Mix together and dispense into small (12 × 75 mm) plastic disposable test tubes with snap-top caps, 0.4 ml per tube. Store frozen for a maximum of 6 months.

2. Prepare ninhydrin reagent as follows:

Ninhydrin (Sigma)   3.5 g
Acetone   50 ml
Butanol   50 ml

Mix the acetone and butanol together, and then add the ninhydrin. Store at room temperature in a brown bottle with a tight seal for a maximum of 6 months.

3. Defrost the sodium hippurate substrate, and heavily inoculate a tube with a pure culture of the organism to be tested. The suspension should be milky.

4. Incubate the capped tubes for 2 hours in a 35° C water bath.

5. Add 0.2 ml ninhydrin reagent and reincubate for an additional 15 minutes. Observe for a change to deep purple, indicating that the hippurate has been hydrolyzed. This test can

also be used to identify other organisms, as mentioned throughout the book.

QUALITY CONTROL

Test *S. agalactiae* ATCC 27956 and *Enterococcus faecalis* ATCC 19433.

EXPECTED RESULTS

The *S. agalactiae* should turn deep purple (positive), and the enterococcus should remain colorless or slightly yellow-pink (negative) after addition of the ninhydrin reagent.

PERFORMANCE SCHEDULE

Test the quality control organisms each time a new batch of reagents is prepared and occasionally thereafter.

Modified from Hwang, M., and Ederer, G.M. 1975. J. Clin. Microbiol. 1:114.

---

occurs when the organism uses oxygen as a terminal electron acceptor. This reaction can be observed on the surface of agar slants or in the surface layer of liquid media loosely capped to allow diffusion of air. Fermentation occurs in the absence of oxygen. During fermentation the organism often produces large amounts of organic acids. Fermentation can be demonstrated by overlaying substrate-containing media with mineral oil or a petroleum jelly (Vaseline) and paraffin combination (vaspar) to exclude oxygen. A special medium containing low concentrations of peptone (oxidative-fermentative [O-F] medium, described in Chapter 29) has been developed for testing this aspect of the metabolism of bacteria. In either case, the final end products are detected by noting a change in the color of the pH indicator incorporated into the medium. It is clear that the pH at which the indicator changes color will substantially affect the outcome of the test. For that reason, the indicator specified for a particular medium formula must not be modified. Many

substrate utilization tests rely on either bromcresol purple, which changes from purple to yellow at pH 6.3, Andrade's acid fuchsin indicator, which changes from pale yellow to pink at pH 5.5, or phenol red, which changes from red to yellow at pH 7.9.

HYDROLYSIS TESTS

An enzyme that breaks down a substrate by adding the components of water to key bonds within the substrate molecule is called a **hydrolyzing** enzyme. The products are measured by some visual reaction. Substrates for such enzymes typically used in differentiating species include sodium hippurate, which can be broken down by *Listeria, Streptococcus uberis, G. vaginalis,* and group B streptococci *(S. agalactiae);* deoxyribonucleic acid (DNA; test described previously), which can be hydrolyzed by certain Enterobacteriaceae, certain species of staphylococci, and other microorganisms; urea (mentioned previously); and esculin, which is hydrolyzed by *L. monocytogenes,* group D streptococci, some viridans streptococci, and some anaerobes, among others. Hydrolysis of starch, casein, xanthine, tyrosine, and lecithin is also used for differentiation of species. Certain mycobacteria can be identified by their ability to hydrolyze the detergent polysorbate-80 (Tween-80). Media for all these reactions can be purchased, either fully prepared or in dry powder form. Use of hydrolysis tests for identification is described throughout the book.

AMINO ACID DEGRADATION

Enzymes formed by some organisms either *deaminate, dihydrolyze,* or *decarboxylate* amino acids, breaking them down into smaller components. The amino acid substrates tested often include lysine, arginine, ornithine, tryptophan (described previously for the indole test), and phenylalanine. Certain mycobacteria can be differentiated by their ability to deaminate pyrazinamide. Some of these amino acid–degrading enzymes are active in the absence of oxygen; therefore, such tests are often carried out in the deep butts of agar tubes or in liquid media with vaspar overlay to prevent the products of more active oxidative enzymes from obscuring the results. One such frequently used system is that of Moeller (Figure 10.10). Amino acids (lysine, arginine, ornithine) are incorporated into broth media containing 0.05% glucose (to allow the organisms to begin growth) and bromcresol purple indicator. Along with each amino acid being tested, a control tube of the glucose-containing broth base without amino acid is inoculated to serve as a standard against which to compare the color of the indicator. In the case of most amino acid degradation reactions, the end products (e.g., ammonia) are alkaline, and the

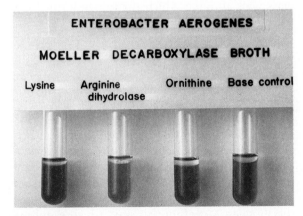

FIGURE 10.10
Moeller's decarboxylase broth media. (From Baron, E.J. 1985. Clinical microbiology, vol.3: Listen, look, and learn. Health and Education Resources, Bethesda, Md.)

color indicator changes to its original color: bromcresol purple turns to purple and phenol red changes to red. Phenylalanine deamination, on the other hand, occurs in air and is measured by the production of a green reactant after the addition of 10% ferric chloride. The media for performing these tests are commercially available.

SINGLE SUBSTRATE UTILIZATION

Many organisms can be recognized by their ability to grow in the presence of only a single compound that supplies enough of the organism's nutrient needs. Substrates that can fulfill this function and are helpful in differentiating microorganisms include citrate, malonate, and acetate. Growth on a substrate-containing agar slant, with or without a pH indicator, is used as the end point of this test. Another variant of this test, used to determine whether a yeast is able to grow with only a single carbohydrate or nitrate, is called an **assimilation** test. Filter paper disks impregnated with the substrate are placed on the surface of a Petri dish of agar without nutrients that has been streaked with a suspension of the yeast being tested so as to obtain confluent growth. The yeast grows around the disks that contain substrates that it can utilize, or assimilate.

NITRATE REACTIONS

Nitrate serves as the source of nitrogen for many bacteria and fungi, but it must be broken down. The first step in nitrate utilization is reduction to nitrite by removal of one oxygen molecule. The Enterobacteriaceae and many other gram-negative bacilli, mycobacteria, and fungi reduce nitrate to nitrite. Certain microorganisms are able to reduce nitrite further to nitrogen by replacing the remaining oxygens with hydrogens. Some *Pseudomonas* species and other nonfermentative gram-negative bacilli possess this capability, as do other species. The classic nitrate reduction test is described in Procedure 10.10. Spot tests using filter paper disks impregnated with nitrate are used for presumptive testing of anaerobes, although that method is less sensitive than broth methods. The ability of yeasts to utilize nitrate can be determined by the same method used for assimilation tests or by a rapid test that uses a swab, described in the mycology section.

TRIPLE SUGAR IRON AGAR AND KLIGLER'S IRON AGAR REACTIONS

Reactions of bacteria in Kligler's iron agar **(KIA)** or triple sugar iron agar **(TSIA)** can be used to direct the initial identification of gram-negative bacilli, particularly members of the Enterobacteri-

PROCEDURE 10.10

CONVENTIONAL NITRATE REDUCTION TEST

PRINCIPLE

Organisms that possess nitrate reductase can reduce nitrate to nitrite. Nitrite combines with an acidified naphthylamine substrate to form a red-colored end product. If the organism has further reduced nitrite to nitrogen gas, the test for nitrite will yield a negative (colorless) result. An additional test for the presence of unreacted nitrate must be performed to validate such a colorless result. Metallic zinc catalyzes the reduction of nitrate to nitrite; thus, with the addition of zinc, a negative test will yield a red color, indicating the presence of unreacted nitrate.

METHOD

1. Grow the organism in 5 ml of nitrate broth (commercially available) for 24 to 48 hours or longer for poorly growing organisms. A small inverted tube, called a Durham tube, may be placed into the broth to trap bubbles of nitrogen gas that may be formed by nitrite-reducing organisms.

2. Prepare the reagents that combine with the nitrite to form colored end products:

   Reagent A

Sulfanilic acid	4 g
Acetic acid (5 M)	500 ml

   Reagent B

N,N-dimethyl-1-naphthylamine	3 ml
Acetic acid (5 M)	500 ml

3. Add 3 drops of reagent A and then 3 drops of reagent B to the suspension of organisms in broth.

4. Wait 30 minutes for the production of a red color, indicating the presence of the nitrate reduction product, nitrite.

5. The presence of unreduced nitrate can be detected by adding a pinch of commercially available zinc powder to the broth if the red color did not develop after the initial reagents were added.

QUALITY CONTROL

Inoculate suspensions of *E. coli* ATCC 25922 and *Acinetobacter calcoaceticus* ATCC 19606 into broth and test.

EXPECTED RESULTS

Nitrate-positive organisms such as all Enterobacteriaceae either yield a red color after addition of reagents A and B or yield no color even after the addition of zinc. Negative organisms such as *Acinetobacter* sp. show no color after addition of reagents A and B but turn red with zinc. This red color after the addition of zinc indicates that nitrate was still present in the broth and that the organism could not reduce it.

PERFORMANCE SCHEDULE

Perform with each new batch of nitrate broth, reagents A or B, and monthly.

aceae. KIA and TSIA can detect three primary characteristics of a bacterium: the ability to produce gas from the fermentation of sugars, the production of large amounts of hydrogen sulfide gas (as visualized by the formation of a black iron-containing precipitate), and the ability to ferment lactose in KIA or lactose and sucrose in TSIA. A small amount of growth from a pure colony is picked onto a straight wire and inoculated to these media by streaking the surface of the slant and stabbing the **butt** of the tube all the way to the bottom. Only the tops of colonies growing on selective agar should be touched, since inhibited flora may still be present and viable. For the same reason, the needle or loop should not be cooled in the agar of any selective medium. Gas formation is usually visualized as bubbles and cracks in the medium, caused by the pressure of the gas formed in the agar. Therefore the inoculating wire must be stabbed down the center of the agar butt of the tube; careless inoculation may allow the wire to form a channel in the agar along the tube's inside glass wall, through which the newly formed gas may escape, preventing its detection. The presence of oxygen in the atmosphere is necessary for the proper reaction to occur on the slant; therefore, caps must be very loose.

Both TSIA and KIA contain a limiting amount of glucose and a 10-fold greater lactose concentration. Enterobacteriaceae and other glucose fermenters first begin to metabolize glucose, since glucose-utilizing enzymes are present constitutively and the bacteria can gain the most energy from using the simplest sugar. All other sugars must be converted to glucose before they enter the Embden-Meyerhof pathway. Glucose utilization occurs both aerobically on the slant, where oxygen is available as a terminal electron acceptor, and in the butt, where conditions are anaerobic. Once a glucose-fermenting bacterium has reduced all the available glucose to pyruvate, it will further metabolize pyruvate via the aerobic Krebs cycle (on the slant) to produce acid end products. The acid in the medium causes the pH indicator, phenol red, to assume a yellow color. Thus, after 6 hours of incubation, both the slant and the butt of a TSIA or a KIA that has been inoculated with a glucose fermenter appear yellow. If the organism cannot ferment glucose, the butt will remain red (indicating no change in pH) or become alkaline (may be indicated by a red color slightly deeper than that of the original medium), demonstrating that the organism is not a member of the Enterobacteriaceae.

After depletion of the limited glucose, an organism that is able to do so will begin to utilize lactose or sucrose. Since there is 10 times as much lactose (and sucrose in TSIA) as glucose in the agar, the organism has enough substrate to continue making acid end products. The slant and butt of the TSIA or KIA remain yellow after 18 to 24 hours incubation. This reaction is called acid over acid (A/A), and the organism is identified as a lactose (or sucrose in TSIA) fermenter. The production of gas will cause the medium to break up or to be pushed up the tube, so that a gas-producing lactose fermenter will give an A/A plus gas reaction.

If the organism being tested cannot use the lactose in the medium, it must produce energy in a less efficient way by using the proteins and amino acids in the medium as nutrient sources. Protein metabolism occurs primarily on the surface of the slant where oxygen is plentiful. The byproducts of peptone breakdown (e.g., ammonia) are alkaline and cause the phenol red indicator to revert back to its original red color. Thus after 18 to 24 hours' incubation on KIA, a nonlactose fermenter shows a red slant; the butt remains yellow because of the early anaerobic glucose metabolism. This reaction is called alkaline over acid (Alk/A or K/A).

Glucose nonfermenters may also produce alkaline products from peptone utilization on the slant. Such reactions are alkaline over alkaline (Alk/Alk or K/K) or alkaline over no change (Alk/NC). Figure 10.11 shows several different possible reactions in TSIA, where occasional sucrose-fermenting nonlactose fermenters exhibit results different from those seen on KIA. Either lactose fermenters or nonlactose fermenters can produce hydrogen sulfide, and the black precipitate may mask the true color of the butt. In this

case the ability to ferment or oxidize glucose can be tested by inoculating the organism into **O-F** (oxidative-fermentative) media. These media are prepared with a low peptone concentration so that the pH change that occurs after metabolism of glucose will not be affected by the alkaline products formed from peptone utilization.

The initial reaction of an inoculum from a pure culture of a bacterial strain in TSIA or KIA gives a microbiologist much information about what the genus might be, and further tests are often chosen based on this reaction. Although these media are used primarily for gram-negative bacilli, the ability of one species of gram-positive bacilli, *Erysipelothrix rhusiopathiae,* to produce hydrogen sulfide is often detected in TSIA. Subtle patterns and colors of reactions, such as a cherry-red slant and very little gas (suggestive of *Serratia* sp.) or a very dark-red slant (suggestive of *Proteus, Providencia,* or *Morganella* sp.) can be recognized by experienced technologists and serve to guide the choice of further tests.

MODIFICATIONS OF CONVENTIONAL BIOCHEMICAL TESTS

Various modifications of conventional biochemicals have been used in recent years to facilitate inoculation of media, to decrease the incubation time, to automate the procedure, or to systematize the determination of species based on reaction patterns. This section briefly describes three types of these systems.

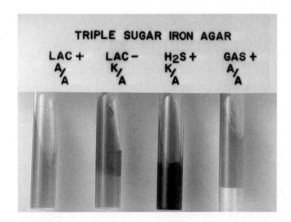

FIGURE 10.11

Examples of inoculated triple sugar iron agar slants. (From Baron, E.J. 1985. Clinical microbiolgoy, vol. 3: Listen, look, and learn. Health and Education Resources, Bethesda, Md.)

PROCEDURE 10.11

RAPID CARBOHYDRATE FERMENTATION REACTION TEST

PRINCIPLE

When carbohydrates are metabolized by organisms with subsequent change in pH, the phenol red indicator changes from red to yellow. The presence of buffers controls the pH change and the heavy inoculum of organisms allows rapid detection of constitutive enzymes.

METHOD

1. Prepare buffered indicator solution as follows:

KH_2PO_4	0.01 g
K_2HPO_4	0.04 g
KCl	0.80 g
Phenol red	0.004 g (or 0.4 ml of a 1% aqueous solution)
Distilled water	100 ml

Adjust the pH to 7.0; sterilize by passing the solution through a 0.2 µm diameter pore-size membrane filter, and store in a tightly capped brown bottle in the refrigerator (4° C).

2. Prepare 20% stock carbohydrate solutions as follows:

Peptone	10 g
Meat extract	3 g
NaCl	5 g
Distilled water	1000 ml

Add 20 g of the appropriate carbohydrate (e.g., glucose, sucrose, available from Difco Laboratories) to 100 ml of the peptone broth solution, adjust to pH 7.0, and filter sterilize as above. One portion of the peptone broth should be retained without carbohydrate to serve as a control. These solutions may be stored frozen in small aliquots.

3. Place 0.1 ml of the indicator buffer solution in a plastic or glass disposable culture tube (10×75 mm diameter) for each carbohydrate to be tested and for one control. Make an extremely heavy suspension of a pure culture of the organism to be tested in each of the indicator buffer tubes (should be milky).

4. Add 1 drop (approximately 0.04 ml) of the appropriate carbohydrate stock solution to each suspension. Add 1 drop of peptone broth without carbohydrate to one suspension to serve as a negative control.

5. Cap the tubes and incubate them in a water bath or heating block at 35° C for up to 4 hours, observing periodically.

QUALITY CONTROL

Test a fresh subculture of an organism that is positive for all carbohydrates prepared. For example, use *Neisseria sicca* ATCC 9913 as a positive control for glucose, maltose, fructose, and sucrose.

EXPECTED RESULTS

Positive organisms show a change in indicator from red to yellow. Many reactions are positive within 30 minutes; questionable reactions may require up to 24 hours for full development. The peptone broth control tube should remain red and serves as an internal negative control for each test.

PERFORMANCE SCHEDULE

Perform with each new batch or reagents prepared.

Modified from Brown, W.J. 1974. Appl. Microbiol. 27:1027; and Hollis, D.G., Sottnek, F.O. Brown, W.J. et al. 1980. J Clin. Microbiol. 12:620.

MODIFIED CONVENTIONAL BIOCHEMICAL TEST SYSTEMS THAT USE SMALL VOLUMES

Several conventional methods can be made to yield more rapid results by heavily inoculating a very small volume of substrate. Used successfully for rapid identification of *Neisseria* sp., the method of Kellogg et al.[6] uses this principle. A modification of the Kellogg broths (Procedure 10.11) has been used for rapid determination of carbohydrate reactions of fastidious gram-negative bacilli.[5] The reagents can be used to test most bacteria, including *Neisseria* sp.[1]

Many tests, including hippurate hydrolysis, nitrate reduction, urease production, decarboxylations, and deaminations, have been modified for rapid results. For these tests, small volumes of media are inoculated with heavy suspensions of the organism to be tested. Media manufacturers produce many such reagents. In several cases, commercial suppliers (including Remel, Austin Biological Laboratories, Key Scientific Products, and others) produce reagent-impregnated paper disks or filter paper strips to be eluted in a small volume of water or saline to form the test substrate. All these systems obviate the need for the technologist to prepare these media in the usual way. The test organism can then be suspended directly in the substrate. Such small volume, singly performed tests are extremely useful for differentiating two similar species that differ in only one characteristic or for ruling out an organism that is characterized by a single parameter. One example is the use of a rapid urease test that can help eliminate certain nonpathogens from further biochemical testing as possible stool pathogens, many of which (except for *Yersinia enterocolitica*) are urease negative. The nitrate and niacin tests for mycobacteria have been adapted to this format. Other tests have also been adapted for rapid determination with filter paper reagents.

MULTITEST SYSTEMS

The simplest multitest system consists of a conventional-type format that can be inoculated once to yield more than one result. By combining reactants, for example, one substrate can be used to determine indole and nitrate results; indole and motility results; motility, indole, and ornithine decarboxylase; or other combinations. These systems are commercially available (powder or prepared media).

In a widely used type of identification system, conventional biochemicals have been put up in smaller volumes and packaged so that they can be inoculated easily with one manipulation instead of several. When used in conjunction with a computer-generated data base (described later), the biochemical patterns generated can be used to predict species identification with much more precision than obtainable using conventional methods with fewer parameters. Several manufacturers produce conventional biochemicals in microdilution trays (Figure 10.12) that are shipped to the user in a frozen state and maintained frozen until they are thawed, inoculated, and used. These systems require overnight incubation, as do conventional biochemicals. The addition of reagents for the demonstration of end products of reactions is necessary. The Sceptor system (BBL Microbiology Systems), UniScept 20E (API), and Sensititre (Radiometer, Inc./Sensititre Microbiology Systems) are also based on a microdilution array, but the substrates are dried in the trays, which are rehydrated with a suspension of the organism during inoculation. The Minitek system (also BBL Microbiology Systems) incorporates a microbroth format and allows the user to determine which reactions are to be tested by arbitrarily adding substrate-impregnated filter paper disks to wells in a plastic tray (Figure 10.13). Many different groups of bacteria, including Enterobacteriaceae, anaerobes, and *Neisseria* sp., can be tested in the versatile Minitek system. If heavy inocula are used to inoculate the Minitek wells, results may be available within 4 hours.

The API 20E for identification of gram-negative bacilli, API 20S for identification of streptococci, API 20C for yeast identification, and API 20A for identification of anaerobes (bioMérieux Vitek, Inc.) incorporate dried reagents in plastic cupules into which a suspension of the test organism is placed (Figure 10.14). With a heavy inoculum, results may be read after 4 to 6 hours' incubation in some cases. The API Rapid E for Enterobacteriaceae, Rapid NFT for nonfermentative gram-negative bacilli, and Rapid Strep for streptococcal identification systems (also bioMérieux Vitek, Inc.) use the same format, although several chromogenic substrates (described later) are incorporated as differential reactants to allow more

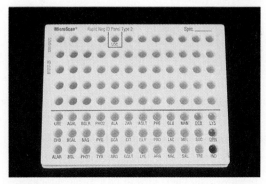

FIGURE 10.12

Microbroth dilution panel used for rapid (2 hour) identifications and MIC determinations of gram-negative bacilli (MicroScan Division of Baxter Healthcare Corp., Sacramento, Calif.). When the organisms metabolize the substrates in the wells, they produce fluorescent end products that can be detected by a special reader in the instrument.

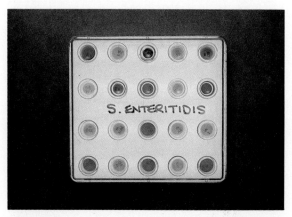

FIGURE 10.13

The Minitek system (Becton-Dickinson Microbiology Systems), which employs filter paper disks impregnated with substrates.

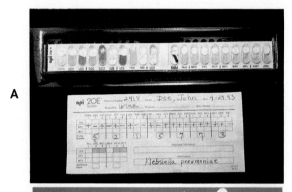

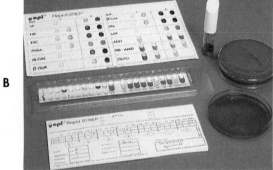

FIGURE 10.14

A, Miniaturized biochemical panel (API; bioMérieux Vitek, Inc., Hazelwood, Mo.) and data recording form. The number generated by coding the biochemical reactions (as shown on the sheet below the panel) is matched to organism identification in an extensive data base.
B, Combination biochemical and enzymatic substrate panel for rapid identification of streptococci (API Rapid Strep; bioMérieux Vitek, Inc.). (Courtesy bioMérieux Vitek, Inc., Hazelwood, Mo.).

rapid results (4 hours). Because of an extensive computer-generated data base incorporating the results of 21 test results (including the oxidase reaction that is performed separately), use of the API system for identification of Enterobacteriaceae has facilitated the recognition of several of the new species of Enterobacteriaceae and has allowed laboratories to designate species to a more

precise level (based on biotype) than was previously possible in a cost-effective manner.

Innovative formats for conventional biochemicals are found in several other systems. The Enterotube II (for lactose fermenters) and Oxi-Ferm (for nonfermenters), distributed by Becton Dickinson Microbiology Systems, are contained within a large tube. A colony is picked onto the end of a long inoculating needle that is drawn through a series of substrate-containing agar compartments in the plastic tube (Figure 10.15). The Enteric-Tek for Enterobacteriaceae, Uni-Yeast-Tek for yeast, Anaerobe-Tek for identification of anaerobes, and Uni-N/F-Tek for nonfermentative gram-negative bacilli (Remel) employ substrate-containing media in pie-shaped wedges in a circular plastic tray (Figure 10.16).

In recent years, several multitest systems have been adapted to yield results more rapidly, often without requiring overnight incubation. The Micro-ID system (now Organon Teknika) was one of the first to exploit the principle that bacterial enzymes can act in the absence of actual viable growth or multiplication. The detection of **preformed enzymes** relies on the inoculation of substrate-containing solutions with an extremely heavy suspension of the organism, resembling the turbidity of skim milk (bacteria ~ 1 × 10^9/ml). Enough of the enzyme is present in this volume of organisms to cause the substrate reaction to occur, even if the organisms are no longer alive. The Micro-ID system uses several substrates that detect preformed enzymes in addition to conventional tests for identification of Enterobacteriaceae. The test organism suspension is inoculated to plastic wells, the tray is incubated, and the entire tray is tipped to allow reactants in a separate chamber to mix with the suspension (Figure 10.17). Results are available after 4 hours' incubation. Tests of this sort yield results after several hours of incubation instead

FIGURE 10.15

The Enterotube II (Becton-Dickinson Microbiology Systems) miniaturized identification system. (From Baron, E.J. 1985. Clinical microbiology, vol. 3: Listen, look, and learn. Health and Education Resources, Bethesda, Md.)

FIGURE 10.16

The Uni-N/F-Tek plate, an example of an agar-based miniaturized identification system (Remel).

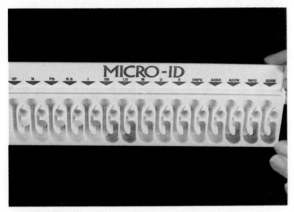

FIGURE 10.17

The Micro-ID (Organon-Teknika) miniaturized identification system. (From Baron, E.J. 1985. Clinical microbiology, vol. 3 Listen, look, and learn. Health and Education Resources, Bethesda, Md.)

of the overnight incubation required by many conventional methods.

Newer biochemical test systems may also use chromogenic substrates. Chromogenic substrates are acted on by enzymes to form colored products; a separate visual indicator system is unnecessary. Because most enzymes being detected by their activity on chromogenic substrates are preformed, results are available within 4 hours for most systems using these substrates. Identification systems that use many chromogenic substrates include the RapID ANA anaerobic identification panel, RapID SS/U for urinary tract bacteria, RapID STR for streptococci, and the RapID N/H panel for identification of *Neisseria* and *Haemophilus* sp., produced by Innovative Diagnostics (Figure 10.18). Rapid Strep, Rapid E, Rapid NFT, An-Ident, Staph-Ident, and Quad FERM+ systems by bioMérieux Vitek, RIM-N (Austin Biologicals), HNID panel (Baxter/Mi-

croScan), and several other *Neisseria* and *Haemophilus* identification systems also use chromogenic substrates.

Most systems that include susceptibility testing as well as identification parameters on the same tray must be incubated overnight to allow enough growth of the organism for accurate test results. Changes in the technology of microbiology are being contemplated, however, that will shorten incubation time by relying on parameters other than growth for measurement of an organism's reaction to an antimicrobial substrate, as further discussed in Chapter 14. Sensititre (Radiometer/Sensititre Microbiology Systems, Westlake, Ohio) and the autoSCAN-Walk-Away (Baxter Diagnostics, Sacramento, Calif.) gram-negative identification system microdilution trays employ fluorogenic substrates that change wavelength and fluoresce when they are acted on by bacterial enzymes. These fluorescent markers of growth can be detected by a fluorometer with greater sensitivity than classical determinants of growth, such as color change caused by pH change or turbidity, and thus can identify metabolic activity faster than conventional growth-dependent methods, although they require instrumentation for detection of results.

AUTOMATED MICROBIAL IDENTIFICATION SYSTEMS

Although they are discussed in more detail in Chapter 12, the most widely used automated identification systems are mentioned briefly here. Systems that use a microdilution tray format have been automated to allow hands-off or partially manual reading of results. The MicroScan (Baxter-American MicroScan), Pasco (Difco Laboratories), Sensititre (Radiometer, Inc.), MicroMedia Systems (Beckman Instruments), and Sceptor systems

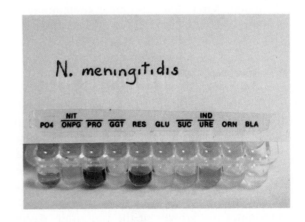

FIGURE 10.18

The RapID N/H (Innovative Diagnostics) miniaturized identification system.

TABLE 10.1

GENERATION AND USE OF GENUS-IDENTIFICATION DATA BASE PROBABILITY: PERCENT POSITIVE REACTIONS FOR 100 KNOWN STRAINS

ORGANISM	BIOCHEMICAL PARAMETER			
	LACTOSE	SUCROSE	INDOLE	ORNITHINE
Escherichia	91	49	99	63
Shigella	1	1	38	20

TABLE 10.2

GENERATION AND USE OF GENUS-IDENTIFICATION DATA BASE PROBABILITY: PROBABILITY THAT UNKNOWN STRAIN X IS MEMBER OF KNOWN GENUS BASED ON RESULTS OF EACH INDIVIDUAL PARAMETER TESTED

ORGANISM	BIOCHEMICAL PARAMETER			
	LACTOSE	SUCROSE	INDOLE	ORNITHINE
X	%+	%+	%−	%+
Escherichia	0.91	0.49	0.01	0.63
Shigella	0.01	0.01	0.62	0.20

Probability that *X* is *Escherichia* = $0.91 \times 0.49 \times 0.01 \times 0.63 = 0.002809$

Probability that *X* is *Shigella* = $0.01 \times 0.01 \times 0.62 \times 0.20 = 0.000012$

(BBL Microbiology Systems) employ readers that interface with the data base for computer-assisted identification. The Sceptor system includes an automated filling device for rehydrating the wells with a suspension of the test organism. API-compatible systems (UniScept Plus and Aladin, bioMérieux Vitek, Inc.) also incorporate an automated reader that interfaces with their extensive data base. The Aladin instrument adds reagents, reads results, and discards trays without operator assistance.

Automated bacterial identification systems usually use growth directly or pH indicators to detect fermentation and oxidation products of bacterial metabolism. After 3 to 6 hours' incubation, the pattern of substrate utilization, as measured by light-scatter photometry, is used to determine the identification of a test organism. The AutoMicrobic System (bioMérieux Vitek, Inc.) determines the growth of the test strain within tiny substrate-containing wells on a plastic card by measuring turbidity and color change. Once the card has been inoculated, the instrument performs all readings without operator assistance; results are available within 4 to 6 hours' incubation. Cards are available for identification of gram-negative fermenters and nonfermenters, yeast, staphylococci, streptococci, and bacilli, some of which require longer incubation periods for final results. The Quantum II system (Abbott Laboratories) uses a plastic cartridge with several wells containing lyophilized substrates that is manually inoculated in one step. The instrument monitors turbidity over time, yielding results based on kinetic patterns of growth and metabolism. Separate reagent cartridges are available for identification of yeasts and Enterobacteriaceae.

COMPUTER-ASSISTED DATA BASE SYSTEMS

The principles on which computer-assisted data base systems for identification of species are based are simple. The first step in developing a data base is to accumulate many organisms of

TABLE 10.3

SELECTED COMMERCIALLY AVAILABLE KITS FOR THE IDENTIFICATION OF ENTEROBACTERIACEAE

PRODUCT NAME	DESCRIPTION	SUBSTRATE/TEST	OPERATION	COMMENTS
API 20E (bioMérieux Vitek, Inc.)	Twenty miniature cupules containing dehydrated substrates are in plasticized strip. After addition of standardized inoculum (McFar. and 0.5) and incubation, color changes are read visually. Reagents must be added to some cupules before reading.	Hydrolysis of *o*-nitrophenyl-β-galactopyranoside (ONPG) Arginine dihydrolase Lysine decarboxylase Ornithine decarboxylase Citrate utilization Production of hydrogen sulfide (H_2S) Urea hydrolysis Tryptophan deaminase Formation of indole Production of acetoin (Voges-Proskauer test) Liquefaction of gelatin Fermentation of: Glucose Mannitol Inositol Sorbitol Rhamnose Sucrose Melibiose Amygdalin Arabinose	Each cupule is manually inoculated with Pasteur pipette. Some cupules are overlaid with oil to provide conditions of reduced oxygen tension. Strip is placed in humid chamber and incubated either 5 h for rapid test or 18 to 24 h for overnight result. Seven-digit number is derived from scoring seven sets of three reactions each. Number is then found in code book that reveals identification of bacterium, probability of its being correct, and aberrant test results.	This is largest commercially available microbiological data base in world. Many reports indicate >90% agreement with conventional methods for both 18-h and 5-h system.
API Rapid E (bioMérieux Vitek, Inc.)	This system is essentially identical to API 20E except that substrates are not buffered and microtubes are smaller.	Hydrolysis of ONPG Lysine decarboxylase Ornithine decarboxylase Urea hydrolysis Phenylalanine deaminase Esculin Citrate utilization Malonate utilization Indole production Voges-Proskauer test	Operation is identical to API 20E except initial incubation period is 4 h.	Results are available in 4 or 18 h. This test is acceptable alternative to API 20E.

Continued.

TABLE 10.3

SELECTED COMMERCIALLY AVAILABLE KITS FOR THE IDENTIFICATION OF ENTEROBACTERIACEAE—cont'd

PRODUCT NAME	DESCRIPTION	SUBSTRATE/TEST	OPERATION	COMMENTS
		Fermentation of: Arabinose Xylose Adonitol Rhamnose Cellobiose Melibiose Sucrose Trehalose Raffinose Glucose		Adonitol
R/B Enteric (Remel)	Fourteen determinations are available through use of four constricted Beckford tubes containing slanted agar media. Two tubes provide presumptive identification of Enterobacteriaceae based on eight tests. Remaining two tubes, called "Expanders," provide additional tests.	R/B 1 Phenylalanine deaminase Lactose fermentation H$_2$S production Glucose fermentation Lysine decarboxylase R/B 2 Indole production Ornithine decarboxylase R/B Expander (1) Citrate utilization Rhamnose fermentation R/B Expander (2) DNAse production Raffinose fermentation Sorbitol fermentation Arabinose fermentation	Agar tube is stabbed with long needle, and then agar surface is streaked with same needle. Tubes are loosely capped and incubated overnight at 35° C. Color changes are read visually and compared with chart provided in kit.	Results of collaborative study indicate >90% agreement with conventional methods.
Enterotube II (Becton-Dickinson Microbiology Systems, Inc.)	Cigar-shaped plastic tube with divided compartments containing conventional biochemical substrates in agar.	Urea hydrolysis H$_2$S production Gas production Indole production Lysine dihydrolase Ornithine decarboxylase Hydrolysis of ONPG Phenylalanine deamination	Wire is touched to tip of colony to be inoculated, and all 15 chambers are inoculated at once by drawing wire through hole running down center of tube; 24 or 48 h incubation is required. Reagents are added to some chambers using needle and	One of earliest multicomponent systems and still one of easiest to inoculate and read. Results compare well with other more complex systems.

System	Description	Substrates	Procedure	Comments
		Citrate utilization Fermentation of: Glucose Adonitol Lactose Arabinose Sorbitol Dulcitol	syringe. Results are determined by visual interpretation.	
GN MicroPlate (Biolog, Inc.)	96-well microplate contains carbon source utilization substrates and oxidation-reduction indicator.	Ninety-five reactions incorporate most carbon sources used by microbes, including carbohydrates, polymers, esters, amides, acids, alcohols, aromatics, and phosphorylated chemicals.	A 15-20 ml turbid suspension of bacteria from an agar purity plate is inoculated to the wells of microplate, and plate is incubated 4-24 h in appropriate atmosphere.	Large inoculum size requires 1 extra day to obtain pure culture.
Micro-ID (Organon Teknika)	Kit consists of 15 tests in plastic tray. Reaction chamber has inoculation part at top. Five chambers have both substrate and detection disk; other five have combination substrate-detection disk. Strip is sealed during incubation.	Voges-Proskauer (production of acetoin) Nitrate reduction Phenylalanine deaminase H_2S production Indole production Ornithine decarboxylase Lysine decarboxylase Malonate utilization Urea hydrolysis Esculin hydrolysis Hydrolysis of ONPG Fermentation of: Arabinose Adonitol Inositol Sorbitol	Heavy suspension (McFarland No. 1 standard) is prepared from the colony to be identified. Each chamber is inoculated with 0.2 ml, and strip is then incubated in upright position. Care must be taken so the substrate strips are moistened, but five separate detection strips remain dry. After 5 h of incubation, strip is rotated 90 degrees to moisten detection strips. Color reactions are read visually.	One of first "rapid" identification systems. Identification is based on constitutive enzyme activity and is not growth dependent. Numerous studies indicate >90% agreement with traditional methods.
Minitek (Becton-Dickinson Microbiology Systems, Inc.)	This kit contains multiwelled plastic tray, reagent disks, and assorted reagents and accessories.	Thirty substrate disks are included. Manufacturer suggests following for Enterobacteriaceae: Arginine Citrate Esculin H_2S Indole Lysine Malonate ONPG Phenylalanine deaminase	Operator selects substrates to be tested and adds one disk to each of the 12 wells. Inoculum is prepared to density of McFarland No. 0.5 standard, and 50 µl are added to each well. Disks are overlaid with oil, and results are read using color comparison chart after 18 to 24 h of incubation.	Greater than 90% agreement with conventional systems. System can be used for nonfermenters and anaerobes, as well as Enterobacteriaceae.

Continued.

TABLE 10.3

SELECTED COMMERCIALLY AVAILABLE KITS FOR THE IDENTIFICATION OF ENTEROBACTERIACEAE—cont'd

PRODUCT NAME	DESCRIPTION	SUBSTRATE/TEST	OPERATION	COMMENTS
Microdilution identification systems	All microdilution products have capability for identification of Enterobacteriaceae, as well as other organisms. Substrates are included in trays either dried or frozen. Panels for identification only or combination identification-antimicrobial susceptibility panels are available.	Urea Voges–Proskauer test Most microdilution products include 20 tests similar to those in API 20E strip. There are minor differences.	Standardized inoculum is added either in very small quantity (3 to 5 µl) with disposable multipronged inoculator or as 50 to 100 µl aliquot with semiautomated inoculation device.	Greater than 90% agreement with traditional methods. One advantage is combination with antimicrobial susceptibility test.
Enteric-Tek (Remel)	Unlike most kits described, Enteric-Tek is a compartmented wheel with central well and 11 surrounding wedge-shaped chambers containing agar media. Fourteen tests can be performed with one wheel.	Indole production Tryptophan–deaminase H_2S production Citrate utilization Malonate utilization Lysine decarboxylase Ornithine decarboxylase Urea hydrolysis Fermentation of: Glucose Lactose Rhamnose Adonitol Sorbitol Arabinose	One drop of inoculum is added to each chamber with Pasteur pipette. To provide anaerobic conditions, medium in central well and lysine and ornithine wells should be stabbed. Wheel is incubated right side up to 18 h at 35° C. Color changes are noted. Spot indole test can be performed from center well.	Agreement is >90% with conventional methods.
Quantum II (Abbott Labs)	This is multipurpose instrumental system using plastic cartridge containing 20 chambers and dual-wavelength photometer.	Lysine decarboxylase Ornithine decarboxylase Urea hydrolysis Citrate utilization Malonate utilization Arginine dihydrolase Indole production	Four or five isolated colonies are suspended in sterile H_2O and adjusted to McFarland No. 0.5 standard; 200 µl of inoculum are added to each chamber. Chamber is sealed and incubated for 4 to 5 h before reading.	Enterobacteriaceae identifications are more accurate than those of oxidase-positive gram-negative bacilli.

RapID SS/U (Innovative Diagnostics, Inc.)	Ten wells contain dehydrated chromogenic and conventional substrates. Reagents are added to visualize some reactions. Most reaction substrates are proprietary.	Growth in acetamide
		Polymyxin B susceptibility
		Fermentation of:
		    Lactose
		    Arabinose
		    Xylose
		    Adonitol
		    Rhamnose
		    Sucrose
		    Glucose
		    Inositol
		    Mannitol
		    Sorbitol
		Hydrolysis of an amino acid arylamide
		Hydrolysis of ONPG
		Hydrolysis of several p-nitrophenyl-β-D-glycosides
		Hydrolysis of p-nitrophenylphosphate
		Urea hydrolysis
		Indole production
		Hydrolysis of several amino acid–β-naphthylamides
			Suspension of organism is introduced in one corner of device, which is tipped to distribute inoculum evenly. After 2 h incubation, reactions are read visually, and reagents are then added to some wells for determination of additional results.	Limited to identification of common uropathogens, such as *E. coli, Klebsiella, Proteus* sp., other Enterobacteriaceae, staphylococci, enterococci, and *Candida albicans.*

* Modified from Howard, B.J., Klaas, J. II, Rubin, S.J., et al 1987. Clinical and pathogenic microbiology. Mosby, St. Louis.

known species. Each strain is subjected to an identical battery of biochemical and enzymatic tests. The reactions are recorded as positive (+) or negative (−), and the cumulative results of each test are expressed as a percentage of each genus or species that possesses that characteristic. For example, suppose that 100 different known *Escherichia* strains and 100 known *Shigella* strains are tested in four biochemicals, yielding the results illustrated in Table 10.1. Now an unknown organism, *X*, is tested in the same four biochemicals, yielding results as follows: lactose (+), sucrose (+), indole (−), and ornithine (+). The results of testing the known strains are now converted to the percentage probability that the unknown strain *X* is a member of one of the known genera based on each separate test result (Table 10.2). If *Escherichia* sp. are 91% lactose positive, the probability that *X* is an *Escherichia* based on lactose alone is 0.91. If *Shigella* sp. are 38% indole positive, then the probability that *X* is a *Shigella* based on indole alone is 0.62 (1.00 [all *Shigella*] − 0.38 [percent of positive *Shigella*] = 0.62 [percent of *Shigella* that are indole negative]). The probabilities are then multiplied to achieve a calculated likelihood that *X* is one or the other of the two genera. Ultimately, *X* is more likely to be an *Escherichia*, with a probability of 357:1 (1 divided by 0.0028). This is still a very unlikely probability, but we have tested only four parameters, and the indole, a very important one, was atypical. As more parameters are added to the formula, the importance of just one test decreases and the overall pattern prevails. With typical organisms being tested for 20 or more reactions, a computer must be used to generate the probabilities. The more organisms in the data base, the more precise are the genus or species designations derived. All commercial suppliers of multicomponent biochemical test systems (Table 10.3) provide their customers with a computer, a computer-derived code book, or access to a telephone inquiry center for matching profile numbers to species. By adding many organisms to the data base, unusual patterns can be recognized. In some cases, new species or unusual species involved in an unsuspected epidemic have been discovered in this way.

REFERENCES

1. Brown, W. J. 1974. Modification of the rapid fermentation test for *Neisseria gonorrhoeae*. Appl. Microbiol. 27:1027.
2. Carlone, G.M., Valadez, M.J., and Pickett, M.J. 1983. Methods for distinguishing gram-positive from gram-negative bacteria. J. Clin. Microbiol. 16:1157.
3. Faruki, H., and Murray, P. 1986. Medium dependence for rapid detection of thermonuclease activity in blood culture broths. J. Clin. Microbiol. 24:482.
4. Halebian, S., Harris, B., Finegold, S.M., et al. 1981. Rapid method that aids in distinguishing gram-positive from gram-negative anaerobic bacteria. J. Clin. Microbiol. 13:444.
5. Hollis, D.G., Sottnek, F.O., Brown, W.J., et al. 1980. Use of the rapid fermentation test in determining carbohydrate reactions of fastidious bacteria in clinical laboratories. J. Clin. Microbiol. 12:620.
6. Kellogg, D.S. , Jr., Holmes, K.K., and Hill, G.A. 1976. Cumitech 4. Laboratory diagnosis of gonorrhea, American Society for Microbiology, Washington, D.C.
7. von Graevenitz, A., and Bucher, C. 1983. Accuracy of the KOH and vancomycin tests in determining the Gram reaction of nonenterobacterial rods. J. Clin. Microbiol. 18:983.

BIBLIOGRAPHY

Balows, A., Hausler, W.J., Jr., Herrmann, K.L., et al., editors. Manual of clinical microbiology, ed. 5. American Society for Microbiology, Washington, D.C.

Blazevic, D.N., and Ederer, G.M. 1975. Biochemical tests in diagnostic microbiology, John Wiley & Sons, New York.

D'Amato, R.F., Bottone, E.J., and Amsterdam, D. 1991. Substrate profile systems for the identification of bacteria and yeasts by rapid and automated approaches. p. 128. In Balows, A., Hausler, W.J., Jr., Herrmann, K.L., et al., editors: Manual of clinical microbiology, ed. 5. American Society for Microbiology, Washington, D.C.

Davis, B.D., Dulbecco, R., Eisen, H.N., et al. 1980. Microbiology, including inmunology and molecular genetics, ed. 3. Harper & Row, New York.

Difco manual, ed. 10. 1984. Difco Laboratories, Detroit.

Estevez, E.G. 1984. Bacteriologic plate media: review of mechanisms in action. Lab. Med. 15:258.

MacFaddin, J. F. 1985. Media for isolation-cultivation-identification-maintenance of medical bacteria, Williams & Wilkins, Baltimore.

Oxoid manual, ed. 5. 1982. Oxoid Ltd., Basingstoke, U.K.

Pincus, D.H., Salkin, I.F., and McGinnis, M.R. 1988. Rapid methods in medical mycology. Lab. Med. 19:296. (Entire issue devoted to automation in clinical microbiology.)

Stager, C.E., and Davis, J.R. 1992. Automated systems for identification of microorganisms. Clin. Microbiol. Rev. 5:302.

Summanen, P., Baron, E.J., Citron, D.M., et al. 1992. Wadsworth anaerobic bacteriology manual, ed. 5. Star, Belmont, Calif.

Washington, J.A., II, editor. 1985. Laboratory procedures in clinical microbiology, ed. 2. Springer-Verlag, New York.

11

NONTRADITIONAL METHODS FOR IDENTIFICATION AND DETECTION OF PATHOGENS OR THEIR PRODUCTS

The microbiologist has traditionally sought to isolate pathogenic organisms in pure culture in an artificial environment outside the human host. In this way the morphology, biochemical activity, and antimicrobial susceptibility pattern of each organism could be examined and evaluated. With viral diseases, identification of the pathogen required recognition of a particular cytopathic effect in a predefined cell culture, a criterion that demanded a high degree of sophistication in the laboratory and often a prolonged incubation period. Only by definitive identification could the cause of an infection or disease be ascertained. A second and less satisfactory method for diagnosis of disease has been the demonstration (by showing a significant rise in titer) of a specific humoral antibody response to the organism in question. For efficient serologic diagnosis, therefore, the clinician must be able to narrow down the possible causes of disease so that the appropriate antigens can be chosen to test against the patient's serum specimens, since testing sera against all possible antigens would be prohibitive. It often takes 2 weeks or more for antibodies to appear. As more patients survive with compromised immune systems such that their antibody production is severely curtailed or absent, the use of rising serologic titers for diagnosis has become less applicable. An even greater problem is that it has been shown that patients may exhibit a nonspecific antibody rise (cross-reacting antibodies) in response to certain antigenic stimuli, confounding the interpretation of serologic tests.

Microbiologists now have many new techniques, unthought of in the days of Pasteur, to aid in the laboratory diagnosis of infectious diseases. The ideal is for rapid, specific diagnostic tests, obviating the classical waiting period for growth of microorganisms, and for tests that detect antigens of pathogens without waiting for an antibody response. This chapter briefly describes several of these new methods in a general sense. Only those methods that lend themselves to visual interpretation are discussed here; diagnostic microbiological methods that require an automated system or instrument for interpretation of results are discussed in Chapter 12.

PARTICLE AGGLUTINATION

LATEX AGGLUTINATION

Antibody molecules can be bound in a random alignment to the surface of latex (polystyrene) beads (Figure 11.1). Since the number of antibody molecules bound to each latex particle is large, the potential number of antigen-binding sites exposed is also large. Antigen present in a solution being tested binds to the combining sites of the antibody exposed on the surfaces of the latex beads, forming cross-linked aggregates of latex beads and antigen. The size of the latex bead (0.8 μm or larger) enhances the ease with which the agglutination reaction is recognized. Levels of bacterial polysaccharides detected by latex agglutination have been shown to be as low as 0.1 ng/ml.[16] The pH, osmolarity, and ionic concentration of the solution influence the amount of binding that occurs, so conditions under which latex agglutination procedures are carried out must be carefully standardized. Additionally, some constituents of body fluids, such as rheumatoid factor, have been found to cause false-positive reactions in the latex agglutination systems available. To counteract this problem, it is recommended that all specimens be pretreated by boiling or with ethylenediaminetetraacetic acid (EDTA) before testing, as outlined in Chapter 16. A similar pretreatment protocol is included as a standard step in the procedure in some commercial systems. Commercial systems should be used precisely, following the manufacturer's recommendations to ensure accurate results. Either a solid, particulate antigen, such as whole bacteria, or soluble antigen, such as capsular polysaccharide, can cause particle agglutination. Microorganisms or their antigenic determinants for which commercial latex agglutination reagents are available are shown in the box on p. 126. A measure of the widespread adaptation of this technology is the fact that at least a dozen products are available for identification of *Staphylococcus aureus* by particle agglutination (latex agglutination or coagglutination, next section). Some of the more widely used individual products are discussed in those chapters that deal with the organisms being sought. Latex tests for the detection of several bacterial toxins, including enterotoxins of *Vibrio cholerae, Staphylococcus aureus, Clostridium perfringens,* and *Escherichia coli;* the staphylococcal toxic shock syndrome toxin, and a *Clostridium difficile*–related product (glutamate dehydrogenase) in stool are available commercially.

COAGGLUTINATION

Similar to latex agglutination, coagglutination uses antibody bound to a particle to enhance visi-

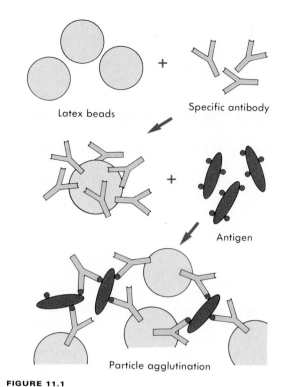

FIGURE 11.1

Alignment of antibody molecules bound to the surface of a latex particle and latex agglutination reaction.

bility of the agglutination reaction between antigen and antibody. In this case the particles are killed and treated *S. aureus* organisms (Cowan I strain), which contain a large amount of an antibody-binding protein, protein A, in their cell walls. In contrast to latex particles, these staphylococci bind only the base of the heavy-chain portion of the antibody, leaving both antigen-binding ends free to form complexes with specific antigen (Figure 11.2). Several commercial suppliers have prepared coagglutination reagents for identification of streptococci, including Lancefield groups A, B, C, D, F, G, and N; *Streptococcus pneumoniae; Neisseria meningitidis; N. gonorrhoeae;* and *Haemophilus influenzae* types A to F. The coagglutination reaction is highly specific but may not be as sensitive for detecting small quantities of antigen as is latex agglutination.

LIPOSOME-ENHANCED LATEX AGGLUTINATION

Phospholipid molecules form small, closed vesicles under certain controlled conditions. These vesicles, consisting of a single lipid bilayer, are called *liposomes.* Molecules bound to the surface of liposomes act as agglutinating particles in a reaction. By combining liposomes containing reactive molecules on their surfaces and latex particles that harbor antibody-binding sites on their surfaces, reagents are created that have the potential to transform a rather weak antigen-antibody par-

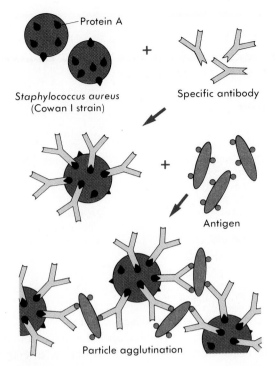

FIGURE 11.2
Coagglutination.

ticle agglutination reaction into a stronger, more easily visualized reaction (Figure 11.3). Liposomes have yet to reach their full potential as diagnostic reagents in clinical microbiology.

LECTIN ASSAYS

Certain proteins or glycoproteins produced as natural biological structural components by various plants and animals, called lectins, have the capability of binding sugars and carbohydrates to form

stable complexes.[5] Lectins therefore can bind to surface components of bacterial and fungal cells that contain these receptors. Although the exact nature of the binding reaction is unknown, various lectins have been found to bind specifically to species of bacteria and fungi (Table 11.1). Lectins can be conjugated to fluorescent and enzymatic markers to create stable diagnostic reagents. Lectin reagents are not widely available at this time but are potentially useful.

ENZYME-LINKED IMMUNOSORBENT ASSAY (ELISA)

ELISA systems were first developed during the 1960s by investigators searching for a substitute for radioimmunoassay procedures. The basic ELISA detection system consists of antibodies bonded to enzymes that remain able to catalyze a reaction yielding a visually discernible end product while attached to the antibody. Furthermore, the antibody-binding sites remain free to react with their specific antigen.[3] The use of enzymes as labels has several advantages. The enzyme itself is not changed during activity; it can catalyze the reaction of many substrate molecules, greatly expanding the reaction and thus enhancing the possibility of detection. Enzyme-conjugated substrates are quite stable, unlike radioactive substrates, and can be stored for relatively long time periods. In addition, the formation of a colored end product allows direct observation of the reaction or measurement with a simple instrument. There are many aspects of ELISA testing to consider. Excellent discussions

FIGURE 11.3

Diagram of liposome–latex agglutination reaction.

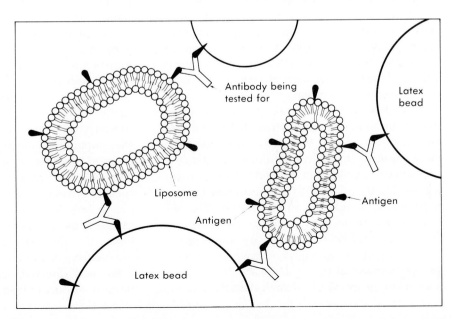

Infectious Agents for Which Latex Particle Agglutination Tests are Available Commercially

Bacteria

Campylobacter species *(coli, fetus, jejuni, laridis)*
Escherichia coli
Haemophilus influenzae type b
Mycoplasma pneumoniae
Neisseria meningitidis
Neiserria gonorrhoeae
Proteus sp.
Rickettsia sp.
Salmonella sp.
Shigella sp.
Staphylococcus aureus
Streptococcus pneumoniae
Streptococcus pyogenes
Streptococcus agalactiae
Other beta-hemolytic streptococci

Fungi

Candida albicans
Coccidioides immitis
Cryptococcus neoformans
Histoplasma capsulatum
Sporothrix schenckii

Parasites

Toxoplasma gondii
Trichinella spiralis

Viruses

Adenovirus
Cytomegalovirus
Herpes simplex virus
Human immunodeficiency virus (HIV)
Rotavirus

of methodology and possible future developments are presented by Conroy et al.[4] and by Meier and Hill.[11] The use of monoclonal antibodies (discussed later in this chapter) has helped increase the specificity of currently available ELISA assays. New ELISA tests are continually developed for detection of etiological agents, their products, or antibodies. In some instances, such as detection of respiratory syncytial virus, human immunodeficiency virus (HIV), and certain adenoviruses, ELISA assays may even be more sensitive than current culture methods.[8,17] ELISA tests have been developed for visual reading of results as well as for automated spectrophotometric reading of end points. The instruments used for such readings are discussed in Chapter 12. ELISA methodology is one of the fastest-growing technologies in diagnostic microbiology. Individual test systems are discussed as they relate to specific etiological agents.

TABLE 11.1

EXAMPLES OF LECTINS KNOWN TO BIND TO SPECIFIC MICROBES

LECTIN	MICROBE
Wheat germ agglutinin (*Triticum vulgaris*)	*Neisseria gonorrhoeae*
Soybean agglutinin (*Glycine max*)	*Bacillus anthracis*
Hairy vetch lectin (*Vicia villosa*)	*Trypanosoma rangeli*
Concanavalin A	*Leishmania donovani* promastigotes
Snail agglutinin (*Cepaea hortensis*)	*Streptococcus agalactiae*

SOLID-PHASE IMMUNOASSAY (SPIA)

Most ELISA systems developed for detection of infectious agents consist of antibody directed against the agent in question firmly fixed to a solid matrix, either the inside of the wells of a microdilution tray or the outside of a spherical plastic or metal bead, or some other solid matrix. Such systems are called solid-phase immunosorbent assays **(SPIA)**. If antigen is present in the fluid to be tested, stable antigen-antibody complexes form when the fluid is added to the matrix. Washing steps to remove nonspecifically adsorbed molecules are very important in ELISA tests. A second antibody against the antigen being sought is then added to the system. This antibody has been complexed to an enzyme, such as alkaline phosphatase or horseradish peroxidase, that catalyzes a reaction that yields a colored end point. If the antigen is present on the solid matrix, it binds the second antibody, forming a sandwich with antigen in the middle. After washing has removed unbound labeled antibody, the addition of the substrates for the enzyme causing the color change completes the reaction, and the visually detectable end point appears wherever the enzyme is present (Figure 11.4). Because of the expanding nature of the reaction, even minute amounts of antigen (<1 ng/ml) can be detected. The system just described requires a specific enzyme-labeled antibody for each antigen tested. It is simpler to use an indirect assay that uses a second unlabeled antibody to bind to the antigen-antibody complex on the matrix. A third antibody, labeled with enzyme and directed against the nonvariable Fc portion of the unlabeled second antibody, can

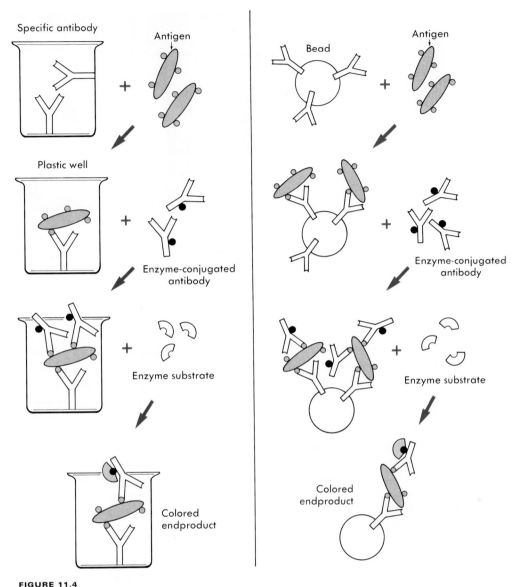

FIGURE 11.4
Principle of solid-phase enzyme immunosorbent assay (SPIA).

then be used as the detection marker for several different antigen-antibody complexes.

MEMBRANE-BOUND SPIA
The flow-through and large surface area characteristics of nitrocellulose, nylon, or other membranes can be exploited to enhance the speed and sensitivity of ELISA reactions.[7] The presence of absorbent material below the membrane can serve to pull the liquid reactants through the membrane and help to separate nonreacted components from the antigen-antibody complexes bound to the membrane and simplify the washing steps. Membrane-bound SPIA assays are available for detection of several viruses, group A beta-hemolytic streptococci antigen directly from throat swabs (Figure 11.5), group B streptococcal antigen in vaginal secretions, and several antibodies in serum. In addition to clinical laboratories, these assays are expected to become more prevalent for home testing systems.

FLUOROGENIC SUBSTRATES

A fluorophore is a compound that absorbs light of a short, excitatory wavelength and emits light of a longer wavelength. The emitted, fluorescent light can be seen visually and can be measured quantitatively by special photometers called *fluorometers*.[6] Fluorophores are used in biological reactions by binding them to a substrate, which effectively inhibits their fluorescence. When the

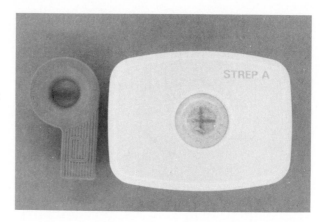

FIGURE 11.5
Membrane-bound SPIA assay for detection of group A streptococcal cell wall antigen. (Courtesy Abbott Diagnostics, Abbott Park, Ill.)

substrate is acted on, such as by the metabolic enzymes of microorganisms, the fluorophore is released and fluoresces under ultraviolet or other short-wave light. Substrate-fluorophore combinations are stable and allow the detection of tiny amounts of reactants. The first practical test using this technology was the MUG test (4-methylumbelliferyl-β-D-glucuronide) for rapid identification of *Escherichia coli*. The enzyme β-glucuronidase, produced by *E. coli* and a few species of *Salmonella* and *Shigella*, breaks the bond holding MUG together and releases the potent fluorophore 4-methylumbelliferone. By observing the fluorescence, microbiologists can identify the organism, often within 30 minutes. Commercially available MUG tests have been found to be specific and convenient. Other fluorophores, including luminol, 7-methyl-coumarin amide, and naphthylamine, are being investigated as diagnostic reagents. Quantitative measurements of fluorescence inhibition by bacteria in the presence of antimicrobial agents is available as a rapid alternative method for susceptibility testing.[9] As with other rapid susceptibility test systems, this method needs to correlate well with standard methods. Fluorophores can be coupled to most nucleic acid probes and immunoassay reagents in the same way enzymes are coupled. The potential exists for developing sensitive detection systems using fluorescence.

COUNTERCURRENT IMMUNOELECTROPHORESIS (CIE)

With some exceptions (e.g., *Streptococcus pneumoniae* serotypes 7 and 14), most bacterial antigens are negatively charged in a slightly alkaline environment, whereas antibodies are neutral. This principle is exploited by CIE assays, in which solutions of antibody and sample fluid to be tested are placed in small wells cut into a slab of agarose (a gelatin-like matrix through which molecules can diffuse readily) on a glass surface (Figure 11.6). A paper or fiber wick is used to connect the two opposite sides of the agarose to troughs of buffer, formulated for each antibody-antigen system. When an electrical current is applied through the buffer, the negatively charged antigen molecules migrate toward the positive electrode and thus toward the wells filled with antibody. The neutrally charged antibodies are carried toward the negative electrode by the flow of the slightly alkaline buffer. At some point between the wells, a zone of equivalence occurs, and the antigen-antibody complexes form a visible precipitin band. A well-visualized set of precipitin bands is seen in a simple immunodiffusion gel in Figure 11.7. The entire procedure usually takes about 1 hour. Any antigen for which antisera are available can be tested by CIE. The sensitivity appears to be less than that of agglutination, detecting approximately 0.01 to 0.05 mg/ml antigen, which trans-

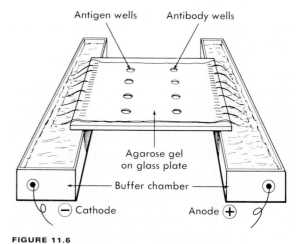

FIGURE 11.6

Apparatus for performing countercurrent immunoelectrophoresis (CIE).

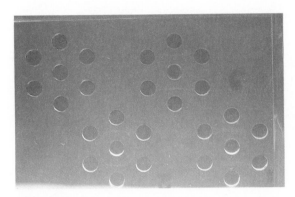

FIGURE 11.7

Precipitin bands formed between *B. dermatitidis* antigen in the center well and antibodies found in patient serum and control antisera in the surrounding wells. (Courtesy R. Glenn Hessel, MD, Department of Pathology, Northwestern University, Chicago, Ill.)

lates to about 10^3 organisms per milliliter of fluid. Bands are often difficult to see, and the agarose gel may require overnight washing in distilled water to remove nonspecific precipitin reactions. Testing positive and negative controls is especially critical, since sera may contain nonspecifically reacting agents that form nonstable complexes in the gel. CIE is more expensive than agglutination-based tests because of the initial capital outlay and the large quantities of antigen and antibody that must be used.

LIMULUS AMEBOCYTE LYSATE (LAL) ASSAY

Within the hemolymph (bloodlike circulating fluid) of the horseshoe crab, *Limulus polyphemus,* are numerous circulating cells called *amebocytes.* Levin and Bang found in 1964 that the lysate of these amebocytes would gel in the presence of minute amounts of lipopolysaccharide (endotoxin) from the cell walls of gram-negative bacteria. The test for gelation of this material, known as the *Limulus* amebocyte lysate (LAL) assay, has been used by industry to detect gram-negative bacterial contamination in a wide range of products, including injectables and other parenterally administered products. This test has been modified for use with body fluids from patients to detect circulating endotoxin, a tangible sign of gram-negative bacterial infection. The gelation of the lysate is specific for and extremely sensitive to (<0.1 ng per milliliter of test fluid) endotoxin. Kits for performing the test are commercially available. Fluids that have been tested by the LAL assay include cerebrospinal fluid, plasma and serum, joint fluid, urine, and cervical and urethral

discharge from patients with gonorrhea. Problems of positive reactions in the absence of any illness (caused either by the extreme sensitivity of the test or by contamination of test containers), particularly for plasma and serum, have precluded its routine use in most clinical laboratory situations. Performance of the LAL assay requires meticulous attention to technique, since any contaminating endotoxin (such as is found in almost all tap water and often in washed glassware) yields a false-positive result.

MONOCLONAL ANTIBODIES: PREPARATION AND UTILIZATION

One of the greatest drawbacks to any of the systems that use antibodies as reagents has been the requirement for large quantities of rather pure, high-avidity, high-affinity antibody molecules. Previous technologies thus required immunization of either many small animals or several large animals with the antigen being sought, repeated bleedings, and subsequent purification of the antibody elicited. Animals often reacted with different antibody responses to the same antigen, resulting in a lack of uniform reagents and necessitating continual retesting of the antibody for reactivity in a given system. Additionally, some antigens, such as *Neisseria meningitidis* group B, were poor immunogens, and good antibodies were impossible to obtain in any quantity. The ability to create an immortal cell line, producing large quantities of a completely characterized and highly specific antibody, known as a **monoclonal antibody,** has revolutionized the field of immunological testing. Monoclonal antibodies are produced by the offspring (clones) of a single hybrid cell, the product of fusion of an antibody-producing plasma cell and an immortal malignant antibody-producing myeloma cell from a plasma cell precursor. One technique for the production of such a clone of cells, called hybridoma cells, is illustrated in Figure 11.8.

A mouse is immunized with the antigen for which an antibody is to be created. The animal responds by producing many antibodies to the antigenic determinants injected. The mouse's spleen, which contains antibody-producing plasma cells, is removed and emulsified so that antibody-producing cells can be separated and placed into individual wells of a microdilution tray. These cells cannot remain viable in cell culture for very long. They must be fused together with cells that are able to survive and multiply in tissue culture, that is, the continuously propagating or immortal cells

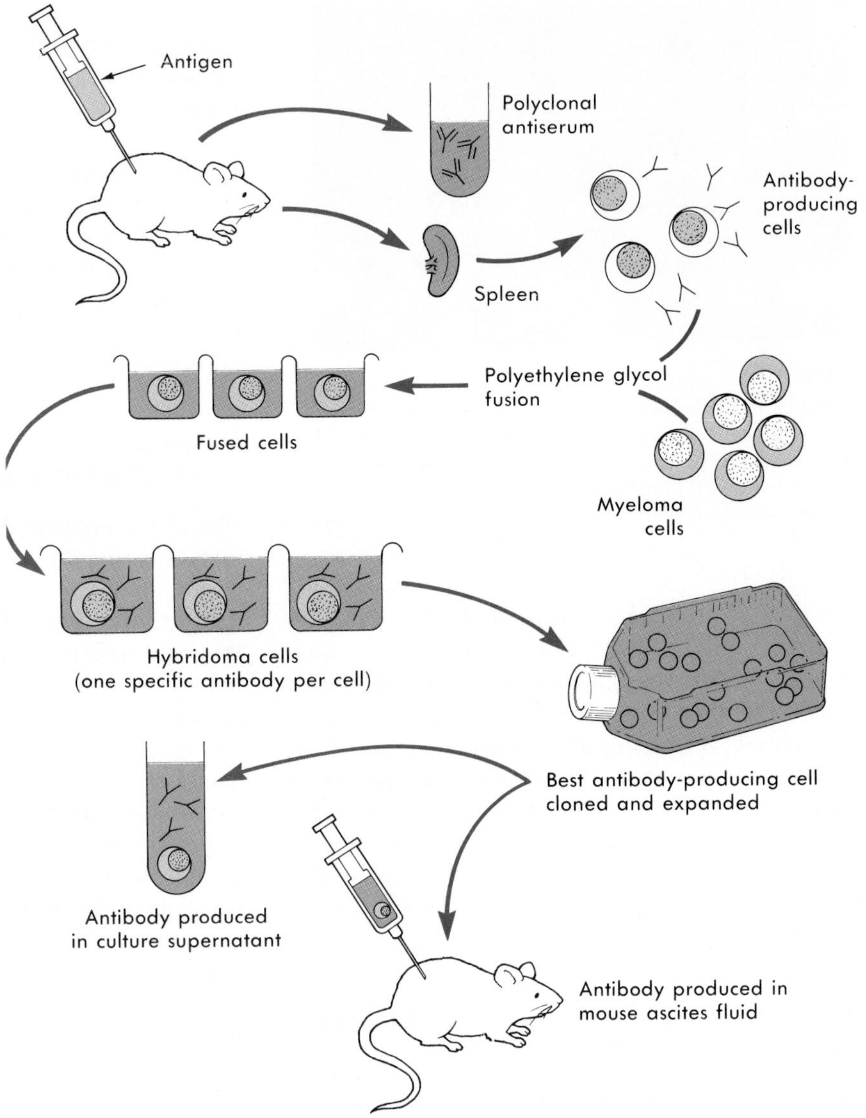

FIGURE 11.8

Production of a
monoclonal antibody.

of multiple myeloma (malignant tumor of antibody-producing plasma cells). The special myeloma tumor cells used for hybridoma production,
however, possess a very important defect. They
are deficient in the enzyme hypoxanthine phos-
phoribosyltransferase. This defect leads to their
inability to survive in a medium containing hypoxanthine, aminopterin, and thymidine (HAT
medium). Antibody-producing spleen cells, however, possess the enzyme. Thus, fused hybridoma
cells survive in the selective medium and can be
recognized by their ability to grow indefinitely in
the medium. Unfused antibody-producing lym-
phoid cells die after several multiplications in
vitro because they are not immortal, and unfused
myeloma cells die in the presence of the toxic enzyme substrates. The only surviving cells will be
true "hybrids." The growth medium supernatant
from the microdilution tray wells in which the

hybridoma cells are growing is then tested for the
presence of the desired antibody. Many such cell
lines are usually examined before a suitable antibody is found, since it must be specific enough to
bind only the type of antigen to be tested, but not
so specific that it binds only the antigen from the
particular strain with which the mouse was first
immunized. When a good candidate antibody-
producing cell is found, the hybridoma cells are
either cultured in large numbers in cell cultures in
vitro, or they are reinjected into the peritoneal
cavities of many mice, where the cells multiply
and produce large quantities of antibody in the ascitic fluid that is formed. Ascitic fluid can be removed from mice many times over the animal's
lifetime. Since all the cells are derived from a single cell producing one antibody molecule type,
the antibody is called monoclonal.

Monoclonal antibodies have been successfully used in commercial systems for the detection of numerous infectious agents. For example, a monoclonal antibody to a protein of *Chlamydia trachomatis* has been conjugated to a fluorescent dye, allowing the visual detection, under fluorescence microscopy, of the elementary bodies and inclusions present in clinical material from patients infected with chlamydiae. Monoclonal antibodies to the *N. meningitidis* group B capsular polysaccharide have been bound to latex particles to create a reagent that can detect the presence of this organism in cerebrospinal fluid, urine, and other body fluids. Many other commercial products using monoclonal antibodies are available.

GENETIC PROBES: PREPARATION AND UTILIZATION

Potentially even more specific than antibody reagents are nucleic acid probes, discrete sequences of single-stranded deoxyribonucleic acid (DNA) or ribonucleic acid (RNA) that form strong covalently bonded hybrids with the specific complementary strand of nucleic acid.[15] Methods are available that allow determination of nucleic acid sequences from cultures of clones of a known bacterial species that constitute a single gene or locus or that are specific for the whole species or a gene that is common to all pathogenic organisms within the species. An example of a method for preparation of a DNA probe for a segment of the genome of a virus is illustrated in Figure 11.9. The nucleic acid from a prototype virus is released by lysing the virions. Bacterial endonuclease restriction enzymes are used to cut the DNA into small segments, which are inserted into a circular strand of plasmid DNA. The plasmid has been chosen for its ability to multiply to large numbers once it has entered the cytoplasm of its bacterial host cell. The expanded plasmids are released from the bacteria and isolated based on their molecular weight and charge. The particular fragments of DNA that match those found in the virus are then isolated using the same restriction endonucleases, purified, and labeled with either digoxigenin, a chemiluminescent marker; a radioisotope, often ^{32}P; or biotin, a small molecule that can be covalently bound to the DNA without destroying its ability to hybridize with complementary DNA.

The biotinylation method of labeling nucleic acid probes was developed in 1981 by Langer et al.,[10] and digoxigenin was developed as recently as 1990.[14a] Bacterial or eukaryotic organism nucleic acid can be isolated and cloned as well. The

material to be tested (either a culture of unknown microorganism or clinical material that may contain the organisms being sought) is usually affixed to a solid matrix, typically some sort of filter.[2] After the material suspected of containing the gene or nucleic acid sequence being sought has been applied, the filter is treated to render the double-stranded DNA single stranded. The probe, a small piece of labeled single-stranded DNA, is then allowed to react with the material on the filter. If complementary sequences are present, the probe DNA forms strong bonds and remains on the filter, even when the filter is washed extensively to remove unbound DNA and nonspecifically bound DNA (Figure 11.10). The labeled DNA probe is now treated to allow visualization (Figure 11.11). If the label is radioactive, the filter is covered with a piece of film; the film develops a dark spot wherever the bound probe is present on the filter underneath. With biotin-labeled probes, a second protein, avidin, which binds tightly to biotin, is added to the matrix. Avidin can be attached to several different marker materials, including enzymes such as horseradish peroxidase, alkaline phosphatase, fluorescent dyes, radioisotope-labeled molecules, or electron-dense markers such as ferritin. Thus the exact location of the bound DNA probe can be determined by the visualization of the avidin-bound marker. Figure 11.12 illustrates the use of a biotin-avidin–horseradish peroxidase probe to detect the antigen of human papillomavirus in infected cells on a filter.

Advantages of the nucleic acid probe systems include the ability of the DNA probe to detect sequences among a mixture of many other genes and molecules, the stability of DNA (allowing the material to remain on the filter for a long period before the test must be performed), and the ability of the probe to detect DNA from nonviable material. For example, a drop of fecal material may be placed on a filter and allowed to dry, and the filter then can be carried or mailed to a distant laboratory for testing at some future date. Conversely, because growth is not necessary, the material can be tested immediately. The biotin-avidin system obviates the need for radioactive materials, but it is not as sensitive. Digoxigenin-labeled probes (Genius kit; Boehringer Mannheim Biochemicals, Indianapolis), on the other hand, are almost as sensitive as isotopic systems, without the problems associated with radioactivity.[6a] Problems with cross-reactivity and sensitivity still exist.[1] Many workers are actively seeking ways to improve this technology, including the use of fluorogenic markers. One company (Gen-Probe, Inc.,

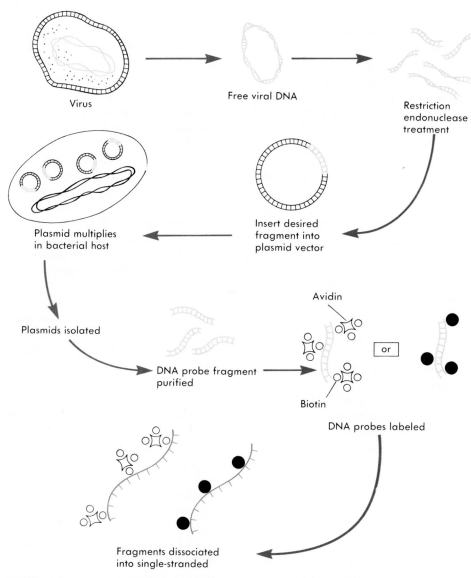

FIGURE 11.9

Production of a radioactive or biotin-avidin-labeled single-stranded nucleic acid probe for a specific segment of DNA from a specific virus. The isolated viral DNA is inserted into a plasmid vector, which is expanded by multiplication within a bacterial host.

San Diego) has developed a new approach using ribosomal RNA detection that has significantly increased sensitivity, can give quantitative results, and does not require a solid testing matrix.

NUCLEIC ACID HYBRIDIZATION

A further development of this same technology is the use of DNA or nucleic acid hybridization to determine the relatedness of two different bacterial clones, primarily for taxonomic grouping. Clones from a single colony of a known bacterium are grown in a medium containing radiolabeled substrate(s) that is (are) incorporated into the bacterial DNA during growth. These known bacteria are then lysed to release their DNA, which is treated with endonucleases to cleave it into small fragments. These fragments are then heated or chemically treated to cause them to dissociate and become single stranded. The bacteria to be compared with the known strain are cultured in unlabeled media, attached to a filter matrix, and treated so that their double-stranded DNA also dissociates into single strands. The known labeled fragments of single-stranded DNA are allowed to react with the DNA of the unknown strain on the filter, and unbound DNA is washed off. The labeled DNA pieces bind (or hybridize) only to DNA

FIGURE 11.10

Detecting genetic material using a labeled nucleic acid probe.

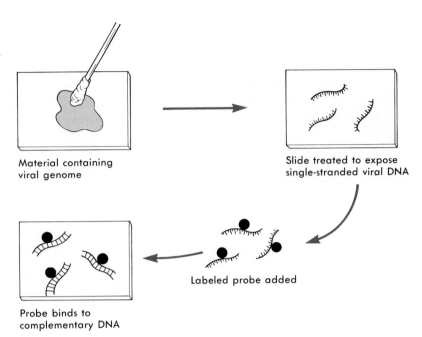

Material containing
viral genome

Slide treated to expose
single-stranded viral DNA

Labeled probe added

Probe binds to
complementary DNA

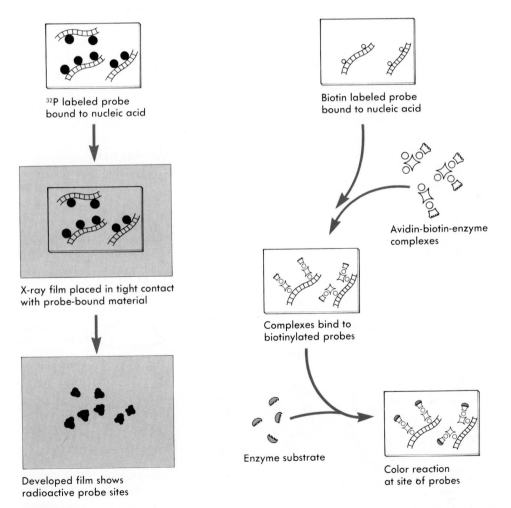

³²P labeled probe
bound to nucleic acid

X-ray film placed in tight contact
with probe-bound material

Developed film shows
radioactive probe sites

Biotin labeled probe
bound to nucleic acid

Avidin-biotin-enzyme
complexes

Complexes bind to
biotinylated probes

Enzyme substrate

Color reaction
at site of probes

FIGURE 11.11

Detection of labeled nucleic acid probe using radiography or a colorimetric enzymatic detection system.

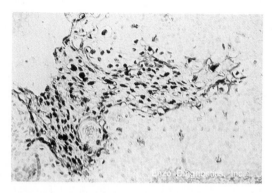

FIGURE 11.12

Biopsy of vaginal condylomata tested with human papillomavirus type 16/18 probe conjugated to horseradish peroxidase (Enzo Diagnostics) and counterstained with hematoxylin and eosin. Infected cells exhibit reddish-brown stained nuclei, indicating presence of HPV 16 or 18 DNA. (Courtesy Enzo Diagnostics, Farmingdale, N.Y.)

that is an exact match, so the amount of labeled DNA left on the filter after washing defines how closely related is the genetic material of the two strains. If both clones are derived from the same original colony, there should be 100% relatedness. Workers have defined a species as those bacteria with very similar morphological and biochemical characteristics that exhibit ≥ 70% genetic relatedness by DNA-DNA hybridization (Figure 11.13).

DNA hybridization technology has been used extensively to determine the relatedness of bacterial clones (strains). In this manner, many species of bacteria have been characterized and given new designations. As a consequence, several his-

torically characterized species have been broken down into different new species, such as some groups within the genera *Streptococcus* and *Enterobacter,* and other species have been combined; for example, all *Salmonella* sp. have been shown to be so closely related genetically that they belong to a single species. By using these new tools, scientists continue to redefine many bacterial taxa that are familiar to microbiologists and physicians. Confusion over changing nomenclature probably will continue for some years to come.

SOUTHERN BLOT ASSAYS

Very specific DNA sequences can be detected using hybridization techniques with a technique known as the Southern blot. DNA to be hybridized is first cleaved by restriction endonucleases, and then the pieces are separated on the basis of size and charge by agarose gel electrophoresis. These fragments are transferred (adsorbed) to a nitrocellulose membrane that is laid over the gel. When this membrane is allowed to react with labeled probe, only the fragment containing the specific sequence of DNA that hybridizes the probe is detected. This method is difficult for most routine clinical laboratories, but a variation (the Western blot that detects antibody instead of DNA) is used as a confirmatory test for HIV infection in diagnostic laboratories.

WESTERN BLOT ASSAYS

The DNA (or RNA) of a particular etiologic agent is treated with endonucleases, or the protein com-

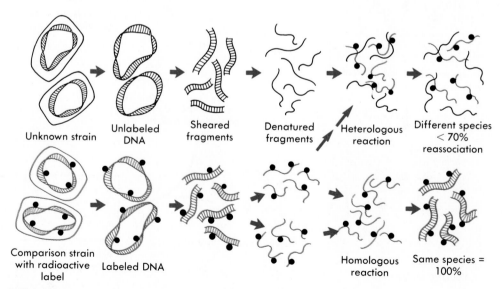

FIGURE 11.13

DNA-DNA hybridization to determine the amount of relatedness between the genomes of two bacterial strains.

ponents of an agent are treated with proteinases to create fragments of different size. The nucleic acid or protein fragments are separated by agarose gel electrophoresis, and the patterns are then blotted onto nitrocellulose as in the Southern hybridization assay. The filter paper containing specific nucleic acids or proteins is then allowed to react with antiserum from a patient or animal suspected of containing antibodies against the agent. If present, antibodies will bind to the protein or nucleic acid against which they were created, and they can then be detected by visual methods for the detection of antibodies (e.g., enzyme-labeled probes, fluorescent markers). Very specific assays for the presence of antibodies have been developed using Western blot methodology.

POLYMERASE CHAIN REACTION

The newest of these techniques to be used in the clinical microbiology laboratory is the polymerase chain reaction (**PCR**). This technique is based on the principle that a large amount of specific DNA can be synthesized from a single piece of DNA as long as a portion of the sequence in the sample or target DNA being sought is known. Short pieces (usually 10 to 30 bases long) of DNA that exactly complement a part of the DNA (or RNA) to be amplified, one at each end of the nucleotide target, are called primers for the reaction, and with a special enzyme called *taq* polymerase, which is not destroyed by the high temperature of the process, these components make the process work. Primers for the polymerase chain reaction are usually synthesized using a nucleotide synthesizer. These are mixed with the sample, the *taq* polymerase, and excess nucleotides, and the entire mixture is cycled multiple times in a special heating block (thermocycler) to make literally millions upon millions of copies of the specific target DNA, if it is present in the sample. A typical PCR run consists of 30 cycles run as follows: 1 minute at 94° C, 1 minute at 55° C, and 30 seconds at 72° C, with 5 minutes at 72° C after the final cycle. The presence of specific DNA can be detected at the end of the PCR run by several methods, ranging from gel electrophoresis to the use of a probe.[13,14]

This method of testing is potentially sensitive enough to detect a single microbe in any patient sample. Since the entire test can be completed in a part of 1 day, it can greatly speed the detection of numerous microbes. Initial test development has focused on finding HIV, chlamydiae, and mycobacteria in clinical specimens, with future development going in many directions, including the ability to determine reliably whether or not a blood specimen contains any bacteria shortly after receipt of the specimen by the laboratory. Two other alternative strategies using molecular diagnostic techniques are under active development as sensitive tests for rapid detection of microbes. They are a transcript-based amplification system, in which a complementary copy of RNA (cDNA) is the target for replication, and a replicatable RNA reporter system based on use of the Q-β-replicase enzyme, using RNA directly as the detection target molecule.[12]

FUTURE USES OF NEW TECHNOLOGY

By using the technological methods described, either singly or in combination, microbiologists will be able to diagnose many infections without the time-consuming requirement of culturing the pathogen in the laboratory. The development of nucleic acid probes for gene-size sequences of DNA and the ability to create infinite amounts of antibody of high specificity offer to students of infectious diseases the promise of exquisitely sensitive and specific reagents. It is currently possible to detect the presence of the gene that codes for resistance to a particular antibiotic in nonviable bacteria that have been affixed to a filter. Even the ability of certain colonies of bacteria to produce a particular toxin can be detected with DNA probe methods.

Recently, microbiologists have begun to detect particular nucleic acid sequences in clinical material deposited as spots on filter paper (called *dot-blot assays*) directly. Presence of organisms as diverse as *Mobiluncus* sp., mycobacteria, and cytomegalovirus have been detected with dot-blot assays. Detection of organisms such as mycobacteria by direct testing of patient specimens using PCR is now possible using commercially produced kits. The future of microbiology is certain to include many non-growth-dependent, rapid testing methods for the presence of pathogens in clinical specimens. Many of these methods will make use of the technological advances described, often combined with instruments to automate the processes further. Microbiologists, who have traditionally taken great pride in the fact that their judgment and interpretation are still necessary for laboratory diagnosis of infectious diseases, will have to integrate results obtained by nontraditional approaches with traditional methods for the worthwhile goal of better patient care through more rapid diagnosis. These new methods will need to be performed as rigorously as more traditional methods and need to be competitive with older tests in accuracy, ease of performance, clinical usefulness, and overall patient cost.

REFERENCES

1. Ambinder, R.F., Charache, P., Staal, S., et al. 1986. The vector homology problem in diagnostic nucleic acid hybridization of clinical specimens. J. Clin. Microbiol. 24:16.
2. Carrow, E., and Folds, J.D. 1987. Recombinant DNA methods in diagnostic microbiology. Lab. Management, March, p. 31.
3. Carter, J.H. 1984. Enzyme immunoassays: practical aspects of their methodology. J. Clin. Immunol. 7:64.
4. Conroy, J.M., Stevens, R.W., and Hechemy, K.E. 1991. Enzyme immunoassay. In Balows, A., Hausler, W.J., Jr., Herrmann, K.L., et al., editors. Manual of clinical microbiology, ed. 5. American Society for Microbiology, Washington, D.C.
5. Doyle, R.J., and Keller, K.F. 1986. Lectins in the clinical microbiology laboratory. Clin. Microbiol. Newsletter 8:157.
6. Grist, R. 1987. Fluorogenics: a new application of an old technology. Clin. Microbiol. Newsletter 9:57.
6a. Gustaferro, C.A., and Persing, D.H. 1992. Chemiluminescent universal probe for bacterial ribotyping. J. Clin. Microbiol. 30:1039.
7. Hendry, R.M., and Herrmann, J.E. 1984. Immobilization of antibodies on nylon for use in enzyme-linked immunoassay. J. Immunol. Methods 67:21.
8. Jackson, J.B., and Balfour, H.H., Jr. 1988. Practical diagnostic testing for human immunodeficiency virus. Rev. Clin. Microbiol. 1:124.
9. Jorgensen, J.H. 1987. Instrument systems which provide rapid (3- to 6-hour) antibiotic susceptibility results. p. 85. In Jorgensen, J.H., editor. Automation in clinical microbiology, CRC Press, Boca Raton, Fla.
10. Langer, P.R., Waldrop, A.A., and Ward, D.C. 1981. Enzymatic synthesis of biotin-labeled polynucleotides: novel nucleic acid affinity probes. Proc. Natl. Acad. Sci. U.S.A. 78:6633.
11. Meier, F.A., and Hill, H.R. 1987. Automation of antigen detection in infectious disease diagnosis. p.
101. In Jorgensen, J.H., editor. Automation in clinical microbiology, CRC Press, Boca Raton, Fla.
12. Pfaller, M.J., and Butcher, P.D. 1991. New strategies in microbial diagnosis. J. Hosp. Infect. 18(suppl A):147.
13. Persing, D.H. 1991. Polymerase chain reaction: Trenches to benches. J. Clin. Microbiol. 29:1281.
14. Peter, J.B. 1991. The polymerase chain reaction: amplifying our options. Rev. Infect. Dis. 13:166.
14a. Pollard-Knight, D., Read, C.A., Downes, M.J., et al. 1990. Nonradioactive nucleic acid detection by enhanced chemiluminescence using probes directly labeled with horseradish peroxidase. Anal. Biochem. 185:84.
15. Tenover, F.C. 1988. Diagnostic deoxyribonucleic acid probes for infectious diseases. Clin. Microbiol. Rev. 1:82.
16. Tilton, R.C. 1987. Microbial antigen detection. p. 693. In Wentworth, B.B., editor. Diagnostic procedures for bacterial infections, ed. 7. American Public Health Association, Washington, D.C.
17. Welliver, R.C. 1988. Detection, pathogenesis, and therapy of respiratory syncytial virus infections. Rev. Clin. Microbiol. 1:27.

BIBLIOGRAPHY

Berry, A.J., and Peter, J.B. 1984. DNA probes for infectious disease. Diagn. Med. 7:62.
de Macario, C. E. and Macario, A.J.L. 1983. Monoclonal antibodies for bacterial identification and taxonomy. Am. Soc. Microbiol. News 49:1.
Edberg, S.C. 1987. Nucleic acid probes. p. 715. In Wentworth, B.B., editor. Diagnostic procedures for bacterial infections, ed. 7. American Public Health Association, Washington, D.C..
Persing, D.H., Smith, T.F., Tenover, F.C., and White, T.J., editors. 1993. Diagnostic molecular microbiology: principles and applications. American Society for Microbiology, Washington, D.C.
Tompkins, L.S. 1985. DNA methods in clinical microbiology. In Lennette, E.H., Balows, A., Hausler, W.J., Jr, and Shadomy, H.J., editors. Manual of clinical microbiology, ed. 4. American Society for Microbiology, Washington, D.C.

12

PRINCIPLES OF AUTOMATED METHODS FOR DIAGNOSTIC MICROBIOLOGY

Since the first automated system for antimicrobial susceptibility testing was developed in the 1960s,[18] the acceptance and use of instrumentation for clinical microbiology have become realities. More than half of all hospital and reference microbiology laboratories now use an automated method for processing blood cultures. Many other instruments for the identification and susceptibility testing of microorganisms, the detection of antigens and microbial products in human clinical material, and the measurement of antimicrobial agents in body fluids are being used in laboratories throughout the United States and Europe. Since the field is expanding rapidly, it is inevitable that all systems will not be represented; however, some of the more commonly used instruments with applications directly related to diagnostic microbiology are characterized in this chapter.

TABLE 12.1

AUTOMATED AND SEMIAUTOMATED MICROBIOLOGY SYSTEMS THAT MEASURE GROWTH OR METABOLISM*

PRINCIPLE	APPLICATION(S)	INSTRUMENT(S)
Nephelometry or colorimetry*	Microbial identification; AST†	Aladin, Autoreader (bioMérieux Vitek, St. Louis, Mo.) Autoscan-4 and Walkaway (Baxter Diagnostics, W. Sacramento, Calif.) Autosceptor (Becton-Dickinson, Sparks, Md.) Eagle Esteem (MSI MicroMedia Systems, Cleveland, Ohio) Pasco (Difco Laboratories, Detroit, Mich.) Vitek System (bioMérieux Vitek)
Carbon source utilization	Microbial identification	Biolog (Biolog, Hayward, Calif.)
Fluorescent substrate metabolism	Microbial identification; AST	Walkaway (Baxter Diagnostics) Sensititre (Radiometer/Sensititre, Westlake, Ohio)
Radiometric or spectrophotometric CO_2 detection	Blood and body fluid cultures; mycobacterial detection and identification	BACTEC (Becton-Dickinson)
Colorimetric CO_2 detection	Blood cultures	BacT/Alert (Organon-Teknika)
Fluorometric CO_2 detection	Blood cultures	BACTEC 9240 (Becton-Dickinson)
Gas production or gas utilization detection	Blood cultures	ESP (Difco Laboratories)
Video image analysis	AST	Aladin (bioMérieux Vitek)

*Avantage (Abbott Laboratories, Irving, Texas) and Autobac II (Organon Teknika, Durham, N.C.), although no longer marketed or supported, are still in use in some laboratories.

†*AST,* Antimicrobial susceptibility testing.

BASIC PRINCIPLES EMPLOYED BY COMMON AUTOMATED SYSTEMS FOR DETECTION AND IDENTIFICATION OF VIABLE PATHOGENS (TABLE 12.1)

TURBIDITY AS AN INDICATOR OF GROWTH

Pasteur initially exploited the fact that microbes would multiply in liquid medium until their presence could be detected visually as an increase in turbidity. Turbidity is the ability of particles in suspension to refract and deflect light rays passing through the suspension, such that the light is reflected back into the eyes of the observer. The optical density (OD), a measurement of turbidity, is determined in a spectrophotometer instrument, which compares the amount of light that passes through the suspension (the percent transmittance) to the amount of light that passes through a control suspension without particles. A photoelectric sensor, or photometer, converts the light that impinges on its surface to an electrical impulse, which can be quantified. A second type of turbidity measurement is obtained by **nephelometry,** or light scatter. In this case, the photometers are placed at angles to the suspension,

and the scattered light, generated by a laser or incandescent bulb, is measured. The amount of light scattered depends on the number and size of the particles in suspension. Turbidity measurements are used to determine whether an organism is present in a clinical specimen (such as in the detection of clinically significant urinary tract infections by the Vitek System [bioMérieux Vitek, St. Louis, Mo.]) or to determine whether a particular strain of bacterium or yeast can grow in the presence of specific growth inhibitors, including antimicrobial agents. A classic 1975 paper by Thornsberry et al.[41] describes parameters to examine in evaluating an automated susceptibility testing system. A recent review discusses the current state of the art.[38] Systems use either lyophilized or frozen substrates. Most automated susceptibility testing systems that generate results within 8 hours to overnight make use of turbidometric or nephelometric measurements of growth.[1]

If the photometers are aimed horizontally, accurate measurement of the turbidity of a suspension requires that the particles be re-suspended before measurement. To circumvent this problem, turbidity in the medium can be measured from the bottom. Susceptibility testing systems that employ microdilution plates have adopted this method. For example, the MicroScan system (Baxter Diagnostics, W. Sacramento, Calif.) uses fiberoptics to deliver an equal amount of light to the bottom of each microdilution well; the light transmitted is determined by photometric measurements from above (Figure 12.1). Other systems may use a single light source and move all wells of the microdilution plate through its path. For descriptions of systems that may be in use but are no longer manufactured, see the previous edition of this book.[5] Rapid results generation (within 4 hours) relies on fluorogenic or other detection systems, as discussed later in this chapter.

Certain systems quantitate the number of microorganisms in the specimen by turbidity measurements. The Vitek System, for example, pulls body fluid such as urine into the growth wells on a plastic card (Figure 12.2) by suction (another instrument, Figure 12.3, is necessary to fill the cards). The wells are arranged in such a way that the body fluid is serially diluted as it fills successive wells. After a relatively short incubation period, the turbidity in the wells is measured and

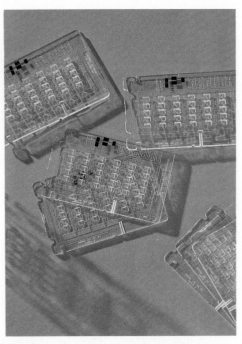

FIGURE 12.2

Plastic cards containing dried substrate in tiny wells for use in Vitek system. The wells are filled with urine or a suspension of an organism by vacuum suction.

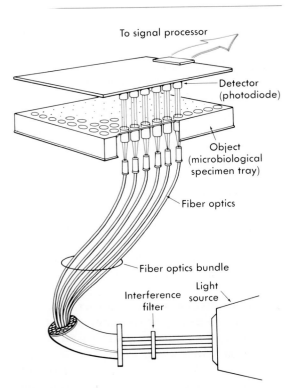

FIGURE 12.1

Diagram of delivery of light to microdilution wells by fiberoptics, as utilized by MicroScan identification and susceptibility testing systems. (MicroScan, Baxter Diagnostics, W. Sacramento, Calif.)

compared to wells containing no growth substances. The ability of an organism to grow in only certain wells is then used to determine the number of colony forming units (CFU) present in the initial specimen. An integrated computer system analyzes and interprets the data and generates final results (Figure 12.3). In systems used commonly today, separate trays are often used for susceptibility testing, based on turbidity, and identification, based on chromogenic substrate utilization (colorimetric recognition).

COLORIMETRIC AND PATTERN RECOGNITION METHODS FOR MICROBIAL IDENTIFICATION

Several systems use a miniaturized modification of conventional biochemical testing that relies on the color changes of pH indicators in media to indicate the presence of metabolic end products. The change in wavelength of light transmitted through the growth cuvette or well is measured by a photoelectric cell. Detection of microbial metabolism by measuring colored end products or indicators is called **colorimetry.** Filters are often used to absorb out certain wavelengths of light for detection of color changes. Vitek System cards with growth substance–containing wells are inoculated either with clinical specimens (primarily urine) or with bacterial or yeast suspensions made from pure cultures for identification. Other systems are useful only for pure cultures. The UniScept AutoReader and Aladin

FIGURE 12.3

Filling module and incubator modules *(left),* computer for data entry and overview *(center),* and printer *(right)* components of the Vitek System.(Courtesy bioMérieux Vitek, St. Louis, Mo.)

(bioMérieux Vitek) overnight-incubation systems identify organisms based on color changes that indicate substrate utilization in transparent plastic cupules. The AutoReader measures turbidity and color changes using filtered light in the standard way. The Aladin instrument detects growth patterns and color changes in the reagent cupules by an innovative computer-assisted video image analysis system.[11,33] Color changes are assessed by measuring filtered light, and susceptibility testing results are interpreted using the patterns of dark and light images, which correspond to density of microbial growth, projected onto a grid on a video screen. The instrument adds reagents, reads results, and discards the trays without manual intervention.

The majority of similar systems use a microdilution tray format. The WalkAway system (Baxter Diagnostics, MicroScan Division, W. Sacramento, Calif.) vies with Vitek and Aladin for total automation by adding reagents automatically within the instrument. The remaining microdilution tray readers are only semiautomated; some assistance from the technologist is required, either for adding reagents or for interpreting results, or both.[16,38]

Aladin and UniScept panels (bioMérieux Vitek), WalkAway panels (MicroScan), and autoSceptor (Becton-Dickinson Diagnostic Instrument Systems, Sparks, Md.) systems are distributed dry, which increases shelf life and allows room-temperature storage. Dry panels must be rehydrated, however, a task that is easily automated; several systems are available for instrument-assisted distribution of a suspension into wells or cupules. Pasco (Difco Laboratories, Detroit, Mich.) and Esteem (MSI/MicroMedia Systems, Cleveland, Ohio) panels are frozen and thus receive a much smaller inoculum of organism suspension but have the added inconvenience of requiring freezer space for storage.

FLUOROPHORE-LABELED OR OXIDATION-REDUCTION SUBSTRATE METABOLISM AS AN INDICATOR OF MICROBIAL GROWTH AND SUBSTRATE UTILIZATION

Two commercial systems, Sensititre (Radiometer America, Westlake, Ohio), and MicroScan (some panels) utilize substrate-fluorophore complexes. When microbial metabolic processes break down the substrates, the fluorogenic markers are freed to assume their fluorescent configuration. Alternatively, pH changes can be measured by changes in fluorescence of certain markers (fluorometric reactions). The fluorophore may become fluorescent, or native fluorescence may be quenched at the designated pH. Ultraviolet light is beamed at the suspension and a special kind of photometer, a fluorometer, is used to measure the resulting fluorescence. These systems offer potentially greater sensitivity than is achievable with conventional substrate metabolism, which depends on a more dramatic color change; susceptibilities and identification of Enterobacteriaeceae and nonfermenters have been measured at 4 and 5 hours with good agreement with subsequent overnight results.* Rapid susceptibility results measured fluorometrically correlate with conventional methods in many cases, but problems have been observed for *Pseudomonas,* staphylococci (probably because of poor growth), and enterococci in general, and for the combinations of ampicillin and *Enterobacter* species, nitrofurantoin, piperacillin, cephalosporins, and tetracycline tested on Enterobacteriaceae.[32,48] So far, the promise of rapid and reliable susceptibility results has not been realized completely.

The Biolog identification system (Biolog, Hayward, Calif.) employs 95 substrate-containing wells and a control well in a microdilution tray format for microbial identification. Each substrate consists of a single carbon source.[28] A large quantity of a pure culture of the organism to be identified is necessary because a turbid suspension must be inoculated into each well. After several hours to overnight, the organism grows in those wells that contain a utilizable carbon source and generates electrons, which reduce the oxidation-reduction potential indicator (tetrazolium violet) to a purple color. The ability to use certain substrates as carbon sources and thus the pattern of purple wells is distinctive for each species and can be read by a spectrophotometric microplate reader. A computer-assisted data base program generates the identification.[26] Biolog does not have antimicrobial susceptibility testing capability. Evaluation of the automated system is still ongoing.

DETECTION OF MICROBIAL GROWTH BY MEASURING CARBON DIOXIDE AS A PRODUCT OF METABOLIC ACTIVITY

The BACTEC system (Becton-Dickinson) measures the production of carbon dioxide by metabolizing organisms. Either radioactive carbon dioxide gas produced as the final end product of metabolism of ^{14}C-labeled substrates (glucose,

amino acids, and alcohols) is measured in an ionization chamber or, in a later configuration of the system, unlabeled CO_2 is quantitated by infrared spectrophotometry. Blood (or sterile body fluid) for routine culture is inoculated into bottles that contain the substrates. The media are incubated and often agitated, preferably on a rotary shaker (also supplied by the manufacturer). At predetermined time intervals thereafter, the bottles are placed into the monitoring module where they are automatically moved past a detector. The detector inserts two needles through a rubber septum seal at the top of each bottle (Figure 12.4) and draws out the gas that has accumulated above the liquid medium, replacing this **headspace gas** with fresh gas of the same mixture (aerobic or anaerobic). Any level of CO_2 (above a preset baseline that covers the metabolism of cellular elements in the blood) is considered to be suspicious for microbial growth. A computerized data-handling section of the instrument allows recording of patient data and collates results. Microbiologists retrieve suspicious bottles and continue processing the culture. Recent modifications have further automated the incubation and measuring device.

Modifications of the basic principle have increased the capabilities of the BACTEC system. The instrument has been used successfully to detect the presence of *Mycobacterium tuberculosis, Mycobacterium avium-intracellulare* complex, and other mycobacteria in clinical specimens.[29] A protective hood is placed over the module to prevent dissemination of aerosols. Sterile body fluids are inoculated directly, and urine and sputum

FIGURE 12.4

Detector module for the BACTEC NR infrared blood culture system (Becton-Dickinson). The needles are inserted through each bottle's septum to withdraw headspace gas and analyze it for the presence of CO_2. Fresh gas is then injected into the bottles and the needles are heat-sterilized before the next bottle is moved automatically into position.(Courtesy Becton-Dickinson Diagnostic Instrument Systems, Sparks, Md.)

*References 9, 12, 16, 24, 31, 48.

specimens are inoculated after decontamination and concentration into BACTEC bottles containing ^{14}C-labeled growth factors utilized only by mycobacteria, such as palmitic acid. Detection of products of metabolism of these substrates by mycobacteria may be possible within 5 to 10 days of inoculation. Because of certain technical considerations, it is still recommended that standard cultures on solid media be inoculated along with BACTEC cultures. If growth is detected radiometrically, a subculture from the initial medium into a second medium that contains p-nitro-α-acetyl-amino-β-hydroxy propiophenone (NAP) can differentiate *M. tuberculosis* and *Mycobacterium bovis,* which are susceptible to the agent, from other mycobacteria, which exhibit radiometric evidence of growth within 3 days. Mycobacterial susceptibility testing has also been performed acceptably using the radiometric BACTEC system; this provides results more rapidly than conventional methods. The use of the BACTEC instrument for mycobacteriology is discussed further in Chapter 42.

Another method for detecting production of CO_2 and other gases is to detect changes in pressure caused by evolution or utilization of gas during microbial metabolism in a closed system. The recently introduced ESP system (Difco Laboratories) attaches a very sensitive pressure detector to blood culture bottles of any configuration and monitors the gas for changes in pressure. Use of a pH indicator in an innovative chamber at the bottom of the bottle allowed Organon Teknika, Durham, N.C., to develop BacT/Alert, the first automated, noninvasive blood culture system. Dissolved CO_2 generated by microbial metabolic activity crosses a membrane into a chamber at the bottom of the bottle where a pH indicator changes from green to yellow. Light reflected from the bottom is measured photometrically, and changes in reflectance are monitored by the computer analysis package. The bottle remains in one place during incubation and monitoring; therefore monitoring takes place at intervals continuously during incubation. The benefit is a shortening of the time necessary for detection of a positive compared to other automated systems that monitor each bottle at greater time intervals.[46] Growth is detected based on changes in absorbance, rate of change, and an overall threshold for color. Although this system currently has only a few media formulas, it has performed essentially as well or better than the BACTEC system.[46] Barcode bottle identification and data handling capability are useful features for laboratories. Two modifica-

tions of the in-bottle indicator chamber using fluorogenic pH indicators have been produced (BACTEC 9240 and the Baxter blood culture system), and a system that uses a fluorescence quenching reagent in the broth (Vital, bioMérieux Vitek) is in development. Each of these systems incorporates noninvasive bottle monitoring in contrast to the infrared spectrophotometry of needle-aspirated headspace gas used by the earlier BACTEC configurations.

BIOLUMINESCENCE ASSAYS FOR VIABLE ORGANISMS

The glowing cold light produced in firefly tails is the end product of a chemical reaction, the conversion of the substrate luciferin to oxyluciferin and light, catalyzed by the enzyme luciferase, and driven by the dephosphorylation of adenosine triphosphate (ATP). The light generated by this reaction can be measured directly with a luminometer, similar to a photometer. The amount of light (measured in photons) produced by the reaction taking place in an excess of luciferin is directly proportional to the amount of ATP present in the solution. Unlike a photometric or fluorometric system, however, there is no need for an exogenous light source. ATP is present within all living cells, and by selectively releasing the ATP from bacterial cells only, the number of colony forming units in a clinical specimen can be estimated. Systems have employed this principle for screening urine specimens for bacteria, but none is currently available.[20,34] Industrial applications have been more successful.

COLORIMETRIC FILTRATION FOR URINE SCREENING

Another technology used to rapidly determine whether a urine specimen contains significant numbers of bacteria is the Bac-T-Screen bacteriuria detection device (Figure 12.5; bioMérieux Vitek). With this system, a measured amount of urine is drawn through a paper filter by vacuum suction. Particles such as bacteria and white blood cells adhere to the filter. A stain is then passed through the filter imparting color, the depth of which is dependent on the number and type of stainable particles. The filter paper is then manually inserted into another section of the instrument, where a photometer compares the color to a preset standard. The filter may also be examined visually. Urines that stain at a level below the control can be considered negative and need not be processed further. This system has shown excellent correlation with clinical diagnoses of uri-

FIGURE 12.5
Bac-T-Screen bacteriuria detection instrument.(Courtesy bioMérieux Vitek, St. Louis, Mo.)

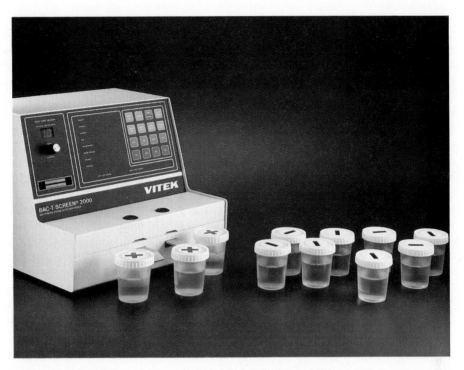

nary tract infection, in part because of the positive staining characteristics of white blood cells present in urines from infected patients.[6]

ELECTRICAL IMPEDANCE AS AN INDICATOR OF MICROBIAL GROWTH

A great theoretical advantage lies in being able to detect the presence of multiplying yeast or bacteria by measuring changes in the flow of an electric current passing through the medium in which a specimen is being incubated (**impedance** changes); however, such systems have not worked well in practice. Impedance devices produced in Scotland by Malthus Instruments and Bactomatic (Princeton, N.J.) are used currently for monitoring bacterial contamination in industrial processing applications.

REPLICA PLATING SYSTEM FOR BACTERIAL IDENTIFICATIONS AND SUSCEPTIBILITIES

The replica plating system (Cathra Repliplates; Automed, Shoreview, Minn.) makes use of the Steers-Foltz replicator (Chapter 14), a device that can deposit a standardized small inoculum of a suspension of bacteria in a discrete position on the surface of an agar plate. Inocula from approximately 35 separate bacterial suspensions can be deposited simultaneously on each of any number of agar plates, containing media for assessing substrate utilization, as well as agar dilution susceptibilities. By recording, storing, and interpreting the results with the computer-assisted data system in-

cluded with the product, technologists can perform rapid and cost-effective identifications and susceptibilities; this is useful primarily for laboratories that must process large numbers of isolates daily.[31]

AUTOMATED MICROBIOLOGY SUPPORT SYSTEMS

A number of instruments have been developed to aid microbiologists by automating certain tasks; often such automation increases reproducibility and consistency. For example, an automated agar sterilizer, Petri dish filler, liquid media dispenser, and colony counter are marketed by New Brunswick Scientific Co., Edison, N.J. Once the medium has been prepared, an automated plate inoculator (Spiral System Instruments, Bethesda, Md.) may be used to distribute the inoculum in a spiral pattern, producing isolated colonies near the outer edge. The Spiral System can be used for determining colony counts of suspensions of organisms and for quantitative susceptibility testing, achieved by placing a graded concentration of antibiotic concentrically onto an agar plate and then streaking organisms from the center to the edge of the agar.[15] The organism's line of growth is inhibited at a defined distance from the edge of the plate, depending on the concentration of antibiotic that is inhibitory. This manufacturer also markets a colony counter that uses computerized data processing.

Several companies produce devices that fill microdilution trays or make serial twofold dilutions in microdilution trays, which can be used for bacterial susceptibility testing and identification systems or for the new immunoassay methods (see section titled Immunological Detection Methods for Antigen or Antibody). Additional instruments have found niches in the practice of microbiology. Automatic Gram-staining devices can pay for themselves by reducing technologist time and reagent costs. Even pipetting has been automated with electrical vacuum-suction devices. Use and quality control procedures for some of these devices have been recently described.[4]

ANTIGEN AND MICROBIAL COMPONENTS OR PRODUCT DETECTION

In addition to identifying, quantifying, and determining susceptibilities of microorganisms, microbiologists are increasingly diagnosing disease by identifying microbial constituents (antigens) or products of microbial metabolism in specimens obtained from the infected host. Antimicrobial agents with sufficient toxicity to require therapeutic monitoring also require quantitation. Structural components of microbes, which are directly related to their genome, can also be detected and measured for a highly specific fingerprint that helps to identify the organism without requiring production of metabolic end products. Widely used methods for such studies that require instrumentation are described here.

GAS-LIQUID CHROMATOGRAPHY

In **gas-liquid chromatography (GLC)** a liquid sample is passed through a matrix that differentially separates its constituents. The volatile components of the sample are carried along with a flow of specially prepared heated gas through a long, narrow column packed with resin or some other material that differentially slows down the rate of travel of the components based on their sizes, molecular weights, or charges. As the various components reach the end of the column, their presence is detected by a change in the temperature or ionization potential, and their relative amounts are plotted on a recording chart as peaks. GLC applications for diagnostic microbiology are increasing. GLC was first used by microbiologists to identify anaerobic bacteria by separating and identifying the metabolic end products (volatile fatty acids and nonvolatile organic acids) of carbohydrate fermentation or mino acid degradation. Different species of anaerobic bacteria yield different types or quantities of end products (see Chapter 35). The GLC patterns obtained from culture supernatants are often definitive.[17,40] With the development of newer, more sensitive column materials, such as very long fused–silica capillary columns, whole cell fatty acids can now be analyzed to aid in identification of aerobic bacteria, anaerobes, and mycobacteria (Figure 12.6).[2,14,21,30,42,43,45] An extensive database and identification software are available (MIDI Microbial Identification System; Newark Del.).[35,45]

Other uses proposed for GLC include the

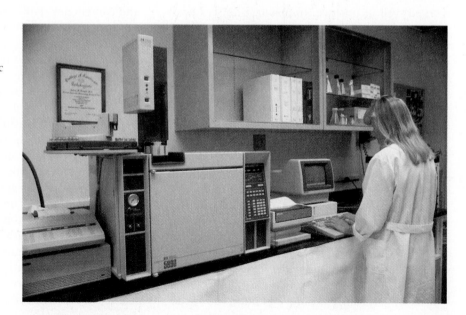

FIGURE 12.6

Hewlett-Packard GLC instrument with automatic injector (tower at *left*) and computer model *(far right)* for analysis of whole cell fatty acids for microbial identification using the MIDI software package.

rapid detection of infection involving anaerobic organisms (with information on the specific organisms involved) directly from infected body fluids. Bacterial vaginosis (see Chapter 20) can be diagnosed in research settings by detecting an increased ratio of succinic to lactic acid in GLC profiles prepared from the foul-smelling vaginal discharge.[36] Abdominal fluid has been evaluated for direct chromatographic analysis.[37] Recently, analysis of headspace gas has been useful for rapid identification of *Clostridium difficile*.[10] Any gas-liquid chromatograph instrument can be used for anaerobic microbiological assays, since the level of sensitivity required is not great. Overviews of chromatographic methods and quality control parameters for clinical laboratories are available.[35,39] Computers may be used to start and stop the GLC analysis, to control the baseline, and to identify and quantify the peaks representing the various end products.

The combination of GLC and **mass spectrometry,** the separation and measurement of ions of a material to determine its structure and probable identification, gives microbiologists a very powerful tool for analysis of microorganisms. When this technology is applied to bacterial constituents or metabolic end products, very specific profiles can be generated that, compared with a data bank of known profiles, will facilitate identification of the substance or the organism. Such systems naturally require sophisticated computerized data handling.[21]

HIGH-PERFORMANCE LIQUID CHROMATOGRAPHY

The column packing material in **high-performance liquid chromatography (HPLC)** consists of special resins that bear a number of ion-exchange groups on their surfaces, such that constituents in the sample undergoing HPLC are preferentially retained in the column based on their affinity for the charged ionic groups. Components with the weakest attraction for the resin will flow through the column fastest and elute first. This process is known as **ion-exchange chromatography,** the same principle as that used to separate IgG from IgM by elution from a column (Chapter 13). HPLC, by using newly developed column packing material, high pressure to force the sample through the column, and liquid buffer carriers at room temperature instead of hot gas, has decreased analysis time and improved resolution compared with GLC. It is currently being used by some medical centers to determine the levels of

antimicrobial agents, including chloramphenicol, cephalosporins, flucytosine, vancomycin, aminoglycosides, quinolones, and others, in body fluids, as well as for other therapeutic drug monitoring. Products of microbial metabolism or cellular constituents may also be detected using HPLC. Its greatest use in microbiology currently is for identification of mycobacteria.[8,35]

IMMUNOLOGICAL DETECTION METHODS FOR ANTIGEN OR ANTIBODY

All of the automated immunological detection methods in use rely on the same principle, the ability of an antigen and its homologous antibody to bind to each other selectively and with high affinity. If either component is labeled in such a way that it can be recognized and quantified, the amount of the other component can be accurately determined. The methods are suitable for detection and quantitation of either antibodies to infectious agents, or antigens of various sorts, not limited to microbial antigens. With the advent of monoclonal technology and the development of nonradioactive labels, the possibilities for sensitivity and broad applicability of immunological methods are practically infinite.

COUNTERIMMUNOELECTROPHORESIS

The simplest method for determining the presence of cross-reacting antigen or antibody is to visually observe the product of the reaction between an antigen and its bound antibody. Originally performed in liquid with whole bacterial cells (agglutination), the procedure was modified by Ouchterlony for performance in a semisolid gel matrix. Antigen and antibody were placed in wells cut into the agarose or agar surface of the matrix, and they slowly diffused toward each other. If they cross-reacted, they bound one another to form a precipitin reaction, visible as a white line in the gel somewhere between the two wells. Counterimmunoelectrophoresis merely adds an electric current to help move the antigen and antibody toward each other more quickly. Since there is really no automation involved in this technique, it is discussed in more depth in Chapter 11.

Counterimmunoelectrophoresis is specific, but it requires enough material to produce a visible reaction. In numerous situations only minute amounts of the substance being sought are present in the patient. In other cases the subject of the assay is a molecule too small to yield a visible precipitin reaction. Therefore more sensitive immunoassays were developed.

NEPHELOMETRY

Antigen-antibody complexes in solution may be too small to see easily with the naked eye, but they do alter the turbidity of the suspension as measured photometrically. **Nephelometry** (described previously in this chapter) measures the increased light scatter of a solution as the complexes form; the kinetics of this change can be determined quite quickly when the photometric results are analyzed by computer. Although more widely used in chemistry laboratories than in microbiology laboratories, nephelometry has been used to measure immune complexes and the presence of antibodies to several substances.

RADIOIMMUNOASSAY

The main difference among various sophisticated immunoassay systems in use today is the choice of label. The first such methods employed radioisotopes, either tritium (^{3}H), iodine (^{125}I), cobalt (^{57}Co), or carbon (^{14}C), to label antigen molecules of the same substance being measured in the assay. **Radioimmunoassay (RIA),** as the method is called, relies on the competitive binding to antibody of labeled antigen, provided by the assay, and unlabeled antigen, present in the unknown patient sample. When all three components are present in the system, an equilibrium exists, dependent on the amount of unlabeled antigen. The more unlabeled (patient) antigen that is added, the less the labeled antigen will be bound to the antibody. When the antigen-antibody complexes are precipitated out of solution and the amount of radioactive label in the precipitate is determined, the unlabeled antigen present in the sample being assayed can be quantified. In practice, a standard curve is first created by adding known amounts of unlabeled antigen to the system. The amount of radiolabel present in the precipitate of the test solution is compared with values obtained from the standard curve to provide quantitative results. Although RIA is an immunoassay procedure, it is used more widely for detection of circulating proteins, hormones, and drugs than for diagnosis of infectious diseases, and thus its primary application seems to be in chemistry laboratories. RIA is being used in a limited way in infectious disease diagnosis to detect antigens of and antibodies to the hepatitis viruses, enteroviruses, and *Legionella,* as well as to determine the presence of immune complexes and to measure the amount of antimicrobial agents in patients' body fluids. RIA techniques are highly specific and sensitive, but they require very expensive instruments and highly skilled technologists. The radioisotopes have a relatively short half-life. In addition, the institution must be licensed and willing to deal with the use and disposal of radioactive substances. Alternatives to radioactive labels are being promulgated, as described later.

ENZYME-LINKED IMMUNOSORBENT ASSAY (ELISA)

By using enzymes instead of isotopes as labels for antigen or antibody, many of the disadvantages of RIA can be overcome. Instead of a radioactive substance, the antigen (or antibody) is bound to an enzyme that can react with a substrate to yield a colored end product. Horseradish peroxidase and alkaline phosphatase (which yield yellow, orange, or blue precipitates depending on the substrates used) are often chosen. The use of biotin-avidin complexes (Chapter 11) for binding enzyme can greatly increase the affinity of the reactants and the speed and intensity of the result. The nonradioactive reagents have a long half-life, and the sensitivity of some assays approaches that of RIA.

ELISA systems are becoming widely used in microbiology laboratories. The basic principles of ELISA assays that can be performed without automation are described in Chapter 11. For most clinical applications, however, the changes being detected are so subtle that computer analysis of results is required. Standard, commercially produced immunoassay kits that rely on competitive binding, removal of free substrate from the bound complexes before measurement, and comparison of results to a standard curve, as in RIA, are available for the detection of antibodies to viruses, bacteria, and parasites. The most commonly used systems use a direct determination of the amount of bound labeled substance after unbound constituents have been removed by a washing step. Sensitive photometers that can determine slight differences in color or fluorometers to detect minute amounts of fluorescence are coupled with computers to automatically compare results to standards. Most of the systems are prepared in microdilution plates, but other configurations are used. In many commercial systems, antigen-antibody complexes are bound to a solid phase, either the inside of the microdilution well, a plastic paddle, a plastic pipette tip, or a plastic bead. This simplifies washing and allows easier manipulation. Table 12.2 lists a sampling of products available

TABLE 12.2

AUTOMATED MICROBIOLOGY SYSTEMS THAT USE ELISA OR FLUORESCENT IMMUNOASSAY TECHNOLOGY*

MATRIX	MANUFACTURER
Plastic bead	Abbott Laboratories (Abbott Park, Ill.)
Microdilution plate	API (bioMérieux Vitek, St. Louis, Mo.)
	Biowhittaker (Walkersville, Md.)
	Granbio, Inc. (Temecula, Calif.)
	INCSTAR Corp. (Stillwater, Minn.)
	Meridian Diagnostics, Inc. (Cincinnati, Ohio)
	Ortho Diagnostic Systems, Inc. (Raritan, N.J.)
	OPUS system, PB Diagnostic Systems, Inc. (Westwood, Mass.)
	Sanofi Diagnostics Pasteur (Chaska, Minn.)
	Seradyn, Inc. (Indianapolis, Ind.)
	Sigma Diagnostics (St. Louis, Mo.)
	Syva Co. (San Jose, Calif.)
	Trend Scientific, Inc. (St. Paul, Minn.)
	Wampole Laboratories (Cranbury, N.J.)
Plastic hollow tip	VIDAS (bioMérieux Vitek)
Plastic paddle	Biowhittaker (Walkersville, Md.)

*Systems are available commercially to detect bacterial antigens (*Chlamydia, Neisseria, Clostridium difficile* toxin), bacterial antibodies *(Borrelia burgdorferi, Bordetella pertussis, Mycoplasma pneumoniae, Helicobacter pylori, Legionella),* fungal antigens *(Cryptococcus, Coccidioides),* parasitic antibodies *(Entamoeba histolytica, Toxoplasma gondii,)* parasitic antigens *(Giardia),* viral antigens (hepatitis A, B, and C, rotavirus, Norwalk virus, adenovirus, HIV-1, and HIV-2), viral antibodies (cytomegalovirus, measles, parvovirus B19, rubella, mumps, Epstein-Barr, herpes simplex, varicella-zoster), and others not listed.

currently that employ automated ELISA technology for detection of microbial antigens or antibodies. In addition to the ELISA readers provided by most manufacturers of the reagents, several companies produce computerized, programmable reader instruments for determining and printing results from any microdilution format assay system (Figure 12.7) and companion instruments such as microdilution plate washers (Figure 12.8). According to Meier and Hill,[27] the use of automated washers can be justified by the improved reproducibility of ELISA assays.

A modification of ELISA assays is widely used for monitoring levels of drugs including antimicrobials. Enzyme-multiplied immunoassay systems use an enzyme bound to a molecule identical to the agent being sought as the third constituent in a competitive assay. The agent, if present in the patient's sample, will bind to antibody, allowing the free enzyme–agent complex to react with its substrate to produce a color change. If the patient's specimen contains only small amounts or none of the agent being tested for, the enzyme-agent complex is free to bind to the antibody. When bound to antibody, the enzyme changes conformation, reducing enzyme activity. Commercial systems that use this and similar technology are used to monitor antibiotic and other chemotherapeutic and toxic agent levels and to detect various proteins with very accurate results.[3]

FLUORESCENT IMMUNOASSAYS (FIAS)
Fluorescent molecules are the newest generation of markers used as substitutes for radioisotope or enzyme labels in several automated systems. In contrast to ELISA systems, **fluorescent immunoassays (FIA)** require only minimal incubation times, no buffered substrate, and fewer washing steps. Current uses in microbiology encompass detection of antibodies to numerous bacteria and viruses, a few parasites, antibodies directed against host tissue (autoantibodies), and some antigens (such as hepatitis virus determinants). Commercially available systems include the TDA (Ames Division of Miles Laboratories, Kankakee, Ill.), which uses a fluorogenic substrate (umbelliferyl-B-D-galactoside) bound to molecules of the agent being assayed to compete with free molecules of the agent for antibody-binding sites. The more competing free agent (from patient serum), the more fluoroescence can be produced by an exogenous enzyme acting on the unbound substrate-agent complex. In another

FIGURE 12.7

Spectrophotometric instrument adapted to read ELISA results in microbroth dilution format.(BioTek Instruments, Winooski, Vt.)

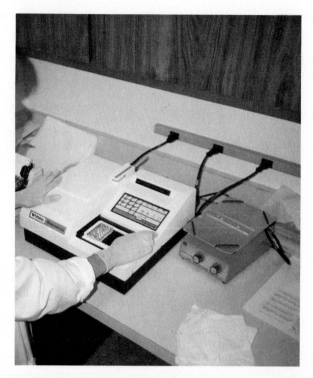

FIGURE 12.8

Semi-automated microdilution plate washer with a second set of detachable reagent reservoir bottles on right. (Courtesy Dynatech Laboratories, Chantilly, Va.)

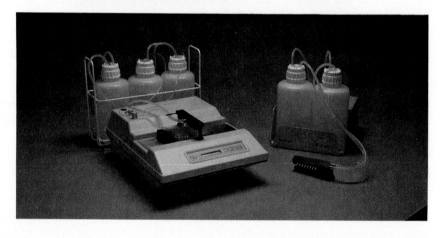

system, the FIAX system (Biowhittaker, Walkersville, Md.), fluorescent-labeled and patient-unlabeled substance compete for antibody (or antigen, if antibody is being assayed) bound to a plastic paddle. This relatively simple system has been formulated to detect antibodies to *Toxoplasma,* rubella virus, *Borrelia burgdorferi,* cytomegalovirus, and herpesvirus, as well as to other infectious agents and proteins (such as immunoglobulin subtypes and C-reactive protein). The TD$_x$ System from Abbott Laboratories (Abbott Park, Ill.) uses a fluorescence polarization format for a competitive therapeutic drug-monitoring assay. A molecule identical to that being assayed is bound to fluorescein, a fluorescent dye. Under polarized light, the amount of fluorescence emit-

ted is dependent on the amount of rotation of the fluorescein-drug complex; the more rotation, the less fluorescence. The conjugated drug and the free drug in the patient sample compete for antibody-binding sites. As the fluorescein complex is bound, it will fluoresce with greater polarization, since its rotation is slowed by the antibody. This sensitive and specific assay is available for monitoring levels of aminoglycosides and vancomycin, as well as many other compounds. The VIDAS enzyme-linked fluorescent immunoassay system (bioMérieux Vitek) uses precoated plastic pipette tiplike receptacles that act as the solid phase. Patient specimens are added to a reagent strip, and the rest of the process proceeds automatically. With its versatile configuration, VIDAS has the capability to perform a number of different test types using a standard technology, some of which include detecting *Clostridium difficile* toxin in stool supernatant, detecting viral antigens, and detecting antibodies. Another potentially very sensitive fluoroimmunoassay system that uses time-resolved immunofluorometry to detect very small amounts of antigens and other proteins, including hepatitis viral proteins, has been developed. With this method, two substances, europium and a diketone (such as isothiocyanatophenyl-DTTA), are chelated together to act as a fluorophore. Because the europium emits fluorescent light for a relatively long time, the fluorometer can measure the light at a time period after all background fluorescence has faded, resulting in very sensitive and specific assay results.

The increasing sensitivity of newer enzyme and fluorescent immunoassays has accelerated the move away from radioactive markers and opened new areas for automation of the diagnosis of infectious diseases. Numerous commercial products are available. It is anticipated that many applications in microbiology will be realized within the next few years.

POSSIBLE FUTURE DIRECTIONS IN CLINICAL MICROBIOLOGY

Every month suggestions for new ways to utilize technology for diagnosis of infectious disease or identification of pathogens are published. A few novel methods that have *not* been adopted for routine use are mentioned here.

Flow cytometry has been explored for detection of bacteria in body fluids such as blood and urine.[25,44] In such methods, bacteria are stained with fluorescent dyes and passed through a detector module in a thin stream of liquid such that only a single organism is measured at any given moment. Based on the size, charge, and staining characteristics of the particles being detected, the number of bacteria per volume of fluid can be determined.

DNA probe systems (Chapter 11) have been automated substantially over the last few years. The current generation uses a fluorescent acridinium ester marker system to detect *Chlamydia* or gonococci in genital tract secretions and mycoplasma DNA in respiratory secretions (Gen-Probe, San Diego, Calif.).[22] Direct detection of other infectious agents in patient specimens is in development.

A final new technology that may soon be au-

tomated and thus practical for the diagnosis of infectious diseases is the **polymerase chain reaction (PCR)** (also see Chapter 11). Once the base pair composition of a specific gene or piece of a gene is known, a tiny amount of the gene in solution can be selectively amplified. Complementary gene sequences with primers for DNA polymerase are used as targets to bind to the sequence in question. If present, the double-stranded primer thus created serves as a template upon which further rounds of replication occur. This reaction has been speeded up dramatically because of the ability to automate the DNA denaturization and subsequent chain elongation steps based on discovery of a heat-stable polymerase, the Taq enzyme of a thermophilic bacterium, *Thermus aquaticus,* and the use of a microprocessor-controlled water bath. Heating the mixture splits the DNA fragments and opens the ends for the next round of polymerase activity; cooling the mixture then activates the enzyme to begin chain elongation again. The average 30-cycle amplification takes approximately 3 hours to complete, but newer enzymes and improved technology promise to revolutionize this field. The technique has been used for detection of tiny amounts of human immunodeficiency virus (HIV) antigen, almost every bacterial, viral, or parasitic agent known, several fungi, and even for identification of organisms in tissue that cannot be grown in culture (see Chapter 41). A practical system for detection of chlamydial DNA by PCR is now available commercially.[23] The current state of the art in nucleic acid–based detection for diagnostic microbiology has been summarized by Wolcott.[47]

COMPUTERIZATION IN MICROBIOLOGY

Very few of the automated systems just described or described in other areas of the book would be possible without computerization. Microprocessor-controlled mechanical functions, such as moving a culture bottle into position for removing headspace gas or shining a beam of light through a solution in a plastic well to determine the turbidity of the solution, are integral to the new microbiology systems. Even more obvious is the computer-controlled recording and integration of data, which are stored and printed, often as directly reportable results. Aside from their important place as the central processing units of almost all automated instruments, however, computers can perform an even more fundamental role in facilitating smooth functioning of the laboratory.

A microbiology computer or a laboratory computer with microbiology functions can perform multiple tasks.[19] Word processing programs can be used in microbiology for a number of administrative functions. Besides the obvious functions of facilitating correspondence and writing manuscripts, a word processor is invaluable for preparing and updating procedure manuals. Quality control recording sheets and records can be generated with such a program, changing parameters as necessary. Ongoing records of quality control discrepancies can be placed into the program and then printed out as a single report when needed. Programs are available for inventory control that can greatly reduce time and anxiety associated with ordering and maintaining supplies. The computer could be programmed to flag certain dates for checking on back orders or initiating inquiries. As supplies are received, they are logged into the system.

Computers can be used to calculate workload units such as those advocated previously by the College of American Pathologists. Raw data can be entered daily as numbers of specimens processed or numbers of tests performed, and the calculations can be analyzed and stored for later retrieval. Cost accounting of procedures can be facilitated by computer-assisted data processing.

Computers can also directly decrease the manual labor involved in logging in new specimens by printing labels for culture media, printing work cards, and notifying the laboratory if a request for processing a particular specimen is repeated within a specified number of days. Barcode labeling of specimens, read by computer-assisted electronic devices, can be used to identify specimens and to order tests correctly. A computer in the accessioning area can even perform automatic billing of tests. The maintenance of records of results, both identifications and susceptibilities, is critical for epidemiological surveillance and for monitoring patient progress. Myriad computer systems have capabilities for data storage and retrieval. Reports can be generated that catalogue susceptibility results of isolates sorted by source, species, doctor, ward, and any other parameter. Unusual susceptibility patterns or those that do not match the supposed identification of the isolate can be flagged. Results of cultures can be studied in many ways, allowing microbiologists access to information that could be gathered previously only by great effort. Data analysis and storage by microcomputers have been outlined by Buck.[7]

In addition to handling management functions, the computer can aid in reporting results. Results immediately accessible to the patient units will have much greater impact on patient care, as well as cut down on communication lags and errors. It is estimated that systems in operation at several large institutions have saved at least 1 hour per physician per day. Physicians may even request tests by computer, which gains them time and reduces transcription errors. Laboratory computers that interface with hospital computers can deliver patient demographic data directly to the laboratory and provide direct billing capability. Epidemiological data entered in the laboratory could be retrieved by nurse epidemiologists in another location for more efficient operation. In the ultimate computerized laboratory, results generated by instruments automatically would be updated to the patient's charts, even during the middle of the night. "Intelligent" information processing systems would interpret the information and pass on only clinically relevant and reliable results. Physicians could use the teaching capabilities of interactive computers to make diagnoses and determine therapy. Such complex use of information systems is already in progress.[13]

REFERENCES

1. Amsterdam, D. 1989. Instrumentation for antimicrobic susceptibility testing: yesterday, today, and tomorrow. Diagn. Microbiol. Infect. Dis. 9:167.
2. Alexander, H. 1987. Gas-liquid chromatography as an aid in identification of glucose-nonfermenting gram-negative bacilli. Microbiol. News 9:25.
3. Anhalt, J.P. 1992. Assays for antimicrobial agents in body fluids. pp. 1192-1198. In Isenberg, H.D., editor. Clinical microbiology procedures handbook. American Society for Microbiology, Washington, D.C.
4. Baron, E.J., editor. 1992. Instrument maintenance and quality control. Section 12. In Isenberg, H.D., editor. Clinical microbiology procedures handbook. American Society for Microbiology, Washington, D.C.
5. Baron, E.J. and Finegold, S.M. 1990. Principles of automated methods for diagnostic microbiology, pp. 142-156. In Bailey & Scott's Diagnostic Microbiology, ed 8. Mosby, St. Louis.
6. Baron, E.J., Tyburski, M., Almon, R., et al. 1988. Visual and clinical analysis of Bac-T-Screen urine screen results. J. Clin. Microbiol. 26:2382.
7. Buck, G.E. 1987. The role of microcomputers for data analysis and storage. p. 177-187. In Jorgenson, J.H., editor. Automation in clinical microbiology, CRC Press, Boca Raton, Fla.
8. Butler, W.R, Thibert, L., and Kilburn, J.O. 1992. Identification of *Mycobacterium avium* complex strains

and some similar species by high-performance liquid chromatography. J. Clin. Microbiol. 30:2698.

9. Colonna, P., Nikolai, D., and Bruckner, D. 1990. Comparison of MicroScan autoSCAN-W/A, Radiometer Sensititre and Vitek systems for rapid identification of gram-negative bacilli. Abstr. C-157, p. 370. Abstr. 90th Annual Meeting Amer. Soc. Microbiol. Amer. Soc. Microbiol., Washington, DC.

10. Cundy, K.V., Willard, K.E., Valeri, L.J., et al. 1991. Comparison of traditional gas chromatography (GC), headspace GC, and the Microbial Identification Library GC system for the identification of *Clostridium difficile.* J. Clin. Microbiol. 29:260.

11. D'Amato, R.F., Isenberg, H.D., McKinley, G.A., et al. 1988. The novel application of video image processing to biochemical and antimicrobial susceptibility testing. J. Clin. Microbiol. 26:1492.

12. Doern, G.V., Staneck, J.L., Needham, C., et al. 1987. Sensititre Autoreader for same-day breakpoint broth microdilution susceptibility testing of members of the family Enterobacteriaceae. J. Clin. Microbiol. 256:1481.

13. Evans, R.S. 1991. Application of artificial intelligence in clinical microbiology. Clin. Microbiol. Newsletter 13:25.

14. Ghanem, F.M., Ridpath, A.C., Moore, W.E.C., and Moore, L.V.H. 1991. Identification of *Clostridium botulinum, Clostridium argentinense,* and related organisms by cellular fatty acid analysis. J. Clin. Microbiol. 29:1114.

15. Hill, G.B., and Schalkowsky, S. 1990. Development and evaluation of the spiral gradient endpoint method for susceptibility testing of anaerobic gram-negative bacilli. Rev. Infect. Dis. 12:200.

16. Hogan, P., Isenberg, H.D., and Staneck, J.L. 1992. Introduction to automated microbial identification and susceptibility testing systems. Section 12.4. In Isenberg, H.D., ed. Clinical microbiology procedures handbook. American Society for Microbiology, Washington, D.C.

17. Holdeman, L.V., Cato, E.P., and Moore, W.E.C, editors. 1977. Anaerobe laboratory manual, ed 4. Virginia Polytechnic Institute and State University, Blacksburg, Va.

18. Isenberg, H.D., Reichler, A., and Wiseman, D. 1971. Prototype of a fully automated device for determination of bacterial antibiotic susceptibility in the clinical laboratory. Appl. Microbiol. 22:980.

19. Jorgensen, J.H. 1987. Use of laboratory computer systems to facilitate reporting of instrument-generated microbiology results. p. 169-176. In Jorgensen, J.H., editor. Automation in clinical microbiology, CRC Press, Boca Raton, Fla.

20. Koenig, C., Tick, L.J., and Hanna, B.A. 1992. Analyses of the FlashTrack DNA probe and UTIscreen bioluminescence tests for bacteriuria. J. Clin. Microbiol. 30:342.

21. Larsson, L. 1987. Gas chromatography and mass spectrometry. p. 153-166. In Jorgensen, J.H., editor. Automation and clinical microbiology, CRC Press, Boca Raton, Fla.

22. Limberger, R.J., Biega, R., Evancoe, A., et al. 1992. Evaluation of culture and the Gen-Probe PACE 2 assay for detection of *Neisseria gonorrhoeae* and *Chlamydia trachomatis* in endocervical specimens

transported to a state health laboratory. J. Clin. Microbiol. 30:1162.

23. Loeffelholz, M.J., Lewinski, C.A., Silver, S.R., et al. 1992. Detection of *Chlamydia trachomatis* in endocervical specimens by polymerase chain reaction. J. Clin. Microbiol. 30:2847.

24. Lyznicki, J., Lester, S., Springer, D., et al. 1991. Comparison of MicroScan and Sensititre panels for the rapid identification and susceptibility testing of gram-negative rods. Abstr. C-208, p. 376. Abstr. 91st General Meeting Amer. Soc. Microbiol. Amer. Soc. Microbiol., Washington, D.C.

25. Mansour, J.D., Robson, J.A., Arndt, C.W., et al. 1985. Detection of *Escherichia coli* in blood using flow cytometry. Cytometry 6:186.

26. McLaughlin, J.C., Barron, W.G., Merlin, T.L., et al. 1991. A comparison of Biolog, MicroScan, and Vitek AMS for the identification of infrequently isolated human gram-negative bacterial pathogens. Abstr. C-210, p. 377. Abstr. 91st General Meeting Amer. Soc. Microbiol. Amer. Soc. Microbiol., Washington, D.C.

27. Meier, F.A., and Hill, H.R. 1987. Automation of antigen detection in infectious disease diagnosis. p. 101-120. In Jorgensen, J.H., editor. Automation in clinical microbiology, CRC Press, Boca Raton, Fla.

28. Miller, J.M. and Rhoden, D.L. 1991. Preliminary evaluation of Biolog, a carbon source utilization method for bacterial identification. J. Clin. Microbiol. 29:1559.

29. Morgan, M.A., and Roberts, G.D. 1987. Radiometric detection, identification, and antimicrobial susceptibility testing of mycobacteria. p. 31-38. In Jorgensen, J.H., editor. Automation in clinical microbiology, CRC Press, Boca Raton, Fla.

30. Moss, C.W., and Nunez-Monteil, O.I. 1982. Analysis of short-chain acids from bacteria by gas-liquid chromatography with a fused-silica capillary column. J. Clin. Microbiol. 15:308.

31. Murray, P.R. 1987. Rapid automated identification systems. p. 53-67. In Jorgensen, J.H., editor. Automation in clinical microbiology, CRC Press, Boca Raton, Fla.

32. Nolte, F.S., Krisher, K.K., Beltran, L.A., et al. 1988. Rapid and overnight microdilution antibiotic susceptibility testing with the Sensititre Breakpoint AutoReader system. J. Clin. Microbiol. 26:1079.

33. O'Hara, C.M., Rhoden, D.L., and Miller, J.M. 1990. Agreement between visual and automated UniScept API readings. J. Clin. Microbiol. 28:452.

34. Pezzlo, M. 1987. Instrument methods for detection of bacteriuria. p. 15-29. In Jorgensen, J.H., editor. Automation in clinical microbiology, CRC Press, Boca Raton, Fla.

35. Sasser, M. and Wichman, M.D. Identification of microorganisms through use of gas chromatography and high-performance liquid chromatography. pp. 111-146. In Balows, A., Hausler, W.J., Jr., Herrmann, K.L., et al., editors. Manual of Clinical Microbiology, ed 5. American Society for Microbiology, Washington, D.C.

36. Spiegel, C.A., Amsel, R., Eschenbach, D., et al. 1980. Anaerobic bacteria in non-specific vaginitis. N. Engl. J. Med. 303:601.

37. Spiegel, C.A., Malangoni, M.A., and Condon, R.E. 1984. Gas-liquid chromatography for rapid

diagnosis of intraabdominal infection. Arch. Surg. 119:28.

38. Stager, C.E. and Davis, J.R. 1992. Automated systems for identification of microorganisms. Clin. Microbiol. Rev. 5:302.

39. Strong, C.A. and Ward, K.W. 1992. Packed-column gas-liquid chromatography. Section 12.8. In Isenberg, H.D., editor. Clinical microbiology procedures handbook. American Society for Microbiology, Washington, D.C.

40. Summanen, P., Baron, E.J., Citron, D.M., et al. 1993. Wadsworth anaerobic bacteriology manual, ed 5. Star Publishing, Belmont, Calif.

41. Thornsberry, C., Gavan, T.L., Sherris, J.C., et al. 1975. Laboratory evaluation of a rapid, automated susceptibility testing system: report of a collaborative study. Antimicrob. Agents Chemother. 7:466.

42. Tisdall, P.A., DeYoung, D., Roberts, G.D., et al. 1982. Identification of clinical isolates of mycobacteria with gas-liquid chromatography: a 10-month follow-up study. J. Clin. Microbiol. 16:400.

43. Tunér, K., Baron, E.J., Summanen, P., and Finegold, S.M. 1992. Cellular fatty acids in *Fusobacterium* species as a tool for identification. J. Clin. Microbiol. 30:3225.

44. Van Dilla, M.A., Langlois, R.G., Pinkel, D., et al. 1983. Bacterial characterization by flow cytometry. Science 220:620.

45. Welch, D.F. 1991. Applications of cellular fatty acid analysis. Clin. Microbiol. Rev. 4:422.

46. Wilson, M.L., Weinstein, M.P., Reimer, L.G., et al. 1992. Controlled comparison of the BacT/Alert and BACTEC 660/730 nonradiometric blood culture systems. J. Clin. Microbiol. 30:323.

47. Wolcott, M.J. 1992. Advances in nucleic acid–based detection methods. Clin. Microbiol. Rev. 5:370.

48. York, M.K., Brooks, G.F., and Fiss, E.H. 1992. Evaluation of the autoSCAN-W/A rapid system for identification and susceptibility testing of gram-negative fermentative bacilli. J. Clin. Microbiol. 30:2903.

BIBLIOGRAPHY

Beckmann, E., and Connolly, P. 1990. Flow cytometry: introduction and microbiological applications. Clin. Microbiol. Newsletter 12:105.

Miller, J.M. 1991. Evaluating biochemical identification systems. J. Clin. Microbiol. 29:1559.

Jorgensen, J.H., editor. 1987. Automation in clinical microbiology, CRC Press, Boca Raton, Fla.

13

DIAGNOSTIC IMMUNOLOGICAL PRINCIPLES AND METHODS

Infectious diseases can be definitively diagnosed in only three ways: (1) by documenting the presence in the patient of an agent known to cause the disease, either by visualizing the agent directly in clinical material obtained from the patient, by detecting antigens or genetic material specific for the agent, or by cultivating the agent in the laboratory (within an animal host or in vitro); (2) by detecting a specific product of the infectious agent in clinical material obtained from the patient, a product that could not have been produced without the agent's presence; and (3) by detecting an immunological response specific to the infecting agent in the patient's serum. Detection of the agent by visualization or cultivation is the focus of much of this text, covered extensively in Parts Two and Three. Detection of products or antigens of the agent is addressed in Chapter 11 and again as this diagnostic method relates to individual pathogens. Methods for documenting the presence of an immune response are described in this chapter.

GENERAL FEATURES OF THE IMMUNE RESPONSE

CATEGORIES OF IMMUNE RESPONSES

The human specific immune responses are simplistically divided into two categories. Cell-mediated immune responses, which are carried out by special lymphocytes of the T (thymus-derived) class, include production of activator cytokines that induce other cells to produce antibodies, to attack and kill pathogens readily, or activate cells to multiply and become aggressive, as well as the direct attacking and killing of pathogens (or host cells damaged or infected by pathogens) by the T lymphocytes themselves. Although diagnosis of certain diseases may be aided by measuring the cell-mediated immune response to the pathogen, such tests entail skin tests performed by physicians or in vitro cell function assays, performed by specially trained immunologists. These tests are usually not within the repertoire of clinical microbiology laboratories. Antibody-mediated immune responses are those produced by specific proteins generated by lymphocytes of the B (bone marrow–derived) class. Because these proteins have immunological function and because they fold into a globular structure in the active state, they are called **immunoglobulins.** Immunoglobulins, also called **antibodies,** since they are produced against a particular foreign structure or "body," attach to offending agents or their products and thus aid the host in removing the agents. These agents need not be pathogenic microorganisms, as the immune system is also effective against some tumor cells, transplanted (nonself) tissue, pollen, and other foreign substances. This chapter will deal only with the detection of antibodies produced against infectious agents, although many of the techniques employed may be used for the detection of other types of antibodies.

Antibodies are either secreted into the blood or lymphatic fluid (and sometimes other body fluids) by the B lymphocytes, or they remain attached to the surface of the lymphocyte or other effector cells. Because the mediators of this category of immune response chiefly circulate in the blood, this type of immunity is also called **humoral immunity.** For purposes of determining whether an antibody has been produced against a particular agent by a patient, the patient's **serum** (or occasionally the plasma) is tested for the presence of the antibody. The study of the diagnosis of disease by measuring antibody levels in serum is thus called *serology.*

CHARACTERISTICS OF ANTIBODIES

Antibodies are produced against a specific, small, chemically and physically defined substance called an antigen or an antigenic determinant, recognized as foreign by the host. By a genetically determined mechanism, normal humans are able to produce antibodies specifically directed against almost all the antigens with which they might come into contact throughout their lifetimes. Antigens may be part of the physical structure of the pathogen, or they may be a chemical produced and released by the pathogen. One pathogen may contain or produce many different antigens that the host will recognize as foreign, so that infection with one agent may cause a number of different antibodies to appear. *Streptococcus pyogenes* is an example of a pathogen that induces production of several different antibodies (Figure 13.1). In addition, some antigenic determinants of a pathogen may not be available for recognition by the host until the pathogen has undergone a physical change. For example, until a pathogenic bacterium has been digested by a human polymorphonuclear neutrophil, certain antigens deep within the cell wall are not "visible" to the host immune system. Once the bacterium is broken down, these new antigens are revealed and antibodies can be produced against them. For this reason, a patient may produce different antibodies at different times during the course of a single disease. The immune response to an antigen also matures with continued exposure, and the antibodies produced against it become more specific and more **avid** (able to bind more tightly).

Antibodies function by attaching to the surface of pathogens and making the pathogens more amenable to ingestion by phagocytic cells **(opsonizing** antibodies), by binding to and blocking surface receptors for host cells, or by attaching to the surface of pathogens and contributing to their destruction by the lytic action of complement (complement-fixing antibodies). Although humans produce several different classes of antibodies differentiated by their structures, routine diagnostic serologic methods are usually used to measure only two antibody classes, **immunoglobulin M (IgM)** and **immunoglobulin G (IgG).** The basic structure of an antibody molecule comprises two mirror images, each made up of two identical protein chains. At the terminal ends are the antigen-binding sites, which specifically attach to the antigen against which the antibody was produced. De-

FIGURE 13.1

S. pyogenes contains many antigenic structural components and produces several antigenic enzymes, each of which may elicit a specific antibody response from the infected host.

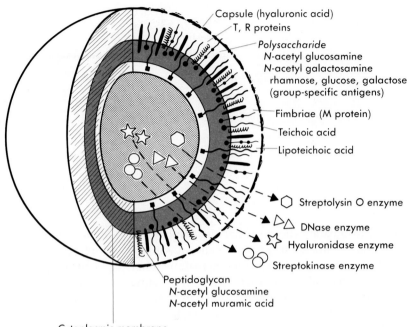

Capsule (hyaluronic acid)
T, R proteins
Polysaccharide
N-acetyl glucosamine
N-acetyl galactosamine
rhamnose, glucose, galactose
(group-specific antigens)
Fimbriae (M protein)
Teichoic acid
Lipoteichoic acid
Streptolysin O enzyme
DNase enzyme
Hyaluronidase enzyme
Streptokinase enzyme
Peptidoglycan
N-acetyl glucosamine
N-acetyl muramic acid
Cytoplasmic membrane

pending on the specificity of the antibody, antigens of some similarity, but not identical, to the inducing antigen may also be bound. The complement-binding site is found in the center of the molecule in a structure that is similar for all antibodies of the same class. IgM is produced as a first response to many antigens, although the levels remain high only transiently. Thus presence of IgM is usually indicative of recent or active exposure to antigen or infection. On the other hand, IgG antibody may persist long after infection has run its course. The IgM antibody type (Figure 13.2) consists of five identical proteins with the basic antibody structures linked together at the bases, leaving 10 antigen-binding sites available. The second antibody

class, IgG, consists of one basic antibody molecule with two binding sites (Figure 13.3). The differences in the size and conformation between these two classes of immunoglobulins result in differences in activities and functions.

FEATURES OF HUMORAL IMMUNE RESPONSE USEFUL IN DIAGNOSTIC TESTING

Normal humans produce both IgM and IgG in response to most pathogens. The larger number of binding sites on IgM molecules can help to clear the offending pathogen more quickly, even though each individual antigen-binding site may not be the most efficient for attaching the antigen. Over time, the cells that were producing IgM switch to producing IgG, often more specific for the antigen (more avid). The IgG has only two binding sites, but it can also bind complement. When IgG has bound to an antigen, the base of the molecule may be left projecting out in the environment. Structures on the base attract and bind phagocytic cell membranes, increasing the chances of engulfment and destruction of the pathogen by the host cells. In most cases IgM is produced by a patient only after the first interaction with a given pathogen and is no longer detectable within a relatively short period afterward. For serologic diagnostic purposes, one important difference between IgG and IgM is that IgM cannot cross the placenta of pregnant women. Therefore any IgM detected in the serum of a newborn baby must have been produced by the baby itself. Other differences between IgG and IgM are ex-

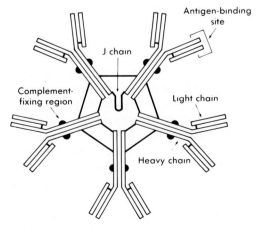

Antigen-binding site
J chain
Complement-fixing region
Light chain
Heavy chain

FIGURE 13.2

Structure of immunoglobulin M.

FIGURE 13.3

Structure of immunoglobulin G.

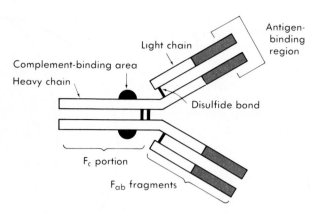

ploited to separate them so that the presence of one class of antibody in a patient's serum will not interfere with testing for the presence of the other class. A second encounter with the same pathogen will usually induce only an IgG response. Because the B lymphocytes retain memory of this pathogen, however, they can respond more quickly and with larger numbers of antibodies than at the initial interaction. This response is called the **anamnestic response.** Because the B cell memory is not perfect, occasional clones of memory cells will be stimulated by an antigen that is similar but not identical to the original antigen; thus the anamnestic response may be polyclonal and nonspecific. For example, reinfection with cytomegalovirus may stimulate memory B cells to produce antibody against Epstein-Barr virus, which they encountered previously, in addition to antibody against cytomegalovirus. The relative humoral responses over time are diagramed in Figure 13.4.

INTERPRETATION OF SEROLOGIC TESTS

A central dogma of serology is the concept of "rise in titer." The **titer** of antibody is the reciprocal of

the highest dilution of the patient's serum in which the antibody is still detectable. Patients with large amounts of antibody have high titers, since antibody will still be detectable at very high dilutions of serum. Serum for antibody levels should be drawn during the acute phase of the disease (when it is first discovered or suspected) and again during convalescence (usually at least 2 weeks later). These specimens are called **acute** and **convalescent sera.** For some infections, such as Legionnaires' disease, hepatitis, and others, titers may not rise until months after the acute infection or may never rise.

Patients with intact humoral immunity will develop increasing amounts of antibody to a disease-causing pathogen over several weeks. If it is the patient's first encounter with the pathogen and the specimen has been obtained early enough, no antibody will be present at the onset of disease. In the case of a second encounter the patient's serum will usually contain measurable antibody during the initial phase of disease and the level of antibody will quickly increase, because of the anamnestic response. For most pathogens, an increase in the patient's titer of two doubling dilutions (e.g., from a positive result at 1:8 to a positive result at 1:32) is considered to be diagnostic of current infection. This is called a fourfold rise in titer. Accurate results useful for diagnosis of many infections are achieved only when acute and convalescent sera are tested concurrently in the same system, since variables inherent in the procedures and laboratory error can easily result in differences of one doubling (or twofold) dilution in the results obtained from even the same sample tested at the same time. If the infecting or disease-causing agent is extremely rare and people without disease or prior immunization would have no chance of developing an immune response, such as the rabies virus or the toxin of botulism, the presence of specific antibody in a single serum specimen can be diagnostic.

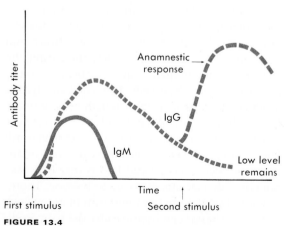

FIGURE 13.4

Relative humoral response to antigen stimulation over time.

With most diseases, there exists a spectrum of responses in infected humans, such that a person may develop antibody from subclinical infection or after colonization by an agent without actually having disease. In these cases the presence of antibody in a single serum specimen or a similar titer of antibody in paired sera may merely indicate past contact with the agent and cannot be used to accurately diagnose a recent disease. On the other hand, patients may respond to an antigenic stimulus by producing antibody that can cross-react with other antigens. These antibodies are nonspecific, and thus they may cause misinterpretation of serologic tests. Therefore for the vast majority of serologic procedures for diagnosis of recent infection, testing both acute and convalescent sera is the method of choice. Except for detecting the presence of IgM, the testing of a single serum can be recommended only in certain cases, such as for diagnosis of recent infection with *Mycoplasma pneumoniae* and viral influenza B, when high titers may indicate recent infection. Unfortunately, a certain proportion of infected individuals may never show a rise in titer, necessitating the use of other diagnostic measures. Cumitech 15, by Chernesky and others,[1] contains an excellent discussion on the collection and interpretation of paired sera for immunological testing. Because the delay inherent in testing paired acute and convalescent sera results in diagnostic information that arrives too late to influence initial therapy, increasing numbers of early serologic testing assays are being evaluated. More sensitive and specific measurements for IgM are often keys to these newer methods.

The prevalence of antibody to an etiological agent of disease in the population relates to the number of people who have come into contact with the agent, not the number who have actually developed disease. For most diseases, only a small proportion of infected individuals actually develop symptoms; another group develops antibodies that are protective without experiencing actual disease. There are a number of circumstances when serum is tested only to determine whether a patient is "immune," that is, has antibody to a particular agent either in response to a past infection or to immunization. These tests can be performed with a single serum sample. Correlation of the results of such tests with the actual immune status of individual patients must be performed to determine the level of detectable antibody present that corresponds to actual immunity to infection or reinfection in the host. For example, sensitive tests can detect the presence of very tiny amounts of antibody to the rubella virus.

Certain people, however, may still be susceptible to infection with the rubella virus with such small amounts of circulating antibody, and a higher level of antibody may be required to ensure protection from disease. Depending on the etiological agent, even low levels of antibody may protect a patient from pathological effects of disease, although they may not prevent reinfection. For example, a person previously immunized with killed poliovirus vaccine who becomes infected with pathogenic poliovirus will experience multiplication of the virus in the gut and virus entry into the circulation, but damage to the central nervous system will be blocked by humoral antibody in the circulation. As more sensitive testing methods are developed, and these types of problems become more common, microbiologists will need to work closely with clinicians to develop guidelines for interpreting serologic test results as they relate to the status of individual patients.

PRINCIPLES OF SEROLOGIC TEST METHODS

Antibodies can be detected in many different ways. In some cases, antibodies to an agent may be detected in more than one way, but the different antibody detection tests may not be measuring the same antibody. For this reason, the presence of antibodies to a particular pathogen as detected by one method may not correlate with the presence of antibodies to the same agent detected by another test method. This concept will become clearer after antibody detection methods have been discussed.

Slide agglutination tests are the easiest to perform and in some cases are the most sensitive tests currently available. Either artificial carriers, such as latex particles or treated red blood cells, or biological carriers, such as whole bacterial cells, can carry on their surface an antigen that will bind with antibody produced in response to that antigen when it was introduced to the host. Reagents for many of the following tests are commercially available.

DIRECT WHOLE PATHOGEN AGGLUTINATION FOR ANTIBODY DETECTION

The most basic tests for antibody are those that measure the antibody produced by a host to determinants on the surface of a bacterial agent in response to infection with that agent. Specific antibodies bind to surface antigens of the bacteria in a thick suspension and cause the bacteria to clump together in visible aggregates (Figure 13.5). Such antibodies are called agglutinating antibod-

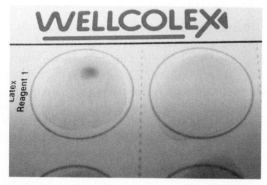

FIGURE 13.5

Color-coded particle agglutination system (Wellcolex, Murex Diagnostics) used for serotyping of *Salmonella* and *Shigella* species recovered from feces. Combinations of particles of different colors, carrying specific antigen receptors, create agglutination end products whose color depends on the serotype of the organism in the suspension. (Courtesy Murex Diagnostics, Norcross, Ga.)

ies, and the reaction is called bacterial **agglutination.** Electrostatic and other forces influence the formation of aggregates in solutions, so that certain conditions are usually necessary for good results. Since most bacterial surfaces exhibit a negative charge, they tend to repel each other. Performance of agglutination tests in sterile physiological saline (0.85% sodium chloride in distilled water), which has free positive ions present in the solution, will enhance the ability of antibody to cause aggregation of bacteria. Bacterial agglutination tests can be performed on the surface of glass slides or in tubes. Tube agglutination tests are often more sensitive, since a longer incubation period (allowing more antigen and antibody to interact) can be used. The small volume of liquid used for slide tests requires a rather rapid reading of the result, before the liquid has evaporated. Examples of bacterial agglutination tests are the tests for antibodies to *Francisella tularensis* and *Brucella* species, usually part of a panel of so-called "febrile agglutinin" tests. Bacterial agglutination tests are often used to diagnose diseases in which the bacterial agent is difficult to cultivate in vitro. Diseases commonly diagnosed by this technique include tetanus, yersiniosis, leptospirosis, brucellosis, and tularemia. The reagents necessary to perform many of these tests are commercially available, singly or as complete systems. Agglutination tests for certain diseases such as typhoid fever have become less useful with the ability of most laboratories to culture and identify the causative agent. Furthermore, the typhoid febrile agglutinin test (called the Widal test) is often positive in patients with infections caused by other bacteria, because of cross-reacting antibodies or previous immunization against typhoid. Appro-

priate specimens from patients suspected of having typhoid fever should be cultured for the presence of salmonellae (stool, urine, or blood, as appropriate); clinicians and microbiologists should not rely on febrile agglutinins for diagnosis of this disease.

Whole cells of parasites, including *Plasmodium, Leishmania,* or *Toxoplasma gondii,* have also been used for direct detection of antibody by agglutination. In addition to using the actual infecting bacteria or parasites as the agglutinating particles for the detection of antibodies, certain bacteria may be agglutinated by antibodies produced against another infecting agent. Many patients infected with one of the rickettsiae produce antibodies that can agglutinate bacteria of the genus *Proteus.* Tests for these cross-reacting antibodies are called the **Weil-Felix tests.** Although these antibodies are also produced in response to bacterial infection, their presence has been used traditionally as presumptive serologic evidence of rickettsial disease. As newer, more specific serologic methods of diagnosing rickettsial disease become more widely available, the use of the *Proteus* agglutinating tests should be discontinued.

PARTICLE AGGLUTINATION TESTS
Numerous serologic procedures have been developed for the detection of antibody via the agglutination of an artificial carrier particle with antigen bound to its surface. As noted in Chapter 11, similar systems employing artificial carriers coated with antibodies are commonly used for detection of microbial antigens. Antigens of streptococci, *Cryptococcus neoformans,* agents of bacterial meningitis, and other etiological agents can be detected using latex, treated erythrocytes, and staphylococcal particle agglutination systems. The size of the carrier enhances the visibility of the agglutination reaction, and the artificial nature of the system allows the antigen bound to the surface to be extremely specific. Complete systems for the use of latex or other particle agglutination tests are commercially available for the accurate and sensitive detection of antibody to cytomegalovirus, rubella virus, varicella-zoster virus, the heterophile antibody of infectious mononucleosis, teichoic acid antibodies against staphylococci, antistreptococcal antibodies, mycoplasma antibodies, and others. Latex tests for antibodies to *Coccidioides, Sporothrix, Echinococcus,* and *Trichinella* are available, although they are not widely used because of the uncommon occurrence of the corresponding infection or its limited geographical distribution. Use of tests for *Candida* antibodies

has not yet shown results reliable enough for diagnosis. Results of particle agglutination tests are dependent on several factors, including the amount and avidity of antigen conjugated to the carrier, the time of incubation together with patient's serum (or other source of antibody), and the microenvironment of the interaction (including pH, tonicity, and protein concentration).[5] Commercial tests have been developed as systems, complete with their own diluents, controls, and containers. For accurate results, they should be used as units, without modifications. If tests have been developed for use with cerebrospinal fluid, for example, they should not be used with serum unless the product insert or the technical representative has certified such usage.

Treated animal red blood cells have also been used as carriers of antigen for agglutination tests, called indirect **hemagglutination** or passive hemagglutination tests, since it is not the antigens of the blood cells themselves but the passively attached antigens that are being bound by antibody. The most widely used of these tests are the microhemagglutination test for antibody to *Treponema pallidum* (MHA-TP, so-called because it is performed in a microtiter plate), the hemagglutination treponemal test for syphilis (HATTS), the passive hemagglutination tests for antibody to extracellular antigens of streptococci, and the rubella indirect hemagglutination tests, all of which are available commercially. Certain reference laboratories, such as the Centers for Disease Control, also perform indirect hemagglutination tests for antibodies to some clostridia, *Pseudomonas pseudomallei, Bacillus anthracis, Corynebacterium diphtheriae, Leptospira,* and the agents of several viral and parasitic diseases.

FLOCCULATION TESTS FOR ANTIBODY DETECTION

In contrast to the aggregates formed when particulate antigens bind to specific antibody, the interaction of soluble antigen with antibody may result in the formation of a precipitate, a concentration of fine particles, usually visible only because the precipitated product is forced to remain in a defined space within a matrix. Two variations of the **precipitin test** are widely used for serologic studies. In **flocculation tests** the precipitin end product forms macroscopically or microscopically visible clumps. The Venereal Disease Research Laboratory test, known as the **VDRL,** is the most widely used flocculation test. Patients infected with pathogenic treponemes, most commonly *T. pallidum,* the agent of syphilis, form an antibody-like protein called **reagin** that binds to the test antigen, cardiolipin-lecithin–coated cholesterol particles, causing the particles to flocculate. Since reagin is not a specific antibody directed against *T. pallidum* antigens, the test is not highly specific, but it is a good screening test, detecting more than 99% of cases of secondary syphilis. Conditions and infections other than syphilis can cause a patient's serum to yield a positive result in the VDRL test, called a "biologic false positive" test. The VDRL is the single most useful test available for testing cerebrospinal fluid in cases of suspected neurosyphilis, although it may be falsely positive in the absence of this disease. The performance of the VDRL test requires scrupulously clean glassware and exacting attention to detail, including numerous daily quality control checks.[3] In addition, the reagents must be prepared fresh each time the test is performed, patients' sera must be inactivated by heating for 30 minutes at 56° C before testing, and the reaction must be read microscopically. For all these reasons it is being supplanted in many laboratories by a qualitatively comparable test, the rapid plasma reagin, or **RPR** test.

The RPR test is commercially available as a complete system containing positive and negative controls, the reaction card, and the prepared antigen suspension. The antigen, cardiolipin-lecithin–coated cholesterol with choline chloride, also contains charcoal particles to allow for macroscopically visible flocculation. Sera are tested without heating, and the reaction takes place on the surface of a specially treated cardboard card, which is then discarded (Figure 13.6). The RPR test is not recommended for testing cerebrospinal fluid. All procedures are standardized and clearly described in product inserts, and these procedures should be adhered to strictly. The failure of the RPR test to detect minimally reactive sera in one statewide proficiency testing survey was shown to be due to deficiencies in technique used, rather than deficiencies in the reagents or the test system itself. An excellent set of guidelines for acceptable test performance was published by Neimeister and coworkers.[4] Overall, the RPR appears to be a more specific screening test for syphilis than the VDRL, and it is certainly easier to perform. Several modifications have been made, such as the use of dyes to enhance visualization of results.

COUNTERCURRENT IMMUNOELECTROPHORESIS FOR ANTIBODY DETECTION

The second variation of the classic precipitin test has been widely used to detect small amounts of antibody. This test takes advantage of the net

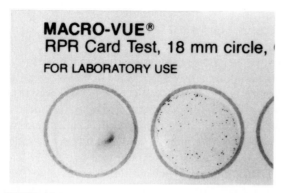

FIGURE 13.6

Rapid plasma reagin test for syphilis (Becton Dickinson Microbiology Systems).

electric charge of the antigens and antibodies being tested in a particular test buffer. Because the antigen and antibody being sought migrate toward one another in a semisolid matrix under the influence of an electrical current, the method is called countercurrent immunoelectrophoresis, or simply **counterimmunoelectrophoresis (CIE).** The principles of this test were outlined in Chapter 11; the same methodology is used to identify specific antigen or antibody. When antigen and antibody meet in optimal proportions, a line of precipitation will appear. Since all variables, such as buffer pH, type of gel or agarose matrix, amount of current, amounts and concentrations of antigen and antibody, size of antigen and antibody inocula, and placement of these inocula, must be carefully controlled for maximum reactivity, CIE tests are difficult to develop and perform. Other methods for detection of antibody to infectious agents are more commonly used in most diagnostic laboratories. A very sensitive, but not very specific, commercially available test employs CIE for detection of antibody to *Entamoeba histolytica* in patients suspected of having invasive, extraintestinal amebic disease, primarily liver abscess.

IMMUNODIFFUSION ASSAYS FOR DETECTION OF ANTIFUNGAL ANTIBODIES

Closely resembling the precipitin test is a method widely used for detecting antibodies directed against fungal cell components, the immunodiffusion assay (ID). Whole cell extracts or other antigens of the suspected fungus are placed in wells in an agarose plate, and the patient's serum and a control positive serum are placed in adjoining wells. If the patient has produced specific antibody against the fungus, precipitin lines will be visible within the agarose between the homologous antigen and antibody wells; their identity to similar lines from the control serum helps establish the results. The type and thickness of the precipitin bands may have prognostic, as well as diagnostic, value. Antibodies against *Histoplasma, Blastomyces, Coccidioides, Paracoccidioides,* and some opportunistic fungi are routinely detected by ID. Immunodiffusion tests usually require at least 48 hours and may require additional steps to develop the bands.

HEMAGGLUTINATION INHIBITION TESTS FOR VIRAL ANTIBODIES

Many human viruses have the ability to bind to surface structures on red blood cells from different species. For example, rubella virus particles can bind to human type O, goose, or chicken erythrocytes and cause agglutination of the red blood cells. Influenza and parainfluenza viruses agglutinate guinea pig, chicken, or human O erythrocytes; many arboviruses agglutinate goose red blood cells; adenoviruses agglutinate rat or rhesus monkey cells; mumps virus agglutinates chicken or goose cells; measles virus binds red blood cells of monkeys; and herpesvirus and cytomegalovirus agglutinate sheep red blood cells. Serologic tests for the presence of antibodies to these viruses exploit the agglutinating properties of the virus particles. Patients' sera that have been treated with kaolin or heparin-magnesium chloride (to remove nonspecific inhibitors of red cell agglutination and nonspecific agglutinins of the red cells) are added to a system that contains the virus suspected of causing disease. If antibodies to the virus are present, they will form complexes and block the binding sites on the viral surfaces. When the proper red cells are added to the solution, all of the virus particles will be bound by antibody, preventing the virus from agglutinating the red cells. Thus the patient's serum is positive for hemagglutination-inhibiting antibodies. As for most serologic procedures, a fourfold increase in such titers is considered diagnostic. The hemagglutination inhibition tests for most agents are performed only at reference laboratories. Procedures for performing such tests are delineated in the *Manual of Clinical Immunology* (American Society for Microbiology). Rubella antibodies, however, are often detected with this method in routine diagnostic laboratories. Several commercial rubella antibody detection systems are available.

NEUTRALIZATION TESTS

To test for certain antibodies, the ability of a patient's serum to block the effect of the antigenic agent can be evaluated. In the case of viruses, antibody that inhibits the infectivity of the virus by blocking the host cell receptor site is called *neutralizing antibody.* The serum to be tested is mixed with a suspension of infectious virus particles of the same type as those with which the patient is suspected of being infected. A control suspension of viruses is mixed with normal serum. The virus suspensions are then inoculated into a cell culture system that supports growth of the virus. The control cells will display evidence of virus infection. If the patient's serum contains antibody to the virus, that antibody will bind the virus particles and prevent them from invading the cells in culture. The antibody has "neutralized" the infectivity of the virus. These tests are technically demanding and time-consuming and are performed only in some laboratories that routinely perform virus cultures.

Antibodies to bacterial toxins and other extracellular products that display measurable activities can be tested in the same way. The ability of a patient's serum to neutralize the erythrocyte-lysing capability of streptolysin O, an extracellular enzyme produced by *S. pyogenes* during infection, has been used for many years as a test for previous streptococcal infection. After pharyngitis with streptolysin O–producing strains, most patients show a high titer of the antibody antistreptolysin O **(ASO).** Streptococci also produce the enzyme deoxyribonuclease B (DNase B) during infections of the throat, skin, or other tissue. A neutralization test that prevents activity of this enzyme, the *anti-DNase B test,* has also been used extensively as an indicator of recent or previous streptococcal disease. The use of particle agglutination (latex or indirect hemagglutination) tests for the presence of antibody to many of the streptococcal enzymes has replaced the use of these neutralization tests in many laboratories.

COMPLEMENT FIXATION FOR ANTIBODY DETECTION

One of the classic methods for demonstrating the presence of antibody in a patient's serum has been the **complement fixation test (CF).** This test consists of two separate systems, the first (the indicator system) consisting of a combination of sheep red blood cells, complement-fixing antibody (IgG) raised against the sheep red blood cells in another animal, and an exogenous source of complement (usually guinea pig serum). When these three components are mixed together in optimum concentrations, the anti-sheep erythrocyte antibody will bind to the surface of the red cells and the complement will then bind to the anti-

gen-antibody complex, ultimately causing lysis of the red cells. For this reason the anti-sheep red cell antibody is also called *hemolysin.* The second system consists of the antigen suspected of causing the patient's disease and the patient's serum. For the CF test these two systems are tested in sequence (Figure 13.7). The patient's serum is first added to the putative antigen; then the limiting amount of complement is added to the solution. If the patient's serum contains antibody to the antigen, the resulting antigen-antibody complexes will bind all of the complement added. In the next step the sheep red blood cells and the hemolysin (indicator system) are added. Only if the complement has not been bound by a complex formed with antibody from the patient's serum, will the complement be available to bind to the sheep cell–hemolysin complexes and cause lysis. A positive result, meaning the patient does possess complement-fixing antibodies, is revealed by failure of the red cells to lyse in the final test system. Lysis of the indicator cells indicates lack of antibody and a negative CF test.

This test has been used over the years for the detection of many types of antibodies, particularly antiviral and antifungal, although it requires many manipulations, at least 48 hours for both stages of the test to be completed, and often yields nonspecific results. Many new systems both for improved recovery of pathogens or their products and for more sensitive and less demanding procedures for detection of antibodies, including particle agglutination, indirect fluorescent antibody tests, and ELISA procedures, described in this chapter and throughout the text, have gradually been introduced to replace the CF test. At this time CF tests are performed chiefly for diagnosis of unusual infections; these tests are done primarily in reference laboratories. This test is still probably the most common method for diagnosis of infection caused by some fungi, respiratory viruses, parainfluenza, and arboviruses, as well as for diagnosis of Q fever. Laboratories without experience in performing these tests should not adopt complement fixation tests for routine diagnostic testing when other less demanding procedures are available.

ENZYME-LINKED IMMUNOSORBENT ASSAYS

Enzyme-linked immunosorbent assay (ELISA) technology is among the most widely used detection methods for biological products, including antigens, antibodies, hormones, peptides, and toxins. ELISA methods for antibodies to infectious agents are sensitive and specific. As described

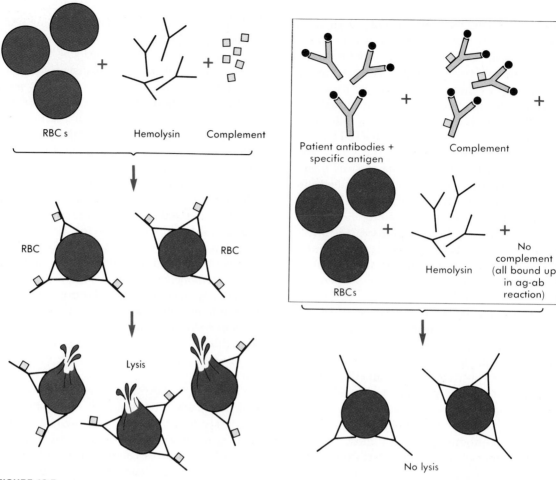

FIGURE 13.7

Complement fixation test.

more thoroughly in Chapter 11, the presence of specific antibody is detected by the ability of a second antibody conjugated to a colored or fluorescent marker to bind to the target antibody, which is bound to its homologous antigen. Various enzyme-substrate systems and the use of avidin-biotin to bind marker substances were also discussed in Chapter 11. The antigen to which the antibodies bind, if they are present in patients' sera, is either attached to the inside of wells of a microtiter plate, adherent to a filter matrix, or bound to the surface of beads or plastic paddles. Although the sensitivity of certain particle-agglutinating tests is greater than ELISA tests for antibody at the current time, ELISA methodology is being improved rapidly. Currently, most ELISA test procedures require several incubation periods and washes. However, the complete automation of these procedures is possible, and much automation has already been introduced into commercial systems. Advantages are that tests can be

performed easily on many serum samples at the same time and that the colored or fluorescent end products are easily detected by instruments, removing the element of subjectivity inherent in so many serologic procedures that rely on a technologist's interpretation of a reaction. Disadvantages include the need for some special equipment, the fairly long reaction times (often hours instead of minutes for microdilution format tests), the relative end point of the test (that relies on measuring the amount of a visible end product that is not dependent on the original antigen-antibody reaction itself, but on a second enzymatic reaction as compared with a directly quantitative result, and the requirement for batch processing to ensure that performance of the test is cost-effective. Possibly as a result of the measurement of the end point of a secondary reaction, ELISA methods do not generate a true titer result, another major disadvantage of the system. ELISA results thus may not correlate well with titers, a problem for those who are accustomed to interpreting serologic results as titers.

A general discussion of the principles of automated ELISA readers currently available is found in Chapter 12. Commercial microdilution or solid phase matrix systems are available for the detection of antibody specific for hepatitis virus antigens, herpes simplex viruses I and II, respiratory syncytial virus (RSV), cytomegalovirus, human immunodeficiency virus (HIV), rubella virus (both IgG and IgM), mycoplasmas, chlamydiae, *Borrelia burgdorferi*, *E. histolytica*, *T. gondii*, and many other agents.

The introduction of membrane-bound ELISA components has improved sensitivity and ease of use dramatically. Slot-blot and dot-blot assays force the target antigen through a membrane filter, where the antigen becomes affixed in the shape of the hole (a dot or a slot). Several antigens can be placed on one membrane. When test serum is layered onto the membrane, specific antibodies, if present, will bind to the corresponding dot or slot of antigen. Addition of a labeled second antibody and subsequent development of the label allows visual detection of the presence of antibodies based on the pattern of antigen sites. Cassette-based membrane-bound ELISA assays, designed for testing a single serum, can be performed rapidly (often within 10 minutes). These systems have gained popularity for detecting antigens (rotavirus, RSV, *Streptococcus pyogenes*, and others), but tests for antibodies to *Helicobacter pylori*, *Toxoplasma gondii*, and some other infectious agents are available. Accuracy of the results of tests using each of these formats is variable.

Antibody capture ELISAs are particularly valuable for detecting IgM in the presence of IgG. Anti-IgM antibodies are fixed to the solid phase, and thus only IgM antibodies, if present in the patient's serum, are bound. In a second step, specific antigen is added in a sandwich format to which a second antigen-specific labeled antibody is finally added (see Figure 11.4). Toxoplasmosis, rubella, and other infections are diagnosed using this technology, usually in research settings. The recent advances in ELISA methods, including membrane-fixed antibodies, fluorescent and enzymatic labels, antibody capture formats, monoclonal antibodies (discussed more completely in Chapter 11), and others have brought ELISA methodology to the forefront of serologic assay techniques in routine use today.

INDIRECT FLUORESCENT ANTIBODY TESTS AND OTHER IMMUNOMICROSCOPIC METHODS

A widely applied method for the detection of diverse antibodies is that of indirect fluorescent antibody determination (IFA). For tests of this type, the antigen against which the patient makes antibody (such as whole *Toxoplasma* organisms or virus-infected tissue culture cells) is affixed to the surface of a microscope slide. The patient's serum to be tested is diluted and placed on the slide, covering the area in which antigen was placed. If antibody is present in the serum, it will bind to its specific antigen. Unbound antibody is then removed by washing the slide. In the second stage of the procedure, a conjugate of antihuman globulin (which may be directed specifically against IgG or IgM) and a dye that will fluoresce when exposed to ultraviolet light (such as fluorescein) is placed on the slide. This labeled marker for human antibody will bind to the antibody already bound to the antigen on the slide and will serve as a detector for the antibody when viewed under a fluorescence microscope (Figure 13.8). After the slides have been stained, they are covered with a drop of buffered glycerol and a glass coverslip is applied. The fluorescence does not fade appreciably for several days if the slides are kept refrigerated in the dark, but it is best to examine them immediately after staining. Such slides cannot be used as permanent mounts, since the fluorescence fades with time. Commercially available systems include the slides with the antigens, positive and negative control sera, diluent for the patients' sera, and the properly diluted conjugate. As with other commercial products, IFA systems should be used as units, without modifying the manufacturers' instructions. Currently, commercially available IFA tests include those for antibodies to *Legionella* species, *B. burgdorferi*, *T. gondii*, varicella-zoster virus, cytomegalovirus, Epstein-Barr virus capsid antigen, early antigen, and nuclear antigen, herpes simplex viruses I and II, rubella virus, *M. pneumoniae*, *T. pallidum* (the fluorescent treponemal antibody absorption test, **FTA-ABS**), and several rickettsiae. Most of these tests, if performed properly, give extremely specific and sensitive results. Proper interpretation of IFA tests requires experienced and technically competent technologists, and these tests can be performed rapidly and are cost-effective with only a few samples, in contrast to ELISA.

FIGURE 13.8

Indirect fluorescent antibody test for *T. gondii* antibodies. *Toxoplasma* organisms affixed to the slide bind specific antibodies in the patient's serum. Antihuman antibody conjugated with fluorescein binds in turn to the bound patient's antibodies, causing the organisms to fluoresce.

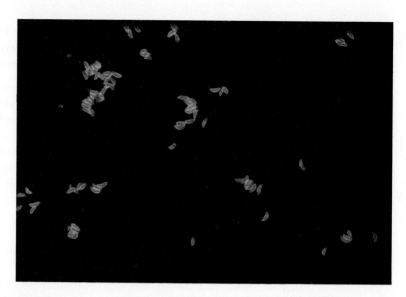

Antibodies may be conjugated to other markers in addition to fluorescent dyes. This technology has been called "colorimetric immunological target detection." Recently, the use of enzyme-substrate marker systems has expanded. Horseradish peroxidase, alkaline phosphatase, and avidin-biotin–conjugated enzyme labels have all been used as visual tags for the presence of antibody. These reagents allow the preparation of permanent mounts, since the reactions do not fade with storage, and require only a standard light microscope for visualizing results.

RADIOIMMUNOASSAY

This automated method of detecting antibodies is usually not performed in a serology laboratory, but in a chemistry laboratory. Radioimmunoassay (RIA) is primarily used to measure antigens (notably certain hormones or proteins) in serum samples. RIA tests were primarily used to detect antibody to hepatitis B viral proteins. As described in Chapter 12, radioactively labeled antibody competes with the patient's unlabeled antibody for binding sites on a known amount of antigen. A reduction in radioactivity of the antigen-patient antibody complex compared with the radioactive counts measured in a control test with no antibody is used to quantitate the amount of patient antibody bound to the antigen. The development of new marker substances, such as enzyme and substrate systems, chemiluminescence, and fluorescence, is allowing development of tests as sensitive as RIA without the hazards of radioactive reagents.

FLUORESCENCE IMMUNOASSAYS

Because of the inconveniences associated with radioactive substances and scintillation counters, fluorescent immunoassays (FIA) were developed. These tests, which use fluorescent dyes or molecules as markers instead of radioactive labels, are based on the same principle as RIA. The primary difference is that in RIA systems the competitive antibody is labeled with a radioisotope and in FIA the antigen is labeled with a compound that will fluoresce under the appropriate light rays. Binding of patient antibody to a fluorescent-labeled antigen can reduce or "quench" the fluorescence, or binding can cause fluorescence by allowing conformational change in a fluorescent molecule. Measurement of fluorescence is thus a direct measurement of antigen-antibody binding, not dependent on a second marker system such as that in ELISA tests. Chapter 12 describes the principles involved in the fluorescence immunoassay for microbiological determinations. Systems are commercially available to measure antibody developed against a number of infectious agents, as well as against self-antigens (autoimmune antibodies).

WESTERN BLOT IMMUNOASSAYS

Requirements for the detection of very specific antibodies have driven the development of the Western blot immunoassay system (Figure 13.9). The method is based on the electrophoretic separation of major proteins of an infectious agent in a two-dimensional agarose gel matrix. A suspension of the organism against which antibodies are being sought is mechanically or chemically dis-

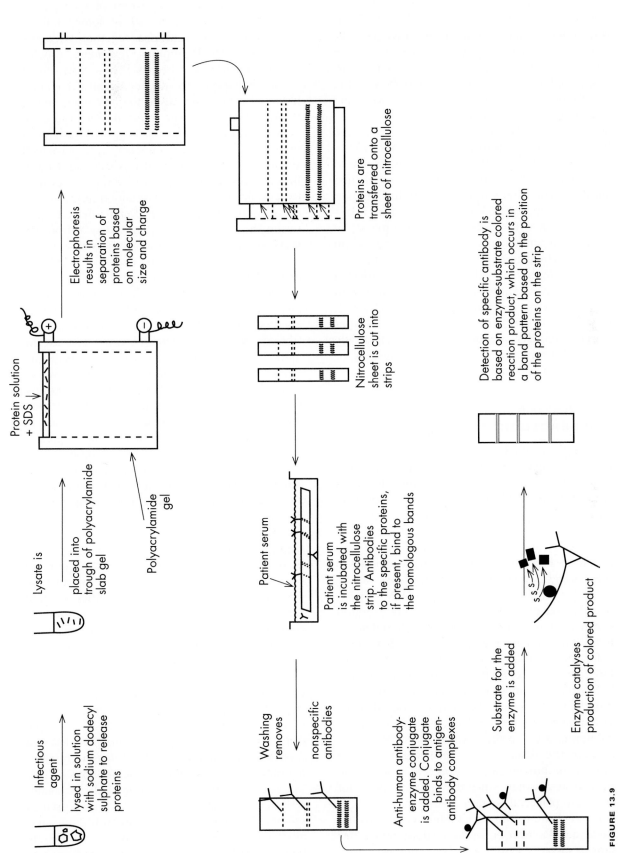

FIGURE 13.9

Diagram of Western blot immunoassay system.

rupted, and the solubilized antigen suspension is placed at one end of a polyacrylamide (polymer) gel. Under influence of an electrical current, the proteins migrate through the gel. Most bacteria or viruses contain several major proteins that can be recognized based on their position in the gel after electrophoresis. Smaller proteins travel faster and migrate farther in the lanes of the gel than do the larger proteins. The protein bands are transferred from the gel to a nitrocellulose or other type of thin membrane, and the membrane is treated to immobilize the proteins. The membrane is then cut into many thin strips, each carrying the pattern of protein bands. When patient serum is layered over the strip, antibodies will bind to each of the protein components represented by a band on the strip. The pattern of antibodies present can be used to determine whether the patient is infected with the agent. Antibodies against microbes with numerous cross-reacting antibodies, such as *T. pallidum, B. burgdorferi,* herpes simplex virus types 1 and 2, and HIV, are identified more specifically using this technology than a method that tests for only one general antibody type.

SEPARATING IgM FROM IgG FOR SEROLOGIC TESTING

Since IgM is usually produced only during a patient's first exposure to an infectious agent, the detection of specific IgM can help the clinician a great deal in establishing a diagnosis, especially for diseases that may have less well-characterized clinical presentations, such as toxoplasmosis, or those for which rapid therapeutic decisions may be required. Pregnant women who are exposed to rubella and develop a mild febrile illness can be tested for the presence of antirubella IgM. If positive, damage to the fetus is possible and the option of elective termination of pregnancy should be offered. An additional reason to measure IgM alone is for the diagnosis of neonatal infections. Since IgG can readily cross the placenta, newborn babies will carry titers of IgG nearly identical to those of their mothers. Accurate serologic diagnosis of infection in neonates requires either demonstration of a rise in titer (which takes time to occur) or the detection of specific IgM directed against the putative agent. This IgM would have to be of fetal origin, since the molecule does not cross the placental barrier. Agents that are difficult to culture or those that adult females would be expected to have encountered during their lifetimes, such as cytomegalovirus, herpesviruses, *Toxoplasma,* rubella virus, or *T. pallidum,* are those

for which specific identification of fetal IgM is most often used. The names of some of these agents have been grouped together with the acronym **STORCH** (syphilis, *Toxoplasma,* rubella, cytomegalovirus, and herpes). These tests should be ordered separately, depending on the clinical illness of a newborn suspected of having one of these diseases. However, infected babies often appear clinically the same as those without any of these infections. There are many cases in which serologic tests can yield false-positive and false-negative results. Other considerations, including patient history and clinical situation, must be employed in the diagnosis of neonatal infection. Culturing the agent from patient material from infants suspected of having herpesvirus or cytomegalovirus infection, for example, is still the most reliable diagnostic method.

Several methods have been developed to measure only the specific IgM in sera that may also contain specific IgG.[6] In addition to using a labeled antibody specific for only IgM as the marker or the IgM capture sandwich assays described previously, the immunoglobulins can be separated from each other by physical means. Centrifugation through a sucrose gradient, performed at very high speeds, has been used in the past to separate IgM, which has a greater molecular weight, from IgG. This method is time-consuming and requires a very expensive ultra-centrifuge. Many laboratories use a commercially available system that follows the principle of ion exchange (Figure 13.10).[2] One commercially available IgM isolation system (Isolab Inc., Akron, Ohio) recovers more than 80% of the IgM present in the original serum, with a small amount of contaminating IgG. Since the serum must be washed through the column with buffer, the IgM solution in the eluate (fluid collected from the column) is more dilute than in the original serum (usually 1:10, depending on the volume of serum originally placed on the column).

Other IgM separation systems available utilize the fact that certain proteins on the surface of staphylococci (protein A) and streptococci (protein G) will bind the Fc portion of IgG. A simple centrifugation then separates the particles and their bound immunoglobulins from the remaining mixture, which contains the bulk of the IgM. The antibody capture method for ELISA tests described previously in this chapter is also used to separate IgG. Other methods use antihuman IgG antibodies to remove IgG from sera containing both IgG and IgM. Different systems perform best under different circumstances and for recovery of

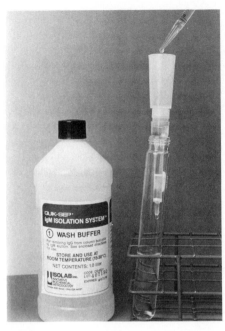

FIGURE 13.10

Separation of IgM from human serum by passing the serum through an ion-exchange column.

specific immunoglobulins. An added bonus of IgM separation systems is that **rheumatoid factor,** IgM antibodies produced by some patients against their own IgG, will often bind to the IgG molecules being removed from the serum, and consequently these IgM antibodies will be removed along with the IgG. Rheumatoid factor can cause nonspecific and interfering results with many serologic tests, and its presence should be taken into account.

REFERENCES

1. Chernesky, M.A., Ray, C.G., and Smith, T.F. 1982. In Drew, W.L., editor. Laboratory diagnosis of viral infections. Cumitech 15. American Society for Microbiology, Washington, D.C.
2. Johnson, R.B., and Libby, R. 1980. Separation of immunoglobulin M (IgM) essentially free of IgG from serum for use in systems requiring assay of IgM-type antibodies without interference from rheumatoid factor. J. Clin. Microbiol. 12:451.
3. Larson, S.A., Hunter, E.F., and Kraus, S.J. 1990. Manual of tests for syphilis, ed 8. Amer. Public Health Assoc., Washington, D.C.
4. Neimeister, R.P., Teschemacher, R., Yankevitch, I.J., and Cocklin, J. 1975. Proficiency testing, trouble shooting and quality control for the RPR test. Am. J. Med. Technol. 41:13.
5. Tinghitella, T.J., and Edberg, S.C. 1991. Agglutination tests and *Limulus* assay for the diagnosis of infectious diseases. pp. 61-72. In Balows, A., Hausler, W.J., Jr., Herrmann, K.L., et al., editors. Manual of clinical microbiology, ed 5. American Society for Microbiology, Washington, D.C.
6. Wiedbrauk, D.L., and Chuang, H.Y.K. 1988. Managing IgM serologies: getting the IgG out. Lab Management 26:24.

BIBLIOGRAPHY

Benjamini, E., and Leskowitz, S. 1991. Immunology: A short course, ed 2. Wiley-Liss, New York.
Nakamura, R.M., Kasahara, Y., and Rechnitz, G.A., editors. 1991. Immunochemical assays and biosensor technology for the 1990s. American Society for Microbiology, Washington, D.C.
Noonan, K.D. 1984. Latex agglutination testing: the importance of the microenvironment. Clin. Microbiol. Newsletter 6:20.
Palmer, D.F. 1980. Complement fixation test. In Rose, N.R., and Friedman, H., editors. Manual of clinical immunology, ed 2. American Society for Microbiology, Washington, D.C.
Proffitt, M.R. 1990. "Blotting" techniques for the diagnosis of infectious diseases. Clin. Microbiol. Newsletter 12:121.
Roitt, I.M., Brostoff, J., and Male, D.K. 1989. Immunology, ed 2. Gower Medical Publishing, London. (Distributed by Mosby, St. Louis.)
Rose, N.R., de Macario, E.C., Fahey, J.L., et al., editors. 1992. Manual of clinical immunology, ed 4. American Society for Microbiology, Washington, D.C.
Thompson, E.J. 1987. Immunoblotting: A rapid and sensitive method for disease diagnosis. Lab. Management. 25:39.

14

METHODS FOR TESTING ANTIMICROBIAL EFFECTIVENESS

Determining appropriate treatment for an infectious disease requires both the isolation of an infectious agent from a patient with disease and determination of susceptibility or resistance to antimicrobial agents used in therapy. Patterns of resistance are constantly changing. Even the pneumococcus, which for decades was invariably susceptible to penicillin G at levels of less than 0.03 μg/ml, is developing increasing resistance to this drug.

Since the susceptibility of many bacteria, mycobacteria, fungi, and viruses to antimicrobial agents cannot be predicted, testing individual pathogens against appropriate antimicrobial agents is often necessary. The appropriate agent (based on activity against the pathogen, including bactericidal activity when indicated, the least toxicity to the host, the least impact on normal flora, appropriate pharmacological characteristics, and most economical drug) can then be chosen, allowing a more certain therapeutic outcome. Table 14.1 lists some commonly encountered organisms and the types of antimicrobial agents that are usually effective against them. The ultimate therapeutic outcome depends on many more variables,

of course. The underlying disease and condition of the host, the use of indicated drainage, debridement, or other surgical procedures, the pharmacological properties of the antimicrobial agent, additional therapeutic agents being given, and other factors will exert a strong influence on the ultimate outcome. In addition, the blood and tissue levels of an antimicrobial agent are dependent on the route of administration (oral versus parenteral).

Clinical microbiologists can only recommend therapeutic agents based on their in vitro activities and their background understanding of "drug-bug" interactions. At times, an antimicrobial agent that shows poor in vitro activity against an organism is used in a patient with good results; the opposite effect may also occur. This occurs because host immune factors play a significant role in the outcome of an infection, with antimicrobial agent therapy being only one important factor in determining success or failure. Although the exact "level" of activity of an antimicrobial agent against an infecting organism may not directly correlate to the outcome of treatment, it is critical that the treating physician accurately know

TABLE 14.1

SELECTED BACTERIA AND GENERAL CATEGORIES OF ANTIMICROBIAL AGENTS USED TO TREAT SYSTEMIC INFECTION CAUSED BY THEM

GRAM-NEGATIVE ORGANISMS OTHER THAN *PSEUDOMONAS AERUGINOSA*	*P. AERUGINOSA*	GRAM-POSITIVE BACILLI AND STAPHYLOCOCCI	ENTEROCOCCI	ANAEROBES	NONENTEROCOCCAL STREPTOCOCCI
Aminoglycosides	Aminoglycosides	Cephalosporins	Penicillin and aminoglycoside combination	β-lactam plus β-lactamase inhibitor	Penicillins
Broad-spectrum penicillins	Broad-spectrum penicillins	Erythromycin	Vancomycin	Broad-spectrum penicillins	
Cephalosporins	Cefoperazone	Penicillins		Cefoxitin	
Cotrimoxazole	Ceftazidime	Vancomycin		Chloramphenicol	
Imipenem	Imipenem			Clindamycin	
Monobactams				Imipenem	
Quinolones				Metronidazole	

whether the microbe being treated is susceptible or resistant to the chosen antibiotic regimen. The clinician must make the final choice based on his or her knowledge of all the pertinent facts. One important concept that must be stressed is that in vitro susceptibility tests cannot be performed on mixed cultures; only pure cultures will yield valid results. With these caveats in mind, we have catalogued the antimicrobial susceptibility testing methods currently in widespread use among microbiologists in the United States.

The term **antibiotic** has been defined as a chemical produced by a microorganism that inhibits the growth of other microorganisms. Antibiotics modified by chemical manipulations are still considered to be antibiotics. An antimicrobial agent is active against microorganisms and may be produced either naturally by microorganisms or synthetically by scientists in laboratories. The term **chemotherapeutic agent** has been used to refer to synthetic and other antimicrobial agents, and also to refer to agents that act against human cells, such as anticancer drugs. The subterms *antiviral agent* and *antifungal agent* are more specific terms that fall into the general category of antimicrobial agents as we refer to them here.

NONAUTOMATED IN VITRO SUSCEPTIBILITY TESTING

Immediately after the discovery of sulfonamides and penicillin, pathogenic organisms were rarely tested for their susceptibility to drugs. Patients were treated empirically, and the organisms were usually susceptible. It was not until resistance began to emerge, soon after the agents were introduced, that microbiologists began to test antimicrobial agents against an infecting organism. The classic method used for in vitro testing was that of broth dilution, which yields a quantitative result for the amount of antimicrobial agent that is needed to inhibit growth of a specific microorganism.

For many types of susceptibility testing, a standard inoculum of bacteria must be used. The number of bacteria in a liquid medium can be determined in four ways:

1. By counting individual cells in a microscopic counting chamber, called a Petroff-Hausser chamber
2. By measuring the optical density of a broth culture of the bacteria and comparing it to the number of colonies found on subculture to solid agar
3. By growing the organisms to stationary phase (when the number of bacteria levels out at approximately 1×10^9/ml)
4. By visually comparing the turbidity of the liquid medium to a standard that represents a known number of bacteria in suspension, the most practical method. Turbidity standards (Figure 14.1) can be prepared by mixing chemicals that precipitate to form a solution of reproducible turbidity. Such solutions, using barium sulfate, were developed by McFarland to approximate numbers of bacteria in solutions of equal turbidity, as determined by colony counts (Table 14.2). They are prepared as detailed in Procedure 14.1.

TABLE 14.2

MCFARLAND NEPHELOMETER STANDARDS

	TUBE NUMBER										
	0.5	1	2	3	4	5	6	7	8	9	10
Barium chloride (ml)	0.05	0.1	0.2	0.3	0.4	0.5	0.6	0.7	0.8	0.9	1
Sulfuric acid (ml)	9.95	9.9	9.8	9.7	9.6	9.5	9.4	9.3	9.2	9.1	9
Approximate cell density ($\times 10^8$/ml)	1.5	3	6	9	12	15	18	21	24	27	30

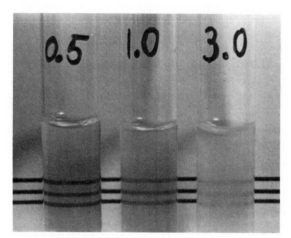

FIGURE 14.1

McFarland standards numbers 0.5, 1, 3.

BACTERIAL BROTH DILUTION METHODS

A complete protocol for performing these tests is found in the National Committee for Clinical Laboratory Standards (NCCLS) publication, M7-A2.[12] We will describe the basics of the procedures, but the reader is referred to the NCCLS publications and other references cited at the end of this chapter for more complete details. For broth dilution methods, decreasing concentrations of the antimicrobial agent(s) to be tested, usually prepared in serial twofold dilutions, are placed in tubes of a broth medium that will support the growth of the test microorganism. The most commonly used broth for these tests is Mueller-Hinton (available commercially) supplemented with the magnesium and calcium cations. Formulas can be found in Appendix A of the eighth edition of this book or in other publications compiling media used in microbiology. For broth dilution susceptibility testing of staphylococci against methicillin, oxacillin, or nafcillin, 2% sodium chloride must be added to the cation-adjusted Mueller-Hinton broth (CAMHB). The antimicrobial agents are prepared in concentrated solution and then diluted to the appropriate concentrations in broth.

The diluents used for the commonly tested antimicrobial agents are shown in Table 14.3. Preparation of the antimicrobial agents must take into consideration that the laboratory standard powder obtained from the manufacturer (addresses can be obtained from pharmaceutical representatives or a local pharmacy), usually at no cost, often does not consist of 100% active compound. The specific activity is the amount, in micrograms per milligram, of active antimicrobial agent in the powder. To make a solution containing 1280 µg/ml rifampin, assuming an activity of 950 µg/mg, a simple calculation is performed. First, determine how much powder will be necessary for the volume of diluent that you desire:

$$\text{Weight (mg)} = \frac{\text{volume (ml)} \times \text{concentration}(\mu g/ml)}{\text{potency }(\mu g/mg)}$$

Example: $\text{Weight (mg)} = \dfrac{20 \text{ ml} \times 1280 \text{ } \mu g/ml}{950 \text{ } \mu g/ml}$

$= 26.9$

Weigh out an amount of powder similar to the desired weight (usually slightly more) on an analytical balance. Then determine the exact amount of diluent necessary to achieve the required concentration:

$$\text{Volume (ml)} = \frac{\text{weight (mg)} \times \text{potency }(\mu g/mg)}{\text{concentration }(\mu g/ml)}$$

Example: $\text{Volume (ml)} = \dfrac{28 \text{ mg} \times 950 \text{ } \mu g/mg}{1280 \text{ } \mu g/ml}$

$= 20.8$

For every 1280 µg of active rifampin desired, 1347 µg of powder of 95% activity must be added to the diluent. The weight of powder greater than 1280 µg includes the inactive components. NCCLS M7-A2 and M100-S4 also describe preparation of antimicrobial agent solutions.[12,16]

To perform the classic broth dilution susceptibility test, a standard inoculum of the microorganism (e.g., organisms 1.5×10^6 colony forming

---

PROCEDURE 14.1

PREPARATION OF MCFARLAND NEPHELOMETER STANDARDS

PRINCIPLE
A chemically induced precipitation reaction can be used to approximate the turbidity of a bacterial suspension.

METHOD
1. Set up 10 test tubes or ampules of equal size and of good quality. Use new tubes that have been thoroughly cleaned and rinsed.

2. Prepare 1% chemically pure sulfuric acid.

3. Prepare a 1.175% aqueous

solution of barium chloride $(BaCl_2 \cdot 2\ H_2O)$.

4. Slowly, and with constant agitation, add the designated amounts of the two solutions to the tubes as shown in Table 14.2 to make a total of 10 ml per tube.

5. Seal the tubes or ampules. The suspended barium sulfate precipitate corresponds approximately to homogenous *E. coli* cell densities per milliliter throughout the range of

standards, as shown in Table 14.2.

6. Store the McFarland standard tubes in the dark at room temperature. They should be stable for 6 months.

From National Committee for Clinical Laboratory Standards. 1991. Performance standards for antimicrobial disk susceptibility tests, ed 4, M2-A4, Villanova, Pa.

units [CFU]/ml, a 1:100 dilution of a suspension of turbidity equal to a McFarland standard 0.5, Table 14.2) is added to an equal volume (often 1 ml) of each concentration of antimicrobial agent and to a tube of the growth medium without antimicrobial agent, which serves as a growth control (Figure 14.2). Notice that adding a bacterial suspension will dilute both the suspension and the concentration of antimicrobial agent in the tube; this must be taken into account during preparation of the inoculum and of the dilutions of antimicrobial agents. An uninoculated tube of medium is incubated to serve as a negative growth control. After sufficient incubation (usually overnight), the tubes are examined for turbidity, indicating growth of the microorganism. The organism will grow in the control tube and in any other tube that does not contain enough antimicrobial agent to inhibit growth. The lowest concentration of the agent that inhibits growth of the organism, as detected by lack of visual turbidity (matching the negative growth control), is designated the minimum inhibitory concentration (MIC). If the concentration of antimicrobial agent represented by the MIC can be readily achieved in the patient's serum by normal routes of delivery, the organism is said to be susceptible to that agent. Most clinicians try to achieve a concentration of antimicrobial agent at the site of infection that is four times (or higher in the case of immunosuppressed patients) the in vitro MIC of the microorganism being tested, although this is not

always feasible or necessary. The level of agent attained in the urine, cerebrospinal fluid, other body fluid, or in an abscess may be very much different, however, and such differences must be considered when therapy is being chosen. If the MIC is above the achievable level or within a range toxic to the host, the microorganism is said to be resistant to the antimicrobial agent. Thus susceptibility and resistance are functions of the site of the infection, the microorganism, and the antimicrobial agent being tested. The breakpoint of an antimicrobial agent is the concentration that can be achieved in the serum or tissue with optimal therapy. One can find several different versions of breakpoints; however, accrediting agencies usually require adherence to guidelines published by the National Committee for Clinical Laboratory Standards.[12,15]

Organisms with MICs at or below the breakpoint are considered "susceptible." An "intermediate" result indicates that the test result is equivocal and that susceptibility of that particular organism cannot be predicted. A "moderately susceptible" category also exists for certain organism-antibiotic combinations. In this case the amount of drug given to the patient can be increased beyond standard dosages to obtain higher levels in the infected site. Certain antimicrobial agents are less active at an acid pH, such as might be found within an abscess. These factors also contribute to the efficacy of treatment. The results of in vitro tests are subject to many variables, such

TABLE 14.3

SOLVENTS AND DILUENTS FOR PREPARATION OF STOCK SOLUTIONS OF ANTIMICROBIAL AGENTS

ANTIMICROBIAL	SOLVENT*	DILUENT*
Amoxicillin, Ticarcillin	Phosphate buffer, pH 6.0, 0.1 M	Phosphate buffer, pH 6.0, 0.1 M
Ampicillin	Phosphate buffer, pH 8.0, 0.1 M	Phosphate buffer, pH 6.0, 0.1 M
Cephalothin†	Phosphate buffer, pH 6.0, 0.1 M	Water
Chloramphenicol and erythromycin	Ethanol	Water
Moxalactam (diammonium salt)‡	0.04 N HCl (let sit for 1.5-2 h‡	Phosphate buffer, pH 6.0, 0.1 M
Nalidixic acid	½ volume of water, then add NaOH 1 M, dropwise to dissolve	Water
Nitrofurantoin§	Phosphate buffer, pH 8.0, 0.1M	Phosphate buffer, pH 8.0, 0.1 M
Rifampin	Methanol	Water (with stirring)
Sulfonamides	½ volume hot water and minimal amount of 2.5 M NaOH to dissolve	Water
Trimethoprim	0.05 N lactic or hydrochloric acid, 10% of final volume	Water (may require heat)

From National Committee for Clinical Laboratory Standards. 1986. Performance standards for antimicrobial susceptibility testing. Second informational supplement. M100-S2: Permission to reproduce this table has been granted by the Committee. NCCLS is not responsible for errors or inaccuracies. The most current edition of the standard is available from NCCLS, 771 E. Lancaster Ave., Villanova, PA, 19085.

* These solvents and diluents are for making *stock solutions* of antimicrobial agents requiring other than water. They can be further diluted as necessary in water or broth. The products known to be suitable for water solvents and diluents are amikacin, azlocillin, carbenicillin, cefamandole, cefonicid, cefotaxime, cefoperazone, cefoxitin, ceftizoxime, ceftriaxone, ciprofloxacin, clindamycin, gentamicin, kanamycin, methicillin, mezlocillin, nafcillin, netilmicin, oxacillin, penicillin G, piperacillin, tetracyclines, tobramycin, trimethoprim (if lactate), and vancomycin.

† Solubize all other cephalosporins and cephamycins (unless manufacturer indicates otherwise) in phosphate buffer, pH 6.0, 0.1 M and further dilute in sterile distilled water. These include cephalothin, cefazolin, and cefuroxime.

‡ The diammonium salt of moxalactam is very stable but is almost pure R isomer. Moxalactam clinical is a 1:1 mixture of R and S isomers. Therefore dissolve the salt in 0.04 N HCl and allow it to react for 1.5 to 2.0 hours to convert it to equal parts of both isomers.

§ Alternatively, dissolve nitrofurantoin in dimethyl sulfoxide (DMSO).

as inoculum size, the rate of growth of the bacterium, the incubation period, the nature of the medium used, and the stability of the antimicrobial agent. By using standard methods, as detailed in NCCLS M7-A2, microbiologists can deliver consistent results that can serve to aid clinicians in their choices of therapeutic regimens. Figure 14.2 illustrates a procedure commonly used to determine the MIC of an antibiotic against a test organism.

The MIC measures the ability of the antimicrobial agent to inhibit multiplication of the organism. Thus organisms in the inoculum may be merely inhibited by the antimicrobial agent and will be able to recommence growing if the antimicrobial influence is removed. In that case the antimicrobial agent is said to be bacteriostatic, or inhibitory. For certain serious infections, such as endocarditis and meningitis, and perhaps osteomyelitis, and in patients who lack a well-functioning immune system, some experts consider it

important to determine the ability of an agent to actually kill the infecting organism.

To measure the ability of the antimicrobial agent to kill the microorganism, the bactericidal activity test is performed using a modification of the broth dilution susceptibility testing system. In reality, macrobroth dilution tests in tubes, as described previously, are rarely performed for MICs alone, but a modification of the procedure is used to determine bactericidal levels for organism-antimicrobial combinations that cannot be tested by commercially available methods. When the initial microorganism suspension is being inoculated into the tubes of broth, a measured portion is removed from the growth control tube immediately after it was inoculated and this aliquot is plated to solid agar for determination of actual CFU in the inoculum. The technologist determines this number by counting the number of colonies present after overnight incubation of the agar plate and multiplying times the dilution factor. For exam-

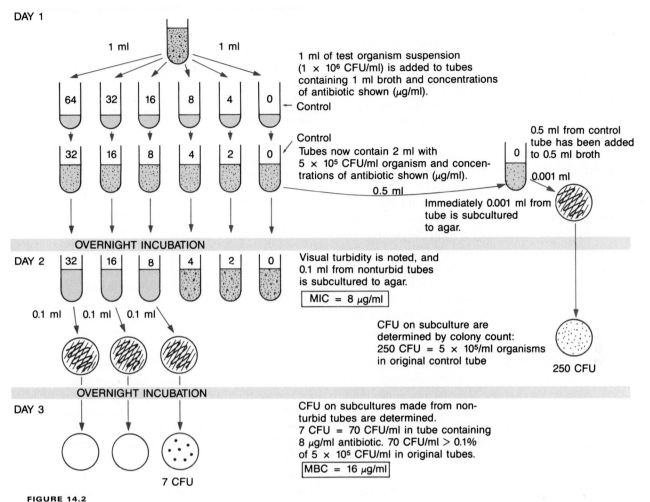

DAY 1

1 ml of test organism suspension
$(1 \times 10^6$ CFU/ml) is added to tubes
containing 1 ml broth and concentrations
of antibiotic shown (μg/ml).

Control
Tubes now contain 2 ml with
5×10^5 CFU/ml organism and concen-
trations of antibiotic shown (μg/ml).

0.5 ml from control
tube has been added
to 0.5 ml broth

0.001 ml

0.5 ml

Immediately 0.001 ml from
tube is subcultured
to agar.

OVERNIGHT INCUBATION

DAY 2

Visual turbidity is noted, and
0.1 ml from nonturbid tubes
is subcultured to agar.

MIC = 8 μg/ml

CFU on subculture are
determined by colony count:
250 CFU = 5×10^5/ml organisms
in original control tube

250 CFU

OVERNIGHT INCUBATION

DAY 3

CFU on subcultures made from nonturbid tubes are determined.
7 CFU = 70 CFU/ml in tube containing
8 μg/ml antibiotic. 70 CFU/ml > 0.1%
of 5×10^5 CFU/ml in original tubes.

MBC = 16 μg/ml

7 CFU

FIGURE 14.2

Determining MIC and MBC for one organism and one antibiotic.

ple, a 0.001 ml calibrated loop can be used to streak a plate for CFU from dilutions made from the growth control tube. If the organism concentration is 5×10^5/ml in the test tubes, there should be around 250 colonies on a subculture plate made from a 1:2 dilution of the growth control tube (0.001 ml of a suspension of 2.5×10^5 organisms contains 2.5×10^2 organisms). The 1:2 dilution is necessary in order to be able to count the colonies on the plate. It is possible to count colonies accurately numbering from 30 to 300 on a subculture plate, but counting a greater number of colonies is time-consuming and often inaccurate.

After the MIC has been determined, a known quantity (often 0.1 ml) of inoculum from each of the tubes of broth that showed no visible turbidity after 22 to 24 hours' incubation is subcultured to solid agar plates. The small amount of the antimicrobial agent that is carried over with this inoculum is easily removed by diffusion into the agar,

and the effect is negated by spreading the inoculum over a large area. The number of colonies that grow on this subculture after overnight incubation is then counted and compared to the number of CFU/ml in the original inoculum. Within those tubes that showed no turbidity, microorganisms were either still viable or they were killed by the antimicrobial agent. Since even bactericidal drugs do not always totally sterilize a bacterial population, the lowest concentration of antimicrobial agent that allowed less than 0.1% of the original inoculum to survive is said to be the minimum bactericidal concentration (MBC), also called the "minimum lethal concentration (MLC)" (Figure 14.2). For our example, the MBC would be the tube from which less than 5 CFU (50 organisms per milliliter in the tube) grew. The technologist must be careful that the original inoculum is great enough to measure a 99.9% lethal effect by the techniques being used in the test. For example, if the original inoculum contained only

10^4 CFU/ml, it would require the subculture of a 1 ml aliquot to yield the number of colonies (greater than 10 CFU) required to demonstrate less than 0.1% survival. One milliliter can be subcultured by pour plate, but it is a cumbersome task. A recent review by Peterson and Shanholtzer,[20] and NCCLS Publication M26-T[17] describe this procedure.

Certain antimicrobial agents are bactericidal in activity; they cause irreversible damage to bacteria. Other agents are known to be bacteriostatic, inhibiting growth by preventing multiplication but not causing death. In some cases drugs that are primarily bacteriostatic may be bactericidal against a given type of organism. The general category of certain antimicrobial agents is shown in Table 14.4. When an agent that has known bactericidal properties exhibits an in vitro MBC that is very much greater than its MIC (requires a great deal more antimicrobial agent to kill the bacteria than to merely inhibit growth; differences of 32 times are sometimes cited) against a bacterial strain, the strain is said to exhibit tolerance. Laboratory techniques play a large part in determining whether a bacterial strain is designated as tolerant, as has been pointed out by Taylor, and others.[20,30] The clinical importance of tolerance, if any, has never been established, and we do not recommend routine testing for it in clinical microbiology laboratories. In addition, truly reproducible MBC determinations are difficult to achieve. MBC testing should be limited to those laboratories proficient in its performance and for those clinical situations in which such results are important (such as osteomyelitis, endocarditis, and immunosuppressed patients).

The MICs and MBCs of a given antimicrobial agent can be determined for any microbe that is able to grow in liquid media. The optimal temperature for such tests is 35° C, since methicillin-resistant *Staphylococcus aureus* only exhibits that resistance at 35° C or below. Many bacteria require a medium different from cation-supplemented Mueller-Hinton broth or require special supplements added to the standard broth to achieve good growth. For example, streptococci are often tested in Todd-Hewitt broth. Certain fastidious bacteria require the addition of horse serum, lysed horse blood, rabbit serum, or special commercially produced nutrient supplements, such as Fildes' extract (Difco) or IsoVitaleX (BBL Microbiology Systems), as recommended by NCCLS M7-A2.[12] Organisms for which standardized procedures have not been agreed upon can be tested for in vitro susceptibility to selected agents. However, in

---

TABLE 14.4

ACTIVITY OF ANTIMICROBIAL AGENTS AGAINST BACTERIA

AGENTS THAT ARE USUALLY BACTERICIDAL	AGENTS THAT ARE USUALLY BACTERIOSTATIC
Aminoglycosides	Chloramphenicol (also
Cephalosporins	bactericidal in some cases)
Cotrimoxazole	Erythromycin
Metronidazole	Nalidixic acid
Monobactams	Sulfonamides
Penicillins	Tetracyclines
Quinolones	
Rifampin	
Vancomycin	

Note: Erythromycin and clindamycin can act as bactericidal or bacteriostatic agents.

---

the absence of standards, results should be interpreted with caution. Susceptibility testing of clinically important unusual organisms for which standard methods are not yet available should be performed only by reference laboratories that have experience in such procedures.

The methods outlined can be used to test many types of bacteria, aerobic and anaerobic, but they are time-consuming and very labor intensive. For each combination of bacterium and antimicrobial agent, a complete set of tubes must be prepared. It would be prohibitive to test more than a few antimicrobial agents against each isolate. Clinicians, however, have to choose from among numerous possible agents; a more practical method had to be adopted. The development of agar dilution and disk diffusion methods, as discussed previously, was one solution to the problem. A more recent innovation has been the adaptation of macrobroth dilution methods to a microbroth format, saving in reagents and technologist time. The use of a microdilution format, which utilizes plastic microdilution trays (Figure 14.3), has made routine reporting of MIC results possible for many laboratories. Although a few large centers still prepare their own microdilution plates to save money and allow flexibility in the antimicrobial agents being tested, most laboratories buy commercially prepared plates. MIC microdilution trays purchased as commercial systems have the advantage of being prepared in great quantities under very strict quality control standards to ensure consistent performance. Their disadvantages include high cost and a predetermined format. Even anaerobic susceptibility test-

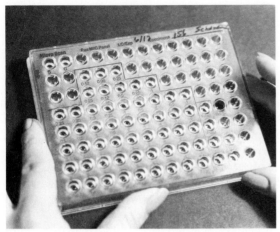

FIGURE 14.3

Plastic microdilution tray used for microbroth dilution antibiotic susceptibility testing. Antimicrobial agents are arranged in linear arrays of serial twofold dilutions.

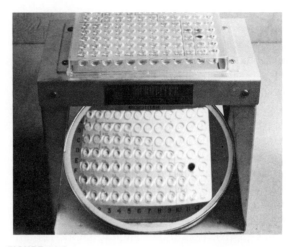

FIGURE 14.4

Reading microbroth dilution results using a magnifying mirror reader.

ing trays are commercially available. All plates are viewed manually on a light box or read from the bottom with a mirror reader (Figure 14.4). Automated readers are described in Chapter 12. Some systems are prepared complete by the manufacturer, shipped frozen to the consumer, and stored in the freezer until thawed and used. Such systems are easy to inoculate, often including a disposable plastic inoculating apparatus that requires only one manipulation to inoculate simultaneously all 96 wells in the microdilution plate. However, they may suffer from degradation of the antimicrobial agents if major temperature variations are encountered during shipping and storage. A second type of commercial microbroth system consists of dried or lyophilized antimicrobials. The organism suspension or another diluent may be used to rehydrate the wells. If any number of such plates must be inoculated, some mechanical or instrumental device is necessary. Because of the small volume of medium in the wells, the incubating plates must be protected from dehydration by sealing them with a plastic tape or stacking them.

Although microdilution plates are prepared using the same growth medium and the same antimicrobial agent concentrations as macrobroth methods, there is one important difference. The total number of bacteria, still 5×10^5 CFU/ml, is considerably decreased (at least tenfold) per microdilution well because of the small volume being tested. These wells usually hold only 0.1 ml, effectively reducing the final inoculum to 5×10^4 CFU per well. This number may not include a sufficient number of colony forming units to allow expression of resistance shared by only a small proportion of the total organisms present in the primary isolate. NCCLS M26-T presents a method for performing MBCs with microdilution trays.[17]

As for all laboratory tests, susceptibility test results must be monitored and evaluated using standard quality control strains of bacteria. By testing strains with known MICs, technologists can uncover breaks in technique or problems with media or reagents. NCCLS recommends routine testing of *Escherichia coli* American Type Culture Collection (ATCC) 25922, *Pseudomonas aeruginosa* ATCC 27853, *S. aureus* ATCC 29213, and *Enterococcus faecalis* ATCC 29212, since their MICs all fall within a measurable range and the results are consistent on repetitive testing. NCCLS M7-A2 discusses performance of quality control procedures, and M100-S4 lists the most recent recommendations.[12,16]

BACTERIAL DISK DIFFUSION METHODS

As more antimicrobial agents became available for treating a wide variety of bacteria, the limitations of the macrobroth dilution method became apparent. Before microdilution technology was available, methods were developed for testing an isolate against more than one antimicrobial agent, such as inoculating the surface of an agar plate with the organism to be tested, placing small glass or metal cylinders into the agar to create tiny wells above the agar surface, and placing suspensions of antimicrobial agents into the cylinders. The antimicrobial agent would diffuse into the medium in a circle around the cylinder, inhibiting the growth of the organism wherever the concentration of drug was high enough. Large zones indicated more effective antimicrobial activity or greater diffusibility of the drug or both. No zone indicated complete resistance. In this crude way, several agents could be tested against one isolate at the same time. This method was modified in 1947 by Bondi and others[3] by incorporating the antimicrobial agents into filter paper disks. Thus a large number of identical disks could be prepared in advance and stored for future use. It was not until the landmark study by Bauer, Kirby, Sherris, and Turck in 1966 that the use of filter paper disks for susceptibility testing was standardized and correlated with MICs using a large number of bacterial strains.[2] The disk diffusion susceptibility test, as the method of Bauer, Kirby, Sherris, and Turck is designated, uses single high antibiotic content disks and yields qualitative results that correlate well with the quantitative results obtained by MIC tests.

These authors chose concentrations of antimicrobial agents to incorporate into the filter paper disks that showed the best results (clearest, within acceptable size, and most consistent zones of inhibition). They then tested many strains of rapidly growing bacteria, such as various Enterobacteriaceae and staphylococci, both by broth dilution and by spreading the bacteria on the surface of Mueller-Hinton agar plates, pressing the disks onto the inoculated plate surfaces, incubating the plates overnight at 35° C, and measuring the zone of inhibition of growth around each disk (Figure 14.5). For each combination of antimicrobial agent and bacterial species tested, a regression analysis was performed, plotting the zone size in millimeters against the $\log_2$ MIC (Figure 14.6). Breakpoint zone diameters were chosen that corresponded to the MICs that fell into the suscepti-

ble and resistant ranges, based on achievable serum levels.

In some cases, because of technical factors, the zone size did not correlate well with the MIC. For this reason, zone sizes that fell within an arbitrarily selected middle range were designated "intermediate." For bacteria whose zones fall in this range, the extrapolated MIC should be considered equivocal and dilution tests may be indicated. The latest modifications of the agar disk diffusion testing procedure have been published in NCCLS M2-A4, the current state of the art in such matters; a "moderately susceptible" category has been established for certain drugs (penicillins and cephalosporins, in particular) in which higher levels of drug may be achieved in those

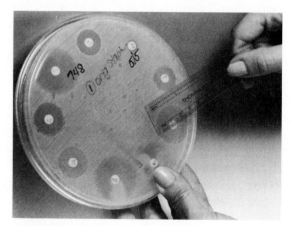

FIGURE 14.5

Measuring zones of inhibition on a disk diffusion susceptibility test plate.

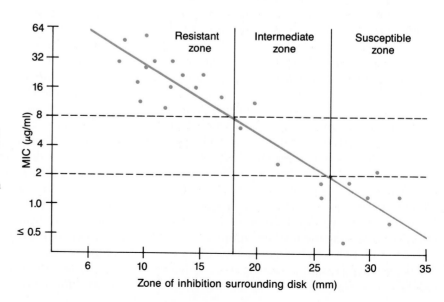

FIGURE 14.6

Example of a regression analysis plot used to determine zone sizes that correspond to susceptibility characteristics of the species of bacterium being tested. In the example the maximum achievable serum concentration is 8 µg/ml. Zone sizes less than or equal to 18 mm are interpreted as resistant; zone sizes more than or equal to 26 mm are interpreted as susceptible.

body sites where the drugs are concentrated, for which a range of dosages might be possible, or for organisms that may require more aggressive therapy.[11]

The agar disk diffusion test is still used by many laboratories in the United States and around the world, because it is easy and inexpensive to perform and standardization is not difficult. The results can also be directly correlated to MIC values. Each time a new antimicrobial agent is developed and released for clinical use, a large number of organisms must be tested in the manner described here to establish interpretive zone sizes for the agent. Unless zone sizes have been specifically established for a species of bacterium, results achieved by the agar disk diffusion method may not be reliable. It must also be remembered that the designations "susceptible" and "resistant" relate to achievable serum levels; antimicrobial agents that are excreted through the kidneys achieve much higher concentrations in urine than are obtainable in serum, such that an isolate from a urinary tract infection that is "resistant" by agar disk diffusion may be "susceptible" to the same antimicrobial agent in the urine. For the same reason, the zone sizes that have been established for antimicrobial agents that are used primarily or exclusively for treatment of urinary tract infections, such as nalidixic acid, cinoxacin, nitrofurantoin, trimethoprim, sulfonamides, and norfloxacin, correspond to levels of antimicrobial agent achievable in urine only. Special criteria have been established for other drugs, such as ceftizoxime, when they are being tested against urinary tract isolates. By the same token, drug levels in spinal fluid are commonly lower than in serum.

Interpretation of the results of the disk diffusion test depends on proper performance of the test. The variables that influence the outcome include the depth, pH, cation content, supplements, and source of the Mueller-Hinton agar; the age and turbidity of the bacterial inoculum; the way in which the inoculum is spread on the plate; the temperature, atmosphere, and duration of incubation; the method of reading results; the antimicrobial content of the disks, their age, and storage conditions; and more. For these reasons, performance of the test must strictly adhere to the NCCLS guidelines.

Laboratories should monitor their performance by periodically testing the recommended quality control strains *S. aureus* ATCC 25923, *E. coli* ATCC 25922, and *P. aeruginosa* ATCC 27853.

For monitoring disks that contain combinations of β-lactam agents plus β-lactamase inhibitors, such as amoxicillin plus clavulanic acid, *E. coli* ATCC 35218 is used.

Trimethoprim-sulfamethoxazole is tested with *E. faecalis* ATCC 29212 to determine that the Mueller-Hinton agar does not itself contain sulfonamide-inhibiting substances. The NCCLS publication M2-A4 describes all aspects of performance of this test.[11] An excellent troubleshooting guideline for tracing sources of error in the disk diffusion test was written by Miller and colleagues.[10]

Enterobacteriaceae, *Aeromonas*, *Acinetobacter*, *Pseudomonas*, staphylococci, enterococci, nonenterococcal streptococci, and *Listeria monocytogenes* can be tested against selected antimicrobial agents by the standard disk diffusion method.[11,16] If the organism is unable to grow satisfactorily after overnight incubation on Mueller-Hinton agar, 5% defibrinated sheep blood may be added to the agar without appreciably affecting the overall results. Modifications of the standard disk diffusion test have been adopted for testing several commonly isolated bacteria with unique growth characteristics: *Haemophilus* species, *Neisseria gonorrhoeae*, and *Streptococcus pneumoniae*. Table 14.5 summarizes some of these considerations, and recent publications by Thornsberry, Neumann, and others[18a,31] detail current recommendations. Detecting resistance of staphylococci to the penicillinase-resistant penicillins (oxacillin, methicillin, nafcillin, cloxacillin, and dicloxacillin) poses special problems. Plates should be incubated at 35° C or slightly lower for at least 24 hours, and zones of inhibition around methicillin and oxacillin disks should be examined carefully, using transmitted light, for a slight haze of growth; such a haze indicates a resistant subpopulation, and the organism should be reported as "resistant." Strains of staphylococci that demonstrate in vitro resistance to methicillin or oxacillin should be considered resistant to all β-lactam agents, including penicillins and cephalosporins, and should be reported as such.

Certain results obtained in vitro are known to be unreliable. Although *Salmonella typhi* is inhibited by aminoglycosides in vitro, these agents are ineffective for treating typhoid fever. Aminoglycosides may inhibit enterococci in vitro, but they are not ever recommended alone for the treatment of enterococcal infections. Disk diffusion susceptibility testing, although appearing deceptively simple, is a demanding test, fraught with pitfalls that, if not circumvented, can render the results less than totally reliable.

TABLE 14.5

NCCLS RECOMMENDATIONS FOR DISK DIFFUSION TESTING OF FASTIDIOUS AEROBIC BACTERIA*

ORGANISM†	MEDIUM SUPPLEMENT	DISK (CONTENT)	ZONE (MM) BREAKPOINT FOR	
			SUSCEPTIBLE	RESISTANT
H. influenzae	*H* Test Medium (HTM)‡ Mueller-Hinton agar 15 µg/ml NAD 15 µg/ml bovine hematin 5 mg/ml yeast extract pH 7.2 to 7.4	Amoxicillin-clavulanic acid (20/10 µg)	≥20	≤19
				≤21
		Ampicillin (10 µg)	≥25	
		Cefotaxime, ceftazidime, ceftriaxone, ceftizoxime (30 µg)	≥26	
		Cefuroxime (30 µg)	≥24	≤20
		Chloramphenicol (30 µg)	≥29	≤25
		Ciprofloxacin (5 µg)	≥21	
		Trimethoprim-sulfamethoxazole (1.25 and 23.75 µg)	≥16	≤10
			(standard breakpoints appear to apply for chloramphenicol, tetracycline, and trimethoprim-sulfamethoxazole)	
S. pneumoniae	Mueller-Hinton agar, 5% sheep blood	Oxacillin (1 µg) (to screen for penicillin resistance)	≥20	≤19
N. gonorrhoeae	GC Agar Base (Difco), 1% IsoVitaleX or Supplement VX	Penicillin (10 U)	≥20	≤19

From Hindler, J. 1983. Strategies for antimicrobial susceptibility testing of fastidious aerobic bacteria. Am. J. Med. Tech. 49:761.

* As modified from NCCLS protocol M2-T4.

† Inoculum preparation: Suspend overnight growth from plate into Mueller-Hinton broth to obtain turbidity equivalent to a McFarland 0.5 standard. Incubation: Overnight at 35° C in CO_2 for *N. gonorrhoeae*. CO_2 is usually not necessary for *H. influenzae* and *S. pneumoniae*.

‡ To make *Haemophilus* test Medium (HTM), first prepare a fresh hematin stock solution by dissolving 50 mg of powder in 100 ml of 0.01 N NaOH with heat and stirring until the powder is thoroughly dissolved. Add 30 ml of the hematin stock solution to 1 L of Mueller-Hinton broth with 5 g of yeast extract also added. After the solution has been autoclaved and cooled, cations are added aseptically, if needed as in CAMHB, and 3 ml of a NAD stock solution (50 mg of NAD dissolved in 10 ml of distilled water; filter-sterilized) also is aseptically added. If sulfonamides or trimethoprim is to be tested. 0.2 IU/ml thymidine phosphorylase should also be aseptically added to the medium.

AGAR DILUTION METHODS

In addition to broth dilution MICs, another method for testing the MICs of a large number of isolates against many concentrations of several antimicrobial agents is that of agar dilution. For this method, concentrations of antimicrobial agents are incorporated into agar plates, one plate for each concentration to be tested. The organisms to be tested are diluted to a slightly greater turbidity than that of a McFarland 0.5 standard. This initial suspension is then further diluted, and an aliquot of each organism is placed into one well of a replicating inoculator device (called a Steers-Foltz replicator, Figure 14.7). This device has metal prongs that are calibrated to pick up a small amount of the bacterial suspension (usually 0.001 ml) and deliver it to the agar surface. At least 25 different strains plus controls can be placed in the wells of the inoculator for delivery to each plate in a single manual movement. In this manner, approximately 1×10^4 CFU are delivered in a discrete drop to the surface of agar plates containing different concentrations of antimicrobial agents. After overnight incubation the organisms will grow on those plates that do not contain enough antimicrobial agent to inhibit them. The lowest concentration of agent that allows no more than one or two CFU or only a slight haze to grow is the MIC. An example of this test is in Figure 14.8. MBC results cannot be determined using this technique. NCCLS publication M7-A2 describes performance of this test in detail.[12]Although agar dilution susceptibility testing is not commonly used by clinical laboratories for testing aerobic organisms, it is highly reliable and can be very cost-effective when used in laboratories performing a

FIGURE 14.7
Steers-Foltz replicator apparatus used for performing agar dilution susceptibility tests.

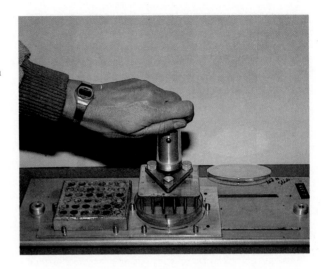

large number of susceptibility tests each day. It is a good research laboratory technique and is also used as a reference method against which other methods are compared. As described in NCCLS publication M11-A,[13] this reference test for anaerobes is performed much like the aerobic test, with several important modifications. Agar plates must be prepared fresh the same day as they are inoculated, using Wilkins-Chalgren agar (available commercially). The bacterial suspensions are brought to the appropriate turbidity in Brucella broth or Schaedler broth before being placed in the inoculator wells. Plates are inoculated quickly and immediately placed into an anaerobic atmosphere for incubation (as described in Chapter 9), and MIC results are read after 48 hours of incubation. It must be noted that this NCCLS procedure is a reference procedure and should not be used for testing susceptibility of clinical isolates for purposes of guiding therapy—not all anaerobes are able to grow on Wilkins-Chalgren agar, and some that do grow do not grow well. Accordingly, the inoculum is relatively small, and the organisms may be susceptible to lower levels of drug than when tested with other media. This procedure may be used with Brucella base blood agar to yield reliable results.[29] The Brucella base blood agar allows better growth and easier interpretation of end points.

AGAR DILUTION SUSCEPTIBILITY TESTING OF MYCOBACTERIA

Although the frequency with which most bacterial populations produce a mutation that leads to resistance to an antimicrobial agent is usually 1 in 10^6, it is 10 times greater for *M. tuberculosis*. Thus primary treatment of mycobacterial disease should include combinations of two or more

agents, and susceptibility testing methods should allow recognition of 1% resistance in the population. The antimicrobial agents tested against these organisms are listed in Table 14.6. Secondary agents are tested when an isolate is found to be resistant to the first-choice (primary) agents. The recommended method, the proportion method, is a modified agar dilution susceptibility test. Performance of this test is detailed in the book by Vestal.[33] Identical methods can be used to test *M. avium* complex, although this organism is unlikely to be susceptible to any one agent alone and different antimicrobial agents are tested. Because of its importance in patients with AIDS, consideration of this group of organisms has become more important (see Chapter 42).

Since isolation of mycobacteria can take as long as several weeks and since primary resistance to the first-choice antimycobacterial agents is becoming more common, we recommend that sus-

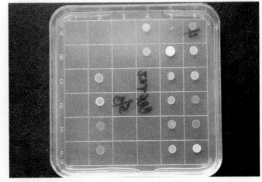

FIGURE 14.8
Results of an antimicrobial agent containing plate inoculated with a Steers-Foltz replicator. Some isolates are growing as seen by a "spot" of growth, whereas others are completely inhibited.

TABLE 14.6

ANTIMICROBIAL AGENTS USED FOR TESTING MYCOBACTERIAL SUSCEPTIBILITIES

PRIMARY DRUGS	SECONDARY DRUGS
Isonaizid	Amikacin
Ethambutol	Ansamycin (for *M. avium-intracellulare*)
Streptomycin	Capreomycin
Rifampin	Cycloserine
Pyrazinamide	Ethionamide
	Kanamycin
	p-Aminosalicylic acid

TABLE 14.7

CONCENTRATIONS OF ANTIMICROBIAL AGENTS USED IN MIDDLEBROOK 7H11 AGAR QUADRANTS FOR SUSCEPTIBILITY TESTING OF MYCOBACTERIA

PLATE	QUADRANT	ANTIMICROBIAL AGENT
1	I	None (control)
	II	Isoniazid (0.2 µg/ml)
	III	Isoniazid (1 µg/ml)
	IV	Ethambutol (10 µg/ml)
2	I	None (control)
	II	Streptomycin (2 µg/ml)
	III	Streptomycin (10 µg/ml)
	IV	Rifampin (1 µg/ml)
3	I	*p*-Aminosalicylic acid (2 µg/ml)
	II	*p*-Aminosalicylic acid (10 µg/ml)

ceptibility testing be performed directly on specimens that are discovered to harbor acid-fast bacilli, as seen on microscopic examination of the concentrate (detailed in Chapter 42). More commonly, however, the presence of mycobacteria will be discovered by isolation of the organisms on culture. The procedures for testing such isolates against primary drugs (indirect susceptibility testing) are included here, as well as modifications for direct testing of positive specimens.

For testing mycobacteria, Middlebrook 7H11 agar containing antimicrobial agents is prepared in quantity, and 5 ml is poured into each section of divided Petri dishes (Felsen quadrant plates), as outlined in Table 14.7. A number of whole Petri plates are poured containing drug-free Middlebrook 7H11 to serve as growth controls for determining CFU. These plates can be stored refrigerated, tightly wrapped in plastic, for a maximum of 4 weeks before they are inoculated. As in prepar-

ing any solution of antimicrobial agent, the specific activity must be taken into account, as described previously. Another method for incorporating antimycobacterial agents into agar is to pour the molten agar into the quadrants over antimicrobial agent–impregnated filter paper disks, as described by Wayne and Krasnow,[34] and outlined in Table 14.8 (Figure 14.9). These plates must be incubated overnight to allow diffusion of the antimicrobial agent throughout the agar in the quadrant. They may then be refrigerated, but it is best to inoculate them within a few days after preparation. This technique of preparing antimicrobial agent solutions using filter paper–impregnated disks as the source of the drug is called *agar disk elution.*

Preparation of the inoculum for indirect my-

FIGURE 14.9
M. tuberculosis susceptibility test (organism is INH-resistant).

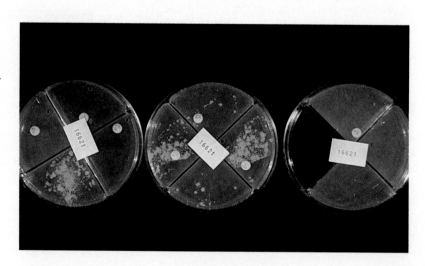

TABLE 14.8

DISK CONTENTS FOR PREPARATION OF DISK-ELUTION MYCOBACTERIAL SUSCEPTIBILITY TEST PLATES USING MIDDLEBROOK 7H11 AGAR

PLATE	QUADRANT	DISK	FINAL CONCENTRATION (μG/ML)
1	I	None (control)	0
	II	Isoniazid (1 μl)	0.2
	III	Isoniazid (5 μl)	1
	IV	Rifampin (5 μl)	1
2	I	None (control	0
	II	Ethambutol (24 μl)	5
	III	Streptomycin (10 μl)	2
	IV	Streptomycin (50 μl)	10
3	I	None (control	0
	II	p-Aminosalicylic acid (10 μl)	2
	III	p-Aminosalicylic acid (50 μl)	10

TABLE 14.9

DILUTION OF CONCENTRATE FOR INOCULA

NO. OF ACID-FAST BACILLI PER OIL IMMERSION FIELD	CONTROL QUADRANT 1	CONTROL QUADRANT 2	DRUG QUADRANTS
Less than 1	Undiluted	10^{-2}	Undiluted
1-10	10^{-1}	10^{-3}	10^{-1}
More than 10	10^{-2}	10^{-4}	10^{-2}

cobacterial susceptibility testing against the primary drugs is outlined in Procedure 14.2. Mycobacteria tend to grow in serpentine fashion, called *cords,* consisting of many individual organisms bound together in parallel rows end to end. It is necessary to attempt to break up these cords as much as possible to determine accurately the number of resistant bacteria in a population. The use of Tween-containing media and glass beads for vortexing bacterial suspensions helps to break up the cords. As for all procedures involving mycobacteria, these manipulations must be carried out within a biological safety cabinet. Only reference laboratories should test secondary drugs.

If acid-fast bacilli are seen on the concentrated direct smear, material from the concentrate can be diluted in Middlebrook 7H9, as shown in Table 14.9, and then inoculated onto quadrants for direct susceptibility testing as outlined in Procedure 14.2. The report of results of drug susceptibility tests should include the following:

1. Type of test—direct or indirect
2. Number of colonies on control quadrant
3. Number of colonies on drug quadrant
4. Concentration of drug in each quadrant

From these data, a rough approximation of the percentage of organisms resistant to the drug may be calculated as follows:

$$\frac{\text{Number of colonies on drug quadrant}}{\text{Number of colonies on control quadrant}} \times 100$$

$$= \% \text{ resistance at that drug concentration}$$

Radiometric growth detection (described in Chapter 12) has recently been modified for detection and susceptibility testing of *Mycobacterium tuberculosis* and other mycobacteria. Such testing yields results in several days, as compared with several weeks required for agar dilution methods. Antimicrobial agents are incorporated into bottles containing radioactive substrates (Becton-Dickinson, Franklin Lakes, N.J.), and growth of the my-

PROCEDURE 14.2

INDIRECT MYCOBACTERIAL SUSCEPTIBILITY TESTING

PRINCIPLE

Antimycobacterial agents can diffuse throughout an agar quadrant; on a solid agar surface quantitative susceptibility testing of mycobacteria can be done by direct inoculation of a known number of organisms.

METHOD

1. Suspend a wooden stickful or a loopful of growth in 5 ml of Middlebrook 7H9 broth (commercially available) containing five glass beads, 1 to 2 mm in diameter. Vortex for 15 seconds. Allow the solution to settle for 10 to 15 minutes after vortexing to discourage aerosol dissemination. Either adjust turbidity directly (as described in 2 below) if culture is fresh, or allow culture to incubate for 7 days.

2. Vortex several times for 15 seconds each to break up cords. Adjust turbidity to that of a McFarland 0.5 standard.

3. Dilute this suspension 1:100 (0.05 ml into 5 ml) and then 1:100 again, to make a final dilution of 1:10,000 in Middlebrook 7H9 broth.

4. Inoculate 0.1 ml of each of the two dilutions (the 10^{-2} and 10^{-4}) onto each quadrant of the drug-containing plates and to two whole plates with drug-free media. Tilt the plates gently to spread the 0.1 ml drop around; do not allow it to pool up around the edges. By spreading the 0.1 ml onto the surface of a whole plate for the growth controls, the actual number of colonies can usually be counted. Leave the plates upright in the safety cabinet with tops slightly ajar for about 10 minutes to allow the inoculum to dry.

5. Tape the plates closed on two sides with clear cellophane tape to prevent accidental opening, and then place the plates, stacked two high, four plates each, into polyethylene plastic bags, sealed with tape. (Do not use Saran Wrap.)

6. Incubate the plates in 5% to 10% CO_2 at 37° C for a maximum of 3 weeks, or until colonies can be counted. Record growth as 4 + (more than 500 CFU), 3 + (200-500 CFU), 2 + (100-200 CFU), 1 + (50-100

CFU), or the actual number if less than 50 CFU.

7. Using the number of CFU on the growth control plates from the 10^{-4} dilution, estimate whether more than 1% of the inoculum grew on any drug-containing quadrant.

QUALITY CONTROL

A strain of *M. tuberculosis* susceptible to all antimicrobial agents (such as H37RV) should be tested concurrently with clinical isolates.

EXPECTED RESULTS

The QC strain should not grow on agar containing antimicrobial agents but should grow on the control agar without antibiotics.

PERFORMANCE SCHEDULE

QC should be performed each time the test is performed.

Modified from Vestal, A.L. 1975. Procedures for the isolation and identification of mycobacteria. DHEW publication no. (CDC) 77-8230. Centers for Disease Control, Atlanta, Ga.

---

cobacteria, indicating resistance, is detected by measuring the amount of radioactive gas produced by the bacteria. Further discussion of radiometric detection of growth of mycobacteria is found in Chapter 42. Although the technology has not been adopted by large numbers of laboratories, those that can use the BACTEC instrument cost-effectively report good results.

The rapidly growing mycobacteria *M. fortuitum* and *M. chelonae* can be tested by microbroth dilution in cation-supplemented Mueller-Hinton broth. The inoculum (McFarland 0.5 turbidity) is prepared from overnight growth in Mueller-Hin-

ton broth plus 0.02% Tween 80. Plates are inoculated as noted previously and incubated in air at 35° C for 72 hours before being read. Antibiotics suitable for testing in this way include aminoglycosides, cefoxitin, doxycycline, erythromycin, sulfonamides, and ciprofloxacin.[31]

ANTIMICROBIAL GRADIENT STRIP METHOD (E-TEST)

A recently described variant of the disk diffusion technique allows the quantitative determination of MICs on agar. The system consists of a plastic strip containing a gradient of antimicrobial agent applied to one side and an interpretive scale printed on the other side (E-test; PDM Epsilometer; AB Biodisk, Solne, Sweden and Piscataway, N. J.). The strip is placed antibiotic side down on the surface of an agar plate that has been inoculated with a lawn of the organism to be tested, in the same manner as for the agar disk diffusion test. Growth of the organism occurs with an elliptical zone of inhibition surrounding the strip relative to the concentration of antibiotic along its length. The value of the MIC is determined by reading the scale at the point where the zone of inhibition intersects the strip (Figure 14.10). The system is versatile and can be adapted to any agar medium and most microorganisms.[1a,3a,8a,22a,22b] It is used currently for anaerobes, *Streptococcus pneumoniae,* and other fastidious and unusual organisms, but additional possibilities are being explored. Quantitative results agree very well with standard agar dilution methods.

FIGURE 14.10

Quantitative gradient susceptibility agar diffusion test (E-test) of one organism against six separate antibiotics (one on each E-test strip) on a large Mueller-Hinton agar plate. (Courtesy AB Biodisk, Piscataway, N. J.)

IN VITRO FUNGAL SUSCEPTIBILITY TESTING METHODS

Because fungi are large cells or hyphae with more variation in size than bacteria, preparation of a standard suspension of known CFU is difficult. For this reason and because of peculiar growth patterns, fungal susceptibility testing is difficult to perform and thus is much less standardized than are bacterial methods. Although possible for yeast, establishing a standardized inoculum for filamentous fungi is beyond the capability of most laboratories. Basically, fungi are tested by the macrobroth dilution method against amphotericin B and 5-fluorocytosine, the two most commonly used antifungal agents. Methods for testing the newer imidazoles, such as fluconazole, ketoconazole, miconazole, and clotrimazole, are described in NCCLS M27-P.[14] Special media are necessary for fungal susceptibility testing; such testing is best left to a reference laboratory.

Disk diffusion tests for testing 5-fluorocytosine have been described by Utz and Shadomy.[32] A good discussion of current methods can be found in the chapter by Shadomy, and other references.[7,26]

IN VITRO VIRAL SUSCEPTIBILITY TESTING METHODS

Although not routinely performed, even by laboratories that perform routine viral cultures, viral susceptibility testing is a procedure that has been described and will undoubtedly become more widely used in the future. As shown by Howell and Miller,[8] routine testing of herpes simplex iso-

lates against acyclovir can be accomplished without much difficulty. With increasing use of antiviral agents, such as amantadine for influenza A and the antiherpetic agents such as acyclovir and ribavirin, the testing of viral isolates for resistance will become necessary to assist clinicians in choices of therapy. Briefly, supernatant from a tissue culture known to be producing infectious virus is inoculated into several wells of freshly passed tissue culture cells, some control wells without drug and some wells containing dilutions of the antiviral agent being tested. After sufficient incubation to produce cytopathic effect in the control wells, the wells containing antiviral agent are examined. The concentration of antiviral agent that eliminates cytopathic effect can be identified. Arbitrary numbers of infected cells can be chosen to approximate different levels of antiviral activity. For herpes cultures, results may be available as quickly as within 24 hours.

SERUM BACTERICIDAL LEVELS

In addition to determining in vitro susceptibility test results for isolates from patients receiving antimicrobial therapy, a laboratory can measure the activity of the patient's own serum (containing an antibacterial agent) against his or her specific pathogen. A serum sample is obtained from the patient at a specific time after an antimicrobial agent dose, serially diluted in doubling dilutions, and antibacterial activity against the infecting microbe determined. The lowest dilution of patient's serum that kills a standard inoculum of the organism is called the serum bactericidal level (SBL). Originally described by Schlichter and MacLean,[24] the test was used to measure the bacterial inhibitory activity of serum from rabbits being treated with penicillin for experimental endocarditis. Still known as the Schlichter test in some laboratories, it has been modified and occasionally is used as a broad guideline to aid clinicians in deciding whether a patient is being given effective therapy for a serious infection, such as endocarditis or sepsis in a severely neutropenic patient.

Current recommendations include testing an inoculum of 10^5 CFU/ml of an actively growing suspension of organisms in both peak (one-half hour after delivery of antibiotic) and trough (immediately before the next antibiotic dose is given to the patient) sera in a diluent containing at least 50% normal, noninhibitory, pooled human serum. We do not recommend using serum as a diluent but prefer removing protein from the patient's sample using a disposable millipore filter (Amicon Corp.) and then using broth medium as the diluent. If a microbroth dilution format is used, the entire contents of wells showing no growth should be subcultured to agar plates (blood, chocolate, or Mueller-Hinton) for evidence of growth.[27] Peak bactericidal serum titers more than or equal to 1:16 and trough titers more than or equal to 1:8 are considered desirable.

Often the results of tests of this type are not clear-cut. Each laboratory must decide how skipped tubes and inconsistent results should be handled. Possible strategies include performing tests in duplicate or repeating the entire procedure. The NCCLS has formed a subcommittee to establish standardized procedures for serum inhibitory and bactericidal testing procedures. M21-T[18] and the review by Stratton[28] contain excellent discussions of this test and include numerous relevant references. We do not recommend the performance of bactericidal testing as a routine procedure for most laboratories since the test is poorly standardized and often difficult to perform in a reproducible manner.[20] If done, the results should always be accompanied by a formal, written consultation from a clinical microbiologist.

IN VITRO SYNERGISM AND ANTAGONISM

When combinations of two or more antimicrobial agents provide more antimicrobial activity against a pathogen than the total additive effect of the agents given separately, the agents are said to exhibit **synergism.** Combinations of antimicrobial agents may be advisable for treating serious infections, for infections in hosts with compromised immune systems, for treating infections caused by organisms known to readily develop resistance to one agent (*M. tuberculosis, P. aeruginosa,* and ente-

rococci), and for certain mixed infections. However, when the infecting organism is known and there is a single agent highly active against it, there is probably no advantage to combination therapy except for organisms that readily develop resistance.

In contrast, when one agent diminishes the activity of a concurrently administered second antimicrobial agent, **antagonism** is said to occur. For instance, if a growth-inhibiting agent such as chloramphenicol is administered with an agent such as penicillin, which requires actively growing bacteria for bactericidal activity, the antagonism is primarily of the nature of slower bactericidal activity; such slowing of bactericidal activity may be significant in meningitis, for example.

Synergism and antagonism of two antimicrobial agents against a particular bacterium can be measured by performing an in vitro test using a checkerboard format. Twofold dilutions of each antibiotic being tested are prepared in large volume and dispensed, one antibiotic in the horizontal rows and the other in the vertical columns of a microbroth dilution tray, beginning with a control well containing no antibiotic. Thus wells in the upper right-hand corner of the tray contain the highest concentrations of both drugs, and the lower left-hand corner well contains no antibiotic. The wells along the left-hand column and the bottom row contain only the individual antibiotic placed in them. After inoculating all wells with the same inoculum of the organism being tested and incubating the tray appropriately, the presence of growth or no growth is recorded. If growth is inhibited by combinations of both antibiotics in concentrations fourfold or more less than that required by either antibiotic alone, the combination is said to be synergistic at those concentrations. Conversely, if the organism is able to grow better at combinations of concentrations fourfold or more higher than those required for inhibition when the agents are used singly, then the combination is said to be antagonistic. No difference in inhibition between either agent used alone and any combination of the two agents is called *indifference.*

If only two agents are being assessed for synergism, time-kill curves may provide the most clinically relevant answer. In such tests the antimicrobial agents are added to broth in fractions of the MIC, which was established in a previous test. For example, drug A is tested at one fourth of its MIC, and drug B is added to drug A at one half, one fourth, and one eighth of its MIC (when tested separately). The definition of synergy is

Basic cephalosporin

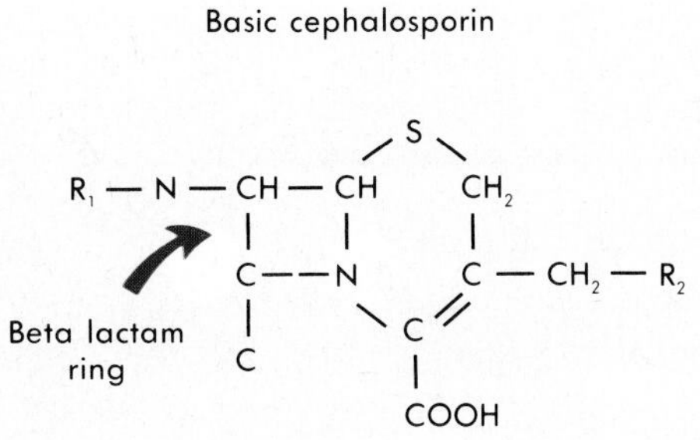

FIGURE 14.11
The β-lactam ring.

that the sum of the fractions for an effective combination is less than unity. The reverse combinations of antimicrobial agents are also tested, as are the antimicrobial agents alone. The inoculum is added as for MIC tests, and samples of the suspensions are removed at intervals over 24 hours for subculture to determine CFU. If there is a decrease of CFU corresponding to 1/100 of the CFU found on subculture from a tube containing the most effective drug tested singly, synergism is said to be demonstrated. Serious infections caused by enterococci, the most penicillin-resistant streptococci, are often treated with combinations of a penicillin and an aminoglycoside. Although time-kill curves of the organism against combinations of the two antimicrobial agents being used would give the most specific information about synergism, a screening procedure has been employed effectively. By determining whether the isolate is resistant to a very high concentration of aminoglycoside, a prediction can be made as to the organism's probable response to such combination therapy. For example, the isolate is tested in a broth dilution system against 2000 µg/ml of streptomycin or gentamicin. If the enterococcus is resistant to this amount, it can be assumed that there will be no synergism. Synergism tests with three or more agents can be performed, but the methods are too cumbersome for routine use. Methods for synergism testing are described in the book edited by Lorian,[9] the review by Norden,[19] discussions by Edberg,[5] as well as in many other publications.

RAPID TESTS FOR ANTIMICROBIAL SUSCEPTIBILITY

Certain bacteria are resistant to some penicillin and cephalosporin antibotics because of their production of β-lactamase enzymes. These enzymes bind to antibiotics that possess a β-lactam ring (Figure 14.11) and, in most cases, ultimately open the ring, inactivating the antibiotic. Although it is not the only resistance mechanism against β-lactam antibiotics, it is a very important one. The presence of β-lactamase, either in body fluid, culture supernatant, or directly within bacterial cells, can be demonstrated with a very rapid test, the "β-lactamase test." The substrate can be a penicillin or a cephalosporin, since both classes of antibiotic may be affected by these enzymes, although one may be affected much more than another. Several methods have been used to detect β-lactamases, as described in the *Manual of Clinical Microbiology*.[25] For example, products of penicillin

breakdown can reduce iodine bound to starch, causing loss of the color of the starch-iodine compound. Hydrolysis of penicillin into acid products by the enzyme has been measured by a color change in a pH indicator (phenol red) from red to yellow.

The most sensitive method, however, is detection of the enzyme by its ability to hydrolyze the β-lactam ring of a chromogenic cephalosporin (a cephalosporin that is a yellow-colored compound in its intact state and red if the β-lactam ring is broken). The most commonly used compound, called nitrocefin (Glaxo), has a very high affinity for most bacterial β-lactamases. The compound is commercially available impregnated onto filter paper disks (Cefinase, BBL Microbiology Systems). A small amount of growth of the organism to be tested is rubbed on the moistened filter paper. Development of a red color within a maximum of 30 minutes indicates presence of β-lactamase (Figure 14.12). The test is most useful for testing *Haemophilus influenzae*, but it can also be used for gonococci, staphylococci, and all anaerobes. *Bacteroides gracilis* and some other anaerobes are often resistant to a number of β-lactam drugs but are nitrocefin negative; this resistance is not mediated by β-lactamases. Resistance of *Bacteroides fragilis* group strains to cefoxitin and imipenem is not usually detectable by the nitrocefin test. *H. influenzae*, gonococci, and other organisms may, of course, be β-lactam–resistant by a mechanism other than β-lactamase production and would thus be β-lactamase–negative. The nitrocefin test is the most sensitive β-lactamase test for all of the aforementioned species.

A second rapid test can be used to detect chloramphenicol resistance among *Haemophilus* species. If the organisms possess chloramphenicol acetyltransferase, the enzyme will modify a substrate (acetyl-S-coenzyme A), the breakdown product of which can induce a chromogenic indicator substance to turn yellow. The reagents for

FIGURE 14.12

Chromogenic cephalosporin (Cefinase; Becton-Dickinson Microbiology Systems, Cockeysville, Md.) impregnated into filter paper disk develops a pink color when hydrolyzed by β-lactamase enzyme present in cell paste rubbed onto the disk. β-lactamase–negative organism (disk on right) does not elicit the reaction. (Photograph by Martin Cohen.)

this test are commercially available as filter paper disks (C.A.T., Remel Laboratories).

Several other strategies can be used to determine susceptibility results more rapidly than is possible with conventional methods. Agar disk diffusion results on plates prepared in the standard manner may be read quite accurately after only 5 to 6 hours' incubation in many cases. Some workers have used tetrazolium dye, a compound that is reduced to a purple-colored compound by metabolizing bacteria, to delineate zones of inhibition on agar disk diffusion plates after 6½ hours of incubation. By increasing the initial inoculum, broth dilution tests may be modified to yield faster results. All of these approaches are not standardized. They should be confirmed by standard methods subsequently. The most radical method for obtaining rapid results, however, is the inoculation of susceptibility testing media with material from the clinical specimen itself. Although results using inocula from body fluids and urines have not correlated well with conventional methods, they do yield general information. Tests performed on positive blood culture broth, however, have given satisfactory results.[4] As detailed in Chapter 4, susceptibility tests may be set up directly or within a few hours of detection of a positive blood culture.

The most accepted methods for obtaining rapid susceptibility results are the automated systems. General principles of such systems are discussed in Chapter 12. New methods being developed include the use of fluorogenic compounds to measure growth, laser imaging of growth in wells, and fluorescence markers as probes for actively growing cells. The ability to deliver susceptibility determinations to the physician within the first few hours of discovery of the infectious process is the ultimate goal of such studies.

MEASUREMENT OF ANTIMICROBIAL AGENT LEVELS IN BODY FLUIDS

The measurement of the amount of antimicrobial agent in the serum or body fluid of a patient is performed for two reasons. First, it is the best method for determining whether an effective level of antimicrobial agent is being achieved. Although serum is usually tested, levels of agents in cerebrospinal fluid or other body fluids may be tested since they do not correlate with serum levels. For some drugs, impaired renal function and other conditions may lead to excessive serum levels with conventional dosage. Second, several agents, notably the aminoglycosides, are therapeutically effective at concentrations very close to those that may be toxic to the host. The administration of these agents should be carefully monitored by determining serum levels. The dosage can usually be adjusted to prevent toxicity while maintaining adequate therapeutic levels. Methods for measuring the concentrations of these agents in body fluids are constantly being improved and refined. Originally, all assays were performed as bioassays, using an indicator strain of bacterium sensitive to the drug being evaluated. Known concentrations of the drug were impregnated into filter paper disks or placed in wells on the surface of an agar plate that had been seeded with an inoculum of an organism known to be sensitive to that drug. The patient's serum or other body fluid was also placed on similar disks or in similar wells in duplicate. By comparing the sizes of zones of inhibition of the bacterial strain around the known amounts of drug (setting up a standard curve) to the zones formed around the patient specimen, the amount of the antimicrobial agent present in the specimen could be determined. This method worked well with single agents but became difficult to perform when a patient was receiving multiple agents. At first, organisms that were resistant to every antimicrobial agent except the one being tested were sought. Eventually, however, technological advances allowed the development of automated methods employing chemical or immunological procedures.

Most antimicrobial agent levels are now performed by gas-liquid chromatography, high-pressure liquid chromatography, fluorescence polarization immunoassay, radioimmunoassay or other type of competitive binding immunoassay, or some other form of immunoassay. Discussion of the principles of some of the instruments involved is found in Chapter 12. The results of tests for levels of antimicrobial agents, also called *therapeutic drug monitoring* **(TDM),** are used to modify treatment being administered to patients with infectious diseases. The same automated test systems, however, are also used to monitor therapy with many other agents, including antiepileptic, antiarrhythmic, antineoplastic, and antidepressant compounds. For this reason, such testing is usually performed in the chemistry, rather than microbiology, section of the laboratory. However, it is important for the clinical microbiologist to remain involved in setting guidelines for interpretation of the results of these tests since knowledge regarding the interactions of microbes and antimicrobial agents is an important area of expertise for the discipline of clinical microbiology and infectious diseases. The immunoassays and chemical assays used for TDM are much more specific and easier to perform than were the bioassays, and they have the advantage of yielding results within hours, rather than requiring overnight incubation. Publications by Anhalt,[1] Edberg and others,[6] Pezzlo,[21] Pfaller,[22] and Saubolle[23] discuss various aspects of TDM testing.

REFERENCES

1. Anhalt, J.P. 1981. Liquid chromatographic assay of antibiotics. Clin. Microbiol. Newsletter 3:159.
1a. Baker, C.N., Stocker, S.A., Culver, D.H., and Thornsberry, C. 1991. Comparison of the E test to agar dilution, broth microdilution, and agar diffusion susceptibility testing techniques by using a special challenge set of bacteria. J. Clin. Microbiol. 29:533.
2. Bauer, A.W., Kirby, W.M.M., Sherris, J.C., et al. 1966. Antibiotic susceptibility testing by a single disc method. Am. J. Clin. Pathol. 45:493.
3. Bondi, A., Spaulding, E.H., Smith, E.D., et al. 1947. A routine method for the rapid determination of susceptibility to penicillin and other antibiotics. Am. J. Med. Sci. 214:221.
3a. Citron, D.M., Ostovari, M.I., Karlsson, A., and Goldstein, E.J.C. 1991. Evaluation of the E test for susceptibility testing of anaerobic bacteria. J. Clin. Microbiol. 29:2197.
4. Coyle, M.B., McGonagle, L.A., Plorde, J.J., et al. 1984. Rapid antimicrobial susceptibility testing of isolates from blood cultures by direct inoculation and early reading of disk diffusion tests. J. Clin. Microbiol. 20:473.
5. Edberg, S.C. 1988. Antibiotic interaction tests: Should they be performed? Clin. Microbiol. Newsletter 10:77.
6. Edberg, S.C., Barry, A.L. and Young, L.S. 1984. Cumitech 20. Therapeutic drug monitoring: antimicrobial agents. American Society for Microbiology, Washington, D.C.
7. Espinell-Ingroff, A., et al. 1992. Collaborative comparison of broth macrodilution and microdilution antifungal susceptibility tests. J. Clin. Microbiol. 30:3138.
8. Howell, C.L., and Miller, M.J. 1984. Rapid method for determining the susceptibility of herpes simplex virus to acyclovir. Diagn. Microbiol. Infect. Dis. 2:77.
8a. Jorgensen, J.H., Howell, A.W., and Maher, L.A. 1991. Quantitative antimicrobial susceptibility testing of *Haemophilus influenzae* and *Streptococcus pneumoniae* by using the E-test. J. Clin. Microbiol. 29:109.
9. Lorian, V., editor. 1991. Antibiotics in laboratory medicine, ed 3. Williams & Wilkins, Baltimore.
10. Miller, J.M., Thornsberry, C., and Baker, C.N. 1984. Disk diffusion susceptibility test troubleshooting guide. Lab. Med. 15:183.
11. National Committee for Clinical Laboratory Standards. 1991. Performance standards for antimicrobial disk susceptibility tests, ed 4, M2-A4. NCCLS, Villanova, Pa. Order from NCCLS, 771 East Lancaster Ave., Villanova, PA 19085.
12. National Committee for Clinical Laboratory Standards. 1988. Methods for dilution antimicrobial susceptibility tests for bacteria that grow aerobically, ed 2, approved standard. M7-A2. NCCLS, Villanova, Pa. Order from NCCLS, 771 East Lancaster Ave., Villanova, PA 19085.
13. National Committee for Clinical Laboratory Standards. 1985. Reference agar dilution procedure for antimicrobial susceptibility testing of anaerobic bacteria; approved standard, M11-A, NCCLS, Villanova, Pa. Order from NCCLS, 771 East Lancaster Ave., Villanova, PA 19085.
14. National Committee for Clinical Laboratory Standards. 1993. Reference method for broth dilution antifungal susceptibility testing for yeasts. M27-P. NCCLS, Villanova, Pa. Order from NCCLS, 771 East Lancaster Ave., Villanova, PA 19085.
15. National Committee for Clinical Laboratory Standards. 1990. Methods for antimicrobial susceptibility testing of anaerobic bacteria, ed 2, approved standard, M11-A2, NCCLS, Villanova, Pa. Order from NCCLS, 771 East Lancaster Ave., Villanova, PA 19085.
16. National Committee for Clinical Laboratory Standards. 1992. Performance standards for antimicrobial susceptibility testing; fourth informational supplement. M100-S4, NCCLS, Villanova, Pa. Order from NCCLS, 771 East Lancaster Ave., Villanova, PA 19085.
17. National Committee for Clinical Laboratory Standards. 1993. Methods for determining bactericidal activity of antimicrobial agents; tentative guideline. M26-T, NCCLS, Villanova, Pa. Order from NCCLS, 771 East Lancaster Ave., Villanova, PA 19085
18. National Committee for Clinical Laboratory Standards. 1993. Methodology for the serum bactericidal test; tentative guideline. M21-T, NCCLS, Villanova, Pa. Order from NCCLS, 771 East Lancaster Ave., Villanova, PA 19085.
18a. Neumann, M.A., Sahm, D.F., Thornsberry, C., and McGowan, J.E. Jr. 1991. New developments in antimicrobial agent susceptibility testing: a practical guide. Cumitech 6A. Coordinating editor, McGowan, J.E. Jr. Amer. Soc. Microbiol., Washington, D.C.
19. Norden, C.W. 1982. Problems in determination of antibiotic synergism in vitro. Rev. Infect. Dis. 4:276.
20. Peterson, L.R., and Shanholtzer, C.J. 1992. Tests for bactericidal effects of antimicrobial agents. Clin. Micro. Rev. 5:420.

21. Pezzlo, M. 1983. Assay of antimicrobial agents. Am. J. Med. Technol. 49:565.

22. Pfaller, M.A. 1987. Immunoassays for measurement of antimicrobial agents in body fluids. p 121-137. In Jorgensen, J.H., editor. Automation in clinical microbiology, CRC Press, Boca Raton, Fla.

22a. Sanchez, M.L., Barrett, M.S., and Jones, R.N. 1992. The E-test applied to susceptibility tests of gonococci, multiply-resistant enterococci, and Enterobacteriaceae producing potent β-lactamases. Diagn. Microbiol. Infect. Dis. 15:459. (Note: entire issue of journal consisted of papers from an E-test symposium.)

22b. Sanchez, M.L. and Jones, R.N. 1992. E test, an antimicrobial susceptibility testing method with broad clinical and epidemiologic application. The Antimicrobic Newsletter 8:1.

23. Saubolle, M.A. 1987. Are assays for serum levels of antifungal agents routinely needed? Clin. Microbiol. Newsletter 9:113.

24. Schlichter, J.G., and MacLean, H. 1947. A method for determining the effective therapeutic level in the treatment of subacute bacterial endocarditis with penicillin: a preliminary report. Am. Heart J. 34:209.

25. Schoenknecht, F.D., Sabath, L.D., and Thornsberry, C. 1985. Susceptibility tests: special tests. In Lennette, E.H., Balows, A., Hausler, W.J. II, and Shadomy, H.J., editors. Manual of clinical microbiology, ed 4. American Society for Microbiology, Washington, D.C.

26. Shadomy, S., Espinel-Ingroff, A., and Cartwright, R.Y. 1985. Laboratory studies with antifungal agents: susceptibility tests and bioassays. In Lennette, E.H., Balows, A., Hausler, W.J., Jr., and Shadomy, H.J., editors. Manual of clinical microbiology, ed 4. American Society for Microbiology, Washington, D.C.

27. Shanholtzer, C.J., Peterson, L.R., Mohn, M.L., Moody, J.A., and Gerding, D.N. 1984. MBCs for *Staphylococcus aureus* as determined by macrodilution and microdilution techniques. Antimicrob. Agents Chemother. 26:214.

28. Stratton, C.W. 1988. Serum bactericidal test. Clin. Microbiol. Rev. 1:19.

29. Summanen, P. Baron, E.J., Citron, D.M., et al. 1993. Wadsworth anaerobic bacteriology manual, ed 5, Star Publishing, Belmont, Calif.

30. Taylor, P.C., Schoenknecht, F.D., Sherris, J.C., et al. 1983. Determination of minimum bactericidal concentrations of oxacillin for *Staphylococcus aureus:* influence and significance of technical factors. Antimicrob. Agents Chemother. 23:142.

31. Thornsberry, C., Swenson, J.M., Baker, C.N., et al. 1988. Methods for determining susceptibility of fastidious and unusual pathogens to selected antimicrobial agents. Diagn. Microbiol. Infect. Dis. 9:139.

32. Utz, C., and Shadomy, S. 1977. Antifungal activity of 5-fluorocytosine as measured by disc diffusion susceptibility testing. J. Infect. Dis. 135:970.

33. Vestal, A.L. 1975. Procedures for the isolation and identification of mycobacteria DHEW Publication No. (CDC) 77-8230, Centers for Disease Control, Atlanta, Ga.

34. Wayne, L.G., and Krasnow, I. 1966. Preparation of tuberculosis susceptibility testing medium by means of impregnated discs. Am. J. Clin. Pathol. 45:769.

BIBLIOGRAPHY

Atlas, R.M. 1993. Handbook of Microbiologic Media. CRC Press, Boca Raton, Fla. (Detailed formulae and instructions for media preparation.)

Baron, E.J, and Finegold, S.M. 1990. Bailey and Scott's Diagnostic Microbiology. ed 8, Mosby, St. Louis. (Appendices for detailed formulae of media and stains.)

Inderlied, C.B., and Hindler, J.A. 1987. Clinical significance of antimicrobial susceptibility testing. ASCP Check Sample Microbiology No. MB 87-7 (MB-168) 30:1-9. American Society for Clinical Pathology, Chicago, Ill.

Thornsberry, C., Swenson, J.M., Baker, C.N., et al. 1988. Methods for determining susceptibility of fastidious and unusual pathogens to selected antimicrobial agents. Diagn. Microbiol. Infect. Dis. 9:139.

Willett, H.P. 1988. Antimicrobial agents. pp. 128-160. In Joklik, W.K., Willett, H.P., and Amos, D.B., editors. Zinsser's Microbiology, ed 19. Appleton & Lange, Norwalk, Conn.

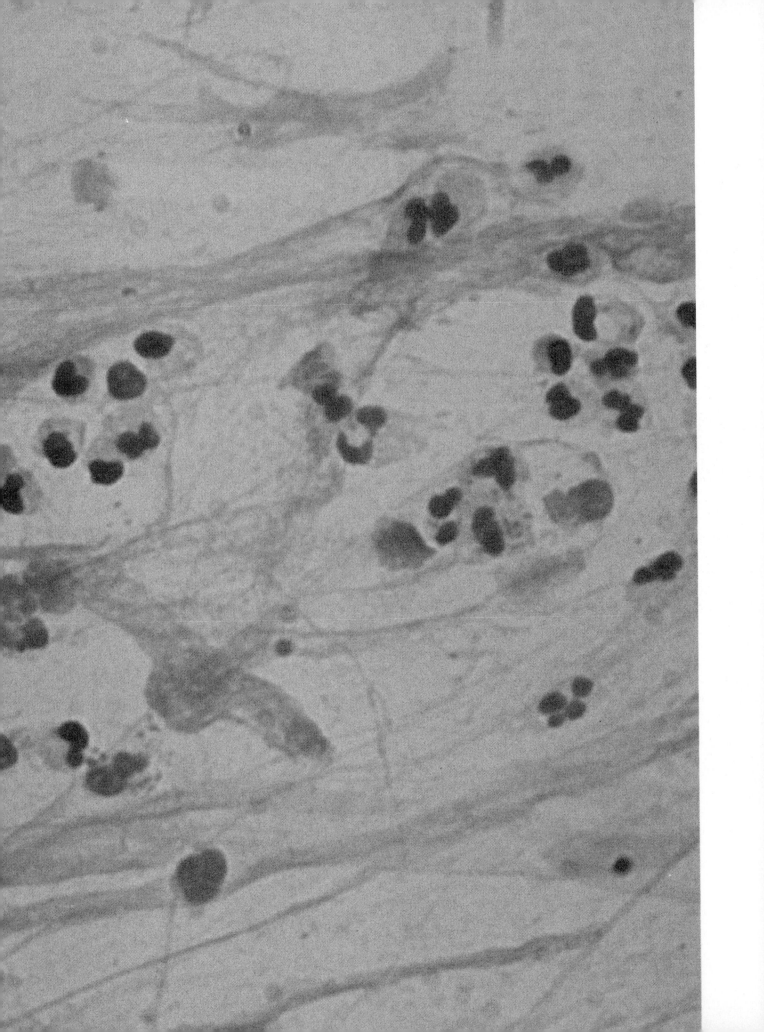

PART THREE

ETIOLOGICAL AGENTS

RECOVERED FROM CLINICAL

MATERIAL

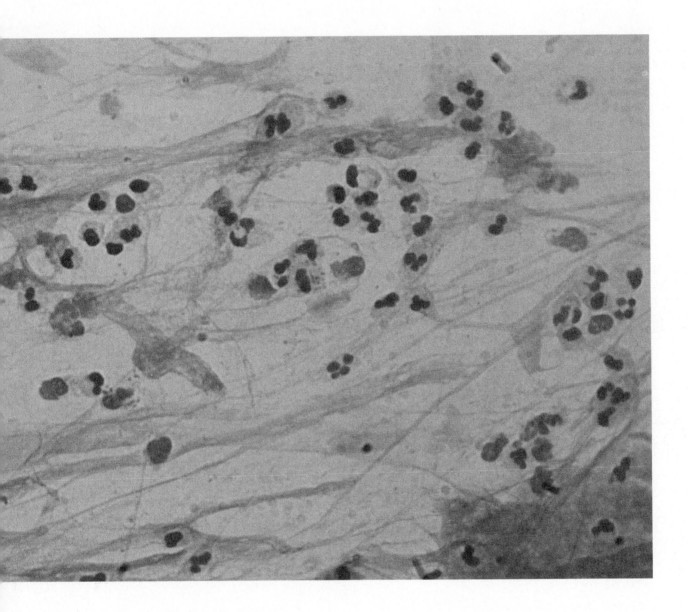

The quality of the specimen received in the microbiology laboratory determines the clinical utility of the results. This Gram stain of expectorated sputum illustrates the lack of squamous epithelial cells and the presence of mucus strands and numerous polymorphonuclear neutrophils that suggest that this was an adequate and potentially diagnostically useful specimen.

15

MICROORGANISMS ENCOUNTERED IN THE BLOOD

Microorganisms present in the circulating blood, whether continuously or transiently, are a threat to every organ in the body. Invasion of the blood-stream by microorganisms, aside from the problems of disseminating infection and affecting the function of implanted foreign bodies (e.g., artificial heart valves and joints), may have serious immediate consequences, including shock, multiple organ failure, disseminated intravascular coagulation (DIC), and death. Thus bacteremia constitutes one of the most serious situations in infectious disease. The expeditious detection and identification of blood-borne pathogens is one of the most important functions of the microbiology laboratory. Positive blood cultures may help provide a clinical diagnosis, as well as a specific etiological diagnosis. Pathogens of all four major groups of microbes—bacteria, fungi, viruses, and parasites—may be found in circulating blood during the course of disease. Bacteria, on the other hand, may also be recovered from peripheral blood in the absence of disease. After any loss of the integrity of the capillary endothelial cells, a few bacteria may enter the bloodstream and live for a short time until the primary defense systems of a healthy host (phagocytic cells in the liver and spleen, polymorphonuclear leukocytes, and antibody and complement) destroy them. Eukaryotic parasites may also be found transiently in the bloodstream as they migrate to the environment of choice. Their presence, however, cannot be considered consistent with a state of good health. Table 15.1 shows the most common clinically significant microorganisms isolated from blood during a 1-year period at one major medical center. The kinds of microorganisms isolated from blood cultures vary according to the patient population served by an institution and the microbial detection system used, but it can be generalized that the numbers of fungi and coagulase-negative staphylococci have increased while the numbers of clinically significant anaerobic isolates have decreased over the past decade.[2,7,19,23]

PARASITIC INFECTIONS

PATHOGENIC MECHANISMS

Parasites that infect cellular components may cause disease by disrupting the normal function of the infected cells. Invasion of the phagocytic cells of the reticuloendothelial system, as occurs in visceral leishmaniasis, renders those cells less able to fulfill their role as nonspecific immune responders. Proliferation of infected mononuclear

TABLE 15.1

BACTERIA AND FUNGI ISOLATED FROM BLOOD CULTURES MONITORED FROM MAY 1990 TO APRIL 1991 AT BARNES HOSPITAL, ST. LOUIS, MO.

ORGANISM	NUMBER OF PATIENTS
Aerobic bacteria	
Escherichia sp.	156
Klebsiella sp.	95
Pseudomonas sp.	94
Enterobacter sp.	63
Proteus sp.	32
Other Enterobacteriaceae	31
Acinetobacter sp.	24
Xanthomonas sp.	19
Other gram-negative rods	21*
Coagulase-negative staphylococci	285
Staphylococcus aureus	173
Viridans streptococci	122
Enterococcus sp.	120
β-hemolytic streptococci	50
Streptococcus pneumoniae	45
Other gram-positive organisms	22†
Anaerobic bacteria	
Bacteroides sp.	26
Other gram-negative anaerobes	5
Clostridium sp.	21
Peptostreptococcus sp.	4
Other gram-positive anaerobes	26‡
Fungi	
Candida albicans	52
Candida tropicalis	25
Torulopsis sp.	25
Cryptococcus sp.	10
Other *Candida* sp.	8

Modified from Murray, P.R., Traynor, P., and Hopson, D. 1992. Critical assessment of blood culture techniques: analysis of recovery of obligate and facultative anaerobes, strict aerobic bacteria, and fungi in aerobic and anaerobic blood culture bottles. J. Clin. Microbiol. 30:1462.
*Includes *Aeromonas* sp. (4), *Haemophilus* sp. (4), *Moraxella* sp. (3), *Neisseria* sp. (3), and others.
†Includes *Corynebacterium* sp. (13), *Listeria* sp. (5), and others.
‡Includes *Propionibacterium* sp. (10), *Actinomyces* sp. (7), *Lactobacillus* sp. (6), and *Bifidobacterium* sp. (3).

phagocytes in the liver, spleen, and bone marrow leads to enlargement of the liver and spleen and contributes to the anemia and leukopenia seen in patients with this infection. Primary growth of *Trypanosoma cruzi* in the human host occurs in various host cells that are initially penetrated. When the parasites mature to the trypomastigote stage and burst out of the host cells, a generalized parasitemia results. Although able to multiply in any organ, the organisms prefer to invade muscle (including heart muscle) and nerve tissue; the cycle then repeats itself. Damage to the host is probably due both to its own inflammatory response attempting to localize or destroy infected tissue and to rupture of infected host cells during the release of trypomastigotes. The most important consequences of the disease are congestive heart failure and cardiac arrhythmias relating to cardiac muscle and its conducting tissue involvement, and dilation of the esophagus and colon, related to neuronal involvement.

Tachyzoites of the parasite *Toxoplasma gondii* may be found in circulating blood. They invade cells within lymph nodes and other organs, including lung, liver, heart, brain, and eye. The resulting cellular destruction accounts for the manifestations of toxoplasmosis. Malarial parasites invade host erythrocytes and hepatic parenchymal cells. Cerebral malaria is especially dangerous. The significant anemia and subsequent tissue hypoxia that may result from destruction of red cells by the parasite and vascular trapping of normal erythrocytes by the infected red cells (which are less flexible and tend to clog small capillaries) are a major cause of morbidity. The host's immunological response is to remove the parasites and damaged red blood cells; the immune response may also have deleterious effects.

Although microfilariae are seen in peripheral blood during infection with *Dipetalonema, Mansonella, Loa loa, Wuchereria,* or *Brugia,* it is the adult worm eliciting the inflammatory response, primarily in lymphatics, that causes the distinctive lesions of filariasis related to lymphatic obstruction.

DETECTION

Parasites in the bloodstream are usually detected by direct visualization. Those parasites for which diagnosis is routinely dependent on observation of the organism in peripheral blood smears include *Plasmodium, Trypanosoma,* and *Babesia.* Patients with malaria or filariasis may display a periodicity in their episodes of fever that allows the physician to time the collection of blood for visual examination for parasites for optimal results. Further discussion of life cycles of bloodborne parasites and techniques for examination of blood for their presence can be found in Chapter 45. Serologic tests for diagnosis of infection caused by blood-borne parasites are also useful, as will be detailed in Chapter 45.

VIRUS INFECTIONS

PATHOGENIC MECHANISMS

Although many viruses do circulate in the peripheral blood at some stage of disease, the primary pathology relates to infection in the target organ or cells. Those viruses that preferentially infect blood cells are the Epstein-Barr virus and the cytomegalovirus, which invade lymphocytes; human immunodeficiency virus (HIV, which involves only certain T lymphocytes and perhaps macrophages); and other human retroviruses that attack lymphocytes. The pathogenesis of viral diseases of the blood is the same as that of viral diseases of any organ; by diverting the cellular machinery to create new viral components or by other means, the virus may prevent the host cell from performing its normal function. The cell may be destroyed or damaged by viral replication, and immunological responses of the host may also contribute to the pathology of the resulting disease.

DETECTION

Although many viral diseases have a viremic stage, recovery of virus particles or detection of circulating viruses has been used routinely in the diagnosis of only a few diseases. Detection of viral antigen by immunological methods has gained widespread use for diagnosis of hepatitis B. By using an enzyme-linked immunosorbent assay (ELISA) procedure, the surface antigen (HbsAg), core antigen (HbcAg), and early antigen (HbeAg) can be measured quantitatively in serum. Separation of the buffy coat cells from blood using a high molecular weight substrate gradient allows this virus-rich component to be inoculated into tissue culture for isolation of cytomegalovirus. ELISA, particle agglutination, and polymerase chain reaction are used commonly for detection of HIV virus antigens or genes. Chapter 43 will discuss recovery of viruses from blood in greater detail.

FUNGAL INFECTIONS

PATHOGENIC MECHANISMS

Fungemia is usually a serious condition, occurring primarily in immunosuppressed patients and in those with serious or terminal illness.[25] *Candida albicans* is the most common species by a vast margin, but *Malassezia furfur* can often be isolated in patients, particularly neonates, receiving lipid-supplemented parenteral nutrition. With the exception of *Histoplasma,* which multiply in leukocytes, the fungi do not invade blood cells, but their presence in the blood usually indicates a focus of infection elsewhere in the body. The large size and sterol-containing cell walls of molds make them particularly insensitive to the primary host defenses, antibody and phagocytic cells. Fungi in the bloodstream can be carried to all organs of the host, where they may grow, invade normal tissue, and elaborate toxic products. Fungi gain entrance to the circulatory system via loss of integrity of the gastrointestinal or other mucosa, through damaged skin, secondary to involvement of the lung or other organ, or by means of intravascular catheters. Several outbreaks of systemic *Rhizopus* infection were traced to the use of bandage material contaminated with spores of the fungus. Systemic fungal diseases that begin as pneumonia disseminate from the lungs, which serve as the portal of entry. Arthroconidia of *Coccidioides immitis* and microconidia of *Histoplasma capsulatum* and *Blastomyces dermatitidis* are ingested by alveolar macrophages in the lung. These macrophages carry the fungi to nearby lymph nodes, usually the hilar nodes. The fungi multiply within the node tissue and ultimately are released into the circulating blood, from which they go on to seed other organs or are destroyed by the body's defenses.

CULTURE

Even though all disseminated fungal disease is preceded by fungemia, recovery of the microorganisms from blood cultures has not been accomplished readily until recently. One reason for this may be that cultures were not often taken at the fungemia stage because presence of the disease had not yet been recognized. However, introduction of better methods for isolation of fungi from blood, including the lysis-centrifugation system described later, has resulted in increasing reports of recovery of fungi from peripheral blood and greater awareness by physicians to order fungal blood cultures.[25]

Many fungi, particularly yeast, can be recovered in standard blood culture media if the bottle has been vented and agitated to allow sufficient oxygen in the atmosphere for fungal growth. Fungi may grow slowly and poorly in the media that best support bacterial growth, however, and the ideal temperature for the dimorphic fungi (30° C) is lower than that recommended for optimal bacterial isolation. Until recently, the best medium for the recovery of fungi from blood has been a biphasic system. Optimal isolation of fungi in blood cultures is achieved with either agitated incubation of a recently introduced

commercial biphasic system, such as the Becton-Dickinson Microbiology Systems Septi-Chek (Figure 15.1) or by using the lysis-centrifugation system, the Wampole Isolator (described later).[17,25]

Blood specimens for detection of fungemia are collected in the same manner as for bacterial culture. Manufacturers of media for automated blood culture systems (discussed later) have developed specific media for isolation of fungi. These new formulas have dramatically increased the numbers of fungi isolated from patients with fungemia and have shortened the incubation time required for detection of the fungi.[11,17,18,19,25]

ANTIGEN DETECTION

Because many patients with systemic or invasive fungal infections are immunosuppressed and do not produce appreciable antibodies, detection of circulating fungal antigens or other cell wall com-

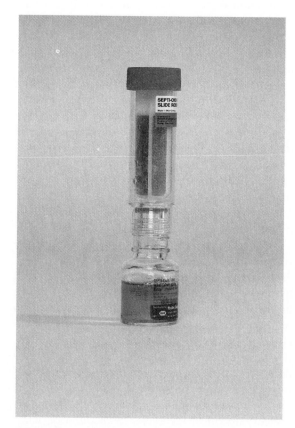

FIGURE 15.1

Becton-Dickinson (formerly Roche) Septi-Chek pediatric size biphasic blood culture bottle. The medium-containing base bottle is inoculated with blood, and the top piece containing agar paddles is added in the laboratory. The agar is inoculated by tipping the bottle to allow the blood-containing medium to flow over the agar.

ponents may be a more fruitful diagnostic method than searching for the presence of antibodies. This procedure might also give positive results early in the illness, before there has been time for an antibody response. This concept is being explored for diagnosis of several of the common fungal diseases, and tests employing it have been marketed for aiding diagnosis of at least two mycotic diseases. The presence of mannan or galactomannan antigen, polysaccharides found in *C. albicans* cell walls, has been noted in the serum of some severely ill patients with invasive disease. A few commercial latex agglutination and ELISA tests are available, but their utility has not been validated. However, latex agglutination and ELISA tests for detection of the polysaccharide capsular antigen of *C. neoformans* in serum or cerebrospinal fluid (Meridian Diagnostics and others) are accepted for diagnostic and prognostic use. Quantitation of this antigen can be followed sequentially to determine the effectiveness of antifungal therapy.[9]

BACTERIAL INFECTIONS

PATHOGENIC MECHANISMS

Bacteria in the bloodstream, called **bacteremia,** may indicate the presence of a focus of disease, such as intravascular infection, pneumonia, or liver abscess, or may merely represent transient release of bacteria into the bloodstream, such as occurs after vigorous toothbrushing by a normal person. Septicemia or sepsis indicates a situation in which bacteria or their products (toxins) are causing harm to the host. Unfortunately, clinicians often use the terms interchangeably. Signs and symptoms of septicemia may include fever or hypothermia, chills, hyperventilation and subsequent respiratory alkalosis, skin lesions, change in mental status, and diarrhea. More serious manifestations include hypotension or shock, disseminated intravascular coagulation (DIC), and major organ system failure.

Shock is the gravest complication of septicemia. In septic shock the presence of bacterial products and the host's responding defensive components act to shut down major host physiological systems. Manifestations include drop in blood pressure, increase in heart rate, impairment of function of vital organs (brain, kidney, liver, and lungs), acid-base problems, and bleeding problems. Gram-negative bacteria contain a substance in their cell walls, endotoxin, that has a strong effect on several physiological functions. This substance, a lipopolysaccharide (LPS) com-

prising part of the cell wall structure (Figure 15.2), may be released during the normal growth cycles of bacteria or after the destruction of bacteria by host defenses. Endotoxin (or the lipoidal disaccharide core of the LPS, lipid A) has been shown to mediate a number of systemic reactions including a drop in circulating granulocytes, a febrile response, and the activation of complement and certain blood-clotting factors.

Although most gram-positive bacteria do not contain endotoxin, many produce exotoxins, and in any case the effects of their presence in the bloodstream may be equally devastating to the patient. The disastrous complication of disseminated intravascular coagulation, in which numerous small blood vessels become clogged with clotted blood and bleeding may result from depletion of coagulation factors, can occur with septicemia involving any circulating pathogen, including parasites, viruses, and fungi, although it is most often a consequence of gram-negative bacterial sepsis. The syndrome characterized by fever, acute respiratory distress, shock, renal failure, intravascular coagulation, and tissue destruction, known as "septic shock," can be initiated by either exotoxins or endotoxins, but it is mediated by the same mechanism—activated mononuclear cells producing cytokines such as tumor necrosis factor and interleukins.

ORGANISMS COMMONLY ISOLATED FROM BLOOD

The commonest portals of entry for septicemia are the genitourinary tract (25%), respiratory tract (20%), abscesses (10%), surgical wound infections (5%), biliary tract (5%), miscellaneous sites (10%), and uncertain sites (25%). The organisms most commonly isolated from blood are gram-positive cocci, including coagulase-negative staphylococci, *Staphylococcus aureus*, and *Enterococcus* sp., organisms likely to be inhabitants of the hospital environment and thus to colonize the skin, oropharyngeal area, and gastrointestinal tract of hospitalized patients (see Table 15.1). Factors contributing to initiation of such infections are immunosuppressive agents, widespread use of broad-spectrum antibiotics that suppress the normal flora and allow the emergence of resistant strains of bacteria, invasive procedures that allow bacteria access to the interior of the host, more extensive surgical procedures, and prolonged survival of debilitated and seriously ill patients.

With the ever-increasing use of intravenous catheters, intraarterial lines, and vascular prostheses, organisms found as normal (or hospital-ac-

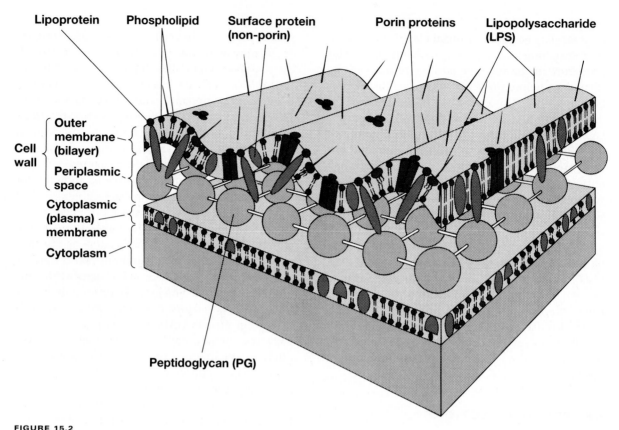

FIGURE 15.2

Diagram of gram-negative cell wall and outer membrane structure. (Modified from Baron, E. J., Chang, R. S., Howard, D. H. et al. 1994. Medical Microbiology: A short course. John Wiley & Sons, New York.)

quired) inhabitants of the human skin are able to gain access to the bloodstream and to find a surface on which to multiply.[8] Heart valves and vascular endothelium may represent such surfaces. When endocarditis occurs, colonies of bacteria embedded in fibrin (vegetations) occur on a heart valve. Constant shedding of fibrin particles and bacteria into the bloodstream is one of the usual consequences of endocarditis. In such a setting, *S. epidermidis* and other coagulase-negative staphylococci have been increasingly detected. The isolation of coagulase-negative staphylococci with the same biochemical and antibiotic susceptibility patterns from more than one blood culture from a patient should alert the clinician to the possibility of a clinically significant isolate. *S. epidermidis* is the most common etiological agent of prosthetic valve endocarditis, with *S. aureus* the second most common. *S. aureus* is an important cause of septicemia without endocarditis, in association with other foci such as abscesses, wound infections, and pneumonia, as well as sepsis related to indwelling intravascular catheters.

Immunocompromised hospitalized patients tend to become colonized with bacteria that are relatively antibiotic resistant, such as *Serratia*

marcescens, *Enterobacter* species, *Enterococcus*, and *Pseudomonas aeruginosa*. Another such bacterium, much less commonly recognized (perhaps because it is relatively fastidious), is *Corynebacterium jeikeium*. This organism should be suspected if a diphtheroid-like bacterium isolated from blood displays an unusually broad antibiotic resistance pattern, often being sensitive only to vancomycin. *C. jeikeium* is extremely virulent and must not be dismissed as a contaminant.

The primary causes of infectious endocarditis are the viridans streptococci, comprising several species. These organisms are normal inhabitants of the oral cavity or gastrointestinal tract, often gaining entrance to the bloodstream because of gingivitis, periodontitis, or dental manipulation. Heart valves, especially those that have been previously damaged, present convenient surfaces for attachment of these bacteria. The resulting vegetations ultimately seed bacteria into the blood at a slow but constant rate. Identification of these streptococci may be useful because certain species (such as *Streptococcus anginosus*) may be associated with increased frequency of metastatic abscess formation. *Streptococcus sanguis* and *Streptococcus mutans* are most frequently isolated in streptococ-

Common Agents of Infectious Endocarditis

Viridans streptococci*
Nutritionally deficient streptococci
Enterococci*
Streptococcus bovis
*Staphylococcus aureus**
Staphylococci (coagulase-negative)
Enterobacteriaceae
Pseudomonas species (usually in drug abusers)
Haemophilus species, particularly *H. aphrophilus*
Unusual gram-negative bacilli (e.g., *Actinobacillus,*
   Cardiobacterium, Eikenella)
Yeast species
Other (including polymicrobial infectious endocarditis)

---

* Most common species associated with native valve
  endocarditis in non-drug abusing adults.

---

cal endocarditis. Common agents of infectious endocarditis are listed in the box above.

Streptococcus bovis is normal gastrointestinal flora but may be found in the oral cavity as well. The presence of *S. bovis,* a nonenterococcal group D streptococcus, is significantly associated with large bowel carcinoma; clinicians should be aware of this association when this organism is isolated from blood. Identification of this group D streptococcus as *S. bovis* is also important because infection with this organism (including endocarditis) will respond to penicillin G, whereas enterococcal endocarditis requires a more aggressive therapeutic approach.

Septicemia occurs in 25% to 30% of patients with pneumococcal pneumonia and is often the most reliable way to establish the diagnosis. The prognosis is poorer in patients with pneumococcal pneumonia complicated by sepsis. Because the organisms produce an autolytic enzyme, early subculture of blood culture broths (before 18 hours) is necessary. Some automated systems may register positive results early during incubation, but when the bottle is examined and subcultured hours later, the autolysed nonviable pneumococci will not be detected, resulting in a pseudo false-positive culture result. *Streptococcus pneumoniae* grows extremely well in biphasic media, as described later.

Haemophilus influenzae infections may be accompanied by septicemia. The organism requires an enriched medium such as chocolate agar, and 10% CO_2 enhances growth. *H. parainfluenzae* can also be recovered from blood, particularly as an etiological agent of endocarditis. This organism grows as large clumps of filamentous rods, which may appear branched in the original stains from

broth. Other *Haemophilus* species, including *H. aphrophilus* and *H. paraphrophilus,* may also cause bacteremia and endocarditis. Major arterial embolization is relatively common in endocarditis involving *Haemophilus* species. All of the *Haemophilus* species grow sparsely in blood culture broth and often do not exhibit visually discernible turbidity. For this reason, blind stains or subcultures are essential for manual methods. Unless such subcultures are made to a medium such as chocolate agar and incubated under 5% to 10% CO_2, the organism can be overlooked. *H. aphrophilus* may grow as clumps that adhere to the walls of the bottle. *Listeria monocytogenes* is a cause of septicemia in immunosuppressed patients, alcoholics, and pregnant women. The bacterium may resemble a diphtheroid or coccus on Gram stain but grows on 5% sheep blood agar as a translucent, weakly β-hemolytic colony, resembling group B streptococci. *Listeria* must be distinguished from diphtheroids and from groups B and D streptococci. It displays a characteristic tumbling motility at room temperature but not at 35° C.

Anaerobic organisms also cause septicemia. *Bacteroides fragilis* is the most common anaerobe isolated from blood. *Clostridium perfringens,* however, can cause rapidly fatal disease and must be presumptively identified by Gram stain appearance and gas and hemolysis in the bottle so that clinicians can be alerted to the possibility of the presence of this virulent bacterium in the bloodstream. In recent studies, however, *C. perfringens* was judged to be a significant isolate (representing true infection) only 50% of the time.[14,30] The isolation of *Clostridium septicum* may be indicative of malignancy, particularly in the cecum or elsewhere in the gastrointestinal tract of the patient.

Capnocytophaga ochracea and *C. sputigena* may cause bacteremia in immunocompromised patients, particularly those with oral lesions.[4] An organism associated with fulminant septicemia, often in splenectomized or immunocompromised patients, although rarely isolated, is *Capnocytophaga canimorsus* (formerly CDC group DF-2). This organism is part of the normal oral flora of dogs; most reported cases of *C. canimorsus* infection are associated with dog contact or dog bites, and rarely with cats. The *Capnocytophaga* are fastidious fermentative gram-negative bacilli that will grow in conventional blood culture media. Subcultures may require as long as 4 days before colonies are visible. These organisms are described in Chapter 30. Almost every recognized bacterial species has been implicated as a cause of bacteremia. Certain organisms that require special techniques for isolation are mentioned here. Other bacteria that may grow in routine culture media include *Rothia dentocariosa, Campylobacter* species (especially *Campylobacter fetus*), *Cardiobacterium hominis, Chromobacterium* species, *Actinobacillus actinomycetemcomitans, Eikenella corrodens, Flavobacterium* species, and *Mycoplasma*. If these less common organisms are suspected by the clinician, the laboratory should be alerted to hold the blood cultures for an extended period of time past the first week and to make blind subcultures to several media, including more supportive media such as buffered charcoal yeast extract.

Although *Francisella tularensis*, the agent of tularemia, and *Yersinia pestis*, the agent of plague (rarely isolated), may grow in routine blood culture media, diagnosis by blood culture is not recommended for two reasons: (1) These organisms are fastidious in their growth requirements and often require prolonged incubation time for detection. The serious nature of the diseases, however, demands a more rapid method of diagnosis. Serologic methods (IgM determination or antigen detection in lesions) are more reliable and rapid. If an organism resembling *Y. pestis* or *F. tularensis* is isolated from blood, the physician should be notified, and the organism should be forwarded to the nearest public health facility that can handle such an isolate. (2) These pathogens require Biosafety Level 3 practices (extremely hazardous) and special containment facilities for their manipulation. In fact, if very small coccobacilli are observed on Gram stain of blood culture broth, all manipulations should be performed in a biological safety cabinet.

TIMING AND QUANTITY OF BLOOD
FOR CONVENTIONAL CULTURE

Conditions in which bacteria are only transiently present in the bloodstream include manipulation of infected tissues, instrumentation of contaminated mucosal surfaces, and surgery involving nonsterile sites. These backgrounds may also lead to significant septicemia. Incidental, transient bacteremia may occur spontaneously or with such minor events as chewing food. Bacteria can be found intermittently in the blood of patients with undrained abscesses. During early stages of typhoid fever, brucellosis, and leptospirosis, bacteria are continuously present in the bloodstream. The causative agents of meningitis, pneumonia, pyogenic arthritis, and osteomyelitis are often recovered from blood during the early course of these diseases. In the case of transient seeding of the blood from a sequestered focus of infection such as an abscess, bacteria are released into the blood approximately 45 minutes before the febrile episode. At least one study showed that perioperative fevers within the first 3 days are unlikely to be due to septicemia; cultures during that time are likely to be not cost-effective.[26]

In septic shock, bacterial endocarditis, and other endovascular infections, organisms are released into the bloodstream at a fairly constant rate, making the timing of cultures unimportant. Many studies have shown that the likelihood of recovering a pathogen increases as the volume of blood cultured increases, provided that there is adequate dilution of such blood to negate the effect of normal antibacterial substances and antimicrobial agents that may be present. When a patient's condition requires institution of therapy as rapidly as possible, there is little time for the collection of multiple blood culture samples. A generally accepted compromise is to collect 40 ml of blood at one time, 20 ml from each of *two separate venipunctures from two different sites, using two separate needles and syringes*, before the patient is given antimicrobial therapy. For initial evaluation of fever of unknown origin, four separate blood cultures, two drawn on each of 2 days, will detect most causative agents.[1]

The large study reported by Weinstein et al.[30] revealed that 99.3% of 500 episodes of septicemia were identified by the first two blood cultures, comprising 30 ml total volume of blood cultured. In another study, in 98% of patients with infective endocarditis who had not received antimicrobial agents, two blood cultures established the diagnosis. If therapy can be delayed, three separate blood collections of 10 to 20 ml each, and an additional blood culture or two taken on the second day (if necessary), will detect most etiological agents of endocarditis, even for patients who have received prior antibiotic therapy. This presumes use of setups adequate for growth of the organism involved, which often entails extending the incubation to 28 days. An excellent discussion on the rational use of blood cultures was presented by Aronson and Bor.[1]

It is not safe to take large samples of blood from children, particularly young infants. Fortunately, infants with more serious disease usually yield more than 10 colony forming units (CFU) of bacteria per milliliter of blood. For infants and small children, only 1 to 5 ml of blood can usually be drawn for bacterial culture.[5] Quantities less than 1 ml may not be adequate to detect pathogens, however, since blood specimens from

---

PROCEDURE 15.1

DRAWING BLOOD FOR CULTURE

PRINCIPLE
Organisms found in circulating blood can be enriched in culture for isolation and further studies. Blood for culture must be obtained aseptically. Once removed from the circulation, unclotted blood must be diluted in growth media.

METHOD
Note: Universal Precautions requires that phlebotomists wear gloves for this procedure.

1. Choose the vein to be drawn by touching the skin before it has been disinfected.

2. Cleanse the skin over the venipuncture site in a circle approximately 5 cm in diameter with 70% alcohol, rubbing vigorously. Allow to air dry.

3. Starting in the center of the circle, apply 2% tincture of iodine (or povidone-iodine) in ever-widening circles until the entire circle has been saturated with iodine. Allow the iodine to dry on the skin for at least 1 minute. The timing is critical; a watch or timer should be used.

4. If the site must be touched by the phlebotomist after preparation, the phlebotomist must disinfect the gloved fingers used for palpation in identical fashion.

5. Insert the needle into the vein and withdraw blood. Do not change needles before injecting the blood into the culture bottle.[12,13]

6. After the needle has been removed, the site should be cleansed with 70% alcohol again, because many patients are sensitive to iodine.

QUALITY CONTROL
Blood culture results should be monitored regularly. If more than 3% of blood cultures are judged on clinical grounds to yield contaminants, aseptic phlebotomy technique may need to be reemphasized. Another factor to consider when contamination rates seem high is that laboratory handling of the specimen may be the source, particularly for systems that require manipulation after phlebotomy, such as the Isolator and biphasic systems.

---

septic children may yield less than 5 CFU/ml of the organism.[16,24,31]

HOW TO DRAW BLOOD FOR CONVENTIONAL CULTURE
Since the media into which blood specimens are placed have been developed as enrichment broths to encourage the multiplication of even one bacterium, it follows that these media will enhance the growth of any stray contaminating bacterium, such as a normal inhabitant of human skin. Because of the increasing incidence of true infections caused by bacteria that normally are nonvirulent indigenous microflora of a healthy human host, interpretation of the significance of growth of such bacteria in blood cultures has become increasingly difficult.[3,21,24] To help clinicians determine whether a *Staphylococcus epidermidis,* other coagulase-negative staphylococcus, *Corynebacterium* species, or *Propionibacterium acnes* is a true pathogen or an inadvertently introduced skin contaminant, blood cultures are drawn from separate sites, even when more than one blood culture is to be drawn at one time, such as before instituting emergency therapy. Recovery of identical bacteria of these types from multiple cultures is considered to be indicative of infection. Quantitation of bacteremia, as is possible with the lysis-centrifugation system, may be helpful.[31]

Another strategy is to reduce the risk of introducing contaminants into blood culture media by careful skin preparation. Procedure 15.1 details the steps necessary for drawing blood for culture. A publication from the National Committee for Clinical Laboratory Standards describes in more detail the mechanics of collection of blood by venipuncture.[20] Laboratories that recover "contaminants," as determined by clinical evaluation of the patients' conditions, at rates greater than 3% should suspect improper phlebotomy techniques and should institute measures to reeducate the phlebotomists in proper skin preparation methods. It is less desirable to draw blood through a vascular shunt or catheter, since these prosthetic devices are difficult to decontaminate completely. It is also recommended to draw blood below an existing intravenous line, if possible, since blood above the line will be diluted with the fluid being infused.

If the blood for culture is not being inoculated directly into broth media, it must be transported with an anticoagulant. Heparin, ethylenediaminetetraacetic acid (EDTA), and citrate have been found to be inhibitory to a number of organisms; thus sodium polyanethol sulfonate (SPS, Liquoid) in concentrations of 0.025% to 0.03% is the best anticoagulant for blood.

HOW TO CULTURE INTRAVENOUS CATHETER TIPS

When colonization of an indwelling catheter is suspected of being the focus for septicemia, the catheter tip may be cultured to determine its status. As first reported by Maki and co-workers, the number of colony forming units of bacteria on the catheter directly relates to whether it is the source of the infection. The skin around the catheter is carefully disinfected with an iodine preparation, and the catheter is removed. A short section (approximately 5 cm [2 in]) including the area directly beneath the skin is aseptically cut off and sent to the microbiology laboratory in a sterile container without liquid. This section of catheter is rolled across the surface of an agar plate with sterile forceps (as shown in Figure 15.3). After overnight incubation the colonies are counted. A positive culture (more than or equal to 15 CFU) correlates well with the catheter tip serving as the source of infection.[15] Some workers will report positive results with greater than 5 CFU.

CONVENTIONAL BLOOD CULTURE MEDIA, ADDITIVES, AND DILUTION FACTORS

The diversity of bacteria that are recovered from blood requires an equally diverse armamentarium of media to enhance the growth of these bacteria. Basic blood culture media contain a nutrient broth and an anticoagulant. Numerous different broth formulations are available, including those that can be prepared by the laboratory in-house and those that are commercially prepared. Most blood culture bottles available commercially contain trypticase soy broth, brain-heart infusion broth, supplemented peptone, or thioglycolate broth. More specialized broth bases include Columbia or *Brucella* broth. Growth of cell wall–deficient bacteria may be enhanced by the addition of osmotic stabilizers such as sucrose, mannitol, or sorbose to create a hyperosmotic (hypertonic) medium. Media without osmotic additives are known as isotonic. The hypertonic bottles are difficult to inspect visually for evidence of bacterial growth, since red cells in the media become partially lysed and no longer sediment to the bottom, giving a muddy appearance to the broth. Hypertonic media should be used only for specific problems, since they can inhibit a number of species of bacteria. The anticoagulant in blood culture media must not harm the bacteria and

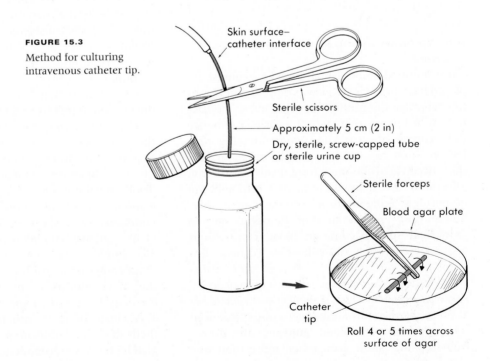

FIGURE 15.3
Method for culturing intravenous catheter tip.

must prevent clotting of the blood, which would entrap bacteria and prevent their detection. The most commonly used preparation in blood culture media today is 0.025% to 0.05% SPS. In addition to its anticoagulant properties, SPS is also anti-complementary, antiphagocytic, and interferes with the activity of some antimicrobial agents, notably aminoglycosides.[22] SPS, however, may inhibit the growth of a few microorganisms such as some strains of *Neisseria* species, *Gardnerella vaginalis*, *Streptobacillus moniliformis*, and all strains of *Peptostreptococcus anaerobius*. The addition of 1.2% gelatin has been shown to counteract this inhibitory action of SPS; however, recovery of other organisms is decreased.

The addition of penicillinase to blood culture media for inactivation of penicillin has been largely superseded in recent years by the availability of a resin-containing medium that inactivates most antibiotics nonselectively by adsorbing them to the surface of the resin particles.[22] Resin-containing media may enhance isolation of staphylococci, particularly when patients are receiving bacteriostatic drugs. The BACTEC system offers several resin-containing media. Preliminary studies have shown one medium (BACTEC Plus), which allows inoculation of a larger volume of blood (10 ml per bottle), to yield enhanced recovery of agents of septicemia in comparison with the Wampole Isolator system and B-D Septi-Chek system.[10,11]

In addition to volume of blood cultured and type of media chosen, the dilution factor for the blood in the medium must be considered. To conserve space and materials, it is desirable to combine the largest feasible amount of blood from the patient (usually 10 ml) with the smallest amount of medium that will still encourage the growth of bacteria and dilute out or inactivate the antibacterial components of the system. For this purpose, a 1:5 ratio of blood to unmodified medium has been found to be adequate.

INCUBATION CONDITIONS FOR CONVENTIONAL BLOOD CULTURES

The atmosphere in commercially prepared blood culture bottles is usually at a low oxidation-reduction potential, allowing most facultative and some anaerobic organisms to multiply. To encourage the growth of strict aerobes, such as yeast and *Pseudomonas aeruginosa*, transient venting of the bottles with a sterile, cotton-plugged needle may be necessary. Agitation also yields improved growth.[19]

DETECTING GROWTH IN CONVENTIONAL BLOOD CULTURES

Most clinical laboratories report a positive blood culture rate of greater than 8%. Approximately 3% to 5% of the positive blood cultures will yield anaerobic bacteria, whereas more than 7% may yield fungi in some medical centers. Restricted patient populations may influence the rate of positive cultures reported from smaller hospitals and clinics. The optimal recovery of agents of septicemia may involve routine inoculation of two aerobic bottles and addition of an anaerobic bottle for selected clinically defined conditions.[7,14,19]

After 6 to 18 hours of incubation, most bacteria are present in numbers large enough to detect by blind subculture or acridine orange viable stain (discussed in Chapter 8). Constant agitation of the bottles during the first 24 hours of incubation will enhance the growth of most bacteria.[17] Agitation does not allow the red cells to settle in the media, however, so that gross visual detection of growth is not possible soon after the initial incubation with agitation.

Blind subcultures from conventional bottles after the first 6 to 12 hours of incubation are performed by aseptically removing a few drops of the well-mixed medium and spreading this inoculum onto a chocolate blood agar plate, which is incubated in 5% to 10% CO_2 at 35° C for 48 hours. Bottles are then reincubated for 5 to 7 days unless the patient's condition requires special consideration, as discussed later. Growth of anaerobic bacteria can be detected in stationary bottles by visual inspection with such success that blind anaerobic subcultures are not recommended. After 48 hours of incubation, a second blind subculture or acridine orange stain may be performed.

Bottles should be examined visually at least daily. Growth is usually indicated by hemolysis of the red cells, gas bubbles in the medium, turbidity, or appearance of small colonies in the broth, on the surface of the sedimented red cell layer (Figure 15.4), or occasionally along the walls of the bottle. When macroscopic evidence of growth is apparent, a Gram stain of a drop of medium will be more rewarding. Methanol fixation of the smear preserves bacterial and cellular morphology, which may be especially valuable for detecting gram-negative bacteria among red cell debris. Examination of the positive broth under phase microscopy will reveal details of morphology and motility that may aid the clinician in making an early decision regarding treatment. As soon as a morphological description can be tentatively as-

FIGURE 15.4
Many colonies of *Staphylococcus aureus* on layer of settled red blood cells in blood culture bottle.

ples must be streaked for isolated colonies. Restreaking positive bottles a second time at the end of the incubation period after the original subculture to detect further organisms is of questionable value and is not recommended.

Because of the seriousness of septicemia, even preliminary results should be telephoned to the physician and the patient care unit immediately.[27] A number of rapid tests for identification and presumptive antimicrobial susceptibilities can be performed from the broth blood culture, if a unimicrobic infection is suspected (based on microscopic evaluation). A suspension of the organism that approximates the turbidity of a 0.5 McFarland standard, achieved directly from the broth, or by centrifuging the broth and resuspending the pelleted bacteria, can be used to inoculate Mueller-Hinton agar plates for the Bauer-Kirby susceptibility test or for performing dilution antimicrobial susceptibilities. These suspensions may also be used to set up preliminary biochemical tests such as coagulase, thermostable nuclease, esculin hydrolysis, bile solubility, or antigen detection by fluorescent-antibody stain or agglutination procedures for gram-positive bacteria and oxidase and rapid biochemicals such as API Rapid E, Minitek, or Vitek-ID for gram-negative bacteria. Presumptive results must be verified with conventional procedures using pure cultures. Chapter 4 includes some discussion of rapid methods of handling positive blood cultures. All isolates from blood cultures should be stored for an indefinite period, preferably by freezing at −70° C in 10% skim milk. Storing an agar slant of the isolate under sterile mineral oil at room temperature is a good alternative to freezing. It is often necessary to compare separate isolates from the same patient or isolates of the same species from different patients, sometimes even months after the bacteria were isolated.

signed to a pathogen detected in blood, the physician should be given all of the available information. Determining the clinical significance of an isolate is the physician's responsibility. If no organisms are seen on microscopic examination of a bottle that appears positive, subcultures should be performed anyway.

PROCEDURES FOR HANDLING POSITIVE BLOOD CULTURES

Subcultures from blood cultures suspected of being positive, whether proved by microscopic visualization or not, should be made to a variety of media so as to be able to support the growth of most bacteria including anaerobes. Initial subculture may include chocolate blood agar, 5% sheep blood agar, MacConkey agar (if gram-negative bacteria are seen), and supplemented anaerobic blood agar. The incidence of polymicrobic bacteremia or fungemia ranges from 3% to 20% of all positive blood cultures. For this reason, sam-

CULTURING SUSPECTED CONTAMINATED BANKED BLOOD, PLASMA, OR OTHER INFUSED FLUIDS

When an infusion is suspected of being the source of bacteremia or fungemia, it must be cultured in a manner similar to that used for blood. Since contamination is usually random and rare, the routine monitoring of intravenous fluids and banked blood is not rational. However, in cases of suspected contamination, the fluid, plasma, or other blood product is aseptically aspirated into a syringe and 5 to 10 ml is aseptically introduced into each of two blood culture bottles. One bottle is incubated at room temperature to enhance the growth of psychrophilic (cold-loving) gram-negative bacilli, which have been implicated as the cause of fatal transfusion reactions, and the other bottle is incubated at 35°C. The normal visual and cultural methods used to detect positive cultures are then followed. The putative contaminated blood product bag, tubing, and all isolates should be saved until epidemiological investigations are complete.

A number of infections have been documented to be transmitted by blood transfusions or needle contact, including malaria, trypanosomiasis, toxoplasmosis, cytomegalovirus disease, AIDS, hepatitis, and bacterial endocarditis. Gram-negative bacteria are most often implicated in contaminated intravenous fluids, although the possibility exists for any organism to multiply in a suitable environment and subsequently be injected into a compromised host. Recently, several episodes of *Yersinia enterocolitica* bacteremia, shock, and even death following transfusion of packed red blood cells were reported.[6] *Yersinia* can multiply to large numbers in red cells stored at refrigerator temperatures without showing visible evidence of their presence.

SPECIAL PROBLEMS AND UNUSUAL MICROORGANISMS

Special handling may be required for recovery of *Brucella* species from blood; septicemia occurs primarily during the first 3 weeks of illness. Best recovery is obtained with *Brucella* or trypticase soy broth; the use of biphasic media may enhance growth or the Isolator system may allow release of intracellular bacteria. The bottles should be continuously vented and incubated in 10% CO_2 at 37° C for at least 4 weeks. Blind subcultures should be performed at 4 days and weekly thereafter onto *Brucella* blood agar plates incubated as described. *Brucella* species may grow slowly, so cultures must be incubated for a minimum of 3 weeks. Cultures should be handled in a biological safety cabinet. The use of the lysis-centrifugation method or an automated method may enhance recovery. In some cases, only bone marrow cultures are positive.

Visualization in direct preparations is diagnostic for 70% of cases of febrile disease caused by *Borrelia* species. The organisms may be seen in direct wet preparations of a drop of anticoagulated blood diluted in saline as long, thin, unevenly coiled spirochetes that seem to push the red blood cells around as they move. Thick and thin smears of blood, prepared as for malaria testing (described in Chapter 45) and stained with Wright's or Giemsa stain, are also quite sensitive for the detection of *Borrelia*.

Leptospirosis can be diagnosed by isolating the causative spirochete from blood during the first 4 to 7 days of illness. Media with up to 14% (vol/vol) rabbit serum, such as Fletcher's, polysorbate 80 (Tween 80)-albumin, or Ellinghausen, McCullough, Johnson, Harris (EMJH) semisolid medium, are recommended (described in Chapter 32). After adding 1 to 3 drops of fresh or SPS-anticoagulated blood to each of several tubes with 5 ml of culture medium, the cultures are incubated for 5 to 6 weeks at 28° C to 30° C in air in the dark. Leptospires will grow 1 to 3 cm below the surface, usually within 2 weeks. The organisms remain viable in blood with SPS for 11 days, allowing for transport of cultures from distant locations. Direct darkfield examination of peripheral blood is not recommended, since many artifacts are present that may resemble spirochetes.

One of the agents of rat-bite fever, *Streptobacillus moniliformis,* may be recovered in blood cultures. Concentrations of SPS greater than 0.025% will inhibit the growth of this organism, however, so the addition of 1.2% gelatin to commercial media is recommended. These organisms will grow in conventional media after several days of incubation, visible as "puffball" colonies on the surface of the red cell layer. Cell wall–defective forms may require the use of hypertonic media for recovery.

Mycobacterium avium-intracellulare complex organisms are isolated often from the blood of patients suffering from AIDS; isolation of *M. tuberculosis* from these patients is also not uncommon. By the time the infection is discovered, the bacteria are usually circulating in great numbers. The use of special media, such as Middlebrook 7H9 broth with 0.05% SPS or brain-heart infusion broth with 0.5% polysorbate 80, with or without a Middlebrook 7H11 agar slant, is recommended. The newer methods of detection, biphasic Septi-Chek, automated systems, and lysis-centrifugation (described later), have been shown to shorten significantly the time required for mycobacterial cultures to become positive. Blood can be held in Isolator tubes for prolonged periods and not decrease substantially the recovery of *M. avium-intracellulare*, which allows blood to be drawn at a site distant from the laboratory.[28]

Streptococcus adjacens and *S. defectivus* are unable to multiply without the addition of 0.001% pyridoxal hydrochloride (also called thiol or vitamin B_6). These streptococci are known as "nutritionally variant" or "satelliting" streptococci. Although human blood introduced into the blood culture medium will provide enough of the pyridoxal to allow the organisms to multiply in the bottle, standard sheep blood agar plates may not support their growth. Subculturing the broth to a 5% sheep blood agar plate and overlaying a streak of *Staphylococcus aureus* to produce the supplement will generally demonstrate colonies of the streptococci growing as tiny "satellites" next to the streak (see Figure 26.1). Some commercial media may be supplemented with enough pyridoxal (0.001%) to support growth of nutritionally variant streptococci.

NONCONVENTIONAL METHODS FOR DETECTING BACTEREMIA

ANTIMICROBIAL REMOVAL DEVICE

Introduction of blood into a vial containing anticoagulant, a cationic resin, and a polymeric adsorbent resin will remove antimicrobial agents present in the blood before the blood is inoculated into culture media. The vial is commercially available (ARD, Becton-Dickinson Microbiology Systems). Blood must be rotated in the vial with the resins for 15 minutes, centrifuged to remove the resins, and then inoculated into the media of choice. This device can remove up to 100 μg of most antibiotics. Although there are studies that imply no significant advantage with this system, most workers have found increased and faster isolation of pathogens from the blood of patients being treated with antimicrobial agents. Negative aspects include inhibition of the growth of some strains of gram-negative bacilli.

AUTOMATED BLOOD CULTURE SYSTEMS

More than half of all hospital microbiology laboratories use an automated blood culture system. Most laboratories use the BACTEC (Johnston Laboratories/Becton-Dickinson) described in Chapter 12 for the radiometric or spectrophotometric detection of CO_2 produced by microbial metabolism and released into the airspace above the fluid medium. A small percent of laboratories use the BacT/Alert (Organon Teknika, Durham, N.C.), which detects CO_2-derived pH changes in the medium using a colorimetric indicator. The blood cultures are always agitated during incubation to facilitate the growth of aerobic and facultative organisms. Resin-containing broth media are also available for the BACTEC; enhanced recovery of fungi and fastidious bacteria have been reported.[10,11,16,29] No further manipulations are required to use this medium after the blood has been introduced into the bottle.

Problems encountered with the BACTEC system include false-positive findings caused by inappropriate threshold values, false-positive findings caused by carryover of organisms from positive bottles to adjacent culture bottles by inadequately sterilized sampling needles, false-negative findings caused by failure of certain microorganisms to metabolize enough substrate, and the need for special disposal of the radioactive media (for the radiometric system) at the end of the incubation period. Regulations governing the disposal of the media vary among states. The BacT/Alert system, which relies on colorimetric detection of pH changes in the medium caused by microbial growth, circumvents some of these problems by being noninvasive (no carryover), nonradioactive, and monitoring against individual bottle baselines. The system currently has only a few media available, none of which contains resins.

Advantages of automated systems are more rapid detection time for many pathogens, the ability to monitor growth without visual inspection or subculture, the automated handling of large numbers of blood culture bottles, and the potential to tie the system into a management information processing data base for monitoring results and generating statistical reports for both historical and epidemiological studies. Modified radiometric BACTEC media have been used successfully for relatively rapid detection of circulating mycobacteria. Other commercial vendors are developing automated blood culture systems based on pressure changes in positive blood cultures and fluorogenic detection of pH changes. Microbiologists will soon have numerous choices among automated blood culture systems.

LYSIS-CENTRIFUGATION

The lysis-centrifugation system commercially available is the Isolator (Wampole Laboratories, Cranbury, N.J). The Isolator consists of a stoppered tube containing saponin to lyse blood cells, polypropylene glycol to decrease foaming, SPS as an anticoagulant, EDTA to chelate calcium ions and thus inhibit the complement cascade and coagulation, and a small amount of an inert fluorochemical (Fluorinert, 3M Co., St. Paul, Minn.) to cushion and concentrate the microorganisms during 30-minute centrifuga-

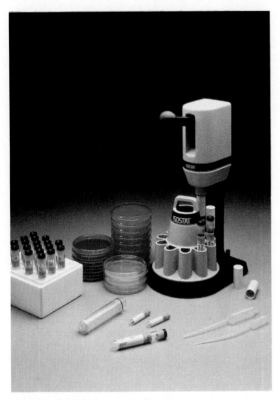

FIGURE 15.5

Lysis-centrifugation blood culture system (Isolator system, Wampole Laboratories) uses vacuum-draw collection tubes with lysing agent and a special apparatus (Isostat) to facilitate removal of the supernatant without the use of needles. (Courtesy Wampole Laboratories, Cranbury, N.J.)

tion at 3000 × *g* (Figure 15.5). After centrifugation, the supernatant is discarded, the sediment containing the pathogens is vigorously vortexed, and the entire sediment is plated to solid agar. Benefits of this system include the more rapid and greater recovery of filamentous fungi, the presence of actual colonies for direct identification and susceptibility testing after initial incubation, the ability to quantify the colony forming units present in the blood, rapid detection of polymicrobial bacteremia, dispensing with the need for a separate antibiotic-removal step, the ability to choose special media for initial culture setup based on clinical impression (such as direct plating onto media supportive of *Legionella* species or *Mycobacterium* species), and possible greater recovery of intracellular microorganisms caused by lysis of host cells. Shortcomings of the system seem to be a relatively high rate of plate contamination, necessitating manipulation inside a laminar flow biosafety cabinet, and failure of the system to

detect certain bacteria, such as *Streptococcus pneumoniae*, *Listeria monocytogenes*, *Haemophilus influenzae*, and anaerobic bacteria, as well as conventional systems.[10,17,18]

SELF-CONTAINED SUBCULTURE

Recent modifications of the biphasic blood culture medium are the Septi-Chek system (Becton-Dickinson) and the Vacutainer Agar Slant system (Becton-Dickinson), consisting of a conventional blood culture broth bottle with an attached chamber containing a slide coated with agar or several types of agars. Special formulas for isolation of fungi and mycobacteria are also available. To subculture, the entire broth contents are allowed to contact the agar surface by inverting the bottle, a simple procedure that does not require opening the bottle or using needles. The large volume of broth that is subcultured and the ease of subculture allow faster detection time for many organisms than is possible with conventional systems. The Septi-Chek system appears to enhance the recovery of *Streptococcus pneumoniae*, but such biphasic systems do not efficiently recover anaerobic isolates.[17,18,29]

NONCULTURE METHODS FOR DETECTING BACTEREMIA

Circulating antigens can be detected by latex agglutination procedures, available for some yeasts, group B streptococci, *H. influenzae* type b, *S. pneumoniae*, staphylococcal teichoic acids, and *Neisseria meningitidis*. These techniques are discussed in Chapter 11.

The lysate from amebocytes of *Limulus polyphemus*, the horseshoe crab, will gel in the presence of as little as 0.1 ng of endotoxin. This highly sensitive test, the *Limulus* amebocyte lysate assay, can be used to detect circulating lipopolysaccharides (LPS) in patients with gram-negative bacteremia. Sites of gram-negative bacterial infection other than blood may also cause a positive test result with serum. Although often used as an emergency test to diagnose gram-negative bacterial meningitis, the amebocyte lysate assay is not commonly used to diagnose bacteremia, in part because of its nonspecific nature in blood testing.

Detecting the presence of metabolic end products of microbial activity in serum using gas-liquid chromatography (GLC) has been explored. Some success in detecting products of anaerobic metabolism has been reported. A promising use of GLC may be its ability to detect metabolic products of mycobacteria, organisms whose diagnosis by conventional methods is extremely time-consuming.

Lysis-filtration of blood and subsequent staining of the filter surface, either with traditional stains or with new molecular reagents, may prove to be a very sensitive and rapid method for detection of circulating microorganisms. New methods are being developed for exploiting nucleic acid probe reagents for detection of pathogens in blood.

REFERENCES

1. Aronson, M.D., and Bor, D.H. 1987. Blood cultures. Ann. Intern. Med. 106:246.
2. Banerjee, S.N., Emori, T.G., Culver, D.H., et al. 1991. Secular trends in nosocomial primary bloodstream infections in the United States, 1980-1989. Am. J. Med. 91 (Suppl. 3B):86S.
3. Bates, D.W. and Lee, T.H. 1992. Rapid classification of positive blood cultures: prospective validation of a multivariate algorithm. JAMA 267:1962.
4. Bilgrami, S., Bergstrom, S.K., Peterson, D.E., et al. 1992. *Capnocytophaga* bacteremia in a patient with Hodgkin's disease following bone marrow transplantation: case report and review. Clin. Infect. Dis. 14:1045.
5. Campos, J.M. 1989. Detection of bloodstream infections in children. Eur. J. Clin. Microbiol. Infect. Dis. 8:815.
6. CDC 1991. Update: *Yersinia enterocolitia* bacteremia and endotoxin shock associated with red blood cell transfusions—United States, 1991. MMWR 40:176.
7. Dorsher, C.W., Rosenblatt, J.E., Wilson, W.R., and Ilstrup, D.M. 1991. Anaerobic bacteremia: decreasing rate over a 15-year period. Rev. Infect. Dis. 13:633.
8. Freeman, J., Goldmann, D.A., Smith, N.E., et al. 1990. Association of intravenous lipid emulsion and coagulase-negative staphylococcal bacteremia in neonatal intensive care units. N. Engl. J. Med. 323:301.
9. Gade, W., Hinnefeld, S.W., Babcock, L.S., et al. 1991. Comparison of the PREMIER cryptococcal antigen enzyme immunoassay and the latex agglutination assay for detection of cryptococcal antigens. J. Clin. Microbiol. 29:1616.
10. Kelly, M., Roberts, F., Henry, D., et al. 1989. Comparison of BACTEC 660 resin media to Isolator for detection of bacteremia. Abstr. Ann. Meeting, Amer. Soc. Clin. Microbiol., C-228, p. 431.
11. Koontz, F.P., Flint, K.K., Reynolds, J.K., and Allen, S.D. 1991. Multicenter comparison of the high volume (10 ml) NR BACTEC Plus system and the standard (5 ml) NR BACTEC system. Diagn. Microbiol. Infect. Dis. 14:111.
12. Krumholz, H.M., Cummings, S., and York, M. 1990. Blood culture phlebotomy: switching needles does not prevent contamination. Ann. Intern. Med. 113:290.
13. Leisure, M.K., Moore, D.M, Schwartzman, J.D., et al. 1990. Changing the needle when inoculating blood cultures: a no-benefit and high-risk procedure. JAMA 264:2111.
14. Lombardi, D.P. and Engleberg, N.C. 1992. Anaerobic bacteremia: incidence, patient characteristics, and clinical significance. Am. J. Med. 92:53.
15. Mermel, L.A., McCormick, R.D., Springman, S.R., and Maki, D.G. 1991. The pathogenesis and epidemiology of catheter-related infection with pulmonary artery Swan-Ganz catheters: a prospective study utilizing molecular subtyping. Amer. J. Med. 91(Suppl. 3B):197S.
16. Morello, J.A., Matushek, S.M., Dunne, W.M., and Hinds, D.B. 1991. Performance of a BACTEC nonradiometric medium for pediatric blood cultures. J. Clin. Microbiol. 29:359.
17. Murray, P.R. 1991. Comparison of the lysis-centrifugation and agitated biphasic blood culture systems for detection of fungemia. J. Clin. Microbiol. 29:96.
18. Murray, P.R., Spizzo, A.W., and Niles, A.C. 1991. Clinical comparison of the recoveries of bloodstream pathogens in Septi-Chek brain-heart infusion broth with saponin, Septi-Chek tryptic soy broth, and the Isolator lysis-centrifugation system. J. Clin. Microbiol. 29:901.
19. Murray, P.R., Traynor, P., and Hopson, D. 1992. Critical assessment of blood culture techniques: analysis of recovery of obligate and facultative anaerobes, strict aerobic bacteria, and fungi in aerobic and anaerobic blood culture bottles. J. Clin. Microbiol. 30:1462.
20. NCCLS. Procedures for the collection of diagnostic blood specimens by venipuncture, ed 3: Approved standard. 1991. National Committee for Clinical Laboratory Standards, Villanova, Pa.
21. Ringberg, H, Thoren, A, and Bredberg, A. 1991. Evaluation of coagulase-negative staphylococci in blood cultures. Scand. J. Infect. Dis. 23:315.
22. Rodriguez, F., and Lorian, V. 1985. Antibacterial activity in blood cultures. J. Clin. Microbiol. 21:262.
23. Scheckler, W.E., Scheibel, W., and Kresge, D. 1991. Temporal trends in septicemia in a community hospital. Am. J. Med. 91 (Suppl. 3B):90S.
24. St. Geme III, J.W., Bell, L.M., Baumgart, S., et al. 1990. Distinguishing sepsis from blood culture contamination in young infants with blood cultures growing coagulase-negative staphylococci. Ped. 86:157.
25. Telenti, A. and Roberts, G.D. 1989. Fungal blood cultures. Eur. J. Clin. Microbiol. Infect. Dis. 8:825.
26. Theuer, C.P., Bongard, F.S., and Klein, S.R. 1991. Are blood cultures effective in the evaluation of fever in perioperative patients? Am J. Surg. 162:615.
27. Trenholme, G.M., Kaplan, R.L, Karakusis, P.H., et al. 1989. Clinical impact of rapid identification and susceptibility testing of bacterial blood culture isolates. J. Clin. Microbiol. 27:1342.
28. von Reyn, C.F., Hennigan, S., Niemczyk, S., and Jacobs, N.J. 1991. Effect of delays in processing on the survival of *Mycobacterium avium-M. intracellulare* in the Isolator blood culture system. J. Clin. Microbiol. 29:1211.
29. Weinstein, M.P., Mirrett, S., Wilson, M.L., et al. 1991. Controlled evaluation of BACTEC Plus 26 and Roche Septi-Chek aerobic blood culture bottles. J. Clin. Microbiol. 29:879.

30. Weinstein, M.P., Reller, L.B., Murphy, J.R., et al. 1983. The clinical significance of positive blood cultures: a comprehensive analysis of 500 episodes of bacteremia and fungemia in adults. I. Laboratory and epidemiologic observations. Rev. Infect. Dis. 5:35.
31. Yagupsky, P. and Nolte, F.S. 1990. Quantitative aspects of septicemia. Clin. Microbiol. Rev. 3:269.

BIBLIOGRAPHY

Balows, A., and Sonnenwirth, A.C. 1983. Bacteremia: laboratory and clinical aspects. Charles C Thomas, Springfield, Ill.

Balows, A., and Tilton, R.C., editors. 1983. Proceedings of a symposium. Body fluids and infectious diseases: clinical and microbiologic advances. Am. J. Med. 75:1.

Farrar, W.E. 1990. Leptospira species (leptospirosis). p. 1813. In Mandell, G.L., Douglas, R.G., Jr., and Bennett, J.E., editors. Principles and practice of infectious diseases, ed 3. Churchill Livingstone, New York.

Reller, L. B., Murray, P.R., and MacLowry, J.D. 1982. Cumitech 1A. Blood cultures II, American Society for Microbiology, Washington, D.C.

Scheld, W.M., and Sande, M.E. 1990. Endocarditis and intravascular infections. p. 670. In Mandell, G.L., Douglas, R.G., Jr., and Bennett, J.E., editors. Principles and practice of infectious diseases, ed 3. Churchill Livingstone, New York.

16

MICROORGANISMS ENCOUNTERED IN THE CEREBROSPINAL FLUID

ANATOMY OF THE MENINGES AND CHOROID PLEXUS; CEREBROSPINAL FLUID

Diagnosis of an infection involving the central nervous system (CNS) is of critical importance. Most clinicians consider infection here one of the medical emergencies relating to infectious diseases. Therefore an understanding of the basic physiology of the CNS is helpful. The cerebrospinal fluid (CSF), which envelops the brain and spinal cord, has several functions. It cushions the bulk of the brain and provides buoyancy, reducing the effective weight of the brain by a factor of 30. The CSF, through its circulation around the entire brain, ventricles, and spinal cord, carries essential metabolites into the neural tissue and cleanses the organs of wastes. Every 3 to 4 hours, the entire volume of CSF is exchanged. The CSF is produced by the choroid plexus, specialized secretory cells located centrally within the brain in the third and fourth ventricles (fluid-filled cavities of the brain). The fluid travels around the outside areas of the brain within the subarachnoid space, driven primarily by the pressure produced initially at the choroid plexus. The subarachnoid space is the intervening space between the pia mater, the tissue layer immediately covering the brain, and the arachnoid, a membrane loosely covering the brain and part of the spinal cord. At-tachment between these two membranes is mediated by a thin network of fibers. The membranes surrounding the brain, the pia mater and the arachnoid, are collectively called the leptomeninges. All of the membranes surrounding the brain, including the dura, which covers the leptomeninges, are called the meninges. The portion of the arachnoid that covers the top of the brain contains special structures, arachnoid villi, that absorb the spinal fluid and allow it to pass into the blood.[6]

Infection within the subarachnoid space or throughout the leptomeninges is called **meningitis.** With bacterial meningitis, cerebrospinal fluid usually contains large numbers of inflammatory cells (greater than 1000/mm³), primarily polymorphonuclear neutrophils, shows a decreased glucose level relative to the serum glucose level (the normal ratio of CSF to serum glucose is approximately 0.6), and shows an increased protein concentration (normal protein is 15 to 50 mg/dl in adults and as high as 170 mg/dl, with an average of 90 mg/dl, in newborns). Most infectious agents reach the leptomeninges via hematogenous spread, entering the subarachnoid space through the choroid plexus or through other blood vessels of the brain. Meningitis can be either acute, often caused by an encapsulated bacterial species, or chronic, caused by *Mycobacterium* *tuberculosis,* other bacteria, or fungi such as *Cryptococcus neoformans.* Some of the more common causes of chronic meningitis are listed in the box on p. 212. Inflammation of the brain parenchyma, called **encephalitis,** is usually a result of viral infection. Concomitant inflammation of the meninges often occurs with this condition, which is called **meningoencephalitis,** although the cellular infiltrate is more likely to be lymphocytic in this situation. Early in the course of viral encephalitis, or when much tissue damage occurs as a part of encephalitis, the nature of the inflammatory cells found in the cerebrospinal fluid may be no different from that associated with bacterial meningitis; cell counts are typically much lower. In the past, patients who showed symptoms of meningitis and had cerebrospinal fluid samples that failed to grow a bacterial agent in culture were said to be suffering from "aseptic meningitis." The other possibility to consider in the case of a patient who previously received antibiotics, of course, is that of partially treated bacterial meningitis. With the advent of improved technology and availability of viral and other special techniques, many instances of "aseptic meningitis" can now be assigned a viral, leptospiral, or other etiology. Aseptic meningitis may also be due to tumor, cysts, chemicals, sarcoidosis, or other noninfectious causes.

PATHOGENESIS AND EPIDEMIOLOGY OF MENINGEAL INFECTIONS

Most cases of meningitis caused by bacteria, or encephalitis caused by viruses, share a similar pathogenesis. The etiological agent impinges on the mucous membranes of the nasopharynx and oropharynx, attaches with the aid of an adherence mechanism (discussed in Chapter 17), and multiplies at the initial site. The organism then invades or is carried across tissue into the lymphatics or bloodstream, where it disseminates throughout the system during a bacteremic or viremic phase. The microorganism enters the CSF through loss of capillary integrity, or through some other mechanism, and multiplies within the CSF, a site initially free of antimicrobial antibodies or phagocytic cells. Organisms that can adhere to mucosal epithelium and evade destruction by possessing an antiphagocytic capsule are the best pathogens. In patients with impaired host defenses, other etiological agents can gain a foothold and cause disease. Patients with prosthetic devices, particularly central nervous system shunts, are at increased risk of developing meningitis

caused by less "virulent" species, such as organisms that make up the normal skin flora like aerobic and anaerobic diphtheroids and *Staphylococcus epidermidis.* Staphylococci may possess specialized virulence factors that allow them to colonize artificial surfaces. The "slime" or noncapsular polysaccharide produced by some staphylococci seems to help these bacteria adhere to smooth surfaces. Viruses may enter the central nervous system through the bloodstream, but they may also travel along the nerves that lead to the brain (e.g., herpes simplex virus, adenovirus). Infections adjacent to the brain, such as sinusitis and subdural abscess, may also lead to meningitis or may simulate it (parameningeal focus). Patients with meningitis or encephalitis may have a fever, stiff neck, headache, nausea and vomiting, neurological abnormalities, and change in mental status.

Which etiological agents are involved as the cause of any occurrence of bacterial meningitis depends on several underlying host factors, but age of the patient is most prominent. Ninety-five percent of all cases in the United States occur in children less than 5 years old. In children in the United States between the ages of 1 month and 6 years, *Haemophilus influenzae* type b is the most common agent. However, the recent widespread use of a conjugate vaccine in children as young as 2 months has already had significant impact on the incidence of invasive *H. influenzae* type b disease. With continued use of these vaccines, meningitis caused by this organism will become less common. From age 6, patients are more likely to develop meningococcal or pneumococcal meningitis. Ninety-five percent of all cases are due to *H. influenzae* type b, *Neisseria meningitidis,* and *Streptococcus pneumoniae.*

The age-group with the highest prevalence of meningitis is that of the newborn, with a concomitant increased mortality rate (as high as 20%). Organisms causing disease in the newborn are different from those that affect other age-groups; many of them are acquired by the newborn during passage through the mother's vaginal vault. Neonates are likely to be infected with, in order of incidence, group B streptococci, *Escherichia coli,* other gram-negative bacilli, *Listeria monocytogenes,* and other organisms. The prevalence of these organisms is probably due to the immature immune system of neonates, the organisms present in the colonized female vaginal tract, as well as the increased permeability of the blood-brain barrier of newborns. *Flavobacterium meningosepticum* has been associated with nursery outbreaks of meningitis. The organism is presumably acquired nosocomially, since it is a normal inhabitant of water in the environment. Isolation of *Staphylococcus aureus* or other unusual organism from the CSF of a child should alert the clinician to look for congenital anatomical defects that might account for the entry of such an organism. Lack of demonstrable humoral antibody against *H. influenzae* type b in children has been associated with increased incidence of meningitis. Before widespread vaccination, most children developed measurable antibody by age 5. The importance of antibody is also a factor in adults, since military recruits without antibody to *N. meningitidis* are more likely to develop disease. *N. meningitidis* has been associated with epidemic meningitis among young adults in crowded conditions (e.g., military recruits and college dormitory mates). Meningococcal meningitis is endemic in certain areas of central Africa and Nepal. The *Morbidity and Mortality Weekly Report,* published by the Centers for Disease Control, provides physicians with a guide to recommendations for precautions to be taken by travelers.

Factors that predispose adults to meningitis are often the same factors that increase the likelihood that the adult will develop pneumonia or other respiratory tract colonization or infection, since the respiratory tract is the primary portal of entry for many etiological agents of meningitis. Alcoholism, splenectomy, diabetes mellitus, prosthetic devices, and immunosuppression contribute to increased risk. Important causes of meningitis in the adult, in addition to the meningococcus in young adults, include pneumococci, *Listeria monocytogenes,* and, less commonly, *Staphylococcus aureus* and various gram-negative bacilli. The latter organisms lead to meningitis via hematogenous seeding from various sources, including urinary tract infections.

The sequelae of acute bacterial meningitis in children are frequent and serious. Seizures occur in 20% to 30% of patients seen at large urban hospitals, and other neurological changes are common. Acute sequelae include cerebral edema, hydrocephalus, cerebral herniation, and focal neurological changes. Permanent sensorineural deafness can occur in 10% of children who recover from bacterial meningitis. It has also been reported that more subtle physiological and psychological sequelae follow an episode of acute bacterial meningitis. Even though the mortality from such infections has dropped from more than 90% in the preantibiotic era to between 3% and 30% today, the morbidity associated with meningitis today is still significant.[12] The *Haemophilus in-*

Etiological Agents of Chronic Meningitis

Mycobacterium tuberculosis
Cryptococcus neoformans
Coccidioides immitis
Histoplasma capsulatum
Blastomyces dermatitidis
Candida species
Miscellaneous other fungi
Nocardia
Actinomyces
Treponema pallidum
Brucella
Salmonella
Rare parasites—*Toxoplasma gondii,* cysticercus, *Paragonimus westermani, Trichinella spiralis*

fluenzae type b conjugate vaccine is expected to have a major role in reducing post-meningitis sequelae.

Chronic meningitis often occurs in those patients who are immunocompromised in some way, although this is not always the case. The various etiological agents of chronic meningitis are listed in the accompanying box.

Patients experience an insidious onset of disease, with some or all of the following: fever, headache, stiff neck, nausea and vomiting, lethargy, confusion, and mental deterioration. Symptoms may persist for a month or longer before treatment is sought. The CSF usually manifests an abnormal number of cells (usually lymphocytic), elevated protein, and some decrease in glucose content. The pathogenesis of chronic meningitis is similar to that of acute disease.[13]

Brain abscesses may occasionally cause changes in the CSF and clinical symptoms that mimic meningitis. Brain abscesses may also rupture into the subarachnoid space, producing a severe meningitis with high mortality. If anaerobic organisms or viridans streptococci are recovered from CSF cultures, the diagnosis of brain abscess must be entertained; however, CSF culture is typically negative in brain abscess. Immunosuppressed patients and diabetics with ketoacidosis may show a rapidly progressive fungal infection (phycomycosis) of the nasal sinuses or palatal region that travels directly to the brain.

Viral encephalitis, which cannot always be distinguished clinically from meningitis, is common in the warmer months. The primary agents are enteroviruses (coxsackieviruses A and B, echoviruses), mumps virus, herpes simplex virus, and arboviruses (togavirus, bunyavirus, equine encephalitis, St. Louis encephalitis, and other encephalitis viruses). Other viruses, such as measles, cytomegalovirus, lymphocytic choriomeningitis, Epstein-Barr virus, hepatitis, varicella-zoster virus, rabies virus, myxoviruses, and paramyxoviruses, are less commonly encountered. Any preceding viral illness and exposure history are important considerations in establishing a cause by clinical means.

Parasites can cause meningoencephalitis, brain abscess, or other central nervous system infection via two routes. The free-living amebae, *Naegleria fowleri* and *Acanthamoeba* species, invade the brain via direct extension from the nasal mucosa. These organisms are acquired by swimming or diving in natural, stagnating freshwater ponds and lakes. Other parasites reach the brain via hematogenous spread. Toxoplasmosis, caused by an agent that grows intracellularly and destroys brain parenchyma, is a common central nervous system affliction in patients with acquired immunodeficiency syndrome (AIDS; discussed further in Chapter 24). *Entamoeba histolytica* and *Strongyloides stercoralis* have been visualized in brain tissue, and the larval form of *Taenia solium* (the pork tapeworm), called a "cysticercus," can travel to the brain via the bloodstream and encyst in that site. Amebic brain infection and cysticercosis cause changes in the CSF that mimic meningitis.[13]

COLLECTION, TRANSPORT, AND INITIAL HANDLING OF SPECIMENS

CSF is collected by aseptically inserting a needle into the subarachnoid space, usually at the level of the lumbar spine. Three or four tubes of CSF should be collected and immediately labeled with the patient's name. Tube 3 or 4 is used for cell count and differential. If a small capillary blood vessel is inadvertently broken during the spinal tap (lumbar puncture), blood cells picked up from this source will usually be absent from the last tube collected; comparison of counts between tubes 1 and 4 is occasionally needed if such bleeding is suspected. The other tubes can be used for both microbiological and chemical studies. In this way, a larger proportion of the total fluid can be concentrated, facilitating detection of infectious agents present in low numbers, and the supernatant can still be used for other required studies. Microbiologists should periodically monitor the sterile tubes used in lumbar puncture kits prepared at their institutions by processing filter-sterilized negative CSF as a quality control check. CSF collecting tubes have been known to harbor non-

viable bacteria, which can cause false-positive Gram stain results.

The volume of CSF is critical for detection of certain microorganisms, such as mycobacteria and fungi. A minimum of 5 to 10 ml of CSF is recommended for analysis of CSF suspected of containing these pathogens by centrifugation and subsequent culture. When an inadequate volume of CSF is received, the physician should be consulted regarding the order of priority of laboratory studies.[9] Processing too little specimen lowers the sensitivity of the testing, which leads to false-negative results. This is more harmful to patient care than performing an additional lumbar puncture to obtain the needed amount of sample.

CSF should be hand-delivered immediately to the laboratory. Specimens should never be refrigerated. If they are not being rapidly processed, they should be incubated or left at room temperature. One exception to this rule involves CSF for viral studies. These specimens may be refrigerated for as long as 24 hours after collection or frozen at −70° C if a longer delay is anticipated until they are inoculated. CSF for viral studies should never be frozen at temperatures above −70° C. Certain agents, such as *Streptococcus pneumoniae,* may not be detectable after an hour or longer unless antigen detection methods are used. Prompt examination by microbiologists can often determine the etiological agent within as short a time as 30 minutes, helping the clinician to direct or correct therapy. CSF is one of the few specimens handled by the laboratory for which information promptly relayed to the clinician can directly affect therapeutic outcome. Such specimens should be processed immediately upon receipt in the laboratory, and all results should then be reported to the physician. Initial processing of CSF for bacterial, fungal, or parasitic studies includes centrifugation of all specimens greater than 1 ml in volume for at least 15 minutes at $1500 \times g$.[9] Specimens in which cryptococci or mycobacteria are suspected must be handled differently. Discussions of techniques for culturing CSF for fungi are found in Chapter 44, and methods for isolating mycobacteria are detailed in Chapter 42. The supernatant is removed to a sterile tube, leaving approximately 0.5 ml fluid in which to suspend the sediment before it is inoculated to media or examined visually. This mixing of the sediment is critical. Laboratories that use a sterile capillary pipette to remove portions of the sediment from underneath the supernatant will miss a significant number of positive specimens. The sediment must be thoroughly mixed after the supernatant has been removed. Forcefully aspirating the sediment up and down into a sterile pipette several times will adequately disperse the organisms that remained adherent to the bottom of the tube after centrifugation. The supernatant can be used to test for the presence of antigens or for chemistry evaluations (e.g., protein, glucose, lactate). Even if the supernatant has no immediate use, it should be kept as a safeguard. Later, supernatants from culture-negative CSF can be pooled, filter-sterilized, and used as a diluent for other tests or quality control. Communication between physician and microbiology laboratory is essential, since the results of hematological and chemical tests directly relate to the probability of infection. Among 555 cerebrospinal fluids from patients older than 4 months of age tested at the University of California-Los Angeles, only two showed normal cell count and protein in the presence of bacterial meningitis.[7] Thus the diagnosis of acute bacterial meningitis can be excluded in patients with normal fluid parameters in almost all cases, precluding further expensive and labor-intensive microbiological processing beyond a standard smear and culture (which must be included in all cases).

VISUAL DETECTION OF ETIOLOGICAL AGENTS OF MENINGITIS

DIRECT WET PREPARATION

Amebas are best observed by examining thoroughly mixed sediment as a wet preparation under phase-contrast microscopy (Chapter 8). If a phase-contrast microscope is not available, observing under light microscopy with the condenser closed slightly is an alternative technique. Amebas are recognized by typical slow, methodical movement in one direction by advancing pseudopodia. The organisms may require a little time under the warm light of the microscope before they begin to move. They must be distinguished from motile macrophages, which occasionally occur in CSF. Following a suspicious wet preparation, a trichrome stain (Chapter 45) can help differentiate amebas from somatic cells (Figure 16.1) The pathogenic amebas can be cultured on a lawn of *Klebsiella pneumoniae* or *Escherichia coli*, as outlined in Chapter 45.

INDIA INK STAIN
Cryptococcus neoformans, because of the large polysaccharide capsule, can often be visualized by the India ink stain, although latex agglutination testing for capsular antigen is more sensitive and extremely specific. Therefore it is recommended that antigen testing be used in place of an India ink stain if accurate diagnosis is

critical. Furthermore, strains of *C. neoformans* that infect patients with AIDS may not possess detectable capsules, so culture is also essential. To perform the India ink preparation, a drop of CSF sediment is mixed with one-third volume of India ink (Pelikan Drawing Ink, Block, Gunther, and Wagner; available at art supply stores). The India ink can be protected against contamination by adding 0.05 ml thimerosal (Merthiolate, Sigma Chemical Co.) to the bottle when it is first opened. After mixing the CSF and ink to make a smooth suspension, a coverslip is applied to the drop and the preparation is examined under high-power magnification (400×) for characteristic encapsulated yeast cells, which can be confirmed by examination under oil immersion. The inexperienced microscopist must be careful not to confuse white blood cells with yeast. The presence of encapsulated buds, smaller than the mother cell, is diagnostic.

QUELLUNG REACTION
A test rarely performed today is the Quellung reaction, in which spinal fluid sediment is mixed with antisera against the suspected organism (*H. influenzae* type b, *N. meningitidis*, or *S. pneumoniae*) and a drop of saturated aqueous methylene blue dye. If the organism is present, the antibody combining with the capsular polysaccharide will cause apparent capsular swelling, which can be visualized under the microscope. This test requires considerable experience and is frequently not reproducible between reference laboratories.

STAINED SMEAR OF SEDIMENT
The Gram stain must be performed on all CSF sediments. False-positive smears have resulted from inadvertent use of contaminated slides. Therefore use of alcohol dipped and flamed, or autoclaved slides (Figure 16.2) is recommended. After thoroughly mixing the sediment, a heaped drop is placed on the surface of a sterile or alcohol-wiped slide. The sediment should never be spread out on the slide surface, since this increases the difficulty of finding small numbers of microorganisms. The drop of sediment is allowed to air dry, is heat- or methanol-fixed, and is stained by either Gram or acridine orange stain (Chapter 8). The acridine orange fluorochrome stain may allow faster examination of the slide under high-power magnification (400×) and thus a more thorough examination. The brightly fluorescing bacteria will be easily visible. All suspicious smears can be restained by the Gram stain

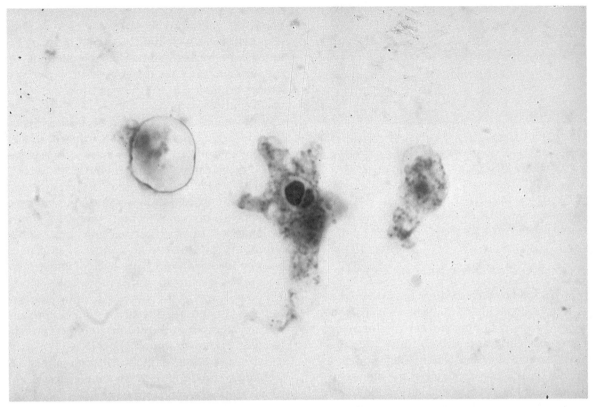

FIGURE 16.1

Trichrome-stained preparation of *Naegleria fowleri* isolated from cerebrospinal fluid and grown on a lawn of Enterobacteriaceae.

FIGURE 16.2

Stainless steel autoclavable glass microscope slide holder, suitable for autoclaving cleaned slides.

(directly over the acridine orange stain) to confirm morphology of any organisms seen. An alternative procedure using a cytospin centrifuge (Cytospin, Shandon-Southern) to prepare slides for staining has also been found very useful.[1,10] Use

of this method for preparing smears for staining concentrates microbes up to a thousandfold and offers a sensitivity for Gram staining approaching that of antigen detection methods (Figure 16.3). Based on demographic and clinical patient data and Gram stain morphology, the etiology of the majority of cases of bacterial meningitis can be presumptively determined within the first 30 minutes after receiving the specimen. Acid-fast smears are handled differently (Chapter 42).

SYPHILIS SEROLOGY

Diagnosis of neurosyphilis is based on several parameters, including the number and types of white and red cells present in the patient's CSF, patient history and epidemiological factors, levels of protein and glucose, and the VDRL test. The VDRL test is the only useful test for detecting antibodies directed against *Treponema pallidum* in CSF, although new technological developments promise better methods in the future. A positive CSF VDRL is still not diagnostic in itself, since patients with successfully treated neurosyphilis may retain a positive CSF VDRL even years later.

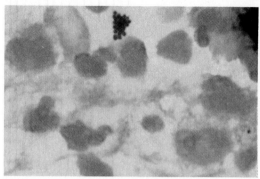

FIGURE 16.3

Gram-stained smear from CSF prepared by a cytocentrifuge. Note the numerous white cells and the excellent morphology of the *Staphylococcus aureus.*

DETECTION OF ANTIGEN IN CSF

Reagents and complete systems for the rapid detection of antigen in CSF have been available for several years. Countercurrent immunoelectrophoresis (CIE), although the first widely accepted method for rapid antigen detection from CSF, has been largely replaced by the more sensitive and simpler techniques of latex agglutination and coagglutination; however, CIE may be used in some laboratories for detection of antigen to *Listeria monocytogenes,* for which no latex tests are available. All of the commercial systems use the principle of an antibody-coated particle that will bind to specific antigen, resulting in macroscopically visible agglutination. The soluble capsular polysaccharide produced by the most common etiological agents of meningitis and the group B streptococcal polysaccharide are well suited to serve as bridging antigens. The systems differ in that certain antibodies are polyclonal and others are monoclonal, and not all systems detect all antigens. With the exception of latex tests for cryptococcal antigen, the use of these reagents is somewhat controversial, and they should be used as an adjunct to standard procedures.[4]

Reagents for the detection of the polysaccharide capsular antigen of *Cryptococcus neoformans* are available commercially. CSF specimens that yield positive results for cryptococcal antigen should be tested with a second latex agglutination test for rheumatoid factor. A positive rheumatoid factor test renders the cryptococcal latex test uninterpretable, and the results should be reported as such, unless the rheumatoid factor antibodies have been inactivated. The commercial test systems incorporate rheumatoid factor testing in their protocol. Undiluted specimens that contain large amounts of capsular antigen may yield a false-negative reaction caused by a prozone phe-

nomenon. Patients with AIDS may have an antigen titer in excess of 100,000, requiring many dilutions to reach an end point. Such dilutions are usually done since the test is used for following a patient's response to treatment, as well as for initial diagnosis.

In general, the commercial systems have been developed for use with CSF, urine, or serum, although results with serum have not been as useful diagnostically as those with CSF. Soluble antigens from *Streptococcus agalactiae* and *Haemophilus influenzae* may concentrate in the urine. Urine, however, seems to produce a higher incidence of nonspecific reactions than either serum or CSF. The manufacturers' directions must be followed for performance of antigen detection test systems for different specimen types. Although some of the systems require pretreatment of samples, usually heating for 5 minutes, not all manufacturers recommend such a step. The reagents, however, may yield false-positive or cross-reactions unless specimen pretreatment is performed. Interference by rheumatoid factor and other substances, more often present in body fluids other than CSF, has also been reported. The method of Smith and co-workers[11] has been shown to effectively reduce a substantial portion of nonspecific and false-positive reactions, at least for tests performed with latex particle reagents. This pretreatment, called rapid extraction of antigen procedure (REAP; Procedure 16.1), is recommended for those laboratories that use commercial body fluid antigen detection kits. Certain commercial systems have such an extraction procedure included in their protocols.

Using a direct antigen detection system, confirmation of identification of organisms seen on Gram stain, as well as the detection of an etiological agent in smear-negative specimens or those from previously treated patients, can be accomplished. The systems are not substitutes for properly performed smears and cultures because they provide less than 100% sensitivity and specificity, but they provide an additional diagnostic test to aid clinicians in initial management of serious meningitis.[2,5] The standard procedures and empirical therapeutic decisions of clinicians served by a laboratory will determine whether these types of tests are cost effective.[3,8]

CULTURE FOR ETIOLOGICAL AGENTS OF CSF INFECTION

After vortexing the sediment and preparing smears, several drops of the sediment should be

---

PROCEDURE 16.1

RAPID EXTRACTION OF ANTIGEN PROCEDURE (REAP)

PRINCIPLE

Removal of nonspecific cross-reactive material can improve the specificity of direct antigen detection particle agglutination tests. Ethylene-diaminetetraacetic acid (EDTA) complexes cross-reactive materials, and they are removed from the reaction mixture by centrifugation.

METHOD

1. Pipette 0.05 ml fluid to be tested (CSF, serum, or urine) into a 1.5 ml plastic conical microcentrifuge tube.

2. Add 0.15 ml of 0.1 M EDTA (Sigma Chemical) to the microcentrifuge tube, close the cap tightly, and vortex the tube.

3. Heat in a dry bath (available from instrument supply companies) for 3 minutes at 100° C.

4. Centrifuge the tubes for 5 minutes at $13,000 \times g$ in a tabletop microcentrifuge. Be certain that the instrument used achieves the required centrifugal force.

5. Remove the supernatant with a capillary pipette, and use one drop of this solution as the test sample in the antigen detection test, following the manufacturer's instructions for performance of the test.

QUALITY CONTROL

Test the reagents using filter-sterilized known-negative CSF.

EXPECTED RESULTS

Agglutination should not occur.

PERFORMANCE SCHEDULE

Test reagents once when each new batch is prepared and after every positive test result.

Modified from Smith, L.P., Hunter, K.W., Jr., Hemming, V.G., et al. 1984. J. Clin. Microbiol. 20:981.

---

inoculated to each medium. Routine bacteriological media should include a chocolate agar plate, 5% sheep blood agar plate, and an enrichment broth, usually thioglycolate without indicator. The plates should be incubated at 37° C in 5% to 10% CO_2 for at least 72 hours. If a CO_2 incubator is not available, a candle jar (Chapter 9) can be used. The broth should be incubated in air at 37° C for at least 5 days. The broth cap must be loose to allow free exchange of air. If organisms morphologically resembling anaerobic bacteria are seen on the Gram stain or if a brain abscess is suspected, an anaerobic blood agar plate may also be inoculated. If the Gram stain revealed large, regular gram-negative rods, a MacConkey plate should be added to the initial media. These media will support the growth of almost all bacterial pathogens and several fungi. The authors of Cumitech 14 recommend that penicillinase (Difco Laboratories or BBL Microbiology Systems) be added to culture media if patients have received prior antibiotic therapy.[9]

The symptoms of chronic meningitis that prompt a physician to request fungal cultures are the same as those for tuberculous meningitis, which should always be sought by the laboratory if chronic or fungal meningitis is suspected. Two drops of the well-mixed sediment should be inoculated to a non–blood-containing medium (brain-heart infusion with gentamicin and chloramphenicol) plate or slant and a brain-heart infusion or inhibitory mold agar slant or plate. Addition of a third medium such as Mycosel or Mycobiotic (selective agar, Chapter 9) is recommended. The material can be deposited on the agar surface without spreading for isolation. Fungal media should be incubated in air at 30° C for 4 weeks. If possible, it is best to inoculate two sets of media and incubate one set at 30° C and one set at 35° C. Centrifugation may not effectively sediment mycobacteria (Chapter 42); other methods are suggested. Cultures positive for fungi or mycobacteria are evaluated in a biological safety cabinet.

Free-living amebae can be co-cultivated on artificial media if they are supplied with a living nutrient, such as *Klebsiella pneumoniae* or *Escherichia coli*. This procedure should be tried if amebic meningoencephalitis is suspected (Chapter 45).

CSF may be inoculated directly to tissue culture for detection of viral agents. Diagnosis of viral encephalitides is often accomplished by isolation of the virus from a throat culture, feces, or blood, so these specimens should be submitted in addition to CSF for identification of an etiological agent. Methods for cultivation of viruses from such specimens are discussed in detail in Chapter

43. Briefly, 0.25 ml of CSF per tube is inoculated into tubes of primary monkey kidney, Hep-2 continuous cell line, and human fetal diploid foreskin fibroblast cell cultures. These culture systems are incubated at 35° C in air for varying amounts of time, depending on the cell culture system.

OTHER DIAGNOSTIC TECHNIQUES

Certain agents of encephalitis or meningitis can be detected only by histopathological examination of brain or meningeal biopsy material. Direct smears (touch preparations) can be stained with fluorescein or enzyme-conjugated antibodies for rapid detection of herpesviruses, varicella-zoster, rabies, and *Toxoplasma gondii*. Conventionally stained smears can be used to visualize *T. gondii* and other parasites, as well as the characteristic inclusions of cytomegalovirus and herpesvirus. These diagnostic procedures are usually performed by anatomical pathologists rather than clinical microbiologists.

Other new approaches are being explored for diagnosis of meningitis. ELISA procedures are being manufactured for detection of *H. influenzae, S. pneumoniae,* and *N. meningitidis* antigens. An ELISA system has been shown to reliably detect antigens of *M. tuberculosis* in CSF. Because of the slow growth of this organism and the lack of any rapid procedures for its detection, such tests are highly desirable. New molecular techniques such as PCR can revolutionize the diagnosis of these slow growing pathogens. Commercial PCR-based systems are currently available. Modifications of gas-liquid chromatography are under active investigation. Mass spectrometry has also been evaluated for rapid detection of etiological agents of disease in CSF. Although reports are promising, the future role of these methods remains to be defined.

REFERENCES

1. Chapin-Robertson, K., Dahlberg, S.E., Edberg, S.C. 1992. Clinical and laboratory analyses of cytospin-prepared Gram stains for recovery and diagnosis of bacteria from sterile body fluids. J. Clin. Microbiol. 30:377.
2. College of American Pathologists. 1987. Rapid microbial detection survey, Set 1D-B. Final critique. The College, Chicago.
3. Dagbjartsson, A., and Ludvigsson, P. 1987. Bacterial meningitis: diagnosis and initial antibiotic therapy. Pediatr. Clin. North Am. 34:219.
4. Gerber, M.A. 1985. Critical appraisal of the clinical relevance of rapid diagnosis in pediatrics. Diagn. Microbiol. Infect. Dis. 4:39S.
5. Granoff, D.M., Murphy, T.V., Ingram, D.L., et al. 1986. Use of rapidly generated results in patient management. Diagn. Microbiol. Infect. Dis. 4:157S.
6. Greenlea, J.E. 1985. Anatomic considerations in central nervous system infections. In Mandell, G.L., Douglas, R.G., Jr., and Bennett, J.E., editors. Principles and practice of infectious diseases, ed 2. John Wiley & Sons, New York.
7. Hayward, R.A., Shapiro, M.F., and Oye, R.K. 1987. Laboratory testing on cerebrospinal fluid, a reappraisal. Lancet 1:1.
8. McCracken, G.H., Nelson, J.D., Kaplan, S.L., et al. 1987. Consensus report: Antimicrobial therapy for bacterial meningitis in infants and children. Pediatr. Infect. Dis. 6:501.
9. Ray, C.G., Wasilauskas, B.L. and Zabransky, R.J. 1982. Laboratory diagnosis of central nervous system infections. In McCarthy, L.R., editor. Cumitech 14, American Society for Microbiology, Washington, D.C.
10. Shanholtzer, C.J., Schaper, P.J., and Peterson, L.R. 1982. Concentrated Gram stain smears prepared with a cytospin centrifuge. J. Clin. Microbiol. 16:1052.
11. Smith, L.P., Hunter, K.W., Jr., Hemming, V.G., et al. 1984. Improved detection of bacterial antigens by latex agglutination after rapid extraction from body fluids. J. Clin. Microbiol. 20:981.
12. Swartz, M.N. 1984. Bacterial meningitis: more involved than just the meninges. N. Engl. J. Med. 311:912.
13. Wilhelm, C., and Ellner, J.J. 1986. Chronic meningitis. Neurol. Clin. 4:115.

BIBLIOGRAPHY

Conly, J.M., and Ronald, A.R. 1983. Cerebrospinal fluid as a diagnostic body fluid. Am. J. Med. 75:102.

Force, R.W., Lugo, R.A., and Nahata, M.C. 1992. *Haemophilus influenzae* type B conjugate vaccines. Ann. Pharmacother. 26:1429.

McGee, Z., and Baringer, J.R. 1990. Acute meningitis. pp. 741-755. In Mandell, G.L., Douglas, R.G., Jr., and Bennett, J.E., editors. Principles and practice of infectious diseases, ed 3. Churchill Livingstone, New York.

Neu, H.C. 1985. CNS infection: first things first. Hosp. Pract. 20:69.

17

MICROORGANISMS ENCOUNTERED IN THE RESPIRATORY TRACT

GENERAL CONSIDERATIONS, ANATOMY, AND NORMAL STATE OF RESPIRATORY TRACT

One of two major connections between the interior of the body and the outside environment (along with the gastrointestinal tract) is the respiratory tract, through which the body acquires fresh oxygen and removes unneeded CO_2. The respiratory tract begins with the nasal or oral passages, which serve to humidify inspired air, and extends past the nasopharynx and oropharynx to the trachea and then into the lungs. The trachea divides into bronchi, which subdivide into bronchioles, the smallest branches of which terminate in the alveoli. Several mechanisms nonspecifically

protect the respiratory tract from infection: the nasal hairs and convoluted passages and mucous lining of the nasal turbinates; secretory IgA and nonspecific antibacterial substances (lysozyme) in respiratory secretions; the cilia and mucous lining of the trachea; and reflexes such as coughing, sneezing, and swallowing. These mechanisms prevent foreign objects or organisms from entering the bronchi and gaining access to the lungs, which remain sterile in the healthy host. Once particles that have escaped the mucociliary sweeping activity enter the alveoli, alveolar macrophages ingest them and carry them to the lymphatics. In addition, normal flora of the nasopharynx and oropharynx help to prevent colonization of the upper respiratory tract by pathogenic microorganisms.

FLORA OF RESPIRATORY TRACT

Those bacteria that can be isolated as part of the indigenous flora of healthy hosts are listed in the box at right, as well as many species that may cause disease under certain circumstances but that are often isolated from the respiratory tracts of healthy persons. Under certain circumstances, for unknown reasons—perhaps because of previous damage by a viral infection, loss of some host immunity, or physical damage to the respiratory epithelium (e.g., from smoking)—these colonizing organisms go on to cause disease. Organisms isolated from normally sterile sites in the respiratory tract by methods that avoid contamination with normal flora should be definitively identified and reported to the clinician.

Certain microorganisms are considered to be etiological agents of disease if they are present in any numbers in the respiratory tract because they possess virulence factors that are expressed in every host. These are listed in the box on p. 222. Many selective media have been designed to isolate these organisms, even in the presence of normal flora. This chapter discusses such methods, as well as strategies to follow for handling of specimens received from the respiratory tract.

PATHOGENIC MECHANISMS USED BY AGENTS OF RESPIRATORY TRACT INFECTIONS

Microorganisms primarily cause disease by a limited number of pathogenic mechanisms. Since these mechanisms relate to respiratory tract infections, they will be discussed briefly. For any organism to cause disease, it must first be able to gain a foothold within the respiratory tract in order to grow to sufficient numbers to produce symptoms. Therefore most etiological agents of respiratory tract disease must first adhere to the mucosa of the respiratory tract. The presence of normal flora and the overall state of the host affect the ability of microorganisms to adhere. Surviving or growing on host tissue without causing overt harmful effects is called **colonization.** Except for those microorganisms that are breathed directly into the lungs, all etiological agents of disease must first colonize the respiratory tract to some degree before they can cause harm.

Bacteria that possess specific adherence fac-

Organisms Commonly Present in Nasopharynx and Oropharynx of Healthy Humans

Rarely pathogens
  Nonhemolytic streptococci
  Staphylococci
  Micrococci
  Corynebacterium species
  Coagulase-negative staphylococci
  Neisseria species, other than N. *gonorrhoeae* and N. *meningitidis*
  Lactobacillus species
  Veillonella species
  Spirochetes

Possible pathogens
  Acinetobacter species
  Viridans streptococci
  β-Hemolytic streptococci
  Streptococcus pneumoniae
  Staphylococcus aureus
  Corynebacterium diphtheriae
  Neisseria meningitidis
  Cryptococcus neoformans
  Mycoplasma species
  Haemophilus influenzae
  Haemophilus parainfluenzae
  Branhamella (Moraxella) catarrhalis
  Candida albicans
  Herpes simplex virus
  Enterobacteriaceae
  Mycobacterium species
  Pseudomonas species
  Filamentous fungi
  Klebsiella ozaenae
  Eikenella corrodens
  Bacteroides species
  Peptostreptococcus species
  Actinomyces species

---

tors include *Streptococcus pyogenes*, whose gram-positive cell wall contains lipoteichoic acids and certain proteins (M protein and others), visible as a thin layer of fuzz surrounding the bacteria. Other bacteria that possess lipoteichoic acid adherence complexes are *Staphylococcus aureus* and certain viridans streptococci. Many gram-negative bacteria (which do not have lipoteichoic acids), including Enterobacteriaceae, *Legionella* species, *Pseudomonas* species, *Bordetella pertussis*, and *Haemophilus* species, adhere by means of proteinaceous fingerlike surface structures called **fimbriae** (Figure 17.1).

Fimbriae are often also called **pili,** although this is technically the term for a similar structure that participates in sexual interaction rather than just adherence. Viruses possess either a hemagglutinin (influenza and parainfluenza viruses) or

FIGURE 17.1

Fimbriae of *Escherichia coli,* extending as thin projections from the cell periphery. (Courtesy A. Ryter, Institut Pasteur. From Boyd, R.F. 1984. General microbiology. Mosby, St. Louis, Mo.)

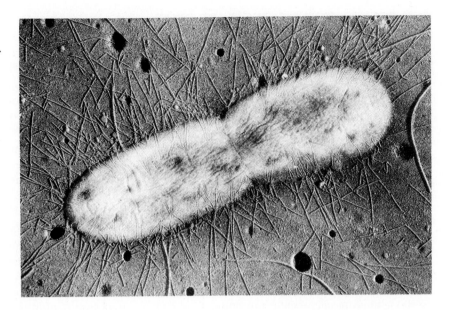

other proteins that mediate their epithelial attachment.

The production of extracellular toxin was one of the first pathogenic mechanisms discovered among bacteria. *Corynebacterium diphtheriae* is a classic example of a bacterium that produces disease through the action of an extracellular toxin. Once the organism colonizes the upper respiratory epithelium, it elaborates a toxin that is disseminated systemically, adhering preferentially to central nervous system cells and muscle cells of the heart. Systemic disease characterized by myocarditis as well as peripheral neuritis and local disease leading to respiratory distress can follow. Growth of *C. diphtheriae* causes necrosis and sloughing of the epithelial mucosa, producing a "diphtheritic (pseudo) membrane," which may extend from the anterior nasal mucosa to the bronchi, or may be limited to any area between—most often the tonsillar and peritonsillar areas. The membrane may cause sore throat and interfere with respiration and swallowing. Although nontoxic strains of *C. diphtheriae* can cause local disease, it is much milder than the version mediated by toxin. The toxin gene is carried into the bacteria by phage (Chapter 34), therefore not all strains are toxigenic.

Some strains of *Pseudomonas aeruginosa* produce a toxin very similar to diphtheria toxin. Whether this toxin actually contributes to the pathogenesis of respiratory tract infection with *P. aeruginosa* has not been established. *Bordetella pertussis,* the agent of whooping cough, also produces toxins. The role of these toxins in production of disease is not clear. They may act to inhibit the activity of phagocytic cells or to damage cells of the respiratory tract. *Staphylococcus aureus* and β-hemolytic streptococci produce extracellular enzymes that act to damage host cells or tissues. Extracellular products of staphylococci aid in production of tissue necrosis and destruction of phagocytic cells, contributing to the commonly seen phenomenon of abscess formation associated with infection caused by this organism. Although *S. aureus* can be recovered from throat specimens, it has not been proven to cause infection there (as pharyngitis). Enzymes of streptococci, including hyaluronidase, allow rapid dissemination of the bacteria. Many other etiological agents of respiratory tract infection also produce extracellular enzymes and toxins.

In addition to adherence and toxin production, pathogens cause disease by merely growing in host tissue, interfering with normal tissue function and attracting host immune effectors, such as neutrophils and macrophages. Once these cells begin to attack the invading pathogens and repair the damaged host tissue, an expanding reaction ensues with more nonspecific and immunological factors being attracted to the area, increasing the amount of host tissue damage. Respiratory viral infections usually progress in this manner, as do many types of pneumonias, such as those caused by *Streptococcus pneumoniae, S. pyogenes, Staphylococcus aureus, Haemophilus influenzae, Neisseria meningitidis, Moraxella catarrhalis, Mycoplasma pneumoniae, Mycobacterium tuberculosis,* and most gramnegative bacilli. Aspiration of minor amounts of oropharyngeal material, as occurs often during sleep, plays an important role in the pathogenesis of many types of pneumonia.

The ability to evade host defense mechanisms

Respiratory Tract Pathogens

Definite respiratory tract pathogens
Corynebacterium diphtheriae (toxin-producing)
Neisseria gonorrhoeae
Mycobacterium tuberculosis
Mycoplasma pneumoniae
Chlamydia trachomatis
Chlamydia (TWAR) *pneumoniae*
Bordetella pertussis
Legionella species
Pneumocystis carinii
Nocardia species
Histoplasma capsulatum
Coccidioides immitis
Cryptococcus neoformans (may also be recovered from
   patients without disease)
Blastomyces dermatitidis
Viruses (respiratory syncytial virus (RSV),
   adenoviruses, enteroviruses, herpes simplex virus
   (HSV), influenza and parainfluenza virus,
   rhinoviruses

Rare respiratory tract pathogens
Francisella tularensis
Bacillus anthracis
Yersinia pestis
Pseudomonas pseudomallei
Coxiella burnetti
Chlamydia psittaci
Brucella species
Salmonella species
Pasteurella multocida
Klebsiella rhinoscleromatis
Varicella-zoster virus (VZV)
Parasites

is another virulence mechanism that certain respiratory tract pathogens possess. *S. pneumoniae*, *N. meningitidis*, *H. influenzae*, *Klebsiella pneumoniae*, mucoid *P. aeruginosa*, *Cryptococcus neoformans*, and others possess polysaccharide capsules that serve both to prevent engulfment by phagocytic host cells and to protect somatic antigens from being exposed to host immunoglobulins. The capsular material is produced in such abundance by certain bacteria, such as pneumococci, that soluble polysaccharide antigen particles can bind host antibodies, blocking them from serving as opsonins. Proof that the capsular polysaccharide is a major virulence mechanism of *H. influenzae*, *S. pneumoniae*, and *N. meningitidis* was established when vaccines consisting of capsular antigens alone were shown to protect individuals from disease.

Some respiratory pathogens evade the host immune system by multiplying within host cells. *Chlamydia trachomatis*, *C. psittaci*, and all viruses replicate within host cells. They have evolved methods for being taken in by the "nonprofessional" phagocytic cells of the host to achieve their required environment, which is protected from host humoral immune factors and other phagocytic cells until the host cell becomes sufficiently damaged that it is recognized as foreign and attacked. A second group of agents of respiratory tract disease is actively taken up by phagocytic host cells, usually macrophages, within which they are able to multiply. *Legionella, Listeria,* some *Salmonella* species, *Pneumocystis carinii,* and *Histoplasma capsulatum* are some of these more common intracellular parasites.

Mycobacterium tuberculosis is the classic representative of an intracellular pathogen. In primary tuberculosis the organism is carried to an alveolus in a droplet nucleus, a tiny aerosol particle containing a few tubercle bacilli (the minimum infective dose is quite small). It is phagocytized there by alveolar macrophages, which carry it to the nearest lymph node, usually in the hilar or other mediastinal chains. In the node the organisms slowly multiply within macrophages, destroying them and being taken up by other phagocytic cells in turn. Within the protected environment of the macrophages, which are somehow prevented from accomplishing lysosomal fusion by the bacteria, tubercle bacilli multiply to a critical mass, which spills out of the destroyed macrophages through the lymphatics into the bloodstream, producing mycobacteremia and carrying tubercle bacilli to many parts of the body. In most cases the host immune system reacts sufficiently at this point to kill the bacilli; however, a small reservoir of live bacteria may be left in areas of normally high oxygen concentration, such as the apical portion of the lung. These bacilli are walled off, and years later an insult to the host, either immunological or physical, may cause breakdown of the focus of latent tubercle bacilli, allowing active multiplication and disease (secondary tuberculosis). In certain patients with primary immune defects of some type, the initial bacteremia seeds bacteria throughout a host that is unable to control them, leading to disseminated or **miliary** tuberculosis. Growth of the bacteria within host macrophages and histiocytes in the lung causes an influx of more effector cells, including lymphocytes, neutrophils, and histiocytes, eventually resulting in granuloma formation, then tissue destruction and cavity formation. The lesion is characteristically a semisolid, amorphous tissue mass resembling semisoft cheese, from which it receives the name **caseation necrosis.** The infected tissue, or organisms from it, can extend into bronchioles and bronchi, from which bacteria are disseminated via respiratory secretions by coughing, to be inhaled by the next victim, and also to other portions of the patient's own lungs through aspiration.

UPPER RESPIRATORY TRACT INFECTIONS

Microorganisms that cause disease in the upper respiratory tract, the nasopharynx, and the pharynx must first come into contact with the mucosal epithelium, adhere to the epithelium, and multiply locally. Organisms such as *N. meningitidis, B. pertussis, C. diphtheriae,* and many viruses are carried in with small droplets as aerosols. They may be passed in secretions from other infected persons, either by touching intimate objects, or from person to person through crowded conditions or poor hygiene. *Mycoplasma* species and *Streptococcus pyogenes* are often passed in this way. Although a large number of different bacteria and fungi may be isolated from specimens taken from the upper respiratory tract, very few organisms have been shown to cause true disease here.

ETIOLOGICAL AGENTS OF NASOPHARYNGEAL AND OROPHARYNGEAL INFECTIONS

The most commonly sought bacterial pathogen is *S. pyogenes,* or the group A β-hemolytic streptococcus, the agent that leads to poststreptococcal (immunologically mediated) sequelae such as acute rheumatic fever and glomerulonephritis. The organism not uncommonly causes pyogenic infections of the tonsils (suppurations), sinuses, and middle ear, or cellulitis as secondary pyogenic sequelae after an episode of pharyngitis. Accordingly, streptococcal pharyngitis is usually treated to prevent both the suppurative and nonsuppurative sequelae, as well as to decrease morbidity. Cases of pharyngitis caused by groups B, C, and G and by nonhemolytic members of these groups of streptococci (including group A) have been reported. Although bacteria other than group A streptococci may cause pharyngitis, this occurs less often, and laboratories should not routinely look for other agents except in special circumstances, and after discussion with the clinician involved.

Although much less common than streptococcal pharyngitis, *C. diphtheriae* can still be isolated from patients with sore throat, as well as more serious systemic disease. Other *Corynebacterium species* (*C. ulcerans* and *C. pyogenes*) can cause pharyngitis with or without membrane formation or associated rash, but such infections are rare. *Ar-canobacterium haemolyticum* (formerly *Corynebacterium*) may cause as many as 10% of pharyngotonsillitis in patients 10 to 30 years old.[18a] *N. meningitidis,* too, can be recovered from the nasopharynx of carriers, although finding it here is not significant unless the patient has recently been exposed to a patient with *N. meningitidis* meningitis. *Neisseria gonorrhoeae* can cause an exudative pharyngitis that is especially prevalent among men who engage in oral-genital sex. In disseminated gonococcal disease, the organism may be isolated from throat cultures as well. Gonococcal pharyngitis is not always symptomatic. Although *H. influenzae, S. aureus,* and *S. pneumoniae* are frequently isolated from nasopharyngeal and throat cultures, they have not been shown to cause pharyngitis. Carriage of any of these organisms, as well as *N. meningitidis,* may have clinical importance for some patients or their contacts. In patients with cystic fibrosis, throat cultures may yield etiological agents of pneumonia that cannot be isolated from sputum because of overgrowth by other bacteria. However, extrapolating results of throat cultures to implicate specific organisms as pathogens for pneumonia is only a "best guess," and close communication between clinical microbiologists and physicians must ensue to clarify the relevance of such findings. Cultures of specimens obtained from the anterior nares often yield *S. aureus.* The carriage rate for this organism is especially high among health care workers, and even 20% of the general population can be colonized with this microbe. *Mycoplasma pneumoniae* may, rarely, cause pharyngitis and can be isolated from nasopharyngeal swabs. Such a swab may also be the most easily collected specimen for isolation of *Chlamydia trachomatis* from suspected cases of pneumonia, especially in very young babies. A number of viruses cause exudative pharyngitis or rhinitis, including rhinoviruses, adenoviruses, respiratory syncytial virus, influenza and parainfluenza viruses, coronaviruses, coxsackieviruses, cytomegalovirus, and Epstein-Barr virus. Nasopharyngeal washings or tracheal secretions are the specimens of choice for these cultures. In addition, *Haemophilus influenzae* (and occasionally other organisms) causes a serious epiglottitis that can rapidly lead to airway obstruction if not treated. The combined effects of certain anaerobic oral flora causes an exudative pharyngitis known as Vincent's angina that may be characterized by a membrane-type lesion similar to that of diphtheria (the foul odor noted in patients with Vincent's angina helps to differentiate the two syndromes).

Peritonsillar abscess is a common feature of this syndrome. Evidence suggests that *Fusobacterium necrophorum* is a key pathogen in this setting, but other anaerobes may also be involved.

Immunosuppressed patients, including very young babies, may develop oral candidiasis, called **thrush.** Oral thrush can extend to produce esophagitis, a common finding in AIDS and other immunosuppressed patients. Oral mucositis or pharyngitis in the granulocytopenic patient may be caused by Enterobacteriaceae, *S. aureus,* or *Candida* species. It is manifested by erythema, sore throat, and possibly exudate or ulceration. Herpes simplex virus can cause painful lesions in the mouth and the oropharynx, another condition that is also prevalent among immunosuppressed patients.

A rare form of chronic, granulomatous infection of the nasal passages, including the sinuses and occasionally the pharynx and larynx, is rhinoscleroma. Associated with *Klebsiella rhinoscleromatis,* the disease is characterized by nasal obstruction appearing over a long period of time, caused by tumorlike growth with local extension. Another species, *Klebsiella ozaenae,* can also be recovered from upper respiratory tract infections. This organism may contribute to another infrequent condition called **ozaena,** characterized by a chronic, mucopurulent nasal discharge (often foul-smelling, caused by secondary low-grade anaerobic infection).

COLLECTION AND TRANSPORT OF SPECIMENS

Although the general concepts of collection and transport of specimens were covered in Chapter 7, a few specific comments are appropriate here. Either cotton-, Dacron-, or calcium alginate–tipped swabs are suitable for collection of most upper respiratory tract microorganisms. Note that all swabs, even including calcium alginate–tipped ones, can be toxic to some *Chlamydia* species. If the swab remains moist, no further precautions need to be taken for specimens that are cultured within 4 hours of collection. After that period, some kind of transport medium to maintain viability and prevent overgrowth by contaminants should be used. Swabs for detection of group A streptococci only are the exception. This organism is highly resistant to desiccation and will remain viable on a dry swab for as long as 48 to 72 hours. Throat swabs of this type can be placed in glassine paper envelopes for mailing or transport to a distant laboratory. Throat swabs are adequate for recovery of adenoviruses and herpesviruses, *Corynebacterium diphtheriae, Mycoplasma, Chlamydia,*

Candida spp., and *Haemophilus* spp. Nasopharyngeal swabs are better suited for recovery of respiratory syncytial virus, parainfluenza virus, *Bordetella pertussis, Neisseria* species, and the viruses causing rhinitis. Although modified Stuart's transport medium, as is often used in commercially produced swab systems, will preserve most viruses for a short time, nasopharyngeal swabs should ideally be transported in veal infusion broth or other protein-containing fluid if they are not being cultured within a few hours. They should be refrigerated, not frozen. Recovery of *C. diphtheriae* is enhanced by culturing both throat and nasopharynx. Transport media should be used as discussed in Chapter 7. Aspirated nasopharyngeal secretions collected in a soft rubber bulb are the best specimens for *Bordetella pertussis.* Specimens for *B. pertussis* ideally should be inoculated directly to fresh culture media at the patient's bedside. If this is impossible, transport for less than 2 hours in 1% casamino acid medium is acceptable.

DIRECT VISUAL EXAMINATION OF UPPER RESPIRATORY TRACT SPECIMENS

A Gram stain of material obtained from upper respiratory secretions or lesions can do very little to help with diagnosis. Yeastlike cells can be identified, helpful in identifying thrush, and the characteristic pattern of fusiforms and spirochetes of Vincent's angina may be visualized. Plain Gram's crystal violet (allowed to remain on the slide for 1 minute before rinsing with tap water) can be used to identify the agents of Vincent's angina, as well as the Gram stain. However, if only crystal violet is used, remember to make the smear very thin because everything will be intensely gram positive, making a thick smear impossible to read. For other causes of pharyngitis, Gram stains are unreliable. Direct smears of exudate from membrane-like lesions to differentiate diphtheria from other causes of such membranes are not reliable and are not recommended. Fungal elements, including yeast cells and pseudohyphae, may be visualized with a 10% potassium hydroxide (KOH) preparation, with calcofluor white fluorescent stain, or with a periodic acid-Schiff stained slide. Direct examination of material obtained from the nasopharynx of suspected cases of whooping cough using a fluorescent antibody stain (Chapter 30) has been shown to yield some early positive results.[1]

Direct fluorescent antibody smears have been used to identify group A β-hemolytic streptococci in throat specimens. This technology has largely been supplanted by direct detection of antigen, for which a large number of commercial products are available. Direct antigen detection systems will be discussed later in this chapter. Fluorescent antibody stain reagents are also commercially available for detection of herpes simplex antigen, influenza virus (rarely used), adenovirus, parainfluenza virus, and respiratory syncytial virus (RSV). Particularly for RSV, direct fluorescent antibody stain methods have been shown to be superior to culture.

CULTURING UPPER RESPIRATORY TRACT SPECIMENS

Classically, throat swabs have been plated onto 5% sheep blood agar (trypticase soy base). Other animal blood cells will serve to provide the necessary nutrients required by the usual pathogens being sought (primarily *Streptococcus pyogenes*), but sheep blood does not support growth of the β-hemolytic bacteria *Haemophilus haemolyticus* and *Haemophilus parahaemolyticus*. Thus these organisms will not be mistaken for possible β-hemolytic streptococci if sheep blood agar is used. The use of a sheep blood agar plate for isolation of *S. pyogenes* has the drawback of requiring overnight growth for colony formation and further manipulations of β-hemolytic growth for definitive identification. If sufficient numbers of pure colonies are not available for identification, a subculture requiring additional growing time is necessary. New selective agars (such as Strep Selective Agar, mentioned in Chapter 9) have been developed that suppress the growth of almost all normal flora and β-hemolytic streptococci except for groups A and B and *Arcanobacterium haemolyticum*. By placing a 0.04 unit differential bacitracin filter paper disk (Taxo A, BBL or Bacto Bacitracin, Difco Laboratories) directly on the area of initial inoculation, presumptive identification of *S. pyogenes* can be made after overnight incubation (Figure 17.2; all group A and a very small percentage of group B streptococci are susceptible). If preferred, direct antigen detection tests (coagglutination or latex agglutination) or PYR test (Chapter 26) can also be carried out from the more isolated β-hemolytic colonies that appear after overnight incubation on selective agar. A new probe test was released for commercial use in 1992 (Gen-Probe, Inc., San Diego, Calif.) that provides highly sensitive and specific results following a 4-hour incubation step in selective media.

Cultures for *Corynebacterium diphtheriae* should be plated onto sheep blood agar or streptococcal selective agar, as well as onto special media

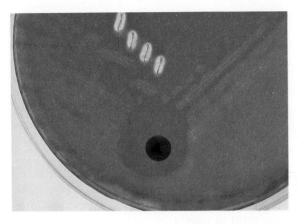

FIGURE 17.2

Primary streptococcal selective agar (BBL Microbiology Systems) culture plate from a throat specimen that contained many group A β-hemolytic streptococci. Growth of the bacteria is inhibited in zone surrounding the 0.04-unit bacitracin disk.

for recovery of the agent of diphtheria, since streptococcal pharyngitis is in the differential diagnosis of diphtheria and since dual infections do occur. These special media include a Loeffler's agar slant and a cystine-tellurite agar plate. Identification of the organism is discussed in Chapter 34. Recovery of this organism is enhanced by culturing specimens from both throat and nasopharynx of potentially infected patients.

Freshly prepared Bordet-Gengou agar is the medium of choice for isolation of *Bordetella pertussis*. A modified medium designed for isolation of *Legionella* species, buffered charcoal yeast extract agar with 3 μg/ml lincomycin and 80 μg/ml anisomycin has also been shown to yield good recovery of *B. pertussis*, as reported by workers at the Centers for Disease Control.[5] Jones and Kendrick[7] have also reported excellent recovery on a blood-free medium containing charcoal, heart infusion base, 2% agar, and 0.3 U/ml penicillin. Stauffer and others[16] showed best recovery of *B. pertussis* when this medium was supplemented with 40 μg/ml cephalexin. The charcoal medium developed for *Legionella* is more readily available and is a useful substitute. Specimens should be plated directly onto media, if possible, because the organisms are extremely delicate. Technologists should use both fluorescent antibody smears and cultures to achieve maximum detection. The yield of positive isolations from clinical cases of pertussis seems to vary from 20% to 98%, depending on the stage of disease, previous treatment of the patient, and laboratory techniques. Specimens received in the laboratory for isolation of *N. meningitidis* (for detection of carriers) or for *N. gonorrhoeae* should be plated to a selective medium, either modified Thayer-Martin or Martin-Lewis agar. After 48 hours of incubation in 5% to 10% CO_2, typical colonies of *Neisseria* species will be visible.

Clinical specimens from cases of epiglottitis (swabs obtained by a physician) should be plated to sheep blood agar, chocolate agar (for recovery of *Haemophilus* species), and a streptococcal selective medium if desired. Since *Staphylococcus aureus*, *Streptococcus pneumoniae* and β-hemolytic streptococci may, rarely, cause this disease, their presence must be sought.

NONCULTURE METHODS FOR DETECTING AGENTS OF PHARYNGITIS

One of the major improvements in selected areas of clinical microbiology in recent years has been the development of rapid methods for antigen detection that obviate the need for culture. Identification of group A streptococcal antigen in throat specimens is a particular application that has been a part of this development. At least 40 commercial products are available, employing latex agglutination, enzyme immunoassay and gene probe technologies, that allow detection of group A streptococcal antigen within as short a time as 10 minutes. Although the specific procedures vary with the products, several generalizations can be made. Throat swabs are incubated in some type acid reagent or enzyme to extract the group A–specific carbohydrate antigen (the gene probe requires preincubation in broth media followed by nucleotide extraction). Dacron swabs seem to be most efficient at releasing antigen, although other types of swabs may yield acceptable results. Following the recommendations of the manufacturer of the extraction kit is mandatory to achieve optimal results. In most laboratory comparisons between a rapid method and conventional culture methods for detecting the presence of group A streptococci in throat swabs, the commercial kits have shown good sensitivity (>90%) and exquisite specificity. There have been doubts expressed, however, about their overall ability to detect all clinically important streptococcal infections. Such systems are discussed further in Chapter 26.

In those situations in which rapid results are important (e.g., outpatient clinics), information about the presence of other organisms is not required, and the number of positive results allows the test to be cost effective. A commercial screening test for group A streptococci in throats may be the method of choice in selected situations. Rapid nonmicroscopic detection methods for most other agents of upper respiratory tract infections are not as well developed as those for *Streptococcus pyogenes*.

ACUTE LOWER RESPIRATORY TRACT INFECTIONS

The etiological agents of community-acquired pneumonia vary with the age and state of health of the patient. Pneumonia among patients with various types of immunosuppressive factors is discussed in Chapter 24. Among previously healthy patients 2 months to 5 years old, viruses, including respiratory syncytial (RSV), parainfluenza and influenza, and adenoviruses, are the most common etiological agents of lower respiratory tract disease. In addition to pneumonia, which indicates involvement of the lung parenchyma, viruses cause bronchiolitis and bronchitis. The latter infections may manifest as croup. Presumably, growth of viruses in host cells disrupts the function of the latter and encourages the influx of nonspecific immune effector cells that exacerbate the damage. Damage to host epithelial tissue by virus infection is known to predispose patients to secondary bacterial infection, as reviewed by Mills.[11] Children suffer less commonly from bacterial pneumonia, usually caused by *H. influenzae*, pneumococci, or *S. aureus*. Neonates may acquire lower respiratory tract infections with *C. trachomatis* or *P. carinii* (which likely indicates an immature immune system or an underlying immune defect). The most common etiological agent of lower respiratory tract infection among adults younger than 30 years old is *Mycoplasma pneumoniae*, which is transmitted via close contact. Contact with secretions seems to be more important than inhalation of aerosols for becoming infected. After they contact respiratory mucosa, *Mycoplasma* organisms are able to adhere to and colonize respiratory mucosal cells. Both a protein adherence factor and gliding motility may be determinants of virulence. Once situated in their preferred site between the cilia of respiratory mucosal cells, *Mycoplasma* organisms multiply and somehow destroy ciliary function. Cytotoxins produced by *Mycoplasma* organisms may account for the cell damage they inflict. The recently described species *C. pneumoniae* (originally called *Chlamydia* TWAR) has been seen as the third most common agent of lower respiratory tract infection in young adults, after mycoplasmas and influenza viruses; it also affects older individuals. Chlamydiae are intracellular parasites; thus their ability to disrupt cellular function and cause respiratory disease is similar to that of viruses.

Lower respiratory tract infections in older patients are most commonly due to bacterial infection, with *Streptococcus pneumoniae* most prevalent (causing 80% of all community acquired bacterial pneumonia). Aspiration pneumonia occurs in the community setting and involves primarily oral anaerobes and viridans streptococci but may also involve *Staphylococcus aureus* or gram-negative rods such as *Klebsiella pneumoniae*, other Enterobacteriaceae, and *Pseudomonas* species, particularly in patients with recent hospital or nursing home experience. *Haemophilus influenzae, Legionella* species, *Acinetobacter, Moraxella catarrhalis, Chlamydia pneumoniae*, meningococci, and other agents may also be implicated.[10,12] In the hospital setting, pneumonia involves many of these same agents. Aspiration pneumonia with infection caused by gram-negative bacilli or staphylococci is probably the major type of hospital-acquired pneumonia. It is followed by pneumococcal disease. *Legionella* has been implicated in a number of hospital outbreaks, but the problem is typically specific to a given institution.

Aspiration of oropharyngeal contents, often not overt, plays an important role in the pathogenesis of many different types of pneumonia. Aided by gravity and often by loss of some host nonspecific protective mechanism, the organisms reach lung tissue, where they multiply and attract host inflammatory cells. Other mechanisms include inhalation of aerosolized material and hematogenous seeding. The buildup of cell debris and fluid contributes to the loss of lung function and thus to the pathology. Agents with polysaccharide capsules, such as pneumococci, *Klebsiella* species, *Haemophilus* species, certain *S. aureus*, and cryptococci, seem more able to avoid phagocytosis and thus are more likely to cause pneumonia in patients without adequate humoral immunity in the form of organism directed specific antibody.

Adults may suffer from viral pneumonia caused by influenza, adenovirus, cytomegalovirus, parainfluenza, varicella, rubeola, or respiratory syncytial virus, particularly during epidemics. After viral pneumonia, especially influenza, secondary bacterial disease caused by β-hemolytic streptococci, pneumococci, *Staphylococcus aureus, Moraxella catarrhalis, Haemophilus influenzae*, and *Chlamydia pneumoniae* is more likely. Unusual

causes of acute lower respiratory tract infection include *Actinomyces* and *Nocardia* species; the agents of plague, tularemia, melioidosis (*Pseudomonas pseudomallei*), *Brucella, Salmonella, Coxiella burnetti* (Q fever), *Bacillus anthracis*, and *Pasteurella multocida. Paragonimus westermani, Entamoeba histolytica, Ascaris lumbricoides*, and *Strongyloides* species (the latter may cause fatal disease in immunosuppressed patients) may, on rare occasion, be recovered from sputum. A high index of suspicion by the clinician is usually a prerequisite to a diagnosis of parasitic pneumonia in the United States. Psittacosis should be ruled out as a cause of acute lower respiratory tract infection in patients who have had recent contact with birds. *Histoplasma capsulatum, Blastomyces dermatitidis, Paracoccidioides brasiliensis, Coccidioides immitis*, and *Cryptococcus neoformans*, and, occasionally, *Aspergillus fumigatus* may cause acute pneumonia. In the immunosuppressed patient, these fungi, as well as other *Aspergillus* species, *Candida*, zygomycetes, and other fungi, can cause life threatening pulmonary infection.

As noted, pneumonia secondary to aspiration of gastric or oral secretions is rather common. It happens most often during a loss of consciousness as might occur with anesthesia, or a seizure, or after alcohol or drug abuse, but other patients, particularly among geriatric groups, may also develop aspiration pneumonia. Neurological disease or esophageal pathology and periodontal disease or gingivitis are other important background factors. The oral black-pigmenting *Prevotella* and *Porphyromonas* species, *Prevotella oris, P. buccae, P. disiens, Bacteroides gracilis*, fusobacteria, and anaerobic and microaerobic streptococci are the most common agents. However, as noted, *Staphylococcus aureus*, various Enterobacteriaceae, and *Pseudomonas* may also be acquired in this way. The anaerobic agents possess many factors that may enhance their ability to produce disease, such as extracellular enzymes and capsules. It is their presence, however, in an abnormal site within the host producing lowered oxidation-reduction potential secondary to tissue damage that contributes most to their pathogenicity.

Nonaspiration pneumonia in hospitalized patients is associated most commonly with *Pseudomonas aeruginosa, Staphylococcus aureus, Klebsiella* species, and other Enterobacteriaceae. Some of these pneumonias are secondary to sepsis, and some are related to contaminated inhalation therapy equipment. Other aspects of pneumonia in hospitalized patients, as well as other groups at increased risk of developing respiratory disease, are discussed in Chapter 24.

CHRONIC LOWER RESPIRATORY TRACT INFECTIONS

Mycobacterium tuberculosis is the most likely etiological agent of chronic lower respiratory tract infection, but fungal infection and anaerobic pleuropulmonary infection may also run a subacute or chronic course. Mycobacteria other than *M. tuberculosis* (MOTT) may also cause such disease, particularly *M. avium-intracellulare* and *M. kansasii*. Although possible causes of acute, community-acquired lower respiratory tract infections, fungi and parasites are more commonly isolated from patients with chronic disease. *Actinomyces* and *Nocardia* may also be associated with gradual onset of symptoms. *Actinomyces* is usually associated with an infection of the pleura or chest wall, and *Nocardia* may be isolated along with an infection caused by *M. tuberculosis*. Agents of chronic disease in compromised hosts are discussed in Chapter 24. The pathogenesis of many of the infections caused by agents of chronic lower respiratory tract disease is characterized by the requirement for some breakdown of cell-mediated immunity in the host or the ability of these agents to avoid being destroyed by host cell–mediated immune mechanisms. This may be by an effect on macrophages, by the ability to mask foreign antigens, sheer size, or by some other factor, allowing microbes to grow within host tissues without eliciting an overwhelming local immune reaction.

LABORATORY DIAGNOSIS OF LOWER RESPIRATORY TRACT INFECTIONS

COLLECTION AND TRANSPORT OF SPECIMENS
Lower respiratory tract secretions will be contaminated with upper respiratory tract secretions, especially saliva, unless they are collected using some invasive technique. For this reason, sputum is among the least clinically relevant specimens received for culture in microbiology laboratories. It is both one of the most numerous and time-consuming specimens. When hospital laboratories began to reject some sputum samples based on visual examination of predominant cell types (as discussed later in this chapter), the numbers of acceptable specimens received by the laboratory increased significantly. It is possible, therefore, for microbiologists and clinicians to influence the quality of specimens received for culture. Good sputum samples depend on thorough health care

worker education and patient understanding throughout all phases of the collection process. Patients should be instructed to provide a deep coughed specimen. The material should be expelled into a sterile container, with an attempt to minimize contamination by saliva. Specimens should be transported to the laboratory immediately, since even a moderate amount of time at room temperature can result in loss of some infectious agents.

Patients who are unable to produce sputum may be assisted by respiratory therapy technicians, who can use postural drainage and thoracic percussion to stimulate production of acceptable sputum. As an alternative, an "aerosol-induced specimen" may be collected. Particularly useful for obtaining material suitable for isolation of the agents of mycobacterial or fungal disease, and now recognized for its high diagnostic yield in cases of *Pneumocystis carinii* pneumonia, aerosol-induced specimens are collected by allowing the patient to breathe aerosolized droplets of a solution containing 15% sodium chloride and 10% glycerin for approximately 10 minutes, or until a strong cough reflex is initiated. The lower respiratory secretions obtained in this way appear watery, resembling saliva, although they often contain material directly from alveolar spaces. These specimens are usually adequate for culture and should be accepted in the laboratory without prescreening. Obtaining such a specimen may obviate the need for a more invasive procedure such as bronchoscopy or needle aspiration in many cases.

Another specimen, exclusively for isolation of acid-fast bacilli, that may be collected from patients who are unable to produce sputum, particularly young children, is the gastric aspirate. Before the patient wakes up in the morning, a nasogastric tube is inserted into the stomach and contents are withdrawn (on the assumption that acid-fast bacilli from the respiratory tract were swallowed during the night and will be present in the stomach). The relative resistance of mycobacteria to acidity allows them to remain viable for a short period. Gastric aspirate specimens must be delivered to the laboratory immediately so that the acidity can be neutralized. Because of the presence of mycobacteria in tap water, acid-fast smears from gastric aspirates are often falsely positive and ordinarily should not be examined. A physician may be able to pass a soft catheter into the tracheobronchial tree and obtain a "tracheal aspirate" by suctioning with a syringe.

Patients with tracheostomies are unable to

produce sputum in the normal fashion, but lower respiratory tract secretions can easily be collected in a Lukens trap (Figure 17.3). These specimens (called **tracheostomy aspirates** or **tracheostomy suction**) should be treated as sputum (without microscopic screening, discussed later in this chapter) by the laboratory. Patients with tracheostomies rapidly become colonized with gram-negative bacilli and other nosocomial pathogens. Such colonization per se does not have clinical relevance, but these organisms may be aspirated into the lungs and cause pneumonia. Thus there can be real confusion for microbiologists and clinicians trying to ascertain the etiological agent of pneumonia in these patients.

When pleural empyema is present, thoracentesis may be used to obtain infected fluid for direct examination and culture. This constitutes an excellent specimen that typically accurately reflects the bacteriology of an associated pneumonia. Laboratory examination of such material is discussed in Chapter 22. Blood cultures, of course, should always be obtained from patients with pneumonia since they will be positive in about 20% of patients requiring hospitalization. Physicians sometimes attempt to obtain more representative specimens via a bronchoscope than are possible with noninvasive techniques. Bronchial secretions are often obtained by instilling a small amount of sterile physiologic saline into the bronchial tree and withdrawing the fluid when purulent secretions are not visualized. Such "bronchial washing" specimens will still be contaminated with upper respiratory tract flora such as viridans streptococci and *Neisseria* species. Recovery of potentially pathogenic organisms from bronchial washings should be attempted, however, since such specimens may be more diagnostically relevant than sputa.

A deeper sampling of desquamated host cells and secretions is also obtained via bronchoscopy by bronchoalveolar lavage (BAL). Lavages are especially suitable for detecting *Pneumocystis* cysts and fungal elements. The collection methods and handling of the specimen are well described by Bartlett and others.[2] Recent publications document the value of this technique, with quantitative culture, for diagnosis of most major respiratory tract pathogens, including bacterial pneumonia.[8,18] Various authors have found significant correlation between acute bacterial pneumonia and greater than 10^3 to 10^4 bacterial colonies per milliliter of BAL fluid.[13]

The specimen obtained by only moderately invasive means that is best suited for microbiolog-

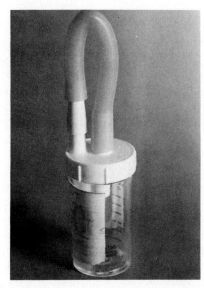

FIGURE 17.3
Tracheal secretions received in the laboratory in a Lukens trap.

ical studies, particularly in aspiration pneumonia, is obtained via a protected catheter bronchial brush as part of a bronchoscopy examination. A small brush that holds 0.01 to 0.001 ml of secretions is placed within a double cannula. The end of the outermost tube or cannula is closed with a displaceable plug made of absorbable gel. Once the cannula has been inserted to the proper area, the inner cannula is pushed out, dislodging the protective plug as it is extruded. Then the brush is extended beyond the inner cannula. The specimen is collected by "brushing" the involved area, and the brush is withdrawn into the inner cannula, which is withdrawn into the outer cannula to prevent contamination as it is removed. The contents of the bronchial brush may be suspended in 1 ml of broth solution with vigorous vortexing and then inoculated onto culture media by technologists using a 0.01 ml calibrated inoculating loop. Some workers have stated that specimens obtained via double lumen protected catheters are suitable for anaerobic as well as aerobic cultures. Colony counts of greater than 1000 organisms per milliliter in the broth diluent (or 10^6/ml in the original specimen) have been considered to correlate with infection, as reported by Pollock and associates.[14] However, other studies question the validity of this procedure, even when cultures are done quantitatively. Furthermore, the tiny sample and its aeration are problems with the technique. When done, the entire procedure should be rigorously standardized.[14]

Several invasive procedures are used to collect pulmonary specimens when definitive results are necessary, or when anaerobic bacteria are being sought. Percutaneous transtracheal aspirates (TTA), as mentioned in Chapter 7, are obtained by inserting a small plastic catheter into the trachea via a needle previously inserted through the skin and cricothyroid membrane. This invasive procedure, although somewhat uncomfortable for the patient and not suitable for all patients (it cannot be used in uncooperative patients, in patients with bleeding tendency, or with poor oxygenation), guarantees a specimen uncontaminated by upper respiratory tract flora and undiluted by added fluids, provided that care is taken to keep the catheter from being coughed back up into the pharynx. Although this technique is rarely used anymore, anaerobes like *Actinomyces* and those associated with aspiration pneumonia can be isolated from transtracheal aspirate specimens.

For patients with pneumonia, a thin needle aspiration of material from the involved area of the lung may be performed percutaneously, often without fluoroscopy. If no material is withdrawn into the syringe after the first try, approximately 3 ml of sterile saline can be injected and then withdrawn into the syringe. Patients with emphysema, uremia, thrombocytopenia, or pulmonary hyptertension may be at increased risk of complication (primarily pneumothorax or bleeding) from this procedure. The specimens obtained are very small in volume, and protection from aeration is usually impossible. This technique is more frequently used in children than in adults.

Transbronchial biopsies, again as part of a bronchoscopic procedure, are used more for histological diagnosis than for culture. Direct examination of biopsy specimens for agents such as *Pneumocystis* often shows positive findings. The most invasive procedure for obtaining respiratory tract specimens is the **open lung biopsy.** Performed by surgeons, this method is used to procure a wedge of lung tissue. Biopsy specimens are extremely helpful for diagnosis of severe viral infections, such as herpes pneumonia, for rapid diagnosis of *Pneumocystis* pneumonia, and for other hard-to-diagnose or life threatening pneumonias. Ramifications of this and all other specimen collection techniques are discussed in Cumitech 7A, *Laboratory Diagnosis of Lower Respiratory Tract Infections.*[2]

DIRECT VISUAL EXAMINATION OF LOWER RESPIRATORY TRACT SPECIMENS

Lower respiratory tract specimens can be examined by direct wet preparation for some parasites, and by special procedures for *Pneumocystis.* Fungal elements can be visualized under phase microscopy with 10% KOH, under UV light with calcofluor white, or using PAS stained smears. If pneumococci or *H. influenzae* type b is suspected, a portion of purulent material may be mixed with an equal amount of antiserum for performance of the Quellung test after a Gram stain has indicated the presence of bacteria.

For most other evaluations, the specimen must be fixed and stained. Bacteria and yeasts can be recognized on Gram stain. One of the most important uses of the Gram stain, however, is to evaluate the quality of expectorated sputum received for routine bacteriological culture.[9] A portion of the specimen consisting of purulent material is chosen for the stain. One enterprising group showed that the smear can be evaluated adequately even before it is stained, thus negating the need for Gram stain of specimens later judged unacceptable.[20] An acceptable specimen will yield less than 10 squamous epithelial cells per low-power field (100×). The number of white cells may not be relevant, since many patients are severely neutropenic and specimens from these patients will not show white cells on Gram stain examination. On the other hand, the presence of 25 or more polymorphonuclear leukocytes per 100× field, together with few squamous epithelial cells, implies an excellent specimen (Figure 17.4). Only expectorated sputa are suitable for rejection based on microscopic screening. However, it is important to realize that in *Legionella* pneumonia, sputum may be scant and watery with few or no host cells. Such specimens may still often be positive by direct fluorescent antibody stain and culture, and should not be subjected to screening procedures. Respiratory secretions may need to be concentrated before staining. The cytocentrifuge instrument has been used successfully for this purpose, concentrating the cellular material in an easily examined monolayer on a glass slide.[15] As an alternative, specimens are centrifuged, and the sediment is used for visual examinations and cultures. For screening purposes, the presence of ciliated columnar bronchial epithelial cells, goblet cells, or pulmonary macrophages in specimens obtained by bronchoscopy or BAL is indicative of a specimen from the lower respiratory tract.

In addition to the Gram stain, respiratory specimens may be stained for acid-fast bacilli with either the classic Ziehl-Neelsen or the Kinyoun carbolfuchsin stain. Auramine or auramine-rhodamine is also used for detection of acid-fast organisms. Because they are fluorescent, these

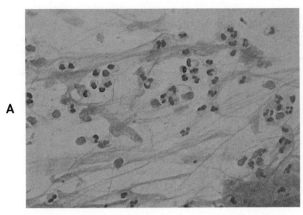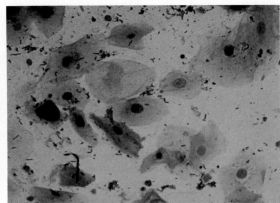

FIGURE 17.4

Gram stain of sputum specimens (400×). **A,** This specimen contains numerous polymorphonuclear leukocytes and no visible squamous epithelial cells, indicating that the specimen is acceptable for routine bacteriological culture. **B,** This specimen contains numerous squamous epithelial cells and rare polymorphonuclear leukocytes, indicating an inadequate specimen for routine sputum culture.

stains are more sensitive than the carbolfuchsin formulas and are preferable for rapid screening. Slides may be restained with the classic stains directly over the fluorochrome stains as long as all of the immersion oil has been removed carefully with xylene. All of the acid-fast stains will also reveal *Cryptosporidium* species if they are present in the respiratory tract, as may occur in immunosuppressed patients.

Immunosuppressed patients are often at risk of infection with *P. carinii.* Although the modified Gomori methenamine silver stain procedure (Chapter 45) has been used traditionally to interpret for recognition of *Nocardia, Actinomyces,* fungi, and parasites, it takes approximately 1 hour of technologist time to perform, is technically demanding, and is not suitable as an emergency procedure. A fairly rapid stain, toluidine blue O, has been used in many laboratories with some success. It stains not only *Pneumocystis* but also *Nocardia asteroides* and some fungi (Chapter 45). A new monoclonal antibody stain for *Pneumocystis* shows great promise as a specific and sensitive stain for *Pneumocystis* (Chapter 45).[17]

Direct fluorescent antibody (DFA) staining has been used for detection of *Legionella* species in lower respiratory tract specimens. Sputum, pleural fluid, aspirated material, and tissue are all suitable specimens. Polyclonal antibody reagents and a monoclonal antibody directed against all serotypes of *Legionella pneumophila* are used, since there are so many different serotypes of legionellae. DFA results should not be relied on in lieu of culture, which remains the gold standard for legionellae because of its high sensitivity.

DFA reagents (commercially available) are also used to detect antigens of a number of viruses, including herpes simplex, cytomegalovirus, adenovirus, and respiratory syncytial virus.[4,6] Commercial suppliers of reagents provide procedure information for each of the tests available. Monoclonal and polyclonal fluorescent stains for *Chlamydia trachomatis* are available and may be useful for staining respiratory secretions of infants with pneumonia.[3]

CULTURE OF LOWER RESPIRATORY TRACT SPECIMENS

Most of the commonly sought etiological agents of lower respiratory tract infection will be isolated on the commonly used media: 5% sheep blood agar, MacConkey agar, and chocolate agar. Because of contaminating oral flora, sputum specimens, specimens obtained by bronchial washing and lavage, tracheal aspirates, and tracheostomy or endotracheal tube aspirates are not inoculated to enrichment broth or incubated anaerobically. Only specimens obtained by percutaneous aspiration (including transtracheal aspiration) and by protected bronchial brush are suitable for anaerobic culture; the latter must be done quantitatively for proper interpretation. Transtracheal and percutaneous lung aspiration material may be inoculated to enriched thioglycolate, as well as to solid media. For suspected cases of Legionnaire's disease, buffered charcoal yeast extract (BCYE) agar and selective BCYE are inoculated. Pertussis media, as mentioned previously in this chapter, may also be inoculated.

Contaminated specimens received for isolation of mycobacteria must be decontaminated and concentrated before they are inoculated to media (see Chapter 42). The radiometric detection system (BACTEC, BBL Microbiology Systems) has been used with good success for more rapid detection of growth of mycobacteria, especially *M. avium-intracellulare*, than can be achieved with conventional cultures. Specimens are inoculated to special media for mycoplasma isolation (Chapter 39).[3] For isolation of the dimorphic fungi, material should be inoculated to brain-heart infusion agar as well as to infusion or Sabouraud's heart infusion agar (SABHI) with sheep blood and Mycobiotic or Mycosel (or some inhibitory mold agar), as discussed in Chapters 9 and 44.

Patients with cystic fibrosis may yield such numerous colonies of mucoid *Pseudomonas aeruginosa* that other growth is obscured.[19] For such specimens, selective agar, mentioned in Chapter 24, should be inoculated. The isolation of *Haemophilus* species in specimens containing large numbers of normal flora can be facilitated by placing a 10-U bacitracin disk in the area of the primary inoculum on the chocolate agar plate (George Tortora, personal communication). This may be particularly valuable with sputum specimens received from patients with cystic fibrosis.

Viruses and chlamydiae require tissue culture techniques,[4] detailed in Chapters 43 and 39, respectively. Specimens for viral culture should be placed in a protein-containing transport medium such as Minimum Essential Medium (MEM) in Eagle's Buffered Salt Solution (commercially available) with amphotericin B or veal infusion broth and refrigerated (never frozen) until they can be cultured. Chlamydia cultures are transported in 2-sucrose phosphate (2SP) or other transport medium (Chapter 39).

NONCULTURE METHODS FOR DIAGNOSING LOWER RESPIRATORY TRACT INFECTIONS

In addition to direct fluorescent antibody staining of lower respiratory tract secretions, several other nonculture techniques have been used. Traditional serology may be valuable for retrospective diagnosis of respiratory tract infections; however, interpretation of results is not standardized and often difficult. An enzyme immunoassay for the detection of *B. pertussis* antibodies (Labsystems Inc., Chicago, Ill.) may be helpful in timely diagnosis of pertussis. Gas-liquid chromatography has been used to analyze fluids for the presence of metabolites of pneumococci and other pathogens (see Chapter 12). Gas-liquid chromatography, electron-capture gas chromatography, and gas-chromatography mass spectroscopy have been evaluated for diagnosis of tuberculosis. Although

the ideas are promising, no practical schemes are in general use for diagnosing lower respiratory tract infection in this way. Elevated pleural fluid lactic acid levels are indicative of bacterial empyema, but there is overlap with malignant pleural effusion and therefore a significant false-positive rate for this test.

Testing clinical material directly for antigens of common pathogens, an approach that has succeeded for detection of group A streptococci in throat swabs, shows promise for diagnosis of pneumonia. The problems of nonspecific reactivity and masking of reactions by mucus and other factors have not been fully resolved. One radioimmunoassay (RIA) procedure, detection of antigens of *L. pneumophila* in urine of patients with legionellosis (DuPont Medical Products, Wilmington, Del.), has been found to be approximately as sensitive as direct fluorescent antibody stains of sputum. Electron microscopy has been used for detection of virus particles, although this procedure is limited to a few very specialized laboratories. DNA hybridization, a promising technique for detection of viral and other microbial gene sequences in clinical material, discussed in Chapter 11, has been used by a number of workers. In situ hybridization procedures utilizing biotinylated or other DNA probes have successfully visualized herpes simplex (HSV), varicella-zoster (VZV), adenovirus, and cytomegalovirus-infected cells in sections from lungs and in cell layers. Currently the easiest to use of the sensitive, rapid techniques are the direct fluorescent antibody (DFA) stains. Reliable DFA stains are available for detection of HSV, VZV, *Chlamydia trachomatis*, and RSV directly from patient specimens.

REFERENCES

1. Anhalt, J.P. 1981. Flourescent antibody procedures and counterimmunoelectrophoresis. In Washington, J.A. II, editor. Laboratory procedures in clinical microbiology, Springer-Verlag, New York.
2. Bartlett, J.G., Ryan, K.J., Smith, T.F., et al. 1987. Cumitech 7A, Laboratory diagnosis of lower respiratory tract infections, American Society for Microbiology, Washington, D.C.
3. Clyde, W.A. Jr., Kenny, G.E., and Schachter, J. 1984. Cumitech 19. In Drew, W.L., editor. Laboratory diagnosis of chlamydial and mycoplasmal infections. American Society for Microbiology, Washington, D.C.
4. Greenberg, S.B., and Krilov, L.R. 1986. Cumitech 21, In Drew, W.L., editor. Laboratory diagnosis of viral respiratory disease. American Society for Microbiology, Washington, D.C.
5. Hayes, P.S., Feeley, J.C., Johnson, S.E., et al. 1985. Use of charcoal yeast extract agar for the isolation of *Bordetella pertussis*, Abstr. Ann. Meet., C188.

American Society for Microbiology, Washington, D.C.

6. Hildreth, S.W., Hall, C.B., and Menegus, M.A. 1983. Respiratory syncytial virus. Clin. Microbiol. Newsletter 5:93.

7. Jones, G.L., and Kendrick, P.L. 1969. Study of a blood-free medium for transport and growth of *Bordetella pertussis.* Health Lab. Sci. 6:40.

8. Kahn, F.W., and Jones, J.M. 1987. Diagnosing bacterial respiratory infection by bronchoalveolar lavage. J. Infect. Dis. 155:862.

9. Lentino, J.R. 1987. The nonvalue of unscreened sputum specimens in the diagnosis of pneumonia. Clin. Microbiol. Newsletter 9:70.

10. Marrie, T.J., Grayston, T., Wang, S.-P., et al. 1987. Pneumonia associated with the TWAR strain of *Chlamydia.* Ann. Intern. Med. 106:507.

11. Mills, E.L. 1984. Viral infections predisposing to bacterial infections. Annu. Rev. Med. 35:469.

12. Parker, R.H. 1983. *Hemophilus influenzae* respiratory infection in adults: recognition and incidence. Postgrad. Med. 73:179.

13. Pisani, R.J., and Wright, A.J. 1992. Clinical utility of bronchoalveolar lavage in immunocompromised hosts. Mayo Clin. Proc. 67:221.

14. Pollock, H.M., Hawkins, E.L., Bonner, J.R., et al. 1983. Diagnosis of bacterial pulmonary infections with quantitative protected catheter cultures obtained during bronchoscopy. J. Clin. Microbiol. 17:255.

15. Shanholtzer, C.J., Schaper, P.J., and Peterson, L.R. 1982. Concentrated Gram stain smears prepared with a cytospin centrifuge. J. Clin. Microbiol. 16:1052.

16. Stauffer, L.R., Brown, D.R. and Sandstrom, R.E. 1983. Cephalexin-supplemented Jones-Kendrick charcoal agar for selective isolation of *Bordetella pertussis:* comparison with previously described media. J. Clin. Microbiol. 17:60.

17. Stratton, N., Hryniewicki, J., Aarnaes, S.L., 1991. Comparison of monoclonal antibody and calcofluor white stains for detection of *Pneumocystis carinii* from respiratory specimens. J. Clin. Microbiol. 29:645.

18. Thorpe, J.E., Baughman, R.P., Frame, P.T., et al. 1987. Bronchoalveolar lavage for diagnosing acute bacterial pneumonia. J. Infect. Dis. 155:855.

18a. Waagner, D.C. 1991. *Arcanobacterium haemolyticum:* biology of the organism and diseases in man. Pediatr. Infect. Dis. J. 10:933.

19. Welch, D.F. 1984. Clinical microbiology of cystic fibrosis. Clin. Microbiol. Newsletter 6:39.

20. Wetterau, L.M., Zeimis, R.T., and Hollick, G.E. 1986. Direct examination of unstained smears for the evaluation of sputum specimens. J. Clin. Microbiol. 24:143.

BIBLIOGRAPHY

Browning, R.J. 1983. Diagnosing pulmonary diseases: a clinician's perspective. Diagn. Med. 6:39; 6:62.

Hilborne, L.H., et al. 1992. Diagnostic applications of recombinant nucleic acid technology: infectious diseases. Lab. Med. 23:89.

Kaufman, L. 1992. Laboratory methods for the diagnosis and confirmation of systemic mycoses. Clin. Inf. Dis. 14(suppl 1):S23.

McDougall, J.K., Beckman, A.M., Galloway, D.A., et al. 1985. Defined viral probes for the detection of HSV, CMV, and HPV. In Kingsbury, D.T., and Falkow, S., editors. Rapid detection and identification of infectious agents. Academic Press, New York.

von Graevenitz, A. 1983. Which bacterial species should be isolated from throat cultures? (Editorial.) Eur. J. Clin. Microbiol. 2:1.

Wakefield, A.E., et al. 1991. DNA amplification on induced sputum samples for diagnosis of *Pneumocystis carinii* pneumonia. Lancet 337:1378.

18

MICROORGANISMS ENCOUNTERED IN THE GASTROINTESTINAL TRACT

ANATOMY AND GENERAL FEATURES OF GASTROINTESTINAL TRACT

We are all connected to the external environment through our gastrointestinal (GI) tract (Figure 18.1). What we swallow enters the GI tract and passes through the esophagus into the stomach, through the duodenum, the jejunum, the ileum, and finally through the cecum and colon to the anus. During passage, fluids and other components are both added to this material (as secretory products of individual cells and as enzymatic secretions of glands and organs) and removed from this material by absorption through the gut epithelium. The duodenum, jejunum, and ileum are collectively called the **small intestine,** and the cecum and colon make up the **large intestine.** The nature of the epithelial cells lining the GI tract varies with each segment; the small intestine is lined with small projections, called *villi,* that greatly increase the surface area. The function of villi is to absorb fluids and nutrients from the intestinal contents. Other cells of the small intestine secrete fluids and metabolites into the lumen. Many cells in the lining of the large intestine are mucus secreting, and there are no villous projections into the lumen. The remaining excess fluid within the GI tract is reabsorbed by the cells lining the large intestine before waste is finally discharged through the

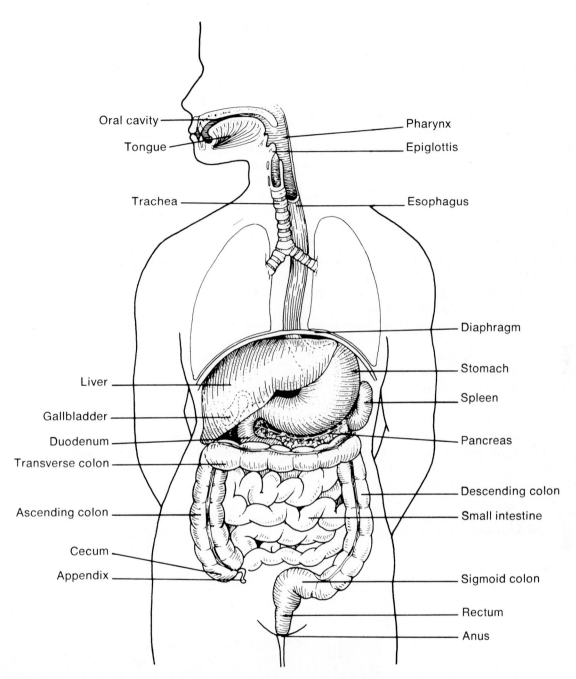

FIGURE 18.1
General anatomy of the gastrointestinal tract. (From Broadwell, D.C., and Jackson, B.S. 1982. Principles of ostomy care. Mosby, St. Louis.)

rectum. The lining of the GI tract is called the *mucosa*. Because of the differing nature of the mucosal surfaces of various segments of the bowel, specific infectious disease processes tend to occur in each segment. The appendix, a long, narrow tube of bowel tissue, extends from the cecum near the ileum. The appendix may become inflamed and even gangrenous when there is stasis within its lumen or when an adjacent inflammatory process occurs. Ap-

pendicitis is the most common initial event leading to intraabdominal abscess in children. Infections of the appendix are discussed in Chapter 21.

The GI tract contains vast, diverse normal flora. Although the acidity of the stomach prevents any significant colonization in a normal host, many species can survive passage through the stomach to become resident within the lower intestinal tract. Normally, the upper small intestine contains only a sparse flora (bacteria, primarily streptococci, lactobacilli, and yeasts, 10^1 to 10^3/ml), but by the time the distal ileum is reached, counts are about 10^6 to 10^7/ml, and Enterobacteriaceae and *Bacteroides* are present. With certain types of pathology (e.g., bleeding or obstructing peptic ulcer) there may be much higher bacterial counts and a more diverse flora in the stomach and duodenum. This has implications for a patient who has a perforated peptic ulcer. Babies usually are colonized by normal human epithelial flora (staphylococci, *Corynebacterium* sp., etc.) and other gram-positive organisms (bifidobacteria, clostridia, lactobacilli, streptococci) within a few hours of birth. Over time, the content of the intestinal flora changes. The normal flora of the adult large bowel, predominantly anaerobic species, including *Bacteroides, Clostridium, Peptostreptococcus, Bifidobacterium, Eubacterium,* and others, is established relatively early in life. The anaerobes outnumber the aerobes, which include primarily *Escherichia coli,* other Enterobacteriaceae, enterococci, and other streptococci, by 1000:1. The number of bacteria per gram of stool within the lumen of the bowel increases steadily as material approaches the sigmoid colon (the last segment). Eighty percent of the dry weight of feces from a healthy human consists of bacteria, which can be present in numbers as high as 10^{11} to 10^{12}/g. After antimicrobial therapy, the large bowel may become colonized with nosocomial pathogens such as *Staphylococcus aureus, Pseudomonas aeruginosa, Candida albicans,* and resistant gram-negative bacilli. This influences the flora of any subsequent intraabdominal infection.

PATHOPHYSIOLOGY OF GASTROINTESTINAL INFECTIONS

ROLE OF NORMAL FLORA IN HOST DEFENSE

The normal flora is an important factor in the response of the host to introduction of a potentially harmful microorganism. It has been shown that whenever a reduction in normal flora occurs, because of antibiotic treatment or some host factor, resistance to GI infection is significantly reduced. The most common example of the protective effect of normal flora is the development of the syndrome **pseudomembranous colitis (PMC).** Caused by the toxins of the anaerobic organism *Clostridium difficile* and occasionally other clostridia and perhaps even *Staphylococcus aureus,* this inflammatory disease of the large bowel (Figure 18.2) seldom occurs except after antimicrobial or antimetabolite treatment has altered the normal flora. Almost every antimicrobial agent and several cancer agents have been associated with the development of PMC. *C. difficile,* usually acquired from the hospital environment, is suppressed by normal flora. When normal flora are reduced, *C. difficile* is able to multiply and produce its toxins. Thus, this syndrome is also known as *antibiotic-associated colitis.* Other microorganisms that may gain a foothold when released from selective pressure of normal flora include *Candida* sp., staphylococci, *Pseudomonas* sp., and various Enterobacteriaceae.

NONSPECIFIC HOST DEFENSE MECHANISMS

Several other factors contribute to host resistance to disease. The acidity of the stomach effectively restricts the number and types of organisms that enter the lower GI tract. Normal peristalsis helps

FIGURE 18.2

Autopsy specimen of colon showing multiple yellow, elevated plaques indicative of the inflammatory exudate of pseudomembranous colitis (PMC).

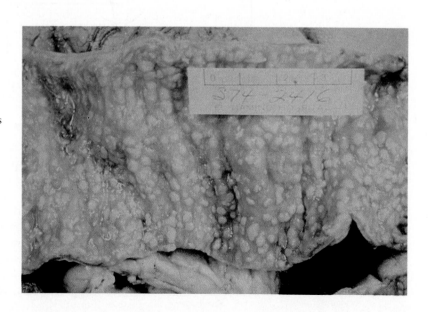

to move organisms toward the rectum, interfering with their ability to adhere to the mucosa. The mucous layer coating the epithelium entraps microorganisms and helps to propel them through the gut. The normal flora prevent colonization by potential pathogens. Secretory immunoglobulin A (IgA) and phagocytic cells within the gut help to destroy etiological agents of disease, as do eosinophils, which are particularly active against parasites. For a microorganism to cause GI infection, it must possess one or more factors that allow it to overcome these host defenses, or it must enter the host at a time when one or more of the innate defense systems is inactive. For example, certain stool pathogens are able to survive gastric acidity only if the acidity has been reduced by bicarbonate or other buffer or by cimetidine, ranitidine, (H$_2$ blockers), or similar medications. Pathogens taken in with milk have a better chance of survival, since milk neutralizes stomach acidity. Organisms such as *Mycobacterium tuberculosis*, *Shigella*, and *C. difficile* (a spore-forming clostridium) are able to withstand exposure to gastric acids and thus require much smaller infectious inocula than do acid-sensitive organisms such as *Salmonella*.

PATHOGENIC MECHANISMS OF AGENTS OF GI INFECTIONS

Etiological agents of GI infection are known to cause disease in only four ways:

1. By producing a toxin that affects fluid secretion, cell function, or neurologic function
2. By growing within or close to intestinal mucosal cells and destroying them, thus disrupting function
3. By invading the mucosal epithelium, causing cellular destruction and occasionally invading the bloodstream and going on to systemic disease
4. By adhering to intestinal mucosa, thus preventing the normal functions of absorption and secretion.

Since the normal adult GI tract receives up to 8 L of fluid daily in the form of ingested liquid, plus the secretions of the various glands that contribute to digestion (salivary glands, pancreas, gallbladder, stomach), of which all but a small amount must be reabsorbed, any disruption of the normal flow or resorption of fluid will have a profound effect on the host.

"Food poisoning" may occur as a result of the ingestion of toxins produced by microorganisms. The microorganisms usually produce their toxins in foodstuffs before they are ingested; thus the patient ingests preformed toxin. Although, strictly speaking, these syndromes are not GI infections but intoxications, they are acquired by ingestion of microorganisms or their products and are considered in this chapter. Particularly in staphylococcal food poisoning and botulism, the causative organisms may not be present in the patient's bowel at all.

GASTROINTESTINAL OR SYSTEMIC DISEASES ASSOCIATED WITH TOXIN-PRODUCING MICROORGANISMS

One of the most potent neurotoxins known is produced by the anaerobic organism *Clostridium botulinum*. This toxin acts to prevent the release of the neurotransmitter acetylcholine at the cholinergic nerve junctions, causing flaccid paralysis. The toxin acts primarily on the peripheral nerves but also on the autonomic nervous system. Patients exhibit descending symmetrical paralysis and ultimately die from respiratory paralysis unless they are given mechanical ventilation. In most cases, adult patients who develop botulism have ingested the preformed toxin in food (home-canned tomato products and canned cream-based foods are often implicated), and the disease is considered to be an intoxication, although *C. botulinum* has been recovered from the stools of many adult patients. Wound botulism, in which the organism grows and produces toxin in an infected wound, also occurs, but it is rare. A relatively recently recognized syndrome, infant botulism, is a true GI infection. Probably the flora of the normal adult bowel usually prevent colonization by *C. botulinum*, whereas the organism is able to multiply and produce toxin in the infant bowel. Infant botulism is not an infrequent condition; babies acquire the organism by ingestion, although the source of the bacterium is not always clear. Since an association has been found with honey and corn syrup, it is recommended that babies under 9 months of age not be fed honey. The effect of the toxin is the same, whether it is ingested in food or produced by growing organisms within the bowel.

Other bacterial agents of food poisoning that produce neurotoxins include *Staphylococcus aureus* and *Bacillus cereus*. Toxins produced by these organisms cause vomiting, independent of other actions on the gut mucosa. Staphylococcal food poisoning is one of the most frequently reported categories of foodborne disease. The organisms grow in warm food, primarily meat or dairy products, and produce the toxin. Onset of disease is usually within 2 to 6 hours of ingestion. *B. cereus* produces two toxins, one of which is preformed, called the *emetic toxin,* because it produces vomiting. The second type, probably involving several enterotoxins (discussed later), causes diarrhea. Often acquired from eating rice, *B. cereus* has also been associated with cooked meat, poultry, vegetables, and desserts. Perhaps the most common cause of food poisoning is from type A *Clostridium perfringens,* which produces toxin in the host after ingestion. It causes a relatively mild, self-limited (usually 24 hours) gastroenteritis, often in outbreaks in hospitals. Meats and gravies are typical offending foods. Clinical symptoms of *B. cereus* and *C. perfringens* foodborne disease occur 8 to 14 hours after ingestion. Investigations of outbreaks of food poisoning caused by preformed toxin ingestion are carried out by public health laboratories and should not be attempted by clinical laboratories.

The second category of toxins, cytotoxins, acts to disrupt the structure of individual intestinal epithelial cells. When these cells are destroyed, they slough from the surface of the mucosa, leaving it raw and unprotected. The secretory or absorptive functions of the cells are no longer performed. The damaged tissue evokes a strong inflammatory response from the host, further inflicting tissue damage. Numerous polymorphonuclear neutrophils and blood are often seen in the stool, and pain, cramps, and an urgent need to defecate (even though little material is present in the bowel) **(tenesmus)** are common symptoms. The term **dysentery** refers to this destructive disease of the mucosa, almost exclusively occurring in the colon. Cytotoxin has not yet been shown to be the sole virulence factor for any etiological agent of GI disease, since most agents produce a cytotoxin in conjunction with another factor.

Escherichia coli strains seem to possess virulence mechanisms of many types.[7] Some strains produce a cytotoxin that destroys epithelial cells and blood cells. Certain strains produce a cytotoxin that affects Vero cells (tested in African green monkey kidney cells) and resembles the cytotoxin produced by *Shigella dysenteriae* (Shiga toxin); such strains of *E. coli* are associated with hemorrhagic colitis and hemolytic uremic syndrome.[3,13] *C. difficile* produces a cytotoxin, the presence of which is a most useful marker for diagnosis of PMC. (The *C. difficile* cytotoxin assay is discussed later in this chapter.) *S. dysenteriae, Staphylococcus aureus, C. perfringens,* and *Vibrio parahaemolyticus* produce cytotoxins that probably contribute to the pathogenesis of diarrhea, although they may not be essential for initiation of disease. Other vibrios, *Aeromonas hydrophila* (a relatively newly described agent of GI disease), and *Campylobacter jejuni,* the most common cause of GI disease in many areas of the United States, have been shown to produce cytotoxins. The role that these toxins actually play in the pathogenesis of the disease syndromes is not yet delineated. With the exception of tests for the toxins (cytotoxin, enterotoxin) of *C. difficile,* tests for cytotoxins are not routinely performed.

Enterotoxins cause an alteration in the metabolic activity of intestinal epithelial cells, resulting in an outpouring of electrolytes and fluid into the lumen. They act primarily in the jejunum and upper ileum, where the most fluid transport takes place. The stool of patients with enterotoxic diarrheal disease involving the small bowel is profuse and watery, and polymorphonuclear neutrophils or blood are not prominent features. The classical example of an enterotoxin is that of *Vibrio cholerae,* called *choleragen* (Figure 18.3). This toxin consists of two subunits, A and B.[14] The A subunit is composed of one molecule of A1, the toxic moiety, and one molecule of A2, which binds an A1 subunit to five B subunits. The B subunits bind the toxin to a receptor (a ganglioside, an acidic glycolipid) on the intestinal cell membrane. Once the toxin binds, it acts on adenylate cyclase enzyme, which catalyzes the transformation of adenosine triphosphate (ATP) to cyclic adenosine monophosphate (cAMP). Increased levels of cAMP stimulate the cell to actively secrete ions into the intestinal lumen. To maintain osmotic stabilization, the cells then secrete fluid into the lumen. The fluid is drawn from the intravascular fluid store of the body. Patients therefore can become dehydrated and hypotensive very rapidly, as mentioned in Chapter 31. Oral fluid replacement with a glucose-containing electrolyte solution adequately treats most patients with cholera, since fluid reabsorption is coupled to glucose transport into the cells. *V. cholerae* must be able to multiply to sufficient numbers to produce toxin in the infected host. To this end, the bacteria possess motility and adherence factors, as discussed later. *V. cholerae* inhabit sea and stagnant water and are spread in contaminated water. They have been isolated from coastal waters of several states, and there have been a few cases of cholera reported that were acquired in the United States in recent years.

Other organisms also produce a choleragen-like enterotoxin. A group of vibrios similar to *V.*

FIGURE 18.3

Diagrammatic representation of the structure and action of cholera toxin.

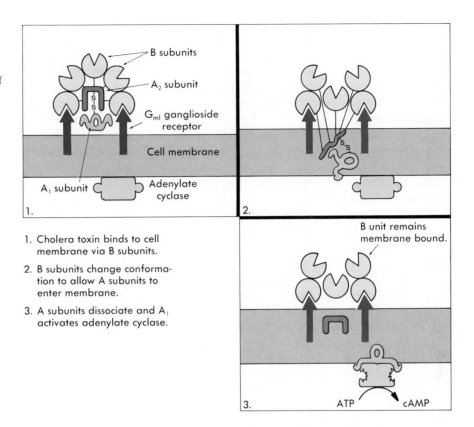

1. Cholera toxin binds to cell membrane via B subunits.

2. B subunits change conformation to allow A subunits to enter membrane.

3. A subunits dissociate and A_1 activates adenylate cyclase.

cholerae but serologically different, known as the *noncholera vibrios,* produce disease clinically identical to cholera, effected by a very similar toxin. The heat-labile toxin (LT) elaborated by certain strains of *E. coli,* called enterotoxigenic *E. coli* (ETEC), is similar to choleragen, sharing cross-reactive antigenic determinants. The enterotoxins of some *Salmonella* sp. (including *S. arizonae*), *Vibrio parahaemolyticus,* the *Campylobacter jejuni* group, *Clostridium perfringens, Clostridium difficile, Bacillus cereus, Aeromonas, Shigella dysenteriae,* and many other Enterobacteriaceae also cause positive reactions in at least one of the tests for enterotoxin. The exact contribution of these enterotoxins to the pathogenicity of most stool pathogens remains to be elucidated.

Certain strains of *E. coli,* in addition to producing a heat-labile toxin similar to choleragen (LT), also produce a heat-stable toxin (ST) with other properties. Although ST also promotes fluid secretion into the intestinal lumen, its effect is mediated by activation of guanylate cyclase, resulting in increased levels of cyclic guanylate monophosphate (GMP), which yields the same net effect as increased cAMP. Tests for ST include enzyme-linked immunosorbent assay (ELISA), immunodiffusion, and the classical *suckling mouse assay,* in which culture filtrate is placed into the stomach of a suckling mouse, with the intestinal contents later measured for fluid volume increase.

Several tests are available for the detection of enterotoxin, although these are not performed routinely in diagnostic microbiology laboratories. The classical test is the *ligated rabbit ileal loop test.* Portions of rabbit ileum are tied off and injected with the culture filtrate being tested; known toxin-positive control material is also injected into separate segments. Within hours the rabbit is sacrificed and the intestinal loops are examined. Those loops that received toxin are found to be distended with fluid that has been secreted into the loop, whereas the loops without enterotoxin are flat with no distention, resembling the control loops. Ten to 12 different fluids along with controls may be tested for enterotoxin in the same rabbit.

Other tests for enterotoxin include the *Chinese hamster ovary (CHO) cell assay* and the *Y-1 adrenal cell assay.* These are tissue culture assays in which a tissue culture consisting of CHO cells or Y-1 adrenal cells, known to be sensitive to the action of enterotoxin, is exposed to test fluids and examined for evidence of damage ("rounding up"). Since enterotoxins are toxigenic, homologous antibodies can be used to identify them specifically. Immunodiffusion, ELISA, and latex agglutination tests are all available to identify specific toxins.

Molecular probes for toxin detection are available for research.

GASTROINTESTINAL INFECTIONS ASSOCIATED WITH DESTRUCTION OF CELL FUNCTION WITHOUT TOXINS

Organisms that multiply within certain cells of the mucosa, disrupting normal function of those cells, can cause diarrheal disease. Viruses probably act in this manner, multiplying within cells of the small bowel. Diarrhea of viral etiology is not associated with the presence of blood and white cells. Hepatitis A, B, and C; rotavirus; Norwalk-like agents; and occasionally adenovirus have been associated with diarrheal symptoms in infected patients, whereas the role of other viruses, such as coxsackievirus and echoviruses, in the etiology of diarrhea has not been established. Viral diarrhea may be the second most common infectious disease among people in the United States (after upper respiratory tract infections). Nursery and day-care center outbreaks of rotavirus diarrhea have been reported. Although young children are at most risk, close adult contacts may also become infected. Rotaviruses and Norwalk-like agents are both visualized by electron microscopy within the absorptive cells at the ends of the intestinal villi, where they multiply and destroy cellular function. The villi become shortened, and inflammatory cells infiltrate the mucosa, further contributing to the pathological condition. Hepatitis viruses have not been found in intestinal cells, but some multiplication may occur there, since diarrhea is a common early symptom of hepatitis. Other viruses, particularly enteroviruses, can be recovered from feces, although the disease-producing activity occurs in other sites. The etiology of diarrhea associated with an overgrowth of *Candida* sp. in the bowel, especially in patients taking broad-spectrum antibiotics, has not been established. It is not caused by *Candida* and probably is similar to other types of antibiotic-associated diarrhea without PMC. Filamentous fungi may occasionally be associated with altered GI function, particularly in immunosuppressed patients.

GASTROINTESTINAL INFECTIONS ASSOCIATED WITH INVASIVE MICROORGANISMS

Certain parasites, particularly *Entamoeba histolytica* and *Balantidium coli,* invade the intestinal epithelium of the colon as a primary site of infection. The ensuing amebic dysentery is characterized by blood and numerous white blood cells, and the patient experiences cramping and tenesmus. Other parasites that are acquired by ingestion, such as *Schistosoma* sp. and *Trichinella,* may cause transient bloody diarrhea and pain during migration through the intestinal mucosa to their preferred sites within the host.

Bacteria that invade the epithelial cells also produce symptoms of dysentery. The most common etiological agent is *Shigella,* although strains of *E. coli* (enteroinvasive) may be associated with an identical syndrome. The bacteria penetrate the superficial layers of the mucosa, rarely passing into deeper host tissues. Therefore, invasiveness is a virulence factor for these organisms. The classical test for invasiveness is the *Sereny test,* in which a suspension of the organism is placed within the conjunctival sac of a guinea pig or rabbit. Within a short time an invasive organism penetrates the epithelial cells, causing a purulent and exudative conjunctivitis. Although tissue culture methods for determining invasiveness have been described, this factor is not tested for in routine clinical laboratories.

Other bacteria not only invade superficial epithelia but also may invade the blood vessels to disseminate systemically. Certain *Salmonella* serotypes, such as *S. typhi* and *S. choleraesuis,* are more likely to do this. *S. typhi* and *S. choleraesuis* are etiological agents of enteric fever, systemic diseases characterized by fever, headache, vomiting, and sometimes constipation (rather than diarrhea). The organisms can be recovered from the blood, stool, and occasionally the duodenal fluid of patients with this syndrome. Invasiveness is also thought to contribute to the pathogenesis of disease associated with species of vibrios, campylobacters, *Yersinia enterocolitica, Plesiomonas shigelloides,* and *Edwardsiella tarda.*[2]

GASTROINTESTINAL INFECTIONS ASSOCIATED WITH ADHERENT MICROORGANISMS

Giardia lamblia has increasingly become more common as an etiological agent of GI disease in the United States. Excreted into fresh water by natural animal hosts such as the beaver, the organism can be acquired by drinking stream water or even city water in some localities, particularly in the Rocky Mountain states, as well as throughout the world. The organism, a flagellated protozoan, adheres to the intestinal mucosa of the small bowel, possibly by means of a ventral sucker, destroying the ability of the mucosal cells to participate in normal secretion and absorption. No evidence indicates invasion or toxin production.

Yet another virulence mechanism for *E. coli* may be adherence. Certain strains have been implicated as etiological agents of malabsorption diarrhea; the strains have no other detectable pathogenic mechanisms but are able to multiply in proximity to the intestinal epithelial cells. This has been documented by electron microscopic studies in both rabbits and humans. *Cryptosporidium* and *Isospora* sp., etiological agents of diarrhea in animals and poultry and more recently recognized as causing human disease, probably also act by adhering to intestinal mucosa and disrupting function. Cryptosporidia are often seen in the diarrhea of patients with acquired immunodeficiency syndrome (AIDS), as well as in travelers' diarrhea, day-care epidemics, and diarrhea in people with animal exposure; cryptosporidia and *Isospora* may cause severe, protracted diarrhea in AIDS patients. Other coccidian parasites, such as microsporidia, produce diarrhea by destroying intestinal cell function.

OTHER INFECTIOUS CAUSES OF GASTROINTESTINAL DISEASE

The curved organism associated with gastritis, called *Helicobacter pylori,* is seen on the surface of gastric epithelial cells of patients with gastritis and peptic ulcer. Recent studies have shown it to be causally associated with these diseases. This organism is recovered from gastric biopsy material obtained endoscopically, but not from stool.

Unusual agents and those that have not been cultured, such as the mycobacteria that may be associated with Crohn's disease and the bacterium associated with Whipple's disease, identified by molecular methods as a new agent, *Trophyrema whippelii,* are also candidates as etiological agents of GI disease. Occasionally, stool cultures from patients with diarrheal disease yield heavy growths of organisms such as enterococci, *Pseudomonas* sp. or *Klebsiella pneumoniae,* not usually found in such numbers as normal flora. Only anecdotal evidence suggests that these organisms actually contribute to the pathogenesis of the diarrhea. Agents of sexually transmitted disease may cause GI symptoms when they are introduced into the colon via sexual intercourse. *Mycobacterium avium-intracellulare* may be transmitted in this way, going on to cause systemic disease in patients with AIDS. Homosexual practices predispose people to infections such as gonococcal and herpetic proctitis, syphilis, and others. Chapter 24 contains further discussion of patients at risk for these and other causes of GI disease. The pathogenesis of infections resulting from *Blastocystis hominis* (a possible coccidian etiological agent of human diarrheal disease) is not well documented, although these organisms are associated with GI symptoms.

COLLECTION AND TRANSPORT OF STOOL SPECIMENS

Specimens that can be delivered to the laboratory within 1 hour may be collected in a clean, waxed, cardboard or plastic container. Stool for direct wet-mount examinations, *Clostridium difficile* toxin assays, concentration for detection of ova and parasites, immunoelectron microscopy for detection of viruses, and ELISA or latex agglutination test for rotavirus must be sent to the laboratory without any added preservatives or liquids. Volume of stool at least equal to 1 teaspoon (5 ml) is necessary for most procedures. If a delay longer than 2 hours is anticipated for stools for bacterial culture, the specimen should be placed in transport medium. For bacterial culture, Cary-Blair transport medium has been shown to preserve best the viability of intestinal pathogens, including *Campylobacter* and *Vibrio* sp. However, the media produced by different manufacturers can vary.[8] Some workers recommend reducing the agar content of Cary-Blair medium from 0.5% to 0.16% for maintenance of *Campylobacter* sp. Buffered glycerol transport medium does not maintain these bacteria. Several manufacturers produce a small vial of Cary-Blair with a self-contained plastic scoop suitable for collecting samples. *Shigella* sp. are delicate; best recovery is obtained by inoculating media (brought to room temperature) directly at the bedside. If direct plating is not convenient, a transport medium of equal parts of glycerol and 0.033 M phosphate buffer (pH 7.0) supports viability of *Shigella* better than Cary-Blair. For this purpose, maintenance of the glycerol transport medium at refrigerator or freezer temperatures yields better results.

For detection of ova and parasites, specimen preservation with a fixative is recommended for visual examination (Chapter 45). Stools for virus culture must be refrigerated if they are not inoculated onto media within 2 hours. A rectal swab, transported in modified Stuart's transport medium, is adequate for recovery of most viruses from feces. Stools received for detection of *Clostridium difficile* cytotoxin should be refrigerated for a maximum of 7 days until they are tested. The stool-free filtrate can be frozen without seriously affecting the qualitative results (although some loss of titer occurs). Repeated freezing and thawing as well as freezing whole stool result in fewer positive results.

If an etiological agent is not isolated with the first culture or visual examination, two additional specimens should be submitted to the laboratory over the next few days. Since organisms may be shed intermittently, collection of specimens at different times over several days enhances recovery. Certain infectious agents, such as *Giardia*, may be difficult to detect, requiring the processing of multiple specimens over many weeks, duodenal aspirates (in the case of *Giardia*), or additional alternative methods.

If stool is unavailable, a rectal swab may be substituted as a specimen for bacterial or viral culture, but it is not as good, particularly for diagnosis in adults. For suspected intestinal infection with *Campylobacter*, the swab must be placed in transport medium immediately to avoid drying. Swabs are not acceptable for detection of parasites, toxins, or viral antigens. The swab should be placed in Cary-Blair transport medium. Further discussion of handling of stool can be found in an article by Gilligan et al.[2]

Other specimens that may be obtained for diagnosis of GI tract infection include duodenal aspirates (usually for detection of *Giardia* or *Strongyloides*), which should be examined immediately by direct microscopy for presence of motile trophozoites, cultured for bacteria, and placed into polyvinyl alcohol (PVA) fixative. The laboratory should be informed in advance that such a specimen is going to be collected so that the specimen can be processed and examined efficiently. The *string test* has proved useful for diagnosis of duodenal parasites such as *Giardia* and for isolation of *Salmonella typhi* from carriers and patients with acute typhoid fever. A weighted gelatin capsule containing a tightly wound length of string is swallowed by the patient, leaving the end of the string protruding from the patient's mouth (and taped to the cheek). After a predetermined time period, during which the capsule reaches the duodenum and dissolves, the string is pulled back out, covered with duodenal contents. This string must be immediately delivered to the laboratory. In the laboratory the technologist strips the mucus and secretions attached to the string with sterile gloved fingers, depositing some material on slides for direct examination, some material into fixative for preparation of permanent stained mounts, and inoculating some material to appropriate media for isolation of bacteria.

Biopsy material and material from mucosal lesions may be obtained from patients during proctoscopy or sigmoidoscopy. Such material is often fixed for histological and pathological examinations. Microbiologists should handle such material as if it were feces, staining or culturing as requested. For detection of *Helicobacter pylori*, biopsy material should be transported to the laboratory without diluent, ground in dextrose-phosphate broth, and inoculated to culture media within 1 hour of collection.[10]

DIRECT DETECTION OF AGENTS OF GASTROENTERITIS IN FECES

A direct wet mount of fecal material, particularly with liquid or unformed stool, is the fastest method for detection of motile trophozoites of *Dientamoeba fragilis*, *Entamoeba*, *Giardia*, and other intestinal parasites (that may not contribute to disease but may alert the microbiologist to the possibility of finding other parasites), such as *Entamoeba coli*, *Endolimax nana*, *Chilomastix mesnili*, and *Trichomonas hominis*. Occasionally the larvae or adult worms of other parasites may be visualized. Experienced observers can also see the refractile forms of cryptosporidia and many types of cysts on the direct wet mount. Examination of a direct wet mount of fecal material taken from an area with blood or mucus, with the addition of an equal portion of Loeffler's methylene blue, is helpful for detection of fecal leukocytes, which occasionally aids in differentiating among the various types of diarrheal syndromes. If present in sufficient numbers, the ova of intestinal parasites can be seen. Under phase-contrast and darkfield microscopy, the darting motility and curved forms of *Campylobacter* may be observed in a warm sample. Water, which will immobilize *Campylobacter*, should not be used. Trained observers working in endemic areas can recognize the characteristic appearance and motility of *Vibrio cholerae*.

Immunoelectron microscopy is a very rapid method for detecting rotavirus and Norwalk-like agents. Details of the procedures are discussed in Chapter 43. For most laboratories, electron microscopy is unavailable, and rotavirus is detected using a solid-phase ELISA procedure or a latex agglutination test. Several manufacturers produce systems for detection of the antigen of rotavirus, including microbroth dilution well systems, plastic bead systems, and latex agglutination systems. For the immunoassay systems, the feces are diluted to suspend the viral antigens, the solid material is removed by filtration or centrifugation, and the filtrate is added to the immunoassay solid phase on which antibody to rotavirus has been adsorbed. After an attachment step, excess material is washed off and an enzyme-conjugated antibody is added. If viral antigens are present, the conjugate will adhere to the solid phase as well. With the addition of the proper substrate, the enzyme effects a visible color change (further discussed in Chapter 13). Instructions for each of the commercial systems vary, but the results have been uniformly good, with detection of rotavirus by ELISA comparable to electron microscopy in sensitivity. The latex agglutination test exhibits similar sensitivity with a much less complex system. This high sensitivity is probably a result of the huge numbers of virus particles shed in the feces of patients with infection. ELISA methods are also available now for detection of antigens of *Giardia lamblia*. Because this organism is often difficult to detect visually, the commercial ELISA may prove to be a welcome diagnostic alternative to direct visualization, even when one considers the added sensitivity of monoclonal fluorescent stain reagents. An accurate, sensitive direct fluorescent antibody (DFA) stain for giardiasis and cryptosporidiosis is now commercially available. ELISA methods have been evaluated for detection of certain bacterial pathogens.[9]

DNA probe technology is available commercially for detection of at least two fecal pathogens, *Campylobacter* sp. and rotavirus. After the cells are lysed in stool, nucleic acids are extracted and fixed to a filter paper matrix, and target sequences are detected by enzyme-labeled probes (SNAP [synthetic nucleic acid probe], Molecular Biosystems, San Diego). This technology is still in its infancy, although the concept is promising. Probes for *Salmonella*, *Shigella*, and *Yersinia* are being evaluated. Disadvantages with probe technology are that the organism itself is not available for susceptibility testing, which is important for certain bacterial pathogens for which susceptibility patterns vary.

Feces may be Gram stained for detection of certain etiological agents. Although gram-negative bacilli cannot be differentiated, sheets of polymorphonuclear neutrophils and many clumps of typical gram-positive cocci are indicative of staphylococcal infection (Figure 18.4). Many thin, comma-shaped, gram-negative bacilli may indicate *Campylobacter* infection (if vibrios have been ruled out), and gram-positive bacilli, large and thin with parallel walls, may be suggestive (but not reliable enough for routine use[1,16]) of overgrowth by *Clostridium difficile* (Figure 18.5). An acid-fast stain (or new DFA from Meridian Diagnostics for *Cryptosporidium*) should be used to detect *Cryptosporidium* sp. and mycobacteria. *Isospora* sp. are also acid fast. Examination of fixed fecal material for parasites by trichrome or other stains is thoroughly covered in Chapter 45. It is recommended that a permanent stained preparation be made from all stool specimens received for detection of parasites. Giardia and cryptosporidia can easily and unequivocally be visualized with a monoclonal antibody fluorescent stain (Meridian Diagnostics). This stain is recommended for specimens from patients who probably have these parasites or for specimens that yield suspicious results by standard methods.

FIGURE 18.4

Gram stain of stool with staphylococcal enterocolitis.

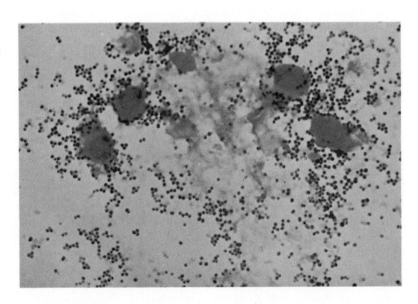

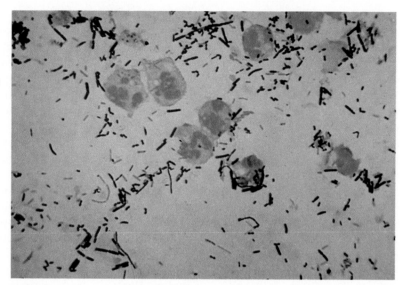

FIGURE 18.5
Gram stain of stool with *Clostridium difficile* pseudomembranous colitis. Note: Gram stain is not recommended as a diagnostic test.

CULTURE OF FECAL MATERIAL FOR ISOLATION OF ETIOLOGICAL AGENTS OF GASTROINTESTINAL DISEASE

As outlined in Chapter 7, fecal specimens for culture should be inoculated to several media for maximal yield, including solid agar and broth. The choice of media is arbitrary and based on the particular requirements of the clinician and the laboratory. Recommendations are given in this section.

CULTURES FOR BACTERIAL ETIOLOGICAL AGENTS

Stools received for routine culture in most clinical laboratories in the United States should be examined for the presence of *Campylobacter, Salmonella,* and *Shigella* sp. under all circumstances. An argument can also be made for performing procedures for detection of rotavirus as part of the standard protocol. If the incidence of *Yersinia enterocolitica* gastroenteritis is high enough in the area served by the laboratory, this agent should also be sought routinely. Special procedures for detection of vibrios, *Clostridium difficile,* Verotoxin-producing *E. coli,* and mycobacteria should be carried out on request. *Vibrio* disease has become increasingly prevalent among travelers, as well as among individuals living in high-risk areas of the United States (seacoasts). We also recommend that detection of *Aeromonas* sp. should be incorporated into the routine stool culture procedures.

Specimens received for detection of the most frequently isolated Enterobacteriaceae and *Salmonella* and *Shigella* sp. should be plated to a supportive medium, a slightly selective and differential medium, and a moderately selective medium. A highly selective medium does not seem to be cost-effective for most microbiology laboratories. Blood agar (tryptic soy agar with 5% sheep blood) is an excellent general supportive medium. Blood agar medium allows growth of yeast species, staphylococci, and enterococci, in addition to gram-negative bacilli. If a gram-positive organism is predominant, it will be detected on the blood agar. For routine cultures, use of an additional selective medium for detection of small numbers of gram-positive organisms does not seem warranted. Another benefit of blood agar is that it will allow oxidase testing of colonies. We recommend that several colonies (that do not resemble *Pseudomonas*) from the third or fourth quadrant be routinely screened for production of cytochrome oxidase. If many *Aeromonas, Vibrio,* or *Plesiomonas* sp. are present, they will be detected by their positive oxidase reaction.

The moderately selective agar should support growth of most Enterobacteriaceae, vibrios, and other possible pathogens; MacConkey agar seems to work very well. Some laboratories use eosin–methylene blue (EMB), which is slightly more inhibitory. All lactose-negative colonies should be tested further, ensuring adequate detection of most vibrios and most pathogenic Enterobacteriaceae. Lactose-positive vibrios *(V. vulnificus),* pathogenic *E. coli,* some *Aeromonas,* and *Plesiomonas* sp. may not be distinctive on MacConkey agar.

The specimen should also be inoculated to a moderately selective agar, such as Hektoen enteric (HE) or xylose-lysine desoxycholate (XLD) media. These media inhibit growth of most Enterobacteriaceae, allowing *Salmonella* and *Shigella* sp.

to be detected. Colony morphologies of lactose-negative, lactose-positive, and H_2S-producing organisms are outlined in Chapter 9. The highly selective brilliant green (BG) or bismuth sulfite (BS) medium may be used for detection of *Salmonella* sp; they should be heavily inoculated, since they are quite inhibitory. *S. typhi* will often display silver metallic colonies with black halos surrounding them on BS (Figure 18.6), whereas salmonellae show pink colonies on BG. Other potential pathogens do not survive on these agars. All these media are incubated at 35° to 37° C in air and examined at 24 and 48 hours for suspicious colonies.

Verotoxin-producing *E. coli* serotype O157:H7 can be detected by screening on a sorbitol-containing selective medium (sorbitol-MacConkey, which contains 1% D-sorbitol instead of lactose), since these strains are unable to ferment sorbitol and 95% of other *E. coli* are sorbitol positive. Sorbitol-negative *E. coli* strains grow as colorless colonies and can be reported as suspicious for *E. coli* O157:H7, whereas sorbitol-fermenting *E. coli* develop a pink to red color on the media (Figure 18.7). Suspicious colonies should be serotyped by slide and tube agglutination tests.[5,13] Final identification of Verotoxin-producing *E. coli* depends on the detection of the toxin by cell culture assay. Alternatively, stool filtrate may be tested for toxin production in Vero cells. Comprehensive and practical methods are described by Karmali,[5] and a simpler approach is presented by Haldane et al.[4]

Cultures for isolation of *Campylobacter jejuni* and *Campylobacter coli* should be inoculated to Campy–blood agar (further discussed in Chapter 31). Brucella broth base has yielded less satisfactory recovery of *Campylobacter* sp. Commercially produced agar plates for isolation of campylobacters are available from several manufacturers. These plates are incubated in a microaerophilic atmosphere at 42° C and examined at 24 and 48 hours for suspicious colonies (Chapter 31). Other campylobacters associated with GI disease, such as *C. laridis, C. hyointestinalis,* and *C. fetus* ss. *fetus,* grow best at 37° C. Tenover and Gebhart[18] recommend that one selective agar be inoculated with fresh stool and one nonselective agar (trypticase-soy blood agar) be inoculated with stool filtrate that has been passed through a cellulose acetate filter of 0.65 or 0.8 μm pore size.[17] The selective agar should be incubated at 42° C and the filtered stool on nonselective agar at 37° C. Although common campylobacters are detected at 48 hours, incubating plates for 72 hours or longer yields better recovery.

Helicobacter pylori, associated with gastritis and possibly with peptic ulcer disease, can be isolated from gastric biopsy tissue by inoculating the specimen onto Skirrow agar or other *Campylobacter* agar containing blood and appropriate antimicrobial agents (cefoperazone is recommended) and incubating microaerophilically at 35° C in high humidity for up to 7 days. The organism may grow on chocolate, modified Thayer-Martin, or gonococcal agar base without supplements from some commercial suppliers; however, not all agars with the same formulas support growth of *H. pylori.* Alternatively, a small piece of tissue may be placed directly into urea broth base or onto a Christensen's urea agar slant. Demonstration of

FIGURE 18.6

Appearance of colonies of *Salmonella typhi* on bismuth sulfite agar.

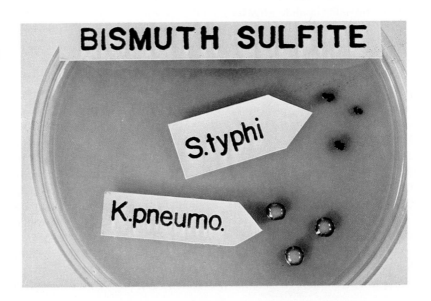

FIGURE 18.7

Appearance of *E. coli* 0157:H7 on sorbitol-MacConkey agar *(right)* and a control strain of *E. coli* that ferments sorbitol *(left)*.

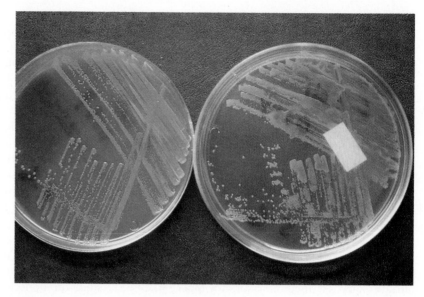

urease activity (often within 30 minutes but occasionally taking as long as 24 hours) is presumptive evidence of the presence of this organism.

Vibrios that are etiological agents of gastroenteritis may be detected by screening all oxidase-positive colonies growing on blood agar. Many vibrios are beta hemolytic. For specific detection of *V. cholerae* and other vibrios, thiosulfate-citrate–bile salts–sucrose (TCBS) agar is recommended. The agar is heavily inoculated with feces, as are the moderately selective media, and incubated in air at 35° to 37° C for up to 48 hours. Either large, flat, yellow or olive-green colonies should be tested further.

A special medium has been developed for recovery of *Yersinia enterocolitica*. Often called CIN agar, this medium is made from a peptone agar base with mannitol, neutral red, crystal violet, and the antimicrobial agents cefsulodin, 3,3′,4′,5-tetrachlorosalicylanilide (irgasan), and novobiocin. CIN agar is available commercially as powder or plates. The fecal specimen is streaked onto the agar, and the plate is incubated at room temperature for 48 hours. Colonies of *Y. enterocolitica* appear bright pink with a red center. Very few other organisms grow on CIN agar, although some *Aeromonas* sp. have been found to grow very well with some formulations of this medium. If *Aeromonas* is suspected as the etiological agent of diarrheal disease, we recommend inoculation of CIN agar at 35° C in addition to other media. A blood agar medium containing ampicillin has been used for selective recovery of *Aeromonas* with good success.

Enrichment broths are often used for enhanced recovery of *Salmonella*, *Shigella*, and *Campylobacter*, although *Shigella* will usually not survive enrichment. Gram-negative broth (Hajna GN) or selenite F broth yields good recovery. Campy-thioglycolate enrichment broth increases the yields of positive cultures for *Campylobacter* sp., although it is not necessary for routine use. Enrichment broths for Enterobacteriaceae should be incubated in air at 35° C for 6 to 8 hours, and then several drops should be subcultured to at least two selective medium plates. A commercial system that allows such broth to be tested for antigen of *Salmonella* or *Shigella* directly has been described; however, the reported sensitivity is lower than desired. Stool would be inoculated to broth only initially; those broths that tested negative could be discarded without subculturing. These systems have not been fully evaluated yet. Enrichment broth for *Campylobacter* is refrigerated overnight or for a minimum of 8 hours before a few drops are plated to *Campylobacter* agar and incubated at 42° C in a microaerophilic atmosphere. *Yersinia enterocolitica* also can be "cold enriched." A swab of fecal material is inoculated to a tube of 5 ml of phosphate-buffered saline (pH 7.2), and the tube is refrigerated for up to 21 days. A few drops are subcultured to CIN agar every 4 days and incubated in air at room temperature for up to 3 days. Such a cold-enrichment procedure is most useful for epidemiological studies, since results may not be known soon enough to benefit an individual patient. Alkaline peptone water has been shown to adequately enrich for vibrios. After 8 hours' incubation, several drops from the broth should be inoculated to TCBS agar and incubated in air at 35°C.

Clostridium difficile may be recovered by plating stool to cycloserine-cefoxitin-fructose egg yolk agar (CCFA). Two or three drops (or 0.1 g) of stool are inoculated to CCFA and streaked for isolation. Plates are incubated anaerobically for 48 hours and examined for the characteristic large (4 mm diameter) yellow colonies with an opalescent sheen.[12] *C. difficile* is one of the most oxygen sensitive of the clinically important clostridia. For this reason, it may fail to grow from material inoculated to plates in the air and incubated in an anaerobic jar or plastic bag. Best recovery is obtained on prereduced plates inoculated with fecal material in an anaerobic chamber. *C. difficile* colonies will fluoresce yellow-green under ultraviolet light, and they exhibit a strong horse manure–like odor. *C. difficile* can be presumptively identified on the basis of colony and Gram stain morphology. Definitive identification is outlined in Chapter 36. Detection of *C. difficile* in stool of patients with diarrhea may be suggestive of PMC; however, this method is not specific, since up to 20% of hospitalized adult patients may be colonized with the organism, and the carriage rate in asymptomatic infants is high. At least one additional test (discussed later) should be performed for a more specific laboratory diagnosis.

Mycobacterium avium-intracellulare complex organisms may be isolated from the stool of patients with AIDS, and intestinal tuberculosis may still be suspected in some unusual cases. Methods for culturing acid-fast bacilli from feces are discussed in Chapter 42. It is not recommended that clinical microbiology laboratories process stool for *M. tuberculosis*, since the procedure is so cumbersome. Reference laboratories are better equipped for such requests.

CULTURES FOR VIRAL ETIOLOGICAL AGENTS

In addition to the viral agents of gastroenteritis, rotavirus, and Norwalk-like agent, whose detection is by latex agglutination, enzyme immunoassay, or electron microscopy, many viral agents of disease are routinely sought in feces. Fresh feces or rectal swabs are inoculated to antibiotic-containing media for 30 minutes before inoculating to tissue culture. Choice of tissue culture is determined by the etiological agents sought, but feces are usually inoculated to primary monkey kidney, human fetal diploid fibroblast, and a human epithelial continuous cell line (Chapter 43).

LABORATORY DIAGNOSIS OF *CLOSTRIDIUM DIFFICILE*-ASSOCIATED DIARRHEA

The definitive diagnosis of *C. difficile*–associated diarrhea is based on clinical criteria combined with laboratory testing.[12] Visualization of characteristic pseudomembrane or plaque (Figure 18.2) on endoscopy is diagnostic for pseudomembranous colitis and, with the appropriate history of prior antibiotic use, meets the criteria for diagnosis of antibiotic-associated pseudomembranous colitis. No one laboratory test will establish that diagnosis unequivocally; three tests are currently available for routine use. Culture on CCFA medium detects the organism in more than 95% of patients with *C. difficile*–associated diarrhea, although this test is not specific, and culture plates cannot be examined until after 48 hours of anaerobic incubation. Detection of cytotoxin by tissue culture is probably most closely associated with disease, although this test also has some false-positive and many false-negative results, and the results also are not available for 48 hours. The tissue culture assay can be performed using any of several cell lines. Although it is possible to maintain a cell culture in the laboratory, cell lines are also available commercially. Two prepared microdilution trays containing tissue culture cells are available for the *C. difficile* toxin assay (Bartels Immunodiagnostics; Baxter Healthcare Systems, Issaquah, Wash.). With the availability of these systems, the cytotoxicity assay is within the ability of most clinical microbiology laboratories. A procedure for passage and maintenance of Hep-2 cells, another recommended cell line for the cytotoxic assay, is detailed in Chapter 43. Currently, several enzyme immunoassay (EIA) tests for toxin A and/or B are available, as is a rapid latex agglutination assay for detection of a *C. difficile*–associated protein. All have shown reasonable correlation with clinical diagnoses of *C. difficile*–associated disease, but with no better sensitivity than cytotoxin testing.[12] We recommend that microbiologists decide which tests best suit the needs of their clinicians and offer more than one test, perhaps in a reflexive panel. Several workers have published in this area,[1,6,11] and current knowledge is summarized in reviews.[12,15]

REFERENCES

1. Gerding, D.N., Olson, M.M., Peterson, L.R., et al. 1986. Prospective case-controlled epidemiologic study of *Clostridium difficile*–associated diarrhea and colitis in adults. Arch. Intern. Med. 146:95.
2. Gilligan P.H., Janda J.M., Karmali M.A., et al. 1992. Laboratory diagnosis of bacterial diarrhea. Cumitech 12A:1. American Society for Microbiology. Washington, D.C.
3. Guerrant, R.L. 1991. Bacterial and protozoal gastroenteritis. N. Engl. J. Med. 325:327.

4. Haldane, D.J.M., Damm, M.A.S., and Anderson, J.D. 1986. Improved biochemical screening procedure for small clinical laboratories for Vero (Shiga-like)-toxin-producing strains of *Escherichia coli* O157:H7. J. Clin. Microbiol. 24:652.

5. Karmali, M.A. 1987. Laboratory diagnosis of verotoxin-producing *Escherichia coli* infections. Clin. Microbiol. Newsletter 9:65.

6. Kelly, M.T., Champagne, S.G., Sherlock, C.H., et al. 1987. Commercial latex agglutination test for detection of *Clostridium difficile*–associated diarrhea. J. Clin. Microbiol. 25:1244.

7. Levine, M.M. 1987. *Escherichia coli* that cause diarrhea: enterotoxigenic, enteropathogenic, enteroinvasive, enterohemorrhagic, and enteroadherent. J. Infect. Dis. 155:377.

8. Mundy, L.S., Shanholtzer, C.J., Willard, K.E., and Peterson, L.R. 1991. An evaluation of three commercial fecal transport systems for the recovery of enteric pathogens. Am. J. Clin. Pathol. 96:364.

9. Pál, T., Páscsa, A.S., Emödy, L., et al. 1985. Modified enzyme-linked immunosorbent assay for detecting enteroinvasive *Escherichia coli* and virulent *Shigella* strains. J. Clin. Microbiol. 21:415.

10. Parsonnet, J., Welch, K., Compton, C., et al. 1988. Simple microbiologic detection of *Campylobacter pylori*. J. Clin. Microbiol. 26:948.

11. Peterson, L.R., Holter, J.J., Shanholtzer, C.J., et al. 1986. Detection of *Clostridium difficile* toxins A (enterotoxin) and B (cytotoxin) in clinical specimens (evaluation of a latex agglutination test). Am. J. Clin. Pathol. 86:208.

12. Peterson, L.R., and Kely, P.J. 1993. The clinical microbiology laboratory in the diagnosis and treatment of *C. difficile*–associated disease. Infect. Dis. Clin. North Am. (in press).

13. Ratnam, S. 1988. Characterization of *Escherichia coli* serotype O157:H7. J. Clin. Microbiol. 26:2006.

14. Ribi, H.O., Ludwig, D.S., Mercer, K.L., et al. 1988. Three-dimensional structure of cholera toxin penetrating a lipid membrane. Science 239:1272.

15. Rolfe, R.D., and Finegold, S.M., editors. 1988. *Clostridium difficile*: its role in intestinal disease, Academic Press, Orlando, Fla.

16. Shanholtzer, C.J., Peterson, L.R., Olson, M.M., and Gerding, D.N. 1983. Prospective study of Gram-stained stool smears in diagnosis of *Clostridium difficile* colitis. J. Clin. Microbiol. 17:906.

17. Steele, T.W., and McDermott, W. 1984. Technical note: the use of membrane filters applied directly to the surface of agar plates for the identification of *Campylobacter jejuni* from feces. Pathology 16:263.

18. Tenover, F.C., and Gebhart, C.J. 1988. Isolation and identification of *Campylobacter* species. Clin. Microbiol. Newsletter 10:81.

BIBLIOGRAPHY

Baron, E.J, and Finegold, S.M. 1990. Bailey and Scott's diagnostic microbiology. ed 8. Mosby, St. Louis. (Appendices provide detailed formulas of media and stains.)

Brenden, R.A., Miller, M.A., and Janda, J.M. 1988. Clinical disease spectrum and pathogenic factors associated with *Plesiomonas shigelloides* infections in humans. Rev. Infect. Dis. 10:303.

Du Pont, H.L., and Pickering, L.K. 1980. Infections of the gastrointestinal tract. In Greenough, W.B., and Merigan, T.C., editors. Current topics in infectious disease. Plenum, New York.

George, W.L., Nakata, M.M., Thompson, J., and White, M.L. 1985. *Aeromonas*-related diarrhea in adults. Arch. Intern. Med. 145:2207.

Gradus, M.S. 1986. Public health criteria for the diagnosis of foodborne illness. Clin. Microbiol. Newsletter 8:85.

Guerrant, R.L. 1985. Principles and definition of syndromes of gastrointestinal infections and food poisoning. In Mandell, G.L., Douglas, R.G., Jr., and Bennett, J.E., editors. Principles and practice of infectious diseases, ed. 2. John Wiley & Sons, New York.

Guerrant, R.L. 1985. Inflammatory enteritides. In Mandell, G.L., Douglas, R.G., Jr., and Bennett, J.E., editors. Principles and practice of infectious diseases, ed. 2. John Wiley & Sons, New York.

Guerrant, R.L., and Hughes, J.M. 1985. Nausea, vomiting, and noninflammatory diarrhea. In Mandell, G.L., Douglas, R.G., Jr., and Bennett, J.E., editors. Principles and practice of infectious diseases, ed 2. John Wiley & Sons, New York.

Holmberg, S.D., Schell, W.L., Fanning, G.R., et al. 1986. *Aeromonas* intestinal infections in the United States. Ann. Intern. Med. 105:683.

Holmberg, S.D., Wachsmuth, I.K., Hickman-Brenner, F.W., et al. 1986. *Plesiomonas* enteric infections in the United States. Ann. Intern. Med. 105:690.

Middlebrook, J.L., and Dorland, R.B. 1984. Bacterial toxins: cellular mechanisms of action. Microbiol. Rev. 48:199.

Nelson, J.D. 1985. Etiology and epidemiology of diarrheal diseases in the United States. Am. J. Med. 78(6B):76.

19

MICROORGANISMS ENCOUNTERED IN THE URINARY TRACT

GENERAL CONSIDERATIONS, ANATOMY, AND NORMAL FLORA

The urinary tract consists of the kidneys, the ureters, the bladder, and the urethra (Figure 19.1). Although the urethra hosts resident microflora that colonize its transitional epithelium, consisting of coagulase-negative staphylococci (excluding *S. saprophyticus*), viridans and nonhemolytic streptococci, lactobacilli, diphtheroids (*Corynebacterium* sp.), nonpathogenic *Neisseria* sp., anaerobic cocci, *Propionibacterium* sp., anaerobic gram-negative bacilli, commensal *Mycobacterium* sp., and commensal *Mycoplasma* sp., all areas of the urinary tract above the urethra in a healthy human are sterile. Potential pathogens, including gram-negative aerobic bacilli (primarily Enterobacteriaceae) and occasional yeasts, are also present as transient colonizers. Because noninvasive methods for collecting urine must rely on a specimen that has passed through a contaminated milieu, special steps must be taken with collection, and interpretation of the culture results of such specimens requires that some discriminatory criteria be used. The use of quantitative cultures for diagnosis of urinary tract infections (UTIs), first popularized by Kass,[8,9] has been modified to include separate quantitative criteria to be applied to different types of patients. Quantitative cultures of urine are still critical for diagnosis of UTI. The classical criterion of greater than 10^5 colony forming units of bacteria per milliliter (CFU/ml)

of urine is highly indicative of infection in patients with acute cystitis, pyelonephritis, and asymptomatic bacteriuria. This definition must be modified, however, for most other patient categories. This chapter discusses various aspects of UTI and its laboratory diagnosis.

EPIDEMIOLOGY AND PATHOGENESIS OF URINARY TRACT INFECTIONS

UTIs are primarily of two types: cystitis, which is infection of the bladder, and pyelonephritis, which is infection of the renal parenchyma. Perinephric abscess, as the name implies, is an abscess adjacent to the kidney. Pyelitis and ureteritis do occur but are not usually diagnosed as such. Urethritis, which occurs often, is generally discussed as a sexually transmitted disease (Chapter 20), although the acute urethral syndrome (see later discussion) is not usually placed in the sexually transmitted category, strictly speaking.

UTIs occur more often in women than men, at least partially because of the short female urethra and its proximity to the anus. Additionally, sexual activity serves to increase the chances of bacterial contamination of the female urethra; pregnancy causes anatomical and hormonal changes that favor development of UTIs; and changes in the genitourinary tract mucosa related to menopause may play a role. Colonization of the introitus by coliforms is a major background

249

FIGURE 19.1

Overview of the anatomy of the urinary tract. (From Potter, P.H., and Perry, A.G. 1985. Fundamentals of nursing. Mosby, St. Louis.)

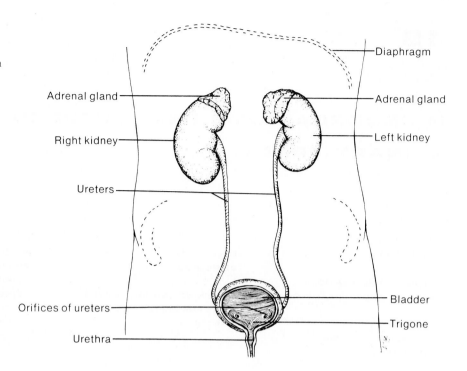

factor for recurrent bladder infection in females. Recent studies have shown that the use of contraceptive foams and gels and a diaphragm results in significantly greater colonization of the vagina than sexual intercourse with no such factors.[7] It is estimated that approximately 20% of all women have a UTI at least once, with the incidence increasing with age. Among *asymptomatic* females of childbearing age or older, a culture yielding greater than 10^5 organisms per milliliter of a single bacterial species probably corresponds to UTI. If a repeat specimen also yields the same result with the same organism, presence of infection is 95% certain. The presence of greater than 10 white blood cells/μl also substantiates UTI.

The incidence of UTI among males is low until after age 60; prostatic hypertrophy then contributes to development of UTI. Bacterial prostatitis is the key background factor for the problem of recurrent cystitis in males. Any anatomical barrier to free flow of urine through the urinary tract contributes to development of UTI; involved are prostatic hypertrophy, neurogenic disorders, tumors, and stones. Previous infection with certain urea-splitting organisms, notably *Proteus* and related species, is often associated with the formation of urinary stones, which further predispose the patient to infection. The introduction of a foreign body into the urinary tract, especially one that remains in place for a time (e.g., Foley catheter), carries a substantial risk of leading to infection, particularly if obstruction is present. As

many as 20% of all hospitalized patients who receive short-term catheterization develop a UTI. Consequently, UTI is the most common nosocomial infection in the United States, and the infected urinary tract is the most frequent source of bacteremia.

Most infections involving the kidneys are acquired by the *ascending route* (bacteria migrate from the bladder to reach the tissues of the kidneys). However, yeast (usually *Candida albicans*), *Mycobacterium tuberculosis, Salmonella* sp., *Leptospira* sp., or *Staphylococcus aureus* in the urine often indicate pyelonephritis acquired via **hematogeneous** spread, or the *descending route*.

Among the bacteria most frequently isolated from patients (overwhelmingly women) with community-acquired acute cystitis are *Escherichia coli, Klebsiella* sp., other Enterobacteriaceae, and *Staphylococcus saprophyticus.* Hospitalized patients are most likely to be infected by *E. coli, Klebsiella* sp., *Proteus mirabilis,* other Enterobacteriaceae, *Pseudomonas aeruginosa,* and enterococci. Other less frequently isolated agents are other gram-negative bacilli, such as *Acinetobacter* and *Alcaligenes* sp., other *Pseudomonas* sp., *Citrobacter* sp., *Gardnerella vaginalis,* beta-hemolytic streptococci, and *Neisseria gonorrhoeae. Trichomonas vaginalis* may occasionally be observed in urinary sediment, and *Schistosoma haematobium* can lodge in the urinary tract and release eggs into the urine.

In most hospitalized patients, UTI is preceded by urinary catheterization or other manipulation of the urinary tract. Although the pathogenesis of catheter-associated UTI is not fully understood, certain possibilities have been discussed. It is certain that soon after hospitalization, patients become colonized with bacteria endemic to the institution, often gram-negative aerobic and facultative bacilli carrying resistance markers. These bacteria colonize patients' skin, gastrointestinal tract, and mucous membranes, including the anterior urethra. With insertion of a catheter, the bacteria may be pushed along the urethra into the bladder or, with an indwelling catheter, may migrate along the track between the catheter and the urethral mucosa, gaining access to the bladder. Some have suggested that nosocomial UTIs caused by certain bacteria, notably *Serratia marcescens,* are more likely to result in secondary bacteremia. Some women are more likely to have chronic or recurrent UTIs. Studies by Svanborg-Eden and de Man[17] have shown stronger binding of *E. coli* (isolated from infected urine) to the genitourinary tract epithelial cells of infection-prone women than to the cells of noninfected control subjects. The organisms themselves may have mechanisms that allow them to adhere well. Mechanisms by which other common etiological agents of UTI produce disease are not yet well understood. Their presence probably induces an inflammatory response, which in turn causes tissue destruction and the characteristic pain, burning, and urinary frequency of acute cystitis. Damage to renal parenchyma that occurs in pyelonephritis results in flank pain, fever, and other systemic symptoms, or sometimes no symptoms until much damage has occurred.

Another urinary tract infection is the **acute urethral syndrome,** studied by Stamm.[16] Women, primarily young, sexually active women, experience dysuria, frequency, and urgency but yield fewer organisms than 10^5 CFU/ml urine on culture. As many as one third of all women who seek medical attention for complaints of symptoms of acute cystitis fall into this group. Although *Chlamydia trachomatis* and *N. gonorrhoeae* urethritis, anaerobic infection, genital herpes, and vaginitis account for some cases of acute urethral syndrome, most of these women are infected with organisms identical to those that cause cystitis, but in numbers less than 10^5 CFU/ml urine. One must use a cutoff of 10^2 CFU/ml, rather than 10^5 CFU/ml, for this group of patients but must insist on concomitant pyuria. The leukocyte esterase test (discussed later) is not sensitive enough for determining pyuria in these patients.[15] Approximately 90% of these women have pyuria, an im-

portant discriminatory feature of infection. Recognition of this large class of patients with UTI that do not fit into classical categories has led to a reexamination of the criteria used to discriminate between urine specimens from patients with infection and those specimens from which culture results indicate only contamination.

Patients with true UTIs whose urine may yield fewer numbers of bacteria than the classical 10^5 CFU/ml include infants and children, males, catheterized patients, patients who have previously received antibacterial agents, patients who consume large amounts of liquid and thus dilute bladder urine, symptomatic patients (usually with pyuria), patients with urinary obstruction that may prevent organisms from being excreted, and patients with pyelonephritis acquired from hematogeneous spread (particularly yeast, *S. aureus,* and *M. tuberculosis* infections). For this reason, good communication between clinicians and microbiologists is essential for proper interpretation of urine culture results. If laboratorians also quantitate polymorphonuclear neutrophils in urine, as they should, the diagnostic value of a urine specimen is increased significantly, just as it is for sputum or other specimens.

COLLECTION AND TRANSPORT OF URINE SPECIMENS

Prevention of contamination by normal vaginal, perineal, and anterior urethral flora is the most important consideration for collection of a clinically relevant urine specimen. The least invasive procedure, the *clean-catch, midstream urine specimen collection,* must be performed carefully for optimal results, especially in females. Good patient education is essential. A method for collection of such a specimen is detailed in Chapter 7. Uncleansed first-void specimens from males were shown to be as sensitive as (but less specific than) midstream urine specimens.[10] Slightly more invasive, urinary catheterization may allow collection of bladder urine with less urethral contamination. There is the risk, however, that urethral organisms will be introduced into the bladder with the catheter. **Suprapubic bladder aspiration,** withdrawal of bladder urine directly into a syringe through a percutaneously inserted needle, ensures a contamination-free specimen. If good aseptic techniques are used, this procedure can be performed with little risk in infants, small children, and adults with full bladders.

Collection of specimens from patients with indwelling catheters requires scrupulous aseptic technique. Health care workers who manipulate a urinary catheter in any way should wear gloves. The catheter tubing should be clamped off above the port to allow collection of freshly voided urine. The catheter port or wall of the tubing should then be cleaned vigorously with 70% ethanol, and urine should be aspirated via a needle and syringe; the integrity of the closed drainage system must be maintained to prevent the introduction of organisms into the bladder. Specimens obtained from the collection bag are inappropriate, since organisms can multiply there, obscuring the true relative numbers.[21] Cultures should be obtained when patients are ill; routine monitoring does not yield clinically relevant data.[3]

Urine, being an excellent supportive medium for growth of most bacteria, must be immediately refrigerated or preserved. Bacterial counts in refrigerated (4° C) urine remain constant for as long as 24 hours. A urine transport tube, B-D Urine Culture Kit (Becton-Dickinson Vacutainer Kits, Rutherford, N.J.), containing boric acid, glycerol, and sodium formate, has been shown to preserve bacteria without refrigeration for as long as 24 hours when greater than 10^5 CFU/ml (100,000 organisms per milliliter) were present in the initial urine specimen.[20] The system may inhibit the growth of certain organisms, and it must be used with 3 to 5 ml of urine. Another preservative system (Sage Products, Carey, Ill.) is also available. Both boric acid products preserve bacterial viability in urine for 24 hours in the absence of antibiotics. A new lyophilized system appears to stabilize microbial populations for 24 hours in the presence of antibiotics.[5] For populations of patients from whom colony counts of organisms of less than 100,000/ml might be clinically significant, plating within 2 hours of collection is recommended. None of the kits has any advantage over refrigeration, except for perhaps convenience or for transport of urine from remote areas in which refrigeration is not practical.

SCREENING AND AUTOMATED TESTS FOR URINARY TRACT INFECTION

As many as 60% to 80% of all urine specimens received for culture by the acute care medical center laboratory may contain no etiological agents of infection or contaminants only. Procedures developed to identify quickly those urine specimens that will be "negative" on culture, thus to circumvent excessive use of media, technolo-

gist time, and the overnight incubation period, are discussed in this section. The Gram stain is the easiest, least expensive, and probably the most sensitive and reliable screening method for identifying urine specimens that contain greater than 10^5 CFU/ml. As outlined by Washington et al.,[19] a drop of well-mixed urine is allowed to air dry. The smear is stained and examined under oil immersion (1000×). Presence of at least one organism per oil immersion field (examining 20 fields) correlates with significant bacteriuria (>10^5 CFU/ml). The Gram stain should not be relied on for detecting polymorphonuclear leukocytes in urine.[1] Most microbiologists are unwilling to adopt this procedure as part of the routine laboratory examination of urine specimens, probably because of the low number of positive results.

Several chemical tests, impregnated onto paper strips, can be used to assess parameters that may indicate UTI. The *Griess test* for presence of nitrate-reducing enzymes, produced by the most common urinary tract pathogens, has been incorporated onto a paper strip that also tests for leukocyte esterase, an enzyme produced by polymorphonuclear neutrophils. These strips are available commercially. Triphenyltetrazolium chloride is reduced by several frequently encountered urinary tract pathogens and can be impregnated into a strip test, but this test has not achieved wide use. The sensitivity of these screening tests is not great enough to recommend their use as a stand-alone test in most circumstances.

Automated methodologies have been developed for screening urine specimens. The detection of bacterial adenosine triphosphate (ATP) by measuring light emitted by the reaction of luciferin-luciferase is discussed in Chapter 12. After somatic cell ATP is removed by selectively disrupting somatic cells enzymatically, the bacterial cells are lysed, and the released ATP is allowed to drive the light-producing reaction. Photons are measured, correlating with the numbers of bacteria present in the sample. The luminescent tests are somewhat expensive and do take time; they are not used widely today.

Several instruments have been developed for the "hands-off" detection and identification of uropathogens. The principles employed by these systems are described in Chapter 12. The MS-2 system (Abbott Laboratories, Abbott Park, Ill.) uses a cuvette system and nephelometry to detect bacteria in urine. The AutoMicrobic System (bioMérieux Vitek Systems, Hazelwood, Mo.) is the most labor-free screening system. Once the urine is introduced into the tiny wells of a plastic substrate card by vacuum suction and the card is placed into the instrument, it is incubated and examined periodically by measuring the amount of light that passes through the individual wells. The wells contain substrates that can be utilized by only certain etiological agents of UTI. After 6 to 8 hours, microorganisms grow in the appropriate substrate wells and are detected by increased turbidity. The internal computer issues a colony count and a preliminary identification of the agent or agents identified. In this system as well, low counts of bacteria or yeast are considered negative urine cultures.

The Bac-T-Screen Bacteriuria Detection Device (bioMérieux Vitek) uses a different approach (Chapter 12, Figure 12.5). Urine is forced through a filter paper, which retains microorganisms, somatic cells, and other particles. A dye is then added to the filter paper to visualize the particulate matter that has adhered. The intensity of color relates to the number of particles. This procedure, which takes approximately 1 minute, has been shown to detect greater than 90% of all positive urine specimens, even if 100 organisms per milliliter are considered to be significant.[1,12,15] The detection of somatic cells and particles as well as bacteria and yeast probably accounts for the increased sensitivity of this system at low numbers of CFUs.[1] Organisms likely to be associated with false-negative results include enterococci and *Pseudomonas aeruginosa;* the reason for this is not known. A manual filtration method using the same reagents as Bac-T-Screen is the Filtracheck-UTI (Meridian Diagnostics, Cincinnati) (Figure 19.2). Another promising recently introduced manual system (Qualture, Future Medical Technologies International, W. Palm Beach, Fla.) combines filtration with differential media to quantitate and identify presumptively uropathogens, with results available within 4 hours.[6]

URISCREEN (bioMérieux Vitek) is a manual screening system that measures the enzyme catalase in urine. Approximately 1.5 to 2.0 ml of urine is added to a tube containing dehydrated substrate. Hydrogen peroxide is added to the urine and the solution is mixed gently. The formation of bubbles above the liquid surface is interpreted as a positive test. Catalase enzyme is present in both somatic cells and most etiological agents of UTI, with the exception of streptococci.[14]

Screening of urine can also be accomplished simply by enumerating polymorphonuclear neutrophils (PMNs) in uncentrifuged specimens, which correlates fairly well with the number of

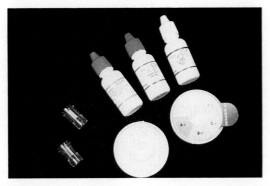

FIGURE 19.2

Manual filtration and staining system for screening urine specimens for presence of inflammatory cells or microorganisms consistent with urinary tract infection (FiltraCheck-UTI; Meridian Diagnostics, Cincinnati). Urine is diluted in small plastic vials *(left)* and dropped into the center of the circular plastic holder. The urine wicks through the filter, leaving particles on the surface to be stained with the reagents provided. The result is interpreted using a color scale (in position on the filter holder, *right*).

PMNs excreted per hour, the best indicator of the host's state. Patients with more than 400,000 PMNs excreted into the urine per hour are likely to be infected, and the presence of more than 8 PMNs/mm³ correlates well with this excretion rate and with infection.[3,4,9] This test can be performed using a hemacytometer, but it is not easily incorporated into the work flow of most microbiology laboratories. The standard urinalysis (usually done in hematology or chemistry sections) includes an examination of the *centrifuged* sediment of urine for enumeration of PMNs, results of which do not correlate well with either PMN excretion rate or the presence of infection.

Given the importance of the 10^2 CFU/ml count and the PMN count, no screening test should be used indiscriminately. In general, there is a cost advantage to screening only in laboratories receiving many culture-negative specimens; accordingly, education of physicians is one important approach. Good medical practice and cost-effective microbiology dictate that only urine from patients with symptoms of UTI plus a selected group expected to have asymptomatic bacteriuria (e.g., patients in the first trimester of pregnancy) should be cultured. Other situations in which patients with no symptoms of UTI might be cultured include bacteremia of unknown source, urinary tract obstruction, follow-up after removal of an indwelling urinary catheter, and follow-up of previous therapy. One must also remember that screening may delay culture results, especially if one of the rapid, automated instruments for urine culturing is used.

Developed in 1974, the *antibody-coated bacteria test* is used to localize the site of infection to the bladder (cystitis) or renal tissue (pyelonephritis) using a noninvasive technique. Previous methods required collection of urine directly from the renal pelvis (an invasive procedure), bilateral ureteral catheterization after bladder washout, or direct tissue biopsy. Since simple bladder infection can usually be treated effectively by administration of a single large dose of an appropriate antimicrobial agent, it is important to be able to differentiate such an infection from one that requires more prolonged therapy. Incidentally, many clinicians use response to single-dose treatment as a means of distinguishing between upper and lower tract infection. The antibody-coated bacteria test consists of washing the centrifuged sediment of urine containing bacteria and then staining the washed bacteria with fluorescein-conjugated antihuman globulin. The bacterial cells are again washed to remove nonspecifically bound conjugate, and the bacteria are examined under ultraviolet light. If at least one fourth of the bacteria are fluorescent, indicating the presence of human immunoglobulin on their surface, the test is considered positive and the patient is considered to have a more deep-seated infection than cystitis.[18] Although not difficult to perform, lack of standardization among laboratories and lack of uniformly available reagents have caused confusion in interpretation of results among different laboratories. Furthermore, there may be both false-positive and false-negative results. This test seems to be useful in research settings but, in its present state, is not warranted for general testing.

Since most clinical laboratories receive more urine specimens than specimens of any other type, it is important that laboratories develop an efficient, cost-effective way of handling them. It seems that some form of preliminary screening system to rule out negative cultures, combined with a standard method for quantifying and identifying significant etiological agents of UTI, will be of most benefit to the patients being served. Strategies employed by laboratories in hospitals that must cut costs without sacrificing patient-relevant quality are discussed in Chapter 4.

METHODS FOR CULTURING URINE AND INTERPRETIVE CRITERIA

ROUTINE CULTURE

Once it has been determined that a urine specimen should be cultured for isolation of the com-

mon agents of UTI, a measured amount of urine is inoculated to each of the appropriate media. The urine should be mixed thoroughly before plating. The plates can be inoculated using disposable sterile plastic tips with a displacement pipetting device, calibrated to deliver a constant amount, but this method is somewhat cumbersome. Most often, microbiologists use a **calibrated loop** designed to deliver a known volume, either 0.01 or 0.001 ml of urine. We recommend the larger volume to detect lower numbers of organisms for specimens collected at surgery and for specific patient groups discussed previously, such as symptomatic, pyuric female outpatients. These loops, made of platinum, plastic, or other material, can be obtained from various laboratory supply companies. They must be inserted into the urine in a cup *vertically* or they will pick up more than the desired volume of urine. A widely used method is described in Procedure 19.1.

The choice of which media to inoculate depends on the patient population served and the microbiologist's preference. The use of a 5% sheep blood agar plate and a MacConkey agar plate allows detection of most gram-negative bacilli and staphylococci. Enterococci and other streptococci, however, may be obscured by heavy growth of Enterobacteriaceae. The addition of a selective plate for gram-positive organisms, such as Columbia colistin–nalidixic acid agar (CNA) or phenylethyl alcohol agar, may add discriminatory capability but also adds cost to the procedure. Many European laboratories use cystine-lactose electrolyte-deficient (CLED) agar.

A recent study showed that incubation for a minimum of 24 hours is necessary for the detection of uropathogens.[11] Thus, some specimens inoculated late in the day could not be read accurately the next morning. If specimens were incubated another night, detection of fungi and accurate quantitation of cultures yielding less than 10^4 of potential pathogens were ensured.

SPECIMENS FROM PATIENTS WITH INDWELLING CATHETERS

The number of patients in hospitals and nursing homes with long-term indwelling urinary catheters continues to increase. These patients ultimately develop bacteriuria, which predisposes them to more severe infections. York and Brooks[21] suggest that urine be cultured quantitatively by allowing a 0.05 ml drop of mixed urine to run down the center of each half of a biplate containing blood agar and CNA, then estimating CFU by comparing growth patterns with photographs. Additional urine is plated to a support-

PROCEDURE 19.1

INOCULATING URINE WITH A CALIBRATED LOOP

PRINCIPLE
The number of microorganisms per milliliter recovered on urine culture can aid in the differential diagnosis of UTI. Plastic or wire loops, available commercially, have been calibrated to deliver a known volume of liquid when handled correctly, thus enabling the microbiologist to estimate numbers of organisms in the original specimen based on CFU of growth on cultures.

METHOD
1. Flame a wire calibrated inoculating loop. Allow it to cool without touching any surface. Alternatively, aseptically remove a plastic calibrated loop from its package.

2. Mix the urine thoroughly and remove the top of the container. If the urine is in a small-diameter tube, the surface tension will alter the amount of specimen picked up by the loop. A quantitative pipette should be considered if the urine cannot be transferred to a larger container.

3. Insert the loop vertically into the urine (Figure 19.3)

to allow urine to adhere to the loop.

4. Spread the loopful of urine over the surface of the agar plate, as shown in Figure 19.4. A standard quadrant streaking technique is also acceptable.

5. Without reflaming, insert the loop vertically into the urine again for transfer of a loopful to a second plate. Repeat for each plate.

6. Incubate plates for at least 24 hours at 35° to 37° C in air. Colonies are counted on each plate. The number of CFUs is multiplied by 1000 (if a 0.001 ml loop was used) or by 100 (if a 0.01 ml loop was used) to determine the number of microorganisms per milliliter in the original specimen.

7. Reincubate plates with no growth or tiny colonies for an additional 24 hours before discarding plates, since antimicrobial treatment or other factors may inhibit initial growth.

8. It is recommended to store the loop (handle down) in a test tube taped to the wall,

rather than flat on the bench, to prevent bending, which would destroy the calibration.

QUALITY CONTROL
Calibrated loops should be tested for proper volume delivery by using the loop to add dye to a measured volume of distilled water, then reading the optical density of the suspension spectrophotometrically, as described in *Clinical Microbiology Procedures Handbook* and *Cumitech 2A*.[2,4]

EXPECTED RESULTS
The loops should dispense the correct volume ±10%.

PERFORMANCE SCHEDULE
Wire loops should be tested monthly because they tend to bend and become encrusted and lose accuracy. Plastic loops may be tested once for each new lot number, although once a manufacturer's product has been determined to be accurate, it is unlikely to change much.

ive and selective differential agar biplate and streaked to obtain isolated colonies. All bacterial species in numbers greater than 10^2 CFU/ml, with the exception of normal skin or genital flora, are identified, and susceptibilities are performed on gram-negative bacilli and *Staphylococcus aureus*. Other methods that allow detection and identification of small numbers of organisms and separation of species in mixed culture are acceptable for specimens from patients with indwelling catheters.

INTERPRETING CULTURE RESULTS
Interpretive criteria have changed over the years.[4] Growth of 10^4 CFU/ml or less of two or more

probable pathogens (organisms not considered part of the indigenous flora, as listed previously) from voided urine is usually considered indicative of contamination unless clinical information indicates a chronic or recurrent infection.[13] For such cultures, the organisms are minimally identified (e.g., lactose-positive gram-negative bacilli, coagulase-negative staphylococci) and quantified. Antimicrobial susceptibility tests are not performed. Plates from such cultures are held, however, since situations occur in which growth of two or more organisms may be significant (e.g., patients with obstruction), and the physician always should have the option of calling the laboratory to request more definitive studies. Specimens

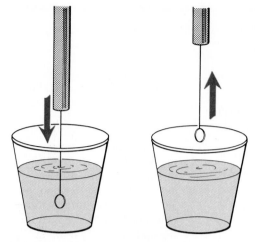

FIGURE 19.3

Method for inserting a calibrated loop into urine to ensure that the proper amount of specimen adheres to the loop.

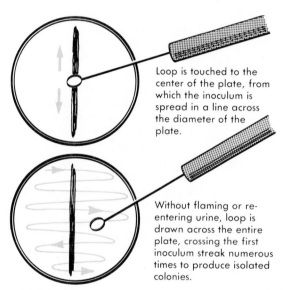

Loop is touched to the center of the plate, from which the inoculum is spread in a line across the diameter of the plate.

Without flaming or re-entering urine, loop is drawn across the entire plate, crossing the first inoculum streak numerous times to produce isolated colonies.

FIGURE 19.4

Method for streaking with calibrated urine loop to produce isolated colkonies and countable colony forming units (CFUs) of organisms.

from patients with long-term indwelling catheters can yield an average of five species, each in numbers greater than 10^5 CFU/ml.[4] Because of the polymicrobic nature of infection in these patients, all species except skin and genital commensals should be evaluated. Urine samples obtained by straight catheterization, by suprapubic aspirate, by cystoscopy, or at surgery should be representative of bladder urine, and all isolates should be identified and tested for antimicrobial susceptibilities. For surgically obtained urine (from kidney or ureter), a few drops should be incubated in broth for detection of very small numbers of organisms.

Growth of greater than 10^4 CFU/ml of a single potential pathogen is usually considered to be clinically significant, and a full workup should be performed.[4,13] Growth of less than 10,000 CFU/ml of one potential pathogen may indicate infection. A CFU count more than 10^3/ml has been shown to be significant in males,[10] and in symptomatic women with pyuria, greater than 10^2 CFU/ml of a potential pathogen should be reported. The organisms are identified, but antimicrobial susceptibility studies are not performed unless the clinician notifies the laboratory to do so. A pure culture of *S. aureus* is considered to be significant regardless of the number of CFUs, and antimicrobial susceptibility tests are performed. The presence of yeast in any number is reported to physicians, and pure cultures of a yeast may be identified as to species. In all urines, regardless of the extent of final workup, all isolates should be enumerated (e.g., "three different organisms present at 10^3 CFU/ml"), and those present in numbers greater than 10^4 CFU/ml should be described morphologically. Pezzlo[13] describes interpretation of urine cultures in great detail.

UNUSUAL ETIOLOGICAL AGENTS OF UTI

Culture methods for isolation of mycobacteria, *Haemophilus influenzae, Leptospira, Gardnerella vaginalis, Trichomonas vaginalis,* and other more rarely encountered agents of UTI are discussed in appropriate other areas of this book. *Salmonella* sp. may be recovered during the early stages of typhoid fever; their presence should be immediately reported to the physician. If anaerobes are suspected, a percutaneous bladder tap should be done by the physician unless urine can be obtained from the upper urinary tract (e.g., from a nephrostomy tube). Most important for detecting such agents is the clinician communicating to the laboratory that such an agent is suspected. However, the laboratory can exert some initiative here as well. In patients with "sterile pyuria," Gram stain may reveal unusual organisms with distinctive morphology (e.g., *H. influenzae,* anaerobes). Presence of any organisms on smear that do not grow in culture is an important clue to the cause of the infection. The laboratory can then take the action necessary to optimize chances for recovery, whether it includes inoculating a special medium or examining a spun sediment under phase-contrast microscopy.

REFERENCES

1. Baron, E.J., Tyburski, M.B., Almon, R., and Berman, M. 1988. Visual and clinical analysis of

Bac-T-Screen urine screen results. J. Clin. Microbiol. 26:2382.

2. Bostick, C.C., and Thomson, R.B., Jr. 1992. Pipetters and calibrated loops. Section 12.17. In Isenberg, H.D., editor. Clinical microbiology procedures handbook, vol. 1. American Society for Microbiology, Washington, D.C.

3. Breitenbucher, R.B. 1984. Bacterial changes in the urine samples of patients with long-term indwelling cathethers. Arch. Intern. Med. 144:1585.

4. Clarridge, J.E., Pezzlo, M.T., and Vosti, K.L. 1987. Laboratory diagnosis of urinary tract infections. In Weissfeld, A.S., coordinating editor. Cumitech 2A. American Society for Microbiology, Washington, D.C.

5. Dorn, G.L. 1991. Microbial stabilization of antibiotic-containing urine samples by using the FLORA-STAT urine transport system. J. Clin. Microbiol. 29:2169.

6. Friedman, M.P., Danielski, J.M., Day, T.E., et al. 1991. Rapid isolation and presumptive diagnosis of uropathogens by using membrane filtration and differential media. J. Clin. Microbiol. 29:2385.

7. Hooton, T.M., Hillier, S., Johnson, C., et al. 1991. *Escherichia coli* bacteriuria and contraceptive method. J.A.M.A. 265:64.

8. Kass, E.H. 1956. Asymptomatic infections of the urinary tract. Trans. Assoc. Am. Physicians 69:56.

9. Kass, E.H. 1960. The role of asymptomatic bacteriuria in the pathogenesis of pyelonephritis. In Quinn, E.L., and Kass, E.H., editors. Biology of pyelonephritis. Little, Brown, Boston.

10. Lipsky, B.A., Ireton, R.C., Fihn, S.D., et al. 1987. Diagnosis of bacteriuria in men: specimen collection and culture interpretation. J. Infect. Dis. 155:847.

11. Murray, P.R., Traynor, P., and Hopson, D. 1992. Evaluation of microbiological processing of urine specimens: comparison of overnight versus two-day incubation. J. Clin. Microbiol. 30:1600.

12. Murray, P.R., Niles, A.C., Heeren, R.L., and Pikul, F. 1988. Evaluation of the modified Bac-T-Screen and FiltraCheck-UTI urine screening systems for detection of clinically significant bacteriuria. J. Clin. Microbiol. 26:2347.

13. Pezzlo, M. 1992. Urine culture procedure. Section 1.17. In Isenberg, H.D., editor. Clinical microbiology procedures handbook, vol. 1. American Society for Microbiology, Washington, D.C.

14. Pezzlo, M.T., Amsterdam, D., Anhalt, J.P., et al. 1992. Detection of bacteriuria and pyuria by URISCREEN, a rapid enzymatic screening test. J. Clin. Microbiol. 30:680.

15. Pfaller, M., Ringenberg, B., Rames, L., et al. 1987. The usefulness of screening tests for pyuria in combination with culture in the diagnosis of urinary tract infection. Diagn. Microbiol. Infect. Dis. 6:207.

16. Stamm, W.E. 1980. Causes of the acute urethral syndrome in women. N. Engl. J. Med. 303:409.

17. Svandborg-Eden, C., and de Man, P. 1987. Bacterial virulence in urinary tract infection. Infect. Dis. Clin. North Am. 1:731.

18. Thomas, V.L. 1983. The antibody-coated bacteria test: uses and findings. Lab. Management 21:39.

19. Washington, J.A., II, White, C.M., Laganiere, M., and Smith, L.H. 1981. Detection of significant bacteriuria by microscopic examination of urine. Lab. Med. 12:294.

20. Weinstein, M.P. 1983. Evaluation of liquid and lyophilized preservatives for urine culture. J. Clin. Microbiol. 18:912.

21. York, M.K., and Brooks, G.F. 1987. Evaluation of urine specimens from the catheterized patient. Clin. Microbiol. Newsletter 9:76.

BIBLIOGRAPHY

Edberg, S.C. 1981. Methods of quantitative microbiological analyses that support the diagnosis, treatment, and prognosis of human infection. C.R.C Crit. Rev. Microbiol. 8:339.

Fihn, S., and Stamm, W.E. 1983. Management of women with acute dysuria. In Rund, D.A., and Wolcott, B.W., editors. Emergency medicine annual, vol. 2. Appleton-Century-Crofts, East Norwalk, Conn.

Gabre-Kidan, T., Lipsky, B.A., and Plorde, J.J. 1984. *Haemophilus influenzae* as a cause of urinary tract infections in men. Arch. Intern. Med. 144:1623.

Kunin, C.M. 1987. Detection, prevention, and management of urinary tract infections, ed. 4. Lea & Febiger, Philadelphia.

Platt, R. 1983. Quantitative definition of bacteriuria. Am. J. Med. 75:44.

Sobel, J.D. 1987. Pathogenesis of urinary tract infections: host defenses. Infect. Dis. Clin. North Am. 1:751.

Stamm, W.E., and Turck, M. 1983. Urinary tract infection. Year Book, Chicago.

20

GENITAL AND SEXUALLY TRANSMITTED PATHOGENS

NORMAL AND PATHOGENIC FLORA OF THE GENITAL TRACT

The lining of the normal human genital tract is a mucosal layer made up of transitional, columnar, and squamous epithelial cells. A variety of species of commensal bacteria colonize these surfaces, causing no harm to the host except under abnormal circumstances and helping to prevent the adherence of pathogenic organisms. Normal urethral flora include coagulase-negative staphylococci and corynebacteria, as well as various anaerobes. The vulva and penis, especially the area underneath the prepuce (foreskin) of the uncircumcised male, may harbor *Mycobacterium smegmatis* along with other gram-positive bacteria. The flora of the female genital tract (Figure 20.1) varies with the pH and estrogen concentration of the mucosa, which depends on the host's age. Prepubescent and postmenopausal women harbor primarily staphylococci and corynebacteria, the same flora present on surface epithelium, whereas women of reproductive age may harbor large numbers of facultative bacteria such as Enterobacteriaceae, streptococci, and staphylococci,

as well as anaerobes such as lactobacilli, anaerobic non-spore-forming bacilli and cocci, and clostridia. The numbers of anaerobic organisms remain constant throughout the monthly cycle. Many women carry group B beta-hemolytic streptococci *(Streptococcus agalactiae),* which may be transmitted to the neonate as it passes through the birth canal and may cause devastating systemic disease.[47] Whether a colonized infant will become ill cannot be predicted. Although yeasts (acquired from the gastrointestinal tract) may be transiently recovered from the female vaginal tract, they are not normal flora of the genital tract. Lactobacilli are the predominant organisms in secretions from normal, healthy vaginas (Figure 20.2). Recent studies have shown that hydrogen peroxide–producing lactobacilli are most associated with a healthy state.[17,18,21,30]

Infections of the genital tract in normal females before or after childbearing age are often a result of irritation or a foreign body and are caused by the same organisms that cause skin wound infections. Genital tract infections caused by sexually transmitted agents in children (pread-

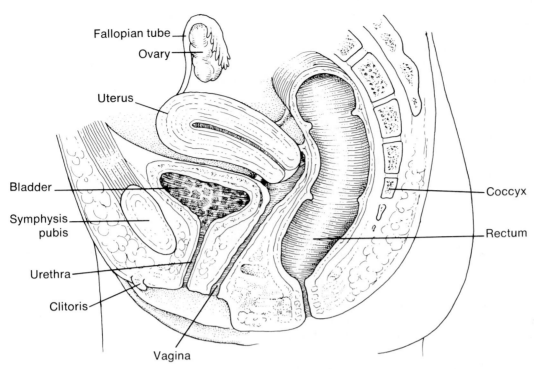

FIGURE 20.1

Diagrammatic representation of the female genital tract. (From Broadwell, D.C., and Jackson, B.S. 1982. Principles of ostomy care. Mosby, St. Louis.)

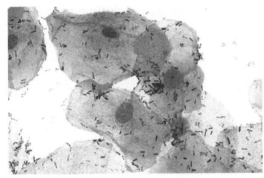

FIGURE 20.2

Predominance of lactobacilli in Gram stain from healthy vagina.

olescents) are most often the result of sexual abuse. The laboratory should treat specimens from such patients with extreme care, carefully identifying and documenting all isolates, since medicolegal implications exist. Cultures should always be obtained, especially for *Chlamydia*, because antigen detection methods are less sensitive in children than in adults.[19]

Chlamydia trachomatis, Gardnerella vaginalis, Neisseria gonorrhoeae and *meningitidis, Treponema pallidum, Ureaplasma urealyticum, Mycoplasma hominis,* other mycoplasmas, human papillomavirus,

herpesvirus, and other organisms may be acquired as people engage in sexual activity. Some of these organisms may exist in the genital tract in the absence of any noticeable pathology. *C. trachomatis,* however, is the most prevalent agent of mucopurulent cervicitis (and one of the most common sexually transmitted diseases [STDs]) in the United States. It is clearly associated with pelvic inflammatory disease, subsequent infertility, and preterm births.[9,30] Although the role of *M. hominis* in the pathogenesis of genital infection is unclear, it has been associated with infections in females, such as pelvic inflammatory disease, pyelonephritis, and postpartum fever, and with morbidity and stillbirths in infants.* *U. urealyticum,* but not *M. hominis,* has been associated causally with the acute urethral syndrome in women and with reproductive failure.[43] *U. urealyticum* may be related to nongonococcal urethritis in males, and it is clearly associated with chorioamnionitis, premature birth, morbidity, and even mortality of the babies of infected females.[7,31,35]

Human papillomavirus (HPV) is present in epidemic proportions among sexually active Americans and may be the most prevalent STD

*References 11, 18, 24, 31, 35, 43.

today. It has been linked unequivocally with cervical carcinoma, and an association with penile malignancy is being explored.[22,33] The syndrome of **bacterial vaginosis** (BV), consisting of copious, malodorous vaginal discharge, is important in the etiology of premature labor and low birth weight.[11,17,24,29,46] Thus, diagnosis of this entity has taken on greater importance. A partial list of organisms recovered in culture from the genital tract that have been implicated as pathogens, at least under some circumstances, are listed in the box at right.

In addition, various other agents cause genital tract disease but are not routinely isolated from clinical specimens. These include adenovirus,[16] coxsackievirus,[49] molluscum contagiosum virus (a member of the poxvirus group), the human papillomaviruses of genital warts (condylomata acuminata; types 6, 11, and others) and those associated with cervical carcinoma (predominantly types 16 and 18, but numerous others are also implicated), *Calymmatobacterium granulomatis, Treponema pallidum,* and ectoparasites such as scabies and lice. Infections with more than one agent may occur. Such dual or concurrent infections should always be considered.

Homosexual practices and increasingly common heterosexual practices of anal-genital intercourse have required that other gastrointestinal and systemic pathogens also be considered etiological agents of STDs. The intestinal protozoa *Giardia lamblia, Entamoeba histolytica,* and *Cryptosporidium* sp. are significant causes of STD, especially among homosexual populations. In the same group of patients, fecal pathogens such as *Salmonella, Shigella, Campylobacter,* and *Microsporidium* are often transmitted sexually. Oral-genital practices probably allow *N. meningitidis* to colonize and infect the genital tract. Viruses shed in secretions or present in blood (cytomegalovirus; hepatitis B, possibly C and E; other non-A, non-B hepatitis viruses; human T-cell lymphotropic virus type I [HTLV-I]; and human immunodeficiency virus [HIV] of acquired immunodeficiency syndrome [AIDS]) are increasingly being spread by sexual practices. The spectrum of diseases (not previously considered to be sexually transmitted) that accompanies AIDS is discussed in Chapter 24.

Certain pathogens are usually sought in specimens from lesions on surface epithelium of or near the genital tract. These include herpes simplex virus, *Haemophilus ducreyi,* and *T. pallidum. C. granulomatis* is recovered from biopsy material taken from beneath the characteristic lesion. Collection of specimens from lesions is discussed later.

Genital Tract Pathogens Recovered in Culture

Bacteria
Chlamydia trachomatis
Gardnerella vaginalis
Haemophilus ducreyi
Mycoplasma hominis
Mycoplasma genitalium
Neisseria gonorrhoeae
Neisseria meningitidis
Ureaplasma urealyticum

Fungi
Candida species
Other yeast

Viruses
Cytomegalovirus
Herpes simplex virus
Human papillomavirus

Protozoans
Trichomonas vaginalis

COLLECTION AND TRANSPORT OF SPECIMENS FROM THE URETHRA AND VAGINA

Urethral discharge may occur in both males and females infected with certain pathogens, such as *Neisseria gonorrhoeae* and *Trichomonas vaginalis.* Because the discharge is usually less profuse and may be masked more easily in the female by vaginal secretions, the presence of infection is more likely to be asymptomatic in females. *Ureaplasma urealyticum* and *Chlamydia trachomatis* can also be isolated from male urethral discharge.

A urogenital swab designed expressly for collection of such specimens should be used. These swabs are made of cotton or rayon that has been treated with charcoal to adsorb material toxic to gonococci, wrapped tightly over one end of a thin wooden stick that has also been treated to remove resins and toxic chemicals (Chapter 7). Certain wooden sticks may be toxic to mycoplasmas and chlamydiae; plastic sticks are generally less toxic. Rayon-tipped swabs on thin wire may also be used for collection of specimens for isolation of mycoplasmas and chlamydiae. Calcium alginate swabs are generally more toxic for herpesvirus, gonococci, and mycoplasma than are treated cotton swabs, but dacron swabs are least toxic and thus recommended for viral specimens. Rare lots of calcium alginate swabs have been toxic for chlamydiae.

The swab is inserted approximately 2 cm into the urethra and rotated gently before withdrawing. Since chlamydiae are intracellular pathogens, it is important to remove epithelial cells (with the swab) from the urethral mucosa. Use of a cytobrush may enhance recovery of endocervical cells, (see following discussion). Separate samples for cultivation of gonococci and chlamydiae or ureaplasma are required. When profuse urethral discharge is present, particularly in males, the discharge may be collected externally without inserting a sampling device into the urethra. A few drops of urine and concentrated voided urine have also been used successfully for detection of chlamydiae in males.[26,28]

Since *T. vaginalis* may be present in urethral discharge, material for culture should be collected by swab as just described, but a swab should also be placed into a tube containing 0.5 ml of sterile physiological saline and hand delivered to the laboratory immediately. Direct wet mounts and cultures can be performed from this specimen. Commercial media for culture of *Trichomonas* are available. The first few drops of voided urine may also be a suitable specimen for recovery of *Trichomonas* from infected males, if it is inoculated into culture media immediately. Alternatively, material may be smeared onto a slide for later performance of a fluorescent antibody stain. Plastic envelope cultures have recently gained popularity.[5]

Organisms that cause purulent vaginal discharge (vaginitis) include *T. vaginalis,* yeast, gonococci, and rarely, beta-hemolytic streptococci. The same organisms that cause purulent infections in the urethra may also infect the epithelial cells in the cervical opening (os), as can herpes simplex virus. Mucus is removed by gently rubbing the area with a cotton ball. A special swab (the urethral swab just described) is inserted into the cervical canal and rotated and moved from side to side for 30 seconds before removal. A small, nylon-bristled brush (cytology brush or cytobrush) may be used to ensure collection of cellular material, but its use is associated with discomfort and bleeding. Some controversy exists over whether the cytobrush results in better specimens, at least for detection of *Chlamydia*.[20] The cytology brush is contraindicated in pregnant patients. Swabs are handled as urethral swabs for isolation of *Trichomonas* and gonococci. Chlamydiae cause a mucopurulent cervicitis with much discharge. Endocervical specimens are obtained after the cervix has been exposed by insertion of a speculum, which allows visualization of vaginal and cervical architecture, and after ectocervical mucus has been adequately removed. The speculum is moistened with warm water, since many lubricants contain antibacterial agents. Because the normal vaginal secretions contain great quantities of bacteria, care must be exercised to avoid or minimize contaminating swabs for culture by contact with these secretions. Swabs of Bartholin gland exudate are not recommended, since vaginal flora contamination is not possible to exclude. Infected Bartholin glands should be aspirated with needle and syringe after careful skin preparation, and cultures should be evaluated for anaerobes and aerobes.

In addition to cervical specimens, which are particularly useful for isolation of herpes, gonococci, *Mycoplasma,* and chlamydiae, vaginal discharge specimens may be collected. Organisms likely to cause vaginal discharge include *Trichomonas,* yeast, and the agents of bacterial vaginosis (BV), presumably combinations of *Gardnerella vaginalis* and various anaerobic bacteria. The absence of inflammatory cells in the vaginal discharge is one sign of BV, formerly called *nonspecific vaginitis.* Swabs for diagnosis of BV are dipped into the fluid that collects in the posterior fornix of the vagina.

Swabs collected for isolation of gonococci may be transported to the laboratory in modified Stuart's or Amies charcoal transport media, held at room temperature until inoculated to culture media. Good recovery of gonococci is possible if swabs are cultured within 12 hours of collection. Material that must be held longer than 12 hours should be inoculated directly to one of the commercial systems designed for recovery of gonococci, described later in this chapter. Swabs for isolation of chlamydiae and *Mycoplasma* are best transported in sucrose buffer with antibiotics (2-SP), purchased or prepared as described in the last edition of this book.[3] If the specimens are not going to be inoculated to cell culture within several hours, they should be refrigerated but *never frozen.* It is important to vortex the swab thoroughly in the transport medium and then remove the swab to prevent toxicity.

DIRECT MICROSCOPIC EXAMINATION

In addition to culture, urethral discharge may be examined by Gram stain for the presence of gram-negative intracellular diplococci (Figure 20.3), usually indicative of gonorrhea in males. After inoculation to culture media, the swab is rolled over the surface of a glass slide, covering an area of at least 1 cm². If the Gram stain is characteristic, cultures of urethral discharge need not be performed. In homosexual males, however, *Neisseria meningitidis* has been increasingly isolated from such specimens, so that the assumption that

FIGURE 20.3
Gram-negative intracellular diplococci; diagnostic for gonorrhea in urethral discharge and presumptive for gonorrhea in vaginal discharge.

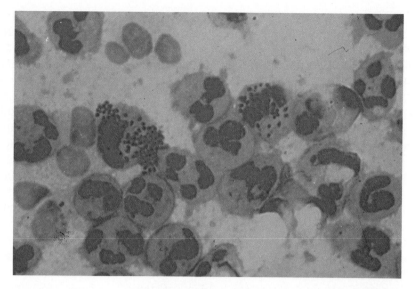

gram-negative diplococci are gonococci may no longer be valid for that patient population. Urethral smears from females may also be examined, but presumptive diagnosis of gonorrhea from vaginal smears is reliable only if the microscopist is experienced, since normal vaginal flora such as *Veillonella* or occasional gram-negative coccobacilli may resemble gonococci. If extracellular organisms resembling *Neisseria gonorrhoeae* are seen, the microscopist should continue to examine the smear for intracellular diplococci for a longer period than if no suspicious bacteria are seen. Presumptive diagnosis can be useful when decisions are to be made regarding immediate therapy, but confirmatory cultures or an alternative antigen detection method should always be performed on specimens from females. Some strains of *N. gonorrhoeae* are sensitive to the amount of vancomycin present in selective media. If suspicious organisms seen on smear fail to grow in culture, reculture on chocolate agar without antibiotics may be warranted.

Fluorescein-conjugated monoclonal antibody reagents are quite sensitive and specific for visualization of the inclusions of *Chlamydia trachomatis* in cell cultures and elementary bodies in urethral and cervical specimens containing cells (Chapter 39). The reagents are available commercially in complete collection and test systems (Figure 20.4), but the relatively greater technologist time required for this method limits its usefulness for laboratories that receive many specimens, except as a confirmatory test for other antigen detection systems with borderline results.[40] The swab is rolled over the surface of a marked glass slide, and the material is fixed with acetone. In some studies

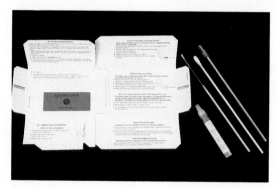

FIGURE 20.4

Commercially produced collection kit for genital tract specimens for direct fluorescent antibody stain for *Chlamydia trachomatis* (Syva, Palo Alto, Calif.). Physician has a choice of swab or cytobrush, and the fixative is included in a small ampoule *(right).* Detailed instructions are printed in the kit.

the sensitivity of visual detection of chlamydiae with these newer reagents has been similar to that of culture, although such comparative results are obtained only by technologists experienced in fluorescent techniques and when the slides are examined quite thoroughly.[25,40] False-positive results should not occur if at least 10 morphologically compatible fluorescing organisms are seen on the entire smear. No direct visual methods exist for detection of mycoplasma at this time, but nucleic acid probes have been evaluated.

Direct microscopic examination of a wet preparation of vaginal discharge provides the simplest rapid diagnostic test for *Trichomonas vaginalis* when such a specimen is available, although use of culture or a monoclonal fluorescent antibody stain detects more positive specimens.[42] The plastic envelope method combines direct visualization with culture.[5] Motile trophozoites of *Trichomonas* can be visualized in a routine wet preparation performed by a proficient technologist in two thirds of cases, and budding cells and pseudohyphae of yeast can also be easily identified. The addition of 10% potassium hydroxide (KOH) to a separate preparation serves two functions: by dissolving host cell protein, it enhances the visibility of fungal elements; and by causing the discharge to become alkaline, it elicits the fishy, aminelike odor associated with BV.[11]

Bacterial vaginosis, characterized by a foul-smelling discharge, can be diagnosed microscopically or clinically.[11,23,34] The discharge is primarily sloughed epithelial cells, many of which are completely covered by tiny, gram-variable rods and coccobacilli. These cells are called "clue cells" (Figure 20.5). *Gardnerella vaginalis* has been associated with the syndrome historically, but the synergistic activity of various anaerobic organisms, including *Prevotella* sp., peptostreptococci, and sometimes *Mobiluncus* sp. (curved, motile rods), and of mycoplasmas seems to contribute to the pathology.[8,29,37] Although *G. vaginalis* can be cultured on a human blood bilayer plate, we do not recommend culture for diagnosis of BV. A clinical diagnosis of BV is best made using three or more of the following criteria: homogeneous, gray discharge; clue cells seen on wet mount or Gram stain; amine or fishy odor elicited by the addition of a drop of 10% KOH to the discharge on a slide or on the speculum; and pH greater than 4.5.[11]

The absence of a predominance of lactobacilli on Gram stain, present in the normal vagina (Figure 20.2), is another sign of bacterial vaginosis.[17] A grading system for Gram stains of vaginal discharge has been developed by Nugent et al. (Procedure 20.1).[34] Using these criteria, the sensitivity of Gram stain for diagnosis of bacterial vaginosis (diagnosed clinically) was 93% to 97%, better than culture for *G. vaginalis* (sensitivity 95%).[11] The Gram stain is more specific than the wet mount for detection of clue cells and the smear can be saved and reexamined later. Additionally, *Mobiluncus* species, often present in this syndrome, can be detected best on Gram stain because of their distinctive curved morphology.[37] With the validation of the Gram stain as the preferred method, diagnosis of BV should not be made by culture.[39]

DIRECT INOCULATION AND LABORATORY HANDLING OF URETHRAL AND VAGINAL SPECIMENS

Samples for isolation of gonococci may be inoculated directly to culture media, obviating the need for transport. Several commercially produced systems have been developed for this purpose, and many clinicians inoculate standard plates directly if convenient access to an incubator is available. Modified Thayer-Martin medium is most often used, although New York City (NYC) medium has the added advantage of supporting the growth of mycoplasma as well as gonococci. A modification of NYC medium has been shown to yield comparable results for growth of *Neisseria gonorrhoeae* but

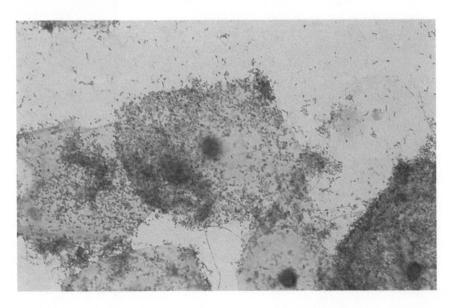

FIGURE 20.5

"Clue cells" in vaginal discharge, suggestive of bacterial vaginosis (BV).

PROCEDURE 20.1

SCORING VAGINAL GRAM STAINS FOR BACTERIAL VAGINOSIS

1. Roll the swab of vaginal discharge over the surface of a slide.

2. Allow the smear to air dry, fix with methanol, and stain.

3. Scores are assigned as follows:

Organism morphotype	Number/oil immersion field	Score	Organism morphotype	Number/oil immersion field	Score
Lactobacillus-like	>30	0	*Gardnerella/Bacteroides*-like	>30	4
(parallel-sided gram-positive rods)	5-30	1	(tiny, gram-variable coccobacilli and	5-30	3
	1-4	2	rounded, pleomorphic	1-4	2
	<1	3	gram-negative rods	<1	1
	0	4	with vacuoles)	0	0
Mobiluncus-like	>5	2			
(curved gram-negative	<1-4	1			
rods)	0	0			

4. Add up total score and interpret as follows:

Score	Interpretation
0-3	Normal
4-6	Intermediate, repeat test later
7-10	Bacterial vaginosis

From Nugent, R.P., Krohn, M.A., and Hillier, S.L. 1991. J. Clin. Microbiol. 29:297.

costs as much as 70% less.[1] These media contain numerous supplements, however, and are not suitable for preparation in most routine clinical laboratories. Excellent recovery of gonococci is the rule when specimens are inoculated directly to any of these media or their modifications in self-contained incubation systems such as the Jembec plates (Figure 20.6). Media are commercially available. The swab containing material is rolled across the agar with constant turning to expose all surfaces to the medium. The Jembec plate, which generates its own increased carbon dioxide atmosphere by means of a sodium bicarbonate tablet, is inoculated in a **W** pattern. It may be cross-streaked with a sterile loop in the laboratory (Figure 20.7).

Specimens must be inoculated to other media in addition to gonococcal selective agar for isolation of yeast, streptococci, mycoplasmas, *Gardnerella vaginalis,* and anaerobic bacteria associated with BV. Yeast grow well on Columbia agar base with 5% sheep blood and colistin and nalidixic acid (CNA), as do most strains of *Gardnerella,* al-

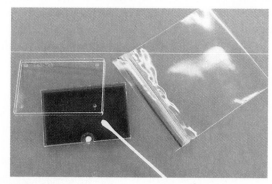

FIGURE 20.6

Jembec plate containing modified Thayer-Martin medium in a plastic, snap-top box with a self-contained CO_2-generating tablet, all sealed inside a zip-lock plastic envelope after inoculation.

though more selective media are available (Chapter 40). Most yeast and the streptococci also grow on standard blood agar; thus the addition of special fungal media such as Sabouraud's brain-heart infusion agar (SABHI) is unwarranted.

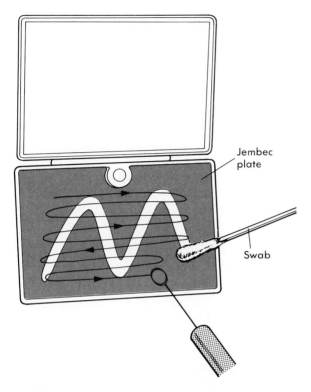

FIGURE 20.7
Method of cross-streaking Jembec plate after original specimen has been inoculated by rolling the swab over the surface of the agar in a **W** pattern.

Culture of the vaginal canal within 6 to 10 weeks of delivery can reliably predict the presence of group B streptococci at delivery.[14] Rapid antigen tests for group B streptococci in vaginal secretions obtained during labor are being evaluated for predicting those mothers likely to give birth to infants colonized with group B streptococci, although the use of these tests is very controversial.[12,13,36] For culture, group B streptococci are detected best by inoculating a swab obtained from the perianal area onto CNA agar and into Todd-Hewitt enrichment broth with blood, gentamicin and nalidixic acid for subculture to agar the next day (Chapter 26). Hillier[16a] found that vaginal swabs detected 14% fewer positive patients than concurrently collected perianal swabs. Cultures on plates without enrichment broth yielded the organism from approximately two thirds of all positive patients.

Trichomonas vaginalis may be cultured in Diamond's medium (available commercially) or plastic envelopes inoculated with discharge material; direct wet preparations may be examined for the presence of motile trophozoites; or a direct fluorescent antibody (DFA) stain (Meridian Diagnostics, Cincinnati) may be used. Culture techniques are most sensitive.[5,42] A8 agar or a commercially available biphasic *Mycoplasma* culture system (Mycotrim, Irvine Scientific, Santa Ana, Calif.) can be used to culture mycoplasmas and *Ureaplasma urealyticum,* although commercially prepared media are not as sensitive as fresh media.[7,48] *Mycoplasma genitalium* may not grow on commercial media because of the presence of thallium acetate.[45]

INFECTIONS OF FEMALE PELVIS AND MALE INTERNAL GENITAL ORGANS

Pelvic inflammatory disease is often caused by the same organisms that cause cervicitis or by organisms that make up the normal flora of the vaginal mucosa. Because of the profuse normal flora of the vaginal tract, species of which may be involved in peritoneal infection, specimens must be collected in such a way as to prevent vaginal flora contamination. Aspirated material collected by needle and syringe represents the best specimen. If this cannot be obtained at the time of surgery or laparoscopy, collection of intrauterine contents using a protected suction curetting device (Pipelle, Unimar, Inc., Wilton, Conn.) or double-lumen sampling device (Uterine Sampling Device, Medi-Tech, Watertown, Mass.) inserted through the cervix is also acceptable. Culdocentesis, after decontamination of the vagina by povidone-iodine, is satisfactory but rarely practiced today. Aspirated material should be placed into an anaerobic transport container. The presence of either mixed anaerobic flora, gonococci, or both can be rapidly detected from a Gram stain. Direct examination with fluorescent monoclonal antibody stain may also detect chlamydiae. All such specimens should be inoculated to media that allow the recovery of anaerobic, facultative, and aerobic bacteria, gonococci, fungi, mycoplasmas, and chlamydiae. All material collected from normally sterile body sites in the genital tract should be inoculated to chocolate agar and placed into a suitable broth such as chopped meat medium or thioglycolate, in addition to the other types of media noted. If only specimens obtained on routine swabs inserted through the cervix are available, cultures should be performed only for detection of gonococci and chlamydiae.[10a]

Infections of the male prostate, epididymis, and testes are usually bacterial. In younger men, chlamydiae predominate as the cause of epididymitis and possibly of prostatitis. Urine or discharge collected via the urethra is the specimen of choice unless an abscess is drained surgically or by needle and syringe. Urine (the first few milliliters of voided urine) may be collected before and after

PROCEDURE 20.2

COLLECTION OF MATERIAL FROM SUSPECTED HERPETIC LESIONS

PRINCIPLE

Herpesvirus is best recovered from the base of active lesions in the vesicular stage. The older the lesion, the less likely it will yield viable virus.

METHOD

1. Open the vesicles with a small-gauge needle or Dacron-tipped swab.

2. Rub the base of the lesion vigorously with a small cotton-tipped or Dacron-tipped swab to recover infected cells.

3. Place the swab into viral transport medium (Chapter 43), and refrigerate until inoculated to culture media. Specimens in media may be stored at −70° C for extended periods without loss of viral yield.

4. If large vesicles are present, material for culture may be aspirated directly by needle and syringe.

5. Material from another lesion can be applied directly to a glass slide for a Tzanck preparation (histology) with Wright-Giemsa or fluorescent antibody stain for detection of multinucleated giant cells.

---

prostatic massage to try to pinpoint the anatomical site of the infection. Cultures are inoculated to support the growth of anaerobic, facultative, and aerobic bacteria, as well as gonococci. In patients with suspected AIDS or other immunosuppression-related systemic cytomegalovirus disease, semen or urine can be cultured for the virus. The specimen is transported immediately to the virology laboratory for culture. Refrigeration may preserve the virus for several hours; specimens that must be held longer should be placed into viral transport media (Chapter 43).

SPECIMENS OBTAINED FROM SKIN AND MUCOUS MEMBRANE LESIONS

External genital lesions are usually either vesicular or ulcerative. Causes of lesions can be determined by physical examination, histological examination, or microscopic examination or culture of exudate. *Since any genital lesion may be highly contagious, all manipulations of lesion material should be carried out by clinicians wearing gloves.*

EXAMINATION OF VESICULAR LESIONS

Vesicles in this area are almost always attributable to viruses, and herpes simplex is the most common cause. Material from the base of a vesicle may be spread onto the surface of a slide and examined for the typical multinucleated giant cells of herpes or stained by immunofluorescent antibody stains for viral antigens. Additionally or alternatively, the material is transported for culture of the virus, as outlined in Procedure 20.2.

Several commercial fluorescein-conjugated monoclonal and polyclonal antibodies directed against herpetic antigens of either type 1 or 2 are available. When fluorescent-antibody-stained lesion material containing enough cells is viewed under ultraviolet light, the diagnosis can be made in 70% to 90% of patients. Laboratories that routinely process genital material for herpes should be using immunofluorescent staining reagents when a rapid answer is desired; otherwise, culture, which is generally positive in 2 days, is the method of choice. Nonfluorescent markers such as biotin-avidin-horseradish peroxidase or alkaline phosphatase have also been conjugated to these specific antibodies, often allowing earlier detection of herpes-infected cells in tissue culture monolayers. Such reagents have been developed for use directly on clinical material, although their sensitivity is not great enough to forego culture if a definitive diagnosis is necessary.

EXAMINATION OF ULCERATIVE LESIONS

Ulcerative lesions are usually one of the following: a syphilitic chancre (caused by *Treponema pallidum*), the chancroid of *Haemophilus ducreyi*, later stages of herpetic vesicles, the ulcer of granuloma inguinale (also known as donovanosis, caused by *Calymmatobacterium granulomatis*), an infected traumatic ulcer, or the result of some other rare infectious disease or even a noninfectious process. Macroscopically, lesions of syphilis are usually clean, with an even, indurated edge.[44] Only 30% of syphilitic chancres are tender. Lesions of chancroid are usually ragged, necrotic, and painful and are often found in pairs where material from the first lesion autoinoculated a second area in close contact ("kissing lesions"). The beefy red, nonpainful lesion of donovanosis is characterized by a white border. These lesions can usually be differentiated from the painful, erythematous-bordered

PROCEDURE 20.3

COLLECTION OF MATERIAL FOR DARKFIELD EXAMINATION FOR SYPHILIS

PRINCIPLE

Numerous motile treponemes are present in a fresh lesion (chancre) of syphilis. Their characteristic motility can aid in presumptive diagnosis of this infection before a serologic response can be detected. Darkfield microscopy is necessary because the organisms are too thin to be visible under phase microscopy.

METHOD

1. Cleanse the area around the lesion with a sterile gauze pad moistened with sterile saline.

2. Abrade the surface of the ulcer with a sterile, dry gauze pad until some blood is expressed.

3. Continue to blot the blood until there is no further bleeding, then squeeze the area until serous fluid is expressed.

4. Touch the surface of a very clean glass slide or coverslip to the exudate, and immediately cover the material on the slide with a coverslip or place the inoculated coverslip face down onto a slide. If only minimal material was expressed, a drop of sterile physiological saline may be added.

5. Immediately examine the material under darkfield microscopy (400×), high dry magnification) for the presence of motile spirochetes. Treponemes are very long (8-10 μm), slightly longer than a red blood cell, and consist of 8 to 14 tightly coiled, even spirals (Figure 20.8). They may bend slightly in the middle as they slowly move about. Once characteristic forms are seen, they should be verified by examination under oil immersion magnification (1000×).

QUALITY CONTROL

Proper setup of the microscope can be ascertained by making a wet preparation of a scraping of the inside of your cheek (mucosal surface).

EXPECTED RESULTS

Numerous thin, fusiform and spirochetal organisms should be seen darting about.

PERFORMANCE SCHEDULE

Test a cheek cell scraping each time the darkfield procedure is performed. NOTE: All slides and materials should be handled as infectious, since syphilitic lesions may contain large numbers of infectious treponemes.

---

lesions of herpes, which usually begin as vesicles and occur in groups. Material from lesions suggestive of syphilis should be examined by darkfield microscopy (Chapter 8). Collection of material for darkfield microscopy is outlined in Procedure 20.3.

All lesions suspected of infectious etiology may be Gram stained in addition to the procedures described. The smear of lesion material from a patient with chancroid may show many small, pleomorphic gram-negative rods and coccobacilli arranged in chains and groups, characteristic of *H. ducreyi* (Figure 20.9). However, culture has been shown to be more sensitive for diagnosis of this agent. Material collected on cotton or Dacron swabs may be transported in modified Stuart's medium until it is inoculated to culture media. A special agar, consisting of chocolate agar enriched with 1% IsoVitaleX (BBL Microbiology Systems) and vancomycin (3 μg/ml), has yielded good isolation if cultures are incubated in 5% to 7% carbon dioxide in a moist atmosphere, such as a candle jar. *H. ducreyi* grows best at 33° C.[32] Inoculation of a duplicate sample from each patient to a second agar, fetal bovine serum agar with vancomycin, may increase chances for recovery.[15]

Diagnosis of granuloma inguinale is achieved by staining a crushed preparation of a small piece of biopsy tissue from the edge of the base of the ulcer with Wright's or Giemsa stain and finding characteristic Donovan bodies (bipolar staining rods intracellularly within macrophages). Cytologists or pathologists usually examine such specimens rather than microbiologists. No acceptable media for isolation of *C. granulomatis* are available.

EXAMINATION OF MATERIAL FROM BUBOES

Buboes, swollen lymph glands that occur in the inguinal region, are often evidence of a genital tract infection. Buboes are common in patients with primary syphilis, genital herpes, lymphogranuloma venereum, and chancroid. Patients

FIGURE 20.8

Appearance of fluorescent antibody-stained *Treponema pallidum.* Organisms appear to bend only near the center.

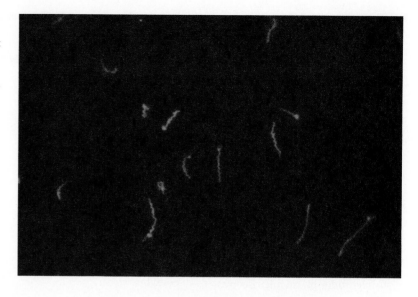

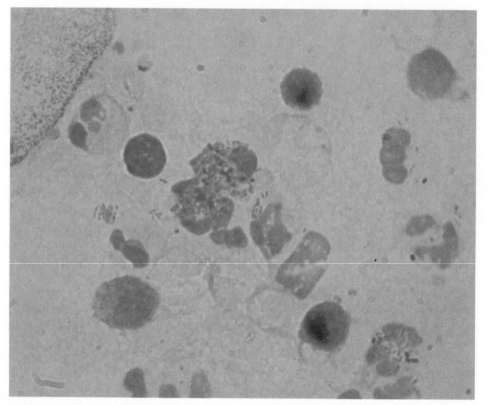

FIGURE 20.9

Gram stain of genital lesion showing typical "school-of-fish" formation of pleomorphic gram-negative bacilli characteristic of *Haemophilus ducreyi* infection.

with AIDS may show generalized lymphadenopathy. Other diseases that are not sexually transmitted, such as plague, tularemia, and lymphoma, can also produce buboes. Those diseases are not discussed further here. Material from buboes may be aspirated for microscopic examination and culture, as described in Procedure 20.4.

INFECTIONS OF NEONATES AND HUMAN PRODUCTS OF CONCEPTION

Certain infectious agents are known to cause fetal infection and even abortion. *Listeria monocytogenes,* although usually causing only mild flulike symptoms in the mother, can cause extensive disease and abortion of the fetus if infection occurs late in

PROCEDURE 20.4

ASPIRATION OF BUBO MATERIAL

To be done by physician

PRINCIPLE
The etiological agent of swollen inguinal lymph glands is often visible in material aspirated from the gland itself.

METHOD
1. Wipe overlying skin with an iodophore, allowing it to remain on the skin for at least 1 min.

2. Vigorously wash off the iodophore with 70% alcohol, allowing it to dry.

3. Introduce a 21-gauge needle with an attached 5-ml syringe through the skin into the node.

4. If the node is nonfluctuant, as in syphilis, the needle is plunged gently in various directions within the node while continuous suction is applied to the syringe.

5. Material aspirated from nonfluctuant nodes is examined under darkfield microscopy for spirochetes.

6. Material from fluctuant nodes is obtained by applying suction. If no pus is aspirated, the needle is left in place, and a syringe containing at least 1 ml of nonbacteriostatic physiological sterile saline is exchanged for the initial

empty syringe. This saline is injected into the node and withdrawn. Rigorous aseptic technique is important.

7. Pus is cultured on chocolate and 5% sheep blood agars, in special media for *Haemophilus ducreyi* (incubated at 33° C), and in broth such as thioglycolate and is Gram stained.

8. Lesion material is also placed into sucrose buffer for chlamydial culture and examined by fluorescein-conjugated monoclonal chlamydial antibody stain.

---

the pregnancy, (Chapter 34). A recent California outbreak traced to contaminated Mexican-style cheese resulted in several deaths among pregnant women. The organism can be isolated from the placenta and from tissues of the fetus. Other agents that cross the placental barrier to infect a developing fetus include *Toxoplasma gondii,* rubella, cytomegalovirus, herpes simplex virus,[41] parvovirus, and *Treponema pallidum.*

Infections that can be acquired by infants as they pass through an infected birth canal include herpes, cytomegalovirus (CMV) infection, gonorrhea, group B streptococcal sepsis,[13] chlamydial conjunctivitis and pneumonia, and *Escherichia coli* or other neonatal meningitis. Laboratory diagnosis of these infections is accomplished by direct detection or culturing for the agents when possible and by performing serologic tests, especially for the presence of specific immunoglobulin M (IgM) directed against the agent in question. Specific IgM tests are commercially available for the TORCH agents (*Toxoplasma,* rubella, cytomegalovirus, and herpes) (Chapter 13). Unless serum IgM is separated from immunoglobulin G (IgG), the IgM tests for neonatal rubella, CMV, and herpes are unreliable.

CMV infection is best diagnosed by culture of the virus from urine. The most reliable diagnostic tests for neonatal herpes infection are demonstration of the antigen in clinical specimens by direct

fluorescent antibody stain or recovery of the virus in tissue culture. Group B streptococcal antigen is reliably detected in serum or cerebrospinal fluid from infected neonates using rapid latex agglutination tests, but urine, the least invasive and most common specimen sent to the laboratory, yields specificities ranging from 81% to 98%. False-positive results seem to be caused partly by collection of urine from bags in infants with perianal colonization or by gastrointestinal absorption of antigen from swallowed organisms.[2,38,47]

NONCULTURE METHODS FOR DIAGNOSIS OF GENITAL TRACT DISEASES

SEROLOGIC TESTS FOR SYPHILIS

Diagnosis of syphilis rests on demonstration of the organism in a lesion, a difficult and insensitive test, or demonstration of specific antibodies to the treponemes. Although a complete discussion of diagnosis of syphilis is beyond the scope of this text, certain tests are mentioned here. The classical serologic tests for syphilis measure the presence of two types of antibodies: "treponemal" and "nontreponemal." Treponemal antibodies are produced against the antigens of the organisms themselves, whereas nontreponemal antibodies, often called **reaginic antibodies,** are produced by infected patients against components of mammalian cells. These antibodies, although almost always produced by patients with syphilis, are also produced by patients with other infectious diseases, such as leprosy, tuberculosis, chancroid, leptospirosis, malaria, rickettsial disease, trypanosomiasis, lymphogranuloma venereum (LGV), measles, chickenpox, hepatitis, and infectious mononucleosis; noninfectious conditions such as drug addiction; autoimmune disorders, including rheumatoid disease; and such nondiseases as old age, pregnancy, and recent immunization. When a positive serologic test for syphilis occurs in a patient without syphilis, it is called a *biological false-positive (BFP) test.*

The two most widely used nontreponemal serologic tests are the **VDRL (Venereal Disease Research Laboratories)** test and the **RPR (rapid plasma reagin)** test. Each of these tests is a flocculation (or agglutination) test, in which soluble antigen particles are coalesced to form larger particles that are visible as clumps when they are aggregated by antibody. The principles of these tests are discussed in Chapter 13.

Specific treponemal serologic tests include the **FTA-ABS (fluorescent treponemal antibody absorption)** test and the **MHA-TP (microhemagglutination)** test. Another test, the **TPI** (*Treponema pallidum* **immobilization**) test, is rarely performed today, since it requires the maintenance of viable treponemes by passage in rabbit testicles.

The FTA-ABS test is performed by overlaying whole treponemes fixed to a slide with serum from patients suspected of having syphilis because of a previous positive VDRL or RPR test. The patient's serum is first absorbed with non–*T. pallidum* treponemal antigens (sorbent) to reduce nonspecific cross-reactivity. Fluorescein-conjugated antihuman antibody reagent is then applied as a marker for specific antitreponemal antibodies in the patient's serum. This test should not be used as a primary screening procedure. The MHA-TP test uses treated erythrocytes from a turkey or other animal that are coated with treponemal antigens. The presence of specific antibody causes the red cells to agglutinate and form a flat mat across the bottom of the microdilution well in which the test is performed. The serologic tests for syphilis can be used to determine quantitative titers of antibody, which are useful for following response to therapy. The relative sensitivity of each test is shown in Table 20.1. To confirm that a positive nontreponemal test result is caused by syphilis rather than one of the other infections or biological false-positive conditions mentioned, a specific treponemal test should be performed.[44] Enzyme-linked immunosorbent assay (ELISA) tests for syphilis antibodies are available, and Western blot methods are being developed.[6] Although not widely used yet, they should offer a sensitive and specific alternative to existing methods.

OTHER SEROLOGIC TESTS FOR GENITAL PATHOGENS

As mentioned in the chapters that deal with individual pathogens, many conventional serologic tests for antibodies to genital pathogens are in use. Complement fixation and microimmunofluorescence are both used with excellent sensitivity for detection of antibodies to chlamydiae and mycoplasmas, although ELISA tests are probably more sensitive. ELISA tests are also primary choices for detection of antibodies to herpes and cytomegaloviruses, and a latex particle agglutination test (CMVScan; Becton Dickinson Microbiology Systems, Cockeysville, Md.) for CMV is sensitive and simpler. Serologic tests for gonorrhea by antibody detection are not satisfactory in

---

TABLE 20.1

SENSITIVITY OF FREQUENTLY USED SEROLOGIC TESTS FOR SYPHILIS

TEST*	STAGE		
	1	2	LATE
Nontreponemal (reaginic tests)			
Venereal Disease Research Laboratories (reaginic) test (VDRL)	70%	99%	1%†
Rapid plasma-reagin card test (RPR) or automated reagin test (ART)	80%	99%	0%
Specific treponemal tests			
Fluorescent treponemal antibody absorption test (FTA-ABS)	85%	100%	98%
Treponema pallidum hemagglutination assay (MHA-TP)	65%	100%	95%
T. pallidum immobilization test (TPI)	50%	97%	95%

* Percentage of patients with positive serologic tests in treated or untreated primary or secondary syphilis.
† Treated late syphilis.
Reproduced from Tramont, E. 1985. *Treponema pallidum* (syphilis). In Mandell, G.L., Douglas, R.G., Jr., and Bennett, J.E., editors: Principles and practice of infectious diseases, ed. 2. John Wiley & Sons, New York.

that they do not distinguish between past and current infection or between gonococcal and meningococcal infection. For patients whose serum contains antibody to HIV-1, as detected by ELISA tests, a Western blot nucleic acid hybridization test is usually used to confirm the diagnosis.

ANTIGEN DETECTION TESTS FOR GENITAL PATHOGENS

In addition to the fluorescent microscopic methods mentioned and described in more detail in the chapters relating to individual pathogens, several ELISA and latex agglutination systems are currently available for detection of antigens of etiological agents involved in STDs without culture.

Numerous commercial ELISA systems for detection of antigens of *Chlamydia trachomatis* are available and have been evaluated. ELISA systems are unable to assess specimen quality, however, they are an important parameter when an intracellular organism is being sought. Sensitivities vary greatly, depending on the patient population, skill of clinicians, and test system. Overall, sensitivity seems to be similar to that of the DFA stain.[10,20,25,26,40] Although no single test (culture, DFA, or ELISA) will detect all infections, laboratories are urged to perform at least one chlamydial test on cervical material. The public health consequences of this infection are too great to bypass any opportunity to detect infected patients.[9]

A *Neisseria gonorrhoeae* ELISA test exists as well (Gonozyme, Abbott Laboratories, Abbott Park, Ill.). The Gonozyme test has been shown to be similar to culture for detection of gonococcal urethritis and especially useful for screening asymptomatic females, although less sensitive than culture. It is not useful for detection of pharyngeal or rectal infection.

Systems are also being perfected for direct detection of herpes antigen by latex agglutination and membrane-fixed solid-phase immunoassay systems, although none is as sensitive as culture, which is still the method of choice for herpes.[41] Polyclonal and monoclonal antibodies conjugated to either a fluorescent stain or to a biotin-avidin-enzyme marker are available for staining clinical material. They have been evaluated favorably, particularly for diagnosis of herpes meningoencephalitis from brain biopsy tissue.

OTHER TESTS

A nucleic acid hybridization assay (PACE 2, Gen-Probe, San Diego) uses a chemiluminescent marker for hybridized sequences of chlamydial and gonococcal ribosomal ribonucleic acid (RNA)

and labeled homologous deoxyribonucleic acid (DNA) probe. Test results are available within 2 hours, and results in some populations are much more sensitive than could be obtained by culture.[27] Early reports indicate sensitivity and specificity at least equal to that of immunoassay methods. This hybridization system is an attractive alternative because it does not require a radioactive marker and because the use of ribosomal nucleic acid as target allows greater sensitivity than DNA-based hybridization assays. DNA-hybridization immunoperoxidase assays for herpes nucleic acid in cell culture and direct specimen material are also available. Such systems are not as sensitive or specific as culture as yet, and results require more labor than fluorescent antibody stain methods.

Systems for detection and typing of HPV DNA by hybridization have been released recently (Life Technologies, Inc., Gaithersburg, Md.; Digene Diagnostics, Silver Spring, Md.). The relationship between certain types of these viruses and cervical cancer is established, but the role of the virus in malignancy is not clear.[4,22,33] Viral genomes can be identified by DNA hybridization studies in specimens from visible genital warts, but those warts can often be treated without resorting to expensive and laborious laboratory procedures. In situ hybridization to detect HPV DNA in cervical smears is also available today, but the practice has not gained universal acceptance.

BV can be diagnosed by an increased succinic/lactic acid ratio in vaginal secretions, as detected by direct gas liquid chromatography. The Gram stain, however, is equally sensitive and more readily available.

We expect that rapid, reliable, nonculture antigen detection methods will eventually be available for diagnosis of all significant STDs so that patients can be diagnosed and treated very soon after being examined. The rapid, specific therapeutic intervention thus made possible will contribute greatly to minimizing spread of STDs and decreasing the morbidity these infections currently produce.

REFERENCES

1. Anstey, R.J., Gun-Munro, J., Rennie, R.P., et al. 1984. Laboratory and clinical evaluation of modified New York City medium (Henderson formulation) for the isolation of *Neisseria gonorrhoeae*. J. Clin. Microbiol. 20:905.
2. Ascher, D.P., Wilson, S., Mendiola, J., and Fischer, G.W. 1991. Group B streptococcal latex agglutination testing in neonates. J. Pediatr.119:458.

3. Baron, E.J. and Finegold, S.M. 1990. Bailey & Scott's diagnostic microbiology, ed. 8. Procedure 19.1. Preparation of sucrose buffer (2-SP) for transport of *Chlamydia* and *Mycoplasma* specimens. p. 267. Mosby, St. Louis.

4. Bauer, H.M., Ting, Y., Greer, C.El, et al. 1991. Genital human papillomavirus infection in female university students as determined by a PCR-based method. J.A.M.A. 265:472.

5. Beal, C., Goldsmith, R., Kotby, M., et al. 1992. The plastic envelope method, a simplified technique for culture diagnosis of trichomoniasis. J. Clin. Microbiol. 30:2265.

6. Byrne, R.E., Laska, S., Bell, M., et al. 1992. Evaluation of a *Treponema pallidum* Western immunoblot assay as a confirmatory test for syphilis. J. Clin. Microbiol. 30:115.

7. Cassell, G.H., Waites, K.B., Watson, H.L., et al. 1993. *Ureaplasma urealyticum* intrauterine infections: role in prematurity and disease in newborns. Clin. Microbiol. Rev. 6:69.

8. Catlin, B.W. 1992. *Gardnerella vaginalis*: characteristics, clinical considerations, and controversies. Clin. Microbiol. Rev. 5:213.

9. Centers for Disease Control. 1991. Pelvic inflammatory disease: guidelines for prevention and management. M.M.W.R. 40(RR-5):1.

10. Clark, A., Stamm, W.E., Gaydos, C., et al. 1992. Multicenter evaluation of the AntigEnz Chlamydia enzyme immunoassay for diagnosis of *Chlamydia trachomatis* genital infection. J. Clin. Microbiol. 30:2762.

10a. Eschenbach, D.A. Personal communication.

11. Eschenbach, D.A., Hillier, S., Critchlow, C., et al. 1988. Diagnosis and clinical manifestations of bacterial vaginosis. Am. J. Obstet. Gynecol. 158:819.

12. Granato, P.A., and Petosa, M.T. 1991. Evaluation of a rapid screening test for detecting group B streptococci in pregnant women. J. Clin. Microbiol. 29:1536.

13. Greenspoon, J.S., Fishman, A., Wilcox, J.G., et al. 1991. Comparison of culture for group B streptococcus versus enzyme immunoassay and latex agglutination rapid tests: results in 250 patients during labor. Obstet. Gynecol. 77:97.

14. Greenspoon, J.S., Wilcox, J.G., and Kirschbaum, T.H. 1991. Group B streptococcus: the effectiveness of screening and chemoprophylaxis. Obstet. Gynecol. Surv. 46:499.

15. Greenwood, J.R., and Robertson, C.A. 1983. *Haemophilus ducreyi*. Microbiol. No. MB 83-8. Check sample. Am. Soc. Clin. Pathol., vol 26, no 8. Chicago.

16. Harnett, G.B., Phillips, P.A., and Gollow, M.M. 1984. Association of genital adenovirus infection with urethritis in men. Med. J. Aust. 141:337.

16a. Hillier, S.L. Personal communication.

17. Hillier, S.L. Krohn, M.A., Klebanoff, S.J., and Eschenbach, D.A. 1992. The relationship of hydrogen peroxide producing lactobacilli to bacterial vaginosis and genital microflora in pregnant women. Obstet. Gynecol. 79:369.

18. Hillier, S.L., Krohn, M.A., Rabe, L.K., et al. 1993. Normal vaginal flora, H_2O_2-producing lactobacilli and bacterial vaginosis in pregnant women. Clin. Infect. Dis. 16(suppl. 4):S273.

19. Janda, W.M. 1991. Sexually transmitted diseases in children: the role of the microbiology laboratory. Clin. Microbiol. Newsletter 13:9.

20. Kellogg, J.A., Seiple, J.W., Klinedinst, J.L., and Levisky, J.S. 1992. Comparison of cytobrushes with swabs for recovery of endocervical cells and for Chlamydiazyme detection of *Chlamydia trachomatis*. J. Clin. Microbiol. 30:2988.

21. Klebanoff, S.J., Hillier, S.L., Eschenbach, D.A., and Waltersdorph, A.M. 1991. Control of the microbial flora of the vagina by H_2O_2-generating lactobacilli. J. Infect. Dis. 164:94.

22. Koutsky, L.A., Galloway, D.A., and Holmes, K.K. 1988. Epidemiology of genital human papillomavirus infection. Epidemiol. Rev. 10:122.

23. Krohn, M.A., Hillier, S.L., and Eschenbach, D.A. 1989. Comparison of methods for diagnosing bacterial vaginosis among pregnant women. J. Clin. Microbiol. 27:1266.

24. Kurki, T., Sivonen, A., Renkonen, O.-V., et al. 1992. Bacterial vaginosis in early pregnancy and pregnancy outcome. Obstet. Gynecol. 80:173.

25. Lefebvre, J., Laperriere, H., Rousseau, H., and Masse, R. 1988. Comparison of three techniques for detection of *Chlamydia trachomatis* in endocervical specimens from asymptomatic women. J. Clin. Microbiol. 26:726.

26. Leonardi, G.P., Seitz, M., Edstrom, R., et al. 1992. Evaluation of three immunoassays for detection of *Chlamydia trachomatis* in urine specimens from asymptomatic males. J. Clin. Microbiol. 30:2793.

27. Limberger, R.J., Biega, R., Evancoe, A., et al. 1992. Evaluation of culture and the Gen-Probe PACE 2 assay for detection of *Neisseria gonorrhoeae* and *Chlamydia trachomatis* in endocervical specimens transported to a state health laboratory. J. Clin. Microbiol. 30:1162.

28. Mahony, J.B., Luinstra, K.E., Sellors, J.W., et al. 1992. Confirmatory polymerase chain reaction testing for *Chlamydia trachomatis* in first-void urine from asymptomatic and symptomatic men. J. Clin. Microbiol. 30:2241.

29. Majeroni, B.A. 1991. New concepts in bacterial vaginosis. Am. Fam. Pract. 44:1215.

30. Martius, J., Krohn, M.A., Hillier, S.L., et al. 1988. Relationship of vaginal *Lactobacillus* species, cervical *Chlamydia trachomatis*, and bacterial vaginosis to preterm birth. Obstet. Gynecol. 71:89.

31. Moller, B.R. 1983. The role of mycoplasmas in the upper genital tract of women. Sex. Transm. Dis. 10(suppl.):281.

32. Morse, S.A. 1989. Chancroid and *Haemophilus ducreyi*. Clin. Microbiol. Rev. 2:137.

33. Moscicki, A.-B., Palefsky, J.M., Gonzales, J., et al. 1992. Colposcopic and histologic findings and human papillomavirus (HPV) DNA test variability in young women positive for HPV DNA. J. Infect. Dis. 166:951.

34. Nugent, R.P., Krohn, M.A., and Hillier, S.L. 1991. Reliability of diagnosing bacterial vaginosis is improved by a standardized method of Gram stain interpretation. J. Clin. Microbiol. 29:297.

35. Oriel, J.D. 1983. Role of genital mycoplasmas in nongonococcal urethritis and prostatitis. Sex. Transm. Dis. 10(suppl.):263.

36. Park, C.H., Hixon, D.L., Spencer, M.L., et al. 1992. The ICON strep B immunoassay for rapid detection of

group B streptococcal antigen. Lab. Med. 23:543.

37. Roberts, M.C., Hillier, S.L., Schoenknecht, F.D., and Holmes, K.K. 1985. Comparison of Gram stain, DNA probe, and culture for the identification of species of *Mobiluncus* in female genital specimens. J. Infect. Dis. 152:74.

38. Sanchez, P.J., Siegel, J.D., Cushion, N.B., and Threlkeld, N. 1990. Significance of a positive urine group B streptococcal latex agglutination test in neonates. J. Pediatr. 116:601.

39. Schreckenberger, P.C. 1992. Diagnosis of bacterial vaginosis by Gram-stained smears. Clin. Microbiol. Newsletter 14:126.

40. Schwebke, J.R., Stamm, W.E., and Handsfield, H.H. 1990. Use of sequential enzyme immunoassay and direct fluorescent antibody tests for detection of *Chlamydia trachomatis* infections in women. J. Clin. Microbiol. 28:2473.

41. Siegel, D., Golden, E., Washington, A.E., et al. 1992. Prevalence and correlates of herpes simplex infections. J.A.M.A. 268:1702.

42. Smith, R.F. 1986. Detection of *Trichomonas vaginalis* in vaginal specimens by direct immunofluorescence assay. J. Clin. Microbiol. 24:1107.

43. Stamm, W.E., Running, K., Hale, J., and Holmes, K.K. 1983. Etiologic role of *Mycoplasma hominis* and *Ureaplasma urealyticum* in women with the acute urethral syndrome. Sex. Trans. Dis. 10(suppl.):318.

44. Tramont, E. 1990. *Treponema pallidum* (syphilis). p. 1794. In Mandell, G.L., Douglas, R.G., Jr., and Bennett, J.E., editors. Principles and practice of infectious diseases, ed. 3. Churchill Livingstone, NY.

45. Tully, J.G., Taylor-Robinson, D., Rose, D.L., et al. 1983. *Mycoplasma genitalium,* a new species from the human urogenital tract. Int. J. Syst. Bacteriol. 33:387.

46. Watts, D.H., Krohn, M.A., Hillier, S.L., and Eschenbach, D.A. 1992. The association of occult amniotic fluid infection with gestational age and neonatal outcome among women in preterm labor. Obstet. Gynecol. 79:351.

47. Weisman, L.E., Stoll, B.J., Cruess, D.F., et al. 1992. Early-onset group B streptococcal sepsis: a current assessment. J. Pediatr. 121:428.

48. Wood, J.C., Lu, R.M., Peterson, E.M., et al. 1985. Evaluation of Mycotrim-GU for isolation of *Mycoplasma* species and *Ureaplasma urealyticum.* J. Clin. Microbiol. 22:789.

49. Zbitnew, A., and Gagne, L. 1992. Coxsackie B5 isolation from genital specimens in a pregnant patient. Clin. Microbiol. Newsletter 14:64.

BIBLIOGRAPHY

Baron, E.J., Cassell, G., Duffy, L.B., et al. 1993. Laboratory diagnosis of female genital tract infections. In Baron, E.J., coordinating editor. Cumitech 17A. American Society for Microbiology, Washington, D.C.

Holmes, K.K., Mårdh, P.-A., Sparling, P.F., and Wiesner, P.J., editors. 1990. Sexually transmitted diseases, ed. 2. McGraw-Hill, New York.

Infectious Disease Society of America. 1992. General guidelines for the evaluation of new anti-infective drugs for the treatment of sexually transmitted diseases. p. S96. In Beam, T.R., Jr., Gilbert, D.N., and Kunin, C.M., editors: Guidelines for the evaluation of anti-infective drug products. Clin. Infect. Dis. 15.

Lugo-Miro, V.I., Green, M., and Mazur, L. 1992. Comparison of different metronidazole therapeutic regimens for bacterial vaginosis. J.A.M.A. 268:92.

21

MICROORGANISMS ENCOUNTERED IN WOUNDS, ABSCESSES, SKIN, AND SOFT TISSUE LESIONS

Infections in the categories covered in this chapter are frequently encountered clinically, and specimens from them are typically sent to the laboratory for culture. A wide variety of bacteria, fungi, and viruses may be involved. Parasites *(Trichinella, Taenia solium, Toxoplasma)* may be involved in myositis and others, such as cysticercus, in soft tissues, but these are not sought in the usual processing of specimens in the clinical laboratory; biopsy and serologic procedures are typically employed. The bacterial flora involved in bite infections and clenched fist injuries originate in the oral cavity of humans and animals and thus are often polymicrobic and include anaerobes.[8] Tissue obtained by debridement or biopsy (Chapter 22) may be an important source of information on the infections discussed in this chapter. Venereal diseases of certain types produce skin and soft tissue lesions; these are discussed in Chapter 20. Other agents of skin and soft tissue infections are also discussed elsewhere in the book (e.g., *Bacillus anthracis, Corynebacterium diphtheriae,* mycobacteria, viruses, fungi).

Some agents involved in the infections covered in this chapter are fastidious; accordingly, careful attention to specimen transport and culture techniques, including the use of specialized media, is important. The potential for introduction of indigenous and transient skin flora into the specimen is great; therefore, proper specimen collection is critical for laboratory diagnosis of these infections (Chapter 7). In some cases the amount of material available for direct examination and culture will be very small. The use of more than one swab (a swab or similar device such as Accu-CulShure[3] may be the only practical collection device in such cases) and careful specimen transport help ensure reliable bacteriological results. Biopsy, if feasible, provides the best material. Discussion between the microbiologist and the clinician results in better specimen collection and transport and should provide the microbiologist with important information that will be of great assistance in selecting optimal culturing and processing procedures. For example, information on a patient's exposure to sea water or ingestion of raw shellfish would alert the laboratory to look for *Vibrio vulnificus* or *Vibrio alginolyticus.* Information that the specimen is from a necrotic palatal lesion in a patient with diabetic ketoacidosis would indicate the need to look carefully for *Aspergillus* or zygomycetes. Clinical evidence of a necrotizing fasciitis would indicate the likely presence of group A streptococcus, *Staphylococcus aureus,* or various anaerobic bacteria. An excellent reference for more information is the *Cumitech* by Simor et al.[13]

WOUND INFECTIONS, ABSCESSES

Wound infections and abscesses occur as complications of surgery, trauma, or disease that may interrupt a mucosal or skin surface. The nature of the infecting flora depends on the underlying problem and the location of the process. In the case of wound infections following appendectomy or other lower bowel surgery, indigenous flora of the lower gastrointestinal tract are involved, primarily *Escherichia coli,* streptococci, *Bacteroides fragilis* group, and other anaerobic gram-negative rods including *Bilophila* sp., *Peptostreptococcus* sp., anaerobic non-spore-forming rods, and *Clostridium* sp.[2] In the case of postoperative wound infection following a perforated peptic ulcer in a patient hospitalized for 2 weeks before the perforation and receiving antimicrobial therapy for an unrelated process (e.g., pneumonia), nosocomial pathogens, such as *S. aureus, Klebsiella, Enterobacter,* and *Pseudomonas aeruginosa,* are likely to be involved, in addition to oral flora (streptococci of the viridans group, oral anaerobes) and flora related to the underlying peptic ulcer disease. In the presence of obstruction or bleeding, the stomach is often colonized with *E. coli* and other elements typical of the colonic flora, including the *B. fragilis* group.

Use of H_2-receptor blockers (e.g., ranitidine), and antacids also promotes floral overgrowth in the upper gastrointestinal tract. Thus, knowledge of the indigenous flora of a viscus or an area regarding how this flora may be modified by disease or antimicrobial agents and regarding the environmental flora permits one to make an educated guess as to the likely etiological agents and even their antimicrobial susceptibility patterns. The clinician uses this approach, along with certain clinical features that may indicate the presence of one organism or another, to choose empirical therapy before availability of microbiological data; information from Gram stain or other direct examination is taken into account as well. The microbiologist uses this type of information to be certain that appropriate media, culture conditions, and so forth are employed and to help in interpreting the Gram stain. The organisms generally encountered are listed in the box at above right.

Organisms Generally Encountered in Wound Infections and Abscesses

Facultative gram-positive cocci
Staphylococcus aureus
Streptococcus pyogenes
"milleri" group streptococci *(S. anginosus, S. constellatus, S. intermedius)*
Microaerobic streptococci
Enterococci

Facultative gram-negative bacilli
Proteus, Morganella, Providencia
Other Enterobacteriaceae
Escherichia coli

Aerobic microbes
Pseudomonas sp.
Candida sp.

Anaerobic bacteria
Bacteroides sp.
Prevotella and *Porphyromonas* sp.
Fusobacterium sp.
Clostridium sp.
Peptostreptococcus sp.
Non-spore-forming anaerobic gram-positive rods

Since anaerobic bacteria are involved in many or most infections of this type, collection of specimens so as to avoid indigenous flora and specimen transport under good anaerobic conditions (which will not interfere with recovery of even obligate aerobes) are particularly important. Blood cultures, including special fungal blood cultures, should be obtained because bacteremia or fungemia is encountered in some patients with wound infection or abscess. Unusual organisms associated with postsurgical wound infection include *Mycoplasma hominis, Mycobacterium chelonae, Mycobacterium fortuitum,* fungi,[11] and even *Legionella* sp. These organisms should not be overlooked.

Quantitative or semiquantitative reporting of culture results is recommended. This provides some information on the relative importance of the various organisms present in a mixed infection (the typical situation with wounds and abscesses). It may also provide information on the infection's response to therapy. Quantitative culture results are particularly important for burn wound infections, in which the number of organisms is indicative of the severity of disease.[13] Chapter 22 covers this topic in more detail.

EXTREMITY INFECTIONS RELATED TO VASCULAR AND NEUROLOGICAL PROBLEMS (INCLUDING DECUBITUS ULCER)

The classical patient with one of these common infections has diabetes mellitus, poor arterial circulation (often both large-vessel and small-vessel disease), and peripheral neuropathy. Such patients are often seen. These individuals traumatize their feet readily (often just by virtue of wearing a new pair of shoes) without being aware of it (loss of sensation because of the neuropathy). The traumatized area develops an ulcer that does not heal readily because of the poor vascular supply and often becomes infected.[1] The infections tend to be chronic and difficult to clear up, particularly since these patients may also have poor vision and therefore may not recognize the problem and may not seek medical attention until the process has gone on for some time. These infections manifest primarily as purulent discharge and necrotic tissue in the base of the ulcer, often with a foul odor. Extension to the underlying bone to produce a difficult-to-manage osteomyelitis occurs often. Periodically, an acute cellulitis and lymphangitis may be associated with the chronic low-grade infection. This infection may make control of the patient's diabetes difficult. Peripheral vascular disease unrelated to diabetes mellitus may also predispose to this problem, but this is usually less difficult to manage because no associated neuropathy is present. Venous insufficiency, particularly when it leads to stasis ulcers, also predisposes to infection, again primarily of the lower extremities (in this case, often in the area of the calf or lower leg, rather than the foot). Infections related to poor blood supply often involve *S. aureus* and *Streptococcus pyogenes*. Those with open ulcers often become colonized with Enterobacteriaceae and *P. aeruginosa,* which may or may not play a role in the infection. Less well appreciated is that anaerobes are frequently involved in the infectious process, particularly in diabetic patients or others with peripheral vascular disease. The poor blood supply contributes to anaerobic conditions. Various anaerobes may be recovered, including the *Bacteroides fragilis* group, *Prevotella* and *Porphyromonas, Peptostreptococcus* sp., and less frequently, *Clostridium* sp.[1,13]

Another common type of infection in this general category, especially with the elderly or very ill bedridden patient, is infected **decubitus ulcer** (pressure sore) (Figure 21.1). Anaerobic conditions are present in such lesions because of necrosis of tissue. Because most of these lesions

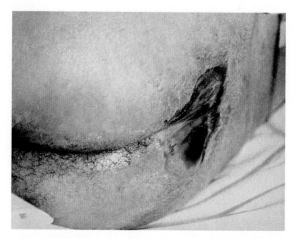

FIGURE 21.1
Sacral decubitus ulcer.

are located in proximity to the anus or on the lower extremities and because so many of these patients are relatively helpless, the ulcers become contaminated with bowel flora, which leads to chronic infection. This contributes to further death of tissue and extension of the decubitus ulcer. Bacteremia is a possible complication; the *B. fragilis* group is often involved, along with clostridia and enteric bacteria. The ulcers themselves yield a variety of anaerobes and nonanaerobes characteristic of the colonic flora; nosocomial pathogens such as *S. aureus* and *P. aeruginosa* may also be recovered.[3]

Proper collection and transport of specimens are important factors in laboratory diagnosis of this type of infection. Specimen collection is particularly difficult because many of these lesions are open and therefore readily colonized by nosocomial pathogens that may not be involved in the infection. In the event of an underlying osteomyelitis or a collection of pus in the subcutaneous tissues, attempts should be made to obtain a bone biopsy (for histology as well as smear and culture) or to aspirate the pus by needle and syringe, in both cases traversing only normal skin to get to the underlying infected area, if possible. Bacteremia does not usually accompany this type of infection, with the major exception of infected decubitus ulcers, unless an acute lymphangitis is part of the clinical picture. An acceptable specimen can be obtained by curette or by swabbing of the base after thorough debridement or cleansing of the ulcer surface.[6]

PYODERMA, CELLULITIS

This category includes infections of the skin and subcutaneous tissue, including impetigo, folliculitis, furuncles, carbuncles, paronychia, cellulitis, erysipelas (Figure 21.2), and similar lesions. *S. aureus* and group A streptococci are the most common infecting organisms, with *Candida* and *P. aeruginosa* (ecthyma gangrenosum) seen less frequently. Various non-spore-forming anaerobes are encountered more often than had been realized in the past.[5] Bacteremia is not a common accompaniment of this type of infection. Diabetic patients and those with underlying hematological (especially plasma cell) malignancy are more likely to have positive results from fine-needle aspiration of cellulitis sites.[4,10]

VESICLES, BULLAE

This type of fluid-filled lesion characteristically involves certain organisms predictably, so if the laboratory is aware of the nature of the lesion, the microbiologist can anticipate the flora and use appropriate techniques to ensure recovery of the agent(s). The organisms seen in these lesions are listed in the box at above right.

The material in the blisterlike lesion varies from serous fluid to serosanguineous or hemorrhagic fluid. Large bullae permit withdrawal of 0.5 to 1 ml of fluid by needle and syringe aspiration. Some vesicles are quite tiny, so one must use a swab for specimen collection. The clinician can usually readily anticipate whether the lesion is viral or bacterial in nature and may even be able to suspect a particular organism; recovery of the agent is facilitated if this information is imparted to the microbiologist. Bullous lesions are caused by bacteria and are often associated with sepsis, so blood cultures are mandatory. In the case of an immunosuppressed patient with acute leukemia, for example, *P. aeruginosa* would be a common cause of bullous lesions (a variety of ecthyma gangrenosum, Figure 21.3). One may see bronzed

---

Organisms Seen in Vesicles and Bullae

Staphylococcus aureus
Group A streptococci
Pseudomonas aeruginosa
Vibrio vulnificus and other vibrios
Other gram-negative bacilli
Clostridium sp.
Varicella-zoster virus
Herpes simplex virus
Other viruses

---

skin with bullous lesions in gas gangrene (Figure 21.4), an entity with a very distinctive clinical picture; *Clostridium perfringens* and other clostridia are the key pathogens.[7] Gram stain of the fluid from such lesions typically reveals the etiological agent and provides the clinician with additional valuable information on which to base initial therapy.

DRAINING SINUSES, FISTULAS

The most common situation in which one finds *draining sinuses* is a deep-seated infection that spontaneously drains itself externally, most often a chronic osteomyelitis. Unfortunately, this type of drainage does not usually cure the underlying process, and such sinuses themselves tend to be chronic. The organisms most often involved in sinuses with an underlying osteomyelitis are *S. aureus*, various Enterobacteriaceae, *P. aeruginosa*, anaerobic gram-negative bacilli, anaerobic gram-positive cocci, and occasionally other anaerobes. In the case of actinomycosis (with or without bone involvement, Figures 21.5 and 21.6), one

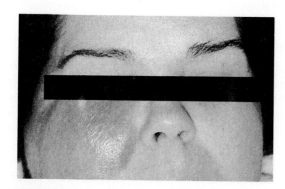

FIGURE 21.2
Erysipelas caused by *Streptococcus pyogenes*.

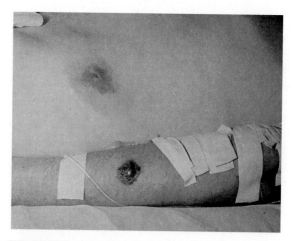

FIGURE 21.3
Ecthyma gangrenosum caused by *Pseudomonas aeruginosa*.

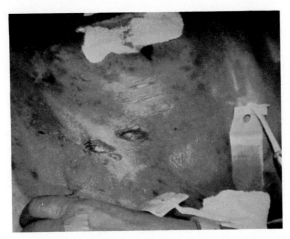

FIGURE 21.4

Gas gangrene, abdominal wall. Note bronze discoloration and fluid-filled blisters (bullae).

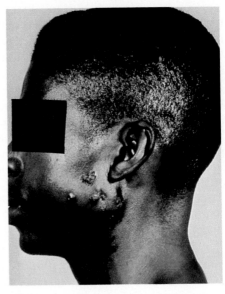

FIGURE 21.6

Actinomycosis, side view. Note sinuses in skin of face and neck.

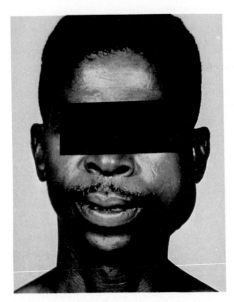

FIGURE 21.5

Actinomycosis. Note "lumpy" jaw.

would expect to recover *Actinomyces* sp., *Propionibacterium propionicum*, *Prevotella* or *Porphyromonas* and other non-spore-forming anaerobes, and *Actinobacillus actinomycetemcomitans*. With other types of draining sinuses, such as that associated with recurrent breast abscess of the subareolar type, the organisms involved depend on the nature of the underlying process. In the example of the type of breast abscess cited, non-spore-forming anaerobic bacteria are frequently isolated.

Other situations in which chronic draining sinuses are found typically include tuberculosis and atypical mycobacterial infection, *Nocardia* infection, and certain infections associated with im-

planted foreign bodies. Curettings or biopsy from the debrided cleansed sinus constitute the best specimen.

Fistulas (abnormal communications between a hollow viscus and the exterior) are difficult management problems. They also pose problems that are often insurmountable in terms of collection of meaningful specimens, since the viscus that has the abnormal communication to the skin surface often has its own profuse indigenous flora. Examples are perirectal fistulas and enterocutaneous fistulas from the small bowel to the skin in association with Crohn's disease or chronic intraabdominal infection. When the bowel is involved, only cultures for specific key organisms such as mycobacteria or *Actinomyces* are meaningful. One should always attempt to rule out specific underlying causes, such as tuberculosis, actinomycosis, and malignancy; biopsy is most useful.

SUPPURATIVE LYMPHADENITIS

Lymphadenopathy may be seen in a variety of infectious processes, but the clinical microbiology laboratory is not usually involved directly in diagnosis of the underlying problem except for serologic testing (e.g., in infectious mononucleosis syndromes). In the case of suppurative breakdown of a lymph node, pus aspirated from the node is submitted for direct examination and culture. Agents that may cause suppuration in a lymph node include group A streptococci, *S. aureus, Mycobacterium tuberculosis, Mycobacterium scrofulaceum, Mycobacterium haemophilum, Yersinia pestis, Francisella tularensis, Pseudomonas pseudomallei, Pseudomonas mallei, Sporothrix, Coccidioides,* and very rarely, *Chlamydia trachomatis* and *Rochalimaea henselae.*

MYOSITIS

Involvement of muscle occurs with a variety of infectious agents. The nature of the pathological process is variable, sometimes involving extensive necrosis of muscle, as in gas gangrene or clostridial myonecrosis, *necrotizing cutaneous myositis* or *synergistic nonclostridial anaerobic myonecrosis,* anaerobic streptococcal myonecrosis, myonecrosis caused by *Bacillus* sp., or myonecrosis caused by *Aeromonas.* Focal collections of suppuration in muscle (staphylococcal or other pyomyositis) are sometimes seen.[7,9] Muscle involvement may also manifest as a vasculitis in muscle tissue *(V. vulnificus),* and sometimes as an inflammatory myositis presenting clinically as a benign myalgia, but occasionally with myoglobinuria secondary to muscle breakdown (rhabdomyolysis), as occurs in influenza and other viral infections, rickettsial infections, infective endocarditis, bacteremia, *Legionella* infection, toxoplasmosis, trichinosis, and cysticercosis. Abscess in the psoas muscle may involve *M. tuberculosis, S. aureus,* or various facultative or anaerobic gram-negative bacilli. Serious vascular problems may lead to death of muscle resulting from loss of blood supply; such muscle may become secondarily infected (infected vascular gangrene), but this is a low-grade process.

Blood cultures should always be drawn from patients with significant myonecrosis. Transport of material (tissue is always better than pus, which is, in turn, better than a swab) should be under anaerobic conditions. An extra set of plates could be set up for early examination (in 12 to 18 hours) by use of a single-plate anaerobic pouch system to rapidly detect *Clostridium perfringens* presumptively, if the laboratory does not have an anaerobic chamber. Similarly, *Bacteroides* bile esculin agar plates should be set up in the case of synergistic nonclostridial anaerobic myonecrosis to permit rapid presumptive identification of the *B. fragilis* group (18 to 24 hours). Very early presumptive identification requires examination of plates at frequent intervals and much sooner than the usual 48-hour interval used conventionally. Organisms producing myositis or

---

Organisms Producing Myositis or Other Muscle Pathology

Clostridium perfringens
C. novyi
C. septicum
C. bifermentans
C. histolyticum
C. sordellii
C. sporogenes
Bacillus sp.
Aeromonas sp.
Peptostreptococcus sp.
Microaerobic streptococci
Bacteroides sp.
Enterobacteriaceae
Staphylococcus aureus
Group A streptococci
Pseudomonas mallei
P. pseudomallei
Vibrio vulnificus
Mycobacterium tuberculosis
Salmonella typhi
Legionella sp.
Rickettsia
Viruses
Trichinella
Taenia solium
Toxoplasma

---

other muscle pathology are listed in the box above.

SERIOUS SOFT TISSUE AND SKIN INFECTIONS

This category is used for a miscellaneous collection of infections, some extensive and some localized, but all serious or potentially serious. Anthrax and cutaneous diphtheria are discussed elsewhere, as noted previously. An often fatal but fortunately rare infection known as *mucormycosis* or **phycomycosis** is found in patients with diabetic ketoacidosis or with hematological neoplasms or other conditions requiring corticosteroid or cytotoxic therapy.[12] Necrotic lesions of the palate or nasal mucous membranes are often noted first, but lesions may also spread to the brain or involve other deep organs early in the illness. The fungi most frequently involved are *Mucor, Rhizopus, Absidia,* and *Rhizomucor,* but occasionally other fungi of the order Mucorales may be involved; *Aspergillus* sp. and *Pseudallescheria boydii* may produce similar clinical pictures. Material should be obtained by biopsy of the necrotic tissue for direct examination and culture.

Necrotizing fasciitis is a serious infection that is more common than mucormycosis but still occurs relatively infrequently. The basic pathology is infection involving the fascia overlying muscle groups, often with concomitant involvement of the overlying soft tissue. At the fascial level, no barrier exists to spread of infection, so fasciitis may extend widely and rapidly to involve huge areas of the body in short periods. This process, once known as hospital gangrene, typically involves group A streptococci or *S. aureus*. Necrotizing fasciitis also frequently involves anaerobic bacteria, especially *Bacteroides* and *Clostridium* sp. Many or most cases of perineal gangrene or phlegmon (Fournier's gangrene) and of necrotizing dermogenital infection are forms of necrotizing fasciitis, although some involve only the more superficial soft tissues and thus are not as dangerous.

Progressive bacterial synergistic gangrene is usually a chronic gangrenous condition of the skin most often encountered as a postoperative complication, particularly after abdominal surgery requiring retention sutures or after thoracic surgery. The lesions may be extensive and, with involvement of the abdominal wall, may lead to evisceration. As the name suggests, this is typically a mixed infection with microaerobic streptococci and *S. aureus*. At times, other organisms may be present, including anaerobic streptococci, *Proteus*, and other facultative and anaerobic bacteria. If only the purulent or necrotic material from the central portion of the wound is cultured, the microaerobic streptococcus will be missed, since this infection occurs very infrequently and the clinician may not recognize the usually distinctive clinical features. Consequently, the nature of the lesion may not be appreciated. Cultures should be taken from the advancing outer edge of the lesion.

Chronic undermining ulcer, or **Meleney's ulcer,** is a slowly progressive infection of the subcutaneous tissue with associated ulceration of portions of the overlying skin. The causative organism is classically a microaerobic streptococcus, but anaerobic streptococci and, occasionally, other organisms may be involved.

Anaerobic **cellulitis** is an acute or subacute anaerobic infection of soft tissue that occurs much more frequently than the other infections in this category. Considerable amounts of gas are usually present in the subcutaneous tissue. It is most often found in the extremities and is particularly common in diabetic patients. It may also involve the neck, the abdominal wall, the perineum, or connective tissue in other areas. Anaerobic cellulitis also may occur as a postoperative problem. Although the onset and spread of this lesion are not usually rapid and patients do not show impressive systemic effects at first, it is not an illness to be taken lightly. The organisms are almost always mixed aerobic and anaerobic. The aerobes include *E. coli*, alpha-hemolytic or nonhemolytic streptococci, and *S. aureus* predominantly, but group A streptococci and other Enterobacteriaceae are encountered as well. The anaerobes are typically found in greater numbers and in more variety than the aerobes; *Peptostreptococcus* sp., *B. fragilis* group strains, *Prevotella* and *Porphyromonas*, other anaerobic gram-negative bacilli, and clostridia are seen. Bacteremia is not usually present. Good anaerobic transport and culture techniques are important.

BURN INFECTIONS

Infection of burn wounds is essentially universal, may be associated with bacteremia, may carry a significant mortality, and interferes with the acceptance of skin grafts. Many organisms are capable of infecting the eschar of a burn. Those most often encountered are various streptococci, *Staphylococcus aureus*, *S. epidermidis*, Enterobacteriaceae, *Pseudomonas* sp., other gram-negative bacilli, *Candida*, and *Aspergillus*. Anaerobes, including clostridia and *Bacteroides*, have been recovered occasionally but are probably more frequently involved than has been appreciated to date.

BITE INFECTIONS

Human bites (Figure 21.7), including clenched fist injuries, yield, among the aerobic or facultative flora, alpha-hemolytic streptococci, *S. aureus*, group A streptococci, and *Eikenella corrodens*, in that order of frequency. Anaerobes recovered include, in order of frequency, *Peptostreptococcus*, pigmented *Prevotella* and *Porphyromonas*, *Prevotella oris*, *P. buccae*, and *Fusobacterium nucleatum*. In infected animal bite wounds (Figure 21.8), the most frequently encountered aerobic and facultative bacteria are alpha-hemolytic streptococci, *S. aureus*, *Pasteurella* sp., and *Enterobacter cloacae*. Predominant anaerobes in the animal bites are anaerobic gram-positive cocci, *Fusobacterium* sp., and anaerobic gram-negative rods.[8]

Oral and nasal fluids from dogs yield Centers for Disease Control (CDC) group EF-4, *Weeksella zoohelcum*, *Pasteurella* sp., and *Staphylococcus inter-*

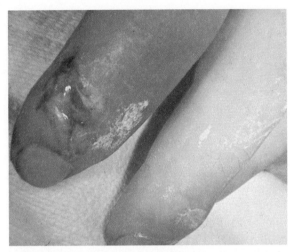

FIGURE 21.7
Human bite infection.

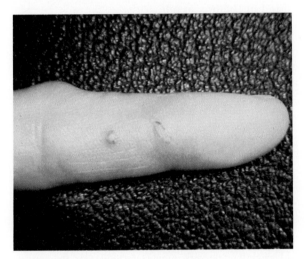

FIGURE 21.8
Animal bite infection caused by *Pasteurella* sp.

medius, with much smaller numbers of *S. aureus*. *Simonsiella* is found in the oral cavity of most dogs and is also found in the oral cavity of humans, cats, and other animals. The oral flora of snakes contains various gram-negative bacilli, including *Pseudomonas*, *Klebsiella*, *Proteus*, and *E. coli*. Clostridia may also be recovered from snakebite wounds.

Capnocytophaga canimorsus (former CDC group DF-2) and CDC group DF-3 have been responsible for several types of serious infections, including bacteremia, endocarditis, and meningitis. Most of these patients had a history of dog bite, and most had underlying diseases that impair host defense mechanisms. The organisms grow slowly on blood and chocolate agar (they do best with a heart infusion agar base), and growth may be enhanced

by rabbit serum and increased carbon dioxide tension. *C. canimorsus* is often recovered from blood cultures. It is typically resistant to aminoglycosides and is susceptible to penicillin G.

Bite wound infections usually involve relatively small lesions and minimal exudate, so a swab technique with anaerobic transport usually is needed. Surrounding skin should be thoroughly disinfected before the specimen is obtained.

UNCOMMON BUT IMPORTANT CAUSES OF SKIN AND SOFT TISSUE INFECTION

Many organisms that may cause infection with serious consequences are encountered infrequently or rarely. These include *Clostridium tetani* (tetanus), *C. botulinum* (wound botulism), *Francisella tularensis* (tularemia), various mycobacteria, *Erysipelothrix rhusiopathiae* (erysipeloid), *Actinomyces*, *Nocardia*, *Vibrio alginolyticus* (cellulitis in marine wound infection), *V. vulnificus* (wound infection), *Sporothrix schenckii* (sporotrichosis), and dematiaceous fungal agents of phaeohyphomycosis, chromoblastomycosis, and mycetoma. Various other systemic fungi, such as *Coccidioides immitis* and *Blastomyces dermatitidis*, may involve skin and subcutaneous tissue either as a primary inoculation infection or, more often, as part of the process of dissemination.

MISCELLANEOUS SKIN AND SOFT TISSUE INFECTIONS OF MILD TO MODERATE SEVERITY

Erythrasma is a very mild and superficial skin infection; the causative organism is *Corynebacterium minutissimum*. The clinical picture is distinctive, with infected lesions fluorescing coral red under long-wave ultraviolet light. Skin scrapings may be cultured in media containing serum, but imprint smears of the lesion should reveal gram-positive pleomorphic rods, precluding the necessity for culture. Numerous superficial mycoses involve the skin, nails, and hair. They are discussed in Chapter 44, but it is well to be aware that under unusual circumstances, organisms such as *Pityrosporum orbiculare* (also called *Malassezia furfur*) or *Trichosporon* can cause systemic infection. *Mycobacterium marinum* causes "swimming pool" or "fishbowl" granuloma. A similar clinical picture, sometimes accompanied by lymphadenitis, is caused by *Mycobacterium haemophilum*.

Hidradenitis suppurativa is a chronic, troublesome infection of the obstructed apocrine

Organisms Involved in Systemic Infection with Cutaneous Lesions

Viridans streptococci
Staphylococcus aureus
Enterococci
Group A and other beta-hemolytic streptococci
Neisseria gonorrhoeae
N. meningitidis
Haemophilus influenzae
Pseudomonas aeruginosa
P. mallei
P. pseudomallei
Listeria monocytogenes
Vibrio vulnificus
Salmonella typhi
Mycobacterium tuberculosis
M. leprae
Treponema pallidum
Leptospira
Streptobacillus moniliformis
Bartonella bacilliformis
Rochalimaea henselae
Rickettsia
Candida sp.
Cryptococcus neoformans
Blastomyces dermatitidis
Coccidioides immitis
Histoplasma capsulatum

glands in the axillae, genital, or perianal areas with intermittent discharge of pus (often foul smelling), draining sinuses at times, and scarring. Subareolar breast abscess probably involves this type of pathology. Secondary infection involves anaerobic gram-negative rods, anaerobic cocci, staphylococci, streptococci, and various aerobic and facultative gram-negative rods. Infected pilonidal cysts almost invariably yield anaerobic bacteria, sometimes to the exclusion of other types of bacteria. The anaerobes involved include the *B. fragilis* group, *Prevotella*, *Fusobacterium*, anaerobic gram-positive cocci, and clostridia. Various nonanaerobes, especially Enterobacteriaceae, enterococci, and other streptococci, may also be recovered. Infected "inclusion" or "sebaceous" cysts often yield anaerobic cocci, and *Bacteroides* or *Propionibacterium* species; staphylococci and streptococci are also often found in such cysts.[6]

SKIN LESIONS AS PART OF SYSTEMIC INFECTION

Cutaneous manifestations of systemic infection such as bacteremia or endocarditis may be im-

portant clues for the clinician and may present an opportunity for direct or cultural demonstration of the presence of a particular organism. For example, one may be able to scrape petechiae from patients with meningococcemia and demonstrate gram-negative diplococci. In other patients the skin lesion represents a more impressive type of metastatic infection. The lesions of ecthyma gangrenosum (see Figure 21.3) in *Pseudomonas aeruginosa* sepsis are prominent. In *Vibrio vulnificus* sepsis, dramatic-appearing cutaneous ulcers with necrotizing vasculitis or bullae may be found. In some patients, skin lesions may actually represent a noninfectious complication of a local or systemic infection. Examples would be the rashes of scarlet fever and toxic shock syndrome. Various organisms that may be involved in systemic infection with cutaneous lesions are listed in the box at left.

REFERENCES

1. Amin, N. 1988. Infected diabetic foot ulcers. Am. Fam. Physician 37:283.
2. Baron, E.J., Bennion, R., Thompson, et al. 1992. A microbiological comparison between acute and complicated appendicitis. Clin. Infect. Dis. 14:227.
3. Baron, E.J., Väisänen, V.-L., McTeague M., et al. 1993. Comparison of Accu-CulShure® system and swab placed in B-D Port-a-Cul® tube for culture collection and transport. Clin. Infect. Dis.16 (suppl. 4): S325.
4. Epperly, T.D. 1986. The value of needle aspiration in the management of cellulitis. J. Fam. Pract. 23:337.
5. Finch, R. 1988. Skin and soft-tissue infections. Lancet 1:164.
6. Finegold, S.M., Baron, E.J., and Wexler, H.M. 1992. A clinical guide to anaerobic infections. Star, Belmont, Calif.
7. George, W.L. 1989. Other infections of skin, soft tissue, and muscle. p. 486. In Finegold, S.M. and George, W.L., editors. Anaerobic infections in humans. Academic Press, San Diego.
8. Goldstein, E.J.C. 1992. Bite wounds and infection. Clin. Infect. Dis. 14:633.
9. Isaacs, R.D., Paviour, S.D., Bunker, D.E., and Lang, S.D.R. 1988. Wound infections with aerogenic *Aeromonas* strains: a review of twenty-seven cases. Eur. J. Clin. Microbiol. Infect. Dis. 7:355.
10. Kielhofner, M.A, Brown, B., and Dall, L. 1988. Influence of disease process on the utility of cellulitis needle aspirates. Arch. Intern. Med. 148:2451.
11. Paparello, S.F., Parry, R.L., MacGillivray, D.C., et al. 1992. Hospital-acquired mucormycosis. Clin. Infect. Dis. 14:350.
12. Sapico, F.L, and Bessman, A.N. 1992. Infections in the diabetic patient. Infect. Dis. Clin. Pract. 1:339.
13. Simor, A.E., Roberts, F.J., and Smith, J.A. 1988. Infections of the skin and subcutaneous tissues. In Smith J.A., coordinating editor. Cumitech 23. American Society for Microbiology, Washington, D.C.

BIBLIOGRAPHY

Bonner, J.R., Coker, A.S., Berryman, C.R., et al. 1983. Spectrum of *Vibrio* infections in a gulf coast community. Ann. Intern. Med. 99:464.

Churchill, M.A., Geraci, J.E., and Hunder, G.G. 1977. Musculoskeletal manifestations of bacterial endocarditis. Ann. Intern. Med. 87:754.

Davison, A.J., and Rotstein, O.D. 1988. The diagnosis and management of common soft-tissue infections. Can. Assoc. Gen. Surg. 31:333.

Finegold, S.M. 1977. Anaerobic bacteria in human disease. Academic Press, New York.

Fleisher, G., Ludwig, S., and Campos, J. 1980. Cellulitis: bacterial etiology, clinical features, and laboratory findings. J. Pediatr. 97:591.

Swartz, M.N. 1990. Skin and soft tissue infections. p. 796. In Mandell, G.L., Douglas, R.G., Jr., and Bennett, J.E., editors. Principles and practice of infectious diseases, ed 3. Churchill Livingstone, New York.

22

MICROORGANISMS ENCOUNTERED IN SOLID TISSUE, BONE, BONE MARROW, AND BODY FLUIDS

Any body tissue or sterile body fluid site can be invaded and infected with etiological agents of disease from all four categories of microbes: bacteria, fungi, viruses, and parasites. This chapter discusses the types of microorganisms likely to be recovered from specimens of tissue, bone marrow, bone, and normally sterile body fluids. Bacteria, fungi, and viruses can be identified by microbiological techniques practiced in routine clinical laboratories, and the following discussion focuses on those agents. Because even one colony of a potential pathogenic microorganism may be significant, and because these specimens often undergo more handling in the laboratory than do other types of specimens, we recommend that all manipulations of the specimens, transferral of material, and inoculation of media be performed by a technologist wearing gloves and working within a biological safety cabinet.

In most cases the detection of parasites in tissue sections is accomplished by observing characteristic morphological structures in stained histological sections; therefore, examination of many tissues for parasites is usually done by anatomical pathologists. Exceptions exist, particularly when body fluids are examined or when tissue is homogenized in the laboratory rather than being processed to make histological sections. For example, pulmonary infection with *Pneumocystis*

carinii is often diagnosed by microbiologists using special stains such as direct immunofluorescence. Occasionally a wet preparation of pleural fluid may reveal *Entamoeba histolytica*. *E. histolytica* may also be identified in material from the wall of an amebic liver abscess. Other tissues may be cultured for parasites, such as *Leishmania* and trypanosomes.

STERILE BODY FLUIDS

In response to infection, fluid may accumulate in any body cavity. Infected solid tissue often presents as a phlegmon or cellulitis or with abscess formation (Chapter 21). Areas of the body from which fluids are typically sent for microbiological studies (in addition to blood and cerebrospinal fluids, Chapters 15 and 16) include the thorax (thoracentesis fluid, usually called pleural fluid), abdominal cavity (paracentesis fluid, usually called ascitic fluid or peritoneal fluid), joints (synovial fluid), and pericardium (pericardial fluid). Techniques for laboratory processing of all sterile body fluids are similar. Clear fluids may be concentrated by centrifugation or filtration, whereas purulent material can be inoculated directly to media. Any body fluid received in the laboratory that is already clotted must be homogenized to release trapped bacteria and minced or cut to release fungal cells. Either processing such specimens in a motorized tissue homogenizer or grinding them manually in a mortar and pestle or glass tissue grinder allows better recovery of bacteria. Hand grinding is often preferred, since motorized grinding has the potential to generate considerable heat and thereby kill microorganisms in the specimen. Grinding may lyse fungal elements; therefore it is not recommended with specimens processed for fungi. Small amounts of whole material from a clot should be aseptically cut with a scalpel and placed directly onto media for isolation of fungi.

PLEURAL FLUID

Pleural fluid is a collection of fluid in the pleural space, normally found between the lung and the chest wall. The fluid usually contains few or no cells and has a consistency similar to that of serum, but with a lower protein content. When excess amounts of this fluid are present, it is called an *effusion* or *transudate* and is often the result of cardiac, hepatic, or renal disease. Pleural fluid that contains numerous white blood cells and other evidence of an inflammatory response (an *exudate*) is usually caused by infection, but malignancy, pulmonary infarction, or autoimmune diseases in which an antigen-antibody reaction initiates an inflammatory response may also be responsible. Such fluid is the usual specimen submitted to microbiology laboratories. This material is collected from the patient by needle aspiration (thoracentesis) and submitted to the laboratory as "pleural fluid," "thoracentesis fluid," or "empyema fluid." Exudative pleural effusions that contain numerous polymorphonuclear neutrophils, particularly those that are grossly purulent, are called **empyema** fluids. Empyema usually occurs secondary to pneumonia, but other infections near the lung (e.g., subdiaphragmatic infection) may seed microorganisms into the pleural cavity.

The bacteria that may be recovered from pleural fluid include those associated with pneumonia, such as *Streptococcus pneumoniae, Staphylococcus aureus, Haemophilus influenzae,* Enterobacteriaceae, *Pseudomonas,* and anaerobes. Anaerobic organisms in pleural fluid or empyema occur secondary to aspiration pneumonia or its complications, such as lung abscess. Less often, other streptococci, *Mycobacterium tuberculosis* or nontuberculous mycobacteria, *Actinomyces* sp., *Nocardia* sp., fungi, and rarely, mycoplasma or viruses may be recovered from cultures of empyema fluid.

Pleural fluid, as with other body fluids from sterile sites, should be transported to the laboratory in a sterile tube or vial that excludes oxygen. From 1 to 5 ml of specimen is adequate for isolation of most bacteria, but the larger the specimen, the better, particularly for isolation of *M. tuberculosis* and fungi; at least 5 ml should be submitted for each of these latter two cultures. Anaerobic transport vials are available from several sources, as mentioned in Chapter 7. These vials are prepared in an oxygen-free atmosphere and are sealed with a rubber septum or short stopper through which the fluid is injected. Fluid should never be transported in a syringe capped with a sterile rubber stopper because this method is unsafe (Chapter 3). The use of syringes should be curtailed whenever possible because recapping and removal of needles is not permitted under universal precautions requirements for hospital safety. Most clinically significant anaerobic bacteria survive adequately in nonanaerobic transport containers (e.g., sterile screw-capped tubes) for short periods if the specimen is frankly purulent and of adequate volume.

Specimens received in anaerobic transport vials should be inoculated to routine aerobic and anaerobic media as quickly as possible. Gram stains, preferably concentrated by some method, such as with a Cytospin centrifuge (Figure 22.1) should always be examined.[3] If long, thin, gram-positive branching rods are seen, a second smear should be prepared and stained with the modified acid-fast stain for *Nocardia.* Non-acid-fast branching rods are usually *Actinomyces* sp.; other possibilities are discussed in Chapter 34.

Specimens for recovery of only fungi or mycobacteria may be transported in sterile, screw-capped tubes. At least 5 to 10 ml of fluid is re-

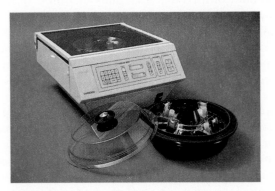

FIGURE 22.1

Photograph of a cytocentrifuge (Cytospin, Shandon Lipshaw, Pittsburgh) used for making concentrated smears of material from body fluid specimens.

quired for adequate recovery of small numbers of organisms. Those specimens that are thin enough are concentrated by centrifugation at 1500× *g* for at least 15 minutes. The supernatant is aseptically decanted or aspirated with a sterile pipette, leaving approximately 1 ml liquid in which to mix the sediment thoroughly. Vigorous vortexing or drawing the sediment up and down into a pipette several times adequately resuspends the sediment. This procedure should be done in a biological safety cabinet. The suspension is used to inoculate media and prepare smears. Smears can be further concentrated up to 1000-fold using a Cytospin centrifuge.[3]

Specimens for fungi should be examined by direct wet preparation or by preparing a separate smear for periodic acid-Schiff (PAS) staining in addition to Gram stain. Either 10% potassium hydroxide (KOH) or calcofluor white is recommended for visualization of fungal elements from a wet preparation. In addition to hyphal forms, material from the thoracic cavity may contain spherules of *Coccidioides* or budding yeast cells. Specimens are inoculated to the primary media that support the growth of fungi, including brain-heart infusion agar supplemented with sheep blood and inhibitory mold agars. Although *Legionella* sp. are seldom recovered from pleural fluid, special media for isolation of these organisms may be inoculated if clinically indicated. The physician should notify the laboratory to culture the specimen for *Legionella*.

PERITONEAL FLUID

The peritoneal cavity contains or abuts the liver, pancreas, spleen, stomach and intestinal tract, bladder, and fallopian tubes and ovaries. The kidneys occupy a retroperitoneal position. Within the healthy human peritoneal cavity is a small amount of fluid that maintains moistness of the surface of the peritoneum. Normal peritoneal fluid may contain as many as 300 white blood cells per milliliter, but the protein content and specific gravity of the fluid are low. Agents of infection gain access to the peritoneum through a perforation of the bowel, through infection within abdominal viscera, by way of the bloodstream, or by external inoculation (as in surgery or trauma). On occasion, as in pelvic inflammatory disease (PID), organisms travel through the natural channels of the fallopian tubes into the peritoneal cavity. In primary peritonitis, no apparent focus of infection is evident. Secondary peritonitis involves a rupture of a viscus or other known source of infection. During an infectious

or inflammatory process, increased amounts of fluid accumulate in the peritoneal cavity. The fluid, often called *ascites* or **ascitic fluid,** contains an increased number of inflammatory cells and an elevated protein level.

The organisms likely to be recovered from specimens from patients with primary peritonitis vary with the patient's age. The most common etiological agents in children are *Streptococcus pnuemoniae* and group A streptococci, Enterobacteriaceae, other gram-negative bacilli, and staphylococci. In adults, *Escherichia coli* is the most common bacterium followed by *S. pneumonia* and group A streptococci. Polymicrobic peritonitis is unusual in the absence of bowel perforation or rupture. Among sexually active young women, *Neisseria gonorrhoeae* and *Chlamydia trachomatis* are common etiological agents of peritoneal infection, often in the form of a perihepatitis (inflammation of the surface of the liver, called Fitz-Hugh-Curtis syndrome). Tuberculous peritonitis occurs infrequently in the United States today. It is more likely to be found among persons recently arrived from South America, Southeast Asia, or Africa. Fungal causes of peritonitis are not common, but *Candida* sp. may be recovered from immunosuppressed patients and patients receiving prolonged antibacterial therapy. *Coccidioides immitis* is an infrequent cause of peritonitis found in patients who live in or have visited endemic areas such as the southwestern United States.

Secondary peritonitis is a sequel to a perforated viscus, surgery, traumatic injury, loss of bowel wall integrity because of destructive disease (e.g., ulcerative colitis, ruptured appendix, carcinoma), obstruction, or a preceding infection (liver abscess, salpingitis, septicemia, etc.). The nature, location, and etiology of the underlying process govern the agents to be recovered from peritoneal fluid. With PID as the background, gonococci, anaerobes, or chlamydia are isolated. With peritonitis or intraabdominal abscess, anaerobes generally are found in peritoneal fluid, usually together with Enterobacteriaceae and enterococci or other streptococci. In patients whose bowel flora has been altered by antimicrobial agents, more resistant gram-negative bacilli and *Staphylococcus aureus* may be encountered. Since anaerobes outnumber aerobes in the bowel by 1000-fold, it is not surprising that anaerobic organisms play a prominent role in intraabdominal infection, perhaps acting synergistically with facultative bacteria. The organisms likely to be recovered include *E. coli*, the *Bacterioides fragilis* group, enterococci and other streptococci, *Bilophila* sp., other anaerobic gram-negative bacilli, anaerobic gram-positive cocci, and clostridia.

Specimens are collected by percutaneous needle aspiration (paracentesis) or at the time of surgery and 10 ml should be inoculated into blood culture broth bottles (aerobic and anaerobic) immediately at the bedside. Any delay results in decreased detection of true positive cultures.[15a] If gonococci or chlamydiae are suspected, additional aliquots should be sent to the laboratory for smears and appropriate cultures. Transport should be in an anaerobic vial. Ten ml of fluid is recommended for the diagnosis of peritonitis. If a large volume of clear serosanguineous fluid is received, it may be concentrated by centrifugation at $1500 \times g$ for 15 minutes or longer. Bacteriological media should include chocolate or other medium for recovery of gonococci and the occasional *Haemophilus* sp. Appropriate procedures for isolation of fungi, *Chlamydia,* and viruses should be used when such tests are appropriate.

PERITONEAL DIALYSIS FLUID

More than 5000 patients with end-stage renal disease are maintained on chronic ambulatory peritoneal dialysis (CAPD). In this treatment, fluid is injected into the peritoneal cavity and subsequently removed, which allows exchange of salts and water and removal of various wastes in the absence of kidney function. The average incidence of peritonitis in these patients is up to two episodes per year per patient. Peritonitis is best diagnosed clinically by the presence of cloudy dialysate with or without abdominal pain.[12,13] Although white blood cells are usually plentiful (leukocytes >100/ml is usually indicative of infection), the number of organisms is usually too low for detection on Gram stain of the peritoneal fluid sediment unless a concentrating technique is used; fungi are more readily detected. Most infections originate from the patient's own skin flora; *Staphylococcus epidermidis* and *S. aureus* are the most common etiological agents, followed by streptococci, aerobic or facultative gram-negative bacilli, *Candida* sp., *Corynebacterium* sp., and others. The oxygen content of peritoneal dialysate is usually too high for development of anaerobic infection. Among the gram-negative bacilli isolated, *Pseudomonas* sp., *Acinetobacter* sp., and the Enterobacteriaceae are frequently seen.

Fluid is usually received in the laboratory in a sterile tube or urine cup; less than 10 ml is unacceptable. Almost all patients with peritonitis have cloudy peritoneal dialysis fluid, with a white cell count greater than 100/mm³.[14] If fluid is received in the original bag, the bag should be entered only once with a sterile needle and syringe to withdraw fluid for culture. Fluid may be processed directly by inoculation into blood culture bottles (at least 20 ml, 10 ml in each of two culture bottles, should be cultured).[15] For other culture methods, the fluid should be concentrated, either by centrifugation, as for other clear, sterile body fluids, or by filtration. A recent study has shown that lysis of leukocytes before concentration enhanced recovery significantly.[9] Filtration through a 0.45 μm pore size membrane filter allows a greater volume of fluid to be processed and usually yields better results. Since the numbers of infecting organisms may be quite low (less than one organism per 10 ml fluid), a large quantity of fluid must be processed. Sediment obtained from at least 50 ml fluid has ben recommended.[1] Gram stains or acridine orange stains should be performed, even though the yield may be low. If the specimen is filtered, the filter should be cut aseptically into three pieces, one of which is placed on chocolate agar for incubation in 5% carbon dioxide, one on MacConkey agar, and the other on a blood agar plate for anaerobic incubation. Commercial filtration-culture systems have be evaluated.[11] Sediment should be inoculated to aerobic media and thioglycolate or similar broth, although anaerobic media are usually unnecessary.

One study had documented the sensitivity of culturing the entire contents of a patient's dialysis exchange bag.[4] Processing dialysis fluid in this way may yield more positive cultures than does processing small amounts of dialysate.

PERICARDIAL FLUID

The heart and contiguous major blood vessels are surrounded by a protective tissue, the pericardium. The area between the epicardium (the membrane surrounding the heart muscle) and the pericardium, the pericardial space, normally contains 15 to 20 ml of clear fluid. If an infectious agent is present within the fluid, the pericardium may become distended and tight, and eventually interference with cardiac function and circulation can ensue (called *tamponade*). Agents of pericarditis (inflammation of the pericardium) are usually viruses. Parasites, bacteria, certain fungi, and noninfectious causes are also associated with this disease. Inflammation of the heart muscle itself, myocarditis, may accompany pericarditis. The pathogenesis of disease involves the host inflammatory response contributing to fluid buildup and cell and tissue damage. The most common etiological agents of pericarditis and myocarditis are enteroviruses, primarily coxsackieviruses A and B. Echoviruses, adenoviruses, influenza viruses, and other viruses play a lesser role. Among nonviral agents, *Mycoplasma pneumoniae, Chlamydia, Mycobacterium tuberculosis, Staphylococcus aureus, Streptococcus pneumonia,* Enterobacteriaceae, other aerobic gram-negative bacilli, anaerobic bacteria, *Coccidioides immitis, Aspergillus* sp., *Candida* sp., *Cryptococcus neoformans, Histoplasma capsulatum, Entamoeba histolytica,* and *Toxoplasma gondii* may be found. Other bacteria, fungi, and parasitic agents have been recovered from pericardial effusions, however, and all agents should be sought. Patients who develop pericarditis resulting from agents other than viruses are often compromised in some way. An example is infective endocarditis when a myocardial abscess develops and then ruptures into the pericardial space. Collection of pericardial fluid is obviously hazardous because the sample is immediately adjacent to the beating heart. It is performed by needle aspiration with electrocardiographic monitoring or as a surgical procedure. The laboratory personnel should be alerted in advance so that the appropriate media, tissue culture media, and stain procedures are available immediately. Fluid should be processed in the same way as other sterile body fluids, as discussed earlier.

JOINT FLUID

Infectious arthritis may involve any joint in the body. It usually occurs secondary to hematogeneous spread of bacteria or, less often, fungi, or it may occur as a direct extension of infection of the bone. It may also occur after injection of material, especially corticosteroids, into joints or after insertion of prosthetic material (e.g., total hip replacement). Although infectious arthritis usually occurs at only one site (monoarticular), a preexisting bacteremia or fungemia may seed more than one joint to establish polyarticular infection, particularly when multiple joints are diseased, such as in rheumatoid arthritis. Knees and hips are the most frequently affected joints. In addition to active infections associated with viable microorganisms within the joint, sterile, self-limited arthritis caused by antigen-antibody interactions may follow an episode of infection, such as meningococcal meningitis. When an etiological agent cannot be isolated from an inflamed joint fluid specimen, either the absence of viable agents or inadequate transport or culturing procedures can be blamed. For example, even under the best circumstances, *Borrelia burgdorferi* is isolated from the joints of less than 20% of patients with Lyme disease.

Nonspecific test results, such as increased white blood cell count, decreased glucose, or elevated protein, may seem to implicate an infectious agent, but they are not conclusive. A role has been postulated for the persistence of bacterial L-forms (cell-wall–deficient forms) in joint fluid after systemic infection, but such theories have not been proved.

Overall, *Staphylococcus aureus* is the most common etiological agent of septic arthritis, accounting for approximately 70% of all such infections. In adults less than 30 years old, however, *Neisseria gonorrhoeae* is isolated most frequently. *Haemophilus influenzae* has been the most common agent of bacteremia in children less than 2 years of age, and consequently it has been the most frequent cause of infectious arthritis in these patients, followed by *S. aureus.* The widespread use of *H. influenzae* type B vaccine should contribute to a change in this pattern. Streptococci, including groups A and B, pneumococci, and viridans streptococci, are prominent among bacterial agents associated with infectious arthritis in patients of all ages. *Bacteroides,* including *B. fragilis,* may be recovered, as may *Fusobacterium necrophorum,* which usually involves more than one joint in the course of sepsis. Among people living in certain endemic areas of the United States and Europe, infectious arthritis is a prominent feature of Lyme disease. Some of the more frequently encountered etiological agents of infectious arthritis are listed in the box on p. 289.

These agents act to stimulate a host inflammatory response, which is initially responsible for the pathology of the infection. Arthritis is also a symptom associated with infectious diseases caused by certain agents, such as *Neisseria meningitidis,* group A streptococci (rheumatic fever), and *Streptobacillus moniliformis,* in which the agent cannot be recovered from joint fluid. Presumably, antigen-antibody complexes formed during active infection accumulate in a joint, initiating an inflammatory response that is responsible for the ensuing damage.

Infections in prosthetic joints are usually associated with somewhat different etiological agents than those in natural joints. After insertion of the prosthesis, organisms that gained access during the surgical procedure slowly multiply until they reach a critical mass and produce a host response. This many occur long after the initial surgery; approximately one half of all prosthetic joint infections occur more than 1 year after surgery. Skin flora are the most common etiological agents, with *Staphylococcus epidermidis,* other

Most Frequently Encountered Etiological Agents of Infectious Arthritis

Bacterial
Staphylococcus aureus
Beta-hemolytic streptococci
Streptococci (other)
Haemophilus influenzae
Haemophilus sp. (other)
Bacteroides sp.
Fusobacterium sp.
Neisseria gonorrhoeae
Pseudomonas sp.
Salmonella sp.
Pasteurella multocida
Moraxella osloensis
Kingella kingae
Moraxella catarrhalis
Capnocytophaga sp.
Corynebacterium sp.
Clostridium sp.
Peptostreptococcus sp.
Eikenella corrodens
Actinomyces sp.
Mycobacterium sp.
Mycoplasma sp.
Ureaplasma urealyticum
Borrelia burgdorferi

Fungal
Candida sp.
Cryptococcus neoformans
Coccidioides immitis
Sporothrix schenckii

Viral
Hepatitis B
Mumps
Rubella
Other viruses (rarely)

coagulase-negative staphylococci, *Corynebacterium* sp., and *Propionibacterium* sp. predominating. However, *Staphylococcus aureus* is also a major pathogen in this infectious disease. Alternatively, organisms may reach joints during hematogenous spread from distant infected sites.[10]

Specimens are collected by aspiration with a sterile needle and syringe. The specimen should be injected into an anaerobic transport vial to preserve viability of anaerobes, which are probably present more often than has been appreciated. It may also prove beneficial to inoculate one bottle of blood culture medium with some of the initial specimen, particularly if the specimen cannot be delivered to the laboratory immediately. One must always remember to send some of the specimen to the laboratory in a container other than a blood culture bottle, since putting the sample into

a blood culture bottle dilutes it, making the preparation of a smear for Gram stain useless. This portion of the culture in the blood culture bottle is processed as a blood culture, facilitating the recovery of small numbers of organisms and diluting out the effects of antibiotics. Citrate or sodium polyanetholesulfonate may be used as an anticoagulant. If the fluid contains a clot, it must be homogenized or ground up to release organisms, which tend to become trapped in clots. Gonococci may be recovered from joint fluids that have been incubated in hyperosmotic medium, such as sucrose-containing broth. Since these specimens are from normally sterile sites, selective media are inadvisable because they may inhibit the growth of some of the organisms being sought. Very purulent specimens are plated directly to several agars, including one that supports the growth of fastidious organisms, such as chocolate agar, and an enriched broth such as thioglycolate. Material should also be inoculated onto primary anaerobic media (Chapter 35). If fungi or mycobacteria are suspected, appropriate media for their isolation should also be inoculated. Methods for isolation of mycoplasma and viruses are described in Chapters 39 and 43. Direct Gram stains, KOH, or calcofluor white preparations for fungi and acid-fast stain for mycobacteria are also performed.

BONE MARROW ASPIRATION OR BIOPSY

Diagnosis of certain diseases, including brucellosis, histoplasmosis, blastomycosis, tuberculosis, and leishmaniasis, can sometimes be made only by detection of the organisms in the bone marrow. *Brucella* sp. can be isolated on culture, as can fungi, but parasitic agents must be visualized in smears or sections made from bone marrow material. Bone marrow is typically aspirated from the interstitium of the iliac crest. It is usually not necessary to process this material for routine bacteria, since blood cultures are equally useful for these microbes, and false-positive cultures for skin bacteria *(Staphylococcus epidermidis)* are frequent. It is best to use the marrow obtained for detection of specific microbes noted earlier. Some laboratories report good recovery from bone marrow material that has been injected into a Pediatric Isolator tube (Wampole, Cranbury, N.J.; Chapter 15) as a collection and transport device. The lytic agents within the Isolator tube are thought to lyse cellular components, presumably freeing intracellular bacteria for enhanced recovery. Clotted specimens must be homogenized or ground up to release trapped microorganisms; specimens are inoculated to the same media as for other sterile body fluids. A special medium for enhancement of growth of *Brucella* sp. and incubation under 10% carbon dioxide may be needed. A portion of the specimens may be inoculated directly to fungal media (Chapter 44). Sections are also made from biopsy material (bone) for fixation, staining, and examination (usually by anatomical pathologists) for the presence of mycobacterial, fungal, or parasitic agents.

BONE BIOPSY

A small piece of infected bone is occasionally sent to the microbiology laboratory for determination of the etiological agent of osteomyelitis (infection of bone). It has been shown that cultures taken from open wound sites above infected bone or material taken from a draining sinus leading to an area of osteomyelitis may not reflect the actual etiological agent of the underlying osteomyelitis. Patients develop osteomyelitis from hematogeneous spread of an infectious agent, invasion of bone tissue from an adjacent site of infection (e.g., joint infection, dental infection), breakdown of tissue caused by trauma or surgery, or lack of adequate circulation followed by colonization of a skin ulceration with microorganisms. Parasites or viruses are rarely, if ever, etiological agents of osteomyelitis. Once established, infections in bone may tend to progress toward chronicity, particularly if effective blood supply to the affected area is lacking.

Staphylococcus aureus, seeded during bacteremia, is the most common etiological agent of osteomyelitis among people of all age groups. The toxins and enzymes produced by this bacterium, as well as its ability to adhere to smooth surfaces and produce a protective glycocalyx coating, seem to contribute to its pathogenicity. Among young persons, osteomyelitis is usually associated with a single agent. Such infections are usually of hematogeneous origin. Other organisms that have been recovered from hematogeneously acquired osteomyelitis include *Salmonella* sp., *Haemophilus* sp., Enterobacteriaceae, *Pseudomonas* sp., *Fusobacterium necrophorum,* and yeasts. *S. aureus* or *Pseudomonas aeruginosa* is often recovered from drug addicts. A human bite may lead to infection with *Eikenella corrodens,* whereas a preceding animal bite may lead to *Pasteurella multocida* osteomyelitis. Stepping on a nail or other sharp object while wearing tennis shoes has been associated with osteomyelitis caused by *P. aeruginosa.*[7]

Bone biopsies from infections that have spread to a bone from a contiguous source or that are associated with poor circulation, especially in diabetic patients, are likely to yield multiple isolates. *Mycobacterium tuberculosis* is presently an infrequent cause of osteomyelitis. Gram-negative bacilli are increasingly common among hospitalized patients; a break in the skin (surgery or intravenous line) may precede establishment of gram-negative osteomyelitis. Breaks in skin from other causes (bite wound, trauma) also may be the initial event that leads to underlying bone infection. Poor oral hygiene may lead to osteomyelitis of the jaw with *Actinomyces* sp., *Capnocytophaga* sp., and other oral flora, particularly anaerobes. Pigmented *Prevotella* (formerly *Bacteroides melaninogenicus* group) and *Porphyromonas, Fusobacterium,* and *Peptostreptococcus* sp. are often involved. Pelvic infection in the female may lead to mixed aerobic and anaerobic osteomyelitis of the pubic bone.

Patients with neuropathy in the extremities, notably diabetic patients, who may have poor circulation as well, are subjected to trauma that they cannot feel simply by walking. They develop ulcers on the feet that do not heal, become infected, and may eventually progress to involve underlying bone. These infections are usually polymicrobial, involving anaerobic and aerobic bacteria.[2] *Prevotella* or *Porphyromonas;* other gram-negative anaerobes, including the *Bacteroides fragilis* group; *Peptostreptococcus* sp.; *Staphylococcus aureus;* and group A and other streptococci are frequently encountered.

Bone removed at surgery or by percutaneous biopsy is sent to the laboratory in a sterile container. It is very difficult to break up bones; however, most bone in this setting is quite soft and necrotic, and grinding in a mortar and pestle may break off some pieces. It is sometimes possible to scrape off aseptically small shavings from the most necrotic-looking areas, which can be inoculated to media. Pieces should be placed directly into media for recovery of fungi. Small bits of bone can be ground with sterile broth to form a suspension for bacteriological and mycobacterial cultures. If anaerobes are to be recovered, all manipulations are best performed in an anaerobic chamber. If such an environment is unavailable, microbiologists should work quickly within a biosafety cabinet to inoculate prereduced anaerobic plates and broth with material from the bone.

TISSUE

Pieces of tissue are removed from patients during surgical or needle biopsy procedures or may be

Infectious Agents in Tissue Requiring Special Media

Actinomyces sp.
Brucella sp.
Legionella sp.
Afipia felis and *Rochalimaea (Bartonella) henselae* (cat-scratch disease bacilli)
Listeria monocytogenes (not fastidious)
Systemic fungi
Mycoplasma
Mycobacteria
Viruses

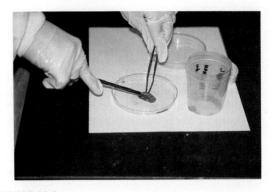

FIGURE 22.2

Photograph of mincing a piece of tissue for culture using a sterile forceps and scissors. Note: perform this procedure in a biosafety cabinet.

collected at autopsy. Any agent of infection may cause disease in tissue, and laboratory practices should be adequate to recover bacteria, fungi, and viruses and to detect the presence of parasites. Fastidious organisms (e.g., *Brucella* sp.), and agents of chronic disease (e.g., systemic fungi and mycobacteria) may require special media and long incubation periods for isolation. Some agents that require special supportive or selective media are listed in the box above.

The conditions required for the isolation of these agents are discussed in the appropriate chapters. This chapter does not attempt to delineate all possible etiological agents that can be recovered from tissue specimens but does discuss general techniques for laboratory handling of such material. The clinician is responsible for informing the laboratory of any suspected etiological agent, since the appropriate procedures can then be initiated once the specimen reaches the laboratory. Often the physician consults the clinical microbiologist about possible agents to search for in advance of removing the tissue. Optimal diagnostic services are rendered when such dialogue is a continuing part of the practice of clinical and laboratory medicine. Tissue specimens are obtained after careful preparation of the skin site. It is critical that biopsy specimens be collected aseptically and submitted to the microbiology laboratory in a sterile container. A wide-mouthed, screw-capped bottle or plastic container is recommended. Anaerobic organisms survive within infected tissue long enough to be recovered from culture. A small amount of sterile, nonbacteriostatic saline may be added to keep the specimen moist. Since homogenizing with a tissue grinder can destroy some organisms by the shearing forces generated during grinding, it is often best to mince larger tissue specimens into small pieces suitable for culturing using a sterile scissors and forceps (Figure 22.2). NOTE: *Legionella* sp. may be inhibited by saline; a section of lung

should be submitted without saline for *Legionella* isolation.

If anaerobic organisms are a particular concern, a small amount of tissue can be placed into a loosely capped, wide-mouthed plastic tube and sealed into an anaerobic pouch system, which also seals in moisture enough for survival of organisms in tissue until the specimen is plated. The surgeon should take responsibility for seeing that a second specimen is submitted to anatomical pathology for histological studies. Formaldehyde-fixed tissue is not very useful for recovery of viable microorganisms, although some organisms can be recovered after very short periods. Therefore, an attempt may rarely be made to subculture from tissue in formalin if that is the only specimen available. Material from draining sinus tracts should include a portion of the tract's wall, obtained by deep curettage. Tissue from infective endocarditis should contain a portion of the valve and vegetation if the patient is undergoing valve replacement.

In some instances, contaminated material may be submitted for microbiological examination. Specimens such as tonsils, autopsy tissue, or similar material may be surface cauterized with a heated spatula or blanched by immersing in boiling water for 5 to 10 seconds to reduce surface contamination. The specimen may then be dissected with sterile instruments to permit culturing of the specimen's center, which will not be affected by the heating. Alternatively, larger tissues may be cut in half with a sterile scissors or blade and the interior portion cultured for microbes.

Because surgical specimens are obtained at great risk and expense to the patient, and because supplementary specimens cannot be obtained easily, it is important that the laboratory save a portion of the original tissue (if enough material is available) in a small amount of sterile broth in the

PROCEDURE 22.1

QUANTITATIVE BACTERIOLOGICAL CULTURE OF TISSUE

1. Cut a small piece of tissue, several cubic millimeters, aseptically onto a small, preweighed piece of sterile aluminum foil.

2. Determine the weight of the tissue by subtracting the weight of the aluminum foil from the total weight.

3. Place the specimen and 1 ml sterile nutrient broth in a sterile tissue grinder, and macerate the specimen.

4. Prepare six serial dilution tubes of 4.5 ml sterile physiological saline. Make 1:10 serial dilutions of the macerated specimen by adding 0.5 ml of the original suspension to the first dilution tube, vortexing vigorously, and then adding 0.5 ml of this dilution to the second tube of saline. Repeat this procedure until all six dilutions have been made. Discard 0.5 ml from the last tube. This process can be done in an anaerobic chamber or alternatively in a biosafety cabinet.

5. Inoculate 0.1 ml from each of the six dilutions onto a blood agar plate, a Columbia colistin–nalidixic acid agar plate, an anaerobic blood agar plate (if indicated), and into thioglycolate broth. Spread the inoculum on the plates with a sterile glass spreading rod or a loop. Alternatively, a 0.01 or 0.001 calibrated loop can be used for sampling. The loop is not quite as accurate as actual dilutions but is simpler to use in the clinical laboratory setting. Additional plates may be inoculated after the results of the Gram stain are known, but not all dilutions need to be plated on selective media.

6. Incubate plates in 5% to 10% carbon dioxide overnight, and count the colonies of bacteria on the plate that contains 30 to 300 CFUs.

7. Calculate the number of CFUs per gram of tissue with the formula:

Number of CFU × Reciprocal of dilution (10^{-3} or 10^{-4}, etc.) × 10 (the inoculation of only 0.1 ml per plate) ÷ Weight of tissue

For example, for tissue that weighed 0.002 g, 68 CFUs were observed on the plate that received the 10^{-3} dilution of suspension:

$$\frac{68 \times 10^3 \times 10}{0.02} = \frac{6.8 \times 10^5}{0.02}$$

$$= 3.4 \times 10^7 \text{ CFU/g}$$

Modified from a method published by Loebl, E.C., Marvin, J.A., Heck, E.L., et al. 1974. Am. J. Clin. Pathol. 61:20.

---

refrigerator and at $-70°$ C (or, if necessary, at $-20°$ C) for at least 4 weeks in case additional studies are indicated. If the entire tissue must be ground up for culture, a small amount of the suspension should be placed into a sterile tube and refrigerated. Refrigeration is also used to recover *Listeria monocytogenes* from tissue (cold enrichment).

Tissue should be manipulated within a laminar flow biological safety cabinet by an operator wearing gloves. Processing tissue within an anaerobic chamber is even better. The microbiologist should cut through the infected area (often discolored) with a sterile scalpel blade. One half of the specimen can then be used for fungal cultures and the other for bacterial cultures. Both types of agents should be sought in all tissue specimens. Clinicians may need to be reminded that some sample should also be sent to surgical pathology for histological examination. Specimens should be cultured for viruses or acid-fast bacilli when such tests are requested. If material is to be cultured for parasites, it is finely minced or teased before inoculation into broth (Chapter 45). Direct examination of stained tissue for parasites is performed by anatomical pathologists. Imprint cultures of tissues may yield bacteriological results identical to homogenates,[6] and may help differentiate microbial infection within the tissue's center from surface colonization (growth only at the edge) when specimens are cut in half before processing.

Additional media can be inoculated for incubation at lower temperatures, which may facilitate recovery of certain systemic fungi and mycobacteria (Chapters 44 and 42). It is possible to obtain reliable microbiological information from tissues after a body has been embalmed. Bacteria and fungi often survive the embalming fluids.

Occasionally it is important to ascertain the number of organisms present per gram of tissue. Particularly for burn patients, greater than 10^5 colony forming units (CFUs) per gram of tissue is considered by some clinicians to be indicative of infection, whereas less than that number may indicate only colonization. Although rarely re-

PROCEDURE 22.2

ISOLATION OF *LISTERIA* SP. *FROM TISSUE*

1. Add 5 to 10 ml of ground tissue to each of two flasks of infusion broth.

2. Incubate one flask at 35° C for 24 hours, and inoculate a drop of this to a blood agar plate and a tellurite blood agar plate. Incubate these in a candle jar or carbon dioxide incubator for at least 48 hours, along with the original broth flask.

3. Store the second flask in a refrigerator at 4° C. If the 35° C subcultures (step 2) are unsuccessful, subculture material from the refrigerated flask at weekly intervals for at least 1 month.

4. Identify isolates as detailed in Chapter 34.

PROCEDURE 22.3

COLLECTION OF POSTMORTEM SPECIMENS FOR MICROBIOLOGICAL STUDIES

GENERAL TENETS

1. Use fresh, sterile instruments for each tissue sample collected.

2. Change to fresh sterile gloves if the current gloves become wet.

3. Sear outside of tissue to dryness before tissue is resected with sterile scalpel or scissors. Handle tissue aseptically from collection to delivery to the laboratory.

4. Bring material for cultures to microbiology laboratory as soon as possible after collection.

PREPARATION OF BODY

The following steps are critical if microbiological data are significant for determination of cause of death or adequacy of treatment before death:

1. Shave anterior trunk

2. Thoroughly scrub body with povidone-iodine (5 minutes) followed by 70% alcohol (5 minutes).

3. Autopsy technicians and pathologists scrub for 10 minutes with disinfectant soap. Personnel don sterile masks, gowns, gloves, and boots, just as for surgery.

4. Drape body in sterile plastic drapes and cover skin with adhesive plastic sterile drape.

BLOOD CULTURES

Make usual Y-shaped incision, but draw blood cultures first before any additional manipulation. Reflect skin and subcutaneous tissue over thorax and sear third intercostal space next to the sternum on the left side with a red-hot spatula. If seared tissue is not completely moisture-free, it must be seared a second time. Insert a sterile 18-gauge needle with attached 30 ml syringe into the heart (right ventricle). Withdraw as much blood as possible, up to 30 ml. Thoroughly and vigorously wipe the rubber septa of the blood culture containers with 70% alcohol. Inject 5 to 10 ml blood into each of two blood culture bottles (aerobic and anaerobic) for bacterial cultures and 10 ml into Isolator tube for fungal and acid-fast bacterial cultures.

COLLECTION OF MATERIAL FROM PERITONEAL CAVITY

Peritoneal fluid should be collected on a swab immediately after entering the peritoneal cavity. If intraabdominal or other abscess is encountered, the outside of the abscess should be seared to dryness with a red-hot spatula, if feasible, and a sterile needle and syringe should be used to collect pus through intact cavity walls. Vigorously scrub the rubber tip of an anaerobic transport vial with 70% alcohol, and inject aspirated material suspected of harboring anaerobes through the rubber septum into the anaerobic transport vial. All other fluid may be collected in a similar manner (preferred) or on standard transport swabs, with the exception of urine. For collection of urine, sear the bladder surface and collect bladder urine with a needle and syringe. Expel the urine into a sterile urine cup.

PROCEDURE 22.3—cont'd

COLLECTION OF POSTMORTEM SPECIMENS FOR MICROBIOLOGICAL STUDIES

COLLECTION OF TISSUE SPECIMENS FOR CULTURE

Avoid sectioning any large vessels before adequate tissue specimens for culture are obtained. Be certain that the gloves of the prosector are dry and cannot drip onto the tissue. Sear a large area of the surface of the tissue to complete dryness, using a 5 × 5 cm red-hot bent spatula tip. Using fresh sterile scissors and forceps for each tissue specimen obtained, cut a 1 cm³ block of tissue from the center of the seared area,

and place it immediately into a sterile Petri dish or sterile screw-cap urine cup, quickly covering it.

ABSCESSES: SPECIAL CONSIDERATIONS

In addition to the aspirated material, the abscess wall may be the only site from which certain organisms, such as *Nocardia* and *Actinomyces*, may be recovered. Submit a small (1 cm²) section of the abscess wall.

COLLECTION OF MATERIAL FROM THORACIC CAVITY

Change to fresh, dry, sterile gloves, and collect fluid from the pericardium, pleural cavity, along with any other fluid collections by needle and syringe (or on swabs, if necessary). Collect purulent or necrotic material with a needle and syringe, as described for abdominal abscess. Submit such material in an anaerobic transport vial. Tissue specimens are collected as previously described.

quired, a laboratory may be asked to perform quantitative cultures of tissue. Procedure 22.1 describes one method.

Tissue may be inoculated to virus tissue culture media for isolation of viruses. Brain, lung, spinal fluid, and blood are most useful. Tissue may be examined by immunofluorescence for the presence of herpes simplex virus, varicella-zoster virus, cytomegalovirus, or rabies viral particles. Lung tissue should be examined by direct fluorescent antibody test for *Legionella* sp. Fluorescent reagents are also available for detection of agents of plague, anthrax, and certain *Bacteroides* sp., although most laboratories do not routinely perform such procedures.

The tissues of all fetuses, premature infants, and young babies who have died from an infectious process should be cultured for *Listeria* (Procedure 22.2). Specimens of the brain, spinal fluid, blood, liver, and spleen are most likely to contain the organism. The isolation procedure is given in detail by Seeliger and Cherry.[16]

AUTOPSY CULTURES

Several concepts about the microbiology of autopsy specimens are important. Most internal organs (other than lungs) of previously uninfected patients remain sterile for approximately 20 hours after death. A significant portion of positive necropsy cultures therefore result from contamination from the autopsy room or autopsy personnel. It is estimated that only half of all autopsies

should yield positive cultures and that 75% of all tissues obtained at necropsy should be sterile. Procedures modified from studies by De Jongh et al.[5] and Silver and Sonnenwirth[17] are outlined in Procedure 22.3.

REFERENCES

1. Buggy, F.P. 1986. Culture methods for continuous ambulatory peritoneal dialysis-associated peritonitis. Clin. Microbiol. Newsletter 8:12.
2. Centers for Disease Control. 1991. The prevention and treatment of complications of diabetes: a guide for primary care practitioners. Department of Health and Human Services, Public Health Service, Atlanta.
3. Chapin-Robertson, K., Dahlberg, S.E., Edberg, S.C. 1992. Clinical and laboratory analyses of Cytospin-prepared Gram stains for recovery and diagnosis of bacteria from sterile body fluids. J. Clin. Microbiol. 30:377.
4. Dawson, M.S., Harford, A.M., Garner, B.K., et al. 1985. Total volume culture technique for the isolation of microorganisms from continuous ambulatory peritoneal dialysis patients with peritonitis. J. Clin. Microbiol. 22:391.
5. De Jongh, D.S., Loftis, J.W., Green, G.S., et al. 1968. Postmortem bacteriology: a practical method for routine use. Am. J. Clin. Pathol. 49:424.
6. Fung, J.C., Sun, T., Kilius, I., and Gross, S. 1983. Print cultures for postmortem microbiology. Ann. Clin. Lab Sci. 13:83.
7. Jacobs, R.F., McCarthy, R.E., and Elser, J.M. 1989. *Pseudomonas* osteochondritis complicating puncture wounds of the foot in children: a 10-year evaluation. J. Infect. Dis. 160:657.
8. Loebl, E.C., Marvin, J.A., Heck, E.L., et al. 1974. The method of quantitative burn wound biopsy cultures

and its routine use in the care of the burned patient. Am. J. Clin. Pathol. 61:20.

9. Ludlam, H.A., Price, T.N., Berry, A.J., and Phillips, I. 1988. Laboratory diagnosis of peritonitis in patients on continuous ambulatory peritoneal dialysis. J. Clin. Microbiol. 26:1757.

10. Maderazo, E.G., Judson, S., and Pasternak, H. 1988. Late infections of total joint prostheses. Clin. Orthop., April(229):131.

11. Males, B.M., Walshe, J.J., Garringer, L., et al. 1986. Addi-Chek filtration, BACTEC, and 10-ml culture methods for recovery of microorganisms from dialysis effluent during episodes of peritonitis. J. Clin. Microbiol. 23:350.

12. Males, B.M., Walshe, J.J., and Amsterdam, D. 1987. Laboratory indices of clinical peritonitis: total leukocyte count, microscopy, and microbiologic culture of peritoneal dialysis effluent. J. Clin. Microbiol. 25:2367.

13. Rubin, J. Rogers, W.A., Taylor, H.M., et al. 1980. Peritonitis during continuous ambulatory peritoneal dialysis. Ann. Intern. Med. 92:7.

14. Rubin, S.J. 1984. Continuous ambulatory peritoneal dialysis: dialysate fluid cultures. Clin. Microbiol. Newsletter 6:3.

15. Ryan, S., and Fessia, S. 1987. Improved method for recovery of peritonitis-causing microorganisms from peritoneal dialysate. J. Clin. Microbiol. 25:383.

15a. Runyon, B.A., Antillon, M.R., Akriviadis, E.A., and McHutchison, J.G. 1990. Bedside inoculation of blood culture bottles with ascitic fluid is superior to delayed inoculation in the detection of spontaneous bacterial peritonitis. J. Clin. Microbiol. 28:2811.

16. Seeliger, H.P.R., and Cherry, W.B. 1957. Human listeriosis: its nature and diagnosis. U.S. Government Printing Office, Washington, D.C.

17. Silver, H., and Sonnenwirth, A.C. 1969. A practical and efficacious method for obtaining significant postmortem blood cultures. Am. J. Clin. Pathol. 52:433.

BIBLIOGRAPHY

Baron, E.J. and Finegold, S.M. 1990. Bailey and Scott's diagnostic microbiology, ed 8. Mosby, St. Louis. (Appendices include detailed formulas for media and stains.)

Brewer, N.S., and Weed, L.A. 1976. Diagnostic tissue microbiology methods. Hum. Pathol. 7:141.

Minshew, B.H. 1983. Are quantitative wound cultures worthwhile? Clin. Microbiol. Newsletter 5:51.

O'Toole, W.F., Saxena, H.M., Golden, A., and Ritts, R.E. 1965. Studies of postmortem microbiology using sterile autopsy technique. Arch. Pathol. 80:540.

Steigbiegel, R.T., and Cross, A.S. 1984. Infections associated with hemodialysis and chronic peritoneal dialysis. In Remington, J.S., and Swartz, M.N., editors. Current clinical topics in infectious diseases, ed. 5. McGraw-Hill, New York.

Young, E.J., and Sugarman, B. 1988. Infections in prosthetic devices. Surg. Clin. North Am. 68:167.

23

INFECTIONS OF THE HEAD AND NECK

GENERAL CONSIDERATIONS AND ANATOMY

Infections of the head and neck can be life-threatening and have their own specific bacteriology.[3] The structure and physiology of the head and neck allow both a rapid response to injury and a potential site for rapid development and spread of infection. This area has a rich vascular supply that promotes rapid healing after trauma or surgery and permits rapid mobilization of host defenses. Conversely, the extensive venous drainage may facilitate the spread of infection throughout the body, especially to the central nervous system. The mastoids and sinuses are unique air-filled cavities within the head (Figure 23.1). These structures, as well as the eustachian tube, the middle ear, and the respiratory portion of the pharynx, are lined by respiratory epithelium. The clearance of secretions and contaminants depends on normal ciliary activity and mucous flow. The maxillary sinuses are close to the roots of the upper teeth so that dental infections can extend into these sinuses. The fascia (firm, fibrous covering of muscles) of the neck behave as cylinders surrounding the important muscles, and other, deeper fasciae surround the important structures of the neck (pharynx, trachea, major salivary glands). These fasciae communicate; they create neck spaces that tend to confine infection to specific areas within the neck. Unfortunately, spread to the central area of the chest cavity (mediastinum) is not always prevented.

The delicate intraocular structures are enveloped in a tough collagenous coat. The eyes themselves are enclosed within the bony orbits, which are four-sided pyramids. Three of the four walls of the orbit are contiguous with the paranasal (facial) sinuses. The roof of the orbit is contiguous with the frontal sinus, the floor of the orbit forms the top wall of the maxillary sinus, and the ethmoid sinuses run along the medial aspect of the orbit. Thus, sinus infections may extend directly to the periocular orbital structures. The conjunctival membranes line the eyelids and extend onto the surface of the eye itself. The eyelashes tend to prevent entry of foreign material into the eye. The lids blink 15 to 20 times per minute, during which time secretions of the lacrimal glands and goblet cells wash away bacteria and foreign matter. Lysozyme and immunoglobulin A (IgA) are secreted locally and serve as part of the eye's natural defense mechanisms.

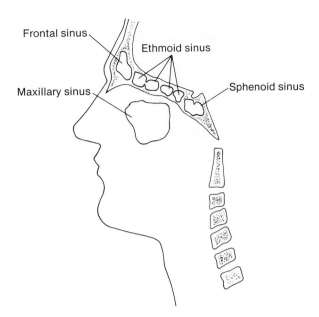

FIGURE 23.1

Location of the paranasal sinuses. (From Milliken, M.E., and Campbell, G. 1985. Essential competencies for patient care. Mosby, St. Louis.)

MICROBIAL FLORA OF THE HEAD AND NECK

Rather sparse indigenous flora exist in the conjunctival sac. *Staphylococcus epidermidis* and *Lactobacillus* sp. are the most frequently encountered organisms; *Propionibacterium acnes* may also be present. *Staphylococcus aureus* is found in less than 30% of people, and *Haemophilus influenzae* colonizes 0.4% to 25%. *Moraxella catarrhalis*, various Enterobacteriaceae, and various streptococci (*Streptococcus pyogenes*, *Streptococcus pneumoniae*, other alpha-hemolytic and gamma-hemolytic forms) are found in a very small percentage of subjects. The normal flora of the external ear canal are also rather sparse, similar to flora of the conjunctival sac qualitatively except that pneumococci, *P. acnes*, *S. aureus*, and Enterobacteriaceae are encountered somewhat more often. *Pseudomonas aeruginosa* is found on occasion. *Candida* sp. (non–*C. albicans*) are common.

The normal respiratory tract flora is discussed in Chapter 17. In addition to the organisms listed there, the oral flora may contain *Leptotrichia buccalis*, *Selenomonas*, *Wolinella*, *Stomatococcus mucilaginosus*, *Campylobacter* sp., and the following organisms that would be more likely to be involved in infection: *Clostridium perfringens* (found irregularly), *Eikenella corrodens*, *Capnocytophaga*, *Actinobacillus* sp., *actinomycetemcomitans*, *Haemophilus aphrophilus*, and closely related species. Two protozoa, *Entamoeba gingivalis* and *Trichomonas tenax*, are found as normal flora but are rarely involved in infectious processes. We must also be concerned with indigenous flora of animals in the case of bite infections (Chapter 21). *Pasteurella*

multocida is usually involved in infection following dog and cat bites. A particularly virulent organism encountered in relation to dog bites, especially in immunocompromised patients, is a fastidious gram-negative bacillus now called *Capnocytophaga canimorsus* (DF-2).

PATHOGENIC MECHANISMS

Much of the discussion of pathogenic mechanisms in Chapter 17 is pertinent to this chapter as well. *Capnocytophaga*, which includes organisms formerly called *Bacteroides ochraceus* and DF-1, has been shown to release lysosomal enzymes from human polymorphonuclear leukocytes. A sonic extract of these bacteria inhibits migration of phagocytes and suppresses the proliferation of human fibroblasts. One species of *Capnocytophaga* (*C. sputigena*) shows relatively weak endotoxin activity. Cell envelopes of *Capnocytophaga* exhibit several types of immunomodulating activity and evoke dermatoxic reactions on rabbit skin. The pathogenicity of *Clostridium perfringens* is discussed in Chapter 36.

EYE INFECTIONS

CONJUNCTIVITIS

Chlamydia trachomatis is responsible for one of the most important types of conjunctivitis, trachoma, one of the leading causes of blindness in the world. *C. trachomatis* acquired by the neonate during passage through an infected vaginal canal is also one of the causes of acute conjunctivitis in the newborn.

Bacterial conjunctivitis is the most common type of infectious conjunctivitis.[7] In adults, the most common organisms cultured are *Streptococcus pneumoniae, Staphylococcus aureus,* and *Staphylococcus epidermidis.* However, there is some dispute as to the significance of isolates of the latter two organisms, since they are often recovered from noninfected eyes. In children the most common causes of bacterial conjunctivitis are *Haemophilus influenzae, S. pneumoniae,* and perhaps *S. aureus.*[4] *S. pneumoniae* and *H. influenzae* (especially subsp. *aegyptius*) have been responsible for epidemics of conjunctivitis. Gonococcal conjunctivitis may be quite destructive. With the common practice of instilling antibiotic drops into the eyes of newborns in the United States, the incidence of gonococcal and chlamydial conjunctivitis has dropped dramatically. Diphtheritic conjunctivitis may occur in conjunction with diphtheria elsewhere in the body. *Moraxella lacunata* produces a localized angular conjunctivitis with little discharge from the eye. Distinctive clinical pictures may also occur with conjunctivitis caused by *Mycobacterium tuberculosis, Francisella tularensis, Treponema pallidum,* and *Yersinia enterocolitica.* Other bacteria also have been isolated from patients with conjunctivitis. Fungi may be responsible for this type of infection as well, often in association with a foreign body in the eye or an underlying immunological problem.

Viruses are an important cause of conjunctivitis; 20% of such infections resulted from adenoviruses in one large U.S. study. Adenoviruses types 4, 3, and 7A are common. Worldwide, enterovirus 70 and coxsackievirus A24 are responsible for outbreaks and epidemics of acute hemorrhagic conjunctivitis.[2] Puerto Rico had an epidemic caused by coxsackievirus A24 in the late 1980s.[1] Physicians in the continental United States need to be aware of this illness.

KERATITIS

Keratitis, or inflammation of the cornea, may be caused by a variety of infectious agents, usually only after some type of trauma produces a defect in the ocular surface. Keratitis should be regarded as an emergency situation, since corneal perforation and loss of the eye can occur within 24 hours when organisms such as *Pseudomonas aeruginosa, Staphylococcus aureus,* or herpes simplex virus (HSV) are involved. The most common symptom of keratitis is pain, usually accompanied by some decrease in vision, but typically with no discharge from the eye. Bacteria account for 65% to 90% of corneal infections.

In the United States the most common infecting organisms are *S. aureus, Streptococcus pneumoniae, P. aeruginosa, Moraxella,* and HSV. The first three organisms listed account for more than 80% of all bacterial corneal ulcers. A toxic factor known as exopeptidase has been implicated in the pathogenesis of corneal ulcer produced by *S. pneumoniae.* With *P. aeruginosa,* proteolytic enzymes are responsible for the corneal destruction. The gonococcus may cause keratitis in the course of inadequately treated conjunctivitis. *Acinetobacter,* which may look identical microscopically to the gonococcus and is resistant to penicillin and many other antimicrobial agents, can cause corneal perforation. Many other organisms, including mycobacteria, *Treponema pallidum, Nocardia, Chlamydia trachomatis,* several viruses other than HSV, and many fungi, may cause keratitis. Fungal keratitis is usually a complication of trauma.

These organisms can be seen in scrapings from the lesion. A previously rare etiological agent of corneal infections has become more common (although still unusual) in users of soft and extended-wear contact lenses. *Acanthamoeba* sp., free living amebae, can survive in improperly sterilized cleaning fluids and be introduced into the eye with the contact lens. Other bacterial and fungal causes of infections in such patients have also been traced to inadequate cleaning of lenses. The organisms can often be recovered from the contact lens cleaning and storage containers.

ENDOPHTHALMITIS

Surgical trauma, nonsurgical trauma (infrequently), and hematogeneous spread from distant sites of infection are the background factors in endophthalmitis. The infection may be limited to specific tissues within the eye or may involve all the intraocular contents. Pain, especially on movement of the eye, is a prominent aspect of the clinical picture. Bacteria are the most common infectious agents responsible for endophthalmitis. Bacterial endophthalmitis develops suddenly and progresses rapidly.

After surgery or trauma, evidence of the disease is usually found within 24 to 48 hours but can be delayed for several days. Postoperative infection involves primarily bacteria from the ocular surface microflora. *Staphylococcus aureus* is responsible for 50% of cases of endophthalmitis after cataract extraction. *S. aureus* and *Pseudomonas* usually lead to fulminating infection early in the postoperative period. Any bacterium, including those considered to be primarily saprophytic, may cause endophthalmitis. In hematogeneous endophthalmitis, a septic focus elsewhere is usually evident before onset of the intraocular infection. *Bacillus cereus* has caused endophthalmitis after transfusion with contaminated blood and in narcotic addicts. Endophthalmitis associated with meningitis may involve various organisms, including *Haemophilus influenzae*, streptococci, and *Neisseria meningitidis*. *Nocardia* endophthalmitis may follow pulmonary infection with this organism. Endophthalmitis may be seen with miliary tuberculosis and in syphilis.

Mycotic infection of the eye has increased significantly over the past 3 decades because of increased use of antibiotics, corticosteroids, antineoplastic chemotherapy, addictive drugs, and hyperalimentation. Fungi generally considered to be saprophytic are important causes of endophthalmitis. Fungi recovered from postoperative infection include *Volutella* sp., *Neurospora sitophila*, *Monosporium apiospermum*, *Candida parapsilosis*, *Trichosporon cutaneum*, *Paecilomyces lilacinus*, *Cephalosporium*, and *Torulopsis glabrata*. Endogenous mycotic endophthalmitis is most often caused by *Candida albicans*. Patients with diabetes and underlying disease are most at risk. Other causes of hematogeneous ocular infection include *Aspergillus*, *Cryptococcus*, *Coccidioides*, *Sporothrix*, and *Blastomyces; Histoplasma* is rarely involved. Viral causes of endophthalmitis include herpes simplex I, varicella (herpes) zoster virus (VZV), cytomegalovirus, and measles viruses. The most common parasitic cause is *Toxocara*. *Toxoplasma gondii* is a well-known cause of chorioretinitis. Thirteen percent of patients with cysticercosis have ocular involvement. *Onchocerca* usually produces keratitis, but intraocular infection also occurs.

PERIOCULAR INFECTIONS

Included in this category are infections of the eyelids, lacrimal apparatus, and orbit. Lid infections include blepharitis (inflammation of the lid margins); hordeolum, or the common sty; and chalazion. Most eyelid infections result from bacteria, especially *Staphylococcus aureus* and *Staphylococcus epidermidis*. Other organisms that may be encountered include *Proteus mirabilis* and *Moraxella*, molluscum contagiosum, and HSV.

The three infections of the lacrimal apparatus are canaliculitis (chronic inflammation of the lacrimal canals), dacryocystitis (inflammation of the lacrimal sac), and dacryoadenitis (inflammation of the main lacrimal gland). Canaliculitis is usually caused by *Actinomyces* or *Propionibacterium*

propionicum (formerly *Arachnia), Pityrosporum pachydermatis*, or *Fusobacterium*. Infection of the lacrimal sac may involve *Aspergillus*, *Candida albicans*, or *Actinomyces*. Other causes are *Streptococcus pneumoniae*, mixed bacterial infection, and *Chlamydia trachomatis*. Dacryoadenitis often involves pyogenic bacteria such as *S. aureus* and streptococci, but gonococci and *Cysticercus cellulosae* have been reported. Chronic infections of the lacrimal gland occur in tuberculosis, syphilis, leprosy, and schistosomiasis. Acute inflammation of the gland may occur in the course of mumps and infectious mononucleosis.

Orbital cellulitis is an acute infection of the orbital contents most often caused by bacteria. It is a serious infection because it may spread posteriorly to produce cavernous sinus thrombosis. Most cases involve spread from contiguous sources such as the paranasal sinuses. In children, bloodborne bacteria, notably *Haemophilus influenzae*, may lead to orbital cellulitis. Intrauterine infections may be a background factor in newborns. *S. aureus* is the most common etiological agent; *Streptococcus pyogenes* and *S. pneumoniae* are also common. Anaerobes may be involved secondary to chronic sinusitis, primarily in adults. Mucormycosis of the orbit is a serious, invasive infection seen particularly in diabetic patients with poor control of their disease, patients with acidosis from other causes, and patients with malignant disease receiving cytotoxic and immunosuppressive therapy. *Aspergillus* may produce a similar infection in the same settings but also can cause mild, chronic infections of the orbit.

Newer surgical techniques involving the ocular implantation of prosthetic or donor lenses have resulted in increasing numbers of iatrogenic infections. Isolation of *Propionibacterium acnes* may have clinical significance in such situations, in contrast to many other sites in which it is usually considered to be a contaminant.

EAR, NOSE, AND THROAT INFECTIONS

PHARYNGITIS, TONSILLITIS, PERITONSILLAR ABSCESS

The various forms of pharyngitis and tonsillitis, including diphtheria, are discussed in Chapter 17. The predominant organisms in peritonsillar abscess are non–spore-forming anaerobes, including *Fusobacterium* (especially *F. necrophorum*), *Bacteroides* (including the *B. fragilis* group), and anaerobic cocci. *Streptococcus pyogenes* and viridans streptococci may also be involved. Vincent's angina, or anaerobic tonsillitis, involves pseudomembrane formation on tonsillar surfaces; the infection is relatively rare today, but it is a very serious disease because it is often complicated by septic jugular thrombophlebitis, bacteremia, and widespread metastatic infection. Multiple anaerobes, especially *F. necrophorum*, are implicated in this syndrome.

DENTAL AND ORAL INFECTIONS

Oral infection with HSV, gonococci, and *Candida* is discussed in Chapter 17. Oral lesions may be encountered in secondary syphilis (see Chapter 20 for a discussion of darkfield preparations and syphilis serology). Although a large spirochete seems to be a key pathogen in acute necrotizing ulcerative gingivitis (Vincent's gingivitis), clinical microbiology laboratories are not asked to do bacteriological studies for diagnosis of this condition. The three dental problems in which help may be requested from clinical laboratories are root canal infections, with or without periapical abscess; orofacial odontogenic infections, with or without osteomyelitis of the jaw; and perimandibular space infections. The bacteriology is similar in all these situations, involving primarily anaerobic bacteria and streptococci except for perimandibular space infections, which may also involve staphylococci and *Eikenella corrodens* in about 15% of patients. The streptococci are microaerobic or facultative and are usually alpha-hemolytic; they are usually found in 20% to 30% of infections of the preceding types of dental infection. Members of the *Bacteroides fragilis* group are found in root canal infections, orofacial odontogenic infections, and bacteremia secondary to dental extraction in 5% to 10% of patients. Anaerobic cocci (both *Peptostreptococcus* and *Veillonella*), pigmented *Prevotella* and *Porphyromonas*, the *Prevotella oralis* group, and *Fusobacterium* are found in about 20% to 50% of the three conditions mentioned, as well as in postextraction bacteremia. Infection with *Actinomyces israelii* may complicate any oral surgery. Bacterial stomatitis is rarely seen now in the United States. It is characterized by superficial (and sometimes deep) tissue necrosis with pseudomembrane formation and a fetid odor. Anaerobes are clearly involved, but bacteriological studies are unsatisfactory; spirochetes may be important. Oral bacteria are clearly important in other dental processes such as caries, periodontal disease (pyorrhea), and localized juvenile periodontitis, but clinical laboratories are not involved in culturing in such cases.

SALIVARY GLAND INFECTIONS

Acute suppurative parotitis is seen in very ill patients, especially those that are dehydrated, malnourished, elderly, or recovering from surgery. It is associated with painful, tender swelling of the parotid gland; purulent drainage may be evident at the opening of the duct of the gland in the mouth. *Staphylococcus aureus* is the major pathogen, but on occasion viridans streptococci and oral anaerobes may play a role. A chronic bacterial parotitis has been described that usually involves *S. aureus*. Less often, other salivary glands may be involved with a bacterial infection; this is usually because of ductal obstruction. The mumps virus is traditionally the major viral agent involved in parotitis; however, since the advent of childhood vaccination, infection with mumps virus is rarely diagnosed. Influenza virus and enteroviruses may also cause this syndrome. Diagnosis of viral parotitis is usually done serologically. Infrequently, *Mycobacterium tuberculosis* may involve the parotid gland in conjunction with pulmonary tuberculosis.

SINUSITIS

Acute sinusitis is usually caused by bacterial infection; it tends to be self-limited, usually lasting 1 to 3 weeks. Most often it follows a common cold or other viral upper respiratory infection. Complications include local extension into the orbit, skull, meninges, or brain and development of chronic sinusitis. Most studies of the microbiology of acute sinusitis have dealt with maxillary sinusitis because it is the most common type and the only one really accessible for puncture and aspiration. Bacterial cultures are positive in about three fourths of patients. In a study involving young adults, *Haemophilus influenzae* was recovered from 50% and *Streptococcus pneumoniae* from 19% of patients. *Streptococcus pyogenes* and *Moraxella catarrhalis* were also found, in addition to normal skin flora such as *Propionibacterium acnes*. Anaerobes were considered pathogens in only 2% of cases.[5] Among children, *S. pneumoniae*, *H. influenzae*, and *M. catarrhalis* are most common.[9] Rhinovirus is found in 15% of patients, influenza virus in 5%, parainfluenza virus in 3%, and adenovirus in less than 1%. Bacteria, particularly anaerobes, are more frequently involved in chronic sinusitis in adults. A report by Tinkelman and Silk[8] has shown *M. catarrhalis* to be an important agent in chronic sinusitis in children. The primary problems are inadequate drainage, impaired mucociliary clearance, and mucosal damage. Ordinarily, surgery or drainage is required for successful management.

EXTERNAL EAR INFECTIONS

Otitis externa is similar to skin and soft tissue infections elsewhere, but there is the unique problem of a narrow, tortuous ear canal. Several types of external otitis exist. Acute localized disease occurs in the form of a pustule or furuncle and typically results from *Staphylococcus aureus.* Erysipelas caused by group A streptococci may involve the external ear canal and the soft tissue of the ear itself. Acute diffuse otitis externa (swimmer's ear) is related to maceration of the ear from swimming or hot, humid weather. Gram-negative bacilli, particularly *Pseudomonas aeruginosa,* play an important role. A severe, hemorrhagic (malignant) external otitis caused by *P. aeruginosa* is very difficult to treat and has occasionally been related to hot tub use.

Chronic otitis externa results from the irritation of drainage from the middle ear in patients with chronic, suppurative otitis media and a perforated eardrum. Rarely, this condition may be caused by tuberculosis, syphilis, yaws, or leprosy. Malignant otitis externa is a necrotizing infection that spreads to adjacent areas of soft tissue, cartilage, and bone. It may progress to a life-threatening situation by spreading into the central nervous system or vascular channels. *P. aeruginosa,* in particular, and anaerobes are frequently associated with this process, which is seen in diabetic patients with small blood vessel disease of the tissues overlying the temporal bone; the poor local perfusion of tissues results in a milieu for invasion by bacteria. On occasion, external otitis extends into the cartilage of the ear; this usually requires surgical intervention. Certain viruses may infect the external auditory canal, the soft tissue of the ear, or the tympanic membrane; influenza A virus is a suspected but not an established cause. VZV may cause painful vesicles within the soft tissue of the ear and the ear canal. *Mycoplasma pneumoniae* is a cause of bullous myringitis (a painful infection of the eardrum with hemorrhagic bullae); the ear canal itself may be involved in this case as well.

MIDDLE EAR INFECTION (OTITIS MEDIA),
MASTOIDITIS

Acute otitis media may begin as a viral infection, but bacterial infection typically supervenes. In children (in whom the disease is most common), pneumococci (33% of cases) and *Haemophilus influenzae* (20%) are the usual etiological agents;

group A streptococci (8%) are the third most frequently encountered agents. Other organisms, encountered in only 1% to 3% of cases, include *Moraxella catarrhalis, Staphylococcus aureus,* gram-negative enteric bacilli, and anaerobes. Viruses, chiefly respiratory syncytial virus (RSV) and influenza virus, have been recovered from middle ear fluid of 4% of children with acute or chronic otitis media. *Chlamydia trachomatis* and *Mycoplasma pneumoniae* have occasionally been isolated from middle ear aspirates.

Chronic otitis media and its complication, mastoiditis, yield a predominantly anaerobic flora, with *Peptostreptococcus* sp., *Bacteroides fragilis* group, *Prevotella melaninogenica* (pigmented anaerobic gram-negative rod) and *Porphyromonas,* other *Prevotella* sp., and *Fusobacterium nucleatum* as the principal pathogens; less frequently present are *S. aureus, Pseudomonas aeruginosa, Proteus* sp., and other gram-negative facultative bacilli. *Mycobacterium tuberculosis* may be an unusual cause of chronic otitis media and mastoiditis.

NECK SPACE INFECTIONS

Infections of the deep spaces of the neck are potentially serious because they may spread to critical structures such as major vessels of the neck or to the mediastinum, leading to mediastinitis, purulent pericarditis, and pleural empyema.[3] One of these infections, Ludwig's angina (Figure 23.2), may lead to airway obstruction. The oral flora is responsible for these infections. Accordingly, the predominant organisms are anaerobes, primarily

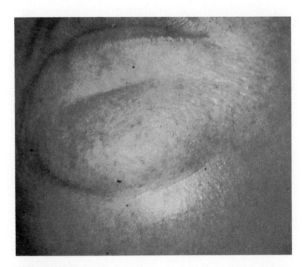

FIGURE 23.2
Ludwig's angina.

Peptostreptococcus; various *Bacteroides, Prevotella, Porphyromonas, Fusobacterium* sp., and *Actinomyces.* Streptococci, chiefly of the viridans variety, are also important. *Staphylococcus aureus* and various aerobic gram-negative bacilli may be recovered, particularly from patients developing these problems in the hospital.

COLLECTION AND TRANSPORT OF SPECIMENS

Purulent material from the surface of the lower conjunctival sac and inner canthus of the eye is collected on a sterile swab for cultures of conjunctivitis. Both eyes should be cultured separately. Chlamydial cultures are taken with a dry calcium alginate swab; this is placed in 2-SP transport medium. An additional swab may be rolled across the surface of a slide, fixed with methanol, and sent for direct fluorescent antibody (DFA) stain.

In the patient with keratitis, an ophthalmologist should obtain scrapings of the cornea with a heat-sterilized platinum spatula. Multiple inoculations with the spatula are made to blood agar, chocolate agar, an agar for fungi, thioglycolate broth, and an anaerobic blood agar plate. Other special media may be used if indicated. For culture of HSV and adenovirus, corneal material is transferred to viral transport media.

Cultures of endophthalmitis specimens are inoculated with material obtained by the ophthalmologist from the anterior chamber of the eye, the vitreous, wound abscesses, and wound dehiscences. Lid infection material is collected on a swab in a conventional manner. For microbiological studies of canaliculitis, concretions from the lacrimal canal should be transported under anaerobic conditions. Aspiration of fluid from the orbit is contraindicated in patients with orbital cellulitis. Since sinusitis is the most common background factor; an otolaryngologist's assistance in obtaining material from the maxilary sinus by antral puncture is helpful. Blood cultures should be obtained. Tissue biopsy is essential for microbiological diagnosis of mucormycosis; cultures are usually negative, and the diagnosis is made by histological examination.

For laboratory diagnosis of external otitis, the external ear should be cleansed with a mild germicide such as 1:1000 aqueous solution of benzalkonium chloride to reduce the contaminating skin flora before obtaining the culture. Material from the ear, especially that obtained after spontaneous perforation of the eardrum or by needle aspiration of middle ear fluid (tympanocentesis),

should be collected by an otolaryngologist, using sterile equipment and a sterile swab designed for microbiological use. Cultures from the mastoid are generally taken on swabs during surgery, although actual bone is preferred; they should be transported anaerobically.

The problem in collecting oral and dental infection material is to avoid or minimize contamination with oral flora. For collection of material from root canal infection, the tooth is isolated by means of a rubber dam. A sterile field is established, the tooth is swabbed with 70% alcohol, and after the root canal is exposed, a sterile paper point is inserted, removed, and placed into semisolid, nonnutritive anaerobic transport medium. Alternatively, needle aspiration can be used if sufficient purulent material is present. Completely defining the flora of such infections is beyond the scope of routine clinical microbiology laboratories.

Specimens from neck space infections can usually be obtained with a syringe and needle or by biopsy during a procedure by the surgeon. Transport must be under anaerobic conditions.

DIRECT VISUAL EXAMINATION

All material submitted for culture should always be smeared and examined directly by Gram stain or other appropriate technique. In bacterial conjunctivitis, polymorphonuclear leukocytes predominate; in viral infection the host cells are primarily lymphocytes and monocytes. Specimens in which chlamydia are suspected can be stained immediately with monoclonal antibody-conjugated fluorescein for detection of elementary bodies or inclusions. Using histological stains, basophilic intracytoplasmic inclusion bodies are seen in epithelial cells. Cytologists and anatomical pathologists usually perform these tests. Direct examination of conjunctivitis specimens using histological methods (Tzanck smear) may reveal multinucleated epithelial cells typical of herpes group virus infections, but immunoelectron microscopy and immunofluorescent techniques are more sensitive and specific and are recommended for diagnosing viral conjunctivitis. A reliable DFA stain is available for both HSV and VZV, which is the best test for a rapid diagnosis of these viral infections. In the patient with keratitis, scrapings are examined by Gram, Giemsa, periodic acid-Schiff (PAS), and methenamine silver stains. If *Acanthamoeba* or other amebae are suspected, a direct wet preparation should be examined for motile trophozoites, and a trichrome stain should be added to the reg-

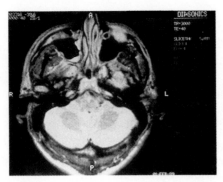

FIGURE 23.3
Computed tomography of head of patient with right maxillary sinusitis. Note fluid level and mucosal thickening of sinus cavity *(left)*. (From Finegold, S.M., Baron, E.J., and Wexler, H.M. 1992. A clinical guide to anaerobic infections. Star Publishing, Belmont, Calif.)

imen. For this diagnosis, however, culture is by far the most sensitive detection method. Transmission electron microscopy may establish an unusual viral etiology. In the patient with endophthalmitis, material is also examined by Gram, Giemsa, PAS, and methenamine silver stains. When ophthalmic specimens are submitted in large volumes of fluid, they must be concentrated by centrifugation before further studies are performed. Material aspirated from the maxillary sinus, from middle ear or mastoid infections, from neck space infections, and from bite infections is also examined directly for bacteria and fungi. The calcofluor white fluorescent or PAS stains reveal fungal elements. Methenamine silver stains have the added efficiency of staining most bacterial, fungal, and several parasitic species.

CULTURING

Because of the constant washing action of the tears, the number of organisms recovered from cultures of certain eye infections may be relatively low. Unless the clinical specimen is obviously purulent, it is recommended that a relatively large inoculum and a variety of media be used to ensure recovery of an etiological agent. Conjunctival scrapings placed directly onto media yield the best results. At a minimum, one should use blood agar and chocolate agar plates incubated under increased carbon dioxide tension. Since potential pathogens may be present in an eye without causing infection, it may be very helpful to the clinician, when only one eye is infected, to culture both eyes. When *Moraxella lacunata* (Morax-Axenfeld bacillus) is suspected, Loeffler's medium may prove useful; the growth of the organism often leads to proteolysis and pitting of the medium,

although nonproteolytic strains are available. If diphtheritic conjunctivitis is suspected, Loeffler's or cystine-tellurite medium should be used. For more serious eye infections such as keratitis, endophthalmitis, and orbital cellulitis, one should always include, in addition to the media just noted, a reduced anaerobic blood agar plate, a medium for fungi, and a broth such as thioglycolate broth. Blood cultures are also essential in the case of serious eye infections.

Cultures of material for *Chlamydia* and viruses should be inoculated to appropriate media from transport broth. Cycloheximide-treated McCoy cells for *Chlamydia* isolation and human embryonic kidney, primary monkey kidney, and Hep-2 cell lines for virus isolation should be inoculated. For recovery of some coxsackieviruses, suckling mice may be required, although this is rarely done by clinical laboratories.

Other infections discussed in this chapter, such as peritonsillar abscess, oral and dental infections, chronic sinusitis, chronic otitis media and mastoiditis, and neck space infections, usually involve anaerobic bacteria. The anaerobes involved typically originate in the oral cavity and are often more delicate than anaerobes isolated from other clinical material. Very careful attention must be paid to providing optimal techniques of anaerobic cultivation, as well as transport.

NONCULTURAL DETECTION METHODS

Although acute and convalescent serologic tests for viral agents might be used in the event of epidemic conjunctivitis, these are not generally done because the infections are self-limited. Reliable DFA stains for VZV and HSV are commercially available. Enzyme-linked immunosorbent assay (ELISA) tests and DFA staining are now possible for detection of *Chlamydia trachomatis*. An immune dot blot test also may be useful for this purpose.[6] It is anticipated that the direct antigen tests should perform well, particularly since so many eyes have been partially treated before culture. An ELISA test of aqueous humor is available for diagnosis of *Toxocara* infection.

REFERENCES

1. Centers for Disease Control. 1988. Acute hemorrhagic conjunctivitis caused by coxsackie A24 variant—Puerto Rico. M.M.W.R. 37:123.
2. Chou, M.Y., and Malison, M.D. 1988. Outbreak of acute hemorrhagic conjunctivitis due to coxsackie A24 variant—Taiwan. Am. J. Epidemiol. 127:795.

3. Chow, A.W. 1992. Life-threatening infections of the head and neck. Clin. Infect. Dis. 14:991.

4. Gigliotti, F., Williams, W.T., Hayden, F.G., et al. 1981. Etiology of acute conjunctivitis in children. J. Pediatr. 98:531.

5. Jousimies-Somer, H.R., Savolainen, S., and Ylikoski, J.S. 1988. Bacteriologic findings of acute maxillary sinusitis in young adults. J. Clin. Microbiol. 26:1919.

6. Mearns, G., Richmond, S.J., and Storey, C.C. 1988. Sensitive immune dot blot test for diagnosis of *Chlamydia trachomatis* infection. J. Clin. Microbiol. 26:1810.

7. Seal, D.V., Barrett, S.P., and McGill, J.I. 1982. Aetiology and treatment of acute bacterial infection of the external eye. Br. J. Ophthalmol. 66:357.

8. Tinkelman, D.G., and Silk, H.J. 1989. Clinical and bacteriologic features of chronic sinusitis in children. Am. J. Dis. Child. 143:938.

9. Wald, E.R., Milmoe, G.J., Bowen, A., et al. 1981. Acute maxillary sinusitis in children. N. Engl. J. Med. 304:749.

BIBLIOGRAPHY

Atlas, R.M. 1993, Handbook of microbiologic media. CRC Press, Boca Raton, Fla. (Includes detailed formulas and instructions for media preparation.)

Bartlett, J.G., and Gorbach, S.L. 1976. Anaerobic infections of the head and neck. Otolaryngol. Clin. North Am. 9:655.

Baum, J.L. 1978. Ocular infections. N. Engl. J. Med. 299:28.

Brook, I. 1987. Microbiology of human and animal bite wounds in children. Pediatr. Infect. Dis. J. 6:29.

Brook, I., and Finegold, S.M. 1979. Bacteriology of chronic otitis media. J.A.M.A. 241:488.

Farr, B., and Gwaltney, J.M., Jr. 1989. Acute and chronic sinusitis. In Pennington, J.E., editor. Respiratory infections: diagnosis and management, ed 2. Raven Press, New York.

Finegold, S.M. 1977. Anaerobic bacteria in human disease. Academic Press, New York.

Hirst, L.W., Thomas, J.V., and Green, W.R. 1985. Conjunctivitis; keratitis; endophthalmitis; and periocular infections. In Mandell, G.L., Douglas, R.G., Jr., and Bennett, J.E., editors: Principles and practice of infectious diseases. John Wiley & Sons, New York.

Klein, J.O. 1989. Otitis media. In Pennington, J.E., editor. Respiratory infections: diagnosis and management, ed 2. Raven Press, New York.

Newman, M.G., and Goodman, A.D., editors. 1984. Guide to antibiotic use in dental practice. Quintessence, Chicago.

Summanen, P., Baron, E,J., Citron, D.M., et al., 1992. Wadsworth anaerobic bacteriology manual, ed 5. Star, Belmont, Calif.

24

INFECTIONS IN THE VULNERABLE HOST

The risk of developing an infectious disease is not the same for all individuals. It varies because of inherent host factors, such as age or disease, environmental factors such as living in a nursing home or in the general community, manipulations performed as part of medical practice such as cancer chemotherapy or organ transplantation, or combinations of these factors (see the box on p. 306). Exposure to certain agents, such as viruses, may be related to occupation or life-style. One of the ironies of modern medicine is that as treatments are improved to prolong life, we all become more vulnerable to infectious diseases. This chapter focuses on some of the patient categories who are at a greater risk of acquiring infections than are normal hosts and on some likely etiological agents of these infections. In most cases, diagnosis of infectious disease is handled in the same way for compromised patients as for any infected host, as described elsewhere in this book. When special considerations are appropriate, they are mentioned in this chapter. We attempt to highlight the most important settings for these infections, and the microbes causing them.

INFECTIONS RELATED TO AGE

INFECTIONS IN NEONATES

As a fetus develops within the uterus, it is protected from most environmental influences, including infectious agents. The human immune system does not become wholly competent until several months after birth. Immunoglobulins that cross the placental barrier, primarily immunoglobulin G (IgG), serve to protect the newborn from many infections until the infant begins to produce immunoglobulins of his or her own in response to antigenic stimuli. This unique environmental niche, however, does expose the vulnerable fetus to pathogens present in the mother. Infections in infants are of two unique types: **congenital,** acquired by the fetus as a result of a maternal infection that crosses the placenta; and *neonatal,* acquired during passage through the birth canal or immediately postpartum (see the box on p. 306).

The most common agents of congenital infection in the United States are *Toxoplasma gondii,* rubella, cytomegalovirus, and herpes simplex virus (HSV), the so-called TORCH agents. Congenital disease may also be caused by other microbes. Of these, *Treponema pallidum* (syphilis) and *Listeria monocytogenes* (neonatal listeriosis or abortion) probably occur most often.

Factors Contributing to Increased Risk of Acquiring Infectious Disease

Host factors
- Age: very young or very old
- Anatomical defects: congenital or acquired, such as meningocele or benign prostatic hypertrophy
- Disease: cystic fibrosis, diabetes, cancer, sickle cell anemia, viral or other concomitant infections
- Immune system abnormalities
- Poor nutritional state

Environmental factors
- Iatrogenic causes: antibiotic or myelosuppressive therapy, foreign implants, surgery, access conduits such as catheters
- Tissue damage: trauma or surgery
- Life-style: sexual behavior, travel, eating habits, alcohol, drugs
- Occupation
- Proximity to other infected hosts, animal or human

Etiological Agents of Congenital and Neonatal Infections

Viruses
*Cytomegalovirus
Enteroviruses
*Hepatitis B virus
*Herpes simplex virus
Human immunodeficiency virus
Measles virus
Parvovirus
*Rubella virus
Vaccinia virus
Varicella-zoster virus

Bacteria
Bacteroides fragilis
Borrelia
Brucella
Campylobacter
**Chlamydia*
**Escherichia coli*
Francisella
Gardnerella
Haemophilus influenzae
Klebsiella
Leptospira
**Listeria*
Mycobacterium
**Mycoplasma* and *Ureaplasma*
Neisseria gonorrhoeae
Salmonella
Staphylococcus
**Streptococcus*, group B
**Treponema*

Fungi
**Candida*
Coccidioides

Parasites
Leishmania
Plasmodium
Pneumocystis
**Toxoplasma*
Trypanosoma

Modified from Peter, J.B., Cherry, J.D., and Bryson, Y.J. 1982. Diagn. Med. 5:61.
* More frequently recognized agents of infection.

Suspected congenital infection can be diagnosed culturally or serologically. Because maternal IgG crosses the placenta, serologic tests are often difficult to interpret. If the presence of IgG is the only available test, serum samples from both mother and neonate should be tested simultaneously, and the infant's titers must be observed to rise over several weeks, indicating that antigenic stimulation of active antibody production occurred as a result of infection. For culturable agents, the most definitive diagnoses involve recovery of the pathogen in culture. Rubella, HSV, varicella-zoster virus (VZV), enteroviruses, and cytomegalovirus (CMV) can be cultured easily, as can most bacterial agents. Nasal and urine specimens offer the greatest yield for viral isolation, whereas blood, cerebrospinal fluid, and material from a lesion can also be productive. Systemic neonatal herpes without lesions may be difficult to diagnose unless tissue biopsy material is examined, since the viruses may not be present in cerebrospinal fluid or blood. Bacteria and fungi can be isolated from lesions, blood, and other normally sterile sites.

Determination of the presence of fetal immunoglobulin M (IgM) directed against the agent in question establishes the serologic diagnosis of congenital infection. Until recently, ultracentrifugation was required for separation of IgM from IgG, the only definitive means of preventing false-positive results caused by maternal IgG or fetal rheumatoid factor. Ion-exchange chromatography columns, antihuman IgG, and bacterial proteins that bind to IgG specifically are now commercially available for removing cross-reactive IgG and rheumatoid factor to obtain more homogeneous IgM for differentiation of fetal antibody. Indirect fluorescent antibody and enzyme-linked immunosorbent assay (ELISA) test systems are commercially available for detection of IgM against *T. gondii*, rubella, CMV, HSV, and VZV. Interference by rheumatoid factor is still a consideration in most commercial IgM test systems. Our ability to detect viral inclusions in tissue, conjunctival scrapings, and vesicular lesions, traditionally performed with Giemsa stain, has been improved by availability of monoclonal and polyclonal fluorescent antibody reagents, described in the chapters that discuss individual agents.

Unfortunately, the incidence of congenital syphilis, whose U.S. levels were at an all-time low in 1978, has been rising steadily over the last 8 years and dramatically since 1989.[2] If an infant is born to a mother with both a positive Venereal Disease Research Laboratories (VDRL) and fluorescent treponemal antibody absorption (FTA-ABS) test, the diagnosis of congenital syphilis is presumptive and treatment is begun. With VDRL-positive mothers who were adequately treated for their disease in the past, diagnosis of a congenitally infected baby as a result of maternal reinfection may be more difficult. Congenital syphilis can be diagnosed by either darkfield examination of lesions (not available in many laboratories) or serologically. Since antitreponemal IgM tests can be unreliable, infants should be tested sequentially over several months. The Sabin-Feldman dye test has been the standard against which all other serologic tests for toxoplasmosis are measured; however, it is performed in only a few specialized laboratories, and newer methods have been developed with equal or better sensitivity. Commercial latex agglutination and ELISA tests have been shown to be adequate for detection of IgG against *Toxoplasma*, a useful screening test for women before they become pregnant. A fourfold rise in titer during pregnancy is indicative of a newly acquired infection. Laboratory serologic diagnosis of congenital disease, however, requires demonstration of specific IgM in fetal serum by either conventional ELISA, indirect immunofluorescent stain, or a reverse-capture IgM ELISA, the most sensitive and specific test available today. Because of the serious consequences of congenital toxoplasmosis, including deafness, microcephaly, and low IQ,[16] detection of high-risk pregnant women is very important so that prenatal therapy can be instituted.[4] For the parasitic agents of congenital infection other than *T. gondii*, visual evidence of the agent in fetal circulation is diagnostic.

Hepatitis B can be transmitted to fetuses of chronically infected mothers, and a significant number of such infected infants will themselves become chronic carriers of hepatitis B. Laboratory diagnosis of congenital hepatitis requires demonstration of hepatitis B surface antigen (HBsAg) in fetal serum, usually accomplished by performing an ELISA or a radioimmunoassay test.

Neonatal infections are often related to a difficult labor, premature birth, premature rupture of membranes, maternal genital infection, or other event. Meningitis with Enterobacteriaceae, usually *Escherichia coli*, sepsis with group B streptococci or *Listeria*, or herpes infections (often systemic) are the most common types encountered.[14] Infections with *Haemophilus influenzae, Gardnerella vaginalis, Bacteroides fragilis*, or other anaerobic bacteria; sepsis caused by mycoplasma that infect the genital tract; ophthalmia neonatorum caused by *Neisseria gonorrhoeae;* and inclusion conjunctivitis and pneumonia caused by *Chlamydia trachomatis* also occur, although less often. Diagnosis of these infections is usually based on recovery of the agent, but direct examination of conjunctival smears by Giemsa, direct fluorescent antibody (DFA), or Gram stain can allow a much more rapid determination of the etiology of purulent discharge from the infant's eye. Direct antigen detection tests are available for gonococci, HSV, respiratory syncytial virus, VZV, CMV, *Chlamydia trachomatis*, and group B streptococci. The appropriate specimens (cerebrospinal fluid, serum, pus, tracheal aspirate, etc.) should be examined immediately with these reagents. For all other agents, a culture of blood, nasopharyngeal secretions, or stool produces the greatest yield. Routine body surface cultures of infants in intensive care have not been shown to be helpful for predicting subsequent disease.[6]

INFECTIONS IN THE ELDERLY

Pneumonia is the most common infection in aged patients; it is the fourth leading cause of death in patients over 75 years in the United States. Factors such as decreased mucociliary function, decreased cough reflex, decreased level of consciousness, periodontal disease, and decreased general mobility probably contribute to a greater incidence of pneumonia in aged patients. Such patients have been found to be more frequently colonized with gram-negative bacilli than are younger people, perhaps because of poor oral hygiene, decreased saliva, or decreased epithelial cell turnover. When these patients are institutionalized, their oral colonization rate with resistant bacteria is higher, and their chances of acquiring gram-negative bacterial pneumonia are enhanced. Mortality from bacterial pneumonia, often after respiratory viral infections, particularly influenza, is a major cause of death in elderly patients. The bacteria associated with pneumonia in these patients are *Streptococcus pneumoniae, Legionella* sp., oral anaerobes, *Staphylococcus aureus*, the Enterobacteriaceae, and *Pseudomonas* sp. (especially in hospitalized patients). Reactivation tuberculosis also needs to be considered in patients over 75 years old. The incidence of atypical mycobacterial disease is low, and the species of organisms involved may vary depending on the locality.

Urinary tract infections pose significant problems for older patients. Women may have cystocele, loss of pelvic muscle tone, and changes in urinary tract mucosa, which predispose them to infection. In men, prostatic obstruction is the most common condition contributing to urinary tract infection. Important factors predisposing to infection include obstructive uropathy and manipulation of the urinary tract (e.g., catheterization).

Gram-negative bacteremia resulting from another focus of infection (often in the urinary tract) is an important problem infection seen in elderly patients. Other infections likely to be encountered among older patients include peritonitis or intraabdominal abscess, often initiated by gastrointestinal tract disorders such as cholecystitis, tumor-causing obstruction, diverticulosis, or peptic ulcer. Circulating bacteria may attach to and infect heart valves, leading to endocarditis. The loss of some immune functions with age may predispose older patients to serious VZV infections (including pneumonia), infectious arthritis, and meningitis. Impaired circulation and diminished sensation predispose to skin and soft tissue infections. The specimens required for diagnosis and the handling of those specimens are no different from those used to diagnose infections in any patient, as described in other chapters.

INFECTIONS IN PATIENTS WITH UNDERLYING DISEASE

Diseases that alter the body's normal metabolism can affect the host's "steady state," rendering the patient more vulnerable to infection. In addition, various diseases affect the immune system itself, preventing functional activity that serves to protect a normal host from many etiological agents. A few of the more notable chronic, debilitating diseases that predispose individuals to infection are mentioned here.

CYSTIC FIBROSIS
Patients with cystic fibrosis may present as young adults with chronic respiratory tract disease, as well as the more common presentation in children that may also include gastrointestinal

problems and stunted growth. A very mucoid *Pseudomonas* sp., characterized by production of copious amounts of extracellular capsular polysaccharide, can be isolated from the sputum of almost all patients with cystic fibrosis older than 18 years, becoming more prevalent with increasing age after 5 years.[7] Even if cystic fibrosis has not been diagnosed, isolation of a mucoid *Pseudomonas aeruginosa* of this type from sputum should alert the clinician to the possibility of such an underlying disease. Microbiologists should always report this unusual morphological feature if it is encountered. In addition to the mucoid *Pseudomonas*, cystic fibrosis patients are likely to harbor *Staphylococcus aureus*, *Haemophilus influenzae*, and *Pseudomonas cepacia*. Respiratory syncytial virus and *Aspergillus* are also important pathogens in this population.[13] Sputum specimens from patients known to have cystic fibrosis should be inoculated to selective agar, such as mannitol salt for recovery of *S. aureus* and selective horse blood–bacitracin, incubated anaerobically and aerobically, for recovery of *H. influenzae* that may be obscured by the mucoid *Pseudomonas* on routine media. The use of a selective medium for *P. cepacia* is also recommended when these patients have a serious acute infection.

DIABETES MELLITUS
Diabetes mellitus, especially in older patients, seems to predispose those affected to several infectious diseases. Peripheral neuropathy, occasionally combined with lack of adequate peripheral circulation (via both small and large vessels) leads to tissue necrosis and the frequently encountered diabetic foot ulcer (Chapter 21). Chronic soft tissue infections, often of mixed bacterial etiology, including anaerobic and facultative bacteria (including *S. aureus*), can lead to osteomyelitis (Chapter 22). Diabetic patients are also at risk of developing various other forms of anaerobic cellulitis and fasciitis (Chapter 21), synergistic necrotizing cellulitis caused by mixed aerobic and anaerobic bacteria, malignant otitis externa, and streptococcal cellulitis. Acidosis predisposes patients to fungal infection. Rhinocerebral mucormycosis and candidiasis occur relatively frequently in patients with diabetes.

SICKLE CELL DISEASE
Circulating antibodies of patients with sickle cell disease seem to display impaired ability to opsonize certain bacteria, particularly *Streptococcus pneumoniae*, *Haemophilus influenzae*, and *Salmonella* sp. These patients have defective complement ac-

tivity and lack a functional spleen, which may also contribute to their increased risk for infection. Primary pneumococcal sepsis, peritonitis, and meningitis are common in patients with sickle cell disease. *Salmonella* is the cause of 80% of osteomyelitis cases in patients with sickle cell disease; this is usually secondary to sepsis. Other gram-negative bacilli account for the other 20%. A recently recognized problem, infection with parvovirus B19 (the agent of erythema infectiosum, or fifth disease), can lead to acute aplastic anemic crisis in these patients.[15]

CONGENITAL IMMUNODEFICIENCY DISEASES

Some children are born with defects of the immune system, either cellular or humoral, or with combined immunodeficiency syndromes that involve components of each of the immune effector systems. Additional defects include lack of certain complement components. Infections that are easily handled by the normal host become problems in the immunodeficient host. For example, patients with antibody deficiencies are often plagued by infections caused by encapsulated bacteria such as pneumococci, meningococci, and *Haemophilus influenzae*. Patients who lack certain elements of the complement cascade are more likely to contract infections with meningococci and gonococci. Children with chronic granulomatous disease, a disease with polymorphonuclear neutrophil (PMN) dysfunction, have gram-negative bacterial and staphylococcal infections. A complete discussion of the infectious consequences of inherited immunological defects is found in other texts.

NEOPLASTIC DISEASES AND BONE MARROW TRANSPLANT RECIPIENTS

Patients with cancer are at high risk to become infected, and the nature of the malignancy often determines the etiological agent(s). Table 24.1 contains a list of etiological agents associated with certain malignancies. Hodgkin's disease affects the cell-mediated immune response that normally helps to protect the body from intracellular pathogens such as herpes, mycobacteria, *Salmonella*, *Brucella*, *Nocardia*, and *Listeria*. Even before chemotherapy, patients with Hodgkin's-type lymphoma are likely to become infected with these and other agents normally handled by the cellular immune system. Treatment with certain chemotherapeutic agents and with corticosteroids also impairs cellular immunity. Other malignancies may affect the humoral immune response, which contributes to protection against agents such

---

TABLE 24.1

INFECTIOUS AGENTS FREQUENTLY ASSOCIATED WITH CERTAIN MALIGNANCIES

MALIGNANCY (SITES AND TYPES OF INFECTIONS)	PATHOGENS
Acute nonlymphocytic leukemia (pneumonia, oral lesions, cutaneous lesions, urinary tract infections, hepatitis, most often sepsis without obvious focus)	Enterobacteriaceae *Pseudomonas* Staphylococci *Corynebacterium jeikeium* *Candida* *Aspergillus* *Mucor* Hepatitis C and other non-A, non-B
Acute lymphocytic leukemia (pneumonia, cutaneous lesions, pharyngitis, disseminated disease)	Streptococci (all types) *Pneumocystis carinii* Herpes simplex Cytomegalovirus Varicella-zoster virus
Lymphoma (disseminated disease, pneumonia, urinary tract, sepsis, cutaneous lesions)	*Brucella* *Candida* (mucocutaneous) *Cryptococcus neoformans* Herpes simplex virus (cutaneous) Herpes zoster virus Cytomegalovirus *Pneumocystis carinii* *Toxoplasma gondii* *Listeria monocytogenes* Mycobacteria *Nocardia* *Salmonella* Staphylococci Enterobacteriaceae *Pseudomonas* *Strongyloides stercoralis*
Multiple myeloma (pneumonia, cutaneous lesions, sepsis)	*Haemophilus influenzae* *Streptococcus pneumoniae* *Neisseria meningitidis* Enterobacteriaceae *Pseudomonas* Herpes varicella-zoster virus *Candida* *Aspergillus*

---

as viruses, *Pneumocystis*, *Giardia*, staphylococci, yeasts, and encapsulated bacteria.

Cancer patients are likely to be undergoing treatment for their disorder, and a discussion of infection in such hosts must consider the effects of therapy as a necessary consequence of the disease. Although a few infectious agents are associated with certain carcinomas irrespective of therapy, such as *Streptococcus bovis* and *Clostridium septicum* bacteremia in patients with bowel malignancy, most infections in cancer patients are at least partially a result of the loss of immunological effector cells, either as a part of the disease process or as a consequence of chemotherapy.

Allogeneic bone marrow transplantation is undertaken for patients (particularly young persons) with hematological malignancies and occasionally for other indications, such as aplastic anemia. Destruction of host immune effectors predisposes the patient to gram-negative infection from endogenous bowel flora and gram-positive infections caused by skin flora. Subsequent graft-versus-host disease can lead to systemic viral infections. After immune function has returned via donor tissue, the transplant patient is at risk for infection associated with encapsulated bacteria, fungi, and systemic viral disease.[19]

A critical factor that determines whether an immunosuppressed patient will have increased risk of infection is the number of circulating granulocytes, primarily PMNs. Patients with severe granulocytopenia (greatly decreased numbers of circulating granulocytes, PMNs <500/mm³ and particularly those with PMNs <100/mm³) are the most likely to develop infections.[23] Gram-negative facultative organisms and staphylococci are the most common etiological agents. Gram-negative anaerobic organisms and fungi are encountered less often. The authors of one major study found that routine surveillance cultures of nose, throat, urine, and stool were not helpful for predicting those patients at risk for septicemia or for determining empirical therapy once the patients became febrile.[10] Follow-up daily blood cultures drawn on febrile patients after they had been started on empirical therapy, however, were helpful for identifying an etiological agent and ultimately for guiding therapy. By drawing such blood cultures into resin-containing media to remove antibiotics from the milieu or into a cell-lysing system such as the Isolator tube, microbiologists should increase the chances of recovering an organism from a patient receiving antibiotics. Persistent fever in the granulocytopenic patient taking antibiotics suggests the presence of a fungal infection, particularly *Candida* or *Aspergillus*, and less often, *Mucor*. Diagnosis of such fungal infections is difficult, and direct visual examination of specimens should always accompany culture, which may be negative, in cases of *Aspergillus* and *Mucor*.[12]

INFECTIONS IN HOSPITALIZED PATIENTS (NOSOCOMIAL INFECTIONS)

Patients are at increased risk for acquiring infections merely by being hospitalized. More than 2 million people (5% to 6% of all hospitalized patients) acquire a nosocomial infection (an infection that occurs at least 72 hours after admission that was not present at admission) each year. The cost of increased antibiotics, increased length of hospital stay, and loss of work caused by nosocomial infections is staggering. Factors that contribute to the risk of acquiring a nosocomial infection include the poor state of health of many patients; the use of immunosuppressive therapy; extensive surgery; invasive diagnostic tests; use of indwelling catheters in veins, arteries, and the bladder; indwelling tubes in the respiratory and gastrointestinal tracts; infusion of contaminated intravenous fluids; contaminated respiratory therapy equipment; and the widespread use of extended-spectrum antibiotics, which leads to the prevalence of antibiotic-resistant strains of bacteria in the hospital environment. Protein malnutrition, which may affect surgical and other hospitalized patients, contributes to such patients' increased risk of infection.[5]

Pneumonia is the leading cause of death among patients with nosocomial infections (as high as 50% mortality in intensive care unit patients), although urinary tract infections are the most common type of nosocomial infection. Etiological agents recovered from nosocomial urinary tract infections include *Escherichia coli*, enterococci, *Pseudomonas aeruginosa*, *Klebsiella* sp., and *Proteus* sp. Urinary tract, surgical wound, and lower respiratory tract infections account for more than 70% of all nosocomial infections. Hospitalized patients are often colonized in the respiratory tract, gastrointestinal tract, and the skin by endemic hospital strains of bacteria within several days of admission.[11] The nosocomial infection rate is highest among patients admitted to the surgery service, where most patients experience a breakdown of skin barriers as a result of surgery. The lowest rates are found among neonates and pediatric patients. Surgical wound infections are most often caused by *Staphylococcus aureus*, *E. coli*, other Enterobacteriaceae, coagulase-negative staphylococci, anaerobes, *P. aeruginosa*, and enterococci. Intraabdominal and burn infections often lead to bacteremia secondary to the primary infection.[5] The most common isolates in these settings are *Bacteroides* sp., *Serratia* sp., *S. aureus*, *Acinetobacter* sp., *Streptococcus agalactiae*, and *Providencia* sp.

Nosocomial pneumonia is a risk for any hospitalized patient, particularly for intubated patients. Organisms associated with these infections can be very hospital specific, but overall the most common include *Klebsiella* sp., other Enterobacteriaceae, *S. aureus,* anaerobes, *Streptococcus pneumoniae, P. aeruginosa,* and *Legionella.* Other agents have been associated with nosocomial outbreaks, including influenza virus. Viruses such as respiratory syncytial virus, adenovirus, and influenza A are often implicated as causes of nosocomial pneumonia among hospitalized children. Hospitalized patients are at increased risk of aspirating upper respiratory flora into the lungs (e.g., because of anesthesia, stroke, or depressed sensorium). "Aspiration pneumonia," as it is called, is usually caused by mixed anaerobic oral bacteria and aerobic streptococci. In the hospital setting, *S. aureus* and various gram-negative bacteria may also be involved. Material introduced into the lungs of patients via contaminated fluids or equipment during respiratory therapy has led to infections with *P. aeruginosa, Acinetobacter* sp., *Serratia* sp., and other gram-negative bacteria. Bronchoalveolar lavage or collecting material with a protected specimen brush technique is useful for diagnosis of nosocomial pneumonias (with quantitative culture), although transtracheal aspiration may be necessary for optimal recovery of anaerobes.

Antibiotic treatment, by altering normal flora and thus removing the protective effect of such flora, can select for multiply resistant strains of bacteria or allow fungal agents to multiply within the susceptible host. Such colonizing organisms may become involved in a variety of hospital-acquired infections. Alteration of the gastrointestinal flora may allow overgrowth of toxin-producing *Clostridium difficile.* This may lead to a syndrome called antibiotic-associated pseudomembranous colitis (PMC), definitively diagnosed by visualizing (by sigmoidoscopy or colonoscopy) the typical pseudomembrane or plaques made up of necrotic epithelial and pus cells on the mucosal surface of the lower gastrointestinal tract. Chapter 18 contains further information about this syndrome and its diagnosis.

INFECTIONS IN PATIENTS WITH GRAFTS, SHUNTS, INTRAVENOUS CATHETERS, OR PROSTHETIC DEVICES

Production of an artificial opening in the surface epithelium and then insertion of a plastic catheter into a blood vessel of a patient compromises the skin's ability to exclude pathogens and allows such agents direct access to the bloodstream. Devices such as intravenous catheters, heparin locks, hemodialysis shunts, total parenteral nutrition catheters, and arterial catheter lines expose patients to increased risk of infection. The agents present on the skin surface near the site of access are the most prevalent isolates from nosocomial infections; included are coagulase-negative staphylococci and *Staphylococcus aureus,* gram-negative bacilli, *Candida albicans,* other *Candida* sp., *Torulopsis glabrata,* and *Corynebacterium jeikeium. S. aureus* is the most common etiological agent of infected hemodialysis shunts, and *Staphylococcus epidermidis* is most frequently associated with peritonitis in patients undergoing ambulatory peritoneal dialysis.

S. epidermidis is the most common etiological agent of infected intravenous catheters, heparin locks, and cerebrospinal fluid shunts. Other skin flora, including diphtheroids, micrococci, *Propionibacterium acnes,* and colonizing gram-negative bacilli, are common causes of infection associated with catheters and shunts. Patients receiving parenteral (intravenous) nutrition via a Broviac, Hickman, or other catheter are at risk of developing infections from *Candida* sp., including *T. glabrata,* staphylococci, or gram-negative bacilli. It is important to culture the blood and the catheter tip to document catheter-associated septicemia (Chapter 15). Semiquantitative cultures of catheter tips have been validated as useful for diagnosis of catheter-associated bacteremia.[3] Contaminated intravenous fluids have been documented as a cause of serious nosocomial sepsis on occasion. The most common agents have been gram-negative rods, including *Citrobacter freundii, Enterobacter agglomerans, Pseudomonas aeruginosa, Pseudomonas cepacia, Klebsiella* sp., and *Serratia* sp.

Prosthetic devices often present a surface on which bacteria and fungi can adhere and multiply. Accordingly, a small percentage of patients given vascular grafts, prosthetic heart valves, cardiac pacemaker devices, and joint prostheses become infected with bacteria or fungi. The time until presentation of infection may vary from 1 week to several years. Staphylococci again predominate as the most important etiological agents of such infections.[22] The role of extracellular "slime" produced by coagulase-negative staphylococci as an adherence mechanism is controversial. Prosthetic heart valve infections may also be caused by diphtheroids, gram-negative bacilli, *C. albicans,* and *Aspergillus* sp., particularly in the early postoperative period. In addition, enterococci and viridans streptococci can cause late-appearing (after 60 days) infections in patients with prosthetic valves. *Mycobacterium chelonae* and fungi have been rare causes of porcine valve infection caused by contaminated valves implanted at surgery.

INFECTIONS IN DRUG ABUSERS

Drug abusers are at increased risk of acquiring infectious disease for several reasons: they may inject contaminated foreign material or their own (cutaneous or oral) flora into their bloodstreams along with the drug; they may be malnourished or inattentive to proper hygiene; they may have lapses of consciousness and thus be at risk of aspirating oral secretions into their lungs; and they may contract viral infections (e.g., hepatitis B, acquired immunodeficiency syndrome [AIDS]) via contaminated needles.

Sites of injection of drugs are often infected, most often with *Staphylococcus aureus*, which is carried on the skin of most drug abusers. Streptococci are also common etiological agents of cutaneous and subcutaneous infections, and severe necrotizing fasciitis and pyomyositis after introduction of mixed aerobic and anaerobic oropharyngeal flora have been seen occasionally in "skin poppers." Injection of microorganisms directly into the bloodstream often results in endocarditis among drug abusers. The most common agents include *S. aureus*, streptococci, enterococci, *Pseudomonas aeruginosa*, *Serratia marcescens*, other gram-negative bacilli including *Eikenella corrodens*, and *Candida* sp. Staphylococcal involvement of the tricuspid valve is particularly common.

Drug addicts, because of drug-induced stupor and depressed cough reflex, may aspirate oropharyngeal flora into the lungs, causing lung abscess, aspiration pneumonia, and empyema. Organisms associated with these syndromes are black-pigmented anaerobes (now referred to as *Prevotella*) and other anaerobic gram-negative rods, *Peptostreptococcus* sp., microaerobic and facultative streptococci, fusobacteria, *S. aureus*, and various gram-negative rods. General poor health and depressed respiratory defense mechanisms predispose such patients to the more common agents of pneumonia, including *Streptococcus pneumoniae*, *Haemophilus influenzae*, viruses, and mycoplasma. As mentioned earlier, many addicts are positive for hepatitis B surface antigen, and many develop hepatitis (either hepatitis B or C) at least once. Sexually transmitted diseases (STDs) are common

among drug abusers, particularly women, who may turn to prostitution to support their habit. Those drug abusers who are homosexual or bisexual or who develop AIDS are at even greater risk of acquiring infections, as outlined next.

INFECTIONS IN HOMOSEXUAL MALES

Homosexuality per se does not predispose a person to infectious disease, and homosexual women are at no greater risk than heterosexuals. Many practices among some homosexual men, however, including oral-genital, anal-genital, and oral-anal sexual activities, as well as interacting with multiple partners over a relatively short period, predispose these men to several infectious diseases. As might be suspected, acquisition of the standard STDs is enhanced by participating in sexual activity with increased numbers of partners. Gonorrhea, syphilis, and genital herpes infection are more prevalent among homosexual men than heterosexuals of similar demographic groups. In addition to the usual genital sites of infection, gonococci and herpes are often isolated from oropharyngeal lesions and as causes of proctitis. Oral-genital practices may predispose some patients to unusual oral presentation of STDs. Frequent isolation of *Neisseria meningitidis* from the oropharynx, anus, and urethra of homosexual men is probably also a consequence of oral-genital contact.

Anal-genital and anal-oral sexual practices predispose participants to several infections that are usually transmitted by the fecal-oral route. Non-B hepatitis (as well as type B) is prevalent among homosexual males. A group of agents has been found to comprise the most common causes of inflammatory, diarrheal, or ulcerative bowel disease in homosexual men, collectively called the "gay bowel syndrome." Diarrhea is usually caused by *Entamoeba histolytica*, *Giardia lamblia*, *Cryptosporidium*, *Microsporidium*, *Salmonella* sp., *Shigella* sp., or *Campylobacter jejuni*. Painful lesions may be caused by herpes, gonococci, meningococci, *Chlamydia* (both lymphogranuloma venereum [LGV] and non-LGV strains), *Haemophilus ducreyi*, *Calymmatobacterium granulomatis*, or *Candida* sp. The primary chancre of syphilis may occur within the rectum. In addition to hepatitis virus, CMV, human immunodeficiency virus type 1 (HIV-1), and human T-cell lymphotropic virus type 1 (HTLV-1) may be transmitted sexually. Homosexual males are one of the largest risk groups in the United States for contracting AIDS, a devastating disease of the cell-mediated immune system marked by infection and destruction of a subclass of T lymphocytes and other cells by the HIV retrovirus.

INFECTIONS IN PATIENTS WITH ACQUIRED IMMUNODEFICIENCY SYNDROME (AIDS)

Since 1981, when the first cluster of cases of pneumonia caused by the rare pathogen *Pneumocystis carinii* and Kaposi's sarcoma was reported among young homosexuals, more than 100,000 Americans have developed AIDS. It is estimated that during the 1990s, more than 1 million persons worldwide will develop AIDS. Although the number of cases among homosexuals has leveled off, cases among drug abusers and their contacts, including children of drug abusers, have continued to increase. Because a primary site of cellular destruction is the host immune system, AIDS patients have numerous infections. Patients often develop pneumonia, meningoencephalopathy and neurological dysfunction, mucocutaneous candidiasis, mycobacterial and other systemic infections, and diarrhea.[8] Common infectious complications of AIDS are listed in the box at right. Other infectious organisms reported in AIDS patients include *Legionella* sp., *Campylobacter jejuni,* other unusual *Campylobacter* sp., *Rochalimaea quintana* and *R. henselae* (Figure 24.1), *Mycobacterium haemophilum, Salmonella* sp., *Isospora belli, Microsporidium,* VZV, *Coccidioides* sp., and other fungi.[9,17,18]

The HIV of AIDS, a retrovirus, has been better studied since its discovery in 1983 than any other virus, but understanding of the pathogenesis of the disease is not complete. The virus, an enveloped ribonucleic acid (RNA) virus that carries its own reverse transcriptase, adheres preferentially to helper T (T4 surface antigen) lymphocytes, to macrophages, and perhaps to certain other cells (e.g., those in brain tissue), from which it enters the host cell and either transforms it or destroys it. HIV can alter its antigenic composition to avoid immune destruction while it undermines the host's immunological capabilities, thus the preponderance of opportunistic infections in patients with AIDS. It is known that the virus itself contributes directly to the neurological symptoms often seen in AIDS patients, but it also predisposes to infection by other pathogens in the central nervous system.

Prevention has focused on alteration of lifestyles (by education) and treatment or screening

Common Infectious Agents and Syndromes Associated with AIDS

Bacterial
Mycobacterium avium-intracellulare complex
M. tuberculosis
Other mycobacteria

Fungal
Esophageal and vulvovaginal candidiasis
Disseminated aspergillosis, histoplasmosis
Cryptococcosis

Parasitic
Chronic cryptosporidiosis
Pneumocystis carinii pneumonia
Strongyloidosis, intestinal and disseminated
Toxoplasmosis (pneumonia or central nervous system infection)

Viral
Disseminated cytomegalovirus infection
Chronic (>1 month) or disseminated herpes simplex infection
Progressive multifocal leukoencephalopathy (JC virus)
Condylomata acuminata

of blood products to prevent transmission. The complexity and antigenic instability of the virus, even within a single host, suggests that an effective vaccine will not be developed quickly. Amelioration of some symptoms has been achieved with azidothymidine (AZT), but this drug is not without side effects, including bone marrow suppression. Numerous other treatments are in developmental stages.

EPIDEMIOLOGY OF AIDS
New York, New Jersey, Washington, D.C., Florida, and California have the highest incidence of AIDS in the United States. Homosexual and bisexual men, intravenous drug abusers, immigrants from Haiti, recipients of transfused blood and blood products, hemophiliac patients who received factor VIII concentrate, and sexual contacts and children of people within risk groups are at risk of acquiring AIDS. Males greatly outnumber females. Recently, increasing numbers of patients in the United States with only heterosexual exposure are developing AIDS. This form of exposure is the primary mode of transmission in underdeveloped countries, such as those in central Africa, where AIDS is a major health crisis and there are no sex-related differences in prevalence. Reasons for these epidemiological differences are not understood. The overall long-term mortality of AIDS approaches 100%. Patients with Kaposi's sarcoma

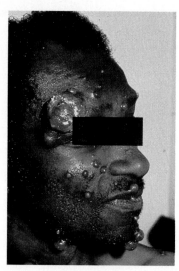

FIGURE 24.1

Appearance of lesions of bacillary angiomatosis on face of man infected with HIV and *Rochalimaea henselae*. The lesions cleared dramatically after the patient received doxycycline. (Reproduced from Mui, B.S., Milligan, M.E., and George, W.L. 1990. Response of HIV-associated disseminated cat-scratch disease to treatment with doxycycline. Am. J. Med. 89:229-231.)

alone have the best prognosis for survival; those with an opportunistic infection have the worst prognosis. Although cases have occurred in health care workers, occupational transmission has been rarely documented. Involvement of children is directly related to congenital spread, subsequent sexual abuse, or drug abuse.

In addition to patients with infectious diseases or neoplasms suggestive of an underlying immune disorder, in some patients with HIV antibody the primary evidence of disease is lymphadenopathy. These patients may also complain of malaise, weight loss, and fever. These symptoms are indicative of the AIDS-related complex (ARC) and signify a prodrome of AIDS. HIV has been cultured from peripheral blood and occasionally from saliva of patients with ARC. Recently it has been shown that another retrovirus, HIV-2, can also cause AIDS. Although the prevalence of this virus is very low in the United States, early protective measures, including screening the blood supply for HIV-2 antibody, are being instituted to prevent an additional nationwide epidemic.

ELISA and Western blot tests for detection of the HIV antibody and ELISA tests for HIV-1 are used to screen all donated blood. This screening is highly sensitive and specific but does miss those patients with recent acquisition of infection who have not yet produced detectable antibodies.

Through screening, individuals with no symptoms are identified. The ultimate fate of such persons is not yet known, although some have gone on to develop ARC or AIDS.

Literature concerning all aspects of AIDS is readily available; books, newsletters, and reviews abound. Readers are referred to the publications listed in the bibliography of this chapter as an initial resource for more information.

LABORATORY DIAGNOSIS OF HUMAN IMMUNODEFICIENCY VIRUS (HIV) INFECTION

Diagnostic testing for the HIV agent encompasses antibody and antigen detection systems. ELISA tests (at least eight commercial tests are approved by the Food and Drug Administration [FDA]) are the first screening test of choice for HIV antibodies. These tests usually yield less than 1% false-negative results but may show false-positive results in as many as 25% of some patient groups. The Western blot confirmatory test helps to identify false-positive results, although a "gold standard" of criteria for a positive test has not been determined. Patients who have been infected recently and produce only IgM may test negative in these tests. Immunofluorescent antibody (IFA) tests are used with reliability by the California State Health Department and other centers as a supplement to ELISA. Radioimmunoprecipitation assay (RIPA) is a relatively new test for HIV that uses radioactively labeled viral proteins that bind to specific serum antibodies. The pattern of antibodies is then determined. RIPA is available only as a research test, although it is probably more specific and sensitive than the commercially available tests. Double-antibody sandwich ELISA tests for HIV antigen are also available. Commercially available antigen detection by ELISA is not yet sensitive enough for routine use, although specific tests for the p24 antigen (most prevalent in patients with HIV infection) are promising. Amplification methods, however, improve the sensitivity of antigen tests. Polymerase chain reaction (PCR) systems show much promise for detection of even small amounts of viral deoxyribonucleic acid (DNA) in patient specimens. Field trials of commercial PCR tests were nearing completion in 1993. Once the viral target nucleic acid is amplified, it can be detected easily using hybridization techniques. Finally, culture of the virus from peripheral blood monocytes in a cocultivation system seems to detect most positive reactions. Detection of either reverse transcriptase activity or HIV antigen denotes a positive culture result. Cultures are expensive and time-consuming but may be the only positive test in some stages of HIV infection.

LABORATORY DIAGNOSIS OF INFECTIONS ASSOCIATED WITH AIDS

Standard methods for diagnosis of infections should be used for this purpose in AIDS patients. Some additional measures that have been useful in diagnosing opportunistic infections in AIDS patients are mentioned here.

Patients with bacteremia caused by *Mycobacterium avium-intracellulare* complex or *Mycobacterium tuberculosis* often have many circulating bacteria, and blood cultures are often positive, particularly with a lysis-centrifugation system (Wampole Isolator) or when blood is cultured directly using a radiometric procedure developed for that purpose (BACTEC, Becton-Dickinson). Blood is allowed to remain at room temperature for approximately 1 hour in the Isolator tube before centrifugation and plating. Presumably this period allows the white cells to lyse, releasing viable mycobacteria that would not have been detected if the white cells remained intact.

Several procedures, including open lung biopsy, transbronchial biopsy, and bronchoalveolar lavage, are useful in diagnosing *Pneumocystis carinii* pneumonia, although induced sputum is often positive in AIDS patients, particularly if a DFA stain is used on the sample.[21] A rapid stain such as toluidine blue O, or preferably, a monoclonal fluorescent antibody for detection of *P. carinii*, should be used on respiratory specimens. A silver stain reveals fungal elements in addition to *Pneumocystis*. Other agents of pulmonary infection can be detected by standard methods. Acid-fast stains of stool, as well as sputum, may be helpful in detecting mycobacteria and also *Cryptosporidium*. A special stain was recently developed by the Centers for Disease Control that detects *Microsporidium* directly in stool specimens.[20]

Bone marrow cultures have proved valuable for diagnosis of infections in febrile AIDS patients.[1] One must carefully examine all specimens from AIDS patients with an open mind, since unusual pathogens and unusual presentations of infectious diseases have been found.

REFERENCES

1. Bishburg, E., Eng, R.H., Smith, S.M., and Kapila, R. 1986. Yield of bone marrow culture in the diagnosis of infectious diseases in patients with acquired immunodeficiency syndrome. J. Clin. Microbiol. 24:312.
2. Centers for Disease Control. 1991. Summary of notifiable diseases, United States, 1991. M.M.W.R. 40:46.
3. Collignon, P.J., Soni, N., Pearson, I.Y., et al. 1986. Is semiquantitative culture of central vein catheter tips useful in the diagnosis of catheter-associated bacteremia? J. Clin. Microbiol. 24:532.
4. Daffos, F., Forrestier, F., Capella-Pavlosky, M., et al. 1988. Prenatal management of 746 pregnancies at risk for congenital toxoplasmosis. N. Engl. J. Med. 318:271.
5. Deitch, E.A. 1988. Infection in the compromised host. Surg. Clin. North Am. 68:181.
6. Evans, M.E., Schaffner, W., Federspiel, C.F., et al. 1988. Sensitivity, specificity, and predictive value of body surface cultures in a neonatal intensive care unit. J.A.M.A. 259:248.
7. Gilligan, P.H. 1991. Microbiology of airway disease in patients with cystic fibrosis. Clin. Microbiol. Rev. 4:35.
8. Janoff, E.N., and Smith, P.D. 1988. Perspectives on gastrointestinal infections in AIDS. Gastroenterol. Clin. North Am. 17:451.
9. Kiehn, T.E. 1992. *Mycobacterium haemophilum:* a new opportunistic pathogen. Clin. Microbiol. Newsletter 14:81.
10. Kramer, B.S., Pizzo, P.A., Robichaud, K.J., et al. 1982. Role of serial microbiologic surveillance and clinical evaluation in the management of cancer patients with fever and granulocytopenia. Am. J. Med. 72:561.
11. Larson, E.L., McGinley, K.J., Foglia, A.R., et al. 1986. Composition and antimicrobic resistance of skin flora in hospitalized and healthy adults. J. Clin. Microbiol. 23:604.
12. McGowan, K.L. 1987. Practical approaches to diagnosing fungal infections in immunocompromised patients. Clin. Microbiol. Newsletter 9:33.
13. McGowan, K.L. 1988. The microbiology associated with cystic fibrosis. Clin. Microbiol. Newsletter 10:9.
14. Peter, J.B., Cherry, J.D., and Bryson, Y.J. 1982. Improving diagnosis of congenital infections. Diagn. Med. 5:61.
15. Sergeant, G.R., Topley, J.M., Mason, K., et al. 1981. Outbreak of aplastic crisis in sickle cell anemia associated with parvovirus-like agent. Lancet ii:595.
16. Sever, J., Ellenberg, J.H., Ley, A.C., et al. 1988. Toxoplasmosis: maternal and pediatric findings in 23,000 pregnancies. Pediatrics 82:181.
17. Slater, L.N., Welch, D.F., and Min, K.-W. 1992. *Rochalimaea henselae* causes bacillary angiomatosis and peliosis hepatis. Arch. Intern. Med. 152:602.
18. Spach, D.H., Callis, K.P., Paauw, D.S., et al. 1993. Endocarditis caused by *Rochalimaea quintana* in a patient infected with human immunodeficiency virus. J. Clin. Microbiol. 31:692.
19. Tutschka, P.J. 1988. Infections and immunodeficiency in bone marrow transplantation. Pediatr. Infect. Dis. J. 7:S22.
20. Weber, R., Bryan, R.T., Owen, R.L., et al. 1992. Improved light-microscopic detection of microsporidia spores in stool and duodenal aspirates. N. Engl. J. Med. 326:161.
21. Wolfson, J.S., et al. 1990. Blinded comparison of a

direct immunofluorescent monoclonal antibody staining method and a Giemsa staining method for identification of *Pneumocystis carinii* in induced sputum and bronchoalveolar lavage specimens of patients infected with human immunodeficiency virus. J. Clin. Microbiol. 28:2136.

22. Young, E.J., and Sugarman, B. 1988. Infections in prosthetic devices. Surg. Clin. North Am. 68:167.

23. Young, L.S. 1988. Antimicrobial prophylaxis in the neutropenic host: lessons of the past and perspectives for the future. Eur. J. Clin. Microbiol. Infect. Dis. 7:93.

BIBLIOGRAPHY

Centers for Disease Control. 1987. Update: acquired immunodeficiency syndrome—United States. M.M.W.R. 36:522.

Inderlied, C.B., and Young, L.S. 1985. Clinical microbiology of acquired immune deficiency syndrome. J. Med. Technol. 2:167.

Jackson, J.B., and Balfour, H.H., Jr. 1988. Practical diagnostic testing for human immunodeficiency virus. Clin. Microbiol. Rev. 1:124.

Mandell, G., Douglas, R.G., Jr., and Bennett, J.E., editors. 1985. Principles and practice of infectious diseases, ed 2. Part IV: Special problems. Section A: Nosocomial infections; Section B: Infections in special hosts. John Wiley & Sons, New York.

Ostrow, D.G. 1984. Homosexuality and sexually transmitted diseases. In Holmes, K.K., Mårdh, P.-A., Sparling, P.F., and Wiesner, P.J., editors. Sexually transmitted diseases. McGraw-Hill, New York.

Quinn, T.C., and Holmes, K.K. 1984. Proctitis, proctocolitis, and enteritis in homosexual men. In Holmes, K.K., Mårdh, P.-A., Sparling, P.F., and Wiesner, P.J., editors. Sexually transmitted diseases. McGraw-Hill, New York.

Rankin, J.A., Coliman, R., and Daniele, R.P. 1988. Acquired immune deficiency syndrome and the lung. Chest 94:155.

Remington, J.S., and Klein, J.O., editors. 1983. Infectious diseases of the fetus and newborn infant. Saunders, Philadelphia.

Sen, P., Kapila, R., Chmel, H., et al. 1982. Superinfection: another look. Am. J. Med. 73:706.

Steigbigel, R.T., and Cross, A.S. 1984. Infections associated with hemodialysis and chronic peritoneal dialysis. In Remington, J.S., and Swartz, M.N., editors. Current clinical topics in infectious diseases. McGraw-Hill, New York.

Yoshikawa, T.T. 1983. Geriatric infectious diseases: an emerging problem. J. Am. Geriatr. Soc. 31:34.

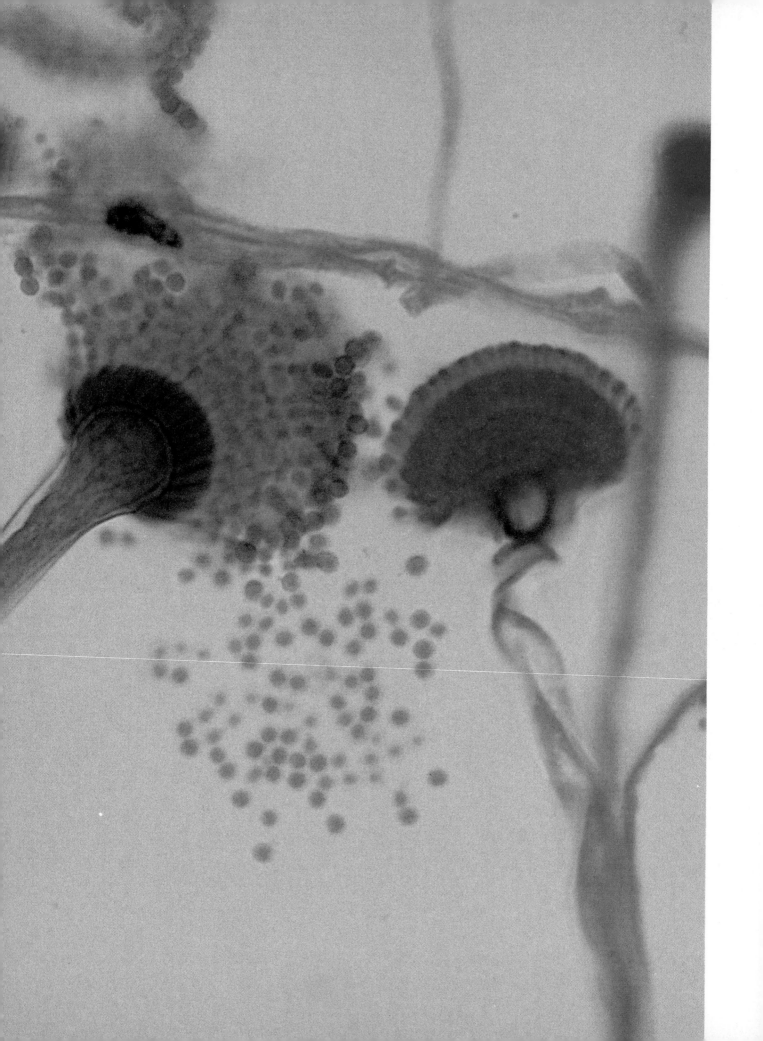

PART FOUR

METHODS FOR IDENTIFICATION

OF ETIOLOGICAL AGENTS OF

INFECTIOUS DISEASE

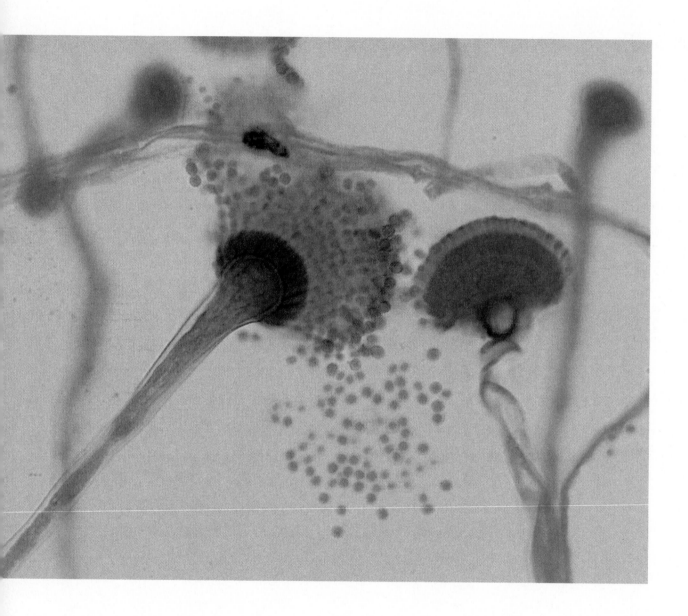

Morphological
characteristics are often the
most important differential
criteria for species
identification, particularly
among parasites and fungi,
such as these conidiophores
of *Aspergillus* species
stained with lactophenol
cotton blue.

25

MICROCOCCACEAE: STAPHYLOCOCCI, MICROCOCCI, AND STOMATOCOCCI

The genera *Staphylococcus, Micrococcus, Stomatococcus,* and *Planococcus* are members of the family Micrococcaceae. The majority of planococci and micrococci are free-living saprophytes, but the natural habitat of staphylococci and *Stomatococcus* species is the surface epithelium of primates and other mammals. Their presence as endogenous flora allows many species of staphylococci the opportunity to cause infection when host defenses break down. Staphylococci, *Stomatococcus* sp., and micrococci have been isolated from clinically significant sources. *Staphylococcus aureus* (usually a coagulase-positive staphylococcus, described later in this chapter) has been recognized historically as a virulent and important human pathogen; its capacity to produce human disease has not diminished with the introduction of antibiotics.[33,37] During the last several years, coagulase-negative staphylococci have surfaced as important pathogens, preying primarily on patients with some sort of prosthetic or indwelling device.[2,28] They are currently the most frequent agent of nosocomial bacteremia (see Chapter 15). *Micrococcus* species and *Stomatococcus mucilaginosus* are opportunistic pathogens, usually seen only in immunocompromised hosts.

Specimens that may harbor clinically important organisms of the Micrococcaceae family come from almost any source. The organisms are hardy and do not require special collection procedures other than those outlined in Chapter 7.

MORPHOLOGY AND PRELIMINARY IDENTIFICATION STRATEGIES

Micrococcaceae are spherical cocci with gram-positive cell walls; those recovered from humans are nonmotile (with the exception of *Micrococcus agilis,* rarely encountered clinically), aerobic or facultatively anaerobic, and, except for stomatococci, are catalase positive. During binary fission, Micrococcaceae divide along both longitudinal and horizontal planes, forming pairs, tetrads, and ultimately irregular clusters (Figure 25-1). The Greek word *staphyle,* meaning a "bunch of grapes," is the descriptive stem for the genus name. Very old cells may lose their ability to retain crystal violet and thus may be more easily decolorized in a Gram stain than young cells.

Micrococcaceae will grow on most laboratory media that support gram-positive organisms, yielding circular, opaque, smooth colonies. The

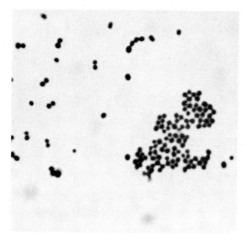

FIGURE 25.1

Appearance of Micrococcaceae in Gram stain.

FIGURE 25.2

Commercial particle agglutination system (Staphaurex, Murex Diagnostics) for identification of *Staphylococcus aureus* based on presence of protein A or clumping factor on bacterial cell surfaces. Agglutination indicates a positive reaction. (Courtesy Murex Diagnostics, Norcross, Ga.)

growth is usually **butyrous** (like butter), and colonies can be emulsified easily in water to form a smooth suspension. Hemolysis on blood agar and color of the colonies are variable. Since yeast colonies can resemble Micrococcaceae and may be catalase positive, all suspect colonies should be examined microscopically by wet mount or Gram stain before further tests are performed. Mannitol salt agar is sometimes used to isolate staphylococci from clinical material. This medium contains a high salt concentration, mannitol, and phenol red pH indicator. *S. aureus* yields colonies surrounded by a yellow halo caused by the acidification of the medium induced by products of mannitol fermentation. Other staphylococci (particularly *S. saprophyticus*) may also ferment mannitol and thus resemble *S. aureus* on mannitol salt agar.

Once an organism isolated from a clinical specimen has been characterized as a gram-positive, catalase-positive coccoid bacterium, it is further identified in a series of steps, the first of which involves the coagulase test. *S. aureus* is identified on the basis of the presence of the enzyme **coagulase,** which binds plasma fibrinogen, causing the organisms to agglutinate or plasma to clot. A rapid screening slide test, the slide coagulase test, for the production of **clumping factor** (cell-bound coagulase) that is positive with more than 95% of strains of *S. aureus* is described in Chapter 10, Procedure 10.2. Commercial particle agglutination tests are also used commonly in clinical laboratories instead of the slide coagulase test for rapid differentiation of *S. aureus* from other staphylococci (Figure 25.2). Some detect only clumping factor, and others detect both clumping factor and protein A (a cell-wall protein found primarily in *S. aureus*). Problems with inter-

pretation of the commercial particle agglutination tests are discussed later.

Isolates that do not produce clumping factor must be tested for the ability to produce extracellular coagulase **(free coagulase).** This enzyme will cause clotting of plasma after 1 to 4 hours of incubation at 35° C to 37° C. The tube coagulase test for free coagulase is described in Procedure 25.1. Coagulase plasma with citrate is not recommended for use in this test because citrate-utilizing organisms may yield false-positive results. A positive coagulase test is sufficient for naming an isolate *S. aureus* in most cases, although *S. lugdunensis, S. schleiferi,* and *S. intermedius* may also produce coagulase or clumping factor. Table 25.1 shows some tests useful for differentiating these coagulase-positive staphylococci from *S. aureus,* which is inherently more virulent and much more likely to be penicillin-resistant.

STAPHYLOCOCCUS AUREUS

CHARACTERISTICS AND EXTRACELLULAR PRODUCTS

S. aureus is recovered from a variety of infections including skin lesions, such as furuncles and carbuncles, abscesses, wound infections, pneumonia, osteomyelitis, and others. From these sites organisms can invade the bloodstream and seed metastatically, appearing in the urine or forming abscesses in various body organs, or producing septic shock or endocarditis. The organism can also be recovered from the anterior nares, perineum, and other skin sites from as many as 10% to 15% of healthy people and a significantly greater percentage of people in the hospital set-

PROCEDURE 25.1

TUBE COAGULASE TEST

PRINCIPLE
Only *S. aureus,* the most
virulent species, and *S.
intermedius* (rarely isolated from
humans) produce coagulase
enzyme able to clot rabbit
plasma, differentiating them
from the rest of the
Micrococcaceae.

METHOD
1. Prepare coagulase reagent,
   rabbit plasma with EDTA
   (Difco Laboratories and BBL
   Microbiology Systems) in
   0.5 ml amounts in 13 × 100
   mm glass tubes. The tubes
   can be prepared in large
   numbers and refrigerated for
   10 days, or frozen at −20° C
   for several months.

2. Emulsify a visible portion of
   growth from isolated
   colonies (grown on
   nonselective medium) in the
   plasma by rubbing the
   material on the side of the

tube while holding the tube
at an angle. Straighten the
tube, causing the plasma
level to cover the site of
inoculation.

3. Incubate the suspension for
   1 to 4 hours at 35° C to 37°
   C and observe for the
   presence of a gel or clot that
   cannot be re-suspended by
   gentle shaking (Figure 25.3).
   A ropy, opaque precipitate
   in the plasma is not a clot. If
   no clot forms after 4 hours,
   the tube should be
   incubated at room
   temperature overnight. Rare
   isolates require such
   extended incubation.

       Some isolates may
   demonstrate a clot at 4
   hours that subsequently
   lyses and is not detectable at
   the 24-hour inspection.
   Thus all assays should be
   examined at 4 to 6 hours
   and reincubated if negative.

4. Organisms that fail to clot
   the plasma within 24 hours
   are considered coagulase
   negative and must be
   identified by other methods.

QUALITY CONTROL
Inoculate coagulase tubes with
fresh subcultures of *S. aureus*
ATCC 25923 and *S. epidermidis*
ATCC 14990 and proceed as
mentioned previously.

EXPECTED RESULTS
S. aureus should clot the plasma
in 4 hours; *S. epidermidis* should
not clot the plasma even after
overnight incubation.

PERFORMANCE SCHEDULE
Test each new lot of coagulase
plasma when it is received,
each prepared batch once it has
been rehydrated, and weekly
thereafter.

ting. This **carrier** state can serve as a reservoir for
infection of hospitalized patients, but most carriers do not disseminate the organism and are not a
risk to others. Although it is also an important
source of food poisoning, *S. aureus* is usually not
isolated from the patient suffering from this intoxication, and clinical laboratories usually do not
attempt to culture pathogens from food.

  Because of its structure, *S. aureus* is uniquely
suited to be a human pathogen. The cell wall is
composed of tightly cross-linked peptidoglycan
and teichoic acid moieties that protect the organism from lysis under harsh osmotic conditions and
probably aid in attachment of the bacteria to mucosal cell receptor sites. Certain strains produce a
polysaccharide capsule that helps to protect them
from phagocytosis by polymorphonuclear neutrophils (PMNs). In addition, most strains possess
a cell wall protein, protein A, that binds the Fc
segment of IgG, preventing antibody-mediated
phagocytosis by PMNs. In the bloodstream, aggre-

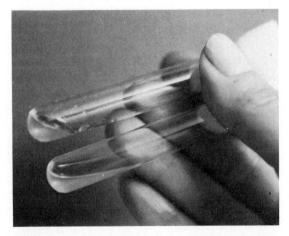

FIGURE 25.3

Tube coagulase test. Presence of clot (tube on *top*) indicates
a positive coagulase result.

TABLE 25.1

SIMPLE PRESUMPTIVE DIFFERENTIATION AMONG ORGANISMS THAT RESEMBLE *STAPHYLOCOCCUS AUREUS* ISOLATED FROM HUMAN CLINICAL SPECIMENS

SPECIES	CLUMPING FACTOR (SLIDE COAGULASE)	TUBE COAGULASE	PYR*	THERMOSTABLE NUCLEASE	VOGES-PROSKAUER	ORNITHINE DECARBOXYLASE
S. aureus	+	+	–	+	+	–
S. intermedius	–	+(rare)	+	+	–	–
S. lugdunensis	+	–	+	–	+	+
S. schleiferi	+	–	+	+	+	–

* Pyroglutamyl β-naphthylamide aminopeptidase (PYR test, described in Chapter 10, Procedure 10.6).

gates of IgG bound to protein A on staphylococcal surfaces will fix complement, causing complement-mediated tissue damage to the host.

S. aureus produces a number of enzymes and toxins that contribute substantially to its ability to cause disease. In addition to coagulase, S. aureus produces the enzymes phosphatase, thermostable deoxyribonuclease and ribonuclease, lipase, gelatinase, protease, and fibrinolysin. Another enzyme, hyaluronidase, may contribute to the spread of infection involving S. aureus through tissues. Among the toxins elaborated by S. aureus are alpha, beta, gamma, and delta toxins and leukocidin, which act on the red and white blood cell membranes of some species. The role of these toxins in the pathogenesis of disease is unclear.

Toxins that have been clearly shown to contribute to certain disease syndromes are enterotoxins, exfoliative toxin, and the toxins associated with toxic shock syndrome. All three types of staphylococcal toxins are now included within a newly recognized class of immune system modulators, the **superantigens.** These agents can stimulate T cells independently of other mediators, causing systemic effects including release of cytokines such as interleukin-2, γ-interferon, and tumor necrosis factors.[40] Some of the previously unexplained systemic effects of staphylococcal disease are now thought to be mediated by these superantigens.[40]

S. aureus produces at least six different enterotoxins, some of which are implicated in food poisoning and antibiotic-induced pseudomembranous colitis. In the case of food poisoning, toxins are elaborated by the organisms growing in food and then are ingested by the patient. Staphylococcal scalded skin syndrome, often affecting newborns, is caused by an exfoliative toxin called **exfoliatin** (which cleaves the middle layers of the epidermis, allowing the surface skin to peel). The most recently described toxin-mediated staphylococcal disease is **toxic shock syndrome,** a systemic disease characterized by fever, hypotension, and multiorgan involvement but negative blood cultures.[35] The disease is caused by toxins, one of which is called toxic shock syndrome toxin-1 (TSST-1), which is elaborated by certain S. aureus strains growing in a localized focus of infection or colonization. Many cases have occurred in menstruating young females who were vaginal carriers of S. aureus and who used tampons. Any site, however, such as an abscess or wound, may harbor the toxin-producing organisms. The clinical laboratory may be asked to isolate S. aureus from suspected sites of infection; a reverse passive latex agglutination test for TSST-1 toxin is available commercially (Oxoid USA, Columbia, Md.).

IDENTIFICATION

S. aureus is coagulase positive, which is its most distinguishing characteristic. Species other than aureus that may be positive in the rapid slide coagulase test or the commercial slide particle agglutination tests include S. lugdunensis, intermedius, and schleiferi.[17] One clue that such an isolate may not be S. aureus is penicillin susceptibility; more than 95% of S. aureus isolated from clinical specimens are penicillin resistant. These coagulase-positive species likely to be encountered in a clinical laboratory may be differentiated from S. aureus using a short series of tests (see Table 25.1).[17,22,32] Staphylococcus intermedius, one coagulase-positive species, is an important agent of dog-bite wound infections; it was isolated four times more frequently than S. aureus from the oral cavity of 135 dogs.[34] This species has probably been misidentified as S. aureus when isolated from such infections. Penicillin-susceptible, coagulase-positive isolates from blood and other sterile body fluids should be chosen for further characterization. In addition to occasional positive tests for non–*S. aureus* strains, commercial slide particle agglutination tests for clumping factor or protein A also yield a small percentage of false-negative results, particularly with methicillin-resistant *S. aureus*.[23] Methicillin-resistant strains of *S. aureus* are often slide coagulase-negative as well, suggesting that a common cell wall structure contributes to both clumping factor expression and methicillin resistance. The rapid commercial tests can be used in most institutions with confidence if microbiologists confirm the identity of methicillin-resistant strains by tube coagulase. However, the traditional coagulase plasma slide test for clumping factor (Chapter 10) yields approximately the same results and can be performed as rapidly and is less expensive than tests using commercial reagents. Tube coagulase should be used to confirm all slide test negative results on clinically significant isolates.

An occasional coagulase-negative isolate will resemble *S. aureus* (hemolytic, yellow-pigmented colony), prompting further identification procedures. Production of thermostable nuclease is an important characteristic used to verify coagulase-positive *S. aureus*, although strains of *S. delphini, schleiferi, intermedius,* and *hyicus* (primarily nonhuman species) are also positive.[22] One method requires that suspensions of overnight broth cultures of the organisms be boiled for 15 minutes and an aliquot placed into wells punched into thermostable nuclease test agar (Remel, Lenexa, Kan. or Edge Biologicals, Memphis, Tenn.). A pink halo surrounding the well after 1 to 4 hours' incubation at 35° C to 37° C indicates presence of the nuclease.[39] Another method used 1 ml of the culture supernatant from blood culture bottles showing positive cocci in clusters for a rapid modification of the thermostable nuclease test described in the preceding sentences.[24] The supernatant was boiled for 15 minutes and placed into 3-mm diameter wells cut into the surface of thermonuclease agar plates. Correlation with standard methods was 100%, and only 2 hours' incubation was needed before results could be interpreted. Another very similar rapid method for performing a thermostable nuclease test directly from blood culture broth was described in Chapter 10. The source of the test agar was critical for reliable results; thermonuclease (not DNase) agar must be used.

Strains of *S. aureus* may be identified for epidemiological purposes by phage typing, but molecular methods are being used more frequently today.[26] Characterizing strains by phage type, biotype, plasmid profile, and genetic restriction analysis may help to delineate the path of spread of strains among hospitalized patients, the environment, and the attending medical staff (see Chapter 6). Gene restriction site polymorphism analysis and a miniaturized commercial identification system have been used for epidemiological studies with staphylococci.[5,26]

SUSCEPTIBILITY

Most strains of *S. aureus*, even those acquired in the community, are penicillin resistant. In most cases this resistance is attributable to β-lactamase production due to genes located on extrachromosomal plasmids. Some staphylococci that are penicillin resistant are also resistant to the newer β-lactamase–resistant semisynthetic penicillins such as methicillin, oxacillin, and nafcillin. This resistance is due primarily to the presence of an unusual penicillin-binding protein in the cell wall of resistant strains; the genetic determinants are chromosomal, probably carried on a transposon.[20]

Clinically significant methicillin-resistant *S. aureus* (MRSA) is being isolated with greater frequency in the United States, often posing problems as causes of nosocomial infections.[3] These strains, although they may exhibit susceptibility to cephalosporins and other β-lactam agents, such as β-lactam/β-lactamase inhibitor combinations and imipenem, in vitro, cannot be treated effectively with these agents or β-lactamase–resistant penicillins. Numerous hospitals have begun active screening for MRSA, since infections caused by these strains significantly affect patients' morbidity and the hospitals' reimbursements. Infection with MRSA is likely to be more severe and require longer hospitalization, with incumbent increased costs, than infection with a methicillin-susceptible strain.[3] Ongoing strategies to prevent nosocomial transmission include treatment of wound sites and carriage sites (nares, for example) with mupirocin ointment and cohorting of caregivers and patients.[3,19,37a] Vigorous attempts to identify carriers should involve enrichment cultures in broth, because plate cultures alone may fail to detect as many as 60% of infected or colonized individuals.[19] An alternative strategy is to inoculate mannitol-salt agar instead of blood agar and incubate the plates for 7 days to detect the organism.

Strategies for susceptibility testing and reporting of *S. aureus* are described in Chapter 14. In particular, the use of oxacillin (1 μg) disks, the addition of 2% NaCl to cation-supplemented

Mueller-Hinton broth or 4% NaCl to Mueller-Hinton agar for dilution tests, incubation at temperatures no greater than 35° C, and incubation for a full 24 hours before reading results have been shown to increase the chances of detecting methicillin-resistant *S. aureus*. A screening test for methicillin-resistant staphylococci incorporating 6 µg/ml oxacillin in 4% NaCl-supplemented Mueller-Hinton agar, inoculated with a spot of a McFarland 0.5 turbidity suspension of the organism to be screened, and incubated for 24 hours at 35° C will accurately detect resistant strains as those that grow on the agar. One of these additional procedures for detecting resistance is sometimes used with automated susceptibility systems, some of which may have problems reliably detecting methicillin-resistant strains of staphylococci.[19,20] All staphylococci are still susceptible to vancomycin, and many or most strains are susceptible to rifampin (which should not be used alone). Certain *S. aureus* strains exhibit in vitro tolerance (minimum bactericidal concentration [MBC] at least 32 times greater than minimum inhibitory concentration [MIC]) to some antimicrobial agents because of clones of bacteria that are killed more slowly than the majority of cells in the population. There is no definitive evidence that tolerance is of clinical significance.

Treatment of catheter-related sepsis will be more successful if the implicated catheter is removed; duration of therapy should be more than 10 days but in the absence of other complications, 2 weeks appears to be adequate.[30]

NONCULTURE METHODS FOR DIAGNOSING STAPHYLOCOCCUS AUREUS INFECTIONS

Tests for staphylococcal antigen have not been shown to aid in diagnosis of disease. However, some workers believe that detection of circulating antibodies to capsular polysaccharide or to the teichoic acid component of the *S. aureus* cell wall is useful for monitoring the course of infection and for differentiating long-standing or deep-seated infections from lesser infections.[1] Although numerous methods have been used to detect such antibodies successfully, many of these methods are too sensitive to differentiate between patients with more serious or less serious infections.[1,7] Use of more than one serologic test can more effectively predict disease state.[36] Diagnostic applications of tests for capsular polysaccharide and teichoic acid antibody, including development and evaluation of commercial kits, are ongoing.[1,7,18,38]

COAGULASE-NEGATIVE STAPHYLOCOCCI

CHARACTERISTICS AND EXTRACELLULAR PRODUCTS

Coagulase-negative staphylococci vary in pathogenic potential.[28] As is true for many other resident human flora of relatively low virulence, these organisms are more likely to cause infection in compromised hosts, particularly cancer patients. Coagulase-negative staphylococci are a common cause of bacteremia in neutropenic patients; of infections related to indwelling catheters, shunts, and prosthetic devices; and of urinary tract infections in young, sexually active women.

Staphylococcus epidermidis is known to cause infection of native heart valves and intravascular prostheses, including intravenous catheters and artificial heart valves.[2] It can cause peritonitis in patients receiving peritoneal dialysis. *S. epidermidis*, usually introduced via a break in the patient's skin, also causes infections in prosthetic joints, central nervous system shunts, and has been isolated from subcutaneous infections.[2]

S. saprophyticus is isolated most frequently from uncomplicated urinary tract infections in nonhospitalized patients, notably sexually active young women. This species may also be involved in recurrent infection and in stone formation. The incidence of *S. saprophyticus* urinary tract infection varies considerably among the patient populations served by institutions and in different geographical areas.

Species that are less commonly implicated as pathogens include *S. cohnii, S. haemolyticus, S. hominis, S. lugdunensis, S. saccharolyticus, S. schleiferi, S. simulans,* and *S. warneri,* although recent articles have emphasized that *S. warneri* and *S. lugdunensis* may be among the more virulent species.[12,16,21] *S. lugdunensis* has been isolated in pure culture from a number of infectious processes in compromised patients, particularly those with prosthetic devices.[12,13,22] Those species rarely isolated from patients are *S. auricularis, S. capitis, S. carnosus, S. lentus,* and *S. xylosus.* Coagulase-negative staphylococci isolated primarily or only from animal sources to date include *S. arlettae, S. caprae, S. caseolyticus, S. chromogenes, S. equorum, S. felis, S. gallinarum, S. hyicus* (coagulase-variable), *S. kloosii, S. lentus, S. muscae,* and *S. sciuri.*[12,16,22,28]

Morphologically similar to *S. aureus,* coagulase-negative staphylococci may possess many of the same virulence properties. Although they do not produce the number of extracellular products that *S. aureus* does, many species produce hemolysins and some seem to possess an antiphagocytic capsule. In addition, the production of slime may correlate with pathogenicity, particularly in foreign body–associated infection in which slime may mediate bacterial adherence.[6,9]

Presently all diagnostic methods for disease caused by coagulase-negative staphylococci require isolation of the bacteria.

IDENTIFICATION

The majority of coagulase-negative staphylococci isolated from clinically significant sources are *S. epidermidis*.[2] These isolations may merely reflect its prevalence as the most common strain on human skin. Since any of the other species can cause disease, coagulase-negative staphylococci cannot arbitrarily be reported as *S. epidermidis* without further biochemical tests. For those laboratories that do not identify this group to species, a report of "coagulase-negative staphylococci" is sufficient.

Isolates obtained from the urinary tract, usually from nonhospitalized patients, may be *S. saprophyticus*. *S. saprophyticus* is resistant to the antibiotic novobiocin, as are several other species rarely isolated from patients. The authors recommend that for coagulase-negative staphylococcal isolates from sources other than blood, body fluids, or other sterile sites, a *presumptive* identification of *S. saprophyticus* may be reported based on novobiocin resistance, as indicated by a zone of inhibition of less than 12 mm diameter surrounding a 5-μg novobiocin disk, in an agar disk diffusion test performed in the same way as the disk diffusion susceptibility test.[15] The 12 mm zone diameter has shown better discrimination than the previously recommended zone of 16 mm. A same-day broth disk elution novobiocin test has been described.[14] Phosphatase production can also be used to distinguish between *S. saprophyticus* (negative) and *S. epidermidis* (positive).[29]

Table 25.2 shows some of the biochemical reactions used to differentiate coagulase-negative staphylococci. Many other reactions may be used to identify these organisms, but the tests require specialized media and may be difficult to perform. Commercial identification kits, including API Staph-Ident and DMS Staph Trac (BioMerieux Vitek), ATB 32 Staph (API System, S.A.), American MicroScan Staph ID system (American Hospital Supply/Baxter), Sceptor (Becton-Dickinson Microbiology Systems), and Vitek GPI (Vitek Microbiology Systems), have been shown to identify coagulase-negative staphylococci with accuracies varying from 65% to 95%, depending on the species studied.[4,5,28] A simplified scheme for identification of some of these organisms using susceptibility to five antimicrobial agents, PYR test, adherence to glass, and synergistic hemolysis has been developed recently by Hebert and others.[15] Epidemiological studies for relatedness among coagulase-negative staphylococci are performed with increasing frequency. Biotyping and molecular methods, particularly restriction fragment length polymorphism, appear to be most useful.[5,27,28,32]

SUSCEPTIBILITY

Unlike *S. aureus*, coagulase-negative staphylococci, with one exception, do not have predictable patterns of susceptibility. Unlike the other coagulase-negative staphylococci, *S. saprophyticus* is usually susceptible to most antimicrobial agents with the exception of the urinary tract agent nalidixic acid (which is active only against gram-negative bacilli). Strains associated with nosocomial infections are likely to be multiply resistant. Infections caused by coagulase-negative staphylococci resistant to penicillinase-resistant penicillins in vitro, like those caused by *S. aureus* with the same resistance patterns, may fail to respond to cephalosporins and other β-lactams. Rapid chromogenic cephalosporin test results (which detect only the presence of β-lactamase) should not be reported for coagulase-negative strains, since most resistance is chromosomal and a negative β-lactamase result is likely to be misleading. Strains exhibiting oxacillin resistance (tested in the same manner as for *S. aureus*) should be reported as cephalosporin resistant (and resistant to other β-lactam agents such as imipenem and the β-lactam/β-lactamase inhibitor combinations) regardless of in vitro susceptibilities to cephalosporins.

STOMATOCOCCUS SPECIES

Several recent reports have focused on recovery of *S. mucilaginosus*, an encapsulated Micrococcaceae species that occurs as part of the normal oral flora and acts as an opportunistic pathogen, from clinically important sites. The organism has been recovered from compromised patients, particularly drug abusers, as an agent of endocarditis and septicemia.[8,25,31] Stomatococci morphologically (colony and microscopic appearance) resemble staphylococci but may adhere to the agar because of their capsule. They are known as "sticky

TABLE 25.2

SELECTED CHARACTERISTICS OF COAGULASE-NEGATIVE STAPHYLOCOCCUS SPECIES ISOLATED FROM HUMANS

SPECIES	PYR HYDROLYSIS	CLUMPING FACTOR	UREASE	ACID PRODUCED AEROBICALLY FROM		ZONE OF INHIBITION*		
				MANNITOL	SUCROSE	BACITRACIN (10 U) <11 mm	POLYMYXIN B (300 U) <10 mm	NOVOBIOCIN (5 µg) <12 mm
capitis†	−	−	−†	+	+d	−	−	−
cohnii‡	−	−	−‡	+/+w	−	+/−	−	+
epidermidis	−	−	+	−	+	−	+	−
haemolyticus	+	−	−	+/−	+	+/−	−	−
hominis	−	−	+	−	+d	−	−	−
hyicus	−	+/−	+/−	+/−	+	+	+	−
intermedius	+	+d	+	−	+	−	−	−
lugdunensis	+	−	+/−	−	+	−	+/−	−
saprophyticus	−	−	NT	+/−	−	−/+	−	+
schleiferi	+	+	+	−	−	−	−	−
simulans	+	−	−	+	+	−	−	−
warneri	+/−	−	+	+/−	+	−	−	−
xylosus		−	+	+/+w	+	−/+	−	+

+ = 90% strains positive; − = 90% strains negative; +/− = variable, results more often positive; −/+ = variable, results more often negative; w = weak; d = delayed; NT = not tested.
* Cotton swab is rinsed in McFarland 1.0 suspension in trypticase soy broth, pressed against side of tube to remove excess moisture, and streaked onto trypticase soy agar in one direction. After agar has dried, disks are placed on surface of agar. Plates are incubated at 35° C for 24 hours.
† *S. capitis* subsp. *ureolyticus* is urease positive.
‡ *S. cohnii* subsp. *ureolyticum* is urease positive.

PROCEDURE 25.2

BACITRACIN AND FURAZOLIDONE SUSCEPTIBILITY

PRINCIPLE

All coagulase-negative staphylococci are resistant to 0.04 U bacitracin, whereas micrococci and stomatococci are susceptible. Staphylococci and *Stomatococcus mucilaginosus* are susceptible to furazolidone (100 µg disk), whereas *Micrococcus* spp. are resistant.[15] The disks are available from commercial suppliers (Becton-Dickinson; Difco; others).

METHOD

1. Growth from the colony to be tested is inoculated into 1 ml of trypticase soy broth to a turbidity equal to that of a McFarland 0.5 standard.

2. Using a cotton swab saturated with the test suspension, the microbiologist must streak a small (90 mm diameter), sheep blood–supplemented Mueller-Hinton or trypticase soy agar plate in three directions, exactly as if preparing the inoculum for agar disk diffusion susceptibility testing.

3. Place a 0.04 U bacitracin-impregnated filter paper disk (Taxo A, BBL Microbiology Systems, or Bacto Group A Differential Disk, Difco) and a 100 µg furazolidone-impregnated filter paper disk (BBL) on the surface of the plate.

4. Incubate the plate at 35° C in air for 18 hours and observe for zones of inhibition around the disk. If an organism fails to grow, it is probably a micrococcus.

QUALITY CONTROL

S. epidermidis ATCC 14990 and a *Micrococcus luteus* strain should be tested as above.

EXPECTED RESULTS

The micrococcus should yield a zone of inhibition around the bacitracin disk, usually greater than 7 mm, whereas the staphylococcus should grow up to the disk. Some *S. aureus* may also be susceptible, so only coagulase-negative strains should be tested by bacitracin. The *M. luteus* should show no zone of inhibition around the furazolidone disk, but the staphylococcus will show a zone of inhibition greater than 6 mm. Any zone determines that the organism being tested is not a *Micrococcus* sp.

PERFORMANCE SCHEDULE

Each new lot of bacitracin and furazolidone disks should be tested, and control organisms should be run with each test.

Modified from Falk, D. and Guering, S.J. 1983. J. Clin. Microbiol. and Hebert, G.A., et al. 1988. J. Clin. Microbiol.

TABLE 25.3

DIFFERENCES BETWEEN *STOMATOCOCCUS*, *STAPHYLOCOCCUS*, AND *MICROCOCCUS* GENERA

GENUS	CATALASE	MODIFIED OXIDASE	RESISTANT TO FURAZOLIDONE (100 µg disk)	RESISTANT TO BACITRACIN (0.04 U disk)
Micrococcus	+	+	+	−
Staphylococcus	+	−	−	+
Stomatococcus	−	−	−	−

staph" in some laboratories. They are catalase negative or weakly positive because of pseudo-catalase. Inability of stomatococci to grow in media containing 5% NaCl and poor growth on Mueller-Hinton agar further distinguish them from all other Micrococcaceae (Table 25.3).[8,25,31] Stomatococci are able to hydrolyze esculin but do not grow on the bile-esculin agar used for entero-cocci. Most isolates are susceptible to erythro-mycin, and all are susceptible to vancomycin; vancomycin is the drug of choice for initial ther-apy of systemic infections, although susceptibility tests should be performed.[25] Ciprofloxacin resis-tance may contribute to environmental selection

PROCEDURE 25.3

MODIFIED OXIDASE TEST

PRINCIPLE

The cytochrome oxidase system of micrococci (and the rare *Staphylococcus caseolyticus, sciuri,* and *lentus*) contains cytochrome C, which yields a colored end product in the presence of the modified oxidase reagent. Disks impregnated with this reagent are available commercially (Remel, Lenexa, Kan.).

METHOD

1. A visible amount of a colony from a 15- to 24-hour-old 5% sheep blood agar plate is smeared onto a piece of filter paper as for the standard oxidase test (Chapter 10). Colonies from other media or colonies older than 24 hours may give aberrant results.

2. A drop of modified oxidase reagent (6% tetramethylphenylene diamine hydrochloride [Eastman Kodak] in dimethyl sulfoxide [DMSO, Sigma Chemical]) is added to the bacteria on the filter paper.

3. Oxidase-positive organisms turn dark blue within 2 minutes. All oxidase-positive organisms are micrococci, with the exception of rarely seen *S. caseolyticus, sciuri,* and *lentus.*

QUALITY CONTROL

Test *Staphylococcus epidermidis* ATCC 14990 and a *Micrococcus luteus* strain as described previously.

EXPECTED RESULTS

The micrococcus should yield a blue color on the filter paper within 2 minutes; the staphylococcus spot should remain colorless.

PERFORMANCE SCHEDULE

Test the reagent with each new batch made and weekly thereafter.

Modified from Faller A. and Schleifer K. H. 1981. J. Clin. Microbiol.[11]

---

of this commensal oral microbe as a pathogen in immunocompromised patients.[25]

MICROCOCCI

Rarely identified as causes of infection, micrococci are saprophytic, often pigmented members of the Micrococcaceae. Micrococci become pathogens during accidental introduction into a susceptible host. They may cause endocarditis and other diverse syndromes. It appears that they are less virulent than the staphylococci.

Species include *M. agilis* (may be motile), *M. kristinae, M. luteus, M. lylae, M. nishinomiyaensis, M. roseus, M. sedentarius,* and *M. varians.* Hospital microbiology laboratories should not attempt identification of species of these bacteria. Isolates of micrococci that are unequivocally implicated as causes of infection should be sent to a reference laboratory for further studies.

DIFFERENTIATION AMONG COAGULASE-NEGATIVE STAPHYLOCOCCI, MICROCOCCI, AND *STOMATOCOCCUS MUCILAGINOSUS*

Catalase-positive, coagulase-negative organisms may be either staphylococci or micrococci. Catalase-negative organisms that resemble staphylo-

cocci by Gram stain and colony morphology may be *S. mucilaginosus,* described previously. The micrococci are not lysed by lysostaphin (an enzyme produced by certain staphylococci), they are resistant to the antibiotic furazolidone (Procedure 25.2), susceptible to 0.04 units of bacitracin (Procedure 25.2), and contain cytochrome C, as determined by a positive modified oxidase test (Procedure 25.3).[10,11,15] Differentiation of coagulase-negative staphylococci from micrococci (considered to be less virulent) is outlined in Table 25.3.[10,11,15,22] Using a combination of these tests, microbiologists should easily be able to differentiate staphylococci and micrococci likely to be found in clinical material.

REFERENCES

1. Abramson, C., Bergdoll, M.S., and Wheat, L.J. 1987. Immunoserology of staphylococcal disease. In Abramson, C., editor. Cumitech 22, American Society for Microbiology, Washington, D.C.
2. Archer, G.L. 1984. *Staphylococcus epidermidis:* the organism, its diseases, and treatment. In Remington, J.S., and Swartz, M.N., editors. Current clinical topics in infectious diseases, ed 5, McGraw-Hill, New York.
3. Baron, E.J. 1992. The detection, significance, and rationale for control of methicillin-resistant *Staphylococcus aureus.* Clin. Microbiol. Newsletter 14:129.
4. Brun, Y., Bes, M., Boeufgras, J.M., et al. 1990.

International collaborative evaluation of the ATB 32 Staph gallery for identification of the *Staphylococcus* species. Zentralbl. Bakteriol. 273:319.

5. Chesneau, O., Aubert, S., Morvan, A., et al. 1992. Usefulness of the ID32 Staph System and a method based on rRNA gene restriction site polymorphism analysis for species and subspecies identification of staphylococcal clinical isolates. J. Clin. Microbiol. 30:2346.

6. Christensen, G.D., Parisi, J.T., Bisno, A.L., et al. 1983. Characterization of clinically significant strains of coagulase negative staphylococci. J. Clin. Microbiol. 18:258.

7. Christensson, B., Boutonnier, A., Ryding, U., and Fournier, J.-M. 1990. Diagnosing *Staphylococcus aureus* endocarditis by detecting antibodies against *S. aureus* capsular polysaccharide types 5 and 8. J. Infect. Dis. 163:530.

8. Coudron, P.E., Markowitz, S.M., Mohanty, L.B., et al. 1987. Isolation of *Stomatococcus mucilaginosus* from drug user with endocarditis. J. Clin. Microbiol. 25:1359.

9. Davenport, D.S., Massanari, R.M., Pfaller, M.A., et al. 1986. Usefulness of a test for slime production as a marker for clinically significant infections with coagulase-negative staphylococci. J. Infect. Dis. 153:332.

10. Falk, D., and Guering, S.J. 1983. Differentiation of *Staphylococcus* and *Micrococcus* spp. with the Taxo A bacitracin disk. J. Clin. Microbiol. 18:719.

11. Faller, A., and Schleifer, K.H. 1981. Modified oxidase and benzidine tests for separation of staphylococci and micrococci. J. Clin. Microbiol. 13:1031.

12. Freney, J., Brun, Y., Bes, M., et al. 1988. *Staphylococcus lugdunensis* sp. nov. and *Staphylococcus schleiferi* sp. nov., two species from human clinical specimens. Int. J. System. Bacteriol. 38:168.

13. Glenn, D., Hinnebusch, C., and Colonna, P. 1992. Isolation and identification of an uncommon pathogen, *Staphylococcus lugdunensis,* from a femoral graft. Abstr. General Meeting, Amer. Soc. Microbiol, C-486, p. 498. American Society for Microbiology, Washington, D.C.

14. Harrington, B.J., and Gaydos, J.M. 1984. Five-hour novobiocin test for differentiation of coagulase negative staphylococci. J. Clin. Microbiol. 19:279.

15. Hebert, G.A., Crowder, C.G., Hancock, G.A., et al. 1988. Characteristics of coagulase-negative staphylococci that help differentiate these species and other members of the family Micrococcaceae. J. Clin. Microbiol. 26:1939.

16. Herchline, T.E., and Ayers, L.W. 1991. Occurrence of *Staphylococcus lugdunensis* in consecutive clinical cultures and relationships of isolation and infection. J. Clin. Microbiol. 29:419.

17. Hinnebusch, C., Glenn, D., and Bruckner, D.A. 1992. Potential misidentification of *Staphylococcus* species when using rapid identification tests that detect clumping factor. Abstr. General Meeting, Amer. Soc. Microbiol., C-485, p. 498. American Society for Microbiology, Washington, D.C.

18. Jacob, E., Durham, L.C., Falk, M.C., et al. 1987. Antibody response to teichoic acid and peptidoglycan in *Staphylococcus aureus* osteomyelitis. J. Clin. Microbiol. 25:122.

19. Jolly, J. and Goldberg, M. 1989. Methicillin resistance in staphylococci: an evaluation of conditions for detection. Med. Lab. Sci. 46:2.

20. Jorgensen, J.H. 1991. Mechanisms of methicillin resistance in *Staphylococcus aureus* and the methods for laboratory detection. Infect. Control Hosp. Epidemiol. 12:14.

21. Kamath, U., Singer, C., and Isenberg, H.D. 1992. Clinical significance of *Staphylococcus warneri* bacteremia. J. Clin. Microbiol. 30:261.

22. Kloos, W.E. and Lambe, D.W., Jr. 1991. *Staphylococcus.* pp. 222-237. In Balows, A., Hausler, W.J., Jr., Herrmann, K.L., et al., editors. Manual of clinical microbiology, ed 5. American Society for Microbiology, Washington, D.C.

23. Lairscey, R., and Buck, G.E. 1987. Performance of four slide agglutination methods for identification of *Staphylococcus aureus* when testing methicillin resistant staphylococci. J. Clin. Microbiol. 25:181.

24. Madison, B.M., and Baselski, V.S. 1983. Rapid identification of *Staphylococcus aureus* in blood cultures by thermonuclease testing. J. Clin. Microbiol. 18:722.

25. McWhinney, P.H.M., Kibbler, C.C., Gillespie, S.H., et al. 1992. *Stomatococcus mucilaginosus:* an emerging pathogen in neutropenic patients. Clin. Infect. Dis. 14:641.

26. Monzon-Moreno, C., Aubert, S., Morvan, A., and El Solh, N. 1991. Usefulness of three probes in typing isolates of methicillin-resistant *Staphylococcus aureus* (MRSA). J. Med. Microbiol. 35:80.

27. Pennington, T.H., Harker, C., and Thomson-Carter, F. 1991. Identification of coagulase-negative staphylococci by using sodium dodecyl sulfate-polyacrylamide gel electrophoresis and rRNA restriction patterns. J. Clin. Microbiol. 29:390.

28. Pfaller, M.A., and Herwaldi, L.A. 1988. Laboratory, clinical, and epidemiological aspects of coagulase-negative staphylococci. Clin. Microbiol. Rev. 1:281.

29. Pickett, D.A., and Welch, D.F. 1985. Recognition of *Staphylococcus saprophyticus* in urine cultures by screening colonies for production of phosphatase. J. Clin. Microbiol. 21:310.

30. Raad, I.I., and Sabbagh, M.F. 1992. Optimal duration of therapy for catheter-related *Staphylococcus aureus* bacteremia: a study of 55 cases and review. Clin. Infect. Dis. 14:75.

31. Relman, D.A., Ruoff, K., and Ferraro, M.J. 1987. *Stomatococcus mucilaginosus* endocarditis in an intravenous drug abuser. J. Infect. Dis. 155:1080.

32. Roberson, J.R., Fox, L.K., Hancock, D.D., and Besser, T.E. 1992. Evaluation of methods for differentiation of coagulase-positive staphylococci. J. Clin. Microbiol. 30:3217.

33. Sheagren, J.N. 1984. *Staphylococcus aureus,* the persistent pathogen. N. Engl. J. Med. 309:1368.

34. Talan, D.A., Staatz, D., Staatz, A., et al. 1989. *Staphylococcus intermedius* in canine gingiva and canine-inflicted human wound infections: laboratory characterization of a newly recognized zoonotic pathogen. J. Clin. Microbiol. 27:78.

35. Todd, J.K. 1988. Toxic shock syndrome. Clin. Microbiol. Rev. 1:432.

36. Verbrugh, H.A., Nelson, R.D., Peterson, P.K. et al. 1983. Serology of *Staphylococcus aureus* infections using multiple antigens and serial serum samples. J. Infect. Dis. 148:608.

37. Waldvogel, F.A. 1990. *Staphylococcus aureus* (including

toxic shock syndrome). pp. 1489-1510. In Mandell, G.L., Douglas, R.G., Jr., and Bennett, J.E., editors. Principles and practices of infectious diseases, ed 3. Churchill Livingstone, New York.

37a. Wenzel, R.P., Nettleman, M.D., Jones, R.N., and Pfaller, M.A. 1991. Methicillin-resistant *Staphylococcus aureus:* implications for the 1990s and effective control measures. Am. J. Med. 91 (suppl. 3B):221S.

38. Wheat, J., Kohler, R.B., Garten, M., and White, A. 1984. Commercially available (ENDO-STAPH) assay for teichoic acid antibodies. Arch. Intern. Med. 144:261.

39. Zarzour, J.Y., and Belle, E.A. 1978. Evaluation of three test procedures for identification of *Staphylococcus aureus* from clinical sources. J. Clin. Microbiol. 7:133.

40. Zumia, A. 1992. Superantigens, T cells, and microbes. Clin. Infect. Dis. 15:313.

BIBLIOGRAPHY

Hershow, R.C., Khayr, W.F., and Smith, N.L. 1992. A comparison of clinical virulence of nosocomially acquired methicillin-resistant and methicillin-sensitive *Staphylococcus aureus.* Infect. Control Hosp. Epidemiol. 13:587.

Hindler, J.A., and Inderlied, C.B. 1985. Effect of the source of Mueller-Hinton agar and resistance frequency on the detection of methicillin-resistant *Staphylococcus aureus.* J. Clin. Microbiol. 21:205.

Panlilio, A.L., Culver, D.H., Gaynes, R.P., et al. 1992. Methicillin-resistant *Staphylococcus aureus* in U.S. hospitals, 1975-1991. Infect. Control Hosp. Epidemiol. 13:582.

Scand. J. Infect. Dis. Suppl. 41, 1983. Entire issue devoted to *S. aureus* organisms and disease.

26

STREPTOCOCCI AND RELATED GENERA

Members of the genus *Streptococcus* are catalase-negative gram-positive cocci that tend to grow in chains in liquid media. Streptococci form large quantities of lactic acid as the end product of carbohydrate metabolism, and they are facultatively anaerobic. They are not only normal flora of humans and animals, but some species also are the etiological agents of several devastating diseases, including pneumococcal pneumonia, meningitis, sepsis, bacterial endocarditis, streptococcal exudative pharyngitis, cellulitis, wound infection, and visceral abscesses. Acute rheumatic fever and poststreptococcal glomerulonephritis are secondary sequelae of infection caused by *Streptococcus pyogenes* and occasional other species. *Enterococcus* sp., also normal animal mucous membrane flora, are the agents of urinary tract and wound infections and are important in bacterial endocarditis, as well as being important secondary pathogens in patients receiving antibiotics. Other related genera have recently been implicated in human disease. *Leuconostoc* sp. have been reported as the etiological agents of septicemia and meningitis.[13,25] *Lactococcus garviae* and *Pediococcus* sp. have been recovered from patients with sepsis.[9,35] Vancomycin resistance, characteristic of *Leuconostoc* and *Pediococcus* sp., may be important in the pathogenesis of

disease associated with these organisms. *Aerococcus viridans,* the single species, is an opportunistic pathogen, probably occurring naturally in the environment. It has been isolated from patients with bacteremia, endocarditis, meningitis, and osteomyelitis. Both *Aerococcus* and *Gemella* sp. resemble streptococci on gross inspection of colonies, although in Gram stains made from cultures grown in broth, they tend to form packets and tetrads and no chains. The genus *Gemella,* a member of the normal upper respiratory flora, is involved in endocarditis, wounds, and abscesses.

The streptococci and related species can be identified in the laboratory by several different processes. The methods most often used in clinical microbiology laboratories involve preliminary grouping of isolates based on the morphology and hemolysis of colonies growing on blood agar. In the United States, 5% sheep blood agar is usually used. Originally used primarily for cultures of inflamed throats because it inhibits growth of hemolytic *Haemophilus* sp., whose presence would complicate the recognition of hemolytic streptococci, 5% sheep blood agar is now used routinely for initial cultivation of almost all types of clinical specimens. Horse blood, which exhibits similar hemolytic reactions, is typically used for initial processing in Europe. In discussions about epidemiology and pathogenesis in this chapter, clinically important streptococci are grouped according to hemolytic patterns or species. Beta-hemolytic streptococci include *S. pyogenes, S. agalactiae,* and several other species. Alpha-hemolytic streptococci include *S. pneumoniae* and the viridans group of species. *Enterococcus* sp., now recognized as a separate genus, are primarily alpha hemolytic or nonhemolytic (and occasionally beta hemolytic). These organisms differ from the streptococci by their resistance to salt and broader antibiotic resistance.

The second major criterion used for identification is the carbohydrate or other antigen group present in the cell wall. These antigens are defined serologically, as originally described by Rebecca Lancefield. Methods for determining serologic groups are discussed later.

EPIDEMIOLOGY AND PATHOGENIC MECHANISMS

STREPTOCOCCUS PNEUMONIAE

S. pneumoniae, a bile-sensitive alpha-hemolytic streptococcus, is found as normal nasopharyngeal and oropharyngeal flora in as many as 15% of children and approximately 5% of adults. It is also found in the upper respiratory tract of animals, but not in the inanimate environment. Organisms are presumably passed from person to person via respiratory secretions and aerosols. The organisms are surrounded by an antiphagocytic capsule composed of polysaccharide antigens useful in strain typing; more than 80 types have been described. *S. pneumoniae,* also called the "pneumococcus," is the second most common cause of bacterial meningitis, often preceded by pneumonia. Pneumococcal pneumonia is the most common type of community-acquired bacterial pneumonia; the fatality rate of uncomplicated pneumococcal pneumonia is still relatively high (5% to 7%), even with prompt institution of appropriate therapy. In addition to pneumonia and meningitis, *S. pneumoniae* is an etiological agent of otitis media, purulent sinusitis, and, occasionally, peritonitis, especially in young patients with nephrotic syndrome.

Pneumococci cause disease in the presence of a predisposing host condition, often a preceding viral respiratory tract infection. Individuals with other compromising conditions of the respiratory tract, such as chronic obstructive pulmonary disease, silicosis, anthracosis, treatment with anesthetic agents, and alcoholism, are at increased risk. The pathogenesis of pneumonia is discussed in Chapter 17. For *S. pneumoniae,* the capsular polysaccharide plays a key role in allowing the establishment of infection. The organisms enter the alveoli via the bronchial tree in the face of impaired host defenses that normally include the cough reflex, ciliary movement, and secretory immunoglobulin. Once within alveoli, they resist phagocytosis by macrophages, multiply, and induce an inflammatory response that impairs lung function. Patients usually experience an abrupt onset of fever after a shaking chill. Sputum is often purulent and may appear rusty colored.

The best evidence that capsular polysaccharide is the primary virulence factor is that antibody to the capsular antigen is protective. Recovery from pneumococcal pneumonia in a nonimmune individual is accompanied by a rise in capsular antibody titer, whereas presence of circulating antibodies prevents establishment of the disease. This fact has spurred development of a pneumococcal vaccine, containing antigens of 23 serotypes of capsular polysaccharide, the most frequently recognized etiological serotypes in the United States and Europe. Patients with chronic respiratory tract disease, those without spleens, immunosuppressed patients, elderly patients, and others deemed to be at increased risk have all been cited as those for whom vaccine is warranted, although not all will respond optimally to it. *S. pneumoniae* produces other factors that may play a role in virulence, including pneumolysin O, an oxygen-sensitive toxin that is cytolytic for cells, and a neuraminidase, an enzyme that degrades surface structures of host tissue. The cell wall of pneumococci contains C substance, a teichoic acid that reacts with a certain serum protein (C-reactive protein, CRP), resulting in the activation of some nonspecific host immune responses.

STREPTOCOCCUS PYOGENES

Most streptococci that contain cell wall antigens of Lancefield group A (discussed later) are known as *S. pyogenes.* Members of this species are almost always beta hemolytic. They are found in the respiratory tract of humans and are always considered to be potential pathogens. Some persons, particularly children, however, do carry the organism without signs of illness. The numbers of organisms isolated from such asymptomatic carriers are usually low. Because the number of colony forming units (CFUs) seen on primary culture plates depends on many factors, including the presence of competing normal oral flora, the skill with which the specimen was obtained, the way in which the culture medium was inoculated, culture medium incubation conditions, and the culture medium itself, the likely presence or absence of streptococcal infection should not be assessed entirely on the numbers of colonies recovered. Infection is spread by aerosols and respiratory secretions, although foodborne and milkborne epidemics do occur. Members of this species are among the etiological agents of pharyngitis, cellulitis, scarlet fever, erysipelas, **pyoderma** (a purulent skin infection), puerperal fever (infrequently seen today), and other purulent (or suppurative) infections. *S. pyogenes* may also be involved alone or with other bacteria, particularly *Staphylococcus aureus,* in impetigo (a form of pyoderma). Infections that extend from the pharynx into the paranasal sinuses, tonsils, and other parts of the respiratory tract can lead to abscesses, pneumonia, otitis media, and other suppurative processes. A recent upsurge in serious invasive group A streptococcal disease has worried many clinicians.[40] With the decline of rheumatic fever in the United States, treatment of streptococcal pharyngitis is often initiated for prevention of these other sequelae.

In terms of human morbidity and mortality worldwide, however, the role of *S. pyogenes* in the subsequent development of acute rheumatic fever and poststreptococcal glomerulonephritis is more important. Both diseases are most prevalent among children. Rheumatic fever and subsequent valvular heart disease are problems of major importance in developing nations throughout the world. In the United States, rheumatic fever occurs infrequently, although mini-epidemics were reported in several areas, including Utah, Pennsylvania, and Ohio, beginning in 1985. Acute rheumatic fever and poststreptococcal glomerulonephritis are considered to be "nonsuppurative," since the organism itself and a purulent inflammatory response are not present in the affected organs (heart, joints, blood vessels, kidneys). Although the pathogenesis of these diseases has not been entirely clarified, it seems certain that they are autoimmune phenomena. It is believed that cross-reactive antibodies, originally directed against streptococcal cell membranes, specifically bind to myosin in human heart muscle cells; other cross-reactive antibodies bind to components of the glomerular basement membrane, forming immune complexes at the affected site. These antigen-antibody complexes attract host reactive cells and enzymes that ultimately cause the damage. Patients with previous rheumatic heart disease (as a result of rheumatic

fever) are at significantly increased risk of developing cardiac malfunction and endocarditis at a later time. Patients who develop streptococcal glomerulonephritis are also at risk of developing later renal failure.

Not all infections with *S. pyogenes* lead to nonsuppurative sequelae. Acute rheumatic fever occurs only after upper respiratory tract infection. The development of this disease depends on host factors, such as histocompatibility-linked antigen (HLA; chromosome constituent) type, immunoglobulin secretory status, and immune responsiveness, and organism factors, such as adherence mechanisms and cell wall carbohydrate components. Glomerulonephritis, on the other hand, occurs after pharyngitis or after suppurative skin infection (pyoderma). Acute glomerulonephritis is more often associated with a limited number of serotypes. These serotypes are defined by antisera against a protein component of the cell wall, the M protein, which is also mentioned next in connection with virulence.

S. pyogenes produces various extracellular products and toxins that probably enhance virulence (see Figure 13.1). Erythrogenic toxin, elaborated by scarlet fever–associated strains, is responsible for the characteristic rash. Group A streptococci also produce deoxyribonuclease (DNase); hyaluronidase (also known as spreading factor), which breaks down host cell connective tissue; and streptokinase, an enzyme that dissolves clots. NADase, proteinases, and other enzymes are secreted. Mucoid strains possess a large hyaluronic acid capsule that acts to inhibit phagocytosis. The two hemolysins, streptolysin O, an oxygen-labile enzyme, and streptolysin S, oxygen stable, can lyse human and other erythrocytes, as well as the cell membranes of polymorphonuclear neutrophils, platelets, and other cells. However, pathogenic strains of *S. pyogenes* exist that do not produce hemolysins, which makes them extremely difficult to recognize on primary blood agar culture plates. Pyrogenic exotoxins seem to induce fever and encourage development of acute respiratory distress syndrome, renal failure, and tissue necrosis.[40] The cell wall lipoteichoic acids serve to effect adherence of the organism to host cell respiratory epithelium, an essential first step in the development of infection (Chapter 17). In association with the hyaluronic acid capsule, cell wall M protein serves to prevent phagocytosis, another essential virulence mechanism. The M protein is the major virulence factor of the bacteria, since strains of *S. pyogenes* that lack M protein cannot cause disease. New theories suggest that streptococcal pyrogenic exotoxins and some M antigens can act as activators of T cells and systemic immunomodulators (known as **superantigens** when acting in this manner). The development of shock is likely to be associated with these streptococcal products.

STREPTOCOCCUS AGALACTIAE

These beta-hemolytic streptococci possess Lancefield group B capsular polysaccharide antigens and are normal flora in the genitourinary tract of humans and other mammals. They are the only streptococci in which specific physiological characteristics (ability to hydrolyze hippurate; positive CAMP test, Procedure 26.1) are associated with a single Lancefield group. Group B streptococci are important etiological agents of bovine mastitis, from which they can be transmitted in milk. They are destroyed by pasteurization. From their carriage site in the human vagina, they can colonize neonates; *S. agalactiae* is currently the most common etiological agent of neonatal sepsis and meningitis, accounting for more than one third of all cases (*Escherichia coli* still accounts for another third).[19,30] Postpartum fever and sepsis, not surprisingly, are also frequently associated with *S. agalactiae*. This organism has been found to be the etiological agent in cases of sepsis in nonparturient women and in men and in joint infection, osteomyelitis, urinary tract infection, and wound infection. *S. agalactiae* is associated with endocarditis, pneumonia, and pyelonephritis in immunosuppressed patients.

OTHER BETA-HEMOLYTIC STREPTOCOCCI

Non–group B beta-hemolytic streptococci can be divided into groups based initially on colony size. Small or "minute" colony types (<1 mm diameter) have been placed into the "*Streptococcus milleri*" group, which also includes nonhemolytic, regular-sized–colony forming streptococci of the species *S. constellatus*, *S. intermedius*, and *S. anginosus*.[7,44] These organisms were briefly grouped in one species, *S. anginosus* (1987-1991), but have since been reseparated.[2,10,43-45] All group F streptococci appear to be either *S. constellatus* or *S. anginosus*. The "*milleri*" group are involved in the production of urogenital tract infections, abscesses, and other purulent infections in soft tissues and internal organs. *S. intermedius* is associated particularly with brain and liver abscesses.[17] Production of hyaluronidase, DNase, and other extracellular enzymes may contribute to their pathogenicity.[7,17] "*S. milleri*" sp. recovered from throats do not possess human immunoglobulin receptors

and are not thought to be associated with pharyngitis.[6,27] Large colony forming, beta-hemolytic streptococci of serologic group C (*S. equisimilis, S. zooepidemicus, S. equi*) may cause a severe pharyngitis, often followed by bacteremia and metastatic infection.[4,41] One species of group C (*S. zooepidemicus*) has been associated with poststreptococcal glomerulonephritis, acquired after drinking milk from a cow with mastitis.[1] Certain group C strains recovered from throats were found to possess human antibody receptors, thought to be a virulence factor for these organisms.[27] Group C and G streptococci may cause suppurative infection in humans, including postpartum sepsis, erysipelas, pharyngitis,[6] and bacteremia. Endocarditis is rare. Beta-hemolytic, large colony forming group G streptococci isolated from human wound infections are related to group L and group C strains.[8] They are included in the species *S. dysgalactiae* by some authors. Species of groups C and G may also colonize the human gastrointestinal tract and vagina. Acute glomerulonephritis in a human caused by a group G streptococcus has been reported.[16] Group G animal strains, now called *S. canis*, are pathogens in dogs and cows.[8]

Large colony forming, beta-hemolytic strains not containing group antigens are rarely etiological agents of human disease. Those possessing group antigens other than A, B, C, and G have been isolated only from animals or environmental sources. They are carried as normal flora of the pharynx, vagina, or skin of wild and domestic animals.

VIRIDANS AND NONENTEROCOCCAL GROUP D STREPTOCOCCI

Species of streptococci that are not beta hemolytic (with rare exceptions), do not possess Lancefield group B or D cell wall antigens, are not *S. pneumoniae* (which are bile soluble and inhibited by the antipneumococcal agent optochin), and cannot grow in broth containing 6.5% sodium chloride are considered to belong to the general category of "viridans streptococci." Viridans was derived from the Latin word *viridis*, meaning "green." Many species in this group are alpha hemolytic and produce a "green" discoloration in blood agar, probably from the production of hydrogen peroxide. Numerous characteristics of this group of streptococcal species are variable, including cell wall carbohydrate group antigens and biochemical and morphological characteristics. For this reason, one should not test for serogroups other than B and D (to rule out non–beta-hemolytic members of these groups) or F (to identify alpha-hemolytic or nonhemolytic "*S. milleri*" group). Many streptococci in the viridans group are able to grow better under anaerobic or **microaerobic** conditions than when incubated in air; thus they are also called *microaerobic streptococci*. We recommend that isolates be identified presumptively only when deemed to be important clinically, such as isolates from blood or abscesses. Coykendall[7] reviewed the current taxonomy of this group of streptococci, but his designations are already being replaced based on new data.[2,26]

The viridans streptococci are normal inhabitants of the oral, respiratory, and gastrointestinal mucosa of humans and animals. Most are opportunistic pathogens and have generally been thought to be of low virulence. Viridans streptococci, however, are the major etiological agents of bacterial endocarditis in the United States. Patients who develop streptococcal endocarditis usually possess a damaged heart valve (from previous rheumatic fever or other cause). Gingival disease or dental manipulations, including dental prophylaxis, are often predisposing factors in the development of endocarditis.

Viridans streptococci are able to adhere to epithelial and endothelial cells, and adherence is probably a key factor in their ability to cause disease. *Streptococcus mutans* has been definitively established as a major etiological agent of dental caries in addition to being a cause of endocarditis. Extracellular polysaccharides, called dextrans, serve as attachment mediators for tooth surfaces as well as heart valves. The viridans streptococci are also associated (often in combination with anaerobic or other bacterial species) with brain abscess, perioral abscess, aspiration pneumonia, liver abscess, and other suppurative infections. Some species of viridans streptococci produce extracellular enzymes and toxins similar to those produced by beta-hemolytic streptococci. The role of these products in virulence is not known. In all series, *S. sanguis*, *S. mitis*, and *S. mutans* are the most common etiological agents of endocarditis. Since the organisms are normal oral flora, isolation of viridans streptococci from a single blood culture may not be indicative of an ongoing infectious process, although it usually is. Cultivation of the same organism from more than one blood culture is highly suggestive of endocarditis, abscess, or other serious infection.

One currently accepted species of streptococcus isolated from humans possesses group D cell wall teichoic acid antigens and is unable to grow in media with a high salt concentration. *S. bovis*, which could be renamed *S. inulinaceus* based on taxonomic studies, may be responsible for as many as one third of all cases of viridans streptococcal endocarditis. Human strains of *S. equinus* have been found to be vancomycin resistant and were moved to the genus *Pediococcus*. Other *S. bovis* strains, which include some of those formerly called *S. equinus*, are isolated only from cows. Since clinicians are familiar with the implication of isolating *S. bovis* from blood, the name should be retained. The presence of *S. bovis* bacteremia, associated with endocarditis or not, is almost always indicative of a loss of integrity of the gastrointestinal mucosa. A strong correlation exists between isolation of *S. bovis* from blood and colon cancer. Detection of this organism may be the first evidence of such a malignancy, an impelling reason for identification of streptococci (at least those isolated from blood) to species.[38]

NUTRITIONALLY VARIANT STREPTOCOCCI

A group of streptococci, first described in the early 1960s, has been found to be significant etiological agents of endocarditis.[36] These organisms, called "nutritionally deficient," "pyridoxal dependent," "satelliting streptococci," or "thiol requiring," were thought to be variants of conventional viridans species. They are now known to be separate species, *S. defectivus* and *S. adjacens*.[3] Nutritionally variant streptococci require vitamin B_6, or pyridoxal, for growth. Consequently, standard 5% sheep blood agar plates do not support growth of these organisms. Commercial lots of chocolate agar vary with respect to their ability to support growth.[33] Subcultures from blood culture broth media, in which human erythrocytes provide the necessary growth factor, must be made to plates known to support growth of such organisms. It was shown that addition of pyridoxal to all blood agar plates would inhibit growth of other frequently encountered bacteria.[34] Alternatively, subcultures can be made to a blood agar plate cross-streaked with an *S. aureus* strain (Figure 26.1). *S. aureus* provides the necessary growth factor that allows development of satelliting colonies of the streptococci in a zone surrounding the staphylococcus. For this reason, the nutritionally variant streptococci have also been called the *satelliting streptococci*.

ENTEROCOCCI

Enterococci possess group D teichoic acid antigen, are able to grow in 6.5% sodium chloride and in 40% bile, and exhibit resistance to some beta-lactam agents, such as penicillin G. They are part

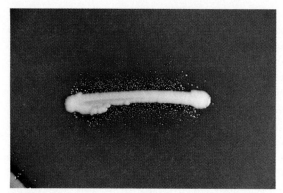

FIGURE 26.1

Nutritionally variant streptococci growing as satelliting colonies surrounding growth of *Staphylococcus aureus.*

of the normal fecal flora of all warm-blooded animals, including humans.[28] *Enterococcus faecalis* and *E. faecium,* the most common human isolates among the enterococci, are encountered as etiological agents of urinary tract infections, wound infections, and intraabdominal abscesses. They must possess some virulence factors, but these have not been characterized. Superinfection with enterococci, including bacteremia, is a relatively common occurrence in patients being treated with a third-generation cephalosporin, since they are resistant to these agents as well as to penicillin. Enterococci are the third most common etiological agents of nosocomial infections in the United States. Other enterococci, *E. dispar, E. durans, E. flavescens, E. gallinarum, E. avium, E. casseliflavus, E. malodoratus, E. mundtii, E. raffinosus, E. pseudoavium, E. solitarius,* and *E. hirae,* have been isolated less frequently from clinically significant human infections.[12]

RELATED GENERA

Leuconostoc species are vancomycin-resistant gram-positive cocci that resemble streptococci morphologically and biochemically, making recognition difficult. They have been isolated from sterile body fluids in patients with septicemia and meningitis and from wound and urinary tract infections.[25] Another vancomycin-resistant genus, *Pediococcus,* is normal fecal flora but is also involved in human infection.[35] Recently, streptococci of Lancefield group N have been transferred to the new genus *Lactococcus.* At least one species, *L. garviae,* has been implicated in human disease. The lactococci are susceptible to vancomycin; are nonmotile, catalase negative, PYR negative (except for *L. garviae*), and alpha hemolytic or nonhemolytic; and otherwise resemble viridans streptococci or, in the case of PYR-positive strains, enterococci (see later discussion of PYR test).

Catalase-negative *Aerococcus* and *Gemella,* whose colonies may be mistaken for streptococci, resemble gram-positive cocci in tetrads and clusters rather than in chains. Since most of these groups have been described recently, their virulence mechanisms are not known. Most are isolated from blood, but they have been implicated in urinary tract infections and wound infections.[37] As microbiologists continue to characterize fully streptococcus-like organisms isolated from infectious processes more often, the extent of their involvement will be better understood.

LABORATORY IDENTIFICATION OF STREPTOCOCCI

Streptococci grow on standard laboratory media containing blood or blood products, such as trypticase soy agar with 5% sheep blood and chocolate agar, as well as the selective agars Columbia agar with colistin and nalidixic acid (CNA) and phenylethyl alcohol agar (PEA). Blood culture media support the growth of streptococci. For those streptococci that require pyridoxal, subculture may require a staphylococcal cross-streak or media containing the necessary growth factors, as discussed earlier. The identification scheme recommended here is not likely to produce definitive species identification of all strains, but it should suffice for most medically important strains encountered in a clinical laboratory. When identification of unusual strains, such as those isolated from normally sterile sites or clearly implicated as pathogens, is required, the isolate should be sent to a reference laboratory that uses both conventional biochemicals and newer methods (molecular, enzymatic).

INITIAL CHARACTERIZATION OF STREPTOCOCCI AND RELATED GROUPS

Because most laboratories incubate all blood agar plates (except the anaerobic plates) in 5% to 10% carbon dioxide, colony morphology of streptococci from that environment is described here. The streptococci are facultative anaerobes and grow very well anaerobically, often better than aerobically. Obligate anaerobes that resemble streptococci belonging to the genus *Peptostreptococcus* are discussed in Chapters 35 and 38. Visualization of beta-hemolytic streptococcal colonies from throat specimens and recovery of group A beta-hemolytic streptococci are enhanced by anaerobic incubation. Detection of group A beta-hemolytic streptococci in pharyngeal exudate is discussed separately.

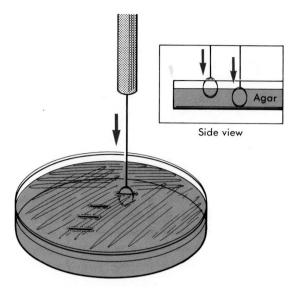

FIGURE 26.2

Stabbing the inoculating loop vertically into the agar after streaking the blood agar plate allows subsurface colonies to display hemolysis caused by streptolysin O.

Identification of streptococci should be initiated by careful examination of colony morphology and hemolytic pattern on 5% sheep blood agar. Hemolysis is enhanced by stabbing the inoculating loop into the agar several times (Figure 26.2). Colonies can then grow throughout the depth of the agar, producing subsurface oxygen-sensitive hemolysins (streptolysin O). Streptococci grow on the agar surface as translucent to milky, circular, small (≤1 mm diameter) colonies with a shiny surface. Colony variants are common, however, including rough, umbonate, dull

matte, and spreading morphologies. *Aerococcus* colonies are gray-white, convex, and circular. All streptococci, leuconostocs, and lactococci are catalase negative. Occasional enterococci and *Pediococcus* and *Aerococcus* sp. produce a pseudocatalase that yields a very weak positive test. *S. pneumoniae, Aerococcus viridans,* and many species of viridans streptococci, enterococci, and *S. bovis* show a zone of alpha hemolysis surrounding the colony. Alpha hemolysis, macroscopically appearing as a green discoloration of the medium surrounding the colony, is characterized by the presence of intact erythrocytes, visible by observing the medium surrounding the colony through the low power of a microscope (Figure 26.3). An unusual type of alpha hemolysis produced by some viridans streptococci is called "wide-zone alpha hemolysis." Colonies are surrounded by a zone of incomplete hemolysis of erythrocytes (Figure 26.4), but a wider zone farther out consists of completely lysed red cells, as seen in true beta hemolysis (Figure 26.5). Wide-zone alpha hemolysis macroscopically resembles beta hemolysis. If the presence of hemolysis is uncertain, it may be detected by moving the colony aside with a loop and examining the area of medium directly beneath the original colony site. Plates are always examined by holding them in front of a light source (Figure 26.6).

Microscopically, streptococci are gram positive and round or oval shaped, occasionally forming elongated forms that resemble pleomorphic

FIGURE 26.3

Alpha hemolysis surrounding the elliptical subsurface colony of an isolate of viridans streptococcus growing in a pour plate of 5% sheep blood agar, as seen by examining the colony microscopically under low power (100× magnification). Notice that the red blood cells are intact.

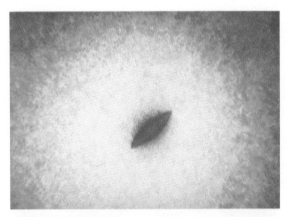

FIGURE 26.4

Wide-zone alpha hemolysis surrounding the elliptical subsurface colony of a streptococcus growing in a pour plate of 5% sheep blood agar, as seen by observing the colony microscopically under low power (100× magnification). Notice that a narrow zone of intact red blood cells, seen as a red haze, immediately surrounds the colony and that a wider zone containing no intact red cells extends farther out from the colony.

FIGURE 26.5

Beta hemolysis surrounding the elliptical subsurface colony of a beta-hemolytic streptococcus growing in a pour plate of 5% sheep blood agar, as seen by observing the colony microscopically under low power (100× magnification). Notice that there are no intact red blood cells in the vicinity of the colony.

corynebacteria or lactobacilli. Streptococci may occasionally appear gram negative if cultures are old or if the patient has been treated with antibiotics. Differentiation between streptococci and lactobacilli may be difficult, especially if the lactobacilli are catalase-negative strains. A Gram stain of growth just outside the zone of inhibition surrounding a 10 U penicillin disk placed on a blood agar plate inoculated with a lawn of the organism reveals long bacilli if the organism is a *Lactobacillus* and spherical forms if it is a *Streptococcus*. Additionally, growth in thioglycolate broth usually induces streptococci to form long chains of cocci (Figure

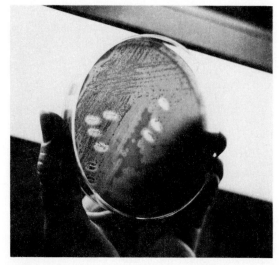

FIGURE 26.6

Observing hemolysis on sheep blood agar plates by viewing the plates in front of a light source.

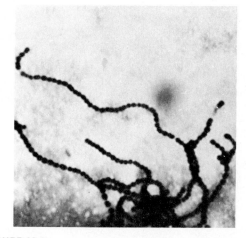

FIGURE 26.7

Chains of streptococci seen in Gram stain prepared from broth culture.

26.7). Growth in broth should always be used for determination of cellular morphology. *Aerococcus, Gemella,* and *Pediococcus* sp. grow in thioglycolate as large, spherical cocci arranged in tetrads and pairs and as individual cells rather than in chains. Lactobacilli grow as rods and filaments and occasionally shorter coccobacillary forms. *Leuconostoc* occasionally elongate to form coccobacilli, but cocci are the primary morphology. Organisms that fail to grow in thioglycolate usually grow in Lactobacilli MRS broth (Difco).

A screening test for vancomycin susceptibility should be performed to differentiate nonstreptococcal isolates. The organism is spread onto the surface of a blood agar plate using a heavy suspension on a swab or using a few colonies picked up on a loop or swab. A 30 μg vancomycin disk is placed on the agar surface, and the plate is incubated overnight in 5% carbon dioxide. Two strains can be tested on one plate. All streptococci, aerococci, gemellas, lactococci, and most enterococci (a few isolates resistant to vancomycin have been described recently) are susceptible to vancomycin (any zone of inhibition), whereas pediococci, leuconostocs, and many lactobacilli are typically resistant (grow right up to the disk).

Serologic grouping of cell wall components has classically been used to separate streptococci into species. Recent deoxyribonucleic acid (DNA) homology studies have shown that this is not possible. However, it is still a very useful practice to aid in the identification of clinical isolates and thus in the management of infected patients. Several reliable and convenient commercial methods for serogrouping are available. The original "Lancefield precipitin test," rarely performed in routine clinical laboratories today, is described in

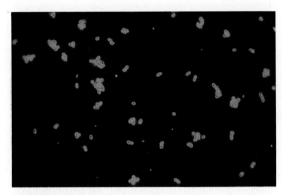

FIGURE 26.8

Group A streptococci stained with fluorescein-conjugated anti–group A antibody, examined with ultraviolet light under oil immersion (1000× magnification). Notice that the organisms stain as rings, with dark centers.

the American Society for Microbiology *Manual of Clinical Microbiology,* fifth edition (see Bibliography). For unusual group antigens and for group D antigens, the Lancefield test in some form is still used in reference laboratories. One must remember that Lancefield group D is teichoic acid and not carbohydrate and thus requires much more aggressive methods for release from the cell wall. Commercial systems often fail to detect group D when it is present. Fluorescent antibody stains for groups A and B streptococci have also been used for identification (Figure 26.8), although they are infrequently used in routine clinical laboratories today because of the availability of commercially produced reagents for rapid serogrouping of beta-hemolytic streptococci.

IDENTIFICATION OF BETA-HEMOLYTIC STREPTOCOCCI (Figure 26.9)

Beta hemolysis is characterized by complete lysis of erythrocytes, as determined by examining the medium microscopically (Figure 26.5). Large-size (>0.5 mm diameter) beta-hemolytic colonies may rarely be *Enterococcus faecalis* or *E. durans* (group D) but they are usually groupable with Lancefield antisera to cell wall carbohydrates of groups A, B, C, or G. Colonies of *Streptococcus pyogenes* are transparent to translucent, convex, entire, circular, and shiny and are surrounded by a rather wide zone of beta hemolysis. They are Lancefield group A and PYR positive. The PYR test, hydrolysis of L-pyrrolidonyl–β-naphthylamide, described in Chapter 10, is positive for *S. pyogenes,* all *Enterococcus* sp., *L. garviae,* nutritionally variant *S. defectivus* and *S. adjacens,* and *Gemella* sp. Modifications of the PYR test, including a broth test that can be read within 4 hours of incubation, and even more rapid chromogenic substrates and fluorescent systems are commercially available. Beta-hemolytic enterococci are distinguished from *S. pyogenes* (both PYR positive) by having a more diffuse zone of hemolysis that is confined to the area below the colony and usually by a larger colony than that of *S. pyogenes.*

Group B streptococci, *S. agalactiae,* yield larger, more translucent to opaque, whitish gray, soft, smooth colonies surrounded by a much smaller zone of beta hemolysis. The colonies of *S. agalactiae* resemble those of *Listeria* sp., which are catalase positive. *S. agalactiae* strains are positive in the CAMP test (Procedure 26.1) and able to hydrolyze hippurate (Chapter 10). Named for the originators, Christie, Atkins, and Munch-Peterson,[5] the CAMP test requires overnight incubation, but it is much less expensive than serologic typing. If an appropriate staphylococcus is unavailable, a commercial CAMP factor–impregnated filter paper disk can be used. A rapid spot CAMP reagent that yields results in 20 min-

FIGURE 26.9

Flow chart for presumptive identification of beta-hemolytic streptococci from humans.

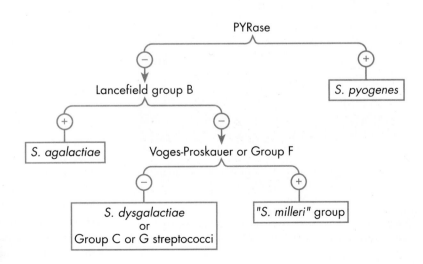

PROCEDURE 26.1

CAMP TEST

PRINCIPLE

Group B streptococci produce a proteinlike compound called the "CAMP factor" that is able to act synergistically with the beta toxin produced by some strains of *S. aureus* to produce even more potent hemolysis.

METHOD

1. Inoculate a streak of beta toxin–producing *S. aureus* down the center of a 5% sheep blood agar plate.

2. Inoculate straight lines of the isolates to be tested at right angles to the staphylococcal streak, stopping just before the staphylococcal line is reached. Several different isolates can be tested on one plate.

3. Incubate plates at 35° C to 37° C overnight in air or for 6 h in 5% to 10% CO_2. Air incubation increases the specificity of the test, since fewer non–group B streptococci are positive in an air atmosphere.

4. Observe for an arrowhead-shaped zone of enhanced hemolysis at the juncture between positive streptococci and the staphylococcus (Figure 26.10). Approximately 95% of group B streptococci, as well as rare strains of other groups, are CAMP positive.

QUALITY CONTROL

Inoculate *S. agalactiae* ATCC 27956 and *S. pyogenes* ATCC 19615 as just described.

EXPECTED RESULTS

The *S. agalactiae* should display enhanced hemolysis in an arrowhead pattern (Figure 26.10), and the *S. pyogenes* should not.

PERFORMANCE SCHEDULE

Test both quality control strains each time a CAMP test is performed.

Modified from Washington, J.A., II, editor. 1985. Laboratory procedures in clinical microbiology, ed 2. Springer-Verlag, New York.

---

utes was reported to be as accurate as the overnight test.[32]

The minute-colony beta-hemolytic streptococci are all likely to be "*S. milleri*" group.[7,44,45] A positive Voges-Proskauer test and a negative PYR

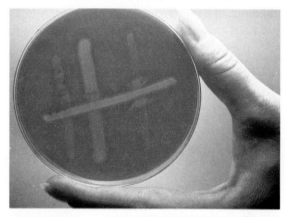

FIGURE 26.10

The CAMP test for presumptive identification of group B streptococci. The *S. agalactiae* (group B streptococcus) streaked down the right side of the plate exhibits the enhanced zone of hemolysis in an "arrowhead" shape near the *S. aureus* streak. The beta-hemolytic streptococcal isolates streaked down the left and center of the test plate are not group B.

test identifies a beta-hemolytic streptococcus as a member of the "*S. milleri*" group. Beta-hemolytic, large-colony-type streptococci groups C and G may be identified by serogroup and further characterized biochemically. Group C streptococci can be differentiated into species and subspecies by carbohydrate fermentation (Table 26.1). Numerous systems are available commercially that yield satisfactory serogroup identifications by latex agglutination or coagglutination methods as compared with the Lancefield precipitin standard. Some of these methods require an extraction step, which may take an hour or longer, whereas others do not. Identification can usually be accomplished with one or two isolated colonies, shortening the laboratory turnaround time by at least 24 hours.

IDENTIFICATION OF ALPHA-HEMOLYTIC AND NONHEMOLYTIC STREPTOCOCCI AND RELATED GENERA

Vancomycin-susceptible, alpha-hemolytic "streptococci" are preliminarily and presumptively identified as shown in Figure 26.11. An occasional *Lactococcus*, *Aerococcus*, or *Gemella* is misidentified as *Enterococcus* using this scheme, so important isolates should be examined more thor-

TABLE 26.1

IDENTIFICATION OF GROUP C STREPTOCOCCI

| | ACID FROM | |
SPECIES	TREHALOSE	SORBITOL
S. equi	−	−
S. zooepidemicus	−	+
S. equisimilis	+	−

oughly. The colony morphology of *S. pneumoniae* is usually distinctive, with very mucoid and glistening colonies that tend to dip down in the center and resemble a doughnut after increased incubation time because of the action of autolytic enzymes (Figure 26.12). Suspicious colonies must be tested for either bile solubility (Chapter 10) or susceptibility to optochin (Procedure 26.2).

Once it has been determined that an alpha-hemolytic streptococcus is not *S. pneumoniae*, it is treated as an enterococcus or viridans streptococcus. Vancomycin-susceptible, nonhemolytic (sometimes called "gamma"-hemolytic) colonies

are usually enterococci or "viridans" streptococci. Alpha-hemolytic and nonhemolytic colonies are screened as *Enterococcus* species by the PYR test. Colonies of enterococci and lactococci are usually larger and less hemolytic on blood agar than those of true streptococci. All enterococci are PYR positive, and clinically important strains are further characterized biochemically (Table 26.2). Additionally, all enterococci can grow in the presence of 6.5% NaCl in trypticase soy broth, a characteristic shared by aerococci and some lactococci but rarely by viridans streptococci. Nonenterococcal alpha-hemolytic and nonhemolytic streptococci are serogrouped. Group D are *S. bovis*.[7,38] If a suspected *S. bovis* does not type, it should be subcultured to bile-esculin agar for verification. Unlike *Enterococcus* sp., *S. bovis* cannot grow in 6.5% NaCl or at 10° C. Because of the associations of *S. bovis* with gastrointestinal malignancy, blood culture isolates of viridans streptococci should be identified enough to rule out *S. bovis*. Ruoff et al.[38] found that the API Rapid Strep system effectively differentiated *S. bovis* from occasional isolates of bile-esculin-positive *S. salivarius*. The ability to hydrolyze esculin in the presence of bile is a positive characteristic among all group D streptococci and entero-

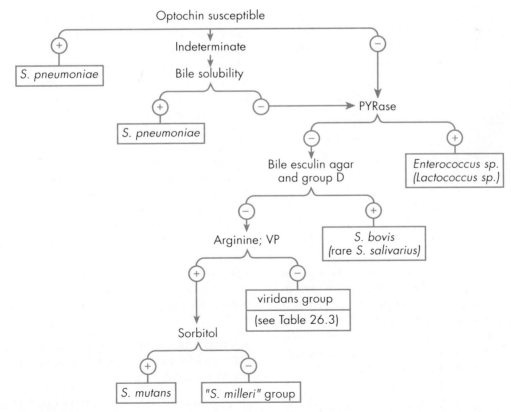

FIGURE 26.11

Flow chart for presumptive identification of alpha-hemolytic streptococci from humans.

PROCEDURE 26.2

OPTOCHIN SUSCEPTIBILITY TEST

PRINCIPLE

S. pneumoniae is susceptible to the antibacterial agent ethylhydrocupreine hydrochloride (optochin), which is impregnated in a filter paper disk. Performance of the test is similar to that of the disk diffusion susceptibility test.

METHOD

1. Streak one quadrant or one half of a 5% sheep blood agar plate with an inoculum from a pure isolate of the organism to be tested.

2. Place an optochin disk (Taxo P, BBL Microbiology Systems, or Bacto Optochin disk, Difco Laboratories) in the center of the inoculum.

3. Incubate overnight at 35° C in 5% to 10% CO_2 or a candle jar. Observe for zones of inhibition surrounding the disk. Zones ≥14 mm surrounding a 6 mm–diameter disk and zones ≥16 mm surrounding a 10 mm–diameter disk are considered positive, presumptive identification of *S. pneumoniae*. Zone sizes of 6 to 14 mm (6 mm disk) and 10 to 16 mm (10 mm disk) are equivocal, and those isolates should be tested for bile solubility.

QUALITY CONTROL

Test *S. pneumoniae* ATCC 27336 and a viridans streptococcus (laboratory strain) as just described.

EXPECTED RESULTS

The pneumococcus should yield a positive test result, and the viridans streptococcal species should grow right up to and under the optochin disk.

PERFORMANCE SCHEDULE

Test each new batch of optochin disks and monthly thereafter.

Modified from Facklam, R.R., and Carey, R.B. 1985. Streptococci and aerococci. In Lennette, E.H., Balows, A., Hausler, W.J., Jr., and Shadomy, H.J., editors. Manual of clinical microbiology, ed 4. American Society for Microbiology, Washington, D.C.

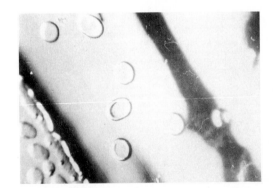

FIGURE 26.12
Colony morphology of pneumococci after 24 hours' incubation (20×).

cocci, although 5% to 10% of the viridans group may also hydrolyze esculin in the presence of bile. This test, *bile esculin,* is performed by inoculating a slant or plate of bile-esculin agar (available commercially), incubating in air for as long as 4 days, and observing for a blackening of the agar, indicating the presence of the end product of esculin hydrolysis combined with iron in the medium.

The *"S. milleri"* group is identified to species based on a new format of enzymatic tests and use of fluorogenic substrates (Procedure 26.3). Other nonenterococcal alpha-hemolytic and non-hemolytic streptococci (viridans group) can be identified by the tests listed in Table 26.3. It is critical that for conventional tests, bromcresol purple broths and biochemical reagents be prepared according to methods described by Facklam.[10a] Esculin hydrolysis is performed on slants of esculin medium without bile, which are available commercially (dry powder is also available commercially). The arginine test is described in Procedure 26.3. Detection of extracellular polysaccharide formation is outlined in Procedure 26.4. Commercial biochemical systems and rapid kits are available in the United States, but none of them has been found to identify accurately all strains of viridans streptococci.[22] Newer systems currently available in Europe appear to be more reliable for identification of this group of bacteria.

RAPID METHODS AND COMMERCIAL SYSTEMS FOR IDENTIFICATION OF STREPTOCOCCI

Beta-hemolytic streptococci have classically been identified based on their serologic group rather

TABLE 26.2

DIFFERENTIATION AMONG SELECTED *ENTEROCOCCUS* SP.

| SPECIES | YELLOW PIGMENT | MOTILITY | ACID PRODUCED FROM ||||||| ARGININE DIHYDROLASE |
			MANNITOL	SORBITOL	SORBOSE	ARABINOSE	RAFFINOSE	SUCROSE	LACTOSE	
*E. faecalis**	–	–	+	+	–	–	NA	NA	+	+
E. avium	–	–	+	+	+	+	–	NA	NA	–
E. faecium	–	–	+	+/–	+	+	NA	NA	+	+
E. raffinosus	–	–	+	+	+	+	+	NA	+	–
E. gallinarum	–	+	+	+/–	–	NA	+/–	+/–	NA	+
E. hirae	–	–	–	–	–	+	NA	NA	+	+
E. casseliflavus	+	+	+	+/–	–	NA	–	NA	+	+
E. durans	–	–	–	–	–	–	–	–	–	+
E. solitarius	–	–	+	+/–	–	+	NA	NA	+	+
E. mundtii	+	–	+	+/–	–	–	+	NA	NA	–
E. malodoratus	–	–	+	+	+	–	+	NA	NA	–
E. pseudoavium	–	–	+	+	+	–	–	NA	NA	–

Modified from Facklam, R.R., and Collins, M.D. 1989. J. Clin. Microbiol. 27:731.

NA, Not applicable; +, 90% strains positive; –, 90% strains negative; +/–, variable (more often positive).

* An asaccharolytic variant of *E. faecalis* has been described.

PROCEDURE 26.3

RAPID TESTS FOR PRESUMPTIVE IDENTIFICATION OF "S. MILLERI" GROUP

PRINCIPLE

S. anginosus, S. intermedius, and *S. constellatus* are causes of purulent infections in humans; can be nonhemolytic, beta hemolytic (minute-colony morphology), and rarely alpha hemolytic; and may possess Lancefield antigens A, C, F, or G. Differentiation from other streptococci is based on biochemical reactions and enzymatic tests. In addition to the reactions described here, all tests can be carried out using tablets available from Key Diagnostics, Round Rock, Texas.

METHODS

A. Voges-Proskauer (VP) test
1. Place 0.2 ml MR-VP broth (available commercially) into a small (10 × 75 mm) tube, and inoculate with a heavy suspension of the organism to be tested.
2. Incubate for 5 h at 35° C.
3. Add 1 drop each of 0.5% aqueous creatine (Sigma and other suppliers), 5% α-naphthol in absolute ethanol (Sigma; VP reagent A as described in Chapter 28), and 40% aqueous potassium hydroxide (Sigma; VP reagent B as described in Chapter 28).
4. Shake or vortex thoroughly and observe for 15 min for development of a pink color.
B. Arginine hydrolysis test
1. Place 0.2 ml Moeller decarboxylase broth base and Moeller base with arginine (available commercially) into small (10 × 75 mm) tubes, and inoculate each with a heavy suspension of the organism to be tested.
2. Overlay each tube with mineral oil, and incubate for 5 h at 35° C.
3. Observe for color change. Purple indicates hydrolysis of arginine. Organisms that do not hydrolyze arginine are still able to ferment glucose in the medium, creating acid by-products and a yellow color change in the pH indicator.
C. Sorbitol fermentation test
1. Place 0.2 ml sorbitol-containing phenol red carbohydrate broth (available commercially) into small (10 × 75 mm) tubes, and inoculate each with a heavy suspension of the organism to be tested.
2. Overlay each tube with mineral oil, and incubate for 5 h at 35° C.
3. Observe for color change. Yellow or orange indicates sorbitol fermentation; red or dark orange indicates negative results.

QUALITY CONTROL

Inoculate *Streptococcus anginosus* ATCC 12393, *S. mitis* ATCC 6249, and *Enterococcus faecalis* ATCC 19433 and test as just described.

EXPECTED RESULTS

S. anginosus is positive for acetoin production (VP positive, pink or red), able to hydrolyze arginine (yellow in base broth and purple in arginine broth), and negative for sorbitol fermentation (red or orange). *S. mitis* should yield a negative VP reaction, negative arginine hydrolysis (yellow in both Moeller tubes), and negative sorbitol fermentation. *E. faecalis* should yield a positive VP test, positive arginine hydrolysis, and positive sorbitol (yellow).

PERFORMANCE SCHEDULE

Test all quality control strains each time new reagents are prepared.

Modified from Ruoff, K.L., and M.J. Ferraro. 1986. J. Clin. Microbiol. 24:495.

---

than biochemically. Numerous commercial systems using either latex agglutination or coagglutination are available (Pharmacia Diagnostics, Wellcome Diagnostics, Adams Scientific, Meridian Diagnostics, API Analytab Products, Diagnostics Products, BBL Microbiology Systems, Difco Laboratories, and others). As long as non–"S. milleri" (Voges-Proskauer-negative) strains are tested, reliable results should be achieved with all commercial systems. The PYR test, which is faster and less expensive than serologic typing, should be used to identify *S. pyogenes* and the rare beta-hemolytic enterococcus. Group B can be identified economically and quickly by a rapid CAMP test.[32]

Identification of *S. pneumoniae* can be made rapidly with the Quellung test, although this is rarely performed today. Addition of specific capsular antibody to a wet preparation of sample

TABLE 26.3

PRESUMPTIVE DIFFERENTIATION AMONG SOME VIRIDANS STREPTOCOCCI LIKELY TO BE ISOLATED FROM CLINICAL SPECIMENS

SPECIES	VOGES-PROSKAUER	ARGININE DIHYDROLASE	UREA	ESCULIN HYDROLYSIS	ACID FROM INULIN	ACID FROM MANNITOL	DEXTRAN	LEVAN
S. bovis I	+	−	−	+	+	+	+	+(rare)
S. bovis II	+	−	−	+	−	−	−	−
S. gordonii	−	+	−	+	+	−	+	−
S. "milleri" group								
anginosus	+	+	−	+	−	−	−	−
constellatus	+	+	−	−	−	−	−	−
intermedius	+	+	−	−/+	−	−	−	−
S. mitis	−	−	−	−	−	−	−	−
S. mutans	+	−	−	+	+	+	+	−
S. oralis	−	−	−	−/+	−	−	+	−
S. parasanguis	−	+	−	−/+	−	−	+	−
S. salivarius	+/−	−	+/−	+	+/−	−	−	+
S. sanguis	−	+	−	+/−	+	−	+	−
S. sobrinus	−	−	−	+	−	+	+	−
S. vestibularis	+/−	−	+	+/−	−	−	−	−

Data from various sources.[2,7,26,45]
+, 90% strains positive; −, 90% strains negative; +/−, variable (more often positive); −/+, variable (more often negative).

containing the organism, either from an isolated colony or in the actual clinical material (sputum or cerebrospinal fluid), results in an antigen-antibody response, visible as apparent capsular swelling (Figure 26.13). The capsules become more visible and refractile because of antibody attachment. Omniserum (available from Statens Serum Institute or DANSerum Inc.) reacts with the most common serotypes of pneumococci. Difco produces a less inclusive serum reagent. Latex reagents are available for pneumococci, as described for detection of pneumococcal meningitis (Wellcome Diagnostics, Pharmacia Diagnostics, BBL Microbiology Systems, and others). A coagglutination reagent (Phadebact by Pharmacia Diagnostics) and latex agglutination reagent (Pneumoslide, BBL Microbiology Systems) are also available for rapid identification from culture.

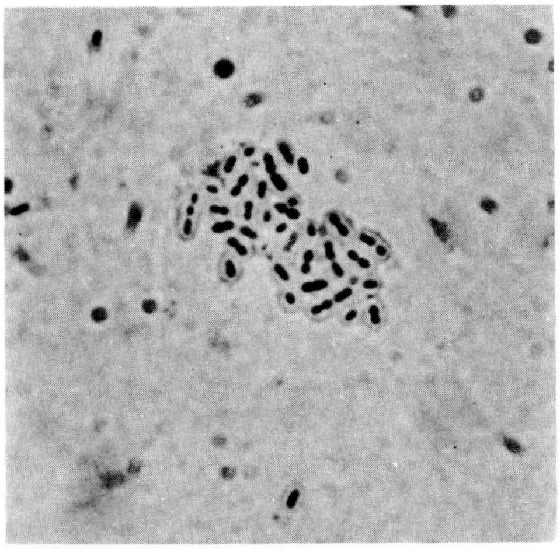

Viridans streptococcal taxonomy is evolving, and new designations often rely on nontraditional or technically demanding parameters. Thus no commercial system in the United States is completely reliable for definitive identification of all important clinical isolates at the time of this writing.[14,22]

FIGURE 26.13
Quellung reaction with *Streptococcus pneumoniae*, showing apparent capsular swelling. This test is only of historical interest now.

SCREENING METHODS FOR SPECIFIC STREPTOCOCCAL SPECIES

DETECTION OF *STREPTOCOCCUS PYOGENES* IN THROAT SWAB SPECIMENS

Detection of *S. pyogenes* in throat cultures is important for clinicians. When specimens are positive, therapy can be initiated to prevent sequelae and the spread of infection to others. In addition, prompt initiation of therapy can significantly shorten the duration of symptoms of pharyngitis.

PROCEDURE 26.4

TEST FOR PRODUCTION OF EXTRACELLULAR POLYSACCHARIDE (GLUCAN)

A. Broth method
1. Prepare 5% sucrose broth by mixing together two separate solutions, autoclaved before mixing, as follows:

  Thioglycolate broth base (Difco), 28.5 g
  K_2HPO_4, 10 g
  Sodium acetate, 12 g
  Distilled water, 500 ml

 Mix together in a 1000 ml flask:

  Sucrose, 50 g
  Distilled water, 500 ml

 Prepare each solution in a separate 1000 ml flask. Autoclave the two solutions at 121° C for 15 min. Cool both to 55° C, and aseptically mix them by pouring the contents of one flask into the other. Dispense aseptically in 5 ml amounts into 16 × 125 mm sterile glass test tubes with screw caps. Store refrigerated.
2. Inoculate a pure culture of the organism; incubate in air at 35° C for up to 1 week, and observe for gelation of the medium. Glucan production is indicated by formation of complete or partial gel in the broth (observed by gently tipping the tube back and forth). If the broth is merely thickened, this is not glucan production but may suggest slime produced by *Streptococcus bovis*.

B. Colony morphology on sucrose agar
1. Prepare 5% sucrose agar as follows:

  Brain-heart infusion agar (Difco), 40 g
  Sucrose, 50 g
  Distilled water, 1000 ml

 Mix together, autoclave for 15 min at 121° C, cool to 55° C, and pour into 90 mm–diameter Petri plates, at least 20 ml per plate. Allow to set, and store tightly wrapped in polyethylene bags in the refrigerator. Plates are good for several weeks.
2. Inoculate plates with pure cultures of the organisms to be tested. Incubate in air for a maximum of 4 days. Examine for colony morphology. *S. sanguis* and *S. mutans*, both glucan producers, form very dry, adherent, refractile colonies that are difficult to move with a loop. Levan producers, such as *S. salivarius*, yield opaque, very gummy (gumdrop-like), but nonadherent colonies. Growth can be scraped up and lifted (in a glob) from the agar surface. Organisms that produce no extracellular polysaccharides, such as *S. mitis* and *Gemella morbillorum*, produce mucoid, nonadherent colonies.

Modified from Rubin. 1984. Lab. Med. 15:171, using data from original work by Facklam.[10a]

Since pharyngitis caused by streptococci cannot always be distinguished from that caused by viruses (the most common etiological agents) on clinical grounds, microbiological procedures are necessary. Many commercial kits for rapid detection of group A streptococcal antigen, using latex agglutination, coagglutination, or enzyme-linked immunosorbent assay (ELISA) technologies, are currently available. These reagents have been reported to be very specific.[11] Sensitivity, however, depends on several variables, and reports have ranged from approximately 60% sensitivity to greater than 95%, with most recent evaluations showing 50% sensitivity.[31,42] Membrane-bound ELISA test systems are probably the most sensitive currently. False-negative results have been reported.[24] It has been found that personnel must be carefully trained in the correct interpretation of these tests.

We recommend that two swabs be collected from each throat to be tested. If the first swab yields a positive result by a direct antigen method, the second swab can be discarded (avoiding the cost of plates for these specimens). During streptococcal season, as many as 25% of patients are likely to test positive in a direct antigen test. For those specimens in which the rapid antigen test yielded a negative result, a blood agar or selective streptococcal blood agar plate should be inoculated with the second swab. Selective streptococcal agar detects the most positive results and should be used unless organisms other than *S. pyogenes* are being sought. None of the current rapid antigen detection products is sensitive enough to use alone, prompting the suggestion by several prominent pediatricians of an algorithm that includes both rapid tests and cultures, depending on severity of illness, risk factors for rheumatic fever, and social factors.[31]

A new technology, using a novel reflective surface for enhanced detection of antigen-antibody reactions (Biostar, Boulder, Colo.), may be sensitive enough to obviate the need for follow-up cultures.[19a] Additional studies are in progress.

Considerations for the adult patient with pharyngitis are different from those for children. One approach advocates treating certain adult patients based on clinical signs and symptoms alone and, for those patients who require laboratory testing, treating only those with positive rapid test results and performing no cultures.[31] The availability of rapid tests for antigen detection has generated significant controversy; no one philosophy is generally accepted. Streptococcal selective media can be used for presumptive identification of *S. pyogenes* in throat specimens after overnight incubation by placing a bacitracin disk on the initial inoculum. Most streptococci other than *S. pyogenes* are resistant to this antibiotic. This method is not suitable for nonselective media.

Other considerations for identification of *S. pyogenes* from throats are collection and transport of the specimen. Unlike most bacteria, streptococci are very resistant to drying. Swabs may be placed in paper, or material may even be transferred to filter paper for transport without significant loss of recovery.

DETECTION OF GROUP B STREPTOCOCCI IN VAGINAL SWAB SPECIMENS AND SPECIMENS FROM INFECTED NEONATES

For diagnosis of neonatal sepsis and meningitis caused by group B streptococci, several commercial antigen detection kits are available that have good sensitivity and specificity. Developed for use on serum, urine, or cerebrospinal fluid, the best results have been achieved with cerebrospinal fluid so far. Latex agglutination procedures appear to be the most sensitive. False-positive results in urine have been problematic.[20,39]

Because neonates acquire group B infection during passage through a colonized vagina, much interest has been generated recently in detection of *Streptococcus agalactiae* in the vagina before delivery. Several studies have reported successful prevention of neonatal disease by treating (with penicillin or ampicillin) colonized mothers during labor and delivery.[19]

Enrichment broths are superior to direct plating for detection of this organism in contaminated specimens.[30] Direct extraction and latex particle agglutination for antigen detection have not been sensitive enough for use alone as a screening test.[19] Solid-phase immunoassays (ICON Strep B, Hybritech; QUIDEL Group B Strep Test) for direct detection from swab specimens have shown some promise in clinical trials.[18,29]

SEROLOGIC DIAGNOSIS OF GROUP A STREPTOCOCCAL INFECTION

Individuals with disease caused by *Streptococcus pyogenes* produce antibodies against many of the extracellular products mentioned at the beginning of this chapter. For determining whether a patient has recently had streptococcal infection, several tests are usually performed. The diagnostic usefulness of such tests is likely to increase, especially with the decline of rheumatic fever in the United States. They should be used only for diagnosis of nonsuppurative sequelae of streptococcal infection. The most common of these is the **ASO,** determination of presence and titers of antistreptolysin O antibodies. Anti–DNAse B, antistreptokinase, and antihyaluronidase have all been used to diagnose streptococcal infection retrospectively. Pharyngitis seems to be followed by rises in antibody titers against all three antigens, whereas the ASO titers of patients with pyoderma are frequently not above normal. Anti–DNAse B titers, however, do increase after pyoderma, and thus this test is the most sensitive. Commercial products are available for detection of these antibodies, using latex agglutination or whole red blood cell agglutination methodologies (Table 26.4). LEAP Strep was not sensitive enough for routine screening in one evaluation.[21] Streptozyme, which detects a mixture of antibodies, has consistently shown the highest sensitivity for detection of ASO but yielded low sensitivity for detection of anti–DNase B in at least one study.[23] False-positive results are associated with its use. No commercial system has been shown to detect accurately all streptococcal antibodies likely to be present in serum of patients with prior infection, although at least one newer test system appears to be useful for detection of ASO antibodies.[15] Serum obtained as long as 2 months after infection usually demonstrates increased antibodies. As with other serologic tests, an increasing titer over time is most useful for diagnosing previous streptococcal infection.

TABLE 26.4

PRODUCTS AVAILABLE FOR DETERMINING ANTISTREPTOCOCCAL ANTIBODY TITERS

TEST	MANUFACTURER
ASO EIA	V-Tech, Inc.
ASO Latex Test	ICL Scientific
ASO Quantum	Sclavo
ASO Test	American Dade
Rapitex ASO Kit	Behring Diagnostics
Rheumagen ASO	biokit USA
SerImmSure	Analytab Products
Strep-phile	BioDiagnostics Systems
Streptonase	Wampole Laboratories
Streptozyme	Wampole Laboratories

TREATMENT OF STREPTOCOCCAL INFECTIONS

Penicillin is still the drug of choice for most streptococcal infections. With the exception of rare *Streptococcus pneumoniae* and the enterococci, most streptococci, including nonenterococcal group D organisms, are susceptible to penicillins, which can be given in high dosage. For penicillin-allergic patients, erythromycin or vancomycin may be used. Erythromycin resistance has been reported. Approximately 5% of all *S. pneumoniae* strains are now being reported to be relatively resistant (minimum inhibitory concentrations [MICs], 0.12 to 1.0 μg/ml) to penicillin. In these instances, penicillin therapy results in treatment failures for patients with pneumococcal meningitis. Less than 0.1% of pneumococci are penicillin resistant, but these isolates may be recovered from seriously ill patients. For these reasons, all clinically important isolates of *S. pneumoniae* should be tested for penicillin susceptibility using an oxacillin disk screening test and MICs (Chapter 14).

The enterococci are uniquely antibiotic resistant. They are resistant to penicillin, aminoglycosides, cephalosporins, clindamycin, and numerous other agents. Synergism between a penicillin and an aminoglycoside (usually gentamicin) has been repeatedly demonstrated, however, and this combination is the therapy of choice. Vancomycin has been effective until recently, when several outbreaks of vancomycin-resistant enterococci were reported.[28] The rising incidence of *E. faecium* infections has also served to increase the number of vancomycin-resistant isolates reported, since this species is inherently more resistant than *E. faecalis*. High-level resistance to streptomycin, indicating in vivo resistance to the combination of penicillin and streptomycin, can be detected by testing the enterococcus isolate against 2000 μg of streptomycin. Growth in the presence of this quantity of antibiotic indicates that synergism with a beta-lactam agent will be absent. Continuous-infusion ampicillin therapy may be useful, but ampicillin resistance has also been encountered.

REFERENCES

1. Barnham, M., Thornton, T.J., and Lange, K. 1983. Nephritis caused by *Streptococcus zooepidemicus* (Lancefield group C). Lancet 1:8331.
2. Beighton, D., Hardie, J.M., and Whiley, R.A. 1991. A scheme for the identification of viridans streptococci. J. Med. Microbiol. 35:367.
3. Bouvet, A., Grimont, F., and Grimont, P.A.D. 1989. *Streptococcus defectivus* sp. nov. and *Streptococcus adjacens* sp. nov., nutritionally variant streptococci from human clinical specimens. Int. J. Syst. Bacteriol. 39:290.
4. Bradley, S.F., Gordon, J.J., Baumgartner, D.D., et al. 1991. Group C streptococcal bacteremia: analysis of 88 cases. Rev. Infect. Dis. 13:270.
5. Christie, R., Atkins, N.E., and Munch-Peterson, E. 1944. A note on the lytic phenomenon shown by group B streptococci. Aust. J. Exp. Biol. 22:193.
6. Cimolai, N., Elford, R.W., Bryan, L., et al. 1988. Do the beta-hemolytic non-group A streptococci cause pharyngitis? Rev. Infect. Dis. 10:587.
7. Coykendall, A.L. 1989. Classification and identification of the viridans streptococci. Clin. Microbiol. Rev. 2:315.
8. Devriese, L.A., Hommez, J., Kilpper-Balz, R., et al. 1986. *Streptococcus canis* sp. nov.: a species of group G streptococci from animals. Int. J. Syst. Bacteriol. 36:422.
9. Elliott, J.A., Collins, M.D., Pigott, N.E., and Facklam, R.R. 1991. Differentiation of *Lactococcus lactis* and *Lactococcus garviae* from humans by comparison of whole-cell protein patterns. J. Clin. Microbiol. 29:2731.
10. Ezaki, T., Facklam, R.R., Takeuchi, N., et al. 1986. Genetic relatedness between the type strain of *Streptococcus anginosus* and minute-colony-forming beta-hemolytic streptococci carrying different Lancefield grouping antigens. Int. J. Syst. Bacteriol. 36:345.
10a. Facklam, R.R. 1977. Physiological differentiation of viridans streptococci. J. Clin. Microbiol. 5:184.
11. Facklam, R.R. 1987. Specificity study of kits for detection of group A streptococci directly from throat swabs. J. Clin. Microbiol. 25:504.
12. Facklam, R.R., and Collins, M.D. 1989. Identification of *Enterococcus* species isolated from human infections by a conventional test scheme. J. Clin. Microbiol. 27:731.
13. Facklam, R., Hollis, D., and Collins, M.D. 1989. Identification of gram-positive coccal and coccobacillary vancomycin-resistant bacteria. J. Clin. Microbiol. 27:724.

14. Freney, J., Bland, S., Etienne, J., et al. 1992. Description and evaluation of the semiautomated 4-hour Rapid ID 32 Strep method for identification of streptococci and members of related genera. J. Clin. Microbiol. 30:2657.

15. Gerber, M.A., Caparas, L.S., and Randolph, M.F. 1990. Evaluation of a new latex agglutination test for detection of streptolysin O antibodies. J. Clin. Microbiol. 28:413.

16. Gnann, J.W., Jr., Gray, B.M., Griffin, F.M., Jr., et al. 1987. Acute glomerulonephritis following group G streptococcal infection. J. Infect. Dis. 156:411.

17. Gossling, J. 1988. Occurrence and pathogenicity of the *Streptococcus milleri* group. Rev. Infect. Dis. 10:257.

18. Granato, P.A., and Petosa, M.T. 1991. Evaluation of a rapid screening test for detecting group B streptococci in pregnant women. J. Clin. Microbiol. 29:1536.

19. Greenspoon, J.S., Wilcox, J.G., and Kirschbaum, T.H. 1991. Group B streptococcus: the effectiveness of screening and prophylaxis. Obstet. Gynecol. Surv. 46:499.

19a. Harbeck, R.J., Teague, J., Crossen, G.R., et al. 1993. Novel, rapid optical immunoassay technique for detection of group A streptococci from pharyngeal specimens: comparison with standard culture methods. J. Clin. Microbiol. 31:839.

20. Harris, M.C., Deuber, C., Polin, R.A., and Nachamkin, I. 1989. Investigation of apparent false-positive urine latex particle agglutination tests for the detection of group B streptococcus antigen. J. Clin. Microbiol. 27:2214.

21. Heath-Fracica, L.A., and Estevez, E.G. 1987. Evaluation of a new latex agglutination test for detection of streptococcal antibodies. Diagn. Microbiol. Infect. Dis. 8:25.

22. Hinnebusch, C.J., Nikolai, D.M., and Bruckner, D.A. 1991. Comparison of API Rapid STREP, Baxter MicroScan Rapid Pos ID panel, BBL Minitek differential identification system, IDS RapID STR system, and Vitek GPI to conventional biochemical tests for identification of viridans streptococci. Am. J. Clin. Pathol. 96:459.

23. Hostetler, C.L., Sawyer, K.P., and Nachamkin, I. 1988. Comparison of three rapid methods for detection of antibodies to streptolysin O and DNase B. J. Clin. Microbiol. 26:1406.

24. Inzana, T.J., and Thompson, C. 1990. Failure of rapid antigen detection tests to identify group A streptococci in throat swabs. Clin. Microbiol. Newsletter 12:111.

25. Isenberg, H.D., Vellozzi, E.M., Shapiro, J., et al. 1988. Clinical laboratory challenges in the recognition of *Leuconostoc* spp. J. Clin. Microbiol. 26:479.

26. Kilian, M., Mikkelsen, L., and Henrichsen, J. 1989. Taxonomic study of viridans streptococci: description of *Streptococcus gordonii* sp. nov. and emended descriptions of *Streptococcus sanguis* (White and Niven 1946), *Streptococcus oralis* (Bridge and Sneath 1982), and *Streptococcus mitis* (Andrewes and Horder 1906). Int. J. Syst. Bacteriol. 39:471.

27. Lebrun, L., Guibert, M., Wallet, P., et al. 1986. Human Fc(gamma) receptors for differentiation in throat cultures of group C *"Streptococcus equisimilis"* and group C *"Streptococcus milleri."* J. Clin. Microbiol. 24:705.

28. Mortensen, J.E., and LaRocco, M. 1992. Enterococci: an old bug has learned new tricks. Clin. Microbiol. Newsletter 14:57.

29. Park, C.H., Hixon, D.L., Spencer, M.L., et al. 1992. The ICON Strep B immunoassay for rapid detection of group B streptococcal antigen. Lab. Med. 23:543.

30. Persson, K.M.-S., and Forsgren, A. 1987. Evaluation of culture methods for isolation of group B streptococci. Diagn. Microbiol. Infect. Dis. 6:175.

31. Radetsky, M., Solomon, J.A., and Todd, J.K. 1987. Identification of streptococcal pharyngitis in the office laboratory: reassessment of new technology. Pediatr. Infect. Dis. 6:556.

32. Ratner, H.B., Weeks, L.S., and Stratton, C.W. 1986. Evaluation of spot CAMP test for identification of group B streptococci. J. Clin. Microbiol. 24:296.

33. Reimer, L.G., and Reller, L.B. 1981. Growth of nutritionally variant streptococci on common laboratory and 10 commercial blood culture media. J. Clin. Microbiol. 14:329.

34. Reimer, L.G., and Reller, L.B. 1983. Effect of pyridoxal on growth of nutritionally variant streptococci and other bacteria on sheep blood agar. Diagn. Microbiol. Infect. Dis. 1:273.

35. Riebel, W.J., and Washington, J.A. 1990. Clinical and microbiologic characteristics of Pediococci. J. Clin. Microbiol. 28:1348.

36. Roberts, R.B., Krieger, A.G., Schiller, N.L., et al. 1979. Viridans streptococcal endocarditis: the role of various species, including pyridoxal-dependent streptococci. Rev. Infect. Dis. 1:955.

37. Ruoff, K.L. 1990. *Gemella:* a tale of two species (and five genera). Clin. Microbiol. Newsletter 12:1.

38. Ruoff, K.L., Miller, S.I., Garner, C.V., et al. 1989. Bacteremia with *Streptococcus bovis* and *Streptococcus salivarius:* clinical correlates of isolates. J. Clin. Microbiol. 27:305.

39. Sanchez, P.J., Siegel, J.D., Cushion, N.B., and Threlkeld, N. 1990. Significance of a positive urine group B streptococcal latex agglutination test in neonates. J. Pediatr. 116:601.

40. Stevens, D.L. 1992. Invasive group A streptococcus infection. Clin. Infect. Dis. 14:2.

41. Turner, J.C., Hayden, G.F., Kiselica, D., et al. 1990. Association of group C β-hemolytic streptococci with endemic pharyngitis among college students. J.A.M.A. 264:2644.

42. Wegner, D.L., Witte, D.L., and Schrantz, R.D. 1992. Insensitivity of rapid antigen detection methods and single blood agar plate culture for diagnosing streptococcal pharyngitis. J.A.M.A. 267:695.

43. Whiley, R.A., Beighton, D., Winstanley, T.G., et al. 1992. *Streptococcus intermedius, Streptococcus constellatus,* and *Streptococcus anginosus* (the *Streptococcus milleri* group): association with different body sites and clinical infections. J. Clin. Microbiol. 30:243.

44. Whiley, R.A., Fraser, H., Hardie, J.M., and Beighton, D. 1990. Phenotypic differentiation of *Streptococcus intermedius, Streptococcus constellatus,* and *Streptococcus anginosus* strains within the *"Streptococcus milleri* group." J. Clin. Microbiol. 28:1497.

45. Winstanley, T.G., Magee, J.T., Limb, D.I., et al. 1992. A numerical taxonomy study of the *"Streptococcus milleri"* group based upon conventional phenotypic tests and pyrolysis mass spectrometry. J. Med. Microbiol. 36:149.

BIBLIOGRAPHY

Auckenthaler, R., Hermans, P.E., and Washington, J.A., II. 1983. Group G streptococcal bacteremia: clinical study and review of the literature. Rev. Infect. Dis. 5:196.

Bisno, A.L. 1990. Nonsuppurative poststreptococcal sequelae: rheumatic fever and glomerulonephritis. p. 1528. In Mandell, G.L., Douglas, R.G., Jr., and Bennett, J.E., editors. Principles and practice of infectious diseases, ed 3. Churchill Livingstone, New York.

Facklam, R.R., and Washington, J.A., II. 1991. *Streptococcus* and related catalase-negative gram-positive cocci. p. 238. In Balows, A., Hausler, W.J., Jr., Herrmann, K.L., et al., editors: Manual of clinical microbiology, ed. 5. American Society for Microbiology, Washington, D.C.

Klein, J.O., Dashefsky, B., Norton, C.R., and Mayer, J. 1983. Selection of antimicrobial agents for treatment of neonatal sepsis. Rev. Infect. Dis. 5(suppl 1):S55.

Krisher, K., and Cunningham, M.W. 1985. Myosin: a link between streptococci and heart. Science 227:413.

Musher, D.M. 1992. Infections caused by *Streptococcus pneumoniae:* clinical spectrum, pathogenesis, immunity, and treatment. Clin. Infect. Dis. 14:801.

Ruoff, K.L. 1990. Update on nutritionally variant streptococci *(Streptococcus defectivus* and *Streptococcus adjacens).* Clin. Microbiol. Newsletter 12:97.

Senitzer, D., and Freimer, E.H. 1984. Autoimmune mechanisms in the pathogenesis of rheumatic fever. Rev. Infect. Dis. 6:832.

27

AEROBIC GRAM-NEGATIVE COCCI (NEISSERIA AND MORAXELLA CATARRHALIS)

The family Neisseriaceae currently contains the genera *Acinetobacter, Kingella, Moraxella, Branhamella,* and *Neisseria.* Recent taxonomic studies support removal of *Moraxella, Branhamella, Acinetobacter,* and *Kingella indologenes* from this family. Accordingly, those discussed in this chapter include *Neisseria* and *Moraxella (Branhamella) catarrhalis.* Other genera within the family are discussed in Chapters 29 and 30. The nomenclature of these organisms became more confusing with the latest edition of *Bergey's Manual of Systematic Bacteriology,* in which *Branhamella* was classified as a subgenus of *Moraxella.* We chose to discuss the *Branhamella* sp. in this chapter because their morphology and biochemical identification parameters most closely resemble those of *Neisseria.* However, we refer to the pathogenic *Branhamella catarrhalis* using the currently more accepted term *Moraxella catarrhalis.*

These organisms are oxidase-positive gram-negative cocci (with the exception of *N. elongata*). They are normal flora of the respiratory, alimentary, and genitourinary mucosal surfaces of humans. Although any of the normally commensal, *Neisseria* sp. can cause disease in a debilitated host; the organisms most often responsible for disease in this group are *N. gonorrhoeae* (gonococcus), the agent of gonorrhea, and *N. meningitidis* (meningococcus), the agent of meningococcal meningitis and septicemia. The pathogenesis of infections caused by the *Neisseria* sp. is discussed in the sections that relate to the diseases they cause, particularly Chapters 16 and 20. Recently, *M. catarrhalis* has been recognized more frequently as a cause of respiratory tract disease. The importance of this rests with *M. catarrhalis* usually being resistant to ampicillin, a frequently used agent for treatment of respiratory tract infections.

MORPHOLOGY

Organisms appear as kidney bean–shaped diplococci on Gram stain. In smears made from clinical material, groups of *Neisseria* and *Moraxella* are often seen within polymorphonuclear neutrophils, sometimes in large numbers (Figure 27.1). This morphological appearance is known as GNID (gram-negative intracellular diplococci) and helps the microbiologist to identify a true infection. *N. elongata,* which has been isolated from throats of patients with pharyngitis, is rod shaped, and extends into long forms under antibiotic pressure.[11]

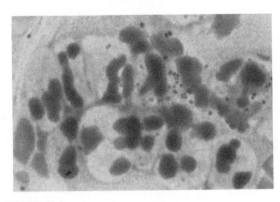

FIGURE 27.1

Methanol-fixed gram stain of bronchial washings showing numerous intracellular and extracellular *Moraxella catarrhalis.* (Courtesy Jay Packer.)

GENERAL GROWTH CHARACTERISTICS

Neisseria and *Moraxella* grow best on enriched media in a humid atmosphere of 5% to 10% carbon dioxide (CO_2). *N. gonorrhoeae* requires the added nutrients of chocolate agar or special enrichments, whereas *N. meningitidis,* other *Neisseria* sp., and *M. catarrhalis* can grow on blood agar. The commensal *Neisseria* sp. and *Moraxella* are able to grow at room temperature, but some strains of the pathogenic *Neisseria* are very sensitive to temperature variations. For this reason, clinical specimens sent to the laboratory for recovery of these organisms must be guarded from the cold and inoculated to growth media as soon as possible after collection. Although holding media at refrigerator temperature does not inhibit recovery, culturing *Neisseria* on media brought to room temperature facilitates luxuriant growth.[5]

NEISSERIA GONORRHOEAE

SPECIMEN COLLECTION

Specimens to be collected for the isolation of *N. gonorrhoeae* include material from genital sources such as the urethra, cervix, vagina, and anal canal, as well as specimens from sites of extragenital infection, such as the oropharynx, skin lesions, inflamed joints, blood, and pelvic inflammatory disease. Sterile body fluids are collected in blood culture bottles or in syringes; they should be transported immediately to the laboratory in transport vials.

Neonates may become infected during passage through the birth canal of an infected mother; a particularly virulent conjunctivitis, gonococcal ophthalmia neonatorum, is often the result. With the requirement that silver nitrate, erythromycin, or penicillin drops be instilled in the eyes of all infants at birth, this disease has become quite uncommon. Swabs of discharge from the eye may be collected for culture.

The use of nontoxic cotton swabs (treated with charcoal to absorb toxic fatty acids), in the absence of gynecological lubricant, spermicides, or douches, allows the best recovery of organisms. Immediate plating is best to ensure optimal recovery of the gonococcus, since virtually all swabs have been shown to have some toxicity toward this organism. Specimens that are not inoculated immediately to culture media should be transported and held at room temperature in a protective carrying medium that prevents or minimizes overgrowth of nonpathogens and drying of the swab-containing specimen, preferably Amie's charcoal transport medium. It is better to refrigerate genital specimen swabs in transport media for a maximum of 3 hours than to incubate the swabs. The best method for culture and transport of *N. gonorrhoeae* is to inoculate agar immediately after specimen collection and to place the medium in an atmosphere of increased CO_2 for transport. Specimens can be inoculated to selective or nonselective agar plates, depending on the source of the specimen, and then placed immediately into a candle extinction jar (see next section). Specialized packaged media consisting of selective agar in plastic trays that contain a CO_2-generating system (Jembec plates, Neigon plates) are also available.

CULTURE MEDIA AND METHODS

Since gonococci must often be recovered from sites that contain large numbers of organisms from the commensal flora, such as the genital tract, and even from sites that may harbor nonpathogenic *Neisseria* sp., such as the oropharynx, special selective media have been developed to aid in the detection of *N. gonorrhoeae.* The first of these was Thayer-Martin medium, chocolate agar with an enrichment supplement and the antibiotics colistin (to inhibit gram-negative bacilli), nystatin (to inhibit yeast), and vancomycin (to inhibit gram-positive bacteria). Modifications of this medium include the addition of trimethoprim lactate to inhibit swarming of *Proteus* sp. (modified Thayer-Martin medium, MTM) and the substitution of anisomycin, an antifungal agent with a longer half-life, for nystatin, combined with an increased vancomycin concentration (Martin-Lewis medium). A transparent medium containing buffered proteose-peptone with many supplements and the same antibiotics as Thayer-Martin medium (New York City medium, NYC) has been employed with similar results for the recovery of gonococci.[7] From 5% to 30% of gonococci are inhibited by the concentrations of antibiotics, particularly vancomycin, in the selective media typically used, so the use of a nonselective medium is warranted in suspect cases that are culture negative and for specimens not contaminated with normal flora (e.g., joint fluid from suspected septic arthritis).

The increased CO_2 necessary for the growth of gonococci can be attained by incubating inoculated media in an incubator that regulates the flow of extraneously provided gas from a commercially purchased gas cylinder or by incubating inoculated media in a closed glass jar (e.g., commercial mayonnaise jar) in which a small, white, unscented candle has been allowed to burn to extinction, called a candle jar (Chapter 9), thus removing atmospheric oxygen (to approximately 3%).

Chapter 20 discusses sites to culture and techniques to apply for maximal recovery of *N. gonorrhoeae* from genital areas. Chapter 17 mentions methods of obtaining material for recovery of gonococci from the oropharynx. Chapter 15 discusses special methods for recovery of gonococci from blood in patients with disseminated disease. The recovery of *N. gonorrhoeae* from normally sterile body fluids does not require any special methods other than those routinely used for such specimens, except that they should be plated onto enriched chocolate agar, in addition to other media, and incubated for 48 hours at 35° to 37° C in a humidified atmosphere of 5% to 10% CO_2. Joint fluids rarely yield isolates, even in definite cases of gonococcal arthritis.

PRELIMINARY IDENTIFICATION

After 24 to 48 hours of incubation, colonies of *N. gonorrhoeae* appear as small (0.5 to 3 mm), translucent, grayish, convex, shiny colonies with margins that may be smooth or irregular. They often lift off the agar surface as whole colonies and may appear mucoid or sticky. Since the organisms produce an active autolytic enzyme, they must not be exposed to room air for too long. Colonies on chocolate or selective agar that resemble those of *N. gonorrhoeae* can be screened for the presence of the enzyme indophenol oxidase either by removing a portion of a colony to a piece of filter paper for the Kovacs' test (Chapter 10) or by dropping the fresh reagent (tetramethyl-*p*-phenylenediamine dihydrochloride) directly onto the colony on the agar surface.

Oxidase-positive colonies turn purple, darkening within 10 seconds (Figure 27.2). These colonies must be subcultured immediately, since the oxidase reagent rapidly renders them nonviable. A Gram stain of the oxidase-positive organism shows typical gram-negative diplococci, with adjacent sides flattened. It has been recommended that the isolation of oxidase-positive, gram-negative diplococci on selective agar such as Thayer-Martin medium from genital sites of infection may be reported presumptively as *N. gonorrhoeae*. However, several recent reports have documented the increasing incidence of gonorrhea-like infections in genital sites caused by *N. meningitidis* and misidentification of other *Neisseria* as *N. gonorrhoeae*, indicating that the reporting of gonococcal disease based on presumptive criteria is no longer valid.[4,6] Definitive biochemical identification methods and newer identification techniques are discussed in later sections. It cannot be assumed that an organism with diplococcal morphology that grows on a selective medium is a pathogenic *Neisseria* sp., since other organisms, including *Kingella denitrificans* (most often), *Moraxella catarrhalis*, and *Neisseria cinerea*, occasionally grow on these media.

COLONY MORPHOLOGY VARIATION

Gonococcal colonies are very diverse, and phase variation is expressed by the presence of at least five distinct types of colony morphology that can be distinguished.[10] Types T1 and T2, small, bright, reflective colonies typical of fresh isolates from cases of gonorrhea, differ from the larger, flatter, nonreflecting colonies of types T3, T4, and T5 that develop from fresh isolates after in vitro passage (Figure 27.3). The T1 and T2 colony types, which are virulent for human volunteers,

FIGURE 27.2

Colonies of *Neisseria* sp. on Thayer-Martin agar after addition of oxidase reagent. Oxidase-positive organisms turn purple within 10 seconds.

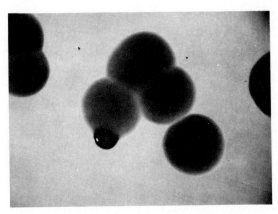

FIGURE 27.3

Colony morphology variants of *Neisseria gonorrhoeae*. The large colonies (type T3 or T4) represent nonpiliated organisms, whereas the small colonies (T1 or T2) indicate that the organisms possess pili and are more virulent. (Courtesy of D.S. Kellogg Jr., Centers for Disease Control.)

possess multiple hairlike projections extending from the cell membrane, called fimbriae or pili. The T3, T4, and T5 organisms fail to cause infection in human urethras and do not have fimbriae. These proteinaceous structures are thinner and shorter than flagella, which they resemble (Figure 27.4). The pili function in adherence of bacteria to mucosal cell surfaces and in the conjugal transfer of genetic material between bacteria. It seems likely that virulence, ability to adhere to host surfaces because of possession of pili, and colony morphology of the gonococci are all interrelated.

NUTRITIONALLY VARIANT GONOCOCCI

Another feature of *N. gonorrhoeae* that deserves mention is the phenomenon of *auxotypes*. Certain strains, or auxotypes, of gonococci require the addition of particular defined nutrients to grow on artificial media. Those strains that require arginine, hypoxanthine, and uracil (AHU-dependent auxotypes) grow slowly and are quite small. They are usually sensitive to penicillin, resistant to the bactericidal activity of human serum, and more likely to cause asymptomatic urethritis in males and disseminated gonococcal disease in females. Although at least 30 auxotypes have been described, the methods are not yet readily available for routine epidemiological studies.

ANTIBIOTIC RESISTANCE

Penicillin-resistant *N. gonorrhoeae* have been causing disease in the United States since the mid-1970s, but the incidence of such strains has now increased in many areas so that penicillin is no longer considered the agent of first choice for treatment of this disease. The ability to hydrolyze the β-lactam ring of penicillin-like antibiotics is mediated by a bacterial plasmid harboring the gene for the enzyme β-lactamase (or penicillinase). Penicillinase-producing *N. gonorrhoeae* (PPNG) can be easily detected using a rapid test for hydrolysis of a chromogenic cephalosporin, nitrocefin, as described in Chapter 14. We recommend that all isolates of *N. gonorrhoeae* be tested by this method and reported as "β-lactamase pos-

FIGURE 27.4

Neisseria gonorrhoea with pili extending from their surfaces. (Courtesy Chuen-mo To and C.C. Brinton Jr., University of Pittsburgh.)

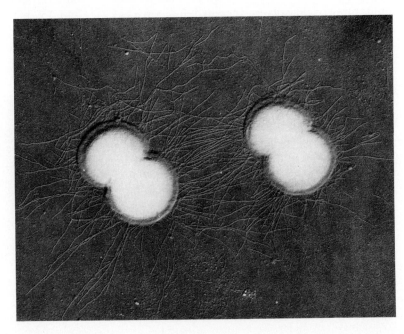

itive" or "β-lactamase negative." Penicillin resistance can occur in non-β-lactamase-producing strains as well. These strains are termed chromosomally mediated resistant *N. gonorrhoeae* (CM-RNG). Disk diffusion on GC agar base plus 1% IsoVitaleX is recommended; a zone 27 mm or greater surrounding a 10 U penicillin disk is considered susceptible.

NEISSERIA MENINGITIDIS

SPECIMEN COLLECTION

N. meningitidis is a normal inhabitant of the human nasopharynx, with carriage rates as high as 44% in some populations. The presence of meningococci in the upper respiratory tract of a normal host has not been shown to predispose the host to meningococcal disease. Thus, isolation of the organism from nasopharyngeal cultures taken from persons with no disease or known exposures has no clinical significance. However, close contacts (persons who eat and sleep together) of a patient with known meningococcal meningitis have as much as 1000 times greater risk of developing meningococcal disease than someone without any known contact. Therefore, carriage of *N. meningitidis* by a close contact of a patient with meningitis has significance; in this particular setting the carrier should be given prophylaxis, often rifampin. Since the organism is not normal oropharyngeal flora, it may be significant if recovered from sputum in large numbers with polymorphonuclear neutrophils in the absence of other pathogens, especially if found intracellularly. It is definitely significant if recovered from a transtracheal aspirate or material from percutaneous needle aspirate of the lung infiltrate from a patient with pneumonia. Definitive diagnosis of meningococcal bacteremia, meningitis, or other serious disease is usually accomplished by isolation of the organism from a normally sterile body fluid, such as blood, cerebrospinal fluid, synovial fluid, or pleural fluid, or from aspirates of petechiae. Other specimens may also yield meningococci, including material from the eyes, genital tract, and anus. Methods for isolation of the *Neisseria* sp. from blood are discussed in Chapter 15. Handling of other specimens for recovery of meningococci must minimize exposure to air and cool temperatures and prevent dehydration.

CULTURE MEDIA AND METHODS

Body fluids for isolation of all bacteria are treated as outlined in Chapter 22. Any volumes of clear fluid greater than 1 ml are centrifuged at room temperature at 1500 to $2800 \times g$ for 15 or 10 minutes, respectively. The supernatant is removed to a sterile tube and may be used for direct detection of soluble antigens; the sediment is vigorously vortexed and inoculated onto an enriched supportive medium, such as enriched chocolate agar, then into a supportive broth medium, such as thioglycolate; and a direct Gram stain is prepared from the last drop of sediment. Purulent fluids may be handled directly without concentrating. All media are immediately incubated in a moist atmosphere of 5% to 10% CO_2 at 36° to 37° C for at least 5 days before discarding as negative.

In contrast to the gonococcus, *N. meningitidis* strains are able to grow on blood agar as well as on chocolate and selective agar, although both organisms grow better in increased CO_2 concentration. Meningococci grow more rapidly and with larger colonies than gonococci. Colonies are typically smooth, round, moist, and uniform and homogeneous in appearance. A greenish cast underneath and a slight tan color of the colonies are typical. Specimens from sites with contaminating normal flora may be inoculated to Thayer-Martin or NYC media for selective isolation.

PRELIMINARY IDENTIFICATION

The same techniques used for preliminary identification of gonococci are useful for the meningococci. A positive oxidase test combined with typical gram-negative diplococcal morphology, particularly if the colony is growing well on selective media, defines those isolates that require further identification, either with the biochemical tests described next or with the use of one of the new techniques discussed later. As with gonococci, it is important that these organisms be fully identified.

SEROLOGIC CLASSIFICATION

Nine serogroups of *N. meningitidis* (A, B, C, D, X, Y, Z, 29E, W135) have been defined on the basis of differing capsular polysaccharides. The most important serogroups in the United States are A, B, C, and Y. However, local patterns can differ, since high numbers of W135 have been seen in Chicago. Interestingly, group B polysaccharide is a very poor immunogen for humans and animals, leading to difficulty in developing an effective vaccine against this strain and limiting the development of serologic reagents prepared in animals for laboratory diagnostic tests. Only since the development of techniques for the preparation of monoclonal antibodies has a consistently good immunological reagent for identification of group B meningococci become widely available.

TABLE 27.1

CHARACTERISTICS OF HUMAN *NEISSERIA* SP., *MORAXELLA CATARRHALIS*, AND *KINGELLA DENITRIFICANS*[a]

SPECIES	PIGMENT[b]	SUPEROXOL[c]	ACID PRODUCED FROM GLUCOSE	MALTOSE	FRUCTOSE	SUCROSE	LACTOSE (ONPG)
N. gonorrhoeae	−	+	+	−	−	−	−
N. meningitidis	−	−	+	+	−	−	−
N. lactamica	−	−	+	+	−	−	+
N. cinerea	−	−	−[h]	−	−	−	−
N. polysacchareae	−	−	+	+	−	−	−
N. kochii	−	+	+	−	−	−	−
N. flavescens	+	−	−	−	−	−	−
N. sicca	d	−	+	+	+	+	−
N. subflava							
Biovar *subflava*	+	−	+	+	−	−	−
Biovar *flava*	+	−	+	+	+	−	−
Biovar *perflava*	+	−	+	+	+	+	−
N. mucosa	+	−	+	+	+	+	−
M. catarrhalis	−	−	−	−	−	−	−
K. denitrificans	−	−	+	−	−	−	−

From Knapp, J.S. 1988. Clin. Microbiol. Rev. 1:41.

[a] *ONPG*, *o*-Nitrophenyl-β-D-galactopyranoside; *DNase*, deoxyribonuclease; *MTM*, modified Thayer-Martin medium; *ML*, Martin-Lewis medium; *NYC*, New York City medium. +, Most strains (≥90%) positive; −, most strains (≥90%) negative; d, some strains positive, some strains negative.

[b] Pigment observed in colonies on nutrient agar. Strains of *N. cinerea* and *N. lactamica* are yellow-brown and yellow pigmented when growth is harvested on a cotton applicator or smeared on filter paper.

[c] All *Neisseria* sp. and *M. catarrhalis* give a positive catalase test with 3% H_2O_2; *N. gonorrhoeae* strains give strong reactions with 30% H_2O_2 (superoxol), whereas other species are negative.

[d] Some strains may be inhibited by 5% sucrose; reactions may be obtained on a starch-free medium containing 1% sucrose. Strains of *N. gonorrhoeae*, *N. meningitidis*, and *N. kochii* do not grow on this medium.

[e] Results for tests in 0.1% (wt/vol) nitrite; *N. gonorrhoeae* and strains of some other species that are negative in 0.1% nitrite can reduce 0.01% (wt/vol) nitrite.

RAPID DETECTION OF *N. MENINGITIDIS* ANTIGEN IN BODY FLUIDS

The sometimes rapidly fulminant course of meningococcal meningitis has spurred the development of several rapid methods for the detection of capsular polysaccharide antigen in body fluids (urine, serum, cerebrospinal fluid) before culture results are available. Although counterimmuno-electrophoresis (CIE) was the first of these methods to be developed, it has been shown to be less sensitive and more time-consuming than particle agglutination techniques such as latex and coagglutination. These reagents are able to detect minute amounts of antigen, often in the absence of viable organisms caused by previous treatment or dilution. The latex particles have been the most sensitive in direct assays for antigen, although a thoroughly studied Gram stain is still the most definitive rapid diagnostic method. Because of a small but significant number of false-positive and false-negative results in the commercially available rapid antigen tests, clinicians rarely rely exclusively on such data.

OTHER *NEISSERIA* SP.

Neisseria cinerea has recently been recognized as a pathogen associated with bacteremia, conjunctivitis, nosocomial pneumonia, and proctitis.[2,4,13] *N. cinerea* is part of the normal oropharyngeal and genital tract flora. Colonies often have a slightly different morphological appearance, with a faint yellow pigmentation as compared with *N. gonorrhoeae*. The recognition of this species largely results from its biochemical resemblance to and consequent confusion with *N. gonorrhoeae*. Rare strains of *N. cinerea* have been found to metabolize glucose when tested by more than one system. To prevent misidentification, all strains of glucose and oxidase-positive gram-negative diplococci recovered from noninhibitory media (e.g., chocolate agar) should be plated to trypticase soy or

POLYSACCHARIDE FROM ≥1% SUCROSE[d]	REDUCTION OF		DNase	EXTRA CO₂ NEEDED[f]	GROWTH ON		
	NO₃	NO₂[e]			MTM, ML, OR NYC MEDIUM	CHOCOLATE, BLOOD AGAR AT 22° C	NUTRIENT AGAR AT 35° C
−	−	−	−	VI	+[g]	−	−
−	−	d	−	I	+	−	+
−	−	d	−	d	+	−	+
−	−	+	−	d	−[i]	−	+
+	−	d	−	d	+	−	+
−	−	−	−	No	+	−	+
+	−	−	−	I	+	−	+
+	−	+	−	No	−	+	+
−	−	+	−	No	−	+	+
−	−	+	−	No	−	+	+
+	−	+	−	No	−[k]	+	+
+	+	+	−	No	−	+	+
−	+	−	+	No	d	+	+
−	+	−	−	I	+	−	−

[f] Extra CO₂: VI, very important; I, important for growth; No, not needed for growth.

[g] ≥90% of vancomycin-susceptible strains of *N. gonorrhoeae* may not grow on Thayer-Martin or MTM medium.

[h] Some strains of *N. cinerea* may give a weak reaction in glucose in some rapid tests for the detection of acid from carbohydrates.

[i] Some strains of *N. cinerea* have been isolated on gonococcal selective medium but are colistin susceptible and will not grow when subcultured on selective media. Colistin-resistant mutants of *N. cinerea* have not been described.

[j] Strains of *N. subflava* biovars give consistent patterns of acid production when tested in appropriate media.

[k] Some strains of *N. subflava* biovar *perflava* grow on gonococcal selective media in primary culture, are colistin resistant, and grow on selective media on subculture.

Mueller-Hinton agar, which support growth of *N. cinerea* but not *N. gonorrhoeae,* and to selective agar such as MTM, on which growth of *N. cinerea* is inhibited. As another differential test, *N. cinerea* is susceptible to colistin, in contrast to *N. gonorrhoeae* and *N. meningitidis. Neisseria mucosa,* also normal flora in the human respiratory tract, has been isolated in association with meningitis, endocarditis, cellulitis, and most recently, neonatal conjunctivitis.[6] Colonies are mucoid and opaque, and the organism is able to grow on nutrient agar.

Certain strains of *Neisseria* exhibit biochemical characteristics of *N. gonorrhoeae* and serologic characteristics of *N. meningitidis.* These strains have been termed *N. gonorrhoeae* subsp. *kochii.* Originally isolated from conjunctivitis specimens from patients in Egypt, a similar strain has been described from a vaginal specimen in Canada.[9,11]

Neisseria polysacchareae, a recently described commensal in the nasopharynx, resembles *N. meningitidis* morphologically. It may be isolated from selective media, such as Thayer-Martin agar; most strains produce acid from glucose and maltose, whereas sucrose oxidation is variable. This nonpathogenic species is distinguished by its production of large amounts of polysaccharides, visible as precipitation after 3 days of growth, when grown on 1% or 5% sucrose agar or in 1% or 5% sucrose broth.[1]

Other commensal *Neisseriae* include *N. flavescens, N. lactamica, N. sicca,* and the *N. subflava* biovars: *subflava, flava,* and *perflava.* Knapp[11] suggests that *N. flavescens* is primarily of historical importance. Except for *N. lactamica* and *N. flavescens,* these organisms do not grow on gonococcal selective media, and colonies are usually opaque and often yellow pigmented. They can also be distinguished from pathogenic species biochemically.

MORAXELLA CATARRHALIS

Former *Branhamella catarrhalis,* the only species in the genus, morphologically and biochemically resembles the *Neisseria* sp. (Table 27.1), but it is now classified as a member of the *Moraxella* genus on the basis of deoxyribonucleic acid (DNA) homology studies. Typical colonies of *Moraxella catarrhalis* are whiter, more raised, and drier in appearance than *Neisseria* sp. Although a normal commensal of the upper respiratory tract, *M. catarrhalis* has been isolated as the causative agent of disease, including septicemia, meningitis, endocarditis, otitis media, conjunctivitis, sinusitis, laryngitis, pneumonia, and bronchitis. Most patients are elderly, and many have predisposing pulmonary disease.[8] Any gram-negative diplococci recovered in large numbers from the lower respiratory tract of a patient with bronchitis or pneumonia should be identified. *M. catarrhalis* is asaccharolytic, usually β-lactamase positive, and the only DNase-positive species within the group.

DEFINITIVE BIOCHEMICAL IDENTIFICATION

TRADITIONAL BIOCHEMICAL IDENTIFICATION

The extent to which identification of isolates resembling *Neisseria* sp. is carried out depends on the source of the specimen. An isolate from a child must be identified unequivocally because of the medicolegal ramifications of those results. Isolates from genital sites of adult patients at risk of sexually transmitted disease can be identified presumptively; this is a good setting for one of the new, rapid methods mentioned later. We recommend complete identification of isolates from normally sterile body fluids.

Isolated bacterial colonies on selective media to be characterized by biochemical confirmatory tests must be subcultured to a nonselective medium, such as chocolate agar, for purification and amplification. Subcultures are incubated for 24 hours (a maximum of 48 hours) before inoculation to carbohydrates. Cystine trypticase soy agar (CTA) base, pH 7.3, with the addition of 1% filter-sterilized carbohydrate solutions, has been used traditionally. Interpretable results may be obtained with a very heavy inoculum scraped from a plate on a cotton swab, which is swirled vigorously in the top ¼ inch of the medium. CTA carbohydrates, as with all carbohydrates, are incubated in air. To overcome some problems associated with the growth-dependent CTA medium test, detection of carbohydrate oxidation patterns may be enhanced by inoculating an extremely heavy suspension of the organism into a small volume of buffered, low-peptone substrate, described in Procedure 10.10.

Table 27.1 shows the reactions used to differentiate members of *Neisseria* and *Kingella* that may be confused. The superoxol test, production of gas bubbles from 30% H_2O_2, is always positive for *N.*

gonorrhoeae, a useful rapid screening and adjunct test from selective media. However, caution is advised when using this test, since about 50% of *M. catarrhalis* and *N. lactamica* also are positive with this test. *Kingella,* being a true rod, elongates and forms filaments under the influence of penicillin. This can be visualized by staining the growth at the margin of the zone of inhibition of a 10 U penicillin disk placed on the surface of an agar plate streaked with the organism being tested.

COMMERCIAL BIOCHEMICAL SYSTEMS

Several commercial identification systems that employ biochemical or enzymatic substrates are available for identification of *Neisseria* sp. and *M. catarrhalis* or of *N. gonorrhoeae* alone (RapID NH, Innovative Diagnostic Systems, Atlanta; Gonochek II, E-Y Laboratories, San Mateo, Calif.; RIM-Neisseria Kit, Remel Laboratories, Lenexa, Kan.; Minitek, Becton-Dickinson Microbiology Systems, Franklin Lakes, N.J.; Quad-FERM+, bioMérieux Vitek, Hazelwood, Mo.; Neisseria-Haemophilus Identification card, bioMérieux Vitek; Identicult-Neisseria, Adams Scientific, Inc., W. Warwick, R.I.; HNID Panel, MicroScan Division of Baxter/American Scientific Products, W. Sacramento). These products usually use the ability of preformed enzymes to act on substrates (conventional or chromogenic) to form a colored end product as the indicator system. A heavy inoculum of the organism is necessary for proper reactions, although viable organisms may not be necessary. More reliable identifications result from systems that combine enzymatic and other substrates. Manufacturers' instructions must be followed exactly; several systems were developed only for strains isolated on selective media and should not be used to test other gram-negative diplococci.

NONBIOCHEMICAL DETECTION AND IDENTIFICATION METHODS

IDENTIFICATION METHODS

Monoclonal fluorescent antibody stains are commercially available (Bartels Baxter, W. Sacramento; Syva Co., Palo Alto, Calif.) for *N. gonorrhoeae* culture confirmation. Because of many cross-reacting organisms in urethral or other discharge, these stains are not particularly useful for direct examination of such specimens. Once an oxidase-positive organism morphologically consistent with gonococci has been isolated on selective media, however, fluorescent stain may provide rapid identification; equivocal reactions must be confirmed by some other method. Good fluorescent reagents are not available for other members of Neisseriaceae.

Antibody-based particle-agglutination identification systems available include GonoGen I (New Horizons Diagnostics, Columbia, Md.), Phadebact (Karo-Bio, Huddinge, Sweden), and Meritec-GC (Meridian Laboratories, Cincinnati). The measurement of radioactive end products of metabolism from labeled substrates in fluid medium has also been used as a method for identification of *Neisseria* sp. Use of the BACTEC system was detailed in Chapter 12.

DIRECT DETECTION OF GONOCOCCI IN SPECIMENS

A commercial enzyme-linked immunosorbent assay (ELISA) system (Gonozyme, Abbott Laboratories, N. Chicago) has been shown to be very sensitive and specific for the detection of gonococcal antigen in the urethral and endocervical discharge from suspected cases of gonorrhea, although no more sensitive than a Gram stain of male urethral discharge. The test is much more rapid than culture and is suitable for large-scale screening programs among sexually active people.[3] The ELISA tests have not been particularly useful for rapid diagnosis of meningitis, perhaps because of the time required to perform the assays (often 4 to 6 hours). Studies are exploring ways to speed up the end point determination, which may allow the ELISA systems to serve better as diagnostic tests for antigen in cerebrospinal fluid.

Gen-Probe (San Diego) has developed a liquid-probe system with a chemiluminescent detection system for direct detection of gonococci in genital specimens. Tests have shown its sensitivity to be equal to that of the ELISA system, with somewhat greater specificity. Clinical studies have shown it to perform well when testing high-risk patients.[12]

The Limulus test for endotoxin may be useful for detection of meningitis caused by meningococci, especially if organisms are present at greater than $10^3/cm^3$. The gel formation of the lysate is nonspecific, however, and will occur in the presence of endotoxin from any organism, such as gram-negative bacteria. Surprisingly, this assay has been found to be useful as a diagnostic test for gonococcal infection in women, as well as in men. Material collected on swabs from the endocervical canal of women with gonorrhea, including early cases and asymptomatic cases, may yield positive Limulus amebocyte lysate tests.

REFERENCES

1. Boquete, M.T., Marcos, C., and Saez-Nieto, J.A. 1986. Characterization of *Neisseria polysacchareae* sp. nov. (Riou, 1983) in previously identified noncapsular strains of *Neisseria meningitidis.* J. Clin. Microbiol. 23:973.
2. Boyce, J.M., Taylor, M.R., Mitchell, E.B. Jr., et al. 1985. Nosocomial pneumonia caused by a glucose-metabolizing strain of *Neisseria cinerea.* J. Clin. Microbiol. 21:1.
3. Demetriou, E., Sackett, R., Welch, D.F., et al. 1984. Evaluation of an enzyme immunoassay for detection of *Neisseria gonorrhoeae* in an adolescent population. J.A.M.A. 252:247.
4. Dossett, J.H., Appelbaum, P.C., Knapp, J.S., et al. 1985. Proctitis associated with *Neisseria cinerea* misidentified as *Neisseria gonorrhoeae* in a child. J. Clin. Microbiol. 21:575.
5. Evans, K.D., Peterson, E.M., Curry, J.I., et al. 1986. Effect of holding temperature on isolation of *Neisseria gonorrhoeae.* J. Clin. Microbiol. 24:1109.
6. Gini, G.A. 1987. Ocular infection in a newborn caused by *Neisseria mucosa.* J. Clin. Microbiol. 25:1574.
7. Greenwood, J.R., Voss, J., Smith, R.F., et al. 1986. Comparative evaluation of New York City and modified Thayer-Martin medium for isolation of *Neisseria gonorrhoeae.* J. Clin. Microbiol. 24:1111.
8. Hager, H., Verghese, A., Alvarez, S., et al. 1987. *Branhamella catarrhalis* respiratory tract infections. Rev. Infect. Dis. 9:1140.
9. Hodge, D.S., Ashton, F.E., Terro, R., et al. 1987. Organism resembling *Neisseria gonorrhoeae* and *Neisseria meningitidis.* J. Clin. Microbiol. 25:1546.
10. Kellogg, D.S., Jr., Peacock, W.L., Jr., Deacon, W.E., et al. 1963. *Neisseria gonorrhoeae.* I. Virulence genetically linked to clonal variation. J. Bacteriol. 85:1274.
11. Knapp, J.S. 1988. Historical perspectives and identification of *Neisseria* and related species. Clin. Microbiol. Rev. 1:41.
12. Panke, E.S., Yang, P.A., Leist, P., et al. 1991. Comparison of Gen-Probe DNA probe test and culture for the detection of *Neisseria gonorrhoeae* in endocervical specimens. J. Clin. Microbiol. 29:883.
13. Southern, P.M., Jr. and Kutscher, A.E. 1987. Bacteremia due to *Neisseria cinerea:* report of two cases. Diagn. Microbiol. Infect. Dis. 7:143.

BIBLIOGRAPHY

Atlas, R.M. 1993, Handbook of microbiologic media. CRC Press, Boca Raton, Fla. (Includes detailed formulas and instructions for media preparation.)
Morello, J.A., editor. 1989. Perspectives on pathogenic Neisseriae. Clin. Microbiol. Rev. Suppl. 2:S1. (Entire issue devoted to Neisseriae.)
Evangelista, A.T., and Beilstein, H.R. 1993. Laboratory diagnosis of gonorrhea. In Abramson, C., coordinating editor. Cumitech 4A. American Society for Microbiology, Washington, D.C.

28

ENTEROBACTERIACEAE

Members of the family Enterobacteriaceae are the most frequently encountered bacteria isolated from many types of clinical specimens. They are gram-negative bacilli, able to grow readily on all supportive media (blood and chocolate agars), and form distinctive colonies on selective agars depending on their metabolic capabilities and the components of the media (Chapter 9). Some species are classical pathogens, assumed to be the etiological agents of disease when they are encountered in clinical specimens, independent of the numbers of colony-forming units detected *(Salmonella, Shigella,* and *Yersinia pestis)*. Other species can colonize humans without evoking any pathological response, such as those Enterobacteriaceae that colonize the human intestinal tract.

These "commensal" organisms, if they are introduced into a susceptible site (peritoneal cavity, lung of an immunosuppressed host, cerebrospinal fluid, etc.), are able to cause disease. Still others, such as *Escherichia coli,* are associated with certain syndromes based on the presence of virulence factors that are not possessed by all strains or even by the same strain at all times. Members of this ubiquitous family of similar organisms have been mentioned in association with each of the anatomical sites of infectious disease covered in previous chapters, including blood, cerebrospinal fluid, respiratory tract, gastrointestinal tract, urinary tract, soft tissue, sterile body fluids, and a variety of loci in immunocompromised hosts. The Enterobacteriaceae are even associated with sexually transmitted diseases as one of the infections encountered in the rectum of homosexual males.

Identification of the Enterobacteriaceae to species level is of importance epidemiologically as well as clinically. The susceptibility patterns of many species are predictable, allowing the clinician to choose therapy based on an isolate's identification. The presence of an epidemic or small outbreak, caused by a particularly virulent strain of Enterobacteriaceae or caused by a breakdown in technique among health care workers, is often first detected by microbiologists who have noticed a cluster of isolates of an unusual biotype. The discovery of an extremely rare biotype in a cluster of blood cultures led a Milwaukee microbiologist to discover the source of a nationwide outbreak of septicemia associated with contaminated intravenous fluids. Chapter 6 discusses the importance of species differentiation for epidemiological purposes.

Within the last several years the number of recognized species of Enterobacteriaceae has grown from 26 to more than 100 (approximately 90 species have been named), and more species are being validated monthly.[7] This rapid expansion is somewhat overwhelming to clinical microbiologists, who are not taxonomists and generally wish only to convey some meaningful information to clinicians concerning bacterial species isolated from clinical material. Most microbiology laboratories are able to identify the newly recognized species of Enterobacteriaceae, primarily because of the availability of commercially produced multibiochemical and enzymatic parameter test systems. This chapter focuses on the 13 genera of Enterobacteriaceae that are most likely to be encountered in clinical specimens. Although we advocate the definitive identification of isolates that can be causally related to disease, particularly

from systemic sites of infection such as blood, cerebrospinal fluid, tissue, or other sterile body fluids, it is not practical for routine clinical laboratories to maintain all the unusual, rarely used, or technically demanding media or test reagents necessary for thorough studies of rarely isolated organisms. Such isolates should be forwarded to a reference laboratory for further studies when necessary for patient care.

EPIDEMIOLOGY AND PATHOGENESIS OF INFECTIONS CAUSED BY ENTEROBACTERIACEAE

Enterobacteriaceae are the predominant facultative flora in the human bowel, a site from which they can easily be disseminated. Lapses in personal hygiene, especially during periods of diarrheal disease, can contribute to the fecal-oral route of transmission of the agents of gastroenteritis and related diseases. Countries with poor sanitation systems are more likely to have environmental reservoirs of the pathogenic Enterobacteriaceae, from which disease is maintained in the population. Hospitalized patients quickly become colonized on the skin or in the respiratory tract with strains endemic to the hospital, allowing these organisms easy access to the host under any compromising situation.

In addition to their presence as normal human fecal flora, Enterobacteriaceae are found in natural habitats worldwide, including soil, water, plants of all types, fish, insects, and other animals. Their importance in human disease is at least equaled by their importance as causes of disease in poultry, domestic livestock, fish, and several vegetable and plant crops, with major economic consequences. Although the genera that have not been associated with human disease are listed here (Table 28.1), they are not discussed in detail.

It is increasingly possible that any species of Enterobacteriaceae may be isolated from clinical specimens for three primary reasons: (1) more immunocompromised patients survive long enough to become infected, (2) technical proficiency among microbiologists is increasing, and (3) commercial identification systems are expanding their data bases. Only by identifying unusual clinical isolates can microbiologists discover previously unrecognized agents of human disease.

ENTEROBACTERIACEAE ASSOCIATED WITH GASTROENTERITIS AND ENTERIC FEVERS
Several Enterobacteriaceae have been associated with gastroenteritis and foodborne disease (Chap-

TABLE 28.1

SELECTED GENERA AND SPECIES OF ENTEROBACTERIACEAE

GENUS	SPECIES	COMMENTS
Budvicia*	aquatica	Found in drinking and surface water; isolated from human feces
Buttiauxella	agrestis	Found in water, not associated with human disease
Cedecea*	davisae	Isolated from humans, primarily respiratory tract and wounds
	lapagei	
	neteri	
	(others)	
Citrobacter	amalonaticus	Isolated from humans, wound infections, urine, sepsis; fecal flora; also isolated from environment
	diversus	
	freundii	
	(others)	
Edwardsiella	hoshinae	Isolated from humans and animals; associated with diarrhea, wound infections, sepsis
	ictaluri	
	tarda	
	(others)	
Enterobacter	aerogenes*	Very common, normal fecal flora; isolated from wounds, respiratory tract, urine, blood; E. agglomerans also in animals, others found in environment; E. cloacae most clinically significant; E. agglomerans and E. sakazakii yellow pigmented
	agglomerans*	
	amnigenus*	
	asburiae	
	cloacae*	
	dissolvens	
	gergoviae*	
	intermedium	
	nimipressuralis	
	sakazakii*	
	taylorae*	
	(others)	
Erwinia	amylovora	Associated with diseases of plants
	(others)	
Escherichia	blattae	Isolated from humans and animals, E. coli most common human pathogen; E. blattae found only in cockroaches; E. hermannii and E. vulneris isolated from wounds; others normal stool flora
	coli*	
	hermannii*	
	vulneris*	
Ewingella*	americana	Isolated from humans, respiratory tract, blood; no environmental source known
Hafnia*	alvei	Isolated from humans, respiratory tract, other sources; no environmental source known
	(others)	
Klebsiella	oxytoca*	Normal stool flora, some strains found in environment; isolated from respiratory tract, urine, wounds, blood; K. terrigena found only in water
	planticola*	
	pneumoniae*	
	pneumoniae subsp. ozaenae	
	pneumoniae subsp. rhinoscleromatis*	
	terrigena	
Kluyvera*	ascorbata	Probably normal stool flora; isolated from human infections, respiratory tract, blood, urine
	cryocrescens	
Leclercia*	adecarboxylata	Isolated from humans, respiratory tract, blood, urine, wounds; also found in food and water
Leminorella*	grimontii	Isolated from human feces; H_2S positive, not known to be pathogenic for humans
	richardii	
Moellerella*	wisconsensis	Isolated from human stool and natural water; not known to be pathogenic for humans; resistant to colistin
Morganella*	morganii	Normal stool flora; isolated from human infections (urine, blood, others)
	(others)	
Obesumbacterium	proteus	Isolated from brewery yeast; not known to be pathogenic for humans

TABLE 28.1

SELECTED GENERA AND SPECIES OF ENTEROBACTERIACEAE—cont'd

GENUS	SPECIES	COMMENTS
Pantoea	*agglomerans* *dispersa* (others)	Similar to *Enterobacter agglomerans;* primarily plant pathogens; yellow pigment produced
Pragia	*fontium*	Found in drinking water and stools from asymptomatic humans; most strains produce H$_2$S
Proteus	*mirabilis** *myxofaciens* *penneri** *vulgaris**	Most species normal fecal flora; isolated from human infections, urine, wounds, blood, others; *P. myxofaciens* isolated from moths only; *P. penneri* resembles *P. vulgaris,* including spreading growth but is indole negative and chloramphenicol resistant
Providencia	*alcalifaciens** *heimbachae* *rettgeri** *rustigianii** *stuartii**	Most species are normal fecal flora; isolated from human infections, urine, wounds, blood; *P. rustigianii* not known to be pathogenic for humans
*Rahnella**	*aquatilis*	Isolated from water; only one human isolate reported; not known to be pathogenic to humans
*Salmonella**	(enterica) >2000 serovars	Widely distributed in humans and animals; cause of gastroenteritis and enteric fever; genus *Arizona* now included in *Salmonella;* nomenclature still undecided
*Serratia**	*ficaria* *fonticola* *grimesii* *marcescens* *odorifera* *plymuthica* *rubidaea* (others)	Isolated from humans, water, rarely animals; some species important human pathogens; isolated from respiratory tract, wounds, blood, urine, others; *S. ficaria, odorifera, plymuthica, rubidaea* not known to be human pathogens; *S. fonticola* probably not true member of *Serratia* genus
*Shigella**	*boydii* *dysenteriae* *flexneri* *sonnei*	Human reservoir only, not normal flora; cause of gastroenteritis and bacterial dysentery; genetically identical to *E. coli*
*Tatumella**	*ptyseos*	Isolated from humans, respiratory tract, urine, blood; unlike other Enterobacteriaceae, *Tatumella* has polar flagella; susceptible to penicillin; poor grower, better at 25° C
Xenorhabdus	*luminescens* *nematophilus*	Isolated from nematodes only; grow only at 25° C, not 35° C; no human isolates
*Yersinia**	*aldovae* *enterocolitica* *frederiksenii* *intermedia* *kristensenii* *pestis* *philomiragia* *pseudotuberculosis* *ruckeri*	Isolated from humans, animals, environment; some important human pathogens; agents of plague, gastroenteritis, other infections; isolated from stool, blood, urine, wounds; *Y. ruckeri* not found in humans, cause of "red mouth" disease of fish; *Y. ruckeri* probably not true member of genus *Yersinia*
Yokenella regensburgei		Isolated from human feces, wounds, respiratory tract, knee joint

*Isolated from clinical specimens.

ter 18). One member, the *Salmonella* sp., is the etiological agent of most foodborne gastroenteritis in the United States caused by this family. More than 50,000 cases were reported to the Centers for Disease Control (CDC) in 1987. By 1990 the number had dropped somewhat, with 20 cases per 100,000 individuals in the U.S. population becoming infected. *Salmonella* sp. are naturally found in poultry, but disease can be transmitted in a variety of foods. A major outbreak occurred in Chicago in the spring of 1985, when contaminated milk from the dairy supplying a large food store chain infected many individuals. Ultimately, more than 5000 individuals contracted *Salmonella* gastroenteritis, resulting in several fatalities.

Salmonella sp. are able to invade the intestinal mucosa. Some strains produce an enterotoxin. *S. typhi*, *S. paratyphi* A (the paratyphoid bacilli), and *S. cholerae-suis* are more likely to invade systemically, entering the bloodstream and causing serious febrile disease (typhoid or enteric fever). These strains are usually carried only in human hosts, passed among individuals by fecal contamination of food or water. Because *Salmonella* sp. are susceptible to gastric acid, it was thought that a large inoculum (approximately 10^5 organisms) must be ingested. However, recent outbreaks of foodborne disease have shown that in some populations the infective dose can be extremely low, similar to that of shigellosis.[5] Very young, very old, and other compromised hosts are most likely to become seriously ill. One third of all cases of salmonellosis occur in children less than 5 years old.

The incidence of typhoid fever in the United States is quite low, about 400 cases reported each year between 1980 and 1990, and in approximately half of these the disease was acquired during foreign travel. Domestically acquired typhoid fever, however, is often transmitted by a chronic carrier of *S. typhi*. After recovering from typhoid fever, some individuals carry the bacterium asymptomatically for long periods. Other individuals are asymptomatic carriers without ever having disease. The silent carriers contribute to continued episodes of infections.

Shigella sp. are carried primarily by humans and are not disseminated in nature. The organism is able to resist gastric acidity, allowing a very small inoculum (as few as 10 organisms) to cause disease. Shigellosis also occurs primarily in children, although the incidence is not as high as that of salmonellosis. In 1987 approximately 24,000 cases of shigellosis were reported to the CDC. *Shigella* sp. produce an enterotoxin, but it is not as important to virulence as is their invasiveness. The organisms invade the epithelial cells of the large bowel, causing a dysentery-like syndrome (Chapter 18). *Shigella* sp. rarely invade systemically.

Escherichia coli is also an important cause of gastrointestinal disease. It can be associated with nursery outbreaks of diarrhea and with travelers' diarrhea (turista, Delhi belly, Montezuma's revenge, and other amusing epithets), an important source of morbidity to Americans traveling to other countries. Virulence mecha-

nisms of *E. coli* include production of an enterotoxin similar to that of *Vibrio cholerae*, invasiveness, production of cytotoxins, and adherence (Chapter 18). In previous years, nursery outbreaks of *E. coli*–associated gastroenteritis were associated with particular serotypes of the organism, called **enteropathogenic *E. coli*** (EPEC). Serotypes may still be associated with epidemics or geographical regions, but the serotype is not the only significant factor; it does serve as a marker for a particular strain. Toxin production is carried on a plasmid, a transmissible genetic element that can be easily transferred from strain to strain among and between species. For this reason, serotyping of single isolates of *E. coli* from pediatric patients with diarrhea is not recommended. For epidemiological reasons and when outbreaks are suspected, serotyping is still a valuable tool. Local health departments are better equipped to handle such situations than are most clinical laboratories.

In 1982 a particular serotype of *E. coli* was recognized as causing a serious diarrheal illness acquired from eating undercooked meat. The serotype, *E. coli* O157:H7, is associated with the hemolytic-uremic syndrome (HUS) in very young persons and institutionalized older adults. This syndrome has a high mortality rate in these populations. In early 1993 a large outbreak of this diarrhea occurred on the west coast of the United States and was associated with eating hamburgers at a fast-food restaurant chain. Several hundred people and several deaths were associated with eating the undercooked meat. This strain of *E. coli* does not readily ferment D-sorbitol, so colonies grown on MacConkey-sorbitol agar can be recognized with subsequent confirmation using specific antisera. *E. coli* O157:H7 is now found more frequently as a cause of either bloody or nonbloody diarrhea than either *Salmonella* or *Shigella* in some areas.[6]

Tests for toxin production and tests using genetic probes to detect the toxin gene sequences (although not universally available) more accurately identify etiological agents (**enterotoxigenic *E. coli*, ETEC**). A particle agglutination test for detection of heat-labile and heat-stable toxins of *E. coli* is available commercially but is not widely used in clinical laboratories. The gastrointestinal disease associated with invasive or traditional cytotoxic *E. coli* can resemble shigellosis, and the disease mediated by ETEC can resemble cholera. Additionally, serious hemorrhagic colitis and HUS have been associated with infection with verocytotoxin-producing (a specific cytotoxin) *E. coli,* such as the O157:H7 strain.[6] Other mechanisms of pathogenicity also occur in still other types of *E. coli.*

The fourth genus of Enterobacteriaceae containing organisms that are major etiological agents of diarrheal disease is *Yersinia,* which also includes the etiological agent of human plague *(Y. pestis). Y. enterocolitica* and, less often, *Y. pseudotuberculosis* and *Y. intermedia*[1] have only recently been implicated as significant causes of gastroenteritis. The organisms possess the ability to invade the intestinal mucosa and to produce a heat-stable enterotoxin similar to that of some strains of *E. coli.* Virulence factors shared among other members of *Yersinia* sp., such as the ability to survive intracellularly, may also contribute to pathogenesis. Symptoms that accompany *Y. pseudotuberculosis* and occasionally *Y. enterocolitica* infections may mimic those of appendicitis, although the symptoms are actually caused by mesenteric lymphadenitis. The organisms can invade the epithelium to cause systemic disease. *Y. enterocolitica* has also been associated with exudative pharyngitis.

Although the incidence of diarrheal disease associated with *Yersinia* is much higher in Europe and Canada than in the United States, more isolations are occurring in the United States over time, aided by increased awareness of the organism and better laboratory techniques designed especially for its detection (Chapter 18). Certain serotypes, primarily 03 and 09, have been recovered from patients with diarrhea in Europe and Scandinavia. These serotypes are less common in the United States, where 08 predominates. The incidence of 03 isolations, however, seems to be increasing, at least in the New York area, with an accompanying greater risk of acquiring disease via human-to-human transmission.[2]

Gastroenteritis has also infrequently been associated with *Edwardsiella tarda,* a member of the Enterobacteriaceae that resembles salmonellae in its pathogenic mechanisms, as well as morphologically. The organism has also been known to invade through the intestinal epithelium, causing systemic disease. Reported cases of *Edwardsiella*-associated gastroenteritis are uncommon.

YERSINIA PESTIS

The agent of human plague, *Y. pestis,* causes disease among many wild animal species, including prairie dogs, squirrels, rabbits, chipmunks, and others. The organism is transmitted among these animals by fleas and other ectoparasites, which also spread the organism to carnivores that prey on the smaller rodents. Humans acquire the disease when bitten by an infected flea or by handling the carcass of an infected animal. Rarely, infection can be spread from person to person via infected aerosol droplets from respiratory secretions of patients who have plague pneumonia (the Black Death). Most cases occur among young individuals in the southwestern and western United States; Navajos are at higher risk because of intense rodent plague in areas of the reservation in Arizona. The incidence increased steadily beginning in 1965, with 19 cases reported in 1982 and 40 cases reported in the peak year of 1983. Cases dropped considerably after 1983, with only 12 cases reported in 1987 and 12 in 1992. This probably reflects a natural cycle of plague within the wild rodent population.

Once the organisms penetrate the skin, through small breaks or via flea bites, or are breathed in, they are taken up by macrophages and polymorphonuclear neutrophils, within which they can multiply. During this first intracellular phase, the bacteria are stimulated by their immediate environment (lack of Ca^{++}, temperature, ionic concentration) to produce a protein and a lipoprotein antigen, called V and W antigens, that render the organism less susceptible to intracellular killing by other reticuloendothelial cells once they escape the original cell through lysis of that cell. The genetic determinants that code for these antigens are located on plasmids. *Y. pestis* also possesses an envelope antigen (F-1) that may contribute to resistance to phagocytosis. Virulent *Y. pestis* produces three factors, a bacteriocin called pesticin I, coagulase, and fibrinolysin. Produced together, these three factors are plasmid mediated; coagulase and fibrinolysin contribute to virulence by promoting invasiveness. Other factors associated with virulence include the ability of virulent strains to absorb hemin and basic aromatic dyes to form colored colonies (related to iron transport and storage), the ability to synthesize purines, and production of a toxin.

Primary disease is usually of the "bubonic" variety, characterized by infected and swollen lymph nodes and fever. Sepsis and sometimes pneumonia follow within several days. Septicemic and pneumonic plague may also occur without the lymphadenopathy stage. Untreated human plague is 50% to 60% fatal. Streptomycin and tetracycline are the antimicrobial agents of choice. If meningitis develops, chloramphenicol should be added to the regimen.

OTHER INFECTIONS ASSOCIATED WITH MEMBERS OF ENTEROBACTERIACEAE

Enterobacteriaceae, particularly *E. coli,* are the most common cause of urinary tract infections. Urinary tract infections are the largest category of nosocomial infections (Chapter 24); they may lead to secondary sepsis, shock, and death. Nosocomial infections associated with Enterobacteriaceae also include pneumonia, surgical wound infections, and catheter-related sepsis. The species most likely to be recovered from urine of infected individuals are those that are able to adhere to urinary epithelium. Attachment-mediating, fingerlike projections composed of protein, called **fimbriae,** are often found on strains that are able to adhere not only to urinary tract epithelial cells, but also to many other human and animal epithelial cell types (see Figure 17.1). Fimbriae are also found among gastroenteritis-producing strains, especially in animals. The ability to colonize the gastrointestinal tract, resisting the action of peristalsis and other factors, is mediated at least in part by fimbriae.

Most Enterobacteriaceae may be involved in bacterial pneumonia, but *Klebsiella pneumoniae* seems especially suited to this role. The copious polysaccharide capsular material produced by this species probably aids the organism to escape phagocytosis. Chapter 17 contains a discussion of those factors that contribute to the pathogenesis of pneumonia. Elderly patients and those compromised by influenza, alcoholism, and other chronic diseases are most likely to contract community-acquired *Klebsiella* pneumonia. It is a significant cause of hospital-acquired pneumonia as well.

Neonatal meningitis has been associated with certain strains of *E. coli,* most of which possess the K1 capsular antigen. Capsular polysaccharide may contribute to the pathogenesis of this disease. *Citrobacter* sp., also encapsulated, have been associated with neonatal meningitis as well. It is postulated that the organisms can somehow invade through the infant's intestinal epithelium to cause systemic disease. *E. coli* is no more likely to be the etiological agent of adult meningitis than are any other members of the Enterobacteriaceae. The role of capsular polysaccharide as a factor in the development of meningitis is discussed in Chapter 16.

Serratia sp. are sometimes associated with especially severe hospital-acquired infections, including pneumonia and surgical wound infections.[9] The organisms recovered from blood may be resistant to the bactericidal activity of serum and complement. Nosocomially derived *Serratia* strains are often resistant to multiple antibiotics. Rarely, strains of *S. marcescens* display a red, non-water-soluble pigment (Figure 28.1). These pigmented strains are less likely to be resistant and are more likely to be community acquired.

Most members of the Enterobacteriaceae have been associated with various infections, particularly among hospitalized and otherwise compromised hosts. The *Proteus, Providencia,* and *Morganella* sp. are most often recovered from urine of patients with urinary tract infections, although

FIGURE 28.1
Pigmented *Serratia marcescens* sp.

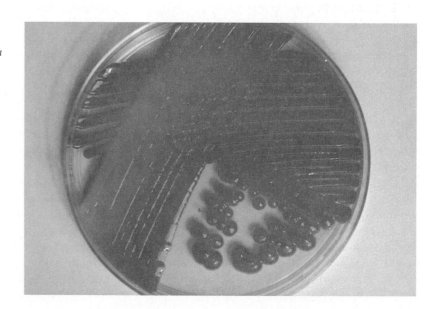

they can also be found associated with pneumonia, wound infections, and other syndromes. Those strains able to hydrolyze urea have been associated with the development of renal calculi in infected patients.

A virulence factor possessed by all Enterobacteriaceae (as well as all gram-negative bacteria) is the cell wall lipopolysaccharide **endotoxin.** Biological activity of this component, released during growth and breakdown of gram-negative bacterial cells, is discussed in Chapter 15. The major pathological consequence of gram-negative septicemia, shock, is mediated by endotoxin. Patients may experience disseminated intravascular coagulopathy, respiratory tract problems, cellular and tissue injury, fever, and other debilitating problems. Multiple organ failure may result. Although many species of Enterobacteriaceae possess flagella, and species other than those mentioned possess polysaccharide capsules and fimbriae, the effect of these factors on the virulence of a particular strain has not been completely elucidated.

TAXONOMY OF ENTEROBACTERIACEAE

With increasing use of multifactorial biotyping systems routinely by clinical microbiology laboratories, unusual and rare biotypes of bacteria are being recognized and sent to reference centers such as the CDC for collection and characterization. The application of deoxyribonucleic acid (DNA) homology technology to the study of the relatedness of two bacterial clones has been the major factor leading to the ability to assign definitively such morphologically or biochemically distinct strains of bacteria to species and occasionally to new genera. As described and illustrated in Chapter 11, the more closely related two strains of bacteria are, the greater the chromosomal homology shared between them, as demonstrated by the extent to which single strands of DNA from one strain realign with single strands of DNA from the other strain under carefully controlled experimental conditions (called *stringency* conditions). The recombined DNA, possessing one strand from each of two test strains, is called a *hybrid,* and the analysis to determine relatedness is called *DNA-DNA hybridization.* Arbitrarily, clones of bacteria that share 70% or more of their chromosomal sequences are considered to belong to the same species. Definition of genera is not so easily determined. In addition to DNA-DNA hybridization studies, relatedness of bacteria can be determined by differences in the melting temperature of DNA,

which is directly related to the number of guanine and cytosine bonds compared with the number of adenine and thymine bonds. The guanine and cytosine content (%G + C) can be used to begin to characterize strains of bacteria, but DNA-DNA hybridization is the most definitive arbitrator.

Ribonucleic acid (RNA) typing *(ribotyping)* has also been used to determine relatedness among bacterial strains.

Table 28.1 lists the genera and species of Enterobacteriaceae that have been validly published. With the exception of *Erwinia amylovora*, most previous *Erwinia* sp. have been renamed as *Enterobacter* sp. Several newer species have not yet been isolated from humans, although in the future they may be found to be associated with human disease. The identification of these species is possible with many of the commercial identification systems (Chapter 10). Laboratories that do not use such systems should refer to the publication by Farmer et al. (see Bibliography) and subsequent descriptions of newly characterized species for biochemical tests used to identify these isolates. Tests described in this chapter and in Chapter 10 are often used in this text to characterize the new species as well. The bulk of discussion in this chapter, however, focuses on the 13 genera most likely to be encountered in clinical specimens. The basic differential characteristics of the most frequently isolated Enterobacteriaceae are given in Table 28.2.

LABORATORY IDENTIFICATION OF ENTEROBACTERIACEAE

ISOLATION OF ENTEROBACTERIACEAE FROM CLINICAL MATERIAL

Material from any site can harbor members of the family Enterobacteriaceae. Specimens of fecal material, in which these organisms are normal flora, are inoculated to selective and differential agars for detection of those organisms known to be the etiological agents of gastroenteritis to the exclusion of the nonpathogenic, commensal fecal Enterobacteriaceae. Strategies for cultivation of stool are discussed in Chapter 18, and an overview of culture media to inoculate initially is presented in Chapter 7. A screening agar for verotoxin-producing *E. coli,* such as sorbitol-MacConkey agar (SMAC), detects the most common serotype of verocytotoxin-producing strains, O157:H7. Other toxin-positive strains must be detected using Vero cells in a tissue culture cytotoxin assay, which is beyond the scope of most clinical laboratories.

TABLE 28.2

IDENTIFICATION OF CLINICALLY IMPORTANT ENTEROBACTERIACEAE*

SPECIES	INDOLE PRODUCTION	METHYL RED	VOGES-PROSKAUER	CITRATE (SIMMONS')	HYDROGEN SULFIDE (TSI)	UREA HYDROLYSIS	PHENYLALANINE DEAMINASE	LYSINE DECARBOXYLASE	ARGININE DIHYDROLASE	ORNITHINE DECARBOXYLASE	MOTILITY (36° C)	GELATIN HYDROLYSIS (22° C)	D-GLUCOSE, GAS	LACTOSE FERMENTATION	SUCROSE FERMENTATION	D-MANNITOL FERMENTATION	DULCITOL FERMENTATION	ADONITOL FERMENTATION	D-SORBITOL FERMENTATION	L-ARABINOSE FERMENTATION	RAFFINOSE FERMENTATION	L-RHAMNOSE FERMENTATION	D-XYLOSE FERMENTATION	MELIBIOSE FERMENTATION	DNase, 25° C	ONPG†
Escherichia coli	98	99	0	1	1	1	0	90	17	65	95	0	95	95	50	98	60	5	94	99	50	80	95	75	0	95
Shigella serogroups A, B, and C	50	100	0	0	0	0	0	0	5	1	0	0	2	0	0	93	2	0	30	60	50	5	2	50	0	2
Shigella sonnei	0	100	0	0	0	0	0	0	2	98	0	0	0	2	1	99	0	0	2	95	3	75	2	25	0	90
Salmonella, most serotypes	1	100	0	95	95	1	0	98	70	97	95	0	96	1	1	100	96	0	95	99	2	95	97	95	2	2
Salmonella typhi	0	100	0	0	97	0	0	98	3	0	97	0	0	1	0	100	0	0	99	2	0	0	82	100	0	0
Salmonella paratyphi A	0	100	0	0	10	0	0	0	15	95	95	0	99	0	0	100	90	0	95	100	0	100	99	95	0	0
Citrobacter freundii	5	100	0	95	80	70	0	0	65	20	95	0	95	50	30	99	55	0	98	100	30	99	100	50	0	95
Citrobacter diversus	99	100	0	99	0	75	0	0	65	99	95	0	98	35	45	100	50	98	99	100	0	100	100	0	0	96
Edwardsiella tarda	99	100	0	1	100	0	0	100	0	100	98	0	100	0	0	99	0	0	9	99	0	0	0	0	0	0
Klebsiella pneumoniae	0	10	98	98	0	95	0	98	0	0	0	0	97	98	99	99	30	90	99	99	99	99	99	99	0	99
Klebsiella oxytoca	99	20	95	95	0	90	1	99	0	0	0	0	97	100	100	99	55	99	99	98	100	100	100	99	0	100

TABLE 28.2

IDENTIFICATION OF CLINICALLY IMPORTANT ENTEROBACTERIACEAE*—cont'd

SPECIES	INDOLE PRODUCTION	METHYL RED	VOGES-PROSKAUER	CITRATE (SIMMONS')	HYDROGEN SULFIDE (TSI)	UREA HYDROLYSIS	PHENYLALANINE DEAMINASE	LYSINE DECARBOXYLASE	ARGININE DIHYDROLASE	ORNITHINE DECARBOXYLASE	MOTILITY (36° C)	GELATIN HYDROLYSIS (22° C)	D-GLUCOSE, GAS	LACTOSE FERMENTATION	SUCROSE FERMENTATION	D-MANNITOL FERMENTATION	DULCITOL FERMENTATION	ADONITOL FERMENTATION	D-SORBITOL FERMENTATION	L-ARABINOSE FERMENTATION	RAFFINOSE FERMENTATION	L-RHAMNOSE FERMENTATION	D-XYLOSE FERMENTATION	MELIBIOSE FERMENTATION	DNase, 25° C	ONPG†
Enterobacter aerogenes	0	5	98	95	0	2	0	98	0	98	97	0	100	95	100	100	5	98	100	100	96	99	100	99	0	100
Enterobacter cloacae	0	5	100	100	0	65	0	0	97	96	95	0	100	93	97	100	15	25	95	100	97	92	99	90	0	99
Hafnia alvei	0	40	85	10	0	4	0	100	6	98	85	0	98	5	10	99	0	0	0	95	2	97	98	0	0	90
Serratia marcescens	1	20	98	98	0	15	0	99	0	99	97	90	55	2	99	99	0	40	99	0	2	0	7	0	98	95
Proteus mirabilis	2	97	0	65	98	98	98	0	0	99	95	90	96	2	15	0	0	0	0	0	1	1	98	0	50	0
Proteus vulgaris	98	95	0	15	95	95	99	0	0	0	95	91	85	2	97	0	0	0	0	0	1	5	95	0	80	1
Providencia rettgeri	99	93	0	95	0	98	98	0	0	0	94	0	10	5	15	100	0	100	1	0	5	70	10	5	0	5
Providencia stuartii	98	100	0	93	0	30	95	0	0	0	85	0	0	2	50	10	5	5	1	1	7	0	7	0	10	10
Providencia alcalifaciens	99	99	0	98	0	0	98	0	0	1	96	0	85	0	15	0	98	98	98	1	1	0	1	0	0	1
Morganella morganii	98	97	0	0	5	98	95	0	0	98	95	0	90	1	0	0	0	0	0	0	0	0	0	0	0	5
Yersinia enterocolitica	50	97	2	0	0	75	0	0	0	95	2	0	5	5	95	98	0	0	99	98	5	1	70	1	5	95
Yersinia pestis	0	80	0	0	0	5	0	0	0	0	0	0	0	0	0	97	0	0	50	100	0	1	90	20	0	50
Yersinia pseudotuberculosis	0	100	0	0	0	95	0	0	0	0	0	0	0	0	0	100	0	0	0	50	15	70	100	70	0	70

From Farmer, J.J., III, Davis, B.R., Hickman-Brenner, F.W., et al. 1985. J. Clin Microbiol. 21:46.

* Each number gives the percentage of positive reactions after 2 days of incubation at 36° C. The vast majority of these positive reactions occur within 24 hours. Reactions that become positive after 2 days are not considered.

† ONPG, o-Nitrophenyl—D-galactopyranoside.

Most other clinical specimens, including material from wounds, respiratory tract secretions, aspirations from abscesses, sterile body fluids, urine specimens, tissues, and blood cultures, are inoculated to at least one supportive medium, such as blood or chocolate agar, that allows the growth of all Enterobacteriaceae, as well as to one selective agar for gram-negative bacilli, usually MacConkey agar. Discussion of initial selection of media for isolation of etiological agents of disease from each of these sites is found in the appropriate chapters of this text. Descriptions of colony morphologies of isolates on various selective and differential media typically used for cultivation of Enterobacteriaceae are found in Chapter 9. Except for the most fastidious or biologically damaged strains, all Enterobacteriaceae grow on MacConkey agar, yielding pink or reddish colonies if they are able to ferment lactose and colorless, translucent colonies if they are slow or nonlactose fermenters (Figure 28.2). Growth on MacConkey agar is one of the first criteria used to establish the identification of an unknown gram-negative bacillus. Gram-positive organisms should not grow on MacConkey agar, although pinpoint colonies of enterococci and yeast may appear after 48 hours.

GENERAL PHENOTYPICAL FEATURES SHARED BY ALL OR MOST ENTEROBACTERIACEAE

The Enterobacteriaceae are straight-sided gram-negative bacilli that grow well on most laboratory media. The Enterobacteriaceae are facultative anaerobes; all grow in air or anaerobically. Carbon dioxide (CO_2) does not appreciably enhance

growth. If they are motile, with the exception of *Tatumella ptyseos,* they possess peritrichous flagella. *T. ptyseos* has polar flagella. All species of *Klebsiella* and *Shigella* are nonmotile. Some species of *Escherichia, Salmonella,* and *Yersinia,* among clinically important Enterobacteriaceae, are nonmotile as well. Selected strains within other species may also be nonmotile.

All Enterobacteriaceae ferment glucose and are oxidase negative, and almost all are able to reduce nitrates to nitrites. Performance of the nitrate test is outlined in Chapter 10. Except for *Shigella dysenteriae* type 1 and a species of *Xenorhabdus,* all Enterobacteriaceae are catalase positive. An additional clue to the identification of an unknown organism as an Enterobacteriaceae is luxuriant growth on most routine media incubated at 35° C in air within 24 hours. Colonies on blood agar are usually gray to white, shiny, entire, convex, and opaque. Hemolysis may be present. Most of the remaining *Proteus* sp. display a recognizable "swarming" pattern that resembles waves of growth spreading from the original inoculum (Figure 28.3) because of their active motility. Enterobacteriaceae typically yield large (at least 1 to 2 mm in diameter after 24 hours) colonies on blood agar and MacConkey agar plates. A few strains of Enterobacteriaceae isolated from clinical specimens, however, may not grow well on MacConkey agar, especially if the organisms have been damaged previously by antibiotic treatment.

The first test that should be performed on a gram-negative bacillus isolated from clinical material is the oxidase test, as described in Chapter 10. The oxidase test must be performed with

FIGURE 28.2

Colonies of lactose-fermenting *(pink)* and non-lactose-fermenting *(colorless)* gram-negative bacilli on MacConkey agar.

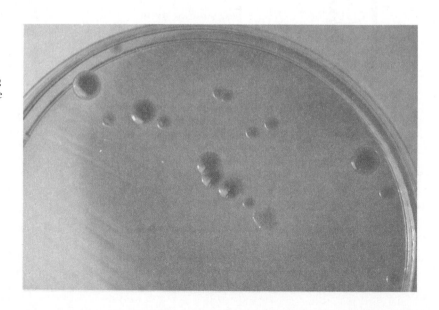

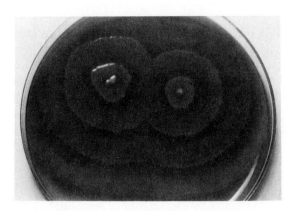

FIGURE 28.3

Appearance of "swarming" *Proteus* sp. on blood agar. Note successive waves of growth extending from a central inoculum.

colonies growing on plates that do not contain dyes that may obscure the results. Plates yielding mixed cultures from clinical material may often contain colonies of more than one species of Enterobacteriaceae. If several colony morphologies can be differentiated only on MacConkey agar, each morphotype should be subcultured to a portion of a blood agar plate for performance of an oxidase test. Oxidase-negative, luxuriantly growing, facultatively anaerobic bacilli (initially isolated aerobically in most cases) are likely to be members of the Enterobacteriaceae.

The ability to utilize glucose anaerobically with formation of acidic end products (fermentation) is tested by inoculating portions of a colony to a Kligler's iron agar (KIA) or triple sugar iron agar (TSIA) slant, with the former test being preferred (Chapter 10). These three preliminary tests (utilization of glucose, oxidase reaction, growth on MacConkey) place the unknown organism into one of several categories, helping to determine which tests are necessary for further identification. Enterobacteriaceae are fermentative, oxidase-negative, MacConkey-positive bacilli. Organisms grouped as nonfermentative gram-negative bacilli or other non-Enterobacteriaceae are further identified, as described in Chapters 29 to 31. More rarely, isolates fall into categories described in Chapter 40. Initial reactions in TSIA or KIA can be used to guide identification of the 13 more frequently isolated genera of Enterobacteriaceae (Figure 28.4).

Many laboratories no longer rely on TSIA for initial testing of gram-negative bacilli, but instead inoculate a commercial identification system (Chapter 10) with a suspension made from one isolated colony. Depending on the system, results are available within 4 hours or after overnight in-

cubation, at which time the oxidase reaction can be correlated with the biochemical reactions for definitive identification. Because these systems usually include more differential parameters than can be used cost-effectively as individual components, the identifications that result are more specific. The immense data bases used by the commercial systems (Chapter 10) contribute to the detection of unusual and new biotypes.

A few basic biochemical parameters, however, have classically been employed by microbiologists to categorize the Enterobacteriaceae (and other bacteria) into species groups. Even if laboratories are using commercial identification systems, with the exception of the totally automated instruments (Chapter 12), technologists can still observe visual changes in indicators that correlate to individual biochemical reactions. Technologists are urged to try to identify isolates based on manually observed biochemical patterns before the organisms are identified by comparing the numerical code generated by the pattern of results with the company-furnished list of profile numbers and species identifications. Attempting to identify species by recognizing their biochemical patterns is challenging and contributes to the continued maintenance of cognitive abilities.

Toward this end, the performance of some of the classical biochemical tests used to identify Enterobacteriaceae is outlined in this chapter. These reactions are used not only to identify Enterobacteriaceae, but also are often necessary to identify more fastidious, unusual, or relatively metabolically inactive microorganisms (not only bacteria) that cannot be identified with commercially available systems. A rudimentary flow chart that begins to differentiate frequently isolated Enterobacteriaceae on the basis of a few biochemical reactions is shown in Figure 28.5. Excellent commentaries on the biochemical bases of these tests and their performance are found in the American Society for Microbiology (ASM) *Manual of Clinical Microbiology*, Washington's laboratory manual, and the books by MacFaddin and by Howard et al. (see Bibliography). Media for performing these tests are all available from commercial suppliers, either already prepared or in dry powder form for in-house preparation. The results listed in Table 28.2 are based on conventional tests. Results of individual tests performed as part of a miniaturized or automated biochemical identification system cannot with complete reliability be compared with the results obtained through conventional methods. Identifications that result from miniature systems are based on

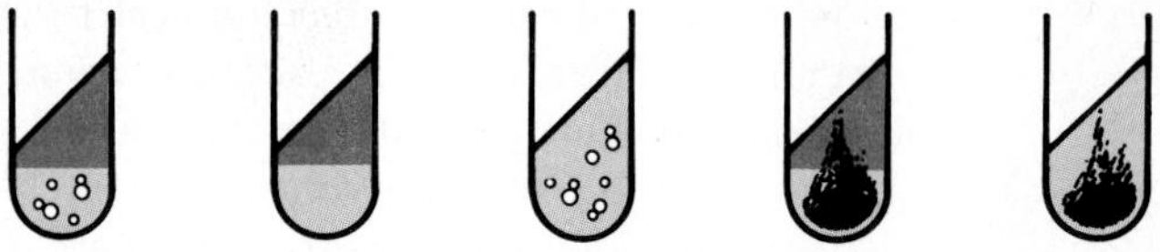

Alk/A gas	Alk/A no gas	A/A gas	Alk/A H₂S	A/A H₂S	Possible species
+	+	+	−	−	Escherichia coli Hafnia alvei
+	+	−	−	−	Morganella morganii Providencia alcalifaciens P. rettgeri P. stuartii Serratia sp.
+	−	+	−	−	Enterobacter aerogenes E. cloacae
+	−	+	+	+	Citrobacter sp.
+	+	−	+	−	Salmonella sp.
−	+	−	−	−	Shigella sp.
−	+	+	−	−	Yersinia sp.
−	−	+	−	−	Klebsiella sp.
−	−	−	+	+	Proteus mirabilis P. vulgaris
−	−	−	+	−	Edwardsiella tarda

*Some species show slight gas.

FIGURE 28.4
Chart showing basic reactions of Enterobacteriaceae on Kligler's iron agar (KIA).

patterns generated from the entire system and not on results of individual tests. Identifications are based on the "best fit."

UREASE TEST
An organism's ability to hydrolyze urea to ammonia and CO_2 (Chapter 10) is determined classically by heavily inoculating a urea-containing agar slant (Christensen urea agar, available commercially) or a urea-containing, buffered broth of the type available commercially. (Rustigian and Stuart urea broth, which is highly buffered and available commercially, yields reliable results.) Organisms should be well dispersed in broth for best results. The media are incubated at 37° C, preferably in a water bath or small incubator. A significant number of Yersinia, Enterobacter, and Citrobacter sp. strains may be urease negative when incubated at 37° C but positive when the urease test is incu-

bated at room temperature. Strongly urease-positive organisms begin to produce alkaline products, which turn the phenol red indicator red to purple within minutes. The heavier the inoculum, the more rapid is the positive result. Tubes or broths should be examined at 10 minutes, 30 minutes, 1 hour, and several hours after inoculation. Several suppliers, including Key Scientific Products, E-Y Laboratories, and Difco Laboratories, produce a rapid urease test reagent containing substrates impregnated on filter paper or in tablets (Chapter 10). Particularly useful for identification of Proteus and Providencia sp. are the filter paper disks produced by Remel Laboratories, which test for urease and phenylalanine deaminase. In addition to differentiation of Enterobacteriaceae, the classical urease test is also used to differentiate between certain Brucella sp., to aid in identification of encapsulated yeasts, and as an additional

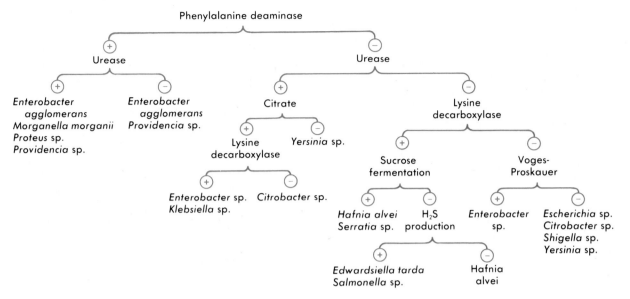

FIGURE 28.5

Flow chart for initial identification of Enterobacteriaceae based on basic biochemical reactions.

test for identification of certain gram-negative coccobacilli.

DECARBOXYLATION AND DIHYDROLATION OF AMINO ACIDS

Differentiation among many species of Enterobacteriaceae requires determining the organism's ability to break down the amino acids lysine, arginine, and ornithine, forming an amine end product and CO_2. Lysine and ornithine are decarboxylated, whereas arginine is broken down first by dihydrolation and then by decarboxylation. The amine end product of these reactions is alkaline, causing a shift in the pH indicators bromcresol purple and cresol red to purple (Chapter 10). The ability to decarboxylate lysine is a prominent feature of most species of *Salmonella, Klebsiella, Enterobacter aerogenes, Hafnia alvei,* and *Serratia marcescens.*

Definitive tests for lysine and ornithine decarboxylases and arginine dihydrolase and to confirm the organism's ability to ferment glucose are carried out in broth media developed by Moeller, called *Moeller's decarboxylase medium* (Chapter 10). Performance of the test entails inoculation of Moeller's medium, which contains glucose, bromcresol purple, a nitrogen source, cresol red indicator, the enzyme activator pyridoxal, and the amino acid to be tested. In addition to tubes containing each amino acid, a control tube containing only the basal medium must be inoculated each time the test is performed to ascertain that the organism does not form alkaline end products in the absence of an available amino acid and to confirm the organism's ability to ferment glucose.

The tubes are inoculated with a small amount of growth from isolated colonies; one must be certain that the inoculum is emulsified in the broth below its surface. All tubes are overlayed with sterile mineral oil or vaspar to a depth of 1 cm to prevent introduction of oxygen. Tubes are incubated at 35° to 37° C for up to 14 days, although results can usually be read in 2 days. A more rapid result is possible if the amount of broth is reduced and the inoculum is extremely turbid. A change in the medium's color to purple indicates a positive test and the organism's ability to utilize the amino acid to form alkaline end products (purple = positive). A yellow color indicates that only the glucose was utilized, with the formation of acid. The control tube should turn yellow if the organism is a glucose fermenter. A gray color may indicate reduction of the indicator rather than alkaline end products. If the test is critical, the addition of more bromcresol purple solution (0.01% aqueous solution) to the broth below the seal may allow detection of a pH shift. Additional incubation of tubes displaying intermediate results after 4 days often allows interpretation of results later. Other bacteria in addition to those within the Enterobacteriaceae may be inoculated to Moeller's broths for detection of decarboxylase and dihydrolase enzymes. The test is particularly useful in the identification of *Aeromonas, Plesiomonas, Vibrio* sp., and nonfermentative gram-negative bacilli.

Lysine iron agar (LIA) detects hydrogen sulfide (H_2S) formation, lysine decarboxylation, and phenylalanine deamination (positive for *Proteus* and *Providencia*) and is often used to differentiate H$_2$S-producing nonpathogens isolated on selective agar such as Hektoen and xylose-lysine-deoxycholate (Chapter 9) from etiological agents of gastroenteritis. *Citrobacter* sp., often confused with *Salmonella* on initial isolation, are lysine negative. LIA contains glucose, lysine, ferric ammonium citrate, and sodium thiosulfate for H$_2$S detection and bromcresol purple pH indicator. Organisms are inoculated in the same manner as for KIA, except that it is helpful to stab the butt of LIA several times to ensure an adequate inoculum. After 18 hours' incubation at 35° C with a *loose cap,* the tube is examined. The formation of a black precipitate in the butt is indicative of H$_2$S production, although LIA is less sensitive than either TSI or KIA for detection of H$_2$S. Organisms that are able to decarboxylate lysine (a reaction that takes place anaerobically in the butt) produce alkaline end products that result in a purple color. Bacteria that do not decarboxylate lysine ferment the small amount of glucose, yielding acid by-products and causing the bromcresol purple to turn the butt yellow. Even in the presence of a large amount of black precipitate, the yellow-colored butt can be seen by observing the tube with light behind it. Deamination of phenylalanine is detected by a change in the color of the slant from purple to reddish. LIA is often used as a basic preliminary screening medium to detect fecal pathogens because it incorporates several parameters into one medium. With the exception of *Salmonella paratyphi A,* most salmonellae and *Edwardsiella tarda* are lysine decarboxylase positive.

METHYL RED AND VOGES-PROSKAUER TESTS

Organisms that utilize glucose may do so by producing abundant acidic end products such as formate and acetate from the intermediate pyruvate metabolite, or they may metabolize the carbon compounds to acetoin and butanediol, which are more neutral in pH. The methyl red and Voges-Proskauer (MR and VP) tests detect the presence of the end products of these two divergent metabolic pathways. The same substrate broth, MR/VP broth, is used for both tests. Tubes of broth (5 ml per tube), containing peptones, glucose, and buffer, are available commercially or can be prepared in the laboratory from powder base. The organism is inoculated below the surface of the medium in the tube, and the tube is incubated with a loose cap at 35° to 37° C for at least 48 hours. A longer incubation period may

be necessary, especially for the demonstration of a positive VP reaction. The time necessary to demonstrate a positive reaction can be shortened by decreasing the amount of MR/VP broth to 0.5 ml for each test. With this modification, the test can be interpreted accurately and with good sensitivity after only 24 to 48 hours. To perform the test, the substrate broth is divided into two equal aliquots, one of which is used for the MR test and the other for the VP test (Procedure 28.1). The subtle nature of this reaction demands that all reagents and substrates are controlled. As for all the tests described in this section, the MR/VP test is sometimes used in the identification of organisms other than Enterobacteriaceae.

INDOLE TEST

The ability of an organism to degrade the amino acid tryptophan can be detected by testing for indole, the product of tryptophanase (tryptophan deamination of indolepyruvic acid). Reaction of indole with an aldehyde results in a colored end product. The rapid indole test (Chapter 10) is performed with dimethylaminocinnamaldehyde reagent, which turns blue or blue-green in the presence of indole. The classic test described here (Procedure 28.2) calls for *p*-dimethylaminobenzaldehyde as the indicator substrate. When the indole test described here is used to test bacteria other than Enterobacteriaceae, the more sensitive Ehrlich's reagent (Chapter 29) should be used to detect indole.

CITRATE UTILIZATION TEST

Certain organisms are able to utilize a single substrate as a sole carbon source. The ability to use citrate can help to differentiate among members of the Enterobacteriaceae. The medium, Simmons' citrate agar, contains buffers, salts, cations, citrate, and bromthymol blue as the pH indicator. The medium can be purchased prepared or as dry powder. Since the reaction requires oxygen, the organism is inoculated to the surface of an agar slant of the medium. A very light inoculum, usually picked with a straight wire, is necessary to prevent false-positive reactions because of carryover of substrates from previous media. The tube is incubated for 24 hours or up to 4 days at 35° to 37° C with a *loose cap.* Growth of the organism on the slant, with change of the color indicator from green to blue, or turquoise blue, is evidence of a positive test, indicating that the organism was able to grow and produce acetate and other alkaline carbonate end products. A rare citrate-positive organism may be able to utilize the substrate

PROCEDURE 28.1

METHYL RED AND VOGES-PROSKAUER TESTS

PRINCIPLE

The presence of certain metabolic enzymes can be used to differentiate organisms based on end products of glucose metabolism detected with various color indicator reagents. Acid products produced by enzymes of the mixed acid fermentation pathways are detected by development of a red color by the methyl red indicator (indicating pH <4.5). Acetoin and butanediol, products of the butanediol fermentation pathway, yield a pink or red color in the presence of α-naphthol in the relatively alkaline environment. Most Enterobacteriaceae, except *Hafnia* and some *Serratia* sp., demonstrate either one or the other metabolic pathway, but rarely both pathways.

METHOD

1. Prepare methyl red (MR) reagent as follows:
   Methyl red (Mallinckrodt), 0.1 g
   Ethyl alcohol (95%), 300 ml
   Distilled water, 200 ml
   Dissolve the methyl red in the alcohol before adding the distilled water. Store in a brown bottle in the refrigerator.

2. Prepare Voges-Proskauer (VP) reagent A as follows:
   α-Naphthol (Sigma Chemical Co.), 5 g
   Ethyl alcohol (absolute), 100 ml
   Dissolve the α-naphthol in a small amount of ethyl alcohol, and bring the volume to 100 ml in a volumetric flask or cylinder. The alcohol should be almost colorless. Store in a brown bottle in the refrigerator.

3. Prepare VP reagent B as follows:
   Potassium hydroxide (KOH), 40 g
   Distilled water, 100 ml
   Weigh out the KOH very quickly, since it is hygroscopic and will become caustic when moist. Add less than 100 ml water to the pellets in a flask in a cold water bath to prevent overheating. Bring the volume to 100 ml in a volumetric flask or cylinder. Store this reagent in the refrigerator in a polyethylene bottle or one that has been specially treated for storage of caustic chemicals.

4. From a tube of inoculated MR/VP broth base that has been incubated for at least 48 hours, remove 2.5 ml to another tube for the VP test.

5. To the 2.5 ml of organism suspension, add 0.6 ml (6 drops) of VP reagent A (α-naphthol). Then add 0.2 ml (2 drops) of VP reagent B (40% KOH). Gently shake the tube and allow it to sit for as long as 15 minutes.

6. Observe for the formation of a pink to red color, indicating the presence of acetoin (acetyl methyl carbinol). A pink color in the medium is indicative of a positive test. A negative test appears colorless or yellow.

7. While the VP reaction is developing, the remaining 2.5 ml of substrate is tested for acidity by adding 0.5 ml (5 drops) of the MR reagent and observing the color. If the indicator remains red, the MR test result is positive. A change to yellow indicates that the pH of the medium is greater than 6.0, a negative MR test. If the reagent remains orange, the test must be repeated after a longer incubation period.

QUALITY CONTROL

E. coli ATCC 25922 and *Klebsiella pneumoniae* ATCC 13883 should be inoculated to MR/VP broths and tested.

EXPECTED RESULTS

Escherichia coli should be MR positive and VP negative, whereas *K. pneumoniae* is MR negative and VP positive.

PERFORMANCE SCHEDULE

Quality control organisms must be used to test each batch of reagents and media prepared before they are used.

Modified from MacFaddin, J. 1980. Biochemical tests for identification of medical bacteria, ed 2. Williams & Wilkins, Baltimore.

PROCEDURE 28.2

INDOLE TEST

PRINCIPLE

Tryptophanase enzymes degrade the tryptophan in peptones and other components of media to indole, skatole, and indoleacetic acid. When combined with certain aldehydes, indole yields a red-colored product. Organisms are grown in a medium rich in tryptophan, and the medium is then tested for the presence of the indole end product. Either Kovacs' reagent (described here) or Ehrlich's reagent (Chapter 29) provides the aldehyde indicator.

METHOD

1. Prepare Kovacs' reagent as follows:
      p-Dimethylaminobenz-
         aldehyde (Sigma
         Chemical Co.), 10 g
      Isoamyl or isobutyl
         alcohol (absolute),
         150 ml
      Concentrated
         hydrochloric acid, 50
         ml
   Dissolve the p-dimethyl-aminobenzaldehyde in the alcohol, and slowly add the acid while constantly stirring the mixture. The reagent should be pale colored and should be stored refrigerated in a brown bottle. If it turns brown or if quality control organisms fail to give correct reactions, discard the reagent and make a fresh lot. The aldehyde may require gentle heating to go into solution.

2. Inoculate the organism into a broth that contains tryptophan, and incubate for at least 24 hours. Most commercial peptone, pancreatic enzymatic casein hydrolysate, or tryptone (a pancreatic digest of casein) broths contain enough tryptophan for use in this test. Note that acid hydrolysate of casein broths that do not contain tryptone are not suitable for the indole test. A specific medium, Indole Nitrate Medium (BBL Microbiology Systems), also performs well for this test.

3. To perform the test, remove 2 ml of the broth suspension to a second tube. Add 0.5 ml (5 drops) of Kovacs' reagent to the broth. Gently shake the tube, and observe for a pink color in a ring around the interface between the broth and the alcohol reagent, which rises to the surface (Figure 28.6).

4. If the test is negative, the remaining broth may be reincubated for an additional 24 hours, and the test may be repeated.

QUALITY CONTROL

Inoculate *Escherichia coli* ATCC 25922 and *Enterobacter cloacae* ATCC 23355 as just described.

EXPECTED RESULTS

The *E. coli* is indole positive and yields a red ring in the lower portion of the alcohol phase layer above the medium. The *E. cloacae* is indole negative, showing either only a cloudy ring or a slightly yellow-colored ring at the alcohol interface.

PERFORMANCE SCHEDULE

Quality control strains should be tested with each new batch of media and reagents.

---

without producing enough alkaline reaction to change the pH indicator. Luxuriant growth on the slant without a blue color may indicate a positive test, but the test should be repeated with a minimal inoculum.

MOTILITY TEST

Another important differential test among Enterobacteriaceae and many other bacteria is motility. The fastest and most direct method for detection of motility is to examine microscopically (under hih the oil immersion objective [1000×]) a wet preparation of the organism from the initial isolation plates or broth incubated at room temperature for 2 to 4 hours (Chapter 8). After the organisms have settled, the directed and purposeful movement of a single bacterium or several bacte-

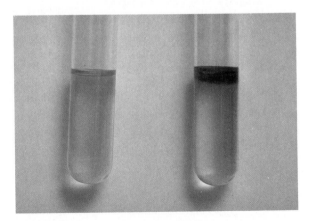

FIGURE 28.6

Indole test. Positive result is visible as a red to pink ring around the top of the tube where the reagents float. A negative test is seen as no change in the reagents' color.

ria, as opposed to the flowing movement of all bacteria within a field, is indicative of motility. Motility must be distinguished from brownian motion, which usually involves all visible organisms equally and appears as a shaking motion rather than true motility.

Motility can be encouraged by growing bacteria in semisolid media. Good commercial motility semisolid media are available from Difco Laboratories, BBL Microbiology Systems, and other commercial media suppliers. The media are prepared as butt tubes with a depth of 5 cm or greater. The organism is inoculated by stabbing a straight wire carrying the inoculum once vertically into the center of the agar butt to a depth of approximately 2 cm. After overnight incubation, motility is evident as a haze of growth extending into the agar from the stab line. It is not recommended to include a colored growth indicator in motility medium, such as tetrazolium salts, since it may be inhibitory to some bacteria. Because a much larger sample of a bacterial population can be tested with the semisolid motility agar than can be visualized microscopically, these media are probably more sensitive for detection of motility. Solid and semisolid motility detection media are read after overnight incubation at the temperature preferred by the organism being tested. For example, *Yersinia enterocolitica* is motile at room temperature but not at 37° C. Organisms other than Enterobacteriaceae for which semisolid motility media are typically used include *Listeria monocytogenes* and nonfermentative gram-negative bacilli. The nonfermentative bacteria show enhanced motility when nitrate is incorporated into the semisolid media because of their more active aerobic metabolism.

PHENYLALANINE DEAMINASE TEST

Of the well-characterized Enterobacteriaceae, only *Enterobacter agglomerans* (20% of strains), *Enterobacter sakazakii* (50% of strains), *Morganella morganii, Proteus* sp., *Providencia* sp., *Rahnella aquatilis,* and *Tatumella ptyseos* are able to deaminate the amino acid phenylalanine. The organism to be tested is grown overnight on an agar slant containing phenylalanine, salts, yeast extract, and buffers. Phenylalanine agar is available prepared or in dry powder form. The product of phenylalanine deamination, phenylpyruvic acid, is detected by a colored end product that forms when phenylpyruvic acid combines with ferric chloride solution. Procedure 28.3 describes performance of the test.

SUBSTRATE FERMENTATION TESTS

Many tests used to differentiate among members of the Enterobacteriaceae (as well as among other microorganisms) determine the organism's ability to utilize a carbohydrate with the production of acid metabolic end products. In addition to detecting acid, carbohydrate fermentation media can be modified to detect the formation of gas by placing a very small, inverted glass tube **(Durham tube)** into the broth. If gas is produced during fermentation of the carbohydrate, it is collected in the inverted tube and is visible as a bubble (Figure 28.7). Most Enterobacteriaceae produce gas during glucose fermentation, but *Shigella, Salmonella typhi, Providencia stuartii,* most *Yersinia* sp., and certain *E. coli* sp. do not produce gas (called **anaerogenic**).

The carbohydrates are usually prepared as individual aqueous solutions and filter sterilized to prevent their destruction by the heat of the autoclave. The heart infusion or other broth base containing a pH indicator, which may be 0.0018% (wt/vol) phenol red, 1% Andrade's acid fuchsin, or 0.0016% (wt/vol) bromcresol purple (originally dissolved in 95% ethanol, 1.6 g/100 ml ethanol), is made up, autoclaved, and allowed to cool before the sterile carbohydrate solutions are aseptically added to a final concentration of 1% carbohydrate. For most applications, bromcresol purple yields satisfactory and consistent results. A variety of sugars, including lactose, sucrose, maltose, glucose, and others, and various alcohols, including mannitol, sorbitol, dulcitol, and others, can be added to carbohydrate broth base for fermentation tests. All generally used carbohydrate broths are available commercially from several media suppliers.

ONPG TEST

Enterobacteriaceae are preliminarily grouped according to their ability to ferment lactose, a β-galactoside. *Salmonella, Shigella, Proteus, Providencia,* and *Morganella* are usually considered to be nonlactose fermenters. Other organisms, including certain *Escherichia coli, Shigella sonnei, Hafnia alvei, Serratia marcescens,* and some *Yersinia* sp., may appear to be nonlactose fermenters because they lack the enzyme (permease) that actively transports lactose across the cell membrane, making it available to intracellular β-galactosidase. True nonlactose fermenters do not possess β-galactosidase. If β-galactosidase but not permease is produced by a species, strains eventually are able to ferment lactose that slowly diffuses across the cell membrane, although this process may be

PROCEDURE 28.3

PHENYLALANINE DEAMINASE TEST

PRINCIPLE

Deamination of phenylalanine yields phenylpyruvic acid. Deamination of tryptophan yields indolepyruvic acid. These alpha keto acids react with ferric chloride to form a green end product (phenylpyruvic acid) or an orange-brown end product (indolepyruvic acid).

METHOD

1. Prepare ferric chloride reagent as follows:
   Ferric chloride ($FeCl_3$), (Sigma Chemical Co.) 12 g
   Distilled water, 94.6 ml
   Concentrated HCl (37%), 5.4 ml
   Dissolve $FeCl_3$ in water, and slowly add the hydrochloric acid while working under a fume hood. Store the reagent in a brown bottle in the refrigerator.

2. After overnight incubation of the test organism on an agar slant containing phenylalanine, add 0.5 ml (5 drops) of the reagent to the growth on the slant's surface, and gently rotate the tube so that the reagent covers the entire surface. Shaking the tube back and forth may intensify the reaction. Observe for an immediate development of a green color, indicating the presence of phenylpyruvic acid and a positive test result. Within 10 minutes the color will fade, so observation must occur immediately after the reagent has been added. Adding additional reagent usually regenerates the color. Certain species may deaminate phenylalanine so rapidly that the test is positive within 4 hours of incubation.

QUALITY CONTROL

Test *Proteus vulgaris* ATCC 33420 and *Escherichia coli* ATCC 25922 as described.

EXPECTED RESULTS

The *P. vulgaris* yields a positive test (green slant), and the *E. coli* slant remains yellow after addition of the ferric chloride.

PERFORMANCE SCHEDULE

Test quality control organisms with each new batch of media or reagents.

Modified from MacFaddin, J. 1980. Biochemical tests for identification of medical bacteria, ed 2. Williams & Wilkins, Baltimore.

---

too slow for detection with carbohydrate fermentation broths. A rapid test for the presence of β-galactosidase (determination of whether an organism has the capability of utilizing lactose) is the ONPG (*o*-nitrophenyl-β-D-galactopyranoside)

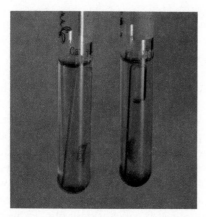

FIGURE 28.7

Fermentation of glucose with the production of gas, as seen by the bubble that has collected in the inverted Durham tube on the right. The organism in the tube on the left does not produce gas.

test (Procedure 28.4). Organisms other than Enterobacteriaceae, particularly *Neisseria*, can also be differentiated with the ONPG test.

Because the reagents are difficult to prepare, we recommend that laboratories purchase ONPG-impregnated filter paper disks from Difco Laboratories or Remel Laboratories or the tablets produced by Key Scientific Products. To perform the tests, an *extremely heavy* inoculum of the organism to be tested is suspended in a small volume of saline or buffer, and the reagents are eluted from the disk or tablet into the liquid. After several hours of incubation, a yellow color is indicative of hydrolysis of the substrate by β-galactosidase, or a positive test. The medium remains colorless when the test is negative. Manufacturers' recommendations and strict quality control practices should be followed.

GELATIN HYDROLYSIS TEST

Enzymes that degrade proteins, often important virulence factors, are produced by relatively few members of the Enterobacteriaceae. *Serratia, Proteus, Xenorhabdus*, some *Yersinia*, and several of the unnamed groups produce proteinases. Gelati-

PROCEDURE 28.4

ONPG (o-NITROPHENYL-β-D-GALACTOPYRANOSIDE) BROTH TEST FOR β-GALACTOSIDASE

PRINCIPLE

β-Galactosidase acts on the substrate **ONPG** in the same way as the enzyme hydrolyzes lactose to form galactose and glucose, although the end product of ONPG hydrolysis is a visible (yellow) product, orthonitrophenol.

METHOD

1. Prepare peptone water as follows:
   Peptone (Difco), 1 g
   NaCl, 0.5 g
   Distilled water, 100 ml
   Autoclave at 121° C for 15 minutes.

2. Prepare sodium phosphate buffer as follows:
   Na_2HPO_4, 13.8 g
   Distilled water, 90 ml
   Add the sodium phosphate to the distilled water in a 100 ml volumetric flask. Dissolve it by warming at 37° C. Slowly add approximately 15 ml of 5 N NaOH (or approximately 6 g) until the pH stabilizes at 7.0. Bring the volume to 100 ml with distilled water. Because of its causticity, NaOH should be handled carefully, and manipulations should be performed in a fume hood.

3. Prepare ONPG solution as follows:
   ONPG (Sigma), 0.6 g
   Sodium phosphate buffer (step 2), 100 ml
   Add the ONPG to the buffer, sterilize this by filtration, and store solution in a sterile brown glass bottle, refrigerated at 4° to 10° C.

4. Prepare ONPG broth as follows: Aseptically add 25 ml of ONPG solution to 75 ml peptone water. Aseptically dispense this solution into sterile 13×100 mm plastic snap-capped tubes, 0.5 ml per tube. Store refrigerated for 1 month or frozen at −20° C for up to 6 months. If the solution appears yellow, discard it.

5. Inoculate one tube of ONPG broth with a heavy suspension of the organism to be tested, taken from a KIA or TSIA slant or from the surface of a fresh subculture plate. Incubate the tubes for a minimum of 1 hour (preferably overnight) at 37° C.

6. Examine tubes for color change. A change to yellow indicates a positive result (presence of β-galactosidase), and no color change indicates a negative result (absence of the enzyme). If the test organism has not changed color after 1 hour of incubation, incubation should be continued for additional time up to the full 24 hours. Tubes may be checked at hourly intervals. If the organism produces yellow pigment, the test cannot be performed, and the longer test for utilization of lactose must be substituted.

QUALITY CONTROL

Inoculate control organisms *Escherichia coli* ATCC 25922 and *Proteus mirabilis* ATCC 29245 as in step 5.

EXPECTED RESULTS

The *E. coli* should be positive (yellow), and the *P. mirabilis* should be negative (clear).

PERFORMANCE SCHEDULE

Quality control strains should be tested each time new reagents are received and monthly thereafter.

Modified from MacFaddin, J. 1980. Biochemical tests for identification of medical bacteria, ed 2. Williams & Wilkins, Baltimore; and from Hendrickson, D.A. 1985. Reagents and stains. In Lennette, E.H., Balows, A., Hausler, W.J., Jr., and Shadomy, H.J., editors. Manual of clinical microbiology, ed 4. American Society for Microbiology, Washington, D.C.

cient growth, which may take as long as 30 days, the tubes are placed in the refrigerator for 30 minutes or until the uninoculated control tube resolidifies. If the test organism made gelatinase, either one or both of the inoculated tubes will remain liquid, indicating breakdown of the gelatin present in the medium. A positive control should always be tested with each assay. The gelatinase test is also used in identification schemes for many other organisms in addition to Enterobacteriaceae.

DNase TEST

Extracellular nucleases are produced primarily by the same species that produce the protease gelatinase. The detection of these enzymes is most sensitive at room temperature. Tests for DNase (deoxyribonuclease) detection are outlined in Chapter 10. Commercially produced agars, either prepared as plates or as slants, are available from several sources, as are dry powder media (Figure 28.8).

RAPID CATALASE SUPPLEMENTAL TEST

A recently described rapid catalase test may help to separate genera of Enterobacteriaceae that are morphologically difficult to differentiate. Chester and Moskowitz[3] observed the time for bubbles to appear after placing a single colony from a fresh agar culture into a drop of 3% hydrogen peroxide. Genera were divided into two groups. Those in which bubbles appeared *immediately* were *Cedecea, Hafnia, Morganella, Proteus, Providencia, Serratia,* and *Yersinia*. Those from which bubbles were

delayed (usually 1 second or longer, but any slight delay was considered "delayed") included *Citrobacter, Edwardsiella, Escherichia, Enterobacter, Klebsiella, Kluyvera, Salmonella, Shigella,* and *Tatumella*. This test may be useful for differentiating similar-appearing strains of *Yersinia* from *Escherichia coli* and *Shigella* and differentiating *Serratia* from *Enterobacter* sp.

SEROLOGICAL CHARACTERIZATION OF ENTEROBACTERIACEAE

GENERAL ANTIGENIC FEATURES

In addition to biochemical and morphological features, the Enterobacteriaceae can also be differentiated serologically, based on their possession of at least three classes of antigenic determinants. The cell wall structure of a typical Enterobacteriaceae is illustrated in Chapter 15 (Figure 15.2). Not all species have been antigenically characterized, but those that have been intensely studied include *Escherichia coli, Shigella,* and most extensively, *Salmonella*. The first category of antigens is associated with the lipopolysaccharide, or endotoxin moiety of the cell wall, called "O" antigens or **somatic** (body) antigens. These antigens are heat stabile.

Among *E. coli*, certain O antigens are associated with specific virulent phenotypes, such that *E. coli* O111 and O125, for example, are often found as etiological agents of infantile diarrhea; *E. coli* O112 is often an invasive strain that causes a dysentery-like syndrome. Serotypes O1, O26, O91, O113, O121, O145, O157, and others have been isolated from patients with serious hemorrhagic colitis and hemolytic-uremic syndromes caused by verocytotoxin-positive *E. coli*.

Shigella sp. have classically been grouped according to O serotypes. Since the *Shigella* have now been shown to be identical to *Escherichia* by DNA hybridization studies, no new serotypes will be added. For the convenience of clinicians and microbiologists, the name *Shigella* will continue to be used for those organisms that can be distinguished biochemically from the *E. coli* group (Table 28.2). Classically, the serotypes of *Shigella* correspond to species as follows: *S. dysenteriae,* serotype A; *S. flexneri,* serotype B; *S. boydii,* serotype C; and *S. sonnei,* serotype D.

The second category of antigens are the capsular polysaccharide antigens, called "K" antigens. These antigens are heat labile and are known to be produced by organisms that include *Klebsiella, Salmonella,* and *E. coli*. The K antigens of some strains of *Salmonella (S. typhi, S. paratyphi C)* are

FIGURE 28.8

DNase (deoxyribonuclease) test agar containing methyl green indicator. Organisms that produce DNase are identified by production of a clear zone (hydrolysis of DNA and subsequent breakdown of the indicator) surrounding growth of the organism after overnight incubation.

called "Vi" antigens. Certain *Citrobacter* sp. may also possess Vi antigens, although they can be differentiated from the salmonellae biochemically. The presence of capsular antigens may inhibit the ability of antiserum to detect the underlying O antigen. By heating the organism suspension to boiling for 30 minutes or longer, technologists may be able to detect the masked O antigens. Nomenclature for *E. coli* antigens associated with virulence became confused when certain noncapsular antigens, important for adherence of the organisms to intestinal epithelium, were named "K" antigens before their composition was known. These antigens, called K88 and K99, are now known to be protein, analogous to fimbriae; they are not capsular polysaccharide and should not be called "K." Under new classification schemes, fimbriae-associated K antigens have been renamed as "F" antigens.

The third category of antigens possessed by Enterobacteriaceae are flagellar, or "H" antigens. These antigens are protein in nature and heat labile. *Salmonella* sp. often produce two different antigenic types of flagella, or H antigens, called "phase 1" and "phase 2." Only motile strains display H antigens, but certain salmonellae may express only one of their two phases of antigens at a given time. These strains can be encouraged to express the second phase of their H antigens by cultivation in special media.

SEROLOGICAL CHARACTERIZATION OF *SALMONELLA* SP.

Salmonella sp. are characterized according to their O and H antigens and to the Kauffmann-White scheme. Many more than 2000 different serovars of *Salmonella* exist, based on the combined O, Vi, and H antigens that each serovar expresses. In addition to their alpha-numerical serovars, each serovar is named for the town in which it was first isolated or for some other factor with which it is associated. For practical purposes, these organisms are treated as if the name is a species; for example, *Salmonella* subgroup I 1,4,5,12:i:1,2, representing the O:phase 1:phase 2 antigens, is called *Salmonella typhimurium*.

Serotyping of isolates of Enterobacteriaceae is important for epidemiological purposes (Chapter 6). The unabridged list of all currently accepted serovars can be found in *Bergey's Manual of Systematic Bacteriology*. Clinical laboratories often perform a very abbreviated, preliminary serological grouping on isolates of *Salmonella* using commercially available polyvalent antisera designated A, B, C_1, C_2, D, E, F, G, H, and Vi. Some of the more com-

mon etiological agents of gastrointestinal disease and enteric fever can be classified by their agglutination in antisera belonging to one of the pools. For example, *Salmonella paratyphi A* is agglutinated by antiserum A, *S. typhimurium* is agglutinated by antiserum B, and *S. typhi* is agglutinated by antiserum D, although if it is Vi positive the group D antiserum will not agglutinate until after the organism suspension has been boiled.

The *slide agglutination test* is performed as follows. A suspension of the organism in physiological saline must be tested for autoagglutination before antiserum is added. To test for autoagglutination, mix 1 drop of the organism suspension with 1 drop of saline (instead of antiserum) and rotate manually for 1 to 2 minutes. No agglutination or clumping should be present. Biochemically suspicious organism isolates that fail to agglutinate in the polyvalent antisera should be boiled by placing a heavy suspension of the organisms in 0.5 ml sterile physiological saline, in a small screw-capped glass tube, and into a beaker of boiling water for 15 to 30 minutes. Some isolates may require prolonged (up to 1 hour) boiling. After cooling, the suspension is retested for specific agglutination. All isolates that biochemically resemble *Salmonella* but fail to agglutinate in any typing sera should be sent to a reference laboratory for definitive identification. The widespread use of commercial biotyping systems with many parameters (Chapter 10) has allowed almost every laboratory, no matter how small, to identify definitively (by biotype) isolates that resemble stool pathogens recovered from patients with gastroenteritis. It is important to notify both the physician caring for the patient and public health workers (or the hospital epidemiologist) so that the patient may be appropriately isolated if he or she is in the hospital, or so that the patient may be instructed as to hygienic practices if at home to prevent spread of the disease.

DNA homology studies have shown all *Salmonella* to be genetically identical. A new species name, *Salmonella enterica*, was proposed but did not gain widespread acceptance.[8] Microbiologists continue to refer to *Salmonella* by their serovar names.

TREATMENT OF INFECTIONS CAUSED BY ENTEROBACTERIACEAE

The great diversity among species of Enterobacteriaceae, as well as their propensity to pass genetic elements containing resistance factors indiscriminately among each other, makes it difficult to make specific statements about antimicrobial resistance. Multiple resistance is common among hospital strains of Enterobacteriaceae, those most involved in serious nosocomial infections. However, a few generalizations can be made. Environmental strains of Enterobacteriaceae are usually susceptible to aminoglycosides, imipenem, quinolones, and third-generation cephalosporins. In larger hospitals, however, these bacteria may be highly resistant to most available antimicrobial agents. A recent study examining new species of Enterobacteriaceae showed *Klebsiella terrigena, Klebsiella planticola, Enterobacter amnigenus, Enterobacter intermedium, Proteus penneri, Rahnella aquatilis, Serratia fonticola, Serratia plymuthica*, and *Buttiauxella agrestis* to be resistant to chloramphenicol, which is rarely used for treatment today.[4]

Several genera of Enterobacteriaceae display patterns of resistance that aid in their identification. Technologists can check the biochemically derived identification with the known susceptibility pattern as a form of quality assurance. *Klebsiella pneumoniae* and *Citrobacter diversus* are resistant to ampicillin and carbenicillin; most *Enterobacter* sp. and *Hafnia* are resistant to ampicillin and cephalothin, although newer cephalosporins are more active against *Enterobacter*. Swarming *Proteus* sp. are usually resistant to tetracycline and nitrofurantoin, and *Proteus, Morganella*, and *Serratia* are resistant to colistin. *Providencia* and *Serratia* are multiply resistant, and *Escherichia coli*, particularly community-acquired strains, in contrast, are often susceptible to many antibiotics tested.

Treatment of nosocomial infections associated with any of the Enterobacteriaceae usually consists of a β-lactam or quinolone, alone or, more often, in combination with another agent such as an aminoglycoside. Some newer agents, such as cefoperazone, ceftazidime, cefotaxime, or ceftriaxone, may be effective against *Proteus, Citrobacter, E. coli*, and *Klebsiella*. For nosocomially acquired infections, the best policy is to start empirical therapy based on the prevailing susceptibility patterns for known isolates in the institution, changing therapy if results of in vitro susceptibility tests indicate the necessity for change. For this reason, it is very important that the laboratory have available statistics concerning the percentage of susceptible isolates, categorized by species and antimicrobial agent. These statistics should be distributed to all clinicians and updated regularly. With seriously ill patients, the best agents should be used in combination until data are available regarding the specific susceptibility pattern of the etiological agent in question.

Treatment of Enterobacteriaceae-associated gastroenteritis varies with the severity of disease. Many infections, particularly with *Salmonella*, are self-limiting and should not be treated. Treatment may encourage development of the carrier state. If treatment is warranted, trimethoprim-sulfamethoxazole is the drug of choice. Quinolones are also effective. Chloramphenicol may be necessary for life-threatening illness. Susceptibility tests are important, however, since multiply resistant strains of *Salmonella* are being reported more frequently. *Shigella* are usually ampicillin resistant. Trimethoprim-sulfamethoxazole, the quinolones, and furazolidone may be useful agents for treatment of bacillary dysentery. *Yersinia* sp. are susceptible to chloramphenicol, aminoglycosides, tetracycline, and trimethoprim-sulfamethoxazole. *Y. enterocolitica* and *Y. pseudotuberculosis* septicemia are associated with high fatality rates despite therapy; these syndromes require aggressive treatment with an aminoglycoside, ampicillin, or both. As stated previously, streptomycin or tetracycline is the preferred agent for treatment of plague. For severe disease, chloramphenicol can be added.

REFERENCES

1. Agbonlahor, D.E. 1986. Characteristics of *Yersinia intermedia*–like bacteria isolated from patients with diarrhea in Nigeria. J. Clin. Microbiol. 23:891.
2. Bottone, E.J. 1981. *Yersinia enterocolitica*. CRC Press, Boca Raton, Fla.
3. Chester, B., and Moskowitz, L.B. 1987. Rapid catalase supplemental test for identification of members of the family Enterobacteriaceae. J. Clin. Microbiol. 25:439.
4. Freney, J., Husson, M.O., Gavini, F., et al. 1988. Susceptibilities to antibiotics and antiseptics of new species of the family Enterobacteriaceae. Antimicrob. Agents Chemother. 32:873.
5. Hedberg, C.W., Korlath, J.A., D'Aoust, J.Y., et al. 1992. A multistate outbreak of *Salmonella javiana* and *Salmonella oranienburg* infections due to consumption of contaminated cheese. J.A.M.A. 268:3203.
6. Karmali, M.A. 1989. Infection by verocytotoxin-producing *Escherichia coli*. Clin. Microbiol. Rev. 2:15.
7. Miller, J.M., and Farmer, J.J., III. 1987. Recent additions to the Enterobacteriaceae. Clin. Microbiol. Newsletter 9:173.
8. Minor, L.L., and Popoff, M.Y. 1987. Designation of *Salmonella enterica* sp. nov., nom. rev., as the type and only species of the genus *Salmonella*. Int. J. Syst. Bacteriol. 37:465.
9. Yannelli, B., Schoch, P.E., and Cunha, B.A. 1987. *Serratia marcescens*. Clin. Microbiol. Newsletter 9:157.

BIBLIOGRAPHY

Akhurst, R.J. 1986. *Xenorhabdus nematophilus* subsp. *beddingii (Enterobacteriaceae)*: a new subspecies of bacteria mutualistically associated with entomopathogenic nematodes. Int. J. Syst. Bacteriol. 36:454.

Atlas, R.M. 1993, Handbook of microbiologic media. CRC Press, Boca Raton, Fla. (Includes detailed formulas and instructions for media preparation.)

Balows, A., Hausler, W.J., Jr., Herrmann, K.L., et al. 1991. Manual of clinical Microbiology, ed 5. American Society for Microbiology, Washington, D.C.

Baron, E.J, and Finegold, S.M. 1990. *Bailey and Scott's diagnostic microbiology,* ed 8. Mosby, St. Louis. (Appendices include detailed formulas of media and stains.)

Brubaker, R.R. 1984. Molecular biology of the dread Black Death. ASM News 50:240.

Centers for Disease Control. 1991. Summary of notifiable diseases United States. M.M.W.R. 40:1.

Clarridge, J.E., and Weissfeld, A.S. 1987. *Enterobacteriaceae* infections. p. 233. In Wentworth, B.B., Baselski, V.S., Doern, G.V., et al., editors. Diagnostic procedures for bacterial infections, ed 7. American Public Health Association, Washington, D.C.

Farmer, J.J., III, Davis, B.R., Hickman-Brenner, F.W., et al. 1985. Biochemical identification of new species and biogroups of *Enterobacteriaceae* isolated from clinical specimens. J. Clin. Microbiol. 21:46.

Hendrickson, D.A. 1985. Reagents and stains. In Lennette, E.H., Balows, A., Hausler, W.J., Jr., and Shadomy, H.J., editors. Manual of clinical microbiology, ed. 4. American Society for Microbiology, Washington, D.C.

Howard, B.J., Klaas, J., II., Rubin, S.J., et al., editors. 1987. Clinical and pathogenic microbiology. Mosby, St. Louis.

Karmali, M.A. 1987. Laboratory diagnosis of verotoxin-producing *Escherichia coli* infections. Clin. Microbiol. Newsletter 9:65.

Kosako, Y., Sakazaki, R., Huntley-Carter, G.P., and Farmer, J.J., III. 1987. *Yokenella regensburgei* and *Koserella trabulsii* are subjective synonyms. Int. J. Syst. Bacteriol. 37:127.

MacFaddin, J. 1980. Biochemical tests for identification of medical bacteria, Williams & Wilkins, Baltimore.

Owen, R.J., Ahmed, A.U., and Dawson, C.A. 1987. Guanine-plus-cytosine contents of type strains of the genus *Providencia*. Int. J. Syst. Bacteriol. 37:449.

Washington, J.A., II., editor. 1985. Laboratory procedures in clinical microbiology, ed. 2. Springer-Verlag, New York.

29

NONFERMENTATIVE GRAM-NEGATIVE BACILLI AND COCCOBACILLI

The nonfermentative gram-negative bacilli are found in nature as inhabitants of soil and water and as harmless parasites on the mucous membranes of humans and animals. These bacteria can cause disease by colonizing and subsequently infecting immunocompromised individuals or by gaining access to normally sterile body sites through trauma. Less than one fifth of all gram-negative bacilli isolated from clinical specimens received in a routine clinical microbiology laboratory are likely to be nonfermentative bacilli, and the predominant species among those strains is *Pseudomonas aeruginosa*. Although nonfermenters make up a small percentage of the total isolates recovered, they generally require more effort for their identification and often require specialized tests.

Pseudomonas is a common gram-negative etiological agent of nosocomial infections in many hospitals. Former members of the genus have been placed in new genera recently; further taxonomic changes are likely in the future. The genera *Moraxella* and *Acinetobacter,* covered in this chapter, are included in the family Neisseriaceae.[1] Other genera in that family are discussed in Chapter 30, since they are able to ferment glucose. *Moraxella urethralis* has been removed from the genus *Moraxella* and the family Neisseriaceae and placed into a new genus, *Oligella*.[22] The genus *Achromobacter* has no validity at this time. Many organisms that were called "*Achromobacter xylosoxidans*" are now called *Alcaligenes xylosoxidans*.[13,20,23] In addition to the named bacteria covered in this chapter, there are several unnamed species, including Centers for Disease Control (CDC) Groups IVc-2, EO-2, and others (often similar to *Alcaligenes* or *Pseudomonas*) that have been recovered from clinical specimens. The definitive identification of all the various nonfermentative gram-negative bacilli that might be encountered in a clinical laboratory is beyond the scope of this text. Readers are referred to the excellent references by Gilardi,[7] Pickett et al.,[25] and Clark et al.,[3] for detailed information and identification schemes.

EPIDEMIOLOGY AND PATHOGENESIS OF INFECTIONS CAUSED BY NONFERMENTERS

PSEUDOMONAS AERUGINOSA

The most frequently isolated nonfermentative bacillus found in clinical specimens, *P. aeruginosa*, has one of the broadest ranges of infectivity among all pathogenic microorganisms. This species can cause disease in plants, insects, fish, amphibians, reptiles, birds, and mammals. It is a significant cause of burn wound infections and nosocomial infections, and an unusual mucoid variant is almost universally found to colonize the respiratory tract of patients with cystic fibrosis. Particularly destructive infections of the eye (keratitis, endophthalmitis) are associated with *P. aeruginosa;* contact lens wearers are one group at increased risk for such infections. *P. aeruginosa* is the third most frequently isolated etiological agent in hospital-acquired pneumonia. The organisms are so well adapted to survival in harsh environments that they can exist on the minimal organic nutrients present in distilled water. Underchlorinated water used to fill hot tubs is the vehicle of *P. aeruginosa*–associated folliculitis, which initially appeared across the United States during the late 1970s. Strains of *P. aeruginosa* have even been isolated from the disinfectant soap solutions used by nurses and physicians to wash their hands. When hospital environments are sampled, *P. aeruginosa* is found in drains, water faucets, sinks, cleaning solutions, medicines, and other sites, including the flowers in patients' rooms.

Although *P. aeruginosa* possesses many possible virulence factors, one must remember that it is a significant human pathogen only in compromised patients. Those strains that cause pneumonia appear to possess fimbriae that mediate adherence to respiratory epithelial cells. The mucoid polysaccharide alginate, particularly pronounced in strains isolated from patients with cystic fibrosis, seems to protect the bacteria from phagocytosis.[19] The ability of *P. aeruginosa* strains to destroy tissue may be related to the production of various extracellular enzymes, including proteases, elastase, and pigments that may inhibit growth of other bacteria (called **pyocins**) and hemolysins. In addition, virulent strains produce an exotoxin, called *exotoxin A*, which inhibits protein synthesis much as does the toxin of *Corynebacterium diphtheriae*. Exotoxin A also seems to promote tissue destruction and aggravate the host's inflammatory response, further contributing to tissue damage. As with all other nonanaerobic gram-negative bacteria, *P. aeruginosa* possesses endotoxin.

OTHER PSEUDOMONADS, *SHEWANELLA (ALTEROMONAS), CHRYSEOMONAS, COMAMONAS, FLAVIMONAS, METHYLOBACTERIUM,* AND *XANTHOMONAS*

Other pseudomonads that have been isolated from human clinical sources include species that are saprophytic inhabitants of the soil and water, infecting humans as opportunistic pathogens. The number of isolations of *Pseudomonas cepacia* in association with *Pseudomonas aeruginosa* in respiratory secretions from patients with cystic fibrosis has recently increased dramatically. The presence of this organism seems to be associated with decreased pulmonary function. Many strains of *P. cepacia* produce protease and lipase, and several strains also produce biologically active endotoxin. Most other pseudomonads and many other nonfermenters are isolated more frequently from respiratory specimens from patients with cystic fibrosis than from those with other types of pneumonia. These patients are particularly at risk for pulmonary infection caused by these agents.

Numerous species of *Pseudomonas* and *Xanthomonas* (formerly *Pseudomonas*) *maltophilia* have been implicated as the etiological agents of human disease.[21,25] Many more species can be recovered from the environment, including soil, water, and plants. Most of these species probably do not possess any particular virulence properties but infect humans opportunistically.

Specific syndromes, however, can be attributed to *Pseudomonas mallei,* the agent of glanders, and *P. pseudomallei,* the agent of melioidosis, both considered to be "true" pathogens. Glanders, a disease of horses and occasionally other animals such as goats, sheep, donkeys, and dogs, is transmitted to humans by contact with infected animals. The disease, rarely seen in the United States, may present as a systemic septicemic infection, a local suppurative infection with lymphadenopathy, acute pneumonia, or chronic disease. Melioidosis, found primarily among returning Vietnam veterans and Southeast Asian immigrants, is usually a pneumonia-like illness, although it may be a systemic, febrile illness or an acute or chronic suppurative infection. Although *P. pseudomallei* can be found in the soil, *P. mallei* has been isolated only from animals that are susceptible to infection with this agent. The etiological *Pseudomonas* sp. can be recovered from blood, aspirated lesion material, or respiratory secretions from patients with glanders or melioidosis. *P. pseudomallei* and *P. mallei* both produce extracellular toxins, and *P. mallei* has an endotoxin-like material that may contribute to pathogenesis.

The relatively new genera *Shewanella* (originally named *"Alteromonas"*) and *Comamonas* include former *Pseudomonas* sp. that were found to differ from *P. aeruginosa* (the type strain) by cell wall component structures and genetic composition. They are opportunistic pathogens similar to the pseudomonads. The new genera *Chryseomonas* and *Flavimonas* produce a yellow pigment, but unlike most pseudomonads, with the major exception of *Xanthomonas maltophilia*, they are oxidase negative.[7,11] Virulence factors have not been studied for these species, although they have been isolated from human clinical specimens.

Methylobacterium extorquens, most recently called *Pseudomonas mesophilica* or *Vibrio extorquens*, are pink-pigmented, nonfermenting bacteria of low virulence. They are involved in catheter-related sepsis, peritoneal dialysis infection, and other rare infections in immunocompromised hosts.[12]

ACINETOBACTER, AGROBACTERIUM, ALCALIGENES, FLAVOBACTERIUM, OCHROBACTRUM, PSYCHROBACTER, SPHINGOBACTERIUM, AND *WEEKSELLA* SP.

These nonfermenters are found in soil and water. They infect humans when they have been introduced into a traumatic wound or by opportunistic colonization and infection of compromised hosts. In addition, *Acinetobacter* sp. have been isolated as the etiological agent of pneumonia in healthy individuals, although they are primarily found as nosocomial pathogens.[16] *Alcaligenes*, which has recently undergone taxonomic changes to include former *Achromobacter xylosoxidans*, is most often recovered from sputum, cerebrospinal fluid, and blood of hospitalized patients.[13,17,20,23]

Although no definite virulence factors have been elucidated for any of these organisms, *Flavobacterium* sp., particularly *F. meningosepticum*, are associated with outbreaks and sporadic cases of neonatal meningitis and septicemia. These bacteria must possess some mechanism, such as neurotropism, that allows them to produce such a syndrome. The flavobacteria also produce proteases and gelatinase, which may contribute to virulence.[8] *Sphingobacterium multivorum*, formerly known as *Flavobacterium multivorum* and CDC Group IIk-2, is rarely isolated from humans.[5] *Weeksella zoohelcum* has been isolated from human wounds, especially those resulting from dog bites.[10] *Agrobacterium tumefaciens* (synonym *A. radiobacter*) has been isolated from sputum, blood, mucous membrane sites, and wounds. Cases of catheter-related sepsis, osteochondritis, and a few other infections in compromised hosts have been attributed to *Ochrobactrum anthropi*.[2,9] *Psychrobacter immobilis* has been reported as a cause of meningitis and ocular infections in infants[15,22].

EIKENELLA, MORAXELLA, AND *OLIGELLA* SP.

These genera are normal inhabitants of human mucous membranes of the mouth, upper respiratory tract, and genitourinary tract. From these sites, they can cause disease in immunosuppressed hosts. As with the other nonfermenters, infection may follow deposition in normally sterile tissues. *Eikenella corrodens* is a frequently recovered etiological agent of human bite wound infections and occasionally an important pathogen in many other types of infection. *Oligella urethralis*, formerly *Moraxella urethralis*, is a commensal in the human genitourinary tract and is an uncommon pathogen.[22] The species *Oligella ureolytica*, formerly CDC group IVe, is also isolated from urine. Its pathogenicity has not been documented. The virulence factors that contribute to pathogenesis of disease caused by these nonfermentative bacteria are not well characterized.

LABORATORY IDENTIFICATION

GENERAL CONSIDERATIONS

Once one has determined that a gram-negative bacillus isolated from clinical material is not a member of the Enterobacteriaceae by using criteria listed in Chapter 28 (Enterobacteriaceae are oxidase negative; are able to ferment glucose, as detected by production of a yellow butt in triple sugar iron agar [TSIA] or Kligler's iron agar [KIA]; and grow well on MacConkey and other media) and that the isolate is not *Pseudomonas aeruginosa* (as detailed in the next section), a set of preliminary tests must be performed to categorize the isolate so that it can be identified by the most expedient method. These tests include determining the organism's ability to grow on MacConkey agar, the oxidase reaction, and determining whether the organism is able to metabolize carbohydrates or the circumstances under which the organism produces acid from the carbohydrates glucose and maltose. An isolate's ability to utilize carbohydrates and whether it does so oxidatively (in the presence of air) or fermentatively (in the absence of air) or in both ways is one of the most important characteristics needed to separate the nonfermenters from Enterobacteriaceae.

Low-peptone media, called *oxidative-fermentative media (O-F media)*, allow detection of acid production by weak metabolizers for which TSIA and

KIA reactions are slow. Preparation, inoculation, and interpretation of test results for one O-F medium are described in Procedure 29.1.

O-F reactions are interpreted as shown in Figure 29.1. Fermentation takes place in the medium without oxygen (overlaid tubes). Organisms that are able to ferment a carbohydrate are also able to oxidize it (open tubes). These organisms are still called "fermenters." Organisms that produce acid products and yellow reactions only in open tubes are "oxidizers," and organisms that do not utilize glucose, either fermentatively or oxidatively, are called "nonoxidizers." O-F media may be purchased from most media manufacturers.

Based on their reaction in KIA or TSIA (or O-F glucose), ability to grow on MacConkey agar, and oxidase reaction, all unusual non-Enterobacteriaceae may be placed preliminarily into a category requiring particular additional tests for identification (Figure 29.2). The ability of these organisms to utilize carbohydrates other than glucose is also usually tested in O-F medium; standard methods (Chapter 10) are used for performance of most other biochemical tests required to identify these organisms. Since indole production may be an important characteristic, a more sensitive method for extracting indole is often used for the nonfermenters (Procedure 29.2). The rapid spot indole test that uses *p*-dimethylaminocinnamaldehyde reagent (Chapter 10) may also be satisfactory. Key differential characteristics of frequently recovered nonfermenters and related strains are shown in Table 29.1. Other clinically significant gram-negative bacilli that grow on conventional media, other than the nonfermenters mentioned in this chapter, are discussed in Chapters 30, 31, and 40.

In addition to biochemical metabolism, another important taxonomic feature of organisms classified as nonfermenters is the arrangement of their flagella. Performance of a flagella stain is not easy; it is an art that must be developed with experience. Flagella are fragile and tend to break off from the surface of bacteria during manipulation. Many individual bacteria within a culture do not produce flagella, and the stain sometimes fails to work. Procedure 29.3 outlines a flagella stain that should yield excellent results if all the steps are followed carefully.

IDENTIFICATION OF *PSEUDOMONAS AERUGINOSA*

As noted earlier, by far the most frequently encountered nonfermenter is *P. aeruginosa*. These organisms can grow on most routine laboratory media, including MacConkey agar. On 5% sheep blood agar, colonies are flat with a feathered edge, rough or "ground glass" appearing, and beta hemolytic (Figure 29.4). Colonies can grow to 3 to 5 mm in diameter within 48 hours, incubation at 35° to 37° C in air, and growth is equally good at room temperature. The mucoid (alginate-producing) strains isolated primarily from patients with cystic fibrosis may be blue-green or pale-yellow-green, and they are extremely mucoid. This mucoid material may be produced in such quantities that it slides across the surface of the plate when the plate is tipped (Figure 29.5). *P. aeruginosa* exhibits a particularly striking characteristic odor, which resembles overripe grapes (or corn

FIGURE 29.1

Interpretation of oxidative-fermentative (O-F) utilization of glucose test.

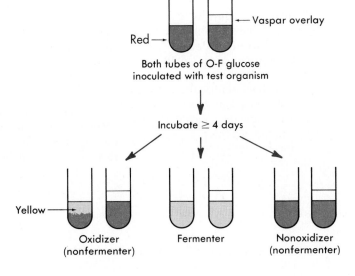

PROCEDURE 29.1

OXIDATIVE-FERMENTATIVE (O-F) TEST FOR CARBOHYDRATE UTILIZATION

PRINCIPLE

The small amount of acid produced by certain fermenters and oxidizers may be masked by larger amounts of alkaline products of protein metabolism. Carbohydrate fermentation reactions can thus be determined by testing such organisms in a medium with minimal proteins.

METHOD

1. Prepare O-F base medium as follows:

   Casitone (Difco), 2.0 g
   Agar, 3.0 g
   Phenol red, 0.03 g
   Distilled water, 1000 ml

   Combine all ingredients, which are available from Difco Laboratories and standard chemical supply companies, or purchase O-F base from commercial suppliers of media. Adjust pH to 7.3, and divide the solution into several smaller flasks with known volume, one flask for each carbohydrate being prepared. For example, if O-F glucose, maltose, sucrose, lactose, and mannitol are to be made, divide the O-F base into five flasks containing 200 ml each. Autoclave the flasks at 121° C for 15 min.

2. Prepare 10% aqueous carbohydrate solutions by adding 2 g carbohydrate (glucose, lactose, etc.) to 20 ml distilled water. Immediately sterilize the carbohydrate solutions by passing them through a 0.2 µm membrane filter. Autoclave sterilization destroys some of the carbohydrates. These solutions can be refrigerated for several months, although they may require redissolving before use.

3. When the O-F base has cooled to 55° C after autoclaving, aseptically add 20 ml of the sterile 10% carbohydrate solution to the 200 ml flask of O-F base, for a final concentration of 1% carbohydrate. The correctly prepared medium is red, but it appears somewhat pale orange while it is still hot.

4. Dispense the 1% carbohydrate O-F media into 16 × 125 mm screw-cap test tubes, 5 ml per tube. Allow the tubes to solidify upright (forming a semisoft butt), tighten caps, and refrigerate. The medium should be usable for several months. If desired, the base can be dispensed without carbohydrates, which can then be added to melted and cooled tubes of base as needed (0.5 ml 10% carbohydrate to 5 ml base).

5. For testing the nature of glucose utilization, two tubes of 1% O-F glucose are inoculated. Material from isolated colonies is picked up with a straight wire and stabbed at least four times into each medium. Stabs should extend approximately 5 mm deep into the surface layer of the butt.

6. The medium in one of each of the two identical tubes is overlayed with sterile melted petrolatum or melted paraffin combined with an equal volume of petroleum jelly (vaspar) approximately 1 cm deep to prevent oxygen from reaching the inoculum. Sterile mineral oil is not recommended for overlaying the medium, since some lots are acidic and can contribute to false-positive results.

7. The tubes are incubated at 35° C for as long as 4 days and examined daily for production of acid, as indicated by a change in the phenol red indicator from red to yellow.

QUALITY CONTROL

Test *Pseudomonas aeruginosa* ATCC 27853 and *Alcaligenes faecalis* ATCC 8750 in O-F glucose and maltose. *Xanthomonas maltophilia* can be used as a positive control for maltose.

EXPECTED RESULTS

P. aeruginosa is oxidative for glucose (yellow reaction in nonoverlaid glucose and no color change in the overlaid tube) and nonfermentative, nonoxidative for maltose (no change in either of the maltose tubes). *A. faecalis* shows no change or a red color, indicating alkaline end products in the four tubes, because it is asaccharolytic. *X. maltophilia* is oxidative (yellow in nonoverlaid tube) for maltose.

PERFORMANCE SCHEDULE

Test the quality control organisms each time a new batch of media is prepared and every 2 months thereafter.

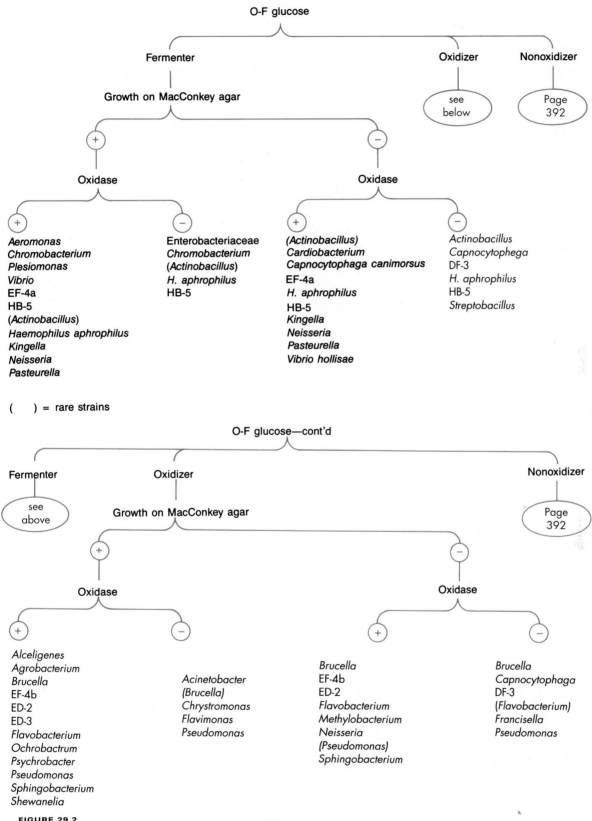

FIGURE 29.2

Initial characterization of gram-negative bacilli based on utilization of glucose in oxidative-fermentative (O-F) media, growth on MacConkey agar, and oxidase reaction.

continued

PROCEDURE 29.2

XYLENE EXTRACTION FOR DEMONSTRATION OF INDOLE PRODUCTION

PRINCIPLE

Small amounts of indole produced from the breakdown of tryptophan by tryptophanase must be extracted with xylene before they can be detected.

METHOD

1. Prepare Ehrlich's reagent as follows:
   Paradimethylaminoben-zaldehyde (Sigma Chemical Co.), 2 g
   Ethyl alcohol (95%), 190 ml
   Hydrochloric acid (concentrated), 40 ml
   Store refrigerated in a brown bottle.

2. Add 1 ml of xylene to a 48 hour culture of organisms in tryptone or trypticase broth or other appropriate medium. Shake the tube well, and allow it to stand for a few minutes until the solvent rises to the surface.

3. Gently add about 0.5 ml of the reagent down the sides of the tube so that it forms a ring between the medium and the solvent. If indole has been produced by the organisms, and since it is soluble in solvent, indole will be concentrated in the solvent layer, and on addition of the reagent, a brilliant *red ring* will develop just below the solvent layer. (see Figure 28.6) If no indole is produced, no color will develop.

QUALITY CONTROL

Test *Escherichia coli* ATCC 25922 and *Pseudomonas aeruginosa* ATCC 27853.

EXPECTED RESULTS

E. coli is strongly indole positive (red ring), and *P. aeruginosa* is negative (no color or pale-yellow ring).

PERFORMANCE SCHEDULE

Test quality control organisms each time fresh reagents are made and monthly thereafter.

Modified from Bohme. 1906. Zentralbl. Bakt., Orig. 40:129.

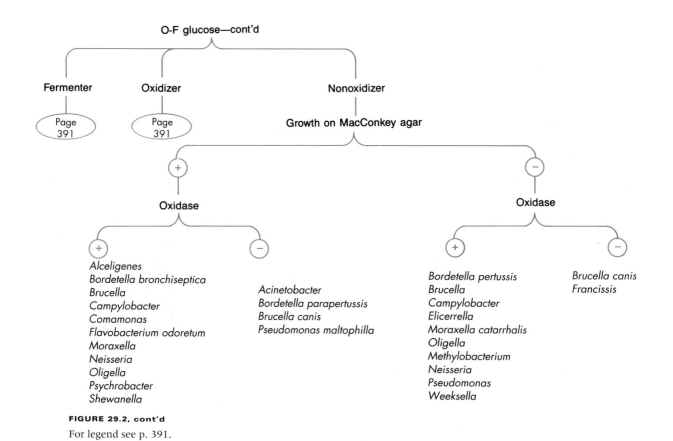

FIGURE 29.2, cont'd

For legend see p. 391.

TABLE 29.1

KEY DIFFERENTIAL CHARACTERISTICS OF FREQUENTLY RECOVERED NONFERMENTERS AND RELATED STRAINS

GENUS	CATALASE	OXIDASE	INDOLE	PIGMENT	UREASE	H₂S IN KIA	SUSCEPTIBLE TO PENICILLIN	FLAGELLA	SUSCEPTIBLE TO POLYMYXIN	ARGININE DIHYDROLASE	OXIDATIVE/ FERMENTATIVE
Acinetobacter	+	–	–	–	–/+	–	–	N	+	–	O/–
Agrobacterium	+	+	–	–	+R	–/+	–	P	+	–	O
Alcaligenes	+	+	–	–	–/+	–	–	P	+	–	–/O
Bordetella	+	+	–	–	+R	–	–	P	+	–	–
Chryseomonas	+	–	–	Y	–/+	–	–	Po	+	+	O
Comamonas	+	+	–	–	–	–	–	PT	+	–	–/O
Eikenella	–	+	–	–/Y	–	–	+	N	+	–	–
Flavimonas	+	–	–/+	Y	–/+	–	–	Po	+	–	O
Flavobacterium	+	+	–/+	–/Y	–/+	–	–	NM	–	–	O/–
Kingella	–	+	–	Pink	–	–	+	N	+	ND	F
Methylobacterium	+	+	–	–	–/+	–	–	Po	–/+	–	O
Moraxella	+	+	–	–	–/+	–	+	N	+	–	–
Ochrobactrum	+	+	–	–	+R	–	–	P	+	–	O
Oligella	+	+	–	–	+R/–	–	+/–	N/P	+	–	–
Pasteurella	+	+	+/–	–	+/–	–	+	N	+	–	F
Pseudomonas	+	+	–	+/–	+/–	–	–	Po/P+	+/–	+/–	O/–
Shewanella	+	+	–	+	–/+	+	–	Po	+	–	O
Sphingobacterium	+	+	–	Y	+	–	–	NM	–	–	O
Weeksella	+	+	+	Tan	+R/–	–	+	NM	+/–	–	–

Modified from G. Gilardi (personal communication).

+, >90% of strains positive; –, >90% of strains negative; +/–, variable (most strains positive); –/+, variable (most strains negative); *NM*, nonmotile; *N*, none; *P*, peritrichous; *Po*, polar; *PT*, polar tuft; *Y*, yellow; *R*, rapid urease; *ND*, not done; *O*, oxidative; *F*, fermentative.

* H₂S, Hydrogen sulfide; *KIA*, Kligler's iron agar.

PROCEDURE 29.3

FLAGELLA STAIN

PRINCIPLE
Tannic acid and fuchsin coat flagella, lending them enough thickness to be observed microscopically.

METHOD
1. Prepare solution A as follows:

   Basic fuchsin (certified for flagella stain; Aldrich Chemical Co.), 0.6 g

   Ethyl alcohol (95%), 50 ml

   Add the fuchsin to the alcohol, and allow to stir overnight on a magnetic stirring plate to dissolve the fuchsin.

2. Prepare solution B as follows:

   Sodium chloride, 0.37 g
   Tannic acid (Sigma Chemical Co.), 0.75 g
   Distilled water, 50 ml

3. Combine solutions A and B and mix thoroughly. Adjust the pH to 5.0 with 1.0 N NaOH. Place the stain into a tightly capped brown bottle and let it remain for 3 days in the refrigerator (4° C) before using it to allow it to cure. It is stable in the refrigerator for 1 month and in the freezer for at least 1 year. The stain deteriorates rapidly at room temperature. Use only the clear supernatant for the staining procedure; do not disturb the sediment that settles in the bottom of the bottle. Before performing each staining procedure, draw a small amount of the stain from the bottle and allow it to warm to room temperature. Discard the stain that is left over after the procedure.

4. Clean slides by soaking them in 3% concentrated hydrochloric acid in 95% ethanol for 4 days. Rinse thoroughly in tap water, allowing the water to run through the slides in a beaker for 10 minutes, and then rinse with several changes of distilled water. Lean the slides against an upright surface to drain and air dry.

5. Flame a clean slide in the blue flame portion of a gas burner. Do not allow carbon to be deposited on the slide. Enclose a rectangular area with a thick, wax pencil line.

6. Place a large drop of distilled water at one end of the slide, within the wax-penciled area.

7. Pick up a small amount of growth from the tops of several young colonies on an agar plate or from a freshly inoculated slant on the end of an inoculating needle. Touch the needle to the drop of water on the slide in several places, allowing a small amount of the organism to disperse in the water. Tip the slide slightly to allow the drop of water to run toward the opposite end of the slide. Allow the suspension to air dry, and do not heat fix before staining.

8. Apply 1 ml of the stain to the dry slide, and observe the slide for the formation of a fine, metallic-red precipitate, which requires 5 to 15 minutes. Each new batch of stain should be tested for optimal timing with a known flagella-producing organism.

9. Rinse the slide gently in tap water, holding the slide horizontally. Do not tilt to drain the water, and allow the slide to air dry.

10. Examine under oil immersion (1000×) for flagella (Figure 29.3). It is possible that only a few of the bacteria in the suspension retained intact flagella; thorough examination is often required.

QUALITY CONTROL
Test a fresh culture of *Pseudomonas aeruginosa* ATCC 27853.

EXPECTED RESULTS
The organism should display polar flagella on at least 10% of cells seen.

PERFORMANCE SCHEDULE
Test a quality control organism each time a flagella stain is performed.

Modified from Clark. 1976. J. Clin. Microbiol. 3:632.

FIGURE 29.3

Bacteria stained with the flagella stain demonstrating peritrichous flagella.

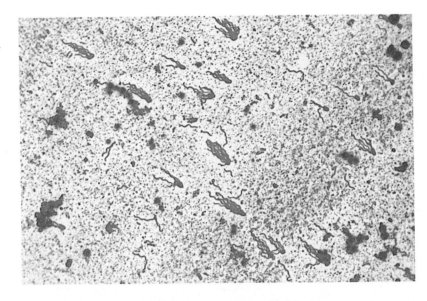

FIGURE 29.4

Strong beta hemolysis exhibited by most strains of *Pseudomonas aeruginosa*.

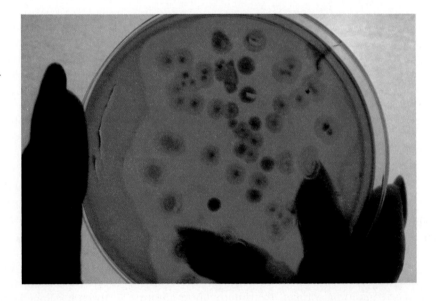

tortillas). Most strains of *P. aeruginosa* produce pyocyanin, a turquoise blue or blue, water-soluble, chloroform-extractable pigment visible on uncolored media, such as Mueller-Hinton agar (Figure 29.6), and especially visible on *Pseudomonas* P agar (Difco Laboratories) and Tech agar (BBL Microbiology Systems). No other bacteria produce this pigment, so its detection is sufficient for identification of an isolate as *P. aeruginosa*. Some strains of *P. aeruginosa* produce other pigments, pyoverdin (yellow), pyorubrin (red), and pyomelanin (brown), which may mask the pyocyanin. Such strains, which appear yellow-green, blue-green, red, purple, or brown, must be further characterized. Occasional strains of *P. aeruginosa* fail to produce pyocyanin. These strains, however, are able to grow at 42° C, are able to dihydrolyze

arginine, are motile by means of polar flagella, and are unable to produce acid in O-F lactose or sucrose. *P. aeruginosa* and other pseudomonads isolated from humans often produce water-soluble fluorescent pigments; pyoverdin is also one of these. Fluorescent pigment-producing strains fluoresce under *short-wave* ultraviolet light. A standard Wood's lamp emits light at 365 nm; the fluorescence of pseudomonads is best observed at 254 nm.

Several selective agars have been developed for isolation of *P. aeruginosa* from "contaminated" specimens, such as stool. It has been shown that development of nosocomial infection with *P. aeruginosa* is often preceded by gastrointestinal colonization; thus detection of colonized persons at risk of developing *Pseudomonas* infection may

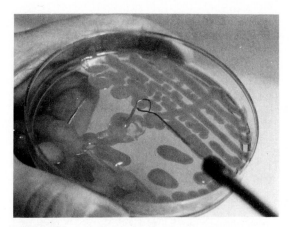

FIGURE 29.5

Abundant mucoid material produced by a strain of *Pseudomonas aeruginosa* isolated from a patient with cystic fibrosis.

FIGURE 29.6

Two different pyocyanin-producing strains of *Pseudomonas aeruginosa* and a control plate with no growth.

aid in prevention of subsequent morbidity. Agar containing *cetrimide* (cetrimethylammonium bromide), available as Pseudosel agar (BBL Microbiology Systems); *irgasan* (2,2,2-trichloro-2-hydroxydiphenyl ether), available as irgasan agar (Difco Laboratories); and noncommercially available agar containing 30 µg/ml *C-390* (9 chloro-9-[diethylaminophenyl]-10-phenylacridan), available from Norwich-Eaton Pharmaceuticals, have been used successfully.[26] Since C-390 inhibits the growth of all other bacteria, it may be used in an identification scheme. Isolates that grow on trypticase soy agar containing 30 µg/ml C-390 can be presumptively identified as *P. aeruginosa.*

For practical purposes, isolates on blood agar that exhibit the characteristic colonial morphology and odor, that are beta hemolytic, and that are oxidase positive can be presumptively identified as *P. aeruginosa.* When the susceptibility pattern also agrees with that generally exhibited by *P. aeruginosa,* which is resistant to most antimicrobial agents except for aminoglycosides and colistin (polymyxin B), the identification is more certain. All *P. aeruginosa* strains are oxidase positive, motile by means of polar flagella, able to oxidize glucose in O-F carbohydrate base, unable to oxidize maltose, unable to decarboxylate lysine or ornithine, and able to dihydrolyze arginine. As with all pseudomonads, they are catalase positive. In addition to growth on C-390, all *P. aeruginosa* are able to grow at 42° C, an ability shared by several other *Pseudomonas* species but no other species that produce fluorescent pigment, such as *P. fluorescens* and *P. putida.* Fluorescent pigment may be enhanced by growth on *Pseudomonas* F agar (Difco Laboratories), GNF agar (Flow Laboratories), or Flo agar (BBL Microbiology Systems). Fluorescence test media should be incubated at room temperature; some strains of *P. fluorescens* and *P. putida* will not produce fluorescent pigments at 35° to 37° C. In KIA or TSIA, *P. aeruginosa* yields an alkaline slant over a red or unchanged butt. Growth on the slant surface often appears metallic. Table 29.2 lists key characteristics used to differentiate some more frequently isolated species of clinically important pseudomonads.

IDENTIFICATION OF OTHER *PSEUDOMONAS* AND SIMILAR SPECIES

Xanthomonas (Pseudomonas) maltophilia is second to *Pseudomonas aeruginosa* in frequency of isolation. As the name implies, this species is able to oxidize maltose as well as glucose. *X. (P.) maltophilia* differs from the genus *Pseudomonas;* it possesses peritrichous flagella and is oxidase negative, which accounts for its proposed transfer to the genus *Xanthomonas,* a group of yellow-pigment-producing plant pathogens. Although most microbiologists have accepted the new name, the taxonomic status of *X. (P.) maltophilia* is still uncertain at this time.

Colonies of *X. (P.) maltophilia* on blood agar are rough, lavender-green, and have an ammonia-like odor (Figure 29.7). They may be slightly alpha hemolytic. The organism is oxidase negative (rare strains may be oxidase positive), arginine dihydrolase negative, ornithine decarboxylase negative, but lysine decarboxylase positive (may be delayed). It does not produce pyoverdin, and most strains are unable to grow on cetrimide. *X. (P.) maltophilia,* rare strains of *Pseudomonas cepacia* and *P. paucimobilis,* and some strains of *P. mallei* and *P. gladioli* (synonym, *P. marginata;* very rarely

TABLE 29.2

SOME CHARACTERISTICS OF FREQUENTLY ENCOUNTERED OR CLINICALLY IMPORTANT *PSEUDOMONAS* AND RELATED SPECIES (GIVEN AS PERCENTAGE OF POSITIVE REACTIONS)

SPECIES (NUMBER OF STRAINS)	OXIDASE	GROWTH AT 42°C	GAS FROM NITRATE	ACID IN O-F† GLUCOSE	ACID IN O-F LACTOSE	ACID IN O-F FRUCTOSE	LYSINE DECARBOXYLASE	ARGININE DIHYDROLASE	ORNITHINE DECARBOXYLASE	GELATIN HYDROLYSIS
P. aeruginosa* (91)	100	100	62	98	0	90	0	99	0	47
Xanthomonas (P.) maltophilia (645)	2	50	0	100	86	99	99	0	0	100
P. cepacia (208)	93	60	0	100	99	100	91	0	66	74
P. fluorescens (217)	100	0	4	100	11	99	0	99	0	100
P. putida (310)	100	0	0	100	14	99	0	99	0	0
P. stutzeri (171)	100	90	100	100	0	94	0	0	0	0
P. putrefaciens (72)	100	76	0	97	0	53	0	0	99	94
P. paucimobilis (168)	90	0	0	100	0	100	0	0	0	0
Comamonas acidovorans (106)	100	9	0	0	0	100	0	0	0	2
P. alcaligenes (52)	100	54	0	0	0	0	0	8	0	2
Ochrobactrum anthropi (52)	100	10	96	98	0	98	0	40	92	0
Shewanella putrefaciens (13)	100	0	0	100	15	100	0	0	0	100
Flavimonas oryzihabitans (101)	0	26	0	100	0	100	0	0	0	5
P. diminuta (44)	100	23	0	0	0	0	0	0	0	61
P. pickettii (56)	100	41	93	100	57	100	0	0	0	60
P. pseudoalcaligenes (76)	100	75	5	17	0	100	0	36	0	1
P. pseudomallei (7)	100	100	100	100	0	100	0	100	0	100
P. vesicularis (47)	100	0	0	58	0	0	0	0	0	38
Methylobacterium extorquens (22)	100	0	0	18	0	41	0	0	0	0

Data from Gilardi, G.L. 1984. Clin. Microbiol. Newsletter 6:111, 127; updated in January 1987 (personal communication).

* Apyocyanogenic strains.

† O-F, Oxidative-fermentative.

FIGURE 29.7

Xanthomonas (Pseudomonas) maltophilia growing on 5% sheep blood and MacConkey agars.

FIGURE 29.8

Colonies of *Pseudomonas pseudomallei* after 96 hours of incubation on blood agar.

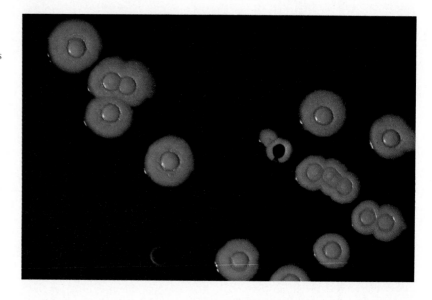

isolated) are the only named oxidase-negative pseudomonads (several unnamed strains are oxidase negative). *Chryseomonas luteloa* (formerly CDC group Ve-1) and *Flavimonas oryzihabitans* (formerly CDC group Ve-2), both pseudomonad-like, are oxidase negative.

The pseudomallei group includes organisms that grow on a variety of organic compounds as single sources of carbon and energy. Several species in the group are not human pathogens, but *P. pseudomallei, P. mallei, P. cepacia, P. pickettii,* and *P. gladioli* have been isolated from clinical specimens. *P. pseudomallei* grows slowly on blood agar, forming colonies that may be wrinkled at first, becoming umbonate with time (Figure 29.8). Colonies exhibit varying degrees of hemolysis. Growth is characterized by a putrid odor fol-

lowed by an aromatic, pungent odor. The organisms possess polar tufts of flagella and are able to grow at 42° C and on MacConkey agar. The patient's clinical history usually suggests melioidosis, which aids the microbiologist in identifying the etiological agent.

P. mallei is the only nonmotile species in the genus *Pseudomonas*. This species, which may be suspected because of a clinical picture suggestive of glanders, is unable to grow at 42° C and may not grow on blood agar. Colonies on brain-heart infusion agar (with added glycerol) are grayish white and translucent after 48 hours' incubation, later becoming yellowish and opaque. A Gram stain reveals coccal forms and occasional filaments and pleomorphic forms that may branch. The CDC considers this or-

FIGURE 29.9

Dry, wrinkled colonies of *Pseudomonas stutzeri* on blood agar after 48 hours of incubation.

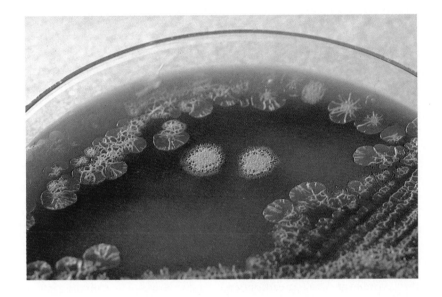

ganism to be too dangerous for routine laboratory study.[3]

The third species in the pseudomallei group is *P. cepacia.* Next to *P. aeruginosa, P. cepacia* is the pseudomonad most frequently isolated from patients with cystic fibrosis and is an important nosocomial pathogen. *P. cepacia* may produce diffusible pigments, appearing as yellow or yellow-green, rough, serrate-edged colonies. It does not fluoresce. Strains are motile and able to oxidize glucose, lactose, maltose, and mannitol. *P. cepacia* is resistant to polymyxin B.

In addition to *P. aeruginosa,* two other species may produce fluorescent pigments, *P. fluorescens* and *P. putida.* Whereas *P. aeruginosa* has one polar flagellum, these two species have polar tufts of flagella. Many strains grow best at 30° C, and none grows at 42° C. Both species are resistant to carbenicillin. *P. stutzeri,* isolated from wounds, urogenital sources, and other sites, is distinguished by production of yellow to tan, flat, dry, and wrinkled colonies that resemble those of *P. pseudomallei* (Figure 29.9).

Shewanella (formerly *Pseudomonas*) *putrefaciens* (divided into two biovars based on oxidation of sucrose and maltose and requirement for Na⁺) is the only nonfermenter known to produce hydrogen sulfide (H_2S) in the butt of a TSIA or KIA slant. For this reason, an isolate may be mistaken for an H_2S-producing Enterobacteriaceae strain unless the oxidase result is checked. These strains are oxidase positive.

Other pseudomonads and similar organisms have been isolated from clinical material, including *Comamonas acidovorans, Comamonas testosteroni, Pseudomonas alcaligenes, P. diminuta, P. mendocina, P. paucimobilis, P. pickettii (P. thomasii), P. pseudoal-*

caligenes, P. putida, and *P. vesicularis.** *Methylobacterium extorquens,* recognized by its pink to salmon-colored colonies, grows poorly on routine blood agar at 37° C. The organism is isolated more frequently when cultures are inoculated onto buffered-charcoal yeast extract agar, Sabouraud's dextrose agar, or trypticase soy agar and incubated at 30° to 35° C for at least 72 hours. The reader is referred to references by Gilardi[7] and Clark et al.[3] for methods for definitive identification of these organisms. If a clinically significant isolate must be identified, it should be sent to an appropriate reference laboratory.

GENUS *ALCALIGENES* (INCLUDING ORGANISMS FORMERLY CALLED *ACHROMOBACTER* SP.)

The *Alcaligenes* sp. of clinical importance have been isolated from blood, respiratory secretions, infected wounds, urine, and several other sources. They are usually opportunistic pathogens, infecting patients with impaired host defenses. The genus *Alcaligenes* consists currently of three species (with two subspecies), *A. faecalis* (also called "*A. odorans*" by CDC), *A. xylosoxidans* subsp. *xylosoxidans* (formerly called *Achromobacter xylosoxidans* and briefly known as *Alcaligenes denitrificans* subsp. *xylosoxidans*), *A. xylosoxidans* subsp. *denitrificans,* and the recently described *A. piechaudii* (Table 29.3).[13] *Alcaligenes* belong to the family Alcaligenaceae, which also includes *Bordetella* (Chapter 30) based on genetic similarity. All *Alcaligenes* are oxidase positive and catalase positive, possess peritrichous flagella (which distinguishes them from pseudomonads), and are able to utilize acetate as a sole source of carbon. On blood agar

---

* References 7, 8, 13, 14, 20, 25.

TABLE 29.3

ALCALIGENES SP.

	NITRATE → NITRITE	NITRITE → GAS	ACID IN O-F* XYLOSE
A. xylosoxidans subsp. *denitrificans*	+	+	−
A. xylosoxidans subsp. *xylosoxidans*	+	+/−	+
A. piechaudii	+	−	−
A. faecalis	−	+	−

Data from Clark, W.A., Hollis, D.G., Weaver, R.E., et al. 1984. Identification of unusual pathogenic gram-negative aerobic and facultatively anaerobic bacteria. Centers for Disease Control, U.S. Department of Health and Human Services, Atlanta; and Kiredjian, M., Holmes, B., Kersters, K., et al. 1986. Int. J. Syst. Bacteriol. 36:282.

+, >90% of strains positive; −, >90% of strains negative; +/−, variable (most strains positive).

* O-F, Oxidative-fermentative.

the colonies are flat, spreading, and rough with a feathery edge. *Alcaligenes* sp. do grow on Mac-Conkey agar. *A. faecalis* is biochemically quite inert, whereas *A. xylosoxidans* sp. are able to reduce nitrate to nitrite. In contrast to the other species, *A. xylosoxidans* subsp. *xylosoxidans* is saccharolytic. The strains of *A. faecalis* previously called *A. odorans* produce a distinctive sweet odor, reminiscent of fresh apple cider; colonies may be slightly alpha hemolytic on sheep blood agar. *A. xylosoxidans* subsp. *xylosoxidans* oxidizes xylose, and some strains may oxidize glucose weakly.

GENUS ACINETOBACTER

Found as a major constituent of the flora of soil, water, and sewage and within the hospital environment, *Acinetobacter* is also found to colonize the skin of hospitalized patients. It may be involved in wound infections, particularly if there has been soil or water contamination of the wound. After *Pseudomonas aeruginosa*, *Acinetobacter* sp. are the most frequently isolated nonfermenters in clinical laboratories. It is possible that *Acinetobacter* sp. are actually normal flora of the skin and mucous membranes of humans. They have been implicated as etiological agents of pneumonia (both community and hospital acquired) and urinary tract infection. They may also be involved in sepsis and other infections.[16] The genus has recently been expanded to include species *A. baumannii*, *A. calcoaceticus*, *A. lwoffii*, *A. haemolyticus*, *A. johnsonii*, and *A. junii*, in addition to several unnamed species.[1,6,22]

Colonies on blood agar are convex, gray to white, 2 to 3 mm in diameter, and variably hemolytic. They are oxidase negative and catalase positive and may initially be confused with Enterobacteriaceae, since colonies resemble non-lactose-fermenting Enterobacteriaceae on Mac-Conkey agar. *Acinetobacter*, however, cannot grow in the butt of a TSIA or KIA and cannot reduce nitrate. Differentiation from *X. (P.) maltophilia*, which is also usually oxidase negative, may be based on the lysine decarboxylase reaction. *Acinetobacter* cannot utilize lysine in Moeller's decarboxylase media, whereas *X. (P.) maltophilia* is lysine decarboxylase positive. *A. baumannii*, *A. calcoaceticus*, *A. haemolyticus*, and *A. lwoffii* are clinically important. Although species were differentiated based on phenotypical characteristics (Table 29.4), recent studies suggest that *Acinetobacter* sp. of the main glucose-oxidizing groups could not be differentiated and should be called *A. calcoaceticus-baumannii* complex.[6] On Gram stain, the cells of *Acinetobacter* are very plump, almost coccoid rods that tend to resist alcohol decolorization; they may be mistaken for *Neisseria* sp., particularly when found intracellularly. Similar to most gram-negative bacilli but unlike other *Moraxella*-like organisms, *Acinetobacter* sp. are usually resistant to penicillin in vitro.

GENERA MORAXELLA AND OLIGELLA

Moraxella atlantae, *M. lacunata*, *M. nonliquefaciens*, *M. osloensis*, and *M. phenylpyruvica* are important isolates from human clinical material. All are normal flora of the mucous membranes of humans and other animals. They have been isolated as the etiological agents of nosocomial infections, wound infections, pneumonia, and other types of opportunistic infections. *Moraxella* sp. are gram-negative coccobacilli, and in the case of *Moraxella* (formerly *Branhamella*) *catarrhalis*, the cells are true cocci. For this reason and because all *Moraxella* are oxidase positive, *M. (B.) catarrhalis* will always be identified initially as a *Neisseria* sp. The complete discussion of this organism therefore has been included in Chapter 27 with the

TABLE 29.4

ACINETOBACTER,* MORAXELLA, AND OLIGELLA SP.

SPECIES	GROWTH AT 42° C	GELATIN HYDROLYSIS	UREASE	CITRATE	GROWTH AT 37° C	ACID IN O-F GLUCOSE	GROWTH IN 3% NACL
Acinetobacter calcoaceticus	−	−	+/−	+	+	+	NA
A. baumannii	+	−	NA	+	+	+	NA
A. haemolyticus	−	+	NA	+	+	+/−	NA
A. junii	+	−	NA	+	+	−	NA
A. johnsonii	−	−	NA	+	−	−	NA
A. lwoffii	−	−	−	−	+	−	NA
Moraxella osloensis	−	−	−	−	+	−	−
M. lacunata	−	+/−	−	−	+	−	+
M. phenylpyruvica	+/−	−	+	−	+	−	+
Oligella urethralis	+	−	−	+/−	+	−	+
O. ureolytica	−	−	+†	−/+	+	−	+

+, >90% of strains positive; −, >90% of strains negative; +/−, variable (most strains positive); −/+, variable (most strains negative); *NA,* not applicable.
*Only named species are listed.
†Rapid reaction.

Neisseria sp., although *M. (B.) catarrhalis* is, according to deoxyribonucleic acid (DNA) homology studies, a true subgenus of the genus *Moraxella.*

Moraxella and *Oligella* are nonpigmented and are unable to utilize carbohydrates. Whereas *Moraxella* is nonmotile, *Oligella ureolytica* is usually motile. *Moraxella* and *Oligella urethralis,* but not *O. ureolytica,* are susceptible to penicillin. Most strains grow slowly on MacConkey agar, resembling non-lactose-fermenting Enterobacteriaceae. *M. lacunata,* isolated most frequently from eye infections, fails to grow on MacConkey agar but may produce pitting of blood agar. The organism is able to liquefy serum, so depressions are formed on the surface of Loeffler's serum agar slants. Strains of *M. nonliquefaciens* may be very mucoid. *M. phenylpyruvica* is able to hydrolyze urea and deaminate phenylalanine. Other characteristics are shown in Table 29.4.

Oligella urethralis (formerly *Moraxella urethralis*) and *O. ureolytica,* newly named for species called CDC group IVe, are asaccharolytic, oxidase-positive, catalase-positive, small, coccoid bacilli. *O. ureolytica* is strongly urease positive, with reactions detected within 30 minutes. *Bordetella* sp. also display such rapid urease reactions. Penicillin susceptibility helps differentiate *O. urethralis* from similar species.

Two unnamed species groups, CDC EO(eugonic oxidizer)-2 and EO-3, are similar to *Moraxella* and *Oligella* sp. in being oxidase and catalase positive and able to grow on MacConkey agar.[22] Unlike *Moraxella,* however, they are able to oxidize glucose, and most strains hydrolyze urea slowly. EO-2 exhibits the unusual staining characteristic of a dark rim around each very coccoid cell, often called "O-shaped" cells. EO-3 produces a yellow pigment.

GENUS EIKENELLA

Eikenella corrodens, the only species within the genus, is usually characterized by the ability to pit ("corrode") the agar. The organisms are found as normal flora in the human mouth and gastrointestinal tract, and they are common etiological agents of human bite wounds and clenched fist wounds. They have also been isolated from patients with meningitis, endocarditis, aspiration pneumonia, osteomyelitis, brain and intraabdominal abscess, and periodontitis. Because some isolates initially grow only anaerobically, the organism was first thought to be an anaerobe and named *"Bacteroides corrodens."* An obligately anaerobic organism that greatly resembles *Eikenella* had been named *Bacteroides corrodens* and later *Bacteroides ureolyticus. Eikenella* is facultatively anaerobic and grows best in a humidified, capnophilic atmosphere.

E. corrodens is oxidase positive but is unable to utilize carbohydrates. It is catalase, urease, indole, and arginine dihydrolase negative but lysine decarboxylase and nitrate positive. Blood is required for growth, and chocolate agar is the most supportive medium. Colonies are yellowish and opaque and consist of three distinct zones of growth: a moist, shiny central zone; a refractile circle of growth with a pearly sheen, resembling a mercury drop; and an outer flat, rougher spreading zone. The colonies are quite tiny, however, and because they often are sunk into small craters in the agar (pitting), they must be viewed under magnification. The colonies produce a bleachlike odor. In broth the organisms grow in discrete granules, often adherent to the sides of the tube. Noncorroding colonies may also be seen, either as morphological variants among the initial colonies of an isolate or as a homogeneous strain. Fresh isolates require hemin for growth. Gram stains reveal a small, even, straight-sided gram-negative bacillus. Because they are so resistant to clindamycin, a clindamycin-impregnated filter paper disk (5 μg clindamycin) may be used to help isolate the organism in mixed culture.

GENERA *FLAVOBACTERIUM*, *SPHINGOBACTERIUM*, AND *WEEKSELLA*

So named because of their typical yellow pigment (Figure 29.10), the flavobacteria are ubiquitous inhabitants of soil and water, and they have been found in food and from sources within the hospital.[8] Although the flavobacteria grow best at 25° to 30° C, clinical isolates grow at 35° C on initial isolation. The most clinically significant species is *F. meningosepticum,* an etiological agent of neonatal meningitis and sepsis. Episodes of this syndrome are likely to occur as small epidemics, affecting several babies in the same nursery. Although many infants are asymptomatically colonized with the organism, mortality can be 50% among those who contract disease.

Other species, *F. breve, F. indologenes, F. odoratum,* and others, have only rarely been isolated from humans.[4] *Sphingobacterium multivorum* and *S. spiritovorum,* former flavobacteria, were placed into a new genus based on cell wall structure. All species have been implicated as the etiological agents of sepsis, wound infections, endocarditis, and pneumonia.[5,8] Flavobacteria, in addition to producing yellow pigmentation, usually grow on MacConkey agar (except for some unnamed strains), are catalase and oxidase positive and nonmotile, and (except for *F. odoratum*) are able to produce acid from glucose in O-F carbohydrate medium. *F. odoratum* produces a sweet, fruity odor similar to that of *Alcaligenes odorans (faecalis),* by which it can be recognized. All flavobacteria are resistant to penicillins and polymyxin. Because of extensive proteolytic enzyme production, colonies on blood agar are usually surrounded by a lavender-green discoloration. Gram stains of colonies reveal long, thin, filamentous gram-negative bacilli with occasional swollen ends. They are nonfermentative; some species are oxidizers. Indole, produced by many strains, is detected more sensitively using a tryptophan-based medium.[24] Gelatinase is produced by most strains. The extracellular proteins and gelatinase may account in part for the virulence of some species.

Weeksella zoohelcum, formerly CDC group IIj, is nonsaccharolytic and nonmotile, unable to grow on MacConkey agar, and urease positive. Growth on rabbit blood agar may yield sticky colonies.[3] This species is almost exclusively associated with animal bite wounds and scratches.[10]

GENERA *AGROBACTERIUM, OCHROBACTRUM,* AND *PSYCHROBACTER*

Only one species of *Agrobacterium, A. tumefaciens,* formerly *A. radiobacter* (from the Greek *radius,* meaning "ray"; star-shaped aggregates of bacteria are observed occasionally), has been recovered from human clinical material. Normally an invasive plant pathogen, *A. tumefaciens* has been isolated from wound infections, sputum, urine, and blood. Whether this organism is a true pathogen or only a saprophytic contaminant has not been determined. *Agrobacterium* is oxidase variable (by Kovacs' method), rapidly urease positive, able to hydrolyze esculin, and motile by means of peritrichous flagella, unlike *Alcaligenes* and certain *Pseudomonas* sp. with which it may be confused.

Ochrobactrum anthropi (formerly CDC group Vd) is catalase, nitrate, and oxidase positive; able to grow on MacConkey and hydrolyze urea; and a glucose oxidizer. As with *Agrobacterium* and *Psychrobacter, O. anthropi* is indole negative.

FIGURE 29.10

Yellow-pigmented colonies of *Flavobacterium* sp. on blood agar.

Psychrobacter can be differentiated by its Gram-stain morphology, which resembles the gram-negative diplococci of *Neisseria* sp., and by its ability to grow in nutrient broth. Some strains are reported to smell like roses (phenylethyl alcohol–like), some strains are nitrate negative, and none can hydrolyze urea.[7,15]

NONCONVENTIONAL METHODS FOR IDENTIFICATION OF NONFERMENTATIVE BACILLI

Numerous identification package systems, rapid methods, and automated methods are available for identification of nonfermenters. Because of the complexity and diversity of this group of organisms, no one system has proved to identify all nonfermenters accurately. Determination of whole-cell fatty acid composition using gas-liquid chromatography (GLC) and a commercial data base available from Microbial ID, Inc., Newark, Del., (Chapter 12) has been very successful.[22] In addition, a novel method of identification using carbon source utilization detected by a colored end point oxidation-reduction reaction (Biolog, Inc., Hayward, Calif.) has been developed.

Several manufacturers produce miniature conventional biochemical or enzymatic identification systems (Chapter 10). The Uni N/F Tek (Flow Laboratories), Oxi-ferm (Roche Diagnostics), Minitek (BBL Microbiology Systems), RapID NF Plus (Innovative Diagnostics), and two API systems, API Rapid NFT and the API 20E (bioMérieux Vitek), have all been used to identify nonfermenters. With those species that make up the great majority of the isolates seen in routine clinical microbiology laboratories, each of the systems performs adequately. The unusual or rare species are identified with difficulty even in reference laboratories; the commercial systems should not be relied on for definitive identifications of such organisms. An example of the results of an early evaluation found that 72% of 229 clinical isolates were correctly identified by one commercial system.[18] A more recent evaluation of the RapID NF Plus system concluded that most (90%) of the nonfermenters likely to be encountered in clinical laboratories were identified correctly.[14]

The makers of automated identification and identification/susceptibility testing systems have incorporated substrates and data bases for identification of nonfermenters. Most manufacturers of MIC/ID systems (bioMérieux Vitek; BBL Microbiology Systems; MicroScan, American Scientific Products; Micro-Media Systems) each include gram-negative nonfermenter identification capability. For all these systems, approximately 90% of the clinically relevant nonfermenters tested are identified accurately. Microbiologists should retain a slight skepticism, however, and correlate the results of the identification system of choice with the morphological characteristics, both cellular and colonial; growth characteristics; motility; and other traits (smell, hemolysis, speed of growth) of the organism being examined.

TREATMENT OF INFECTIONS CAUSED BY NONFERMENTATIVE GRAM-NEGATIVE BACILLI

Pseudomonads, particularly *P. aeruginosa,* are notoriously resistant to many frequently used antimicrobial agents. However, they are usually susceptible to aminoglycosides, the drugs of choice. Much literature details treatment of *Pseudomonas* infections; treatment choices vary with the clinical situation and the site of infection. Certain medical centers experience a high rate of aminoglycoside-resistant *P. aeruginosa* isolates. New β-lactam antimicrobials, including aztreonam, cefotaxime, ceftriaxone, ceftazidime, imipenem, and others, have shown promise, although resistance can occur during a course of therapy. If the organism is susceptible, ciprofloxacin, ofloxacin, piperacillin, mezlocillin, or carbenicillin may be chosen. For serious infections, high dosage of the β-lactam drug and combination therapy (addition of an aminoglycoside) are recommended. It is beyond the scope of this text to discuss the many ramifications of therapy of *Pseudomonas* infections.

Most strains of *Alcaligenes* are susceptible to some aminoglycosides, carbenicillin, rifampin, piperacillin, cefoperazone, cefamandole, trimethoprim-sulfamethoxazole, and colistin. For *Alcaligenes* sp., in vitro susceptibility test results should be used to guide therapy. *Acinetobacter* sp. are usually susceptible to trimethoprim-sulfamethoxazole, carbenicillin, cefotaxime, and ceftizoxime. Aminoglycosides and imipenem are also usually effective. The new quinolones are extremely active against *Acinetobacter* sp.[27] For serious infections, combinations of an aminoglycoside and a β-lactam drug are recommended. *Moraxella* sp. are susceptible to penicillin, ampicillin, chloramphenicol, and aminoglycosides in vitro. Cumulative susceptibility data from clinical isolates of *Ochrobactrum* indicate general susceptibility to trimethoprim-sulfamethoxazole and variable resistance to all other agents tested.[2] *Eikenella corrodens* has been successfully treated with penicillin, ampicillin, carbenicillin, and tetracycline. Newer agents such as moxalactam and third-generation cephalosporins, as well as the quinolones, are also effective against *Eikenella* in vitro.

Flavobacteria are multiply resistant, although most isolates are susceptible to trimethoprim-sulfamethoxazole, rifampin, moxalactam, cefotaxime, and clindamycin. In vitro bactericidal susceptibility tests should be used to guide therapy of patients with serious infections. Treatment with erythromycin has not been generally successful, although in vitro test results indicate susceptibility. Most strains are resistant to aminoglycosides, penicillins, and chloramphenicol. Limited susceptibility data available for *Agrobacterium* sp. indicate most strains tested were susceptible to carbenicillin, gentamicin, piperacillin, and tetracycline and resistant to penicillin, chloramphenicol, cephalosporins, amikacin, and tobramycin. Both clinical isolates of *Psychrobacter* were resistant to penicillin but susceptible to cephalothin, chloramphenicol, gentamicin, and erythromycin in vitro.[15]

REFERENCES

1. Bouvet, P.J.M., and Grimont, P.A.D. 1986. Taxonomy of the genus *Acinetobacter* with the recognition of *Acinetobacter baumannii* sp. nov., *Acinetobacter haemolyticus* sp. nov., *Acinetobacter johnsonii* sp. nov., and *Acinetobacter junii* sp. nov. and emended descriptions of *Acinetobacter calcoaceticus* and *Acinetobacter lwoffii*. Int. J. Syst. Bacteriol. 36:228.
2. Cieslak, T.J., Robb, M.L., Drabick, C.J., and Fischer, G.W. 1992. Catheter-associated sepsis caused by *Ochrobactrum anthropi*: report of a case and review of related nonfermentative bacteria. Clin. Infect. Dis. 14:907.
3. Clark, W.A., Hollis, D.G., Weaver, R.E., et al. 1984. Identification of unusual pathogenic gram-negative aerobic and facultatively anaerobic bacteria. Centers for Disease Control. U.S. Department of Health and Human Services, Atlanta.
4. Dees, S.B., Moss, C.W., Hollis, D.G., et al. 1986. Chemical characterization of *Flavobacterium odoratum*, *Flavobacterium breve*, and *Flavobacterium*-like groups IIe, IIh, and IIf. J. Clin. Microbiol. 23:267.
5. Freney, J., Hansen, W., Ploton, C., et al. 1987. Septicemia caused by *Sphingobacterium multivorum*. J. Clin. Microbiol. 25:1126.
6. Gerner-Smidt, P., Tjernberg, I., and Ursing, J. 1991. Reliability of phenotypic tests for identification of *Acinetobacter* species. J. Clin. Microbiol. 29:277.
7. Gilardi, G.L. 1991. *Pseudomonas* and related genera. p. 429. In Balows, A., Hausler, W.J., Jr., Herrmann, K.L., et al., editors. Manual of clinical microbiology,

ed. 5. American Society for Microbiology, Washington, D.C.
8. Gilardi, G.L. 1986. *Flavobacterium*. Clin. Microbiol. Newsletter 8:143.
9. Holmes, B., Popoff, M., Kiredjian, M., and Kersters, K. 1988. *Ochrobactrum anthropi* gen. nov., sp. nov. from human clinical specimens and previously known as group Vd. Int. J. Syst. Bacteriol. 38:406.
10. Holmes, B., Steigerwalt, A.G., Weaver, R.E., et al. 1986. *Weeksella zoohelcum* sp. nov. (formerly group IIj), from human clinical specimens. Syst. Appl. Microbiol. 8:191.
11. Holmes, B., Steigerwalt, A.G., Weaver, R.E., et al. 1987. *Chryseomonas luteola* comb. nov. and *Flavimonas oryzihabitans* gen. nov., comb. nov., *Pseudomonas*-like species from human clinical specimens and formerly known, respectively, as groups Ve-1 and Ve-2. Int. J. Syst. Bacteriol. 37:245.
12. Kaye, K.M., Macone, A., and Kazanjian, P.H. 1992. Catheter infection caused by *Methylobacterium* in immunocompromised hosts: report of three cases and review of the literature. Clin. Infect. Dis. 14:1010.
13. Kiredjian, M., Holmes, B., Kersters, K., et al. 1986. *Alcaligenes piechaudii*, a new species from human clinical specimens and the environment. Int. J. Syst. Bacteriol. 36:282.
14. Kitch, T.T., Jacobs, M.R., and Appelbaum, P.C. 1992. Evaluation of the 4-hour RapID NF Plus method for identification of 345 gram-negative nonfermentative rods. J. Clin. Microbiol. 30:1267.
15. Lloyd-Puryear, M., Wallace, D., Baldwin, T., and Hollis, D.G. 1991. Meningitis caused by *Psychrobacter immobilis* in an infant. J. Clin. Microbiol. 29:2041.
16. Lyons, R.W. 1983. *Acinetobacter calcoaceticus*. Clin. Microbiol. Newsletter 5:87.
17. Mandell, W.F., Garvey, G.J., and Neu, H.C. 1987. *Achromobacter xylosoxidans* bacteremia. Rev. Infect. Dis. 9:1001.
18. Martin, R., Siavoshi, F., and McDougal, D.L. 1986. Comparison of Rapid NFT system and conventional methods for identification of nonsaccharolytic gram-negative bacteria. J. Clin. Microbiol. 24:1089.
19. May, T.B., Shinabarger, D., Maharaj, R., et al. 1991. Alginate synthesis by *Pseudomonas aeruginosa*: a key pathogenic factor in chronic pulmonary infections of cystic fibrosis patients. Clin. Microbiol. Rev. 4:191.
20. Morrison, A.J., Jr., and Boyce, K., IV. 1986. Peritonitis caused by *Alcaligenes denitrificans* subsp. *xylosoxydans*: case report and review of the literature. J. Clin. Microbiol. 24:879.
21. Morrison, A.J., Jr., and Shulman, J.A. 1986. Community-acquired bloodstream infection caused by *Pseudomonas paucimobilis*: case report and review of the literature. J. Clin. Microbiol. 24:853.
22. Moss, C.W., Wallace, P.L., Hollis, D.G., and Weaver, R.E. 1988. Cultural and chemical characterization of CDC groups EO-2, M-5, and M-6, *Moraxella (Moraxella)* species, *Oligella urethralis*, *Acinetobacter* species, and *Psychrobacter immobilis*. J. Clin. Microbiol. 26:484.
23. Peel, M.M., Hibberd, A.J., King, B.M., et al. 1988. *Alcaligenes piechaudii* from chronic ear discharge. J. Clin. Microbiol. 26:1580.
24. Pickett, M.J. 1989. Methods for identification of flavobacteria. J. Clin. Microbiol. 27:2309.

25. Pickett, M.J., Hollis, D.G., and Bottone, E.J. 1991. Miscellaneous gram-negative bacteria. p. 410. In Balows, A., Hausler, W.J., Jr., Herrmann, K.L., et al., editors. Manual of clinical microbiology, ed 5. American Society for Microbiology, Washington, D.C.

26. Robin, T., and Janda, J.M. 1984. Enhanced recovery of *Pseudomonas aeruginosa* from diverse clinical specimens on a new selective agar. Diagn. Microbiol. Infect. Dis. 2:207.

27. Rolston, K.V.I., and Bodey, G.P. 1986. In vitro susceptibility of *Acinetobacter* species to various antimicrobial agents. Antimicrob. Agents Chemother. 30:769.

BIBLIOGRAPHY

Palleroni, N.J. 1984. Genus *Pseudomonas* Migula *1894, 237*[AL]. p. 141. In Krieg, N.R., and Holt, J.G., editors. Bergey's manual of systematic bacteriology, vol. 1, ed 9. Williams & Wilkins, Baltimore.

30

GRAM-NEGATIVE FACULTATIVELY ANAEROBIC BACILLI AND AEROBIC COCCOBACILLI

The gram-negative organisms discussed in this chapter are usually normal flora of mucous membranes of mammals, and some species cause severe diseases. Three genera, *Pasteurella, Brucella,* and *Francisella,* are primarily agents of zoonoses and infect humans only by accident, when the humans come into close contact with the preferred animal hosts, with insects that have acquired the organisms from the animals, or from environmental contamination. Six other genera, *Haemophilus, Bordetella, Kingella, Cardiobacterium, Capnocytophaga,* and *Actinobacillus,* are found among the normal flora of the human respiratory tract and occasionally other mucosal surfaces. *Capnocytophaga canimorsus* and *C. cynodegmi* are also carried by animals as normal oral flora, particularly in dogs and rarely in cats. DF-3, named **dysgonic** fermenter by the Centers for Disease Control (CDC) because the organism grows poorly in carbohydrate-containing media, may also be normal flora of animals. Organisms designated by the CDC as HB-5 (recently named *Pasteurella bettae*) seem to be normal flora of the human genitourinary mucosa. All the organisms discussed in this chapter are morphologically **pleomorphic,** occurring as short bacilli and coccobacilli or exhibiting chains or filamentous forms. *Cardiobacterium, Capnocytophaga, Pasteurella,* HB-5, *Kingella,* and *Actinobacillus* sp. will grow on routine laboratory media, but the other genera may require special growth factors.

GENUS *FRANCISELLA*

EPIDEMIOLOGY AND PATHOGENESIS OF *FRANCISELLA* INFECTION

Francisella consists of *F. tularensis,* the agent of human and animal tularemia, and two other related species, *F. novicida,* a rare human pathogen that is genetically identical to *F. tularensis* (it has been proposed to reclassify this organism as a biogroup of *F. tularensis*), and *F. philomiragia. F. novicida* was isolated from lymph nodes and blood, respectively, of two human patients whose disease resembled tularemia (R.E. Weaver, CDC), although *F. novicida* is primarily an animal pathogen. *F. philomiragia* has been isolated from several patients, many of whom were immunocompromised or were victims of near-drowning incidents. The organism is present in animals and ground water.

F. tularensis is carried by many species of wild rodents, rabbits, beavers, and muskrats in North America. Humans become infected by handling the carcasses or skin of infected animals, through insect vectors (primarily deerflies and ticks in the United States), by being bitten by carnivores that have themselves eaten infected animals, or by inhalation. The capsule of *F. tularensis* may be a factor in virulence, allowing the organism to avoid immediate destruction by polymorphonuclear neutrophils. The organism is extremely invasive, being one of only a few infectious agents reported to penetrate intact skin, still a debatable property. In addition to invasiveness, the organisms are intracellular parasites, able to survive in the cells of the reticuloendothelial system, where they reside after a bacteremic phase. Granulomatous lesions may develop in various organs. Patients have high fevers, headache, lymphadenopathy, and an ulcerative lesion at the site of inoculation; other clinical presentations may also occur. Pneumonia and rhabdomyolysis, evidenced by elevated creatine phosphokinase levels and hemoglobinuria without red blood cells, are common complications. Pneumonia may be seen as a primary infection. Tularemia is primarily a disease of adult men who hunt, since they have the most contact with wild animals and are most likely to be exposed to the insect vectors.

LABORATORY IDENTIFICATION

F. tularensis is a Biosafety level 2 pathogen, a designation that requires technologists to wear gloves and to work within a biological safety cabinet when handling clinical material that potentially harbors this agent. For cultures the organism is designated Biosafety level 3; a mask, recommended for handling all clinical specimens, is very important for preventing aerosol acquisition with *F. tularensis.* Because tularemia is one of the most common laboratory-acquired infections, however, most microbiologists do not attempt to work with infectious material from suspected patients. We recommend that specimens be sent to reference laboratories or state or other public health laboratories that are equipped to handle *Francisella.* Fluorescent antibody stains are available commercially for direct detection of the organism in lesion smears, but such procedures are best performed by reference laboratories. *F. tularensis* is strictly aerobic and requires enriched media (containing cysteine and cystine) for primary isolation. Commercial media for cultivation of the organism are available (glucose cystine agar, BBL Microbiology Systems; cystine heart agar, Difco Laboratories); both require the addition of 5% sheep or rabbit blood. *F. tularensis* may also grow on chocolate agar and the nonselective buffered charcoal–yeast extract agar used for isolation of legionellae. It does not grow on Mac-Conkey agar, and growth is not enhanced by carbon dioxide (CO_2). The organisms, which grow slowly, requiring 2 to 4 days for maximal colony formation, are weakly catalase positive and oxidase negative. Some strains may require up to 2 weeks to develop visible colonies. *F. tularensis* has been isolated from blood with the BACTEC 460 blood culture system, although positive growth indices were not detected until a minimum of 3 days after inoculation.[22] Isolation of this organism has not been reported with other automated systems.

Colonies are transparent, mucoid, and easily emulsified. Although carbohydrates are fermented, isolates should be identified serologically (by agglutination) or by a fluorescent antibody stain. Ideally, they should be sent to a reference laboratory for characterization. Diagnosis of disease in patients is usually accomplished serologically by whole-cell agglutination (febrile agglutinins, Chapter 13) or by newer enzyme-linked immunosorbent assay (ELISA) techniques.

F. novicida and *F. philomiragia* do not require cysteine or cystine for isolation, although they resemble *F. tularensis* by being small, coccobacillary rods that grow poorly or not at all on MacConkey agar. These two species differ from *F. tularensis* biochemically; *F. philomiragia* is oxidase positive, and most strains produce hydrogen sulfide (H_2S) in triple sugar iron agar (TSIA) medium, produce gelatinase, and grow in 6% sodium chloride (NaCl) (no strains of *F. tularensis* share either characteristic). *F. novicida* is oxidase negative and may grow on MacConkey agar and in 6% NaCl. All three strains tested are able to oxidize glucose and sucrose in oxidative-fermentative (O-F) media.

TREATMENT OF *FRANCISELLA* INFECTIONS

The organism is susceptible to aminoglycosides, and streptomycin is the drug of choice. Gentamicin is a possible alternative; tetracycline and chloramphenicol have also been used, although these two agents have been associated with a higher rate of relapse after treatment.

GENUS *BRUCELLA*

EPIDEMIOLOGY AND PATHOGENESIS OF *BRUCELLA* INFECTION

The agents of brucellosis, *Brucella* sp., are normal flora of the genital and urinary tracts of many animals, including goats, pigs, cows, and dogs. Most humans acquire disease through ingestion of con-

taminated milk or through occupational exposure; the disease is particularly common among slaughterhouse workers. Farmers and veterinarians are also at increased risk. The primary virulence factor for *Brucella* seems to be the organism's ability to survive intracellularly. This characteristic, as well as granuloma and abscess formation, probably accounts for the chronic and relapsing nature of the febrile disease. Patients often have gradual onset of nonspecific systemic symptoms, including fever, malaise, chills, sweats, fatigue, and mental changes. Localized lesions can occur in almost any organ or bone; they may be granulomatous or suppurative and destructive. An infectious, mononucleosis-like illness may occur. Pulmonary, cardiovascular, hepatic, skeletal, genitourinary, and central nervous system localization are also possible complications.

COLLECTION AND TRANSPORT OF SPECIMENS

Brucella sp. are usually isolated from blood cultures or bone marrow cultures during the acute illness. Organisms may also be recovered from joint fluid of patients with arthritis. Blood cultures may require prolonged incubation times (Chapter 15). Bone marrow cultures are handled as discussed in Chapter 22. Material from infected lymph nodes, other tissue, cerebrospinal fluid, and urine may also yield *Brucella*. The laboratory should be notified that brucellosis is suspected so that supplemented media that better support growth of the organism can be inoculated. Although most strains grow on chocolate and blood agar and some strains grow on MacConkey agar, *Brucella* agar (Gibco Laboratories, BBL Microbiology Systems) or some type of infusion base agar is recommended. The addition of 5% heated horse or rabbit serum enhances growth on all media. Cultures should be incubated in 5% to 10% CO_2 in a humidified atmosphere, such as a candle extinction jar. Blood cultures are held for 4 weeks, and inoculated plates are incubated for 3 weeks before being discarded as negative. It has been found that organisms may be present in blood culture broths without visible evidence and without an increase in the growth index, as measured on the BACTEC radiometric blood culture instrument. For this reason, weekly blind subcultures to *Brucella* broth are recommended for optimal recovery of *Brucella* sp. from blood cultures. The use of biphasic blood culture bottles (Castañeda type with an agar slant and broth or B-D Septi-Check) is a more cost-effective option, and the bottles are less likely to become contaminated during repeated subcultures. Lysis-centrifugation is an excellent method for culturing blood and bone marrow. *Brucella* is considered to be a type 3 biohazard and should be handled with all appropriate precautions. Brucellosis is one of the most prevalent laboratory-acquired infections. A recent report describes an outbreak involving eight microbiology laboratory workers who acquired *B. melitensis* from aerosols produced during routine subcultures (on the bench) of a stock strain. The same isolate infected a CDC worker subsequently, although correct biosafety practices were followed.[23]

LABORATORY IDENTIFICATION

Serologic tests are used for diagnosis more often than are cultures (*Brucella* is one of the "febrile agglutinins"), but only a positive culture yields a definitive diagnosis. Both immunoglobulins M and G (IgM, IgG) are quantified initially; IgM is then inactivated by 2-mercaptoethanol to reveal rising antibody titers of IgG without IgM interference. Serological diagnosis should be performed by an experienced reference laboratory.

On culture, colonies appear small, convex, smooth, translucent, and slightly yellow and opalescent after at least 48 hours of incubation. The colonies may become brownish with age. The organisms are catalase positive, and most strains are oxidase positive. Other nonfermentative gram-negative coccobacilli that may be confused with *Brucella* are *Bordetella, Moraxella, Kingella, Acinetobacter* sp., and others (Chapter 29). *Brucella*, however, is nonmotile, urease and nitrate positive, and strictly aerobic. The most rapid test for presumptive identification of *Brucella* is the particle agglutination test with antismooth *Brucella* serum, available from Difco Laboratories. Agglutination is carried out as described in Chapter 13. This test cannot be performed on rough colony variants.

Six species of *Brucella* may be isolated from humans: *B. melitensis, B. abortus, B. suis, B. canis, B. ovis* (rare), and *B. neotomae* (also rare). The species are differentiated by the rapidity with which they hydrolyze urea, their relative ability to produce H_2S gas, their requirements for CO_2, and their susceptibility to the aniline dyes thionine and basic fuchsin, as briefly described in Procedure 30.1.[21] Performance of the urease and H_2S detection tests is also described in Procedure 30.1. Differential characteristics are listed in Table 30.1. For determination of CO_2 requirement, identical plates of *Brucella* agar or brain-heart infusion agar should be given equal inocula (e.g., with a calibrated loop) of a broth suspension of the organism to be tested. One plate should be incubated in a candle jar and the other plate in air, both in the same incubator. Most strains of *B. abortus* and all strains of *B. ovis* do not grow in air but show growth in the candle jar. Isolates of *Brucella* should be sent to state or other reference laboratories for confirmation or definitive identification, since most clinical laboratories lack the necessary media and containment facilities. The tests are variable enough to necessitate very stringent quality control measures employing known strains of *Brucella*. Because of the difficulty of isolating the organism, many cases of brucellosis are diagnosed serologically by the *Brucella* **febrile agglutinin** test. Titers of 1:80 or greater are always considered to be significant; however, a fourfold increase in titer between acute and convalescent sera is considered to be more definitive. An ELISA procedure has also been described.

TREATMENT OF BRUCELLOSIS

For initial therapy, tetracycline combined with an aminoglycoside, streptomycin or gentamicin, is probably best. Trimethoprim-sulfamethoxazole has also been effective for treatment of brucellosis in adults. Prolonged therapy is often required.

GENUS *BORDETELLA*

EPIDEMIOLOGY AND PATHOGENESIS OF *BORDETELLA* INFECTION

Bordetella sp. are obligate parasites of animals and humans, residing on the mucous membranes of the respiratory tract. The organisms colonize the ciliated epithelial cells, adhering closely to the surface; the adherence may be mediated by fimbriae. Mechanisms of pathogenesis of etiological agents of respiratory tract infections are discussed in Chapter 17. Bordetellae seem to be the only bacteria that adhere only to ciliated mucosal cells. Other virulence factors associated with Bordetellae, particularly *B. pertussis*, include a capsule that surrounds the cells of virulent strains, production of dermonecrotic toxins, production of a filamentous hemagglutinin, production of extracellular enzymes (adenylate cyclase), and production of other biologically active substances, including endotoxin (Chapter 15), tracheal cytotoxin, and an exotoxin (pertussis toxin) that acts on various cells of the immune system and other tissues.[7,12]

There are four species, *B. pertussis, B. bronchiseptica, B. avium,* and *B. parapertussis; B. avium* is found in turkeys, and *B. bronchiseptica* is found naturally in animals such as rabbits, dogs, swine, and many other wild and domestic animals. Ex-

PROCEDURE 30.1

SPECIAL TESTS FOR SPECIES DIFFERENTIATION AMONG *BRUCELLA*

1. Urease activity is determined by inoculating a large loopful of growth from a broth suspension made from isolated colonies to the surface of a Christensen's urea agar slant (Chapter 10).

2. Examine the slant immediately, and incubate the slant at 37° C, examining at 15 minutes, 1 hour, 2 hours, and 24 hours.

3. Urease activity is evidenced by a change in color on the slant surface from yellow to pink to bright pinkish purple.

4. Production of H_2S is determined by placing a strip of lead acetate–impregnated filter paper (available from Remel Laboratories and other suppliers) suspended in the atmosphere above an agar slant culture (trypticase soy agar or serum-supplemented agar) of the organism being tested. The paper must not contact the agar or the tube's sides. This can be accomplished by bending the paper over the edge of the tube mouth before lightly screwing on the cap, leaving a length of the strip inside the tube.

5. The tube is incubated at 37° C in air or CO_2, as required, and the paper strip is examined and changed daily. A complete blackening of the entire strip after the first day, or slight blackening of the strip on subsequent days, is indicative of H_2S production.

6. Dye inhibition tests can be carried out quite easily using dye-impregnated tablets available from Key Scientific Products, Round Rock, Texas.

7. Test and control organisms are inoculated onto a *Brucella* agar plate to yield sparse but confluent growth, and dye tablets are placed in a manner similar to that of agar disk diffusion testing. The plates are incubated at 37° C in the appropriate atmosphere for 2 days in 5% to 10% CO_2, and the zone of inhibition around the dye disks is measured.

QUALITY CONTROL

Because strains of *Brucella* must be maintained and manipulated to assess the performance characteristics of these media adequately, we recommend that only reference laboratories with extensive experience with this genus perform the tests just described.

Modified from Corbel, M.J., and Brinley-Morgan, W.J. 1984. Genus *Brucella* (Meyer and Shaw 1920). In Krieg, N.R., and Holt, J.G., editors. Bergey's manual of systematic bacteriology, vol 1. Williams & Wilkins, Baltimore. Additional information available in the reference by Pickett et al.[21] and product material supplied by Key Scientific Products, Round Rock, Texas.

TABLE 30.1

CHARACTERISTICS OF *BRUCELLA* SP.

SPECIES	CO₂ REQUIRED FOR GROWTH	TIME TO POSITIVE UREASE	H₂S PRODUCED	INHIBITION BY DYE	
				THIONINE*	FUCHSIN*
B. abortus	+/−	2 h (rare 24 h)	+ (most strains)	+	−
B. melitensis	−	2 h (rare 24 h)	+	−	−
B. suis	−	15 min	+/−	−	+ (most)
B. ovis	+	≥7 days	−	−	+ (most)
B. neotomae	−	15 min	+	+	+
B. canis	−	15 min	−	−	+

+, >90% of strains positive; −, >90% of strains negative; +/−, variable results.
*Dye tablets (Key Scientific Products, Inc.).

cept for *B. avium,* all species can cause respiratory disease in humans, although the disease associated with *B. pertussis,* "pertussis" or "whooping cough," is most severe.[19] *B. bronchiseptica* has been isolated from infected wounds and from the respiratory tract of humans. The incidence of infections in immunocompromised hosts seems to be increasing.[27] Incidence of whooping cough in the United States had steadily declined since 1950, when vaccine became widely used, leveling at approximately 1900 cases reported annually. An increase has been seen since 1981, however, and more than 4500 cases were reported in 1990. Fatalities still occur. Recent epidemiological studies in New York and Wisconsin have shown recovery of *B. pertussis* from children and adults with respiratory tract disease that does not resemble classical whooping cough.[19] The organism may be more prevalent than currently assumed, since very few laboratories use methods that allow its detection unless requested by a clinician who suspects the disease.

Laboratory diagnosis is difficult; all tests used currently, including cultures, immunofluorescence, and serology, lack sensitivity in previously immunized individuals.[11]

COLLECTION AND TRANSPORT OF SPECIMENS

Specimens, either nasopharyngeal washings or a nasopharyngeal swab (calcium alginate on a wire handle), should be collected within the first week of paroxysmal coughing. The swab is bent to conform to the nasal passage and held against the posterior aspect of the nasopharynx. If coughing does not occur, another swab is inserted into the other nostril to initiate the cough. The swab is left in place during the entire cough, removed, and

immediately inoculated onto a selective medium such as Bordet-Gengou agar containing 20% sheep blood cells and 2.5 µg/ml methicillin instead of penicillin or the recently evaluated Jones-Kendrick charcoal agar containing 40 µg/ml cephalexin. Stauffer et al.[24] have found cephalexin to be superior to methicillin for inhibition of normal respiratory flora and Jones-Kendrick agar to be acceptable for transport and cultivation of *Bordetella* sp. from respiratory tract specimens. This agar remains effective even after 2 months of storage in tightly wrapped plastic bags in the refrigerator. Other cephalexin-containing media have also shown good yields.[11,14,25,29] Buffered charcoal–yeast extract agar also supports growth of *Bordetella* sp. A recent study advocates Regan-Lowe media for best recovery of *B. pertussis* from nasopharyngeal swabs.[17] A fluid transport medium may be used for swabs; it must be held for less than 2 hours. Half-strength Regan-Lowe agar enhanced yields when used as a transport and enrichment medium.[14] Cold casein hydrolysate medium and casamino acid broth (available commercially) have been found to be effective transport media, particularly for preparation of slides for direct fluorescent antibody staining. Sensitivity of culture approaches 60% in the best of hands.

LABORATORY IDENTIFICATION

Plates are incubated at 35° to 37° C in a humidified atmosphere; they do not require increased CO_2. If grown on Bordet-Gengou agar with 20% blood, colonies show a diffuse zone of hemolysis. Colonies grow within 7 days, appearing small, smooth, convex, with a pearly luster, resembling mercury drops (Figure 30.1). The colonies are

FIGURE 30.1

Colonies of *Bordetella pertussis* resembling mercury drops on Bordet-Gengou agar after 48 hours of incubation.

PROCEDURE 30.2

DIRECT FLUORESCENT ANTIBODY STAIN FOR *BORDETELLA PERTUSSIS*

PRINCIPLE

The presence of even one bacterium of *B. pertussis* is indicative of disease; therefore direct visualization of the organism, specifically stained by an immunofluorescent reagent, is pathognomic.

METHOD

1. Prepare casamino acid transport and holding medium. Add 1 g casamino acids (commercially available) to 100 ml distilled water in acid-washed glassware. Adjust pH to 7.2 and dispense in 0.5 ml amounts in acid-washed, screw-capped, 12 × 75 mm tubes. Tighten caps and autoclave 20 minutes. This medium can be stored for 1 year at refrigerator temperature.

2. Immediately after collection, insert nasopharyngeal swab or material obtained by aspiration into cool casamino acid medium. Incubate the specimen for 1 hour at 35° C.

3. With a sterile capillary pipette, place a heaped drop of the well-mixed medium onto each of two circles of a clean glass slide suitable for

fluorescent staining. Allow the smears to air dry, and reapply material for a total of three applications. Heat fix the slide. Remove a known positive and a known negative control slide (prepared from stock cultures) from the freezer at the same time, following an identical procedure for each slide.

4. Place a drop of fluorescein-conjugated anti–*B. pertussis* antibody to cover one of the smears completely, and place a drop of fluorescein-conjugated normal rabbit serum over the second control smear. Incubate at room temperature in a humidified chamber in the dark for 30 minutes; placing the slide on a broken wooden applicator stick in a Petri dish with a moist gauze pad works well.

5. Rinse the slides two times in phosphate buffered saline, pH 7.5 (commercially available), allowing them to remain in the second buffer rinse for 10 minutes, and then rinse in distilled water.

6. Mount with glycerol buffered mounting fluid, pH

8.5 (commercially available), and examine under ultraviolet light for bright fluorescing coccobacilli with bright rims and darker centers. The patient control smear should show no fluorescence, and the known positive smear should yield brightly fluorescing short bacilli and coccobacilli.

QUALITY CONTROL

Obtain a *B. pertussis* culture (from your local health department or large university laboratory), pass it on the agar used for specimens, and prepare slides from suspensions of colonies. Stain one control slide each time a patient specimen is examined.

Modified from Anhalt, J. 1981. In Washington, J.A., Jr., editor: Laboratory procedures in clinical microbiology. Springer-Verlag, New York; and from Parker and Linneman. 1980. In Lennette, E.H., Balows, A., Hausler, W.J., Jr., and Truant, J.P., editors: Manual of clinical microbiology, ed 3. American Society for Microbiology, Washington, D.C.

---

mucoid and tenacious. A Gram stain of the organism reveals minute, faintly staining coccobacilli singly or in pairs. The use of a 2-minute safranin O counterstain or a 0.2% aqueous basic fuchsin counterstain enhances their visibility. *B. bronchiseptica* and *B. parapertussis* grow on MacConkey agar; *B. pertussis* will not.

A direct fluorescent antibody stain is available for detection of *B. pertussis* in smears made from nasopharyngeal material. The procedure is described in Procedure 30.2. In experienced hands, this reagent has been as much as 100% sensitive for early diagnosis of cases of whooping cough, al-

though sensitivities as low as 20% have also been reported.[25,29] Slides may be prepared, heat fixed, and mailed to a reference laboratory for evaluation. This reagent may also be used to identify presumptively organisms growing on agar. The fluorescent antibody identification method is much faster than conventional biochemicals.

Whole-cell agglutination reactions in specific antiserum may also be used for species identification (Chapter 13). Abbreviated biochemical reactions of the three known species of *Bordetella* are shown in Table 30.2.

An enzyme immunoassay for detection of an-

TABLE 30.2

CHARACTERISTICS OF *BORDETELLA* SP.

SPECIES	GROWTH ON HEART INFUSION AGAR	UREASE	NITRATE REDUCTION	MOTILITY
B. pertussis	−	−	−	−
B. parapertussis	+	+ (18 h)	−	−
B. bronchiseptica	+	+ (4 h)	+	+

+, All strains positive; −, all strains negative.

tibody to *B. pertussis* was recently shown to be valuable in diagnosing culture-negative individuals.[11,25] Commercial variations of this test have recently become available (Labsystems-USA, Chicago). Preliminary trials have yielded promising results for early diagnosis of pertussis. As for many other serological tests, a rising titer is the best diagnostic indicator.

TREATMENT OF PERTUSSIS

Adequate microbiological therapy may not alter the course of the disease. Erythromycin does eradicate organisms from the respiratory tract, however, and should be given to all patients with pertussis, as well as to contacts prophylactically.

GENUS *HAEMOPHILUS*

The following three genera, *Haemophilus*, *Actinobacillus*, and *Pasteurella*, are all members of the family Pasteurellaceae, a group of organisms that are pleomorphic, gram-negative, nonmotile bacilli and coccobacilli that are able to reduce nitrates and utilize carbohydrates either fermentatively or with oxygen as the terminal electron acceptor. All three genera live on animals, including humans, primarily on mucosal surfaces. They may be differentiated by their requirements for nicotinamide adenine dinucleotide (NAD), as well as by other criteria. *Actinobacillus actinomycetemcomitans* was placed briefly in the genus *Haemophilus*.

EPIDEMIOLOGY AND PATHOGENESIS OF *HAEMOPHILUS* INFECTION

Carried in the normal respiratory tract and occasionally the genital tract of humans, pigs, sheep, and most other vertebrates, the organisms cause upper respiratory tract infections; suppurative infections, primarily in areas adjacent to the respiratory tract; and systemic infections, particularly meningitis. In humans, *Haemophilus influenzae* has

been the most common etiological agent of acute bacterial meningitis in young children. With the recent widespread use of an effective polysaccharide vaccine for babies and now for neonates, the incidence of invasive disease, particularly meningitis, is dropping. Expectations are that *H. influenzae* meningitis and sepsis will become increasingly less common.[4] *H. influenzae*, the most important human pathogen among the *Haemophilus* sp., is also associated with acute, contagious conjunctivitis ("pinkeye" [biogroup aegyptius, formerly called *H. aegyptius*]), acute and chronic otitis media, epiglottitis, cellulitis, and other infections. *H. influenzae* has even been isolated from the urine of patients with urinary tract infections.[8] A recently recognized serious disease in children, Brazilian purpuric fever, preceded by conjunctivitis, has been associated with *H. influenzae* biogroup aegyptius.[3]

H. ducreyi, not found as part of normal genital flora of humans or known to be harbored by any animals, is the cause of chancroid, a widespread venereal disease, found most often in tropical parts of the world. Other *Haemophilus* sp., which include *H. parainfluenzae*, *H. haemolyticus*, *H. aphrophilus*, *H. paraphrophilus*, and *H. segnis*, are normal flora of humans only, occasionally causing upper and lower respiratory tract infections. These organisms may also occasionally enter the bloodstream from a mucosal focus, disseminating to cause systemic disease, including endocarditis, metastatic abscesses (especially brain abscess), septic arthritis, and osteomyelitis. They have also been known to infect tissue adjacent to the mucosa, causing jaw infections, dental abscesses, appendicitis,[26] and, when traumatically inoculated into healthy tissue, bite wound infections. *H. influenzae* and *H. parainfluenzae* are associated with invasive disease in humans more often than are the other species. *H. parainfluenzae* tends to grow in tangled clumps or strands of filamentous bacterial forms; these may easily

TABLE 30.3

BIOTYPES OF *HAEMOPHILUS INFLUENZAE*

BIOTYPE	SYNDROME (SITE OF INFECTION)	INDOLE	UREASE	ORNITHINE
I	Sepsis, meningitis	+	+	+
II	Eye (United States), bacteremia, ear, lower respiratory tract	+	+	−
III	Eye (Europe), respiratory tract	−	+	−
IV	Ear (rare), respiratory tract, genital tract	−	+	+
V	Ear, lower respiratory tract	+	−	+
VI	Upper respiratory tract (rare)	−	−	+

Modified from Kilian, M., and Biberstein, E.L. 1984. Genus *Haemophilus* (Winslow, Broadhurst, Buchanan, Krumwiede, Rogers and Smith 1917). In Kreig, N.R., and Holt, J.G., editors. Bergey's manual of systematic bacteriology, vol. 1. Williams & Wilkins, Baltimore.
+, >90% of strains positive; −, >90% of strains negative.

break off the main colony. Particularly in patients with sepsis or endocarditis, metastatic abscess formation may occur. The polysaccharide capsule is clearly a virulence factor for *H. influenzae*, since anticapsular antibody is protective against acquisition of meningitis. Several immunological reagents have been developed for the rapid identification of *H. influenzae* capsular polysaccharide type b (the most common pathogenic agent), as described later in this discussion and in Chapter 16.

Nonencapsulated strains, however, are often recovered from specimens taken from patients with infections other than meningitis, including pneumonia, cellulitis, otitis media, and soft tissue infections. Pathogenic mechanisms of the *Haemophilus* sp. are not well understood. Since the organisms enter the bloodstream so readily after colonization of the respiratory epithelium, they may possess some invasive properties. Their ability to multiply in host tissues and evoke an inflammatory response contributes to the pathology of pyogenic *Haemophilus* disease. The presence of a specific plasmid was significantly associated with severe disease among strains of *H. influenzae* biogroup aegyptius isolated from children with Brazilian purpuric fever.[3] The nature of a possible virulence factor on this plasmid, however, is not known.

Recent studies on identification of *H. influenzae* have shown that certain biotypes of that organism are more frequently associated with particular syndromes, as outlined in Table 30.3. Biotyping can be easily accomplished, and microbiologists should routinely identify significant *H. influenzae* isolates to biotype. As more laboratories procure such definitive identifications, our understanding of the epidemiology and pathogenesis of infection with this prevalent microorganism will expand.

COLLECTION AND TRANSPORT OF SPECIMENS

Most specimens may harbor *Haemophilus* sp., and it is primarily for isolation of this organism that most clinical material received for routine culture in microbiology laboratories is inoculated to chocolate agar and incubated in 5% to 10% CO_2. Blood, cerebrospinal fluid, middle ear exudate collected by tympanocentesis, joint fluids, respiratory tract specimens, swabs from inflamed conjunctivae, and abscess drainage are a few of the many types of specimens likely to yield *Haemophilus* sp.

Most swabs routinely used for collection of material for culture, including calcium alginate, cotton, and Dacron, are suitable for collection of specimens for detection of most *Haemophilus* sp. Modified Stuart's transport medium or Amie's charcoal transport medium adequately maintains the organism's viability until the specimen is inoculated to growth media. For recovery of *H. ducreyi* from genital ulcers, however, special measures are necessary because of the bacterium's fastidious nature. The ulcer should be cleansed with sterile gauze moistened with sterile saline. A physician uses a cotton swab premoistened with phosphate buffered saline to collect material from the ulcer base. This swab must be plated to media (described in Chapter 20 and mentioned in the following section) within 10 minutes of collection for maximal yield. A Gram stain of material from a chancroid lesion may display the characteristic "school of fish" arrangement, seen as numerous gram-negative coccobacilli in groups resembling fish (see Figure 20.9).

MEDIA AND INCUBATION CONDITIONS FOR ISOLATION OF *HAEMOPHILUS* SP.

Chocolate agar supplies the factors necessary for growth of most *Haemophilus* sp. These factors are *hemin* (called "X factor") and *nicotine adenine dinucleotide* (*NAD;* called "V factor"). *H. influenzae, H. haemolyticus,* and *H. ducreyi* require X factor, whereas *H. aphrophilus* requires X factor only for initial isolation. All clinically significant species except *H. aphrophilus* and *H. ducreyi* require V factor. Chocolate agar, horse blood agar, and rabbit blood agar all contain adequate quantities of both growth factors and support the growth of *Haemophilus* sp. that require both factors. Either rabbit blood or horse blood is used for visualization of hemolysis. Sheep blood that has not been chocolatized contains enough NADase (produced by the sheep blood cells) to destroy the NAD normally present, although sheep blood agar does contain adequate hemin for growth of *Haemophilus* that require hemin. *Haemophilus* sp. that require both X and V factors can grow on sheep blood agar if the V factor is exogenously provided. Several other bacterial species produce V factor as a metabolic by-product. Therefore, tiny colonies of *Haemophilus* may be seen growing very close to colonies of bacteria that produce V factor on sheep blood agar. This phenomenon, known as *satelliting* (Figure 30.2), often occurs around hemolytic staphylococci because they produce sufficient V factor. Pneumococci, neisseriae, and many other organisms also produce enough NAD to allow satellite colonies of *Haemophilus* sp. to form in their vicinity. With the exceptions of *H.*

FIGURE 30.2

Satelliting colonies of *Haemophilus influenzae* surrounding the *Staphylococcus aureus* streaks on the right. The *Haemophilus* was streaked for confluent growth on the blood agar plate, but colonies appear only close to the *Staphylococcus* growth. Chocolate agar *(plate on left)* provides all nutrients necessary for growth of *H. influenzae.* (Courtesy Pete Rose.)

aphrophilus and *H. ducreyi,* all clinically significant *Haemophilus* sp. require V factor.

Media that adequately support growth of *H. ducreyi* are Mueller-Hinton base chocolate agar, supplemented with 1% IsoVitaleX (BBL Microbiology Systems) and 3 µg/ml vancomycin, and heart infusion base agar with 10% fetal bovine serum and 3 µg/ml vancomycin. For best recovery, material from suspected chancroid lesions should be inoculated to two plates, one of each different agar formula; both are incubated in a candle jar at 35° to 37° C for 7 days.

H. influenzae, often untypable strains, are significant etiological agents of lower respiratory tract infection, particularly among elderly patients and young individuals with cystic fibrosis. The number of isolates of *Haemophilus* sp. is increased when selective media such as horse blood–bacitracin are employed for recovery of these organisms. Such media are particularly important for isolating *Haemophilus* from respiratory secretions of patients with cystic fibrosis, in whom the organisms may be etiological agents of disease, but they are overgrown and their presence obscured by the coexisting mucoid *Pseudomonas aeruginosa.* Anaerobic incubation of cultures from patients with cystic fibrosis also enhances isolation of *Haemophilus* by inhibiting growth of *Pseudomonas.* Most strains of *Haemophilus* are able to grow aerobically and anaerobically. Growth is usually stimulated by 5% to 10% CO_2 so that incubation in a candle extinction jar or CO_2 incubator is recommended. The organisms usually grow within 24 hours, but some strains, particularly *H. ducreyi,* may require as long as 7 days.

LABORATORY IDENTIFICATION OF CLINICALLY SIGNIFICANT *HAEMOPHILUS* SP.

Haemophilus species are often not visible in broth, especially in blood culture media, since they do not grow to turbidity. For this reason, blind subcultures to chocolate agar or examinations of smears (acridine orange fluorescent stain is recommended) are necessary for their detection. *H. parainfluenzae* may appear as filamentous, intertwining strands of gram-negative bacilli, whereas the other species usually are seen as coccobacilli with long filamentous forms occasionally interspersed (Figure 30.3). Gram stains from colonies growing on solid agar are more typical, showing pleomorphic coccobacilli.

Colonies on chocolate agar are transparent, moist, smooth, and convex with a distinct odor, variously described as "mousy" or "bleachlike." X and V factor requirements are carried out using

FIGURE 30.3

Gram stain of cerebrospinal fluid containing many polymorphonuclear leukocytes and *Haemophilus influenzae.* Note both coccobacilli and long filamentous forms.

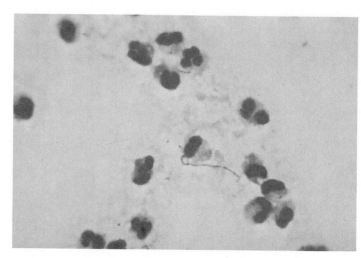

the porphyrin test (Procedure 30.3). This test detects the presence of enzymes that convert δ-aminolevulinic acid (ALA) into porphyrins or protoporphyrins. The more traditional method for detection of X and V factor requirements, which involves inoculating the organism to media without blood and placing X and V factor–impregnated filter paper strips on the plate, incubating overnight, and examining for growth around the appropriate strip, is not as reliable. Many X factor–requiring organisms are able to carry over enough X factor from the primary medium to yield false-negative tests with the filter paper method.[9]

A commercial ALA reagent impregnated on a filter paper disk is available from Remel Laboratories. The test is performed in 4 hours. The technologist must be careful not to overmoisten the disk, since false-negative results can occur because of dilution of the end product.[9] An easy way to perform this test is to place the filter paper disk directly on the primary plate, where the moisture from the agar is enough to moisten the disk. The entire plate is incubated during testing.

In addition to the porphyrin test, *Haemophilus* isolates may be identified using biochemical parameters (Table 30.4). Although conventional media are acceptable for performance of these tests, several new rapid enzymatic biochemical systems have been developed for identification of *Haemophilus,* including the Minitek System (BBL Microbiology Systems), the RapID NH (Innovative Diagnostics), HNID (American MicroScan), and others.

Those species that require V factor have been grouped into biotypes, but the biotypes have not yet been associated with particular clinical syndromes. *H. segnis* is the only oxidase-negative species among this group. *H. aphrophilus* can be confused with a closely related organism, *Actinobacillus actinomycetemcomitans.* As discussed later, *A. actinomycetemcomitans* is catalase positive, does not require V factor, and cannot ferment lactose or sucrose. For these and other reasons, the genus name *Actinobacillus* has been retained. *H. aphrophilus* yields the opposite reactions in those tests. Colony morphology of *H. aphrophilus* on a clear medium can be used to aid in differentiating it from *A. actinomycetemcomitans.* The colony of *H. aphrophilus* is round and convex and displays an opaque zone near the center (Figure 30.4), whereas *A. actinomycetemcomitans* colonies appear to have a central star-shaped structure (Figure 30.5).

RAPID IDENTIFICATION OF *HAEMOPHILUS INFLUENZAE* TYPE B

The clinical significance of *H. influenzae* type b has prompted workers to develop rapid immunological methods for detection of its capsular polysaccharide directly in clinical specimens, such as cerebrospinal fluid, serum, and urine. The polysaccharide concentrates in the urine, which is thus an easy and useful specimen to test. The organism, once isolated, can also be rapidly identified with the same reagents. Coagglutination reagents produced by Pharmacia Biotechnology and latex agglutination reagents (BBL Microbiology Systems, Wellcome Diagnostics, Wampole Laboratories) have been shown to be very sensitive and specific for detecting the presence of *H. influenzae* type b in cerebrospinal fluid and slightly less sensitive when used with other body fluids (Chapter 16). Because of false-negative tests, however, and rare cross-reactive antigens, such tests should not be used as the sole diagnostic modality.

PROCEDURE 30.3

PORPHYRIN TEST FOR X FACTOR REQUIREMENT

PRINCIPLE
Non-hemin-requiring *Haemophilus* sp. can use δ-aminolevulinic acid (ALA) as a substrate for the synthesis of porphyrin (another name for hemin). Porphyrin fluoresces reddish orange under ultraviolet light and can thus be detected.

METHOD
1. Prepare ALA substrate as follows:
   δ-Aminolevulinic acid (Sigma Chemical Co.) (2 mol/L), 0.34 g
   MgSO$_4$ (0.08 mol/L), 0.0096 g
   Phosphate buffer (0.1 mol/L, pH 6.9), 1000 ml
   The MgSO$_4$ may be prepared in a more concentrated solution and diluted to reach the desired final concentration. Mix thoroughly, dispense into small, plastic snap-top tubes (12 × 75 mm), 0.5 ml per tube, and freeze at −20° C until use.
2. Thaw the substrate, and add a very heavy loopful of the organism to be tested, making a heavy suspension.
3. Incubate the tubes at 37° C for 4 hours, and then take the tubes into a darkroom and shine an ultraviolet light source of 360 nm wavelength (long wave) on the tube's contents.
4. Presence of a brick-red to orange fluorescence indicates that porphyrins were produced and that the organism does not require hemin (X factor).

QUALITY CONTROL
Test *H. influenzae* ATCC 43065 and *H. parainfluenzae* ATCC 7901 in the same manner as test organisms.

EXPECTED RESULTS
H. parainfluenzae, the positive control, shows red fluorescence in the tube; *H. influenzae,* the negative control, should display no fluorescence.

PERFORMANCE SCHEDULE
Test fresh subcultures of the quality control strains each time the test is performed.

Modified from Kilian, M., and Biberstein, E.L. 1984. Genus *Haemophilus* (Winslow, Broadhurst, Buchanan, Krumwieded, Rogers and Smith 1917). In Krieg, N.R., and Holt, J.G., editors. Bergey's manual of systematic bacteriology, ed 9. Williams & Wilkins, Baltimore.

In addition to rapid detection and identification of the organism, the increasing incidence of ampicillin resistance necessitates that microbiologists perform a rapid test for detection of β-lactamase. Because of widespread chloramphenicol resistance found in other parts of the world, and occasional chloramphenicol-resistant isolates seen in the United States, microbiologists may want to perform a rapid test for detection of chloramphenicol resistance as well. These tests, produced commercially as reagent-impregnated filter paper disks, are discussed fully in Chapter 14. A special susceptibility testing medium, *Haemophilus* testing medium (HTM), is recommended for both disk diffusion and broth dilution methods.

TREATMENT OF *HAEMOPHILUS* INFECTIONS
During meningitis, certain antimicrobial agents that would normally be unable to do so can penetrate the blood-brain barrier, allowing agents such as ampicillin to be used for treatment of this disease. However, because of the recent increase in resistance of *H. influenzae* to ampicillin and the possibility of chloramphenicol resistance, the current recommended initial therapy for meningitis is a combination of ampicillin and chloramphenicol pending the results of in vitro susceptibility tests. Ceftriaxone or cefotaxime is a popular alternative. Chloramphenicol achieves excellent levels in most tissues, including bone, and is the drug of choice for serious infections caused by susceptible *Haemophilus* sp., alone or in combination with ampicillin. Once an isolate has been shown to be susceptible to ampicillin, this agent may be used alone to continue treatment. Patients with less severe disease caused by *H. influenzae* being treated as outpatients are usually given oral amoxicillin–clavulanic acid, amoxicillin, ampicillin, or trimethoprim-sulfamethoxazole. Other species of *Haemophilus* may often be successfully treated with tetracycline in addition to the agents mentioned. Treatment for chancroid involves either

TABLE 30.4

CHARACTERISTICS OF *HAEMOPHILUS* SP. ISOLATED FROM HUMANS

SPECIES	REQUIRES V FACTOR FOR GROWTH	ALA→ PORPHYRINS	UREASE	INDOLE	ACID FROM GLUCOSE	ACID FROM LACTOSE	BETA HEMOLYSIS†	CATALASE
*H. influenzae**	+	−	+/−	+/−	+	−	−	+
H. parainfluenzae	+	+	+/−	−	+ (some gas)	−	−	+/− (slow)
H. haemolyticus	+	−	+	+/−	+ (some gas)	−	+	+
H. aphrophilus	−	+ (weak)	−	−	+ (gas)	+	−	−
H. paraphrophilus	+	+	−	−	+ (gas)	+	−	−
H. segnis	+	+	−	−	+ (weak)	−	−	+/− (slow)
H. ducreyi	−	−	−	−	+/−	−	−	−
H. parahaemolyticus	+	+	+	−	+ (some gas)	−	+	+/−
H. paraphrohaemolyticus‡	+	+	+	−	+	−	+	+/−

Modified from Kilian, M., and Biberstein, E.L. 1984. Genus *Haemophilus* (Winslow, Broadhurst, Buchanan, Krumwiede, Rogers and Smith 1917). In Krieg, N.R., and Holt, J.G., editors. Bergey's manual of systematic bacteriology, vol 1. Williams & Wilkins, Baltimore.

+, >90% of strains positive; −, >90% of strains negative; +/−, variable results; some gas, some strains produce gas.

*Includes *H. influenzae* biogroup aegyptius.

†Rabbit or horse blood agar.

‡Requires CO_2 for growth; may not be a separate species.

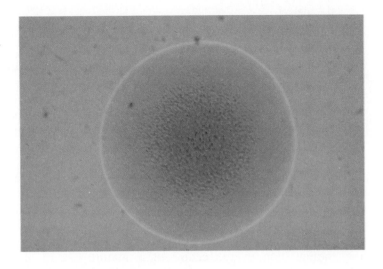

FIGURE 30.4
Colony of *Haemophilus aphrophilus* on heart infusion agar. Note opaque central area.

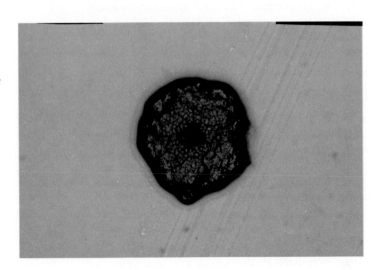

FIGURE 30.5
Colony of *Actinobacillus actinomycetemcomitans* on heart infusion agar. Note the central star-shaped structure.

oral erythromycin or ceftriaxone. Trimethoprim-sulfamethoxazole is a frequently used alternative. Rifampin is the antimicrobial agent of choice for prophylaxis of close contacts of a patient with *H. influenzae* meningitis.

GENUS *ACTINOBACILLUS*

EPIDEMIOLOGY AND PATHOGENESIS OF *ACTINOBACILLUS* INFECTION

A. lignieresii, A. equuli, A. ureae, and *A. suis* are found as pathogens and as part of the normal oral flora of many domestic animals, including cows, horses, sheep, and pigs. *A. actinomycetemcomitans* is normal flora of human oral mucosa and is not found in other animals. This genus is very similar phenotypically to *Pasteurella* and can be distinguished only biochemically. Humans are infected with the animal species through traumatic inoculation of the organism, such as by a bite wound or

other wound.[13,20] Infections associated with *A. actinomycetemcomitans* are probably all of endogenous origin. This species is often isolated along with *Actinomyces* or *Propionibacterium* (formerly *Arachnia*) *propionicum* from actinomycotic lesions of the jaw or abscesses within the thoracic cavity. Failure to eradicate the *Actinobacillus* may account for persistence of actinomycosis, even after the *Actinomyces* or *P. propionicum* has been cleared. *A. actinomycetemcomitans* has recently been associated with a destructive periodontitis that occurs in adolescents. This species, as well as the closely related *Haemophilus aphrophilus*, is occasionally isolated as the etiological agent of endocarditis.[13] The organism probably possesses very low virulence, acting as a pathogen only opportunistically.

LABORATORY IDENTIFICATION

A. actinomycetemcomitans, the most important species clinically, grows on blood and chocolate agars; unlike the other species of *Actinobacillus,* it does not grow on MacConkey agar. Growth is enhanced in a candle jar or a CO_2 incubator, but colonies may require up to 7 days to achieve maximal size. Growth in broth media is often barely visible, with no turbidity produced. Microcolonies may be seen as tiny puffballs growing on the blood cell layer in blood culture bottles, or alternatively, growth may appear as a film or as tiny puffs on the sides of a bottle or tube. On blood agar, the medium surrounding the colonies exhibits a slight greenish tinge. The colonies themselves are rough and sticky. A characteristic finding is the presence of a four- to six-pointed star-like configuration in the center of a mature colony growing on clear medium (e.g., brain-heart infusion agar); this star pattern can be easily visualized by examining the colony under low power (100×) with a standard microscope (Figure 30.5). On Gram stain, the organisms are gram-negative bacilli, often displaying coccoid forms. The coccoid shapes may be found at the end of bacilli, giving the overall appearance of the dots and dashes of Morse code. *Actinobacillus* sp. may resemble *Haemophilus* sp. and *Pasteurella* sp.; their differentiation is outlined in Table 30.5. Key characteristics of *Actinobacillus* sp. likely to be encountered are shown in Table 30.6. *A. ureae* is more similar to *Pasteurella* than *Actinobacillus* and is presented in Table 30.7 with other pasteurellae.

TREATMENT OF *ACTINOBACILLUS* INFECTIONS
Most members of the genus are susceptible to cephalosporins, aminoglycosides, tetracycline, and chloramphenicol. Penicillin and ampicillin resistance has been reported, and the organisms are often resistant to erythromycin and clindamycin.[13]

GENUS *PASTEURELLA*

EPIDEMIOLOGY AND PATHOGENESIS OF *PASTEURELLA* INFECTION
Pasteurella sp. are normal flora of the respiratory and gastrointestinal tracts of many species of domestic and wild animals and birds. They are etiological agents of hemorrhagic septicemia, gastrointestinal infections, and respiratory tract infections in many animals. Humans acquire *Pasteurella* primarily through animal exposure. Most human infections are associated with animal bites, but many respiratory tract and intraabdominal infections are associated with possible inhalation of the organism or with no known source of acquisition. *Pasteurella multocida* (which now in-

cludes three subspecies, subsp. *multocida,* subsp. *septica,* and subsp. *gallicida*), the most common human pathogen among the Pasteurellae, can be isolated from animal bite wounds, respiratory secretions from patients with pneumonia, cerebrospinal fluid, abscesses, biopsy specimens from patients with osteomyelitis, blood, joint fluid from septic arthritis, and many other sites. Other species of *Pasteurella* (at least 17 more) are more often associated with animal disease, although they have been reported from human cases of endocarditis, pneumonia, septicemia, and primarily infected wounds, occasionally involving osteomyelitis. Some species of *Pasteurella, haemolytica, ureae,* and *aerogenes* have been shown to be similar to the genus *Actinobacillus* by deoxyribonucleic acid (DNA) homology studies.

Virulence of *Pasteurella* is related to the polysaccharide capsule that allows the organism to resist phagocytosis. Although some strains produce a cytotoxin, its role in the pathogenesis of disease has not been elucidated. *P. multocida* appears to be the most virulent of the species. Virulence may be enhanced by the ability of the organism to utilize free iron, a factor in the pathogenicity of several bacterial strains.

LABORATORY IDENTIFICATION
Since *Pasteurella* sp. grow well on routinely inoculated laboratory media, including blood agar, they are easily isolated from all sites in which they may be present. Growth is not appreciably enhanced by increased concentrations of CO_2. Colonies are small and translucent and may be smooth or rough. A brown discoloration of the medium may develop around colonies of *P. multocida.* The colonies on blood agar have a musty or "mushroom" smell, which is recognizable to experienced microbiologists. All species should be oxidase positive by the Kovacs' test; several subcultures may be necessary to demonstrate this. They are nitrate and catalase positive, spot indole positive, and glucose fermenters. A quick screening procedure used for presumptive identification of *Pasteurella* sp. takes advantage of their penicillin susceptibility, unusual for gram-negative bacteria. The surface of a Mueller-Hinton blood agar plate is inoculated with a suspension of the organism equal to a McFarland 0.5 turbidity, using a cotton swab and streaking in three directions, as for the disk diffusion susceptibility test. A filter paper disk containing 2 U of penicillin (used for susceptibility testing) is placed in the center of the plate, which is then incubated overnight in air. A zone of inhibition around the disk is indicative of susceptibil-

TABLE 30.5

DIFFERENTIAL CHARACTERISTICS OF MORPHOLOGICALLY SIMILAR GRAM-NEGATIVE BACILLI AND COCCOBACILLI

GENUS/SPECIES	OXIDASE	CATALASE	CELL SHAPE COCCOID	CELL SHAPE FUSIFORM	INDOLE	NITRATE → NITRITE	FERMENTATION OF GLUCOSE	MALTOSE	SORBITOL	SUCROSE	LACTOSE
Actinobacillus actinomycetemcomitans	+/-	+	+	-	-	+	+	+	-	-	-
Capnocytophaga gingivalis/ochracea/sputigena	-	-	-	+	-	+/-	+	+	-	+	+/-
Cardiobacterium hominis	+	-	-	-	+	-	+	+	+	+*	-
Capnocytophaga canimorsus/cynodegmi	+ (weak)	+	-	+	-	-	+	+	-	-	+
Eikenella	+	-	-	-	-	+	-	-	-	-	-
Haemophilus aphrophilus	-/+ (weak)	-	+/-	-	-	+	+	+	-	+	+
HB-5 (Pasteurella bettae)	+ (weak)	-	+/-	-	+ (weak)	+	+	-	-	-	-
Kingella	+	-	-	-	-	-/+	+	+	-	-	-
Pasteurella multocida	+	+	+/-	-	+/-	+	+	-/+	+/-	+	-
Suttonella indologenes	+	-	-	-	+	-	+	+	-	+	-

+, >90% of strains positive within 2 days; -, >90% of strains negative; +/-, variable (most strains positive); -/+, variable (most strains negative).

*Positive results may require extended incubation period.

TABLE 30.6

CHARACTERISTICS OF *ACTINOBACILLUS* SP.

SPECIES	OXIDASE	UREASE	ESCULIN HYDROLYSIS	FERMENTATION OF			
				TREHALOSE	MELIBIOSE	SALICIN	MANNOSE
A. actinomycetemcomitans	+	−	−	−	−	−	+
A. hominis	+	+	+/−	+	+	+/−	−
A. lignieresii	+/−	+	−	−	−	−	+
A. equuli	+/−	+	−	+	+	−	+
A. suis	+	+	+	+	+	+	+
A. ureae	+/−	+	−	−	−	−	+/−

+, >90% of strains positive; −, >90% of strains negative; +/− = variable results.

FIGURE 30.6

Susceptibility of *Pasteurella multocida* to a 2 U penicillin disk.

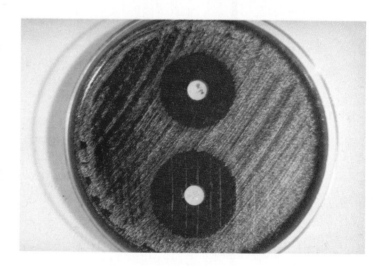

ity; most other gram-negative bacilli are resistant to penicillin (Figure 30.6).

The organism is a small, coccoid or rod-shaped bacillus that often exhibits bipolar staining. All species are nonmotile. Differentiation among species can be achieved using biochemical parameters listed in Table 30.7. Only some strains of *P. dagmatis* produce gas from glucose. Isolates of this species (previously called "new species" or "gas") are often associated with dog and cat bites and may go on to cause systemic disease.

Most biochemical identification systems available commercially identify *Pasteurella*, at least to genus. *Pasteurella* should be suspected when any isolate is associated with an animal bite. Again, it is important for clinicians to inform the laboratory of the source of a wound specimen and pertinent clinical information so that the microbiologist can perform the appropriate tests early in the identification protocol.

TREATMENT OF *PASTEURELLA* INFECTIONS

Penicillin is the antimicrobial agent of choice for all *Pasteurella* infections. Tetracycline and chloramphenicol are also effective, but aminoglycosides, erythromycin, and clindamycin are not recommended. Cephalothin and the β-lactam/β-lactamase inhibitor combinations are also effective for therapy of wound infections caused by *Pasteurella*.

GENUS *KINGELLA*

EPIDEMIOLOGY AND PATHOGENESIS OF *KINGELLA* INFECTION

The two true species of *Kingella*, *K. kingae* and *K. denitrificans*, are normal oropharyngeal flora of humans. Although the pathogenicity of *K. denitrificans* is unknown, *K. kingae* is an important pathogen in children. The organism has been found as the etiological agent of bacteremia and skin lesions, and it seems to be particularly asso-

TABLE 30.7

CHARACTERISTICS OF *PASTEURELLA* AND RELATED SPECIES OF POSSIBLE CLINICAL SIGNIFICANCE

SPECIES	BETA HEMOLYSIS	GROWTH ON MACCONKEY AGAR	INDOLE	UREASE	ORNITHINE DECARBOXYLASE	GAS FROM GLUCOSE	FERMENTATION OF			
							MALTOSE	MANNITOL	SORBITOL	SUCROSE
P. multocida	–	+/–	+/–	–	+/–	–	+/–	+/–	+/–	+
P. gallinarum	–	–	–	–	–	–	+	–	+/–	+
P. bettii	–	+/–	–	–	–	+/–	+/–	–	–	–
P. dagmatis	–	–	+	+	–	+/–	+	–	–	+
P. canis	–	–	+/–	–	+	–	–	–	–	+
P. pneumotropica	–	+/–	+	+	+	+/–	+/–	+	+	+
P. haemolytica	+/–	+	–	–	+	–	+	+/–	+	+
P. aerogenes	–	+	+/–	+	+/–	+/–	+	+/–	+/–	+
Actinobacillus ureae	–	+/–	+/–	+	–	–	+/–	+	+/–	+

Modified from Carter, G.R. 1984. Genus *Pasteurella* (Trevisan 1887). In Krieg, N.R., and Holt, J.G., editors. Bergey's manual of systematic bacteriology. vol. 1. Williams & Wilkins, Baltimore; and Sneath, P.H.A., and Stevens, M. 1990. *Actinobacillus rossii* sp. nov., *Actinobacillus seminis* sp. nov., nom. rev., *Pasteurella bettii* sp. nov., *Pasteurella lymphangitidis* sp. nov., *Pasteurella mairi* sp. nov., and *Pasteurella trehalosi* sp. nov. Int. J. Syst. Bacteriol. 40:148.
+, >90% of strains positive; –, >90% of strains negative; +/–, variable results.

ciated with septic arthritis and other osteoarticular diseases.[10,15,16,28] Recovery of the organism from joint fluids is enhanced by inoculating the specimen into blood culture broth.[28] The lysis-centrifugation system should also be attempted for joint fluids. Pathogenic mechanisms have not been well characterized. The previously named species *K. indologenes* is more similar to *Moraxella* sp. and has been renamed *Suttonella indologenes*. It has been associated (rarely) with eye infections.

LABORATORY IDENTIFICATION

Colonies of *K. kingae* on 5% sheep blood agar are small with a clear zone of hemolysis. Initial growth may require several days of incubation. Increased CO_2 seems to enhance growth. Colony morphology may vary; one morphotype pits the agar and assumes a "fried egg" appearance with a thin, spreading haze of growth surrounding the central colony. Gram stain reveals short, plump bacilli with squared-off ends that may form chains.

K. denitrificans produces nonhemolytic colonies resembling those of *K. kingae*. One distinguishing feature is the ability of *K. denitrificans* to grow on Thayer-Martin medium. Because it is also oxidase positive and able to produce acid from glucose, *K. denitrificans* may be confused with *Neisseria gonorrhoeae*, a closely related genus.[5] Differential tests include nitrate reduction (*N. gonorrhoeae* is unable to reduce nitrate, whereas most *K. denitrificans* are nitrate positive) and inhibition of growth of *N. gonorrhoeae* by amylase. All 52 *K. denitrificans* isolates tested by Odugbemi and Arko[18] were not inhibited from growing up to a 13 mm–diameter filter paper disk impregnated with 0.05 ml diluted human saliva or α-amylase (Sigma Chemical Co.) when inoculated to the surface of plates made from GC agar base with IsoVitaleX (BBL Microbiology Systems).

All *Kingella* sp. are oxidase positive and non-motile and able to ferment glucose. Most *Kingella* sp. are catalase negative, which helps to differentiate them from *Moraxella* sp. and *Neisseria* sp. In contrast to *Kingella*, *Suttonella indologenes* is indole positive, as the name suggests. None of the species produces acid in the butt of a TSIA tube; they may not grow at all in that medium. O-F carbohydrate media (Chapter 29) may need to be supplemented with 5% horse serum to allow sufficient growth.

TREATMENT OF *KINGELLA* INFECTIONS

β-Lactam agents appear to be efficaceous for treatment of infections caused by *K. kingae*.[10] The organism is also susceptible to gentamicin and chloramphenicol, both of which have been used successfully to treat septic arthritis. Limited information is available on susceptibilities of the other two species.

GENUS *CAPNOCYTOPHAGA*

EPIDEMIOLOGY AND PATHOGENESIS OF *CAPNOCYTOPHAGA* AND RELATED ORGANISM INFECTION

The *Capnocytophaga* are capnophilic (CO_2 loving) gram-negative bacilli, found as normal oral flora in humans (*C. ochracea, C. sputigena, C. gingivalis*) and animals (*C. canimorsus, C. cynodegmi*).[6] The human strains are often found in significantly larger numbers within the gingival pockets of patients with periodontal disease. The organisms produce proteases that are able to cleave secretory immunoglobulins, which are postulated to have a protective effect in preventing oral disease. An additional possible virulence factor is production by *Capnocytophaga* of a heat-stable factor that alters polymorphonuclear neutrophil chemotactic activity. Interestingly, young patients with severe periodontitis also often display a neutrophil dysfunction that may spontaneously revert with the effective treatment of their periodontitis. The role of the extracellular products of *Capnocytophaga* in virulence of the organism has not yet been fully elucidated.

In addition to periodontal disease, human *Capnocytophaga* is associated with sepsis or systemic infection, usually in leukemic and/or granulocytopenic patients.[1] However, there are several reports of osteomyelitis, septic arthritis, endocarditis, and soft tissue infections in normal hosts from whom *Capnocytophaga* has been isolated. The portal of entry is thought to be the oral cavity. Many patients with predisposing immunosuppression had bleeding gums or oral mucosal erosion before they developed sepsis.

C. canimorsus, probably normal oral flora in dogs and possibly cats, has been isolated from numerous patients with severe infection, often following a dog bite.[2] Patients with underlying immunocompromising disorders are more likely to develop infection. The organism is usually isolated from blood or cerebrospinal fluid, but it may be missed in other specimens because of its fastidious nature and slow growth rate.

C. cynodegmi and DF-3, an organism resembling these species, have been isolated from blood, wounds, and other sources, again associated with animal bites.

LABORATORY IDENTIFICATION

The organism grows well on trypticase soy agar with 5% sheep blood, the standard blood agar plate used by most microbiology laboratories for routine cultures, as well as on chocolate or trypticase soy agar without blood. After 48 to 72 hours of incubation at 37° C in 5% CO_2 or anaerobically, colonies are 1 to 5 mm in diameter, opaque, shiny, and nonhemolytic, with a pale-beige or yellowish color. The color may be more apparent if growth is scraped from the surface with a loop and examined against a white background. The organism does not grow in air. Occasionally, gliding motility of the organism may be observed as outgrowths from the colonies or as a haze on the surface of the agar (Figure 30.7). The gliding motility is best observed on media containing 3% agar. The organism fails to grow on MacConkey agar.

Gram stain reveals gram-negative, fusiform-shaped bacilli with one rounded end and one tapered end and occasional filamentous forms (Figure 30.8). The organisms may appear as aggregates. They are oxidase and catalase negative except for *C. canimorsus,* which is positive in both tests. All are indole negative. Some differential features are listed in Table 30.8. The addition of 3% rabbit serum may be necessary for growth of the organism in carbohydrate broths. Alternatively, the rapid enzymatic carbohydrate tests described in Chapter 10 may be used.

Commercially available anaerobic identification systems, such as the RapID ANA (Innovative Diagnostics, Inc.), Minitek System (BBL Microbiology Systems), and API AnIdent (bioMérieux Vitek), and the multiparameter API ZYM enzymatic identification system can be used to differentiate *Capnocytophaga* sp. For most situations, however, a presumptive identification to genus is as beneficial to clinicians as is definitive identification. In addition, all species share the traits of slightly yellow pigment, acid production from glucose, and negative urease and lysine decarboxylase. The organisms differ from morphologically similar organisms, such as *Actinobacillus actinomycetemcomitans, Haemophilus aphrophilus, Eikenella corrodens, Gardnerella vaginalis,* and CDC Group HB-5. Table 30.5 outlines some of the major differential characteristics among these fastidious gram-negative organisms.

DF-3 is catalase, oxidase, and nitrate negative and may be indole positive or negative.

TREATMENT OF *CAPNOCYTOPHAGA* INFECTIONS

Sepsis caused by this organism is severe, especially in immunocompromised patients, and must be approached aggressively. Bactericidal activity has been shown for penicillin, carbenicillin, clindamycin, and erythromycin. All strains are resistant to aminoglycosides. *C. canimorsus* is usually susceptible to penicillin, chloramphenicol, and carbenicillin but may be resistant to aminoglycosides. Presence of β-lactamases in some strains demands that susceptibility tests be performed.[6] In such cases the use of a β-lactam/β-lactamase inhibitor combination may be efficaceous. Quinolones are also active against all strains in vitro. Resistance to colistin may help differentiate *C. canimorsus* from similar organisms, such as *Pas-*

FIGURE 30.7

Colonies of *Capnocytophaga* growing on 5% sheep blood agar. Note haze of growth extending from colonies, evidence of gliding motility of the bacteria.

TABLE 30.8

KEY DIFFERENTIAL CHARACTERISTICS AMONG *CAPNOCYTOPHAGA* SP.*

TEST	C. OCHRACEA	C. SPUTIGENA	C. GINGIVALIS	C. CANIMORSUS	C. CYNODEGMI
Catalase	–	–	–	+	+
Oxidase	–	–	–	+	+
Esculin	+	+	–	–/+	+/–
Galactose	+/–	–	–	+/–	+/–
Lactose	+	+/–	–	+	+
Melibiose	–	–	–	–	+
Raffinose	+/–	–/+	–/+	–	+/–
Sucrose	+	+	+	–	+/–
Nitrate reduced	–/+	+/–	–	–	–/+

Modified from Forlenza, S.W. 1991. Clin Microbiol. Newsletter 13:89.

+, >90% of strains positive; –, >90% of strains negative; +/–, variable (most strains positive); –/+, variable (most strains negative).

*Carbohydrate fermentations were performed in buffered, single-substrate medium.

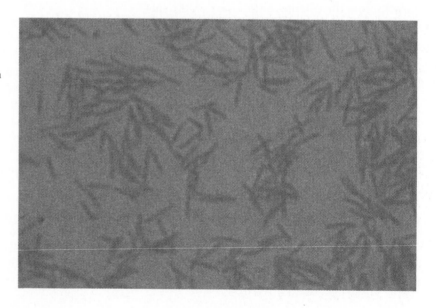

FIGURE 30.8

Gram stain of *Capnocytophaga.* Courtesy Diane Citron, R.M. Alden Research Laboratory, Santa Monica, Calif.)

teurella multocida, EF-4, *Kingella* sp., other *Capnocytophaga, Eikenella,* and *Cardiobacterium hominis,* all of which are susceptible to colistin.

GENUS *CARDIOBACTERIUM*

EPIDEMIOLOGY AND PATHOGENESIS OF *CARDIOBACTERIUM* INFECTION

Cardiobacterium hominis, the only species in the genus, is normal flora in the human mouth and perhaps on other mucous membranes. It is noninvasive and has not been associated with any type of infection other than bacteremia or endocarditis. The incidence of *Cardiobacterium*-related bacteremia, however, is increasing. This may be be-

cause of the increased number of susceptible hosts, or more likely, because of increased awareness among microbiologists and better blood culturing techniques. Only 34 cases of *C. hominis* infection had been reported until 1983. Most patients had prior cardiac valvular disease. No specific virulence factors have been delineated for this organism.

LABORATORY IDENTIFICATION OF *CARDIOBACTERIUM HOMINIS*

As an etiological agent of disease, this organism has been isolated only from blood. It grows well in standard blood culture media and can be subcultured to 5% sheep blood agar, chocolate agar, and trypticase soy agar without blood. The organism does not grow on MacConkey agar. Incubation of blood agar subcultures at 37° C in 5% CO_2 or in air with increased humidity (a candle jar probably provides the best atmosphere) for 48 hours yields colonies that are 1 to 2 mm in diameter, slightly alpha hemolytic, smooth, round, glistening, and opaque. The organism grows equally well at 30° and 37° C. Colonies grown on trypticase soy agar without blood should be observed microscopically under low power (100×) after 24 hours' incubation for typical rough appearance with an internal serpentine pattern of bacilli, some of which stream from the edge of the colony toward adjacent colonies. This appearance is in contrast to the prominent starlike pattern seen in colonies of *Actinobacillus actinomycetemcomitans*, with which *Cardiobacterium* may be confused. *Cardiobacterium* is a pleomorphic gram-negative bacillus that tends to form rosette-like arrangements.

The organism is oxidase positive. *C. hominis* is able to utilize carbohydrates in media with sufficient nutritional factors. The addition of 10% horse serum to standard fermentation media allows sufficient growth of the organism. A key biochemical parameter of *C. hominis* is the weak production of indole. The spot indole test (Chapter 10), performed on a colony grown on blood agar for 72 hours, may be positive even if the conventional test is negative. Alternatively, the more sensitive indole reagent described in Chapter 29, that of Ehrlich, combined with xylene extraction, may be required for demonstration of this property. *C. hominis* is urease, catalase, and nitrate negative, as outlined in Table 30.5.

TREATMENT OF *CARDIOBACTERIUM HOMINIS* INFECTIONS

The organism is susceptible in vitro to penicillin, ampicillin, cephalothin, chloramphenicol, aminoglycosides, and tetracycline. Penicillin is probably the drug of choice, although the organism's identity must be certain, since *A. actinomycetemcomitans* and *H. aphrophilus,* with which *C. hominis* may be confused, are often resistant to penicillin.

HB-5 *(PASTEURELLA BETTAE)*

HB-5 (recently named *Pasteurella bettae*) is isolated from the genitourinary tract of men and women. The organism has been associated with skin infections and abscesses. It is a gram-negative coccobacillus. Growth occurs as nonhemolytic colonies on blood agar after 24 hours of incubation at 37° C in 5% to 10% CO_2. Oxidase is vari-

able, and catalase is not produced. The organisms are nitrate positive and weakly indole positive. A TSIA slant inoculated with HB-5 is uniformly yellow after overnight incubation; glucose is fermented with the production of gas (unlike *Pasteurella, Cardiobacterium, Capnocytophaga,* and DF-3), but neither sucrose nor lactose is fermented.

REFERENCES

1. Bilgrami, S., Bergstrom, S.K., Peterson, D.E., et al. 1992. *Capnocytophaga* bacteremia in a patient with Hodgkin's disease following bone marrow transplantation: case report and review. Clin. Infect. Dis. 14:1045.
2. Brenner, D.J., Hollis, D.G., Fanning, G.R., and Weaver, R.E. 1989. *Capnocytophaga canimorsus* sp. nov. (formerly CDC group DF-2), a cause of septicemia following dog bite, and *C. cynodegmi* sp. nov., a cause of localized wound infection following dog bite. J. Clin. Microbiol. 27:231.
3. Brenner, D.J., Mayer, L.W., Carlone, G.M., et al. 1988. Biochemical, genetic, and epidemiologic characterization of *Haemophilus influenzae* biogroup Aegyptius (*Haemophilus aegyptius*) strains associated with Brazilian purpuric fever. J. Clin. Microbiol. 26:1524.
4. California Morbidity. 1992. Invasive *Haemophilus influenzae* disease in California, 1990. Department of Health Services, Berkeley.
5. Dewhirst, F.E., Paster, B.J., and Bright, P.L. 1989. *Chromobacterium, Eikenella, Kingella, Neisseria, Simonsiella,* and *Vitreoscilla* species comprise a major branch of the beta group *Proteobacteria* by 16S ribosomal ribonucleic acid sequence comparison: transfer of *Eikenella* and *Simonsiella* to the family *Neisseriaceae* (emend.) Int. J. Syst. Bacteriol. 39:258.
6. Forlenza, S.W. 1991. *Capnocytophaga:* an update. Clin. Microbiol. Newsletter 13:89.
7. Friedman, R.L. 1988. Pertussis: the disease and new diagnostic methods. Clin. Microbiol. Rev. 1:365.
8. Gabre-Kidan, T., Lipsky, B.A., and Plorde, J.J. 1984. *Hemophilus influenzae* as a cause of urinary tract infections in men. Arch. Intern. Med. 144:1623.
9. Gadberry, J.L., and Amos, M.A. 1986. Comparison of a new commercially prepared porphyrin test and the conventional satellite test for the identification of *Haemophilus* species that require the X factor. J. Clin. Microbiol. 23:637.
10. Goutzmanis, J.J., Gonis, G., and Gilbert, G.L. 1991. *Kingella kingae* infection in children: ten cases and a review of the literature. Pediatr. Infect. Dis. J. 10:677.
11. Halperin, S.A., Bortolussi, R., and Wort, A.J. 1989. Evaluation of culture, immunofluorescence, and serology for the diagnosis of pertussis. J. Clin. Microbiol. 27:752.
12. Halperin, S.A., and Marrie, T.J. 1991. Pertussis encephalopathy in an adult: case report and review. Rev. Infect. Dis. 13:1043.
13. Kaplan, A.H., Weber, D.J., Oddone, E.Z., and Perfect, J.R. 1989. Infection due to *Actinobacillus actinomycetemcomitans:* fifteen cases and review. Rev. Infect. Dis. 11:46.

14. Kurzynski, T.A., Boehm, D., Rott-Petri, J.A., et al. 1988. Comparison of modified Bordet-Gengou and modified Regan-Lowe media for isolation of *Bordetella pertussis* and *Bordetella parapertussis*. J. Clin. Microbiol. 26:2661.

15. Lacour, M., Duarte, M., Beutler, A., et al. 1991. Osteoarticular infections due to *Kingella kingae* in children. Eur. J. Pediatr. 150:612.

16. Meis, J.F., Sauerwein, R.W., Gyssens, I.C., et al. 1992. *Kingella kingae* intervertebral diskitis in an adult. Clin. Infect. Dis. 15:530.

17. Morrill, W.E., Barbaree, J.M., Fields, B.S., et al. 1988. Effects of transport temperature and medium on recovery of *Bordetella pertussis* from nasopharyngeal swabs. J. Clin. Microbiol. 26:1814.

18. Odugbemi, T., and Arko, R.J. 1983. Differentiation of *Kingella denitrificans* from *Neisseria gonorrhoeae* by growth on a semisolid medium and sensitivity to amylase. J. Clin. Microbiol. 17:389.

19. Orenstein, W.A., Bernier, R.H., and Chen, R.T. 1989. Pertussis: a disease and a vaccine that are not going away. West. J. Med. 150:339.

20. Peel, M.M., Hornidge, K.A., Luppino, M., et al. 1991. *Actinobacillus* spp. and related bacteria in infected wounds of humans bitten by horses and sheep. J. Clin. Microbiol. 29:2535.

21. Pickett, M.J., Nelson, E.L., Hoyt, R.E., and Eisenstein, B.E. 1952. Speciation within the genus *Brucella*. I. Dye sensitivity of smooth *Brucella*. J. Lab. Clin. Med. 40:200.

22. Provenza, J.M., Klotz, S.A., and Penn, R.L. 1986. Isolation of *Francisella tularensis* from blood. J. Clin. Microbiol. 24:453.

23. Staszkiewicz, J., Lewis, C.M., Colville, J., et al. 1991. Outbreak of *Brucella melitensis* among microbiology laboratory workers in a community hospital. J. Clin. Microbiol. 29:287.

24. Stauffer, L.R., Brown, D.R., and Sandstrom, R.E. 1983. Cephalexin-supplemented Jones-Kendrick charcoal agar for selective isolation of *Bordetella pertussis:* comparison with previously described media. J. Clin. Microbiol. 17:60.

25. Steketee, R.W., Burstyn, D.G., Wassilak, S.G.F., et al. 1988. A comparison of laboratory and clinical methods for diagnosing pertussis in an outbreak in a facility for the developmentally disabled. J. Infect. Dis. 157:441.

26. Welch, W.D., Southern, P.M., Jr., and Schneider, N.R. 1986. Five cases of *Haemophilus segnis* appendicitis. J. Clin. Microbiol. 24:851.

27. Woolfrey, B.F., and Moody, J.A. 1991. Human infections associated with *Bordetella bronchiseptica*. Clin. Microbiol. Rev. 4:243.

28. Yagupsky, P., Dagan, R., Howard, C.W., et al. 1992. High prevalence of *Kingella kingae* in joint fluid from children with septic arthritis revealed by the BACTEC blood culture system. J. Clin. Microbiol. 30:1278.

29. Young, S.A., Anderson, G.L., Mitchell, P.D. 1987. Laboratory observations during an outbreak of pertussis. Clin. Microbiol. Newsletter 9:176.

BIBLIOGRAPHY

Corbel, M.J., and Brinley-Morgan, W.J. 1984. Genus *Brucella* (Meyer and Shaw 1920, 173[AL]). p. 377. In Krieg, N.R., and Holt, J.G., editors. Bergey's manual of systematic bacteriology, vol 1, ed 9. Williams & Wilkins, Baltimore.

Friedman, R.L. 1988. Pertussis: the disease and new diagnostic methods. Clin. Microbiol. Rev. 1:365.

Hicklin, A., Verghese, A., and Alvarez, S. 1987. Dysgonic fermenter-2 septicemia. Rev. Infect. Dis. 9:884.

Pickett, M.J., Hollis, D.G., and Bottone, E.J. Miscellaneous gram-negative bacteria. p. 410. In Balows, A., Hausler, W.J., Jr., Herrmann, K.L., et al., editors. Manual of clinical microbiology, ed 5. American Society for Microbiology, Washington, D.C.

31

VIBRIO AND RELATED SPECIES, *AEROMONAS*, *PLESIOMONAS*, *CAMPYLOBACTER*, *HELICOBACTER*, AND OTHERS

Members of the genera *Vibrio, Campylobacter, Aeromonas,* and *Plesiomonas* share a few features: most are oxidase-positive, gram-negative rods; some have curved cellular morphology; and some species from each genus are the etiological agents of diarrheal disease in humans. Some species in these genera are motile by means of polar flagella, although the types of flagella and their arrangements vary. The Vibrionaceae family has classically included the genera *Vibrio, Plesiomonas,* and *Aeromonas.* Renamed species of former Vibrionaceae include *Shewanella* and *Photobacterium.* The Vibrionaceae are found in seawater and fresh water, where they are pathogenic for aquatic animals. The world has been shaken recently by a surprising new pandemic of cholera, devastating parts of South America and the Far East.[6] Recovery of *Vibrio* sp. as the etiological agents of diarrheal disease may require the use of special media, and special atmosphere and temperature of incubation may be required for some *Campylobacter* sp. as well.

Campylobacter sp. differ from vibrios by requiring a microaerobic atmosphere. They display curved cellular morphology with occasional "seagull" forms representing two organisms end to end. Members of the genus are found in domestic animals, in soil and water, and in humans. The closely related *Helicobacter pylori* has been recently identified and detected worldwide as an etiological agent of human gastritis, persistence of which can lead to peptic ulcer and gastric cancer.[3] Two other *Helicobacter* sp., *H. cinaedi* and *H. fennelliae,* are agents of diarrhea and sepsis in homosexual men. Two additional species transferred from *Campylobacter* to the genus *Arcobacter, A. butzleri* and *A. cryaerophilus,* have been isolated from diarrheic stool and from human blood and tissues.[35-37] Both species cause infections in animals, but *A. butzleri* occurs more often in humans. *Arcobacter skirrowii* has not yet been implicated in human infections.[36] The anaerobic species *Wolinella recta* and *W. curva* have been shown to belong with the *Campylobacter* sp. by genetic analysis, as *C. rectus* and *C. curvus.* These organisms are involved primarily in oral infections.

VIBRIO SPECIES

EPIDEMIOLOGY AND PATHOGENESIS OF *VIBRIO* INFECTION

Members of the genus *Vibrio* are natural inhabitants of sea water. All organisms in this group are able to grow in media containing increased salt concentrations, and with the exception of *V. cholerae* and *V. mimicus* (the so-called nonhalophilic vibrios), they require sodium chloride for growth. The sodium ion–requiring members are called **halophilic** vibrios, based on the assumption that the chloride ion is required. Twelve species of potential clinical importance exist among the *Vibrio*. Table 31.1 lists the more common species.[18] *V. carchariae* (also known as *V. harveyi*), *V. metschnikovii,* and *V. cincinnatiensis* have been reported from rare human infections.[20] *V. cholerae,* no longer rare in the United States, is usually acquired during foreign travel, although more than 40 cases of domestically acquired *V. cholerae* infection had been reported in the United States from 1973 to 1990. The situation changed dramatically in association with the newest worldwide pandemic that spread to South America from China in January 1991. The epidemic in Latin America spread rapidly and continues at this time. More than 600,000 cases and 5000 deaths in persons from 20 countries had been reported as of August 1992. More than 100 cases of cholera were reported in 1992 alone in the United States.[6] As long as tourists follow recommended dietary precautions, their risk of acquiring cholera should be low. In most persons from whom a *Vibrio* has been isolated from a domestically acquired infectious process, some association has been found with water or seafood. Although selective media are recommended for recovery of *Vibrio* from stool, the organisms grow well on routine bacteriological media (sheep blood agar) and would probably be detected when isolated from blood, wound cultures, or sources other than stool as long as all gram-negative colonies are screened for oxidase production.

The enterotoxin of *V. cholerae* is one of the best characterized among bacteria (Chapter 18). The action of the toxin is to activate the adenylate cyclase of intestinal mucosal cells, which in turn raises the intracellular level of cyclic adenosine monophosphate (cAMP), ultimately causing an active secretion of electrolytes into the intestinal lumen by altering ion transport across the cell membranes of the gastrointestinal mucosa epithelial cells. The outpouring of ions is followed by a release of water. Patients with cholera have been known to secrete fluid volumes equal to their

TABLE 31.1

CLINICALLY SIGNIFICANT *VIBRIO* SP. AND THEIR ASSOCIATED INFECTIONS (IN ORDER OF RELATIVE FREQUENCY)

TYPE OF INFECTION	ASSOCIATED SPECIES
Gastroenteritis	*V. cholerae* serotype O1
	V. cholerae non-O1
	V. parahaemolyticus
	V. fluvialis
	V. mimicus
	V. furnissii
	V. hollisae
Wound infection	*V. alginolyticus*
	V. vulnificus
	V. damsela
	V. harveyi (carchariae)
Sepsis	*V. cholerae* non-O1
	V. vulnificus
	V. cincinnatiensis
Ear infection	*V. alginolyticus*
	V. cholera non-O1
	V. mimicus
	V. parahaemolyticus

body weight within 4 days. More importantly, the loss of water can dehydrate the patient so quickly that death occurs within 12 hours of onset of disease. In addition to the toxin, adherence to intestinal epithelium and motility are important factors in the pathogenicity of *V. cholerae.* Cholera-like toxin activity has also been observed in strains of *V. cholerae* non-O1 and *V. mimicus.* Other enterotoxins are produced by other vibrios.

Other *Vibrio* sp. produce extracellular cytolytic toxins or cytotoxins and enzymes that may contribute to their virulence.[18,31] Pathogenicity of *V. damsela* is probably enhanced by cytolysin production. *V. fluvialis* has been shown to produce a cytotoxin. *V. parahaemolyticus* can produce disease that resembles dysentery. It appears that *V. parahaemolyticus* virulence may be associated with production of a thermostable, direct hemolysin that hemolyzes human red blood cells in mannitol agar with a high salt content, called the *Kanagawa phenomenon.*[18] Environmental strains of *V. parahaemolyticus* seem to lack this ability. *V. vulnificus,* the lactose-positive halophilic *Vibrio,* appears to have a remarkable invasive potential, perhaps mediated by proteases. A major virulence factor for *V. vulnificus* is an acidic capsular mucopolysaccharide.[31] Primarily seen in patients with underlying hepatic or immunocompromising disease but also in healthy patients, *V. vulnificus* produces a primary septicemia, with 50% mortality in patients with underlying disease. The organisms are assumed to enter the bloodstream via the gastrointestinal tract, usually within 24 to 48 hours after ingestion of raw shellfish, particularly oysters. Infections caused by *V. vulnificus* have been reported primarily from coastal states.

ISOLATION OF *VIBRIO* SP. FROM FECES

Fecal specimens received in microbiology laboratories in the United States are not routinely inoculated to media selective for vibrios because of the low incidence of *Vibrio*-related gastroenteritis in the United States. If a patient with diarrheal disease has a history of travel, shellfish consumption, or other contact with fresh water or sea water, vibrios should be considered. Recent foreign travel is the most important risk factor. Stool should be transported in Cary-Blair transport media if it cannot be plated to selective agar immediately. Plating onto 5% sheep blood agar also adequately supports the growth of *Vibrio* sp. The most frequently used selective agar is thiosulfate citrate–bile salts–sucrose (TCBS) agar, which can be heavily inoculated. Eiken (Japan) produces a highly regarded product.[15a] Media produced by Oxoid Ltd. in England is also regarded highly.

Conventional selective enteric agars (Chapter 9) should be inoculated for isolation of other possible etiological agents. All vibrios grow on nonselective media, but their colonies are not distinctive enough to ensure their detection in mixed flora containing large numbers of other bacteria unless some screening procedure (e.g., oxidase) is used. In very dilute, typical "rice water" stools of cholera, vibrios are present in large numbers, often to the exclusion of much normal flora. An enrichment broth particularly useful for detection of small numbers of *V. cholerae* is alkaline peptone water (pH 8.4). Janda et al.[18] recommend that alkaline peptone water be prepared in 20 ml amounts to retain proper pH. This broth should be subcultured to selective media (TCBS) as soon as possible after 6 hours' incubation (at 35° C) to prevent overgrowth of other bacteria. Longer incubation times are acceptable with lower incubation temperatures (room temperature); however, a recent outbreak strain failed to multiply sufficiently in alkaline peptone water after 6 hours' incubation, and prolonged incubation may be necessary.[15a] Supplementing media with sodium chloride (NaCl) is not necessary, since these organisms do grow in brain-heart infusion broth and on blood agar without added salt.

LABORATORY IDENTIFICATION

After 18 to 24 hours' incubation at 35° C in air, sucrose-fermenting vibrios *(V. cholerae, V. alginolyticus, V. harveyi (carchariae), V. cincinnatiensis V. fluvialis, V. furnissii, V. metschnikovii,* some *V. vulnificus)* appear as medium-sized, smooth, opaque, thin-edged yellow colonies on TCBS agar (Figure 31.1). The other clinically important vibrios and most *V. vulnificus* do not ferment sucrose and appear olive-green (Figure 31.2).

On Gram stain, vibrios are small gram-negative bacilli that may occasionally exhibit slightly curved forms (Figure 31.3). Morphology cannot reliably be used for identification. Oxidase-posi-

FIGURE 31.1

Colonies of *Vibrio cholerae* on thiosulfate citrate–bile salts–sucrose (TCBS) agar.

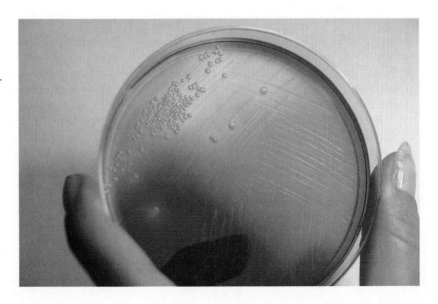

FIGURE 31.2
Colonies of *Vibrio parahaemolyticus* on TCBS agar.

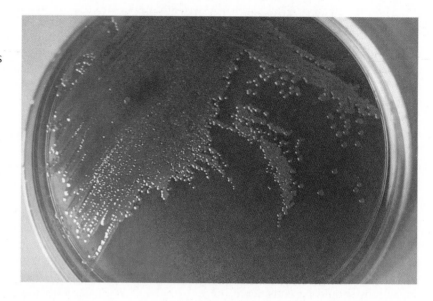

FIGURE 31.3
Gram stain of smear made from *Vibrio cholerae* colony on TCBS agar.

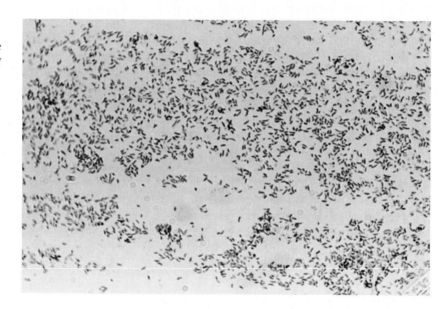

tive fermentative organisms isolated from wound cultures, blood, and other sources may be vibrios, and biochemical tests should be performed. The commercially produced biotyping systems usually identify the most common *Vibrio* sp. correctly; newly described and unusual species may not be reliably identified by such kits, and some systems may have problems separating *Aeromonas* from *Vibrio*. Except for *V. metschnikovii,* all vibrios are oxidase positive. The halophilic vibrios may require the addition of NaCl (up to a total of 3%) to all identification media and prepackaged commercial systems in order to display proper reactions. Systems that use a saline suspension of the inoculum are likely to support the growth of some clinically important vibrios without added salt.

Identification procedures are no different than those used for other gram-negative bacilli. Key differential characteristics of the more frequently isolated, clinically relevant vibrios are listed in Table 31.2.

V. cholera has been divided into serogroups based on somatic "O" antigens; thus, all epidemic strains were assigned originally to serotype "O," subgroup 1 (O1). Epidemic strains have now been identified in non-O1 serogroups. Other grouping systems identify more than 70 non-O1 serogroups.[18] Although no genetic basis exists for a division, *V. cholerae* have been divided into two biovars or biotypes: classical and eltor (formerly El Tor). The eltor biovar is Voges-Proskauer positive, agglutinates chicken red blood cells, and is resistant to a 50 U disk of polymyxin B, whereas the classical biovar yields the opposite reactions. Many recently isolated eltor biovars are non-hemolytic on sheep blood agar, in contrast to the typical beta hemolysis displayed by organisms of this biovar that were isolated several years ago.

V. cholerae of both serotypes, O1 and non-O1, cause cholera-like disease, although serotype O1 is associated more often with epidemic disease. These two serotypes are biochemically identical but differ in their ability to agglutinate in *V. cholerae* O1 antiserum. Non-O1 vibrios have been associated rarely with large outbreaks. In the United States the sporadic, domestically acquired isolate of *V. cholerae* is likely to be nonagglutinable in O1 typing serum, but cases acquired in association with travel to areas affected by the current epidemic are usually the eltor O1 serotype.[6]

Gastroenteritis-associated strains of *V. parahaemolyticus* are usually able to produce a hemolysin that hemolyzes human red blood cells in mannitol agar with a high salt content, called Kanagawa phenomenon. Environmental strains seem to lack this ability. Interestingly, strains of urease-positive *V. parahaemolyticus* are currently being isolated from patients with gastroenteritis in many countries and particularly in the United States, in contrast to the predominance of urease-negative strains recovered before 1980.[18]

Oxidase-positive organisms that may be confused with vibrios include *Aeromonas* sp., *Plesiomonas* sp., and *Pseudomonas* sp.. Key differential characteristics are outlined in Table 31.3. One important test for differentiation of vibrios from *Aeromonas* sp. (discussed in the following section), which they closely resemble, has been susceptibility to the vibriostatic agent 2,4-diamino-6,7-diisopropylpteridine phosphate (O/129). This agent, available in 10 and 150 µg disks (or as powder) from Oxoid USA inhibits the growth of many vibrios at a concentration of 150 µg/ml or greater. However, a recent report indicates that O/129 resistance is common (>90% of isolates) among *V. cholerae* in some parts of India and and Southeast Asia; these strains are increasingly likely to be recovered in the United States.[2] *Aeromonas*, Enterobacteriaceae, and *Pseudomonas* sp. are resistant. O/129-resistant organisms that might be *V. cholerae* (growth on TCBS) should be identified by complete biochemicals and sent to a reference laboratory for confirmation. Of organisms likely to be confused with vibrios, most *Plesiomonas* sp. are susceptible to O/129 at 10 µg, and all are susceptible to O/129 at 150 µg. However, *Plesiomonas*

does not grow on TCBS and fails to grow in 6% salt; all vibrios, except for strains of *V. hollisae*, grow on TCBS, and except for certain *V. mimicus* and *V. cholerae* strains, all are able to grow in media containing 6% salt.

TREATMENT OF *VIBRIO* INFECTIONS

Cholera is best treated with fluid and electrolyte replacement, although tetracycline or furazolidone shortens the course of disease and may decrease bacterial excretion. The quinolones are often the empirical agents of choice until susceptibility results are known. In areas of the world where cholera is endemic, such as Bangladesh, tetracycline resistance is common. The decision of whether to treat a patient with antimicrobial agents depends on demographic, economic, and clinical factors, including the patient's nutritional status. There have been ampicillin-, trimethoprim-, sulfamethoxazole-, and tetracycline-resistant isolates reported from the developed world as well, so a susceptibility test should be performed. The agar disk diffusion test is adequate; Mueller-Hinton agar supports the growth of halophilic and nonhalophilic vibrios; salt should not be added, since it alters the activity of certain antimicrobial agents. Most strains of *V. parahaemolyticus*, *V. alginolyticus*, *V. damsela*, *V. vulnificus*, and *V. hollisae* are susceptible to gentamicin, tetracycline, and chloramphenicol. Resistance to ampicillin, cephalothin, and carbenicillin has been reported for *V. parahaemolyticus*, *V. harveyi (carchariae)*, *V. alginolyticus*, and *V. furnissii*.[20]

AEROMONAS SPECIES

EPIDEMIOLOGY AND PATHOGENESIS OF *AEROMONAS* INFECTION

Aeromonas are ubiquitous inhabitants of natural waters, both fresh and salt, where they infect animals, including amphibians, reptiles, and fish. In humans, they are most often associated with infections of wounds acquired near or in water or with diarrheal disease. The nonmotile species, *A. salmonicida*, and the motile species, *A. hydrophila* group (*A. hydrophila*, *A. caviae*, *A. veronii* biotype sobria), *A. schubertii*, *A. veronii* biotype veronii, *A. media*, *A. jandaei*, and *A. trota*, all have been associated with human disease.[1,13,16,17,30] The fish pathogen, *A. salmonicida*, because of its preferential growth temperature of 23° C, causes rare human infections. *Aeromonas* sp. have been recovered from blood, cerebrospinal fluid, lung tissue, exudate from otitis media, urine, peritoneal fluid, necrotic muscle, infected heart valves, and

TABLE 31.2

DIFFERENTIAL CHARACTERISTICS OF *VIBRIO* SP. LIKELY TO BE RECOVERED FROM CLINICAL SPECIMENS

SPECIES	GROWTH IN NaCl (%)					VOGES-PROSKAUER	LYSINE DECARBOXYLASE
	0	3	6	8	10		
V. alginolyticus	−	+	+	+	+/−	+	+
V. cholerae	+	+	+/−	−	−	+/−	+
V. cincinnatiensis	−	+	+	−	−	+	+
*V. damsela**	−	+	+	−	−	+	+/−
V. fluvialis	−	+	+/−	−	−	−	−
V. furnissii†	−	+	+/−	−	−	−	−
V. harveyi	−	+	+	−	−	+/−	+
V. hollisae	−	+	+/−	−	−	−	−
V. metschnikovii‡	+/−	+	+	+/−	−	+/−	+/−
V. mimicus	+	+	+/−	−	−	−	+
V. parahaemolyticus§	−	+	+	+	−	−	+
V. vulnificus	−	+	+/−	−	−	−	+

Data from Janda, J.M., Powers, C., Bryant, R.G., et al. 1988. Clin. Microbiol. Rev. 1:245.
+, ≥90% positive; −, ≥90% negative; +/−, variable results; *ND,* test not done.
*Urease positive.
†Gas from glucose.
‡Oxidase negative and nitrate negative.
§Rare urease-positive strains.

bone.[15,16] *A. caviae* is less frequently isolated from blood than the other species in the *A. hydrophila* group, suggesting that it is less virulent. *A. schubertii* has been isolated from only extraintestinal infections, but all other strains cause gastroenteritis, which is the most common infection associated with this genus.[16]

Janda and Duffey[17] describe five diarrheal presentations for *Aeromonas*-related gastroenteritis: secretory (acute watery diarrhea, often with vomiting), dysenteric (accompanied by blood and mucus in the stool), chronic (lasting longer than 10 days), choleric ("rice water" stools), and traveler's. Drinking untreated water is one risk factor for acquisition of *Aeromonas*-related gastroenteritis, an increasingly recognized syndrome. Other risk factors for acquisition of gastroenteritis by adults, including current gastrointestinal or liver disease, reduced stomach acidity, and recent antibiotic administration, were reported by George[10] and George et al.[11]

It has been shown that *A. hydrophila* produces a heat-labile enterotoxin (synonymous with the beta hemolysin) and a heat-stable cytotoxic enterotoxin. *Aeromonas* also produces many extracellular enzymes, including protease, amylase, lipase, nuclease, and others.[16,17] The role of these enzymes in the pathogenesis of *Aeromonas* infec-

tions, if any, has not been defined. Adherence mediated by pili may also serve as a virulence factor.[16,17]

LABORATORY IDENTIFICATION

Aeromonas sp. grow on standard laboratory media, including blood agar and MacConkey agar. They would be identified as would any other gram-negative bacillus isolated from a source such as a sterile body fluid, tissue, or wound. On Gram stain the cells are straight-sided, medium-sized, nonpleomorphic gram-negative bacilli. Recognition of *Aeromonas* sp. as etiological agents of diarrhea, however, is more difficult, since many strains ferment lactose and sucrose and would not be distinctive enough to be recognized on MacConkey and Hektoen agars. The selective media typically used to detect *Salmonella, Shigella,* and *Campylobacter* sp. are inhibitory to many *Aeromonas.* Many *Aeromonas* are beta hemolytic on 5% sheep blood agar, allowing one to choose colonies to screen for the presence of oxidase by Kovacs' method. Nonhemolytic colonies should be screened as well. All members of the Vibrionaceae family are oxidase positive except for *V. metschnikovii,* and they can be differentiated from oxidase-positive pseudomonads by their ability to ferment glucose in triple sugar iron agar (TSIA)

ORNITHINE DECARBOXYLASE	ARGININE DIHYDROLASE	INDOLE	ACID FROM			SUSCEPTIBLE TO O/129	
			LACTOSE	SUCROSE	ARABINOSE	10µg	150 µg
+/–	–	+	–	+	–	–	+
+	–	+	+/–	+	–	+/–	+/–
–	–	–	–	+	+	–	+
–	+	–	–	–	–	+	+
–	+	–	–	+	+	–	+
–	+	–	–	+	+	–	+
–	–	+	–	+/–	–	+	+
–	–	+/–	–	–	+	–	+
–	+/–	+/–	+/–	+	–	+	+
+	–	+	+/–	–	–	+	+
+	–	+	–	–	+	–	+
+/–	–	+	+	+/–	–	+	+

TABLE 31.3

KEY CHARACTERISTICS OF OXIDASE-POSITIVE GRAM-NEGATIVE BACILLI

CHARACTERISTIC	AEROMONAS	PLESIOMONAS	VIBRIO	PSEUDOMONAS
Susceptible to O/129				
10 µg	–	+/–	+/–	–
150 µg	–	+	+/–	–
Ferment glucose	+	+	+	–
Acid from				
Insolitol	–	+	–	NA
Mannitol	+	–	+	NA
Gelatin liquefaction	+	–	+	+/–
Growth on TCBS*	–	–	+	–
Requires or stimulated by Na+	–	–	+	–

+, Most strains positive; –, most strains negative; +/–, variable results; NA, not applicable.

*Thiosulfate citrate–bile salts–sucrose agar.

and by the spot indole test (Chapter 10). *Pseudomonas* sp. are all indole negative, whereas many Vibrionaceae are indole positive. If a predominance of non-*Pseudomonas*-appearing oxidase-positive colonies are found on the blood agar plate inoculated with diarrheal stool, the possibility of *Aeromonas* should be further explored. Trypticase soy agar with 5% sheep blood and 10 or 30 µg/ml ampicillin or a 30 µg ampicillin disk on a blood agar plate is also useful for selective isolation of *Aeromonas,* except for *A. trota,* which is susceptible to ampicillin.[10,16] A combination of ampicillin blood agar and cefsulodin-irgasan-novobiocin

(CIN) agar, used to select for *Yersinia enterocolitica,* showed best recovery of strains of *Aeromonas* from feces in one recent study.[21] *A. trota* would not be recovered. Pril-xylose-ampicillin agar has been used successfully for recovery of aeromonads from clinical specimens.[17,29]

Aeromonas are distinguished from the Enterobacteriaceae, which they closely resemble both morphologically and biochemically, by the oxidase test; Enterobacteriaceae are oxidase negative. *Aeromonas* are differentiated from oxidase-positive vibrios by their resistance to the vibriostatic agent O/129 (Table 31.3). Table 31.4

TABLE 31.4

KEY DIFFERENTIAL CHARACTERISTICS OF *AEROMONAS* SP. (ISOLATED FROM HUMANS) AND *PLESIOMONAS SHIGELLOIDES*

GENUS AND SPECIES	ESCULIN HYDROLYSIS	SUSCEPTIBILITY TO AMPICILLIN (10 μg)	VOGES-PROSKAUER*	LYSINE DECARBOX-YLASE	ARGININE DIHYDRO-LASE	ORNITHINE DECARBOX-YLASE	GAS FROM GLUCOSE	MAN-NITOL	ACID FROM ARABI-NOSE	SUCROSE
A. hydrophila	+	–	+	+	+	–	+	+	+	+
A. caviae	+	–	–	–	+	–	–	+	+	+
A. veronii biotype sobria	–	–	+	+	+	–	+	+	–	+
A. veronii biotype veronii	+	–	+	+	–	+	+	+	–	+
A. schubertii†	–	–	+	+	+	–	–	–	–	–
A. jandaei	–	–	+	+	+	–	+	+	–	–
A. trota	–	+	–	+	+	–	+/–	+/–	–	–
P. shigelloides	–	–	–	+	+	+	–	–	–	–

Data from Janda, J.M. Personal communication.

+, ≥90% positive; –, ≥90% negative; +/–, variable results.

*Plus 1% NaCl.

†Indole negative.

outlines differentiation of the *Aeromonas* sp. pathogenic for humans, as well as the closely related *Plesiomonas shigelloides*. The commercial biochemical and enzymatic identification systems routinely used for characterization of Enterobacteriaceae will identify *Plesiomonas* and *Aeromonas*, at least to genus, with the addition of the oxidase test result to the profile number generated. At least one widely used system (API 20E, bioMérieux Vitek, Inc.) does not distinguish between *Aeromonas* and certain *Vibrio* sp., the O/129 test and growth on TCBS are therefore critical.

TREATMENT OF *AEROMONAS* INFECTIONS

Aeromonas-associated diarrhea is usually self-limited and normally does not require treatment. For protracted, bloody, or chronic diarrhea and for more severe infections, several antimicrobial agents are effective. The organisms are susceptible to quinolones, gentamicin, trimethoprim-sulfamethoxazole, third-generation cephalosporins, and chloramphenicol. They are resistant to streptomycin, penicillins, and most first- and second-generation cephalosporins.

PLESIOMONAS

The one species of *Plesiomonas*, *P. shigelloides*, resembles *Aeromonas hydrophila*, although it is not beta hemolytic under normal conditions. *Plesiomonas* also maintains a water habitat. Although primarily associated with gastroenteritis, the organism has been isolated as the etiological agent of meningitis, bacteremia, septic arthritis, cholecystitis, and endophthalmitis.[5] The pathogenesis of diarrhea associated with *Plesiomonas* has not been conclusively established, although epidemiological studies and other lines of evidence strongly suggest a causal relationship.[5,14] Disease is associated with eating raw shellfish and foreign travel. Virulence factors include an enterotoxin, a beta hemolysin related to iron uptake, and adherence to intestinal epithelial cells.[5,9,23]

Gram-stained colonies may reveal long filamentous forms as well as regular bacilli. Although the colonies are not distinctive from Enterobacteriaceae on MacConkey agar, they are oxidase positive and may be detected in the same manner as *Aeromonas* colonies. Both lactose-fermenting and non-lactose-fermenting strains exist, exhibiting different colony morphologies on MacConkey agar. *P. shigelloides* is identified biochemically, as outlined in Table 31.4.

As with *Aeromonas*, only patients with underlying disease or very severe infections should be treated. Patients treated with tetracycline

or trimethoprim-sulfamethoxazole have recovered from *Plesiomonas* infection. Chloramphenicol, trimethoprim-sulfamethoxazole, cephalothin, and aminoglycosides are active against *Plesiomonas* in vitro. Most isolates have been resistant to penicillin, ampicillin, and carbenicillin.

CAMPYLOBACTER AND RELATED SPECIES

EPIDEMIOLOGY AND PATHOGENESIS OF *CAMPYLOBACTER* INFECTION

Campylobacter sp. are microaerobic inhabitants of the gastrointestinal tracts of various animals, including poultry, dogs, cats, sheep, and cattle, as well as the reproductive organs of several animal species (Table 31.5). When fecal samples from chicken carcasses chosen at random from butcher shops in the New York City area were tested for *Campylobacter*, 83% of the samples yielded more than 10^6 colony-forming units per gram of feces. *Campylobacter* sp. produce two primary syndromes in humans: systemic disease characterized by fever and, most often, gastroenteritis. Extraintestinal disease, including meningitis, endocarditis, and septic arthritis, is being recognized increasingly, particularly in patients with acquired immunodeficiency syndrome (AIDS). The *Campylobacter* sp. associated with the most human infections (*jejuni* and *coli*) are usually transmitted via contaminated food, milk, or water. Outbreaks have been associated with contaminated drinking water and consumption of improperly pasteurized milk, among other sources. In contrast to other agents of foodborne gastroenteritis, including *Salmonella* and staphylococci, *Campylobacter* does not multiply in food.[7]

Other campylobacters have been isolated from patients who drank untreated water, were compromised in some way, or were returning from international travel. *C. jejuni* subsp. *doylei* has been isolated from children with diarrhea and from gastric biopsies from adults.[33] *Campylobacter* sp., usually *jejuni*, have been recognized as the most common etiological agent of gastroenteritis in the United States. Ratios of recovery of *Campylobacter* versus *Salmonella* range from 2:1 to 46:1.[7] *C. jejuni*, *lari*, *coli*, *hyointestinalis*, and *upsaliensis* and *Arcobacter butzleri* are all associated with human enteritis.[22,36-38] *C. hyointestinalis* disease probably does not occur often, but the California State Microbial Disease Laboratory reports three isolates, including one from blood.[15a] *Campylobacter* sp. seem to possess several virulence factors. The production of bloody diarrhea, characterized by the presence of numerous polymorphonuclear neutrophils in patients with gastroenteritis and the

TABLE 31.5

CAMPYLOBACTER AND *ARCOBACTER* SP. ASSOCIATED WITH HUMAN DISEASE

GENUS AND SPECIES	PREDOMINANT HOST(S)	SYNDROME IN HUMANS
C. coli	Pigs	Enteritis
C. concisus	Humans	Oral disease
C. curvus	Humans	Oral infections
C. fetus subsp. fetus	Cows, sheep	Sepsis, meningitis, abscesses, other
C. hyointestinalis	Pigs	Proctitis, diarrhea
C. jejuni	Animals, birds	Enteritis
C. jejuni subsp. doylei	Humans	Diarrhea
C. lari	Seagulls	Enteritis
C. rectus	Humans	Oral infections
C. sputorum	Humans	?Diarrhea, abscess
C. upsaliensis	Dogs, humans	Gastroenteritis, sepsis
A. cryaerophilus	Cows, pigs	Diarrhea, sepsis
A. butzleri	Animals, humans	Diarrhea, sepsis

appearance of certain strains of *Campylobacter* in the blood of patients after ingestion of contaminated material, are indicative of the organism's invasive capacity.[38] *Campylobacter* sp. *(coli, jejuni, lari)* are known to produce a cytotoxin, a cytotonic factor, and an enterotoxin, which may account for those cases characterized by watery diarrhea.[19] The actual importance of toxin production in development of disease is still being investigated.[38]

Although 97% of patients with diarrhea from *C. jejuni* cease excreting the organism within 4 to 7 weeks even without treatment, the remaining patients (now asymptomatic) continue to carry the organism for long periods (1 to several years). Carriers of *C. jejuni* are much more common in tropical countries than in the United States. Increased prevalence of infection in tropical countries results in an endemic level of mild disease, probably because of herd immunity.

C. concisus is isolated from the human gingival crevice of patients with periodontal disease, but the organism's role in this syndrome is not clear. *Arcobacter* (formerly *Campylobacter*) *butzleri* is found at times, but *A. cryaerophilus* has been isolated only rarely from patients with gastroenteritis.[4,22,35-37]

ISOLATION OF *CAMPYLOBACTER* SP.
FROM BLOOD CULTURES

C. fetus subsp. *fetus*, *C. lari*, *C. jejuni*, *Helicobacter cinaedi*, and *C. upsaliensis* have been documented

as agents of sepsis[19,27] and grow in most blood culture media, although they may require as long as 2 weeks for growth to be detected. Subcultures from broths must be incubated in a microaerobic atmosphere, or the organisms will not multiply. Turbidity is often not visible in blood culture media; therefore blind subcultures or microscopic examination using acridine orange stain is necessary. The presence of *Campylobacter* sp. in blood cultures is detected effectively by carbon dioxide (CO_2) monitoring. Isolation from sources other than blood or feces (extremely rare) is ideally accomplished by inoculating the material (macerated tissue, wound exudate) to a nonselective blood or chocolate agar plate and incubating the plate at 37° C in a CO_2-enriched, microaerobic atmosphere. Selective agars containing a cephalosporin, rifampin, and polymyxin B may inhibit growth of some strains and should not be used for isolation from normally sterile sites.

ISOLATION OF *CAMPYLOBACTER* SP.
FROM FECES

The most common agents of gastroenteritis, *C. jejuni* and *C. coli*, are able to grow well at 42° C and are resistant to cephalosporin, characteristics useful for their initial isolation. The number of colonies does not increase at this temperature, but the colonies appear sooner and are larger, and the growth of most normal fecal flora is inhibited. Feces may be transported to the laboratory directly for inoculation onto media. If a delay of longer than 2 hours is anticipated, material should be placed either in Cary-Blair transport medium or in campy thio, a thioglycolate broth base with 0.16% agar and vancomycin (10 mg/L), trimethoprim (5 mg/L), cephalothin (15 mg/L), polymyxin B (2,500 U/L), and amphotericin B (2 mg/L). The same antimicrobial agents are incorporated into *Brucella* agar base with 10% sheep blood to produce campy-BAP, one of the selective agars that is useful for cultivation of these strains. Several media manufacturers produce commercial plates for *Campylobacter* isolation. Since these *Campylobacter* are resistant to cold, refrigeration of fresh stools for as long as 24 hours before plating will probably not result in false-negative cultures. The stool should be plated onto a selective agar and incubated at 42° C in a microaerobic atmosphere. The atmosphere can be generated in several ways, including commercially produced gas-generating envelopes meant to be used in conjunction with plastic bags or plastic jars. Evacuation and replacement in plastic bags or anaerobic jars with an atmosphere of 10% CO_2, 5% O_2, and the balance N_2 is the most cost-effective method, although it is somewhat labor intensive. Plates should be held for 72 hours before being discarded as negative.

Isolation of *C. fetus* subsp. *fetus, C. lari, C. hyointestinalis, C. upsaliensis, Helicobacter fennelliae, H. cinaedi, Arcobacter butzleri,* and *A. cryaerophilus* from stool may be attempted by inoculating a selective medium without cephalosporins and incubating for up to 7 days at 37° C microaerobically. Skirrow's medium, containing blood agar base, 5% lysed horse blood, vancomycin (10 mg/L), polymyxin (2500 U/L), and trimethoprim (5 mg/L) would adequately support growth of the less common species. Media with cephalosporins may be inhibitory. Blood-free media with charcoal and reducing agents is also very useful for recovery of fastidious campylobacters.

Filtration of feces is an excellent method for isolation of *Campylobacter* sp. Either a 0.65 or a 0.8 μm pore-size cellulose acetate filter can be used. The stool is diluted if necessary, and several drops of filtrate are placed directly onto nonselective agar and incubated at 37° C in a microaerobic atmosphere. In a modification, a filter is placed onto the agar surface, and a drop of stool is placed on the filter. The plate is incubated upright. After 30 to 60 minutes at room temperature or 37° C, the filter is removed, and the plates are reincubated. *Campylobacter* and *Helicobacter* sp. are able to move through the filter to produce colonies on the agar surface.

Plates should be examined at 48 hours and after 5 days for characteristic colonies, which are gray to pinkish or yellowish gray, slightly mucoid-looking colonies; some colonies may exhibit a tailing effect along the streak line (Figure 31.4). However, other colony morphologies are also frequently seen.

LABORATORY IDENTIFICATION

Suspicious-looking colonies seen on selective media incubated at 42° C may be presumptively identified as *Campylobacter* sp., usually *jejuni* or *coli,* with a few simple tests. A wet preparation of the organism may be examined under phase or darkfield microscopy for characteristic darting motility and curved forms. On Gram stain the organisms are small, curved or seagull-winged, faintly staining gram-negative rods (Figure 31.5). Almost all the pathogenic *Campylobacter* sp. are oxidase and catalase positive. For most laboratories, reporting of such isolates from feces as *"Campylobacter* sp." should suffice.

Most *Campylobacter* sp. are asaccharolytic, unable to grow in 3.5% NaCl, although strains of *Arcobacter* appear more resistant to salt,[22,36] and except for *Arcobacter cryaerophilus,* are unable to grow in air. Growth in 1% glycine is variable. Susceptibility to nalidixic acid and cephalothin, an important differential characteristic among species (Table 31.6), is determined by inoculating a 5% sheep blood agar plate or a Mueller-Hinton agar plate with a McFarland 0.5 turbidity suspension of the organism as for agar disk diffusion susceptibility testing, placing 30 μg disks on the agar surface, and incubating microaerobically at 37° C. Other tests useful for identifying these species are the rapid hippurate hydrolysis test (Chapter 10,

FIGURE 31.4

Colonies of *Campylobacter jejuni* after 48 hours' incubation on Campy blood agar in a microaerobic atmosphere. Note that some colonies appear elongated as they extend along the streak line.

FIGURE 31.5

Gram stain appearance of *Campylobacter jejuni* smeared from primary isolation plate. Note seagull-shaped and curved forms.

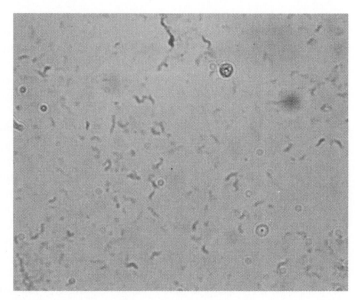

Procedure 10.9), production of hydrogen sulfide (H_2S) in TSIA butts, nitrate reduction,[33] and hydrolysis of indoxyl acetate (Procedure 31.1).[24,28] Indoxyl acetate disks are available commercially. The indoxyl acetate test is positive for *C. jejuni, C. coli, C. jejuni* subsp. *doylei, C. upsaliensis, H. fennelliae, H. cryaerophilus,* all *Arcobacter* sp., *C. curvus,* and *C. rectus.* The species *C. cinaedi, C. concisus, C. fetus* subsp. *fetus, C. hyointestinalis, C. lari, C. mucosalis, C. sputorum, Wolinella succinogenes,* and *Helicobacter pylori* are negative.[28,36] Identification of an isolate as *C. jejuni, C. coli,* or *H. pylori* is also accomplished by detecting C-19 cyclopropane fatty acid methyl ester as a component of the cell wall using gas-liquid chromatographic methods. Cellular fatty acid analysis can also help differentiate among other species. This method is not available to most clinical microbiology laboratories. Several commercial products are available for species identification, including particle agglutination methods and nucleic acid probes.

TREATMENT OF *CAMPYLOBACTER* INFECTIONS
Erythromycin is the drug of choice for those patients with complicated gastroenteritis (hyperpyrexia, dehydration, bacteremia). For patients with bacteremia, therapy is continued for 4 weeks. Gastroenteritis is usually treated for 10 days. Tetracycline is an alternative choice in the United States, although 15% to 20% of strains are resistant; children under 7 years should be given clindamycin. The gastroenteritis-associated *Campylobacter* sp. are susceptible to gentamicin, other aminoglycosides, chloramphenicol, cefotaxime, and ciprofloxacin.

HELICOBACTER SPECIES

EPIDEMIOLOGY AND PATHOGENESIS OF *HELICOBACTER* INFECTION
Among homosexual populations, *H. cinaedi* and *H. fennelliae* are transmitted sexually, as are other agents of gastroenteritis.[27] The organisms cause proctitis, gastroenteritis, and sepsis.

Recent studies have amassed evidence for the role of *H. pylori* as an etiological agent of gastritis and gastric ulcer.[3,25,32] The organism is acquired early in life in underdeveloped countries, but the mode of transmission is unknown. In industrialized nations, antibody surveys show that approximately 50% of adults older than 60 years are infected. Gastritis incidence increases with age. *H. pylori* colonizes the mucous layer of the antrum and fundus but does not invade the epithelium. Virulence factors include a vacuole-promoting cytotoxin and the potent urease enzyme, which creates an alkaline microenvironment that may allow continued growth despite stomach acidity. The host mounts an inflammatory response, which accounts for much of the tissue damage (ulcer formation).[3]

DETECTION OF *HELICOBACTER* IN GASTRIC TISSUES
Tissue biopsy material should be transported to the laboratory in a sterile container. It may be refrigerated for up to 5 hours before processing. Smears and cultures should be examined for presence of the organisms. The Warthin-Starry or other silver stain and Giemsa stains are used by pathologists for examination of biopsy specimens. One potential problem is that of sampling error.

TABLE 31.6

DIFFERENTIAL CHARACTERISTICS OF CLINICALLY RELEVANT CAMPYLOBACTER, HELICOBACTER, AND ARCOBACTER SP.

GENUS AND SPECIES	GROWTH AT 25°C	GROWTH AT 42°C	HIPPURATE HYDROLYSIS	CATALASE	H₂S IN TRIPLE SUGAR IRON AGAR	INDOXYL ACETATE HYDROLYSIS	NITRATE TO NITRITE	SUSCEPTIBLE TO 30 µg DISK CEPHALOTHIN	SUSCEPTIBLE TO 30 µg DISK NALIDIXIC ACID
C. coli	–	+	–	+	–	+	+	–	+
C. concisus	–	+	–	–	+	–	+	ND	+
C. curvus*	–	+	–	+	+	+	+	+	–
C. fetus subsp. fetus	+	–/+	–	+	–	–	+	+	–
C. hyointestinalis	+/–	+	–	+	+	–	+	+	+
C. jejuni	–	+	+	+	–	+	+	+	+
C. jejuni subsp. doylei	–	+/–	+	+/– or weak +	–	+	–		
C. lari	–	+	–	+	–	–	+	ND	–
C. rectus*	–	Slight +	–	–/+	+	+	+	+	–/+
C. sputorum	–	+	–	–/weak +	+	–	+	+/–	+
C. upsaliensis	–	+	–	+	–	+	+	+	+
H. cinaedi	–	–/+	–	+	–	–/+	+	+	–
H. fennelliae	–	–	–	+	–	+	–	–	
H. pylori†	–	+	–	–/weak +	–	–	+/–	+	
A. butzleri‡,§	+	–	–	+	–	+	+	–	+/–
A. cryaerophilus§	+	–	–	+/–	–	+	+	–	+/–

+, Most strains positive; –, most strains negative; +/–, variable (more often positive); –/+, variable (more often negative); ND, test not done.
*Anaerobic, not microaerobic.
†Strong and rapid positive urease.
‡Grows at 40°C.
§Aerotolerant, not microaerobic; except for a few strains, A. cryaerophilus cannot grow on MacConkey agar, whereas A. butzleri grows on MacConkey agar.

PROCEDURE 31.1

HYDROLYSIS OF INDOXYL ACETATE

PRINCIPLE
Certain *Campylobacter* and related species produce a bacterial esterase that hydrolyzes indoxyl acetate to produce a blue end product. Commercially prepared disks are also available.

METHOD
1. Prepare indoxyl acetate disks as follows:
   a. Dissolve 1 g indoxyl acetate (Sigma Chemical) in 10 ml acetone.
   b. Place filter paper disks (0.25-inch [0.64 cm] diameter, available from Difco and BBL) on a dry surface such as wax paper (not plastic), and

add 0.5 ml of the indoxyl acetate solution to each disk.
   c. Allow the disks to air dry, and store desiccated in an amber bottle at 4° C. Disks are stable for at least 1 yr.
2. To test an isolate, place several colonies on the surface of an indoxyl acetate disk and moisten the disk with a drop of sterile water. Observe for up to 10 min for a blue color, indicating hydrolysis of the substrate.

QUALITY CONTROL
Test a fresh subculture of *Campylobacter jejuni* (ATCC

29428) on one disk, and simply add water to a second disk.

EXPECTED RESULTS
The *C. jejuni,* as a positive control, should yield a blue color reaction within 10 min. The uninoculated disk should remain the original color, as a negative control.

PERFORMANCE SCHEDULE
Perform a quality control check when disks are first prepared and monthly thereafter.

Modified from Mills, G.K., and Gherna, R.L. 1987. J. Clin. Microbiol. 25:1560.

Squash preparations of biopsy material can be Gram stained with good results; the 0.1% basic fuchsin counterstain enhances recognition of the bacteria's typical morphology. Parsonnet et al.[26] report good recovery in culture using the following technique. Tissue is rinsed in dextrose-phosphate broth, blotted onto a sterile gauze or filter paper to remove excess broth, and pressed onto the surface of a sterile glass slide. After the smear has been prepared, the tissue is ground in dextrose phosphate broth and inoculated to media. Nonselective agar media, including chocolate agar and *Brucella* agar with 5% sheep blood, have been useful for isolation of *H. pylori* from gastric antral biopsy specimens. Selective agar, such as Skirrow's and modified Thayer-Martin agars, also support growth.[3,8] Several agars must be evaluated before a laboratory determines what is best in the particular circumstances.[8]

Incubation up to 1 week in a humidified, microaerobic atmosphere at 35° to 37° C may be required before growth is visible as small, translucent, circular colonies. *H. pylori* is identified presumptively by typical cellular morphology and by positive results for oxidase, catalase, and rapid urease tests.

Presumptive evidence of the presence of the organism in biopsy material may be obtained by placing a portion of crushed tissue biopsy material directly into urease broth or onto commercially available urease agar kits. A positive test is considered indicative of the organism's presence. Problems of low sensitivity and the necessity for a 24-hour incubation have precluded universal acceptance of this test.[8,25] Other tests for detection of *H. pylori* infection include the urea breath test and serologic tests for antibodies.[12,32,34] The noninvasive tests will probably be used more widely in the future.

LABORATORY IDENTIFICATION

Colonies of *H. pylori* require several days of incubation in humidified microaerobic conditions. They are small, translucent, and oxidase and catalase positive. The presence of urease detected with a rapid test (Chapter 10) is virtually pathognomic.

TREATMENT OF *HELICOBACTER* INFECTIONS

Therapy is still problematic; current regimens usually combine a bismuth salt, metronidazole or other nitroimidazole, and either amoxicillin or tetracycline. If the infecting strain is resistant to the nitroimidazole, therapy is likely to be unsuccessful in the long term. An alternative therapy for patients with metronidazole-resistant strains is omeprazole and amoxicillin.[3] Relapses occur often. The *Helicobacter* sp. associated with enteritis and proctitis may respond to quinolones, but ideal therapy has not been established.

REFERENCES

1. Abbott, S.L., Cheung, W.K.W., Kroske-Bystrom, S., et al. 1992. Identification of *Aeromonas* strains to the genospecies level in the clinical laboratory. J. Clin. Microbiol. 30:1262.

2. Abbott, S.L., Cheung, W.K., Portoni, B.A., and Janda, J.M. 1992. Isolation of vibriostatic agent O/129–resistant *Vibrio cholerae* non-O1 from a patient with gastroenteritis. J. Clin. Microbiol. 30:1598.

3. Blaser, M.J. 1992. *Helicobacter pylori*: its role in disease. Clin. Infect. Dis. 15:386.

4. Boudreau, M., Higgins, R., and Mittal, K.R. 1991. Biochemical and serological characterization of *Campylobacter cryaerophila*. J. Clin. Microbiol. 29:54.

5. Brenden, R.A., Miller, M.A., and Janda, J.M. 1988. Clinical disease spectrum and pathogenic factors associated with *Plesiomonas shigelloides* infections in humans. Rev. Infect. Dis. 10:303.

6. Centers for Disease Control. 1992. Cholera associated with international travel, 1992. M.M.W.R. 41:664.

7. Centers for Disease Control. 1988. *Campylobacter* isolates in the United States, 1982-1986. M.M.W.R. 37(SS-2):1.

8. Coudron, P.E., and Kirby, D.F. 1989. Comparison of rapid urease tests, staining techniques, and growth on different solid media for detection of *Campylobacter pylori*. J. Clin. Microbiol. 27:1527.

9. Daskaleros, P.A., Stoebner, J.A., and Payne, S.M. 1991. Iron uptake in *Plesiomonas shigelloides*: cloning of the genes for the heme-iron uptake system. Infect. Immun. 59:2706.

10. George, W.L. 1987. *Aeromonas*-associated diarrhea. Clin. Microbiol. Newsletter 9:121.

11. George, W.L., Nakata, M.M., Thompson, J., et al. 1985. *Aeromonas*-related diarrhea in adults. Arch. Intern. Med. 145:2207.

12. Graham, D.Y., Klein, P.D., Evans, D.J., et al. 1987. *Campylobacter pylori* detected noninvasively by the ^{13}C-urea breath test. Lancet 1:1174.

13. Holmberg, S.D., Schell, W.L., Fanning, G.R., et al. 1986. *Aeromonas* intestinal infections in the United States. Ann. Intern. Med. 105:683.

14. Holmberg, S.D., Wachsmuth, I.K., Hickman-Brenner, F.W., et al. 1986. *Plesiomonas* enteric infections in the United States. Ann. Intern. Med. 105:690.

15. Isaacs, R.D., Paviour, S.D., Bunker, D.E., et al. 1988. Wound infection with aerogenic *Aeromonas* strains: a review of twenty-seven cases. Eur. J. Clin. Microbiol. Infect. Dis. 7:355.

15a. Janda, J.M. Personal communication.

16. Janda, J.M. 1991. Recent advances in the study of the taxonomy, pathogenicity, and infectious syndromes associated with the genus *Aeromonas*. Clin. Microbiol. Rev. 4:397.

17. Janda, J.M., and Duffey, P.S. 1988. Mesophilic aeromonads in human disease: current taxonomy, laboratory identification, and infectious disease spectrum. Rev. Infect. Dis. 10:980.

18. Janda, J.M., Powers, C., Bryant, R.G., et al. 1988. Current perspectives on the epidemiology and pathogenesis of clinically significant *Vibrio* spp. Clin. Microbiol. Rev. 1:245.

19. Johnson, W.M., and Lior, H. 1986. Cytotoxic and cytotonic factors produced by *Campylobacter jejuni, Campylobacter coli,* and *Campylobacter laridis.* J. Clin. Microbiol. 24:275.

20. Kelly, M.T., Hickman-Brenner, F.W., and Farmer, J.J.III. 1991. *Vibrio.* p. 384. In Balows, A., Hausler, W.J., Jr., Herrmann, K.L., et al., editors., Manual of clinical microbiology, ed. 5, American Society for Microbiology, Washington, D.C.

21. Kelly, M.T., Stroh, E.M.D., and Jessop, J. 1988. Comparison of blood agar, ampicillin blood agar, MacConkey-ampicillin-tween agar, and modified cefsulodin-irgasan-novobiocin agar for isolation of *Aeromonas* spp. from stool specimens. J. Clin. Microbiol. 26:1738.

22. Kiehlbauch, J.A., Brenner, D.J., Nicholson, M.A., et al. 1991. *Campylobacter butzleri* sp. nov. isolated from humans and animals with diarrheal illness. J. Clin. Microbiol. 29:376.

23. Matthews, B.G., Douglas, H., and Guiney, D.G. 1988. Production of a heat stable enterotoxin by *Plesiomonas shigelloides.* Microb. Pathog. 5:207.

24. Mills, C.K., and Gherna, R.L. 1987. Hydrolysis of indoxyl acetate by *Campylobacter* species. J. Clin. Microbiol. 25:1560.

25. Nedenskov-Sorensen, P., Aase, S., Bjorneklett, A., et al. 1991. Sampling efficiency in the diagnosis of *Helicobacter pylori* infection and chronic active gastritis. J. Clin. Microbiol. 29:672.

26. Parsonnet, J., Welch, K., Compton, C., et al. 1988. Simple microbiologic detection of *Campylobacter pylori.* J. Clin. Microbiol. 26:948.

27. Penner, J.L. 1988. The genus *Campylobacter*: a decade of progress. Clin. Microbiol. Rev. 1:157.

28. Popovic-Uroic, T., Patton, C.M., Nicholson, M.A., and Kiehlbauch, J.A. 1990. Evaluation of the indoxyl acetate hydrolysis test for rapid differentiation of *Campylobacter, Helicobacter,* and *Wolinella* species. J. Clin. Microbiol. 28:2335.

29. Rogol, M., Sechter, I., Grinberg, L., et al. 1979. Pril-xylose-ampicillin agar, a new selective medium for the isolation of *Aeromonas hydrophila.* J. Med. Microbiol. 12:229.

30. Semel, J.D., and Trenholme, G. 1990. *Aeromonas hydrophila* water-associated traumatic wound infections: a review. J. Trauma 30:324.

31. Simpson L.M., White, V.K., Zane, S.F. and Oliver, J.D. 1987. Correlation between virulence and colony morphology in *Vibrio vulnificus.* Infect. Immun. 55:269.

32. Sobala, G.M., Crabtree, J.E., Pentith, J.A., et al. 1991. Screening dyspepsia by serology to *Helicobacter pylori.* Lancet 338:94.

33. Steele, T.W., and Owen, R.J. 1988. *Campylobacter jejuni* subsp. *doylei* subsp. nov., a subspecies of nitrate-negative campylobacters isolated from human clinical specimens. Int. J. Syst. Bacteriol. 38:316.

34. Talley, N.J., Newell, D.G., Ormond, J.E., et al. 1991. Serodiagnosis of *Helicobacter pylori*: comparison of enzyme-linked immunosorbent assays. J. Clin. Microbiol. 29:1635.

35. Vandamme, P., Falsen, E., Rossau, R., et al. 1991. Revision of *Campylobacter, Helicobacter,* and *Wolinella* taxonomy: emendation of generic descriptions and

proposal of *Arcobacter* gen. nov. Int. J. Syst. Bacteriol. 41:88.

36. Vandamme, P., Vancanneyt, M., Pot, B., et al. 1992. Polyphasic taxonomic study of the emended genus *Arcobacter* with *Arcobacter butzleri* comb. nov. and *Arcobacter skirrowii* sp. nov., an aerotolerant bacterium isolated from veterinary specimens. Int. J. Syst. Bacteriol. 42:344.

37. Vandamme, P., Pugina, P., Benzi, G., et al. 1992. Outbreak of recurrent abdominal cramps associated with *Arcobacter butzleri* in an Italian school. J. Clin. Microbiol. 30:2335.

38. Walker, R.I., Caldwell, M.B., Lee, E.C., et al. 1986. Pathophysiology of *Campylobacter* enteritis. Microbiol. Rev. 50:81.

BIBLIOGRAPHY

Morgan, D.R., editor. 1991. Symposium on *Helicobacter pylori*: a cause of gastroduodenal disease. Rev. Infect. Dis. 13(suppl. 8):S655.

32

SPIROCHETES AND OTHER SPIRAL-SHAPED ORGANISMS

Diseases caused by microbes in the Spirochaetaceae family (genera *Treponema, Borrelia, Leptospira*) have been known to exist since prehistoric times. These include pinta, caused by *Treponema carateum,* as well as organisms that have been newly recognized as recently as 1975, such as Lyme disease resulting from an infection with *Borrelia burgdorferi.* Most spirochetes multiply within a living host, having no natural reservoir in the inanimate environment. *Treponema* pathogenic for humans are transmitted from person to person through direct contact, either sexual or otherwise. *Borrelia* sp. pass through an arthropod vector, and *Leptospira* sp. are contracted accidentally by humans who come into contact with water contaminated by animal urine or who are bitten by an infected animal. Although the spirochete *Treponema hyodysenteriae* has been unequivocally established as an agent of swine dysentery, the role of intestinal spirochetes in human gastroenteritis is still largely unknown. This chapter briefly mentions intestinal spirochetosis.

MORPHOLOGY

The spirochetes are all long, slender, helically curved, gram-negative bacilli, with the unusual morphological features of axial fibrils and an outer coating, the sheath. These fibrils, or axial fil-aments, are flagella-like organelles that wrap around the bacteria's cell wall, are enclosed within the outer sheath, and facilitate motility of the organisms. The fibrils are attached within the cell wall by platelike structures located near the ends of the cells, called *insertion disks.* The protoplasmic cylinder gyrates around the fibrils, causing bacterial movement to appear as a corkscrew-like winding. Differentiation of genera within the Spirochaetaceae is based on the number of axial fibrils, the number of insertion disks present (Table 32.1), and biochemical and metabolic features. In general, the spirochetes also fall into genera based loosely on their morphology (Figure 32.1). *Treponema* are slender with tight coils; *Borrelia* are somewhat thicker with fewer and looser coils; and the *Leptospira* resemble the *Borrelia* except for their hooked ends.

TAXONOMIC CONSIDERATIONS

The genera that contain organisms pathogenic for humans are *Treponema,* including the causative agents of syphilis, pinta, yaws, and bejel; *Borrelia,* including vectorborne agents of relapsing fever and Lyme disease; *Leptospira,* which includes the agents of leptospirosis; and the one human intestinal spirochete *Brachyspira,* the type species of which, *B. aalborgii,* may cause intestinal disease.

TABLE 32.1

SPIROCHETES PATHOGENIC FOR HUMANS

GENUS	AXIAL FIBRILS	INSERTION DISKS
Treponema	6-10	1
Leptospira	2	3-5
Borrelia	30-40	2
*Brachyspira**	4	2

*Proposed name.

Leptospira sp., in contrast to the other spirochetes, contain diaminopimelic acid in their cell walls instead of ornithine. *Leptospira* organisms can be cultivated more easily in vitro than the other genera, but the *Borrelia* sp. associated with relapsing fever stain more readily and are thus more easily visualized microscopically. *Borrelia*, except for *B. burgdorferi,* are named primarily for the vector in which they are transmitted.

TREPONEMA SPECIES

Virulent strains of these spirochetes, the agents of syphilis and several other diseases, had not been cultivated in the laboratory except by animal passage in rabbit testicles or mouse footpads until 1981. At that time, Fieldsteel et al. and Norris[5,13] using a very complex tissue culture system of fibroblast cell monolayers, reported in vitro growth of *Treponema pallidum* biotype *pallidum,* the causative agent of syphilis. The organisms were microaerobic, in contrast to the anaerobic non–syphilis-associated treponemes that had been previously cultivated. To cause disease, the spirochetes must attach to host cell membranes, penetrate, and multiply. Inflammatory responses by the host mediate pathology. Recent studies in-

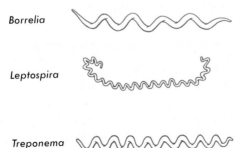

FIGURE 32.1
Species designation of spirochetes based on morphology.

dicate that these organisms are able to escape host defenses by being antigenically relatively inert.

LABORATORY DIAGNOSIS OF SYPHILIS: SEROLOGY AND DARKFIELD EXAMINATION

Cultivation of this organism is still not possible for clinical laboratories, so the diagnosis of syphilis primarily must rest on serologic methods and clinical acumen. Serologic diagnosis of syphilis is discussed in Chapters 13 and 20.

The microbiology laboratory, however, can examine the primary lesion of syphilis, the chancre, or secondary lesions for the presence of motile spirochetes under darkfield illumination. Although the darkfield examination depends greatly on technical expertise and the numbers of organisms in the lesion, it can be highly specific when performed on genital lesions. This test is best performed by laboratories servicing sexually transmitted disease clinics or affected patient populations to allow laboratory personnel to maintain acceptable diagnostic expertise. Since the normal oral cavity contains spirochetes that are usually nonpathogenic but can mimic *T. pallidum* in appearance, one should not perform darkfield microscopy on material from lesions within the oral cavity. Samples from the oral cavity can serve as a positive control, when needed.

Chapter 20 describes the method for specimen collection. Treponemes, if present in material from the lesion, are long (5 to 20 μm), slender, tightly coiled spirochetes, moving very slowly (rarely spiraling) and perhaps bending in the middle (Figure 32.2). It is best to examine the darkfield preparation under the darkfield oil immersion lens (1000×) to substantiate the morphology. Since positive lesions may be teeming with viable spirochetes that are highly infectious, all supplies and patient specimens must be handled with extreme caution and carefully discarded as required for contaminated materials. Although a fluorescent antibody reagent for *T. pallidum* is available, its use on exudate from lesions has not yielded consistently reliable results.

OTHER TREPONEMES

The other species of treponemes associated with human disease include *T. pallidum* biotype *pertenue,* the agent of yaws; *T. pallidum* biotype *endemicus,* the agent of endemic syphilis (bejel); and *T. carateum,* the agent of pinta. These diseases are primarily found in tropical or subtropical regions and are usually spread by human contact. Poor personal hygiene plays a role in their transmission. The serologic tests and the microscopic mor-

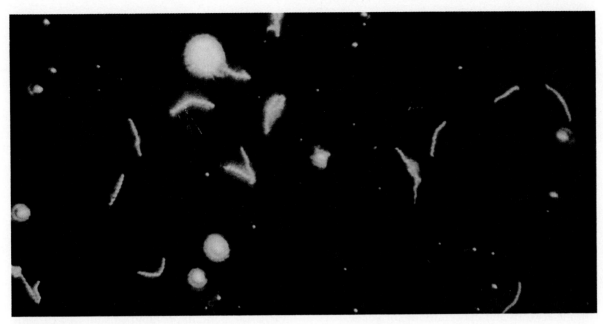

FIGURE 32.2

Appearance of *Treponema pallidum* in darkfield preparation.

phologies of all these organisms, including *T. pallidum* biotype *pallidum*, are indistinguishable from each other; only manifestations of the disease vary.

Another group of treponemes are normal inhabitants of the oral cavity or the human genital tract; these include *T. vincentii*, *T. denticola*, *T. refringens*, *T. macrodentium*, and *T. oralis*. These organisms are cultivable anaerobically on artificial media. Patients with acute necrotizing ulcerative gingivitis, also known as Vincent's gingivitis, a destructive lesion of the gums, show combinations of morphological types of bacteria on methylene blue–stained material from the lesions, which include spirochetes and fusiforms. It appears that oral spirochetes, particularly an unusually large one, may be important in this disease, along with other anaerobes. Chapter 23 further discusses oral infections. Penicillin G, or tetracycline in penicillin-allergic patients, is effective against all treponematoses.

BORRELIA SPECIES

RELAPSING FEVER

Human relapsing fever is caused by more than 15 species of *Borrelia* and is transmitted to humans by the bite of a louse or a tick. Those carried by lice are called *B. recurrentis*. All others that cause disease in the United States are transmitted via tick bites and are named after the species of tick, usually of the genus *Ornithodoros*, from which they

are recovered. Common species in the United States include *B. hermsii*, *B. turicatae*, *B. parkeri*, and *B. bergmanni*.

Although the organisms can be cultured in nutritionally rich media under microaerobic conditions, the procedures are cumbersome and unreliable and are used primarily as research tools. Clinical laboratories rely on direct observation of the organism in peripheral blood from patients for diagnosis. Organisms can be found in 70% of cases when blood specimens from febrile patients are examined. The organisms can be seen either directly in wet preparations of peripheral blood (mixed with equal parts of sterile nonbacteriostatic saline) under darkfield illumination, in which the spirochetes move rapidly, often pushing the red blood cells around, or by staining thick and thin films with Wright's or Giemsa stains, by procedures similar to those used to detect malaria.

Serologic tests for *Borrelia*-caused relapsing fever have not proved reliable for diagnosis because of the many antigenic shifts undergone by the *Borrelia* organisms during the course of disease. Patients may exhibit increased titers to *Proteus* OX K antigens (up to 1:80), but other crossreacting antibodies are rare. Certain reference laboratories, such as those at the Centers for Disease Control (CDC) in Atlanta, may perform special serologic procedures on sera from selected patients. Several antibiotics, including tetracycline, are effective therapeutically.

LYME DISEASE

Lyme disease, named after its discovery as the cause of an epidemic of "pseudojuvenile rheumatoid arthritis" among young boys living in Lyme, Conn., is now the most prevalent tickborne disease in the United States. It is characterized by three stages, not all of which occur in any given patient. The first stage, erythema migrans, is the characteristic red, annular skin lesion with central clearing that first appears at the site of the tick bite but may develop at distant sites as well (Figure 32.3). Patients may experience headache, fever, muscle and joint pain, and malaise during this stage. The second stage, beginning weeks to months after infection, may include arthritis, but the most important features are neurological disorders (meningitis, neurological deficits) and carditis. The third stage is usually characterized by chronic arthritis and may continue for years.

The agent of Lyme disease is *Borrelia burgdorferi*. The spirochete has been recovered from the blood, cerebrospinal fluid, and skin lesions (erythema migrans) of patients, and rarely from joint fluid.[15] The bacteria are transmitted by the tick bite. Several genera of ticks act as vectors in the United States, including *Ixodes pacificus* in California and in other areas *I. dammini*. The ticks' natural hosts are deer and rodents, although they will attach to pets as well as humans; all stages of tick can harbor the spirochete and transmit disease. The nymphal form of the tick is most likely to transmit disease because it is active in the spring and summer when people are dressed lightly and present in the woods habitat. This stage of the tick is very tiny (pinhead size); thus the initial tick bite

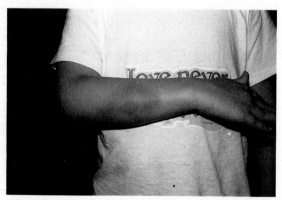

FIGURE 32.3

Appearance of the classic erythema migrans lesion of acute Lyme disease. This particular lesion has a central "target" of erythema not seen in all cases. (Courtesy Richard Dykoski, P.A., Laboratory Service, VA Medical Center, Minneapolis, Minn.)

may be overlooked. Ticks require a lengthy period of attachment, a minimum of 24 hours, before they transmit disease. The use of insect repellent and careful inspection for ticks daily during outdoor activity in high-risk areas are the best preventive strategies. Endemic areas of disease have been identified in 14 states, including Massachusetts, Connecticut, Maryland, Wisconsin, Minnesota, Oregon, and California, as well as in Europe, Russia, Japan, and Australia.

Culture of the spirochete may be attempted, although the yield is low. The periphery of the annular lesion of erythema migrans, blood, and cerebrospinal fluid provide the best specimens for culture. The resuspended plasma from blood, spinal fluid sediment, or macerated tissue biopsy is inoculated into a tube of modified Kelly's medium (BSK II) and incubated tightly capped at 33° C for 1 month.[10] Blind subcultures (0.1 ml) are performed weekly from the lower portion of the broth to fresh media, and the cultures are examined by darkfield microscopy for the presence of spirochetes.

The organism may also be visualized in tissue sections stained with Warthin-Starry silver stain and in blood and cerebrospinal fluid stained with acridine orange, specific fluorescent stain, or Giemsa.[3] Serologic diagnosis is the best routine test available for clinical laboratory performance today. Although several commercial products are available, the reproducibility of these tests is extremely poor, and the interlaboratory variation of test results very high.[6] Classically, patients show a rise in titer of immunoglobulin M (IgM) against the spirochete within 2 to 3 weeks of onset of symptoms and a concomitant rise in immunoglobulin G (IgG) very early in the disease. Both these antibodies can last for several years. Because of antigenic similarity, however, antibodies produced by patients with Lyme disease often cross-react with other spirochetal antigens, leading to false-positive and false-negative tests.[12] Patients with syphilis may harbor antibodies that cross-react with antigens of *B. burgdorferi*. The more than 15% of patients with Lyme disease who fail to develop antibody are also problematic.[4,7] Western blot tests are commercially available (MarDx, Carlsbad, Calif.) and tests for antigen in urine are being developed. Any laboratory test result must be cautiously interpreted along with appropriate patient clinical and exposure information.

Phenoxymethyl penicillin or erythromycin is a first choice drug for therapy during the first stage of the illness. Broad-spectrum cephalosporins, particularly ceftriaxone or cefotaxime, and chloramphenicol have been successful with patients who either fail initial treatment or present in later stages of the disease.[3] Symptomatic treatment failures, particularly in patients with chronic disease, have been reported.

LEPTOSPIROSIS

The leptospires include both free-living and parasitic forms. Pathogenic species are called *Leptospira interrogans,* and most saprophytic leptospires are called *L. biflexa.* The pathogens include more than 180 different serologically defined types, which formerly were designated as species and are now known as serovars, or serotypes, of *L. interrogans.* Physiologically, the saprophytes can be differentiated from pathogens by their ability to grow at 10° C and lower, or at least 5° C lower than the growth temperature of pathogenic leptospires.[9] The organisms are too small to be seen in wet preparations made from fresh blood, and they do not cause motion of erythrocytes as do *Borrelia.* Currently, the most reliable method for laboratory diagnosis of leptospirosis is to culture the organisms from blood or cerebrospinal fluid during the first week of illness or from urine thereafter for several months. A few drops of heparinized or sodium oxalate–anticoagulated blood are inoculated into tubes of Fletcher's or Ellinghausen, McCullough, Johnson, and Harris (EMJH) media. Urine should be inoculated soon after collection, since acidity (diluted out in the broth medium) may harm the spirochetes. One or two drops of undiluted urine and a 1:10 dilution of urine are added to 5 ml medium. The addition of 200 µg/ml of 5-fluorouracil (an anticancer drug) may prevent contamination by other bacteria without harming the leptospires. Tissue specimens, especially from the liver and kidney, may be aseptically triturated (macerated) and inoculated in dilutions of 1:1, 1:10, and 1:100 as for urine cultures.

All cultures are incubated at room temperature or 30° C in the dark for up to 6 weeks. The organisms grow below the surface. Material collected from a few centimeters below the surface of broth cultures should be examined weekly for the presence of growth using a direct wet preparation under darkfield illumination. Leptospires exhibit corkscrew-like motility. A more complete discussion of these methods can be found in the chapter by Alexander.[1] Another species of *Leptospira,* "*L. inadai*" serovar *lyme,* has been recovered from a skin biopsy of a patient with Lyme disease.[14] The

spirochete displayed typical morphology, failed to agglutinate in antisera against any known strains, and showed low deoxyribonucleic acid (DNA) homology with other species and serovars of leptospires. Based on growth at low temperature and low antibody titers in the patient from whom it was recovered, the spirochete was thought to be nonpathogenic.

Serologic diagnosis of leptospirosis is best performed using pools of bacterial antigens containing many serotypes in each pool. Positive results are visualized by examining for the presence of agglutination under darkfield examination. A macroscopic agglutination procedure is more readily accessible to routine clinical laboratories. Reagents are available commercially. Indirect hemagglutination and an enzyme-linked immunosorbent assay (ELISA) test for IgM antibody are also available.

Penicillin or tetracycline may modify the course of the illness, but only if started not later than the fourth day of illness.

MISCELLANEOUS SPIRAL-SHAPED ORGANISMS

The normal gastrointestinal tract of humans is colonized with spirochetes, most of which cause no harm to the host. Since 1981, however, several groups have reported patients with diarrhea, inflammatory cellular responses, and numerous spirochetes found to be invading the intestinal epithelium (into the lamina propria), as well as within macrophages. Many of these patients also had other illnesses, such as Crohn's disease. Hovind-Hougen et al.[8] were able to culture the spirochetes from five patients in vitro and found that they differed from other known genera morphologically.[2] Since growth characteristics and morphology differed from previously named spirochetes, they designated this organism *Brachyspira aalborgii.* Jones et al.[11] found large spirochetes in diarrheal stool from homosexual males. Some of the isolates resembled *B. aalborgii* and others showed different colony morphologies and biochemical characteristics. The role, if any, of these organisms in the etiology of the patients' diarrhea is not known.

Whether intestinal spirochetosis is a real entity and a more common syndrome than has been appreciated until now remains to be elucidated. *Spirillum minus,* the other agent of rat-bite fever (in addition to *Streptobacillus moniliformis,* Chapter 40) is a short, helical organism with only a few spirals that resembles *Campylobacter* more than spirochetes. It does not have an accepted name at this time, and thus its taxonomic position is unclear. Seen almost exclusively in Asia, this organism has not been cultivated in vitro. It is mentioned again in Chapter 40.

REFERENCES

1. Alexander, A.D. 1991. *Leptospira*. In Balows, A., Hausler, W.J., Jr., Herrmann, K.L, et al., editors. Manual of clinical microbiology, ed 5. American Society for Microbiology. Washington, D.C.
2. Antonakopoulos, G., Newman, J., and Wilkinson, M. 1982. Intestinal spirochetosis; an electron microscopic study of an unusual case. Histopathology 6:477.
3. Barbour, A.G. 1988. Laboratory aspects of Lyme borreliosis. Clin. Microbiol. Rev. 1:399.
4. Dattwyler, R.J., Volkman, D.J., Luft, B.J., et al. 1988. Seronegative Lyme disease: discussion of specific T- and B-lymphocyte responses to *Borrelia burgdorferi*. N. Engl. J. Med. 319:1441.
5. Fieldsteel, A.H., Cox, D.L., and Moeckii, R.A. 1981. Cultivation of virulent *Treponema pallidum* in tissue culture. Infect. Immun. 32:908.
6. Golightly, M.G. 1993. Laboratory considerations in the diagnosis and management of Lyme borreliosis. Am. J. Clin. Pathol. 99:168.
7. Grodzicki, R.L., and Steere, A.C. 1988. Comparison of immunoblotting and indirect enzyme-linked immunosorbent assay using different antigen preparations for diagnosing early Lyme disease. J. Infect. Dis. 157:790.
8. Hovind-Hougen, K., Birch-Anderson, A., Henrik-Nielsen, R., et al. 1982. Intestinal spirochetosis: morphological characterization and cultivation of the spirochete *Brachyspira aalborgi* gen. nov., sp. nov. J. Clin. Microbiol. 16:1127.
9. Johnson, R.C., and Harris, V.G. 1967. Differentiation of pathogenic and saprophytic leptospires. I. Growth at low temperatures. J. Bacteriol. 94:27.
10. Johnson, S.E., Klein, G.C., Schmid, G.P., et al. 1984. Lyme disease: a selective medium for isolation of the suspected etiological agent, a spirochete. J. Clin. Microbiol. 19:81.
11. Jones, M.J., Miller, J.N., and George, W.L. 1986. Microbiological and biochemical characterization of spirochetes isolated from the feces of homosexual men. J. Clin. Microbiol. 24:1071.
12. Magnarelli, L.A., Anderson, J.F., and Johnson, R.C. 1987. Cross-reactivity in serologic tests for Lyme disease and other spirochetal infections. J. Infect. Dis. 156:183.
13. Norris, S.J. 1982. In vitro cultivation of *Treponema pallidum;* independent confirmation. Infect. Immun. 36:437.
14. Schmid, G.P., Steere, A.C., Kornblatt, A.N., et al. 1986. Newly recognized *Leptospira* species ("*Leptospira inadai*" serovar *lyme*) isolated from human skin. J. Clin. Microbiol. 24:484.
15. Steere, A.C., Grodzicki, R.L., Craft, J.E., et al. 1984. Recovery of Lyme disease spirochetes from patients. Yale J. Biol. Med. 57:557.

BIBLIOGRAPHY

Alexander, A.D. 1980. Serological diagnosis of leptospirosis. In Rose, N.R., and Friedman, J., editors. Manual of clinical immunology, ed 2. American Society for Microbiology, Washington, D.C.

Atlas, R.M. 1993, Handbook of microbiologic media. CRC Press, Boca Raton, Fla. (Includes detailed formulas and instructions for media preparation.)

Baron, E.J, and Finegold, S.M. 1990. Bailey and Scott's diagnostic microbiology, ed 8. (Appendices include detailed formulas of media and stains.)

Benach, J.L., and Bosler, E.M., editors. 1988. Lyme disease and related disorders. Ann. N.Y. Acad. Sci. 539:1. (Entire issue devoted to Lyme disease.)

Burgdorfer, W. 1985. Spirochetes. In Lennette, E.H., Balows, A., Hausler, W.J., Jr., and Shadomy, H.J., editors. Manual of clinical microbiology, ed 4. American Society for Microbiology, Washington, D.C.

Coleman, J.L., and Benach, J.L. 1987. Isolation of antigenic components from the Lyme disease spirochete: their role in early diagnosis. J. Infect. Dis. 155:756.

Larsen, S.A., Hunter, E.F., and Kraus, S.J. 1990. A manual of tests for syphilis. American Public Health Association, Washington, DC.

Steere, A.C., Grodzicki, R.L., Kornblatt, A.N., et al. 1983. The spirochetal etiology of Lyme disease. N. Engl. J. Med. 308:733.

33

AEROBIC OR FACULTATIVE SPORE-FORMING RODS (*BACILLUS* SPECIES)

The genus *Bacillus* contains many species, all of which are aerobic or facultatively anaerobic gram-positive or gram-variable spore-forming bacilli. *Bacillus* sp. are widely distributed in nature. Because the spores from these microbes can survive even in adverse conditions, *Bacillus* sp. are frequent contaminants in laboratory cultures.

Most *Bacillus* sp. are rarely associated with disease; however, several species have been involved in a variety of clinically significant infections. As a rule, species identification within the genus is not important for clinical purposes. Identification of *Bacillus anthracis* (the agent of anthrax) and *Bacillus cereus* (an agent of food poisoning) are two exceptions.

MORPHOLOGY AND GENERAL CHARACTERISTICS

The vegetative cells of *Bacillus* sp. are straight sided with rounded or square (boxy) ends. They can occur singly or in chains. The single spore may be placed centrally or terminally and can be oval, cylindrical, or rounded. With a few exceptions, all *Bacillus* sp. are motile. Morphologically, the typical *Bacillus* sp. exhibit large, flat colonies on nonselective laboratory media. They are often oxidase positive and frequently beta hemolytic. The *Bacillus* strains that simulate gram-negative bacilli are not typical; they can be very small colonies, and spore formation often fails to occur. *Bacillus* sp. are able to grow on most nonselective agar; they may or may not grow on eosin–methylene blue (EMB) agar. Those with typical Gram stain appearance (Figure 33.1) are easily identified as *Bacillus* sp. Because atypical strains may vary in Gram stain, oxidase, and other reactions and spores may not be evident, they may resemble nonfermentative gram-negative bacilli. When identification is attempted using gram-negative schemes, these organisms cannot confidently be classified in any known gram-negative species or Centers for Disease Control (CDC) taxon. *Bacillus* strains that are strict aerobes may appear as nonfermentative gram-negative bacilli on Kligler's iron agar

451

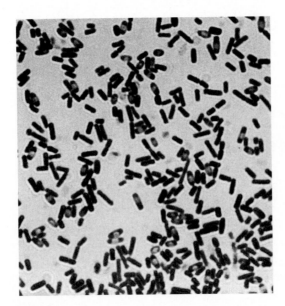

FIGURE 33.1

Gram stain of *Bacillus* sp. visualized under oil immersion (1000×) showing the clear intracellular (subterminal) spores that fail to pick up stain.

(KIA) or triple sugar iron agar (TSIA) as well. Most do not grow on enteric agars, but some show limited growth. However, several gram-negative nonfermenters also fail to grow on such media. Sporulation of the atypical strain may be stimulated by subculture onto esculin agar and may be facilitated on acidified media such as TSIA.

Two other tests may be useful. One is susceptibility to vancomycin; with the exception of *Flavobacterium* sp., no gram-negative nonfermenter is known to be susceptible to vancomycin (30 μg disk), whereas most *Bacillus* sp. are susceptible. However, some strains of *Bacillus* sp. are vancomycin resistant, so only vancomycin susceptibility can be used diagnostically. The second procedure that may be helpful is the potassium hydroxide (KOH) test (Chapter 8), in which gram-negative organisms show a viscous thread. Some *Bacillus* strains, particularly from older cultures that have also lost their gram-positive characteristic, may give a "gram-negative" reaction by this test. Accordingly, the test is of value for identification of *Bacillus* sp. only if no viscous thread is formed.

Treatment of specimens or mixed cultures that may contain *Bacillus* sp. with 50% filter-sterilized ethanol for 1 hour is effective for selecting *Bacillus* sp., just as it is for selecting spore-forming anaerobes (Chapter 35). Serial dilutions of the treated specimens can be directly plated on blood agar plates.

DIFFERENTIATION OF *BACILLUS* SPECIES

Species differentiation within the genus *Bacillus* is based on cell morphology, spore shape and location, whether the organism is obligately aerobic or facultatively anaerobic, action on xylose and arabinose, starch hydrolysis, indole production, nitrate reduction, presence of protein bodies, lecithinase production, susceptibility to gamma phage, capsule production, and animal pathogenicity. Table 33.1 shows differential characteristics of species likely to be encountered in clinical specimens. The three groups are based on size of cell and whether the spore swells the cell wall: group I cells are 0.9 μm wide; group II cells are smaller than group I; and group III spores cause swelling of the cell wall. Identification of a sporadic isolate, besides verifying that the organism is not *B. anthracis,* is not warranted in most circumstances.

BACILLUS ANTHRACIS

B. anthracis is the primary human pathogen in the genus. Anthrax is a rare disease in the United States, so the organism is seldom encountered in the average hospital or public health laboratory. Cases in the United States are most often related to handling of imported wool or goat hair, animal hides, shaving brushes, and so forth, originating principally in Asia, the Middle East, and Africa. Cases may also result from contact with diseased animals, primarily cattle, swine, and horses, and from inadvertent self-inoculation of veterinary anthrax vaccine. During the last decade, an average 2.5 cases of anthrax per year have been reported in the United States.

Anthrax is manifested in humans in three forms:

1. *Cutaneous anthrax* (malignant pustule) is the most common form in the United States and accounts for 95% to 99% of all cases worldwide. Infection is initiated by the entrance of bacilli through an abrasion of the skin. A pustule usually appears on the hands or forearms. The bacilli are readily recognized in the serosanguineous discharge.
2. *Pulmonary anthrax* is also known as woolsorter's disease. The bacilli may be found in large numbers in the sputum. Spores are inhaled during shearing, sorting, or handling of animal hair. If not properly treated, this form progresses to fatal septicemia.
3. *Gastrointestinal anthrax* is the most severe and rarest form of anthrax. The bacilli or spores

TABLE 33.1

IDENTIFICATION OF BACILLUS SP.*

SPECIES	USUAL CELL DIAMETER (μm)	PENICILLIN (10 U)	MOTILITY	WIDE ZONE LECITHINASE	BETA HEMOLYSIS	SPORES SWELL CELLS	VOGES-PROSKAUER	NITRATE REDUCTION	STARCH	DISTINGUISHING CHARACTERISTICS
Group I										
B. anthracis	≥0.9	S	0	27	−	−	85	100		Produces capsule
B. cereus	≥0.9	R	99	100	+	−	84	89		
B. mycoides	≥0.9	R	63	100	−	−	50	100		Rhizoid colony
B. megaterium	≥0.9	V	42	0	−	−	0	10		
B. thuringiensis	≥0.9	R	+	+	+	−	V	+		Insect pathogen that produces toxin crystals
Group II										
B. subtilis	<0.9	S	84	−	V	−	71	89	80	
B. pumilis	<0.9	S	100	−	V	−	78	0	6	
B. licheniformis	<0.9	S	80	−	+	−	83	100	100	
B. firmus	<0.9	S	90	−	V	−	0	70	50	
B. coagulans	<0.9	S	+	−	V	−	V	0		Grows at 55° and 35° C
Group III										
B. circulans	<0.9		+	−	−	+	6	76		Colonies may migrate on agar
B. sphaericus	<0.9		+	−	−	+	0	26		Produces spherical spores
B. laterosporus	<0.9		+	V	−	+	0	90		Produces canoe-shaped spores
B. brevis	<0.9		+	−	−	+	−	26		
B. polymyxa	<0.9		+	−	−	+	100	0		
B. alvei (H$_2$S + Bacillus)	<0.9		+	−	V	+	V	+		
B. stearothermophilus	<0.9		+	−	−	+	V	0		Grows at 65° C, no growth at 35° C

From Howard, B.J., Klaas, J.J., II, Rubin, S.J., et al. 1987. Clinical and pathogenic microbiology. Mosby, St. Louis.

+, >90% positive; −, <10% positive; S, susceptible; R, resistant; V, 10% to 90% positive.

*Percentages are given when available.

are swallowed, thus initiating intestinal infection. The organisms may be isolated from the patient's stools. This form is also usually fatal if not treated.

STRUCTURE AND EXTRACELLULAR PRODUCTS

B. anthracis is a facultative, large, square-ended, nonmotile rod, with an ellipsoidal to cylindrical centrally located spore. The sporangium usually is not swollen. The cells frequently occur in long chains, giving a bamboo appearance, especially on primary isolation from infected tissue or discharge. The chains of virulent forms are usually surrounded by a capsule. Encapsulation occurs also in enriched media and when grown on 0.7% sodium bicarbonate agar under 5% carbon dioxide (CO_2). Avirulent forms are usually unencapsulated. Sporulation occurs in the soil and on culture media but not in living tissue, unless exposed to air.

The colonies of *B. anthracis* are normally large (4 to 5 mm), opaque, white to gray in color, raised, and irregular, with a curled margin. When the margin of the colony is pushed inward and then lifted gently with an inoculating needle, the disturbed portion of the colony stands up resembling beaten egg whites. Comma-shaped colony outgrowths typically occur. On sheep blood agar the colonies are invariably nonhemolytic. Smooth and rough colony forms may be observed, and both of these may be virulent. In broth the bacillus produces a heavy pellicle, with little if any subsurface growth.

Two potent exotoxins are produced. Both toxin molecules depend on a third protein, called *protective antigen (PA)*, for their biological activity. The two exotoxins are known as *lethal factor (LF)* and *edema factor (EF)*. It appears that PA is essential for the toxins to enter into the target cells. The combination of EF and PA increases host susceptibility to infection by suppressing polymorphonuclear neutrophil function.

IDENTIFICATION METHODS

When working with suspected *B. anthracis,* one should use extreme caution, work in a level 3 (containment level) bacteriological safety hood, avoid creating aerosols, and decontaminate all areas thoroughly with a sporicidal germicide such as Alcide (Alcide Corp., Westport, Conn.).

The optimal growth temperature for the organism is 35° C; when grown at 42° or 43° C, the capsule is destroyed and the organism becomes avirulent. Biochemically the organism is characterized as follows:

- Carbohydrate fermentation: glucose, fructose, maltose, sucrose, and trehalose fermented with acid only; arabinose, xylose, galactose, lactose, mannose, raffinose, rhamnose, adonitol, dulcitol, inositol, inulin, mannitol, and sorbitol not fermented
- Gelatin: inverted pine tree growth; slow liquefaction
- Nitrates: reduced to nitrites
- Starch: hydrolyzed
- Voges-Proskauer: positive

Specific identification may be made by use of a gamma bacteriophage (done at the CDC or through state health department laboratories). The pathogenicity of *B. anthracis* may be determined by injecting each of 10 white mice (2 to 3 weeks of age) subcutaneously with 0.2 ml of a saline suspension of the organism. Rabbits or guinea pigs may also be used. Animals usually die 2 to 5 days after inoculation from septicemia; the organism is readily recovered from the animal's heart, blood, spleen, liver, and lungs. Animal inoculation is not generally done in clinical laboratories.

SUSCEPTIBILITY TO ANTIMICROBIAL AGENTS

Penicillin is the drug of choice for treatment. Alternative drugs for the penicillin-sensitive patient are tetracycline, erythromycin, and chloramphenicol.

NONCULTURAL METHODS OF IDENTIFICATION

A fluorescent antibody stain for encapsulated strains is available through the CDC. Both the cell wall and the capsule fluoresce simultaneously. This test would presumably work directly on clinical specimens if performed immediately after obtaining the specimen.

Antibodies to the organism may be detected by an indirect hemagglutination or enzyme-linked immunosorbent test, using the PA component of the toxin as the capture antigen. These tests are available through state health department laboratories.

BACILLUS CEREUS

This organism is the cause of serious infections of various types, usually, but not always, in immunocompromised hosts. Other patient background factors include surgery and other trauma, burns, intravenous drug abuse, implantation of catheters and various prosthetic devices (including heart valves), and use of hemodialysis and peritoneal dialysis. Types of infections described include septicemia, endocarditis, necrotizing pneumonia with or without empyema, meningitis, rapidly destructive ophthalmitis, peritonitis, wound infection, myonecrosis, and osteomyelitis. Myonecrosis has followed surgery and trauma and may simulate clostridial myonecrosis (gas gangrene). Injection of *B. cereus* exotoxin into the skin of rabbits causes increased vascular permeability and necrosis; production of this toxin seems to correlate with the severity of clinical infection. Rabbit skin necrosis is attributed to two factors, an enterotoxin and a hemolysin. Phospholipase C (lecithinase) is produced by almost all strains of *B. cereus*; its pathogenic role is doubtful. The toxin can also be demonstrated to cause fluid accumulation in the rabbit ileal loop model. *B. cereus* produces β-lactamase.

B. cereus has long been known as an important cause of food poisoning. The short-incubation type of food poisoning caused by *B. cereus* is most often associated with oriental rice dishes that have been cooked and held warm for extended periods; it results from preformed toxin. Long-incubation food poisoning caused by *B. cereus* is frequently associated with meat or vegetable dishes; in these cases, toxin is formed in vivo. These entities are discussed in greater detail in Chapter 18.

LABORATORY IDENTIFICATION

B. cereus colonies vary from small, shiny, and compact to the large, feathery, spreading type. A lavender colony with beta hemolysis is seen on sheep blood agar.

Unlike *B. anthracis*, *B. cereus* is resistant to lysis by gamma phage, is usually resistant to penicillin, does not encapsulate on bicarbonate agar, and does not exhibit fluorescence of both cell wall and capsule on fluorescent antibody–stained smears. Lecithinase and hemolysin production are very useful in the provisional identification of *B. cereus*.

SUSCEPTIBILITY TO ANTIMICROBIAL AGENTS

Because of β-lactamase production, *B. cereus* is resistant to penicillin, ampicillin, and many other β-lactam antimicrobial agents. It is also resistant to trimethoprim. *B. cereus* is usually susceptible to gentamicin, clindamycin, erythromycin, vancomycin, and often to chloramphenicol, sulfonamides, and tetracycline.

OTHER *BACILLUS* SPECIES

Several other of the 60 or so species in this genus have been encountered in significant infections, especially *B. subtilis* and *B. licheniformis*. The latter two species have been implicated in food poisoning and in various infections, including septicemia, panophthalmitis, pneumonia, wound infection, and peritonitis. Ocular infection has been produced by a biological insecticide made up of *B. thuringiensis* when the material was inadvertently splashed into the patient's eye. Other *Bacillus* sp. occasionally incriminated as causing an infectious disease include *B. brevis, B. circulans, B. coagulans, B. macerans, B. megaterium, B. pumilus,* and *B. sphaericus.* Allergic reactions in workers manufacturing laundry detergent containing derivatives of *B. subtilis* have occurred. Also, hypersensitivity pneumonitis related to exposure to wood dust contaminated with *B. subtilis* has been described.

MORPHOLOGY AND EXTRACELLULAR PRODUCTS

B. subtilis colonies are usually large, flat, and dull, with a ground-glass appearance (Figure 33.2). Several species produce collagenases. *B. megaterium* produces a filterable hemolysin, and *B. mycoides* produces potent toxins that can be lethal to experimental animals.

SUSCEPTIBILITY TO ANTIMICROBIAL AGENTS

Susceptibility to penicillin G and other older β-lactam antibiotics is variable and may be species related, tending to be high for *B. subtilis* and inter-

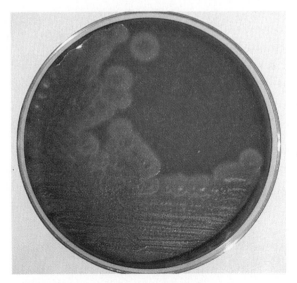

FIGURE 33.2
Bacillus subtilis on blood agar plate (BAP).

mediate for *B. pumilus*. Although clindamycin is generally considered to be an effective drug for treating *Bacillus* infections, one strain of *B. licheniformis* was reported to be resistant to clindamycin and susceptible to penicillin. Vancomycin should continue to be one of the more dependable drugs for therapy of serious infections caused by *Bacillus* sp.

BIBLIOGRAPHY

Bekemeyer, W.B., and Zimmerman, G.A. 1985. Life-threatening complications associated with *Bacillus cereus* pneumonia. Am. Rev. Respir. Dis. 131:466.

Blachman, U., Gilardi, G.L., Pickett, M.J., et al. 1980. *Bacillus* spp. strains posing as nonfermentative gram-negative rods. Clin. Microbiol. Newsletter 2:8.

Brachman, P.S. 1983. Anthrax. In Hoeprich, P.D., editor. Infectious diseases, ed 3. Harper & Row, Philadelphia.

Clarridge, J.E. 1987. Gram-positive bacilli. In Howard, B.J., Klass, J., II, Rubin, S.J., et al., editors. 1987. Clinical and pathogenic microbiology. Mosby, St. Louis.

Editorial. 1983. *Bacillus cereus* as a systemic pathogen. Lancet 2:1469.

Parry, J.M., Turnbull, P.C.B., and Gibson, J.R. 1983. A colour atlas of *Bacillus* species. Wolfe Medical Atlases—19. Wolfe Medical Publications, Ipswich, England.

Pennington, J.E., Gibbons, N.D., Strobeck, J.E., et al. 1976. *Bacillus* species infection in patients with hematologic neoplasia. J.A.M.A. 235:1473.

Tuazon, C.V., Murray, H.W., Levy, C., et al. 1979. Serious infections from *Bacillus* spp. J.A.M.A. 241:1137.

Turnbull, P.C.B., and Kramer, J.M. 1983. Non-gastrointestinal *Bacillus cereus* infections: an analysis of exotoxin production by strains isolated over a two-year period. J. Clin. Pathol. 36:1091.

Turnbull, P.C.B., and Kramer, J.M. 1991. *Bacillus*. In Balows, A., Hausler, W.J., Jr., Herrmann, K.L, et al., editors. Manual of clinical microbiology, ed 5. American Society for Microbiology, Washington, D.C.

34

AEROBIC, NON-SPORE-FORMING, GRAM-POSITIVE BACILLI

Aerobic, gram-positive, non–spore-forming bacilli are taxonomically diverse and are ubiquitous inhabitants of the natural environment. The group comprises many genera (see the box on p. 458). For purposes of this chapter, the discussion is limited to the aerobic *Actinomyces, Arcanobacterium, Corynebacterium, Erysipelothrix, Kurthia, Lactobacillus, Listeria, Nocardia, Oerskovia, Rothia,* and *Rhodococcus.* The remaining genera are not discussed either because they are rarely, if ever, isolated in the clinical laboratory or because they are covered in other chapters.

With the exception of *Corynebacterium diphtheriae,* the aerobic, gram-positive, non–spore-forming bacilli are of low pathogenicity and generally require a major break in the host's immunological defenses to cause significant infection. However, because they constitute a large portion of the normal bacterial flora of patients and the environment, they are frequently isolated in the clinical laboratory. Given their widespread distribution and that some of these organisms may be highly antibiotic resistant, it is not surprising that both epidemic and sporadic infections with certain of

457

Aerobic or Facultative, Gram-Positive, Non–Spore-Forming Bacilli

Actinomadura
Actinomyces
Agromyces
Arcanobacterium
Arthrobacter
Aureobacterium
Brevibacterium
Brochothrix
Caryophanon
Caseobacter
Cellulomonas
Clavibacter
Corynebacterium
Curtobacterium
Dermatophilus
Erysipelothrix
Gardnerella
Jonesia
Kurthia
Lactobacillus
Listeria
Microbacterium
Mycobacterium
Nocardia
Nocardiopsis
Oerskovia
Propionibacterium
Renibacterium
Rhodococcus
Rothia
Streptomyces
Tsukamurella

these organisms are becoming increasingly recognized. *Rhodococcus equi* has become an important pathogen in patients with acquired immunodeficiency syndrome (AIDS), producing serious pulmonary and systemic infections.[18]

Most bacteria in this group are pleomorphic, exhibiting various morphologies that include branching, clubbing, and coccoid forms and palisading. For this reason, the currently preferred term is *coryneform bacteria*.[11] These organisms all grow well on typically used nonselective media, including 5% sheep blood agar. The organisms may be divided into groups on the basis of the catalase test.[7,8,20] Figure 34.1 illustrates a basic strategy for preliminary identification of aerobically-growing gram-positive, non–spore-forming bacilli.

LISTERIA SPECIES

EPIDEMIOLOGY AND PATHOGENESIS OF LISTERIOSIS

Taxonomy of *Listeria* remains in a state of flux. Species currently accepted include *monocytogenes, innocua, ivanovii, seeligeri,* and *welshimeri. L. denitrificans* has been transferred to the new genus *Jonesia. L. grayi* and *L. murrayi* will probably be combined into a single species.[5] *L. monocytogenes* is the major pathogen and the only species often encountered in clinical laboratories. *L. monocytogenes* is ubiquitous in the environment and has been isolated from soil and decaying vegetation and wild and domestic animals. Stool-carriage studies have also documented asymptomatic intestinal carriage of *L. monocytogenes* in 1% to 5% of humans. *L. monocytogenes* is a well-known cause of spontaneous abortions in animals and of basilar meningitis, or "circling disease," in sheep.

Human listeriosis has been a reportable disease since 1986, and although our knowledge of the epidemiology remains incomplete, recent analysis of both epidemic and sporadic disease has provided considerable insight into the epidemiology and transmission of *L. monocytogenes*. Four major epidemics of listeriosis have occurred in North America and one in Switzerland since 1981. The total number of cases ranged from 36 to 142, with case fatality rates of 27% to 44%. The cases were linked to the ingestion of contaminated dairy products in three of the epidemics, contaminated cabbage in one, and no source could be implicated in the fifth outbreak. Studies of dairy products, raw vegetables, and sausages have documented the presence of *L. monocytogenes* in low numbers, suggesting that this organism may be a common contaminant of both processed and unprocessed foods of plant and animal origin. Because the organism can survive and multiply at refrigerator temperatures and in a wide range of pHs, even a small amount of contamination may be significant.[24]

Most human cases are sporadic, and the source of the infection is usually unknown. Active surveillance conducted by the Centers for Disease Control (CDC) in six areas of the United States has provided estimates of at least 1700 serious infections, 450 deaths, and 100 stillbirths caused by listeriosis annually in the United States. Listeriosis is seen almost exclusively in neonates, pregnant women, and immunocompromised individuals. The clinical presentation ranges from sepsis and meningitis in neonates and immunocompromised patients to a nonspecific febrile illness in pregnant

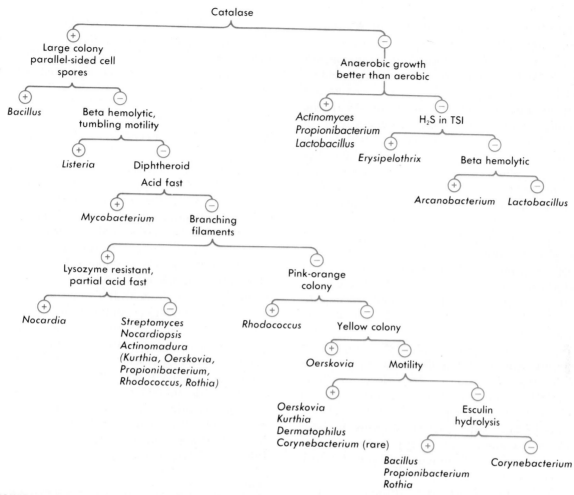

FIGURE 34.1

Flow chart for preliminary identification of aerobically-growing, gram-positive, non–spore-forming bacilli.

women. Transplacental infection occurs often, possibly because of deficiencies in local immunoregulation at the placenta, and results in premature labor, septic abortion, and neonatal listeriosis. The case fatality rate ranges from 3% to 50% with perinatal disease and is estimated at 35% in nonperinatal cases.

Outside of the perinatal period, the most common presentation of listeriosis is meningitis. In persons with meningitis, the cerebrospinal fluid (CSF) is purulent and the CSF protein is usually elevated; however, the CSF glucose level is normal and the Gram stain is negative in greater than 60% of patients. Cultures of blood and CSF are positive in 60% to 75% of patients and remain the most reliable means of diagnosis. The pathogenesis of listeriosis is related to several factors, including host immunity, inoculum size, and possibly strain-specific virulence factors. *L. monocytogenes* is frequently cited as a classical example

of an intracellular pathogen, and the importance of T-cell–mediated immunity to this organism is well established. The available data support the alimentary tract as the major portal of entry of *L. monocytogenes*; however, it is unclear whether disruption of the mucosal barrier is necessary for the organism to cause systemic disease. Although beta hemolysis on blood agar distinguishes *L. monocytogenes* from the nonpathogenic species of *Listeria*, additional virulence factors have been difficult to demonstrate. Currently, it appears that virulence is mediated by action of at least three components: a phospholipase, a hemolysin, and a lipopolysaccharide.

LABORATORY IDENTIFICATION

On Gram stain, *L. monocytogenes* appears as a non–spore-forming, short, gram-positive bacillus. It is somewhat smaller than most gram-positive bacilli, with cells ranging from 0.4 to 0.5 μm × 1 to

2 μm. The cells may be coccobacillary and occur in pairs suggestive of *Streptococcus pneumoniae* or other species of streptococci. *L. monocytogenes* grows well on most laboratory media, yielding small, smooth, translucent gray colonies with a narrow zone of beta hemolysis (Figure 34.2). The colony morphology closely resembles that of group B beta-hemolytic streptococci, with which it is often confused. *L. monocytogenes* also hydrolyzes sodium hippurate and yields a positive reaction in the CAMP test (used for group B streptococci); however, it can readily be distinguished from *Streptococcus agalactiae* by positive tests for catalase and esculin hydrolysis (*S. agalactiae* is negative in both). *Listeria* can also be presumptively identified by observation of motility under direct wet preparation. *Listeria* exhibits a characteristic end-over-end tumbling motility when incubated in nutrient broth for 1 to 2 hours at room temperature. An identical broth culture incubated at 37° C shows considerably less motility. This motility pattern also is illustrated by an "umbrella-shaped" pattern that develops after overnight incubation at room temperature of a culture that has been stabbed into a tube of semisolid agar. Table 34.1 summarizes the basic biochemical and enzymatic tests used to differentiate *L. monocytogenes* from organisms with which it might be confused.

Although *L. monocytogenes* grows readily from CSF, blood, and most other specimens, it may be difficult to isolate from placental and other tissue. Placing the specimen at 4° C for several weeks to months and subculturing at frequent intervals may serve to enhance recovery. This procedure, called *cold enrichment*, is particularly useful for epidemiological studies when specimens are contaminated with other flora. Chapter 22 includes a procedure for handling tissue suspected of containing *Listeria*.

THERAPY

L. monocytogenes is susceptible to a wide variety of antimicrobial agents in vitro, including penicillin, ampicillin, tetracyclines, erythromycin, chloramphenicol, rifampin, sulfamethoxazole and trimethoprim, and the aminoglycosides. Of these, only trimethoprim-sulfamethoxazole (in combination) and the aminoglycosides are bactericidal.

The optimal therapy for listeriosis is controversial and has not been established in controlled clinical trials. Ampicillin is generally considered to be more active than penicillin, and the combination of ampicillin plus an aminoglycoside, the recommended therapy for severe disease, results in synergistic bactericidal activity in vitro. Clinical failures have been observed with chloramphenicol, and most agree that cephalosporins are never effective. The combination of sulfamethoxazole and trimethoprim may be useful in treatment of patients allergic to penicillin. This regimen is bactericidal, achieves adequate levels in serum and CSF, and has documented clinical efficacy.

ERYSIPELOTHRIX RHUSIOPATHIAE

EPIDEMIOLOGY AND PATHOGENESIS

E. rhusiopathiae is the only species in the genus and is a well-known veterinary pathogen of considerable economic importance. *E. rhusiopathiae* is

FIGURE 34.2

Colonies of *Listeria monocytogenes* on sheep blood agar after 48-hour incubation, showing the diffuse zone of beta hemolysis surrounding the colonies.

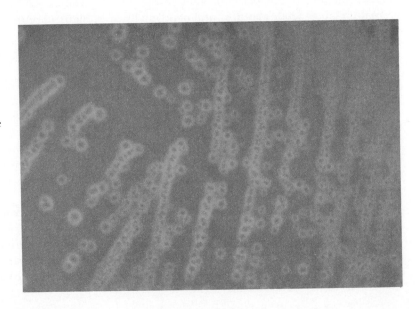

TABLE 34.1

DIFFERENTIATION OF *LISTERIA MONOCYTOGENES* FROM OTHER BACTERIA IT RESEMBLES

ORGANISM	CATALASE	ESCULIN	MOTILITY	BETA HEMOLYSIS	H$_2$S/TSI
L. monocytogenes	+	+	+	+	−
Corynebacterium sp.	+	−	−/+	−/+	−
Propionibacterium sp.	+	+/−	−	+/−	−
Arcanobacterium haemolyticum	−	+/−	−	+*	−
Erysipelothrix rhusiopathiae	−	−	−	−	+
Lactobacillus sp.	−	−	−	−	−
Streptococcus agalactiae	−	−	−	+	−
Enterococcus sp.	−	+	− (+ rare)	−/+	−

+, 90% of strains positive; −, >90% of strains negative; +/−, variable (most strains positive); −/+, variable (most strains negative)

* Delayed

ubiquitous in nature and has been isolated from a variety of environmental sources, including soil, food, and water. Contamination of the environment is thought to occur secondary to excretion of the organism by infected or colonized animals. Human infection is usually related to occupational exposure (veterinarians, farmers) and occurs most often in those who have direct contact with animals or with organic matter in which the organism is typically found. The most common form of human disease is a spreading, cellulitis-like lesion of the hands and fingers known as *erysipeloid*. Rarely a septic form may be observed associated with bacteremia and skin involvement. Bacteremia is not always associated with endocarditis. Most cases of sepsis and endocarditis caused by *E. rhusiopathiae* occur in patients with an occupational risk for cutaneous infection. The crude mortality in these patients is almost twice as high as that in patients with other causes of bacterial endocarditis. Although pathogenicity of *E. rhusiopathiae* for swine has been related to neuraminidase production, surface antigen structure, endothelial adherence, and fibrin deposition on synovial membrane during bacteremia, little is known about the pathogenesis of infection caused by this organism in humans.

LABORATORY IDENTIFICATION

E. rhusiopathiae is a facultatively anaerobic, non-motile, nonsporulating, gram-positive rod. The organism grows well on blood agar, showing alpha hemolysis after prolonged incubation. Older colonies may show a tendency to become gram negative. Aspirated material or biopsy specimens of skin lesions should be inoculated into brain-

heart infusion broth with 1% dextrose and incubated in air or carbon dioxide (CO_2) at 35° C for initial isolation. Subcultures of this broth to blood agar are made at 24-hour intervals.

On primary isolation, the organism may yield two or more different morphological colony types, each containing bacteria with different cellular morphologies, mimicking a polymicrobic infection. *E. rhusiopathiae* smeared from large, rough colonies is a slender, filamentous gram-positive bacillus with a tendency to overdecolorize and become gram negative. A Gram stain made from the smaller, smooth, translucent colony yields small, slender bacilli.[7] *E. rhusiopathiae* may be distinguished from other gram-positive rods, such as *Listeria*, lactobacilli, and *Corynebacterium* sp., by type of hemolysis, presence or absence of motility, and catalase production. However, the organism's primary distinguishing characteristic is that it is the only catalase-negative, gram-positive, aerobic bacillus that produces hydrogen sulfide (H_2S) when inoculated into triple sugar iron agar (TSIA) or Kligler's iron agar (KIA) slant tubes. The important biochemical characteristics for the identification of *E. rhusiopathiae* are listed in Table 34.1.

THERAPY

Clinical isolates of *E. rhusiopathiae* are usually highly susceptible to penicillins, cephalosporins, erythromycin, and clindamycin. Most notably, this organism is usually resistant to vancomycin, as well as sulfonamides, trimethoprim-sulfamethoxazole, and the aminoglycosides. Variable susceptibility is reported for tetracycline and chloramphenicol. The treatment of choice is penicillin. Addition of an aminoglycoside does not appear to be warranted. Given the resistance of *Erysipelothrix* to vancomycin, prompt identification and antimicrobial susceptibility testing of gram-positive bacilli isolated from blood or skin lesions should be performed to minimize the risk of complications from inappropriate therapy.

CORYNEBACTERIUM AND RELATED SPECIES, INCLUDING *RHODOCOCCUS*, *ARCANOBACTERIUM*, AND *ACTINOMYCES PYOGENES*

GENERAL CONSIDERATIONS, EPIDEMIOLOGY, AND PATHOGENESIS

The corynebacteria are pleomorphic, primarily nonmotile, aerobic or facultative, gram-positive bacilli. Preferential synthesis of cell wall components at one end of the cell often results in clubbing and uneven division during binary fission, seen as "Chinese letter forms" (Figure 34.3). The *Corynebacterium* sp. and related genera most likely to be encountered in the clinical laboratory include *C. diphtheriae, C. ulcerans, C. aquaticum, C. pseudodiphtheriticum, C. xerosis, C. renale, C. pseudotuberculosis, C. bovis, C. minutissimum, C. jeikeium, C. matruchotii* (formerly *Bacterionema matruchotii*), *C. urealyticum* (formerly CDC group D2), CDC groups A4 and G2, *Arcanobacterium haemolyticum, Rhodococcus equi,* and *Actinomyces pyogenes*. These organisms are found throughout nature, acting as pathogens and parasites of plants and animals. Although most species are harmless saprophytes and many are part of the normal human skin and mucous membrane flora, *C. diphtheriae,* the agent

of diphtheria, and *C. jeikeium* and *C. urealyticum,* both relatively common causes of infection in compromised hosts, are major pathogens of humans.[35] *A. haemolyticum* is being recognized more often as an agent of infection and pharyngitis,[1,26] *R. equi* is emerging as a major pathogen in patients with AIDS,[18,22] and other species have also been associated with human infection, as outlined in Table 34.2. Certain strains have been placed in new genera that were formerly within *Corynebacterium,* such as *R. equi* (formerly *C. equi*), *A. haemolyticum* (formerly *C. haemolyticum*), and *A. pyogenes* (formerly *C. pyogenes*). *C. xerosis, C. pseudodiphtheriticum, C. minutissimum, C. ulcerans, C. bovis,* and others have also been reported to cause human disease, but rarely.[11,17] *C. aquaticum,* recently associated with serious disease in immunocompromised patients, may initially be confused with *L. monocytogenes* because of its similar motility.[3,6] The leprosy-derived corynebacteria are closely related to each other genetically and closely related immunologically to *Mycobacterium leprae.*[10] These organisms can be cultivated from lesions of leprosy and may contribute synergistically with *M. leprae* in development of disease.

Diphtheria is a well-known toxin-mediated infectious disease caused by *C. diphtheriae*. The disease is an acute, contagious, febrile illness characterized by local oropharyngeal inflammation and pseudomembrane formation and damage to the heart and peripheral nerves caused by the exotoxin produced by the organisms. The exotoxin of *C. diphtheriae* is one of the most well-studied bacterial toxins and acts to inhibit protein synthesis, particularly in the heart muscle and neural cells.

FIGURE 34.3
Pleomorphic forms of *Corynebacterium* sp. seen on Gram stain.

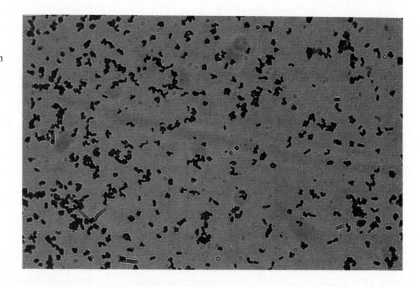

TABLE 34.2

CORYNEBACTERIA ASSOCIATED WITH HUMAN INFECTION

SPECIES	INFECTION(S)
C. diphtheriae	Pharyngeal, nasal, cutaneous*
C. jeikeium (group JK)	Systemic, catheter, cutaneous, endocarditis, pneumonia
C. urealyticum (group D2)	Cystitis, systemic, cutaneous, catheter
C. pseudotuberculosis	Lymphadenitis, pneumonia
C. minutissimum	Associated with erythrasma, bacteremia (rare)
Corynebacterium sp. (leprosy-derived)	Associated with leprosy
C. aquaticum	Meningitis, peritonitis (rare)
Rhodococcus equi	Systemic, pneumonia, osteomyelitis
Arcanobacterium haemolyticum	Wound, pharyngitis, abscess, cutaneous, rash
Actinomyces pyogenes	Cutaneous, endocarditis (rare)

*Although the infection is localized, the effects of the toxin are systemic.

The gene for toxin production is carried into the bacteria by a bacteriophage, which can produce a nonlytic infection, replicating with the bacterial genome. Such an infection that does not lyse the host cell is called *lysogenic*. Although *C. diphtheriae* that are not infected with toxin-producing phage can cause diphtheria-like disease, the disease is usually not as severe. Diphtheria may occur in both endemic and epidemic forms; however, the annual incidence has declined from 200 cases per 100,000 population in 1920 to approximately 0.001 per 100,000 in the 1980s, largely because of vaccination efforts in the United States. Recent endemic foci include skid row inhabitants of Seattle, who had wound infections with *C. diphtheriae*. The toxin can be elaborated from wound colonization as well as from pharyngeal sites of infection.

C. jeikeium and *C. urealyticum* are of particular interest among the nondiphtheriae species of *Corynebacterium*. Both these organisms are resistant to a wide range of antimicrobial agents, such as penicillins, cephalosporins, and aminoglycosides. In addition, both organisms have been recognized as the etiological agent in an increasing number of serious infections among hospitalized and immunocompromised individuals.[11,35]

C. jeikeium causes a variety of infections, including septicemia, meningitis, and peritonitis; however, the most frequent type of infection with this agent is bacteremia or foreign body infection (e.g., intravascular device, CSF shunts). Most often the infections are nosocomial, occurring in patients with compromised host defenses, exposure to broad-spectrum antimicrobials, and an extended hospital stay.[35] Nosocomial transmission of *C. jeikeium* has been well documented; bacteremia caused by this organism can be prevented by the introduction of control measures such as reducing catheter manipulation and wearing sterile gloves.

C. urealyticum has also been reported as a cause of nosocomial bacteremia and pneumonia, but most frequently it is found to cause cystitis, bacteriuria, and pyelonephritis. The urease produced by *C. urealyticum* seems to play an important role in its pathogenicity. Infection with this organism produces an alkaline urine and promotes the formation of struvite (ammonium magnesium phosphate) stones in the bladder.[35] Most patients infected with *C. urealyticum* are elderly and have a history of urological abnormalities, instrumentation, and exposure to broad-spectrum antimicrobial therapy.[35]

Colonization of hospitalized patients with *C. jeikeium* or *C. urealyticum* occurs often; however, only a few patients are colonized by both organisms simultaneously. The sites most frequently colonized are the groin and rectum, with *C. jeikeium* isolated more frequently from males and *C. urealyticum* more frequently from females.[35]

Arcanobacterium haemolyticum has been associated with cutaneous ulcers, abscesses, sepsis, and endocarditis. Both phospholipase D and lecithinase may act as virulence factors.[9] The pharyngitis caused by this organism is often associated with a characteristic maculopapular or scarlatiniform rash, which occasionally may desquamate.[1,26] The disease occurs most often in patients 10 to 30 years old.

ISOLATION OF *CORYNEBACTERIUM DIPHTHERIAE* FROM PATIENTS WITH SUSPECTED DIPHTHERIA

Although the cornyebacteria grow well on most routinely used laboratory media, selective and differential media for *C. diphtheriae* should be used if diphtheria is suspected. Cystine tellurite blood agar reliably supports growth of *C. diphtheriae*. The colonies appear black or gray with a characteristic garliclike odor, usually after 24 hours' incubation in air. Colony morphology differences on cystine tellurite agar have been used to differentiate the three subspecies of *C. diphtheriae*: *mitis*, *intermedius*, and *gravis*. They may occasionally require 48 hours' incubation for initial isolation. Modified Tinsdale agar has been used for isolation of *C. diphtheriae;* colonies are black with dark-brown halos. The halos develop particularly around colonies that have grown into stabbed areas of the agar. This medium must be used within 4 days of preparation and does not support growth of many strains. When small numbers of organisms must be isolated, primary inoculation to a Loeffler slant and overnight incubation at 35° C in air before subculturing to cystine tellurite blood agar may improve recovery. Gram-stained colonies that exhibit coryneform morphology should be streaked to a sheep blood agar plate for purity before biochemical tests are performed. Identification of subspecies of diphtheria isolates is best accomplished biochemically.

LABORATORY IDENTIFICATION
OF CORYNEFORM BACTERIA

Corynebacterium sp. grow well on most typically used media and usually appear as opaque, white, or gray colonies. However, they may resemble nonhemolytic or alpha-hemolytic streptococci, yeast, staphylococci, and commensal *Neisseria* sp. Gram stain morphology is typical of the genus, and no true branching should be observed. Corynebacteria are catalase positive.

Commercially available biochemical systems are reliable for identification of several groups of corynebacteria and should be used for clinically important isolates.[14,15,21,34,38] Selected biochemical reactions of some clinically relevant species are shown in Table 34.3. The lack of pyrazinamidase by *C. diphtheriae, C. ulcerans,* and *C. pseudotuberculosis* may help differentiate these species.[37] The enhancement of growth by lipids (Tween 80, serum, others) by certain coryneform bacteria, such as *C. jeikeium, C. urealyticum, Arcanobacterium,* and *Actinomyces pyogenes,* has been used as an aid to identification.[11] Cell wall fatty acid analysis has been found to be more useful for separating coryneform genera than for species identification.[4] Additionally, although definitive schemes for species identification within *Corynebacterium* and related genera are available, it should be recognized that many clinical isolates cannot be identified to species because of current taxonomic uncertainties.[8,11,12] *A. haemolyticum* displays beta hemolysis on blood agar after the second day of incubation, producing a large zone of hemolysis by day 4. At this point, examination of colonies under 100 × magnification may reveal dark dots in the center.[9] Although *Actinomyces pyogenes* colonies may also exhibit similar mor-

phology, they grow better anaerobically and hydrolyze gelatin, characteristics lacking in *Arcanobacterium.* Gram stain of *Arcanobacterium* shows delicate gram-positive rods with slight branching that breaks into coccal forms after 48 hours (Figure 34.4).[9] Branching is enhanced by anaerobic incubation. The organism inhibits the hemolysin of *Staphylococcus aureus* ATCC 25923 (the reverse CAMP test, Figure 34.5).[2,9]

Rhodococcus equi is usually recognized by its pigment, which may require several days to develop and varies from cream to yellow to coral. These organisms usually form mucoid colonies and are typically acid fast. Extensive testing is necessary to differentiate among the species.[23] Rhodococci are susceptible to vancomycin and erythromycin, which helps to differentiate them from nocardiae. *R. equi* is ONPG (*o*-nitrophenyl-β-D-galactopyranoside) negative, in contrast to nocardia and streptomyces.[29]

IN VITRO TOXIGENICITY TEST

Definitive identification of a strain of *C. diphtheriae* as a true pathogen requires demonstration of toxin production. A patient may be infected with several strains at once, so that testing a pooled inoculum enhances chances of detecting toxin activity. The most useful test is that described originally by Elek,[13] the in vitro toxigenicity test, or *Elek plate.* The test should be performed only in reference laboratories today, since the incidence of diphtheria is quite low and the test poses numerous technical problems because of variability of reagents. Briefly, a filter paper strip impregnated with diphtheria antitoxin is buried just below the surface of a special agar plate before the agar hardens. Strains to be tested and known positive and negative toxigenic strains are streaked on the agar's surface in a line across the plate, at a right angle to the antitoxin paper strip (Figure 34.6). After 24 hours' incubation at 37° C, plates are examined with transmitted light for the presence of fine precipitin lines at a 45-degree angle to the streaks. The presence of precipitin lines indicates that the strain produced toxin that reacted with the homologous antitoxin. Under some circumstances, precipitin lines may take up to 72 hours to develop.

THERAPY FOR PATIENTS WITH INFECTIONS
CAUSED BY CORYNEFORM BACTERIA

C. diphtheriae are usually susceptible to penicillin, erythromycin, and most other antimicrobial agents useful against gram-positive bacteria. In all patients with diphtheria, antibiotic therapy should be initi-

TABLE 34.3

DIFFERENTIATION OF CORYNEBACTERIUM AND RELATED SPECIES

SPECIES	CATALASE	NITRATE	UREASE	DEXTROSE	MALTOSE	SUCROSE	XYLOSE	PIGMENT	PYRAZINAMIDASE	PHOSPHOLIPASE D
C. diphtheriae (mitis)	+	+*	-	+	+	-	-	N	-	-
C. diphtheriae (gravis)	+	+	-	+	+	-	-	N	-	-
C. diphtheriae (intermedius)	+	+	-	+	+	-	-	N	NT	-
C. aquaticum	+	+/-	+/-	+	+	+	+	Y	NT	NT
C. ulcerans	+	-	+	+	+	-	-	N	-	+
C. pseudodiphtheriticum	+	+	+	-	-	-	-	N	+	-
C. pseudotuberculosis	+	+/-	+	+	+	+	-	N, Y	-	+
C. xerosis	+	+	-	+	+/-	+	-	N, T	+	-
C. jeikeium	+	-	-	+/-	+/-	-	-	N	+	-
C. urealyticum	+	-	+	-	-	-	-	N	NT	NT
Oerskovia sp.	+	+/-	+/-	-	-	-	+	Y	NT	NT
Rhodococcus equi	+	-	-	+	+	+	+	P	+	-
Actinomyces pyogenes	-	-	-	+	+	+/-	+	N	+	+
Arcanobacterium haemolyticum	-	-	-	+	+	+/-	-	N	+	+

Compiled from material presented by Coyle, M.B., Harborview Medical Center, Seattle; Thompson, J.S., Gates-Davis, D.R., and Young, D.C. 1983. J. Clin. Microbiol. 18:926; and Coyle, M.B., and Lipsky, B.A. 1990. Clin. Microbiol. Rev. 3:227.

+, >90% of strains positive; −, >90% of strains negative; +/−, variable (most strains positive); NT, not tested; N, none; T, tan; P, orange to pink; Y, yellow.

*C. diphtheriae (mitis, belfanti strain) is nitrate negative.

FIGURE 34.4
Gram stain from colony of *Arcanobacterium haemolyticum* after 48 hours' incubation. Filamentous forms break up into coccal shapes after a few days' incubation. (Photo courtesy Jay Packer.)

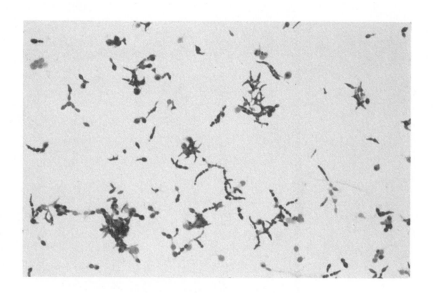

FIGURE 34.5
Arcanobacterium haemolyticum (top horizontal line of growth) inhibits the hemolysis of *Staphylococcus aureus* ATCC 25923 (vertical line of growth), called "reverse CAMP" reaction. In contrast, a group B streptococcus is streaked across the lower half of the plate, exhibiting a true CAMP reaction (enhancement of hemolysis in an arrowhead-shaped area). (Photo courtesy Martin Cohen.)

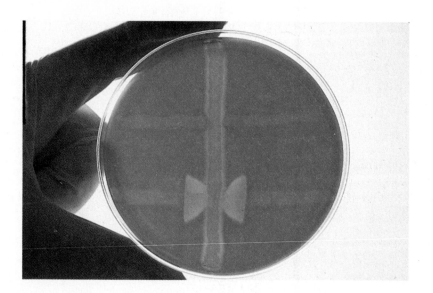

ated. Antitoxin therapy is given to patients from whom toxigenic strains are isolated or if the clinical picture dictates such therapy. The isolates should be tested for susceptibility, however, since erythromycin and tetracycline resistance has been observed. The other corynebacteria are variable in susceptibility. All corynebacteria are resistant to oxacillin, methicillin, and nafcillin. Also, *C. jeikeium* and *C. urealyticum* typically show multiple resistance, usually being susceptible only to vancomycin, which is the drug of choice for treatment of infections with those strains. Some newer antimicrobial agents, such as teicoplanin and the quinolones, are also active against *C. jeikeium* and *C.*

urealyticum. Although synergy between aminoglycosides and such agents has been demonstrated, it is not clear what role these agents will have in the prevention and therapy of infections caused by the *Corynebacterium* sp.[36]

Arcanobacterium infections usually respond to penicillin or erythromycin.[26] Antibiotics effective against *Rhodococcus equi* include vancomycin, chloramphenicol, erythromycin, rifampin, and aminoglycosides. Penicillin, cephalosporins, and tetracyclines are not likely to be effective. Lipophilic drugs such as erythromycin and rifampin that can penetrate phagocytes containing the bacteria are most efficaceous.[29]

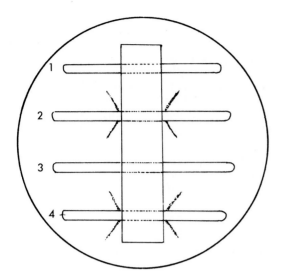

FIGURE 34.6

Diagram of an Elek plate for demonstration of toxin production by *Corynebacterium diphtheriae.*

NOCARDIA SPECIES

GENERAL CONSIDERATIONS, EPIDEMIOLOGY, AND PATHOGENESIS

Nocardia organisms are members of the group of aerobic actinomycetes, which include *Streptomyces, Nocardiopsis, Actinomadura, Dermatophilus,* and other genera that rarely cause human infection. Some of these organisms are discussed briefly later in this chapter. *Dermatophilus* is discussed in Chapter 40. *Nocardia* are inhabitants of the soil and water. They infect humans after inhalation of the organisms and primary establishment of growth in the lungs or by inoculation through breaks in the skin. The infection is usually chronic, but occasionally is fulminant, and is usually seen in immunosuppressed patients. The clinical syndrome of a chronic disease with both respiratory and central nervous system components, particularly if abscess formation is apparent, is suggestive of nocardiosis.

Nocardiosis is most often caused by *Nocardia asteroides* (>90% of patients) and rarely by *N. brasiliensis, N. otitidis-caviarum* (also called *caviae*), or other species. Infections caused by *Nocardia* are primarily suppurative, with necrosis and abscess formation. The incidence is higher in males than in females. Although pneumonic processes are often seen, a lymphocutaneous infection resembling sporotrichosis (Chapter 44) can occur. Spread to other tissues from the primary pulmonary lesions occurs by advancing growth (to produce empyema, chest wall involvement, and draining sinuses) or hematogeneously. Hematogeneous dissemination involving the central ner-

vous system, resulting in single or multiple brain abscesses, is particularly common, occurring in approximately 30% of patients. Osteomyelitis is also being recognized more frequently.[30] The organism seems to be able to resist phagocytosis and therefore persists in the host. Definitive determinants of pathogenicity are not known, but the filamentous stage of growth seems to be more virulent than the stationary coccoid phase.

LABORATORY IDENTIFICATION

Cells of *Nocardia* are pleomorphic, exhibiting branching and coccoid forms. The most distinguishing characteristic of *Nocardia* is its tendency to retain carbolfuchsin stain with mild acid decolorization, called *partial acid-fastness;* this characteristic may be difficult to demonstrate. Growing the organism to be tested, along with a fresh subculture of a known positive control (*N. asteroides*) and a known negative control (*Streptomyces* sp.), for approximately 4 days on Middlebrook 7H11 agar or in litmus milk broth will enhance the acid-fast nature of an isolate.[12] The partial acid-fast stain is performed as outlined in Procedure 34.1.

Nocardia may be first suspected when long, thin, branching gram-positive bacilli are seen in smears of lesion material or sputum. The organism also stains with Gomori methenamine-silver stain (Figure 34.7). Since it grows slowly, *Nocardia* is often overgrown by normal flora and other bacteria on routine microbiological media. Insertion of a glass rod dipped in paraffin to a liquid culture (nonnutritive medium) of a contaminated specimen (e.g., sputum) diluted 1:1 in buffered saline has been shown to enhance recovery of *Nocardia,* which can use paraffin as a sole carbon source.[27,33] A solid medium that uses paraffin as the sole source of carbon has been effective for isolating *Nocardia* sp. and rapidly growing mycobacteria from contaminated clinical specimens.[32] Material from specimens that contain no other flora, such as tissue and abscess drainage, may be inoculated to brain-heart infusion agar, which supports growth of *Nocardia.* Recently, sputum specimens diluted 1:10 in KCl-HCl, pH 2.2, vortexed, and allowed to sit for 10 minutes before inoculation yielded *Nocardia* and *Streptomyces* on buffered charcoal–yeast extract medium (used primarily for isolation of legionellae).[39] The organism grows well on Sabouraud dextrose agar but is inhibited by chloramphenicol. Occasional isolates are first seen on mycobacterial culture media, since *Nocardia* can survive the usual decontamination procedures. Incubating at least one primary isolation medium at 45° C may enhance

PROCEDURE 34.1

PARTIAL ACID-FAST STAIN FOR IDENTIFICATION OF *NOCARDIA*

PRINCIPLE
The nocardiae, because of unusual long-chain fatty acids in their cell walls, can retain carbolfuchsin dye during mild acid decolorization, whereas other aerobic branching bacilli cannot.

METHOD
1. Emulsify a very small amount of the organisms to be stained in a drop of distilled water on the slide. A known positive control and a negative control should be stained along with the unknown strain.

2. Allow to air dry and heat fix.

3. Flood the stain with Kinyoun's carbolfuchsin and allow the stain to remain on the slide 3 minutes.

4. Rinse with tap water, shake off excess water, and decolorize briefly with 3% acid alcohol (940 ml of 95% ethanol and 60 ml of concentrated HCl) until no more red color rinses off the slides.

5. Counterstain with Kinyoun's methylene blue for 30 seconds.

6. Rinse again with tap water. Allow the slide to air dry, and examine the unknown strain compared with the controls. Partially acid-fast organisms show reddish to purple filaments, whereas non–acid-fast organisms are blue only.

QUALITY CONTROL
A known partially acid-fast organism, *N. asteroides,* and a known negative control, *Streptomyces* sp., should be subcultured onto the same medium as the organism to be tested and incubated concurrently. The *Nocardia* should exhibit reddish and purple filaments with occasional areas of blue, whereas the *Streptomyces* stain only blue. No designated strains have been reserved for quality control.

PERFORMANCE SCHEDULE
Quality control strains should be tested each time the stain is performed.

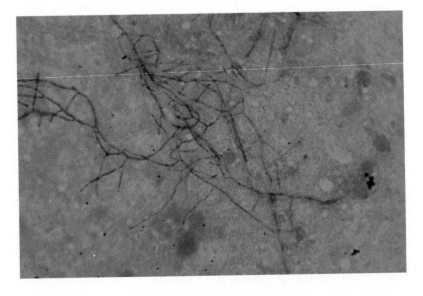

FIGURE 34.7
Nocardia in lesion of lower leg, stained with Gomori methenamine-silver stain.

recovery of *N. asteroides. Nocardia* sp. grow better in 10% CO_2.

Colonies are visible from 3 to 30 days after inoculation, appearing as waxy, bumpy, or velvety rugose forms, often with yellow to orange pigment, but also can appear chalky white (Figure 34.8). The presence of branching filaments and partial acid-fastness differentiates these isolates from mycobacteria. Differentiation from *Streptomyces* sp. is more difficult. A urease-negative species cannot be *Nocardia,* but both *Streptomyces* and *Nocardia* may be urease positive.

FIGURE 34.8
Colony of *Nocardia asteroides* after 10 days of incubation.

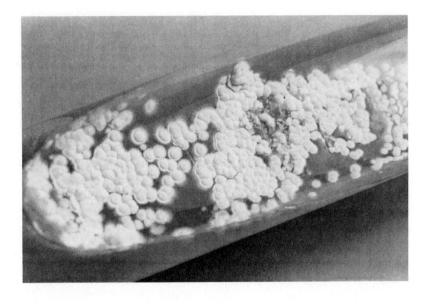

Aerial hyphae of *Nocardia, Streptomyces, Nocardiopsis,* and *Actinomadura* are best displayed on the slightly acidic Sabouraud dextrose agar made from Difco Laboratories' dry powder, according to Leonore Haley of the CDC. Mishra et al.[28] have developed a taxonomic scheme that divides these genera on the basis of biochemical characteristics. Numerous species of *Nocardia* exist, not all of which have been recovered from clinical specimens. A simplified scheme for identification of the most frequently isolated species, developed by McGinnis et al.,[25] is shown in Table 34.4. Readers are referred to the publication by those authors for instructions on preparation of the media used to identify these organisms.

Nocardia sp. are resistant to the action of lysozyme, whereas almost all *Streptomyces* sp. are susceptible. Lysozyme resistance is tested according to Procedure 34.2. This test is subject to variation, however, and may not yield useful results. Another helpful test for differentiating *Nocardia, Nocardiopsis,* and *Streptomyces* from the other aerobic actinomycetes is ONPG, the test for β-galactosidase. The reagents are available commercially (e.g, Key Scientific Products, Round Rock, Texas), since they are usually used for identification of Enterobacteriaceae. The three genera mentioned are positive (able to utilize lactose), whereas other genera of actinomycetes and rapidly growing mycobacteria are negative. A differential agar combining a substrate for β-galactosidase and paraffin as a carbon source can enhance early diagnosis of *Nocardia* infections.[31] When definitive identification of a *Nocardia*-like organism is required, the isolate should be sent to a reference laboratory where cell wall analy-sis by gas-liquid chromatography (GLC) can be performed.

THERAPY

Sulfonamides still appear to be the antimicrobial agents of choice for treatment of nocardiosis. Trimethoprim-sulfamethoxazole has not been shown to be as active as sulfonamides alone against *Nocardia.* Minocycline seems to be a good alternative for sulfonamide-allergic patients, and newer aminoglycosides such as amikacin are very promising. In vitro susceptibility tests can be performed by reference laboratories, and they should be attempted with isolates from patients who appear resistant to treatment.

MISCELLANEOUS OTHER AEROBIC ACTINOMYCETES

Human infections with strains of the genera *Nocardiopsis, Actinomadura, Dermatophilus,* and *Streptomyces* are usually chronic, granulomatous lesions of the skin and subcutaneous tissue, acquired by inoculation. *Mycetoma,* a chronic lesion of the extremities, usually the feet, is characterized by swollen tissue and draining sinus tracts and is usually associated with fungal infection, although it can be caused by members of the actinomycetes. The infection is then called *actinomycetoma.* These organisms are becoming more common as causes of infection in immunocompromised hosts. *Dermatophilus* is discussed further in Chapter 40. All the aerobic actinomycetes are soil saprophytes.

Streptomyces sp., *Actinomadura, Nocardiopsis,* and *Rhodococcus equi* are susceptible to the action of lysozyme, which helps to differentiate them

---

PROCEDURE 34.2

LYSOZYME RESISTANCE FOR DIFFERENTIATING *NOCARDIA* FROM *STREPTOMYCES*

PRINCIPLE

The enzyme lysozyme, present in human tears and other secretions, can break down cell walls of certain microorganisms. Susceptibility to the action of lysozyme can differentiate among certain morphologically similar genera and species.

METHOD

1. Prepare basal broth as follows:
   Peptone (Difco Laboratories), 5 g
   Beef extract (Difco), 3 g
   Glycerol (Difco), 70 ml
   Distilled water, 1000 ml
   Dispense 500 ml of this solution into 16 × 125 mm screw-cap glass test tubes, 5 ml per tube. Autoclave the test tubes and the remaining solution for 15 minutes at 121° C. Tighten caps and store the tubes in the refrigerator for a maximum of 2 months.

2. Prepare lysozyme solution as follows:
   Lysozyme (Sigma Chemical Co.), 100 mg
   HCl (0.01 N), 100 ml
   Sterilize through a 0.45 μm membrane filter.

3. Add 5 ml lysozyme solution to 95 ml basal broth; mix gently, avoiding bubbles; and aseptically dispense in 5 ml amounts to sterile, screw-cap tubes as in step 1. Store refrigerated for a maximum of 2 weeks.

4. Place several bits of the colony of the isolate to be tested into a tube of the basal glycerol broth without lysozyme (control) and into a tube of broth containing lysozyme.

5. Incubate at room temperature for up to 7 days. Observe for growth in the control tube. An organism that grows well in the control tube but not in the lysozyme tube is considered to be susceptible to lysozyme.

QUALITY CONTROL

Test a known susceptible control *(Streptomyces griseus)* and a known resistant control *(Nocardia asteroides)* in the same way. No specific strains have been designated for quality control.

PERFORMANCE SCHEDULE

Test quality control organisms each time the test is performed.

Modified from McGinnis, M.R., D'Amato, R.F., and Land, G.A. 1982. Pictorial handbook of medically important fungi and aerobic actinomycetes. Praeger, New York.

---

from the lysozyme-resistant *Nocardia* sp. A few other species of *Rhodococcus* are also susceptible. Because the biochemical reactions used to identify these organisms are difficult to perform, and because cell wall analysis is usually needed for ultimate species identification, clinically significant strains should be sent to a reference laboratory for testing. Table 34.4 illustrates a simplified scheme for differentiation among some medically important species.

Another group of actinomycetes, the thermophilic actinomycetes, are responsible for hypersensitivity pneumonitis. Clinical laboratories are rarely asked to diagnose such disease, which occurs in farmers, factory workers, and others who are exposed occupationally to spores or fungal elements from which they develop allergy. The species involved, *Thermoactinomyces*, *Saccharomonospora*, and *Micropolyspora*, grow readily on trypticase soy agar with 1% yeast extract or on one-half strength nutrient agar. Hollick[19] presents a practical scheme for identification of the most common etiological agents. Because no effective

therapy exists, patients must prevent disease by altering life-style to avoid exposure to these sensitizing agents.

OTHER AEROBIC NON–SPORE-FORMING BACILLI: *ROTHIA*, *KURTHIA*, *OERSKOVIA*, AND *LACTOBACILLUS*

ROTHIA DENTOCARIOSA

The genus *Rothia* was created in 1967 to accommodate an organism that resembled *Actinomyces* (but grew well aerobically) and resembled *Nocardia* (but did not contain the same cell wall constituents). The only species, *R. dentocariosa*, has caused abscesses and endocarditis and is part of normal human oral flora. It has also been recovered from gingival crevices of patients with periodontal disease and can cause periodontitis in germ-free rats. Unlike other filamentous, gram-positive organisms, *Rothia* contains fructose as a component of the cell wall. The organism grows well on nutrient agar, producing smooth colonies

TABLE 34.4

DIFFERENTIATION OF SELECTED AEROBIC ACTINOMYCETES

SPECIES	ACID FROM CELLOBIOSE	HYDROLYSIS OF					LYSOZYME RESISTANCE
		CASEIN	HYPOXANTHINE	TYROSINE	UREA	XANTHINE	
Actinomadura madurae	+	+	+	+	−	−	−
A. pelletierii	−	+	+/−	+	−	−	−
Nocardia asteroides	−	−	−	−	+	−	+
N. brasiliensis	−	+	+	+	+	−	+
N. otitidis- caviarum (caviae)	−	−	+	+/−	+	+	+
Nocardiopsis dassonvillei	+	+	+	+	+/−	+	−
Streptomyces griseus	+	+	+	+	+	+	−
S. somaliensis	−	+	−	+	+/−	+	−
Streptomyces sp. (other)	+/−	+	+	+	+/−	+/−	−

Modified from McGinnis, M.R., D'Amato, R.F., and Land, G.A. 1982. Pictorial handbook of medically important fungi and aerobic actinomycetes. Praeger, New York.

+, ≥90% of strains positive; −, ≥90% of strains negative; +/−, variable results.

TABLE 34.5

SELECTED KEY FEATURES OF SOME LACTOBACILLI FREQUENTLY ISOLATED FROM CLINICAL SPECIMENS

SPECIES	ARGININE TO AMMONIA	GROWTH AT 45° C	ACID FROM	
			MANNITOL	RAFFINOSE
Lactobacillus acidophilus	−	−	−	+/−
L. casei	−	+	+	−
L. leichmannii	+	−	−	+/−
L. plantarum	−	+/−	+/−	+

+, ≥90% of strains positive; −, ≥90% of strains negative; +/−, variable results.

at 48 hours that become rough or globose on prolonged incubation. Microscopically the cells are coccoid, only becoming branched after several days of incubation in broth. The organism is catalase positive, lactose negative, and produces acetoin from glucose (positive Voges-Proskauer test). *Rothia* is susceptible to most antimicrobial agents, including penicillin, erythromycin, aminoglycosides, and cephalosporins.

KURTHIA AND OERSKOVIA SP.

Both these genera are soil saprophytes, being only rarely implicated in human infection. *Kurthia bessonii* has been isolated from endocarditis. The organism resembles *Bacillus cereus* by forming irregular colonies and by its cellular morphology (large, straight-sided, motile gram-positive bacilli). *Kurthia*, however, does not form spores and is strictly aerobic and nonsaccharolytic.

At least two species of *Oerskovia*, *O. turbata* and *O. xanthineolytica*, can cause human infection. Isolates have been recovered from infected heart valves, blood, wounds, and other sites. Most strains of *Oerskovia* produce a yellow pigment after 48 hours' incubation. Microscopically, the organisms are branching, filamentous bacilli that fragment into motile, coccoid forms after prolonged incubation. Differentiation from *Corynebacterium* sp. is extremely difficult if colonies are not pigmented. The ability of *Oerskovia* to hydrolyze esculin and its motility may be helpful. Only very rare strains of corynebacteria, *C. aquaticum* and a few CDC groups, are motile.

LACTOBACILLUS SP.

The lactobacilli are common inhabitants of the oral cavity, gastrointestinal tract, and female genital tract of humans. They have only rarely been reported to cause serious human infection. Lactobacilli have been isolated from blood, CSF, abscess material, amniotic fluid, and pleural fluid. Endocarditis cases have been reported.[16] The organism is a facultative or strict anaerobe, usually preferring to grow microaerobically; strains are usually identified using anaerobic media and methods (Chapters 35 and 36). The presence of many lactobacilli in the culture or Gram stain of vaginal contents indicates a healthy vagina. When the organism gains access to an unnatural site, it is able to cause suppurative infection. Large amounts of lactic acid are produced as byproducts of metabolism, which helps to maintain the acid pH in which lactobacilli thrive. The absence of hydrogen peroxide–producing endocervical lactobacilli is a sign of bacterial vaginosis. The role of these organisms in protecting the normal vaginal milieu has not been fully elucidated (Chapter 20).

Lactobacillus colonies can assume any morphology on blood agar, ranging from pinpoint, alpha-hemolytic colonies resembling streptococci to large, rough, gray colonies. The microscopic morphology is highly variable as well, with the existence of forms similar to *Bacillus* sp., possibly chaining, as well as coccobacilli, spiral forms, and others. All species are catalase negative. Differentiation from streptococci may be difficult, but the formation of chains of rods in thioglycolate broth and moderate vancomycin resistance (unusual for any gram-positive organism) help to characterize isolates as lactobacilli. Using prereduced, anaerobically sterilized carbohydrate media (Chapter 35), microbiologists can identify some species of lactobacilli. Minimal characteristics of four of the more frequently isolated species that are able to grow aerobically are shown in Table 34.5.

Penicillin, ampicillin, clindamycin, and cephalothin are able to inhibit growth of lactobacilli, but bactericidal activity is not typically seen. The acid environment produced by the organisms may contribute to the decreased activity of certain antimicrobial agents against lactobacilli. Strains of lactobacilli resistant to vancomycin have been isolated from colonized, but not infected, hospitalized individuals; however, these isolates remain susceptible to penicillin. For treatment of patients with endocarditis, it is recommended that combinations of penicillin and aminoglycosides be used, perhaps with addition of rifampin for increased bactericidal activity.[16]

REFERENCES

1. Banck, G., and Nyman, M. 1986. Tonsillitis and rash associated with *Corynebacterium haemolyticum*. J. Infect. Dis. 154:1037.
2. Barksdale, L., Lindr, R., Sulea, I.T., and Pollice, M. 1981. Phospholipase D activity of *Corynebacterium pseudotuberculosis* (*Corynebacterium ovis*) and *Corynebacterium ulcerans*, a distinctive marker within the genus *Corynebacterium*. J. Clin. Microbiol. 13:335.
3. Beckwith, D.G., Jahre, J.A., and Haggerty, S. 1986. Isolation of *Corynebacterium aquaticum* from spinal fluid of an infant with meningitis. J. Clin. Microbiol. 23:375.
4. Bernard, K.A., Bellefeuille, B., and Ewan, E.P. 1991. Cellular fatty acid composition as an adjunct to the identification of asporogenous, aerobic gram-positive rods. J. Clin. Microbiol. 29:83.
5. Boerlin, P., Rocourt, J., and Piffaretti, J.-C. 1991. Taxonomy of the genus *Listeria* by using multilocus enzyme electrophoresis. Int. J. Syst. Bacteriol. 41:59.
6. Casella, P., Bosoni, M.A., and Tommasi, A. 1988. Recurrent *Corynebacterium aquaticum* peritonitis in a patient undergoing continuous ambulatory peritoneal dialysis. Clin. Microbiol. Newsletter 10: 62.
7. Clarridge, J.E. 1984. Aerobic asporogenous gram-positive bacilli. Clin. Microbiol. Newsletter 6:115.
8. Clarridge, J.E. 1986. When, why, and how far should coryneforms be identified. Clin. Microbiol. Newsletter 8:32.
9. Clarridge, J.E. 1989. The recognition and significance of *Arcanobacterium hemolyticum*. Clin. Microbiol. Newsletter 11:41.
10. Cocito, C., and Delville, J. 1983. Properties of microorganisms isolated from human leprosy lesions. Rev. Infect. Dis. 4:649.
11. Coyle, M.B., and Lipsky, B.A. 1990. Coryneform bacteria in infectious diseases: clinical and laboratory aspects. Clin. Microbiol. Rev. 3:227.
12. Crech, T., and Hollis, D.G. 1991. *Corynebacterium* and related organisms. p. 277. In Balows, A., Hausler, W.J., Jr., Herrmann, K.L., et al., editors. Manual of clinical microbiology, ed 5. American Society for Microbiology, Washington, D.C.
13. Elek, S.D. 1949. The plate virulence test for diphtheria. J. Clin. Pathol. 2:250.
14. Freney, J., Duperron, M.T., Courtier, C., et al. 1991. Evaluation of API Coryne in comparison with conventional methods for identifying coryneform bacteria. J. Clin. Microbiol. 29:38.
15. Gavin, S.E., Leonard, R.B., Briselden, A.M, and Coyle, M.B. 1992. Evaluation of the Rapid CORYNE identification system for *Corynebacterium* species and other coryneforms. J. Clin. Microbiol. 30:1692.
16. Griffiths, J.K., Daly, J.S., and Dodge, R.A. 1992. Two cases of endocarditis due to *Lactobacillus* species: antimicrobial susceptibility, review, and discussion of therapy. Clin. Infect. Dis. 15:250.

17. Guarderas, J., Karnad, A., Alvarez, S., and Berk, S.L. 1986. *Corynebacterium minitissimum* bacteremia in a patient with chronic myeloid leukemia in blast crisis. Diagn. Microbiol. Infect. Dis. 5:327.

18. Harvey, R.L., and Sunstrum, J.C. 1991. *Rhodococcus equi* infection in patients with and without human immunodeficiency virus infection. Rev. Infect. Dis. 13:139.

19. Hollick, G.E. 1986. Isolation and identification of thermophilic actinomycetes associated with hypersensitivity pneumonitis. Clin. Microbiol. Newsletter 8:29.

20. Hollis, D.G., and Weaver, R.E. 1981. Gram-positive organisms: a guide to identification. Centers for Disease Control, Atlanta.

21. Kelly, M.C., Smith, I.D., Anstey, R.J., et al. 1984. Rapid identification of antibiotic-resistant corynebacteria with the API 20S system. J. Clin. Microbiol. 19:245.

22. Kunke, P.J. 1987. Serious infection in an AIDS patient due to *Rhodococcus equi.* Clin. Microbiol. Newsletter 9:163.

23. Lasker, B.A., Brown, J.M., and McNeil, M.M. 1992. Identification and epidemiological typing of clinical and environmental isolates of the genus *Rhodococcus* with use of a digoxigenin-labeled rDNA gene probe. Clin. Infect. Dis. 15:223.

24. Linnan, M.J., Mascola, L., Xiao, D.L., et al. 1988. Epidemic listeriosis associated with Mexican-style cheese. N. Engl. J. Med. 319:823.

25. McGinnis, M.R., D'Amato, R.F., and Land, G.A. 1982. Pictorial handbook of medically important fungi and aerobic actinomycetes. Praeger, New York.

26. Miller, R.A., Brancato, F., and Holmes, K.K. 1986. *Corynebacterium hemolyticum* as a cause of pharyngitis and scarlatiniform rash in young adults. Ann. Intern. Med. 105:867.

27. Mishra, S.K., and Randhawa, H.S. 1969. Application of paraffin bait technique to the isolation of *Nocardia asteroides* from clinical specimens. Appl. Microbiol. 18:686.

28. Mishra, S.K., Gordon, R.E., and Barnett, D.A. 1980. Identification of nocardiae and streptomycetes of medical importance. J. Clin. Microbiol. 11:728.

29. Prescott, J.F. 1991. *Rhodococcus equi:* an animal and human pathogen. Clin. Microbiol. Rev. 4:20.

30. Schwartz, J.G., and Tio, F.O. 1987. Nocardial osteomyelitis: a case report and review of the literature. Diagn. Microbiol. Infect. Dis. 8:37.

31. Shawar, R.M., and Clarridge, J.E. 1992. Modified paraffin agar: a selective and differential medium for earlier diagnosis of *Nocardia.* Abstr. 32nd Intersci. Conf. Antimicrob. Agents Chemother. Abstr. 1134, p. 300.

32. Shawar, R.M., Moore, D.G., and LaRocco, M.T. 1990. Cultivation of *Nocardia* spp. on chemically defined media for selective recovery of isolates from clinical specimens. J. Clin. Microbiol. 28:508.

33. Singh, M., Sandhu, R.S., and Randhawa, H.S. 1987. Comparison of paraffin baiting and conventional culture techniques for isolation of *Nocardia asteroides* from sputum. J. Clin. Microbiol. 25:176.

34. Slifkin, M., Gil, G.M., and Engwall, C. 1986. Rapid identification of group JK and other corynebacteria with the Minitek system. J. Clin. Microbiol. 24:177.

35. Soriano, F., Rodriguez-Tudela, J.L., Fernandez-Roblas, R., et al. 1988. Skin colonization by *Corynebacterium* groups D2 and JK in hospitalized patients. J. Clin. Microbiol. 26:1878.

36. Spitzer, P.G., Eliopoulos, G.M., Karchmer, A.W., and Moellering, R.C., Jr. 1988. Synergistic activity between vancomycin or teicoplanin and gentamicin or tobramycin against pathogenic diphtheroids. Antimicrob. Agents Chemother. 32:434.

37. Sulea, I.T., Pollice, M.C., and Barksdale, L. 1980. Pyrazine carboxylamidase activity in *Corynebacterium.* J. Clin. Microbiol. 30:466.

38. Tillotson, G., Arora, M., Robbins, M., and Holton, J. 1988. Identification of *Corynebacterium jeikeium* and *Corynebacterium* CDC group D2 with the API 20 Strep system. Eur. J. Clin. Microbiol. Infect. Dis. 7:675.

39. Vickers, R.M., Rihs, J.D., and Yu, V.L. 1992. Clinical demonstration of isolation of *Nocardia asteroides* on buffered charcoal–yeast extract media. J. Clin. Microbiol. 30:227.

BIBLIOGRAPHY

Ciesielski, C.A., and Swaminathan, B. 1987. *Listeria monocytogenes*—a foodborne pathogen. Clin. Microbiol. Newsletter 9:149.

Fagnant, J.E., Sanders, C.C., and Sanders, W.E., Jr. 1982. Development of and evaluation of a biochemical scheme for identification of endocervical lactobacilli. J. Clin. Microbiol.16:926.

Gellin, B.G., and Broome, C.V. 1989. Listeriosis. J.A.M.A. 261:1313.

Goodfellow, M., Alderson, G., and Lacey, J. 1979. Numerical taxonomy of *Actinomadura* and related organisms. J. Gen. Microbiol. 112:95.

Gorby, G.L., and Peacock, J.E., Jr. 1988. *Erysipelothrix rhusiopathiae* endocarditis: microbiologic, epidemiologic, and clinical features of an occupational disease. Rev. Infect. Dis. 10:317.

Haley, L.D., and Calloway, C.S. 1978. Isolation and identification of some aerobic actinomycetes. In Laboratory methods in medical mycology. U.S. Department of Health, Education, and Welfare Publication CDC 78-8361, Washington, D.C.

McGowan, J.E. 1988. JK coryneforms: a continuing problem for hospital infection control. J. Hosp. Infect. 11(suppl. A):358.

Middlebrook, J.L., and Dorland, R.B. 1984. Bacterial toxins: cellular mechanisms of action. Microbiol. Rev. 48:199.

Reboli, A.C., and Farrar, W.E. 1989. *Erysipelothrix rhusiopathiae:* an occupational pathogen. Clin. Microbiol. Rev. 2:354.

Schwartz, B., Hexter, D., Broome, C.V., et al. 1989. Investigation of an outbreak of listeriosis: new hypotheses for the etiology of epidemic *Listeria monocytogenes* infections. J. Infect. Dis. 159:680.

Slack, J.M., and Gerencser, M. 1975. Actinomyces, filamentous bacteria: biology and pathogenicity. Burgess, Minneapolis.

Smego, R.A., Jr., and Gallis, H.A. 1984. The clinical spectrum of *Nocardia brasiliensis* infection in the United States. Rev. Infect. Dis. 6:164.

Soriano, F., and Fernandez-Roblas, R. 1988. Infections caused by antibiotic-resistant *Corynebacterium* group D2. Eur. J. Clin. Microbiol. Infect. Dis. 7:337.

35

PROCESSING CLINICAL SPECIMENS FOR ANAEROBIC BACTERIA: ISOLATION AND IDENTIFICATION PROCEDURES

GENERAL CONSIDERATIONS

Anaerobic bacteria lack cytochromes and thus are unable to use oxygen as a terminal electron acceptor. They do use many other energy-generating metabolic pathways. Many of their enzymes are more sensitive to oxygen than those of aerobic bacteria. Some anaerobes are able to grow or survive in the presence of small amounts of oxygen, whereas others are rapidly killed. The reasons for this spectrum of oxygen sensitivity vary among the genera and species.[40]

ANAEROBIC BACTERIA AS NORMAL FLORA

Anaerobic bacteria are the predominant flora colonizing the skin and mucous membranes of the body. Table 35.1 summarizes the relative incidence in various body sites. Knowledge of the normal flora can be used to predict what types of organisms will be isolated from infections at these sites, since anaerobes spread from colonized sites to adjacent areas to cause infection. Although it is estimated that 400 to 500 different types or species of anaerobic bacteria may be present as indigenous flora on the body, relatively few of these are isolated frequently from properly collected specimens from infectious processes. Table 35.2 summarizes the incidence of the different genera or groups of anaerobes isolated from clinical specimens.

ANAEROBIC BACTERIA AS ETIOLOGICAL AGENTS OF INFECTION

Anaerobic infections typically contain polymicrobic mixtures of aerobic and anaerobic bacteria. Isolation of five or more anaerobes from a specimen does not imply that the specimen was "contaminated." Anaerobic bacteria produce many different virulence factors that are responsible for tissue destruction and inhibition of host defenses. In some instances, anaerobic bacteria provide growth factors for other organisms present in the infected site. The facultative organisms help to maintain a low oxidation-reduction (redox) potential by reducing oxygen, and they produce en-

TABLE 35.1

INCIDENCE OF ANAEROBES AS NORMAL FLORA OF HUMANS

GENUS	SKIN	UPPER RESPIRATORY TRACT*	INTESTINE	EXTERNAL GENITALIA	URETHRA	VAGINA
Gram-negative bacteria						
Bacteroides	0	±	2	±	±	±
Prevotella	0	2	2	1	±	1
Porphyromonas	0	1	1	U	U	±
Fusobacterium	0	2	1	U	U	±
Veillonella	0	2	1	0	U	1
Gram-positive bacteria						
Peptostreptococcus	1	2	2	1	±	1
Clostridium	±	±	2	±	±	±
Actinomyces	0	1	1	0	0	±
Bifidobacterium	0	1	2	0	0	±
Eubacterium	0	1	2	U	U	1
Lactobacillus	0	1	1	0	±	2
Propionibacterium	2	1	±	±	±	1

Modified from Summanen, P., Baron, E.J., Citron, D.M., et al. 1993. Wadsworth anaerobic bacteriology manual, ed 5. Star Publishing, Belmont, Calif.

U, unknown; *0*, not found or rare; ±, irregular; *1*, usually present; *2*, present in large numbers.

*Includes nasal passages, nasopharynx, oropharynx, and tonsils.

TABLE 35.2

INCIDENCE OF ANAEROBES IN CLINICAL SPECIMENS*

ORGANISM	NO.	PERCENT OF TOTAL
Bacteroides fragilis group	141	34.8
B. fragilis	77	19.0
B. thetaiotaomicron	12	3.0
B. vulgatus	10	2.4
B. distasonis	10	2.4
B. ovatus	6	1.5
B. uniformis	3	0.7
Unidentified	23	5.7
Pigmented *Prevotella/Porphyromonas*	26	6.4
Other *Bacteroides* and *Prevotella* sp.	45	11.1
Fusobacterium sp.	32	7.9
Peptostreptococcus	117	28.9
Clostridium sp.	9	2.2
Non–spore-forming gram-positive rods	20	4.9
Gram-negative cocci	15	3.7

Modified from Goldstein, E.J.C., and Citron, D.M. 1988. J. Clin. Microbiol. 26:2361.

*Excluding *Propionibacterium acnes*.

zymes that inactivate oxygen radicals. Thus the mixture of organisms frequently acts synergistically to produce infection. Anaerobes may be isolated from virtually any body site. Table 35.3 summarizes the incidence of anaerobic involvement in infections at different sites.

RECENT TAXONOMIC CHANGES

Historically, anaerobic bacteria were classified according to morphological appearance and phenotypical characteristics. As a result, genetically diverse organisms were grouped within a genus. Use of biochemical methods such as gas-liquid

TABLE 35.3

INCIDENCE OF ANAEROBIC BACTERIA IN SELECTED INFECTIONS

TYPE OF INFECTION	INCIDENCE (%)
Bacteremia	3-10
Ocular infections	38
Corneal ulcers	7
Central nervous system	
Brain abscess	89
Subdural empyema	29
Epidural abscess	39
Head and neck	
Chronic sinusitis	52
Acute sinusitis	7
Chronic otitis media	50
Cholesteatoma	92
Neck space infections	100
Head and neck surgical infections	95
Dental, oral, and facial infections	86-100
Thoracic	
Aspiration pneumonia	62-100
Lung abscess	85-93
Abdominal	
Intraabdominal (general)	81-94
Appendicitis with peritonitis	92
Liver abscess	52
Biliary tract	45
Obstetrical-gynecological	
Pelvic abscess	88
Vulvovaginal abscess	75
Septic abortion	67
Pelvic inflammatory disease	25-48
Soft tissue and miscellaneous	
Nonclostridial crepitant cellulitis	75
Pilonidal abscess	88
Bite wounds	
Human	50-56
Dog	30-41
Diabetic foot ulcers	95
Soft tissue abscess	60
Osteomyelitis	40
Breast abscess, nonpuerperal	53-79
Perirectal abscess	77

Modified from Sunnamen, P., Baron, E.J., Citron, D.M., et al. 1993. Wadsworth anaerobic bacteriology manual, ed 5. Star Publishing, Belmont, Calif.

chromatography (GLC) for determination of end products of glucose metabolism, guanine plus cytosine content determination, and gel electrophoresis of cellular proteins provided more reliable tools for classifying anaerobes. More recently, molecular biological techniques have been developed that allow for classification based on genetic similarity.[44] This has resulted in taxonomic changes, especially within the genus *Bacteroides*.[45-47] Other genera, such as *Eubacterium* and *Clostridium*, also include many genetically diverse species and will likely be further subdivided and renamed. The specific taxonomic changes that have occurred are described in subsequent chapters.

Although these changes appear to complicate identification, they actually allow for grouping of organisms based on similarities in metabolic activity, ecological niches, and virulence. Also, these changes help to explain the differences observed with pathogenic or nonpathogenic organisms isolated from different sites and types of infections. Fortunately for clinical microbiologists, most of these taxonomic changes have been described for organisms associated with periodontal disease and, to a much lesser extent, for organisms isolated from clinical specimens.

INITIAL PROCESSING PROCEDURE

Specimens for anaerobic culture may be processed on the bench using anaerobic jars and pouches, in an anaerobic chamber, or with the prereduced anaerobically sterilized (PRAS) system (Chapter 9). Collection procedures must exclude contamination from areas colonized with anaerobes, and transport methods must exclude air to protect the anaerobes (Chapter 7). Initial processing includes a direct examination and inoculation of the specimen onto appropriate media, as represented in Figure 35.1.

DIRECT EXAMINATION

The specimen is inspected for characteristics that strongly indicate the presence of anaerobes, such as (1) foul odor and (2) sulfur granules (see Figures 36.14 and 36.15) associated with *Actinomyces* sp., *Propionibacterium propionicum* (formerly *Arachnia*), or *Eubacterium nodatum*. The presence of blood, purulence, or other distinguishing characteristics of the specimen are also noted. Specimens containing large quantities of pigmented *Prevotella* or *Porphyromonas* may fluoresce brick red under long-wave ultraviolet (UV) light.

The microscopic examination always includes a Gram stain (Chapter 8). The Gram stain reveals the types and relative numbers of microorganisms and host cells present and serves as a quality control measure for the adequacy of anaerobic techniques. Results of the Gram stain may also indicate the need for additional media, such as egg yolk agar for clostridia. Some anaerobes have characteristic cell morphology on Gram stain, such as boxcar-shaped cells of *Clostridium perfringens* (Figure 35.2) or thin, pointed-ended cells of *Fusobacterium* sp. (Figure 35.3). Because the organisms found in anaerobic infections usually

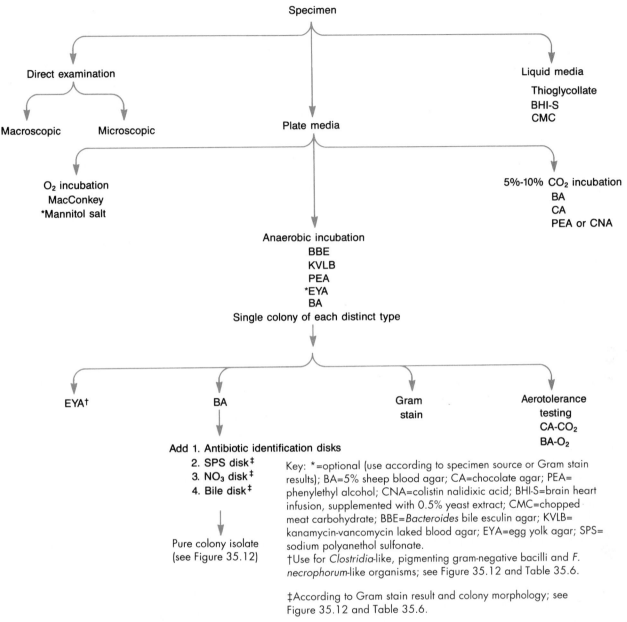

FIGURE 35.1

Initial specimen processing scheme, with suggested primary media and rapid tests. (From Sutter, V.L., Citron, D.M., Edelstein, M.A.C., and Finegold, S.M. 1985. Wadsworth anaerobic bacteriology manual, ed 4. Star Publishing, Belmont, Calif.)

originate from normal flora, knowledge of the flora at various sites can usually predict which organisms will be isolated from the infected site (Table 35.1). For example, a tangled mass of branching or clubbing gram-positive bacilli observed on a Gram stain smear from a jaw is good presumptive evidence for actinomycosis.

A positive Gram stain with a negative culture may indicate (1) poor transport methods, (2) excessive exposure to air during the processing procedure, (3) lack of anaerobic atmosphere in the incubation system, (4) inadequate types of media or old media, or (5) nonviable cells resulting from an-

tibiotic use or presence of other bactericidal factors.

Darkfield or phase-contrast microscopy may aid in detecting poorly staining organisms and spores and in observing motility directly, but it is generally not useful.

SPECIMEN PREPARATION AND INOCULATION

After direct visual examination, the specimen is inoculated onto appropriate anaerobic and aerobic plating media, into a liquid medium, and onto a slide for Gram stain. In preparation for culture, grossly purulent material is vortex mixed in a gas-tight anaerobe transport vial or in a regular tube within an anaerobic chamber to ensure even distribution of the bacteria. Pieces of tissue or bone fragments are homogenized with approximately 1 ml of liquid medium using a tissue grinder to make a thick mixture. This should be done in an anaerobic chamber to minimize aeration. If a chamber is not available, one should process the specimen as rapidly as possible. Swab specimens are vortexed and "wrung out" in about 0.5 ml of broth and treated as a liquid specimen. Using a Pasteur pipette, the prepared specimen is transferred as follows:

- 1 drop per plate for purulent material
- 2 to 3 drops per plate for nonpurulent material
- 0.5 to 1 ml to middle or bottom of liquid medium
- 1 drop evenly spread onto a slide

The plates are then streaked for isolation.

The direct smear is fixed in methanol[31] for 30 seconds, rather than heat fixed, to preserve tissue cell and bacterial cell morphology. Standard Gram stain procedures and reagents are used, except that the counterstain is 0.5% aqueous basic fuchsin, or Kopeloff's method (VPI *Anaerobe Laboratory Manual;* see Bibliography) can be followed.

Enriched and selective media are required for optimal recovery of anaerobes, since many of these organisms are fastidious and most anaerobic infections are mixed with more rapidly growing aerobes. Primary plates should be freshly prepared or used within 2 weeks of preparation. Plates stored for longer periods accumulate peroxides and become dehydrated, which results in growth inhibition.[20] Reduction of media in an anaerobic environment eliminates dissolved oxygen but has no effect on the peroxides. Prereduced, anaerobically sterilized PRAS plate media are available[30] (Anaerobe Systems, San Jose, Calif.). The following supplements are added to basal media:

- For blood agar plates, 5% sheep, horse, or rabbit blood for enrichment and for detecting hemolysis and pigment production (laked rabbit blood best for the latter)
- Vitamin K_1 (final concentration, 10 µg/ml for solid media and 0.1 µg/ml for liquid media) for growth of some pigmented *Prevotella* and *Porphyromonas* sp. (formerly *Bacteroides*)
- Hemin (final concentration, 5 µg/ml) for growth enhancement of the *Bacteroides fragilis* group and other *Bacteroides* and *Prevotella* sp.

The following media are inoculated for the primary isolation of anaerobic bacteria:

1. Nonselective, 5% sheep blood agar supplemented with vitamin K_1 and hemin (BA). Blood agar media may be prepared with Columbia, Schaedler, CDC, or brain-heart infusion base medium supplemented with 0.5% yeast extract. Unsupplemented trypticase soy and heart infusion or brain-heart infusion agar plates do not adequately support the growth of most anaerobes.

2. *Bacteroides* bile esculin (BBE) agar for the selection and presumptive identification of *Bacteroides fragilis* group organisms and *Bilophila wadsworthia.*[28] Rapid presumptive evidence of the *B. fragilis* group is important, since these organisms occur in more than one third of specimens containing anaerobes (Table 35.2), produce several virulence factors, and are more resistant to many antimicrobial agents than are other anaerobes. The medium contains 100 µg/ml of gentamicin, which inhibits most aerobic organisms; 20% bile, which inhibits most anaerobes except for the *B. fragilis* group and a few other species; and esculin, which differentiates esculin-positive and esculin-negative *B. fragilis* group (most are esculin positive). Other non–*B. fragilis* group organisms that may rarely grow on this medium are *Fusobacterium mortiferum* and *F. varium* and gentamicin-resistant strains of Enterobacteriaceae and enterococci. However, their colony size and morphology are unlike those of the *B. fragilis* group or *Bilophila* sp.

3. Kanamycin-vancomycin-laked blood (KVLB) agar for the selection of *Prevotella* and *Bacteroides* sp. The medium contains 75 µg/ml of kanamycin, which inhibits most aerobic, facultative, and anaerobic gram-negative rods except for *Prevotella* and *Bacteroides,* and 7.5 µg/ml of vancomycin, which inhibits most gram-positive organisms. The laked blood allows earlier pigmentation by the pigmented *Prevotella. Porphyromonas* sp. do not grow on this medium because of their susceptibility to vancomycin. Other formulations containing 2 µg/ml of vancomycin permit growth of *Porphyromonas* sp. Some strains of fusobacteria also grow on KVLB, especially when they are present as part of a mixed culture. Yeast and other kanamycin-resistant organisms sometimes grow on this medium; therefore one should perform a Gram stain and aerotolerance for all isolates. BBE and KVLB are available as biplates from several prepared media manufacturers (Figure 35.4).

4. Phenylethyl alcohol sheep blood agar (PEA) for the inhibition of enteric gram-negative bacilli and swarming by some clostridia. Most

FIGURE 35.2

Clostridium perfringens.
Boxcar-shaped cells.
(From Summanen, P.,
Baron, E.J., Citron, D.M.,
et al. 1993. Wadsworth
anaerobic bacteriology
manual, ed 5. Star
Publishing, Belmont,
Calif.)

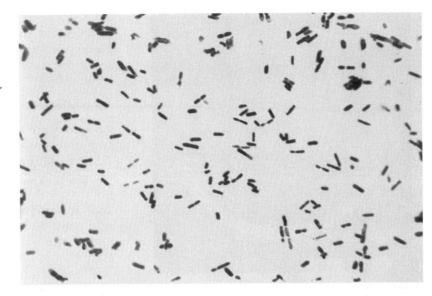

FIGURE 35.3

Fusobacterium nucleatum.
Thin bacillus with pointed
ends.

anaerobes grow on the primary PEA plates, especially in mixed culture. Colony morphology on PEA is similar to that on the blood agar, but it may take an additional 48 hours' incubation to detect the more slowly growing and pigmenting anaerobes. Columbia blood agar with colistin and nalidixic acid (CNA) may be used in place of PEA, although it inhibits fusobacteria, *Veillonella,* the *Bacteroides ureolyticus* group, and some strains of *Prevotella* and *Clostridium perfringens.* It does not inhibit swarming by *Clostridium septicum.*

5. Thioglycolate medium (BBL 135C), supplemented with vitamin K_1 (0.1 µg/ml) and hemin (5 µg/ml), and a marble chip or sodium bicarbonate (1 mg/ml, added before use). This medium must be boiled or steamed to drive off residual oxygen and is used on the same day. Alternatively, PRAS thioglycolate may be used without steaming (Anaerobe Systems). Thioglycolate medium contains a small amount of agar that inhibits convection currents; thus the tube may be incubated aerobically. The broth serves only as a backup source of culture material, which may be needed in the event of a jar failure or poor growth on plates. Other broths, such as chopped meat carbohydrate or chopped meat, are also acceptable. Supplemented brain-heart infusion broth may also be used, but it must be gassed with anaerobic gas after inoculating or incubated with the cap loosened in an anaerobic atmosphere.

If clostridia are suspected, an egg yolk agar (EYA) plate is inoculated for rapid detection of

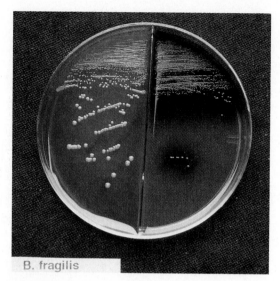

FIGURE 35.4

Bacteroides fragilis growing on biplate of *Bacteroides* bile esculin and laked blood with kanamycin and vancomycin agars. (Courtesy Anaerobe Systems, San Jose, Calif.)

lecithinase-producing or lipase-producing species. If boxcar-shaped cells resembling *C. perfringens* are seen on Gram stain (Figure 35.2), a direct Nagler test (Procedure 35.6) can also be set up along with the primary plates.

INCUBATION OF CULTURES

Inoculated plates are immediately placed into an anaerobic environment (jars, chamber, anaerobic bag; Chapter 9), and are incubated at 35° to 37° C for 48 hours. In general, cultures must not be exposed to oxygen until after 48 hours' incubation, since anaerobes are most sensitive to oxygen during their log phase of growth, and colony morphology often changes dramatically between 24 and 48 hours. Plates incubated in a chamber or anaerobic bag can be observed at 24 hours or earlier without oxygen exposure. At 24 hours, colonies of the *B. fragilis* group are evident on the BA and BBE plates and *Clostridium*-like organisms on the BA and EYA plates. Other rapidly growing anaerobes may also be apparent after 24 hours. Plates showing no growth are incubated anaerobically for at least 5 days before discarding.

The thioglycolate broth culture associated with plates showing no growth is inspected daily for 7 days and, if turbid, subcultured to a BA plate (anaerobic incubation) and a chocolate agar plate (carbon dioxide [CO_2] incubation). If no growth occurs on the primary aerobic plates, growth in the broth, especially in the lower half of the tube, suggests the presence of anaerobes. A final report

of a negative culture should be issued at 7 days, unless special circumstances exist that warrant longer incubation.

EXAMINATION OF PRIMARY PLATES

Anaerobes are usually present in mixed culture with other anaerobes and with facultative bacteria. The combination of selective and differential agar plates yields information that suggests the presence and perhaps the types of anaerobe(s). Primary anaerobic plates should be examined with a hand lens ($\times$ 8) or, preferably, a stereoscopic microscope. Colonies are described from the various media and semiquantitated.

The *B. fragilis* group grows selectively on BBE; therefore, on first inspection, only colonies greater than 1 mm in diameter should be described, enumerated, and subcultured. At 48 hours, *B. fragilis* group colonies are greater than 1 mm in diameter, circular, entire, and raised, with three distinct morphotypes:

1. Low convex, dark gray, friable, and surrounded by a dark gray zone (esculin hydrolysis) and sometimes a precipitate (bile)
2. Glistening, convex, light to dark gray, and surrounded by a gray zone
3. Similar to the second morphotype, but with no gray zone (esculin not hydrolyzed), suggestive of *Bacteroides vulgatus*

Bilophila wadsworthia also grows on BBE, but only after 3 to 5 days. Colonies are usually gray with a black center because of the production of hydrogen sulfide (H_2S). The black center may disappear after exposure to air.

Growth on the KVLB plate suggests the presence of *Prevotella* or *Bacteroides* sp. One must check carefully for pigment, which may take more than 48 hours to appear, and for brick-red fluorescence, which is detected with a long-wave UV light (Wood's lamp).

All colony morphotypes from the nonselective BA plate are characterized and subcultured to purity plates, since facultative and obligately anaerobic bacteria frequently have similar colony appearances. Nevertheless, the distinct colony morphology of certain anaerobes is highly suggestive of their presence in a culture. Pitting colonies present on the anaerobic but not on the CO_2 plate are suggestive of the *Bacteroides ureolyticus*–like group, (Figure 35.5). Large colonies with a double zone of beta hemolysis (Figure 35.6), made up of large gram-positive bacilli, are presumptively *Clostridium perfringens*. Slender, fusiform, gram-

FIGURE 35.5

Bacteroides ureolyticus. Pitting colonies on blood agar. (From Summanen, P., Baron, E.J., Citron, D.M., et al. 1993. Wadsworth anaerobic bacteriology manual, ed 5. Star Publishing, Belmont, Calif.)

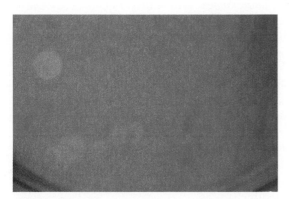

FIGURE 35.6

Clostridium perfringens. Double zone of beta hemolysis on blood agar.

negative bacilli (Figure 35.3) from bread crumb–like or speckled colonies showing greening of the agar around the colonies suggest *Fusobacterium nucleatum*. Dark, pigmented (Figure 35.7) or brick-red fluorescent colonies (Figure 35.8) that are gram-negative coccobacilli are pigmented *Prevotella* sp. or *Porphyromonas*. A "molar tooth" colony (Figure 35.9) of gram-positive branching rods (Figure 35.10) may be *Actinomyces israelii*.

Colonies on the PEA plate are processed further only if they are different from ones growing on the BA plate or if colonies on the BA plate are impossible to subculture because of overgrowth by swarming clostridia, *Proteus,* or other organisms.

SUBCULTURE OF ISOLATES

A single colony of each distinct type described is examined microscopically using a Gram stain and is subcultured to the following (Figure 35.1):

- BA plate to be incubated anaerobically for isolation of the organism and further testing (purity plate)
- Chocolate agar plate to be incubated in CO_2 for aerotolerance testing
- EYA plate may be inoculated if lipase and lecithinase producers *(Clostridium* sp., *Fusobacterium necrophorum, Prevotella intermedia)* are suspected
- Rabbit blood (preferably laked) agar for gram-negative bacilli to better detect pigment production (optional)

One should use a sterile wooden applicator stick to subculture colonies; enough cells can usually be picked up from one colony to inoculate all of the aforementioned plates, a broth, and a slide for Gram stain. If the original colony is not large enough to determine atmospheric requirements directly, use growth from the purity plate for aerotolerance testing. Eight to 12 isolates can be inoculated onto a single, subdivided chocolate agar plate. An additional BA plate incubated in air further defines the atmospheric requirements of facultative organisms if necessary (e.g., microaerobic streptococci). The Centers for Disease Control (CDC) Anaerobe Laboratory recommends using five different atmospheres, although this is usually not necessary for routine identification. Resistance to a 5 µg metronidazole disk may also aid in determining atmospheric requirements of the organism, since only strict anaerobes are resistant to metronidazole.[43] This may be helpful for those microaerobic or fastidious bacteria that may re-

FIGURE 35.7

Prevotella intermedia. Black pigment on sheep blood agar *(left)* versus rabbit blood agar *(right)*. (From Summanen, P., Baron, E.J., Citron, D.M., et al. 1993. Wadsworth anaerobic bacteriology manual, ed 5. Star Publishing, Belmont, Calif.)

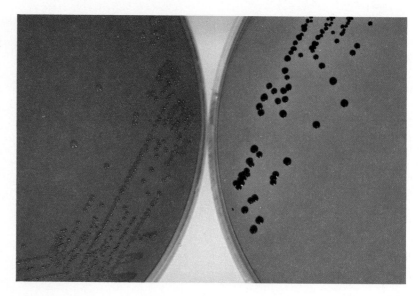

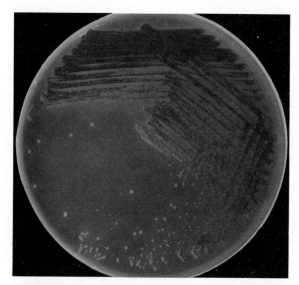

FIGURE 35.8

Pigmented *Prevotella* sp. Red fluorescence under ultraviolet light (365 nm).

quire an anaerobic or reduced oxygen environment for initial growth.

The following antibiotic identification disks are placed on the first quadrant of the purity BA plate (Procedure 35.1): kanamycin, 1 mg (BBL or Anaerobe Systems): colistin, 10 μg; and vancomycin, 5 μg (Figure 35.11) These disks aid in preliminary grouping of anaerobes and serve to verify the Gram stain, but they *do not* imply susceptibility of an organism for antibiotic therapy.

Three other disks may be added to the BA at this time or may be used as needed after isolation of the organism. A sodium polyanetholsulfonate (SPS) disk can be placed near the colistin disk for rapid presumptive identification of *Peptostreptococ-*

cus anaerobius. A nitrate disk may be placed on the second quadrant for subsequent determination of nitrate reduction. A bile disk is added to the second quadrant to detect bile inhibition.

If processing is performed on the open bench, all plates should promptly be incubated anaerobically, since some clinical isolates (for example, *Fusobacterium necrophorum* and some *Prevotella* sp.) may die after relatively short exposure to oxygen. The primary plates are reincubated along with the purity plates for an additional 48 to 72 hours and are again inspected for slowly growing or pigmenting strains.

IDENTIFICATION OF ISOLATES

Pure culture isolates are processed for identification as shown in Figure 35.12. The extent of identification and susceptibility testing of anaerobes varies according to the laboratory's interests and resources. Three levels of identification can be defined. All laboratories should be able to identify colonies presumptively from primary plates as belonging to major groups of anaerobes and to isolate and maintain an anaerobe in pure culture so that it can be sent to a reference laboratory if needed (level I). Most laboratories should be able to use simple tests for further grouping of the anaerobes and to identify certain ones to species level (level II). Reference laboratories and laboratories with sufficient resources should be able to identify most isolates definitively using a variety of techniques (level III). These methods may include PRAS biochemicals, miniaturized biochemical systems (e.g., API, Minitek; Chapter 10), rapid-preformed enzyme detection panels (e.g.,

FIGURE 35.9

Actinomyces israelii. "Molar tooth" colonies.

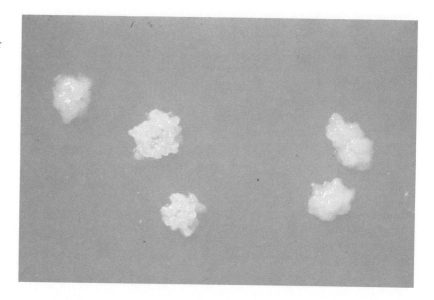

FIGURE 35.10

Actinomyces israelii. Sulfur granule demonstrating gram-positive branching bacilli on microscopic examination.

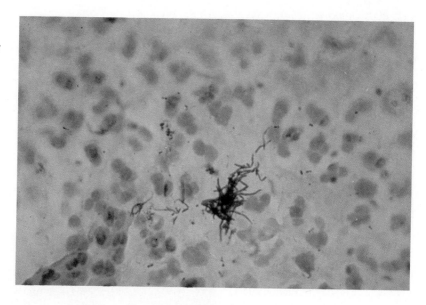

RapID-ANA II, Innovative Diagnostic Systems, Norcross, Ga.; AnIDENT, bioMérieux, Hazelwood, Mo.) or individual tests (Rosco ID Disks, Rosco Diagnostica, Taastrup, Denmark), GLC (Chapter 12), toxin assays, 4-methylumbelliferone derivative substrates,[37] and others.

LEVEL I IDENTIFICATION

As described earlier, information from the primary plates in conjunction with the atmospheric requirements, Gram stain, and colony morphology of a pure isolate provides presumptive identification of many anaerobic organisms. Table 35.4 summarizes the extent to which isolates can be identified using this information. It is also useful to consider the specimen source and expected or-

ganisms from that site to aid in presumptive identification.

LEVEL II IDENTIFICATION TESTS

Preliminary grouping of anaerobes and identification of some are based on colony and cell morphology, Gram reaction, susceptibility to antibiotic identification disks, nitrate reduction (disk test), indole and catalase production, plus several other simple tests, such as glutamic acid decarboxylase production[3] and L-analyl–L-analyl-aminopeptidase production (Carr-Scarborough, Stone Mountain, Ga). The following characteristics are noted, and tests are performed from the purity BA plate on all isolates, as depicted schematically in Figure 35.12.

PROCEDURE 35.1

ANTIBIOTIC IDENTIFICATION DISKS

PRINCIPLE

Most anaerobes have a characteristic susceptibility pattern to colistin (10 μg), vancomycin (5 μg), and kanamycin (1 mg) disks. The pattern generated will usually confirm a dubious Gram stain reaction (with few exceptions, most gram-positive anaerobes are susceptible to vancomycin) and aid in subdividing the anaerobic gram-negative bacilli into groups.

METHOD

1. Reagents
   a. Vancomycin, 5 μg disk (Va)
   b. Kanamycin, 1 mg disk (K)
   c. Colistin, 10 μg disk (Co)
   d. Supplemented blood agar plate (BA)
2. Procedure
   a. Allow the three cartridges of disks to equilibrate to room temperature. A small supply of disks is stored desiccated at 2° to 8° C and a larger supply at −20° C.
   b. Transfer a portion of one colony to a BA plate. Streak the first quadrant several times to produce a heavy lawn of growth, and then streak the other quadrants for isolation.
   c. Place the Co, K, and Va disks in the first quadrant, well separated from each other (Figure 35.11).
   d. Incubate the plates anaerobically for 48 hours at 35° C.
3. Reading and interpretation: observe from a zone of inhibition of growth. A zone of 10 mm or less indicates resistance, and a zone greater than 10 mm indicates susceptibility.

QUALITY CONTROL

Test *Fusobacterium necrophorum*, *Bacteroides fragilis*, and *Clostridium perfringens* as described under Methods. Two isolates can be inoculated per plate.

EXPECTED RESULTS

The disks meet performance standards when the following patterns are obtained:

1. *F. necrophorum* is susceptible to K and Co and resistant to Va.

2. *B. fragilis* is resistant to all three antibiotics.

3. *C. perfringens* is sensitive to Va and K and resistant to Co.

PERFORMANCE SCHEDULE

Test each new lot of disks and weekly thereafter.

FIGURE 35.11

Bacteroides fragilis. Special-potency antibiotic disks, bile-disk resistance, and blank disk for the spot indole test.

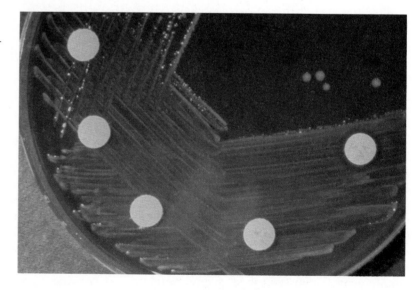

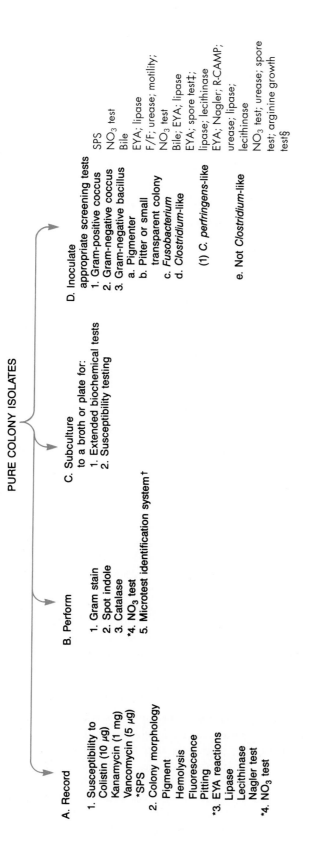

PURE COLONY ISOLATES

A. Record

1. Susceptibility to
   Colistin (10 μg)
   Kanamycin (1 mg)
   Vancomycin (5 μg)
   *SPS
2. Colony morphology
   Pigment
   Hemolysis
   Fluorescence
   Pitting
*3. EYA reactions
   Lipase
   Lecithinase
   Nagler test
*4. NO₃ test

B. Perform

1. Gram stain
2. Spot indole
3. Catalase
*4. NO₃ test
5. Microtest identification system†

C. Subculture
to a broth or plate for:
1. Extended biochemical tests
2. Susceptibility testing

D. Inoculate
appropriate screening tests
1. Gram-positive coccus   SPS
2. Gram-negative coccus   NO₃ test
3. Gram-negative bacillus   Bile
   a. Pigmenter   EYA; lipase
   b. Pitter or small   F/F; urease; motility;
      transparent colony   NO₃ test
   c. *Fusobacterium*   Bile; EYA; lipase
   d. *Clostridium*-like   EYA; spore test‡;
                                 lipase; lecithinase
      (1) *C. perfringens*-like EYA; Nagler; R-CAMP;
                                 urease; lipase;
                                 lecithinase
   e. Not *Clostridium*-like   NO₃ test; urease; spore
                                 test; arginine growth
                                 test§

Key: * =if available at this time; SPS=sodium polyanethol
sulfonate; EYA=egg yolk agar; F/F=formate-fumarate growth test;
R-CAMP=reverse CAMP test
† Use test systems selectively and with appropriate
  organisms.
‡ Perform when spores are not detected on Gram stain.
§ Perform on tiny gram-positive rod.

FIGURE 35.12

Pure culture isolate processing scheme. (From Summanen, P., Baron, E.J., Citron, D.M., et al. 1993. Wadsworth anaerobic bacteriology manual, ed 5. Star
Publishing, Belmont, Calif.)

TABLE 35.4

LEVEL I IDENTIFICATION

ORGANISM	GRAM REACTION	CELL SHAPE	COMMENTS	AERO-TOLER-ANCE	DISTINGUISHING CHARACTERISTICS
Bacteroides fragilis group	−	B	Can be pleomorphic with safety pin feature	−	Grows on BBE; >1 mm in diameter, ± esculin
Pigmented gram-negative bacilli	−	B, CB	Can be very coccoid or *Haemophilus*-like	−	Foul odor; black or brown pigment; some fluoresce brick red
Bacteroides ureolyticus–like	−	B	Thin; some curved	−	May pit agar or spread; transparent colony
Fusobacterium nucleatum	−	B	Slender cell with pointed ends	−	Foul odor; three colony types: bread crumb, speckled, or smooth
Gram-negative bacillus	−	B		−	
Gram-negative coccus	−	C	*Veillonella* cells are tiny	−	
Gram-positive coccus	+	C, CB	Variable size	−	
Clostridium perfringens (presumptive)	+	B	Large boxcar shape; no spores observed; may appear gram negative	−	Double-zone beta hemolysis
Clostridium sp.	+	B	Spores usually observed; may appear gram negative	−+	
Gram-positive bacillus	+	B, CB	No spores observed, no boxcar-shaped cells	−+	
Actinomyces-like	+	B	Branching cells	−+	Sulfur granules on direct examination; "molar tooth" colony

−, Negative; +, positive; −+, some strains positive; *B*, bacillus; *C*, coccus; *CB*, coccobacillus; *BBE, Bacteroides* bile esculin agar.

COLONY MORPHOLOGY Colony features of the isolate are described from a 48-hour pure BA culture and include the shape, edge, elevation, opacity, color, and any other distinguishing characteristics (e.g., irregular, smooth, convex, opaque, white, speckling).

HEMOLYSIS Transmitted light is used to look for hemolysis, especially a double zone of hemolysis (Figure 35.6) and greening of the agar, which is caused by hydrogen peroxide production by the organism after exposure to air.

PIGMENT Few anaerobes produce pigment, but colors common among those that do vary from light tan to black (e.g., pigmented *Prevotella* sp. or *Porphyromonas*) or pink to reddish brown (e.g., *Actinomyces odontolyticus*). Whole or laked rabbit blood is best for detecting pigment.

FLUORESCENCE Long-wave UV light (366 nm) is used to detect fluorescense. A variety of colors may be seen, including brick red, pink, orange, and chartreuse (Figure 35.8). Certain pigmented *Prevotella* sp. and *Porphyromonas* fluoresce coral pink, brick red, or not at all. Fluorescence often disappears or is masked as pigmentation of the colony deepens. A few other *Prevotella* sp. fluoresce yellow-green or coral. Many fusobacteria fluoresce yellow-green. *Veillonella* sp. fluoresce red, but this property is medium dependent and dissipates rapidly on exposure to air. Certain *Clostridium* and *Eubacterium* sp. fluoresce yellow-green, red, or coral.

PITTING The top of the colony usually appears just above the level of a craterlike depression in the agar; this depression has been created by an agarase enzyme produced by the organism. One should hold the BA plate at an angle to the light source to observe this phenomenon. This characteristic may be easier to detect after 4 days of incubation (Figure 35.5).

CATALASE See Chapter 10 and Procedure 10.1. One uses 15% hydrogen peroxide in place of 3%

TABLE 35.5

GROUPING OF ANAEROBES USING THE ANTIBIOTIC IDENTIFICATION DISKS PATTERN

GROUP	VANCOMYCIN (5 µg)	KANAMYCIN (1 mg)	COLISTIN (10 µg)
Gram-positive	S	V	R
Gram-negative	R+	V	V
Bacteroides fragilis group	R	R	R
Prevotella sp.	R	R	V
Porphyromonas sp.†	S	R	R
Fusobacterium	R	S	S
Bacteroides ureolyticus group and gram-negative cocci	R	S	S

S, sensitive; *R*, resistant; *V*, variable.

†*Porphyromonas* is unusual; it is a gram-negative bacillus with a gram-positive disk pattern.

FIGURE 35.13

Spot indole test using *p*-dimethylaminocinnamal-dehyde. Blue indicates positive; colorless or other than blue indicates negative. (From Sutter, V.L., Citron, D.M., Edelstein, M.A.C., and Finegold, S.M. 1985. Wadsworth anaerobic bacteriology manual, ed 4. Star Publishing, Belmont, Calif.)

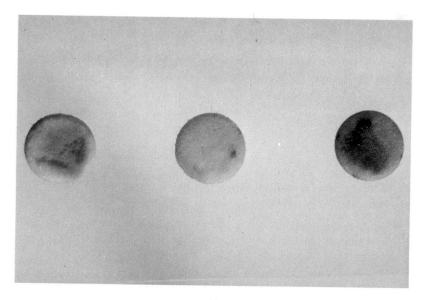

hydrogen peroxide. Most anaerobes are catalase negative; some exceptions include *Bilophila wadsworthia, Actinomyces viscosus,* most species in the *Bacteroides fragilis* group, most *Propionibacterium* sp., and some strains of anaerobic cocci, *Veillonella,* and *Eubacterium lentum.*

SPOT INDOLE   See Chapter 10 and Procedure 10.4, and Figure 35.13. The reagent is *p*-dimethyl-aminocinnamaldehyde, and blue color indicates a positive reaction.

ANTIBIOTIC IDENTIFICATION DISKS   See Procedure 35.1 and Table 35.5. Anaerobes vary in their susceptibility to colistin (10 µg), vancomycin (5 µg), and kanamycin (1 mg), and the reaction patterns aid in verifying the Gram reaction, identifying the

genus *Porphyromonas,* separating *Prevotella* and *Bacteroides* sp. from *Fusobacterium* sp., and subdividing the *Bacteroides* and *Prevotella* (Tables 35.5 to 35.7). Generally, gram-positive organisms are sensitive to vancomycin and resistant to colistin, whereas the gram-negative organisms are resistant to vancomycin. Young cultures of pigmenting *Prevotella* and clostridia may appear gram variable, and some clostridia consistently stain gram negative; *Clostridium* sp. are sensitive to vancomycin, and the *Prevotella* sp. are resistant. A few exceptions do occur, such as the *Porphyromonas* sp. (formerly *Bacteroides asaccharolyticus* group), which are resistant to colistin and sensitive to vancomycin. *Fusobacterium* sp. are susceptible to both kanamycin and colistin, the *Bacteroides fragilis* group are resistant to kanamycin and colistin,

TABLE 35.6

LEVEL II IDENTIFICATION OF GRAM-NEGATIVE ANAEROBES

	CELL CURVED	CELL SHAPE	SLENDER CELLS WITH POINTED ENDS	KANAMYCIN (1 mg)	VANCOMYCIN (5 µg)	COLISTIN (10 µg)	GROWTH IN BILE	INDOLE	CATALASE	PIGMENTED COLONY	BRICK-RED FLUORESCENCE	LIPASE	PITS THE AGAR	REQUIRES FORMATE-FUMARATE	NITRATE REDUCTION	UREASE	MOTILE	STRONG OXIDASE
Bacteroides fragilis group	–	B	–	R	R	R	+	V	V	–	–	–	–	–	–	–	–	–
Pigmented sp.	–	B, CB	–	R	R	V	–	V	–	+*	V	V	–	–	–	–	–	–
Prevotella intermedia	–	B, CB	–	R	R	S	–	+	–	+	+	+	–	–	–	–	–	–
Prevotella loescheii	–	B, CB	–	R	R	Rˢ	–	–	–	+⁻*	+	–⁺	–	–	–	–	–	–
Porphyromonas sp.	–	B, CB	–	R	S	R	–	+	–	+	+⁻†	–	–	–	–	–	–	–
Other *Prevotella* sp.	–	B, CB	–	R	R	V	–	V	–	–‡	–	–	–	–	–	–	–	–
Bilophila sp.	–	B	–	S	R	S	+	–	+	–	–	–	–	–	–	+	+⁻	–
Bacteroides ureolyticus–like	V	B	–	S	R	S	–	–	–	–	–	–	V	+	+	V	V	V
B. ureolyticus	–		–	S	R	S	–	–	–	–	–	–	+⁻	+	+	+	–	
B. gracilis	–	B	–	S	R	S	–	–	–	–	–	–	V	+	+	–	–§	
Wolinella/Campylobacter	V	B	–	S	R	S	–	–	–	–	–	–	V	+	+	–	+	V
Campylobacter concisus	V	B	–	S	R	S	–	–	–	–	–	–	V	+	+	–	+	+
Fusobacterium sp.	–⁺	B	V	**S**	**R**	**S**	V	V	–	–	–	V	–	–	–	–	–	–
F. nucleatum	–	B	+	S	R	S	–	+	–	–	–	–	–	–	–	–	–	–
F. necrophorum	–	B	–	S	R	S	–⁺	+	–	–	–	+⁻	–	–	–	–	–	–
F. mortiferum-varium	–	B	–	S	R	S	+	V	–	–	–	–	–	–	–	–	–	–
Gram-negative cocci	–	C	–	S	R	S	–	–	V	–	V	–	–	–	–	–	V	–
Veillonella	–	C	–	S	R	S	–	–	V	–⁺	–	–	–	–	–	–	+	–

Modified from Summanen, P., Baron, E.J., Citron, D.M., et al. 1993. Wadsworth anaerobic bacteriology manual, ed 5. Star Publishing, Belmont, Calif.

Reactions in **bold type** are key tests; *B*, bacillus; *CB*, coccobacillus; *C*, coccus; *R*, resistant; *S*, sensitive; *V*, variable; +, positive; –, negative; superscripts indicate reactions of occasional strains.

*P. melaninogenica group often requires prolonged incubation before pigment is observed.

†P. gingivalis does not fluoresce.

‡P. bivia produces pigment on prolonged incubation.

§B. gracilis displays a "twitching" motion.

and *Prevotella* sp. are resistant to kanamycin but variable in susceptibility to colistin. The *Bacteroides ureolyticus* group, *Bilophila*, and *Veillonella* have the same pattern as that of *Fusobacterium*. Some unusual or less frequently isolated anaerobic gram-negative bacilli may not conform to these generalizations. In that case, a motility test, a flagellar stain, and GLC may be required to place the organism in the correct genus.

As with any disk diffusion test, many variables affect zone sizes. It is important to inoculate and streak the plate so that a heavy lawn is produced. The organism may appear falsely susceptible if the inoculum is too light. The disks must be placed well apart, since a large zone around one disk may interfere with interpretation of an adjacent disk. The elapsed time between inoculation, streaking, disk application, and incubation should be kept to a minimum. It may be helpful to inoculate the plates in batched groups of 5 to 10 plates before disk placement.

With the addition of some of the following tests, further grouping or definitive identification of some anaerobes is possible.

BILE TEST See Procedure 35.2 and Figure 35.11. The ability to grow in 20% bile is a key for separating the bile-resistant *B. fragilis* group from *Prevotella* sp. and for differentiating *Fusobacterium mortiferum-varium* from most of the other fusobacteria. The test can be performed using bile-containing agar, a bile-tube test, or bile-impregnated disks (Remel, Oxoid, or Anaerobe Systems). The disk test is performed as with the

TABLE 35.7

LEVEL II IDENTIFICATION OF GRAM-POSITIVE ANAEROBES

	CELL SHAPE	SPORES OBSERVED	BOXCAR-SHAPED CELLS	DOUBLE-ZONE BETA HEMOLYSIS	KANAMYCIN (1 mg)	VANCOMYCIN (5 µg)	COLISTIN (10 µg)	INDOLE	SODIUM POLYANETHOLSULFONATE	CATALASE	SURVIVES ETHANOL SPORE TEST	LECITHINASE	NAGLER TEST	STRONG REVERSE-CAMP TEST	ARGININE STIMULATION	UREASE	NITRATE REDUCTION	GROUND-GLASS YELLOW COLONIES ON CCFA* MEDIUM	COMMENTS
Gram-positive cocci	C, CB	–	–	–	V	S	R	V	V	V	–	–	–			–		–⁺	–	
Peptostreptococcus anaerobius	C, CB	–	–	–	**R**ˢ	S	R	–	S	–	–	–	–					–	–	Sweet, putrid odor; may chain
Peptostreptococcus asaccharolyticus	C	–	–	–	S	S	R	+	R	V	–	–	–					–	–	"Indole" odor
Clostridium sp. Nagler positive	B	+⁻	–⁺	–	V	S	R	V		–	+⁻	V	–		–	V		–⁺	V	
C. perfringens	B	–	+	+⁻	S	S	R	–		–	–⁺	+	+	+	–	–	+⁻		
C. baratii	B	+	–	–	S	S	R	–		–	+	+	+ʷ	–	–	–	V		
C. sordellii	B	+	–	–	S	S	R	+		–	+	+	+ʷ	–	–	+⁻	–		
C. bifermentans	B	+	–	–	S	S	R	+		–	+	+	+ʷ	–	–	–	–		
Nagler negative Presumptive																			
C. difficile	B	+⁻	–	–	S	S	R	–		–	+	–	–	–	–		–	+	"Horse stable" odor; fluoresces charteuse
Non–spore forming	B, CB	–	–	–	V	S	R	V		V	–	V	–		V	V	V	–	
Propionibacterium acnes	B, CB	–	–	–	S	S	R	+⁻		+	–	–	–		–	–	+	–	May branch; diphtheroid
Eubacterium lentum	B	–	–	–	S	S	R	–		–⁺	–	–	–		+	–	+	–	Small bacillus

Modified from Summanen, P., Baron, E.J., Citron, D.M., et al. 1993. Wadsworth anaerobic bacteriology manual, ed 5. Star Publishing, Belmont, Calif.

Reactions in **bold type** are key tests; *B*, bacillus; *CB*, coccobacillus; *C*, coccus; *R*, resistant; *S*, sensitive; *V*, variable; +, positive; –, negative; *w*, weak; superscripts indicate reactions of occasional strains.

*Cycloserine-cefoxitin fructose agar.

antibiotic identification disks test, and appearance of any zone denotes susceptibility, whereas growth up to the disk indicates resistance.

NITRATE DISK TEST The principle, reagents, and reactions are the same as those discussed in Chapter 10. One tests tiny indole-negative gram-positive rods (*Eubacterium lentum*–like), curved or straight gram-negative bacilli that form small, transparent colonies or pit the agar (*B. ureolyticus*–like), and gram-negative cocci.

FORMATE AND FUMARATE (F/F) GROWTH STIMULATION TEST See Procedure 35.3. The *B. ureolyticus*–like organisms grow poorly, if at all, in broth medium that is not supplemented with formate and fumarate. These organisms require formate or hydrogen as an electron donor and fumarate or nitrate as an electron acceptor in their metabolic processes. A simple test method is to supplement a thioglycolate broth or comparable medium with formate and fumarate and compare growth in that tube with growth in an unsupplemented

PROCEDURE 35.2

TWENTY-PERCENT-BILE TUBE TEST

PRINCIPLE

Anaerobic bacteria vary in their ability to grow in the presence of 20% bile (equivalent to 2% oxgall). Bile tolerance is most helpful in separating the *Bacteroides fragilis* group from other *Bacteroides* and *Prevotella* sp. and in separating *Fusobacterium mortiferum-varium* from most other clinically significant fusobacteria.

METHOD

1. Reagents
    a. Oxgall (Difco Laboratories): prepare a solution of 40% oxgall (Difco). Sterilize by autoclaving, and store at 2° to 8° C.
    b. Thioglycolate broth (thio)

2. Procedure
    a. Add 0.5 ml of 40% oxgall to a 10 ml tube of freshly steamed or boiled thio or to any broth supporting good growth of the organism. This yields a concentration of 2% oxgall, equivalent to 20% bile. Add the bile to a warm tube, since this facilitates mixing of the bile and thio.
    b. Inoculate both the thio-bile tube and an unsupplemented thio control tube with one or two colonies from a pure BA plate or broth culture of the organism. Do not vigorously mix the thio tubes, since this will introduce too much air into the system.
    c. Incubate the thio tubes aerobically 24 to 48 hours with the caps tightened. If the tubes have been aerated or another broth is used, loosen the caps and incubate anaerobically.

3. Reading and interpretation: compare the growth in the bile-supplemented tube to the growth in the plain tube. Record as "bile inhibited" when no growth or significantly less growth occurs in the bile tube. Record as "bile tolerant" when good growth occurs in both tubes.

QUALITY CONTROL

Test *B. fragilis* (bile tolerant) and any bile-sensitive organism (e.g., *Prevotella oralis* or *P. melaninogenica*) by inoculating each into bile-supplemented and unsupplemented thio broth tubes as described in the Method.

EXPECTED RESULTS

B. fragilis is bile tolerant and therefore grows to the same turbidity in broth containing 20% bile as in non–bile-containing broth. *P. oralis* is inhibited by bile, so it grows poorly, if at all, in the presence of bile.

PERFORMANCE SCHEDULE

Test each new lot of oxgall and each batch of the 2% oxgall solution.

---

broth culture. One should test all gram-negative rods that form small transparent colonies or pit the agar.

UREASE TEST This test is useful for differentiating between the two indole-positive, Nagler-positive clostridia *(C. sordellii, C. bifermentans)*, among the *B. ureolyticus*–like group, and among the *Actinomyces* sp. Broth and disk (Difco, Rosco) tests are available.[35] (See Chapter 10 for the principle and reagents.) This test is incubated aerobically at 37° C. A positive, bright-pink to red color often occurs within 5 to 30 minutes of incubation if a heavy inoculum is prepared.

MOTILITY Motility is occasionally helpful for identification, primarily for unusual gram-negative bacilli. Almost all the clinically isolated gram-negative bacilli belong to the genera *Bacteroides, Prevotella, Porphyromonas, Fusobacterium*, or *Bilophila*, all of which are nonmotile; however, one should test the *B. ureolyticus*–like group because it includes *Wolinella* and *Campylobacter* sp., which are motile. One should also test any other gram-negative bacillus that cannot be identified according to Table 35.6. Motility is optimally determined on young broth or plate cultures by examining a wet-mount preparation.

SPS SUSCEPTIBILITY TEST *Peptostreptococcus anaerobius* is susceptible to a 1 mg SPS disk, whereas the other anaerobic gram-positive cocci are resistant. One should follow the procedure used for the antibiotic identification disks. A zone size of 12 mm or greater is considered susceptible. All anaerobic gram-positive cocci that resemble *P. anaerobius* (large colony, sweetish foul odor, large cocci in chains) are tested.

PROCEDURE 35.3

FORMATE-FUMARATE (F/F) GROWTH STIMULATION TEST

PRINCIPLE

In broth culture, *Bacteroides ureolyticus*, *B. gracilis*, certain *Campylobacter* sp., and *Wolinella* sp. derive energy (adenosine triphosphate, ATP) during the transfer of electrons from formate to fumarate via a formate dehydrogenase-sulfur-molybdenum protein complex and fumarate reductase. Other anaerobes do not require this metabolic pathway for energy production.

METHOD

1. Reagents
   a. Prepare a solution of F/F:
      Sodium formate (Fisher Scientific Co.), 3g
      Fumaric acid (Fisher), 3g
      Distilled water, 50ml
      Sodium hydroxide, 18 pellets
      (1) Combine ingredients, stirring until pellets are dissolved and fumaric acid is in solution.
      (2) Adjust pH to 7.0 with 4 N sodium hydroxide.
      (3) Sterilize by filtration through a 0.2 μm membrane filter. Store at 2° to 8° C.
   b. Thioglycolate broth (thio)
2. Procedure
   a. Add 0.5 ml F/F to a 10 ml freshly steamed or boiled thio tube.
   b. Inoculate both an F/F-supplemented thio and an unsupplemented thio with several colonies from a 48 to 72 hour BA plate culture.
   c. Incubate aerobically with the caps tightened for 24 to 72 hours at 35° C.
3. Reading and interpretation: compare the growth in the F/F-supplemented tube to that in the plain tube. Record as "F/F stimulated" when growth occurs in the supplemented tube and poor to no growth occurs in the plain tube.

QUALITY CONTROL

Test *B. ureolyticus* (requires F/F for growth in thio broth) and *B. fragilis* (does not require F/F for growth in thio broth) by inoculating both into F/F-supplemented and F/F-unsupplemented thio as described under Method.

EXPECTED RESULTS

Performance standards are met when *B. ureolyticus* grows only in the F/F-supplemented broth and *B. fragilis* grows in both the F/F-supplemented and F/F-unsupplemented broth.

PERFORMANCE SCHEDULE

Test each new lot of reagents and each new broth of working solutions.

GLUTAMIC ACID DECARBOXYLASE TEST[3] (CARR-SCARBOROUGH) A tube containing the substrate is inoculated with a heavy suspension of the organism. After 4 hours of aerobic incubation, a change in color from green to blue denotes a positive reaction. Most members of the *B. fragilis* group, *Clostridium perfringens*, *C. sordellii*, and *C. baratii*, are positive. Other anaerobes are negative.

ALN DISK TEST (CARR-SCARBOROUGH) An ALN disk is placed on a slide or empty Petri dish and moistened slightly with distilled water. Colony paste from a pure culture of the isolate is rubbed onto a small area of the disk. Allow the organism to react for 2 minutes and add a drop of cinnamaldehyde reagent. A pink to red color that develops within 30 seconds indicates a positive reaction. No color change is considered negative. Most *Bacteroides*, *Prevotella*, and *Capnocytophaga* sp. produce L-alanyl–L-alanyl-aminopeptidase and are positive in this test. Fusobacteria and the *B. ureolyticus* group are negative.

ETHANOL SPORE TEST See Procedure 35.4. Spores are resistant to the effect of alcohol, whereas vegetative cells are killed. One tests gram-positive bacilli to separate *Clostridium* sp. from other gram-positive bacilli if spores are not seen on a Gram stain smear.

LECITHINASE TEST See Procedure 35.5. The compound lecithin, a normal component of egg yolk, is split by the enzyme lecithinase to release insoluble diglycerides. This results in an opaque halo in egg yolk–containing medium that extends beyond the edge of colony growth (Figure 35.14). It is important to test all clostridia and clostridia-like organisms.

PROCEDURE 35.4

ETHANOL SPORE TEST

PRINCIPLE

Alcohols, such as ethanol, kill vegetative cells in a hydrated environment by coagulating proteins. Spores are not readily hydrated and are usually resistant to ethanol. This test separates spore formers (*Clostridium* and *Bacillus* sp.) from non–spore formers.

METHOD

1. Reagents
   a. 95% ethanol
   b. Chopped meat carbohydrate (CMC) (Carr-Scarborough) or thioglycolate (thio) broth
   c. *Brucella* blood agar plate (BA)
2. Procedure
   a. Inoculate one or two colonies of the isolate into a CMC or thio tube. Incubate 48 h at 37° C, and then incubate an

additional 1 to 2 days at room temperature.
   b. Add 1 ml of the broth culture to 1 ml of 95% ethanol.
   c. Gently mix and allow to stand for 30 to 45 min.
   d. Dip a swab into the alcohol mixture and spread onto a BA plate.
   e. Dip a second swab into the original broth culture; spread this onto a BA; and streak for isolation. This plate serves as a control for purity and viability of the organism.
   f. Incubate both plates in an anaerobic environment at 35° C for 48 h.
3. Reading and interpretation: observe for growth. If growth occurs only on the control plate, the organism is considered a non–spore

former. If growth occurs on both plates, the organism is considered a spore former. If no growth occurs on the control plate, the test is inconclusive. Be certain that the colony morphology on all the plates is the same as the original colony (i.e., the growth is not caused by a contaminant). Also *C. perfringens, C. clostridioforme,* and a few other *Clostridium* sp. may not survive exposure to ethanol, so their identification is based on other characteristics (Chapter 36).

QUALITY CONTROL

The control plate serves as an internal quality control. The media used must have already met quality control performance standards.

LIPASE REACTION   Free fats present in egg yolk are broken down by the enzyme lipase to produce glycerol and fatty acids. The fatty acids appear either as a surface "oil-on-water" or "mother-of-pearl" layer that covers the colony and may extend beyond the colony edge or as a zone of opacity directly beneath the colony (Figure 35.15). All clostridia, pigmented *Prevotella* sp., and fusobacteria are tested for lipase production. One inoculates and incubates the EYA plate as described in the lecithinase test (Procedure 35.5). The plate is held at an angle to the light source to detect the oil-on-water phenomenon more easily. It may be necessary to reincubate the plate for 1 week to detect slow lipase producers (e.g., some strains of *Fusobacterium varium*).

NAGLER TEST   See Procedure 35.6 and Figure 35.16. *Clostridium perfringens, C. baratii, C. bifermentans,* and *C. sordellii* produce an α-lecithinase that is detected by a neutralization test using type A *C. perfringens* antitoxin (Pittman and Moore, Inc., Mundelein, Ill.). Test *C. perfringens*–like and other lecithinase-positive clostridia.

REVERSE CAMP TEST   This test is similar to the CAMP test for identifying group B beta-hemolytic streptococci (Chapter 26), except that the *Clostridium* sp. replaces *Staphylococcus aureus* and a known group B beta-hemolytic *Streptococcus* is used. Although group B streptococci may exhibit some enhanced hemolysis with other clostridia, it is only with *C. perfringens* that the characteristic arrowhead form is demonstrated. Test all *C. perfringens*–like gram-positive bacilli with this test. Only the typical arrowhead of synergistic beta hemolysis (Figure 35-17) indicates a positive test.

ARGININE GROWTH STIMULATION TEST   *Eubacterium lentum* is one of the few anaerobic gram-positive rods whose growth is enhanced by arginine; therefore one should test all small gram-positive bacilli that do not resemble diphtheroids. This test is performed in a manner similar to the formate-fumarate growth test, but uses arginine hydrochloride in a final concentration of 0.5% (wt/vol).

• • •

PROCEDURE 35.5

LECITHINASE TEST

PRINCIPLE
Bacterial lecithinase splits lecithin (a normal component of egg yolk) to insoluble diglycerides, resulting in an opaque halo surrounding a colony growing on egg yolk–containing medium.

METHOD
1. Reagents: commercially prepared egg yolk agar (EYA) plates are available from many sources. If preparing the medium in house, egg yolk emulsion is available from Difco Laboratories, Hana Biologics, and Oxoid USA. Store the plates at 2° to 8° C. Reducing the plates before use is not usually necessary.

2. Procedure
   a. Transfer one colony of a 24 to 72 h plate culture or 2 to 3 drops of a broth culture of the test organism to an EYA plate.
   b. Incubate anaerobically at 35° C for at least 24 h. If negative, reincubate the plate for an additional 24 to 48 h.

3. Reading and interpretation: examine for a white opacity in the medium that surrounds the colony and extends beyond the edge of growth; this is a positive lecithinase test. Some organisms require extended incubation to demonstrate a positive test; however, this is not practical in a clinical laboratory, so 72 h is chosen as an arbitrary maximal incubation time.

QUALITY CONTROL
Test *Clostridium perfringens* (lecithinase positive) and *C. difficile* (lecithinase negative) by streaking each onto one half of an EYA plate as described under Method.

EXPECTED RESULTS
Performance standards are met when both organisms grow on the medium. *C. perfringens* produces a white opaque zone in the medium extending beyond the colony edge (lecithinase positive, Figure 35.14), and *C. difficile* produces no opaque zone in the medium (lecithinase negative).

PERFORMANCE SCHEDULE
Test each new lot of EYA plates.

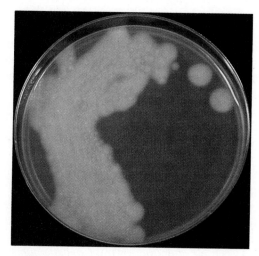

FIGURE 35.14

Clostridium perfringens. Lecithinase reaction.

A few of the preceding tests are incorporated in the Lombard-Dowell (LD) Presumpto 1 plate, which is available from several sources, including Carr-Scarborough and Remel Laboratories. Some fastidious strains of fusobacteria and *Prevotella* do not grow well on the LD basal medium, and some manufacturers supplement it. Good growth is required to interpret results accurately.

LEVEL II GROUP IDENTIFICATION: GRAM-NEGATIVE ORGANISMS
The gram-negative organisms are divided into the following major categories based on microscopic morphology, antibiotic identification disks pattern, or a few other simple tests.

BACTEROIDES FRAGILIS GROUP These gram-negative bacilli grow in 20% bile and are almost always resistant to all three special-potency antibiotic disks. Rare strains of *B. fragilis* are susceptible to colistin.

PROCEDURE 35.6

NAGLER TEST

PRINCIPLE

α-Lecithinase-producing clostridia *(C. perfringens, C. baratii, C. bifermentans, C. sordellii)* are detected by inhibition of the lecithinase reaction by an α-lecithinase antitoxin. This results in a drastic reduction or complete disappearance of the white opaque zone produced by the splitting of lecithin to insoluble diglycerides (positive lecithinase reaction).

METHOD

1. Reagents
   a. Egg yolk plate
   b. *Clostridium welchii (perfringens)* type A antitoxin (Cooper Animal Health, Kansas City, Mo.)
2. Procedure
   a. Swab one half of an egg yolk agar plate with antitoxin.
   b. Allow the liquid to absorb.
   c. Inoculate the test organism on the antitoxin-free half, and then streak once across to the antitoxin side of the plate.
   d. Incubate anaerobically 24 to 48 hours at 37° C.
3. Reading and interpretation: examine for the disappearance or dramatic reduction of the opacity on the antitoxin half of the plate. This denotes neutralization of the type A lecithinase and is a positive Nagler test (Figure 34.16).

QUALITY CONTROL

Test *C. perfringens* (α-lecithinase producer) and a lecithinase-positive (but not α) *C. limosum. C. haemolyticum, C. sardiniensis,* and *C. novyi* type B are also lecithinase producers not of the alpha type, but they are oxygen sensitive and may be difficult to maintain.

EXPECTED RESULTS

Performance standards are met when both organisms grow on the egg yolk–containing agar. Both organisms produce a white, opaque halo in the medium. The opacity in the medium is dramatically decreased or obliterated in the presence of α-lecithinase antitoxin with *C. perfringens.* The opacity remains in the presence of the antitoxin with *C. limosum.*

PERFORMANCE SCHEDULE

Test each new lot of antitoxin and each time the test is performed.

FIGURE 35.15

Clostridium. Positive lipase reaction on egg yolk agar. (From Summanen, P., Baron, E.J., Citron, D.M., et al. 1993. Wadsworth anaerobic bacteriology manual, ed 5. Star Publishing, Belmont, Calif.)

NONPIGMENTED *PREVOTELLA* **SP.** Most bile-sensitive, kanamycin-resistant gram-negative rods belong in this genus. Colistin susceptibility is variable, and almost all strains are catalase and indole negative.

PIGMENTED *PREVOTELLA* **AND** *PORPHYROMONAS* **SP.** Colonies that fluoresce brick red or produce brown-to-black pigment are placed in this group. Some species are coccobacillary.

***BACTEROIDES UREOLYTICUS*-LIKE GROUP** Members of this group reduce nitrate and require formate and fumarate for growth in broth culture. Their disk pattern is the same as the fusobacteria; however, colony morphology is different. *B. ureolyticus, B. gracilis, Wolinella, Campylobacter concisus, C. curva,* and *C. recta* form small, translucent to transparent colonies that may corrode the agar, whereas the *Fusobacterium* colony is generally larger and more opaque.

OTHER *BACTEROIDES* **OR** *PREVOTELLA* **SP.** This term is applied to gram-negative bacilli that do not fit the preceding categories or the *Fusobacterium* sp.

FUSOBACTERIUM **SP.** These gram-negative bacilli are sensitive to kanamycin, and most strains fluoresce a chartreuse color. Different species have characteristic cell and colony morphology.

FIGURE 35.16

Clostridium perfringens. Positive Nagler test on egg yolk agar. Inhibition of lecithinase reaction by antitoxin *(left).*

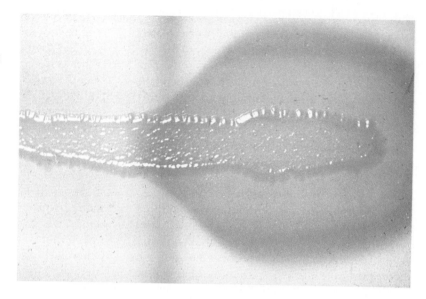

FIGURE 35.17

Clostridium perfringens. Reverse CAMP test with group B beta-hemolytic *Streptococcus (vertical)* and *C. perfringens (horizontal).* (From Summanen, P., Baron, E.J., Citron, D.M., et al. 1993. Wadsworth anaerobic bacteriology manual, ed 5. Star Publishing, Belmont, Calif.)

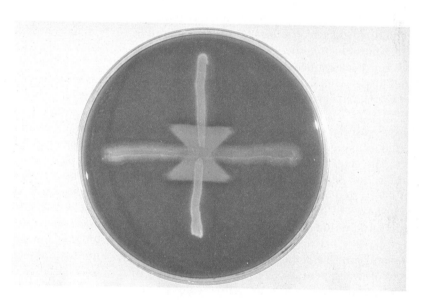

ANAEROBIC GRAM-NEGATIVE COCCI This category is based on Gram stain morphology and includes *Veillonella, Megasphaera,* and *Acidaminococcus.*

LEVEL II GROUP IDENTIFICATION: GRAM-POSITIVE ORGANISMS

The gram-positive organisms are divided into three major categories based on microscopic morphology and the presence of spores.

ANAEROBIC GRAM-POSITIVE COCCI The genera of clinical importance are *Peptostreptococcus* and *Streptococcus.* If a coccus is resistant to metronidazole, it probably is a *Streptococcus* or *Staphylococcus saccharolyticus.*

***CLOSTRIDIUM* SP.** These are the endospore-forming, anaerobic, gram-positive bacilli. If spores are not present on Gram stain, the ethanol spore or heat spore test will separate this group from the non–spore-forming anaerobic bacilli. Some strains of *C. perfringens, C. ramosum,* and *C. clostridioforme* may not produce spores or survive a spore test, so it is important to recognize these organisms using other characteristics. Some clostridia typically stain gram negative although they are susceptible to vancomycin on the disk test. Several species of clostridia grow aerobically, but they produce spores only under anaerobic conditions.

ANAEROBIC GRAM-POSITIVE NON-SPORE-FORMING BACILLI This group includes the genera *Actinomyces, Bifidobacterium, Eubacterium,* anaerobic *Lactobacillus,* and *Propionibacterium.* It is difficult to differentiate this group accurately to the genus level without end product analysis, except for *Propionibacterium acnes* and *Eubacterium lentum.* They can be identified with a few additional tests, as discussed in the next section.

LEVEL II FINAL IDENTIFICATION

Those anaerobes that can be definitively identified using simple tests are shown in Tables 35.6 and 35.7. The results of characteristics in bold type are keys to identification. It is necessary that all stated characteristics be present to identify them definitively. If a characteristic is not present, only a group name can be applied, unless further biochemical testing is carried out.

The *Bacteroides fragilis* group and *Prevotella* sp. require further biochemical tests for definitive identification. Two species of pigmented *Prevotella* sp. can be identified readily. An indole-positive and lipase-positive gram-negative coccobacillus is *P. intermedia.* Lipase-positive, indole-negative pigmented *Prevotella* strains are *P. loescheii;* however, lipase-positive strains of *P. loescheii* are encountered infrequently.

The *Bacteroides ureolyticus* group includes *B. ureolyticus, B. gracilis, Campylobacter concisus,* and *Wolinella* sp. *Wolinella curva* and *W. recta* have been transferred to the genus *Campylobacter.* These organisms are thin, gram-negative rods with round ends; they are susceptible to colistin and kanamycin. The colonies are small, are translucent or transparent, and may produce greening of the agar on exposure to air. Three colony morphotypes exist: smooth and convex, pitting, and spreading. All three can occur in the same culture. The organisms are asaccharolytic, reduce nitrate, and require formate and fumarate for growth in broth culture. The organisms can use nitrate and hydrogen in place of formate and fumarate. The *Wolinella* sp. and *Campylobacter* sp. are motile. *B. ureolyticus* and *B. gracilis* are nonmotile, although *B. gracilis* may demonstrate "twitching" in a wet preparation and "gliding" on agar. *B. ureolyticus* is urease positive, whereas *B. gracilis* is urease negative.

Two *Fusobacterium* sp. can be identified with a few simple tests. *F. nucleatum* is an indole-positive, thin bacillus with pointed ends. The colony fluoresces chartreuse and produces greening of the agar on exposure to air. Three colony morphotypes of *F. nucleatum* exist: speckled, bread crumb–like, and smooth. The colony size varies from less than 0.5 to 2 mm in diameter. The bread crumb colony is white, whereas the speckled and smooth types are gray to gray-white. A lipase-positive *Fusobacterium* is *F. necrophorum.* It is a pleomorphic rod with rounded ends and sometimes forms bizarre shapes. It is indole positive, fluoresces chartreuse, and produces greening of the agar. The colonies are umbonate and range in size from 0.5 to 2 mm in diameter. Lipase-positive strains are beta hemolytic. Lipase-negative strains require further biochemical tests for identification. A bile-resistant *Fusobacterium* may be identified tentatively as *F. mortiferum-varium* group. Other *Fusobacterium* sp. may grow in 20% bile; therefore further testing is required to confirm the presumptive identification.

A gram-negative coccus less than 0.5 mm in diameter that reduces nitrate is *Veillonella* sp. The other gram-negative cocci do not reduce nitrate. Catalase production is variable. The colonies are small and almost transparent.

A gram-positive coccus sensitive to SPS is *Peptostreptococcus anaerobius.* It is a large coccobacillus and often occurs in chains. The colonies vary in size from 0.5 to 2 mm in diameter (usually larger than most anaerobic cocci) and are gray-white and opaque. A sweet, fetid odor is associated with this organism. An indole-positive, gram-positive coccus is *P. asaccharolyticus. P indolicus,* another indole-positive coccus, is an animal strain and is rarely isolated from clinical specimens. Unlike *P. asaccharolyticus, P. indolicus* converts lactate to propionate and is usually nitrate and coagulase positive. Another recently described indole-positive gram-positive coccus is *P. hydrogenalis.* This organism has been isolated from vaginal secretions and can be differentiated from the other indole-producing cocci by glucose fermentation and alkaline phosphatase production.

Four *Clostridium sp.* are fairly simple to identify because they are all Nagler positive. *C. perfringens* is readily recognizable by its microscopic and colony morphology. It is a boxcar-shaped, gram-positive rod with a typical double zone of beta hemolysis surrounding the colony (Figure 35.6). *C. sordellii* and *C. bifermentans* are indole positive. *C. sordellii* is usually urease positive, and *C. bifermentans* is urease negative.

Propionibacterium acnes is an indole positive, catalase positive, pleomorphic (diphtheroid-like), gram-positive bacillus. The colonies are initially small and white and become larger and more yellowish tan with age. Indole-negative strains are identified with more extensive biochemical tests.

Eubacterium lentum is a nitrate-positive, small, gram-positive bacillus whose growth in broth is stimulated by arginine. It forms small, gray, translucent colonies.

LEVEL III IDENTIFICATION

In addition to tests performed at level II, species identification of most anaerobes requires additional biochemical tests and sometimes end product analysis by GLC. Differentiation of genera of anaerobes is shown in tables in Chapters 36 to 38. For level III identification tables and flow diagrams, fermentation reactions are based on use of PRAS liquid media, and preformed enzyme reactions are based on test kits and individual tube tests. Thioglycolate-based media with bromthymol blue indicator *(CDC Laboratory Manual)* give similar results for fermentation reactions, whereas results obtained in other systems may not agree with these tables. Gas chromatographic analysis may be performed from any carbohydrate-containing medium that supports good growth of the organism.

LEVEL III IDENTIFICATION SYSTEMS

Several identification systems are available, including conventional tube tests,[6,9,21,24,25] micromethods,[5,24,25,38] agar plate,[14,15] and enzyme substrate types.* The PRAS system is still the standard method, but other chemical and biochemical tests may be required. Table 35.8 compares features of some of these systems.

CONVENTIONAL BIOCHEMICAL SYSTEMS Biochemical systems test for fermentative, proteolytic, and other metabolic activities of the organism. The substrates are incorporated into PRAS peptone yeast (refer to the VPI *Anaerobe Laboratory Manual* or *Wadsworth Anaerobic Bacteriology Manual)* or thioglycolate broth (refer to the *CDC Anaerobe Manual).* A wide variety of biochemicals is available. Identification is determined by comparing reaction patterns of the test organism to tables of reaction patterns of known organisms. The PRAS II system (Adams Scientific) uses a computer program for interpretation of reaction patterns.[6,25] However, less than 100% agreement with expected results (Table 35.9) occurs with this system. Conventional biochemical tests, including GLC, are often required for the identification of many gram-positive, non–spore-forming bacilli, some *Clostridium* sp., some anaerobic cocci, and

unusual gram-negative bacilli. Additional tests may be required for some organisms. For example, *Prevotella oris* and *P. buccae* have the same fermentation and GLC patterns; α-fucosidase and β-N-acetylglucosaminidase (available in preformed enzyme kits) are used to separate them.

PRAS biochemicals are inoculated by either the open or the closed technique. The open technique uses a special apparatus that allows for a continuous flow of anaerobic gas into the tube during manipulations, whereas the closed technique uses a needle and syringe to inject the inoculum into the tube through a rubber stopper (Carr-Scarborough) or a diaphragm in a Hungate-type screw cap (PRAS II, Adams Scientific). After good growth has occurred, the pH is measured in each tube to test for acid production. Appropriate biochemicals are processed for GLC. (See the *Wadsworth Anaerobic Bacteriology Manual* or VPI *Anaerobe Laboratory Manual* for details.)

The PRAS system is expensive, labor intensive, and not problem free. The major problem is uninterpretable biochemical results that may be caused by (1) an inadequate, nonviable, or mixed inoculum, (2) partial oxidation of tubes, or (3) contamination during inoculation. Repeat testing is required about 10% to 20% of the time. Because proper interpretation requires good growth, fastidious strains may require special supplementation and up to a week of incubation. Careful anaerobic technique is required to maintain an anaerobic atmosphere during manipulations.

MICROMETHODS The microsystems also require growth of the organism for substrate degradation. The API 20A strip contains 16 carbohydrates and tests for indole, urea, gelatin, esculin, and catalase. The indicator is bromcresol purple, which turns yellow at a pH of 5.3 to 5.5. The Minitek system offers a wide choice of biochemical tests using microdilution plates and disks saturated with various substrates. The indicator is phenol red, which turns yellow at pH 6.8. A 24- to 48-hour plate culture of the test organism is harvested and suspended into Lombard-Dowell broth to a density equal to a No. 3 MacFarland turbidity standard for API 20A and No. 5 turbidity for Minitek. After a 24- to 48-hour anaerobic incubation period, the required reagents are added and the tests read. The code number generated is matched to a code number representing a specific organism. A codebook or a more extensive computer data bank is available.

The color reactions in both these systems are not always clear-cut (shades of brown [API] and

---

*References 2, 5, 8, 11, 19, 24, 25, 29, 32, 33, 38, 42, 48, 50.

TABLE 35.8

COMPARISON OF ANAEROBIC IDENTIFICATION SYSTEMS

PARAMETER	PRAS BIO-CHEMICALS	4 HOUR PREFORMED ENZYME KITS				MICROTUBE KITS	
		AnIDENT	RapID-ANA II	MICROSCAN	VITEK	MINITEK	API 20A
Inoculum							
Source	Broth*	Plate†	Plate†	Plate†	Plate	Plate	Plate†
Age	24-48 h	24 h	24-72 h	24-48 h	24-48 h	Not specified	Fresh
Diluent	None	Sterile distilled H$_2$O	Available	Sterile deionized H$_2$O	Sterile saline	Provided (Lombard-Dowell)	Provided (Lombard-Dowell)
Turbidity§							
Size	>2+ in 10 ml	No. 5 in 3 ml	No. 3 in 1 ml	No. 3 in 3 ml	No. 3	No. 5 in 1.5 ml	≥No. 3 in 5 ml
	2-10 drops per tube	About 85 µl per capule	About 0.1 ml per well	50 µl per well	Semiautomatic filling	0.05 ml per well	—
Tests							
Number	5-30	20	18	24	28	20	21
Type	Variable	Fixed	Fixed	Fixed	Fixed	Variable	Fixed
Incubation							
Time	24 h to 1 wk	4 h	4 h	4 h	4 h	48 h	24 h
Atmosphere	Anaerobic	Aerobic	Aerobic	Aerobic	Aerobic	Anaerobic	Anaerobic
Data base	Charts	Codebook	Codebook	Codebook	Computer program	Codebook	Codebook
	Computer assist	Computer assist	Computer assist	Computer assist		Computer assist	Computer assist
Cost ($)/tube or panel‖	0.56	5.00	4.50	4.50	4.50	7.00	5.28

Modified from handout material from Anaerobe Systems (Santa Clara, Calif.) Anaerobe Workshop Lecture Series and product package inserts.

*Peptone yeast glucose, thioglycolate, or any broth that supports growth of isolate can be used. If supplement is required, add supplement to each tube to be inoculated. Inoculum can also be prepared directly from agar plate.

†Inoculum is prepared from growth on nonselective medium. Read package insert for specific media.

‡Selective media may be used, except for those that detect esculin with a ferric ion (e.g., BBE agar). Read package insert for specific media.

§Expressed in terms of McFarland turbidimetric standard, except for prereduced anaerobically sterilized (PRAS), which is a subjective estimate of turbidity.

‖Based on list price of panel; does not include cost of codebook or any other required materials. Except for Minitek and API 20A, it does not include cost of diluent.

TABLE 35.9

SPECIES-LEVEL AGREEMENT OF RAPID IDENTIFICATION SYSTEMS WITH CONVENTIONAL IDENTIFICATION METHODS*

ORGANISMS	AGREEMENT FOR DEFINITIVE IDENTIFICATION (%)				
	RapID-ANA II	AnIDENT	VITEK	API 20A	MINITEK
Gram-negative bacilli	62-86 (2)	74-76 (2)	74 (1)	53-91 (3)	74-80 (2)
Bacteroides/Prevotella	57-98 (2)	81 (1)	77 (1)	67-88 (4)	83 (1)
Bacteroides fragilis group	64-82 (2)	79 (1)	83 (1)	68-84 (1)	90 (1)
Pigmented *Prevotella/Porphyromonas*	46-96 (1)	—	58 (1)	—	—
Other *Bacteroides*	66-100 (2)	89 (1)	65 (1)	67 (1)	—
Fusobacterium	0-33 (2)	58 (1)	27 (1)	0-100 (4)	8 (1)
Gram-positive cocci	72-96 (2)	77-86 (2)	66 (1)	29-80 (3)	55-67 (2)
Gram-positive bacilli					
Clostridium	74-76 (2)	59-63 (2)	64 (1)	66-80 (3)	72-74 (2)
C. perfringens	100 (2)	100 (1)	100 (1)	82-100 (2)	88 (1)
C. difficile	50-67 (2)	9-78 (3)	64 (1)	96 (1)	66-91 (2)
Non–spore-forming	70-81 (2)	44-77 (2)	62 (1)	61-80 (3)	79-80 (2)
Overall					
Correct genus and species	68-87 (2)	67-94 (2)	71 (1)	57-85 (3)	70-76 (2)
Correct genus only	80-94 (2)	90 (1)	77 (1)	69-76 (3)	—

Data from References 2, 5, 6, 8, 19, 25, 32, 33, 38, 42, 48, and 50.

*Number in parentheses is the number of studies used to determine percent agreement.

yellow-orange [Minitek]) and make interpretation of test results for some organisms difficult and unreliable. Addition of indicator after incubation sometimes remedies this problem. Of the few current reports, the percent agreement of final identification between PRAS plus GLC and these micromethods demonstrates that neither system is adequate for identification of many anaerobes without GLC and other tests.

AGAR PLATE SYSTEM The agar plate system developed by the CDC Anaerobe Laboratory also requires growth of the organism. The substrates in the Presumpto I, II, and III four-quadrant plates are incorporated into Lombard-Dowell agar base medium. GLC, motility, and other tests are often required for identification.

PREFORMED ENZYME SYSTEMS Several rapid identification systems are available that do not require growth of the organism in the system, but instead rely on chromogenic and conventional substrate breakdown by enzymes produced during growth on the agar plate. Most kits use two chromogens, ortho-, para-, or β-naphthylamides and ortho- or para-nitrophenyls. When the colorless chromogen-substrate complex is split, the chromogen is detected directly (yellow nitrophenyl) or after the addition of a cinnamaldehyde reagent that complexes with β-naphthylamides to pro-

duce a pink to red color. The chromogenic and conventional substrates of the various kits are similar. Some substrates are available as individual tests (Carr-Scarborough), and others may be available, so a battery of selected tests could be used. These individual substrates require further evaluation and U.S. Food and Drug Administration (FDA) approval (Rosco Diagnostica).

A pure culture of the organism grown on nonselective agar is harvested and suspended in inoculating fluid to the appropriate turbidity (Table 35.8). The panels are inoculated following manufacturer's instructions and incubated aerobically at 35° to 37° C for 4 hours. It is important to adhere strictly to the manufacturer's instructions, since culture medium, inoculum density, correct filling of wells, and incubation time and atmosphere can affect results.

The IDS and API kits have been on the market the longest and are the most frequently used 4-hour systems. The RapID-ANA II has 10 test wells, eight of which are bifunctional (i.e., two tests per well), for a total of 18 tests. The reagent is added to the wells after the first set of reactions has been read. The API AnIDENT panel is a series of 20 substrate wells with a total of 21 tests. The developer is added to nine of the wells. MicroScan and Vitek also produce 4-hour rapid enzyme test kits. These are similar to IDS and API. A codebook or computer access or both are available. The API

ZYM system has almost 40 tests, but no data base is available for anaerobes. The ATB-32A system (bioMérieux Vitek, Hazelwood, Mo.) contains 29 tests, 24 of which are used to generate an eight-digit code; the other five are used as supplemental tests if needed. This system may not be available in the United States.

Although not yet available as a kit, the use of 4-methylumbelliferone derivatives of a variety of preformed enzymes appears to be a potentially useful method for identification.[37] This technique detects preformed enzymes by fluorescense after 15 minutes of aerobic incubation. It has been evaluated for pigmenting gram-negative rods and for the *Bacteroides fragilis* group.

Some advantages and disadvantages of these rapid systems are listed in the box at right.

Table 35.9 is a compilation of several studies showing the agreement of identification between PRAS biochemicals plus GLC with various systems without the use of additional tests. It is important to know the percent agreement of identification at the species level because this is an indicator of accuracy. Several other studies have been done, but few used PRAS and GLC as the reference method and most did not present their data in a standard format, so interpretation for all categories is not possible. Wide ranges of percent agreement for some categories may be caused by differences in the types and numbers of anaerobes selected for testing, the general expertise in identifying anaerobes, the preliminary tests used (e.g., Gram stain, antibiotic disk identification pattern, indole, catalase), previous experience in using the test kits, the number of times the test was repeated, the method and criteria of calculating percent agreement, or inclusion or exclusion of isolates that did not have an interpretation for a code number. The RapID-ANA and API AnIDENT are probably adequate for identifying *Bacteroides fragilis;* most anaerobic gram-positive cocci; *Clostridium perfringens, C. septicum,* and other frequently isolated *Clostridium* sp.; and some but not all of the bile-sensitive gram-negative bacilli (e.g., high agreement with *Prevotella bivia* and *P. disiens*).[11,50] The kits are often not helpful for identifying other *Clostridium* sp., gram-positive non–spore-forming bacilli (except *Propionibacterium acnes*), and most gram-negative cocci. Indole-positive members of the *B. fragilis* group are also problematic. They are more reliably identified with fermentation tests.[9] In all cases, one should use preliminary test results before using a kit, since these may provide a species-level identification, and to confirm the kit-generated identification. For example, the bile

---

Advantages and Disadvantages of Rapid Identification Systems

Advantages

Aerobic incubation
Short incubation time
Direct inoculum from a plate
Easy handling of kits
Minimal storage space
Expanded computer service
Updated data banks
Amenability to taxonomic changes

Disadvantages

Considerable training time is required for familiarity with test reactions.
Interpretation of color reactions is sometimes difficult because of weak reactions (color charts are provided by some manufacturers).
Data bases include only common clinical isolates from human sources.
Misidentifications are possible for organisms not listed in the data base.
Additional testing, including GLC, is sometimes required.

---

reaction is necessary because *B. fragilis* could be identified as *Prevotella oralis* or *P. loeschii,* and vice versa. A few reports discuss the efficacy of the kits for the identification of nonhuman strains.[1]

OTHER IDENTIFICATION METHODS

Serology, GLC of both volatile fatty acid end products and cellular fatty acids, and deoxyribonucleic acid (DNA) probes offer other techniques for the identification of anaerobes.

SEROLOGY

Serologic methods have been used to detect either bacterial cell wall antigens or extracellular toxins. Agglutination,[54] immunodiffusion, immunofluorescence,[22,26,53] immunoperoxidase,[23] and enzyme-linked immunosorbent assay (ELISA) tests[16,34,39] have been developed for many bacteria, including the pigmented anaerobic gram-negative bacilli (especially from the oral cavity), the *Bacteroides fragilis* group and its individual members, and the toxins of *Clostridium perfringens,*[4] *C. botulinum,*[12] and *C. difficile.* Although distinct serologic differences exist among the genera of anaerobic gram-positive cocci, no test has been developed commercially. A latex test for *C. difficile* is commercially available, as are ELISA tests for toxins A and B (Chapter 18). The CDC has antiserum for *Actinomyces* sp.; they accept slides only after prior approval. With the development of monoclonal antibodies for some *B. fragilis* group sp. and *Porphyromonas gingivalis,* perhaps more specific kits will be developed for anaerobes of clinical importance.

GAS-LIQUID CHROMATOGRAPHY (GLC)

GLC is used to measure the end products of metabolism, which provide information for classification of anaerobes at the genus level. Some newer techniques increase the sensitivity for detecting certain end products.[21] Their major drawback has been lack of specificity. GLC is used also to identify long-chain fatty acids in the cell wall.[7,10,17,49,51] This application has been developed for species identification by Hewlett-Packard, and a data base for anaerobes has been created (Microbial Identification System, MIDI, Newark, Del.).

After a series of multistep procedures, including processing, saponification, methylation, and extraction and washing, the sample is injected into a flame ionization detector GLC fitted with a glass capillary column. The chromatogram produced is automatically matched to a best-fit chromatogram in a computer library. In a recent study on clinical isolates of anaerobic gram-negative rods, 72.4% were correctly identified to the species level.[49] Closely related species, such as within the *Bacteroides fragilis* group, had very similar chromatograms that could not be reliably distinguished from each other by this technique.

NUCLEIC ACID PROBES

DNA probes for the *Bacteroides fragilis* group, *B. fragilis,* and *B. thetaiotaomicron* and a multiprobe for the *Bacteroides* sp.–*Fusobacterium nucleatum–F. necrophorum* complex have been developed but are not commercially available.[13,27,36,41,52] The major problem with these techniques is relatively low sensitivity.

REFERENCES

1. Adney, W.S., and Jones, R.L. 1985. Evaluation of the RapID-ANA system for identification of anaerobic bacteria of veterinary origin. J. Clin. Microbiol. 22:980.
2. Appelbaum, P.C., Kaufmann, C.S., Keifer, J.C., and Venbrux, H.J. 1983. Comparison of three methods for anaerobe identification. J. Clin. Microbiol. 18:614.
3. Banks, E.R., S.D. Allen, Siders, J.A., and O'Brien, N.A. 1989. Characterization of anaerobic bacteria by using a commercially available rapid tube test for glutamic acid decarboxylase. J. Clin. Microbiol. 27:361.
4. Bartholomew, B.A., Stringer, M.F., Watson, G.N., and Gilbert, R.J. 1985. Development and application of an enzyme-linked immunosorbent assay for *Clostridium perfringens* type A enterotoxin. J. Clin. Pathol. 38:222.
5. Bate, G. 1986. Comparison of Minitek Anaerobe II, API AnIDENT, and RapID-ANA systems for identification of *Clostridium difficile*. Am. J. Clin. Pathol. 85:716.
6. Beaucage, C.M., and Onderdonk, A.B. 1982. Evaluation of prereduced anaerobically sterilized medium (PRAS II) system for identification of anaerobic microorganisms. J. Clin. Microbiol. 16:570.
7. Brondz, I., and Olsen, I. 1991. Multivariate analysis of cellular fatty acids in *Bacteroides, Prevotella, Porphyromonas, Wolinella,* and *Campylobacter* spp. J. Clin. Microbiol. 29:183.
8. Celig, D.M., and Schreckenberger, P.C. 1991. Clinical evaluation of the RapID-ANA II panel for identification of anaerobic bacteria. J. Clin. Microbiol. 29:457.
9. Citron, D.M., Baron, E.J., Finegold, S.M., and Goldstein, E.J.C. 1990. Short prereduced anaerobically sterilized (PRAS) biochemical scheme for identification of clinical isolates of bile-resistant *Bacteroides* species. J. Clin. Microbiol. 28:2220.
10. Cundy, K.V., Willard, K.E., Valeri, L.J., et al. 1991. Comparison of traditional gas chromatography (GC), headspace GC, and the microbial identification library GC system for the identification of *Clostridium difficile*. J. Clin. Microbiol. 29:260.
11. Dellinger, C.A., and Moore, L.V.H. 1986. Use of the RapID-ANA system to screen for enzyme activities that differ among species of bile-inhibited *Bacteroides*. J. Clin. Microbiol. 23:289.
12. Dezfulian, M., and Bartlett, J.G. 1984. Detection of *Clostridium botulinum* type A toxin by enzyme-linked immunosorbent assay with antibodies produced in immunologically tolerant animals. J. Clin. Microbiol. 19:645.
13. Dix, K.S., Watanabe, M., McArdle, S., et al. 1990. Species specific oligonucleotide probes for the identification of periodontal bacteria. J. Clin. Microbiol. 28:318.
14. Dowell, V.R., Jr., and Hawkins, T.M. 1974. Laboratory methods in anaerobic bacteriology. In CDC laboratory manual, DHEW Pub. No. (CDC) 74-8272, U.S. Government Printing Office, Washington, D.C.
15. Dowell, V.R., Jr., and Lombard, G.L. 1981. Reactions of anaerobic bacteria in differential agar media. U.S. Department of Health and Human Services, Public Health Service, Centers for Disease Control, Atlanta.
16. Ebersole, J.L., Frey, D.E., Taubman, M.A., et al. 1984. Serological identification of oral *Bacteroides* spp. by enzyme-linked immunosorbent assay. J. Clin. Microbiol. 19:639.
17. Ghanem, F.M., Ridpath, A.C., Moore, W.E.C., and Moore, L.V.H. 1991. Identification of *Clostridium botulinum, Clostridium argentinense,* and related organisms by cellular fatty acid analysis. J. Clin. Microbiol. 29:1114.
18. Goldstein, E.J.C., and Citron, D.M. 1988. Annual incidence, epidemiology, and comparative in vitro susceptibilities to cefoxitin, cefotetan, cefmetazole, and ceftizoxime of recent community-acquired isolates of the *Bacteroides fragilis* group. J. Clin. Microbiol. 26:2361.

19. Gulletta, E., Amato, G., Nani, E., and Covelli, I. 1985. Comparison of two systems for identification of anaerobic bacteria. Eur. J. Clin. Microbiol. 4:282.

20. Hanson, C.W., and Martin, W.J. 1976. Evaluation of enrichment, storage, and age of blood agar medium in relation to its ability to support growth of anaerobic bacteria. J. Clin. Microbiol. 4:394.

21. Harpold, D.J., and Wasilauskas, B.L. 1987. Rapid identification of obligately anaerobic gram-positive cocci using high-performance liquid chromatography. J. Clin. Microbiol. 25:996.

22. Holmberg, K., and Forsum, U. 1973. Identification of *Actinomyces, Arachnia, Bacterionema, Rothia,* and *Propionibacterium* species by defined immunofluorescence. Appl. Microbiol. 25:834.

23. Hsu, P.C., Minshew, B.H., Williams, B.L., and Lennard, E.S. 1979. Use of an immunoperoxidase method for identification of *Bacteroides fragilis.* J. Clin. Microbiol. 10:285.

24. Hussain, Z., Lannigan, R., Schieven, B.C., et al. 1987. Comparison of RapID-ANA and Minitek with a conventional method for biochemical identification of anaerobes. Diagn. Microbiol. Infect. Dis. 6:69.

25. Karachewski, N.O., Busch, E.L., and Wells, C.L. 1985. Comparison of PRAS II, RapID-ANA and API 20A systems for identification of anaerobic bacteria. J. Clin. Microbiol. 21:122.

26. Kasper, D.L., Fiddian, A.P., and Tabaqchali, S. 1979. Rapid diagnosis of *Bacteroides* infections by indirect immunofluorescence assay of clinical specimens. Lancet 1:239.

27. Kuritza, A.P., Getty, C.E., Shaughnessy, P., et al. 1986. DNA probes for identification of clinically important *Bacteroides* species. J. Clin. Microbiol. 23:343.

28. Livingston, S.J., Kominos, S.D., and Yee, R.B. 1978. New medium for selection and presumptive identification of the *Bacteroides fragilis* group. J. Clin. Microbiol. 7:448.

29. Looney, W.J., Gallusser, A.J.C., and Modde, H.K. 1990. Evaluation of the ATB 32 A system for identification of anaerobic bacteria isolated from clinical specimens. J. Clin. Microbiol. 28:1519.

30. Mangels, J.I., and Douglas, B.P. 1989. Comparison of four commercial *Brucella* agar media for growth of anaerobic organisms. J. Clin. Microbiol. 27:2268.

31. Mangels, J.I., Cox, M., and Lindberg, L.H. 1984. Methanol fixation: an alternative to heat fixation of smears before staining. Diagn. Microbiol. Infect. Dis. 2:129.

32. Marler, L., Allen, S., and Siders, J. 1984. Rapid enzymatic characterization of clinically encountered anaerobic bacteria with the API ZYM system. Eur. J. Clin. Microbiol. 3:294.

33. Marler, L.M., Siders, J.A., Wolters, L.C., et al. 1991. Evaluation of the new RapID-ANA II system for the identification of clinical anaerobic isolates. J. Clin. Microbiol. 29:874.

34. Millar, S.J., Chen, P.B., and Haussman, E. 1987. Monoclonal antibody for identification of *Bacteroides gingivalis* lipopolysaccharide. J. Clin. Microbiol. 25:2437.

35. Mills, C.K., Grimes, B.Y., and Gherna, R.L. 1987. Three rapid methods compared with a conventional detection of urease production in anaerobic bacteria. J. Clin. Microbiol. 25:2209.

36. Moncla, B.J., Strockbine, L., Braham, P., et al. 1988. The use of whole-cell DNA probes for the identification of *Bacteroides intermedius* isolates in a dot blot assay. J. Dent. Res. 67:1267.

37. Moncla, B.J., Braham, P., Rabe, L.K., and Hillier, S. 1991. Rapid presumptive identification of black-pigmented gram-negative anaerobic bacteria by using 4-methylumbelliferone derivatives. J. Clin. Microbiol. 29:1955.

38. Murray, P.R., Weber, C.J., and Niles, A.C. 1985. Comparative evaluation of three identification systems for anaerobes. J. Clin. Microbiol. 22:52.

39. Okuda, K., Ohta, K., Kato, T., et al. 1986. Antigenic characteristics and serological identification of 10 black-pigmented *Bacteroides* species. J. Clin. Microbiol. 24:89.

40. Rolfe, R.D., Hentges, D.J., Campbell, B.J., and Barrett, J.T. 1978. Factors related to the oxygen tolerance of anaerobic bacteria. Appl. Environ. Microbiol. 36:306.

41. Salyers, A.A., Lynn, S.P., and Gardner, J.F. 1983. Use of randomly cloned DNA fragments for identification of *Bacteroides thetaiotaomicron.* J. Bacteriol. 154:287.

42. Schreckenberger, P.C., Celig, D.H., and Janda, W.M. 1988. Clinical evaluation of the Vitek ANI card for identification of anaerobic bacteria. J. Clin. Microbiol. 26:225.

43. Senne, J.E., and McCarthy, L.R. 1982. Evaluation of a metronidazole disk test for the presumptive identification of anaerobes. Am. J. Med. Tech. 48:613.

44. Shaw, H.N., and Collins, M.D. 1983. Genus *Bacteroides:* a chemotaxonomic perspective. J. Appl. Bacteriol. 55:403.

45. Shaw, H.N., and Collins, M.D. 1988. Proposal for reclassification of *Bacteroides asaccharolyticus, Bacteroides gingivalis,* and *Bacteroides endodontalis* in a new genus, *Porphyromonas.* Int. J. Syst. Bacteriol. 38:128.

46. Shaw, H.N., and Collins, M.D. 1989. Proposal to restrict the genus *Bacteroides* (Castellani and Chalmers) to *Bacteroides fragilis* and closely related species. Int. J. Syst. Bacteriol. 39:85.

47. Shaw, H.N., and Collins, M.D. 1990. *Prevotella,* a new genus to include *Bacteroides melaninogenicus* and related species formerly classified in the genus *Bacteroides.* Int. J. Syst. Bacteriol. 39:205.

48. Stenson, M.J., Lee, D.T., Rosenblatt, J.E., and Contezac, J.M. 1986. Evaluation of the AnIDENT system for the identification of anaerobic bacteria. Diagn. Microbiol. Infect. Dis. 5:9.

49. Stoakes, L., Kelly, T., Schieven, B., et al. 1991. Gas-liquid chromatographic analysis of cellular fatty acids for identification of gram-negative anaerobic bacilli. J. Clin. Microbiol. 29:2636.

50. Tanner, A.C.R., Strzempko, M.N., Belsky, C.A., and McKinley, G.A. 1985. API ZYM and API AnIDENT reactions of fastidious oral gram-negative species. J. Clin. Microbiol. 22:333.

51. Tunér, K., Summanen, P., Baron, E.J., and Finegold, S.M. 1992. Cellular fatty acids in *Fusobacterium* species. J. Clin. Microbiol. 30:3225.

52. Van Damme-Jongsten, M., Rodhouse, J., Gilbert, R.J., and Notermans, S. 1990. Synthetic DNA probes for detection of enterotoxigenic *Clostridium perfringens* strains isolated from outbreaks of food poisoning. J. Clin. Microbiol. 28:131.

53. Viljanen, M.K., Linko, L., and Lehtonen, O. 1988. Detection of *Bacteroides fragilis, Bacteroides thetaiotaomicron,* and *Bacteroides ovatus* in clinical specimens by immunofluorescence with a monoclonal antibody to *B. fragilis* lipopolysaccharide. J. Clin. Microbiol. 26:448.

54. Wong, M., Catena, A., and Hadley, W.K. 1980. Antigenic relationships and rapid identification of *Peptostreptococcus* species. J. Clin. Microbiol. 11:515.

BIBLIOGRAPHY

Edelstein, M.A.C. 1989. Laboratory diagnosis of anaerobic infections in humans. In Finegold, S.M., and George, W.L., editors. Anaerobic infections in humans, p. 111. Academic Press, Orlando, Fla.

Engelkirk, P.G., Duben-Engelkirk, J., and Dowell, V.R., Jr. 1992. Principles and practice of clinical anaerobic bacteriology. Star Publishing, Belmont, Calif.

Finegold, S.M., and Edelstein, M.A.C. 1988. Coping with anaerobes in the 80s. In Hardie, J.M., and Borriello, S.P., editors. Anaerobes today, p. 1. John Wiley & Sons, London.

Finegold, S.M., Baron, E.J., and Wexler, H.M. 1992. A clinical guide to anaerobic infections. Star Publishing, Belmont, Calif.

Holdeman, L.V., Cato, E.P., and Moore, W.E.C. 1977. Anaerobe laboratory manual, ed 4. VPI Laboratory, Virgina Polytechnic Institute and State University, Blacksburg, Va.

Rodloff, A.C., Appelbaum, P.C., and Zabransky, R.J. 1991. Cumitech 5A. American Society for Microbiology, Washington, D.C.

Summanen, P., Baron, E.J., Citron, D.M., et al. 1993. Wadsworth anaerobic bacteriology manual, ed 5. Star Publishing, Belmont, Calif.

36

ANAEROBIC GRAM-POSITIVE BACILLI

The anaerobic gram-positive bacilli of human clinical significance consist of one genus of endospore-formers, *Clostridium,* and six genera of non–spore-formers, *Actinomyces, Bifidobacterium, Eubacterium, Lactobacillus, Propionibacterium,* and *Mobiluncus.* They are part of the indigenous flora of the oral cavity, gastrointestinal and genitourinary tracts, and skin. Some *Clostridium* sp. produce potent toxins and can cause life-threatening infections. The non–spore-forming gram-positive bacilli are usually present in chronic infections. Although these species are classified with anaerobes, most *Actinomyces* sp. and many lactobacilli are slow-growing facultative anaerobes. Four species of clostridia are aerotolerant and may be difficult to distinguish from other aerobic bacilli. The aerotolerant clostridia do not form spores under aerobic conditions and produce larger colonies under anaerobic conditions. Differentiation of genera of anaerobic gram-positive bacilli is represented in Table 36.1. They are initially separated by the presence of spores; however, some *Clostridium* sp. rarely form spores and have to be grouped by other characteristics. The non–spore-forming genera are differentiated by their metabolic end products.

TABLE 36.1

DIFFERENTIATION OF GENERA OF ANAEROBIC, GRAM-POSITIVE BACILLI

CHARACTERISTIC	GENUS
Spore forming	*Clostridium*
Non–spore forming:	
1. Produces propionic and acetic acids with or without lactic acid	*Propionibacterium*
2. Produces acetic and lactic acids (acetic > lactic)	*Bifidobacterium*
3. Produces lactic acid as sole major end product	*Lactobacillus*
4. Produces major succinic acid with or without acetic and lactic acid; nonmotile	*Actinomyces*
5. Produces acetic and succinic acid; motile	*Mobiluncus*
6. Not as above: butyric ± other acids, acetic, or no major acids	*Eubacterium*

TABLE 36.2

TAXONOMIC CHANGES OF ANAEROBIC, GRAM-POSITIVE BACILLI

NEW	PREVIOUS
Actinomyces pyogenes	*Corynebacterium pyogenes*
A. georgiae	sp. nov.
A. gerensceriae	sp. nov.
Clostridium argentinense	*C. botulinum* type G, some strains of *C. hastiforme* and *C. subterminale*
C. baratii	*C. baratii, C. perenne,* and *C. paraperfringens*
C. spiroforme	sp. nov.
C. symbiosum	*Fusobacterium symbiosum*
Eubacterium plautii	*Fusobacterium plautii*
E. yurii	sp. nov.
E. yurii ssp. *yurii*	sp. nov.
E. yurii ssp. *margaretiae*	sp. nov.
Lactobacillus oris	sp. nov.
Propionibacterium propionicus	*Arachnia propionica*

SPORE-FORMING BACILLI
CLOSTRIDIUM SPECIES

More than 120 described species make up the genus *Clostridium*. Some are found only in soil and sludge, whereas others colonize the intestinal tract of humans and animals and may be isolated from infections caused by fecal organisms. *C. tetani* and *C. botulinum* produce potent neurotoxins, whereas *C. perfringens* and *C. septicum* produce toxins that can cause myonecrosis under certain circumstances. Some strains of *C. perfringens* produce an enterotoxin that causes a form of food poisoning. Most clostridia are obligate anaerobes, but those that can grow on enriched media in air are *C. tertium, C. histolyticum, C. carnis,* and *C. durum*. Aerotolerant clostridia can be confused with other aerobic bacilli, especially lactobacilli, because they are catalase negative and do not form spores when growing under aerobic conditions. They are different from *Bacillus* sp., which sporulate under aerobic conditions and are usually catalase positive. Spores are difficult to detect in cells of *C. ramosum, C. clostridioforme* and certain other species; however, most strains will withstand heat shock or ethanol treatment, which kill vegetative cells but not spores. *C. clostridioforme* and several other species typically stain gram negative; however, they are susceptible to vancomycin in the disk identification test (Chapter 35), which confirms their gram positivity.

Several new species have been described, but few of these have been isolated from human clin-
ical specimens. All *C. botulinum* type G and some strains of *C. hastiforme* and *C. subterminale* are now included in a new species, *C. argentinense*.[32] *C. perenne* and *C. paraperfringens* are now *C. baratii* (Table 36.2).

NORMAL FLORA AND INFECTIONS

Clostridia are widely distributed in the environment in soil, dust, and water. They are present in large numbers as part of the indigenous flora of the intestinal tract of humans and animals. Most infections are acquired endogenously and are part of a mixed flora with other anaerobes and aerobes.

Clostridial disease syndromes of exogenous origin occur less frequently and include tetanus and rarely gas gangrene (clostridial myonecrosis); intoxications include tetanus, botulism, and *C. perfringens* food poisoning. The pathogenic species can be placed in five categories according to the disease syndromes or intoxications they produce. Some species can be involved in several types of diseases. *C. perfringens* and *C. ramosum* are the most common clinical isolates.

DISEASE SYNDROMES AND INTOXICATIONS

GROUP I: GAS GANGRENE   *C. perfringens* (type A), *C. septicum, C. novyi* type A, *C. bifermentans, C. histolyticum, C. sordellii, C. sporogenes,* and occasionally other clostridia can cause gas gangrene; the first three species are the most important. Infection can be either of endogenous or exogenous origin.

Bacteremia occurs in approximately 15% of the patients with clostridial myonecrosis. A Gram stain of the clinical specimen may confirm the clinical diagnosis of gas gangrene (Figures 36.1 to 36.3). The two important findings are absence or distortion of white blood cells and other cells and the presence of large, relatively short, fat, gram-positive rods with blunt ends and without spores. The changes in the host cells are caused by the alpha toxin produced by *C. perfringens* and by a variety of other histotoxins produced by other clostridia.

GROUP II: TETANUS   *C. tetani* is the etiological agent of tetanus. This organism is present in soil and may contaminate an existing wound, or it may be inoculated into a puncture wound by a contaminated object. When the proper conditions exist, the toxin, tetanospasmin, is produced, and disease may follow.

GROUP III: BOTULISM   The primary agent of botulism is *C. botulinum*; however, there are reports of rare cases caused by *C. butyricum* and *C. baratii* that can produce types E and F botulinal neurotoxins.[1,21,22,33] The three manifestations of disease are food, wound, and infant botulism. Food botulism occurs with the ingestion of preformed toxin in contaminated food. This occurs most frequently with improperly canned vegetables but has also been found in noncanned processed vegetables, such as prepared garlic, especially when they are improperly stored without chemical preservatives. Wound botulism occurs with the introduction of spores or bacteria into a wound

with subsequent infection and toxin production.[11] Infant botulism occurs with *C. botulinum* colonization of the infant's gastrointestinal tract, presumably because the infant's own flora is not yet fully developed, and subsequent toxin production; the toxin is then absorbed to produce the disease.[15] In vivo toxin production and absorption from the gastrointestinal tract has been reported in several patients with adult botulism as well. Because the toxins are extremely potent, cultivation of *C. botulinum* and toxigenicity determinations should be performed by reference laboratories and personnel involved should be immunized. Selective media for this purpose have been described.[23]

GROUP IV: ANTIBIOTIC-ASSOCIATED DIARRHEA AND COLITIS   Most cases are preceded by antibiotic prophylaxis or therapy, with β-lactam agents and clindamycin most frequently implicated, although any antimicrobial agent may be causative. Occasionally, methotrexate and other cytotoxic agents also have been implicated. It can also occur in exacerbations of inflammatory bowel disease, possibly related to therapy, and in complications of bowel obstruction. The primary etiological agent is *C. difficile*. Rare cases may result from *C. perfringens* type C and other clostridia, as well as *Staphylococcus aureus*. Most strains of *C. difficile* produce at least two toxins; cytotoxin B is cytopathic for most tissue culture cell lines and enterotoxin A promotes fluid secretion and destroys intestinal lining cells. *C. difficile* and rarely type A *C. perfringens* also may cause a mild diarrhea without colitis following antimicrobial therapy.[4] Chapter 18 pre-

FIGURE 36.1

Perineal gas gangrene. *Clostridium perfringens* (large, broad gram-positive rods), gram-negative rods, and white blood cells distorted by toxins of *C. perfringens*.

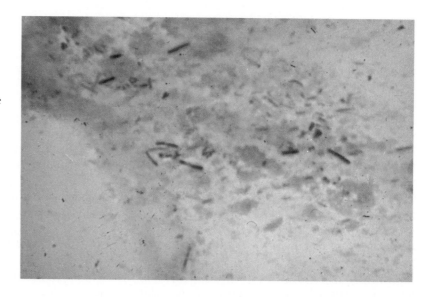

FIGURE 36.2

Clostridium septicum. Note free spores.

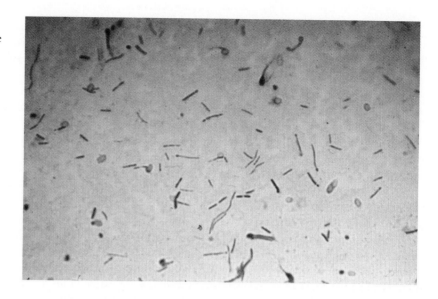

FIGURE 36.3

Intracellular *Clostridium perfringens* in peripheral blood smear.

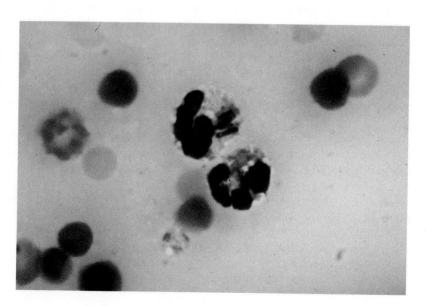

sents details on examination of specimens for *C. difficile* and its toxins.

GROUP V. MISCELLANEOUS *C. perfringens* and other species of clostridia are most frequently isolated from an assortment of infections similar to those caused by other anaerobes. Common sites include intraabdominal infection, postoperative wound infection, infections related to gynecological disease or surgery, and soft tissue infections.[20,24] They are rarely isolated from pleuropulmonary infections and brain abscess. The classical infections involving *C. perfringens*, besides gas gangrene, are gangrenous cholecystitis, with or without visceral gas gangrene, and postabortal sepsis

with intravascular hemolysis. Bacteremia occurs occasionally with *C. perfringens* and other clostridia, especially when blood is obtained from the femoral artery. *C. tertium* bacteremia has been reported in patients with neutropenia and hematological malignancies.[30] *C. septicum* is also associated with colorectal carcinoma and may be isolated from blood or infected wounds.[5,18] Its presence in a culture is often the first sign that a carcinoma may be present. A high mortality rate is associated with *C. septicum* infection. The isolation of clostridia from blood in the absence of clinical signs of disease may be a transient event and not necessarily related to disease, except as noted for *C. septicum*.

VIRULENCE FACTORS

The clostridia produce many virulence factors (e.g., spores), many types of toxins, and other extracellular products (e.g., collagenase, proteases). The spores of *C. botulinum* may survive and grow in improperly canned or processed foods, and *C. perfringens* has been cultured from insufficiently sterilized surgical dressings and bandages, in plaster of Paris for casts, and on the clothing and skin of humans.

The toxins of *C. difficile* have already been mentioned; enterotoxin A appears to be responsible for most clinical findings. The exotoxins of *Clostridium tetani* and *C. botulinum* are among the most potent poisons known. The toxin produced by *C. tetani*, tetanospasmin, becomes bound to gangliosides within the central nervous system; as with strychnine, it suppresses the central inhibitory balancing influences on motor neuron activity, thus leading to intensified reflex response to afferent stimuli and to spasticity and convulsions (Figure 35.4). Botulinal toxin attaches to individual motor nerve terminals, preventing release of acetylcholine and causing passive paralysis. *C. perfringens* may produce at least 10 toxins, including hemolysins, proteases, ribonuclease (RNase), collagenase, hyaluronidase, neuraminidase, and an enterotoxin with activity similar to that of *Vibrio cholerae*. The beta toxin of type C *C. perfringens* is the key factor in necrotizing enterocolitis ("Darmbrand," pig-bel). Several other *Clostridium* sp. produce toxins, some of which are clinically important. The interested reader is referred to Smith's and Willis' excellent books cited in the bibliography.

MORPHOLOGY AND GENERAL CHARACTERISTICS

The clostridia possess no one typical cell shape or size. Some are very large rods, whereas others are slender and short or long and sometimes even coiled. Most strains are gram positive, but a few always appear gram negative (e.g., *C. clostridioforme*), and others may appear gram variable depending on growth medium and age of the culture. Spores are often swollen, showing spindle, "drumstick" (*C. tetani*) (Figure 36.5), and "tennis racket" (*C. botulinum*) forms, and may be located in the cell's central, subterminal, or terminal part. Spores may be apparent on Gram stain or a wet preparation using phase-contrast or darkfield microscopy. It is difficult to detect spores of *C. perfringens*, *C. ramosum*, and *C. clostridioforme*. *C. perfringens* and *C. clostridioforme* are usually differentiated from non–spore-forming gram-positive bacilli by their cell morphology. *C. perfringens* is boxcar shaped (Figure 36.3). *C. clostridioforme* has an elongated football shape with cells often in pairs, and it stains gram negative (Figure 36.6). *C. ramosum* cells are slender and longer than *C. perfringens* cells.

As with cell morphology, the clostridia possess no one typical colony morphology, although generally a large colony (> 2 mm) with irregular edges or swarming growth is probably a *Clostridium* (Figures 36.7 to 36.9). Some *Clostridium* sp., however, form small, convex nonhemolytic colonies with an entire edge, such as *C. ramosum* and *C. clostridioforme*. Some clostridia produce several different-looking colony types, so the culture appears to be mixed. A subculture of a single colony that yields the same pleomorphism should

FIGURE 36.4

Tetanus. Patient exhibits opisthotonos (head and legs bent backward toward each other).

FIGURE 36.5

Clostridium tetani. Note round terminal spores, the "drumstick" appearance.

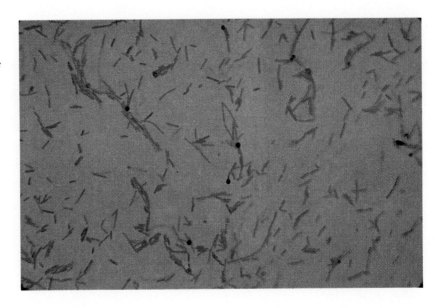

FIGURE 36.6

Clostridium clostridioforme. Note gram-negative cells.

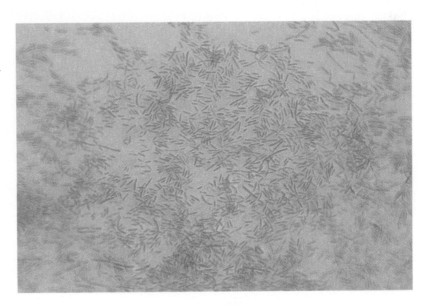

be considered pure. A few *Clostridium* sp. have distinct colony characteristics. *C. perfringens* usually produces a double zone of beta hemolysis (Chapter 35, Figure 35.6). The inner zone shows complete hemolysis, whereas the outer zone may display partial hemolysis. *C. septicum* produces a "Medusa head" colony within a few hours of incubation that spreads over the entire plate in less than 24 hours (Figure 36.7). *C. tetani* and *C. sporogenes*, and rare strains of *C. sordellii* and *C. novyi* may also swarm. *C. difficile* produces a yellow ground-glass colony on the selective medium cycloserine-cefoxitin fructose agar (CCFA)[12] (Figure 36.10). Relatively few organisms other than *C. difficile* grow on the CCFA medium, and their colonies are unlike those of *C. difficile*. *C. difficile*

colonies on blood agar (BA) are usually 2 mm or more in diameter after 24 hours of incubation (Figure 36.11), fluoresce yellow-green, and emit a horse stable odor. Such colonies, which on Gram stain show typical gram-positive rods with subterminal spores, constitute good presumptive evidence of the presence of *C. difficile*. Fluorescense is best observed on BA because CCFA is itself fluorescent. *C. difficile* does not produce spores on CCFA. Most clostridia grow well in broth, and some strains produce abundant gas (e.g., *C. perfringens, C. septicum*) and a foul odor.

IDENTIFICATION: ISOLATION TECHNIQUES

Although clostridia are usually present as part of a mixed flora in infections, their isolation is not dif-

FIGURE 36.7

Clostridium septicum. "Medusa head" colony on blood agar plate.

FIGURE 36.8

C. tetani on blood agar plate. Note irregular edge.

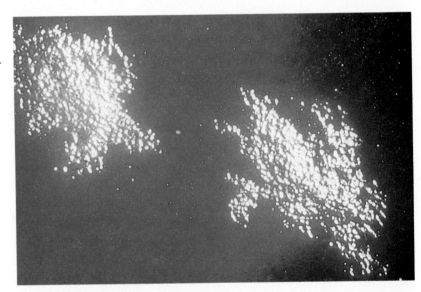

FIGURE 36.9

Colony of *Clostridium sordelli.* (From Summanen, P., Baron, E.J., Citron, D.M., et al. 1993. Wadsworth anaerobe laboratory manual, ed 5. Star Publishing, Belmont, Calif.)

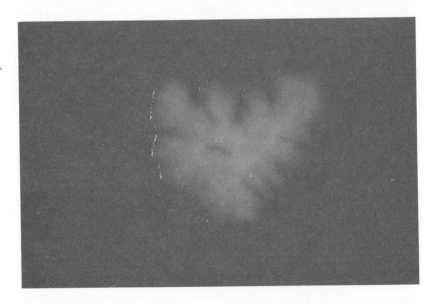

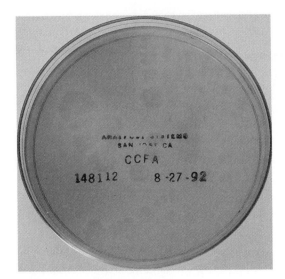

FIGURE 36.10

Clostridium difficile on cycloserine-cefoxitin fructose agar (CCFA). (Courtesy Anaerobe Systems, San Jose, Calif.)

ficult when selective media, such as phenylethyl alcohol (PEA) or colistin and nalidixic acid (CNA) agar, are included as one of the primary culture plates. Growth of some strains of *C. perfringens* and *C. septicum* is inhibited on CNA. In the event these media do not adequately isolate the clostridia, heat shock, ethanol exposure, or other selective agars are alternative isolation methods (Procedure 36.1). One should always split the specimen, so that part is treated and the other is processed in the usual manner. If the specimen amount is too small, one should use the backup broth culture or cell paste from a sweep of the primary BA plate as the inoculum for the spore detection test. Media

with 5% agar or PEA may be useful to minimize swarming of some species. Neomycin (100 μg/ml), McClung-Toabe agar, or cycloserine-containing (500 μg/ml) blood agar or egg yolk agar (EYA) are good selective media. Nonselective media should always be inoculated in parallel with selective media.

GROUPING AND DEFINITIVE IDENTIFICATION

The *Clostridium* sp. can be divided into three groups based on proteolytic and saccharolytic activity. Table 36.3 and Figure 36.12 show test reactions that are useful for definitive identification. Use the flow diagram as a guide only, since atypical strains may give variable reactions.

C. tertium and *C. histolyticum* are the most frequently encountered aerotolerant clostridia. Growth can be very good in air, but a Gram stain of growth from an aerobic plate will not show spores. The Nagler test (Chapter 35) can be used as an adjunct for identification of *C. perfringens, C. baratii, C. sordellii,* and *C. bifermentans. C. perfringens* demonstrates the most complete neutralization of the α-lecithinase in the Nagler test. It is readily differentiated from the other Nagler-positive bacteria by displaying double-zone hemolysis, boxcar-shaped cells, a positive glutamic acid decarboxylase test[2] and a positive reverse CAMP test.[8] It is also nonmotile. *C. baratii* is infrequently isolated from clinical specimens. *C. sordellii* is indole positive and almost always urease positive, whereas *C. bifermentans* is indole positive and urease negative. *C. perfringens* is the most frequent clinical isolate, followed by *C. ramosum, C. septicum, C. clostridioforme* and *C. innocuum. C. difficile* is in-

FIGURE 36.11

Clostridium difficile on blood agar plate.

PROCEDURE 36.1

ISOLATION OF CLOSTRIDIA FROM MIXED CULTURES

Heat shock

PRINCIPLE

Most *Clostridium* sp. produce spores resistant to heating at 80° C for 10 minutes, whereas vegetative cells are usually killed. *Bacillus* sp. also produce heat-resistant spores, so all isolates should be tested for aerotolerance and identified. Some strains of *Clostridium* do not survive exposure to 80° C.

METHOD

1. For a clinical specimen or broth culture from the specimen:
   a. Remove an aliquot of liquid specimen or the broth culture and transfer it to a sterile screw-cap glass tube.
   b. Place the tube in a water bath at 80° C for 10 minutes. The water level should be above the liquid level in the tube. A heating block can also be used.
   c. Transfer 1 or 2 drops of the heated sample onto enriched blood agar (BA) and egg yolk agar (EYA) plates. Streak for isolation.
   d. Incubate the plates anaerobically for 24 to 48 hours.
   e. If growth occurs, identify the colonies as described in this chapter and Chapter 35.
   f. If growth does not occur, the specimen did not contain *Clostridium* sp., or it contained a strain that was sensitive to heating at 80° C.

2. For primary plate:
   a. If the specimen has been inoculated onto agar plates and no broth culture is available or spores were observed on a Gram stain, cell paste from the first quadrant of growth can be used for the spore test.
   b. Roll a swab across the first quadrant of the plate and inoculate a thioglycolate, chopped meat, or prereduced anaerobically sterilized (PRAS) starch broth tube.
   c. Heat immediately at 80° C for 10 minutes.
   d. Incubate the broth 24 to 48 hours.
   e. If growth occurs, subculture 2 or 3 drops of the broth to a BA or EYA plate.
   f. Incubate the plates for 24 to 48 hours, and identify the colonies as described in this chapter and Chapter 35.

QUALITY CONTROL

A specific quality control procedure for clinical specimens is impractical and not required. If equipment, reagents, and media are performing properly, failure to recover a spore-forming organism observed on Gram stain may result from killing of the spores at 80° C. If possible, either repeat the test at a lower temperature, or perform the ethanol spore test.

Ethanol treatment

PRINCIPLE

Ethanol kills vegetative cells, but most spores can withstand exposure for at least 30 minutes.

METHOD

1. The specimen can be a clinical specimen, a broth culture of the specimen, or cell paste from the first quadrant of the primary plate suspended in thioglycolate or chopped meat broth.

2. Add 1 ml of liquid specimen to 1 ml of 95% alcohol.

3. Mix gently and allow to stand for 30 to 45 minutes. Subculture 2 or 3 drops onto BA or EYA medium. Streak for isolation.

4. Incubate the plates anaerobically for 24 to 48 hours.

5. If growth occurs, identify the colonies as described in this chapter and Chapter 35.

QUALITY CONTROL

A specific quality control procedure for clinical specimens is impractical and not required. Determine that equipment, reagents, and media are performing properly. If spore-forming cells were observed on Gram stain and organisms were not recovered, repeat the test with 10 to 15 minutes of ethanol exposure. Some strains of *Clostridium* (e.g., some strains of *C. clostridioforme*) are oxygen sensitive and may die from the oxygen exposure inherent in the procedure.

TABLE 36.3

CHARACTERISTICS OF *CLOSTRIDIUM* SP.

SPECIES	AEROBIC GROWTH	GELATIN HYDROLYSIS	LECITHINASE	NAGLER TEST	LIPASE	INDOLE	UREASE	ESCULIN HYDROLYSIS	FERMENTATION OF GLUCOSE	LACTOSE	MALTOSE	MANNITOL	MANNOSE	MELIBIOSE	STARCH	SUCROSE	XYLOSE	SPORE LOCATION	MOTILITY	FATTY ACID END PRODUCTS FROM PEPTONE-YEAST-GLUCOSE BROTH
Saccharolytic, proteolytic																				
C. bifermentans	−	+	+	+	−	+	−	+⁻	+	−	+⁻	−	−	−	−	−	−	ST	+	A (p ib b iv ic l s)
C. botulinum*	−	V	−	−	+	−	−	V	+	−	V	−	V	−	V	V	−	ST	+⁻	AB (p ib iv v ic l s)
C cadaveris	−	+	−	−	+	−	−	−	+	−	+	−	−	−ʷ	−	−	−	T	+⁻	A p B (l s)
C. difficile	−	+	−	−	−	+	−	+	+	−	−	+⁻	+⁻	−	−	−	−	ST	+⁻	A p ib B iv IC (l)
C novyi type A	−	+	+	−	+	−	−	−	+	−	V	−	−	−	−	−	−	ST	+⁻	A P B
C. perfringens†,§	−	+	+	+	−	−	−	−⁺	+	+	+	−	+	V	+⁻	+	−	ST	−	A B L
C. putrificum	−	+	−	−	−	−	−	−	+	−	−ʷ	−	−	−	−	−	−	T, ST	+	A ib B iv (p ic v l s)
C. septicum	−	+	−	−	−	−	−	+	+	+	+	−	+	−	−	−	−	ST	+⁻	A B (l)
C. sordellii	−	+	+	+	−	+	+	−	+	−	+ʷ	−	−ʷ	−	−	−	−	ST	+⁻	A (p ib iv ic l)
C. sporogenes‡	−	+	−	−	+	−	−	+	V	−	−ʷ	−	−	−	−	−	−	ST	+⁻	A ib B iv v ic
Saccharolytic, nonproteolytic																				
C. baratii	−	−	+	+	−	−	−	+	+	+ʷ	+ʷ	−	+	−ʷ	V	+	−	ST	−	A B L (s)
C. butyricum	−	−	−	−	−	−	−	+	+	+	+	−⁺	+	+	+	+		ST	+⁻	A B (ls)
C. clostridioforme†,§	−	−	−	−	−	−⁺	−	+	+ʷ	V	+ʷ	−	+	−⁺	−	+ʷ	+ʷ	ST	+⁻	A (Ls)
C. innocuum	−	−	−	−	−	−	−	+	+	−	−	+	+	−	−	+	−ʷ	T	−	A B L (s)
C. paraputrificum	−	−	−	−	−	−	−	+	+	+	+	−	+	−	+	+	−	T, ST	+⁻	A B (s)
C. ramosum†,§	−	−	−	−	−	−	−	+	+	+	+	+⁻	+	+⁻	−ʷ	+	−ʷ	T	−	A l (s)
C. tertium	+	−	−	−	−	−	−	+	+	+	+	+	+	+ʷ	+	+	V	T	+⁻	A b L (s)
Asaccharolytic, proteolytic																				
C. hastiforme	−	+	−	−	−	−	−	−	−	−	−	−	−	−	−	−	−	ST, T	+⁻	A p ib B iv (ic s)
C. histolyticum	+	+	−	−	−	−	−	−	−	−	−	−	−	−	−	−	−	ST	+⁻	A (ls)
C. subterminale	−	+	−⁺	−	−	−	−	−	−	−	−	−	−	−	−	−	−	ST	+⁻	A ib b iv (ls)
C. tetani	−	+	−	−	−⁺	+⁻	−	−	−	−	−	−	−	−	−	−	−	T	−⁺	A p B (s)

Reactions: −, Negative; +, positive; *w*, weakly positive; *V*, variable. Superscripts represent rare strain reactions.

Prereduced anaerobically sterilized (PRAS) carbohydrate reactions: +, pH <5.5; w, pH 5.5–5.9; −, pH > 5.9.

Spore location: ST, subterminal; T, terminal.

Fatty acid end products: A, acetic; P, propionic; IB, isobutyric; B, butyric; IV, isovaleric; V, valeric; IC, isocaproic; L, lactic; S, succinic. Capital letters indicate major amount of end product; lower-case letters indicate minor amount. Parentheses indicate variable reaction.

C. botulinum types vary in saccharolytic and proteolytic activity. Send suspected organisms or specimens to local or state public health laboratory.

†Spores rarely observed.

‡*C. sporogenes* and *C. botulinum* types A, B, and F are biochemically similar. Toxin neutralization tests are recommended for definitive differentiation.

§Cell morphology distinctive. (See earlier discussion.)

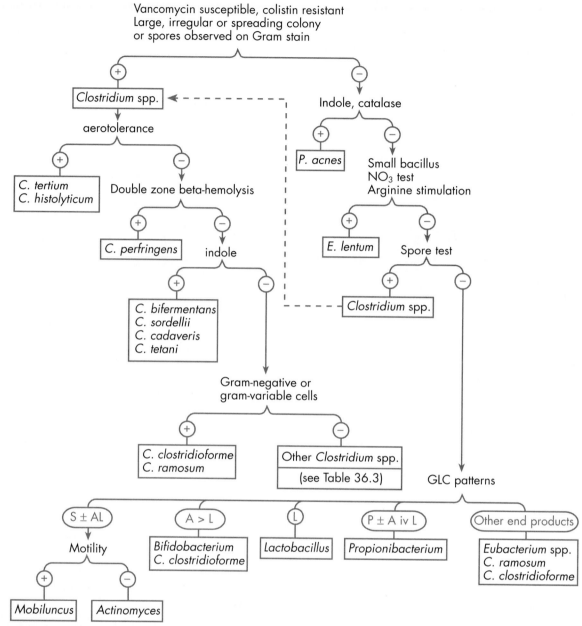

FIGURE 36.12

Flow diagram for identification of gram-positive anaerobic bacilli. Use only as a guide, since atypical strains may have different reactions.

frequently isolated from specimens other than stool. Lipase-positive *C. botulinum* strains and *C. sporogenes* are phenotypically similar, so differentiation traditionally relies on animal toxicity studies. However, recent studies of cell wall fatty acids suggest that these organisms can be identified reliably with this technique.[13] Animal toxicity studies should be performed by a reference laboratory. Production of lecithinase and lipase on EYA, hydrolysis of gelatin, and fermentation of glucose are useful for subdividing this diverse group. *C. novyi* type A is the only *Clostridium* that consis-

tently produces both lecithinase and lipase. Definitive identification relies on end product analysis using gas-liquid chromatography (GLC), carbohydrate fermentation patterns, and various other tests. Characteristics of some *Clostridium* sp. are presented in Table 36.3.

SUSCEPTIBILITY TO ANTIMICROBIAL AGENTS
Penicillin G is the drug of choice to treat most clostridial infections, although β-lactamase has been detected in some strains of *C. butyricum*, *C. ramosum*, and *C. clostridioforme*. Approximately

TABLE 36.4

SIMPLE TESTS FOR PRESUMPTIVE DIFFERENTIATION OF NON–SPORE-FORMING BACILLI

GENUS	AEROBIC GROWTH	CATALASE	NITRATE REDUCTION	CELLS WITH PARALLEL SIDES	CURVED CELLS, MOTILE
Actinomyces	+⁻	−⁺	+⁻	−	−
Bifidobacterium	−⁺	−	−	−	−
Eubacterium	−	−⁺	+⁻	+⁻	−
Propionibacterium	−⁺	+⁻	+⁻	−⁺	−
Mobiluncus	−	−	V	−	+
Lactobacillus	V	−	−	+	−

+, Positive; −, negative, *V*, variable. Superscripts represent reactions of rare strains. See also Table 36.1 for fatty acid end products.

15% of *C. ramosum* strains are highly resistant to clindamycin and cephalosporins, and many strains are resistant to tetracycline and erythromycin. Most strains of *C. difficile* and *C. paraputrificum* and some strains of *C. clostridioforme* and *C. innocuum* are also resistant to clindamycin. All strains of *C. difficile* and *C. innocuum* are resistant to cefoxitin. Some strains of *C. subterminale, C. limosum,* and other species display variable susceptibilities to cefoxitin and third-generation cephalosporins. Chloramphenicol, imipenem, and metronidazole are active against most clostridia. Surgical debridement and drainage are usually essential for treatment of serious clostridial infections. Oral vancomycin, metronidazole, and bacitracin are used to treat *C. difficile* colitis.

NON–SPORE-FORMING BACILLI

The genera of anaerobic, non–spore-forming, gram-positive bacilli encountered in human clinical specimens are *Actinomyces, Bifidobacterium, Eubacterium, Lactobacillus, Propionibacterium,* and *Mobiluncus.* The genus *Arachnia* has been eliminated, and *A. propionica* has been transferred to the genus *Propionibacterium* as *P. propionicus.*[9] Several new species of *Eubacterium* associated with the oral flora have been described. Table 36.2 lists other taxonomic changes. Non–spore-forming bacilli are found as indigenous flora in the upper respiratory tract and oral cavity, the bowel, and the vagina. *Propionibacterium acnes* is the predominant organism on the skin. With the exception of *Actinomyces* sp. and occasional other species, the anaerobic, gram-positive, non–spore-forming bacilli are considered to have low pathogenic potential. They are rarely isolated in pure culture.

DIFFERENTIATION OF GENERA

The genera are identified by the types of volatile and nonvolatile fatty acid end products of metabolism in broth media (Table 36.1). However, a few generalizations can be made based on a limited number of simple tests (Table 36.4):

1. A catalase-positive, non–spore-forming rod is probably *Propionibacterium* sp. or *Actinomyces viscosus.*
2. A nitrate-positive bacterium is probably not a *Lactobacillus* or *Bifidobacterium* sp.
3. An aerotolerant strain is *not* a *Eubacterium* species.
4. A straight bacillus with parallel sides is probably not an *Actinomyces, Bifidobacterium,* or *Propionibacterium* sp.
5. A curved gram-variable or gram-negative motile bacillus susceptible to vancomycin in the disk identification test is probably *Mobiluncus* sp.

Usually, morphological features are not reliable for distinguishing among different gram-positive, non–spore-forming, anaerobic bacilli. Moreover, it may be difficult to distinguish some members of this group of organisms from other groups, since they may appear coccoid at times or gram negative (Figure 36.13) and may resemble some clostridia that do not readily demonstrate spores. Use of antibiotic identification disks helps resolve Gram stain problems, and the use of EYA can be helpful in differentiating some clostridia from these organisms. On EYA, only rare *Eubacterium* sp. produce lecithinase, and only a few strains of *P. acnes* produce lipase. Clostridia, on the other hand, often produce lecithinase or lipase. *Nocardia* can be differentiated from *Actinomyces* and *P. propionicus* by the modified acid-fast stain de-

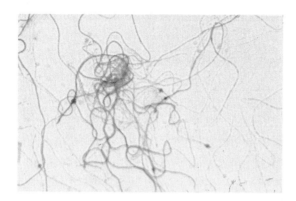

FIGURE 36.13

Eubacterium sp. Gram stain. Some strains appear gram variable or gram negative.

scribed in Chapter 34; *Nocardia* are weakly acid fast.

Whole-cell, long-chain fatty acid, methyl ester analysis by GLC also can be used for differentiation of some genera and species of anaerobic, gram-positive, non–spore-forming bacilli. The data base and software are available from Microbial ID, Inc., Newark, Del. (Chapter 35).

ACTINOMYCES SPECIES

NORMAL FLORA, INFECTIONS, AND VIRULENCE FACTORS

Actinomyces are part of the indigenous flora of the mouth, including tonsillar crypts and dental plaque, and also the female genital tract. They produce infection primarily as endogenous opportunists. *A. israelii* is isolated most frequently as the etiological agent of human actinomycosis, and *A. naeslundii*, *A. odontolyticus*, *A. viscosus*, and *A. meyeri* are found less often. Two new species of *Actinomyces* from dental plaque have been described: *A. georgiae* and *A. gerensceriae*.[17] Actinomycosis is typically a mixed infection with *Actinomyces* plus pigmented and nonpigmented anaerobic gram-negative bacilli, anaerobic cocci, and sometimes *Actinobacillus actinomycetemcomitans*. Pelvic actinomycosis caused by *Actinomyces* or *Eubacterium* has been described in association with the use of intrauterine contraceptive devices.[27] The *Actinomyces* sp. are also associated with periodontal disease. They are infrequently isolated from intraabdominal and central nervous system infections.[29]

Actinomyces sp., particularly *A. israelii,* produce phosphatases and capsular material.

MORPHOLOGY AND GENERAL CHARACTERISTICS

Except for *A. meyeri,* whose rods are usually small and nonbranching, *Actinomyces* cells are usually

branching and beaded. Diphtheroid and coccal forms are sometimes present in young cultures. The microscopic morphology of a Gram stain of "sulfur granules" (Figure 36.14) from actinomycotic pus shows a tangled mass of long filamentous forms displaying a similar morphology (Figure 36.15). Some *Actinomyces* sp. have characteristic colony morphology. *A. israelii* and *A. gerensceriae* produce colonies that are opaque, white, and peaked to pulvinate and may resemble a molar tooth (Chapter 35, Figure 35.9). Some strains may also produce a bread crumb–like, raspberry, or smooth colony; *A. odontolyticus* colonies usually turn a rust-brown color on BA after several days' incubation. Colonies of the other *Actinomyces* sp. generally are smooth, flat to convex, gray-white, and translucent, with entire margins. *A. pyogenes* and sometimes *A. odontolyticus* display beta hemolysis. The *Actinomyces* generally grow slowly, so agar plates should be incubated for a minimum of 5 to 7 days. Many require carbon dioxide (CO_2) for optimal growth and for production of succinate in broth media. Growth in broth of some strains may appear crumby or granular (Figure 36.16) and is usually enhanced by Tween 80 supplementation. Most *Actinomyces* sp. are microaerobic, although many grow better anaerobically. *A. meyeri* is described as an obligate anaerobe, although facultative, phenotypically similar strains are often isolated from clinical specimens.

IDENTIFICATION

The *Actinomyces* sp. produce major succinate and, with the exception of *A. meyeri* and *A. odontolyticus,* major lactate. Most species also produce variable amounts of acetate. Several selective media are available for *Actinomyces,* but no one medium isolates all species[19] (see Summanen, P., et al. in the bibliography). A metronidazole disk placed on a blood plate inhibits other strict anaerobes and may help select for *Actinomyces* and other aerotolerant gram-positive bacilli.

Most *Actinomyces* sp. are nitrate positive, and all are indole negative (Figure 36.17). Other useful simple tests for identification are urease, catalase, and gelatin hydrolysis. *A. naeslundii* and some strains of *A. viscosus* are urease positive, whereas the others are urease negative. *A. viscosus* is catalase positive and may be overlooked because it resembles a *Propionibacterium* sp. or diphtheroid. Unlike *Propionibacterium acnes,* it grows equally well aerobically and anaerobically, usually hydrolyzes esculin, does not produce indole, is nonproteolytic, does not produce a pink sedi-

FIGURE 36.14

Sulfur granules in thoracic empyema fluid; actinomycosis. (Photo courtesy Charles V. Sanders.)

ment in thioglycolate broth, and is more fermentative. The rapid systems aid in preliminary identification of some species,[6] but definitive identification of *Actinomyces* sp. requires conventional biochemical tests (Table 36.5).

SUSCEPTIBILITY TO ANTIMICROBIAL AGENTS

Penicillin G is the drug of choice in the treatment of actinomycosis; however, mixed infections may also harbor β-lactamase-producing organisms, such as *Prevotella* sp., that could inactivate penicillin and cause a therapeutic failure. Tetracycline, clindamycin, and erythromycin are also active

FIGURE 36.16

Actinomyces israelii in thioglycolate broth showing granular growth.

against these organisms. Treatment failures with clindamycin probably result from the presence of clindamycin-resistant bacteria, such as *Actinobacillus actinomycetemcomitans*. Metronidazole is not activite against *Actinomyces* and other non–spore-forming, gram-positive bacilli that are not strict anaerobes.

BIFIDOBACTERIUM SPECIES

NORMAL FLORA AND INFECTIONS

Bifidobacterium sp. are part of the indigenous flora of the gastrointestinal tract and may play a protec-

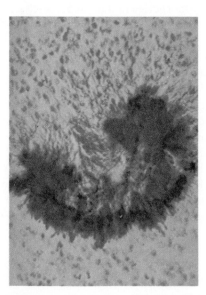

FIGURE 36.15

Sulfur granule, microscopic from clinical actinomycosis. Note aggregate of filamentous bacilli.

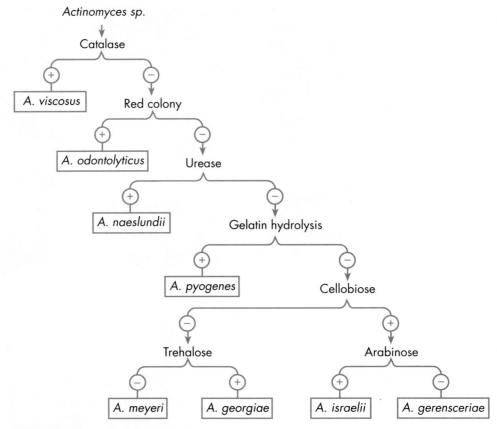

FIGURE 36.17

Flow diagram for identification of *Actinomyces* sp.

tive role in preventing overgrowth by pathogens. Some species are also present as indigenous flora in the oral cavity. These organisms are rarely isolated from clinical specimens, although *B. dentium* (formerly *B. eriksonii*) has been isolated from mixed flora from pulmonary infections.

MORPHOLOGY AND GENERAL CHARACTERISTICS

Bifidobacteria cells may be pleomorphic and may appear clubbed, or branching or may display bifurcated ends. The cells are usually thicker than branching cells of *Actinomyces*. *B. dentium* produces a white, convex, shiny colony with an irregular edge and shows diffuse growth in broth, unlike many *Actinomyces* sp. As with *Lactobacillus* sp., *Bifidobacterium* sp. are acidophilic and grow best on low pH agar (e.g., tomato juice agar). Some bifidobacteria grow aerobically in the presence of CO_2.

IDENTIFICATION

Bifidobacterium sp. produce major amounts of acetic and lactic acids in peptone-yeast-glucose (PYG) broth, with the quantity of acetic acid at least twice that of lactic acid. Strains are generally indole, nitrate, and catalase negative. The bifidobacteria are strongly saccharolytic and produce a low pH in carbohydrate media. Definitive identification requires many biochemical tests and is difficult because of the phenotypical similarity of many of the species (Table 36.5).

EUBACTERIUM SPECIES

NORMAL FLORA AND INFECTIONS

Eubacterium sp. are part of the indigenous flora of the gastrointestinal tract and oral cavity and are usually not particularly pathogenic. In rare cases, *Eubacterium* sp. have caused endocarditis. These bacteria have been isolated from blood, abscesses, dental infections, and wounds, but almost always with other anaerobes or facultative bacteria. *E. nodatum* has been isolated from actinomycosis of the jaw and intrauterine contraceptive device–associated genital tract infections.[16]

MORPHOLOGY AND GENERAL CHARACTERISTICS

The Gram stain appearance of *Eubacterium* sp. varies greatly. They often appear as pleomorphic

TABLE 36.5

CHARACTERISTICS OF GRAM-POSITIVE NON–SPORE-FORMING BACILLI

SPECIES	INDOLE PRODUCTION	CATALASE	UREASE	NITRATE REDUCTION	RED COLONY	STRONG GELATIN HYDROLYSIS	ESCULIN HYDROLYSIS	OXYGEN TOLERANCE	AMYGDALIN	CELLOBIOSE	GLUCOSE	MALTOSE	MANNITOL	SUCROSE	FATTY ACID END PRODUCTS FROM PEPTONE-YEAST-GLUCOSE BROTH
Actinomyces															
A. israelii	−	−	−	+⁻	−	−	+	AM	+	+	+	+	+	+	A L S
A. odontolyticus	−	−	−	+	+	−	V	AM	−	−	+	+⁻	−	+⁻	A S
A. meyeri	−	−	−	−	−	−	−	A	+⁻	−	+	+	−	+	A S
A. pyogenes	−	−	−	−	−	+	−	AF	−	−	+	+⁻	−	−	A L S
A. naeslundii	−	−	+⁻	+	−	−	+	MF	+⁻	+⁻	+	+	−	+	A L S
A. viscosus	−	+	−⁺	+⁻	−	−	V	AMF	−	−	+	+	−	+	A L S
A. georgiae	−	−	−	−⁺	−	−w	+⁻	AM	−	−	+	+	−⁺	+	A S (l)
A. gerensceriae	−	−	−	−⁺	−	−	+	AM	+⁻	+⁻	+	+	+⁻	+	a L S
Bifidobacterium	−	−	V	−	−	−	V	AM	V	V	+	+⁻	−⁺	+	A L [A > L]
B. dentium	−	−	−	−	−	−	+	A	+	+	+	+	+	+	L [A > L]
Eubacterium	V	−							+	V	+⁻	V	−⁺	V	
E. lentum	−	−		+⁻	−		−	A	−	−	−	−	−	−	(a l s)
Lactobacillus	−	−		−⁺	−		V	AM	V	V	+⁻	+⁻	V	+⁻	L
Propionibacterium															
P. acnes	+⁻	+⁻		+⁻	−	+	−	AM	−	−	+	−	−⁺	−	A P (iv L s)
P. avidum	−	+	−	−	+	+	+	AMF	−	−	+	+	−⁺	+	A P (iv s)
P. granulosum	−	+	−	−	−	−⁺	−	AM	−⁺	−	+	+⁻	+⁻	+	A P (iv s)
P. propionicus	−	−	−	+	−	+⁻	−	AF	V	−	+	+	+	+	A P S (L)
Mobiluncus *	−	−	−	V	−	−	−	A	−	−	+⁻	+	−	−⁺	A S (l)

Reactions: −, negative; +, positive; *V*, variable. Superscripts represent rare strain reactions.
PRAS carbohydrate reactions: +, pH <5.5; w, pH 5.6–5.9; −, pH > 5.9.
Oxygen tolerance: *A*, anaerobic; *M*, microaerobic; *F*, facultative.
Fatty acid end products: *A*, acetic; *P*, propionic; *IV*, isovaleric; *L*, lactic; *S*, succinic. Capital letters indicate major amount of end product; lower-case letters indicate minor amount.
Parentheses indicate variable reaction.
*Biochemical tests require supplementation with 2% rabbit serum.

rods or coccobacilli occurring in pairs or short chains, but also as straight uniform or curved rods. *E. alactolyticum* has a seagull-wing shape. *E. nodatum* cells may appear similar to *Actinomyces,* with beading, filaments, and branching. *E. lentum* is a small, straight rod with rounded ends. Colonies of *Eubacterium* are usually not distinctive; they may be raised to convex and may be transparent to translucent. The colony apearance of *E. nodatum,* however, may be raspberry shaped or molar tooth–like and may resemble *Actinomyces israelii.*

IDENTIFICATION

By default, any anaerobic gram-positive bacillus that does not produce end products applicable to the other described genera is classified as *Eubacterium* sp. All members are obligate anaerobes; all, except occasional strains of *E. lentum,* are catalase negative; and most are nonmotile. This genus can be divided into three groups based on fatty acid end product patterns: (1) butyric acid produced, (2) butyric acid not produced, and (3) little or no fatty acids produced. As described in Chapter 35, *E. lentum* can be presumptively identified

FIGURE 36.18

Propionibacterium acnes. Disk identification tests: indole positive and nitrate positive. (From Sutter, V.L., Citron, D.M., Edelstein, M.A.C., and Finegold, S.M. 1985. Wadsworth anaerobic bacteriology manual, ed 4. Star Publishing, Belmont, Calif.)

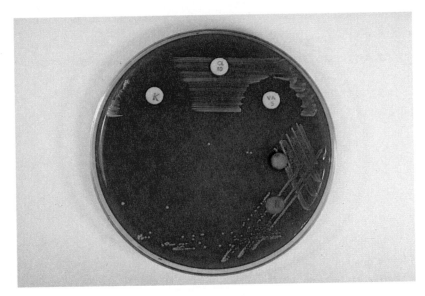

on the basis of growth stimulation by arginine in broth culture, nitrate reduction, and typical microscopic morphology.

LACTOBACILLUS SPECIES

NORMAL FLORA AND INFECTIONS

Lactobacilli are prominent as indigenous flora in the vagina and colon and are present in smaller numbers in the mouth. They are only occasionally involved in human infections, usually pleuropulmonary infections or dental caries, and almost always as part of a mixed bacterial flora. They have been isolated, although rarely, from urinary tract infection, bacteremia, endocarditis, local suppurative infections, and chorioamnionitis.[14]

MORPHOLOGY AND GENERAL CHARACTERISTICS

Most *Lactobacillus* cells are straight and uniform with rounded ends and they may form chains. Some are short and coccobacillary and may resemble streptococci. Most of the species are microaerobic, but they usually form larger colonies when incubated in an anaerobic atmosphere. Colonies on sheep blood or chocolate agar grown in an enriched CO_2 atmosphere are usually small, and sometimes greening is observed. As with the bifidobacteria, they grow best on a low pH medium.

IDENTIFICATION

Lactobacillus species produce major amounts of lactic acid, usually as the sole end product. With rare exception, they are nonmotile and indole-, catalase-, and nitrate-negative. Definitive identi-

cation requires many biochemical tests and may be difficult because of the phenotypic similarity of many of the species.

SUSCEPTIBILITY TO ANTIMICROBIAL AGENTS

Most strains are susceptible to penicillin G and other beta-lactam agents. Some strains may be resistant to vancomycin.[3]

PROPIONIBACTERIUM SPECIES

NORMAL FLORA AND INFECTIONS

Propionibacterium species are the predominant indigenous flora of the skin, but also can be recovered from the gastrointestinal tract, upper respiratory tract, especially the anterior nares, and the urogenital tract. Because of their presence on skin, propionibacteria are often contaminants of specimens such as blood or other body fluids obtained through skin puncture. Accordingly, particular care must be exercised in the preparation of the skin before obtaining these types of specimens, as described in Chapters 7 and 15. Recovery of *P. acnes* within 48 hours or on repeat cultures of blood suggests that the organism may have clinical significance. Various *Propionibacterium* sp., particularly *P. acnes*, have been implicated in infections of corneal ulcers,[26] heart valves and prosthetic devices (e.g., artificial joints), and ventricular shunts.[7,10] These infections may lead to osteomyelitis, bacteremia, endocarditis, and meningitis.[25] Most of these infections are chronic. *P. acnes* plays an important role in acne. Some strains produce lipase, which causes release of free fatty acids from sebum triglycerides; these are irritating and comedogenic. *P. acnes* produces several other enzymes, including hyaluronidase, chondroitin sulfatase, neuraminidase, and acylneuraminic acid. *P. avidum* and *P. granulosum* are infrequently isolated from clinical specimens. *P. propionicus* (formerly *Arachnia propionica*) can cause actinomycosis and lacrimal duct infection. Many strains of propionibacteria enhance cellular and humoral immunity in certain animals and in humans and are sometimes used as adjuvants.

MORPHOLOGY AND GENERAL CHARACTERISTICS

Propionibacterium sp. are often referred to as the "anaerobic diphtheroids" because the bacilli are irregular, pleomorphic, and often club shaped. At times, short, branched, and beaded forms can be observed. Young colonies of *P. acnes* are small and white to gray-white, and older colonies are often yellow. *P. avidum* is beta hemolytic. As with other anaerobic non–spore-forming bacilli, some *Propionibacterium* sp. are aerotolerant but demonstrate better growth anaerobically. They generally grow well in broth.

IDENTIFICATION

When grown in PYG broth, *Propionibacterium* sp. produce a major amount of propionic and acetic acids, with a lesser amount of lactic and isovaleric acids. Except for *P. propionicus* and a few other species, the propionibacteria are catalase positive. When an anaerobic diphtheroid is catalase and indole positive, it can be identified presumptively as *P. acnes* (Figure 36.18).

SUSCEPTIBILITY TO ANTIMICROBIAL AGENTS

P. acnes and the other propionibacteria are very sensitive to penicillin G and most other antibiotics used to treat anaerobic infections. As with most other aerotolerant, anaerobic, gram-positive bacilli, propionibacteria are resistant to metronidazole.[9a,34]

MOBILUNCUS SPECIES

NORMAL FLORA AND INFECTIONS

The genus *Mobiluncus* consists of two species: *M. curtisii* and *M. mulieris*.[31] Unlike the previous anaerobic bacilli, *Mobiluncus* sp. do not appear to be a major component of the indigenous flora in any part of the body. They have been isolated from vaginal cultures of some asymptomatic women but are usually associated with bacterial vaginosis (BV) and other pelvic infections. They have been isolated from rectal cultures from women with BV and from urethral specimens of their male sex partners, but the incidence in nor-

mal humans is not well established. *Mobiluncus* sp. also have been isolated from breast abscess, blood culture, and the chorioamnion of the placenta after preterm births. Because this genus is fastidious and slowly growing, it is difficult to determine if its rare isolation from clinical specimens is caused by inadequate laboratory procedures or if it has little pathogenic potential.

MORPHOLOGY AND GENERAL CHARACTERISTICS

Mobiluncus sp. grow on supplemented BA that supports growth of anaerobes. The colonies are very small after 48 hours of incubation and could be overlooked, especially in a mixed culture. After 3 to 5 days of incubation, the colonies are 1 to 2 mm in diameter, low convex, and translucent. Cells of *M. curtisii* appear gram variable and are usually less than 2 μm in length, whereas *M. mulieris* typically stains gram negative and has cells greater than 2.5 μm in length. These rods are usually curved and motile by means of tufts of two to six subpolar flagella. Electron micrographs of the cell wall reveal a multilayered, gram-positive cell wall lacking an outer membrane. The thinness of the peptidoglycan layer may explain the tendency of these organisms to stain gram negative.

IDENTIFICATION

Mobiluncus sp. grow poorly in unsupplemented broth media. In broth supplemented with rabbit serum or glycogen, however, they ferment a variety of carbohydrates and produce acetate and succinate and sometimes lactate as end products. They are indole, oxidase, and catalase negative. *M. mulieris* was originally differentiated from *M. curtisii* by hippurate hydrolysis and growth stimulation by arginine. However, more recent studies have shown that atypical strains exist in both species and that cell length, as determined on Gram stain, is the most reliable means for differentiation.[28] *M. mulieris* cells are greater than 2.5 μm, and *M. curtisii* cells are less than 2 μm in length. *M. curtisii* is also positive for α-D-galactosidase and arginine dihydrolase, whereas *M. mulieris* is negative for these enzymes, as determined in the RapID-ANA system (Innovative Diagnostic Systems, Norcross, Ga.).

SUSCEPTIBILITY TO ANTIMICROBIAL AGENTS

More than two thirds of *Mobiluncus* sp. are resistant to metronidazole, and occasional strains are resistant to clindamycin. Most are susceptible to the penicillins, cefoxitin, chloramphenicol, and imipenem.

REFERENCES

1. Aureli, P., Fenicia, L., Pasolini, B., et al. 1986. Two cases of type E infant botulism caused by neurotoxigenic *Clostridium butyricum* in Italy. J. Infect. Dis. 154:207.

2. Banks, E.R., Allen, S.D., Siders, J.A., and O'Bryan, N.A. 1989. Characterization of anaerobic bacteria by using a commercially available rapid tube test for glutamic acid decarboxylase. J. Clin. Microbiol. 27:361.

3. Bantar, C.E., Relloso, S., Castell, F.R., et al. 1991. Abscess caused by vancomycin-resistant *Lactobacillus confusus*. J. Clin. Microbiol. 29:2063.

4. Bartlett, J.G. 1992. Antibiotic associated diarrhea. Clin. Infect. Dis. 15:573.

5. Brahan, R.B., and Kahler, R.C. 1990. *Clostridium septicum* as a cause of pericarditis and mycotic aneurysm. J. Clin. Microbiol. 28:2377.

6. Brander, M.A., and Jousimies-Somer, H.R. 1992. Evaluation of the RapID ANA II and API ZYM systems for identification of *Actinomyces* species from clinical specimens. J. Clin. Microbiol. 30:3112.

7. Branger, C., Bruneau, B., and Goullet, P. 1987. Septicemia caused by *Propionibacterium granulosum* in a compromised patient. J. Clin. Microbiol. 25:2405.

8. Buchanan, A.G. 1982. Clinical laboratory evaluation of a reverse CAMP test for presumptive identification of *Clostridium perfringens*. J. Clin. Microbiol. 16:761.

9. Charfreitag, O., Collins, M.D., and Stackebrandt, E. 1988. Reclassification of *Arachnia propionica* as *Propionibacterium propionicus* comb. nov. Int. J. Syst. Bacteriol. 38:354.

9a. Denys, G.A., Jerris, R.C., Swenson, J.M., and Thornsberry, C. 1983. Susceptibility of *Propionibacterium acnes* clinical isolates to 22 antimicrobial agents. J. Clin. Microbiol. 23:335.

10. Dunne, W.M., Jr., Kurschenbaum, H.A., Deshur, W.R., et al. 1986. *Propionibacterium avidum* as the etiologic agent of splenic abscess. Diagn. Microbiol. Infect. Dis. 4:87.

11. Elston, H.R., Wang, M., and Loo, L.K. 1991. Arm abscess caused by *Clostridium botulinum*. J. Clin. Microbiol. 29:2678.

12. George, W.L., Sutter, V.L., Citron, D., and Finegold, S.M. 1979. Selective and differential medium for isolation of *Clostridium difficile*. J. Clin. Microbiol. 9:214.

13. Ghanem, F.M., Ridpath, A.C., Moore, W.E.C., and Moore, L.V.H. 1991. Identification of *Clostridium botulinum*, *Clostridium argentinense*, and related organisms by cellular fatty acid analysis. J. Clin. Microbiol. 29:1114.

14. Griffiths, J.K., Daly, J.S., and Dodge, R.A. 1992. Two cases of endocarditis due to *Lactobacillus* species: antimicrobial susceptibility, review, and discussion of therapy. Clin. Infect. Dis. 15:250.

15. Hatheway, C.L., and McCroskey, L.M. 1987. Examination of feces and serum for diagnosis of infant botulism in 336 patients. J. Clin. Microbiol. 25:2334.

16. Hill, G.B., Ayers, O.M., and Kohan, A.P. 1987. Characteristics and sites of infection of *Eubacterium nodatum*, *Eubacterium timidum*, *Eubacterium brachy*, and other asaccharolytic eubacteria. J. Clin. Microbiol. 25:1540.

17. Johnson, J.L., Moore, L.V.H., Kaneko, B., and Moore, W.E.C. 1990. *Actinomyces georgiae* sp. nov., *Actinomyces gerencseriae* sp. nov., designation of two genospecies of *Actinomyces naeslundii*, and inclusion of *A. naeslundii* serotypes II and III and *Actinomyces viscosus* serotype II in *A. naeslundii* genospecies 2. Int. J. Syst. Bacteriol. 40:273.

18. Kornbluth, A.A., Danzig, J.B., and Bernstein, L.H. 1989. *Clostridium septicum* infection and associated malignancy: report of 2 cases and review of the literature. Medicine 68:30.

19. Kornman, K.S., and Loesche, W.J. 1978. New medium for isolation of *Actinomyces viscosus* and *Actinomyces naeslundii* from dental plaque. J. Clin. Microbiol. 7:514.

20. Lew, J.F., Wiedermann, B.L., Sneed, J., et al. 1990. Aerotolerant *Clostridium tertium* brain abscess following a lawn dart injury. J. Clin. Microbiol. 28:2127.

21. McCroskey, L.M., Hatheway, C.L., Fenicia, L., et al. 1986. Characterization of an organism that produces type E botulinal toxin, but which resembles *Clostridium butyricum* from the feces of an infant with type E botulism. J. Clin. Microbiol. 23:201.

22. McCroskey, L.M., Hatheway, C.L., Woodruff, B.A., et al. 1991. Type F botulism due to neurotoxigenic *Clostridium baratii* from an unknown source in an adult. J. Clin. Microbiol. 29:2618.

23. Mills, D.C., Midura, T.F., and Arnon, S.S. 1985. Improved selective medium for the isolation of lipase-positive *Clostridium botulinum* from feces of human infants. J. Clin. Microbiol. 21:947.

24. Nachman, S., Kaul, A., Li, K.I., et al. 1989. Liver abscess caused by *Clostridium bifermentans* following blunt abdominal trauma. J. Clin. Microbiol. 27:1137.

25. Noble, R.C., and Overman, S.B. 1987. *Propionibacterium acnes* osteomyelitis: case report and review of the literature. J. Clin. Microbiol. 25:251.

26. Perry, L.D., Brisner, J.H., and Klodner, H. 1982. Anaerobic corneal ulcers. Ophthalmology 89:636.

27. Pine, L., Malcolm, G.B., Curtis, E.M., and Brown, J.M. 1981. Demonstration of *Actinomyces* and *Arachnia* species in cervicovaginal smears by direct staining with species-specific fluorescent-antibody conjugate. J. Clin. Microbiol. 13:15.

28. Schwebke, J.R., Lukehart, S.A., Roberts, M.C., and Hillier, S.L. 1991. Identification of two new antigenic subgroups within the genus *Mobiluncus*. J. Clin. Microbiol. 29:2204.

29. Smego, R.A. 1987. Actinomycosis of the central nervous system (review). Rev. Infect Dis. 9:855.

30. Speirs, G., Warren, R.E., and Rampling, A. 1988. *Clostridium tertium* septicemia in patients with neutropenia. J. Infect Dis. 158:1336.

31. Spiegel, C.A., and Roberts, M. 1984. *Mobiluncus* gen. nov., *Mobiluncus curtisii* subsp. *curtisii* sp. nov., *Mobiluncus curtisii* subsp. *holmesii* subsp. nov., and *Mobiluncus mulieris* sp. nov., curved rods from the human vagina. Int. J. Syst. Bacteriol. 34:177.

32. Suen, J.C., Hatheway, C.L., Steigerwalt, A.G., and Brenner, D.J. 1988. *Clostridium argentinense* sp. nov.: a genetically homogeneous group composed of all strains of *Clostridium botulinum* toxin type G and some nontoxigenic strains previously identified as *Clostridium subterminale* or *Clostridium hastiforme*. Int. J. Syst. Bacteriol. 38:375.

33. Suen, J.C., Hatheway, C.L., Steigerwalt, A.G., and Brenner, D.J. 1988. Genetic confirmation of identities of neurotoxigenic *Clostridium baratii* and *Clostridium*

butyricum implicated as agents of infant botulism. J. Clin. Microbiol. 26:2191.

34. Wang, W.L.L., Everett, E.D., Johnson, M., and Dean, E. 1977. Susceptibility of *Propionibacterium acnes* to seventeen antibiotics. Antimicrob. Agents Chemother. 11:171.

BIBLIOGRAPHY

Cato, E.P., George, W.L., and Finegold, S.M. 1986. Genus *Clostridium* Prazmowski 1880. In Sneath, P.H.A., Mair, N.S., Sharpe, M.E., and Holt, J.G., editors: Bergey's manual of systematic bacteriology, vol 2, p. 1141. Williams & Wilkins, Baltimore.

Cummings, C.S., and Johnson, J.J. 1986. Genus I. *Propionibacterium* Orla-Jensen 1909. In Sneath, P.H.A., Mair, N.S., Sharpe, M.E., and Holt, J.G., editors: Bergey's manual of systematic bacteriology, vol 2, p. 1346. Williams & Wilkins, Baltimore.

Dowell, V.R., Jr., and Hawkins, T.M. 1968. Detection of clostridial toxins, toxin neutralization tests, and pathogenicity tests. Centers for Disease Control, Atlanta.

Farrow, J.A.E., and Collins, M.D. 1988. *Lactobacillus oris* sp. nov. from the human oral cavity. Int. J. Syst. Bacteriol. 38:116.

Finegold, S.M. , and George, W.L. 1989. Anaerobic infections in humans. Academic Press, New York.

Gorbach, S.L., and Thadepalli, H. 1975. Isolation of *Clostridium* in human infections: evaluation of 114 cases. J. Infect. Dis. 131(suppl.):S81.

Holdeman, L.V., Cato, E.P., and Moore, W.E.C. 1977. Anaerobe laboratory manual, ed 4. Virginia Polytechnic Institute and State University, Blacksburg, Va.

Moore, L.V.H., Cato, E.P., and Moore, W.E.C. 1987 and 1988. VPI Anaerobe laboratory manual update, supplement to the VPI anaerobe laboratory manual, ed 4. 1977. Virginia Polytechnic Institute and State University, Blacksburg, Va.

Schal, K.P. 1986. Genus *Actinomyces* Harz 1877 and Genus *Arachnia* Pine and Georg 1969. In Sneath, P.H.A., Mair, N.S., Sharpe, M.E., and Holt, J.G., editors: Bergey's manual of systematic bacteriology, vol 2, pp. 1332; 1383. Williams & Wilkins, Baltimore.

Smith, L.D.S. 1975. The pathogenic anaerobic bacteria, ed 2. Charles C Thomas, Springfield, Ill.

Summanen, P., Baron, E.J., Citron, D.M., et al. 1993. Wadsworth anaerobic bacteriology manual, ed 5. Star Publishing, Belmont, Calif.

Willis, A.T. 1969. Clostridia of wound infection. Butterworth, London.

Willis, A.T., and Phillips, K.D. 1988. Anaerobic bacteriology: clinical and laboratory practice. Public Health Laboratory Service, Luton, England.

37

ANAEROBIC GRAM-NEGATIVE BACILLI

BACTEROIDACEAE

The anaerobic gram-negative rods belong to the family Bacteroidaceae (Table 37.1 and box on p. 526). They are obligately anaerobic, non–spore-forming bacilli that are straight, curved, or helical and may be motile or nonmotile. They metabolize carbohydrates, peptones, or metabolic intermediates and produce organic fatty acid end products. Although not members of the Bacteroidaceae, *Campylobacter concisus, C. curvus,* and *C. rectus* (the latter two formerly being *Wolinella*) are included here because they are similar to the *Bacteroides ureolyticus* group and *Wolinella.*

The Bacteroidaceae are prevalent among the indigenous flora of mucous membranes of hu-

mans and animals. They constitute two thirds of the total anaerobic isolates from clinical specimens. Anaerobic or mixed infections result from disruption of a mucosal surface and introduction of anaerobes into a normally sterile site. Predisposing factors include surgical procedures, trauma, peripheral vascular disease, carcinomas, and prior infection with facultative organisms. Unconsciousness is often associated with aspiration of oral secretions, which can develop into aspiration pneumonia and other more serious pulmonary infections.

Infection with gram-negative anaerobic bacilli, as with anaerobic infection generally, is characterized by abscess formation and tissue de-

TABLE 37.1

DIFFERENTIATION OF GENERA OF ANAEROBIC GRAM-NEGATIVE BACILLI

DISTINGUISHING CHARACTERISTIC	GENUS
I. Nonmotile	
A. Major end product: butyric without isoacids	*Fusobacterium*
B. Major end product: lactic acid	*Leptotrichia*
C. Major end product: acetic; H$_2$S produced	*Desulfomonas, Bilophila*
D. Major end product: not as above (A, B, or C)	
1. Pigmented and nonfermentative	*Porphyromonas*
2. Other than D.1	
a. Common clinical isolate	*Bacteroides, Prevotella*
b. Rare clinical isolate	
(1) Strongly fermentative	*Mitsuokella*
(2) Weakly fermentative	*Anaerorhabdus*
II. Motile	
A. Fermentative	
1. Major end product: butyric acid	*Butyrivibrio*
2. Major end product: succinic acid	
a. Single, polar flagellum	
(1) Spiral-shaped cell	*Succinivibrio*
(2) Straight ovoid cell	*Succinimonas*
b. Bipolar tufts of flagella	*Anaerobiospirillum*
3. Major end product: propionic acid	
a. Single polar flagellum	*Anaerovibrio*
b. Tufts of flagella	
(1) Near center concave side	*Selenomonas*
(2) Lateral spiral arrangement	*Centipeda*
B. Nonfermentative	
1. Single polar flagellum	*Campylobacter, Wolinella*
2. Tufts of flagella (gram-positive but stains gram-negative)	*Mobiluncus*
3. Peritrichous flagellation	*Tissierella*

struction. Infections of any type anywhere in the body may involve these organisms (Figure 37.1). Most of the gram-negative bacilli isolated from human infections belong to the genera *Bacteroides, Prevotella, Porphyromonas, Bilophila,* and *Fusobacterium.*

Many of these organisms produce virulence factors that contribute to pathogenesis. These include catalase, superoxide dismutase, collagenase, neuraminidase, deoxyribonuclease (DNase), heparinase, proteinases, and immunoglobulinases.[3] Several produce capsules that inhibit phagocytosis and promote abscess formation.[4] End products of anaerobic metabolism, such as butyrate and succinate, are cytotoxic. Enterotoxin production by some strains of *Bacteroides fragilis* has been recently described.[24]

Since these organisms are prevalent on mucosal surfaces, it is important to avoid contamination with indigenous flora when obtaining a specimen (Chapter 7). Transport must be carried out in an oxygen-free atmosphere.

MORPHOLOGY AND GENERAL CHARACTERISTICS

Cell morphology varies greatly within the family Bacteroidaceae. Some cells may be filamentous, pleomorphic, and vacuolated or may stain irregularly, and motile cells may have peritrichous, polar or bipolar flagella. Cell morphology is useful for differentiation of some genera and species (Table 37.1). For example, *Fusobacterium nucleatum* is a thin bacillus with pointed ends, whereas most other fusobacteria have rounded ends.

Members of this family exhibit a variety of colony types. Some species have a characteristic morphology. For example, the pigmented *Prevotella* sp. and *Porphyromonas* produce a brown to black pigment, and some members of the *Bacteroides ureolyticus* group produce an agarase that pits the agar around the colony. Some strains of *F. nucleatum* produce colonies that resemble bread crumbs, and others have a ground-glass appearance. Members of the *B. fragilis* group have colonies that are more convex than colonies of

Family Bacteroidaceae

Bacteroides fragilis group (Table 37.3)
  B. fragilis
  B. vulgatus
  B. distasonis
  B. caccae
  B. merdae
  B. thetaiotaomicron
  B. uniformis
  B. ovatus
  B. stercoris
  B. eggerthii
Pigmented organisms (Table 37.4)
  Pigmented *Prevotella*
    P. intermedia
    P. nigrescens
    P. corporis
    P. melaninogenica
    P. loescheii
    P. denticola
  Porphyromonas
    P. asaccharolytica
    P. endodontalis
    P. gingivalis
Nonpigmented, bile-sensitive organisms (Table 37.5)
  Prevotella oralis group (pentose nonfermenters)*
    P. oralis
    P. veroralis
    P. buccalis
    P. oulora
  Pentose fermenters
    P. oris
    P. buccae
    P. heparinolytica
    P. zoogleoformans
  Other *Prevotella* sp.
    P. bivia
    P. disiens
Asaccharolytic or weakly fermentative
  Require formate and fumarate as growth factor
    Bacteroides ureolyticus
    B. gracilis
    Wolinella sp.
    Campylobacter concisus
    C. rectus
    C. curvus
  Other
    Bilophila wadsworthia
    Bacteroides putredinis
    B. capillosus
Fusobacterium (Table 37.6)
  F. nucleatum
  F. necrophorum
  F. mortiferum
  F. varium
  Other *Fusobacterium* sp.

*Pentoses usually tested include arabinose and xylose.

enteric gram-negative bacilli grown under anaerobic conditions.

DIFFERENTIATION OF ANAEROBIC GRAM-NEGATIVE BACILLI

The genera in the family Bacteroidaceae are differentiated by their metabolic end products, cell shape, and location of flagella in motile forms (Table 37.1). More recently, genetic techniques such as guanine-cytosine content and presence and types of respiratory quinones have been used in their differentiation.[17-21] Fortunately, many frequently encountered clinical isolates can also be identified with simple tests. These tests are described in Chapter 35 and later in this chapter. Most anaerobic gram-negative bacilli encountered in clinical specimens are in the nonmotile genera. However, a motility test and flagellar stain should be performed on curved or helical bacilli, bacilli that are urease negative and require formate and fumarate for growth in broth media, or if the organism's characteristics do not match the characteristics of the nonmotile genera.

BACTEROIDES, PREVOTELLA, AND PORPHYROMONAS

The anaerobic gram-negative bacilli *Bacteroides*, *Prevotella*, and *Porphyromonas* are nonmotile straight rods with rounded ends that produce a variety of metabolic end products, usually including acetic acid and succinic acid.

TAXONOMIC CHANGES

Some taxonomic changes in the *Bacteroides* result from new chemotaxonomic approaches to categorizing bacteria. Other changes result from investigations into the bacterial flora of specific sites, particularly dental structures. Table 37.2 lists many of these taxonomic changes. It has recently been proposed to limit the genus *Bacteroides* to the *Bacteroides fragilis* group.[19] Three previously unnamed members of the *B. fragilis* group have been named: *B. caccae*, *B. stercoris*, and *B. merdae*.[9] They are all part of the normal bowel flora, and *B. caccae* (formerly "3452A") is occasionally isolated from clinical specimens. The other two are rarely encountered. The indole-positive, asaccharolytic, pigmented *Bacteroides* sp. now belong to the genus *Porphyromonas*.[18] Several new species from animal sources have been described, and these may be important in animal bite wounds. *Wolinella curva* and *W. recta* have been transferred to *Campylobacter*. The taxonomic status of *Bacteroides ureolyticus*

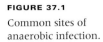
FIGURE 37.1

Common sites of anaerobic infection.

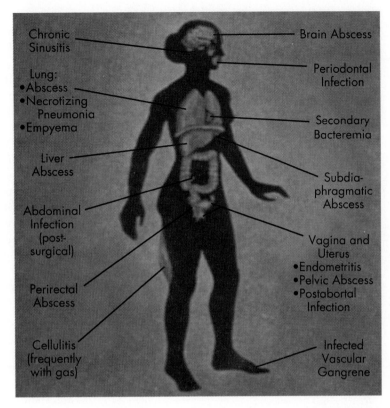

and *B. gracilis* remains uncertain, although they appear more closely related to *Campylobacter* sp. than *Bacteroides*.

NORMAL FLORA, INFECTIONS, AND VIRULENCE FACTORS

The *Bacteroides, Prevotella,* and *Porphyromonas* are found as part of the indigenous flora of the oropharynx, gastrointestinal tract, and female genital tract. The *B. fragilis* group constitutes the predominant flora of the colon. Its members are isolated most frequently from infections occurring below the diaphragm, such as intraabdominal and liver abscesses, decubitus ulcers, and other soft tissue infections and bacteremia. They have been isolated infrequently from pleuropulmonary infection or brain abscess. The *B. fragilis* group makes up one third of all clinical isolates of anaerobes. *B. fragilis* is the most frequently encountered, with *B. thetaiotaomicron* second. Other members of the *B. fragilis* group are present less often. Pigmented *Prevotella* and *Porphyromonas* sp. are not isolated as frequently as the *B. fragilis*

group; however, *Porphyromonas gingivalis* is an important pathogen in periodontal disease, and members of both genera are frequently isolated from the head and neck and from pleuropulmonary infections. *Porphyromonas asaccharolytica* may be isolated from any type of infection. Pigmented gram-negative bacilli also have been isolated from the female genital tract, but less frequently than some other *Prevotella* sp., especially *P. bivia* and *P. disiens*. Nonpigmented *Prevotella* sp. are present in a variety of infections, but also occur less often than the *B. fragilis* group.

Capsules are produced by several species, including *B. fragilis* and occasionally other members of the *B. fragilis* group, *P. asaccharolytica*, and probably *Prevotella melaninogenica, P. intermedia, P. oralis, P. oris,* and *P. buccae*. The *B. fragilis* capsule can be demonstrated by ruthenium red staining, India ink staining, and electron microscopy. *B. fragilis* produces an endotoxin that differs from that of aerobic gram-negative bacilli by structure and weaker biological activity. *Prevotella bivia*, in contrast, produces a potent endotoxin. *B. fragilis*

TABLE 37.2

TAXONOMIC CHANGES IN ANAEROBIC GRAM-NEGATIVE BACILLI

NEW	PREVIOUS
Anaerorhabdus furcosus	*Bacteroides furcosus*
Bacteroides caccae	*Bacteroides "3452A"*
B. forsythus	sp. nov.
B. merdae	*Bacteroides "T4-1"*
B. stercoris	*B. fragilis "subsp. a"*
Bilophila wadsworthia	sp. nov.
Campylobacter curvus	*Wolinella curva*
C. rectus	*W. recta*
Fusobacterium alocis	sp. nov.
F. periodonticum	sp. nov.
F. pseudonecrophorum	sp. nov.
F. sulci	sp. nov.
F. ulcerans	sp. nov.
Mitsuokella dentalis	sp. nov.
Porphyromonas asaccharolytica	*Bacteroides asaccharolyticus*
P. circumdentaria	sp. nov.
P. endodontalis	sp. nov.
P. gingivalis	*Bacteroides gingivalis*
P. salivosa	*B. salivosus*
Prevotella bivia	*Bacteroides bivius*
P. buccae	*B. buccae*
P. buccalis	*B. buccalis*
P. corporis	*B. corporis*
P. denticola	*B. denticola*
P. disiens	*B. disiens*
P. heparinolytica	*B. heparinolytica*
P. intermedia	*B. intermedius*
P. loescheii	*B. loescheii*
P. melaninogenica	*B. melaninogenicus*
P. oralis	*B. oralis*
P. oris	*B. oris*
P. oulora	*B. oulorum*
P. veroralis	*B. veroralis*
P. zoogleoformans	*B. zoogleoformans*
Selenomonas artemidis	sp. nov.
S. dianae	sp. nov.
S. flueggei	sp. nov.
S. infelix	sp. nov.
S. noxia	sp. nov.
Tissierella praeacuta	*B. praeacutus*

also produces many enzymes, such as catalase, neuraminidase, DNase, hyaluronidase, gelatinase, fibrinolysin, superoxide dismutase, and heparinase. Some *Prevotella* sp. produce phospholipase A, neuraminidase, collagenase, hyaluronidase, and fibrinolysin. *Porphyromonas gingivalis* produces an immunoglobulinase, and many other tissue-damaging enzymes. Heparinase is produced by *Prevotella heparinolytica*.

DIFFERENTIATION AND GROUPING

Bacteroides, Prevotella, and *Porphyromonas* sp. are readily divided into three major groups based on bile susceptibility, pigment production, and antibiotic disk identification patterns. The *B. fragilis* group is nonpigmented, bile resistant, and resistant to all three antibiotic identification disks; the pigmented *Prevotella* sp. and *Porphyromonas* typically produce pigment and are bile sensitive. Other *Prevotella* sp. are nonpigmented and bile sensitive. Other gram-negative rods that, at present, remain in the genus *Bacteroides* are nonpigmented and bile sensitive. *Porphyromonas* has a unique antibiotic disk identification pattern; it is resistant to colistin and kanamycin and sensitive to vancomycin. Figure 37.2 illustrates differentiation between anaerobic gram-negative bacilli using key reactions.

BILE-RESISTANT *BACTEROIDES*

This group of *Bacteroides* includes the *B. fragilis* group and two other bile-resistant species not often encountered: *B. eggerthii* and *B. splanchnicus*.

MORPHOLOGY AND GENERAL CHARACTERISTICS Microscopically, the members of the *B. fragilis* group are pale-staining, gram-negative bacilli with rounded ends. The cells are fairly uniform if smears are prepared from young cultures on supplemented agar. Pleomorphism and irregularity of staining are often seen in direct smears of clinical material or from broth cultures (Figure 37.3). Cells of *B. ovatus* are usually more ovoid. Colonies of *B. fragilis* group strains are 1 to 3 mm in diameter, smooth, white to gray, and nonhemolytic on blood agar (Figure 37.4). *B. thetaiotaomicron* may appear whiter than colonies of *B. fragilis*. Colonies of *B. ovatus* are often mucoid. Preliminary grouping procedures for these organisms are discussed in Chapter 35.

IDENTIFICATION The *B. fragilis* group is distinguished from the other two bile-resistant *Bacteroides* sp. by fermentation of sucrose (Table 37.3). Indole, catalase, and esculin are good tests for presumptive identification of the *B. fragilis* group. Further tests for differentiation of species within the *B. fragilis* group and the other bile-resistant *Bacteroides* are noted in Table 37.3.

SUSCEPTIBILITY TO ANTIMICROBIAL AGENTS The species in the *B. fragilis* group are among the most resistant of the anaerobes. Resistance to β-lactam drugs (e. g., penicillins, cephalosporins) is primarily mediated by production of β-lactamases that, although primarily cephalosporinases, also show significant ability to hydrolyze penicillins. Certain β-lactam agents, such as cefoxitin and imipenem, are resistant to these β-lactamases and thus retain activity against the *B. fragilis* group. The broad-spectrum penicillins (ticarcillin, piperacillin) are also susceptible to hydrolysis by many of these enzymes. They are often combined with a β-lactamase inhibitor, such as clavulanate, sulbactam, or tazobactam, which restores their activity against more than 95% of strains. Resistance to tetracycline is common, and 5% to 15% of strains of *B. fragilis* and greater than 30% of the other *B. fragilis* group sp. may be resistant to clindamycin. Resistance to β-lactam agents and clindamycin is variable in different geographical areas and even in institutions within a local area. Metronidazole and chloramphenicol are active against almost all strains of the *B. fragilis* group. *B. thetaiotaomicron, B. ovatus, B. caccae,* and *B. distasonis* are generally much more resistant to antimicrobial agents than *B. fragilis.*

PIGMENTED *PREVOTELLA* AND *PORPHYROMONAS*

Nine of the at least 12 species in this group have been isolated from human clinical material; the others are associated with animals. Those most frequently encountered are *Porphyromonas asaccharolytica, Porphyromonas gingivalis, Prevotella intermedia,* and *Prevotella melaninogenica.* Some strains of *Prevotella bivia* produce pigment after 21 days of incubation on laked rabbit blood agar; however, this is not a practical procedure for most laboratories. Thus, *P. bivia* is discussed with the nonpigmented *Prevotella* sp.

MORPHOLOGY AND GENERAL CHARACTERISTICS Most strains of *Porphyromonas* and occasional strains of *Prevotella* appear as pale-staining coccobacilli (Figure 37.5), whereas other strains appear as pleomorphic bacilli. Young cultures may appear gram variable. The organisms produce a variety of colony types with varying amounts of pigment, and this may be helpful in subdividing the group. Generally, *Prevotella intermedia, Prevotella nigrescens,* a recently described species,[21] and *Prevotella corporis* produce dark brown to black dry colonies; *Porphyromonas* colonies are usually more mucoid and dark brown to black; and the *Prevotella melaninogenica* group produces tan to light brown, smooth colonies. Pigment production may not always be readily detected and may require prolonged incubation (> 5 days), especially with the *P. melaninogenica* group. Pigment production is most reliably demonstrated on laked rabbit blood agar (Figure 35.7). With the exception of *Porphyromonas gingivalis*, colonies fluoresce (Figure 35.8) brick red under ultraviolet light. After pigment is produced, the fluorescence may not

be detectable. Most *Porphyromonas* sp. are difficult to grow in broth and on some selective agars containing 7.5 μg/ml vancomycin (e.g., kanamycin-vancomycin–laked blood [KVLB]). Addition of serum, laked rabbit blood, or hemin may enhance growth in broth. Reduction of the vancomycin concentration to 2 μg/ml allows for growth of most *Porphyromonas* while still inhibiting many gram-positive organisms.

IDENTIFICATION The pigmented anaerobic gram-negative bacilli are all sensitive to bile and form three groups based on the antimicrobial disk identification pattern, indole reaction, and colony morphology (Table 37.4 and Figure 37.6): (1) *Prevotella intermedia, P. corporis,* and *P. nigrescens* are resistant to vancomycin and kanamycin and susceptible to colistin, their colonies are typically dry and black, and *P. intermedia* and *P. nigrescens* are indole positive; (2) the *Prevotella melaninogenica* group (*P. loescheii, P. denticola, P. melaninogenica*) are indole negative, often resistant to all antibiotic identification disks (colistin, kanamycin, vancomycin), and their colonies do not develop dark pigmentation; (3) *Porphyromonas* sp. are indole positive, sensitive to vancomycin, and resistant to kanamycin and colistin, and their colonies are often mucoid, shiny, smooth, and dark brown to black. Some key reactions for differentiation of species are production of indole, lipase, and phenylacetic acid; esculin hydrolysis; fermentation of glucose, sucrose, lactose, and cellobiose; and salt tolerance. Presence of α-fucosidase, β-N-acetyl-glucosaminidase, and arginine dihydrolase are also useful. A rapid method that uses 4-methylumbelliferone derivatives for identification of pigmented gram-negative bacilli was recently described.[13] As noted in Chapter 35, *Prevotella intermedia* and *P. loescheii* are easy to identify when the lipase reaction is positive; *P. intermedia* is indole positive, and *P. loescheii* is indole negative. Lipase-positive strains of *P. loescheii* are rarely encountered.

SUSCEPTIBILITY TO ANTIMICROBIAL AGENTS Many pigmented organisms produce β-lactamase in varying amounts, although in some instances the minimal inhibitory concentration (MIC) of penicillin is within the normally considered susceptible range (2 to 4 U/ml). Penicillin G can no longer be considered the drug of choice without specific susceptibility data for seriously ill patients harboring these organisms. Metronidazole, chloramphenicol, and the β-lactamase inhibitor combinations, such as ampicillin/sulbactam and ticarcillin/clavulanate, are active against these or-

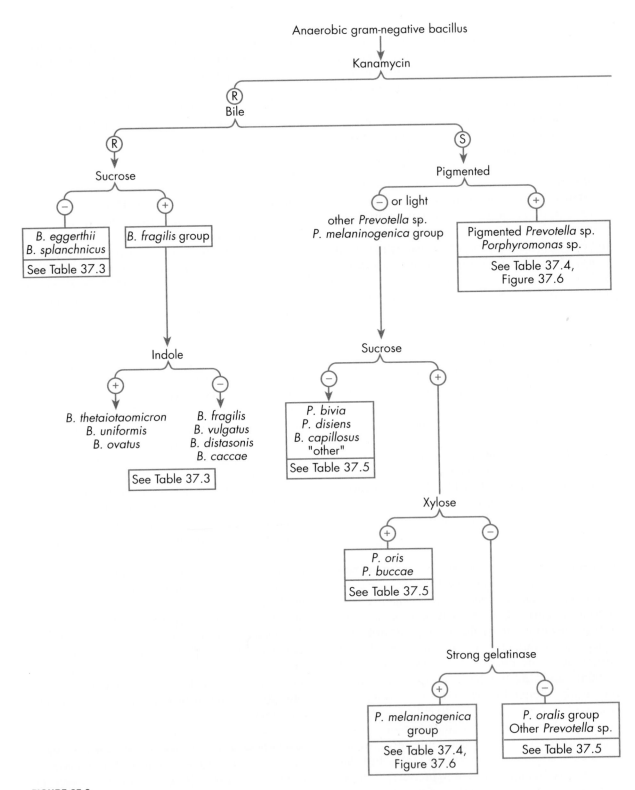

FIGURE 37.2

Flow chart to separate *Bacteroides, Prevotella, Porphyromonas,* and *Fusobacterium.*

ganisms; clindamycin is also usually effective. The broad-spectrum penicillins, cefoxitin, and other second- and third-generation cephalosporins are active against more than 95% of strains in this group.

NONPIGMENTED *PREVOTELLA* SP.

The most frequently isolated members of this group are *P. buccae, P. bivia,* and *P. oralis* (Figure 37.7). *P. oris, P. disiens, P. buccalis,* and *P. veroralis* are less common in clinical specimens. *P. zoogleo-*

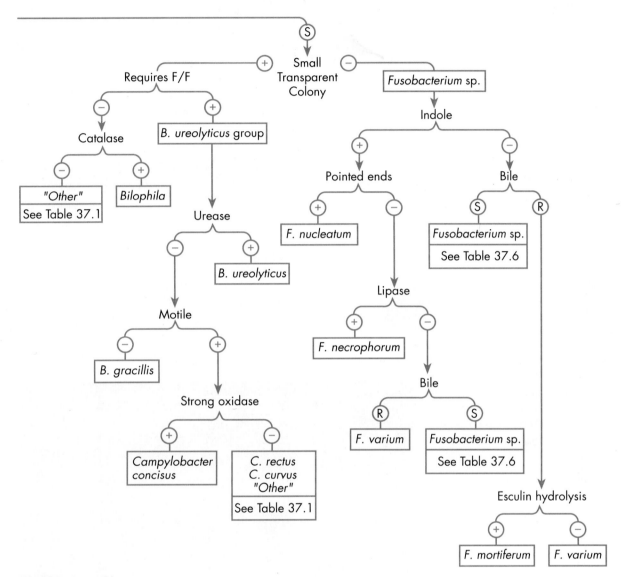

FIGURE 37.2, cont'd
For legend see opposite page.

formans, *P. heparinolytica,* and *P. oulora* are associated with periodontal infections and are rarely isolated from clinical specimens. These organisms are found as normal flora in the oral cavity, gastrointestinal tract, and female genital tract. They are isolated from all types of infections. *P. bivia* and *P. disiens* are often isolated from female genital tract infections, and *P. buccae* and *P. oralis* are found more frequently in pulmonary infections.

MORPHOLOGY AND GENERAL CHARACTERISTICS The cellular morphology of these organisms can be coccobacillary or pleomorphic bacilli. They are bile sensitive, kanamycin resistant, and usually

catalase negative. Many strains of *P. oris* and *P. buccae* exhibit a chartreuse fluorescense under ultraviolet light, and *P. bivia* often fluoresces a coral color. They are somewhat slower growing than members of the *Bacteroides fragilis* group strains, although their colony morphology is generally similar.

IDENTIFICATION This group can be subdivided into four groups (Table 37.5) based on proteolytic and fermentation activity: (1) saccharolytic pentose (arabinose, xylose) fermenters, (2) saccharolytic pentose nonfermenters, (3) saccharolytic and proteolytic, and (4) nonsaccharolytic or weakly

FIGURE 37.3
Bacteroides fragilis. Irregular staining and pleomorphism.

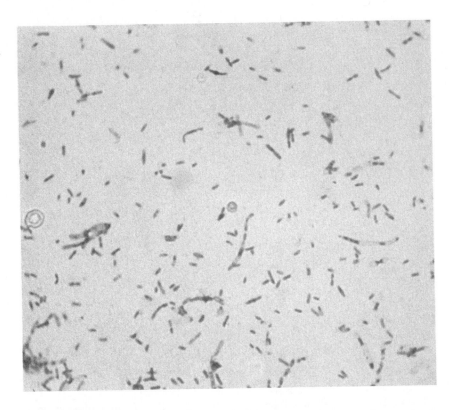

FIGURE 37.4
Bacteroides fragilis. Gray, shiny colony on blood agar.

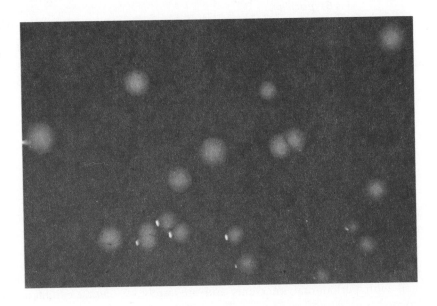

saccharolytic fermenters. Further differentiation requires a combination of conventional biochemical tests and tests for preformed enzymes (Chapter 35).

SUSCEPTIBILITY TO ANTIMICROBIAL AGENTS   As with the pigmented *Prevotella* sp. and *Porphyromonas*, ore than half these strains produce β-lactamase and are thus resistant to penicillins. Rare strains may also be resistant to clindamycin, but most

other antimicrobials typically used to treat anaerobic infections are effective.

ASACCHAROLYTIC OR WEAKLY SACCHAROLYTIC GRAM-NEGATIVE BACILLI

This group of organisms includes *Bacteroides ureolyticus, B. gracilis, Wolinella succinogenes, Campylobacter concisus, C. rectus, C. curvus,* (previously

TABLE 37.3

CHARACTERISTICS OF BILE-RESISTANT *BACTEROIDES* SP.

SPECIES	GROWTH IN 20% BILE	INDOLE	CATALASE	ESCULIN HYDROLYSIS	FERMENTATION OF SUCROSE	MALTOSE	ARABINOSE	CELLOBIOSE	RHAMNOSE	SÁLICIN	TREHALOSE	XYLAN	α-FUCOSIDASE	FATTY ACIDS FROM PEPTONE-YEAST-GLUCOSE BROTH
B. fragilis group														
B. fragilis	+	−	**+**	+	+	+	−	−w	−	−	−	−		A p S pa
B. vulgatus	+	−	−+	−+	+	+	**+**	−	+	−	−	V		A p S
B. distasonis	+	−	**+−**	+	+	+	−+	+	V	+	+	−	−	A p S (ib iv l pa)
*B. merdae**	+	−	−+	+	+	+	−+	−+	+	+	+	−	−	A S (p ib iv)
B. caccae	+	−	−+	+	+	+	+	−+	+−	V	+	−	+	A S (p iv)
B. thetaiotaomicron	+	**+**	**+**	+	+	+	+	+	+	−+	+	−		A p S pa
B. uniformis	w+	+	−+	+	+	+	+	+w	−+	+−	−	V		a p l S (ib iv)
B. ovatus	+	+	+−	+	+	+	+	+	−+	+	+	+		A p S pa (ib iv l)
*B. stercoris**	+	+	−	+	+	+	−+	−	+	−	−	V	V	A p S (ib iv)
Other														
*B. eggerthii**	+	+	−	+	−	**+**	+	−w	w −	−	−			A p S (ib iv l)
*B. splanchnicus**	w+	+	−	+	−	−	+	−	−	−	−	−		A P ib b iv S (l)

Reactions: results in **bold type** are key characteristics; −, negative; −⁺, rare strains positive; +, most strains positive; +⁻, rare strain negative, −ʷ, rare strain weak positive; *V*, variable; *w*, weak positive.

Prereduced anaerobically sterilized (PRAS) carbohydrate reactions: +, pH <5.5; w, pH 5.5–5.7; −, pH > 5.7.

Fatty acid end products: A, acetic; *P,* propionic; *IB,* isobutyric; *IV,* isovaleric; *L,* lactic; *S,* succinic; *PA,* phenylacetic, Capital letters indicate major metabolic products; lower-case letters indicate minor products. Parentheses indicate variable reaction. Isoacids are primarily from carbohydrate-free media (e.g., peptone-yeast) in the case of saccharolytic organisms.

*Rarely isolated; part of the normal bowel flora.

FIGURE 37.5

Pigmented *Prevotella* sp. or *Porphyromonas.* Many coccobacillary forms exist, as well as rods. (From Sutter, V.L., Citron, D.M., Edelstein, M.A.C., and Finegold, S.M. 1985. Wadsworth Anaerobic Bacteriology Manual, ed 4. Star Publishing, Belmont, Calif.)

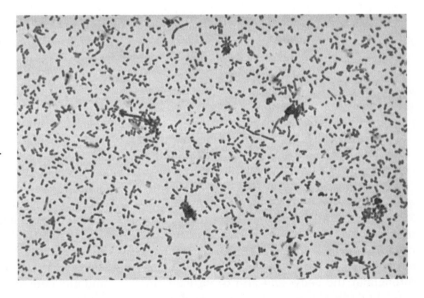

TABLE 37.4

CHARACTERISTICS OF PIGMENTED ANAEROBIC GRAM-NEGATIVE BACILLI

	VANCOMYCIN (5 µg)	COLISTIN (10 µg)	INDOLE	LIPASE	FERMENTATION OF GLUCOSE	LACTOSE	CELLOBIOSE	SUCROSE	ESCULIN HYDROLYSIS	GROWTH 3% SALT	β-NAG*	α-FUCOSIDASE	ARGININE DIHYDROLASE	FATTY ACIDS FROM PEPTONE-YEAST-GLUCOSE BROTH	PHENYLACETIC ACID
Saccharolytic															
Prevotella melaninogenica	R	Rˢ	–	–	+	+	–	+	–⁺			+	+	+	A S (ib v l)	–
P. denticola	R	Rˢ	–	–	+	+	–	+	+			+	–	–	A S (ib iv l)	–
P. loescheii	R	Rˢ	–	–ʷ	+	+	**+**	+⁻				+	+	–	a S (l)	–
Partially saccharolytic															
Prevotella corporis	R	S	–	–	+	–	–	–				+	–	–	A ib iv S (b)	–
P. intermedia	R	S	+	+	+	–	–	+	–			–	+	+	A iv S (p ib)	–
P. nigrescens	R	S	+	–	+	–	–	+	–			–	+	+	A iv S (p ib)	–
Asaccharolytic															
Porphyromonas asaccharolytica	**S**	R	+	–	–	–	–	–	–	G	–	+	+	A p ib B iv (S)	–
P. endodontalis	**S**	R	+	–	–	–	–	–	–	NG	–	–	–	a p ib b iv	–
P. gingivalis	**S**	R	+	–	–	–	–	–	–	NG	+	–	–	a p ib B IV s	+

Reactions: results in **bold type** are key characteristics; *R*, resistant; *Rˢ*, some strains sensitive; *S*, sensitive; –, negative; –⁺, rare strains positive; +, most strains positive; +⁻, rare strain negative; –ʷ, rare strain weak positive; *V*, variable; *G*, growth; *NG*, no growth.
PRAS carbohydrate reactions: +, pH <5.5; w, pH 5.5-5.9; –, pH >5.9.
Fatty acid end products: *A*, acetic; *P*, propionic; *IB*, isobutyric; *B*, butyric; *IV*, isovaleric; *L*, lactic; *S*, succinic; *PA*, phenylacetic. Capital letters indicate major metabolic products; lower-case letters indicate minor products. Parentheses indicate variable reaction. Isoacids are primarily from carbohydrate-free media (e.g., peptone-yeast) in the case of saccharolytic organisms.
*β-*N*-acetyl glucosaminidase.

Wolinella curva and *W. recta*), which require formate/fumarate supplementation for growth, and *Bilophila wadsworthia*, a new genus,[2] which does not have this requirement.

The taxonomic status of these *Bacteroides* sp. and *W. succinogenes* is uncertain because they are very similar to *Campylobacter*, except that their G plus C content is quite different. This characteristic remains an important criterion for placement of an organism into a genus.[16,23] *W. curva* and *W. recta* have been reclassified with *Campylobacter*.

These *Campylobacter* sp. and *B. gracilis* are part of the indigenous oral flora and are present in infections of the oral cavity and sometimes in pulmonary infections. *B. gracilis* has also been isolated from deep-seated, soft tissue infections.[8] *B. ureolyticus* is found on mucous membranes throughout the body, and also in virtually any type of infection. *B. wadsworthia* is present as part of colonic flora and is most frequently isolated from intraabdominal infections secondary to a ruptured appendix. However, because it is slow growing, it may not be as readily detected as other organisms present in this type of infection.

MORPHOLOGY, GENERAL CHARACTERISTICS, AND IDENTIFICATION

These organisms generally have delicate, pale-staining cells. *C. concisus* may be curved. *Campylobacter* sp. and *Wolinella* are motile, and *B. gracilis* may exhibit a "twitching" motility. Their colony morphology is similar: they are slow growing and produce transparent to translucent colonies. Some produce an agarase that creates pitting of the agar (Fig 37.8). Selective media for isolation of *B. ureolyticus* and *B. gracilis* have been described.[6,10] They are sensitive to kanamycin and colistin on the disk identification test. Except for *Bilophila*, they use formate as a hydrogen donor and fumarate or nitrate as hydrogen acceptors. *B. wadsworthia* and *B. ureolyticus* produce urease, and *B. wadsworthia* is very strongly catalase positive. They all reduce nitrate to nitrite or beyond. *B. wadsworthia* is bile resistant and grows on *Bac-*

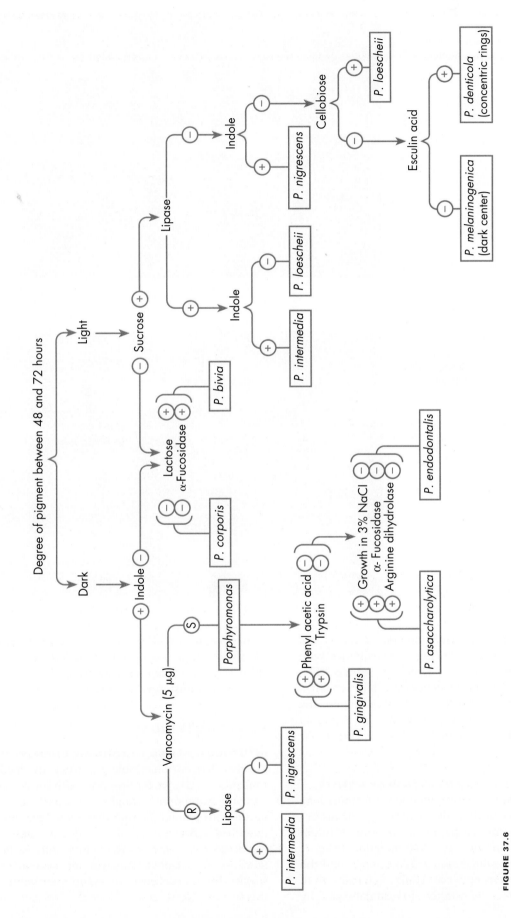

FIGURE 37.6

Flow chart for identification of pigmented, anaerobic, gram-negative bacilli.

FIGURE 37.7

Prevotella oris-buccae on blood agar. Colonies are somewhat mucoid.

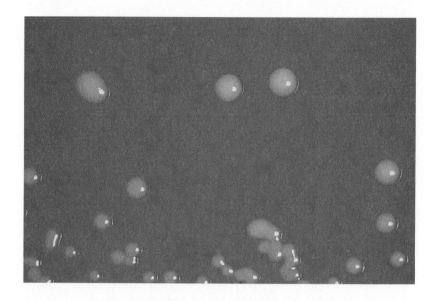

FIGURE 37.8

Bacteroides ureolyticus on blood agar. Note spreading and transparent colonies.

teroides bile esculin (BBE) medium after 4 or 5 days' incubation, producing small colonies with black centers because of hydrogen sulfide (H$_2$S) production. The other genera are bile sensitive. Because these organisms are asaccharolytic, the simplest way to identify them is with the rapid tests already mentioned. These reactions are found in Table 37.5.

SUSCEPTIBILITY TO ANTIMICROBIAL AGENTS

Determination of susceptibility to antimicrobial agents has been difficult for these organisms because they form transparent colonies that result in hazy end points in MIC determination tests. *B. wadsworthia* strains produce β-lactamase, and the others are microaerobic and thus often resistant to metronidazole. Resistance to clindamycin by some strains has been reported. The β-lactamase inhibitor combinations, second- and third-generation cephalosporins, and imipenem are active against most strains.

FUSOBACTERIUM

INTRODUCTION AND TAXONOMIC CHANGES

Fusobacteria are anaerobic gram-negative bacilli that produce major butyric acid without isoacids and are nonmotile, straight, or slightly curved bacilli with pointed or rounded ends. Five newly described species are *F. alocis, F. sulci, F. pseudonecrophorum, F. periodonticum,* and *F. ulcerans.*[1,5,22] *F. nucleatum* may contain several subspecies, but no consensus exists on their names at this time (Table 37.2).

TABLE 37.5

CHARACTERISTICS OF NONPIGMENTED BILE-SENSITIVE ANAEROBIC GRAM-NEGATIVE BACILLI

SPECIES	KANAMYCIN (1 mg)	COLISTIN (10 μg)	GLUCOSE	XYLOSE (PENTOSE)	SUCROSE	LACTOSE	SALICIN	XYLAN	STRONG GELATIN HYDROLYSIS	INDOLE	ESCULIN HYDROLYSIS	α-FUCOSIDASE	β-NAG	ZOOGLEAL MASS	REQUIRES FORMATE/FUMARATE FOR GROWTH	NITRATE REDUCTION	MOTILE	UREASE	STRONG OXIDASE	PITS THE AGAR	FATTY ACIDS FROM PEPTONE-YEAST-GLUCOSE BROTH
Prevotella oralis group																					
P. buccalis	R		+	−	+	+	−	−	−	−	+	+	+	−	−	−	−	−		−	a iv S
P. oulora	R		+	−	+	+	−	−	−	−	+	+	+	−	−	−	−	−		−	a S
P. oralis	R		+	−	+	+	+	−	−	−	+	+	+	−	−	−	−	−		−	A S (l)
P. veroralis	R		+	−	+	+	−	+	−	−	+	+	+	−	−	−	−	−		−	a S
Pentose fermenting																					
Prevotella buccae	R	S	+	+	+	+	+	+−	−	−	+	−	−	−	−	−	−	−		−	A S (p ib b iv l)
P. oris	R	**R**	+	+	+	+	+	−	−	−	+	+	+	−	−	−	−	−		−	A S (p ib iv)
P. heparinolytica	R		+	+	+	+	+	−	−	+	+	+	+	+−	−	−	−	−		−	S (a p iv)
P. zoogleoformans	R		+	+	+	+	V	−	−	+	+	−	−	+	−	−	−	−		−	A P S (ib iv)
Saccharolytic and proteolytic																					
P. bivia	R		+	−	−	+	−	−	+	−	−	+	+	−	−	−	−	−		−	A iv S (ib)
P. disiens	R		+	−	−	−	−	−	+	−	−	−	−	−	−	−	−	−		−	A S (p ib iv)
Weak or nonfermentative																					
Bacteroides capillosus	R		w−	−	−	−	−	−	−	−	+			−	−	−	−	−		−	a s (p l)
B. putredinis	R		−	−	−	−	−	−	+	+	−			−	−	+	−	−		−	a P ib b IV S (l)
Bilophila wadsworthia*	S		−	−	−	−	−	−	−	−	−			−	−	+	−	+		−	
Formate-fumarate requiring																					
Bacteroides gracilis	S		−	−	−	−	−	−	−	−	−			−	+	+	−	−		+−	a S (l)
B. ureolyticus	S		−	−	−	−	−	−	−	−	−			−	+	+	−	+	V	+−	a S
Wolinella sp.	S		−	−	−	−	−	−	−	−	−			−	+	+	+	−	−	−+	S
Campylobacter concisus group	S		−	−	−	−	−	−	−	−	−			−	+	+	+	−	+−	−+	S

Reactions: results in **bold type** are key characteristics; R, resistant; S, sensitive; −, negative; −+, rare strains positive; +−, rare strains negative; w−, most strains weak, rare strains negative; V, variable.

PRAS carbohydrate reactions: +, pH <5.5; w, pH 5.5–5.9; −, pH >5.9.

Fatty acid end products: A, acetic; P, propionic; IB, isobutyric; B, butyric; IV, isovaleric; L, lactic; S, succinic. Capital letters indicate major metabolic products; lower-case letters indicate minor products. Parentheses indicate variable reaction. Isoacids are primarily from carbohydrate-free media (e.g., peptone-yeast) in the case of saccharolytic organisms.

*Other reactions include bile resistant, strongly catalase positive.

TABLE 37.6

CHARACTERISTICS OF FUSOBACTERIUM SP.

SPECIES	MICROSCOPIC FEATURES	INDOLE	GROWTH IN 20% BILE	LIPASE	GAS IN PYG AGAR	FERMENTATION OF			ESCULIN HYDROLYSIS	CONVERTS LACTATE TO PROPIONATE	CONVERTS THREONINE TO PROPIONATE	FATTY ACIDS FROM PEPTONE-YEAST-GLUCOSE BROTH
						GLUCOSE	FRUCTOSE	MANNOSE				
F. nucleatum	**Slender pointed ends**	+	−	−	$-^2$	$-^w$	$-^w$	−	−	−	+	a p B (L s)
F. gonidiaformans	Gonidia forms	+	−	−	4^2	−	−	−	−	−	+	A p B (l s)
F. necrophorum	Round ends	+	$-^+$	$+^-$	4^2	$-^w$	$-^w$	−	−	**+**	+	a p B (l s)
F. naviforme	**Boat shape**	+	−	−	−	w^-	−	−	−	−	−	a B L (p s)
F. varium	Small, round ends	$+^-$	+	$-^+$	4	$+^w$	$+^w$	$+^w$	−	−	+	a p B L (s)
F. mortiferum	**Bizarre, round bodies**	−	+	−	4	$+^w$	$+^w$	$+^w$	+	−	+	a p B (v l a)
F. russii	Round ends	−	−	−	2^-	−	−	−	−	−	−	a B L

Reactions: results in **bold type** are key characteristics; −, negative; $-^+$, rare strains positive; +, most strains positive; $+^-$, rare strain negative; $-^w$, rare strain weak positive; w^-, most strains weak, rare strain negative.

PRAS carbohydrate reactions: +, pH <5.5; w, pH 5.5–5.9; −, pH >5.9.

Gas in peptone-yeast-glucose (PYG) agar deep: −, no gas detected; 2, splits agar horizontally; 3, agar displaced halfway to top of the tube; 4, agar displaced to top of the tube.

Fatty acid end products: A, acetic; P, propionic; B, butyric; V, valeric; L, lactic; S, succinic. Capital letters indicate major metabolic products; lower-case letters indicate minor products. Parentheses indicate variable reaction.

NORMAL FLORA, INFECTIONS, AND VIRULENCE FACTORS

F. nucleatum and *F. necrophorum* are found as indigenous flora in the upper respiratory and intestinal tracts. They may be implicated in infection throughout the body, but with some predilection for the lower respiratory tract, head and neck, periodontium, gingivae, and central nervous system. *F. mortiferum* and *F. varium* are indigenous flora in the intestinal tract and are occasionally isolated from intraabdominal infections. Of the seven fusobacteria encountered in human infections, *F. nucleatum* is isolated most often. It is an important pathogen, particularly in head and neck and lower respiratory infections. *F. necrophorum* may be very virulent in certain types of infections. In postanginal sepsis (Lemierre's syndrome) the infection begins with a membranous tonsillitis and proceeds to septicemia with metastatic infection that can include lung abscess, empyema, liver abscess, osteomyelitis, and purulent arthritis. Other species of fusobacteria (Table 37.6) are isolated infrequently.

The lipopolysaccharide of *Fusobacterium* strains resembles that of nonanaerobic gram-negative rods in structure. With most strains, biological activity is strong. *F. necrophorum* produces a

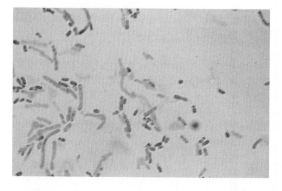

FIGURE 37.9

Fusobacterium mortiferum. Pleomorphic with round swellings. (Courtesy of Mike Cox, Anaerobe Systems, San Jose, Calif.)

leukocidin and hemolyzes red blood cells of humans and other animal species. Phospholipase A and lysophospholipase are produced by *F. necrophorum*.

MORPHOLOGY AND GENERAL CHARACTERISTICS

The fusobacteria are pale-staining gram-negative bacilli with a diversity of cell shapes and colony morphology. *F. nucleatum* is a long, slender rod with tapered ends; occasionally the cells occur in pairs end-to-end (Chapter 35, Figure 35.3). *F. mortiferum* display pleomorphism, with spheroid swellings along irregularly stained filaments and round forms (Figure 37.9). *F. necrophorum* is a slightly pleomorphic bacillus with rounded ends that becomes more pleomorphic with increasing age of the culture. Colonies on blood agar are slightly hemolytic and produce greening of the agar around the colonies after exposure to air. This results from hydrogen peroxide (H_2O_2) production. Colonies of *F. nucleatum* may resemble bread crumbs (Figure 37.10) or appear convex and glistening with a flecked internal structure (Figure 37.11). *F. varium* produces a colony with

an opaque center and translucent, irregular margin that resembles a fried egg, and *F. necrophorum* produces an umbonate colony (Figure 37.12). Some strains of *F. mortiferum* have an umbonate colony, whereas others are not distinctive. *F. nucleatum* and sometimes *F. necrophorum* grow in balls in broth culture.

IDENTIFICATION METHODS

Fusobacteria produce large amounts of butyric acid as a metabolic end product. They are resistant to vancomycin and susceptible to kanamycin and colistin in the antibiotic disk identification test (Chapter 35). This is the same pattern as the *Bacteroides ureolyticus* group, but the colony and

FIGURE 37.10

Fusobacterium nucleatum. Bread crumb–like colonies and greening of agar.

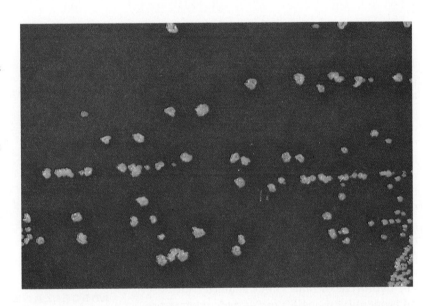

FIGURE 37.11

Fusobacterium nucleatum. Internal speckling in colonies.

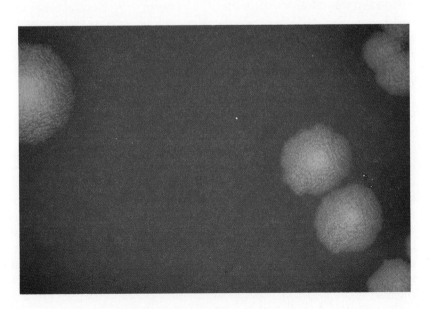

FIGURE 37.12

Umbonate colonies of
Fusobacterium necrophorum
on blood agar.

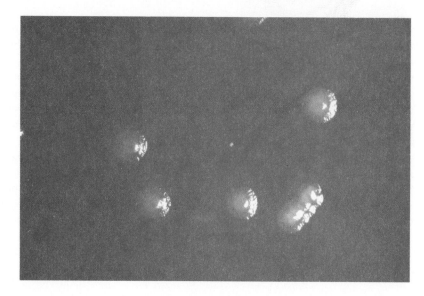

growth characteristics readily distinguish them. The *Bacteroides fragilis* group and *Prevotella* sp. produce alanine aminopeptidase; this differentiates them from fusobacteria, which do not produce this enzyme. This test is available commercially as a rapid disk test (Carr-Scarborough, Stone Mountain, Ga.). *F. nucleatum* is readily identified by microscopic and colonial morphology and a positive test for indole (Chapter 35). *F. necrophorum* is easily identified when the indole and lipase tests are both positive. *F. mortiferum* and *F. varium* and some strains of *F. necrophorum* are bile resistant. *F. mortiferum* is indole negative and hydrolyzes esculin, whereas most strains of *F. varium* produce indole but do not hydrolyze esculin. Other characteristics useful in identifying species of *Fusobacterium* are presented in Table 37.6.

SUSCEPTIBILITY TO ANTIMICROBIAL AGENTS

In general, fusobacteria are more susceptible to antimicrobial agents than are the *Bacteroides fragilis* group. Unfortunately, resistance is beginning to appear in some species. Some strains of *F. nucleatum* and *F. mortiferum* produce β-lactamase. More than 50% of *F. varium* strains are resistant to clindamycin. Fusobacteria produce L-form colonies on exposure to cell wall–active antimicrobial agents. These appear as transparent growth on agar dilution susceptibility tests and make interpretation of MIC end points ambiguous. Metronidazole, imipenem, the β-lactamase inhibitor combinations, and second- and third-generation cephalosporins are active against most strains.

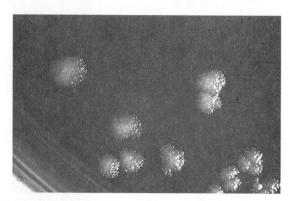

FIGURE 37.13

Raspberry-like colonies of *Leptotrichia buccalis.*

MISCELLANEOUS BACTEROIDACEAE

Several other genera in the family Bacteroidaceae are infrequently encountered in clinical specimens. These genera are listed, with key differential characteristics, in Table 37.1. Motility and cell shape provide an initial means of subdividing the group. Some of these species are associated with dental infections and bacterial vaginosis, and others are isolated from immunocompromised hosts. Several taxonomic changes have occurred within this family. Some *Bacteroides* sp. have been assigned to a new genus, and many new species have been described (Table 37.2). Several new species of *Selenomonas* associated with periodontitis have been described.[14] *Mobiluncus* is included in the table because, although it has a gram-positive cell wall, it usually stains gram variable or gram negative. It is discussed in detail in Chapter 36.

NONMOTILE RODS

Leptotrichia buccalis is part of the normal flora of the oral cavity and is occasionally found in the intestine and vagina. It has been recovered from dental infections, osteomyelitis of the mandible, bite wound infections, infections of the head and neck, and the blood of a patient with a lung abscess.[15] It is a long, straight, or slightly curved gram-negative bacillus that grows end-to-end in pairs. The ends that abut are flattened, whereas the other ends are often pointed. The colonies are 2 to 3 mm in diameter with a characteristic rasberry-like appearance (Figure 37.13). It is unique in producing lactic acid as its sole metabolic end product. *L. buccalis* is susceptible to virtually all antibacterial agents used against anaerobes.

Mitsuokella dentalis has been isolated from infected dental root canals. The bacterium produces a characteristic "water drop" colony and grows poorly.

MOTILE OR CURVED RODS

The various anaerobic vibrios and curved rods are rarely recovered from clinical material; they have been involved in bacteremia, central nervous system, pulmonary, intraabdominal, and soft tissue infections and endophthalmitis.[7] *Selenomonas* sp. have been isolated from dental infections[14] and occasionally from pulmonary infections. *Anaerobiospirillum succiniciproducens* has been isolated from diarrheal stools and from blood.[11,12] *Butyrivibrio* has been isolated from the anterior chamber and vitreous fluid of a patient with traumatic endophthalmitis. *Succinivibrio* sp. and *Desulfovibrio* sp. have been isolated from blood of several patients with underlying gastrointestinal disease. Definitive identification within the group of anaerobic vibrios and curved rods requires a flagella stain, prereduced anaerobically sterilized (PRAS) biochemicals, and fatty acid end product analysis by gas-liquid chromatography (GLC).

REFERENCES

1. Adriaans, B., and Shah, H. 1988. *Fusobacterium ulcerans* sp. nov. from tropical ulcers. Int. J. Syst. Bacteriol. 38:447.
2. Baron, E.J., Downes, J., Summanen, P., et al. 1989. *Bilophila wadsworthia*, gen. nov., sp. nov., a unique gram-negative anaerobic rod recovered from appendicitis specimens and human feces. J. Gen. Microbiol. 135:3405.
3. Briselden, A.M., Moncla, B.J., Stevens, C.E., and Hillier, S.L. 1992. Sialidases (neuraminidases) in bacterial vaginosis and bacterial vaginosis–associated microflora. 1992. J. Clin. Microbiol. 30:663.
4. Brook, I. 1986. Encapsulated bacteria in synergistic infections. Microbiol. Rev. 50:452.
5. Cato, E.P., Moore, L.V.H., and Moore, W.E.C. 1985. *Fusobacterium alocis* sp. nov. and *Fusobacterium sulci* sp. nov. from the human gingival sulcus. J. Clin. Microbiol. 35:475.
6. Eley, A., Clarry, T., and Bennett, K.W. 1989. Selective and differential media for isolation of *Bacteroides ureolyticus* from clinical specimens. Eur. J. Clin. Microbiol. Infect. Dis. 8:83.
7. Johnson, C.C., and Finegold, S.M. 1987. Uncommonly encountered, motile anaerobic gram-negative bacilli associated with infection. Rev. Infect. Dis. 9:1150.
8. Johnson, C.C., Reinhardt, J.F., Edelstein, M.A.C., et al. 1985. *Bacteroides gracilis*, an important anaerobic bacterial pathogen. J. Clin. Microbiol. 22:799.
9. Johnson, J.L., Moore, W.E.C., and Moore, L.V.H. 1986. *Bacteroides caccae* sp. nov., *Bacteroides merdae* sp. nov., and *Bacteroides stercoris* sp. nov. isolated from human feces. Int. J. Syst. Bacteriol. 36:499.
10. Lee, K.E., Baron, E.J., Summanen, P., and Finegold, S.M. 1990. Selective medium for isolation of *Bacteroides gracilis*. J. Clin. Microbiol. 28:1747.
11. Malnick, H., Williams, K., Phil-ebosie, J., and Levy, A.S. 1990. Description of a medium for isolating *Anaerobiospirillum* spp., a possible cause of zoonotic disease, from diarrheal feces and blood of humans and use of the medium in a survey of human, canine, and feline feces. J. Clin. Microbiol. 28:1380.
12. McNeil, M.M., Martone, W.J., and Dowell, V.R., Jr. 1987. Bacteremia with *Anaerobiospirillum succiniciproducens*. Rev. Infect. Dis. 9:737.
13. Moncla, B.J., Braham, P., Rabe, L.K., and Hillier, S.L. 1991. Rapid presumptive identification of black pigmented gram-negative anaerobic bacteria by using 4-methylumbelliferone derivatives. J. Clin. Microbiol. 29:1955.
14. Moore, L.V.H., Johnson, J.L., and Moore, W.E.C. 1987. *Selenomonas noxia* sp. nov., *Selenomonas flueggei*, sp. nov., *Selenomonas infelix*, sp. nov., *Selenomonas dianae*, sp. nov., and *Selenomonas artemidis*, sp. nov. from the human gingival crevice. Int. J. Syst. Bacteriol. 36:271.
15. Morgenstein, A.A., Citron, D.M., Orisek, B., and Finegold, S.M. 1980. Serious human infection with *Leptotrichia buccalis*: report of a case and review of the literature. Am. J. Med. 69:782.
16. Paster, B.J., and Dewhirst, F.E. 1988. Phylogeny of Campylobacters, Wolinellas, *Bacteroides gracilis*, and *Bacteroides ureolyticus* by 16S ribosomal ribonucleic acid sequencing. Int. J. Syst. Bacteriol. 38:56.
17. Shah, H.N., and Collins, M.D. 1983. Genus *Bacteroides*: a chemotaxonomical perspective (a review). J. Appl. Bacteriol. 55:403.
18. Shah, H.N., and Collins, M.D. 1988. Proposal for reclassification of *Bacteroides asaccharolyticus*, *Bacteroides gingivalis*, and *Bacteroides endodontalis* in a new genus, *Porphyromonas*. Int. J. Syst. Bacteriol. 38:128.
19. Shah, H.N., and Collins, M.D. 1989. Proposal to restrict the genus *Bacteroides* (Castellani and Chalmers) to *Bacteroides fragilis* and closely related species. Int. J. Syst. Bacteriol. 39:85.
20. Shaw, H.N., and Collins, M.D. 1990. *Prevotella*, a new genus to include *Bacteroides melaninogenicus* and related

species formerly classified in the genus *Bacteroides*. Int. J. Syst. Bacteriol. 40:205.

21. Shaw, H.N., and Gharbia, S.E. 1992. Biochemical and chemical studies on strains designated *Prevotella intermedia* and proposal of a new pigmented species, *Prevotella nigrescens* sp. nov. Int. J Syst. Bacteriol. 42:542.

22. Slots, J., Potts, T.V., and Mashimo, P.A. 1983. *Fusobacterium periodonticum*, a new species from the human oral cavity. J. Dent. Res. 62:960.

23. Vandamme, P., Falsen, E. Rossau, R., et al. 1991. Revision of *Campylobacter, Helicobacter,* and *Wolinella* taxonomy: emendation of generic descriptions and proposal of *Arcobacter* gen. nov. Int. J. Syst. Bacteriol. 41:88.

24. Van Tassell, R.L., Lyerly, D.M., and Wilkins, T.D. 1992. Purification and characterization of an enterotoxin from *Bacteroides fragilis*. Infect. Immun. 60:1343.

BIBLIOGRAPHY

Collins, M.D., and Shah, H.N. 1986. Reclassification of *Bacteroides praeacutus* Tissier (Holdeman and Moore) in a new genus, *Tissierella,* as *Tissierella praeacuta* comb. nov. Int. J. Syst. Bacteriol. 36:461.

Holdeman, L.V., Cato, E.P., and Moore, W.E.C. 1977. Anaerobe laboratory manual, ed 4. VPI Anaerobe Laboratory and Virginia Polytechnic Institute and State University, Blacksburg, Va.

Holdeman, L.V., Kelley, R.W., and Moore, W.E.C. 1984. Family I Bacteroidaceae Pribram, 1933. In Sneath, P.H.A, Mair, N.S., Sharpe, M.E., and Holt, J.G., editors: Bergey's manual of systematic bacteriology, vol 1. p. 602. Williams & Wilkins, Baltimore.

Jousemies-Somer, H.R., and Finegold, S.M. 1991. Anaerobic gram-negative bacilli and cocci. In Balows, A., Hausler, W.J., Herrmann, K.L., et al., editors: Manual of clinical microbiology, ed 5. p. 538. American Society for Microbiology, Washington, D.C.

Love, D.N., Bailey, G.D., Collings, S., and Briscoe, D.A. 1992. Description of *Porphyromonas circumdentaria* sp. nov. and reassignment of *Bacteroides salivosus* (Love, Johnson, Jones, and Calverley 1987) as *Porphyromonas* (Shah and Collins 1988) *salivosa* comb. nov. Int. J. Syst. Bacteriol. 42:434.

Mayrand, D., Bourgeau, G., Grenier, D., and LaCroix, J-M. 1984. Properties of oral asaccharolytic black-pigmented *Bacteroides*. Can. J. Microbiol. 30:1133.

Moore, L.V.H., Cato, E.P., and Moore, W.E.C. 1988, 1991. Anaerobe laboratory manual update. Supplement to Anaerobe laboratory manual, ed 4. 1977. Virginia Polytechnic Institute, Blacksburg, Va.

Summanen, P., Baron, E.J., Citron, D.M., et al. 1992. Wadsworth anaerobic bacteriology manual, ed 5. Star Publishing, Belmont, Calif.

van Winkelhoff, A.J., van Steenbergen, T.J.M., Kippuw, N., and deGraaff, J. 1986. Enzymatic characterization of oral and nonoral black-pigmented *Bacteroides* sp. Antonie Van Leeuwenhoek 52:163.

38

ANAEROBIC COCCI

The anaerobic cocci are a diverse group of organisms with several genera but no family affiliation (Table 38.1). The anaerobic gram-negative cocci belong to the genera *Veillonella*, *Acidaminococcus*, and *Megasphaera*. The anaerobic gram-positive cocci are in the genera *Peptococcus*, *Peptostreptococcus*, *Streptococcus*, *Ruminococcus*, and *Coprococcus*. Some strains of streptococci and *Staphylococcus saccharolyticus* (formerly *Peptococcus saccharolyticus*) are included with the anaerobic gram-positive cocci because they often require good anaerobic transport and culture techniques for primary isolation. Anaerobic strains of *Streptococcus intermedius*, *S. constellatus*, and *Gemella morbillorum* were previously classified as species in *Streptococcus* and *Peptostreptococcus*.[11]

TAXONOMIC CHANGES

As with other groups of anaerobes, genetic studies are changing the taxonomy of the anaerobic cocci. Most species of *Peptococcus* have been transferred to *Peptostreptococcus*.[6] *P. niger* is the only species retained in *Peptococcus*. Strains previously called *"Gaffkya anaerobia"* are now included in *Peptostreptococcus tetradius*. A new species, *Peptostreptococcus hydrogenalis*, has been described.[5] In deoxyribonucleic acid–ribosomal ribonucleic acid (DNA-rRNA) hybridization studies, *Peptostreptococcus anaerobius* appears to be more closely related to *Eubacterium* sp. or *Clostridium* sp.[10] The taxonomic status of several of the anaerobic streptococci is uncertain.[20] *Streptococcus hansenii* is an obligate

anaerobe and appears to be more closely related to *Clostridium*. *Streptococcus morbillorum* was reclassified as *Gemella*,[11] and *Streptococcus constellatus* and *S. intermedius* are genetically similar to *S. anginosus*[4] (Chapter 26).

NORMAL FLORA, INFECTIONS, AND VIRULENCE FACTORS

The anaerobic cocci are numerically important members of the indigenous flora in the bowel, female genital tract, and oral cavity. They are found in smaller numbers in other sites, including the skin. The anaerobic gram-positive cocci make up about 30% of anaerobes isolated from infections containing anaerobes.[3,7,17] Reflecting differences in their prevalence as normal flora, the gram-positive cocci are isolated more frequently from respiratory tract and female genital tract infections than from intraabdominal infections. They are also found frequently in diabetic foot infections. Virtually all types of infection may contain anaerobic cocci. *Ruminococcus* and *Coprococcus* are part of the normal flora of the colon and are isolated very rarely from clinical specimens.[9] The anaerobic gram-positive cocci most often encountered in infections are *Peptostreptococcus magnus*, *P. asaccharolyticus*, *P. prevotii*, *P. micros*, and *P. anaerobius*. *P. magnus* appears to be the most virulent. It produces collagenase and has been isolated in pure culture from a variety of infections, especially bone and joint infections.[1,12,16] The anaerobic gram-negative cocci are isolated less frequently

TABLE 38.1

ANAEROBIC COCCI

DISTINGUISHING CHARACTERISTIC	GENUS
Gram-negative cocci	
I. Produce propionic and acetic acids	*Veillonella*
II. Produce butyric and acetic acids	*Acidaminococcus*
III. Produce isobutyric, butyric, isovaleric, valeric, and caproic acid	*Megasphaera*
Gram-positive cocci	
I. Require fermentable carbohydrate	
A. Produce butyric acid (plus other acids)	*Coprococcus*
B. Do not produce butyric acid	*Ruminococcus*
II. Do not require fermentable carbohydrate	
A. Lactic acid sole major product	*Streptococcus, Gemella*
B. Not as above	*Peptostreptococcus* or *Peptococcus*

Modified from Sutter, V.L., Citron, D.M., Edelstein, M.A.C., and Finegold, S.M. 1985. Wadsworth anaerobic bacteriology manual, ed 4. Star Publishing, Belmont, Calif.

than the anaerobic gram-positive cocci and appear to be less pathogenic. *Veillonella* is encountered in about 3% of clinical specimens that contain anaerobes,[7] whereas *Acidaminococcus* and *Megasphaera* are rarely isolated.

Virulence factors of anaerobic cocci have not been studied as extensively as for clostridia and anaerobic gram-negative rods. However, lipo-

polysaccharides, hyaluronidase, collagenase, and capsules have been described for the anaerobic cocci.[12,18]

DIFFERENTIATION OF ANAEROBIC COCCI

Differentiation among genera of anaerobic cocci is based on Gram stain reaction and cell wall characteristics, metabolic end products, genetic studies including G plus C content, DNA-RNA homology, DNA-DNA homology, and requirement for a fermentable carbohydrate for growth (Table 38.1).

ANAEROBIC GRAM-POSITIVE COCCI

This group includes one species of *Peptococcus*, *P. niger*, which is rarely encountered in clinical infection; two anaerobic species in the genus *Streptococcus* (*S. intermedius*, *S. constellatus*); *Gemella morbillorum*; and the genus *Peptostreptococcus*, seven species of which may be encountered in infection. *Coprococcus*, *Ruminococcus*, and *Streptococcus hansenii* are excluded from this discussion because they are part of the normal bowel flora and rarely isolated from clinical specimens.

MORPHOLOGY AND GENERAL CHARACTERISTICS

Microscopic morphology of the anaerobic cocci is varied. *Peptostreptococcus anaerobius* (Figure 38.1) and *P. productus* are large coccobacilli that form chains, especially in broth cultures. Other anaerobic cocci, such as *Peptostreptococcus tetradius*, *P. magnus*, and *P. prevotii*, have cells greater

FIGURE 38.1

Peptostreptococcus anaerobius. Large coccobacillary cells in chains, pairs, and singles. (From Sutter, V.L., Citron, D.M., Edelstein, M.A.C., and Finegold, S.M. 1985. Wadsworth anaerobic bacteriology manual, ed 4. Star Publishing, Belmont, Calif.)

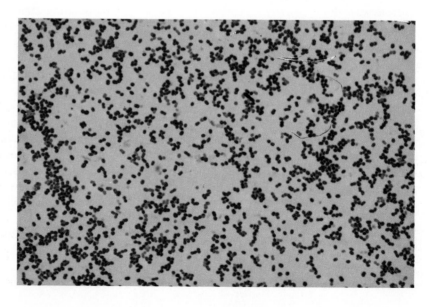

FIGURE 38.2

Peptostreptococcus magnus in pus. Appearance is identical to that of staphylococci. (From Sutter, V.L., Citron, D.M., Edelstein, M.A.C., and Finegold, S.M. 1985. Wadsworth anaerobic bacteriology manual, ed 4. Star Publishing, Belmont, Calif.)

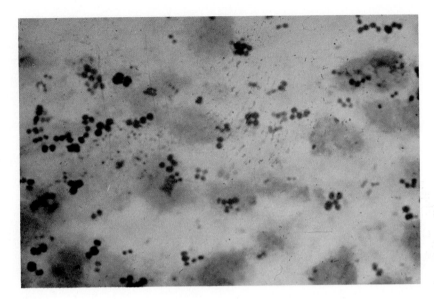

than 0.6 μm in diameter in pairs and clusters and may resemble staphylococcal cells (Figure 38.2). The presence of gram-positive cocci on direct Gram stain, with no staphylococci recovered on aerobic culture after 18 to 24 hours, suggests the presence of one of these species of *Peptostreptococcus*. *P. micros* cells are less than 0.7 μm in diameter and occur in packets and short chains (Figure 38.3); difference in cell size has been used as one characteristic to distinguish between *P. micros* and *P. magnus,* which are similar in their biochemical reactions. The anaerobic strep-tococci vary in size and occur in chains, pairs, or singly.

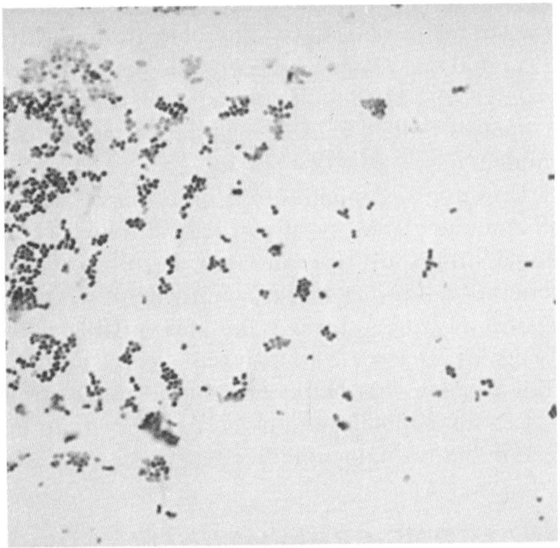

Colony morphology of the gram-positive cocci is variable. *Peptostreptococcus asaccharolyticus* produces gray, translucent colonies about 1 mm in diameter (Figure 38.4). *P. magnus* is similar in appearance. *P. micros* produces smaller, white, opaque colonies with a halo of discoloration surrounding them. *P. anaerobius* typically produces larger and more opaque colonies than other anaerobic cocci (Figure 38.5); it also has a sweet, fetid odor. *Peptococcus niger,* an organism rarely encountered in clinical specimens, produces colonies with a greenish black color, which is caused by hydrogen sulfide (H_2S) production. Most anaerobic cocci grow only to a moderate turbidity in broth media. Growth of some anaerobic cocci in broth is slow and results in balls, clumps, or aggregates of growth rather than diffuse turbidity. Supplementation of the broth with Tween 80 (0.5%) often enhances growth of these strains.

IDENTIFICATION

Preliminary grouping and identification procedures for the anaerobic gram-positive cocci appear in Chapter 35. Tables 38.1 and 38.2, and Figure 38.6 list characteristics that aid in differentiation of the anaerobic gram-positive cocci.

A few cocci can be identified with relatively simple tests. *Peptostreptococcus anaerobius* is susceptible to sodium polyanetholsulfonate (SPS) (Figure 38.5) and produces isocaproic acid as an end product of metabolism. *P. asaccharolyticus, P. hydrogenalis,* and *P. indolicus* are indole positive. *P. indolicus* is rarely isolated from human clinical spec-

FIGURE 38.3

Peptostreptococcus micros. Small cells often form short chains. (From Sutter, V.L. Citron, D.M., Edelstein, M.A.C., and Finegold, S.M. 1985. Wadsworth anaerobic bacteriology manual, ed 4. Star Publishing, Belmont, Calif.).

FIGURE 38.4

Typical colony morphology of anaerobic gram-positive coccus on blood agar *(Peptostreptococcus asaccharolyticus).*

imens; it produces coagulase and usually reduces nitrate. *P. hydrogenalis* ferments glucose and produces alkaline phosphatase; *P. asaccharolyticus* is negative in these tests.

Definitive identification requires end product analysis by gas-liquid chromotography (GLC) and other tests, such as indole, alkaline phosphatase, and urease production; nitrate reduction; and carbohydrate fermentation. Alkaline phosphatase and catalase production and gelatin hydrolysis aid in differentiating *P. magnus* and *P. micros.* As mentioned in Chapter 35, the rapid 4-hour preformed enzyme kits identify the anaerobic gram-positive cocci fairly well without the use of GLC,[14] except that there may be problems accurately separating *P. tetradius* and *P. prevotii.*

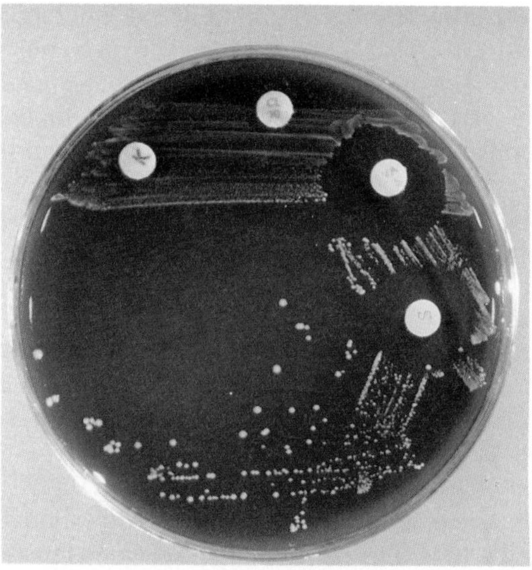

SUSCEPTIBILITY TO ANTIMICROBIAL AGENTS

The drug of choice for anaerobic gram-positive coccal infections has generally been considered to be penicillin G. However, some strains require as much as 8 µg/ml for inhibition, and rare strains may require as much as 32 µg/ml. Infection with the latter strains would best be treated with another, more active agent. Chloramphenicol, the carboxypenicillins and ureidopenicillins, and cefoxitin are active against all strains, although some strains require 16 µg/ml of cefoxitin for inhibition. Metronidazole is active against all but a few strains of obligately anaerobic gram-positive cocci and the "anaerobic" streptococci. Approximately 10% to 15% of strains, especially among *P. magnus* and *P. asaccharolyticus,* are resistant to clindamycin, and 20% to 30% are resistant to erythromycin. Cephalothin is quite active against all anaerobic gram-positive cocci, although occasional strains are resistant to some of the third-generation cephalosporins. Imipenem is active against all strains. Tetracycline has variable activity against anaerobic gram-positive cocci; doxycycline is somewhat better but should not be used unless susceptibility of a given strain from an infection has been documented.[8,15,19]

FIGURE 38.5

Peptostreptococcus anaerobius. Compare colony morphology to Figure 38.4. Note disk identification tests: resistant to kanamycin and colistin and sensitive to vancomycin and sodium polyanetholsulfonate (SPS) (S disk).

ANAEROBIC GRAM-NEGATIVE COCCI

Only three of at least seven species of *Veillonella* are frequently found in human clinical infections.[13] *Megasphaera* and *Acidaminococcus* are rarely encountered.

TABLE 38.2

CHARACTERISTICS OF ANAEROBIC GRAM-POSITIVE COCCI

SPECIES	ISOCAPROIC ACID	BUTYRIC ACID	LACTIC ACID	SODIUM POLYANETHOLSULFONATE	INDOLE	NITRATE	CATALASE	ALKALINE PHOSPHATASE	UREASE	CELLS <0.7 μm IN DIAMETER	GELATIN HYDROLYSIS	FERMENTATION OF GLUCOSE	MALTOSE	SUCROSE	FATTY ACIDS FROM PEPTONE-YEAST-GLUCOSE BROTH
Peptostreptococcus															
P. anaerobius	+	V	–	**S**	–	–	–	–	–		–ʷ	wˉ	wˉ	ˉw	A IC (ib b iv)
P. asaccharolyticus	–	+	–	R	+	–	V	–	–		–	–	–	–	B (A l p)
P. indolicus	–	+	–	R	+	+ˉ					–	–	–	–	A B (p s)
P. prevotii	–	+	–	R	–	–	V	+	–⁺		–	w	wˉ	ˉw	B (A L p)
P. tetradius	–	+	–	R	–	–	V	–	+		–	+	+	+	B L (a p)
P. magnus	–	–	–	R	–	–	+ˉ	–*	–	–	+ˉ	–	–	–	A
P. micros	–	–	–	R	–	–	–	+	–	+	–	–	–	–	A (s)
P. productus	–	–	–	R	–	–	–	V*			–	+	+	+	A l (s)
Staphylococcus saccharolyticus	–	–	–	R	–	+	+				–	+	–	–	A (s)
"Anaerobic" streptococci	–	–	+	R	–	–	–				–	+	+	+	L (a s)

Reactions: Results in **bold type** are key differential tests: –, negative for most strains; +, positive for most strains; V, variable; w, most strains weakly positive. Superscripts indicate reactions of occasional strains.

Prereduced anaerobically sterilized (PRAS) carbohydrate reactions: +, pH <5.5; w, pH 5.5–5.9; –, pH >5.9.

Fatty acid end products: A, acetic; P, propionic; IB, isobutyric; B, butyric; IV, isovaleric; V, valeric; IC, isocaproic; C, caproic; L, lactic; S, succinic. Capital letters indicate major metabolic products; lower-case letters indicate minor products. Parentheses indicate variable reaction. Isoacids are primarily from carbohydrate-free media (e.g., peptone-yeast) in the case of saccharolytic organisms.

*Varies with reporting laboratories.

MORPHOLOGY AND GENERAL CHARACTERISTICS

Veillonella sp. are tiny (less than 0.5 μm in diameter) gram-negative cocci that occur in pairs and packets (Figure 38.7). Unusually large cocci, especially when found in groups or packets, would suggest *Megasphaera* or *Acidaminococcus*. *Megasphaera* typically stains gram variable but has a gram-negative cell wall composition. It could initially be confused with one of the peptostreptococci by Gram stain morphology.

Veillonella sp. produce small, convex, translucent to transparent colonies with an entire edge. These colonies may show red fluorescence under long-wave ultraviolet light, but fluorescence is lost rapidly after exposure of colonies to air.[2] *Veillonella* grows poorly in broth and produces a very fine, granular turbidity. Supplementation of broth with pyruvate may enhance growth for some strains.

IDENTIFICATION

The genera of gram-negative cocci can be differentiated by their end products of metabolism (Table 38.1), and, as noted in Table 38.3, nitrate reduction, catalase production, and glucose fermentation.

SUSCEPTIBILITY TO ANTIMICROBIAL AGENTS

Veillonella is seldom a significant pathogen, and *Acidaminococcus* and *Megasphaera* are rarely encountered in infection, so specific therapy directed against the anaerobic gram-negative cocci is seldom indicated. Penicillin G is quite active against these organisms, but other penicillins, including the carboxypenicillins and ureidopenicillins, are not active against all strains. Metronidazole, chloramphenicol, clindamycin, and cefoxitin, cephalothin, the newer cephalosporins, and imipenem are all active against the gram-negative cocci.

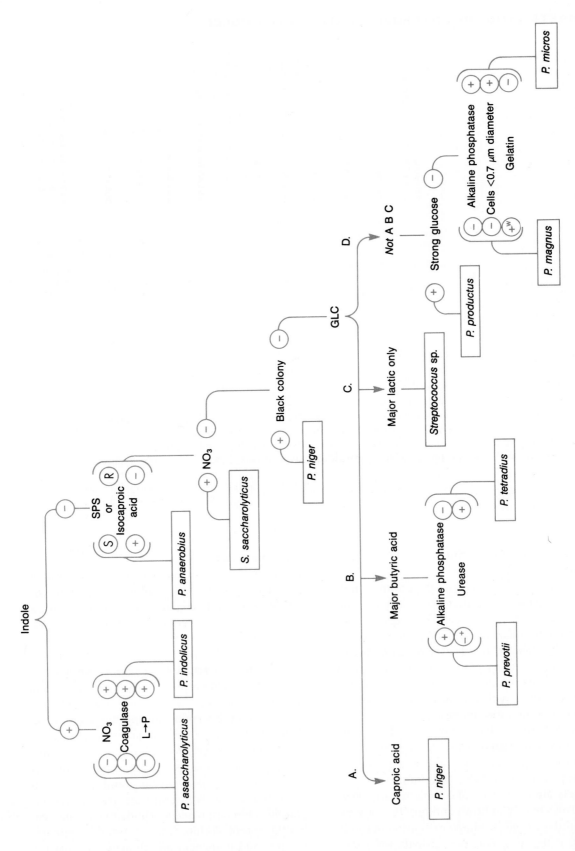

FIGURE 38.6

Flow chart for differentiation of anaerobic gram-positive cocci. Use this flow chart only as a guide; unusual strain variations may give anomalous results. *Reactions:* +, positive; −, negative; L→P, converts lactic acid to propionic acid (detected by GLC); SPS, sodium polyanetholsulfonate; *S*, sensitive; *R*, resistant; +w, most strains positive, some strains weakly positive; −+, most strains negative, some strains positive; GLC, results of end product analysis by gas-liquid chromatography.

TABLE 38.3

CHARACTERISTICS OF ANAEROBIC GRAM-NEGATIVE COCCI

SPECIES	NITRATE REDUCTION	CATALASE	GLUCOSE	FATTY ACIDS FROM PEPTONE-YEAST-GLUCOSE BROTH
Veillonella sp.	+	V	–	A p
Acidaminococcus fermentans	–	–	–	A B
Megasphaera elsdenii	–	–	+	a ib b iv v C

Reactions: See Table 38.2
PRAS carbohydrate reactions: See Table 38.2.
Fatty acid end products: See Table 38.2.

FIGURE 38.7

Veillonella. Small, gram-negative coccus. (From Sutter, V.L., Citron, D.M., Edelstein, M.A.C., and Finegold, S.M. 1985. Wadsworth anaerobic bacteriology manual, ed 4. Star Publishing, Belmont, Calif.)

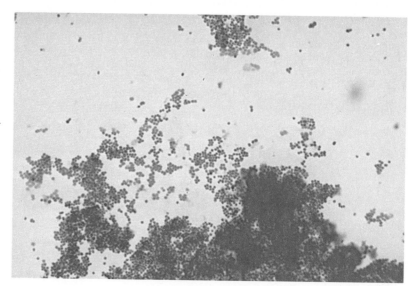

REFERENCES

1. Bourgault, A.M., Rosenblatt, J.E., and Fitzgerald, R.H. 1980. *Peptococcus magnus:* a significant human pathogen. Ann. Intern. Med. 93:244.

2. Brazier, J.S., and Riley, T.V. 1988. UV red fluorescence of *Veillonella* sp. J. Clin. Microbiol. 26:383.

3. Brook, I. 1988. Recovery of anaerobic bacteria from clinical specimens in twelve years at two military hospitals. J. Clin. Microbiol. 26:1181.

4. Coykendall, A.L., Wesbecher, P.M., and Gustafson, K.B. 1987. "Streptococcus milleri," *Streptococcus constellatus,* and *Streptococcus intermedius* are later synonyms of *Streptococcus anginosus.* Int. J. Syst. Bacteriol. 37:222.

5. Ezaki, T., Liu, S.L., Hashimoto, Y., and Yabuchi, E. 1990. *Peptostreptococcus hydrogenalis* sp. nov. from human fecal and vaginal flora. Int. J. Syst. Bacteriol. 40:305.

6. Ezaki, T., Yamamoto, N., Ninomiya, K., et al. 1983. Transfer of *Peptococcus indolicus, Peptococcus asaccharolyticus, Peptococcus prevotii,* and *Peptococcus magnus* to the genus *Peptostreptococcus* and proposal of *Peptostreptococcus tetradius* sp. nov. Int. J. Syst. Bacteriol. 33:683.

7. Goldstein, E.J.C., and Citron, D.M. 1988. Annual incidence, epidemiology, and comparative in vitro susceptibilities to cefoxitin, cefotetan, cefmetazole, and ceftizoxime of recent community acquired isolates of the *Bacteroides fragilis* group. J. Clin. Microbiol. 26:2361.

8. Goldstein, E.J.C., Citron, D.M., Cherubin, C.E., and Hillier, S.L. 1993. Comparative susceptibility of the *Bacteroides fragilis* group and other anaerobic bacteria to meropenem, imipenem, piperacillin, cefoxitin, ampicillin/sulbactam, clindamycin and metronidazole. J. Antimicrob. Chemother. 31:363.

9. Holdeman, L.V., and Moore, W.E.C. 1974. New genus *Coprococcus,* twelve new species, and emended descriptions of four previously described species of bacteria from human feces. Int. J. Syst. Bacteriol. 24:260.

10. Huss, V.A.R., Festl, H., and Schleifer, K.H. 1984. Deoxyribonucleic acid hybridization studies and deoxynucleic acid base compositions of anaerobic, gram-positive cocci. Int. J. Syst. Bacteriol. 34:95.

11. Kilpper-Balz, R., and Schleifer, K.H. 1988. Transfer of *Streptococcus morbillorum* to the genus *Gemella* as *Gemella morbillorum* comb. nov. Int. J. Syst. Bacteriol. 38:442.

12. Krepel, C.J., Gohr, C.M., Edmiston, C.E., and Farmer, S.G. 1991. Anaerobic pathogenesis: collagenase production by *Peptostreptococcus magnus* and its relationship to site of infection. J. Infect Dis. 163:1148.

13. Mays, T.D., Holdeman, L.V., Moore, W.E.C., et al. 1982. Taxonomy of the genus *Veillonella* Prevot. Int. J. Syst. Bacteriol. 32:28.

14. Murdoch, D.A., Mitchelmore, I.J., Nash, R.A., et al. 1988. Preformed enzyme profiles of reference strains of gram-positive anaerobic cocci. J. Med. Microbiol. 27:65.

15. Ohm-Smith, M.J., and Sweet, R.L. 1987. In vitro activity of cefmetazole, cefotetan, amoxicillin-clavulanic acid, and other antimicrobial agents against anaerobic bacteria from endometrial cultures of women with pelvic infections. Antimicrob. Agents Chemother. 31:1434.

16. Phelps, R., and Jacobs, R.A. 1985. Purulent pericarditis and mediastinitis due to *Peptostreptococcus magnus*. J.A.M.A. 254:947.

17. Pien, F.D., Thompson, R.L., and Martin, W.J. 1972. Clinical and bacteriologic studies of anaerobic gram-positive cocci. Mayo Clin. Proc. 47:251.

18. Steffen, E.K., and Hentges, D.J. 1981. Hydrolytic enzymes of anaerobic bacteria isolated from human infections. J. Clin. Microbiol. 14:153.

19. Sutter, V.L., and Finegold, S.M. 1976. Susceptibility of anaerobic bacteria to 23 antimicrobial agents. Antimicrob. Agents Chemother. 10:736.

20. Weizenegger, W.L., Kilpper-Balz, R., and Schleifer, K.H. 1988. Phylogenetic relationships of anaerobic streptococci. Int. J. Syst. Bacteriol. 38:15.

BIBLIOGRAPHY

Holdeman, L.V., Cato, E.P., and Moore, W.E.C. 1977. Anaerobe laboratory manual, ed 4. VPI Laboratory, Virginia Polytechnic Institute and State University, Blacksburg, Va.

Moore, L.V.H., Johnson, J.J., and Moore, W.E.C. 1986. Genus *Peptococcus;* genus *Peptostreptococcus.* In Sneath, P.H.A., Mair, N.S., Sharpe, M.E., and Holt, J.G., editors: Bergey's manual of systematic bacteriology. p. 1082. Williams & Wilkins, Baltimore.

Summanen, P., Baron, E.J., Citron, D.M., et al. 1993. Wadsworth anaerobic bacteriology manual, ed 5. Star Publishing, Belmont, Calif.

39

CHLAMYDIA, MYCOPLASMA, AND RICKETTSIA

Organisms of the genera *Chlamydia, Mycoplasma,* and *Rickettsia* are prokaryotes that differ from most other bacteria in their very small size, their unusual cell wall structure, and, in the case of *Rickettsia* and *Chlamydia,* their obligate intracellular parasitism. *Chlamydia* and *Mycoplasma* genera each include species that cause pneumonia and species that cause venereal disease. *Mycoplasma hominis* is also being recognized as a cause of extragenital infection that can be very difficult to cure. *Rickettsia* sp. typically cause generalized illness with fever and rash. The fastidious nature of these organisms is at least one reason for the parallel development of similar disease syndromes; the organisms must adhere to and multiply within a protective environment, and mucous membranes provide favorable conditions at the first site these pathogens reach in a hospitable host. Isolation of pathogenic chlamydiae and mycoplasma is a possiblity for hospital clinical laboratories. Rickettsiae are extremely difficult to cultivate and too hazardous to work with for routine isolation, so most diagnosis is done by serologic tests. These three genera are treated separately in this chapter.

CHLAMYDIAE

Members of the family Chlamydiaceae (order Chlamydiales) are obligate intracellular bacteria that were once regarded as viruses. *Chlamydia* is divided currently into three distinct species: *C. trachomatis, C. pneumoniae* (TWAR), and *C. psittaci.*

The species share less than 10% homology between each other.[2]

C. psittaci is the agent of pneumonia and systemic disease associated with bird contact; the disease is variously known as **psittacosis** or **ornithosis.** Although many species variants occur, they have not been separated further. *C. pneumoniae* is also implicated as an important cause of lower respiratory tract infections.[5] It is thought to be exclusively a human pathogen with no animal reservoir.

C. trachomatis is associated with four primary syndromes:

1. Endemic trachoma (repetitive conjunctival infection ultimately leading to blindness and associated with types A to C)
2. Sexually transmitted chlamydial disease (including nongonococcal urethritis and cervicitis and associated with types D to K)
3. Inclusion conjunctivitis (caused by the sexually transmitted strains and not leading to blindness)
4. Lymphogranuloma venereum (**LGV,** strains L1 to L3).

In some patients, *C. trachomatis* may infect the entire genital tract and may cause pneumonia in infants and transplant recipients. Table 39.1 outlines some biological differences among the *Chlamydia* sp. Although the chlamydiae possess a cell wall similar to that of gram-negative bacteria, their cell wall is distinctive in lacking a peptidoglycan layer. They do produce an endotoxin-like

TABLE 39.1

DIFFERENTIATION AMONG *CHLAMYDIA* SP.

PROPERTY	C. TRACHOMATIS	C. PSITTACI	C. PNEUMONIAE
Host range	Humans, mice	Birds, humans, lower mammals	Humans
Elementary body morphology	Round	Round	Pear-shaped
Inclusion morphology	Round, vacuolar	Variable, dense	Round, dense
Glycogen inclusions	Yes	No	No
Susceptibility to sulfonamides	Yes	No	No
Deoxyribonucleic acid (DNA) homology	10%	10%	10%
Plasmid DNA	Yes	Yes	No

lipopolysaccharide antigen. Unlike any other bacteria, chlamydiae undergo an unusual developmental cycle (Figure 39.1). The infectious particles, rigidly enveloped structures approximately 0.3 µm in diameter called **elementary bodies,** attach to the surface of a susceptible host cell. They induce their own phagocytosis through a process called endocytosis and are enclosed within the cell interior in a cytoplasmic vesicle, a *phagosome.* The organisms are able to prevent lysosomal fusion, which prevents the cell from emptying lysosomal contents into the phagosome; thus the chlamydiae are able to escape being killed and survive intracellularly. After the initial stage, elementary bodies reorganize to a more metabolically active form, the **reticulate body.** Reticulate bodies multiply by binary fission until approximately 18 to 24 hours after infection, at which point many reticulate bodies reorganize again into infectious elementary bodies. Newly formed reticulate (replicative) and elementary (infectious) bodies are still enclosed within the phagosomic intracellular vesicle, which is visible microscopically as an **inclusion** after it has been highlighted by iodine, Giemsa, or a fluorescent antibody stain. The *C. trachomatis* inclusions, consisting of a collection of organisms, can be stained with iodine, since they consist primarily of glycogen and exist in a rather compact form, in contrast to the diffuse, Giemsa-staining inclusions of *C. psittaci.* At 48 to 72 hours after initial attachment, the cell ruptures to release many infectious elementary bodies. The diseases caused by each of the species and the laboratory diagnosis of those diseases differ in several important ways. Each entity, *C. psittaci, C. trachomatis,* and *C. pneumoniae,* is discussed separately.

Tetracycline, erythromycin, and sulfonamides are usually effective for therapy of chlamydial infections. Some of the new fluoro-quinolone antimicrobial agents, such as ofloxacin, may also be effective therapy.

PSITTACOSIS

C. psittaci is an endemic pathogen of all bird species. Psittacine birds (e.g. parrots, parakeets) are a major reservoir for human disease, but outbreaks have occurred among turkey processing workers and pigeon aficionados. The birds may show diarrheal illness or may be asymptomatic. Humans acquire the disease by inhalation of aerosols. The organisms are deposited in the alveoli; some are ingested by alveolar macrophages and then carried to regional lymph nodes. From there they are disseminated systemically, growing within cells of the reticuloendothelial system. Clinical findings associated with this infection are diverse and include pneumonia, severe headache, changes in mentation, and hepatosplenomegaly.

Diagnosis of psittacosis is almost always by serologic means. Because of hazards associated with working with the agent, only laboratories with biosafety level 3 biohazard containment facilities can culture *C. psittaci* safely. State health departments take an active role in consulting with clinicians about possible cases. Complement fixation has been the most frequently used serologic test to detect psittacosis infection. A more recent test, indirect microimmunofluorescence, is more sensitive but difficult to perform. For this new test, different strains of *C. psittaci* are grown in hens' egg yolk sac cultures. The cultures, rich with elementary bodies, are diluted and suspended in buffer. Next, a dot of each antigen suspension is placed in a geometrical array on the surface of a glass slide. Several identical arrays, each containing one dot of every antigen, are prepared on a single slide. These slides are fixed and frozen until use. At the time of the test, serial dilutions of a patient's serum are placed over the antigen arrays.

FIGURE 39.1

Life cycle of *Chlamydia. EB,* Elementary body. *RB,* Reticulate body. (From Boyd, R.F. 1984. General microbiology. Mosby, St. Louis.)

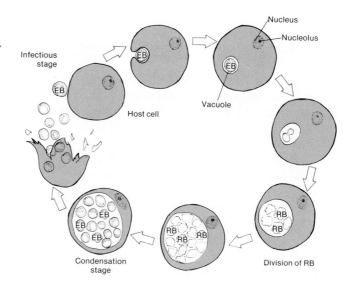

After incubation and washing steps, fluorescein-conjugated antihuman immunoglobulin (either IgG or IgM) is overlaid on the slide. The slide is finally read for fluorescence of particular antigen dots using a microscope. Either a fourfold rise in titer between acute and convalescent serum samples or a single IgM titer of 1:32 or greater in a patient with an appropriate illness is considered diagnostic of an infection.

LYMPHOGRANULOMA VENEREUM

LGV is a sexually transmitted disease that is unusual in Europe and North America but relatively frequent in Africa, Asia, and South America. The disease is characterized by a primary genital lesion at the initial infection site lasting a short time. This lesion is often small and may be unrecognized, especially by female patients. The second stage, acute lymphadenitis, often involves the inguinal lymph nodes, causing them to enlarge and become matted together, forming a large area of groin swelling or bubo. During this stage, infection may become systemic and cause fever or may spread locally, causing granulomatous proctitis. In a few patients (more women than men) the disease progresses to a chronic third stage, causing the development of genital hyperplasia, rectal fistulas, rectal stricture, draining sinuses, and other manifestations. Occasionally, chronic ulcerative or infiltrative lesions of the penis and scrotum develop.

Diagnosis of LGV is established by the isolation of an LGV strain from a bubo or other infected site. However, organism recovery rates of only 24% to 30% are reported. An intradermal skin test of LGV antigen, the Frei test, lacks sensitivity in early LGV and lacks specificity later, since the Frei antigen is only a genus-specific antigen. Moreover, the Frei test can remain positive for many years, limiting its usefulness. Again, because a genus-specific antigen is used, antibody detection by complement fixation is nonspecific. Despite its shortcomings, a complement fixation titer of 1:16 is presumptive evidence of infection. Other serologic tests, such as microimmunofluorescence and neutralization, are still performed only in specialized laboratories.

CHLAMYDIA TRACHOMATIS AS A CAUSE OF NON-LGV DISEASE

This has surpassed gonococcal infections in frequency as a cause of sexually transmitted disease (STD) in the United States. In addition to causing illness similar to that of gonococci, including urethritis, cervicitis, bartholinitis, proctitis, salpingitis, inclusion conjunctivitis (ophthalmia neonatorum), perihepatitis (Fitz-Hugh-Curtis syndrome), epididymitis, and endocarditis, it is a significant cause of pneumonia in neonates. Chlamydiae have been found as one cause of the acute urethral syndrome in women, characterized by symptoms of a urinary tract infection with pyuria but few or no usual bacteria.[16] Both chlamydiae and gonococci are major causes of pelvic inflammatory disease (PID), contributing significantly to the rising rate of infertility and ectopic pregnancies in young women. After only one episode of PID, as many as 10% of women may become infertile because of tubal occlusion. The risk increases dramatically with each additional episode. Chlamydial salpingitis leading to infertility may be asymptomatic. Chlamydial infection can be transmitted to an infant from an infected mother during delivery. Approximately

0.5% of live births are complicated by neonatal chlamydial inclusion conjunctivitis or pneumonia.

Many genital chlamydial infections in both sexes are asymptomatic. It is estimated that as many as 22% of women and 7% of men are asymptomatic carriers. The highest-risk group is patients with gonococcal infection; 20% to 35% of men and 50% of women with gonorrhea are likely to harbor *C. trachomatis* concomitantly. Furthermore, in the United States, 60% of cases of nongonococcal urethritis (NGU) are caused by chlamydiae. The incidence of this disease is even higher in Scandinavian countries. With the diagnostic tests outlined next, we expect that a much clearer epidemiological picture and ultimately better control of this widespread disease are now attainable goals.

Diagnosis of *C. trachomatis* can be achieved by cytology, culture, direct antigen detection, and serologic testing. The organism can be recovered from infected cells of the urethra, cervix, conjunctiva, nasopharynx, and rectum and from material aspirated from the fallopian tubes and epididymis. Cytological examination of cell scrapings for the presence of inclusions was the first method used for diagnosis of *C. trachomatis*; it is still used occasionally for rapid diagnosis of inclusion conjunctivitis in newborns. However, this method is insensitive compared with culture or direct immunofluorescence, the methods of choice.[15]

Isolation of *C. trachomatis* in cell culture remains the most sensitive and specific diagnostic method. Cycloheximide-treated McCoy cells have been shown to produce consistently higher inclusion counts than other cell lines. Centrifugation of the specimen onto the cell monolayer (usually growing on a coverslip in the bottom of a vial (the "shell vial") presumably facilitates adherence of elementary bodies. After 48 to 72 hours' incubation, monolayers are stained with iodine or an immunofluorescent stain and examined microscopically for inclusions. Procedure 39.1 describes a method for isolation of chlamydiae. Despite the high sensitivity of culture, technical problems adversely affect the recovery of *C. trachomatis*, and relatively few laboratories offer cell culture diagnosis.

To circumvent the shortcomings of cell culture, antigen detection methods are commercially available. Direct fluorescent antibody **(DFA)** staining methods employ fluorescein-isothiocyanate–conjugated monoclonal antibodies to either outer membrane proteins or lipopolysaccharides of *C. trachomatis* for the detection of elementary bodies in smears of clinical material (Figure 39.2). The MicroTrak DFA (Syva Co., San Jose, Calif.) is one good commercial system. DFAs achieve at least a 90% sensitivity compared with culture in many evaluations of symptomatic men and high-risk women (sex partners of *Chlamydia*-positive men or women attending STD clinics).[9] In populations with asymptomatic infections, the sensitivity of DFA is decreased, presumably because fewer inclusion-forming units are present in patient specimens.[1,19]

Chlamydial antigen can also be detected by enzyme-linked immunosorbent assays (ELISA). Several commercially available kits are marketed,

FIGURE 39.2

Appearance of fluorescein-conjugated, monoclonal antibody–stained elementary bodies in direct smear of urethral cell scraping from patient with chlamydial urethritis. (Courtesy Syva Co., San Jose, Calif.)

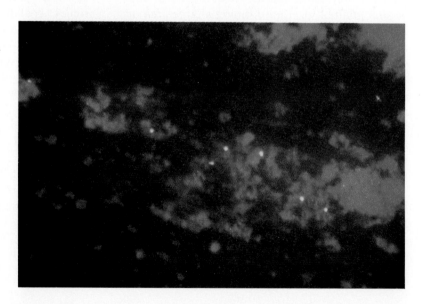

PROCEDURE 39.1

CELL CULTURE METHOD FOR ISOLATION OF CHLAMYDIAE

METHOD

1. Collect a swab or tissue in a vial of sucrose transport medium containing antibiotics with two to four sterile glass beads (Chapter 20). Specimens should be kept refrigerated at all times before transport to the laboratory. If the culture is not inoculated within 24 hours of collection, the specimen should be frozen at −70° C.

2. Vortex the specimen vigorously and remove the swab. The glass beads cause rupture of infected cells and release of elementary bodies. Tissue should be minced and ground to make a cell suspension at a 1:10 dilution using transport medium.

3. Use aseptic technique and aspirate the cell culture

medium above the McCoy cell monolayer in the shell vial (commercially available), and add 0.2 ml of the patient specimen to the vial.

4. Centrifuge the vial for at least 1 hour at 2500 to 3000 × *g* at ambient temperature in a temperature-controlled centrifuge. Do not allow the temperature to increase past 40° C.

5. Aseptically remove the inoculum and add 1 ml fresh maintenance medium containing 1 µg/ml cycloheximide.

6. Recap the vials tightly, and incubate at 35° C for 48 to 72 hours.

7. Remove medium from the vials, fix with methanol or ethanol (according to

manufacturer's instructions), and stain with iodine or fluorescein-conjugated chlamydial antibody.

8. Examine slides at 200 to 400 × magnification for dark-brown (iodine stain; Figure 39.3) or apple-green (fluorescent stain) intracytoplasmic inclusions (Figure 39.4).

QUALITY CONTROL
Laboratories performing chlamydial culture should subscribe to a commercial proficiency testing service to verify their procedures.

FIGURE 39.3
Iodine-stained inclusions in McCoy cell monolayer infected with *Chlamydia trachomatis.* (Courtesy of Ellena Peterson, University of California, Irvine.)

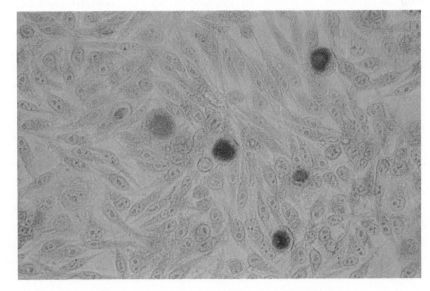

FIGURE 39.4
Fluorescent-antibody-stained inclusions in McCoy cell monolayer infected with *Chlamydia trachomatis*. (Courtesy of Ellena Peterson, University of California, Irvine.)

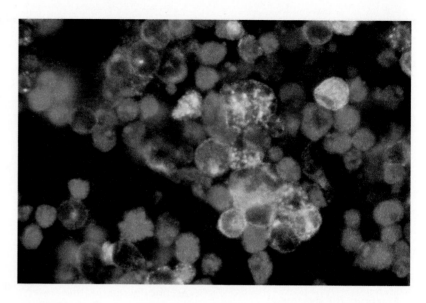

and the Chlamydiazyme assay (Abbott Laboratories, N. Chicago, Ill.) has been extensively evaluated. The median sensitivity of this ELISA is approximately 79% in men with urethritis and 90% in high-risk women. However, only 50% sensitivity has been seen in asymptomatic men. In women with intermediate prevalence of disease (6% to 12%) the sensitivity is approximately 85%.[6,7,18,20] A direct antigen detection method using a ribonucleic acid (RNA)–directed deoxyribonucleic acid (DNA) probe conjugated to a chemiluminescent marker (PACE, Gen-Probe, San Diego, Calif.) is also available. This test is at least as sensitive as DFA.[8a] In low-prevalence populations (<5%), however, culture remains the test of choice.

Serologic testing has limited value for diagnosis of urogenital infections in adults. Most adults with chlamydial infection have had a previous exposure to *C. trachomatis* and are therefore seropositive. Antibodies to a genus-specific antigen can be detected by complement fixation. For type-specific antibodies of *C. trachomatis*, the microimmunofluorescence assay, a tedious and difficult test, is used. A high titer of IgM (≥128) suggests a recent infection; however, not all patients produce IgM.[10] Negative serology can reliably exclude chlamydial infection. Detection of *C. trachomatis*–specific IgM is useful in diagnosis of neonatal infections.

CHLAMYDIA PNEUMONIAE (TWAR)

The TWAR strain of *Chlamydia* was first isolated from the conjunctiva of a child in Taiwan. It was initially considered to be a psittacosis strain because the inclusions produced in cell culture resembled those of *C. psittaci*. Subsequently, the Taiwan isolate (TW-183) was shown to be serologically related to a pharyngeal isolate (AR-39) isolated from a college student in the United States, and thus the new strain was called "TWAR," an acronym for TW and AR (acute respiratory). Only one serotype of the new species, *C. pneumoniae*, has been identified. *C. pneumoniae* has been associated with pneumonia, bronchitis, pharyngitis, sinusitis, and a flulike illness. Infection in young adults is usually mild to moderate; the microbiological differential diagnosis primarily includes *Mycoplasma pneumoniae*. Severe pneumonia may occur in elderly or respiratory-compromised patients. Treatment with tetracycline and erythromycin has been successful. Diagnosis of *C. pneumoniae* infection can be achieved by cell culture isolation and serology. A cell culture procedure similar to that used for *C. trachomatis* but substituting the more sensitive HeLa cell line for McCoy cells has been used. Complement fixation using a genus-specific antigen has been used, but it is not specific for *C. pneumoniae*. A microimmunofluorescence test using *C. pneumoniae* elementary bodies as antigen is more reliable but available only in specialized laboratories. A fourfold rise in either IgG or IgM is diagnostic, and a single IgM titer of 16 or greater or an IgG titer of 512 or greater is suggestive of recent infection. Commercial kits for diagnosing *C. pneumoniae* are not available yet because of the difficulty in developing species-specific detection methods.

MYCOPLASMAS

Members of the class Mollicutes, these organisms are found throughout the animal and plant king-

TABLE 39.2

MYCOPLASMAS RECOVERED FROM HUMANS

GENUS AND SPECIES	HEMADSORPTION	HYDROGEN PEROXIDE PRODUCED	GLUCOSE FERMENTED	HYDROLYSIS OF ARGININE	HYDROLYSIS OF UREA
Mycoplasma					
pneumoniae	+	+	+	–	–
salivarium	–	–	–	+	–
orale	–	–	–	+	–
buccale	–	–	–	+	–
faveium	–	–	–	+	–
lipophilum	–	–	–	+	–
primatum	–	–	–	+	–
fermentans	–	–	+	+	–
hominis	–	–	–	+	–
genitalium	+	+	+	–	–
Ureaplasma					
urealyticum	+/–		–	–	+
Acholeplasma					
laidlawii	–		+	–	–

dom as free-living saprophytes and parasites. First recognized from a case of pleuropneumonia in a cow, the organism was designated "pleuropneumonia-like organism," or PPLO. That term is still used today, particularly for certain mycoplasma media. The smallest organisms able to survive extracellularly, they are highly pleomorphic because they lack a cell wall, which also renders them completely resistant to β-lactam and other cell wall–active drugs. Although some species are normal human respiratory tract flora, *M. pneumoniae* is a major cause of respiratory disease (primary atypical pneumonia, sometimes called "walking pneumonia"). *M. pneumoniae* seldom causes invasive disease.

M. hominis, M. genitalium, and *Ureaplasma urealyticum* are important colonizers (and possibly pathogens) of the human genital tract. *M. hominis* has been associated with sepsis (particularly postpartum), arthritis, and recently with sternal wound infections.[4,17] *U. urealyticum* contributes to chorioamnionitis, perinatal disorders, and premature birth; its association with acute urethral syndrome in women is fairly strong.[2a,8,16] None of the other species shown in Table 39.2 is definitely pathogenic for humans. These organisms colonize the mucosa of the respiratory and genital tracts of humans, growing extracellularly. By an unknown mechanism, infection with mycoplasma causes injury to mucosal cells. The tissue damage leads to impairment of ciliary function and thus impairs the ability of the host to clear the microbes from the site of infection.

MYCOPLASMA PNEUMONIAE

It is estimated that *M. pneumoniae* may cause as many as 50% of all community-acquired pneumonias during noninfluenza periods. The disease is primarily one of children and young adults, spreading from person to person via respiratory tract secretions. The most common syndrome produced is tracheobronchitis, but pneumonia, pharyngitis, rhinitis, ear infection (including bullous myringitis), meningitis, myocarditis, and pericarditis also occur. Very few laboratories perform mycoplasma cultures. Accurate, rapid diagnosis is highly desired, since penicillin and other β-lactam agents are ineffective treatment; erythromycin and tetracycline are the antibiotics of choice. Laboratory diagnosis is usually made serologically. Nonspecific production of cold agglutinins occurs in approximately half of patients with atypical pneumonia. The most widely used serologic tests today are complement fixation (CF) and ELISA, although newly developed indirect fluorescent antibody tests are being used with some success. Although a fourfold rise in titer from acute to convalescent serum is better evidence, CF titers of 1:128 or greater in single serum specimens suggest recent infection. The RNA-directed DNA probe test (Gen-Probe) has been evaluated for detection of *M. pneumoniae* genes in sputum specimens, with sensitivities similar to culture. However, the isotopic based test was withdrawn in 1992, with a new, nonisotope-based test under development. Whether culture

PROCEDURE 39.2

ISOLATION OF *MYCOPLASMA PNEUMONIAE*

1. Place 0.1 to 0.2 ml of liquid specimen, or dip and twirl a specimen received on a swab in a vial of biphasic SP-4 culture medium prepared as follows:

   Mix the ingredients below for the media base:
   Mycoplasma broth base (BBL), 3.5 g
   Tryptone (Difco), 10 g
   Bacto-Peptone (Difco), 5 g
   Dextrose (50% aqueous solution, filter sterilized), 10 ml
   Distilled water, 615 ml
   Stir the solids into boiling water to dissolve them, and adjust the pH to 7.5. Autoclave according to the manufacturer's instructions. Cool to 56° C before adding the following supplements.

   Add the following supplements for each 625 ml of base to make a final volume of 1 L:
   CMRL 1066 tissue culture medium with glutamine, 10× (GIBCO), 50 ml
   Aqueous yeast extract (prepared as described in New York City Medium), 35 ml
   Yeastolate (Difco; 2% solution), 100 ml
   Fetal bovine serum (heat inactivated at 56° C for 30 minutes), 170 ml
   Penicillin G sodium, 1million U, or Ampicillin, 1 mg/ml
   Amphotericin B, 0.5 g
   Polymyxin B, 500,000 U
   To prepare the agar necessary for making biphasic media, add 8.5 g Noble agar (Difco) to the basal medium ingredients before adding the supplements.

   Dispense 1 ml of SP-4 agar aseptically into the bottom of sterile 4 ml screw-capped vials. Allow the agar to set and dispense 2 ml of SP-4 broth above the agar layer in each vial. Seal caps tightly, and store at − 20° C.

   After expressing as much fluid as possible from the swab, remove the swab to prevent contamination.

2. Seal the vial tightly and incubate in air at 35° C for up to 3 weeks.

3. Inspect the vial daily. During the first 5 days, a change in pH, indicated by a color shift from orange to yellow or violet, or increased turbidity is a sign that the culture is contaminated and should be discarded.

4. If either a slight acid pH shift (yellow color) with no increase in turbidity or no change occurs after 7 days' incubation, subculture several drops of the broth culture to Edward-Hayflick agar. Continue to incubate the original broth.

5. If broth that exhibited no changes at 7 days shows a slight acid pH shift at any time, subculture to agar as above. Broths that show no change at 3 weeks are blindly subcultured to agar.

6. Incubate the agar plates in a very moist atmosphere with 5% CO_2 at 35° C for 7 days.

7. Observe the agar surface under 40× magnification after 5 days for colonies, which appear as spherical, grainy, yellowish forms, embedded in the agar, with a thin outer layer ("fried egg,"similar to those shown in Figure 39.5).

8. Definitive identification of *M. pneumoniae* is accomplished by overlaying Edward-Hayflick agar plates showing suspicious colonies with 5% sheep or guinea pig erythrocytes in 1% agar prepared in physiological saline (0.85% NaCl) instead of water. The 1% agar is melted and cooled to 50° C, the blood cells are added, and a thin layer is poured over the original agar surface.

9. Reincubate the plate for 24 hours, and observe for beta hemolysis around colonies of *M. pneumoniae* caused by production of hydrogen peroxide (Figure 39.6). Additional incubation at room temperature overnight enhances the hemolysis. No other species of *Mycoplasma* produces this reaction.

Modified from Clyde, W.A., Jr., Kenny G.E., and Schachter, J. 1984. Cumitech 19: Laboratory diagnosis of chlamydial and mycoplasmal infections. Drew, M.L., coordinating editor. American Society for Microbiology, Washington D.C.

FIGURE 39.5
Colonies of *Mycoplasma hominis* on A7 agar.

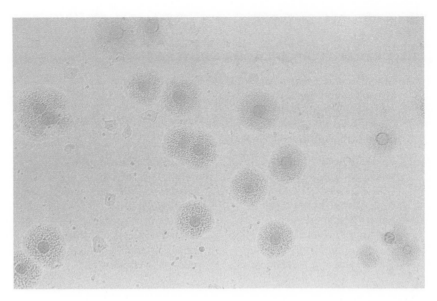

FIGURE 39.6
Mycoplasma pneumoniae colonies with sheep red blood cell overlay.

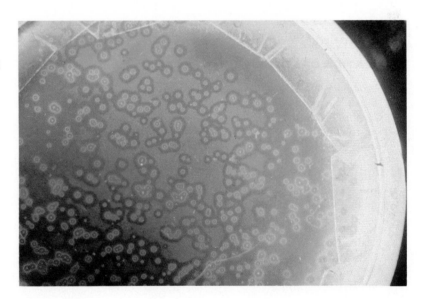

should be the "gold standard" has not been determined, since mycoplasmas can persist in respiratory secretions long after signs of infection have disappeared.

Collection and transport of respiratory tract specimens, including nasopharyngeal and throat swabs and washings, tracheal aspirates, sputa, and lung biopsy specimens, are detailed in Chapters 17 and 22. Procedure 39.2 is a method for isolation of *M. pneumoniae.*

Other methods are available for identifying *Mycoplasma* colonies isolated by Procedure 39.2, including hemadsorption, immunofluorescent stain of colonies, and inhibition by specific antisera impregnated onto filter paper disks.

GENITAL MYCOPLASMAS

Of the species of *Mycoplasma* found in the human genital tract, only *Ureaplasma urealyticum* and *M. hominis* are established pathogens. Under what circumstances they cause disease is still unclear, since many adults are asymptomatically colonized with both species. Ureaplasmas are a cause of nongonococcal urethritis in males, but they also colonize sexually active males without urethritis.[12] In females, ureaplasmas do not seem to cause disease, but chorioamnionitis from these organisms contributes to low birth weight of newborns.[2a] *M. hominis* does not seem to cause illness often in males, but it is a cause of postpartum fevers in women. Additionally, strains of the or-

---

PROCEDURE 39.3

ISOLATION OF *UREAPLASMA UREALYTICUM*

1. Inoculate one *Ureaplasma* agar plate and one *Ureaplasma* broth each with 0.1 ml specimen from transport medium.

2. Incubate broth in tightly sealed test tubes for 5 days. Observe twice daily for a color change in the broth to red with no increase in turbidity. If color change occurs, immediately transfer one loopful to a *Ureaplasma* agar plate, and streak for isolated colonies.

3. Agar plates are incubated in a candle jar or optimally in an anaerobic jar at 35° C. Colonies appear on agar within 48 hours. Plates are inspected in the same way as described for *M. pneumoniae* (Procedure 39.2). *Ureaplasma* colonies appear as small, granular, yellowish spheres that simulate the yolk of fried egg.

4. To identify definitively colonies on *Ureaplasma* agar after 48 hours' incubation, pour a solution of 1% urea and 0.8% $MnCl_2$ in distilled water over the agar surface. *U. urealyticum* stains dark brown because of production of urease (Figure 39.8).

PROCEDURE 39.4

ISOLATION OF *MYCOPLASMA HOMINIS*

1. Inoculate one *M. hominis* agar plate and two *M. hominis* broth tubes, one broth containing phenol red indicator and one without the possibly inhibitory phenol red, each with 0.1 ml specimen from transport media.

2. Incubate broths in tightly sealed test tubes for 5 days. If the phenol red–containing broth changes color to red or violet, both broths are subcultured to *M. hominis* agar. After 48 hours of incubation, transfer 0.1 ml or a loopful of broth from tubes that exhibited no change or only a slight increase in turbidity to *M. hominis* agar, and streak for isolated colonies.

3. *M. hominis* agar plates are incubated in the same manner as *Ureaplasma* cultures. Plates should be observed daily for up to 5 days for colonies.

ALTERNATE METHOD

1. Inoculate specimen onto a prereduced colistin and nalidixic acid (CNA) sheep blood agar or anaerobic blood plate, and incubate anaerobically for 48 to 72 hours.

2. Examine for pinpoint colonies that show no bacteria on Gram stain.

3. Streak suspicious colonies to *M. hominis* agar and incubate as in step 3 above.

---

ganism can be isolated significantly more often from patients with PID than from control subjects. *M. hominis* also causes wound infections, arthritis, and neonatal meningitis and bacteremia.

Specimens are collected on swabs, as described in Chapter 20. Transport media such as modified Stuart's, sucrose phosphate transport medium (2SP, used for chlamydiae), or trypticase soy broth (BBL Microbiology Systems) with 0.5% albumin plus 25 U/ml of nystatin, 100 μg/ml of vancomycin, and 50 μg/ml of streptomycin will preserve viability for up to 48 hours at refrigerator temperatures. Specimens that must be held longer should be frozen at −70° C in a medium containing protein. Culture media may also be used for transport. Culture methods described here (Procedures 39.3 and 39.4) are modified from references 3 and 11.

Shepard's A7-B agar, a satisfactory medium for cultivation of genital mycoplasmas, is available from commercial sources, as is a biphasic medium (Hana Biologics).[14] Calcium chloride and urea incorporated into these media impart a dark-brown color to *Ureaplasma* colonies (Figure 39.7) and re-

FIGURE 39.7
Colonies of *Ureaplasma urealyticum* growing on Shephard A7-B agar visualized under 100× magnification.

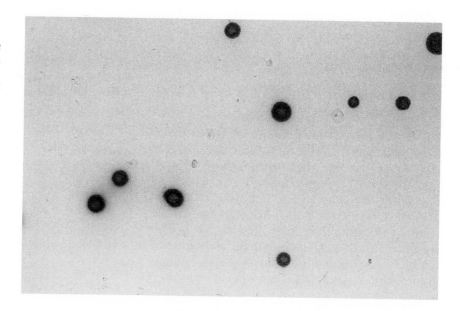

FIGURE 39.8
Colonies of *Ureaplasma urealyticum* after 48 hours' incubation, overlaid with urea–manganese chloride reagent.

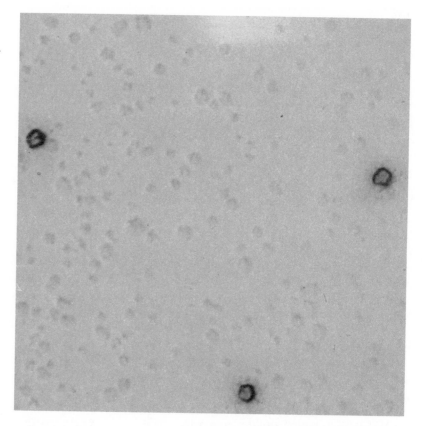

move the need to stain the colonies with magnesium chloride–urea reagent, described in Procedure 39.3. *M. hominis* also grows anaerobically on culture media used for detection of anaerobic bacteria and on sheep blood agar incubated for 72 hours in carbon dioxide (CO_2).

Although serologic tests such as indirect

hemagglutination and metabolism inhibition for genital mycoplasmas are available, they are rarely used. ELISA tests for mycoplasmal antigen or RNA-directed DNA probes (PACE system, Gen-Probe) of sufficient specificity for diagnosis may be available in the future. Nucleic acid dot-blot hybridization methods have been used for detection of *M. genitalium*, which is difficult to grow in vitro.

RICKETTSIA, COXIELLA, AND EHRLICHIA

The Rickettsiaceae are a group of organisms that infect wild animals, with humans acting as accidental hosts in most cases. Most of these organisms are passed between animals by an insect vector. Since 1986, human infection from *Ehrlichia canis* and *E. sennetsu*, *Rickettsia*-like organisms previously thought to infect only animals, has been recognized.[13] All rickettsiae multiply only intracellularly. *Coxiella burnetii*, which is similar to but distinct from the rickettsiae, can survive extracellularly. The rickettsiae cannot even survive outside of their animal hosts or insect vectors. The diseases caused in humans are usually characterized by fever, headache, and rash. With the exception of Q fever, caused by *C. burnetii*, the rickettsiae multiply in the endothelial cells of blood vessels, causing cell injury and death. Damage to cells results from vascular changes and the influx of inflammatory cells that serve to magnify the host's immune response and increase overall morbidity.

Rickettsiae are pleomorphic gram-negative coccobacilli. The organisms multiply by binary fission in the cytoplasm of host cells, which are lysed during release of mature rickettsiae. The rickettsiae, but not *C. burnetii*, infect humans following the bite of an infected arthropod vector. The agents, their vectors, and diseases are shown in Table 39.3. *C. burnetii* is more hardy than the others, resisting desiccation and sunlight. It infects humans by the aerosol route. In contrast to rickettsiae, *Coxiella* is passively phagocytized by host cells and multiplies within vacuoles. The other agents, after being deposited directly into the bloodstream through the bite of a louse, ixodid tick, flea, or mite, induce host cells to engulf them, after which they quickly escape the vacuole to become free in the cytoplasm. *Coxiella* is inhaled into the alveoli, picked up by macrophages, and carried to lymph nodes, from which it disseminates into the bloodstream. Granulomatous hepatitis and endocarditis are possible sequelae.

Early diagnosis of Rickettsiae is usually made on clinical grounds based on the symptoms of fever, rash, and exposure to insect vectors (ticks). The characteristic spread of the rash from the extremities to the trunk helps to distinguish Rocky Mountain spotted fever (RMSF) from meningococcemia. Ehrlichiosis resembles "rashless" RMSF,

TABLE 39.3

VECTORS OF SELECTED DISEASES CAUSED BY RICKETTSIA, COXIELLA, AND EHRLICHIA

AGENT	DISEASE	VECTOR
Rickettsia akari	Rickettsialpox	Mites
R. conorii	Boutonneuse fever	Ticks
R. prowazekii	Epidemic typhus Brill-Zinsser disease	Lice
R. rickettsii	Rocky Mountain spotted fever	Ticks
R. tsutsugamushi	Scrub typhus	Mites
R. typhi	Murine (endemic) typhus	Fleas
Rochalimaea quintana	Trench fever	Lice
Coxiella burnetii	Q fever	Ticks, aerosols
Ehrlichia canis	Ehrlichiosis	Ticks
E. sennetsu	Ehrlichiosis	Ticks

since only 20% of patients infected with *Ehrlichia* develop rash, whereas almost 90% of RMSF patients exhibit a rash.

LABORATORY IDENTIFICATION

Biopsy specimens of skin tissue from the rash of RMSF can be stained directly with a specific immunofluorescent reagent. This test, available at certain state health laboratories and the Centers for Disease Control (CDC), provides a diagnosis as early as 3 to 4 days after symptoms have appeared. The Giménez stain is also used for examination of clinical material. Although the rickettsiae can be cultured in embryonated eggs and in tissue culture, the risk of laboratory-acquired infection is extremely high, limiting availability of culture to a few specialized laboratories. Diagnosis of rickettsial disease and ehrlichiosis is primarily accomplished serologically. The least specific but most widely used test in the United States is the Weil-Felix reaction, the fortuitous agglutination of certain strains of *Proteus vulgaris* by serum from patients with rickettsial disease. Although this test has been clinically useful, false-positive tests are a continuing problem. Table 39.4 outlines the positive reactions associated with each disease. Commercial suppliers produce kits that contain all necessary antigens and human antirickettsial control antisera for performance of the Weil-Felix test. This test is only presumptive, however, and a more specific serologic test, such as CF or direct immunofluorescence, must be used for confirma-

TABLE 39.4

REACTION OF *PROTEUS* STRAINS IN WEIL-FELIX TEST

DISEASE	OX-19	OX-2	OX-K
Brill-Zinsser	V	V	–
Epidemic typhus	+	V	–
Murine typhus	+	V	–
Rickettsialpox	–	–	–
Rocky Mountain spotted fever	+	+	–
Scrub typhus	–	–	V
Q fever	–	–	–
Ehrlichiosis	–	–	–

+, >90% positive agglutination; –, >90% negative agglutination; *V,* variable results.

tion of disease. These tests are performed primarily by reference laboratories, and kits can be obtained from CDC for performing tests for IgG antibody.

The microimmunofluorescent dot test, described previously for diagnosis of psittacosis, has shown excellent sensitivity for detecting antibodies to rickettsiae. By overlaying the antigen dots on the slide with fluorescein-conjugated antihuman IgM after the patient's serum has been allowed to form complexes, an early diagnosis of RMSF can be achieved within 7 to 10 days after onset of symptoms. Both immunofluorescent and a recently developed latex agglutination test method have shown good sensitivity for diagnosis of RMSF within the first week. Q fever, ehrlichiosis, and rickettsialpox do not induce Weil-Felix antibody in infected patients. In those illnesses, specific immunofluorescent tests are valuable.

C. burnetii undergoes an antigenic phase variation during infection. The organisms exist in phase II during initial infection, and humans produce antibody to phase II early in disease. If infection progresses to chronic disease, such as endocarditis or hepatitis, antibodies to phase I are present and can be measured by complement fixation or immunofluorescence.

Except for latex agglutination, immunofluorescence, and direct fluorescent antibody testing for diagnosing RMSF, none of the serologic tests is useful for diagnosing disease in time to influence therapy, since antibodies to rickettsiae other than *R. rickettsii* cannot be reliably detected until at least 2 weeks after the patient has become ill. With newer immunological reagents under develop-

ment, the potential exists for new tests for all the rickettsial diseases.

REFERENCES

1. Baselski, V.S., McNeeley, S.G., Ryan, G., and Robison, M.A. 1987. A comparison of nonculture-dependent methods for detection of *Chlamydia trachomatis* infections in pregnant women. Obstet. Gynecol. 70:47.
2. Campbell, L.A., Kuo, C.-C., and Grayston, J.T. 1987. Characterization of the new *Chlamydia* agent, TWAR, as a unique organism by restriction endonuclease analysis and DNA-DNA hybridization. J. Clin. Microbiol. 25:1911.
2a. Cassell, G.H., Waites, K.B., Watson, H.L., et al. 1993. *Ureaplasma urealyticum* intrauterine infection: role in prematurity and disease in newborns. Clin Microbiol. Rev. 6:69.
3. Clyde, W.A., Jr., Kenny, G.E., and Schachter, J. 1984. Cumitech 19: Laboratory diagnosis of chlamydial and mycoplasmal infections. Drew, W.L., coordinating editor. American Society for Microbiology, Washington, D.C.
4. Gibbs, R.S., Cassell, G.H., Davis, J.K., et al. 1986. Further studies on genital mycoplasms in intra-amniotic infection: blood cultures and serologic response. Am. J. Obstet. Gynecol. 154:717.
5. Grayston, J.T., Kuo, C.-C., Wang, S.-P., and Altman, J. 1986. A new *Chlamydia psittaci* strain, TWAR, isolated in acute respiratory tract infections. N. Engl. J. Med. 315:161.
6. Hammerschlag, M.R., Roblin, P.M., Cummings, C., et al. 1987. Comparison of enzyme immunoassay and culture for diagnosis of chlamydial conjunctivitis and respiratory infections in infants. J. Clin. Microbiol. 25:2306.
7. Hipp, S.S., Yangsook, H., and Murphy, D. 1987. Assessment of enzyme immunoassay and immunofluorescence test for detection of *Chlamydia trachomatis.* J. Clin. Microbiol. 25:1938.
8. Kundsin, R.B., Driscoll, S.G., and Pelletier, P.A. 1980. *Ureaplasma urealyticum* incriminated in perinatal morbidity and mortality. Science 213:474.
8a. Limberger, R.J., Biega, R., Evancoe, A., et al. 1992. Evaluation of culture and the Gen-Probe PACE 2 assay for detection of *Neisseria gonorrhoeae* and *Chlamydia trachomatis* in endocervical specimens transported to a state health laboratory. J. Clin. Microbiol. 30:1162.
9. Lipkin, E.S., Moncada, J.V., Shafer, M.A., et al. 1986. Comparison of monoclonal antibody staining and culture in diagnosing cervical chlamydial infection. J. Clin. Microbiol. 23:114.
10. Mahony, J.B., Chernesky, M.A., Bromberg, K., and Schachter, J. 1986. Accuracy of immunoglobulin M immunoassay for diagnosis of chlamydial infections in infants and adults. J. Clin. Microbiol. 24:731.
11. Mårdh, P.-A. 1984. Laboratory diagnosis of sexually transmitted diseases. In Holmes, K.K., Mårdh, P.-A., Sparling, P.F., and Wiesner, P.J., editors. Sexually transmitted diseases. McGraw-Hill, New York.
12. Oriel, J.D. 1983. Role of genital mycoplasmas in nongonococcal urethritis and prostatitis. Sex. Transm. Dis. 10 (suppl.):263.
13. Peterson, L.R., Sawyer, L.A., Fishbein, D.B., et al. 1989. An outbreak of ehrlichiosis in members of an

army reserve unit exposed to ticks. J. Infect. Dis. 159:562.

14. Phillips, L.E., Goodrich, K.H., Turner, R.M., and Faro, S. 1986. Isolation of *Mycoplasma* species and *Ureaplasma urealyticum* from obstetrical and gynecological patients by using commercially available medium formulations. J. Clin. Microbiol. 24:377.

15. Smith, J.W., Rogers, R.E., Katz, B.P., et al. 1987. Diagnosis of chlamydial infection in women attending antenatal and gynecologic clinics. J. Clin. Microbiol. 25:868.

16. Stamm, W.E., Running, K., Hale, J., and Holmes, K.K. 1983. Etiologic role of *Mycoplasma hominis* and *Ureaplasma urealyticum* in women with acute urethral syndrome. Sex. Transm. Dis. 10 (suppl.):318.

17. Steffenson, D.O., Dummer, J.S., Granick, M.S., et al. 1987. Sternotomy infections with *Mycoplasma hominis*. Ann. Intern. Med. 106:204.

18. Taylor-Robinson, D., Thomas, B.J., and Osborn, M.F. 1987. Evaluation of enzyme immunoassay (Chlamydiazyme) for detecting *C. trachomatis* in genital tract specimens. J. Clin. Pathol. 40:194.

19. Tilton, R.C., Judson, F.N., Barnes, R.C., et al. 1988. Multicenter comparative evaluation of two rapid microscopic methods and culture for detection of *Chlamydia trachomatis* in patient specimens. J. Clin. Microbiol. 26:167.

20. Tjiam, K.H., van Heijst, B.Y., van Zuuren, A., et al. 1986. Evaluation of an enzyme immunoassay for the diagnosis of chlamydial infections in urogenital specimens. J. Clin. Microbiol. 23:752.

BIBLIOGRAPHY

Anonymous. 1983. International symposium on *Mycoplasma hominis*—a human pathogen. Sex. Transm. Dis. 10(suppl.). (Entire issue devoted to human mycoplasmas.)

Baca, O.G., and Paretsky, D. 1983. Q fever and *Coxiella burnetii*: a model for host-parasite interactions. Microbiol. Rev. 47:127.

Balows, A., Hausler, Jr., W.J., Herrmann, K.L., et al. 1991. Manual of clinical microbiology, ed 5, American Society for Microbiology. (Source for procedures, media formulas, and stains.)

Barnes, R.C. 1989. Laboratory diagnosis of human chlamydial infections. Clin. Microbiol. Rev. 2:119.

Baron, E.J, and Finegold, S.M. 1990. Bailey and Scott's diagnostic microbiology, ed 8. Mosby, St. Louis, (Appendices include detailed formulas of media and stains.)

Batteiger, B.E., and Jones, R.B. 1987. Chlamydial infections. Infect. Dis. Clin. North Am. 1:55.

Cassel, G.H., and Cole, B.C. 1981. Mycoplasma as agents of human disease. N. Engl. J. Med. 304:80.

Centers for Disease Control. 1985. *Chlamydia trachomatis* infections: policy guidelines for prevention and control. M.M.W.R. 34 (suppl.):53.

Couch, R.C. 1984. Mycoplasma diseases. In Holmes, K.K., Mårdh, P.-A., Sparling, P.F., and Wiesner, P.J., editors. Sexually transmitted diseases. McGraw-Hill, New York.

Grayston, J.T. 1992. Infections caused by *Chlamydia pneumonia* strain TWAR. Clin. Infect. Dis. 15:757.

LaScolea, L.J., Jr. 1986. Chlamydial infections: the mother-infant connection. Clin. Microbiol. Newsletter 8:77.

Marrie, T.J., 1990. *Coxiella burnetii* (Q fever). In Mandell, G.L., Douglas, R.G., Jr., and Bennett, J.E., editors. Principles and practice of infectious diseases, ed 3. John Wiley & Sons, New York. (Several other chapters in the section on rickettsiosis contain valuable information.)

McMahon, D.K., Dummer, J.S., and Pasculle, A.W. 1990. Extragenital *Mycoplasma hominis* infections in adults. Am. J. Med. 89:275.

Oriel, D., Ridgway, G., Schachter, J., et al., editors. 1986. Chlamydial infections. Cambridge University Press, Cambridge, England. (Entire book devoted to chlamydial topics.)

Philip, R.N., Casper, E.A., Ormsbee, R.A., et al. 1976. Microimmunofluorescence test for the serological study of Rocky Mountain spotted fever and typhus. J. Clin. Microbiol. 3:51.

Raoult, D., and Walker, R.B. 1990. *Rickettsia rickettsii* and other spotted fever group *Rickettsiae* (Rocky Mountain spotted fever and other spotted fevers). In Mandell, G.L., Douglas, D.H., Jr., and Bennett, J.E., editors. Principles and practice of infectious diseases, ed 3. John Wiley & Sons, New York.

Rikihisa, Y. 1991. The tribe *Ehrlicheae* and Ehrlichial diseases. Clin. Microbiol. Rev. 4:286.

Salgo, M.P., Telzak, E.E., Currie, B., et al. 1988. A focus of Rocky Mountain Spotted Fever within New York City. N. Engl. J. Med. 318:1346.

Schachter, J. 1984. Biology of *Chlamydia trachomatis*. In Holmes, K.K., Mårdh, P.-A., Sparling, P.F., and Wiesner, P.J., editors. Sexually transmitted diseases. McGraw-Hill, New York.

Stamm, W.E. 1988. Diagnosis of *Chlamydia trachomatis* genitourinary infections. Ann. Intern. Med. 108:710.

Weissfeld, A.S. 1983. Genital mycoplasma. Clin. Microbiol. Newsletter 5:65.

Walker, D.H. 1989. Rocky Mountain spotted fever: a disease in need of microbiological concern. Clin. Microbiol. Rev. 2:227.

Wyrick, P.B., Gutman, L.T., and Hodinka, R.L. 1988. Chlamydiae. In Joklik, W.K., Willett, H.P., Amos, D.B., and Wilfert, C.M., editors: Zinsser's microbiology, ed 19. Appleton & Lange, Norwalk, Conn.

40

UNCLASSIFIED OR UNUSUAL BUT EASILY CULTIVATED ETIOLOGICAL AGENTS OF INFECTIOUS DISEASE

The microorganisms discussed in this chapter are not placed in any of the previous chapters either because they are unusual enough not to fit into existing classifications or because they are isolated rarely or only in association with a defined syndrome. Although they are able to grow on media easily available to most clinical laboratories, several organisms discussed here require specific isolation techniques for their cultivation, necessitating that the physician communicate with the laboratory the need for special procedures.

BARTONELLA BACILLIFORMIS

EPIDEMIOLOGY AND PATHOGENESIS OF *BARTONELLA BACILLIFORMIS* INFECTION

Bartonella bacilliformis, formerly the only species in the genus *Bartonella,* is classified in the order Rickettsiales, although it is more closely related to *Rochalimaea quintana* than to the *Rickettsia* sp.[21] *Rochalimaea henselae* (see Chapter 41) has been moved recently into the genus *Bartonella. B. bacilliformis* grows within and on the surface of erythrocytes and occasionally in vascular endothelium. It can be cultivated in vitro on nonliving media. The disease bartonellosis is confined to the Andes region of Peru, Ecuador, and Colombia, where the arthropod vector, the sand fly *Phlebotomus,* and susceptible humans both exist. Humans acquire the disease from the bite of the fly and serve as a reservoir for the organism in the region. Two forms of the disease are common, a febrile, systemic disease characterized by anemia, called *Oroya fever,* and a chronic, cutaneous disease characterized by tumorlike lesions of the skin and mucous membranes, called *verruga peruana.* Verruga peruana usually follows an episode of Oroya fever. The bacteria on the surface of and within erythrocytes seem to contribute to the anemia.[26] Patients with partial immunity develop verruga peruana. It is not known exactly how the organisms damage the red blood cells, but an endothelial cell proliferative substance (angiogenic factor) is elaborated by *B. bacilliformis.*[11] Patients with Oroya fever appear to be at increased risk for contracting *Salmonella* infection.

LABORATORY DIAGNOSIS OF BARTONELLOSIS

The organisms can be isolated from blood and from material aspirated from lymph nodes or cutaneous lesions. Organisms grow best at room temperature and do not require increased carbon dioxide (CO_2) in the atmosphere. Brain-heart infusion agar, modified to contain 0.4% agar, with the addition of 5% human, rabbit, or horse blood, should support growth of *B. bacilliformis.* Colonies appear as white puffballs, 1 to 5 mm diameter, after 1 to 2 weeks.[10] The organisms stain gram negative and are visualized best with the 0.2% basic fuchsin counterstain. They are curved or coccoid, with mixtures of rods and granules, which tend to grow in aggregates. In vitro, cells produce a polar tuft of flagella. Biochemically, *B. bacilliformis* is quite inert.

The organisms may also be visualized in blood films from infected patients.[26] They stain red-violet with Giemsa stain, appearing as cocci or bacilli, with occasional curved or ringlike forms. The cells may show polar enlargements, appearing only at one end.

TREATMENT OF BARTONELLOSIS

Patients with the disease seem to respond rapidly to therapy with penicillin, chloramphenicol, tetracycline, or streptomycin.

GENUS *CHROMOBACTERIUM*

EPIDEMIOLOGY AND PATHOGENESIS OF *CHROMOBACTERIUM* INFECTION

Chromobacterium violaceum, instantly recognizable by its purple pigment, is the etiological agent of a particularly devastating and virulent bacteremia, usually acquired in the southeastern United States, particularly in Florida. The organism has been recovered from soil and water in tropical countries. Patients usually acquire the organism from puncture wounds or wounds that come into contact with soil or contaminated water. One patient, however, presumably acquired the infection by swallowing water during a near-drowning accident. Since at least 3 of 12 patients were known to have chronic granulomatous disease, and since the status of other patients is unknown, it has been suggested that an underlying neutrophil dysfunction puts patients at increased risk for developing fulminant *C. violaceum* infections.[19]

The organism is extremely virulent; a mortality rate of 57% occurs in patients with infection acquired in the United States. The infection often presents as a febrile illness, and skin lesions are common. Liver abscess and pulmonary involvement are also prominent features. Virulence factors have not been well studied, but extracellular proteases are produced by strains of *C. violaceum.*

LABORATORY IDENTIFICATION OF *CHROMOBACTERIUM VIOLACEUM*

The organisms are usually recognized by their violet pigment (Figure 40.1), produced on media containing tryptophan, such as 5% sheep blood agar. Although the organisms grow best at 25° C,

FIGURE 40.1

Colonies of violet-pigmented *Chromobacterium violaceum* on blood agar.

they will grow at 37° C. *C. violaceum* is facultatively anaerobic. Cultures may smell of ammonium cyanide, since the organisms produce hydrogen cyanide (HCN). *C. violaceum* is resistant to penicillin and to the vibriostatic agent O/129. Although it is usually oxidase-positive, the purple color may mask the reaction. It is catalase, Voges-Proskauer, and esculin negative. The rarely encountered pigment-negative strains may be confused with *Vibrio* or *Aeromonas* sp., but since they have not been implicated in human disease, they should not pose a problem. Resistance to O/129 distinguishes *Chromobacterium* from vibrios, which are susceptible, and lack of indole production differentiates *Chromobacterium* from *Aeromonas* and *Plesiomonas,* many of which are indole positive. Unlike many other unusual gram-negative bacilli, *C. violaceum* grows well on MacConkey agar.

The organisms are always motile, as demonstrated by wet preparation. On Gram stain, they are medium to long bacilli with rounded ends; the bacilli are occasionally slightly curved. *Chromobacterium* possesses the unusual flagellar arrangement of both a single polar flagellum and lateral flagella.

TREATMENT OF INFECTIONS CAUSED BY *CHROMOBACTERIUM VIOLACEUM*

Most isolates are susceptible in vitro to chloramphenicol, erythromycin, tetracycline, and trimethoprim-sulfamethoxazole. Resistance to ampicillin, cephalothin, carbenicillin, and penicillins has been reported. Those patients who recovered from septicemia in the series reported by Macher et al.[19] were treated with gentamicin,

with or without chloramphenicol. Long-term therapy may be necessary to eradicate all foci of the infection.

GENUS *DERMATOPHILUS*

EPIDEMIOLOGY AND PATHOGENESIS OF *DERMATOPHILUS* INFECTION

Within the same family as the *Actinomyces* sp. and therefore classified as a bacterium, *Dermatophilus congolensis* is the etiological agent of dermatophilosis, also called strawberry foot rot, lumpy wool, mycotic dermatitis, and streptotrichosis, a disease of the skin characterized by acute or chronic development of an exudative, scabbing dermatitis.[15] The disease occurs worldwide in domestic animals, including cattle, sheep, goats, horses, other mammals, and humans. Manifestations of the disease are more severe in tropical regions. The organism has been isolated from both patients and wild animals in New York. A *Dermatophilus* infection of the tongue of a patient with human immunodeficiency virus (HIV) infection had lesions that resembled those of Epstein-Barr–associated "hairy" leukoplakia.[3] As with the fungi, pathogenesis is probably related to the organism's large size and its ability to resist cell-mediated killing. Extracellular enzymes may contribute to pathogenesis.

LABORATORY IDENTIFICATION OF *DERMATOPHILUS*

Biopsy material and touch preparations of the underside of scabs should be stained by Giemsa or methenamine silver. Typical organisms displaying septate hyphal-like forms, tapering and dividing transversely, with production of packets of coccoid spores as a final morphotype, are visible, particularly around hair follicles. The organism is gram positive, but fine structure is not visible with the Gram stain.

Biopsy specimens and tissue scrapings from skin lesions, as well as exudate aspirated from unopened pustules, should be streaked to blood agar, which is incubated in air at 37° C. Brain-heart infusion agar with 5% sheep blood, recommended for primary inoculation of all specimens for isolation of fungi, also supports growth of *D. congolensis.* Colonies develop to a diameter of approximately 5 mm after 2 to 4 days' growth. Colonies are grayish and usually become orange after several days, heaped, rough, glabrous, and adherent to the agar; pitting may be observed with early growth (Figure 40.2). In heavy areas the colonies are beta hemolytic and in later stages may become mucoid. The organisms also grow in

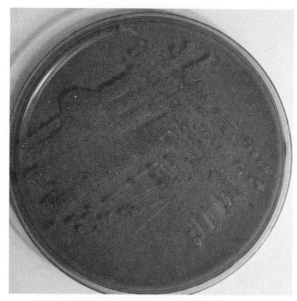

FIGURE 40.2

Colonies of *Dermatophilus congolensis* on 5% sheep blood agar. (Courtesy Morris Gordon.)

brain-heart infusion–peptone broth, but they do not grow on Sabouraud's dextrose agar. A direct wet preparation from a colony shows branching filaments 0.5 to 5 μm in width; branches are at right angles. The filaments divide longitudinally and transversely, finally becoming aggregates of motile coccoid-shaped spores (Figure 40.3). If only coccoid forms are seen on wet preparation from a colony, a younger subculture should be examined for filaments. Because the Gram stain tends to obscure details, a plain methylene blue stain (e.g., Loeffler's methylene blue) should be used. Although cellular morphology is characteristic, biochemical reactions may help confirm the

identification. The organism is catalase and urease positive. *D. congolensis* produces extracellular enzymes, such as protease and gelatinase. All strains are indole, nitrate, and Voges-Proskauer negative.

TREATMENT OF INFECTIONS CAUSED BY *DERMATOPHILUS CONGOLENSIS*

The organism is susceptible in vitro to penicillin, streptomycin, chloramphenicol, tetracycline, erythromycin, and sulfonamides. It is resistant to the antidermatophyte griseofulvin. The treatment of choice appears to be a penicillin in combination with an aminoglycoside.

GENUS *GARDNERELLA*

EPIDEMIOLOGY AND PATHOGENESIS OF *GARDNERELLA* INFECTION

Gardnerella vaginalis is found in the genital and urinary tracts of humans. In **bacterial vaginosis,** (BV), a disease of women characterized by copious, malodorous discharge, *G. vaginalis* is often found in large numbers, where it serves as one indicator organism of the syndrome.[12] BV, formerly called "nonspecific vaginitis," is one of the most common complaints of postpubescent women visiting their gynecologists. The vaginal discharge of this syndrome is usually homogeneous, unaccompanied by pain or itching. Unlike the purulent discharge of gonococcal urethritis, *Trichomonas* vaginitis, and most yeast infections, few neutrophils are found in the discharge of patients with BV. For this reason, the term *vaginosis* rather than "vaginitis" was coined. BV is characterized by a decrease in hydrogen peroxide–producing lactobacilli and an increase in *G. vaginalis*, anaero-

FIGURE 40.3

Methylene blue–stained organisms of *Dermatophilus congolensis* from colony on blood agar plate. (Courtesy Morris Gordon.)

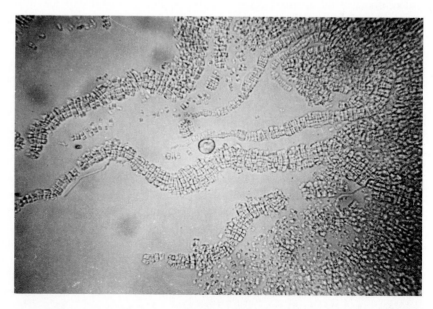

bic gram-negative bacilli, peptostreptococci, and mycoplasmas.[12] Recent studies have emphasized the importance of BV as a risk factor for premature labor, chorioamnionitis, and postpartum endometritis.[14] G. vaginalis is one of the indicator organisms for the BV syndrome, but it is not thought to be an etiological agent. The complete etiology of BV has not been defined. G. vaginalis, however, is involved in certain infections just mentioned, for which BV is a risk factor. The difficulty in cultivating G. vaginalis from blood has probably led to a lack of recognition of its role in postpartum bacteremia and fever, as well as infection in the neonate.

Although male partners of women with BV can be found to harbor the organism in the genital tract, it rarely causes infection. In males, G. vaginalis has been associated with urethritis, balanoposthitis, prostatitis, and urinary tract infections.[4] In addition to women with BV, approximately two thirds of asymptomatic normal control subjects also harbor the organism, although the number of colony forming units is often lower in asymptomatic females. Urinary tract infection with G. vaginalis may occur in women.

LABORATORY IDENTIFICATION
OF GARDNERELLA VAGINALIS

Specimens most likely to be sent to the microbiology laboratory for cultivation of G. vaginalis include cervical, urethral, and vaginal swabs from patients with discharge. It is not recommended, however, to culture these specimens. BV can be detected best by Gram stain[20]; results of cultures are not specific enough to predict reliably the presence of the syndrome. Women with postpartum fever and neonates showing signs of sepsis may be infected with G. vaginalis, which may be isolated from blood. The sodium polyanetholsulfonate (SPS) currently used as an anticoagulant in most commercial blood culture media has been found to inhibit growth of Gardnerella, however, and these media are not suitable for culture unless gelatin is added to counteract the SPS. Good results have been achieved by direct plating of blood onto agar. Although no studies have been done, it seems likely that the lysis-centrifugation blood culture method would allow growth of G. vaginalis plated onto chocolate or other supportive agar, such as Columbia agar with colistin and nalidixic acid (CNA). Chocolate agar can be used to enhance recovery of this organism from urine.

If isolation of G. vaginalis from female genital tract specimens is necessary, such as for an epidemiological study, the use of the semiselective medium human blood Tween™ (HBT) agar is recommended.[30] This medium consists of a base layer of Columbia agar base overlaid with a layer of the same base containing 5% human blood. HBT medium is available commercially. Material from genital discharge is inoculated within 4 to 6 hours of collection onto this medium, and the plates are incubated in 5% to 10% CO_2 or in a candle jar for 48 hours. The organism has been shown to survive poorly in frequently used transport media. Colonies of G. vaginalis on HBT agar are convex, opaque, and gray, surrounded by a diffuse zone of beta hemolysis. On sheep blood agar, these organisms are not hemolytic and are

FIGURE 40.4

Culture of vaginal discharge after 48 hours' incubation. Colonies of *Gardnerella vaginalis* growing in large numbers on colistin and nalidixic acid (CNA) agar that also supports the growth of some normal flora.

barely visible. They grow well on CNA agar, without yielding any hemolysis (Figure 40.4). The organism's ability to hemolyze human erythrocytes forms the basis for the differential nature of the HBT agar. *G. vaginalis* does not grow on Mac-Conkey agar.

The organisms are small, pleomorphic, gram-variable or gram-negative coccobacilli and short rods (Figure 40.5). They possess an unusual laminated cell wall structure with a possible glycocalyx, which may aid adherence to vaginal epithelium. The organism is oxidase and catalase negative. Presumptive identification based on colony morphology and hemolysis on HBT, cellular morphology, and the rapid enzyme tests mentioned here are adequate for isolates from female genital sources. *Gardnerella* hydrolyzes sodium hippurate and ferments starch and raffinose.[4] Several commercial systems for fastidious gram-negative bacterial identification adequately identify this species.

Presumptive diagnosis of BV can be made without performing cultures, as described in Chapter 20. Aside from the distinctive clinical picture, a direct wet mount of vaginal discharge material usually shows "clue cells," squamous epithelial cells covered with tiny bacilli, especially around the periphery, giving the cell a stippled appearance. Discharge material has a characteristic fishy odor, particularly after addition of 10% potassium hydroxide (KOH) (the sniff test), and the pH of the material is greater than 4.5. The Gram stain of vaginal discharge has been validated as the most reliable diagnostic laboratory test for BV.[20]

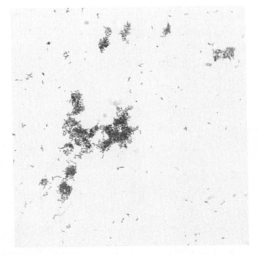

FIGURE 40.5

Gram stain of *Gardnerella vaginalis* from growth on CNA agar.

TREATMENT OF *GARDNERELLA VAGINALIS* VAGINOSIS

Although in vitro susceptibility of *G. vaginalis* to metronidazole is variable, metronidazole has consistently yielded the best cure of BV.[18] Since metronidazole is effective primarily against anaerobic bacteria, the ability of metronidazole to affect this syndrome speaks against a primary causative role for *Gardnerella*. A topical gel is available commercially.[12a] A newer, equally effective therapeutic alternative is a topical clindamycin cream applied once daily.[27] No standardized treatment has been promoted for treatment of systemic infections; all strains of *G. vaginalis* have been susceptible in vitro to ampicillin, carbenicillin, oxacillin, penicillin, and vancomycin. Susceptibility to the latter three agents is usually a characteristic of gram-positive organisms, another enigmatic finding for *Gardnerella*.

GENUS *LEGIONELLA*

EPIDEMIOLOGY AND PATHOGENESIS OF *LEGIONELLA* INFECTION

Species of *Legionella* are distributed worldwide in surface water, soil, mud, lakes, and streams. The organisms are not known to have an animal reservoir, although they are maintained intracellularly in freshwater protozoa that function as their natural host.[9] *Legionella* is the causative agent of Legionnaires' disease, a febrile and pneumonic illness with numerous clinical presentations. *Legionella* was discovered in 1976 by scientists at the Centers for Disease Control (CDC), who were investigating an epidemic of pneumonia among Pennsylvania state American Legion members attending a convention in Philadelphia. There is retrospective serologic evidence of *Legionella* infection as far back as 1947. *Legionella pneumophila*, the type species of the genus, is the most prevalent etiological agent of Legionnaires' disease. At least 40 other species of *Legionella* have now been characterized, isolated from sporadic cases of pneumonia, other outbreaks, and environmental sources. A well-documented outbreak of prosthetic valve endocarditis caused by *L. pneumophila* and *L. dumoffii* has been reported.[29] Many reported cases of non–*L. pneumophila* pneumonia have been nosocomial. The genus name *Tatlockia* is used by some authors for species *micdadei* and *maceachernii*, and the genus name *Fluoribacter* is used for species *dumoffii*, *gormannii*, and *bozemanii*; all three genus names are legitimate currently.

Legionella sp. live in natural waters and can be found in air-conditioning cooling-tower water, air-conditioning condensate, and reservoir waters, often in association with blue-green algae and free-living protozoa.[28] The organisms can also multiply in relatively hot water and can be found in shower heads, whirlpool baths, and institutional potable water systems. They are able to grow at temperatures of 25° to 43° C on artificial media and at even higher temperatures in nature. Patients usually acquire the organism by breathing in aerosols; no person-to-person transmission occurs. Eradication of *Legionella* from institutional water systems requires increased heat (hot water tanks must be heated to >70° C for 72 hours) or increased chlorination (to 2 parts per million of free chlorine). Surgical patients; those who are immunosuppressed, particularly those receiving corticosteroids; and persons who smoke are at increased risk for acquiring Legionnaires' disease.

The disease occurs sporadically, at low levels endemically, or as epidemics among susceptible hosts.[28] It is not a rare infection; an estimated 25,000 to 50,000 cases occur annually in the United States. Usually a seasonal peak occurs during late summer. The disease involves multiple organ systems but is manifested primarily as severe, consolidated pneumonia. Bacteremia probably occurs, especially in the immunosuppressed patient. Another manifestation of disease caused by *L. pneumophila* and occasionally by *L. anisa, L. feeleii,* and *L. micdadei* is Pontiac fever, a multisystem disease with respiratory symptoms, fever, myalgia, and headache but without pneumonia. Pontiac fever is a mild, self-limited disease with no mortality, in contrast to Legionnaires' disease, which has a significant mortality. The clinical manifestations exhibited by infection with a particular species primarily are caused by differences in the host's immune response and perhaps by inoculum size; the same *Legionella* sp. gives rise to different expressions of disease in different individuals.

Pathogenesis of *Legionella* infection has been studied quite extensively. Although the organisms produce several extracellular enzymes, including a phosphatase, a lipase, and nucleases, these have not been shown to contribute significantly to virulence. The primary virulence mechanism of *Legionella* is its ability to inhibit lysosomal fusion with organism-containing phagosomes in host cells. Cell-mediated immune responses are necessary for the host to overcome disease. Serologic evidence exists for the presence of asymptomatic disease, since many healthy people surveyed possess antibodies to *Legionella* sp., and since in a recent hospital outbreak (VA Wadsworth Medical Center) a significant incidence of asymptomatic seroconversion occurred.

LABORATORY DIAGNOSIS OF LEGIONNAIRES' DISEASE BY CULTURE[8]

The general isolation and direct detection of *Legionella* can be done effectively by all laboratories that possess a class II biological safety cabinet. Specimens from which *Legionella* can be isolated include respiratory tract secretions of all types, including sputum and pleural fluid; other sterile body fluids such as blood; and lung, transbronchial, or other biopsy material. Sputum specimens are preferable to bronchial washings, which are likely to contain small amounts of inhibitory local anesthetic and are diluted by saline. Sputum from patients with Legionnaires' disease is usually nonpurulent and may appear bloody or watery. Therefore the grading system used for screening sputum for routine cultures is not applicable. Patients with Legionnaires' disease usually have detectable numbers of organisms in their respiratory secretions, even for quite some time after antibiotic therapy has been initiated. If the disease is present, the initial specimen is often likely to be positive. However, additional specimens should be processed if suspicion of the disease persists. Pleural fluid has not yielded many positive cultures in studies performed in several laboratories, but it may contain organisms. Specimens should be transported without holding media, buffers, or saline, which may be inhibitory to growth of *Legionella*. The organisms are actually very hardy and are best preserved by maintaining specimens in a small, tightly closed container to prevent desiccation and transporting them to the laboratory within 30 minutes of collection. If longer delay is anticipated, specimens should be refrigerated. If one cannot ensure that specimens will remain moist, a small amount (1 ml) of sterile broth may be added.

Material should be inoculated to culture plates and applied to slides for direct fluorescent antibody testing (as described next) as soon as it is received in the laboratory. Technologists should work in a biological safety cabinet. As described in Chapter 17, sputum can best be picked from upper respiratory secretions with a wooden applicator stick, to which the mucus and pus cells will adhere. Sputum specimens that may contain contaminating bacteria should be diluted 1:10 in trypticase soy broth, the suspension should be vortexed with sterile glass beads, and plates should be inoculated with several drops of the diluted specimen. Lesser dilutions are recommended for aspirates. An acid-wash method may enhance recovery (Procedure 40.1).[8]

Sputum and respiratory secretions for direct fluorescent antibody tests are applied to slides

---

PROCEDURE 40.1

ACID-WASH TREATMENT OF CONTAMINATED SPECIMENS FOR RECOVERY OF *LEGIONELLA* SP.

PRINCIPLE
Acid washing may reduce contaminating organisms and increase the yield of *Legionella* organisms.

METHOD
1. Prepare stock solutions before preparing working solution, and discard excess stock solutions:
   0.2 N HCl, 50 ml
   0.2 N KCl, 50 ml

2. Add 5.3 ml of HCl stock solution and 25 ml of KCl stock solution to 100 ml of distilled water.

3. Adjust pH to 2.2 by adding HCl or additional KCl.

4. Filter-sterilize working solution and dispense, 0.9 ml per tube plus sterile glass beads, into small screw-cap tubes.

5. Store refrigerated; stable for 6 months.

6. Add 0.1 ml of specimen to 0.9 ml of acid-wash solution.

7. Vortex vigorously, and allow to sit on bench 4 minutes.

8. Revortex and plate 3 drops per plate to buffered charcoal–yeast extract (BCYE) and BCYE-selective agar.

---

without dilution. As described in Chapter 22, tissues are homogenized before smears and cultures are performed, and clear sterile body fluids are centrifuged for 30 minutes at 4000 × *g*. The sediment is then vortexed and used for culture and smear preparation. Blood for culture of *Legionella* should probably be processed with the lysis-centrifugation tube system and plated directly to buffered charcoal–yeast extract agar (BCYE). A biphasic culture medium containing BCYE agar and yeast extract broth, designed for isolation of *Legionella* from blood, is available from Remel Laboratories.

Specimens should be inoculated to two agar plates for recovery of *Legionella,* at least one of which is BCYE without inhibitory agents. This medium, developed at the CDC, contains charcoal to detoxify the medium, remove CO_2, and modify the surface tension to allow the organisms to proliferate more easily. BCYE is also prepared with ACES buffer and growth supplements cysteine (required by *Legionella*), yeast extract, α-ketoglutarate, and iron. A second medium, BCYE base with polymyxin B, anisomycin (to inhibit fungi), and cefamandole, is recommended for specimens such as sputum that are likely to be contaminated with other flora. These media are commercially available. Several other media, including a selective agar containing vancomycin and a differential agar containing bromthymol blue and bromcresol purple, are also available from Remel and others.

Specimens obtained from sterile body sites may be plated to two media without selective agents, and perhaps also inoculated into the special blood culture broth without SPS. Specimens should always be plated to standard media for recovery of pathogens other than *Legionella* that may be responsible for the disease.

Plates should be incubated in air or in a candle jar at 35° to 37° C in a humid atmosphere. Only growth of *L. gormanii* is stimulated by increased CO_2, so air incubation is preferable to 5% to 10% CO_2, which may inhibit some legionellae. Within 3 to 4 days, colonies should be visible. Plates should be held for a maximum of 2 weeks before discarding. Blood cultures in biphasic media should be held for 1 month. At 5 days, colonies are 3 to 4 mm in diameter, gray-white to blue-green, glistening, convex, and circular and may exhibit a cut-glass type of internal granular speckling (Figure 40.6). A Giménez stain reveals small, coccobacillary organisms with occasional filamentous forms (Figure 40.7). A Gram stain yields thin, gram-negative bacilli.

All *Legionella* sp. are weakly catalase positive; strongly catalase-positive organisms are less likely to be legionellae. Oxidase test results are variable. Another rapid differential screening test is the production of fluorescent pigment, visible under long-wave fluorescent light (long-wave light is generated at 365 nm; a Wood's light generates light of wavelength near 250 nm). *L. pneumophila,*

FIGURE 40.6

Colonies of *Legionella pneumophila* on buffered charcoal–yeast extract (BCYE) agar. (Courtesy Clifford Mintz.)

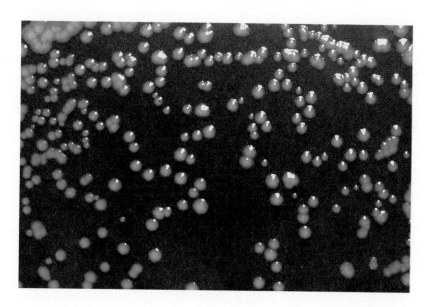

FIGURE 40.7

Giménez stain of *Legionella pneumophila* from a colony grown on BCYE.

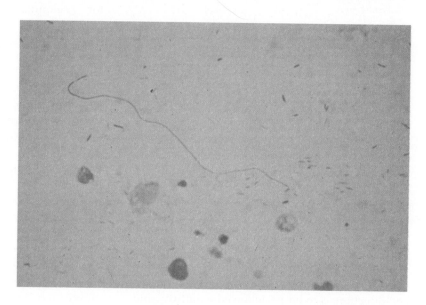

L. micdadei, L. jordanis, L. wadsworthii, L. oakridgensis, and *L. longbeachae* fluoresce pale yellow-green, whereas *L. gormanii, L. dumoffii,* and *L. bozemanii,* fluoresce blue-white. All *Legionella* sp. described so far are able to liquefy gelatin.

The currently recognized species of legionellae are listed in the box on p. 574. Most cases of Legionnaires' disease are associated with *L. pneumophila.* Definitive identification requires the facilities of a specialized reference laboratory. Identification of *L. pneumophila* sp. can be achieved, however, by a monoclonal immunofluorescent stain (Genetic Systems). Emulsions of organisms from isolated colonies are made in 10% neutral formalin, diluted 1:100 (to produce a very thin suspension), and placed on slides for fluorescent antibody staining, as de-

scribed in the next section. Clinical laboratories can probably perform sufficient service to clinicians by indicating the presence of *Legionella* sp. in a specimen. If further identification is necessary, the isolate should be forwarded to an appropriate reference laboratory.

DIRECT DETECTION OF *LEGIONELLA* IN INFECTED CLINICAL MATERIAL

The most rapid means of diagnosing Legionnaires' disease is direct detection of the presence of the bacilli in clinical specimens. Although *Pseudomonas* sp. and other contaminating material may occasionally cause cross-reactive and nonspecific fluorescence, direct fluorescent antibody (DFA) tests should be performed on specimens received in the laboratory for *Legionella* de-

Currently Recognized Species of *Legionella*

L. adelaidensis	*L. longbeachae**
L. anisa	*L. maceachernii**
*L. birminghamensis**	*L. micdadei**
*L. bozemanii**	*L. moravica*
L. brunensis	"*L. nautarum*"
L. cherrii	*L. oakridgensis**
*L. cincinnatiensis**	*L. parisiensis*
*L. dumoffii**	*L. pneumophila**
L. erythra	"*L. quarteiraensis*"
L. fairfieldensis	*L. quinlivanii*
*L. feeleii**	*L. rubrilucens*
"*L. geestiae*"	*L. sainthelensi*
L. gormanii	*L. santicrucis*
L. gratiana	*L shakespearei*
*L. hackeliae**	*L. spiritensis*
*L. israelensis**	*L. steigerwaltii*
L. jamestownensis	*L. tucsonensis**
*L. jordanis**	*L. wadsworthii**
*L. lansingensis**	"*L. worsleiensis*"
"*L. londiniensis*"	

*Isolated from patient specimens. Names in quotations are proposed.

tection. Antisera conjugated with fluorescein are available from several commercial suppliers (BioDx, Biological Products Division of Centers for Disease Control, Litton Bionetics, Genetic Systems, MarDx, Zeus Technologies, others). Specimens are first tested with pools of antisera containing antibodies to several species or serotypes of *L. pneumophila*. Those that exhibit positive results are then reexamined with specific conjugated antisera. The Genetic Systems reagent is a monoclonal antibody directed against a cell wall protein common to *L. pneumophila*. Manufacturers' directions should be followed explicitly, and material from commercial systems should never be divided and used separately. Laboratories should decide which serotypes to test routinely for, based on the prevalence of isolates in their geographical area.

Briefly, material from the specimen is applied directly to two marked circles on an alcohol-cleaned microscope slide. Tissue is applied to the slide by touch preparation, with adequate pressure to express bacteria from spaces. The slide is allowed to air dry, is heat fixed, and then is fixed by overlaying the slide with 10% buffered (pH 7.0) formalin for 10 minutes. The formalin is rinsed from the slide with sterile distilled water, and the slide is allowed to air dry. If it cannot be stained immediately (delay is not acceptable for clinical specimens from living patients), the slide may be frozen at −20° C in the dark for several weeks. One of the two specimen areas is tested against pooled antisera, and the other is tested against specific antiserum in the event of a positive result. The sensitivity of the DFA test is low, since 10^4 to 10^5 bacteria per milliliter of specimen are required for detection. Slides should be examined for 15 to 20 minutes before being declared negative, especially by technologists who rarely perform the test. The organisms appear as brightly fluorescent rods, with a velvety texture (Figure 40.8). Silver stains or even Giemsa stain may reveal organisms in tissue, and *L. micdadei* stains acid fast in tissue sections and occasionally in sputum.[2] Culture of specimens must always be performed as well, since it will detect those organisms that are not included within the usual battery of serotypes, and since culture is more sensitive than DFA.

Two recently developed immunological tests for Legionnaires' disease have shown great promise. Detection of *L. pneumophila* antigen in urine by means of a radioimmunoassay (RIA) (BINAX, South Portland, Me., and DuPont Co.)[1] or enzyme-linked immunosorbent assay (ELISA) test has yielded very specific and sensitive results. ELISA methods are also being used to detect circulating antigens in serum with good results. A genetic probe (Gen-Probe) has been evaluated for direct detection tests.[7] The reagent is being modified to yield more reliable results.

SEROLOGIC AND IMMUNOLOGICAL DIAGNOSIS OF LEGIONNAIRES' DISEASE

Most patients with legionellosis have been diagnosed retrospectively by detection of a fourfold rise in anti-*Legionella* antibody with an indirect fluorescent antibody test. Serum specimens no closer than 2 weeks apart should be tested. Confirmation of disease is accomplished by a fourfold rise in titer to more than 128 or, for unusual cases, a single serum with a titer of more than 256 and a characteristic clinical picture. However, as many as 12% of healthy persons yield titers as high as 1:256, and unfortunately, individuals with Legionnaires' disease may not exhibit serologic titers until as long as 8 weeks after the primary illness or may never display significant antibody titer rises. Commercially prepared antigen-impregnated slides are available from numerous suppliers.

TREATMENT OF LEGIONNAIRES' DISEASE

Erythromycin, 4 g per day intravenously in adults, is the drug of choice for treatment of disease caused by *Legionella*. Penicillins, cephalosporins of all generations, and aminoglycosides are not effective and should not be used. In vitro suscepti-

FIGURE 40.8
Fluorescent antibody–stained *Legionella pneumophila* from a positive control slide.

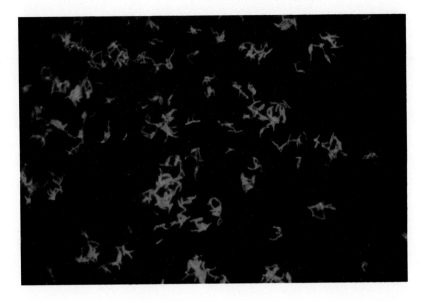

bility studies are not predictive of clinical response and should not be performed for individual isolates. High-dose trimethoprim-sulfamethoxazole and tetracyclines have been effective in some patients. The concomitant administration of rifampin along with erythromycin may prove beneficial, especially in severe cases. For patients who cannot take erythromycin, doxycycline (plus rifampin for moderately and severely ill patients) is recommended. Clinical response usually follows the introduction of effective therapy within 48 hours.

GENUS *PROTOTHECA*

EPIDEMIOLOGY AND PATHOGENESIS
OF *PROTOTHECA* INFECTION

Prototheca are algae without chlorophyll that morphologically resemble green algae. The organisms are worldwide in distribution and are found in water and soil. They enter the body through breaks in the skin, usually associated with minor trauma. The primary syndrome is a chronic, nodular skin lesion, although the organism can cause joint infection, wound infection, and rare disseminated disease in humans. The presence of the organism, which is resistant to killing by phagocytes, results in a granulomatous response. Spontaneous recovery does not occur.

LABORATORY IDENTIFICATION
OF *PROTOTHECA*

Two species of *Prototheca* are associated with human disease, *P. wickerhamii* and *P. zopfii*. The organisms grow very well on Sabouraud's dextrose agar and presumably on any other agar that supports the growth of yeast, such as brain-heart in-

fusion agar and Sabouraud's dextrose with heart infusion. Colonies are opaque, white to tan, smooth, heaped, and moist and resemble those of yeast. Examination under wet mount, however, reveals the characteristic structure of *Prototheca* (Figure 40.9). Cells are spherical, varying in size from tiny (1.5 μg diameter) to large (three times the size of a red blood cell). Within the cells, called *sporangia*, endospores can be seen developing.

For most laboratories, identification of *Prototheca* to genus by observation of typical cellular morphology is sufficient. If definitive identification is required, the API 20C system (Analytab Products, Inc.), a multiwell identification system for yeasts, accurately determines the assimilation pattern and other biochemical parameters of *Prototheca* sp. within 4 days.[22]

The infections caused by this organism must be treated or surgically excised. Most patients respond to amphotericin B, although ketoconazole has been used successfully in at least one patient. *Prototheca* sp. are resistant to flucytosine.

GENUS *SIMONSIELLA*

EPIDEMIOLOGY AND PATHOGENESIS
OF *SIMONSIELLA* INFECTION

These strangely shaped gram-negative organisms are normal oral flora in mammals and other warm-blooded animals. They resemble fat caterpillars and are related genetically to other members of the family Neisseriaceae. Although isolated from gastric aspirate of a neonate, the species was not assumed to be pathogenic in one reported case.[31]

FIGURE 40.9
Prototheca organisms seen in wet mount stained with lactophenol cotton blue.

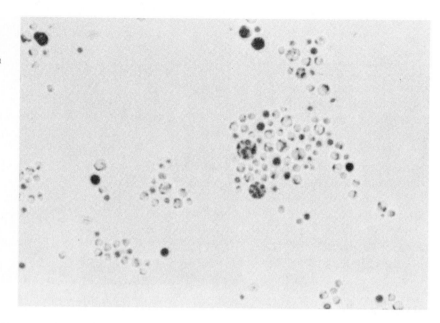

LABORATORY IDENTIFICATION OF *SIMONSIELLA*

Simonsiella grows well on blood agar aerobically. Colonies are translucent, gray, and 2 mm in diameter after overnight incubation. The isolate reported was oxidase and catalase positive.[31] Identification can be made on the basis of Gram stain morphology alone. The cells are arrayed side by side, tightly packed, to form a large structure approximately 3 μm wide by 10 to 20 μm long (Figure 40.10). Recognition of this organism may result in better definition of its role, if any, in infection.

GENUS *STOMATOCOCCUS*

EPIDEMIOLOGY AND PATHOGENESIS OF *STOMATOCOCCUS* INFECTION

Stomatococcus mucilaginosus, a member of the family Micrococcaceae, is normal oral flora in humans. It has been reported as an agent of endocarditis, primarily in intravenous drug abusers; bacteremia; and chronic ambulatory dialysis–associated peritonitis.[6,23,25]

LABORATORY IDENTIFICATION OF *STOMATOCOCCUS*

Stomatococcus is differentiated from other micrococcaceae by its negative or weak catalase reaction. The organism does not grow in the presence of 5% sodium chloride. It hydrolyzes esculin and exhibits a positive *l*-pyroglutamyl aminopeptidase (PYR, as used for preliminary identification of enterococci and *Streptococcus pyogenes*) test. *Stomatococcus* has been characterized as "sticky staphylococci" when it is recognized among normal oral

flora on culture. Important isolates should be sent to a reference laboratory for definitive identification.

TREATMENT OF *STOMATOCOCCUS MUCILAGINOSUS* INFECTIONS

Isolates have been reported susceptible to most antimicrobial agents, including penicillin and cephalosporins. Penicillin is the drug of choice. One strain, isolated from the blood of a drug abuser with endocarditis, was described as relatively resistant to penicillin.[23]

GENUS *STREPTOBACILLUS*

EPIDEMIOLOGY AND PATHOGENESIS OF *STREPTOBACILLUS* INFECTION

Streptobacillus moniliformis, the only species within the genus, is one of the etiological agents of rat-bite fever. Humans acquire the infection via the bite of an infected rat, including laboratory rats, or less often via contaminated milk, food, or water.[13]

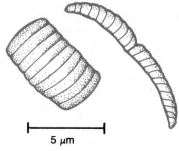

FIGURE 40.10

Diagram of structure of *Simonsiella* sp.

When the organism is acquired by ingestion, the disease is called *Haverhill fever*. The organism is normal oral flora of mice, wild and laboratory rats, and other rodents, although it is also associated with respiratory disease and fatal systemic disease in laboratory animals. Virulence factors have not been defined. The organism is known to spontaneously develop L forms (without cell walls), which may allow persistence of the organism in some sites.

Patients develop acute onset of chills, fever, headache, and severe joint pains. Within the first 2 days, patients exhibit a maculopapular, petechial, or morbilliform rash on the palms, soles of the feet, and extremities. Complications include endocarditis, septic arthritis, and pneumonia.[5,13,17] Another consequence of rat-bite fever with *Streptobacillus* is a false-positive Venereal Disease Research Laboratories (VDRL) test.

LABORATORY IDENTIFICATION
OF *STREPTOBACILLUS*

Organisms may be cultured from blood, material aspirated from infected joints or lymph nodes, or aspirated from the pus within lesions.[17] The organism requires the presence of blood, ascitic fluid, or serum for growth. Growth occurs on blood agar, incubated in a very moist environment with 5% to 10% CO_2, after 48 hours' incubation at 37° C. Colonies are not hemolytic. Addition of 10% to 30% ascitic fluid (available commercially from some media suppliers, such as Difco Laboratories) or 20% horse serum should facilitate recovery of the organism. The organism grows as "fluff balls" or "bread crumbs" near the bottom of the tube of broth or on the surface of the sedimented red blood cell layer in blood culture media. The SPS present in most blood culture media is inhibitory.[5] Colonies grown on brain-heart infusion agar supplemented with 20% horse serum are small, smooth, glistening, colorless or grayish, with irregular edges.

Colonies may also exhibit a fried egg appearance, with a dark center and a flattened, lacy edge. These colonies have undergone the spontaneous transformation to the L form. Stains of L-form colonies yield coccobacillary or bipolar staining coccoid forms; usually a special stain, such as the Dienes stain (performed by pathologists), is required. Acridine orange stain also reveals the bacteria when Gram stain fails because of lack of cell wall constituents.

Gram-stained organisms from standard colonies show extreme pleomorphism, with long filamentous forms, chains, and swollen cells. The carbolfuchsin counterstain or the Giemsa stain may be necessary for visualization (Figure 40.11). With the same stains, organisms may be visualized directly in clinical material for rapid diagnosis.

S. moniliformis does not produce indole and is catalase, oxidase, and nitrate negative, in contrast to organisms with which *Streptobacillus* may be confused, including *Actinobacillus*, *Haemophilus aphrophilus*, and *Cardiobacterium*. In addition, the clinical history of a rat bite is usually elicited. The other putative agent of rat-bite fever, *Spirillum minus* (Chapter 41), cannot be grown in vitro. Serologic diagnosis of rat-bite fever is also useful; most patients develop agglutinating titers to the

FIGURE 40.11

Gram stain of *Streptobacillus moniliformis* from growth in thioglycolate broth with 20% serum. (Courtesy Robert E. Weaver, Centers for Disease Control, Atlanta.)

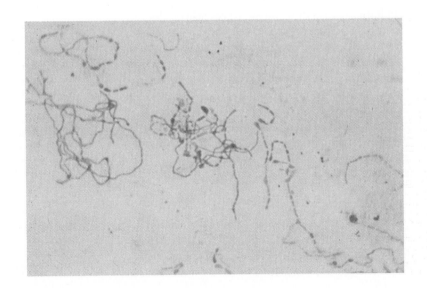

causative organism. The specialized serologic tests are performed only at national reference laboratories, since the disease is extremely rare in the United States.

TREATMENT OF *STREPTOBACILLUS MONILIFORMIS* INFECTIONS

The organism is susceptible to penicillin and streptomycin. Therapy with a combined regimen of a penicillin and an aminoglycoside is probably most effective. The organism is only weakly susceptible to tetracycline in vitro. *S. moniliformis* is resistant to sulfonamides.

THERMOPHILIC BACTERIA

An unusual group of bacteria has been implicated as etiological agents of human disease.[24] Isolated from specimens such as cerebrospinal fluid, blood, and heart valves, these organisms are able to grow at increased temperatures. Called thermophilic bacteria, most isolates grew better at 42° to 50° C than they did at 37° C, the temperature at which they were first isolated. All species were grown on tryptone-glucose–yeast extract agar. None was able to grow at room temperature. No strains were fermentative, and only a few strains were able to use glucose or maltose oxidatively in O-F media (Chapter 29). No organisms grew on Mac-Conkey agar, and none grew on triple sugar iron agar (TSIA) or Kligler's iron agar (KIA) agar slants. All strains were oxidase positive; none was indole positive. There are probably several species among this group, but further studies are needed to delineate them. It is certain that at least several isolates were etiological agents of disease in humans who had no predisposing conditions. All isolates were susceptible to penicillin and cephalosporins, and most were susceptible to aminoglycosides. Since the isolates grow poorly at normal incubator temperatures, they might be dismissed by microbiologists unless an index of suspicion exists. Other thermophilic microorganisms, such as certain actinomycetes (Chapter 34) and *Cytophaga* sp., found in environmental sources, have been implicated as etiological agents of human allergic pneumonitis, although they do not grow within the human body.[16]

UNNAMED BACTERIA

Several unnamed bacterial groups, morphologically similar but not yet definitively characterized, have been implicated as etiological agents of human disease. Because the CDC is the national ref-

erence laboratory, most clinically significant isolates that cannot be identified by other laboratories in the United States are ultimately sent to CDC. Such organisms, designated by letters that reflect their characteristics (EF stands for eugonic fermenter, which means good-growing fermenter; DF stands for dysgonic fermenter, which means poor-growing fermenter; EO stands for eugonic oxidizer) or by other letters (e.g., IIk, IVc-2, HB-5) so that they can be categorized by microbiologists, may be isolated from clinical specimens. Although most individual groups are not discussed in this text, they may be listed in some tables and charts because they resemble organisms being discussed. Identification tables published by the CDC and the *Manual of Clinical Microbiology*, published by the American Society for Microbiology, should be consulted for further details concerning these organisms.

REFERENCES

1. Aguero-Rosenfeld, M.E., and Edelstein, P.H. 1988. Retrospective evaluation of the DuPont radioimmunoassay kit for detection of *Legionella pneumophila* serogroup 1 antigenuria in humans. J. Clin. Microbiol. 26:1775.
2. Bryce, E.A., Barteluk, R.L., Noble, M.A., et al. 1991. Community-acquired *Legionella micdadei* pneumonia presenting as acid fast bacilli in sputum. Clin. Microbiol. Newsletter 13:95.
3. Bunker, M.L., Chewning, L., Wang, S.E., and Gordon, M.A. 1988. *Dermatophilus congolensis* and "hairy" leukoplakia. Am. J. Clin. Pathol. 89:683.
4. Catlin, B.W. 1992. *Gardnerella vaginalis*: characteristics, clinical considerations, and controversies. Clin. Microbiol. Rev. 5:213.
5. Clausen, C. 1987. Septic arthritis due to *Streptobacillus moniliformis*. Clin. Microbiol. Newsletter 9:123.
6. Coudron, P.E., Markowitz, S.M., Mohanty, L.B., et al. 1987. Isolation of *Stomatococcus mucilaginosus* from drug user with endocarditis. J. Clin. Microbiol. 25:1359.
7. Edelstein, P.H. 1986. Evaluation of the Gen-Probe DNA probe for the detection of legionellae in culture. J. Clin. Microbiol. 23:481.
8. Edelstein, P.H. 1984. Legionnaires' disease laboratory manual. National Technical Information Service. Publ. No. PB84-156827, U.S. Department of Commerce, Springfield, Va.
9. Fields, B.S. 1991. The role of amoebae in legionellosis. Clin. Microbiol. Newsletter 13:92.
10. Frothingham, T.E. 1984. Bartonella. p. 781. In Joklik, W.K., Willett, H.P., and Amos, D.B., editors. Zinsser microbiology, ed 18. Appleton-Century-Crofts, Norwalk, Conn.
11. Garcia, F.U., Wojta, J., Broadley, K.N., et al. 1990. *Bartonella bacilliformis* stimulates endothelial cells in vitro and is angiogenic in vivo. Am. J. Pathol. 136:1125.
12. Hillier, S.L., Krohn, M.A., Rabe, L.K., et al. 1993. Normal vaginal flora, H_2O_2-producing lactobacilli and

bacterial vaginosis in pregnant women. Clin. Infect. Dis. 16 (suppl 4):S273.

12a. Hillier, S.L., Lipinski, C., Briselden, A.M., and Eschenbach, D.A. 1993. Efficacy of intravaginal 0.75% metronidazole gel for the treatment of bacterial vaginosis. Obstet. Gynecol. 81:963.

13. Jenkins, S.G. 1988. Rat-bite fever. Clin. Microbiol. Newsletter 10:57.

14. Krohn, M.A., Hillier, S.L., Lee, M.L., et al. 1991. Vaginal *Bacteroides* species are associated with an increased rate of preterm delivery among women in preterm labor. J. Infect. Dis. 164:88.

15. Land, G., McGinnis, M.R., Staneck, J., and Gatson, A. 1991. Aerobic pathogenic *Actinomycetales*. p. 340. In Balows, A., Hausler, W.J., Jr., Herrmann, K.L., et al., editors. Manual of clinical microbiology, ed 5. American Society for Microbiology, Washington, D.C.

16. Liebert, C.A., Hood, M.A., Deck, F.H., et al. 1984. Isolation and characterization of a new *Cytophaga* species implicated in a work-related lung disease. Appl. Environ. Microbiol. 48:936.

17. Lopez, P., Euras, J., Anglada, A., et al. 1992. Infection due to *Streptobacillus moniliformis*. Clin. Microbiol. Newsletter 14:38.

18. Lugo-Miro, V.I., Green, M., and Mazur, L. 1992. Comparison of different metronidazole therapeutic regimens for bacterial vaginosis. J.A.M.A. 268:92.

19. Macher, A.M., Casale, T.B., and Fauci, A.S. 1982. Chronic granulomatous disease of childhood and *Chromobacterium violaceum* infections in the southeastern United States. Ann. Intern. Med. 97:51.

20. Nugent, R.P., Krohn, M.A., and Hillier, S.L. 1991. Reliability of diagnosing bacterial vaginosis is improved by a standardized method of Gram stain interpretation. J. Clin. Microbiol. 29:297.

21. O'Connor, S.P., Dorsch, M., Steigerwalt, A.G., et al. 1991. 16S rRNA sequences of *Bartonella bacilliformis* and cat scratch disease bacillus reveal phylogenetic realtionships with the alpha-2 subgroup of the class *Proteobacteria*. J. Clin. Microbiol. 29:2144.

22. Padhye, A.A., Baker, J.G., and D'Amato, R.F. 1979. Rapid identification of *Prototheca* species by the API 20C system. J. Clin. Microbiol. 10:579.

23. Pinsky, R.L., Piscitelli, V., and Patterson, J.E. 1989. Endocarditis caused by relatively penicillin-resistant *Stomatococcus mucilaginosus*. J. Clin. Microbiol. 27:215.

24. Rabkin, C.S., Galaid, E.I., Hollis, D.G., et al. 1985. Thermophilic bacteria: a new cause of human disease. J. Clin. Microbiol. 21:553.

25. Relman, D.A., Ruoff, K., and Ferraro, M.J. 1987. *Stomatococcus mucilaginosus* endocarditis in an intravenous drug abuser. J. Infect. Dis. 155:1080.

26. Ristic, M., and Kreier, J.P. 1984. Bartonellaceae Gieszczykiewicz 1939, 25[AL]. p. 717. In Krieg, N.R., and Holt, J.G., editors. Bergey's manual of systematic bacteriology, vol 1. Williams & Wilkins, Baltimore.

27. Schmitt, C., Sobel, J.D., and Meriwether, C. 1992. Bacterial vaginosis: treatment with clindamycin cream versus oral metronidazole. Obstet. Gynecol. 79:1020.

28. Stout, J.E., Yu, V.L., Maraga, M.E., et al. 1992. Potable water as a cause of sporadic cases of community acquired Legionnaires' disease. N. Engl. J. Med. 326:151.

29. Tompkins, L.S., Roessler, B.J., Redd, S.C., et al. 1988. *Legionella* prosthetic-valve endocarditis. N. Engl. J. Med. 318:530.

30. Totten, P.A., Amsel, R., Hale, J., et al. 1982. Selective differential human blood bilayer media for isolation of *Gardnerella (Haemophilus) vaginalis*. J. Clin. Microbiol. 15:141.

31. Whitehouse, R.L.S., Jackson, H., Jackson, M.C., and Ramji, M.M. 1987. Isolation of *Simonsiella* sp. from a neonate. J. Clin. Microbiol. 25:522.

BIBLIOGRAPHY

Greenwood, J.R., and Pickett, M.J. 1984. Genus *Gardnerella* Greenwood and Pickett 1980 170[VP]. p. 587. In Krieg, N.R., and Holt, J.G., editors. Bergey's manual of systematic bacteriology, vol 1. Williams & Wilkins, Baltimore.

Lee, T.S., and Wright, B.D. 1981. Fulminating chromobacterial septicaemia presenting as respiratory distress syndrome. Thorax 36:557.

Sneath, P.H.A. 1984. Genus *Chromobacterium* Bergonzini 1881 153[AL]. p. 580. In Krieg, N.R., and Holt, J.G., editors. Bergey's manual of systematic bacteriology, vol 1. Williams & Wilkins, Baltimore.

Starr, A.J., Cribbett, L.S., Poklepovic, J., et al. 1981. *Chromobacterium violaceum* presenting as a surgical emergency. South. Med. J. 74:1137.

Winn, W.C., Jr. 1988. Legionnaires' disease: historical perspective. Clin. Microbiol. Rev. 1:60.

41

NEW, CONTROVERSIAL, DIFFICULT-TO-CULTIVATE, OR NONCULTIVABLE ETIOLOGICAL AGENTS OF DISEASE

In addition to those diseases caused by agents that cannot be grown in culture using existing methods and those that are difficult to isolate, a number of diseases appear to be caused by an infectious agent, although such an agent has either not been identified or not isolated in vitro. Reasons for which an infectious agent might be considered to be associated with a disease include an epidemiological pattern suggestive of dissemination by an agent or a clinical presentation that is similar to that produced by known etiological agents. Once the suggestion of infectious etiology is advanced, tools of science can be used to investigate the problem. Within the last several years, a number of diseases, either newly recognized or described long ago, have been definitely traced to an etiological agent for the first time. For example, *Helicobacter pylori* has now been definitely associated with gastritis, a disease originally not thought to be infectious in nature (see Chapter 31). At least three agents of cat-scratch disease have been isolated and characterized. The diseases mentioned in this chapter are either syndromes in which the etiological agent is difficult to isolate or demonstrate; well-described syndromes for which no etiological agent has been found, although the diseases appear to be of an infectious nature; or diseases with which a newly described agent is associated. The agents mentioned in this chapter may not be routinely sought by clinical microbiology laboratories, depending on their resources.

DISEASES ASSOCIATED WITH AN ETIOLOGICAL AGENT

There are, of course, a number of infectious diseases for which the agent is well known but cannot be cultivated easily, if at all, such as syphilis, leprosy, hepatitis A and B, warts (papillomavirus), and the slow virus diseases of progressive multifocal leukoencephalopathy (JC virus) and subacute sclerosing panencephalitis (measles-like paramyxovirus). Because these diseases are so well characterized, they are discussed in the appropriate chapters of this book.

SPIRILLUM MINUS, AN AGENT OF RAT-BITE FEVER

Spirillum minus (formerly called *S. minor,* an incorrect Latin usage), an agent of rat-bite fever, has never been grown in culture. The organism has been visualized in Giemsa- or Wright-stained blood films and can be seen under darkfield microscopy. *S. minus* is a thick, spiral, gram-negative organism with two or three coils and polytrichous polar flagella. Infection can be transmitted by the bite or scratch of an infected animal. The bite wound heals spontaneously, but 1 to 4 weeks later reulcerates to form a granulomatous lesion at the same time that the patient develops constitutional symptoms of fever, headache, and a generalized, blotchy purplish maculopapular rash. Although the organism rarely can be seen in blood, it may be visualized in darkfield preparations made from the lesion. Diagnosis is definitively made by injection of lesion material or blood into (experimental) white mice or guinea pigs. Differentiation between rat-bite fever caused by *S. minus* ("Sodoku") and that caused by *Streptobacillus moniliformis* (Haverhill fever; Chapter 40) is usually accomplished based on clinical presentations of the two infections and the isolation of the latter organism in culture. The incubation period for *S. minus* is much longer than that of streptobacillary rat-bite fever, which has occurred within 12 hours of the initial bite.

ROCHALIMAEA (BARTONELLA) HENSELAE AND *AFIPIA FELIS* IN THE ETIOLOGY OF CAT-SCRATCH DISEASE

Cat-scratch disease was first recognized by Robert Debre in Paris during the 1930s. The disease is not uncommon, with probably more than 6000 cases

occurring annually in the United States.[2] Patients are usually children (80%), and the primary sites of lesions are the hands, arms and legs, and face and neck. Most cases are acquired following the scratch, bite, or lick of a cat or kitten, usually newly arrived to the household. Patients exhibit regional tender lymphadenopathy, anorexia, fatigue, fever, and headache, usually beginning 2 weeks after the cat scratch or contact. They may display a rash, generalized lymphadenopathy, seizures, and other systemic findings. The disease is usually self-limited; complications such as a suppurative lymph node or encephalitis have been reported; fatalities are rare. A form of cat-scratch disease presenting as a proliferative endothelial cell lesion resembling Kaposi's sarcoma has been recently seen in AIDS patients. A more systemic disease, bacillary peliosis, is seen in AIDS and other immunocompromised patients as well, with lesions in internal organs (bone marrow, liver, and spleen) resembling blood-filled cavities (Figure 41.1).[21,28] Until 1983, clinical diagnosis required that a patient fulfill three of the following four criteria:

1. History of animal contact plus a site of primary inoculation
2. Negative laboratory studies for other causes of lymphadenopathy
3. Characteristic histopathology of the lesion or lymph node
4. A positive skin test

The skin test antigen was prepared from heat-treated pus taken from another patient's lesion. Injection of 0.1 ml of this material intradermally elicited a wheal or papule within 72 hours, with

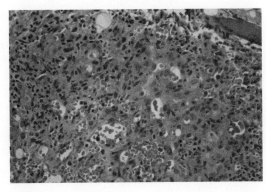

FIGURE 41.1
Cavernous blood-filled sinuses in bone marrow of patient with bacillary peliosis. Warthin-Starry silver stain of this tissue revealed numerous organisms. (Courtesy Dr. Jay Parker, St. Jude Medical and Rehabilitation Center, Fullerton, Calif.)

PROCEDURE 41.1

WARTHIN-STARRY STAIN

PRINCIPLE

Silver ions precipitate on the cell wall of bacteria and other structures; the black color enhances visibility of the organisms.

METHOD

1. Make solutions as follows:
   a. Acidulated water
      Triple distilled water 1000 ml
      1% aqueous citric acid enough to bring triple distilled water to pH 4.0
   b. 1% silver nitrate solution
      Silver nitrate, crystalline 1 g
      Acidulated water 100 ml
   c. 2% silver nitrate solution
      Silver nitrate, crystalline 2 g
      Acidulated water 100 ml
   d. 5% gelatin solution
      Sheet gelatin, high grade 10 g
      Acidulated water 200 ml
   e. 0.15% hydroquinone solution
      Hydroquinone crystals, 0.15 g, photographic quality
      Acidulated water 100 ml
      All chemicals should be available from a standard chemical supply company, such as Sigma Chemical Co. or E. Merck Darmstadt.

2. Fix tissue sections in buffered neutral formalin. Avoid chromate fixatives. This fixative is standard for most pathology laboratories. The tissue is embedded in paraffin; sections are cut at a thickness of 6 μm.

3. Warm the 2% silver nitrate, the 5% gelatin, and the 0.15% hydroquinone in 50 ml beakers in a water bath heated to 54° C.

4. Deparaffinize the (patient's) tissue slide and a known positive control slide through a series of xylene changes with increasing concentrations of water, until the slide is finally placed into triple distilled water for final hydration. These procedures are standard histological methods. Place the slides into a Coplin jar containing the 1% silver nitrate solution, preheated in a 43° C water bath. Allow the silver to impregnate the slides for 30 minutes at 43° C in the water bath.

5. While the slides are being impregnated, prepare the developer as follows:
   Silver nitrate solution (2%) 1.5 ml
   Gelatin solution (5%) 3.75 ml
   Hydroquinone solution (0.15%) 2 ml
   Prepare this developer, immediately before it is used, in a small beaker or flask.

6. Move the slides from the impregnator, lay them flat across a slide holder, and immediately flood them with the warm developer solution. Allow the sections to develop until they are light brown or yellow.

Check the known control periodically, which should show black spirochetes or bacteria against a yellow or light-brown background. When handling the slides, use paraffin-coated or plastic forceps.

7. Wash quickly and thoroughly in hot tap water, approximately 50° C. Rinse in distilled water.

8. Dehydrate in 95% alcohol, then absolute alcohol, and clear in xylene by placing the slides in two jars of each solution for 5 minutes in each jar.

9. Mount with Permount or Histoclad (available from Baxter/American Scientific Products). Observe for characteristic black organisms against a light brown or yellow background.

QUALITY CONTROL

Stain a known positive control slide concurrently with each specimen staining procedure. Although most bacteria will stain black with Warthin-Starry, a thin, spiral organism will best control the procedure.

Modified from Luna, L.G. 1968. Warthin-Starry method for spirochetes and Donovan bodies. In Manual of histologic staining methods of the Armed Forces Institute of Pathology. McGraw-Hill Book Co., New York.

approximately 10% of patients exhibiting false reactions.

Since 1983, when organisms were first visualized in the lesions of cat-scratch disease with the Warthin-Starry silver stain[15] (Procedure 41.1) (Figure 41.2), workers have been trying to cultivate the agent. In 1988 English and colleagues isolated an agent, later named *Afipia felis*, from lesions of cat-scratch disease. Similar bacteria have now been isolated by clinical laboratories.[2] *Afipia* are catalase-negative, oxidase-positive, gram-negative bacilli, motile by means of a

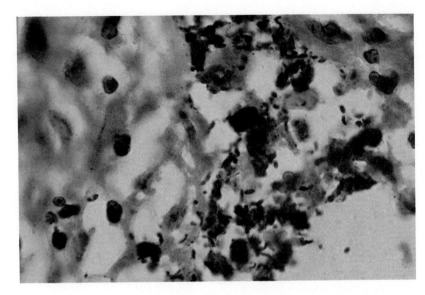

FIGURE 41.2

Cat-scratch disease bacillus in lymph node tissue (Warthin-Starry silver stain). The identification of the organism, *Rochalimaea (Bartonella)* or *Afipia*, is unknown. (Courtesy Dr. Ellen Kahn, North Shore University Hospital, Manhasset, N.Y.)

single flagellum. They grow best at 30° C and have been recovered in blood culture media and a cell culture system. Subcultures will grow on buffered charcoal yeast-extract agar and blood-containing agars. However, another gram-negative bacillus called *Rochalimaea (Bartonella) henselae*, isolated from blood of patients with fever and identified by DNA methods in tissues of patients with bacillary angiomatosis and peliosis, may actually be a more common etiological agent of cat-scratch disease, as well as the agent of bacillary angiomatosis and peliosis.[19-21,28] The original cat-scratch disease skin test antigen was shown to contain DNA homologous to that of *Rochalimaea* sp. but not to that of *Afipia* sp. The complete etiology of these diverse syndromes and the range of infections caused by *Rochalimaea* sp. are rapidly evolving areas of study. The genus *Bartonella* may be more accurate, and a name change has been proposed. Current recommendations for cultivating the agent from blood and tissue involve using the lysis-centrifugation system, freshly made chocolate agar plates or Brucella agar supplemented with Fildes and hemin (Figure 41.3).[23] Fresh agar helps to supply the moisture necessary for growth. Isolation of the organism may require prolonged (several weeks) incubation in a humid atmosphere with 5% to 10% CO_2.[23,28] Treatment with tetracycline or doxycycline has been effective for various *Rochalimaea* syndromes, but disease caused by *Afipia* may not respond.

CALYMMATOBACTERIUM GRANULOMATIS, THE AGENT OF GRANULOMA INGUINALE

Calymmatobacterium granulomatis is the etiological agent of "granuloma inguinale," or "donovanosis," a venereally transmitted disease charac-

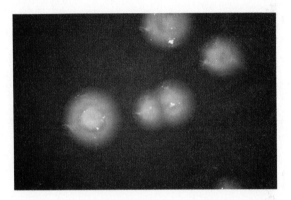

FIGURE 41.3

Colonies of *Rochalimaea (Bartonella) henselae* on sheep blood agar. Note the cauliflower-like appearance. (Courtesy Catherine Nesbit and Linda Mann.)

terized by subcutaneous nodules that enlarge and evolve to form beefy, erythematous, granulomatous, painless lesions that bleed easily. The lesions, which usually occur on the genitalia, have been mistaken for neoplasms. Patients often have inguinal lymphadenopathy. The organism has been isolated only from human lesions, and the disease is confined primarily to the tropics, although it has been reported in the United States.[5] Infectivity of this bacillus must be low, since sexual partners of infected patients often do not themselves become infected.

The organism can be visualized in scrapings of lesions stained with Wright's or Giemsa stain. Subsurface infected cells must be present; surface epithelium is not an adequate specimen. Groups of organisms are seen within mononuclear endothelial cells; this pathognomonic entity is known as a "Donovan body," named after the physician who first visualized the organism in

PROCEDURE 41.2

CULTIVATION OF *C. GRANULOMATIS*

METHOD

1. Prepare medium as follows:
   Peptone (Difco Laboratories) 10 g
   Tryptone (Difco Laboratories) 3 g
   Glucose 3 g
   Commercial sea salts 2 g (any chemical supplier)
   Agar (Bacto agar, Difco Laboratories) 1.2 g
   Distilled water 1000 ml
     Dissolve all ingredients by heating gently. Adjust pH to 7.3. Autoclave at 121° C for 15 minutes and cool to 45° C. Soak fresh eggs in 70% alcohol for 10 minutes before aseptically removing egg yolks. Dilute egg yolks with equal volume of sterile saline (0.85% NaCl) and mix with equal parts of the basal medium. Dispense the medium in sterile 16 × 125 mm tubes, 20 ml per tube. (Commercially available egg yolk [Difco Laboratories] may be suitable as a substitute for fresh egg yolks.)

2. A physician aspirates lymph node material or obtains material from the lesion with a sterile scalpel. The lesion is cleansed by carefully swabbing its surface with a sterile gauze pad soaked in sterile saline to remove surface contamination. A small slice near the edge of the lesion will allow removal of subsurface material. The best material is obtained from under the surface of the lesion near the edge.

3. Inoculate the medium well below the surface with the scraping or aspirated material. Incubate for at least 72 hours at 37° C. Incubating the tubes in a candle jar may help to retain the low oxygen concentration of the medium. Gram stain the broth, using a carbolfuchsin counterstain. *C. granulomatis* is gram-negative

Modified from Dienst R.B. and Brownell G.H. 1984. Genus *Calymmatobacterium Aragao* and *Vianna 1913.* In Krieg, N.R., and Holt, J.G., editors. Bergey's manual of systematic bacteriology, vol 1, Williams & Wilkins, Baltimore.

---

such a lesion. The organism stains as a blue rod with prominent polar granules, giving rise to a "safety pin" appearance, surrounded by a large, pink capsule. The capsular polysaccharide shares several cross-reactive antigens with *Klebsiella* species, fueling speculation that *Calymmatobacterium* is closely related to *Klebsiella.*

Cultivation in vitro is very difficult, but it can be done using media containing some of the growth factors found in egg yolk (Procedure 41.2).[5,6] A medium described by Dienst has been used to culture *Calymmatobacterium* from aspirated bubo material, as described by Dienst and Brownell.[5]

Although gentamicin and chloramphenicol are the most effective for the therapy of granuloma inguinale, tetracycline or ampicillin is usually the drug of first choice. Trimethoprim-sulfamethoxazole or erythromycin (in pregnancy) is usually effective treatment for granuloma inguinale.

HEPATITIS C, D, E, AND OTHER VIRAL HEPATITIS

The viruses of hepatitis A and B have been identified and characterized, but only hepatitis A virus has been grown in culture outside of research facilities. The infectious particle of hepatitis B virus, the Dane particle, is present in the blood and hepatocytes of patients and can be visualized by electron microscopy. Several different antigenic constituents are released during infection. Diagnosis is based on detection of either antibodies to the constituents or the antigens themselves. Newer methods for detecting the surface antigen (HBsAg), an early protein antigen (HBeAg), and antibodies to HBsAg, HBeAg, and hepatitis core antigen (HBc) include radioimmunoassay (RIA) (Chapter 12) and enzyme-linked immunosorbent assay (ELISA) (Chapter 11). Microbiology laboratories usually do not perform these tests.

Coinfection of patients with hepatitis B virus and a defective mutant virus, the "delta agent," now called hepatitis D virus, is responsible for some proportion of the most severe cases of hepatitis. Hepatitis D requires coinfection with hepatitis B in the delta-infected cell for damage to occur. Persons at risk of acquiring hepatitis D are those with chronic asymptomatic hepatitis B infection. Serologic tests for delta antigen and antibody are available in ELISA and RIA formats.

Non-A, non-B hepatitis is the most common transfusion-associated hepatitis. Hepatitis C virus has been identified as the major etiological agent of this type of hepatitis. Hepatitis C virus, characterized as a single-stranded RNA virus, appears to be similar to togaviruses or flaviviruses (see Chapter 43). The virus itself has not been isolated or cultivated in vitro, but serologic tests (ELISA and RIA) are available and used routinely to screen donated blood.[1] Other agents of non-A, non-B hepatitis have not yet been identified. Hepatitis E has recently been identified as the agent of a water-borne hepatitis similar to that of hepatitis A, seen commonly in developing nations.[27] This virus resembles those in the calicivirus family. Immunoassays are available but not used widely yet. It is safe to assume that other agents of viral hepatitis will be discovered in the future.

RHINOSPORIDIOSIS

Rhinosporidiosis is a chronic, celluloproliferative infection of the mucous membranes of the face and other areas, usually in the nose and less often in the conjunctiva, caused by a fungus that produces a spherule-like body similar to that of *Coccidioides immitis*. The disease is much more common in men than in women. The lesions ultimately develop into pedunculated polyps. Most prevalent in India, the infection has been described in the United States. A similar disease occurs in horses, cattle, and other animals. The agent, seen macroscopically in the polyp as tiny white dots, consists of large, thick-walled cysts or sporangia containing many endospores. The endospores, each about the size of a red blood cell, are released through a pore in the wall of the sporangium once it has matured to its full size of 200 to 300 μm. The agent of rhinosporidiosis, called *Rhinosporidium seeberi*, has recently been cultured in mammalian epithelial cell tissue culture, which may be a requirement for growth.[13] The fungus induced cellular proliferation in vitro. The cyst can be stained in biopsy material with mucicarmine; histological studies are probably the best diagnostic tool available currently, although it may now be possible for laboratories to cultivate the agent from tissue biopsy material if appropriate tissue culture methods are used.

LOBOMYCOSIS

Lobomycosis is a presumed fungal infection from which an etiological agent has not been cultured. Found in adult males primarily living in the Amazon River basin and in bottle-nose dolphins off the coast of Florida, the disease is characterized by chronic, progressive, subcutaneous nodules, usually on the face or extremities. The lesions can become verrucous or ulcerative. The organism, *Loboa loboi*, can be seen in scrapings and histological sections from the lesions as a large, thick-walled, spheroidal or oval yeast. The yeasts possess multiple buds and often appear in short chains. The buds are the same size as the mother cells, unlike the multiple buds of *Paracoccidioides brasiliensis*. The yeasts can be visualized with 10% KOH and stain with methenamine silver or periodic acid-Schiff (PAS). Diagnosis is made histologically.

KURU, CREUTZFELDT-JAKOB DISEASE, AND GERSTMANN-STRAUSSLER DISEASE[3]

Kuru, a progressively fatal dementing disease of head-hunting tribes of New Guinea, was first thought to be of sex-linked genetic origin, since female adults were involved more often than males, and children were universally affected. When it was realized that the histological lesions in the brains of patients with kuru resembled those found in sheep with a naturally occurring neurological disease with similar symptoms, scrapie, the search began in earnest for an etiological agent. Scrapie was known to be transmissible. In 1965 kuru-like disease was transmitted to monkeys after intracerebral injection of material from the brains of humans who died from kuru. Brains of humans, monkeys, and sheep suffering from the disease show a characteristic spongiform degeneration of tissue (abnormal spaces in the tissue), with vacuoles in the neurons and extensive glial cell proliferation with almost no inflammatory changes. A similar histological and clinical picture was characteristic of another rare human disease, Creutzfeldt-Jakob disease (CJD). This disease, found in humans throughout the world, consists of a very slowly progressive first stage of intellectual deterioration and behavioral abnormalities, followed by a shorter, rapidly progressive terminal stage characterized by myoclonic contractions of various muscle groups. Patients suffer from ataxia, dementia, and delirium that progresses to stupor, coma, and death. Gerstmann-Straussler disease affects adults, leading ultimately to dementia. There is no treatment for any of these diseases. Another histological feature of the brain of patients and laboratory animals with CJD, shared by patients with Alzheimer's disease (another progressive dementing disease of humans), is the presence of characteristically staining areas called amyloid plaques.

The putative agent of CJD and kuru was first purified by workers led by Stanley Prusiner in 1982. The agent consisted primarily of a single protein of approximately 27 to 30 kilodaltons molecular weight, designated PrP 27-30. This structural component was called a **prion** (proteinaceous infectious agent) by Prusiner. Antibodies to PrP 27-30 react with the "amyloid" plaques found in brains of patients with CJD and with the same structures found in the brains of scrapie-infected experimental animals, all of which were found to be crystalline arrays of prion protein, called scrapie-associated fibrils (SAFs). The gene encoding this protein has been identified in human chromosome 20 and cloned.[7] This finding may negate the theory that prions are infectious agents and suggests that prion protein is an altered product of a normal gene. Although Alzheimer's disease also demonstrates similar-appearing amyloid plaques, the existence of this gene and differences in plaque morphology at the electron microscopic structural level argue against a direct association between prions and Alzheimer's syndrome.

The prion genes of humans and hamsters share 89% homology, and there is no homology between this protein and other known proteins. The nature of the infectious process is still not understood. Perhaps another, yet unknown agent induces the prion gene to produce a modified protein that aggregates to cause symptoms of CJD and kuru. Until the pathogenesis of these and similar diseases are known, the tissues from patients with these syndromes must be handled as though they were infectious.

Prions are extremely resistant to denaturation by proteases, nucleases, heat, ultraviolet light, and chemicals. Phenolic compounds, active against most other infectious agents, have no effect on prions. Full-strength chlorine bleach is effective with increased exposure (more than 1 minute). Dilute (1 N) NaOH will also effectively render the prion noninfectious.[4] It is not known how or even whether they reproduce.

Cases of CJD have been diagnosed in several individuals who ate the brains of wild or domestic animals. Ingestion of brains, even well cooked, therefore, cannot be recommended, since cooking practices would be entirely unable to reduce infectivity of an agent such as a prion. Direct inoculation of infectious material has been known to transmit disease, as has transplantation of

corneas, cadaveric dura mater grafts, and administration of growth hormone from infected humans. Organs and tissues from any patients with degenerative neurological diseases should not be used for transplantation.

Laboratory workers who must handle tissue from the brains of patients with CJD or Gerstmann-Straussler disease should take extra precautions. Double gloves, masks, and gowns should be worn, and all manipulations should take place within a biological laminar flow safety cabinet. A solution of 1 N NaOH should be used to soak all instruments, and even the hands of technologists can be soaked briefly in this alkali if accidental exposure occurs. Chlorine bleach is another effective decontaminating agent.

WHIPPLE'S DISEASE

Whipple's disease, found primarily in middle-aged men, is characterized by the presence of PAS-staining macrophages (indicating mucopolysaccharide or glycoprotein) in almost every organ system.[17] A bacillus is seen in macrophages and affected tissues but it has never been cultured. Patients develop diarrhea, weight loss, arthralgia, lymphadenopathy, hyperpigmentation, often a long history of joint pain, and a distended and tender abdomen. Neurological and sensory changes often occur. It has been suggested that a cellular immune defect is involved in pathogenesis of this disease. Recently the agent of this disease was identified in tissues of five patients by amplifying bacterial DNA sequences by PCR and then sequencing the entire genome.[22] The organism resembles actinomycetes but is not related closely to known genera. It has been tentatively named *Tropheryma whippelii*.[22] Patients usually respond well to long-term therapy with antibacterial agents, including trimethoprim-sulfamethoxazole, tetracycline, and penicillin; tetracycline has been associated with serious relapses, however. Colchicine therapy appears to control symptoms.[17] Without treatment the disease is uniformly fatal.

PARVOVIRUS B19, THE AGENT OF ERYTHEMA INFECTIOSUM (FIFTH DISEASE) AND AN AGENT OF APLASTIC CRISIS

Parvovirus B19, the agent of the relatively mild febrile exanthematous disease called "fifth disease" (because it was the fifth childhood exanthem, after the four others: rubella, measles, scarlet fever, and roseola), has recently been associated with aplastic crisis, adult arthritis, and congenital infection. The rash of fifth disease begins on the cheeks, causing them to resemble slapped cheeks. The rash then moves to the extremities, where it may disappear and recur several times. Headache and fever may be associated with the rash in one fourth of patients. The more serious syndrome of aplastic crisis, characterized by a sudden and significant decrease in red blood cell production, is most common in children with congenital anemias, particularly sickle cell disease. A potentially important cause of birth defects, the virus has been shown to infect the fetus of infected pregnant women. At this time, most infected fetuses spontaneously abort. Arthritis, especially in children suffering from fifth disease and family contacts, has been described.

The parvoviruses are among the smallest DNA viruses, infecting a number of animals. Parvovirus B19 has not been grown in culture, but the viral DNA has been cloned, providing antigen for ELISA tests, which are now commercially available.[24]

MISCELLANEOUS OTHER DISEASES

Recently *Mycobacterium paratuberculosis,* originally known as *M. linda,* has been isolated from tissue biopsy material from the intestinal mucosa of patients with Crohn's disease (Chapter 42), although it is not fully established as a cause.[16] Only with great difficulty can this mycobactin-dependent agent be cultivated in the laboratory. The same mycobacterium species is an important pathogen in cattle, causing Johne's disease. Recent genetic studies have shown that *M. paratuberculosis* and *M. avium* complex are genetically closely related and may ultimately be reclassified to the same species.[16] Other intestinal organisms, including spirochetes, have been cultivated but not yet firmly associated with disease.[9,11,26]

At least two mycoplasma species are newly recognized agents of disease. *M. genitalium* is very difficult to grow, but several DNA probe studies have demonstrated its presence in the genital tract. It has been associated with urethritis in males and may be a pathogen in females as well.[10] Because of its fastidious nature, it is rarely detected without special studies. *M. fermentans* (originally called *M. incognitus*) was first isolated from tissues of patients dying of an acute fatal disease who were concurrently infected with HIV.[8,14,25] The virus has subsequently been postulated to be the cause of death in otherwise healthy humans because it was recovered from virtually every tissue sampled. Its role in disease

is very controversial. Ciprofloxacin has been the most effective drug in vitro, but clinical trials have not been performed. Erythromycin is thought to be effective.[14]

Human herpesvirus 6 has now been determined to be the etiological agent of the childhood exanthem roseola infantum. Like fifth disease, the illness is mild and self-limited and appears to be somewhat contagious. The virus is ubiquitous, and most children become infected during the first year of life, usually after maternal antibodies have disappeared. The rash of roseola occurs within 48 hours after the febrile period, which usually lasts 3 to 5 days. Antibody prevalence is approximately 80% among adults.[12]

DISEASES FOR WHICH NO ETIOLOGICAL AGENT HAS BEEN DEMONSTRATED

KAWASAKI DISEASE

Kawasaki disease occurs in children, usually younger than 8 years old, primarily in Japan, but many cases have been reported in the United States.[29] The symptoms include fever that lasts at least 5 days, congestion of the conjunctivae, dry, red lips and red, fissured or "strawberry" tongue, reddening of the palms of hands and soles of feet with ultimate desquamation, a rash over the trunk and face, and cervical lymphadenopathy. The disease has also been called "mucocutaneous lymph node syndrome." Patients are usually ill for 1 month, with the convalescent phase lasting 6 to 10 weeks. The most important sequela of the disease is cardiopathy, which occurs in 20% of patients. Diagnosis of Kawasaki disease can be made in patients in whom other etiologies are excluded and who meet five of the six criteria, providing also that antibiotic treatment does not ameliorate the fever. The rash and subsequent desquamation resemble that caused by staphylococci, particularly toxic shock syndrome staphylococci, but efforts to isolate an organism from patients with Kawasaki disease have been unsuccessful. Several studies have verified the efficacy of treating Kawasaki patients with intravenous gamma globulin plus aspirin, which dramatically reduced the incidence of the serious sequelae, coronary artery abnormalities.[18]

OTHER DISEASES

It seems reasonable to mention multiple sclerosis in this section. Epidemiological studies point to an infectious agent, and the pattern of pathological findings in the neural tissue resembles that found in a progressive multifocal leukoencephalopathy, a slow virus infection. No definite virus particles have been seen in electron microscopic studies of brain tissue from patients with multiple sclerosis, but these patients do exhibit high levels of serum antibodies against a number of known viruses. The pathogenesis of this relatively common demyelinating disease is unknown, but an infectious agent has not been ruled out.

REFERENCES

1. Alter, H.J., Purcell, R.H., Shih, J.W., et al. 1989. Detection of antibody to hepatitis C virus in prospectively followed transfusion recipients with acute chronic non-A, non-B hepatitis. N. Engl. J. Med. 321:1494.

2. Brenner, D.J., Hollis, D.G., Moss, C.W., et al. 1991. Proposal of *Afipia* gen. nov., with *Afipia felis* sp. nov. (formerly the cat scratch disease bacillus), *Afipia clevelandensis* sp. nov. (formerly the Cleveland Clinic Foundation strain), *Afipia broomeae* sp. nov., and three unnamed genospecies. J. Clin. Microbiol. 29:2450.

3. Brown, P. and Gadjusek, D.C. 1991. The human spongiform encephalopathies: kuru, Creutzfeldt-Jakob disease, and the Gerstmann-Straussler-Scheinker syndrome. Curr. Top. Microbiol. 172:1. (Entire issue devoted to this topic.)

4. Brown, P., Rohwer, R.G., and Gajdusek, D.C. 1984. Sodium hydroxide decontamination of Creutzfeldt-Jakob disease virus. N. Engl. J. Med. 310:727.

5. Dienst, R.B., and Brownell, G.H. 1984. Genus *Calymmatobacterium Aragao and Vianna 1913.* In Krieg, N.R., and Holt, J.G., editors. Bergey's manual of systematic bacteriology, vol 1. Williams & Wilkins, Baltimore.

6. Goldberg, J. 1959. Studies on granuloma inguinale. IV. Growth requirements of *Donovania granulomatis* and its relationship to the natural habitat of the organism. Br. J. Ven. Dis. 35:266.

7. Harrison, P.J. and Roberts, G.W. 1991. "Life, Jim, but not as we know it?": Transmissible dementias and the prion protein. Br. J. Psychiatry 158:457.

8. Hawkins, R.E., Rickman, L.S., Vermund, S.H., and Carl, M. 1992. Association of mycoplasma and human immunodeficiency virus infection: detection of amplified *Mycoplasma fermentans* DNA in blood. J. Infect. Dis. 165:581.

9. Henrik-Nielsen, R., Lundbeck, F.A., Teglbjaerg, P.S., et al. 1985. Intestinal spirochetosis of the vermiform appendix. Gastroenterology 88:971.

10. Hooton, T.M., Roberts, M.C., Roberts, P.L., et al. 1988. Prevalence of *Mycoplasma genitalium* determined by DNA probe in men with urethritis. Lancet 1:266.

11. Jones, M.J., Miller, J.N., and George, W.L. 1986. Microbiological and biochemical characterization of spirochetes isolated from the feces of homosexual men. J. Clin. Microbiol. 24:1071.

12. Levy, J.A., Ferro, F., Greenspan, D., and Lennette, E.T. 1990. Frequent isolation of HHV-6 from saliva and high seroprevalence of the virus in the population. Lancet 335:1047.

13. Levy, M.G., Meuten, D.J., and Breitschwerdt, E.B. 1986. Cultivation of *Rhinosporidium seeberi* in vitro: interaction with epithelial cells. Science 234:474.

14. Lo, S.C. 1990. A newly identified human *Mycoplasma* disease. Infect. Dis. Newsletter 9:73-75.

15. Luna, L.G. 1968. Warthin-Starry method for spirochetes and Donovan bodies. In Manual of histologic staining methods of the Armed Forces Institute of Pathology. McGraw-Hill Book Co., New York.

16. McFadden, J.J., Butcher, P.D., Chiodini, R., and Hermon-Taylor, J. 1987. Crohn's disease–isolated mycobacteria are identical to *Mycobacterium paratuberculosis,* as determined by DNA probes that distinguish between mycobacterial species. J. Clin. Microbiol. 25:796.

17. McMenemy, A. 1992. Whipple's disease, a familial Mediterranean fever, adult-onset Still's disease, and enteropathic arthritis. Curr. Opin. Rheumatol. 4:479.

18. Newburger, J.W., Takahashi, M., Burns, J.C., et al. 1986. The treatment of Kawasaki syndrome with intravenous gamma globulin. N. Engl. J. Med. 315:341.

19. Regnery, R.L., Anderson, B.E., Clarridge, J.E. III, et al. 1992. Characterization of a novel *Rochalimaea* species, *R. henselae* sp. nov., isolated from blood of a febrile, human immunodeficiency virus-positive patient. J. Clin. Microbiol. 30:265.

20. Regnery, R.L., Olson, J.G., Perkins, B.A., and Bibb, W. 1992. Serological response to "*Rochalimaea henselae*": antigen in suspected cat scratch disease. Lancet i. 339(8807):1443-45.

21. Relman, D.A., Loutit, J.S., Schmidt, T.M., et al. 1990. The agent of bacillary angiomatosis: an approach to the identification of uncultured pathogens. N. Engl. J. Med. 323:1573.

22. Relman, D.A., Schmidt, T.M., MacDermott, R.P., and Falkow, S. 1992. Identification of the uncultured bacillus of Whipple's disease. N. Engl. J. Med. 327:293.

23. Schwartzman, W.A., Nesbit, C., and Baron, E.J. 1993. Development and evaluation of a blood-free medium for determining growth curves and optimizing growth of *Rochalimaea henselae.* J. Clin. Microbiol. 31:1882.

24. Schwarz, T.F., Modrow, S., Hottentrager, B., et al. 1991. New oligopeptide immunoglobulin G test for human parvovirus B19 antibodies. J. Clin. Microbiol. 29:431.

25. Stratton, C.W. 1990. AIDS-associated *Mycoplasma*: the rest of the story. Infect. Dis. Newsletter 9:75.

26. Surawicz, C.M., Roberts, P.L., Rompalo, A., et al. 1987. Intestinal spirochetosis in homosexual men. Am. J. Med. 82:587.

27. Tam, A.W., Smith, M.M., Guerra, M.E., et al. 1991. Hepatitis E virus (HEV): molecular cloning and sequencing of the full-length viral genome. Virology 185:120.

28. Welch, D.F., Pickett, D.A., Slater, L.N., et al. 1992. *Rochalimaea henselae* sp. nov., a cause of septicemia, bacillary angiomatosis, and parenchymal bacillary peliosis. J. Clin. Microbiol. 30:275.

29. Wortmann, D.W. 1992. Kawasaki syndrome. Semin. Dermatol. 11:37.

BIBLIOGRAPHY

Cormier, D.P., and Mayo, D.R. 1988. Parvovirus B-19 infections. Clin. Microbiol. Newsletter 10:49.

Gajdusek, D.C., Gibbs, C.J., Asher, D.M., et al. 1977. Precautions in medical care of, and in handling materials from, patients with transmissible virus dementia (Creutzfeldt-Jakob disease). N. Engl. J. Med. 297:1253.

Lohr, J.A. 1990. Kawasaki disease (mucocutaneous lymph node syndrome). p. 2171. In Mandell, G.L., Douglas, R.G., Jr., and Bennett, J.E., editors. Principles and practice of infectious diseases, ed 3. Churchill Livingstone, New York.

Murray, H.W., 1990. *Spirillum minor* (rat-bite fever). p. 1827. In Mandell, G.L., Douglas, R.G., Jr., and Bennett, J.E., editors. Principles and practice of infectious diseases, ed 3. Churchill Livingstone, New York.

Taylor, A.F., Stephenson, T.G., Giese, H.A., and Petterson, G.R. 1984. Rat-bite fever in a college student. Calif. Morbid., No. 13. Department of Health Services, Sacramento, Calif.

42

MYCOBACTERIA

Disease from mycobacteria has been documented in the bones of prehistoric humans and animals. In the eighteenth and nineteenth centuries it was known as the white plague, decimating the populations of Europe. According to the World Health Organization (WHO), tuberculosis still kills 3 million people yearly in underdeveloped countries. *Mycobacterium tuberculosis*, the cause of tuberculosis, is one of 54 recognized species of mycobacteria. Of these, 14 are known to cause disease in humans (Table 42.1).[11] For the first time since 1953, the United States experienced a 1.1% increase in new active cases of tuberculosis in 1986; the increase has continued, and it is thought that it is related to the acquired immune deficiency syndrome (AIDS) epidemic.[26] More than 28,000 cases were reported in the United States in 1992, a 10% increase over 1991. A remarkable increase in infections caused by *M. avium* complex can also be attributed to AIDS patients.[44]

The physician depends on the microbiology laboratory for information that will help him or her to (1) make a definitive diagnosis of tuberculosis or other mycobacterial disease by isolation

College of American Pathologists Extents of Service for Participation in Mycobacterial Interlaboratory Comparison Surveys

1. No mycobacterial procedures performed
2. Acid-fast stain of exudates, effusions, and body fluids, etc., with inoculation and referral of cultures to reference laboratories for further identification
3. Isolation of mycobacteria; identification of *M. tuberculosis* and preliminary identification of the atypical forms such as photochromogens, scotochromogens, nonphotochromogens, and rapid growers; drug susceptibility testing may or may not be performed
4. Definitive identification of mycobacteria isolated to the extent required to establish a correct clinical diagnosis and aid in the selection of safe and effective therapy; drug susceptibility testing may or may not be performed

and identification of the etiological agent, (2) treat the patient effectively by determining susceptibility of the isolated organisms, and (3) monitor the chemotherapeutic response of the patient by periodic examination of specimens for evidence of a decrease in numbers of organisms by smear and colony count in culture. The laboratory must determine what "level of service" it can provide most effectively. "Levels of Service" has been outlined by the American Society of Clinical Pathologists (see box). At whatever level the laboratory chooses to operate, it must be prepared to offer exemplary service.[20]

EPIDEMIOLOGY AND PATHOGENESIS OF MYCOBACTERIAL INFECTIONS

Species of mycobacteria produce a spectrum of infections in humans and animals ranging from localized lesions to dissemination. Although some species cause only human infections, others have been isolated from a wide variety of animals. Many of the species are also found in water and soil. A tentative classification is proposed based on pathogenesis and natural history. Species include (1) obligate pathogens that cause exclusively human infections; (2) facultative pathogens that are found primarily in animals or the environment but produce documented human infections; (3) potentially "opportunistic" pathogenic species that are found in the environment but produce documented human infections; and (4) saprophytic species found in the environment that do not cause human infection (Table 42.1).

MYCOBACTERIUM TUBERCULOSIS

Tuberculosis still accounts for a large number of deaths and great morbidity worldwide. The disease is most common in parts of the developing nations of the world. Infection is spread from person to person by inhalation of airborne **droplet nuclei,** 1 to 5 μm in diameter. They are small enough to avoid being trapped by nasal turbinates or mucociliary membranes and may reach and impinge on the alveolar walls and be taken up by alveolar macrophages.[42] The pathogenesis of tuberculosis was discussed in Chapter 17.

Almost all tuberculosis acquired in the United States today is by aerosol inhalation. With the exception of leprosy, no other mycobacterial diseases are presumed contagious. Patients with cavitary disease are the primary reservoir for dissemination of *M. tuberculosis.* Although the disease incidence among Caucasians is falling, increasing numbers of cases are being discovered among Latin American and Southeast Asian immigrants and refugees. These patients, often infected with antibiotic-resistant strains of *M. tuberculosis,* pose new problems for public health officials and represent new foci for spread of the disease in certain locations, notably larger cities. Elderly patients, alcoholics, and non-Caucasians are more likely to develop disease. In addition, a resurgence of *M. tuberculosis* and other mycobacterial diseases in major metropolitan areas has been caused by the advent of AIDS.

One of the most distressing problems to arise from the resurgence of tuberculosis among AIDS patients and the homeless population is the increasing incidence of multiple drug resistance among *M. tuberculosis* isolates (called MDRTB).[7b] In New York City, the area with the most serious problem, approximately 25% of all new isolates are resistant to at least one antimycobacterial agent. Of tuberculosis cases reported to the Centers for Disease Control in 1991, 14% of isolates were resistant to one or more agents and 6% were resistant to two or more agents.

Inhalation of a single viable organism has been shown to lead to infection, although close contact is usually necessary for acquisition of infection. Of persons who become infected, 15% to 20% develop disease. Disease usually occurs some years after the initial infection, when the patient's immune system breaks down for some reason other than the presence of tuberculosis bacilli within the lung. In a small percentage of infected hosts, the disease becomes systemic, affecting such organs as the kidneys, spleen, bone marrow, central nervous system, and intestinal tract. In most patients,

TABLE 42.1

CURRENTLY RECOGNIZED SPECIES OF THE GENUS *MYCOBACTERIUM* ISOLATED FROM HUMANS

GROUP	OBLIGATORY	FACULTATIVE	POTENTIAL	SAPROPHYTE	
Strict pathogens	*M. africanum* *M. leprae* *M. tuberculosis* *M. ulcerans*	*M. bovis*			
Photochromogens		*M. asiaticum* *M. kansasii* *M. marinum* *M. simiae*			
Scotochromogens		*M. scrofulaceum* *M. szulgai* *M. xenopi*	*M. gordonae* *M. flavescens*		
Nonchromogens*	*M. genavense*	*M. avium* *M. haemophilum* *M. intracellulare* *M. malmoense* *M. shimoidei*	*M. gastri* *M. nonchromogenicum* *M. terrae* *M. triviale*		
Rapid growers		*M. chelonae* *M. fortuitum*	*M. fallax* *M. smegmatis*	*M. agri* *M. aichiense* *M. austroafricanum* *M. aurum* *M. chitae* *M. chubuense* *M. diernhoferi* *M. duvalii* *M. gadium* *M. gilvum* *M. komossense*	*M. neoaurum* *M. parafortuitum* *M. obuense* *M. phei* *M. pulveris* *M. rhodesiae* *M. sphagni* *M. thermoresistibile* *M. tokaiense* *M. vaccae*
Strict animal pathogens	*M. farcinogens* *M. lepraemurium* *M. porcinum*	*M. microti* *M. paratuberculosis* *M. senegalense*			

Modified from Good, R.C. 1985. Opportunistic pathogens in the genus *Mycobacterium*. Annu. Rev. Microbiol. 39:347.
*Some strains of *M. avium* and *M. intracellulare* are pigmented.

however, the disease is restricted to the lungs, with granuloma formation, caseous necrosis, and ultimately cavitary disease (Chapter 17). A definitive diagnosis is made by isolating and identifying the etiological agent, *M. tuberculosis*.

OTHER CULTIVABLE MYCOBACTERIA

M. bovis, the agent of bovine tuberculosis, is an uncommon cause of disease in the United States today, although it used to be one of the primary agents of human intestinal tuberculosis, acquired through ingestion of contaminated milk.[8] An attenuated strain of *M. bovis*, **bacille Calmette-Guérin (BCG),** has been used extensively in many parts of the world to immunize susceptible individuals against tuberculosis. BCG has recently been used as part of a controversial protocol to boost the nonspecific cellular immune response of certain

immunologically deficient patients, particularly those with malignancies. Because mycobacteria are the classic examples of intracellular pathogens and the body's response to BCG hinges on cell-mediated immunoreactivity, immunized individuals are expected to react more aggressively against all antigens that elicit cell-mediated immunity. Rarely, the unfortunate individual's immune system will be so compromised that it cannot handle the BCG, and systemic BCG infection may develop.

Respiratory infection in immunocompetent hosts is usually caused by *M. tuberculosis*, but *M. kansasii* and *M. avium-intracellulare* complex also cause tuberculosis-like disease. Other mycobacteria cause pulmonary disease as well. *M. avium* complex is the most commonly isolated mycobacterial species in the United States today because of its high prevalence in patients suffering from

AIDS (see Chapter 24).[12] In homosexual patients with AIDS the mode of acquisition of *M. avium* complex is believed to be through the gastrointestinal tract.[3,43] The first symptom of mycobacterial disease in many of these patients is protracted diarrhea. Acid-fast stains of feces or intestinal biopsy specimens reveal huge numbers of acid-fast bacilli. The infection usually disseminates in AIDS patients, with organisms being recovered from multiple sites, including blood, sputum, feces, semen, lymph nodes, and internal organs.

Isolation of *M. tuberculosis* or *M. bovis* is always indicative of disease, whereas isolation of other mycobacteria may or may not be clinically significant. *M. gordonae,* for example, one of the more common mycobacteria isolated from clinical specimens, is a normal inhabitant of tap water and rarely causes disease in humans. Table 42.1 lists the "mycobacteria other than tuberculosis **(MOTT)**" that have been implicated as etiological agents of human disease.

The AIDS epidemic has contributed to the recognition of other species of mycobacteria involved in human infection. *M. haemophilum,* a hemin-requiring, slow-growing strain, has been isolated from skin infections such as abscesses and papules, joint fluids from patients with septic arthritis, osteomyelitis, blood, and respiratory cultures from adults.[13b] It can cause systemic infections in immunocompromised adults. *M. haemophilum* is also pathogenic for healthy children, in which it causes cervical or submandibular adenitis.

When BACTEC 13A medium blood cultures from AIDS patients are incubated for extended periods (as long as 10 weeks), they have occasionally yielded a fastidious mycobacterium with the proposed name of *"M. genavense"* (first isolated in Geneva, Switzerland).[7d] This organism seems to be a pathogen, causing disseminated infection in AIDS patients. It can be subcultured only onto solid medium supplemented with the growth factor mycobactin.

Four human cases of disease (breast abscess, pulmonary granuloma, surgical wound, pneumonia) have been attributed to *M. thermoresistibile,* a rapid-growing scotochromogen that can grow at high temperatures.[40a] Although rare in human disease, the organism is present in soil.

CROHN'S DISEASE-ASSOCIATED MYCOBACTERIUM

Isolation of a new species of *Mycobacterium* that is genetically similar to the animal pathogen *M. paratuberculosis* from the bowel mucosa of patients with Crohn's disease has been reported by Chiodini and co-workers (see Bibliography).[23a] This news is particularly interesting, since the etiology of Crohn's disease, a chronic inflammatory bowel disease, previously had not been considered to be infectious. The organism is extremely fastidious, seems to require a growth factor, mycobactin, produced by other species of mycobacteria, such as *M. phlei,* a saprophytic strain, and may take as long as 6 to 18 months for primary isolation. Whether these and other mycobacteria actually contribute to development of Crohn's disease or are simply colonizing an environmental niche in the bowel of these patients remains to be elucidated.[23a]

MYCOBACTERIUM LEPRAE

Possibly second to *M. tuberculosis* in worldwide importance is *M. leprae,* the agent of leprosy (also called Hansen's disease). The organism has not yet been cultivated in vitro. Leprosy is a chronic disease of the skin, mucous membranes, and nerve tissue, with a strong immunological component to the pathology. Understanding of the pathogenesis and epidemiology of the disease is hampered by our inability to grow the organism in culture. In tropical countries, where the disease is most prevalent, it may be acquired from infected humans; however, infectivity is very low. Prolonged close contact and host immunological status play a role in infectivity. However leprosy is acquired, it passes through many stages in the host, characterized by various clinical and histopathological features. Although there are many intermediate stages, the primary stages include a silent phase, during which leprosy bacilli multiply in the skin within macrophages, and an indeterminate phase, in which the bacilli multiply in peripheral nerves and begin to cause sensory impairment. More severe disease states that may follow are called "tuberculoid," in which granulomas and hyperactive tissue are evident in the skin, and "lepromatous," in which the patient's immune response is virtually absent and the bacilli multiply rapidly, often within many organs of the body and the reticuloendothelial system. A patient may recover spontaneously at any stage, and there are several borderline stages defined by subtle differences in the host response to infection. Berlin and Martin[2] have summarized current knowledge about leprosy.

Diagnosis is usually made through histopathological examination of material taken from a skin lesion or from the earlobe of a suspected patient. Although acid-fast bacilli are not seen during the early stages, they are later visualized in large numbers by the Fite stain or by routine Ziehl-Neelsen stain. Other sites from which positive smears may be seen include lymph nodes, bone marrow, and human milk. The stages of leprosy can be additionally pinpointed by a patient's response to "lepromin," a crude suspension of material taken from excised nodules of patients with Hansen's disease. Treatment consists of dapsone or DDS (4,4'-diaminodiphenyl sulfone), often given in combination with rifampin. Excellent results have also been seen with clofazimine, which requires several months to build up active concentrations within the patient's tissues.[33]

LABORATORY SAFETY AND QUALITY CONTROL

Tuberculosis ranks high among laboratory-acquired infections. Therefore laboratory and hospital administrators must provide laboratory personnel with facilities, equipment, and supplies that will reduce this risk to a minimum. All tuberculin-negative personnel should be skin-tested every year. Tuberculin-positive persons should have a chest x-ray examination annually.

FACILITIES AND EQUIPMENT; SAFETY MEASURES

Facilities and equipment

1. A "hot room" devoted only to diagnostic mycobacteriology, maintained under negative air exhaust to adjacent corridors and work areas
2. Laminar flow biological safety cabinets, type 2B, which draw a minimum of 75 linear feet of air per minute across the front opening and exhaust 100% of the air to the outside
3. High-speed, refrigerated centrifuges equipped with bucket covers and safety domes

Safety measures

1. Masks, gowns, and gloves must be worn when working in the "hot room." The threat of MDRTB has prompted the CDC to recommend particulate respirators for health care workers likely to be exposed to *M. tuberculosis*[1]
2. Manipulations resulting in formation of aerosols must be kept to a minimum
3. Hands should be scrubbed thoroughly before leaving the laboratory
4. The increasing use of syringes and needles for BACTEC system inoculations requires ex-

treme precaution to avoid finger sticks. Work slowly and deliberately and do not take your eyes off the needle
5. Substitute egg and agar media in 1 ounce prescription bottles for media in test tubes. Bottles seldom break when dropped, whereas test tubes will always break. Bottles are easier to stack, store, and handle during incubation and examination
6. In case of an accident with formation of aerosols, hold your breath, make certain biological safety cabinets are on, centrifuges are turned off, and leave the room. Return in 30 minutes to cover the spill with 3% Amphyl and paper towels. Allow the Amphyl to stand for 30 minutes. Change clothing and shower if necessary. Return to the accident area for a final cleanup. Skin-test all tuberculin-negative personnel at 3 and 6 months after the accident

QUALITY CONTROL

Quality control in diagnostic mycobacteriology requires the same attention to equipment, media, reagents, stains, and antimycobacterial susceptibility testing procedures as do other areas of the laboratory. Quality control of the decontamination and concentration procedures should be of particular concern, since the accuracy and reliability of these procedures determine whether mycobacteria are isolated from specimens. The slow rate of growth of mycobacteria interferes with prompt detection of deficiencies. Procedure 42.1 detects lethal effects of the decontaminating agents employed (Table 42.2); it also detects deficiencies in media quality. However, two to three weeks will have elapsed before such deficiencies are noted. Therefore routine monitoring with this procedure should help alert the laboratory to impending problems.

SPECIMEN COLLECTION

Acid-fast bacilli may infect almost any tissue or organ of the body (Table 42.3). The successful isolation of the organism depends on the quality of the specimen obtained and the appropriate processing and culture techniques employed by the mycobacteriology laboratory. In suspected mycobacterial disease, as in all other infectious diseases, the diagnostic procedure begins at the patient's bedside. Collection of proper clinical specimens requires careful attention to detail by the attending physician, nurse, or other ward personnel. An essential prerequisite of good specimen

PROCEDURE 42.1

QUALITY CONTROL FOR MYCOBACTERIOLOGY

REAGENTS

1. Media: Routine media used for cultivation of mycobacteria

2. Quality control organism: A recent isolate of *M. tuberculosis* or *M. tuberculosis* strain H37Rv

3. Other materials
   Autoclaved sputum (AFB-negative)
   7H9 liquid medium containing 15% glycerol
   Sterile buffer, pH 7.0
   50-ml plastic, conical centrifuge tubes

PROCEDURE

1. Suspend several colonies of H37Rv in a tube containing 3 ml of 7H9 liquid medium and several plastic or glass beads. Mix vigorously on a test tube mixer; then allow large particles to settle for 15 minutes.

2. Prepare a dilution of approximately 10^6 organisms per ml by adding the above cell suspension drop-by-drop to 1 ml of buffer until a barely turbid suspension occurs. Transfer 0.5 ml of the 10^6 cells per ml dilution to 4.5 ml of glycerol broth to give a suspension of 10^5 cells per ml. Repeat the procedure to make a 10^4/ml and a 10^3/ml suspension.

3. Label 15 3-dram vials for each suspension (10^5, 10^4,

and 10^3). Transfer 0.3 ml of the appropriate suspension to each vial. Store the vials at $-70°$ C to use for future quality control testing.

4. Thaw one vial of each of the three dilutions each time the quality control procedure is performed.

5. Add 2.7 ml autoclaved sputum to each cell suspension to effect a tenfold dilution and inoculate three sets of the media used for primary isolation with each of the three dilutions of sputum. Inoculate 0.1 ml of sputum per bottle.

6. Decontaminate and concentrate the remainder as you do with sputum specimens. Reconstitute the sediments with sterile buffer to 2.6 ml, resuspend vigorously, and inoculate a second set of media with 0.1 ml of each of the concentrated and resuspended samples.

7. Incubate at $35°$ C in 5% to 10% CO_2 for 21 days.

INTERPRETING AND RECORDING RESULTS

The bottle of egg media should have been inoculated with approximately 10^4, 10^3, and 10^2 organisms, respectively. The first dilution should produce semiconfluent growth,

and the second and third dilutions should produce countable colonies in each bottle. Because of the retrospective nature of these determinations, close comparisons must be made between current and previous results to note trends or developing deficiencies. Failures may be due to faulty media, lethal effects of decontamination and concentration procedures, improperly prepared reagents, or overexposure of specimens to these reagents (Table 42.2). Should deficiencies become evident, techniques should be reviewed and attempts made to determine the source of the problem. New batches of media must be substituted for deficient media and the latter rechecked to verify deficiencies. Personnel should be included in all discussions of problems and corrective measures. All deficiencies and corrective actions should be recorded in the appropriate section of the Quality Control records.

---

collection is the use of sturdy, sterile, leakproof containers placed into bags to contain leakage should it occur.

PULMONARY SPECIMENS

Pulmonary secretions may be obtained by any of the following methods: spontaneously produced or induced sputum, gastric lavage, transtracheal aspiration, bronchoscopy, and laryngeal swab-

bing. Sputum, aerosol-induced sputum, bronchoscopic aspirations, and gastric lavage comprise the majority of specimens submitted for examination. Spontaneously produced sputum is the specimen of choice. To raise sputum, patients **must** be instructed to take a deep breath, hold it momentarily, and then cough deeply and vigorously. Patients must also be instructed to cover their mouths carefully while coughing and to discard

Suggested Media for Cultivation of Mycobacteria from Clinical Specimens*

Solid
  Agar-based
    1. Middlebrook 7H10
    2. Middlebrook 7H11
    3. Mitchison's selective 7H11
  Egg-based
    1. Wallenstein
    2. Löwenstein-Jensen (L-J) with RNA
    3. L-J with pyruvic acid
Liquid
  1. BACTEC 12B medium
  2. Middlebrook 7H9 Broth
  3. Septi-Chek AFB

---

*For optimal recovery of mycobacteria, a combination of liquid medium (BACTEC) and a minimum of two solid media is recommended.

tissues in an appropriate receptacle. Saliva and nasal secretions are not to be collected nor is the patient to use oral antiseptics during the period of collection. Sputum specimens must be free of food particles, residues, and other extraneous matter.

The aerosol induction procedure can best be done on ambulatory patients who are able to follow instructions. Aerosol-induced sputums have been collected from children as young as 5 years old. The procedure should be avoided for patients with asthma or other respiratory difficulties. The patient must be questioned closely; if the patient admits to difficulty, the physician should confirm that the patient can be safely induced. A physician should be present or close by when asthmatics are being induced. This procedure should be performed only in properly ventilated rooms by operators wearing particulate respirators and taking all appropriate safety measures to avoid exposure. Aerosol induction can be done at any time during the day, but preferably approximately 2 hours after meals to avoid nausea and dizziness. The patient should be told that the procedure is being performed to induce coughing to raise sputum that the patient cannot raise spontaneously and that the salt solution is irritating. The patient should be instructed to inhale slowly and deeply through the mouth and to cough at will, vigorously and deeply. The patient should be instructed to cover his or her mouth with tissues while coughing and to expectorate into a collection tube. The procedure should be discontinued if the patient fails to raise sputum after 10 minutes

or feels any discomfort. Ten milliliters of sputum should be collected in a 50-ml conical centrifuge tube; if the patient continues to raise sputum, a second specimen should be collected and submitted. Specimens should be delivered promptly to the laboratory and refrigerated if processing is delayed.

When the procedure is completed, the nebulizer should be thoroughly rinsed with filtered distilled water and allowed to soak in 2% acetic acid for 10 minutes. The nebulizer should then be rinsed with sterile filtered distilled water and air dried. The effluent should be collected weekly for bacteriological monitoring. Swabs of the transducer may be submitted as an alternative culture.

Gastric lavage is used to collect sputum from patients who may have swallowed their sputum during the night. The procedure is limited to senile, nonambulatory patients, children less than 3 years old, and patients who fail to produce sputum by aerosol induction. The most desirable gastric lavage is collected at the patient's bedside before the patient arises and before exertion empties the stomach. **Gastric lavage cannot be performed as an office or clinic procedure.** A series of three specimens should be collected within 3 days. The patient should be told that the procedure is being performed to collect sputum that has been swallowed. If the patient appears unable to tolerate the procedure, gastric lavage should not be attempted.

The collector should wear a cap, gown, and particulate respirator mask and stand beside (not in front of) the patient, who should sit up on the edge of the bed or in a chair, if possible. The Levine collection tube is inserted through a nostril, and the patient is instructed to swallow the tube. Small sips of filtered water and deep breaths may aid this process. When the tube is fully inserted, a syringe is attached to the end of the tube and 5 ml of filtered distilled water is inserted through the tube. The syringe is then used to withdraw 20 to 25 ml gastric secretions. The fluid should be expelled slowly down the sides of the 50-ml conical collecting tube. Do not pump the syringe or otherwise create aerosols. The syringe is reattached to the Levine tube while the tube is slowly withdrawn, avoiding motion of the free end. Excess fluid is expelled into the collection tube, and the collection equipment and clothing are bagged in the appropriate hazardous waste containers. The top of the collection tube is screwed on tightly, and the tube is held upright during **prompt** delivery to the laboratory.

TABLE 42.2

SAMPLE RESULTS FOR INTERPRETING QUALITY CONTROL TEST OF DECONTAMINATION AND CONCENTRATION PROCEDURE

| | SPUTUM | | | | | | |
| | UNPROCESSED | | | PROCESSED | | | |
SPUTUM SAMPLE	10^4	10^3	10^2	10^4	10^3	10^2	INTERPRETATION
1	3+	2+	50-100 Col	2+	1+ 2+	Approx. 10	Media and decontamination procedures are acceptable
2	3+	2+	50-100 Col	1+	0	0	Media acceptable; procedure too toxic
3	2+ or 1+	2+ or 1+	0	1+ or 0	1+ or 0	0	One or more of the media being used is not supporting growth of AFB adequately

TABLE 42.3

TYPES OF SPECIMENS AND SPECIES OF MYCOBACTERIA LIKELY TO BE RECOVERED FROM THEM

OCULAR	G.I. TRACT	CUTANEOUS	PULMONARY	GENITAL	VASCULAR (BLOOD AND LYMPH NODES)	NERVOUS
M. fortuitum	M. bovis	M. leprae	M. tuberculosis	M. leprae	M. tuberculosis	M. leprae
M. chelonae	M. avium	M. marinum	M. africanum	M. tuberculosis	M. avium	M. tuberculosis
	M. intracellulare	M. haemophilum	M. kansasii	M. avium	M. intracellulare	
	M. paratuberculosis	M. fortuitum	M. simiae	M. intracellulare	M. kansasii	
		M. chelonae	M. asiaticum	M. xenopi	M. leprae	
		M. ulcerans	M. scrofulaceum		M. scrofulaceum	
		M. smegmatis	M. szulgai		M. fortuitum	
		M. thermoresistibile	M. xenopi		M. chelonae	
			M. avium		M. haemophilum	
			M. intracellulare		"M. genavense"	
			M. malmoense			
			M. fortuitum			
			M. chelonae			
			M. thermoresistibile			

Bronchial lavages, washings, and brushings are collected and submitted by medical personnel. These are the specimens of choice for detecting nontuberculous mycobacteria and other opportunistic pathogens in patients with immune dysfunctions.

URINE SPECIMENS

Urogenital infections show little evidence of decreasing. It is presumed that a relatively long incubation period exists between the initial infection and later urinary tract manifestations. Whereas 2% to 3% of patients with pulmonary tuberculosis exhibit urinary tract involvement, 30% to 40% of patients with genitourinary disease have tuberculosis at some other site. The clinical manifestations of urinary tuberculosis are variable, including frequency of urination (most common), dysuria, hematuria, and flank pain. Definitive diagnosis requires recovery of acid-fast bacilli from the urine.

Early morning voided urine specimens in sterile containers should be submitted daily for at least 3 days. Because of excessive dilution, higher contamination, and difficulty in concentrating, 24-hour urine specimens are undesirable.

GENITAL SPECIMENS

A variety of species of mycobacteria have been isolated from semen, seminal vesicles, prostate gland, and testes specimens; numerous isolations have been made from AIDS patients. Diagnosis requires biopsy of the suspected site for histological examination and culture.

FECAL SPECIMENS

Until recently, fecal specimens were rarely examined for acid-fast bacilli. However, the frequent isolation of *M. avium* complex from feces of AIDS patients has significantly increased the number of specimens submitted for this purpose. Feces should be submitted in a clean, dry container without preservative or diluent. Contamination with urine should be avoided.

TISSUE SPECIMENS

Pus and exudates may be submitted on swabs or on small pieces of bandage, placed in sterile, leakproof containers. Swabs should be placed in a neutral transport medium to avoid desiccation. Tissue biopsy specimens should be submitted in large-mouth tubes or in Petri dishes, bagged to prevent leakage.

OTHER SPECIMENS

Tuberculous meningitis is uncommon but still occurs in both immunocompetent and immunosuppressed patients. Sufficient quantity of specimen is most critical for isolation of acid-fast bacilli from cerebrospinal fluid. There may be very few organisms in the spinal fluid, which makes their detection difficult. At least 10 ml of cerebrospinal fluid is recommended for recovery of mycobacteria.

The discovery of circulating mycobacteria in the blood of patients with AIDS has led to increased requests for such cultures. Best recovery can be achieved by collecting the blood in either a broth or a lysis-centrifugation system. If the blood is allowed to remain in the lysing solution of the Wampole Isolator tube for approximately 1 hour, lysis of phagocytic cells may occur, allowing release and enhanced recovery of intracellular mycobacteria. In patients with AIDS, quantitation of such organisms can be used to monitor therapy and determine prognosis. For this purpose the 1.5 ml Pediatric Isolator tubes can be used, since the entire contents are plated and colony forming units can be determined per milliliter of initial

specimen. The Becton-Dickinson Septi-Chek biphasic blood culture system for mycobacteria is also satisfactory.[13a] Automated radiometric detection of mycobacteria in blood is another alternative for such cultures. Use of this system is discussed later.

SPECIMEN PROCESSING FOR THE RECOVERY OF ACID-FAST ORGANISMS

Specimen processing for the recovery of acid-fast bacilli from clinical specimens involves a number of complex procedures, each of which must be carried out with precision. Specimens from sterile sites can be inoculated directly to media (small volume) or concentrated to reduce volume. Other specimens require decontamination and concentration. A scheme for initial processing is depicted in Figure 42.1. The procedures will be explored in detail in the following sections.

DECONTAMINATION AND CONCENTRATION PROCEDURES

Many techniques for decontaminating and concentrating sputum and other specimens have been discarded because of a failure to fulfill some practical requirement, such as low toxicity to acid-fast bacilli, effective inhibition of contaminants, or ability to completely liquefy mucoid sputum in a reasonable time period. The procedures outlined here, although still suffering some of these disadvantages, yield acceptable numbers of positive cultures *if properly performed*. Because of toxicity of the reagents toward acid-fast bacilli, procedures must be carried out within the stated time limits. The *N*-acetyl–L cysteine–sodium hydroxide method (Procedure 42.2) is the most widely used decontamination and concentration procedure and may be considered the "gold standard."[19]

An alternative method for decontamination and concentration, the Zephiran-trisodium phosphate method (Procedure 42.3), permits a longer exposure of specimens to the decontaminating agents without affecting the recovery of acid-fast bacilli. A modification of this procedure (Procedure 42.4), which reduces the concentrations of the toxic reagents and requires a second wash of the sediment, further enhances survival of these organisms.

A decontaminating procedure utilizing oxalic acid is very useful for treating specimens known to harbor gram-negative rods, particularly *Pseu-*

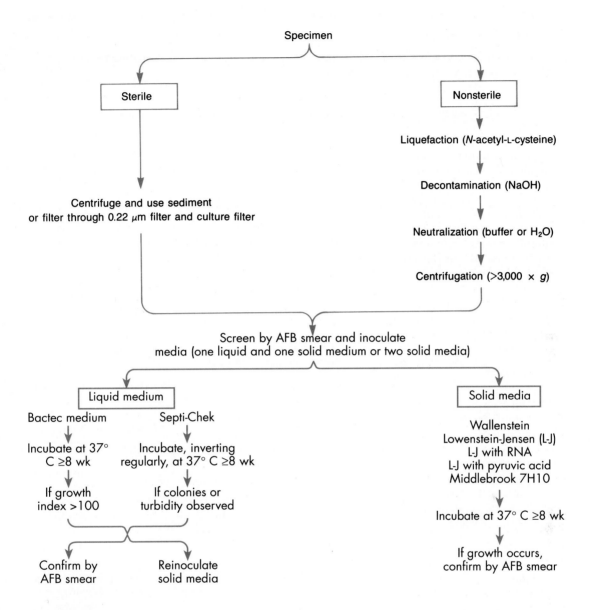

FIGURE 42.1

A flow chart for specimen processing for isolation of mycobacteria.

domonas and *Proteus,* which are extremely troublesome contaminants (Procedure 42.5). Excellent discussions of decontamination and concentration procedures can be found elsewhere.[14,19,27,35]

Many other specimens that contain normal flora require decontamination and concentration; handling procedures for such specimens are described briefly here. Aerosol-induced sputums should be treated as sputums. Gastric lavages should be processed immediately or neutralized with 10% sodium carbonate (check with pH paper to determine that the specimen is at neutral pH) and refrigerated until processed as for sputum. If more than 10 ml of watery-appearing aspirate was obtained, the specimen can be centrifuged at $3600 \times g$ for 30 minutes, the supernatant decanted, and the sediment processed as for sputum.

Urine specimens should be divided into a maximum of four 50-ml centrifuge tubes and centrifuged at $3600 \times g$ for 30 minutes. The supernatant should be decanted, leaving approximately 2 ml of sediment in each tube. The tubes are vortexed to suspend the sediments, and sediments are combined. If necessary, distilled water can be added to make the volume rise to 10 ml. This urine concentrate is then treated as for sputum or with the sputolysin–oxalic acid method.

PROCEDURE 42.2

N-ACETYL-L CYSTEINE-SODIUM HYDROXIDE METHOD FOR LIQUEFACTION AND DECONTAMINATION OF SPECIMENS

PRINCIPLE

Sodium hydroxide, an extremely toxic decontaminating agent, also acts as an emulsifier. The addition of a mucolytic agent, *N*-acetyl-L-cysteine (NALC), makes it possible to reduce the concentration of sodium hydroxide used and also shortens the time required for decontamination, thus aiding the optimal recovery of acid-fast bacilli.

METHOD

1. NALC-sodium hydroxide preparation

   For each day's cultures, add up the total volume of specimens to be treated and prepare an equal volume of the digestant-decontamination mixture, as follows:

   1 N (4%) NaOH 50 ml
   0.1 M (2.94%) trisodium citrate · 3H$_2$O 50 ml
   N-Acetyl-L-cysteine (NALC) powder 0.5 g

   It is useful to use sterile distilled water for preparation of solutions to minimize chances of inadvertently adding acid-fast tap water contaminants to the specimens. Mix, sterilize, and store the NaOH and the citrate in sterile, screw-capped flasks for later use. This solution should be used within 24 hours after the NALC is added.

2. 0.676 (1/15) M phosphate buffer preparation
   Solution A
   Sodium monohydrogen phosphate (anhydrous) 9.47 g
   Distilled water 1000 ml
   Solution B
   Potassium dihydrophosphate 9.07 g
   Distilled water 1000 ml

   Add 50 ml of solution B to 50 ml of solution A and adjust pH to 6.8.

3. Work within a biological safety cabinet and wear gloves. Transfer a maximum of 10 ml of sputum, urine, or other fluid to be processed to a sterile, disposable plastic 50-ml conical centrifuge tube with a leakproof and aerosol-free plastic screw-cap. Tubes with easily visible volume indicator marks are best.

4. Add an equal volume of freshly prepared digestant to the tube, being very careful when pouring digestant not to touch the lip of the specimen container, which might inadvertently transfer positive material to a negative specimen. Tighten the cap completely.

5. Vortex the specimen for approximately a slow count of 15, or for a maximum of 30 seconds, being certain to create a vortex in the liquid and not to merely agitate the material. Check for homogeneity by inverting the tube. If clumps remain, vortex the specimen intermittently while the rest of the specimens are being digested. An extra pinch of NALC crystals may be necessary to liquefy really mucoid sputa.

6. Start a 15-minute timer when the first specimen is finished being vortexed. Continue digesting the other specimens, noting the amount of time that the entire run takes. The digestant should remain on the specimens for a

   maximum exposure of 20 minutes.

7. After 15 minutes of digestion, add enough phosphate buffer to reach within 1 cm of the top, screw the cap tightly closed, and invert the tube to mix the solutions and stop the digestion process. Addition of this solution also reduces the specific gravity of the specimen, aiding sedimentation of the bacilli during centrifugation.

8. Centrifuge all tubes at 3600 × *g* for 15 minutes.

9. Carefully pour off the supernatant into a splash-proof container (Figure 42.2). The lip of the tube may be wiped with an amphyl- or phenol-soaked gauze to absorb drips. Be careful not to touch the lip of any tube to another container. It is helpful to watch the sediment carefully as the supernatant is being decanted, since sometimes a very mucoid sediment can be loose and may pour right out with the supernatant. The technologist who sees the sediment beginning to slip can then stop decanting and use a sterile capillary pipette to remove the supernatant without losing the sediment.

10. Resuspend the sediment in 1 to 2 ml sterile water or buffer (with bovine serum albumin [BSA]).

11. Inoculate the sediment to culture media and prepare slides.

QUALITY CONTROL

See Procedure 42.1.

PROCEDURE 42.3

BENZALKONIUM CHLORIDE (ZEPHIRAN)-TRISODIUM PHOSPHATE METHOD FOR DIGESTION AND DECONTAMINATION

PRINCIPLE
A milder decontaminating solution allows a longer time period for decontamination to occur.

METHOD
1. Prepare Zephiran-trisodium phosphate digestant as follows:
   Trisodium phosphate ($Na_3PO_4 \cdot 2H_2O$) 500 g
   Hot, sterile distilled water 2000 ml
     Add 3.25 ml of Zephiran concentrate (17% benzalkonium chloride, Winthrop laboratories). Mix. Store at room temperature.

2. Mix equal parts of the specimen and digestant, exactly as detailed in Procedure 42.2. Vortex for 30 seconds.

3. Allow the mixtures to stand without agitation for an additional 15 minutes.

4. Centrifuge as in Procedure 42.2 for 30 minutes.

5. Decant supernatant as in Procedure 42.2

6. Resuspend the supernatant in 20 ml of neutralizing buffer (obtained from Difco Laboratories). Although the 1/15 M phosphate buffer described in Procedure 42.2 may be used, the commercial buffer contains components to specifically neutralize the Zephiran.

7. Centrifuge the tubes as done previously for 20 minutes.

8. Decant supernatant fluids as done previously and retain the sediments for smears and cultures.

QUALITY CONTROL
See Procedure 42.1.

Feces are usually submitted for isolation of *M. avium* complex. One or two grams of formed stool or 5 ml of liquid stool is transferred to a 50-ml centrifuge tube and sterile, filtered distilled water is added to make the volume 10 ml. The suspension is vortexed thoroughly. The specimen is then filtered through gauze to remove particles. Add 10 ml of NALC-NaOH to the suspension and allow to stand at room temperature for 45 to 60 minutes. Then add 25 ml phosphate buffer, mix thoroughly, and centrifuge for 20 minutes. The supernatant should be decanted

and the sediment used to prepare slides and inoculate media.

Swabs and small pieces of bandage from pus and wound aspirates should be transferred to a sterile 50-ml centrifuge tube with 10 ml sterile, filtered distilled water. The specimen should be vortexed vigorously and allowed to stand for 20 minutes. The swab or bandage is removed, and the resulting suspension is processed as for sputum.

Large pieces of tissue thought to be contaminated should be finely minced using a sterile scalpel and scissors. This material is transferred to a mortar and pestle for grinding or to a 50-ml centrifuge tube containing 6 to 8 glass beads. The tube is vortexed vigorously and allowed to stand for 20 to 30 minutes to permit the aerosol to settle. As an alternative a Stomacher instrument can be used to homogenize tissue. Homogenate is poured through sterile gauze into a 50-ml centrifuge tube. An equal volume of NALC-NaOH is added; the tube is mixed vigorously and allowed to stand for 20 minutes. Then 25 ml phosphate buffer is added; the tube is mixed thoroughly and centrifuged at $3600 \times g$ for 20 minutes. The supernatant is decanted, and the sediment is suspended in 1 to 2 ml buffer and used to prepare slides and inoculate media. If the tissue is collected aseptically and not thought to be contaminated, it may be processed without treatment with NALC-NaOH.

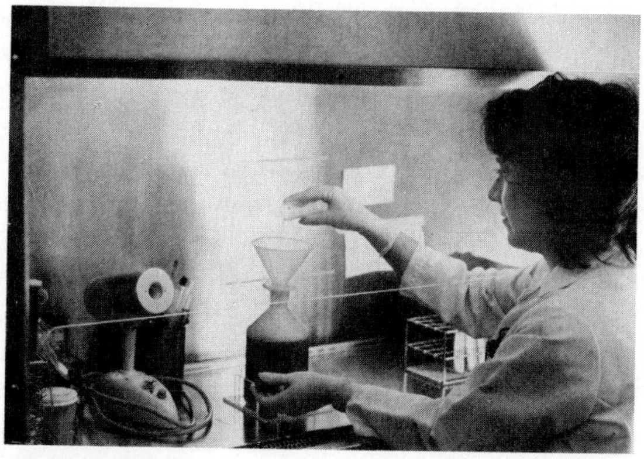

FIGURE 42.2
Supernatant from acid-fast specimen concentrates is carefully decanted into a splash-proof container that can be autoclaved and reused.

PROCEDURE 42.4

MODIFIED ZEPHIRAN-TRISODIUM PHOSPHAGE (TSP) METHOD

PRINCIPLE

By reducing the concentration of reagents and adding a second wash, survival of acid-fast bacilli is enhanced.

METHOD

Note: All reagents should be prepared with filter-sterilized distilled water.

1. Prepare Zephiran-trisodium phosphate (1:5000 Zephiran in 15% TSP) as follows:
   Trisodium phosphate
   $Na_3PO_4 \cdot 2H_2O$) 150 g
   Hot, sterile distilled water 1000 ml
   Concentrated Zephiran chloride 1.1 ml (17%)
   Bromthymol blue, 0.1% aqueous 10.0 ml solution

2. Prepare 1/15 M sterile phosphate buffer, pH 6.6, as follows:
   Stock solution A
   Na_2HPO_4 9.47 g
   Distilled water 1000 ml
   Stock solution B
   KH_2PO_4 9.07 g
   Distilled water 1000 ml

Add 37.5 ml of solution A to 62.5 ml of solution B. Filter-sterilize through a 0.47-μm pore size nitrocellulose filter and autoclave for 15 minutes at 121° C. Aseptically add 15 U/ml penicillin G when the solution reaches room temperature.

3. To liquefy the specimen, add an equal volume of filtered distilled water to a maximum of 10-ml sputum in a 50-ml conical centrifuge tube. Sputolysin (Behring Diagnostics, La Jolla, Calif.) may be used to liquefy thick, mucoid specimens. Vortex until the sputum is thoroughly homogenized; allow to stand 15 minutes.
   Note: Sputolysin is a concentrated sterile solution of dithiothreitol (Cleland's reagent), used to liquefy sputum. Prepare a 1:10 dilution with sterile distilled water and use within 24 hours.

4. Add an equal volume of Zephiran-TSP to the diluted specimen (maximum of 10 ml to 10 ml sputum). Vortex thoroughly for approximately 30 seconds and allow the specimen to stand for 20 minutes.

5. Add 30 ml buffer. Vortex. Centrifuge at 1800 to 2400 $\times g$ for 20 minutes.

6. Decant to dryness, add 40 ml buffer, and vortex for 30 seconds to resuspend the sediment.

7. Centrifuge for 15 minutes as previously, decant the supernatant, and add 1 to 2 ml phosphate buffer to the sediment. Resuspend by vortexing.

8. Prepare slides and inoculate media.

QUALITY CONTROL

See Procedure 42.1.

Tissue specimens too small to be homogenized should be vortexed in 2 ml of buffer in a 50 ml centrifuge tube with 6 to 8 glass beads. The suspension should stand until aerosol settles after mixing. The volume is brought up to 10 ml with sterile, filtered distilled water, and the decontamination process is continued as described for sputum.

SPECIMENS THAT DO NOT REQUIRE DECONTAMINATION

Processing a variety of clinical specimens that do not routinely require decontamination for acid-fast culture is described here. However, should such a specimen initially appear contaminated because of color, cloudiness, or foul odor, perform a Gram stain or acid-fast stain to detect bacteria other than acid-fast bacilli. Specimens found to be contaminated should be processed as described in the previous section.

Cerebrospinal fluid should be handled aseptically and centrifuged for 30 minutes at 3600 $\times g$ to concentrate the bacteria. The supernatant is decanted and the sediment is vortexed thoroughly before preparing the smear and inoculating media. If insufficient quantity of spinal fluid is received, the specimen should be used directly for smear and culture. Because recovery of acid-fast bacilli from cerebrospinal fluid is difficult, additional solid and liquid media should be inoculated if material is available.

Pleural fluid should be collected in sterile anticoagulant (1 mg/ml ethylenediaminetetraacetic acid [EDTA] or 0.1 mg/ml heparin). If the fluid becomes clotted, it should be liquefied with an equal volume of Sputolysin and vigorous mixing. To lower the specific gravity and density of pleural fluid, transfer 20 ml to a sterile 50 ml centrifuge tube and dilute the specimen by filling the tube with distilled water. Invert several times to

PROCEDURE 42.5

SPUTOLYSIN-OXALIC ACID METHOD FOR DECONTAMINATION AND CONCENTRATION

PRINCIPLE

Oxalic acid provides an acid environment that is inhibitory to certain gram-negative rods, such as *Pseudomonas* and *Proteus,* which are difficult to remove from specimens without affecting recovery of mycobacteria.

METHOD

1. Prepare reagents as follows:
   Sputolysin (Behring Diagnostics, La Jolla, Calif.) 1:10 dilution in distilled water
   Oxalic acid 2.5% aqueous solution
   Phosphate buffer (pH 7.0)

2. Add an equal volume of Sputolysin to a maximum of 10 ml of sputum in a 50-ml conical centrifuge tube. Tighten the screw-cap and vortex thoroughly for 30 seconds. Allow to stand for 5 minutes.

3. Add an equal volume of 2.5% oxalic acid. Vortex thoroughly and allow to stand for 20 minutes.

4. Centrifuge at 3600 × *g* for 20 minutes and decant to dryness.

5. Add 40 ml phosphate buffer, resuspend the sediment, and

centrifuge again for 15 minutes.

6. Decant the supernatant, add 1 to 2 ml phosphate buffer to the sediment, and vortex to thoroughly resuspend the sediment.

7. Prepare slides and inoculate media.

QUALITY CONTROL
See Procedure 42.1.

---

mix the suspension and centrifuge at 3600 × *g* for 30 minutes. The supernatant should be removed, and the sediment should be resuspended for smear and culture.

Joint fluid and other sterile exudates can be handled aseptically and inoculated directly to media. Bone marrow aspirates may be injected into Wampole Pediatric Isolator tubes, which help to prevent clotting, and the specimen can be removed with a needle and syringe for preparation of smears and cultures. As an alternative, these specimens are either inoculated directly to media or, if clotted, treated with Sputolysin or glass beads and distilled water before concentration.

Several media have been described for recovery of acid-fast bacilli from blood, including trypticase soy broth, brain-heart infusion broth (fungal medium), 7H11-brain heart infusion biphasic medium, Ficoll-Hypaque isolation of white cell component, and the lysis-centrifugation system. In all patients with AIDS and other immune deficiencies, blood is the specimen of choice for detection of *M. avium* complex.[4,13,43]

CULTURE MEDIA AND INOCULATION METHODS

CONVENTIONAL CULTURE MEDIA

The following solid media are recommended because of the development of characteristic, repro-

ducible colonial morphology, good growth from small inocula, and a low rate of contamination. Media in one ounce prescription bottles are preferred to screw-cap tubes for safety reasons described previously. They also provide a larger surface area for examination of growth and contamination. A minimum of three bottles of at least two different media should be used for each specimen (media are all available from commercial sources). In addition, all aseptically collected specimens from normally sterile sites should be inoculated into 7H9 broth. If a liquid system is being used (described later), one bottle of egg medium may be deleted. All specimens must be processed appropriately before inoculation. It is imperative to inoculate test organisms to commercially available products for quality control. Wallenstein's medium, composed of egg yolk, 2.5% glycerin, malachite green, and water, is an excellent medium for the recovery of MOTT, particularly *M. avium* complex. Löwenstein-Jensen (L-J) medium supplemented with ribonucleic acid and L-J medium supplemented with pyruvic acid are two other preferred media (see box on p. 596). The high rate of contamination and poor recovery of *M. tuberculosis* negate the value of 7H10 or 7H11 as a primary isolation medium. It has no equal, however, for the study of colony morphology, as a selective medium with antibiotics, or as a medium for susceptibility testing. Cultures are incubated at 35° C in the dark in an atmosphere of 5% to 10% CO_2 and high humidity. A second set of cultures of material from lymph nodes or skin lesions suspected of harboring *M. marinum* is incubated at room temperature (or at 30° C if such an incubator is available). Addition of a chocolate or sheep blood agar plate (or the placement of an X-factor [hemin] disk on conventional media) is needed for recovery of *M. haemophilum* from these specimens. Bottle caps should be slightly loose during the first 2 to 3 weeks to allow for evaporation of excess fluids and the entry of CO_2. Caps should be tightened when the surface of the medium is dry, but should be loosened and tightened weekly to allow for gas exchange.

Cultures are examined weekly for growth. Contaminated cultures are discarded and reported as "contaminated, unable to detect presence of mycobacteria," and repeat specimens are requested. If sediment is still available, it may be recultured after enhanced decontamination or by inoculating the sediment to a more selective medium. Most isolates will appear between 3 and 6 weeks; a few isolates will appear after 7 or 8 weeks of incubation. When growth appears, the rate of growth, pigmentation, and colony morphology are recorded. After 8 weeks of incubation, negative cultures (those showing no growth) from smear-negative specimens are reported, and the cultures are discarded.

BROTH CULTURE SYSTEMS

The BACTEC system (Becton-Dickinson Diagnostic Instrument Systems, Sparks, Md.), a recent development for the rapid detection of mycobacteria based on radiometric monitoring, has added a new dimension to diagnostic mycobacteriology. It has reduced the turnaround time for isolation of acid-fast bacilli to approximately 10 days, compared with 17 days for conventional media.[24,30] The BACTEC system is based on the principle that the organisms multiply in the broth and metabolize ^{14}C-containing palmitic acid, producing radioactively labeled $^{14}CO_2$ in the atmosphere that collects above the broth in the bottle. The BACTEC instrument withdraws this CO_2-containing atmosphere and measures the amount of radioactivity present. Those bottles that yield a radioactive index, called a "growth index," greater than 10 are considered to be positive.

The BACTEC culture process may be described briefly as follows: after the specimen has been digested and decontaminated using NALC and NaOH, the sediment is resuspended in sterile phosphate buffer to a total volume of approximately 2 ml. A small quantity (0.4 to 0.5 ml) of this suspension is injected into a BACTEC bottle containing 7H12 medium with added antimicrobial agents. The rest of the sediment is used for conventional smear and culture. Cultures are incubated as noted previously, and BACTEC bottles are incubated in air at 35° C. The bottles are tested for $^{14}CO_2$ production three times weekly for 6 weeks. Bottles that register a growth index greater than 10 are subcultured to media and smeared for acid-fast stain. Broth from bottles registering a growth index greater than 50 are split into two samples, each of which is injected into a fresh bottle of medium. A 5-μg quantity of *p*-nitro-α-acetylamino-β-hydroxypropiophenone (NAP) is added to one of the bottles, and they are both reincubated. These two bottles are then tested on the BACTEC daily for 4 days. An increase in the growth index in the unaltered control bottle without a corresponding increase of the growth index in the bottle containing NAP indicates the presence of *M. tuberculosis* or *M. bovis*, both of which are susceptible to NAP. NAP does not inhibit growth of MOTT. An additional medium containing TCH (thiophene-2-carboxylate hydrazide), to which only *M. bovis* is susceptible, can be used to separate *M. tuberculosis* and *M. bovis*. Definitive identification of mycobacteria other than the tuberculosis complex requires biochemical testing of isolated colonies, of course. The BACTEC system has also been adapted for performance of mycobacterial susceptibility testing. Sediments from acid-fast, smear-positive specimens are inoculated into BACTEC bottles containing special Middlebrook 7H12 with antimicrobial agents and a control broth without antimicrobial agents. The bottles are incubated for a maximum of 2 weeks, and the growth index readings from the antibiotic-containing bottles are compared with those of the control. Once the growth index in the control bottle reaches 20, the test can be reported. A change in the growth index in the antibiotic media that is equal to or greater than that in the control medium indicates resistance, and a greater change in the control indicates susceptibility. Results of trials of this method compared with the conventional direct susceptibility testing method indicate that the BACTEC agreed very well with conventional results, especially when large numbers of organisms were present in the sediment. The radiometric mycobacterial culturing and susceptibility testing systems are continuously improving. At this time, conventional media should be inoculated in addition to BACTEC media for detection and isolation of mycobacteria.

PROCEDURE 42.6

PREPARATION OF SMEARS FOR ACID-FAST STAIN FROM DIRECT OR CONCENTRATED SPECIMENS

METHOD

1. Vortex concentrated sediment, unconcentrated sputum, other purulent material, or stool. Aspirate 0.1 to 0.2 ml into a Pasteur pipette and place 2 to 3 drops on the slide. Place the end of the pipette or a sterile applicator stick parallel to the slide and slowly spread the liquid uniformly to make a thin smear.

2. For cerebrospinal fluid sediment, vortex thoroughly

and apply to the slide in heaped drops. A heaped drop is allowed to air dry, and a second application of sediment is placed on the same spot and allowed to dry. A minimum of three layers, applied to the same 1-cm diameter circle, should facilitate detection of small numbers of bacilli.

3. Fix the smear at 80° C for 15 minutes or for 2 hours at 65° to 70° C on an electric hot plate.

Note: Survival of mycobacteria at this temperature has been reported; handle all specimens with proper precautions.

4. Stain slides by Ziehl-Neelsen or fluorochrome stain.

Modified from Allen, J. 1981. J. Clin. Pathol. 34:719.

A modification of traditional biphasic broth systems has been developed for detection and isolation of mycobacteria. This manual method (Septi-Chek AFB; Becton-Dickinson Microbiology Systems) shortened the average time required for detection of a positive culture from 26 days (using solid media) to 20 days.[13a] This does not yet approach the speed of the BACTEC system (average 13 days), but the Septi-Chek has the advantage of yielding isolated colonies for further studies.

Both liquid systems are used in conjunction with DNA probes for rapid identification of isolates. Sediments of broth cultures are suitable specimens for the probe test so that isolates can be identified even before colonies are available.

MICROSCOPIC EXAMINATION

The mycobacteria possess cell walls that contain mycolic acids, long-chain multiply cross-linked fatty acids. These long-chain mycolic acids probably serve to complex basic dyes, contributing to the characteristic of "acid-fastness" that distinguishes them from other bacteria. Mycobacteria are not the only group with this unique feature. Species of *Nocardia* and *Rhodococcus* are also partially acid fast; *Legionella micdadei*, the causative agent of Pittsburgh pneumonia, is partially acid fast in tissue. Cysts of the genera *Cryptosporidium* and *Isospora* are distinctly acid fast. The mycolic acids and lipids in the mycobacterial cell wall probably account for the unusual resistance of these organisms to the effects of drying and harsh

decontaminating agents, as well as for acid-fastness.

When Gram-stained, mycobacteria usually appear as slender, poorly stained, beaded gram-positive bacilli; sometimes they appear as "gram-neutral" or "gram-ghosts" by failing to take up either crystal violet or safranin. It has been shown that acid-fastness is affected by age of colonies, media on which growth occurs, and ultraviolet light. Rapid-growing species appear to be acid-fast-variable. Three types of staining procedures are used in the laboratory for rapid detection and confirmation of acid-fast bacilli: fluorochrome, Ziehl-Neelsen, and Kinyoun. Smears for all methods are prepared in the same way (Procedure 42.6).

The visualization of acid-fast bacilli in sputum or other clinical material should be considered only presumptive evidence of tuberculosis, since stain does not specifically identify *M. tuberculosis.*[27] The report form should indicate this. For example, *M. gordonae*, a nonpathogenic scotochromogen commonly found in tap water, has been a problem when tap water or deionized water has been used in the preparation of smears or even when patients have rinsed their mouths with tap water before the use of aerosolized saline solution for inducing sputum. However, the incidence of false-positive smears is very low when good quality control is maintained.

FLUOROCHROME STAIN

This is the screening procedure recommended for those laboratories that possess a fluorescent

PROCEDURE 42.7

AURAMINE-RHODAMINE FLUOROCHROME STAIN

PRINCIPLE
The fluorochrome dyes used in this stain complex to the mycolic acids in acid-fast cell walls. Detection of fluorescing cells is enhanced by the brightness against a dark background.

METHOD
1. Heat-fix slides at 80°C for at least 15 minutes, or for 2 hours at 65° to 70° C.

2. Flood slides with auramine-rhodamine reagent and allow to stain for 15 to 20 minutes at room temperature.

3. Rinse with deionized water and tilt slide to drain.

4. Decolorize with acid alcohol (70% ethanol and 0.5% hydrochloric acid) for 2 to 3 minutes.

5. Rinse with deionized water and tilt slide to drain.

6. Flood slides with 0.5% potassium permanganate for 2 to 4 minutes.

7. Rinse with deionized water and air dry.

8. Examine under low power (250×) for fluorescence.

9. Confirm all positive results by Ziehl-Neelsen stain.

QUALITY CONTROL
A series of slides made from a 7H9 suspension of acid-fast organisms, such as *M. tuberculosis* H37Rv, and gram-positive, non–acid-fast organisms, such as *Staphylococcus aureus,* should be prepared in advance and stored in a closed box. One slide from each control set should be heat-fixed and stained along with the specimen slides each time any slides are stained. The mycobacterial slide should display characteristic fluorescence and cell morphology when examined microscopically; the *S. aureus* should not be visible on the slides.

(ultraviolet) microscope (Procedure 42.7). This stain is more sensitive than the conventional carbolfuchsin stains because the fluorescent bacilli stand out brightly against the background; the smear can be initially examined at lower magnifications (250× to 400×), and therefore more fields can be visualized in a short period of time.[5] In addition, a positive fluorescent smear may be restained by the conventional Ziehl-Neelsen or Kinyoun procedure, thereby saving the time needed to make a fresh smear. Screening of specimens with rhodamine or rhodamine-auramine will result in a higher yield of positive smears and will substantially reduce the amount of time needed for examining smears. One drawback associated with the fluorochrome stains is that most strains of rapid growers may not appear fluorescent with these reagents. It is recommended that all negative fluorescent smears be confirmed with Ziehl-Neelsen stain; at least 100 fields should be examined before being reported as negative. It is important to wipe the immersion oil from the objective lens after examining a positive smear, since stained bacilli can float off the slide into the oil and may possibly contribute to a false-positive reading for the next smear examined.

ZIEHL-NEELSEN STAIN
This classic carbolfuchsin stain requires heating the slide for better penetration of stain within the mycobacterial cell wall; hence it is also known as the "hot stain" procedure (Procedure 42.8).[6] At least 100 oil immersion fields should be examined before reporting the smear as negative.

KINYOUN STAIN
Procedure 42.9 is the method of choice for small laboratories. The method is similar to the Ziehl-Neelsen stain but without heat; hence the term "cold stain."[16]

INTERPRETATION AND REPORTING OF THE SMEAR
Smears should be examined carefully by scanning at least 100 fields before reporting a smear as "negative." Typical acid-fast bacilli appear purple to red, slightly curved, short or long rods (2 to 8 μm); they may also appear beaded or banded (*M. kansasii*). For some nontuberculous species, such as *M. avium* complex, they appear pleomorphic, usually coccoid. Results of microscopic examination should be reported "positive for acid-fast bacilli" or "no acid-fast bacilli found." Positive

---

PROCEDURE 42.8

ZIEHL-NEELSEN ACID-FAST STAIN

PRINCIPLE

Heating the slide allows greater penetration of carbolfuchsin into the cell wall. Mycolic acids and waxes complex the basic dye, which then fails to wash out with mild acid decolorization.

METHOD

1. Heat-fix slides on a hot plate as previously described.

2. Flood smear with carbolfuchsin stain reagent and steam the slides gently for 1 minute. This can be accomplished by flaming from below the rack with a gas burner, or by staining the slides directly on a special hot plate. Do not permit the slides to boil or dry out.

3. Allow the stain to remain on the slides for an additional 4 to 5 minutes without heat.

4. Rinse with deionized water and tilt slides to drain.

5. Decolorize with acid alcohol (95% ethanol and 3.0% hydrochloric acid) for 3 minutes.

6. Rinse slides with deionized water and tilt to drain.

7. Flood slides with methylene blue reagent for 1 minute.

8. Rinse with deionized water and allow to air dry.

9. Examine under oil immersion (1000×) for presence of acid-fast bacilli.

QUALITY CONTROL

Prepare control slides in advance, as detailed in Procedure 42.7. Stain one control slide for each slide when slides are stained. Bacilli on the positive smear should display numerous red, small, slightly curved, possibly beaded and tapered ends against a blue background . The negative smear should display numerous blue (non–acid-fast) bacilli.

PROCEDURE 42.9

KINYOUN STAIN

PRINCIPLE

By increasing the concentration of basic fuchsin and phenol, the need for heating the slide is avoided.

METHOD

1. Heat-fix slides on an 85° C hot plate for at least 15 minutes.

2. Flood slides with Kinyoun's carbolfuchsin reagent and allow to stain for 5 minutes at room temperature.

3. Rinse with deionized water and tilt slide to drain.

4. Decolorize with acid-alcohol for 3 minutes and rinse again with deionized water.

5. Redecolorize with acid-alcohol for 1 to 2 minutes or until no more red color runs from the smear.

6. Rinse with deionized water and drain standing water from slide surface by tipping slide.

7. Flood slide with methylene blue counterstain and allow to stain for 4 minutes.

8. Rinse with distilled water and allow to air dry.

9. Examine under high dry (400×) magnification, and confirm acid-fast structures under oil immersion (1000×).

QUALITY CONTROL

Follow the same protocol as for Procedure 42.8.

Modified from Kinyoun, J.J. 1915. Am. J. Public Health 5:867.

TABLE 42.4

ENUMERATION OF ACID-FAST BACILLI IN SMEARS

NO. OF ORGANISMS SEEN	REPORT
1-2 per entire smear specimen	Negative; request another
3-9 per entire smear	Rare (1+)
10 or more per entire smear	Few (2+)
1 or more per oil immersion field	Numerous (3+)

findings should be based only on typical forms, but atypical cells should be noted. When large numbers of typical acid-fast bacilli are seen, it is reasonable to assume that they are *M. tuberculosis*. When atypical rods are seen, they may represent other pathogenic or nonpathogenic mycobacteria or other partially acid-fast organisms. It is desirable to report the number of acid-fast bacilli seen; the criteria recommended by the American Lung Association (Table 42.4) may be used.

When only one or two acid-fast bacilli are seen on the entire smear, it is recommended that they not be reported until confirmation is obtained by examining other smears from the same or another specimen. One study involving sputum and gastric contents found that when more than six acid-fast organisms were seen per high-power field (400×), pathogenic mycobacteria were always recovered from the culture.

STAINING MATERIAL OTHER THAN SPUTUM AND GASTRIC LAVAGES

When repeated specimens of cerebrospinal fluid are examined, a high percentage of specimens from patients with disease will reveal acid-fast bacilli by microscopy or culture. Microscopy is particularly important because of the need to initiate therapy as soon as possible when tuberculous meningitis is suspected. Either the fluorochrome or Ziehl-Neelsen stain can be used for detecting acid-fast bacilli in tissues. The tissue sections should be deparaffinized in three changes of xylene and hydrated before staining with rhodamine-auramine. Fluorochrome stains are ideal for tissue sections in which acid-fast bacilli are usually found in clusters. Fluorochrome positive slides may be restained with Ziehl-Neelsen.

ADDITIONAL CONSIDERATIONS

Species of *Nocardia* and *Rhodococcus* are partially acid fast; therefore less stringent destaining (weaker acid or shorter duration or both) is used to detect these organisms. Red-green color-blind people can use Spengler's method to visualize acid-fast bacilli in 10% formalin-fixed smears or tissue sections. With this method, acid-fast bacilli stain black and non–acid-fast organisms appear yellowish.

The examination of stained smears is considered the least sensitive of the diagnostic methods for tuberculosis; **cultures should be performed on all specimens.** Because of its simplicity and speed, however, the stained smear is an important and useful test, particularly for the detection of smear-positive patients ("infectious reservoirs"), who are the greatest risk to others in their environment.

MACROSCOPIC EXAMINATION

The preliminary identification of mycobacterial isolates depends on their rate of growth, colony morphology, colony texture, and pigmentation. The identification procedures are modified from material published by George Kubica and the Bacteriology Training Section of the Centers for Disease Control (CDC).[18] Quality control organisms should be tested along with unknowns, as listed in Table 42.5. The commonly used quality control organisms can be maintained in broth at room temperature and transferred monthly. In this way they will always be available for inoculation to test media along with suspensions of the unknown mycobacteria being tested.

The first test that must always be performed on colonies growing on mycobacterial media is an acid-fast stain to confirm that the colonies are indeed mycobacteria. At this point, inoculate several colonies of the organism to 5 ml of Middlebrook 7H9 broth (available from several commercial suppliers) and incubate the broth at 35° C for 5 to 7 days with daily agitation (with the cap tightened shut) to enhance growth. If there is sufficient growth on the original L-J medium, the L-J slant can be used for performance of the niacin test, as outlined later. The broth subculture can be used for inoculation of all test media, including biochemical tests as well as pigmentation and growth rate determinations (Procedure 42.10). It is up to the microbiologist to determine whether all tests are inoculated initially at the same time or whether the results of growth rate and pigment production are determined before additional tests are set up. Some observations about colony morphology and permissive incubation temperatures of mycobacteria that may be

TABLE 42.5

CONTROLS AND MEDIA USED AND DURATION OF TESTS

NO.	BIOCHEMICAL TEST	CONTROL ORGANISMS POSITIVE	CONTROL ORGANISMS NEGATIVE	RESULT POSITIVE	RESULT NEGATIVE	MEDIUM USED AND AMOUNT	DURATION OF TEST	REMARKS
1	Niacin	*M. tuberculosis*	*M. avium*	Yellow	No change of color	0.5 ml DH$_2$O	30 min	Room temperature
2	Nitrate	*M. tuberculosis*	*M. avium*	Pink or red	No change of color	0.3 ml 7H9	2 h	37° C bath
3	Urease	*M.kansasii*	*M. avium*	Pink or red	No change of color	0.5 ml DH$_2$O	2 h	37° C bath
4	68° C Catalase	*M. fortuitum* *M. avium*	*M. tuberculosis*	Bubbles	No bubbles	0.5 ml phosphate buffer [pH 7]	30 min	68° C bath
5	S.Q. Catalase	*M. kansasii*	*M. avium*	>45 mm	≤40 mm	Commercial medium	14 d	37° C incubator [with CO$_2$]
6	Tween 80	*M. kansasii*	*M. tuberculosis*	Pink or red	No change of color	1 ml DH$_2$O	5 d or 10 d	37° C incubator [without CO$_2$]
7	Tellurite	*M. avium*	*M. tuberculosis*	Smooth fine black precipitate (smokelike action)	Gray clumps (no smokelike action)	Middlebrook, 7H9 broth	7 + 3 d	37° C incubator [with CO$_2$]
8	Arylsulfatase	*M. fortuitum*	Medium	Pink or red	No change of color	Wayne's arylsulfatase medium	3 d	37° C incubator [without CO$_2$]
9	5% NaCl	*M. fortuitum*	*M. gordonae*	Substantial growth	Little or no growth	Commercial slant	28 d	37° C incubator [with CO$_2$]
10	(TCH)	*M. bovis*	*M. tuberculosis*	No growth	Growth	(TCH) flat and L-J	8 wk	37° C incubator [with CO$_2$]

From Berlin, O.G.W., and Martin, W.J. 1980. Importance of nitrate test in the identification of mycobacteria. Clin Microbiol. Newsletter. 2:4.

PROCEDURE 42.10

DETERMINATION OF PIGMENT PRODUCTION AND GROWTH RATE

PRINCIPLE

Certain mycobacteria produce pigmented chemicals (carotenoids), either dependently or independently of exposure to light. This characteristic, in addition to their doubling time under standard conditions, is useful for initial identification.

METHOD

1. After the broth culture has incubated for 5 to 7 days, adjust the turbidity to that of a McFarland 0.5 standard.

2. Dilute the broth (McFarland 0.5 turbidity) 10^{-4}.

3. Inoculate 0.1 ml of the diluted broth to each of three tubes of Löwenstein-Jensen agar. Completely wrap two of the tubes in aluminum foil to block all light. If the isolate was obtained from a skin lesion, or the initial colony was yellow-pigmented (possible *M. szulgai*), six tubes should

be inoculated. The second set of tubes, two of them also wrapped with aluminum foil, is incubated at 30° C, or at room temperature if a 30° C incubator is not available.

4. Examine the cultures after 5 and 7 days for the appearance of grossly visible colonies. Examine again at intervals of 3 days. Interpretation: Rapid growers produce visible colonies in less than 7 days; slow growers require more than 7 days.

5. When colonies are mature, expose the growth from a foil-wrapped tube to a bright light, such as a desk lamp, for 2 hours. The cap must be loose during exposure, since pigment production is an oxygen-dependent reaction.[39] The tube is rewrapped and returned to the incubator, and the cap is left loose.

6. The three tubes are examined 24 and 48 hours after light exposure. For tubes incubating at 30° C, pigment may require 72 hours for development.

7. Results are interpreted as shown in Figure 42.3.

QUALITY CONTROL

Inoculate suspensions of known organisms to 7H9 broths weekly and test these cultures along with unknowns each time the pigment and growth rate tests are performed. A typical photochromogen control is *M. kansasii*, a typical scotochromogen control is *M. flavescens*, and a typical nonphotochromogen control is *M. avium* complex. The *M. flavescens* can also serve as a quality control organism for rapid growth rate.

found in clinical specimens are listed in Table 42.6.

The rate of growth is an important criterion for determining the initial category of an isolate. Rapid growers will usually produce colonies within 3 to 4 days after subculture. Even a rapid grower, however, may take longer than 7 days to initially produce colonies because of inhibition by the harsh decontaminating procedure. The dilution of the organism used to assess growth rate is critical. Even slow-growing mycobacteria will appear to produce colonies in less than 7 days if the inoculum is too heavy. One organism particularly likely to exhibit false-positive rapid growth is *M. flavescens*. This species therefore serves as an excellent quality control organism for this procedure.

Species of mycobacteria synthesize carotenoids in varying amounts; they are categorized into three groups based on productions of these pigments (Procedure 42.10). Photochromogens produce the yellow pigment when exposed to light (Figure 42.3). To achieve optimum photochromogenicity, the colonies should be young, actively metabolizing, isolated, and well aerated. Although some species, such as *M. kansasii*, turn yellow after a few hours of light exposure, others, such as *M. simiae*, may take a prolonged exposure to light. Scotochromogens produce pigmented colonies even in the absence of light and colonies often become darker with prolonged exposure to light. One member of this group, *M. szulgai*, is peculiar in that it is a scotochromogen at 35° C and nonpigmented when grown at 25° to 30° C. For this reason, all pigmented colonies should be subcultured to test for photoactivated pigment at both 35° C and 25° to 30° C. Nonchromogens are not affected by light. Although commonly nonpigmented, some strains of *M. avium* complex isolated from AIDS patients may exhibit differential colony pigmentation.

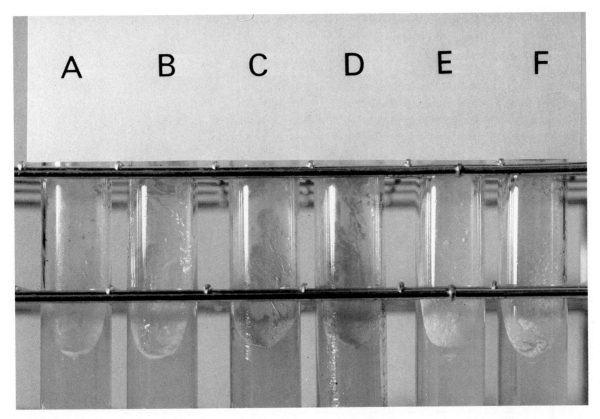

FIGURE 42.3

Initial grouping of mycobacteria based on pigment production before and after exposure to light. In one test system, subcultures of each isolate are grown on two agar slants. One tube is wrapped in aluminum foil to prevent exposure of the organism to light, and the other tube is allowed light exposure. After sufficient growth is present, the wrapped tube is unwrapped, and the tubes are examined together. **Photochromogens** are unpigmented when grown in the dark *(tube A)* and develop pigment after light exposure *(tube B)*. **Scotochromogens** are pigmented in the dark *(tube C)*; the color does not intensify after exposure to light *(tube D)*. **Nonphotochromogens** are nonpigmented when grown in the dark *(tube E)* and remain so even after light exposure *(tube F)*.

BIOCHEMICAL TESTING

Once an organism has been identified on the basis of acid-fast morphology, growth rate, and pigment production, definitive identification is based on a battery of biochemical tests.[14,35] Many of the biochemical tests are qualitative and based on color reactions. Good growth of the isolate is an important prerequisite for performing these tests.[31] A list of commonly used biochemical tests and the strains suggested for quality control testing are summarized in Table 42.5; biochemical profiles of individual species and species groups are summarized in Tables 42.6 to 42.10.

NIACIN TEST

Niacin or nicotinic acid plays an important role in the oxidation-reduction reactions that occur during mycobacterial metabolism. Although all species produce nicotinic acid, *M. tuberculosis* accumulates the largest amount. *M. simiae* and some strains of *M. chelonae* also produce niacin. Niacin

therefore accumulates in the medium in which these organisms are growing. A positive niacin test is preliminary evidence that an organism that exhibits a buff-colored, slow-growing, rough colony may be *M. tuberculosis*. This test is not sufficient, however, for confirmation of the identification. If sufficient growth is present on the initial L-J slant (the egg-base medium enhances accumulation of free niacin), a niacin test can be performed immediately.[17] If growth on the initial culture was scanty, the subculture used for growth rate determination can be used. If this culture yields only rare colonies, the colonies should be spread around with a sterile cotton swab (after the growth rate has been determined) to distribute the inoculum over the entire slant, which is then reincubated until light growth over the surface of the medium is visible. For reliable results, the niacin test should be performed only from cultures on L-J that are at least 3 weeks old and show at least 50 colonies.

TABLE 42.6

DISTINCTIVE PROPERTIES OF CULTIVABLE MYCOBACTERIA ENCOUNTERED IN CLINICAL SPECIMENS[a]

GROUP/ COMPLEX[b]	SPECIES	CLINICAL SIGNIFICANCE[c]	GROWTH RATE[d] AT 45°C	37°C	31°C	24°C	USUAL COLONY MORPHOLOGY[e]	PIGMENTATION[f]	NIACIN	SUSCEPTIBILITY TO TCH[g] (5 µg/ml)	NITRATE REDUCTION	SEMIQUANTITATIVE CATALASE (>45mm)	68°C CATALASE	TWEEN HYDROLYSIS, 5 DAYS	TELLURITE REDUCTION	TOLERANCE TO 5% NaCl	IRON UPTAKE	ARYLSULFATASE, 3 DAYS	GROWTH ON MacCONKEY AGAR	UREASE	PYRAZINAMIDASE, 4 DAYS
TB	*M. tuberculosis*	1	–	S		–	R	N	+	–	+	–	–	–[j]	±	–	–	–	–	+	+
	M. bovis	1	–	S		–	Rt	N	–	+	–	–	–	–	±	–	–	–	–	+	–
	M. africanum	1	–	S	S	–	R	N	–	V	–	–	–	–	–	–	–	–	–	+	–
	M. ulcerans	1	–	–	S	–	R	N	–	–	–	–	+	–	±	–	–	–	–	–	–
Photochromogens	*M. marinum*	2		+±	M	M	S/SR	P	+±	–	–	+	–	+	+±	–	–	+±	–	+	+
	M. kansasii	2		S	S	S	SR/S	P	–	–	+	+	+	+	+±	–	–	–	–	+	–
	M. simiae	3–2		S			S	P	+	–	–	+	+	–	–	–	–	–	–	+	+
	M. asiaticum	3–2		S		S		P	–	–	–	+	+	+	+	–	–	–	–	–	–
Scotochromogens	*M. scrofulaceum*	3–2		S		S	S or R	S	–	–	–	+	+	–	–	–	–	V	–	+	±
	M. szulgai	1		S		S	S	S/P	–	–	+	+	+	+±[h]	+±	–	–	V	–	+	+
	M. gordonae	4		S		S	S	S	–	–	–	+	+	+	+±	–	–	V	–	–	+±
	M. flavescens	4		M		M	S	Si	–	–	+	+	+	+	–	+	–	–	–	+	+
	M. thermoresistibile	1	R	R	R	R	S	S	–	–	+	+	+	+	+±	+	–	+	–	+	
	M. xenopi	3	S	R	R		Sf	S	–	–	V	+		+	+±	–	–	+	+±	–	V
M. avium (complex)	*M. avium*	2	V	S		±	St/R	N	–	–	–	–	+	–	+±	–	–	–	+±	+	+
	M. intracellulare	2	V	S		±	St/R	N	–	–	–	–	+±	–	+±	–	–	–	+±	+	+
	"*M. genavense*"[m]	1		S			St/R	N	–	–	–	–	+±	–	+±	–	–	–		–	+
	M. gastri	4		S		S	S/SR/R	N	–	–	–	+	+	+	+	–	–	+	–	+	–
	M. malmoense	1		S	S	S	S	N	–	–	–	–	+±	+	+±	–	–	–	–	+	+
	M. haemophilum	1		–	S[k]	S	R	N	–	–	–	–	+±	+	+	–	–	+	–	V	+
	M. shimoidei	2		S	–		R	N	–	–	–	+	+	–	+	–	–	–	–	–	+

TABLE 42.6

DISTINCTIVE PROPERTIES OF CULTIVABLE MYCOBACTERIA ENCOUNTERED IN CLINICAL SPECIMENS^a—cont'd

GROUP/COMPLEX^b	SPECIES	CLINICAL SIGNIFICANCE^c	GROWTH RATE^d AT 45°C	37°C	31°C	24°C	USUAL COLONY MORPHOLOGY^e	PIGMENTATION^f	NIACIN	SUSCEPTIBILITY TO TCH^g (5μg/ml)	NITRATE REDUCTION	SEMIQUANTITATIVE CATALASE (>45mm)	68°C CATALASE	TWEEN HYDROLYSIS, 5 DAYS	TELLURITE REDUCTION	TOLERANCE TO 5% NaCl	IRON UPTAKE	ARYLSULFATASE, 3 DAYS	GROWTH ON MacCONKEY AGAR	UREASE	PYRAZINAMIDASE, 4 DAYS
M. terrae (complex)	M. terrae	4		S	S	S	SR	N	−	−	+	+	+	+	−	−	−	−	V	−	V
	M. triviale	4		M	S	S	R	N	−	−	+	+	+	+	−	+	−	±	−	−	V
	M. nonchromogenicum	4		S	S	S	SR	N	−	−	+	+	+	+	−	−	−	−	V	−	V
M. fortuitum	M. fortuitum	4-3	−	R	R	R	Sf/Rf	N	V	−	+	+	V	V	+	V^l	+	+	+	+	+
	M. chelonae	4-3	−	R	R	R	S/R	N	V	−	−	+	+	V	+	+	−	+	+	+	+
	M. phlei	4	R	R	R	R	R	S			+	+	+	+	+	+	+	−	−		
	M. smegmatis	4	R	R	R	R	R/S	N			+	+	+	+	+	V	+	−	−		
	M. vaccae	4	R	R	R	R	S	S			+	+	+	+	+		+	−	−		

^a Plus and minus signs indicate the presence or absence, respectively, of the feature; blank spaces indicate either that the information is not currently available or that the property is unimportant. V, Variable; ±, usually present; ∓, usually absent.

^b For most cinical laboratories, designation to the complex is usually sufficient.

^c Potential clinical significance; 1, present only as pathogens; 2, present usually as pathogens; 3, present commonly as nonpathogens; 4, present usually as nonpathogens.

^d S, Slow; M, moderate; R, rapid.

^e R, Rough, S, smooth; SR, intermediate in roughness; t, thin or transparent; f, filamentous extensions.

^f P, Photochromogenic; S, scotochromogenic; N, nonphotochromogenic; NOTE: M. szulgai is scotochromogenic at 37° C and photochromogenic at 25° C.

^g TCH, Thiophene-2-carboxylic acid hydrazide.

^h Tween hydrolysis may be positive at 10 days.

^i Arylsulfatase, 14 days, is positive.

^j Young cultures may be nonchromogenic or possess only pale pigment that may intensify with age.

^k Requires hemin as a growth factor.

^l M. chelonae subsp. chelonae is negative, M. chelonae subsp. abscessus is positive.

^m Requires mycobactin for growth on solid media.

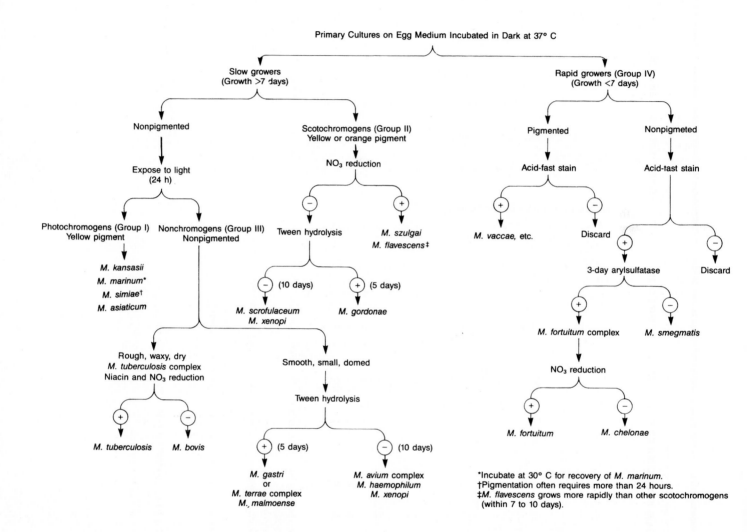

Primary Cultures on Egg Medium Incubated in Dark at 37° C

Slow growers
(Growth >7 days)

Rapid growers (Group IV)
(Growth <7 days)

Nonpigmented

Scotochromogens (Group II)
Yellow or orange pigment

Pigmented

Nonpigmeted

Expose to light
(24 h)

NO_3 reduction

Acid-fast stain

Acid-fast stain

Photochromogens (Group I)
Yellow pigment

Nonchromogens (Group III)
Nonpigmented

⊖

⊕

⊕

⊖

⊕

⊖

M. kansasii
*M. marinum**
M. simiae†
M. asiaticum

Tween hydrolysis

M. szulgai
M. flavescens‡

M. vaccae, etc.

Discard

⊕

⊖

⊖ (10 days)

⊕ (5 days)

3-day arylsulfatase

Discard

M. scrofulaceum
M. xenopi

M. gordonae

⊕

⊖

Rough, waxy, dry
M. tuberculosis complex
Niacin and NO_3 reduction

Smooth, small, domed

M. fortuitum complex

M. smegmatis

Tween hydrolysis

NO_3 reduction

⊕

⊖

⊕

⊖

M. tuberculosis

M. bovis

⊕ (5 days)

⊖ (10 days)

M. fortuitum

M. chelonae

M. gastri
or
M. terrae complex
M. malmoense

M. avium complex
M. haemophilum
M. xenopi

*Incubate at 30° C for recovery of *M. marinum*.
†Pigmentation often requires more than 24 hours.
‡*M. flavescens* grows more rapidly than other scotochromogens
(within 7 to 10 days).

FIGURE 42.4

The identification of mycobacteria. See Tables 42.6 to 42.12 for additional confirmatory tests.

Classic tests for niacin involve cyanogen bromide, which is hazardous to use and presents disposal problems. For this reason, a method that utilizes reagent-impregnated paper strips should be used (Figure 42.5). The results are comparable and the procedure is simple (Procedure 42.11). Control organisms are inoculated at the same time as the test organisms, as listed in Table 42.5.

NITRATE REDUCTION TEST[37]
Many species of mycobacteria are able to reduce nitrate to nitrite (Table 42.11).[1a] The ability of acid-fast bacilli to reduce nitrate is influenced by age of the colonies, temperature, pH, and enzyme inhibitors. Although rapid growers can be tested within 2 weeks, slow growers should be tested after 3 to 4 weeks of luxuriant growth. Commer-

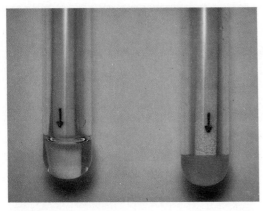

FIGURE 42.5

Niacin test performed with filter paper strips. The positive test *(right)* displays a yellow color, and the negative result remains milky white or clear.

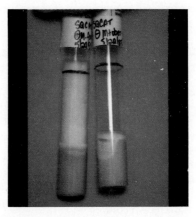

FIGURE 42.6

Semiquantitative catalase test. The tube on the left contains a column of bubbles that has risen past the line indicating 45 mm height (a positive test). The tube on the right is the negative control.

cially available nitrate strips yield acceptable results only with strongly nitrate positive organisms, such as *M. tuberculosis*. This test (Procedure 42.12) may be tried first because of its ease of performance. The *M. tuberculosis* positive control must be strongly positive in the strip test or the test results will be unreliable. If the paper strip test is negative or if the control test result is not strongly positive, the chemical procedure (Procedure 42.13) must be carried out, using strong and weakly positive controls. Steps and reagents used for performing the classic chemical nitrate procedure are outlined in detail in the CDC manual by Vestal.[35]

CATALASE TEST

Most species of mycobacteria, with the exception of certain strains of *M. tuberculosis* complex, produce the intracellular enzyme **catalase,** which splits hydrogen peroxide into water and oxygen. Catalase is measured in two ways:

1. By the relative activity of the enzyme, as determined by the height of a column of bubbles of oxygen (Figure 42.6) formed by the action of untreated enzyme produced by the organism (semiquantitative [S.Q.] catalase test,[21] Procedure 42.14).
2. By the ability of the catalase enzyme to remain active after heating, a measure of the heat stability of the enzyme (heat-stable catalase test[22] Procedure 42.15).

HYDROLYSIS OF TWEEN 80[40]

The commonly-nonpathogenic slow-growing scotochromogens and nonchromogens produce a lipase that is able to hydrolyze Tween 80 (the detergent polyoxyethylene sorbitan monooleate)

into oleic acid and polyoxyethylated sorbitol, whereas pathogenic species do not. Tween 80 hydrolysis (Procedure 42.16) is useful for separating species of photochromogens, nonchromogens, and scotochromogens, as shown in Table 42.12. Because laboratory-prepared media have a very short shelf life, the CDC recommends the use of a commercial Tween 80 hydrolysis substrate (Difco Laboratories or Remel Laboratories) that is stable for up to 1 year.

TELLURITE REDUCTION[15]

Some species of mycobacteria reduce potassium tellurite at variable rates. The ability to reduce tellurite in 3 days distinguishes members of *M. avium* complex from most other nonchromogenic species (Procedure 42.17). All rapid growers reduce tellurite in 3 days.

ARYLSULFATASE TEST[38]

The enzyme arylsulfatase, which splits free phenolphthalein from the tripotassium salt of phenolphthalein disulfate, is present in most mycobacteria. Varying the test conditions helps to differentiate different forms of the enzyme (Procedure 42.18). The 3-day test is particularly useful for identifying the potentially pathogenic rapid growers, *M. fortuitum,* and *M. chelonae.* Slow-growing *M. marinum* and *M. szulgai* are positive in the 14-day test (Figure 42.7).

GROWTH INHIBITION BY THIOPHENE-2-CARBOXYLIC ACID HYDRAZIDE (TCH)[36]

This test is used to distinguish *M. bovis* from *M. tuberculosis,* since only *M. bovis* is unable to grow in the presence of 2 µg/ml TCH (Procedure 42.19).

PROCEDURE 42.11

NIACIN TEST WITH COMMERCIALLY AVAILABLE PAPER STRIPS (AVAILABLE FROM DIFCO LABORATORIES, DETROIT, MICH.)

PRINCIPLE

The accumulation of niacin in the medium caused by lack of an enzyme that converts niacin to another metabolite in the coenzyme pathway is characteristic for *M. tuberculosis* and a few other species. Niacin is measured by a colored end product.

METHOD

1. Add 1 ml sterile distilled water to the surface of the egg-based medium on which the colonies to be tested are growing.

2. Lay the tube horizontal, so that the fluid is in contact with the entire surface. With the pipette used to add the water, scratch or lightly poke through the surface of the agar; this allows niacin in the medium to dissolve in the water.

3. Allow the tube to sit for 15 to 30 minutes at room temperature. It can incubate longer to achieve a stronger reaction.

4. Remove 0.6 ml of the distilled water (which appears cloudy at this point) to a clean, 12 × 75 mm screw-cap or snap-top test tube. Insert a niacin test strip with the arrow down, following manufacturer's instructions.

5. Cap the tube tightly and incubate at room temperature, occasionally shaking the tube to mix the fluid with the reagent on the bottom of the strip.

6. After 20 minutes, observe the color of the liquid against a white background (Figure 42.5). Yellow liquid indicates a positive test. The

color of the strip should not be considered when evaluating results. If the liquid is clear, the test is negative.

7. Discard the strip into alkaline disinfectant (10% NaOH) to neutralize the cyanogen bromide.

8. Negative tests may be repeated on a fresh 3- to 4-week-old subculture of the isolate.

QUALITY CONTROL

Test positive and negative control organisms (Table 42.5) with each performance of the test.

OTHER TESTS

Tests other than those mentioned here are often performed to make more subtle distinctions between species. They include the ability to grow using only sodium citrate or mannitol as a sole source of carbon, used to distinguish between certain rapid growers. *M. fortuitum* and most other rapid growers are able to convert ferric ammonium citrate to iron oxide. The "iron uptake test" is useful for separating *M. chelonae*, which is unable to convert iron in this way (Figure 42.8). Ability to grow on MacConkey agar without crystal violet can help to identify *M. fortuitum*, which is not inhibited by the medium. Pyrazinamide can be deaminated by *M. tuberculosis* and the *M. avium* complex, as well as by *M. marinum*, whereas *M. bovis* and *M. kansasii* do not possess the enzyme pyrazinamidase. The inability of *M. chelonae* to grow in 5% NaCl is yet another test for differentiating this organism from *M. fortuitum*. Most rapid growers and some *M. flavescens* strains will grow on medium containing such high salt concentrations. The urease

test[32] for hydrolysis of urea by mycobacteria can provide valuable information. As it is currently performed, however, results vary among laboratories, and no reliable method can be recommended. It is not cost effective for routine clinical microbiology laboratories to be able to perform all of the procedures necessary for definitive identification of mycobacteria, since excellent reference laboratories are available in every state. With a minimal number of basic procedures, however, the great bulk of all strains isolated can be presumptively identified and those that require further testing can be presumptively identified and forwarded to regional laboratories.

ANTIMICROBIAL SUSCEPTIBILITY TESTING

Antituberculosis antimicrobial susceptibility testing, if performed properly, provides important information to the physician for determining the patient's course of treatment.[25] It is a time-

PROCEDURE 42.12

NITRATE REDUCTION TEST WITH COMMERCIALLY AVAILABLE PAPER STRIPS (AVAILABLE FROM DIFCO LABORATORIES, DETROIT, MICH.)

PRINCIPLE

The presence of the enzyme nitroreductase can be detected by the ability of a suspension of organisms to produce a colored end product from substrates that combine with nitrite, the product of nitroreductase. *M. tuberculosis* and several other species of mycobacteria possess this enzyme.

METHOD

1. Add 1 ml sterile saline to a sterile 13 × 100 ml screw-cap test tube.

2. Emulsify two very large clumps of growth from a 4-week-old culture in the saline. The solution should be very turbid (milky).

3. Using sterile forceps, carefully insert a nitrite test strip according to the package insert instructions. The strip should touch only the fluid at the bottom of the tube, not the sides.

4. Cap the tube tightly and incubate upright for 2 hours at 35° C. Incubation in a water bath will ensure maintenance of adequate temperature.

5. After the first hour, shake the tube gently without tilting.

6. After the 2-hour incubation, tilt the tube six times, wetting the entire strip.

7. Place the tube in a slanted position for 10 minutes with the liquid covering the strip.

8. Observe the top portion of the strip for any blue color change, indicating a positive reaction. A negative reaction (lack of nitroreductase) yields no color change.

QUALITY CONTROL

Test positive and negative control strains (Table 42.5) each time the nitrate test is performed. These strips should be stored refrigerated in the dark. Discolored strips have deteriorated and should not be used.

---

consuming procedure that requires meticulous care in the preparation of the medium, selection of adequate samples of colonies, standardization of the inoculum, use of appropriate controls, and interpretation of results. The susceptibility test may be performed by either the direct or the indirect method. The direct method utilizes a smear-positive concentrate containing more than 50 acid-fast bacilli per 100 oil immersion fields; the indirect method uses a culture as the inoculum source. Isolates of mycobacteria should be tested for susceptibility to primary drugs. In areas where resistant strains of *M. tuberculosis* are commonly seen, susceptibility testing of all isolates is recommended. Laboratories that see very few positive cultures should consider sending isolates to a reference laboratory for testing. Those isolates that must be saved for the future possibility of additional studies (such as susceptibilities if the patient does not respond well to treatment) can be frozen in sterile 10% skim milk in distilled water at −70° C.

With the advent of AIDS, several nontuberculous species have been detected with increasing frequency from patients with immune dysfunctions. Many of these species, including *M. avium* complex, *M. fortuitum*, and *M. chelonae*, are very resistant to all conventional antituberculosis drugs. Therefore new and previously little-used antimycobacterial drugs have been tested in vitro and in vivo. The conventional primary drugs are tested against *M. tuberculosis* complex and some photochromogens and scotochromogens, as outlined in Chapter 14. *M. avium* complex and *M. scrofulaceum* are resistant to these agents. It has been shown that some novel drugs, such as ansamycin (rifabutin), and antileprosy drugs, such as clofazimine (lamprene), are effective against these organisms in vitro. One important consideration when testing isolates of *M. avium-intracellulare* is that colony morphology relates to virulence in this species. Transparent, glossy colonies are more virulent than the opaque, dull, white or yellow colonies often seen on isolation media (Figure 42.9). Therefore isolated colonies of the transparent type should be chosen for in vitro susceptibility testing of these organisms. The rapidly growing species such as *M. fortuitum* and *M. chelonae* are susceptible to amikacin and ciprofloxacin, even though they are resistant to most of the conventional agents. They may exhibit varying susceptibility to erythromycin, cefoxitin, minocycline, and imipenem. Indirect susceptibility testing can be done by agar dilution,

PROCEDURE 42.13

NITRATE REDUCTION TEST USING CHEMICAL REAGENTS

PRINCIPLE

As in the conventional nitrate test, the presence of nitrite (product of the nitroreductase enzyme) is detected by production of a red-colored product on the addition of several reagents. If the enzyme has reduced nitrate past nitrite to gas, then addition of zinc dust (which converts nitrate to nitrite) will detect the lack of nitrate in the reaction medium.

METHOD

1. Prepare the dry crystalline reagent as follows:
   Sulfanilic acid (Sigma Chemical Co., St. Louis, Mo.) 1 part
   N-(1-Naphthyl) ethylenediamine dihydrochloride (Eastman Chemical Co.) 1 part
   1-Tartaric acid (Sigma Chemical Co.) 10 parts
   These crystals can be measured out with any small scoop or tiny spoon, since the proportions are by volume, not weight. The

mixture should be ground in a mortar and pestle to ensure adequate mixing, because the crystals are of different textures. The reagent can be stored in a dark glass bottle at room temperature for at least 6 months.

2. Add 0.2 ml sterile distilled water to a 16 × 125 mm screw-cap tube. Emulsify two very large clumps of growth from a 4-week culture on Löwenstein-Jensen agar in the water. The suspension should be milky.

3. Add 2 ml nitrate substrate broth (Difco Laboratories or Remel, Lenexa, Kan.) to the suspension and cap tightly.

4. Shake gently and incubate upright for 2 hours in a 35° C water bath.

5. Remove from water bath and add a small amount of the crystalline reagent. A wooden stick or a small

spatula can be used to add crystals; the amount is not critical.

6. Examine immediately for a pink to red color, indicating the presence of nitrite, demonstrating the ability of the organism to reduce nitrate to nitrite.

7. If no color results, the organisms may have reduced nitrate beyond nitrite (as in the conventional nitrate test, Chapter 10). Add a small amount of powdered zinc to the negative tube. If a red color develops, that indicates that unreduced nitrate was present in the tube and the organism was nitroreductase-negative.

QUALITY CONTROL

Test the quality control organisms, including the weakly positive control (*M. kansasii*, see Table 42.5), each time the test is performed.

---

disk elution, a modified agar dilution, a radiometric method that uses the BACTEC instrument, and microdilution procedures. With agar dilution, antimycobacterial agents are serially diluted into Middlebrook agar (often in quadrant plates), and dilutions of the organism are inoculated to the surface. Disk elution is a modified agar preparation in which antibiotics elute from disks (obtained from commercial suppliers) that are embedded in the molten agar. The radiometric system involves inoculating either sediments of acid-fast smear-positive specimens or suspensions of organisms grown in culture into BACTEC bottles containing Middlebrook 7H12 B broth with antibiotics. Growth in the liquid medium is monitored radiometrically, and the results are interpreted in 4 to 5 days. An increase in the growth index greater than or equal to the growth index measured in concurrently inoculated control bottles without antibiotics indicates resistance.

Greater growth in the control suggests susceptibility. Results of trials of this method compare well with conventional testing and are achieved much more quickly. A microdilution procedure can be used for rapidly growing mycobacteria. Wells of microbroth dilution plates containing dilutions of the antibiotics to be tested are inoculated in the same manner as for routine susceptibility testing (Chapter 14). Results can usually be read within 2 to 3 days. The E test (Chapter 14) is also a promising method for testing rapidly growing strains.

BRIEF DISCUSSION OF INDIVIDUAL ORGANISMS

M. TUBERCULOSIS COMPLEX AND RELATED SPECIES

This group contains *M. tuberculosis*, *M. bovis*, *M. africanum*, *M. microti*, and the BCG strain. Except for *M. microti*, an animal pathogen, all species in-

PROCEDURE 42.14

SEMIQUANTITATIVE (S.Q.) CATALASE TEST

PRINCIPLE

The amount of catalase present within cells of species of mycobacteria can be estimated by the height of the column of bubbles of oxygen produced during reduction of hydrogen peroxide.

METHOD

1. Prepare 10% Tween 80 solution as follows:
   Tween 80 (Difco Laboratories) 10 ml
   Distilled water 90 ml
     Mix together and autoclave for 10 minutes at 121° C. Swirl solution after autoclaving to disperse Tween 80. Store in refrigerator. This reagent is also available from Remel Laboratories and can be stored for at least 6 months.

2. Inoculate the surface of a 5-ml L-J agar butt tube (available commercially) with 0.1 ml of the broth subculture originally made for preparing the inoculum for the growth rate and pigment tests (described previously). Inoculate tubes with the quality control organisms at the same time.

3. Incubate tubes with caps loose for 2 weeks at 35° C.

4. Prepare hydrogen peroxide reagent fresh for each use:
   30% hydrogen peroxide (Superoxol, Merck Chemical Co.) 1 part
   10% Tween 80 1 part
     Make 1 ml of the hydrogen peroxide reagent for each test to be performed.

5. Place the tubes upright in a rack that is standing in a tray lined with paper towels soaked with disinfectant. The tray must be autoclavable and waterproof. The column of bubbles produced by this test often tends to overflow; the tray will catch the overflow. If the microbiologist is quick enough, he or she can tighten the cap on a vigorously bubbling tube before it flows out the top.

6. Add 1 ml hydrogen peroxide reagent to each tube and to an uninoculated control tube.

7. Allow tubes to stand and column of bubbles to develop for 5 minutes before measuring the height of the column of bubbles above the surface of the medium. As

an alternative, a line can be drawn on the side of each tube at a height of 45 mm above the agar surface before the reagent is added. Cultures whose bubbles rise above the line are considered positive; those with bubble column heights less than 20 mm are considered negative (Figure 42.6).

8. If the height of the column of bubbles is between 20 and 45 mm, the test should be considered equivocal and repeated.

QUALITY CONTROL

Test positive and negative control strains (see Table 42.5) and an uninoculated control tube (above) each time the test is performed.

---

fect humans. DNA hybridization studies show close homology among all species in the complex.

M. tuberculosis. Colonies of human tubercle bacilli generally appear on egg media after 2 to 3 weeks at 35° C; no growth occurs at 25° or 45° C. Growth first appears as small (1 to 3 mm), dry, friable colonies that are rough, warty, granular, and buff-colored. After several weeks, these increase in size (as large as 5 to 8 mm); typical colonies have a flat irregular margin and a "cauliflower" center (Figure 42.10). Because of their luxuriant growth, these mycobacteria are termed **eugonic.** Colonies are easily detached from the medium's

surface but are difficult to emulsify. After some experience with mycobacteria has been gained, one can recognize typical colonies of human-type tubercle bacilli without great difficulty. However, final confirmation by biochemical tests must be carried out in all cases.

Virulent strains tend to orient themselves in tight, "serpentine cords," best observed in smears from the condensation water or by direct observation of colonies on a cord medium. Catalase is produced in moderate amounts but not after heating at 68° C for 20 minutes in pH 7 phosphate buffer; human strains resistant to isoniazid (INH)

TABLE 42.7

DIFFERENTIAL CHARACTERISTICS OF PHOTOCHROMOGENS

PROPERTY	M. KANSASII	M. MARINUM	M. SIMIAE	M. ASIATICUM
Rate of growth	S	S(30° C)	S	S
Colonial morphology	SM/R	SM/R	SM	SM
Pigmentation—dark	NC	NC	NC	NC
—light	Y	Y	Y	Y
Niacin	–	–	+	–
Nitrate reduction	+	–	–	–
Catalase S.Q. > 45 mm	+	–	+	+
Tween hydrolysis, 5-day	+	+	–	+
Urease	+	+	+	–

S, slow; *SM,* smooth; *R,* rough; *NC,* nonchromogenic; *Y,* yellow.

TABLE 42.8

DIFFERENTIAL CHARACTERISTICS OF SCOTOCHROMOGENS

PROPERTY	M. SCROFULACEUM	M. SZULGAI	M. GORDONAE	M. FLAVESCENS	M. XENOPI	M. THERMO-RESISTIBILE
Rate of growth	S	S	S	S	S	R
Colonial morphology	SM	SM/R	SM	SM/R	SM	SM
Pigmentation—dark	OR	37° C, OR; 24° C, NC	OR	OR	OR	OR
—light	OR	37° C, OR; 24° C, OR	OR	OR	OR	OR
Nitrate reduction	–	+	–	+	–	V
Catalase, S.Q. > 45 mm	+	+	+	+	V	+
Tween hydrolysis, 5-day	–	Wk	+	+	–	+
Sodium chloride, 5%	–	–	–	+	–	+
Urease	+	+	V	+	–	+

S, slow; *SM,* smooth; *R,* rough; *NC,* nonchromogenic; *OR,* orange; *Wk,* weak; *V,* variable.

are frequently catalase negative and yield smooth colonies on egg media. Nitrate reduction is positive. The niacin test is useful in that most niacin-positive strains encountered in the diagnostic laboratory prove to be *M. tuberculosis.* Susceptibility to antituberculous drugs must be determined.

M. bovis. Bovine tubercle bacilli are rarely isolated in the United States but remain significant pathogens in other parts of the world. They require a longer incubation period—generally 3 to 6 weeks—and appear as tiny (less than 1 mm), translucent, smooth, pyramidal colonies when grown at 35° C. They adhere to the surface of the medium but are emulsified easily. On the basis of these characteristics, their growth is termed **dys-**gonic. On 7H10 the colonies are rough and resemble those of *M. tuberculosis. M. bovis* grows only at 35° C. It forms serpentine cords in smears from colonies on egg media; the niacin reaction and nitrate reduction tests are negative. *M. bovis* is also susceptible to thiophene-2-carboxylic acid hydrazide (TCH) (unless it is isoniazid-resistant), a useful test to differentiate it from other mycobacteria. The susceptibility of *M. bovis* to the primary antituberculous drugs is similar to that of *M. tuberculosis.*

M. africanum. Rarely isolated in the United States, *M. africanum* resembles *M. tuberculosis.* It is inactive biochemically, showing only positive urease and variably positive ability to grow in the

TABLE 42.9

DIFFERENTIAL CHARACTERISTICS OF NONPIGMENTED SLOW GROWERS

PROPERTY	M. AVIUM* COMPLEX	M. GASTRI	M. TERRAE COMPLEX	M. MALMOENSE	M. HAEMOPHILUM†	"M. GENAVENSE"
Rate of growth	S	S	S	S	S	S
Colonial morphology	SM	SM	SM/R	SM	SM	SM/R
Pigmentation—dark	NC	NC	NC	NC	NC	NC
—light	NC	NC	NC	NC	NC	NC
Niacin	–	–	–	–	–	–
Nitrate reduction	–	–	+	–	–	–
Catalase S.Q. >45 mm	–	–	+	–	–	–
Catalase 68° C	+	–	+	–	–	+
Tween hydrolysis 5-day	–	+	+	+	–	–
Arylsulfatase	–	–	–	–	–	–
Sodium chloride, 5%	–	–	V	–		
Urease	–	+	–	–	–	+

S, slow; *SM*, smooth; *R*, rough; *NC*, nonchromogenic.

M. avium complex includes *M. intracellulare*.

†Growth between 25° and 30° C; requires hemin for growth.

TABLE 42.10

DIFFERENTIAL CHARACTERISTICS OF RAPID GROWERS

PROPERTY	M. FORTUITUM*	M. CHELONAE; M. CHELONAE SUBSP. ABSCESSUS	M. SMEGMATIS‡
Rate of growth	Ra	Ra	Ra
Colonial morphology	SM/R	SM/R	SM/R
Pigmentation—dark	NC	NC	NC
—light	NC	NC	NC
Nitrate reduction	+	–	+
Tween hydrolysis, 5-day	V	–	+
Arylsulfatase	+	–	+
Sodium chloride, 5%	+	+	–
		–/+†	+

Ra, rapid; *SM*, smooth; *R*, rough; *NC*, nonchromogenic; *V*, variable.

M. fortuitum complex includes *M. chelonae* and *M. chelonae* subsp. *abscessus*.

†*M. chelonae* is negative; *M. chelonae* subsp. *abscessus* is positive.

‡Not clinically significant.

presence of TCH. This species has been isolated from pulmonary cultures of humans, as well as from monkeys.

M. ulcerans. *M. ulcerans,* probably one of the ancient species, exhibits discontinuous distribution. It has been reported from parts of Australia, Malaysia, Africa, and Central America. It is associated with skin lesions and is considered the causative agent of Buruli ulceration, a necrotizing ulcer found in African natives. The organism requires several weeks' incubation at 32° C. Unlike *M. marinum,* this organism is a **nonchromogen.** It

is resistant to isoniazid, ethambutol, para-amino-salicylic acid (PAS), and ethionamide, but it is susceptible to rifampin, streptomycin, kanamycin, and cycloserine.

PHOTOCHROMOGENS

Four well-defined photochromogenic mycobacteria display pigment after exposure to light: *M. kansasii, M. marinum, M. simiae,* and *M. asiaticum.* All species in this category are facultative pathogens. *M. kansasii* organisms are responsible for pulmonary disease in humans, often appear-

PROCEDURE 42.15

HEAT STABLE (68° C, pH 7.0) CATALASE TEST

PRINCIPLE
Species of mycobacteria produce catalases that are either resistant to the effect of heat (heat-stable) or that become denatured when heated (heat-labile). This trait can be used for differentiation.

METHOD
1. Prepare 1/15 M phosphate buffer, pH 7.0 as follows:
   Solution A
   Na_2HPO_4 (anhydrous) 9.47 g
   Distilled water 1000 ml
   Solution B
   KH_2PO_4 9.07 g
   Distilled water 1000 ml
   Mix 61.1 ml of solution A with 38.9 ml of solution B and check that the pH is 7.0. Adjust if necessary. Since the suspension will be tested for only a short time, the buffer need not be sterile. Store the two solutions in the refrigerator.

2. Suspend several large clumps of growth from an actively growing agar slant in 0.5 ml of the phosphate buffer in a 16 × 125 mm screw-cap test tube. Prepare control organisms in the same way.

3. Incubate tubes for exactly 20 minutes in a 68° C heating block or water bath. The temperature must be strictly monitored during the entire incubation.

4. Cool the suspensions to room temperature.

5. Add 0.5 ml of the Tween-hydrogen peroxide solution (same as Procedure 42.14) and observe for formation of bubbles. Negative tubes should be held for 20 minutes before reporting as negative. Observe carefully, since even a few tiny bubbles are indicative of a positive result. The tubes must not be agitated in any way, since Tween will bubble if shaken.

QUALITY CONTROL
Test positive and negative control strains (Table 42.5) each time the test is performed.

PROCEDURE 42.16

TWEEN 80 HYDROLYSIS TEST

PRINCIPLE
Intact Tween 80 binds the phenol red indicator, causing it to assume the structure it would normally assume only at a more acid pH. When Tween 80 is hydrolyzed, it no longer binds the indicator, allowing it to assume its normal configuration at pH 7.0, which is visible as pink or red. This pink color, which is the sign of a positive reaction, does not result from a pH shift but results from a breakdown of the Tween 80.

METHOD
1. Prepare Tween 80 hydrolysis substrate from concentrate according to manufacturer's instructions.

2. Inoculate the substrate with one large clump of actively growing mycobacteria from an agar slant. Inoculate control substrate tubes at the same time.

3. Incubate at 35° C.

4. Examine after 1, 5, and 10 days for a change in the substrate from orange to pink. Compare the color to an uninoculated substrate control. Tubes should not be shaken during reading.

QUALITY CONTROL
Test positive and negative control strains (see Table 42.5) and an uninoculated control tube each time the test is performed.

PROCEDURE 42.17

TELLURITE REDUCTION TEST

PRINCIPLE

The enzyme tellurite reductase reduces potassium tellurite to metallic tellurite, which is visualized as a black precipitate. Only *M. avium* complex strains and rapid-growing mycobacteria possess a fast-acting enzyme, which is detected in this test.

METHOD

1. Prepare tellurite stock solution as follows:
   Potassium tellurite 0.1 g
   Distilled water 50 ml
     Dissolve the tellurite and dispense in 2 ml amounts into 12 × 75 mm screw-cap tubes, sterilize by autoclaving, and store refrigerated. The solution should remain stable for 6 months. Commercially prepared 1% potassium tellurite is available.

2. Inoculate commercial 7H9 broth with a heavy inoculum of organisms and incubate for 7 days at 35° C.

3. Add 2 to 3 drops of the tellurite stock solution to each tube, vortex, and reincubate for 3 more days at 35° C.

4. Examine the sedimented cells in each tube. Formation of a black precipitate indicates a positive reaction. There is no color change with a negative reaction.

QUALITY CONTROL

Test a positive and negative control strain (see Table 42.5) each time the test is performed.

TABLE 42.11

NITRATE REDUCTION TEST

GROUP	POSITIVE	NEGATIVE
Slow-growing nonchromogens	*M. tuberculosis*	*M. bovis*
		M. simiae
Slow-growing scotochromogens	*M. szulgai* (weak)	*M. scrofulaceum*
	M. flavescens	*M. gordonae*
Rapid-growing nonchromogens	*M. fortuitum*	*M. chelonae*
Photochromogens	*M. kansasii*	
		M. marinum
		M. asiaticum
		M. simiae

TABLE 42.12

TWEEN 80 HYDROLYSIS TEST

GROUP	POSITIVE	NEGATIVE
Photochromogens	*M. kansasii*	*M. simiae*
	M. marinum	
	M. asiaticum	
Nonphoto-chromogens	*M. gastri*	*M. bovis*
	M. terrae complex	*M. avium* complex
	M. malmoense	*M. xenopi*
		M. simiae
		M. haemophilum
Scotochromogens	*M. szulgai* (slow)	*M. scrofulaceum*
	M. gordonae	*M. xenopi*
	M. flavescens	

ing in white emphysematous men older than 45 years. The disease with its complications is indistinguishable from that caused by *M. tuberculosis*, but it follows a more chronic and indolent course. Other types of disease, including disseminated infection, may occur. The disease is not communicable, in distinct contrast with that caused by the tubercle bacillus. Optimal growth of *M. kansasii* occurs after 2 to 3 weeks at 35° C (slower at 25° C); growth does not occur at 45° C. Colonies are generally smooth, although there is a tendency to develop roughness. They are cream-colored when grown in the dark and be-come a bright lemon yellow if exposed to light (Figure 42.11).

If cultures are grown continuously under light (2 to 3 weeks), bright orange crystals of β-carotene form on the surface of colonies, especially where growth is heavy. Nonchromogenic and scotochromogenic variants of *M. kansasii* occur very rarely. Clinically significant strains of *M. kansasii* are strongly catalase positive, even after heating at 68° C at pH 7 (especially when using the semiquantitative test of Wayne, Procedure 42.14). Low-catalase strains of *M. kansasii* have

PROCEDURE 42.18

ARYLSULFATASE TEST

PRINCIPLE

The rate of activity of the ability of arylsulfatase to break down phenolphthalein disulfate into phenolphthalein (which forms a red color in the presence of sodium bicarbonate) and other salts can help differentiate among certain strains of mycobacteria, particularly rapid growers (see Table 42.10).

METHOD

1. Prepare enzyme substrate stock solution as follows:
   Phenolphthalein disulfate tripotassium, salt (Sigma Chemical Co.) 2.6 g
   Distilled water 50 ml
   a. Mix thoroughly to make a 0.08 M solution. Sterilize through a 0.2-μm pore size membrane filter. Store refrigerated.
   b. Prepare two flasks of 180 ml each of basal Middlebrook 7H9 or Dubos-Tween-albumin broth (from commercial dry powder). Autoclave for 15 minutes at 121° C and cool to room temperature. Aseptically add one 20-ml vial of Middlebrook albumin-dextrose-catalase (ADC) enrichment to each flask.
   c. For the 3-day test, aseptically add 2.5 ml of the enzyme substrate stock to one flask, for a final concentration of 0.001 M phenolphthalein substrate.
   d. For the 14-day test, add 7.5 ml of the substrate stock to the other flask, for a concentration of 0.003 M substrate.
   e. These two reagents should be dispensed in 2 ml amounts into 16 × 125-mm screw-cap test tubes. Store at room temperature. The substrates are also available commercially (ready-made) from Remel Laboratories.

2. Prepare the alkaline indicator solution (6 N sodium carbonate) as follows:
   Na_2CO_3, anhydrous 10.6 g
   Distilled water 100 ml
   Mix together.

3. Inoculate tubes of the 3-day substrate and the 14-day substrate with 0.1 ml of the broth culture originally set up for determination of growth rates and pigment production. As an alternative, a heavy inoculum of material from an agar slant is acceptable.

4. After 3 days of incubation at 35° C, 6 drops of the sodium carbonate are added to the 3-day 0.001 M substrate.

5. After 14 days of incubation, 6 drops of the sodium carbonate solution are added to the 14-day 0.003 M substrate.

6. Observe for an immediate color change to pink or red after addition of carbonate, indicating a positive result for arylsulfatase (Figure 42.7).

QUALITY CONTROL

Test a positive control strain (see Table 42.5) and uninoculated medium as a negative control each time the test is performed.

been described; these were not associated with human pathogenicity. Most strains do not produce niacin, although aberrant strains have been noted; nitrates are reduced; mature colonies show loose cords.

Stained preparations of *M. kansasii* show characteristically *long, banded, and beaded cells* that are strongly acid fast. There is variable susceptibility to INH and streptomycin, moderate susceptibility to rifampin, and resistance to PAS, but *M. kansasii* infections typically respond to conventional antituberculous therapy; however, triple therapy, including rifampin, is desirable.

Primarily associated with granulomatous lesions of the skin, particularly of the extremities, *M. marinum* infection usually follows exposure of

FIGURE 42.7

A positive arylsulfatase test is shown on the left; the tube containing the negative control is on the right.

PROCEDURE 42.19

TCH SUSCEPTIBILITY

PRINCIPLE

The antimycobacterial compound TCH is inhibitory to *M. bovis,* which can be used to distinguish niacin-positive *M. bovis* from *M. tuberculosis.*

METHOD

1. Prepare the TCH agar as follows: Prepare Middlebrook 7H10 agar according to manufacturer's instructions, adding the enrichment when the agar has cooled to 55° C. At this time, also add sufficient filter-sterilized TCH (Aldrich Chemical Co.) to make a final concentration of 5 μg/ml. Chapter 14 discusses methods for making such solutions. Mix the agar thoroughly. Also prepare drug-free control agar. Pour the molten agar into quadrants of Felsen Petri dishes, 5 ml per quadrant. Allow the plates to set.

2. Make dilutions of the 7-day-old broth culture that was originally inoculated for preparation of the growth control, 1:1000 and 1:10,000, in sterile water or saline.

3. Inoculate 0.1 ml of each dilution to one drug-containing and one control agar medium. Incubate for 3 weeks in 10% CO_2 at 35° C and examine amount of growth.

4. Organisms are resistant to TCH if growth in the TCH-containing medium is equal to or exceeds 1% of the growth on the control agar. Results are interpreted much as are those of susceptibility tests, described in Chapter 14.

QUALITY CONTROL

Test positive and negative control strains (Table 42.5) each time the test is performed.

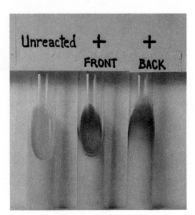

FIGURE 42.8
Iron uptake test for mycobacteria.

the abraded skin to contaminated water. Known best for causing "swimming pool granuloma," this organism also has been implicated in infections related to home aquariums, bay water, and industrial exposures involving water. It grows best at 25° to 32° C, with sparse to no growth at 35° C (corresponding to the reduced skin temperature of the extremities), and it is never isolated from sputum. It may be distinguished from *M. kansasii* by its source, its more rapid growth at 25° C, negative nitrate reduction, and weaker catalase production. The drugs most active against *M. marinum* are amikacin and kanamycin. Tetracyclines are inhibitory, chiefly at concentrations slightly below the expected blood levels.

First isolated from monkeys, *M. simiae* has subsequently been recovered from humans with pulmonary disease. Pigmentation may be erratic. It has a positive niacin reaction, a high thermostable catalase activity, and a negative nitrate test, hydrolyzes Tween 80 slowly (more than 10 days), and is resistant to all first-line antituberculous drugs. It is sensitive to cycloserine and ethionamide.

M. asiaticum grows at 35° C and yields smooth colonies; photochromogenicity may vary with different growth conditions. This species is rarely encountered in the United States. It has been isolated from cases of pulmonary disease in humans and from monkeys. More isolates are needed to better define the characteristics of *M. asiaticum.*

SCOTOCHROMOGENS

The **scotochromogens** are pigmented in the dark (Greek *scotos,* dark), usually a deep yellow to orange, which darkens to an orange or dark red when the cultures are exposed to continuous light for 2 weeks (Figure 42.12). This pigmentation in the dark occurs on nearly all types of media at all stages of growth—characteristics that clearly aid in their identification.

The scotochromogens include the potential pathogens *M. scrofulaceum, M. szulgai,* and *M.*

FIGURE 42.9
Different colony morphologies seen on culture of one strain of *M. avium* complex; the transparent, glossy colonies should be chosen for susceptibility studies because they are known to be more drug resistant.

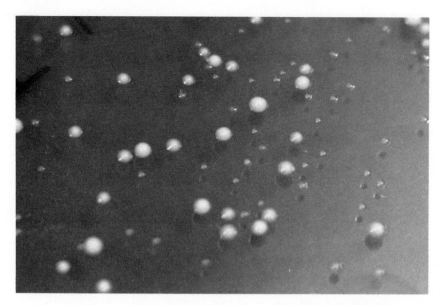

FIGURE 42.10
M. tuberculosis colonies on L-J agar after 8 weeks of incubation.

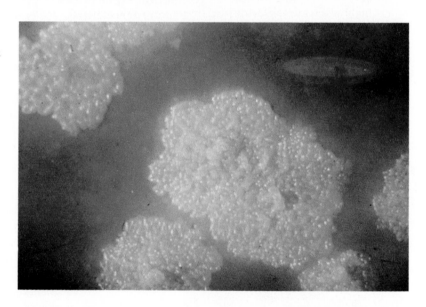

xenopi; the so-called tap water scotochromogen, isolated from laboratory water stills, faucets, soil, and natural waters (now classified as *M. gordonae*); and *M. flavescens*. Since the tap water scotochromogen is rarely associated with human disease, it is important to differentiate it from the potentially pathogenic mycobacteria. The former may contaminate equipment used in specimen collection, as in gastric lavage.

M. thermoresistibile, newly recognized as a cause of rare human infections, is a soil organism that is resistant to high temperatures (up to 57° C). The colonies grow rapidly and turn color after 2 weeks. The organism can be differentiated from *M. flavescens* or *M. gordonae* by its rapid growth and

from *M. phlei* (which also grows at 57° C) by biochemicals (see Table 42.8).

M. scrofulaceum is a slow-growing organism, producing smooth, domed to spreading, yellow colonies in both light and darkness. When exposed to continuous light, the colonies may increase in pigment to an orange or brick red; the aluminum foil shield should remain in place (Procedure 42.10) until visible growth occurs in the unshielded tube; then colonies in the shielded tube should be exposed to continuous light. Initial growth may be inhibited by too much light. The hydrolysis of Tween 80 (and urease activity) separates *M. scrofulaceum* from the tap water organisms in that the latter hydrolyze it within 5 days,

FIGURE 42.11
M. kansasii colonies
exposed to light.

FIGURE 42.12
Scotochromogen
Mycobacterium gordonae
with yellow colonies.

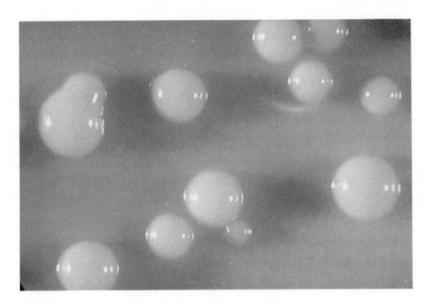

whereas *M. scrofulaceum* remains negative up to 3 weeks. *M. scrofulaceum* is a cause of cervical adenitis and bone and other infections, particularly in children. It is often resistant to INH and PAS.

M. szulgai has been associated with pulmonary disease, cervical adenitis, cutaneous infection, tenosynovitis, and olecranon bursitis. It gives a positive nitrate reduction test. It is relatively susceptible to ethionamide, rifampin, ethambutol, and higher levels of INH.

M. xenopi has been isolated from the sputum of patients with pulmonary disease. Differentiating this organism from *M. avium* complex can be difficult. The optimal temperature for its growth is 42° C, and it fails to grow at 22° to 25° C. Four to

five weeks' incubation is required to produce a typical morphology of tiny dome-shaped colonies of a characteristic yellow color; branching filamentous extensions are seen around colonies on 7H10 agar, resembling a miniature bird's nest. Tween 80 hydrolysis and tellurite reduction tests are negative. The arylsulfatase test is often positive at 3 days. Therapy with standard antituberculous drugs is generally successful.

Growth of *M. gordonae* appears late on L-J and 7H10 media, usually after 2 weeks and frequently after 3 to 6 weeks, as scattered small yellow-orange colonies in both light and darkness. Hydrolysis of Tween 80 characteristically occurs. *M. flavescens,* another nonpathogenic organism, also produces a yellow-pigmented colony in both light and darkness but is considerably more rapid in growth (usually within 1 week) and reduces nitrate as well as hydrolyzing Tween 80.

NONCHROMOGENS

This group of mycobacteria is made up of a heterogeneous variety of both pathogenic and non-pathogenic organisms that do not usually develop pigment on exposure to light. The *M. avium* complex grow slowly at 35° C (10 to 21 days) and 25° C and usually produce thin, translucent, radially lobed to smooth, cream-colored colonies (Figure 42.13). Isolates often undergo rapid-phase variation to a domed, opaque, yellow colony. This colony is commonly the predominant morphotype initially isolated from patients with AIDS. It has been shown that the rough or opaque colony type is less virulent and that its cell wall is more permeable to antibacterial agents. *M. avium* complex contains two phenotypically similar but genetically distinct species, *M. avium* and *M. intracellulare*. These organisms are encountered most commonly as opportunistic pathogens in patients with AIDS and other immune dysfunctions.[44] DNA probe tests performed against both species have shown that more than 90% of isolates from AIDS patients react with only the *M. avium* probe, whereas two thirds of isolates from non-AIDS patients react with only the *M. intracellulare* probe.[9,10]

Another variant of *M. avium* complex bacilli resembles *M. scrofulaceum* in certain biochemical characteristics; the colonies also become darker yellow with age (unlike other *M. avium* strains).

Although biochemical results vary, the majority of isolates of the *M. avium* complex are niacin negative (with rare exceptions), do not reduce nitrate, and produce only a small amount of catalase. Tween 80 is not hydrolyzed in 10 days, but most of these organisms reduce tellurite within 3 days, a useful test to differentiate these potential pathogens from clinically insignificant nonchromogens.

The *M. avium* complex, while producing serious tuberculosis-like endobronchial lesions in non-AIDS patients, causes disseminated disease in the majority of AIDS patients. However, these organisms may also occur in clinical specimens as nonpathogens. The organisms are resistant to most antituberculosis drugs. Newer agents such as ansamycin and clofazimine may be partially active against these organisms. Combinations of drugs, including ethambutol, rifampin, and amikacin, have also had limited success, at least for treatment (but not cure) of the disseminated infection commonly seen in patients with AIDS (Chapter 24). Standard practice has been to use five drugs in combination therapy (if five drugs with in vitro activity can be found).

Mycobacterium gastri ("J" bacillus) has been described as occurring primarily as single-colony isolates from gastric washings. However, it has not been associated with disease in humans. It is closely related to the low catalase-producing strains of *M. kansasii*, from which it may be readily differentiated by the photochromogenic ability of the latter. *M. gastri* may be differentiated from other nonphotochromogens by its ability to hydrolyze Tween 80 rapidly, loss of

FIGURE 42.13

M. avium complex. Smooth colonies on L-J medium.

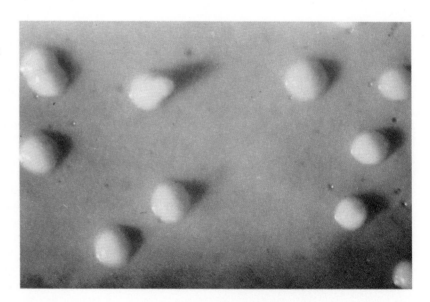

catalase activity at 68° C, and nonreduction of nitrate.

M. malmoense was first described in 1977 in Sweden, where it was found associated with pulmonary disease. Subsequently, it has been found in human pulmonary disease in Australia and Wales. The organism is nonphotochromogenic and grows slowly (2 to 3 weeks at 35° C and up to 6 weeks at 22° C). Colonies are colorless, smooth, glistening, grayish white, opaque, domed, and circular, 0.5 to 1.5 mm in diameter. The organism does not produce niacin, is nitrate negative, hydrolyzes Tween 80 and pyrazinamide, and produces heat-labile catalase. It is resistant to isoniazid, streptomycin, PAS, and rifampin and is susceptible to ethambutol, cycloserine, kanamycin, and ethionamide.

M. haemophilum is a recently described organism isolated from skin lesions.[29] It has also been described as the etiological agent of disseminated infections in AIDS patients.[13b] It requires hemin for growth and may be isolated on chocolate agar, 7H10 agar containing hemolyzed but not whole sheep red blood cells, or L-J medium containing 1% ferric ammonium citrate. Incubation should be at 32° C for a minimum of 2 to 4 weeks. The organism does not grow at 35° C. It should not be expected in specimens of sputum or gastric lavage. It is highly resistant to INH, streptomycin, and ethambutol but is susceptible to PAS.

M. terrae has been called "radish" bacillus. A number of these mycobacteria have been isolated from soil and vegetables as well as from humans, where their pathogenicity remains questionable. The slow-growing (35° C) colonies may be circular or irregular in shape and smooth or granular in texture. They actively hydrolyze Tween 80, reduce nitrate, and are strong catalase producers, but they do not reduce tellurite in 3 days. *M. terrae* is resistant to INH.

M. triviale ("V" bacillus) occurs on egg media as rough colonies that may be confused with *M. tuberculosis* or rough variants of *M. kansasii*. These bacilli have been recovered from patients with previous tuberculous infections but are considered to be unrelated to human infections. There is one report of human infection with this organism. They are nonphotochromogenic, moderate to strong nitrate reducers, and maintain a high catalase activity at 68° C. They hydrolyze Tween 80 rapidly and do not reduce tellurite.

M. shimoidei was first isolated from a patient in Japan in 1975. It has since been isolated from other patients with pulmonary disease. The organism hydrolyzes Tween 80 and produces heat-stable catalase. Colonies are rough, and growth occurs from 35° to 45° C.

Eighteen European patients with AIDS were reported in 1992 to have disseminated infections with a new mycobacterium, named *"M. genavense,"*, after Geneva, where they were isolated. Recently, a Seattle laboratory recovered an additional 17 isolates from blood, bone marrow, and spleen of AIDS patients using BACTEC 13A broth.[7d] The organisms did not grow in Septi-Chek AFB broth or on routine solid medium. A solid medium (Middlebrook 7H11 agar) supplemented with mycobactin J (a growth supplement produced by other mycobacteria), but not mycobactin-supplemented broth, yielded colonies after 2 weeks (range was 5 to 67 days). The organisms resemble *M. simiae* genetically and by cell wall fatty acid analysis. They are nonpigmented and positive for catalase, urease, and pyrazinamidase.

RAPID GROWERS

These mycobacteria are characterized by their ability to grow in 3 to 5 days on a variety of culture media, incubated either at 25° or 35° C. Two members, *M. fortuitum* and *M. chelonae*, are associated with human pulmonary infection, although *M. fortuitum* is also a common soil organism and may frequently be recovered from sputum without necessarily being implicated in a pathological process.

M. fortuitum has been incriminated in progressive pulmonary disease, usually with a severe underlying complication, and has resulted in death. This potential pathogen also grows rapidly (2 to 4 days), is generally nonchromogenic, and may be readily separated from the rapidly growing saprophytes by its positive 3-day arylsulfatase reaction and by growing on MacConkey agar within 5 days, producing a change in the indicator. Both rough and smooth colonies are produced, with increased dye absorption (greening) on L-J medium (Figure 42.14). The niacin test is negative. *M. fortuitum* is usually resistant to PAS, streptomycin, and INH but is susceptible to amikacin and cefoxitin and often the tetracyclines.

M. chelonae (formerly *M. borstelense*) comprises two distinct subspecies: *M. chelonae* subsp. *chelonae*, which fails to grow on 5% NaCl medium, and *M. chelonae* subsp. *abscessus*, which grows on the NaCl medium. *M. fortuitum* and *M. chelonae* are distinguished from each other by the combined use of five tests. *M. fortuitum* is typically nitrate reductase positive, β-glucosidase positive, penicillinase negative, and trehalose negative and produces

FIGURE 42.14
Smooth, multilobate colonies of *M. fortuitum* on L-J medium.

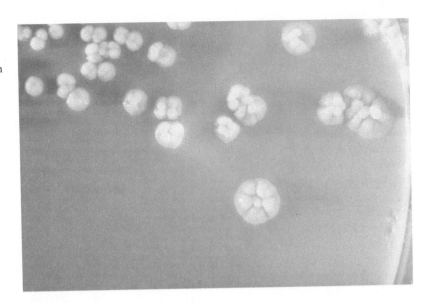

FIGURE 42.15
M. phlei (rapid grower). Rough colonies.

acid from fructose. *M. chelonae* has the opposite reactions. *M. chelonae* subsp. *abscessus* has been isolated from pulmonary, postsurgical wounds, and other infections. There have been several cases of disseminated disease attributed to this species since 1978. *M. chelonae* subsp. *chelonae* is less common, and it has not been isolated from sputum. Infection occurs at the site of an injury that penetrates the skin.[28]

M. smegmatis, M. phlei (Figure 42.15), and *M. vaccae* are considered saprophytes and are nonpathogenic. *M. smegmatis* and *M. phlei* produce pigmented colonies and show filamentous extensions from colonies growing on cornmeal-glycerol agar. The ability of *M. phlei* ("hay bacillus") to produce large amounts of CO_2 has been utilized to stimulate primary growth of *M. tuberculosis* on 7H10 agar plates incubated in CO_2-impermeable

(Mylar) bags. *M. phlei* also produces the growth factor "mycobactin," required by some species of *M. avium* complex and the putative Crohn's disease–associated *Mycobacterium*.

M. fallax resembles *M. tuberculosis*, with cord formation on solid media. The organisms grow rapidly only at 30° C and slowly at 35° C.[23] Several strains have been isolated from humans with pulmonary disease; other strains are found in soil and water. The species is nitrate reductase positive and negative for thermostable catalase, arylsulfatase, and urease.

CROHN'S DISEASE-ASSOCIATED MYCOBACTERIUM

The organism first seen in acid-fast–stained bowel tissue sections from Crohn's disease in 1984 has been isolated in culture and identified using DNA probes. Two variants have been detected so far, an organism similar to *M. paratuberculosis,* the animal pathogen responsible for Johne's disease in cattle, and a wood pigeon–associated strain of *M. avium.*[23a] The *M. paratuberculosis*-like strain may be isolated from cultures but only after prolonged incubation (as long as 18 months). Colonies are brilliant white and mucoid. The organisms exhibit positive thermostable catalase, positive 14-day arylsulfatase, and a weak positive niacin test but are otherwise inactive. Evidence is mounting that these organisms are actually involved in the inflammatory bowel disease process in some cases.

NEW APPROACHES TO MYCOBACTERIAL IDENTIFICATION

New approaches are being explored for identification of the mycobacteria because, as can be appreciated from the number and difficulty of tests necessary and the time required for organisms to grow, the state of the art is somewhat archaic. Thin-layer chromatography and gas-liquid chromatography have seemed logical choices for test systems, because cell wall lipids lend themselves to such procedures. Workers such as Tisdall, de Young, Roberts, Anhalt[34] and Brennan, Heifers, and Ullom[7] have used chromatographic methods for identification of clinical isolates of mycobacteria. Patterns produced by mycobacterial nucleic acids after restriction endonuclease treatment and gel electrophoresis have also proved useful for separating the bacilli into species. Those tests are not yet within the realm of routine microbiology laboratories, although a number of large reference laboratories are beginning to use such methods on a routine basis.

High-performance liquid chromatography (HPLC) of extracted mycobacteria is a very specific and rapid method for identification of species.[7a] Many state health departments and the CDC now use this method routinely. The long chain mycolic acids are separated better by HPLC than by GLC, as they do not withstand the high temperatures needed for GLC. The patterns produced by different species are very reproducible, and a typical identification requires only a few hours.

Many of the MOTT mycobacterial groups have been characterized by serotyping schemes and phage typing. In particular, the *M. avium* complex has been studied and defined serologically and by phage typing. Once serovars are established, more sensitive techniques, such as enzyme-linked immunosorbent assay (ELISA), may

prove useful for rapid identification of clinically relevant strains.[41]

DNA hybridization methods have been used recently for detection of mycobacteria in clinical specimens. DNA-RNA probes (Gen-Probe) have been used for identifying *M. tuberculosis* complex, *M. gordonae, M. kansasii,* and *M. avium* complex isolates within several hours of obtaining good growth. Results have been very encouraging, and numerous laboratories now incorporate such tests as part of their routine procedures.[9,10]

Reports during 1993 have shown that both new technology (PCR) and enhancement of old technology (direct microscopy) have exciting promise for immediate application in the rapid detection of mycobacteria at a level of sensitivity not previously achieved. PCR and routine mycobacterial staining of cytocentrifuge concentrated slides have each detected 100% of patients infected with tuberculosis, based on culture results, in a clinical laboratory setting.[7c,29a] Such developments put rapid, sensitive detection of mycobacteria within the grasp of most laboratories at a time when improved accuracy and speed of tuberculosis diagnosis is urgently needed.

REFERENCES

1. Barenfanger, J. 1993. Making your lab safe against multi-drug–resistant *Mycobacterium tuberculosis.* Clin. Microbiol. Newsletter 15:76.
1a. Berlin, O.G.W., and Martin, W.J. 1980. Importance of nitrate test in the identification of mycobacteria. Clin. Microbiol. Newsletter 2:4.
2. Berlin, O.G.W., and Martin, W.J. 1981. Leprosy or Hansen's disease. Clin. Microbiol. Newsletter 3:135.
3. Berlin, O.G.W., Zakowski, P., Bruckner, D.A., et al. 1984. *Mycobacterium avium:* a pathogen of patients with acquired immunodeficiency syndrome. Diagn. Microbiol. Infect. Dis. 2:213.
4. Berlin, O.G.W., Zakowski, P., Bruckner, D.A., and Johnson, B.L. 1984. New biphasic culture system for isolation of mycobacteria from blood of patients with acquired immune deficiency syndrome. J. Clin. Microbiol. 20:572.
5. Bennedson, J., and Larson, S.O. 1966. Examination for tubercle bacilli by fluorescence microscopy. Scand. J. Respir. Dis. 47:114.
6. Bishop, P.J., and Newman, G. 1970. The history of the Ziehl-Neelsen stain. Tubercle 51:196.
7. Brennan, P.J., Heifers, M., and Ullom, B.P. 1982. Thin-layer chromatography of lipid antigens as a means of identifying nontuberculous mycobacteria. J. Clin. Microbiol. 15:447.
7a. Butler, W.R., and Kilburn, J.O. 1988. Identification of major slowly growing mycobacteria and *Mycobacterium gordonae* by high-performance liquid chromatography of their mycolic acids. J. Clin. Microbiol. 26:50.
7b. Chawla, P.K., Klapper, P.J., Kamholz, S.L, et al. 1992. Drug-resistant tuberculosis in an urban population including patients at risk for human

immunodeficiency virus infection. Am. Rev. Respir. Dis. 146:280.

7c. Clarridge, J.E. III, Shawar, R.M., Shinnick, T.M., et al 1993. Large-scale use of polymerase chain reaction for detection of *Mycobacterium tuberculosis* in a routine mycobacteriology laboratory. J. Clin. Microbiol. 31:2049.

7d. Coyle, M.B., Carlson, L.C., Wallis, C.K., et al. 1992. Laboratory aspects of *"Mycobacterium genavense,"* a proposed species isolated from AIDS patients. J. Clin. Microbiol. 30:3206.

8. Dankner, W.M., Waecker, N.J., Essey, M.A., et al. 1993. *Mycobacterium bovis* infections in San Diego: a clinical epidemiologic study of 73 patients and a historical review of a forgotten pathogen. Medicine 72:11.

9. Drake, T.A., Herron, R.M., Hindler, J.A., et al. 1988. DNA probe activity of *Mycobacterium avium* complex isolates from patients without AIDS. Diagn. Microbiol. Infect. Dis. 11:125.

10. Drake, T.A., Hindler, J.A., Berlin, O.G.W., and Bruckner, D.A. 1987. Rapid identification of *Mycobacterium avium* complex. J. Clin. Microbiol. 25:1442.

11. Good, R.C. 1985. Opportunistic pathogens in the genus *Mycobacterium*. Annu. Rev. Microbiol. 39:347.

12. Good, R.C., and Snider, D.E., Jr. 1982. Isolation of nontuberculous mycobacteria in the United States, 1980. J. Infect. Dis. 146:829.

13. Inderlied, C.B., Baron, E.J., and Gagné, C. 1984. Disseminated mycobacterial disease in a patient with acquired immune deficiency syndrome. ASCP Check Sample No. MB 84-9 (MB-140). American Society of Clinical Pathologists, Chicago.

13a. Isenberg H.D., D'Amato, R.F., Heifets, L., et al. 1991. Collaborative feasibility study of a biphasic system (Roche Septi-Chek AFB) for rapid detection and isolation of mycobacteria. J. Clin. Microbiol. 29:1719.

13b. Kiehn, T.E. 1992. *Mycobacterium haemophilum:* a new opportunistic pathogen. Clin. Microbiol. Newsletter 14:81.

14. Kent, P.T., and Kubica, G.P. 1985. Public Health mycobacteriology. A guide for the level III laboratory. Centers for Disease Control, Atlanta.

15. Kilburn, J.O., Silcox, V.A., and Kubica, G.P. 1969. Differential identification of mycobacteria. V. The tellurite reduction test. Am. Rev. Respir. Dis. 99:94.

16. Kinyoun, J.J. 1915. A note on Uhlenhuth's method for sputum examination for tubercle bacilli. Am. J. Public Health 5:867.

17. Konno, K. 1956. New chemical method to differentiate human-type tubercle bacilli from other mycobacteria. Science 124:985.

18. Kubica, G.P. 1984. Culture examination and identification. Centers for Disease Control, Atlanta.

19. Kubica, G.P., Dye, W.E., Cohen, M.L., and Middlebrook, G. 1963. Sputum digestion and concentration with *N*-acetyl-L-cysteine-sodium hydroxide for culture of mycobacteria. Am. Rev. Respir. Dis. 87:775.

20. Kubica, G.P., Gross, W.M., Hawkins, J.E., et al. 1975. Laboratory services for mycobacterial diseases. Am. Rev. Respir. Dis. 112:773.

21. Kubica, G.P., Jones, W.D., Abbott, V.D., et al. 1966. Differential identification of mycobacteria. I. Tests on catalase activity. Am. Rev. Respir. Dis. 94:400.

22. Kubica, G.P., and Pool, G.L. 1960. Studies on the catalase activity of acid fast bacilli. I. An attempt to subgroup these organisms on the basis of their catalase activity at different temperatures and pH. Am. Rev. Respir. Dis. 81:387.

23. Levy-Frebault, V., Rafidinarivo, E., Prome, J.C., et al. 1983. *Mycobacterium fallax* sp. nov. Int. J. Syst. Bacteriol. 33:336.

23a. McFadden, J., Collins, J., Beaman, B., et al. 1992. Mycobacteria in Crohn's disease: DNA probes identify the wood pigeon strain of *Mycobacterium avium* and *Mycobacterium paratuberculosis* from human tissue. J. Clin. Microbiol. 30:3070.

24. Middlebrook, G., Reggiordo, Z., and Tigertt, W.D. 1977. Automatable radiometric detection of growth of *Mycobacterium tuberculosis* in selective media. Am. Rev. Respir. Dis. 115:1066.

25. Mitchison, D.A. 1979. Basic mechanisms of chemotherapy. Chest 76:771.

26. Morbidity and Mortality Weekly Report. 1987. Tuberculosis and acquired immunodeficiency syndrome—New York City. United States Public Health Service, Centers for Disease Control, Atlanta. 36:785.

27. Rickman, T.W., and Meyer, N.P. 1980. Increased sensitivity of acid fast smears. J. Clin. Microbiol. 11:618.

28. Righter, J., Hart, G.D., and Howes, M. 1983. *Mycobacterium chelonei:* report of a case of septicemia and review of the literature. Diagn. Microbiol. Infect. Dis. 1:323.

29. Ryan, C.G., and Dwyer, B.W. 1983. New characteristics of *Mycobacterium haemophilum*. J. Clin. Microbiol. 18:976.

29a. Saceanu, C.A., Pfeiffer, N.C., and McClean, T. 1993. Evaluation of sputum smears concentrated by cytocentrifugation for detection of acid-fast bacilli. J. Clin. Microbiol 31:2371.

30. Siddiqi, S.H. 1988. Bactec T.B. System Product and Procedure Manual. Becton, Dickinson & Co., Towson, Md.

31. Sommers, H.M. 1978. The identification of mycobacteria. Lab. Med. 9:34.

32. Steadham, J.E. 1979. Reliable urease test for identification of mycobacteria. J. Clin. Microbiol. 10:134.

33. Thangaraj, R.H., and Yawalker, S.J. 1987. Leprosy for medical practitioners and paramedical workers. Ciba-Geigy Limited, Basel, Switzerland.

33a Thoresen, O.F., and Saxegaard, F. 1991. Gen-Probe Rapid Diagnostic System for the *Mycobacterium avium* complex does not distinguish between *Mycobacterium avium* and *Mycobacterium paratuberculosis*. J. Clin. Microbiol. 29:625.

34. Tisdall, P.A., DeYoung, D.R., Roberts, G.D., and Anhalt, J.P. 1982. Identification of clinical isolates of mycobacteria with gas-liquid chromatography: a 10-month follow up study. J. Clin. Microbiol. 16:400.

35. Vestal, A.L. 1975. Procedures for the isolation and identification of mycobacteria. DHEW (CDC 75-8230). Centers for Disease Control, Atlanta.

36. Vestal, A.L., and Kubica, G.P. 1967. Differential identification of mycobacteria. III. Use of thiacetazone, thiophen-2-carboxylic acid hydrazide and triphenyltetrazolium chloride. Scand. J. Respir. Dis. 48:142.

37. Virtanen, A. 1960. A study of nitrate reduction by mycobacteria. Acta. Tuberc. Scand. Suppl. 48:1.

38. Wayne, L.G. 1961. Recognition of *Mycobacterium fortuitum* by means of the 3 day phenolphthalein sulfatase test. Am. J. Clin. Pathol. 36:185.

39. Wayne, L.G. 1964. The role of air in the photochromogenic behavior of *Mycobacterium kansasii.* Am. J. Clin. Pathol. 42:431.

40. Wayne, L.G., Doubaek, J.R., and Russel, R.L. 1964. Classification and identification of mycobacteria. I. Tests employing Tween-80 as substrate. Am. Rev. Respir. Dis. 90:588.

40a. Wolfe, J.M., and Moore, D.G. 1992. Isolation of *Mycobacterium thermoresistibile* following augmentation mammaplasty. J. Clin. Microbiol. 30:1036.

41. Yanagihara, D.L., Barr, V.L., Knisely, C.V., et al. 1985. Enzyme-linked immunosorbent assay of glycolipid antigens for identification of mycobacteria. J. Clin. Microbiol. 21:569.

42. Youmans, G.P. 1979. Tuberculosis. W. B. Saunders, Philadelphia.

43. Young, L.J., Inderlied, C.B., Berlin, O.G.W., and Gottlieb, M.S. 1986. Mycobacterial infections in AIDS patients, with an emphasis on the *Mycobacterium avium* complex. Rev. Infect. Dis. 8:1024.

44. Zakowski, P., Fligiel, S., Berlin, O.G.W., and Johnson, B.L. 1982. Disseminated *Mycobacterium avium-intracellulare* infection in homosexual men dying of acquired immunodeficiency. J.A.M.A. 248:2980.

BIBLIOGRAPHY

Atlas, RM. 1993. Handbook of microbiologic media. CRC Press, Boca Raton, Fla. (Detailed formulae and instructions for media preparation)

Baron, E.J, and Finegold, S.M. 1990. Bailey and Scott's diagnostic microbiology. ed 8. (Appendices for detailed formulae of media and stains.)

Chiodini, R.J. 1989. Crohn's disease and the mycobacterioses: a review and comparison of two disease entities. Clin. Microbiol. Rev. 2:90.

Chiodini, R.J., Van Kruiningen, H.J., Merkal, R.S., et al. 1984. Characteristics of an unclassified *Mycobacterium* species isolated from patients with Crohn's disease. J. Clin. Microbiol. 20:966.

Hastings, R.C., Gillis, T.P., Krahenbuhl, J.L., and Franzblau, S.G. 1988. Leprosy. Clin. Microbiol. Rev. 1:330.

Hernandez, R., Munoz, O., and Guiscafre, H. 1984. Sensitive enzyme immunoassay for early diagnosis of tuberculous meningitis. J. Clin. Microbiol. 20:533.

Krasnow, I. 1978. Primary isolation of mycobacteria. Lab. Med. 9:26.

Lowry, P.W., Jarvis, W.R., Oberle, A.D., et al. 1988. *Mycobacterium chelonae* causing otitis media in an ear-nose-throat practice. N. Engl. J. Med. 319:978.

Sommers, H.M., and Good, R.C. 1985. *Mycobacterium.* In Lennette, E.H., Balows, A., Hausler, W.J., Jr., and Shadomy, H.J., editors. Manual of clinical microbiology, ed 4. American Society for Microbiology, Washington, D.C.

Swenson, J.M., Thornsberry, C., and Silcox, V. 1982. Rapidly growing mycobacteria: testing of susceptibility to 34 antimicrobial agents by broth microdilution. Antimicrob. Agents Chemother. 22:186.

Wallace, R.J., Jr. 1987. Nontuberculous mycobacteria and water: a love affair with increasing clinical importance. Infect. Dis. Clin. North Am. 1:677.

Wallace, R.J., Jr., Nash, D.R., Tsukamura, M., et al. 1988. Human disease due to *Mycobacterium smegmatis.* J. Infect. Dis. 158:52.

Wallace, R.J., Jr., Swenson, J.M., Silcox, V.A., et al. 1983. Spectrum of disease due to rapidly growing mycobacteria. Rev. Infect. Dis. 5:657.

Wayne, L.G., David, H., Hawkins, J.E., et al. 1976. Referral without guilt or how far should a good lab go? ATS News 2:8.

Wayne, L.G., Good, R.C., Krichevsky, M.I., et al. 1981. First report of the cooperative, open-ended study of slowly growing mycobacteria by the International Working Group on Mycobacterial Taxonomy. Int. J. Syst. Bacteriol. 31:1.

Wolinsky, E. 1984. Nontuberculous mycobacteria and associated diseases. pp. 1141-1207. In Kubica, G.P., and Wayne, L.G., editors. The mycobacteria: a sourcebook. Marcel Dekker, New York.

43

LABORATORY METHODS IN BASIC VIROLOGY

GENERAL PRINCIPLES

VIRAL STRUCTURE

Viruses are composed of a nucleic acid *genome* surrounded by a protein coat called a **capsid.** Together the genome and capsid are referred to as the **nucleocapsid** (Figure 43.1). Genomes are either RNA or DNA. Viral capsids are composed of many individual subunits called **capsomeres.** Capsomeres assemble into an *icosahedral-* or *helical*-shaped capsid. Icosahedral-shaped capsids are cubical with 20 flat sides, whereas helical capsids are spiral shaped. Some of the larger viruses have a lipid-containing *envelope* that surrounds the capsid. In addition, many viruses have *glycoprotein spikes* that extend from the surface of the virus, acting as attachment proteins or as enzymes. The entire virus, including nucleic acid, capsid, envelope, and glycoprotein spikes, is called the **virion** or viral particle (see Figure 43.1).

Viruses that cause disease in humans range in size from approximately 20 to 300 nm. Even the largest viruses are not easily seen with a light microscope, since they are less than one fourth the size of a staphylococcal cell (Figure 43.2).

VIRAL REPLICATION

Viruses are strict intracellular parasites, reproducing or replicating only within a host cell. The steps in virus replication, called the *infectious cycle,* in-

FIGURE 43.1

Illustration of viral particle. Enveloped and nonenveloped virions have icosahedral or helical shape. (Modified from Murray P.R., Drew W.L., Kobayashi G.S., Thompson J.H., *Medical microbiology,* 1990, Mosby, St. Louis.)

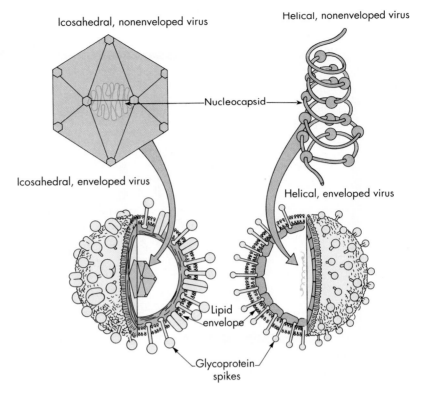

FIGURE 43.2

Relative sizes of viruses, bacteria, and chlamydia. (From Murray P.R., Drew W.L., Kobayashi G.S., Thompson J.H., *Medical microbiology,* 1990, Mosby, St. Louis.)

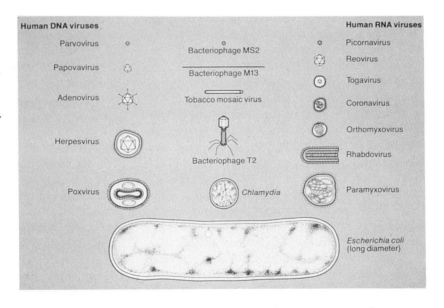

clude attachment, penetration, uncoating, macromolecular synthesis, assembly, and release (Figure 43.3).

To infect a cell, a virus must first recognize and attach to a suitable host. Typically glycoprotein spikes bind to host cell carbohydrate receptors. Viruses recognize and attach to a limited number of host cell types, allowing infection of some tissues but not others. This is referred to as viral **tropism.**

The process by which viruses enter the host cell is called *penetration.* One mechanism of penetration involves fusion of the viral envelope with the host cell membrane. Not only does this internalize the virus but it can also lead to fusion between this and other host cells nearby, forming multinucleated cells called **syncytia.** The detection of syncytia is used by those attempting to identify the presence of virus in cell cultures or clinical specimens.

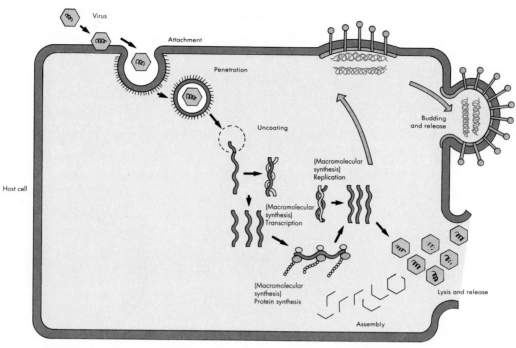

FIGURE 43.3

Illustration of viral infectious cycle. (Modified from Murray P.R., Drew W.L., Kobayashi G.S., Thompson J.H., *Medical microbiology,* 1990, Mosby, St. Louis.)

Uncoating occurs once the virus is internalized. Uncoating is necessary before the viral DNA or RNA is delivered to its intracellular site of replication in the nucleus or cytoplasm.

Macromolecular synthesis includes the production of nucleic acid and protein polymers. Viral transcription leads to the synthesis of mRNA and early and late viral proteins. *Early proteins* are nonstructural components such as enzymes, and *late proteins* are viral structural proteins. Rapid identification of virus in cell culture can be accomplished by detecting early viral proteins in infected cells using immunofluorescence techniques. Replication of viral nucleic acid is necessary to provide genomes for progeny viruses.

During viral assembly, structural proteins, genomes, and, in some cases, viral enzymes are assembled into virions. Viral envelopes are acquired during "budding" from a host cell membrane. Nuclear and cytoplasmic membranes are common areas for viral budding. Acquisition of an envelope is the final step in viral assembly.

Release of intact virions occurs following cell lysis or by budding from cytoplasmic membranes. Release by budding does not result in rapid host cell death as does release by lysis. Detection of virus in cell culture is facilitated by recognition of areas of cell lysis. Detection of virus released by budding is more difficult, since the cell monolayer remains intact. Influenza viruses, which are re-

leased with minimal cell destruction, can be detected in cell culture by an alternative technique called **hemadsorption.** Influenza virus–infected cells contain virally encoded glycoprotein hemagglutinins in the cytoplasmic membrane. Cells infected with influenza adsorb the RBCs added to their outer membranes, uninfected cells do not.

VIRAL CLASSIFICATION

Viruses can be classified according to morphology (type of capsid), type of genome (RNA or DNA), and means of replication. Morphology would include type of capsid, such as icosahedral or helical. For example, enteroviruses have RNA genomes that synthesize additional strands of RNA directly, whereas retroviruses make RNA in a two-step process by first synthesizing DNA, which subsequently makes RNA. Viruses of medical importance to humans comprise six families of DNA viruses and 13 families of RNA viruses (Table 43.1).

VIRAL PATHOGENESIS

Viruses are transmitted from person to person by the respiratory and fecal-oral routes, by trauma or injection with contaminated objects or needles, by tissue transplants (including blood transfusions), and by arthropod or animal bites. Once introduced into a host, the virus infects susceptible cells, frequently in the upper respiratory tract. Lo-

TABLE 43.1

LIST OF DNA AND RNA VIRUSES OF HUMAN IMPORTANCE

DNA VIRUSES		RNA VIRUSES	
FAMILY	**VIRAL MEMBERS**	**FAMILY**	**VIRAL MEMBERS**
Pox	Variola, vaccinia, orf, molluscum contagiosum	Orthomyxovirus	Influenza A, B, C
Herpesvirus	Herpes simplex, varicella-zoster, cytomegalovirus, Epstein-Barr virus, human herpesvirus 6	Paramyxovirus	Parainfluenza, mumps, measles, respiratory syncytial virus
Adenovirus	Types 1 to 8 most common causes of human disease	Togavirus	Western and eastern equine encephalitis, rubella
Papovavirus	Human papillomavirus, JC and BK polyomaviruses	Flavivirus	St. Louis encephalitis, yellow fever, dengue, hepatitis C
Hepadnavirus	Hepatitis B	Bunyavirus	California and La Crosse encephalitis, Rift Valley fever and other hemorrhagic fever agents
Parvovirus	B-19	Calicivirus	Calicivirus (Norwalk agent)
		Coronavirus	Coronavirus
		Reovirus	Rotavirus, Colorado tick fever
		Picornavirus	Polio, coxsackie A, coxsackie B, echovirus, enterovirus 68-71, 72 (hepatitis A), rhinovirus
		Arenavirus	Lymphocytic choriomeningitis, Lassa and other hemorrhagic fever agents
		Rhabdovirus	Rabies
		Filovirus	Ebola and Marburg hemorrhagic fever agents
		Retrovirus	Human T-lymphotropic virus (HTLV-I and II), human immunodeficiency virus (HIV 1 and 2)

cal infection leads to a **viremia,** which inoculates secondary target tissue distant from the primary site. Symptomatic disease ensues. Disease resolves when specific antibody and cell-mediated immune mechanisms halt continued replication of the virus (Figure 43.4). Tissue is damaged by lysis of virus-infected cells or by immunopathological mechanisms directed against the virus but also destructive to neighboring tissue. Some viruses, especially DNA-containing viruses such as the herpes group, remain **latent** in host tissue with no observable clinical impact. Reactivation may occur accompanying immune suppression, resulting in clinically apparent disease.

ANTIVIRAL VACCINES
Control of many viral diseases has been accomplished by vaccination. Since Jenner developed the first vaccine against smallpox nearly 200 years ago, attenuated-live or inactivated-dead viral vaccines have been successively used to prevent yellow fever, poliomyelitis, measles, mumps, rubella, hepatitis B, and influenza.[9,59] Smallpox was eliminated in 1977 by an effective worldwide vaccina-

tion program. Clinical virology laboratories occasionally detect vaccine strains of viruses, as occurs with the attenuated oral polio virus. Throat and rectal swab specimens of vaccinated children or their contacts, who have become infected by the fecal-oral route, may be positive for poliovirus. The vaccine strains do not cause disease in a normal host. Passive, incidental inoculation of siblings and contacts with the vaccine strain acts to vaccinate those who have not been properly immunized or boost antibody levels of those previously immunized.

ANTIVIRAL THERAPY
The availability of specific antiviral chemotherapeutic agents has added importance and urgency to the need for laboratory diagnosis of viral infections. Acyclovir, amantadine, azidothymidine (AZT), ribavirin, and ganciclovir are commonly used antivirals (Table 43.2).[2] Detection of virus to establish an exact etiological diagnosis is usually needed to justify use of these expensive and sometimes toxic agents. Although resistance does occur, routine antiviral testing, available in some

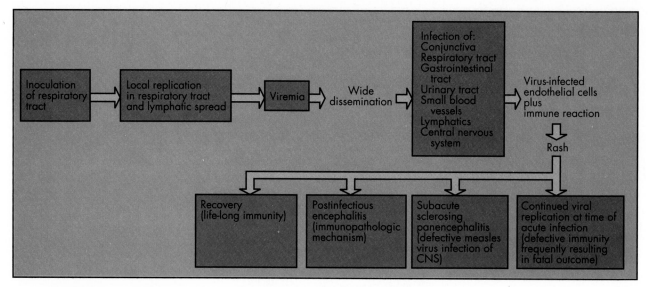

FIGURE 43.4

Viral pathogenesis illustrated by the mechanisms of the spread of measles virus within the body. (Modified from Murray P.R., Drew W.L., Kobayashi G.S., Thompson J.H., *Medical microbiology,* 1990, Mosby, St. Louis.)

TABLE 43.2

ANTIVIRAL CHEMOTHERAPY

ANTIVIRAL	INTENDED VIRUS
Amantadine	Influenza A
Acyclovir	Herpes simplex virus
	Varicella-zoster virus
Zidovudine (AZT)	Human immunodeficiency virus
Ribavirin	Respiratory syncytial virus
Ganciclovir	Cytomegalovirus

reference laboratories, is not yet necessary for most laboratories to perform.

OVERVIEW OF HUMAN VIRAL DISEASES

Specific disease syndromes may be caused by several unrelated viruses. For example, a partial list of viruses that can cause encephalitis includes herpes simplex virus, arboviruses, rabies virus, human immunodeficiency virus, and measles virus. Viral syndromes and the viruses that are important pathogens in each are summarized in Table 43.3. Conversely, individual viruses may cause many different diseases. Herpes simplex virus can cause pharyngitis, genital infection, conjunctivitis, and encephalitis. Human viruses and the diseases they cause are summarized in Table

43.4. Retroviruses and most DNA viruses establish a latent state following primary infection. Some latent viruses also are known to cause cancer. The site of latency and **oncogenic** (cancer causing) potential of human viruses is included in Tables 43.4.

METHODS FOR THE LABORATORY DIAGNOSIS OF VIRAL DISEASES

SETTING UP A CLINICAL VIROLOGY LABORATORY

The demand for clinical virology laboratory services skyrocketed during the 1980s. This growth resulted from the commercial availability of high quality reagents, the use by clinicians of virus-specific antiviral drugs, the development of rapid diagnostic techniques, and the application of virology cell culture procedures to chlamydial culture and *Clostridium difficile* cytotoxin testing. In deciding which virology tests to offer, a clinical laboratory should determine if the test is required for the appropriate care of the patient and if techniques are available that provide an accurate and cost-effective test result. Viral diseases that commonly require laboratory diagnosis include sexually transmitted diseases, diarrhea, respiratory disease in adults and children, aseptic meningitis, congenital disease, and infections in immunocompromised hosts. Viruses detected by a community hospital virology laboratory are shown in Table 43.5.[22,78]

Text continued on p. 645.

TABLE 43.3

LIST OF VIRAL SYNDROMES AND COMMON VIRAL PATHOGENS

VIRAL SYNDROME	VIRAL ETIOLOGY
Infants and children	
Upper respiratory tract infection	Rhinovirus, coronavirus, parainfluenza, adenovirus, respiratory syncytial virus, influenza
Pharyngitis	Adenovirus, coxsackie A, herpes simplex, Epstein-Barr, rhinovirus, parainfluenza, influenza
Croup	Parainfluenza, respiratory syncytial
Bronchitis	Parainfluenza, respiratory syncytial
Bronchiolitis	Respiratory syncytial, parainfluenza
Pneumonia	Respiratory syncytial, adenovirus, influenza, parainfluenza
Gastroenteritis	Rotavirus, adenovirus 40-41, calicivirus, astrovirus
Adults	
Upper respiratory tract infection	Rhinovirus, coronavirus, adenovirus, influenza, parainfluenza
Pneumonia	Influenza, adenovirus
Pleurodynia	Coxsackie B
Gastroenteritis	Norwalk-like virus
All Patients	
Parotitis	Mumps, parainfluenza
Myocardia/pericarditis	Coxsackie and echoviruses
Keratitis/conjunctivitis	Herpes simplex, varicella-zoster, adenovirus
Pleurodynia	Coxsackie B
Herpangina	Coxsackie A
Febrile illness with rash	Echo and coxsackie viruses
Infectious mononucleosis	Epstein-Barr, cytomegalovirus
Meningitis	Echo and coxsackie viruses, mumps, lymphocytic choriomeningitis, HSV 2
Encephalitis	Herpes simplex, togaviruses, bunyaviruses, flaviviruses, rabies, enteroviruses, measles, HIV, JC virus
Hepatitis	Hepatitis A, B, C, non-A, non-B, delta agent, E
Hemorrhagic cystitis	Adenovirus
Cutaneous infection with rash	Herpes simplex, varicella-zoster, enteroviruses, Epstein-Barr, measles, rubella, parvovirus, human herpesvirus 6
Hemorrhagic fever	Ebola, Marburg, Lassa, hantavirus, and other viruses
Acute respiratory failure	Hantavirus

TABLE 43.4

SUMMARY OF HUMAN VIRUSES AND ASSOCIATED DISEASES

Family:	Adenoviridae	Common name:	Adenovirus
Virus:	Adenovirus		
	Characteristics:	Double-stranded DNA genome. Icosahedral capsid, no envelope. Approximately 42 human serotypes	
	Transmission:	Respiratory, fecal-oral, and direct contact (eye)	
	Site of latency:	Replication in oropharynx	
	Disease:	Pharyngitis, pharyngoconjunctival fever, keratoconjunctivitis, pneumonia, hemorrhagic cystitis, disseminated disease, and gastroenteritis in children	
	Treatment:	Supportive	
	Prevention:	Vaccine (adenovirus serotypes 4 and 7) for military recruits	

Continued.

Family: Bunyaviridae   Common name: Bunyavirus
Virus: Arboviruses* including California encephalitis virus and LaCrosse virus. The hantaviruses are also included in
          the Bunyaviridae family. These viruses are carried by rodents, presumably infect humans by the respiratory
          route, and cause hemmorrhagic fever, kidney disease, or acute respiratory failure.
          Characteristics: Segmented, single-stranded RNA genome. Helical capsid with envelope
          Transmission: Mosquito vector
          Disease: Encephalitis
          Treatment: Supportive
          Prevention: Avoid contact with vector. Vector control programs

*Arthropod-borne viruses (arboviruses) were once grouped together because of their common mode of transmission.
Present classification includes three families containing arboviruses: the Togaviridae, Flaviviridae, and Bunyaviridae. The
virus group within togaviridae that includes arboviruses is the alphavirus group. Common arboviruses are referred to as
bunyaviruses, flaviviruses, and alphaviruses.

Family: Coronaviridae   Common name: Coronavirus
Virus: Coronavirus
          Characteristics: Single-stranded RNA genome. Helical capsid with envelope
          Transmission: Unknown. Probably direct contact or aerosol
          Disease: Common cold. Possibly gastroenteritis, especially in children
          Treatment: Supportive
          Prevention: Avoid contact with virus

Family: Hepadnaviridae   Common name: Hepadnavirus
Virus: Hepatitis B virus (HBV)
          Characteristics: Partly double-stranded DNA genome. Icosahedral capsid with envelope. Virion also
                             called Dane particle. Surface antigen originally termed Australia antigen
          Transmission: Humans are reservoir and vector. Spread by direct contact including exchange of body
                             secretions, recipient of contaminated blood products, and percutaneous injection of
                             virus
          Site of latency: Liver
          Disease: Acute infection with resolution (90%). Fulminant hepatitis (1%). Chronic hepatitis
                             (9%) followed by resolution, asymptomatic carrier state, chronic persistent, or chronic
                             active disease
          Oncogenic potential: Liver cancer
          Treatment: Liver transplant for fulminant disease
          Prevention: HBV vaccine. Hepatitis B immune globulin

Family: Herpesviridae   Common name: Herpesvirus
  Characteristics: Double-stranded DNA genome. Icosahedral capsid with envelope. At least six human herpesviruses
                known: HSV-1, HSV-2, VZV, CMV, EBV, and HH6
Virus: Herpes Simplex Virus Types 1 and 2 (HSV-1 and HSV-2)
          Transmission: Direct contact with infected secretions
          Site of latency: Sensory nerve ganglia
          Disease: Predominant virus in parentheses. Gingivostomatitis (HSV-1), pharyngitis (HSV-1),
                                   herpes labialis (HSV-1), genital infection (HSV-2), conjunctivitis (HSV-1), keratitis
                                   (HSV-1), herpetic whitlow (HSV-1 and HSV-2), encephalitis (HSV-1), disseminated
                                   disease (HSV-1 or HSV-2 in neonates)
          Treatment: Acyclovir
          Prevention: Avoid contact
Virus: Varicella-zoster virus (VZV)
          Transmission: Close personal contact, especially respiratory
          Site of latency: Dorsal root ganglia
          Disease: Chicken pox (varicella), shingles (zoster)
          Treatment: Acyclovir
          Prevention: Vaccine
Virus: Cytomegalovirus (CMV)
          Transmission: Close contact with infected secretions, blood transfusions (WBCs), organ transplants,
                                   transplacental
          Site of latency: White blood cells, endothelial cells, cells in a variety of organs
          Disease: Asymptomatic infection, congenital disease of newborn, symptomatic disease of
                                   immunocompromised host, heterophile-negative infectious mononucleosis
          Treatment: Supportive, decrease immune suppression, ganciclovir
          Prevention: Use CMV antibody-negative blood and tissue for transfusion and transplantation,
                                   respectively
Virus: Epstein-Barr Virus (EBV)
          Transmission: Close contact with infected saliva
          Site of latency: B lymphocytes
          Disease: Infectious mononucleosis, progressive lymphoreticular disease, oral hairy leukoplakia
          Oncogenic potential: Burkitt's lymphoma, nasopharyngeal carcinoma
          Treatment: Supportive
          Prevention: Avoid contact
Virus: Human Herpesvirus 6 (HH6)
          Transmission: Most likely close contact or respiratory
          Site of latency: T lymphocytes
          Disease: Exanthem subitum (roseola)
          Treatment: Supportive
          Prevention: Avoid contact

Family: Orthomyxoviridae Common name: Orthomyxovirus
  Characteristics: Segmented, single-stranded, RNA genome. Helical capsid with envelope. Three major antigenic
                types: influenza A, B, and C. Types A and B cause nearly all human disease
Virus: Influenza A
          Transmission: Contact with respiratory secretions
          Disease: Influenza (malaise, headache, myalgia, cough). Primary influenza pneumonia. In children,
                                   bronchiolitis, croup, otitis media
          Epidemiology: Viral subtypes based on hemagglutinin and neuraminidase glycoproteins abbreviated "H" and
                                   "N," respectively (e.g., H1N1 or H3N2). Infects humans and other animals. Antigenic drift,
                                   resulting in minor antigenic change, caused by mutation of viral genome. Drift causes local
                                   outbreaks of influenza every 1-3 years. Antigenic shift, resulting in major antigenic change,
                                   caused by genome reassortment. Shift causes periodic worldwide outbreaks
          Treatment: Supportive. Amantadine for severe disease
          Prevention: Influenza vaccine or amantadine
Virus: Influenza B
          Transmission: Contact with respiratory secretions
          Disease: Similar to "mild" influenza
          Epidemiology: Antigenic drift only, resulting in local outbreaks every 1-3 years
          Treatment: Supportive. Amantadine not effective
          Prevention: Influenza vaccine

Family: Papovaviridae   Common name: Papovavirus
   Characteristics: Double-stranded DNA genome. Icosahedral capsid, no envelope. Includes papilloma viruses, and BK
       and JC polyomaviruses.
Virus: Human Papillomavirus (HPV)
      Characteristics: Contains more than 60 DNA types
      Transmission: Direct contact
      Site of latency: Epithelial tissue
      Disease: Skin warts, benign head and neck tumors, anogenital warts
      Oncogenic potential: Cervical and penile cancer (especially HPV types 16 and 18)
      Treatment: Spontaneous disappearance the rule. Surgical or chemical removal may be necessary
      Prevention: Avoid contact with infected tissue
Virus: Polyomavirus (BK and JC viruses infect humans)
      Transmission: Probably direct contact with infected respiratory secretions
      Site of latency: Kidney
      Disease: Mild or asymptomatic primary infection, virus remains dormant in kidneys.
         Reactivation in immunocompromised patients causes hemorrhagic cystitis (BKV) or
         progressive multifocal leukoencephalopathy (JCV)
      Treatment: Supportive. Decrease immune suppression
      Prevention: Avoid contact with virus. Prevention unlikely until more is learned about mode of
         transmission

Family: Paramyxoviridae   Common name: Paramyxovirus
   Characteristics: Single-stranded, RNA genome. Helical capsid with envelope. No segmented genome like
      orthomyxoviruses
Virus: Measles virus
      Transmission: Contact with respiratory secretions. Extremely contagious
      Disease: Measles, atypical measles (occurs in those with waning "vaccine" immunity), and subacute
         sclerosing panencephalitis
      Treatment: Supportive. Immunocompromised patients can be treated with immune serum globulin
      Prevention: Measles vaccine
Virus: Mumps
      Transmission: Person-to-person contact, presumably respiratory droplets
      Disease: Mumps
      Treatment: Supportive
      Prevention: Mumps vaccine
Virus: Parainfluenza virus
      Transmission: Contact with respiratory secretions
      Disease: Adults: upper respiratory, rarely pneumonia. Children: respiratory including croup,
         bronchiolitis, and pneumonia
      Epidemiology: Four serotypes, disease occurs year-around.
      Treatment: Supportive
      Prevention: Avoid contact with virus
Virus: Respiratory syncytial virus (RSV)
      Transmission: Person to person by hand and respiratory contact
      Disease: Primarily in infants and children. Infants: bronchiolitis, pneumonia, and croup. Children:
         upper respiratory
      Epidemiology: Disease occurs annually late fall through early spring. Nosocomial transmission can occur
         readily
      Treatment: Supportive. Treat severe disease in infants with ribavirin
      Prevention: Avoid contact with virus. Prevent nosocomial transmission wtih isolation and cohorting

Family: Parvoviridae   Common name: Parvovirus
Virus: Parvovirus B-19
   Characteristics: Single-stranded DNA virus. Icosahedral capsid, no envelope. Parvovirus B-19 is the only
      known human parvovirus
   Transmission: Close contact, probably respiratory
   Disease: Erythema infectiosum (fifth disease), aplastic crises in patients with chronic hemolytic
      anemias, and fetal infection and stillbirth
   Treatment: Supportive
   Prevention: Avoid contact

Family: Picornaviridae   Common name: Picornavirus
  Charcteristics: Single-stranded RNA genome. Icosahedral capsid with no envelope
Virus: Enteroviruses
   Poliovirus (3 types)
   Coxsackievirus, Group A (23 types)
   Coxsackievirus, Group B (6 types)
   Echovirus (31 types)
   Enteroviruses (5 types)†
   Transmission: Fecal-oral
   Disease: Predominant virus in parentheses. Polio (poliovirus), herpangina (coxsackie A), pleurodynia
      (coxsackie B), aseptic meningitis (many enterovirus types), hand-foot-mouth disease
      (coxsackie A), pericarditis and myocarditis (coxsackie B), acute hemorrhagic conjunctivitis
      (enterovirus 70), and fever, myalgia, summer "flu" (many enterovirus types)
   Treatment: Supportive
   Prevention: Avoid contact with virus. Vaccination for polio
Virus: Hepatitis A virus (enterovirus type 72)
   Transmission: Fecal-oral
   Disease: Hepatitis with short incubation, abrupt onset, and low mortality. No carrier state
   Treatment: Supportive
   Prevention: Prevent clinical illness with serum immunoglobulin. Vaccine being developed
Virus: Rhinovirus (common cold virus)
   Characteristics: Approximately 90 serotypes
   Transmission: Contact wtih respiratory secretions
   Disease: Common cold
   Treatment: Supportive
   Prevention: Avoid contact with virus

---

†Enteroviruses were originally divided into poliovirus, coxsackie virus, and echovirus groups based on similarity of diseases caused in humans. Genetic diversity of these viruses dictates that newly characterized strains be given enterovirus-type designations because of the inexactness of the old classification system.

Family: Reoviridae   Common name: Reovirus
Virus: Rotavirus
   Characteristics: Segmented, double-stranded, RNA genome. Icosahedral capsid with no envelope
   Transmission: Fecal-oral. Survives well on inanimate objects
   Disease: Gastroenteritis in infants and children 6 months to 2 years
   Epidemiology: Winter-spring seasonality in temperate climates. Nosocomial transmission can occur easily
   Treatment: Supportive, especially fluid replacement
   Prevention: Avoid contact with virus. Vaccine trials under way

Family: Retroviridae   Common name: Retrovirus
  Characteristics: Single-stranded, RNA genome. Icosahedral capsid with envelope. Reverse transcriptase converts
                genomic RNA into DNA
Virus: Human immunodeficiency virus types 1 and 2 (HIV-1 and HIV-2)
          Characteristics: Most disease in humans caused by HIV-1. Infected cells include CD4 (helper) T-
                lymphocytes, monocytes, and some cells of the central nervous system
          Transmission: Sexual contact, blood and blood product exposure, and perinatal exposure
          Site of latency: CD4 T-lymphocytes
          Disease: Asymptomatic infection, acute "flu-like" disease, AIDS-related complex, and AIDS
          Epidemiology: Those at risk of infection are homosexual or bisexual males, intravenous drug abusers,
                sexual contacts of HIV-infected individuals, and infants of infected mothers
          Treatment: AZT. Treat infections resulting from immunosuppression
          Prevention: Avoid contact with infected blood/blood products and secretions. Blood for transfusion
                is screened for antibody to HIV-1 and HIV-2
Virus: Human T-lymphotropic virus (HTLV-1, HTLV-2, and HTLV-5)
          Transmission: Known means of transmission are similar to HIV
          Disease: T-cell leukemia and lymphoma, and tropical spastic paraparesis (HTLV-1)
          Epidemiology: HTLV-1 present in 0.025% of volunteer blood donors in U.S. Blood is now screened for
                antibody to this virus. Rates in areas of Japan and the Caribbean are considerably
                higher
          Oncogenic potential: T-cell lymphoma (HTLV-1)
          Treatment: Supportive
          Prevention: Avoid contact with virus

Family: Rhabdoviridae   Common name: Rhabdovirus
Virus: Rabies virus
          Characteristics: Single-stranded, RNA genome. Helical capsid with envelope
          Transmission: Bite of rabid animal most common. 20% of human rabies cases have no known exposure to
                rabid animal
          Disease: Rabies
          Treatment: Supportive
          Prevention: Avoid contact with rabid animals. Vaccinate domestic animals. Postexposure prophylaxis with
                hyperimmune antirabies globulin and immunization with rabies vaccine

---

Family: Togaviridae and Flaviviridae   Common name: Togavirus and Flavivirus
  Characteristics: Togaviruses and flaviviruses are very similar. Both families contain many arboviruses pathogenic for
                humans, in addition to other viruses. They have the following characteristics: single-stranded RNA
                genome and icosahedral capsid with envelope
Virus: Rubella virus (togavirus)
          Transmission: Respiratory, transplacental
          Disease: Rubella (mild exanthematous disease), congenital rubella
          Treatment: Supportive
          Prevention: Rubella vaccine
Virus: Arboviruses including alphaviruses and flaviviruses*
          Transmission: Arthropod vector, usually mosquito
          Disease: Eastern, western, Venezuelan and St. Louis encephalitis, dengue, and yellow fever
          Treatment: Supportive
          Prevention: Avoid contact with vector. Vector control programs
Virus: Hepatitis C virus (possibly a flavivirus)
          Transmission: Parenteral or sexual
          Disease: Acute and chronic hepatitis (non-A, non-B hepatitis)
          Treatment: Supportive
          Prevention: Avoid contact with virus. Blood supply screened for antibody to hepatitis C virus

---

*Arthropod-borne viruses (arboviruses) were once grouped together because of their common mode of transmission.
Present classification includes three families containing arboviruses: the Togaviridae, Flaviviridae, and Bunyaviridae. The
virus group within togaviridae that includes arboviruses is the alphavirus group. Common arboviruses are referred to as
bunyaviruses, flaviviruses, and alphaviruses.

TABLE 43.5

VIRUSES DETECTED BY A COMMUNITY HOSPITAL VIROLOGY LABORATORY (1986-1988)

VIRUS	NO . ISOLATED FROM	
	CHILDREN	ADULTS
Adenovirus	98	19
Cytomegalovirus	68	191
Enterovirus	229	30
Herpes simplex virus	95	656
Influenza virus	94	71
Parainfluenza virus	161	33
Respiratory syncytial virus	425	2
Rotavirus	307	0
Varicella-zoster virus	5	14
Total	1,482	1,016

Data from Children's Hospital Medical Center of Akron, Akron, Ohio.

FIGURE 43.5
Roller drums used to hold cell culture tubes during incubation. Slow rotation continually bathes cells in medium. (Courtesy Children's Hospital Medical Center of Akron; Akron, Ohio.)

Those working in a clinical virology laboratory must be familiar with cell culture, enzyme immunoassay, and immunofluorescence methods, in addition to other common laboratory techniques. Large equipment needed for a full-service virology laboratory include a laminar flow **biological safety cabinet,** fluorescence microscope, inverted bright field microscope, refrigerated centrifuge, incubator, refrigerator/freezer, and roller drum for holding cell culture tubes during incubation (Figures 43.5 through 43.7)

SPECIMEN SELECTION AND COLLECTION

Specimen selection depends on the specific disease syndrome, viral etiologies suspected, and time of year.[17] Selection of specimen based on disease is complicated, since most viruses enter via the upper respiratory tract, yet they infect tissues and produce symptoms distant from the primary, oropharyngeal site. For example, aseptic meningitis, caused by many enterovirus types, is identified by detecting virus in throat, rectal swab, or CSF specimen. Pharyngitis and gastrointestinal symptoms may not be included in the patient's complaints.

Specimen selection based on virus suspected is complicated by clinical syndromes that are caused by many different viruses. By collecting specimens needed to detect only a specific virus, other important etiologies may be missed. For example, testing smears of nasal secretions from an infant by fluorescence staining for the detection of RSV does not allow for diagnosis of dis-

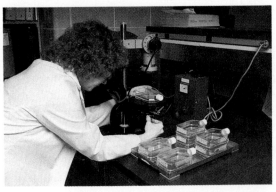

FIGURE 43.6
Inverted microscope used to examine cell monolayers growing attached to the inside surface beneath the liquid medium.

ease caused by influenza, parainfluenza, and adenoviruses.

Appropriate specimen selection dictates that the specimen type and viruses suspected always should be indicated on the requisition. Serum for serologic testing may be necessary, and some viral diseases need to be considered only during certain months because their appearance is seasonal. Table 43.6 suggests specimens for the diagnosis of viral diseases, noting seasonality when important.

Specimens for the detection of virus should be collected as early as possible following the onset of disease. Virus may no longer be present as early as 2 days after the appearance of symptoms. Recommendations for collection of common specimens are summarized in the following.

Throat, nasopharyngeal swab, or aspirate. Nasopharyngeal aspirates are superior, in general, to swabs for recovering viruses, but swabs are considerably more convenient. Throat swabs are acceptable for recovering enteroviruses, aden-

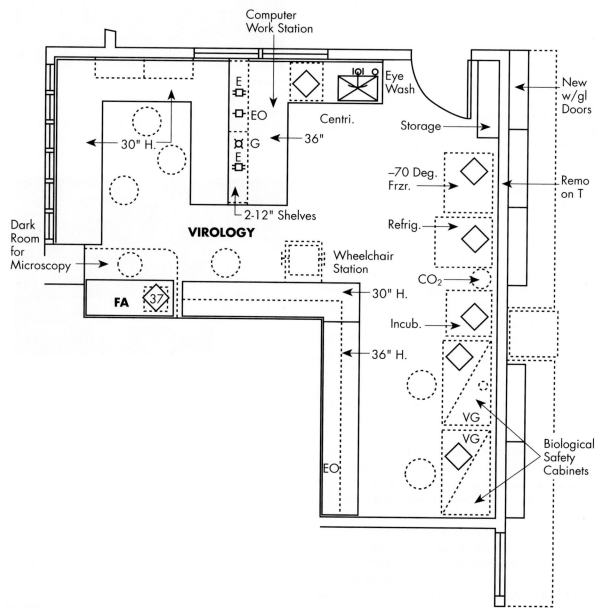

FIGURE 43.7

Floor plan of clinical virology laboratory that includes biological safety cabinet for specimen processing and cell culture handling, dark room for fluorescence microscopy, stand-up and sit-down counter space, computer station, storage areas, and incubator, refrigerator, and freezers.

TABLE 43.6

SPECIMENS FOR THE DIAGNOSIS OF VIRAL DISEASES*

DISEASE CATEGORIES AND PROBABLE VIRAL AGENTS	SEASON OF MOST COMMON OCCURRENCE	THROAT/ NASOPHARYNX	STOOL	CSF	URINE	OTHER
Respiratory syndrome		++++				
Adenoviruses	Y					
Influenza virus	W					
Parainfluenza virus	Y					
Respiratory syncytial virus (RSV)	W					
Rhinoviruses	Y					+++ Nasal
Dermatologic and mucous membrane disease						
Vesicular						+++ (Vesicle fluid or scraping)
Enterovirus	S,F	++	+++			
Herpes simplex†	Y					
Varicella-zoster†	Y	++				
Exanthematous		++				
Enterovirus	S,F	++	+++			
Measles	Y				++	Serum for antibody detection
Rubella	Y				++	Serum for antibody detection
Parvovirus	Y					Serum for antibody detection
Meningoencephalitis		++		+++		
Arboviruses	S,F					
Enteroviruses	S,F		++++			
Herpes simplex	Y					++(Brain biopsy)
Lymphocytic choriomeningitis	Y					Serum for antibody detection
Mumps virus	Y					
Gastrointestinal			++			
Adenoviruses	Y					EM‡, EIA (stool)
Norwalk agent	S					EM (stool)
Rotavirus	W,SP					EIA, Latex agglutination (stool)
Congenital and perinatal		++	+	++		++ (Blood)
Cytomegalovirus	Y				+++	
Enteroviruses	S,F					
Herpes simplex virus	Y					+++ (Vesicle fluid)
Rubella	Y				++	
Eye syndrome						++++ (Conjunctival swab or scraping)
Adenoviruses	Y	++				
Herpes simplex virus	Y					
Varicella-zoster virus	Y					
CMV infection		++			++++	+++ (Blood, tissue)
Cytomegalovirus (CMV)	Y					
Myocarditis, pericarditis, and		++	++++			++ (Pericardial fluid)
pleurodynia						
Coxsackie B	S,F					

Y, year-round; *SP*, spring; *S*, summer; *F*, fall; *W*, winter.

*Specimens indicated beside the disease categories should be obtained in all instances; others should be obtained if the specific virus is suspected.

†Direct fluorescent antibody studies are available for herpes simplex virus and varicella-zoster.

‡EM, electron microscopy.

oviruses, and herpes simplex virus, whereas nasopharyngeal swab or aspirate specimens are preferred for the detection of respiratory syncytial, influenza, and parainfluenza viruses.[34] Rhinovirus detection requires a nasal specimen.[18] Throat specimens are collected by rubbing inflamed or purulent areas of the posterior pharynx with a dry, sterile swab. Avoid touching the tongue, buccal mucosa, or teeth and gums. Nasopharyngeal secretion specimens are collected by inserting a swab through the nostril to the nasopharynx or by using a bulb syringe with 3 to 7 ml of buffered saline. The saline is squirted into the nose by squeezing the bulb and aspirated back by releasing the bulb.

Bronchial and bronchoalveolar washes. Wash and lavage fluid collected during bronchoscopy are excellent specimens for the detection of respiratory viruses, especially influenza and adenoviruses.

Rectal swabs and stool specimens. Stool specimens and rectal swabs of feces are used to detect rotavirus, enteric adenoviruses (serotypes 40 and 41), and enteroviruses. Many agents of viral gastroenteritis do not grow in cell culture and require electron microscopy for detection (see section regarding detection of viruses in patient specimens). In general, stool specimens are preferable to rectal swabs and should be required for rotavirus and enteric adenovirus testing. Rectal swabs are acceptable for detecting enteroviruses in patients suspected of having enteroviral disease such as aseptic meningitis. Rectal swabs are collected by inserting a swab 3 to 5 cm into the rectum to obtain feces. Five to 10 ml of freshly passed diarrheal stool is sufficient for rotavirus and enteric adenovirus detection.

Urine. Cytomegalovirus, mumps, rubella, measles, and adenoviruses can be detected in urine. Virus recovery may be increased by processing multiple (2 or 3) specimens because virus may be shed intermittently or in low numbers. The best specimen is at least 10 ml of a clean-voided, first-morning urine.

Skin and mucous membrane lesions. Enteroviruses, herpes simplex virus, varicella-zoster virus, and, rarely, cytomegalovirus can be detected in vesicular lesions of skin and mucous membranes. Once the vesicle has ulcerated or crusted, detection of virus is very difficult. Fluid and cells from the vesicle should be aspirated with a tuberculin syringe or scraped and collected with a swab.

Sterile body fluids other than blood. Sterile body fluids, especially cerebrospinal, pericar-

dial, pleural, and joint fluids, may contain enteroviruses, herpes simplex virus, influenza viruses, or cytomegalovirus. These specimens are collected aseptically by the physician.

Blood. Viral culture of blood is used primarily to detect cytomegalovirus; however, HSV, VZV, enteroviruses, and adenovirus may occasionally be encountered. Five to 10 ml of anticoagulated blood collected in a vacutainer tube is needed. Heparinized, citrated, or EDTA anticoagulated blood is acceptable for cytomegalovirus detection. Citrated blood should be used in instances when other viruses are being sought.

Bone marrow. Bone marrow for viral culture should be added to a sterile tube with anticoagulant. Heparin, citrate, or EDTA anticoagulants are acceptable. Specimens are collected by aspiration.

Tissue. Tissue specimens are especially useful for detecting viruses that commonly infect the lungs (cytomegalovirus, influenza virus, adenovirus), brain (herpes simplex virus), and small bowel (cytomegalovirus). Specimens are collected during surgical procedures.

Serum for antibody testing. Acute and convalescent serum specimens may be needed to detect antibody to specific viruses. Acute specimens should be collected as soon as possible after the appearance of symptoms. The convalescent specimen is collected a minimum of 2 to 3 weeks after the acute specimen. In both cases, an appropriate specimen is 3 to 5 ml of serum collected by venipuncture.

SPECIMEN TRANSPORT AND STORAGE

Ideally all specimens collected for detection of virus should be processed by the laboratory immediately. Although inoculation of specimens into cell culture at the bedside has been recommended in the past, potential biohazards, sophisticated processing steps, and necessary quality control make this practice impractical today. Specimens for viral isolation should not be allowed to sit at room or higher temperature. Specimens should be placed in ice and transported to the laboratory at once. If a delay is unavoidable, the specimen should be refrigerated, *not frozen,* until processing occurs. Every attempt should be made to process the specimen within 12 to 24 hours of collection. Under unusual circumstances specimens may need to be held for days before processing. For storage up to 5 days, hold specimen at 4° C. Storage for 6 or more days should be at −20° C or preferably −70° C. Specimens for freezing should first be diluted or emulsified in viral transport medium. Significant loss of viral infectivity may occur during prolonged storage, regardless of conditions.[40]

Many types of specimens for the detection of virus can be collected using a swab. Most types of swab material, such as cotton, rayon, and Dacron, are acceptable. Calcium alginate, however, is not acceptable for the detection of herpes simplex virus because calcium alginate may inactivate HSV.[40] If calcium alginate must be used, the specimen should be transferred into viral transport medium by swirling the swab vigorously and then the swab should be discarded immediately.

Viral transport media contain protein, such as serum, albumin or gelatin, to stabilize virus and antimicrobials to prevent overgrowth of contaminating bacteria and fungi. Examples of successful transport media include Stuart's medium, Amies medium, Leibovitz-Emory medium, Hanks balanced salt solution (HBSS), and Eagles tissue culture medium. Transport of specimens containing herpes simplex virus can be accomplished using a Culturette (Becton-Dickinson Microbiology Products, Cockeysville, Md.; modified Stuart's medium), and respiratory and rectal/stool specimens are maintained in modified Stuart's, modified Hanks balanced salt solution, or Leibovitz-Emory medium containing antimicrobials. Penicillin (500 units/ml) and streptomycin (500 to 1000 µg/ml) have been used traditionally; however, a more potent mixture includes vancomycin (20 µg/ml), gentamicin (50 µg/ml), and amphotericin B (10 µg/ml). If serum is added as a protein source, fetal calf serum is recommended because it is less likely to contain inhibitors, such as specific antibody.[39,40]

Blood for viral culture, transported in a sterile tube containing anticoagulant, must be kept at refrigeration temperature (4° C) until processing.

Blood for viral serology testing should be transported to the laboratory in the sterile tube in which it was collected. Serum should be separated from the clot as soon as possible. Serum can be stored for hours or days at 4° C or for weeks or months at −20° C or below before testing. Testing for virus specific IgM should be done before freezing, whenever possible, since IgM may form insoluble aggregates upon thawing.[64]

SPECIMEN PROCESSING

Specimens for viral culture should be processed immediately upon receipt in the laboratory (Table 43.7). This may best be accomplished by combining bacteriology and virology processing responsibilities. Although the threat of cell culture contamination in the past dictated separation of virology procedures, the addition of broad spectrum antimicrobials to cell cultures significantly decreased the possibility of cross-contamination with bacteria. In most laboratories, processing with other microbiology specimens allows viral cultures to be processed 7 days per week. If delays must occur, specimens should be stored in viral transport medium at 4° C as described previously. Fluid specimens that have not been added to viral transport medium and that must be stored for days before processing should be diluted in transport medium (1:2 to 1:5) before storage.

Each specimen for virus isolation should be accompanied by a requisition that provides the following information, in addition to patient identification and demographics: source of specimen, clinical history or viruses suspected, and date and time of specimen collection. If this information is not available, a call should be made to the requesting physician or person caring for the patient for additional details.

Processing viral specimens should occur in a biological safety cabinet whenever possible. This protects specimens from contamination by the processing technologist and protects those in the laboratory from infectious aerosols created when specimens are manipulated. If processing cannot be performed in a biological safety cabinet, it should occur behind a protective Plexiglass shield on the countertop. Latex gloves and a laboratory coat should be worn during manipulation of all patient specimens. Vortexing, pipetting, and centrifugation all can create dangerous aerosols. Vortexing should be done in a tightly capped tube behind a shield. After vortexing, the tube should be opened in a safety cabinet. Pipetting also should be performed behind a protective shield. Pipettes must be discarded into a disinfectant fluid so that the disinfectant reaches the inside of the pipette or into a leakproof biosafety bag for autoclaving or incineration.

Processing virology specimens is not complicated (see Table 43.7). In general, any specimen that might be contaminated with bacteria or fungi or any swab specimen should be added to viral transport medium. Presumably sterile fluid specimens can be inoculated directly. Viral transport medium or fluid specimens not in transport medium should be vortexed just before inoculation to break up virus-containing cells and resuspend the inoculum. Sterile glass beads added to the transport medium help release virus from cells during vortexing. Grossly contaminated or potentially toxic specimens such as minced or

TABLE 43.7

LABORATORY PROCESSING OF VIRAL SPECIMENS

SOURCE	SPECIMEN	PROCESSING*	CELLS FOR COMPREHENSIVE CULTURE†
Blood	Anticoagulated blood	Separate leukocytes (see Procedure 43.1)	PMK, HDF, HEp-2
Cerebrospinal fluid (CSF)	1 ml CSF	Inoculate directly	PMK, HDF, HEp-2
Stool or rectal swab	Pea-sized aliquot of feces	Place in 2 ml of viral transport medium; vortex. Centrifuge at $1000 \times g$ for 15 min. and use supernatant fluid for inoculum	PMK, HDF, HEp-2
Genital, skin	Vesicle fluid or scraping	Scraping: emulsify in viral transport medium Fluid: inoculate directly	PMK, HDF
Miscellaneous	Swab, fluids	Swab: vortex in viral transport medium Fluid: inoculate directly	PMK, HDF, HEp-2
Respiratory tract	Nasopharyngeal or throat swab or washings	Dilute with viral transport medium	PMK, HDF, HEp-2
Tissue	Tissue in sterile container	Mince with sterile scalpel and scissors and gently grind Prepare 20% suspension in viral transport medium. Centrifuge at $1000 \times g$ for 15 min. and use supernatant fluid for inoculum	PMK, HDF, HEp-2
Urine	Fresh refrigerated urine	Inoculate directly if clear. Turbid specimens should be centrifuged at $1000 \times g$ for 15 min. and use supernatant fluid for inoculum	HDF, HEp-2 (if adenovirus suspected)

PMK, Primary monkey kidney; *HDF*, Human diploid fibroblast; *HEp-2*, Human epidermoid.

*All inocula into tissue culture tubes are 0.25 ml volumes.

†Comprehensive culture implies detection of all possible viruses. A more limited cost-effective approach to the use of viral

ground tissue can be centrifuged ($1000 \times g$, 15 minutes) and the virus-containing supernatant used as inoculum. Each viral culture tube is inoculated with 0.25 ml of specimen. If insufficient specimen is available, dilute with viral transport medium to increase volume. Excess specimen can be saved at 4° C in case the culture is contaminated. Contaminated specimens can be reprocessed with antibiotic-containing viral transport medium, if they were not originally handled in this manner, or they can be filtered using a disposable 0.22 μm filter and the filtrate recultured. Table 43.7 represents an overview of processing procedures. Blood for viral culture requires special processing to isolate leukocytes that are then inoculated to cell culture tubes (Procedure 43.1).[38,56,71,83] Rapid shell vial cultures are used to detect many viruses. Processing shell vials is described later. Handling and examining cell cultures once inoculated with specimen also is covered later in this chapter.

DETECTION OF VIRUSES IN PATIENT SPECIMENS

Cytology. The most readily available rapid technique for the detection of virus is cytologic examination for the presence of characteristic viral inclusions. Viral inclusions are intracellular structures formed by aggregates of virus within an infected cell or abnormal accumulations of cellular materials resulting from viral-induced metabolical disruption. Inclusions occur in single or syncytial cells. Syncytial cells are aggregates of cells fused to form one large cell with multiple nuclei (Figure 43.8, *A*). Papanicolaou (Pap) or Giemsa-stained smears are most commonly examined for inclusions or syncytia. Cytology can be used to detect infections with varicella-zoster, herpes simplex, and cytomegaloviruses (Figure 43.8, *A* and *B*). A stained smear of cells from the base of a skin vesicle used to detect VZV or HSV inclusions is called a **Tzanck test.**[73] Less commonly, inclusions characteristic of measles and rabies viruses are detected (Figure 43.8, *C* and *D*). Rabies virus inclu-

Text continues on pg. 655.

PROCEDURE 43.1

PROCESSING BLOOD FOR VIRAL CULTURE: LEUKOCYTE SEPARATION BY POLYMORPHPREP

PRINCIPLE

Detection of most viruses in blood specimens is best accomplished by separating and culturing leukocytes. An exception is enterovirus, which, when infecting newborns, may be found free in serum. Leukocytes were originally added to cell culture by harvesting the buffy coat following concentration by centrifugation (100 × *g*, 15 minutes). This method was found to be inferior to white blood cell concentration using density gradient centrifugation and sedimentation with ficoll-paque/macrodex.[38] Other density gradient methods have been employed, including those that use Plasmagel, LeucoPREP, Sepracell-MN, Mono-Poly Resolving Medium, and Polymorphprep. Leukocyte preparation using Polymorphprep, a mixture of sodium metrizoate and dextran, emerges as one of the best, since mononuclear and polymorphonuclear cells are isolated from red blood cells in a one-step procedure.[56,71,83]

SPECIMEN

Three to 5 ml of anticoagulated blood is optimal. EDTA, citrate, or heparin anticoagulant containing tubes can be used. As few as 2 ml of specimen are acceptable for pediatric patients. Clotted specimens are unacceptable. Specimens should be processed immediately. When processing delay is unavoidable, hold specimen at 2° C to 8° C. Processing must occur within 12 to 24 hours. Loss of virus may occur during any processing delay.

MATERIALS

1. Polymorphprep. Robbins Scientific (Sunnyvale, Calif.). Store at or below 20° C.

2. Phosphate buffered saline (PBS)-10X. Biowhittaker (Walkersville, Md.). Store at room temperature.

3. Sterile phosphate buffered saline-0.5X. Dilute 10X PBS 1:20 with distilled water. Filter sterilize through 0.22 μm filter. Store at 2° C to 8° C.

4. Eagles minimal essential medium (2% fetal bovine serum-FBS). Store at 2° C to 8° C.

5. Sterile sodium chloride (0.9%). Store at 2° C to 8° C.

6. Sterile conical centrifuge tubes.

7. Sterile pipettes.

8. Centrifuge.

METHODS

1. Wear latex gloves and laboratory coat when handling patient specimens. Use biological safety cabinet whenever tube containing specimen is opened. Use biocontainment safety covers during centrifugation procedures.

2. Allow blood specimen and all reagents to come to room temperature.

3. Mix blood specimen well by inverting 5 times. Using a 5 or 10 ml pipette, transfer specimen to a sterile conical centrifuge tube.

4. Centrifuge at 500× *g* for 10 minutes at room temperature (18° C to 22° C).

5. Remove plasma with a pipette.

6. Measure approximate volume of remaining cells. Using a 5 or 10 ml pipette, dilute cells with the same approximate volume of 0.9% sodium chloride. Mix well.

7. Add 3.5 ml of Polymorphprep to a 15 ml conical centrifuge tube with a 5 ml pipette.

8. Carefully layer the entire volume of diluted blood cells over the 3.5 ml of Polymorphprep in the centrifuge tube. Do not mix the blood with the Polymorphprep.

9. Centrifuge the tube containing blood and Polymorphprep at 450 to 500 × *g* for 30 minutes at room temperature (18° C to 22° C).

10. Two leukocyte bands should be visible after centrifugation. The top band at the blood sample/Polymorphprep interface consists of mononuclear cells. The lower band contains polymorphonuclear cells. Erythrocytes are pelleted at the bottom of the tube.

11. The mononuclear and polymorphonuclear bands are collected and transferred to a 15 ml conical centrifuge tube with a pasteur pipette.

Continued on p. 652.

PROCEDURE 43.1

PROCESSING BLOOD FOR VIRAL CULTURE: LEUKOCYTE SEPARATION BY POLYMORPHPREP—cont'd

12. Add 5 ml of 0.5X sterile PBS to the centrifuge tube containing the harvested cells. Mix well by aspirating and expelling repeatedly with a pipette.

13. Centrifuge cells at $400 \times g$ for 10 minutes at room temperature (18° C to 22° C).

14. Aspirate supernatant PBS.

15. Add 5 ml of EMEM with 2% FBS to pelleted cells. Mix well by aspirating and expelling repeatedly with a pipette.

16. Centrifuge as in step 13.

17. Aspirate supernatant EMEM and resuspend in 2.0 ml of fresh EMEM with 2% FBS.

18. Specimen is ready for inoculation to cell culture.

TROUBLESHOOTING

1. If all cells are at the top of the Polymorphprep after centrifugation, remix and overlay on top of a new Polymorphprep layer. It may be necessary to increase the volume of Polymorphprep so proportion is still approximately 1:1 to 1:1.5. Centrifuge at either increased force of gravity for the same length of time, or at the same force of gravity for a longer time.

2. If buffy coat is on top of the erythrocyte layer, take buffy coat and erythrocytes and relayer over a new gradient. Centrifuge for a shorter time at the same force of gravity or at a lesser force of gravity for the same time.

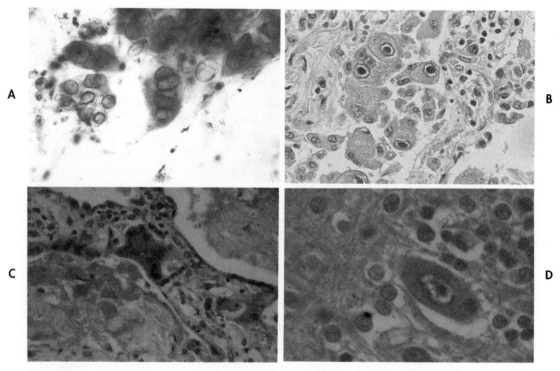

FIGURE 43.8

Viral inclusions. **A,** Pap-stained smear showing multinucleated giant cells typical of herpes simplex or varicella-zoster viruses. **B,** Hematoxylin and eosin-stained lung tissue containing intranuclear inclusion within enlarged CMV infected cells. **C,** Hematoxylin and eosin-stained lung tissue containing multinucleated giant cells infected with measles virus. **D,** Hematoxylin and eosin-stained brain tissue showing oval, eosinophilic rabies cytoplasmic inclusion (Negri body).

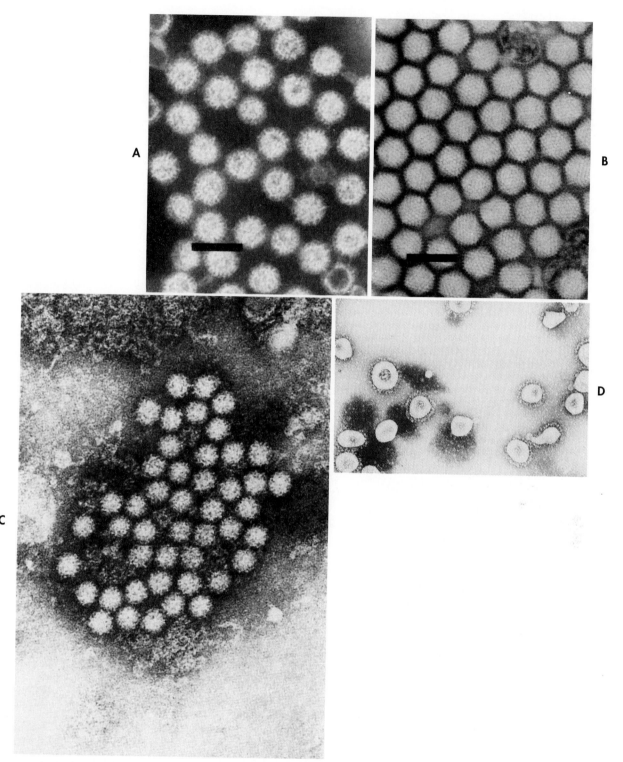

FIGURE 43.9

Electron micrographs of viruses. **A,** Rotavirus. **B,** Adenovirus. **C,** Norwalk agent virus. **D,** Coronavirus. **E,** Calicivirus. **F,** Herpes simplex virus. **G,** Measles virus. **H,** Negatively stained preparation of JC virus from the brain of John Cunningham, the original patient in whom the virus was discovered, passaged in human fetal glial cells. Magnification (248,000×.) (**C** from Howard B.J., Klaas J., Rubin S.J., et al, Clinical and pathogenic microbiology, 1987, Mosby, St. Louis; **D, E, G** from U.S. Department of Health, Education, and Welfare; Public Health Service, Centers for Disease Control, Atlanta Georgia; **H** courtesy Dr. Gabriele M. ZuRhein.)

Continued.

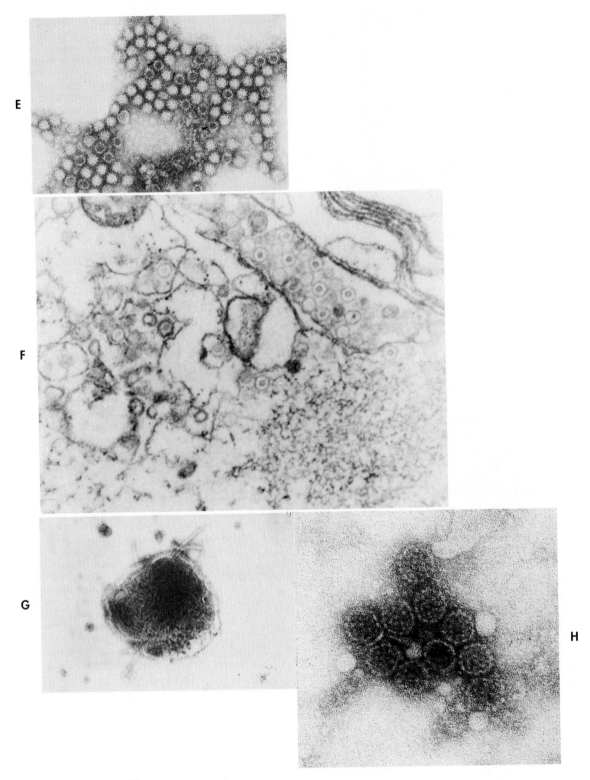

FIGURE 43.9, cont'd

For legend see p. 653.

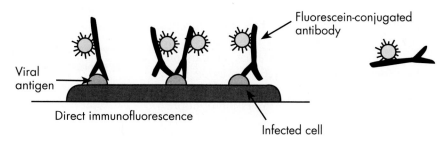

FIGURE 43.10

Illustration of direct immunofluorescent staining method.

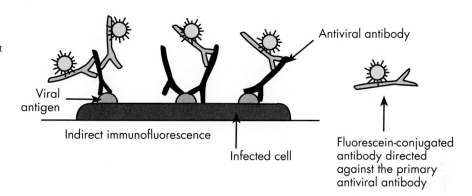

FIGURE 43.11

Illustration of indirect immunofluorescent staining method.

sions in brain tissue are called **Negri bodies.** Cytology is less sensitive than culture but is especially helpful for those viruses that are difficult or dangerous to isolate in the laboratory, such as polyomavirus and rabies virus.

Electron microscopy. Few laboratories use electron microscopy (EM) to detect viruses because it is labor intensive and relatively insensitive. EM is most helpful for the detection of viruses that do not grow readily in cell culture and works best if the titer of virus is at least 10^6 to 10^7 particles per milliliter.[50] Immune EM allows visualization of virus particles present in numbers too small for easy direct detection. The addition of specific antiserum to the test suspension causes the virus particles to form antibody-bound aggregates, which are more easily detected than are single virus particles. EM is still used by some clinical laboratories for the detection of viruses causing gastroenteritis (rotavirus, enteric adenoviruses, Norwalk agent virus, coronaviruses, and caliciviruses) and encephalitis (herpes simplex virus, measles virus, and JC polyomavirus) (Figure 43.9, *A to H*).[57] EM is no longer needed to detect common rotavirus types causing infection in humans because relatively simple and accurate enzyme immunoassay and latex agglutination antigen assays have been developed.

Immunodiagnosis. High quality, commercially available, viral antibody reagents have led to the development of fluorescent antibody, enzyme immunoassay, radioimmunoassay, latex ag-

glutination, and immunoperoxidase tests that detect viral antigen in patient specimens.

Direct and indirect fluorescent antibody methods are used.[52] Direct fluorescent antibody testing involves use of a labeled antibody; the label is usually fluorescein isothiocyanate (FITC), which is layered over specimen suspected of containing homologous virus (Figure 43.10). The indirect fluorescent antibody procedure is a two-step test in which unlabeled antiviral antibody is added to the slide first, followed by a labeled (FITC) antiglobulin that binds to the first-step antibody bound to virus in the specimen (Figure 43.11). Direct immunofluorescence is, in general, more rapid and specific than indirect immunofluorescence, but less sensitive. Increased sensitivity of the indirect test results from the signal amplification that occurs with the addition of the second antibody. Signal amplification decreases specificity by increasing nonspecific background fluorescence. Direct immunofluorescence is best suited to situations in which large quantities of virus are suspected or high quality, concentrated monoclonal antibodies are used, such as for the detection of RSV in patient specimen or the identification of viruses growing in cell culture. Indirect immunofluorescence should be used when lower quantities of virus are suspected, such as detection of respiratory viruses in specimens from adult patients. High quality monoclonal antibodies improve the sensitivity and specificity of the indirect test. Whenever possible, the direct test should be used because it is significantly faster without the second antibody step. A standard procedure for the direct fluorescent antibody test is shown in Procedure 43.2.

Strict criteria for the interpretation of fluorescent patterns must be used. This includes standard interpretation of fluorescence intensity (Table 43.8) and recognition of viral inclusion morphology. Nuclear and cytoplasmic staining patterns are typical for influenza virus, adenovirus, and the herpesviruses; cytoplasmic staining only for respiratory syncytial, parainfluenza, and mumps viruses, and staining within multinucleated giant cells is typical of measles virus (Figure 43.12, *A* to *H*). False-positive staining can occur with specimens containing yeasts, certain bacteria, mucus, or leukocytes. Leukocytes, which contain Fc receptors for antibody, can also cause nonspecific binding of antibody conjugates. To verify one's ability to interpret fluorescent antibody tests, every laboratory should perform viral culture or some alternative detection method in parallel with immunofluorescence until in-house performance has been established.

Immunofluorescent stains that are most useful in the clinical virology laboratory are those for respiratory syncytial virus, influenza and parainfluenza viruses, adenovirus, herpes simplex, varicella-zoster, and cytomegaloviruses. A pool of antibodies can be used to screen a specimen for multiple viruses. A positive screen is then tested with each individual reagent to identify the exact virus that is present. A screening pool has been used successfully for the detection of respiratory viruses in children (Viral Respiratory Kit, Baxter/Bartels Diagnostics, Inc., Issaquah, Wash.). Such pools are much less sensitive when used with specimens from adults.[60,75]

Enzyme immunoassay methods used most frequently in clinical virology are the solid phase enzyme-linked immunosorbent assay (solid phase ELISA) and the membrane bound enzyme-linked immunosorbent assay (membrane ELISA).[68,72] Solid phase ELISA is performed in a small test tube or microtiter tray. Breakaway strips of microtiter wells are available for low volume test runs. The remaining unused wells can be saved for future testing. A solid phase ELISA used for the detection of rotavirus in stool specimens is illustrated in Figure 43.13. Membrane ELISA tests have been developed for low volume testing and situations in which rapid results are needed. They can be performed by those with minimum training and usually require less than 30 minutes to

---

TABLE 43.8

INTERPRETATION OF FLUORESCENCE INTENSITY

INTENSITY	INTERPRETATION
Negative	No apple-green fluorescence
1+	Faint, yet unequivocal apple-green fluorescence
2+	Apple-green fluorescence
3+	Bright, apple-green fluorescence
4+	Brilliant, apple-green fluorescence

---

complete. The membrane method uses a hand-held reaction chamber with a cellulose-like membrane. Specimen and reagents are applied to the membrane. Following a short incubation time, a chromogenic (color) reaction occurs on the surface of the membrane and is read visually. Built-in controls on the same membrane provide convenient monitoring of test procedures. Figure 43.14 illustrates a membrane ELISA used to detect respiratory syncytial virus in a nasal wash specimen. The most useful enzyme immunoassays used for antigen detection are those for respiratory syncytial virus (solid phase and membrane), rotavirus (solid phase and membrane), enteric adenoviruses (solid phase), and herpes simplex virus (solid phase).

Advantages of enzyme immunoassays are the use of nonradioactive, relatively stable reagents and results that can be interpreted qualitatively (positive or negative) or quantitatively (titer or degree of positive reaction). It is important to note that enzyme immunoassays frequently have an indeterminant or borderline interpretative category. This result implies that low levels of viral antigen or background interference has prevented a clear-cut positive or negative result. Such results usually require testing a second specimen to avoid interference or to detect a rise in antigen level. ELISAs are more sensitive than most other assays, including fluorescent immunoassays, and can be easily automated. A drawback with ELISAs is that specimen quality cannot be evaluated, that is, the number of cells cannot be assessed, as can be done microscopically with fluorescent immunoassay. Acellular specimens are clearly inferior since cell-associated virus would be missing.

Radioimmunoassay (RIA), immunoperoxidase staining, and latex agglutination are additional techniques used to detect viral antigen. RIA

DETECTION OF RESPIRATORY SYNCYTIAL VIRUS BY FLUORESCENT ANTIBODY STAINING OF RESPIRATORY SECRETIONS

PRINCIPLE

Respiratory syncytial virus is recognized as a major cause of acute lower respiratory tract disease in infants and young children, commonly presenting as bronchiolitis or pneumonia. Hospitalized children with RSV disease serve as a reservoir for nosocomial transmission of the virus. Isolation or cohorting of infected patients decreases transmission. Conventional culture for the identification of virus requires 4 to 14 days and may be insensitive because of the lability of the virus. Rapid detection of RSV can be accomplished by fluorescent antibody or enzyme immunoassay detection of viral antigen. Direct fluorescent antibody staining for RSV antigen can be performed on upper respiratory wash or swab specimens that contain infected epithelial cells.

SPECIMEN

Nasal washes are the specimen of choice. Three to five ml of saline are quickly squirted into and aspirated back through the nostril. The nasopharyngeal lavage is sent immediately to the laboratory in a sterile container.
Nasopharyngeal swab material can be used for DFA stain but is less likely to contain an abundance of epithelial cells. Swab material should be vigorously extracted in viral transport medium and sent to the laboratory immediately.

MATERIALS

1. Phosphate buffered saline.

2. *N*-acetyl-L-cysteine (NALC). Prepare working stock by diluting 50 mg of NALC in 10 ml of 1.5% sodium citrate. Working stock may be stored for one month at 2° C to 8° C.

3. Anti-RSV monoclonal, FITC conjugated antibody with Evans blue counterstain (Baxter/Bartels Diagnostics, Inc., Issaquah, Wash.).

4. Buffered glycerol mounting medium (pH 8.0 to 8.4).

5. Centrifuge tubes.

6. Pipettes.

7. Microscope slides and coverslips.

8. Humidified incubation chamber.

9. Centrifuge.

10. Fluorescence microscope.

11. RSV control slides. Each control slide with two wells. One well containing RSV infected HEp-2 cells and one containing uninfected HEp-2 cells.

METHODS

1. Wear latex gloves when handling all patient specimens. Perform all specimen handling procedures in laminar flow hood.

2. Examine specimen for the presence of mucous. If sufficient mucous is present to prevent mixing and distribution of specimen to slide, digest with mucolytic agent (NALC) as follows:
   A. Treat specimen with equal volume of stock NALC. Vortex for 20 seconds. Allow to sit at room temperature for 20 minutes.
   B. Vortex for 20 seconds. Add 5 ml PBS to cells, mix, and centrifuge at 380× g for 15 minutes.
   C. Aspirate and discard the supernatant, being careful not to disturb the cell pellet.
   D. Wash cell pellet with 5 ml PBS.
   E. Resuspend cell pellet in 0.2 ml PBS. Cell suspension should be hazy but not opaque. Dilute cell suspension with PBS if necessary.
   F. Place one drop of the suspension into the well of the microscope slide, one patient specimen per slide.

3. If specimen is free of mucus, process as follows:
   A. Centrifuge 1 ml of specimen for 2 minutes in the centrifuge.
   B. Carefully aspirate supernatant and wash twice wtih PBS.
   C. Aspirate supernatant and re-suspend cell pellet in 0.2 ml PBS.
   D. Prepare slide as described above in 2, F.

4. Air dry all slides and then fix for 10 minutes in acetone at 2° C to 8° C. Slides may be held for 24 hours at 2° C to 8° C or for longer periods at −70° C.

5. Stain slides by adding 20 to 30 µl/well of RSV antibody conjugate to each patient slide and to both wells of the control slide. Spread conjugate to cover slide completely.

6. Place slides in humidified chamber at 36° C and incubate for 30 minutes.

Continued on p. 658.

DETECTION OF RESPIRATORY SYNCYTIAL VIRUS BY FLUORESCENT ANTIBODY STAINING OF RESPIRATORY SECRETIONS—cont'd

7. Rinse antiserum from each slide with a gentle stream of PBS and place slides in Coplin jar filled with PBS for 5 minutes.

8. Gently rinse twice in distilled water for 1 to 2 minutes.

9. Drain excess distilled water and air dry. Do not blot.

10. Add coverslip using buffered glycerol as mounting medium.

INTERPRETATION AND QUALITY CONTROL

1. Examine using 200× magnification (10× eyepiece, 20× objective).

2. Examine control slide first. The well containing RSV infected HEp-2 cells should show ≥ 2+ fluorescing inclusions (Figure 43.12, *E*). The well containing uninfected HEp-2 cells should show no fluorescence. If the control slide shows other results, the batch of slides should be discarded and restained with a new control slide.

3. Examine patient slides. Fewer than 3 to 5 cells per 200× field should be considered insufficient cells for an adequate examination, and a new specimen should be

requested. If sufficient cells are seen, examine for typical green fluorescing RSV inclusions. A positive stain will show fluorescing inclusions; a negative stain will have no fluorescing inclusions.

---

has been largely replaced by ELISA because of expensive equipment and disposal procedures for radioactive materials. Immunoperoxidase staining is commonly used to stain histologic sections for virus but is less popular than immunofluorescence staining in clinical virology laboratories. Latex agglutination is an easy and inexpensive method but lacks sensitivity compared with ELISA and fluorescent immunoassays.

Commercially available virus detection systems. The growing need for easy, inexpensive virus detection has resulted in an explosion of commercially available "systems." These systems primarily use fluorescent and ELISA immunoassay methods. Table 43.9 lists viral antigen detection systems that are commercially available.

Nucleic acid probes and polymerase chain reaction assays (PCR). Nucleic acid probes are short segments of DNA that are designed to hybridize with complementary viral DNA or RNA segments. The probe is labeled with a fluorescent, chromogenic, or radioactive tag that allows detection if hybridization occurs. The probe reaction can occur in situ, such as in a tissue thin section, in liquid, or on a reaction vessel surface or membrane.[33,77,82] A DNA probe test used to detect papillomavirus DNA in a smear of cervical cells is illustrated in Figure 43.15.

Nucleic acid probes are most useful in situations in which the amount of virus is relatively abundant, viral culture is slow or not possible, and immunoassays lack sensitivity or specificity. Although direct probe assays have been used for detection of many viruses in research settings, few commercially available reagents exist for probe assays in clinical virology laboratories (Table 43.10). Probe detection of virus suffers most notably from lack of sensitivity when compared with tests that offer amplification of virus before detection, such as culture or the rapid shell vial assay. Probe identification coupled with amplification by PCR, however, is very promising.

The polymerase chain reaction (PCR) is a method that duplicates short DNA segments thousands to a million fold.[45,58] DNA fragments that can be identified with a specific probe, but are too few in number in the original specimen to be detected, can be duplicated (amplified) using PCR. This provides the probe with enough target to readily identify the presence of a specific virus. Probe identification of human immunodeficiency virus following PCR amplification is illustrated in Figure 43.16.

Use of PCR to identify nearly every medically important virus has been attempted in research settings. PCR is a patented technique, so commer-

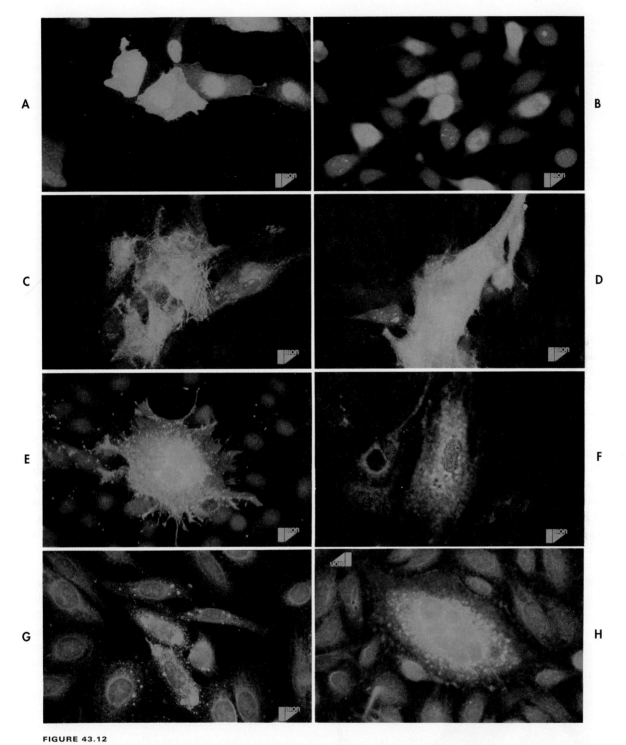

FIGURE 43.12

Fluorescent antibody staining of virus-infected cells. **A,** Influenza virus. **B,** Adenovirus. **C,** Herpes varicella-zoster virus. **D,** Herpes simplex virus. **E,** Respiratory syncytial virus. **F,** Parainfluenza virus. **G,** Mumps virus. **H,** Measles virus. (Courtesy Bion Enterprises Ltd., Park Ridge, Ill.)

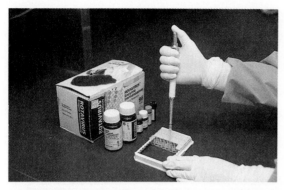

FIGURE 43.13

Solid-phase enzyme immunoassay for the detection of rotavirus with breakaway strips of microtiter wells for small batch testing. (Courtesy Children's Hospital Medical Center of Akron; Akron, Ohio)

cial availability of virus detection tests has been slow to develop. Licensing arrangements of proprietary PCR procedures, or the introduction of other amplification techniques by competitors, should lead to use of PCR or similar nucleic acid amplification techniques in the clinical laboratory setting in the near future.

Conventional cell culture. Viruses are strict intracellular parasites, requiring a living cell for multiplication and spread. To detect virus using living cells, suitable host cells, cell culture media, and techniques in cell culture maintenance are necessary. Host cells, referred to as cell cultures, originate as a few cells and grow into a monolayer on the sides of glass or plastic test tubes. Cells are kept moist and supplied with nutrients by keeping them continuously immersed in cell culture medium (Figure 43.17). Cell cultures are routinely incubated in a roller drum that holds cell

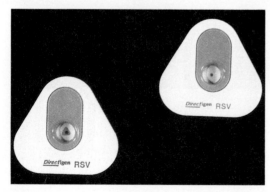

FIGURE 43.14

Positive and negative membrane ELISA for the detection of respiratory syncytial virus. Triangle or triangle and dot indicates a positive test. Dot indicates a negative test. If neither a triangle nor a dot is visible, the test is uninterpretable and must be repeated.

culture test tubes tilted while they slowly revolve ½ to 1 RPM) at 35° C to 37° C (see Figure 43.5). Incubation of cell culture tubes in a stationary rack can be used in place of a roller drum. Rapidly growing viruses, such as HSV, appear to be detected equivalently by both methods. Comparative studies are not available for most viruses. Metabolism of growing cells in a closed tube results in the production of CO_2 and acidification of the growth liquid. To counteract the pH decrease, a bicarbonate buffering system is employed in the culture medium to keep the cells at physiologic pH (pH 7.2). Phenol red, a pH indicator, that is red at physiologic, yellow at acidic, and purple at alkaline pHs is added to monitor adverse pH changes. Once inoculated with specimen, cell cultures are incubated for 1 to 4 weeks, depending on the viruses suspected. Periodically the cells are inspected microscopically for the presence of virus, indicated by areas of dead or dying cells called **cytopathic effect (CPE).**

Two kinds of media, growth medium and maintenance medium, are used for cell culture. Both are prepared with Eagle's minimum essential medium (EMEM) in Earle's balanced salt solution (EBSS) and include antimicrobials to prevent bacterial contamination. Usual antimicrobials added are vancomycin (10 µg/ml), gentamicin (20 µg/ml), and amphotericin (2.5 µg/ml). Growth medium is a serum-rich (10% fetal, newborn, or agammaglobulinemic calf serum) nutrient medium designed to support rapid cell growth. This medium is used for initiating growth of cells in a tube when cell cultures are being prepared in-house or for feeding tubes of purchased cell cultures that have incomplete cell monolayers. Feeding refers to the removal of old medium followed by the addition of fresh culture medium. Maintenance medium is similar to growth medium but contains less serum (0 to 2%) and is used to keep cells in a steady state of metabolism. Fetal, newborn, or agammaglobulinemic calf serum is used to avoid inhibitors, such as specific antibody, and to be free of mycoplasmas present in serum from older animals.

Several kinds of cell cultures are routinely used for isolation of viruses. A cell culture becomes a **cell line** once it has been passed or subcultured in vitro. Cell lines are classified as primary, low passage, or continuous. **Primary cell lines** are those that have been passed only once or twice since harvesting, such as primary monkey kidney cells. Further passage of primary cells results in a decreased receptivity to viral infection. **Low passage cell lines** are those that remain

TABLE 43.9

COMMERCIALLY AVAILABLE VIRUS DETECTION SYSTEMS*

SYSTEM NAME	COMPANY	VIRAL ANTIGENS DETECTED	ANTIBODY ASSAYS AVAILABLE
Acculyza	Analytab Products	Rotavirus, Adenovirus	Yes
MicroTrak EIA	Syva Company	In development	Yes
Prima	Bartels Diagnostics	RSV, Influenza A	Yes
Vidas Assay	BioMerieux	RSV, HSV, Rotavirus, Adenovirus	Yes

*Kits for the detection of individual viral antigens are available from many suppliers.

TABLE 43.10

COMMERCIALLY AVAILABLE DNA PROBES FOR DETECTION OF VIRUSES

COMPANY	VIRUSES	USE
Digene Diagnostics (Silver Spring, Md.)	Human papillomavirus	FDA-approved
	Hepatitis B virus	Research
	Cytomegalovirus	Research
	Epstein-Barr virus	Research
Enzo Diagnostics (Farmingdale, N.Y.)	Human papillomavirus	Research
	Adenovirus	Research
	Cytomegalovirus	Research
	Epstein-Barr virus	Research
	Hepatitis B virus	Research
Oncor (Gaithersburg, Md.)	HIV-1	Research
	HTLV-1	Research
	Human papillomavirus	Research

virus sensitive through 20 to 50 passages. Human diploid fibroblast cells, such as lung fibroblasts, are a commonly used low passage cell line. **Continuous cell lines** such as HEp-2 cells (human epidermoid carcinoma cells) can be passed and re-

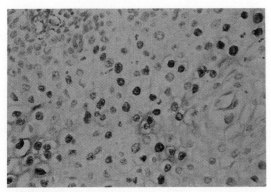

FIGURE 43.15

Smear of cervical cells stained with probe for papillomavirus DNA. Dark-staining cells contain viral DNA. (Courtesy Children's Hospital Medical Center of Akron; Akron, Ohio)

main sensitive to some virus infections indefinitely. Unfortunately most viruses do not grow well in these cells. The majority of clinically significant viruses can be recovered using one cell culture type from each group. A combination frequently used by clinical laboratories is Rhesus monkey kidney cells, MRC-5 diploid fibroblast cells, and HEp-2 cells (Table 43.11).

Inoculated cell cultures should be incubated immediately at 35° C. After allowing virus to adsorb to the cell monolayer for 12 to 24 hours, it is common practice to remove the remaining inoculum and culture medium and replace it with fresh maintenance medium. This avoids most inoculum-induced cell culture toxicity and improves virus recovery. Incubation should be continued for 5 to 28 days depending on viruses suspected (Table 43.11). Maintenance medium should be changed periodically (usually 1 to 2 times weekly) to provide fresh nutrients to the cells. Blind passage of cell cultures is used occasionally to detect viruses that may not produce CPE in the initial culture tube. Cell cultures that show nonspecific

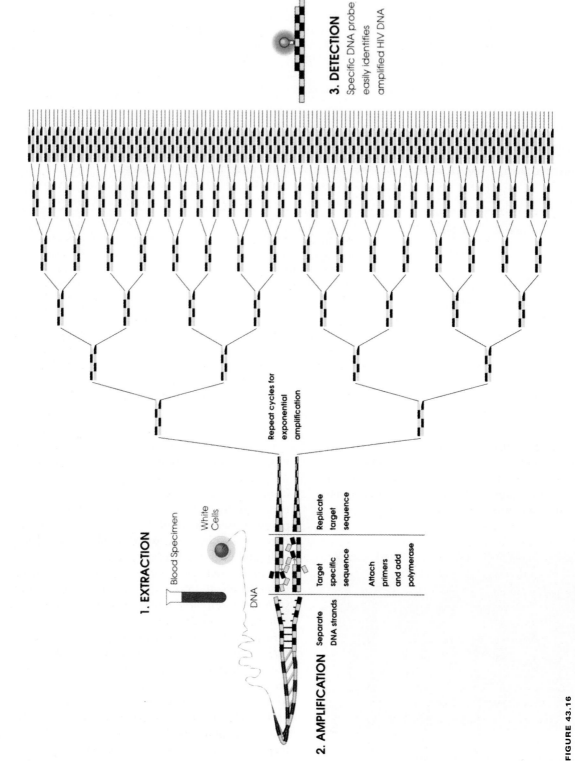

FIGURE 43.16

Probe identification of HIV DNA following PCR amplification. (Courtesy Roche Biomedical Laboratories, Research Triangle Park, N.C.)

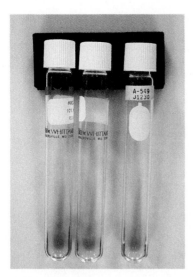

FIGURE 43.17

Cell culture tube showing emblem indicating side that should be incubated upward. Cell monolayer forms on glass surface beneath the culture medium.

or ambiguous CPE are also passed to additional cell culture tubes. Passage in both instances is performed by scraping the monolayer off the sides of the tube with a pipette or disrupting the monolayer by vortexing with sterile glass beads added to the culture tube, followed by inoculating 0.25 ml of the resulting suspension into new cell cultures. Absence of CPE in the subpassages indicates that the original cell culture tube did not contain virus. Blind passage occasionally stimulates productive (CPE-causing) viral infection but is less frequently used today because the added time and expense do not justify detection of a few additional isolates. Toxicity, which causes ambiguous CPE, will be diluted during passage and should not appear in the second cell culture tube.

Viruses are most often detected in cell culture by the recognition of **cytopathic effect (CPE).** Virus-infected cells change their usual morphology and eventually lyse or detach from the glass surface while dying. Viruses have distinct CPEs, just as colonies of bacteria on agar plates have unique morphologies (Figures 43.18, *A* to *K*). CPE may be quantitated as indicated in Table 43.12. Preliminary identification of a virus can sometimes be made based on the cell culture that supports viral replication, how quickly the virus produced CPE, and a description of the CPE (see Table 43.11). Definitive identification of virus detected in cell culture is discussed later in this chapter.

Shell vial culture. The shell vial culture is a rapid modification of a conventional cell cul-

ture.[31,70] Virus is detected more quickly using the shell vial technique because the infected host cell is stained for viral antigens found soon after infection, before the development of CPE. Viruses that normally take days to weeks to produce CPE can be detected within 1 to 2 days. A shell vial culture tube, a 15×45 mm one-dram vial, is prepared by adding a round coverslip to the bottom, covering this with growth medium, and adding appropriate cells (Figure 43.19). During incubation a cell monolayer forms on top of the coverslip. Shell vials should be used 5 to 9 days after cells have been inoculated. Shell vials with the monolayer already formed can be purchased. Specimens are inoculated onto the shell vial cell monolayer by low speed centrifugation. This enhances viral infectivity by poorly understood mechanisms. Coverslips are stained using virus-specific fluorescent conjugates. Typical fluorescing inclusions confirm the presence of virus (Figure 43.20). The shell vial procedure for detecting CMV is presented in detail in Procedure 43.3.

The shell vial culture technique can be used to detect most viruses that grow in conventional cell culture. It is best used for viruses requiring relatively long incubation before producing CPE, such as CMV and VZV. The advantage of shell vial is its speed; most viruses are detected within 24 hours. The disadvantage is that only one type of virus can be detected per shell vial. For example, a specimen that might contain influenza A or B, or adenovirus, would need to be inoculated to three separate shell vials so that each vial could be stained with a separate virus-specific conjugate.

IDENTIFICATION OF VIRUSES DETECTED IN CELL CULTURE

Viruses detected in conventional cell culture by production of CPE may be identified by noting the cell types that support viral replication, time to detection of CPE, and morphology of the CPE (Figure 43.18, *A* to *K*). Experienced virologists are able to presumptively identify most viruses isolated in clinical laboratories based on these criteria. When definitive identification is required, additional testing can be performed. Fluorescent-labeled antisera, available for most viruses, are used commonly for culture confirmation. In addition, acid lability is used to differentiate enteroviruses from rhinoviruses, and neutralization is used to identify viruses with many serotypes for which fluorescent-labeled antisera are not available. Some viruses, such as influenza, parainfluenza, and mumps, which

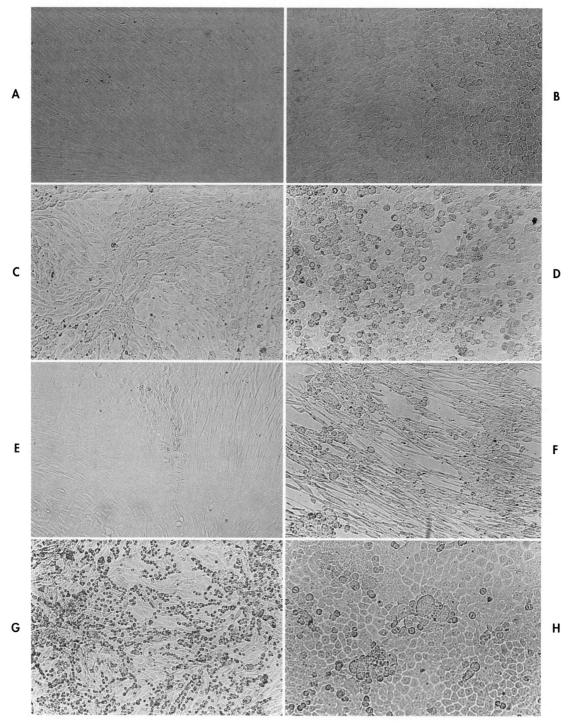

FIGURE 43.18

Cell culture morphology and viral CPE. **A,** Normal human diploid lung fibroblast cells (HDF). **B,** Normal HEp-2 cells. **C,** Normal primary monkey cells (PMK). **D,** HEp-2 cells infected with adenovirus. **E,** HDF cells infected with cytomegalovirus. **F,** HDF cells infected with herpes simplex virus. **G,** PMK cells infected with hemadsorbing virus, such as influenza, parainfluenza, or mumps, plus guinea pig erythrocytes. **H,** HEp-2 cells infected with respiratory syncytial virus. **I,** HDF cells infected with rhinovirus. **J,** PMK cells infected with echovirus. **K,** HDF cells infected with herpes varicella-zoster virus. (From the U.S. Department of Health, Education, and Welfare Public Health Service, Centers for Disease Control, Atlanta.)

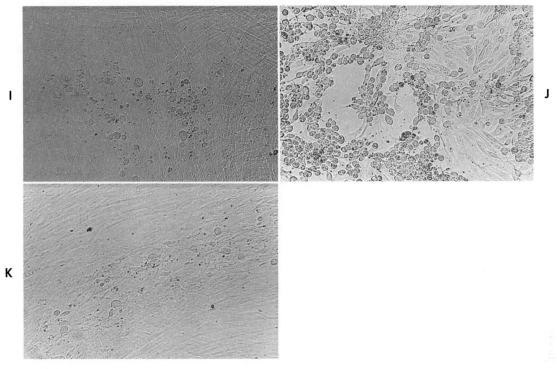

FIGURE 43.18, cont'd

For legend see p. 664.

produce little or no CPE, can be detected by **hemadsorption,** since infected cells contain viral hemadsorbing glycoproteins in their outer membranes.[51] The addition of guinea pig red blood cells to the cell culture tube, followed by a wash to remove nonadsorbed RBCs, results in a ring of RBCs around infected cells (see Figure 43.18, *G*). Cell cultures demonstrating hemadsorption can be stained with fluorescent-labeled antisera to identify the specific hemadsorbing virus present. Detailed procedures for culture confirmation by fluorescent antibody staining, hemadsorption for the detection of influenza and parainfluenza viruses, acid-lability test, and the virus neutralization test are presented in Procedures 43.4 through 43.7.

VIRAL SEROLOGY

Serology was the primary means for the laboratory diagnosis of viral infections until the mid-1970s. At that time, culture and detection of viral antigen became more widely available because of high quality, commercially available reagents, such as cell cultures, and a broad range of immunodiagnostic test kits for the detection of viral antigen directly in patient specimens. The use of viral serology tests has changed to complement this growing menu of

FIGURE 43.19

Preparation of shell vial culture tubes. (Courtesy Bostick C.C.: Laboratory Detection of CMV, Microbiology Tech Sample No. MB-3, p. 5, 1992.)

diagnostic tests. Viral serology is now used primarily to detect immune status and to make the diagnosis of infections in situations in which the virus cannot be cultivated in cell culture or detected by immunoassay.[78]

In most viral infections, IgM is undetectable after 1 to 4 months but detectable levels of IgG remain, as a rule, for the life of the patient. If this patient is infected by an antigenically similar virus or the original strain has remained latent and re-

TABLE 43.11

ISOLATION AND IDENTIFICATION OF COMMON CLINICALLY ENCOUNTERED VIRUSES

VIRUS	PMK	HEp-2	HDF	CPE DESCRIPTION	RATE OF GROWTH (DAYS)	IDENTIFICATION AND COMMENTS
Adenovirus	++*	+++	++	Rounding and aggregation of infected cells in grapelike clusters	2-10	Confirm by FA. Serotype by cell culture neutralization
Cytomegalovirus (CMV)	–	–	++++	Discrete small foci of rounded cells	5-28	Distinct CPE sufficient to identify. Confirm by FA
Enterovirus	++++	±	++	Characteristic refractile angular or tear-shaped CPE; progresses to involve entire monolayer	2-8	Identify by cell culture neutralization test. Stable at pH 3
Herpes simplex (HSV)	±	++++	++++	Rounded, swollen refractile cells. Occasional syncytia, especially with type 2. Rapidly involves entire monolayer	1-3 (may take up to 7)	Distinct CPE sufficient to identify. Confirm by FA
Influenza	++++	–	±	Destructive degeneration with swollen, vacuolated cells	2-10	Detect by hemadsorption or hemagglutination with guinea pig RBCs. Identify by FA
Mumps	+++	±	±	CPE, usually absent. Occasionally syncytia are seen	5-10	Detect by hemadsorption with guinea pig RBCs. Confirm by FA
Parainfluenza	+++	–	–	CPE, usually minimal or absent	4-10	Detect by hemadsorption with guinea pig RBCs. Identify by FA
Respiratory syncytial virus (RSV)	+	+++	+	Syncytia in HEp-2	3-10	Distinct CPE in Hep-2 is sufficient for presumptive identification. Confirm by FA
Rhinovirus	±	–	+++	Characteristic refractile rounding of cells. In PMK. CPE is identical to that produced by enteroviruses	4-10	Labile at pH 3
Varicella-zoster	–	–	++	Discrete foci of rounded, swollen, refractile cells. Slowly involves entire monolayer	5-28	Confirm by FA

FA, Fluorescent antibody.
*Relative sensitivity of cell cultures for recovering the virus: –, none recovered; ±, rare strains recovered; +, few strains recovered; ++++, ≥ 80% of strains recovered.

activates at a later time, these virus-specific IgG and IgM antibody levels again rise. The secondary IgM response may be difficult to detect, however, a significant (fourfold) IgG titer rise is readily apparent (see Figure 13.4).

An immune status test measures whether or not a patient has been infected by a particular virus in the past. A positive result with a sensitive, virus-specific IgG test indicates past infection. Some immune status tests include methods that detect both IgG and IgM, to detect recent or active infections. To help diagnose active disease, two approaches are helpful. Detection of virus-specific IgM in an acute phase specimen collected at least 10 to 14 days after the onset of infection indicates current or very recent disease. Detection of a fourfold antibody titer rise between **acute** and **convalescent sera** also indicates current or recent disease. Acute phase serum should be collected as soon as possible after onset of symptoms. The convalescent specimen should be collected 2 to 3 weeks after the acute specimen. If a single post-acute serum, collected between acute and convalescent times, or convalescent specimen is

TABLE 43.12

QUANTITATION OF CELL CULTURE CPE

QUANTITATION	INTERPRETATION
Negative	Uninfected monolayer
Equivocal (±)	Atypical alteration of monolayer involving few cells
1+	1-25% of monolayer exhibits CPE
2+	25-50% of monolayer exhibits CPE
3+	50-75% of monolayer exhibits CPE
4+	75-100% of monolayer exhibits CPE

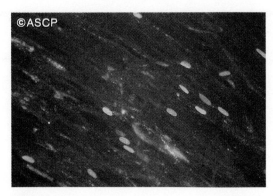

FIGURE 43.20

Typical fluorescing nuclei of human diploid fibroblast cells infected with cytomegalovirus as seen in the shell vial assay. (Courtesy Bostick C.C.: Laboratory Detection of CMV, Microbiology Tech Sample No. MB-3, p. 5, 1992.)

all that is available for testing, an extremely high, virus-specific IgG titer may suggest infection. The exact titer specific for active disease, if known at all, varies with each testing method and virus. In general, titers high enough to be diagnostic are unusual and testing of single specimens should not be performed. A reasonable policy would include using IgM tests when available and performing IgG testing only on paired acute and convalescent specimens. No IgG testing need be performed on the first acute specimen until receipt of the convalescent specimen. This eliminates useless testing of single specimens in those instances when a second sample is never collected or submitted for analysis.

Many serologic methods are or have been routinely used to detect antiviral antibody.[12] Prominent among those used today are **complement fixation (CF), enzyme-linked immunosorbent assay (ELISA),** indirect **immunofluorescence,** anticomplement immunofluorescence (ACIF), and **Western immunoblotting.** Complement fixation is a labor intensive, technically demanding method best fitted to batch testing. The major advantage of CF is the years of experience that have accumulated for interpreting results. As less demanding, easily automated techniques for batch testing are developed, such as ELISA, the need for CF testing is disappearing. Additional advantages of ELISA are that it can be used to detect IgM-specific antibodies free of common interfering factors, in particular through use of an antibody capture technique (Figure 43.21).[68] Indirect immunofluorescence is best used for individual specimens or small batch testing. Immunofluorescence can also be used to detect virus-specific IgM but requires prior separation and elimination of the IgG fraction that, if present, can result in both false-positive and false-negative results. IgM and IgG can be separated by either ion exchange chro-

matography (see Figure 13.10) or immune precipitation (Figure 43.22).[24,28,41,80]

Immunoglobulin G indirect fluorescent antibody testing is subject to false-positive results resulting from Fc receptors that occur in cells infected with virus. IFA testing is performed using virus-infected cells fixed to a microscope slide. When the substrate cells are overlaid with patient serum, the Fc portion of the antibody molecule binds to these receptors. Fluorescent-labeled antiglobulin will attach to both homologous antibody, which is bound to viral antigen, and to Fc bound antibody. Subsequent fluorescence of Fc bound antibody results in a false-positive or falsely elevated reading. To avoid this complication, the anticomplement immunofluorescence (ACIF) test can be used (Figure 43.23).[64] Because fluorescent labeled complement binds only to antigen-antibody complexes, the nonspecific antibody attached by Fc receptors, which is complement free, does not fluoresce. Western immunoblotting (Chapter 13) is also used for viral antibody detection.

False-positive and false-negative results can occur when testing for virus-specific IgM antibodies. False-positive results occur when rheumatoid factor, an anti-IgG IgM-type globulin, combines with homologous or virus-specific IgG present in the patient specimen. Labeled anti-IgM combines with either bound virus-specific IgM or rheumatoid factor, causing falsely positive fluorescence. False-negative IgM tests occur when high levels of strongly binding homologous IgG antibodies prevent binding of IgM molecules, decreasing or eliminating IgM-specific fluorescence (Figure 43.24). Both problems can be eliminated by testing the IgG-free serum fraction.[28,64,80]

Text continued on p. 674

PROCEDURE 43.3

SHELL VIAL CULTURE FOR CYTOMEGALOVIRUS

PRINCIPLE

The shell vial culture for CMV is used to rapidly detect this virus in clinical specimens. Use of the shell vial culture with urine, blood, and bronchoalveolar lavage fluid specimens has demonstrated equivalent or superior sensitivity compared with conventional cell culture.

SPECIMEN

Acceptable specimens include urine, blood, respiratory secretions and washes, tissue, and cerebrospinal fluid. Specimens are collected, transported, and stored according to standard protocols.

MATERIALS

1. Shell vials, 15 × 45 mm one-dram vial with plastic closure (Baxter Diagnostic, Inc.)

2. Circular coverslips, #1, 12 mm diameter (Bellco Glass, Inc.)

3. MRC-5 cells growing in shell vial (5 to 9 days old)

4. In place of steps 1-3, use MRC-5 shell vials purchased from commercial source such as Biowhittaker Inc., Walkersville, Md. or Viromed Laboratories, Inc., Minneapolis, Minn.

5. Centrifuge; sterile 15 ml conical centrifuge tubes

6. Sterile 1 and 5 ml pipettes and cotton-plugged sterile Pasteur pipettes

7. pH paper (3 to 9 pH)

8. Maintenance medium

9. CMV positive control (strain AD 169)

10. Negative control (use maintenance medium with no virus)

11. Anti-CMV early nuclear protein monoclonal antibody

12. FITC-labeled anti-mouse IgG conjugate

13. Phosphate buffered saline (PBS)

14. Methanol, forceps, buffered glycerol mounting medium, and glass microscope slides

METHODS

1. Wear latex gloves and laboratory coat when handling patient specimens. Use biological safety cabinet whenever tube containing specimen is open. Use biocontainment safety covers during centrifugation procedures.

2. Specimen preparation
   a. Urines
      (1) Using a 5 ml pipette, pipette 2 ml of urine into sterile centrifuge tube.
      (2) Using a 1 ml pipette, add 0.5 ml of viral transport medium containing antimicrobials.
      (3) Centrifuge at $1000 \times g$ for 10 minutes at 2° C to 8° C.
      (4) Aspirate supernatant with cotton-plugged sterile Pasteur pipette and transfer to a sterile test tube.
      (5) Adjust pH to 7.0 with sterile cold 0.1N HCl or 0.1 N NaOH. Determine pH with pH paper
      (6) Store at 2° C to 8° C until inoculation into shell vial cultures.
   b. Blood
      (1) Refer to leukocyte separation by Polymorphprep (see Procedure 43.1.). Store leukocytes at 2° C to 8° C until inoculation into shell vial cultures.
   c. Respiratory secretions and washes
      (1) Process as for urines. pH adjustment is not necessary.
      (2) Specimens received in viral transport medium do not require further preparation.
      (3) Store all specimens at 2° C to 8° C until inoculation into shell vial culture.
   d. Tissue
      (1) Mince and grind 5 mm³ piece of tissue gently in approximately 5 ml of viral transport medium. This results in a 10% suspension.
      (2) Centrifuge at $1000 \times g$ for 10 minutes at 2° C to 8° C.
      (3) Aspirate supernatant with cotton-plugged sterile Pasteur pipette and transfer to small sterile test tube.
      (4) Store at 2° C to 8° C until inoculation into shell vial culture.
   e. Cerebrospinal fluid
      (1) CSF and other body fluids are inoculated into shell vial culture undiluted.
      (2) Specimens with less than sufficient volume for proper inoculation (≤ 0.5 ml) should be diluted with enough viral transport medium to inoculate shell vial cultures (0.5 to 0.6 ml).
      (3) Specimens received in viral transport medium are ready to inoculate.
      (4) Store at 2° C to 8° C until inoculation into shell vial culture.

PROCEDURE 43.3

SHELL VIAL CULTURE FOR CYTOMEGALOVIRUS—cont'd

3. Inoculation and incubation of shell vial cultures
   a. Aspirate maintenance medium from a sufficient number of shell vials for all specimens, including one positive and one negative control.
   b. Using a new sterile 1 ml pipette for each specimen, inoculate shell vials as follows:

Specimen	No. Vials	Inoculum/Vial
Urine	1	0.25 ml
Respiratory	1	0.1 ml
Blood	3	all 0.25 ml
All other specimens	2	0.1 and 0.25 ml
Negative control	1	0.1 ml
Positive control	1	0.1 ml

   Inoculate all specimens before inoculating controls. A negative and positive control are used with each batch of shell vial cultures. Inoculate negative control before positive.
   c. Cap vials with sterile plastic lids and label.
   d. Centrifuge vials at $700 \times g$ for 40 minutes at 35° C.
   e. Using a 5 ml pipette, add 2 ml of maintenance medium to each shell vial.
   f. Recap vials and incubate at 36° C ± 1° C for 16 to 24 hours.

4. Staining shell vial coverslips
   a. Wear latex gloves and perform test in laminar flow hood.
   b. Remove lids and aspirate media from shell vials.
   c. Wash coverslips with 1 ml of PBS. Aspirate. Wash again with 1 ml PBS and let stand for five minutes. Aspirate. Additional PBS washes may be necessary for blood cultures, to remove blood cells from cell sheet. If necessary, check cell monolayer using the microscope.
   d. Fix coverslips in 1 ml cold methanol for 10 minutes.
   e. Aspirate methanol. Coverslips should be allowed to dry, but not overdry. A white clearing across coverslips is apparent at the right amount of dryness. Overdrying causes nonspecific fluorescence around the edge. Underdrying causes uneven staining.
   f. Add 150 µl monoclonal antibody, at current working dilution, to each vial by dropping antibody into the middle of the coverslip. Replace shell vial lids and gently rotate to spread stain over entire coverslip. Incubate at 35° C to 37° C for 30 minutes.
   g. After staining, add 1 ml PBS to each vial (do not remove antibody first). Mix. Aspirate entire volume of fluid from vial.
   h. Wash twice with 1 ml PBS. Let the second wash remain on coverslip for five minutes. Aspirate PBS, do not allow coverslips to dry.
   i. Add 150 µl of anti-immunoglobulin conjugate (FITC), at current working dilution, to each vial by dropping antibody onto the middle of the coverslip. Spread as Step 4, f. Incubate at 35° C to 37° C for 30 minutes.
   j. Repeat wash steps g and h. Wash coverslips an additional time with 1 ml distilled/deionized water for 2 minutes.
   k. Using small forceps, remove coverslips from vials and place cell side down onto drop of mounting medium. Wipe tip of forceps with alcohol-soaked gauze between each coverslip.

5. Reading and interpretation of shell vial coverslips
   a. Read entire coverslip using the 20× fluorescence microscope objective. The nuclei of CMV-infected cells exhibit an apple-green fluorescence that may vary in intensity (Figure 43.20). Infected nuclei may be distinguished from artifacts by their regular oval shape and evenly stained, slightly granular appearance. Occasionally the nuclei may appear rounded. The background should stain a dull red or a pale green (dull red if Evans blue counterstain is used with conjugate; pale green if no counterstain is used).

6. Quality control
   a. The negative control coverslip should have no fluorescing nuclei typical of a CMV-infected cell.
   b. The positive control coverslip should have multiple fluorescing nuclei typical of CMV-infected cells.

PROCEDURE 43.4

CULTURE CONFIRMATION BY FLUORESCENT ANTIBODY STAINING

PRINCIPLE

Many viruses can be identified presumptively in cell culture by observing CPE. Definitive identification, or in some cases, typing, of a virus in cell culture can be accomplished by staining a smear of infected cells with a virus-specific antibody conjugate. Results are easier to interpret if there is at least 2+ CPE before collecting cells for staining.

MATERIALS

1. Virus-specific antibody conjugate

2. Phosphate buffered saline (PBS)

3. Glass slides with circled staining areas and coverslips

4. 25 μl pipettes and tips

5. Control slides with homologous virus for positive staining and nonhomologous virus for negative staining

6. Mounting fluid (25% buffered glycerol in PBS)

7. Sterile glass beads

8. Centrifuge

9. Pasteur pipettes

SPECIMEN

Cells can be collected from the surface of the glass test tube by scraping with a 1 ml pipette, adding 3 to 4 sterile glass beads and disrupting by vortexing, or detaching cells with trypsin. Scraping is best to use if there is less than 2+ CPE. Areas of CPE can be circled with a marker on the outside of the culture tube. If only these areas are scraped on the inside, the relative number of virus-infected cells will be greater and detection will be easier in the stained smear. Although trypsin treatment is the traditional method for releasing cells, use of glass beads during vortexing is simple and reliable. The glass bead method will be described.

1. Add 3 to 4 sterile glass beads to the cell culture tube. Vortex vigorously for 15 to 30 seconds. Examine microscopically to ensure that the cells are detached from the glass. Vortex a second time if cells remain on glass.

2. Pellet cells by centrifugation at $900 \times g$ for 10 minutes. Carefully remove supernatant.

3. Wash cells by re-suspending the pellet in PBS and repeat centrifugation.

4. Remove supernatant and re-suspend in 0.05 to 0.1 ml of PBS.

5. Place 25 μl aliquot of the cell suspension onto two wells on clean labeled slides. Prepare enough smears to test with antisera for all viruses under consideration.

6. Air dry at temperatures no higher than 30° C and fix in acetone for 10 minutes. Slides may be stored at −70° C.

Determine working dilution of fluorescent conjugate

1. Prepare serial doubling dilutions of the stock conjugate in PBS.

2. Test each conjugate dilution for its ability to stain smears of homologous and nonhomologous virus-infected cells, as well as smears of normal uninfected cells.

3. The working dilution of the antiserum is the highest dilution producing 3-4+ fluorescence of homologous virus, ≤1+ fluorescence of nonhomologous virus, and no fluorescence of uninfected cells.

METHOD

1. Apply working dilution of conjugate to circled areas on slide. Cover entire circled area (approximately 25 μl of conjugate per circle).

2. Place slides in a moist chamber and cover the chamber to minimize evaporation. Incubate slides at 35° C for 20 minutes. Be careful not to tilt slides, since conjugates will run together. Do not allow conjugate to dry on the slide. This will cause nonspecific staining.

3. Tip slides to drain conjugate from smear. Remove excess reagent with running PBS, followed by a 5-minute PBS wash in a Coplin jar. Agitate slide occasionally during wash.

4. Rinse slides in distilled water to remove PBS salt crystals.

5. Allow slides to air dry at room temperature. Protect slides from bright light.

6. Apply mounting fluid and coverslip.

7. Examine with a fluorescence microscope containing appropriate filters for FITC staining (490 nm UV needed). Use 20× and 40× objectives.

INTERPRETATION OF RESULTS

1. Positive staining cells are characterized by ≥1+ apple-green fluorescent intranuclear and/or cytoplasmic granules (see Figure 43.12).

PROCEDURE 43.4

CULTURE CONFIRMATION BY FLUORESCENT ANTIBODY STAINING—cont'd

2. Negative staining cells are characterized by no or <1+ fluorescence. Uninfected cells may stain red if Evans blue counterstain is used.

QUALITY CONTROL

1. Positive and negative controls should be used each time the stain is performed.

2. The positive control (homologous virus) should exhibit ≥2+ staining.

3. The negative control (nonhomologous virus) should exhibit <1+ staining.

4. If control results are different than expected, stain should be repeated.

PROCEDURE 43.5

HEMADSORPTION OF PRIMARY MONKEY KIDNEY MONOLAYERS TO DETECT INFLUENZA, PARAINFLUENZA, AND MUMPS VIRUSES

PRINCIPLE

Hemadsorption of cell cultures is used to detect or confirm the presence of orthomyxoviruses (influenza A, B, and C) and some paramyxoviruses (parainfluenza 1-4 and mumps). Respiratory syncytial virus is not detected by this method. Cell cultures infected with these viruses may not show CPE; however, infected cells do acquire the ability to hemadsorb guinea pig red blood cells to their outer cell membranes. In addition, the presence of viral hemagglutinin in the culture medium, after release from the cell, can also cause agglutination of the RBCs in the fluid medium.

Hemadsorption is performed on various days during culture incubation depending on the time of year and the viruses suspected. Influenza virus detection is performed on specimens from adult and pediatric patients during the influenza season (November through March). Early hemadsorption (day 2 or 3) decreases time to detection of positive influenza cultures.

Detection of parainfluenza viruses from pediatric patient specimens is performed year-round. Hemadsorption produced by parainfluenza viruses generally occurs later than that produced by influenza. Hemadsorption to detect mumps virus is performed in most clinical settings only if this virus is suspected by the clinician.

SPECIMEN

A culture exhibiting CPE should be hemadsorbed as soon as possible to expedite reporting of a positive result. Perform hemadsorption on CPE-negative cell cultures of respiratory tract specimens or of specimens for mumps virus detection according to the following schedule:

Culture Type	Perform Hemadsorption on Day:
Influenza virus suspected	2 and 7
Parinfluenza virus suspected	3 and 10
Mumps virus suspected	3 and 10

MATERIALS

1. Guinea pig red blood cells. Store in buffer in which cells were received at 2° C to 8° C. Cells expire 10 days after date drawn from animal.

2. Sterile, 10× phosphate buffered saline. Prepare working 1× phosphate buffered saline (PBS) by diluting with distilled water. Filter (0.22 μm) sterilize. Store at 2° C to 8° C.

3. Viral culture maintenance medium with antibiotics.

4. Positive control virus (parainfluenza virus) in tube of current lot of PMK cells.

5. Negative control is uninoculated tube of current lot of PMK cells.

6. Sterile, conical centrifuge tubes.

7. Sterile pipettes.

8. Repeater Eppendorff pipettes.

9. Centrifuge.

Continued on p. 672.

PROCEDURE 43.5

HEMADSORPTION OF PRIMARY MONKEY KIDNEY MONOLAYERS TO DETECT INFLUENZA, PARAINFLUENZA, AND MUMPS VIRUSES—cont'd

PROCEDURE

1. Preparation of guinea pig RBCs
   a. Transfer guinea pig cells to 15 ml conical centrifuge tube. Centrifuge for 5 minutes at $500 \times g$ ($2°$ C to $8°$ C). Aspirate supernatant and buffy coat.
   b. Resuspend RBCs in cold ($2°$ C to $8°$ C) PBS. Wash in cold PBS until supernatant is clear (2 to 3 times)
   c. Aspirate PBS after last wash. Measure remaining volume of cells using graduations on centrifuge tube. Add a sufficient amount of PBS to make a 10% cell solution.
   d. Hemadsorption test is performed using a 0.4% guinea pig cell solution prepared with maintenance medium. Depending on the number of tubes to be tested (0.2 ml/tube), make as follows: 0.2 ml 10% RBCs + 4.8 ml medium (25 tubes) or 0.4 ml 10% RBCs + 9.6 ml medium (50 tubes).

2. Hemadsorption
   a. Wear latex gloves and laboratory coat during procedure. Perform in biological safety cabinet.
   b. Aspirate mainenance medium from all patient and control tubes to be tested.
   c. Add 0.2 ml of the 0.4% RBC suspension to each tube. Be careful not to cross-contaminate tubes. Positive and negative controls should be tested last.
   d. Gently tilt stationary rack from side to side to ensure RBC contact with cell monolayer. Place stationary rack in refrigerator ($2°$ C to $8°$ C) for 30 minutes. Tubes should be placed horizontally so that the erythrocyte suspension covers the monolayer.
   e. Invert tubes quickly, following incubation, to dislodge RBCs lying on the cell sheet. Examine cell culture tubes microscopically with 4× objective for RBCs that adhere to the monolayer (see Figure 43.18, *G*) Culture fluids should be inspected for hemagglutination.

INTERPRETATION

1. Negative: No hemadsorption of guinea pig cells and no hemagglutination of cells in fluid medium. RBCs float freely in medium.
   Positive: Hemadsorption of guinea pig cells to infected PMK cells or hemagglutination of RBCs in fluid medium.
   Hemadsorption is graded on a scale of 1-4+.

   1+ = <25% of cell sheet showing hemadsorption.

   2+ = 25-50% of cell sheet showing hemadsorption.

   3+ = 51-75% of cell sheet showing hemadsorption.

   4+ = 76-100% of cell sheet showing hemadsorption.

2. Viruses that both hemadsorb and hemagglutinate guinea pig erythrocytes are usually influenza virus. Mumps and parainfluenza viruses generally produce only hemadsorption. Identification of virus-causing hemadsorption should be confirmed using virus-specific antibody conjugates. Certain simian viruses that may contaminate the PMK cells hemadsorb at $4°$ C (see following section on quality control). Examine all tubes as soon as possible after removal from the refrigerator because the neuraminidase of the myxoviruses is active at room temperature and will destroy hemagglutinins if tubes are left for extended periods at this temperature.

QUALITY CONTROL

1. Positive control: Parainfluenza virus should exhibit ≥2+ hemadsorption.

2. Negative control: Uninoculated PMK cells should exhibit no hemadsorption. If hemadsorption attributed to an endogenous simian virus occurs, the patient must exhibit more hemadsorption than the negative control to be considered positive.

PROCEDURE 43.6

VIRUS NEUTRALIZATION TEST

PRINCIPLE

The virus neutralization test is based on the ability of specific immune sera to neutralize 100 $TCID_{50}$ (median tissue culture infective dose) of virus and to inhibit its growth in cell culture, as measured by preventing CPE. Neutralization is used to identify or serotype a viral isolate. Neutralization, as a means of viral identification in the clinical virology laboratory, has been largely replaced by the use of fluorescent antibody staining. Neutralization is still useful for confirming the identification of viral groups for which there is no fluorescent-labeled antisera available, such as the rhinoviruses. It is also useful for serotyping viruses containing many serotypes, such as the adenoviruses. Adenovirus neutralization is used as an example of the virus neutralization test procedure.

SPECIMEN

Harvest culture fluids of virus to be neutralized when culture shows 3-4+ CPE. The infectivity titer of the fluid can be roughly determined by selecting arbitrarily the dilution that is most likely to approximate 100 $TCID_{50}$. A 1:10 dilution of a culture showing 3-4+ CPE is usually adequate for adenovirus; whereas, for enteroviruses a 1:100 or 1:1000 dilution should be used. To check this estimate of virus titer, make serial log dilutions of the virus in maintenance medium and inoculate 0.1 ml of each dilution (10^0 to 10^{-4}) to tissue culture tubes. The approximate virus titer can be calculated for the dilutions that exhibit CPE.

MATERIALS

1. Appropriate antisera
2. Appropriate cell culture tubes containing unknown and control viruses showing 3-4+ CPE
3. Water bath (56° C)
4. Maintenance medium
5. Sterile tubes for virus and antisera dilution

METHODS

1. Preparation of neutralization antisera
   a. Antisera should be inactivated at 56° C for 30 minutes before determining working dilution or before use, if working dilution has already been determined.
   b. The working dilution of antisera should contain a minimum of 20 standard antibody units. If this titration has not been done by the supplier, it can be performed by using 100 $TCID_{50}$ of a reference virus per 0.1 ml and an equal volume of serial doubling dilutions of the inactivated antiserum. The highest serum dilution that neutralized 100 $TCID_{50}$ of the reference virus is 1 standard antibody unit. Twenty units are used in the test.
   c. Dilute to the working dilution with maintenance medium.
2. Neutralization
   a. Mix 0.2 ml of diluted virus (approximately 100 $TCID_{50}$ per 0.1 ml) with an equal volume of each antiserum at its working dilution (20 antibody units).
   b. Allow the serum-virus mixture to incubate for 1 hour at room temperature.
   c. Use 0.2 ml of each serum-virus mixture to inoculate duplicate tubes of appropriate cell cultures (HEp-2 for adenovirus).
   d. Incubate all tubes at 35° C and examine daily for 7 days.

INTERPRETATION OF NEUTRALIZATION

1. Inhibition of CPE by specific antisera when control tubes show ≥2+ CPE constitutes identification of the serotype.
2. Complete inhibition is not always possible. There should be at least a 2+ difference in CPE between the positive control and the inhibited serotype.

QUALITY CONTROL

1. Positive control: Duplicate set of tubes inoculated with virus of homologous serotype and antiserum should show no CPE. Neutralization has occurred.
2. Negative control: Duplicate set of tubes inoculated with 0.2 ml of a 1:1 mixture of maintenance medium and diluted virus should show ≥2+ CPE. No neutralization. An uninoculated tube can serve as a cell culture control.

PROCEDURE 43.7

ACID LABILITY TEST TO DIFFERENTIATE RHINOVIRUS FROM ENTEROVIRUS

PRINCIPLE

Rhinoviruses and enteroviruses may have similar patterns of CPE in cell culture. Both can be detected in upper respiratory tract specimens. Sensitivity to an acidic environment differentiates these two groups of viruses. Rhinoviruses are labile at pH 3, whereas enteroviruses are stable at this pH.

SPECIMEN

1. 0.2 ml suspension of test virus

2. 0.2 ml suspension of stock rhinovirus

3. 0.2 ml suspension of stock enterovirus

MATERIALS

1. EMEM prepared without sodium bicarbonate. pH should be approximately 3. Test with pH meter and adjust with 2 N HCl if necessary.

2. EMEM prepared the usual way with sodium bicarbonate. pH should be 7.2 to 7.4. Check with pH meter and adjust if necessary.

3. Sterile pipettes.

4. Human diploid fibroblast cells.

METHODS

1. The neutralization procedure is performed on each unknown virus and the rhinovirus and enterovirus controls.

2. For each virus to be tested, use a 1 ml pipette to add 0.1 ml of the suspension of virus to 0.9 ml of EMEM pH 3.

3. For each virus to be tested, use a 1 ml pipette to add 0.1 ml of the suspension of virus to 0.9 ml of EMEM pH 7.

4. Using a 1 ml pipette, inoculate 0.1 ml of each of the EMEM pH 3 virus suspensions to a separate human diploid fibroblast cell culture tube.

5. Inoculate a second set of culture tubes with 0.1 ml of each of the EMEM pH 7 virus suspensions.

6. Incubate cell culture tubes at 35° C to 37° C. Incubate an uninoculated tube of the same lot of fibroblast cells as a cell culture control. Observe daily for 7 to 10 days for CPE.

INTERPRETATION

1. Control viruses must show appropriate growth patterns. See section on quality control.

2. Presence of CPE in cells inoculated with EMEM pH 3 approximately equal to that in EMEM pH 7 indicates enterovirus.

3. Absence of CPE in EMEM pH 3 and presence of CPE in EMEM pH 7 indicates rhinovirus.

QUALITY CONTROL

1. Rhinovirus control: Presence of CPE in EMEM pH 7. Absence of CPE in EMEM pH 3.

2. Enterovirus control: Presence of CPE in EMEM pH 7 and EMEM pH 3.

PRESERVATION AND STORAGE OF VIRUSES

Clinical virology laboratories must have a method for storing and retrieving viruses. Isolates should be kept as control strains and, in rare instances, for epidemiological investigations. Public health laboratories may use current enterovirus or influenza virus strains for typing. Viruses can be stored by freezing at −70° C or in liquid nitrogen. Freezing at −70° C is more practical for clinical laboratories. A procedure for the preservation and storage of viruses by freezing is described in Procedure 43.8.[37]

USE OF VIRAL DIAGNOSTIC TESTS

In the past, the laboratory diagnosis of viral infections was made using a comprehensive viral culture and serology batteries, both designed to detect any of the many viruses responsible for a disease syndrome. As reagents become more plentiful and expensive, comprehensive approaches to the diagnosis of disease become prohibitive. Virology laboratories should have a menu of individual culture and serology tests and an algorithm for their use, rather than one comprehensive viral culture or serology test battery for use in all situations. Tables 43.13 and 43.14 represent virus detection and serology tests that might be useful in a community clinical virology laboratory.[78]

PROCESSING VIRUS SPECIMENS BASED ON SPECIMEN TYPE

An algorithm for the use of virus detection tests can be based on the type of specimen received or

substrate

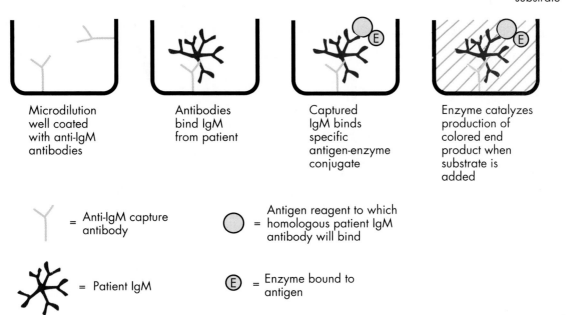

Microdilution
well coated
with anti-IgM
antibodies

Antibodies
bind IgM
from patient

Captured
IgM binds
specific
antigen-enzyme
conjugate

Enzyme catalyzes
production of
colored end
product when
substrate is
added

Y = Anti-IgM capture antibody

○ = Antigen reagent to which homologous patient IgM antibody will bind

✳ = Patient IgM

Ⓔ = Enzyme bound to antigen

FIGURE 43.21

Illustration of antibody-capture ELISA for IgM.

the specific virus suspected. Figure 43.25 summarizes a "specimen" algorithm. Most laboratories receive specimens with little or no clinical data and no specific virus indicated. The Figure 43.25 algorithm is designed for use with these specimens.[1,76]

All lip and genital specimens should be cultured only for herpes simplex virus. Disease by other viruses at lip and external genital sites is unusual and detection of these agents, such as VZV, should be prompted by a special request. Urine specimens require a CMV detection test. This is best done using the shell vial assay. Stool specimens from infants (5 years or younger) in North America should be tested for rotavirus during the fall, winter, and spring. Enteric adenoviruses (serotypes 40 and 41) cause diarrhea in young children and infants throughout the year.[46] Adenovirus gastroenteritis appears to be more common in some geographical areas. Routine testing is necessary only in locations where disease is known to occur. Stool from adults, children, and infants should be tested for enterovirus in summer and fall months as an aid in the diagnosis of aseptic meningitis. Stool for enterovirus should be tested in conjunction with throat and CSF specimens when possible. Respiratory specimens should be divided according to the patient's age and underlying medical condition. Immunocom-

promised patients require a comprehensive virus detection test consisting of cell culture, and shell vial for CMV and other agents when available. Immunocompetent adults should have an influenza virus culture during influenza season (November through April in most areas). Children (younger than 10 years) are susceptible to serious infection caused by influenza, parainfluenza, respiratory syncytial, and adenoviruses and need a full respiratory virus culture. Infants (younger than 2 years) are especially vulnerable to RSV bronchiolitis, which requires hospitalization and occasionally antiviral therapy with ribavirin. A rapid nonculture RSV detection test, such as fluorescent antibody staining or enzyme immunoassay, is appropriate in these situations. Respiratory specimens from newborns with the possibility of congenital or perinatal viral disease should receive a comprehensive virus culture and shell vial for CMV. Blood for viral culture might contain CMV, VZV, adenovirus, or enteroviruses; however, CMV is by far the most common and the one agent whose detection and reporting is clinically useful information. Therefore blood for viral culture should be processed for CMV only, unless other agents are mentioned by the requesting physician. All specimens from immunocompromised hosts and tissues or fluids from presumably sterile sites should be processed for compre-

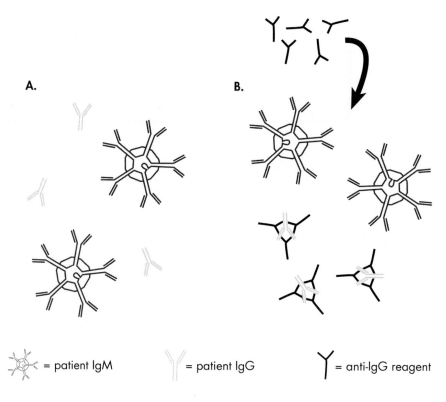

A.

B.

= patient IgM   = patient IgG   = anti-IgG reagent

FIGURE 43.22

Removal of IgG antibodies from human serum. **A,** Patient IgG and IgM in serum. **B,** Patient IgG inactivated by addition of anti-IgG, which binds and precipitates patient IgG. IgM-specific testing can now be performed free of IgG interference.

FIGURE 43.23

Illustration of ACIF technique for antibody detection.

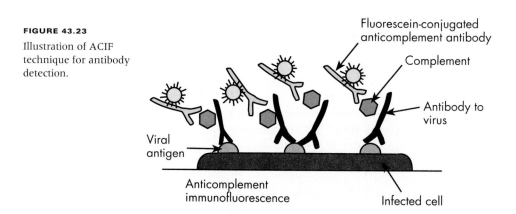

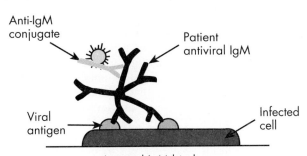

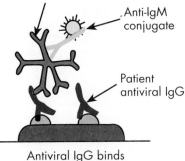

TRUE-POSITIVE REACTION

Antiviral IgM binds
to antigen; anti-IgM
conjugate binds to IgM.

Antiviral IgG binds
to antigen; Rheumatoid
factor binds to IgG;
anti-IgM conjugate binds
to Rheumatoid factor.

FALSE-NEGATIVE REACTION

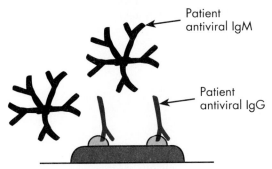

Antiviral IgG binds
to antigen, blocking
all potential antiviral
IgM binding sites.

FIGURE 43.24 FALSE-NEGATIVE REACTION
Illustration of false-positive and false-negative IgM test resulting from the presence of rheumatoid factor
and excess IgG.

TABLE 43.13

MENU OF VIRUS DETECTION TESTS

TEST	FREQUENCY OF REQUEST (%)*
HSV detection (outpatient)	9
Influenza virus detection	2
Enterovirus detection	5
Rapid CMV detection (shell vial)	8
Rotavirus detection	8
RSV detection	14
Pediatric respiratory virus detection	10
Comprehensive virus detection	44
Total	**100**

From Children's Hospital Medical Center of Akron,
Akron, Ohio.
*Frequency of request expressed as percent of all tests.

hensive virus detection. This algorithm represents
a starting point for clinical virology laboratories.
Processing of specimens should be continuously
modified to match the needs of local physicians
and endemic viral diseases.

**PROCESSING VIRUS SPECIMENS BASED
ON REQUESTS FOR SPECIFIC VIRUSES**

Frequently detecting only one virus in a specimen
is advised because the physician recognizes the
patient's clinical condition or that virus is the sole
agent whose detection is clinically useful. Individual viruses are detected as follows.

Arboviruses. Serologic tests for **arboviruses**
are offered by State Public Health Laboratories or
by some commercial laboratories. Diagnosis of arbovirus encephalitis, such as Eastern, Western,
Venezuelan, and St. Louis, and that caused by
California encephalitis and LaCrosse viruses, requires detection of a rise in titer of antibody in
acute and convalescent serum specimens. Detection of virus-specific IgM in cerebrospinal fluid

PROCEDURE 43.8

PRESERVATION AND STORAGE OF VIRUSES BY FREEZING

PRINCIPLE

Viral isolates should be saved for use as quality control strains or epidemiological investigation. Maximal recovery of virus is accomplished by collecting cells and fluid from infected culture tubes.

MATERIALS

1. Sterile glass beads
2. Dimethyl sulfoxide (DMSO)
3. Dry ice and acetone
4. Freezer tubes

METHODS

1. Viruses other than CMV and VZV.
   a. Select tubes showing ≥2+ CPE or hemadsorption.

RBCs may be eluted following hemadsorption by placing tubes in a 37° C water bath for 15 minutes. Wash to remove RBCs.
   b. Add 3 sterile glass beads to cell culture tube. Tighten cap. Vortex vigorously behind shield for 30 seconds. Examine microscopically to ensure that cells have been disrupted and removed from glass surface.
   c. Add cells and medium to freezer tube. Snap freeze by swirling the tube in a slurry of dry ice and acetone.
   d. Store at −70° C.

2. CMV and VZV
   a. Use tubes showing ≥3+ CPE. Harvest as noted previously except add 10% DMSO and glass beads. For example, add 0.2 ml of DMSO to 2 ml of medium in cell culture tube. Mix and proceed as described previously.

3. Thaw in a 37° C water bath. Inoculate to appropriate cell culture.

TABLE 43.14

MENU OF VIRAL SEROLOGY TESTS

TEST	FREQUENCY OF REQUEST (%)*
Arbovirus antibodies	1
CMV immune status	28
CMV IgM antibodies	9
CMV IgG antibodies	9
EBV IgM antibodies	11
EBV IgG antibodies	9
HSV IgM antibodies	3
HSV IgG antibodies	2
Rubella immune status	11
Measles immune status	15
Varicella immune status	2
Total	**100**

From Children's Hospital Medical Center of Akron, Akron Ohio.
*Frequency of request expressed as percent of all tests.

has been used for the diagnosis of LaCrosse virus disease.[5,6,7,53]

Cytomegalovirus. CMV can be detected in clinical specimens using conventional cell culture, shell vial assay, or antigenemia immunoassay.[25,31] CMV produces CPE in diploid fibroblast cells in 3 to 28 days, averaging 7 days. Shell vial for CMV has a sensitivity equivalent to conventional cell culture but takes only 16 hours to complete. The antigenemia immunoassay uses monoclonal antibody in an indirect immunoperoxidase stain to detect CMV protein in peripheral blood leukocytes. The antigenemia assay requires only 3 to 5 hours to complete; however, the clinical utility of a positive result must be established before it can replace culture or shell vial.

Enteroviruses. Enteroviruses can be detected only by conventional cell culture.[15,61] Although most enteroviruses grow in primary monkey kidney cells, some strains grow faster in diploid fibroblast, buffalo green monkey kidney, or rhabdomyosarcoma cell lines.[42] Most enteroviruses are detected during June through December. Mean time to detection in cell culture is 4 days. An efficient cell culture procedure is summarized in Figure 43.26

FIGURE 43.25

Algorithm for processing virus specimens based on specimen type.

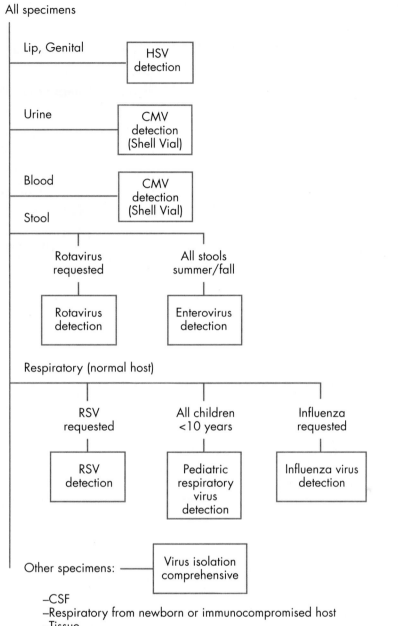

Epstein-Barr virus. Serology tests are used to help diagnose EBV-associated disease, including infectious mononucleosis.[49,55] Isolation of EBV (in cultured B-lymphocytes) is not routinely performed in clinical laboratories. Detection of heterophile antibody, antibody to viral capsid antigen, early antigen, and Epstein-Barr nuclear antigen is used as indicated in Figure 43.27 and Table 43.15.

Hepatitis viruses. Disease or asymptomatic carriage caused by hepatitis A, B, C, and D viruses is detected using serology tests (Table 43.16).[13,21,36] Hepatitis B virus requires detection

of antigen and antibody to multiple antigens in order to classify disease type (Figure 43.28 and Table 43.17).

Herpes simplex virus. HSV grows rapidly in most cell lines.[11,27,63] MRC-5 or mink lung fibroblast cell lines are recommended. Fifty percent of genital HSV isolates will be detected within 1 day and 100% within 3 to 5 days. Cultures should be examined daily and finalized if negative after 5 days of incubation.

Human immunodeficiency virus and other retroviruses. HIV-1 and HIV-2, and HTLV-1 immune status tests (ELISA) are used to screen

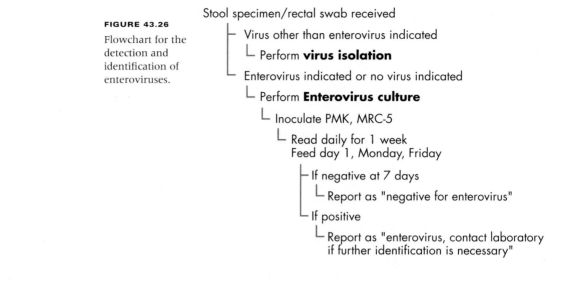

FIGURE 43.26

Flowchart for the detection and identification of enteroviruses.

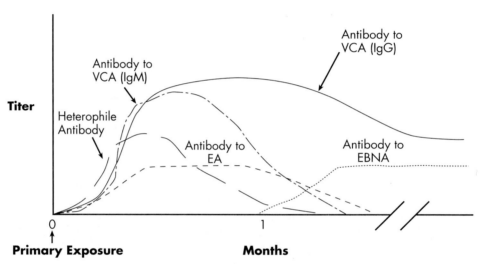

FIGURE 43.27

Illustration of time-course of immune response to EBV infection.

units of blood for donation.[3,10,44,74] Units with a positive HIV-1 screening test are confirmed as positive with the western blot test. The western blot identifies antibody specific for several HIV antigens (Figure 43.29). Antibody to HIV p24 and either gp41 or gp160 most often confirms HIV infection. Units containing antibody to HIV or HTLV are discarded because of the risk of transferring latent virus to the recipient. Diagnosis of infection may require detection of virus or viral antigen.[4] PCR testing for HIV antigen is available and useful for the newborn population, who may have maternal HIV antibody confounding interpretation of serology tests, and for all patients who may not produce detectable antibody for months following primary infection.[54]

Influenza A and B viruses. Influenza viruses cause sporadic outbreaks each winter, epi-

demics every few years, and pandemics at greater intervals. These viruses are detected by using conventional cell culture, shell vial culture, membrane EIA, or direct staining of respiratory tract secretions.[26,62,67] Conventional cell culture using primary monkey kidney cells is superior. Median time to detection, using hemadsorption at day 2 or 3, is 3 days.[51] FA staining is used for confirmation and typing of isolates as A or B. All influenza viruses will be detected by the end of 1 week's incubation. Detection of influenza virus is outlined in Figure 43.30.

Pediatric respiratory viruses. Influenza, parainfluenza, respiratory syncytial, and adenoviruses should be sought in specimens from hospitalized infants and young children (less than 10 years).[32] Influenza and parainfluenza viruses are detected in PMK cells by CPE and

TABLE 43.15

INTERPRETATION OF EPSTEIN-BARR VIRUS SEROLOGY TEST RESULTS

CLINICAL SITUATION	HETEROPHILE ANTIBODY	IgG-VCA	IgM-VCA	EA	EBNA
No past infection	usually –	–	–	–	–
Acute infection	usually +	+	+	+	–
Convalescence phase	+/–	+	+ or –	+ or –	+
Past infection	usually –	+	–	– or wk+	+
Chronic or reactivation	not useful	+	–	+	+

TABLE 43.16

SEROLOGIC TESTS FOR HEPATITIS VIRUSES

DISEASE	VIRUS	SEROLOGY TEST
Hepatitis A	Enterovirus 72	Hepatitis A virus–specific IgG and IgM
Hepatitis B	Hepadnavirus	Hepatitis B surface-antigen (HBsAg)
		Hepatitis B e-antigen (HBeAg)
		Anti-HBsAg
		Anti-HBeAg
		Anti-HB core antigen (Anti-HBcAg)
Hepatitis C	Flavivirus	Antibody to hepatitis C virus
Hepatitis D	Delta agent (hepatitis D virus)	Antibody to delta agent
		Delta antigen

FIGURE 43.28

Illustration of time-course of antigenemia and immune response in a patient who recovers from acute hepatitis B infection.

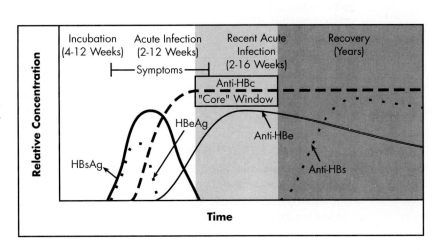

hemadsorption. Fluorescent staining is used for confirmation and typing. Adenovirus and RSV are detected in HEp-2 cell culture and confirmed, if necessary, by fluorescence staining. Specimens from infants or young children sent for RSV detection should be tested by a rapid nonculture RSV test.[35,43,79] Fluorescent antibody staining, conducted by experienced personnel, is equivalent to culture in sensitivity and is best used for single specimens or small batches. Conventional ELISA is also very accurate and is recommended for large batches of specimens. Membrane ELISA is less sensitive than culture but very rapid (less than 1 hour). It is convenient for stat testing. Negative membrane ELISA results should be confirmed by an alternate method. Figure 43.31 describes procedures for the detection of pediatric respiratory viruses.

TABLE 43.17

SEROLOGIC PROFILES FOLLOWING TYPICAL HEPATITIS B VIRUS (HBV) INFECTION

HBsAg	ANTI-HBsAg	HBeAg	ANTI-HBeAg	ANTI-HBcAg	MOST LIKELY INTERPRETATION
−	−	−	−	−	No (or very early) exposure to HBV
+	−	+/−	−	−	Early acute HB
+	−	+	−	+	Acute or chronic HB
+	−	−	+	+	Chronic HBV carrier state
−	−	−	+	+	Early recovery phase from acute HB
−	+	−	+	+	Recovery from HB with immunity
−	+	−	−	−	Distant HBV infection or HB vaccine

FIGURE 43.29

Western blot detecting specific HIV antibody. Lane 1 is the high positive control, lane 2 the low positive control, lane 3 the negative control, lanes 4-8 are positive sera, and lane 9 is an indeterminate serum. Numbers at left refer to approximate molecular weights of HIV antigens. Sera must react with specific, multiple HIV antigens before they are considered positive. (Western Blot test courtesy Hodinka, R.L., Children's Hospital of Philadelphia.)

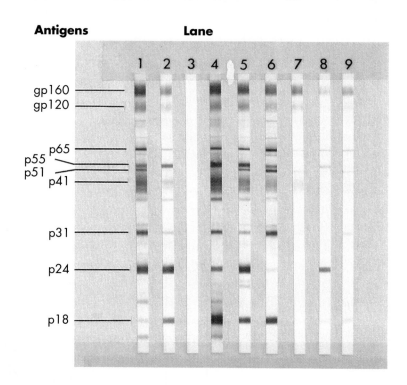

Gastroenteritis viruses. Electron microscopy (EM) can be used to identify all viral agents known to cause gastroenteritis. EM, however, is labor intensive and not broadly available in clinical virology laboratories.[69] Immunoassays for rotaviruses and enteric adenoviruses types 40 and 41 are commercially available.[20,29,47,48,65] Their use was discussed earlier in this chapter.

TORCH testing. TORCH is an acronym for *Toxoplasma*, rubella, cytomegalovirus, and herpes simplex virus.[66] Testing for these agents is appropriate during early pregnancy, because transplacental infection followed by congenital defects can occur, and postnatally because some congenital disease of newborns can be treated to prevent serious permanent damage to the infant. A blan-

ket request for TORCH tests should be avoided whenever possible, especially with specimens from newborns (Chapter 13). Clinical presentation usually is characteristic for one or two of the TORCH agents, and tests only for these etiologies should be pursued. Table 43.18 suggests laboratory tests for the diagnosis of the viral constituents of TORCH.

Varicella-zoster virus. Varicella-zoster virus causes chicken pox (varicella) and shingles (zoster). Varicella (a vesicular eruption) represents the clinical presentation of a primary VZV infection. VZV, a DNA-containing virus, establishes latency in the dorsal nerve root ganglion. Months to years later, during periods of relative immune suppression, VZV reactivates to cause

FIGURE 43.30

Flowchart for the detection and identification of influenza viruses.

Respiratory tract specimen received

├─ Other viruses in addition to influenza virus indicated

└ Perform **virus isolation**

└─ Influenza virus indicated

└ Inoculate PMK

└ Read daily for 1 week
Hemadsorb day 3 and day 7
Feed day 3

├ If negative at 1 week

Report as "negative for influenza virus"

└ If positive

Type by FA and report as "influenza A (or B) virus detected"

FIGURE 43.31

Flowchart for the detection and identification of pediatric respiratory viruses.

Respiratory tract specimen received

├─ RSV indicated (RSV rapid test plus cell culture-PMK and HEp-2)

├ RSV detection positive

└ Report RSV
Discard culture tubes
<1% have second virus

└ RSV detection negative

└ Incubate cell culture tubes

└ Read days 1-7, 10
Hemadsorb day 3 and 10
Feed twice/week

├ If negative at 10 days

└ Report as negative for RSV, adenovirus, influenza, and parainfluenza viruses.

└ If positive

Report as RSV, adenovirus, influenza, or parainfluenza. Type Influenza and parainfluenza if appropriate.

├─ No RSV request

├ Virus other than pediatric respiratory viruses requested

└ Perform **virus isolation**

└ No virus other than pediatric respiratory virus requested

└ Perform virus culture as above for "RSV detection negative"

zoster. Zoster is a modified or limited form of varicella, localized to a specific dermatome, the cutaneous area served by the infected nerve ganglion. Virus is present in the vesicular fluid and in the cells at the base of the vesicle. Material for virus detection should be collected from newly formed vesicles. Once the vesicle has opened and crusted over, detection is unlikely. Virus can be detected by staining cells from the base of the vesicles or by culturing cells and vesicular fluid. The Tzanck test, which is a smear of cells from the base of the vesicle stained by the Giemsa, Pap, or other suitable method, detects typical multinucleated giant cells and inclusions (see Figure 43.8, *A*).[73] The FA test, using FITC labeled anti-VZV antibody, detects virus in infected cells (see Figure 43.12, *C*). Traditionally, diploid fibroblast cell culture (e.g, MRC-5) has been used to detect VZV, which re-

LABORATORY DIAGNOSIS OF THE VIRAL AGENTS OF TORCH INFECTIONS*

VIRUS	CULTURE	SEROLOGY	
		IgM	IgG
CMV	**X**	X	Hold†
HSV	**X**	X	Hold
Rubella	X	**X**	Hold

*Bold indicates preferred test.
†Hold until acute and convalescent can be tested together.

quires up to 28 days before producing visible CPE. The shell vial assay reduces detection time to 48 hours and significantly increases sensitivity, identifying virus that fails to produce CPE in conventional cell culture. A comparison of FA of vesicle cells, conventional cell culture, and shell vial shows FA to be the most sensitive method for diagnosis.[30]

Immune status testing. Immune status tests are many and varied (Table 43.19). Rubella is used to test women of childbearing age. A positive result (presence of antibody) indicates past infection or immunization and implies that congenital infection will not occur during subsequent pregnancies. Absence of antibody implies susceptibility to infection and should prompt rubella vaccination if the woman is not pregnant. Varicella and measles immune status assays are used most commonly to test health care workers. Those with no antibody must avoid diseased patients. CMV immune status is most useful for organ transplant donors and recipients and premature babies hospitalized in newborn intensive care nurseries who are likely to receive blood transfusions. Transplant recipients are susceptible to life-threatening CMV infection. Knowing the CMV status of the donor and recipient enables the physician to better diagnose and treat this disease. Newborns whose mothers were never infected with CMV are susceptible to serious primary CMV infection that can be transmitted in white blood cells during blood transfusion. CMV-negative babies should receive only CMV-negative blood.

Serology panels. Special situations exist in which testing for antibody to individual viruses is less helpful than using a battery of antigens to test for antibody to many viruses. The use of a combination of serologic tests to diagnose a clinical syndrome (Table 43.19) may be useful in situations when the viruses under consideration cannot be cultured, specimens of infected tissue are not available (brain tissue), antiviral agents have been administered, or the patient is convalescing and isolation of virus is unlikely.[12]

INTERPRETATION OF LABORATORY TEST RESULTS

Interpretation of laboratory test results must be based on knowledge of the normal viral flora in the site sampled, the clinical findings, and the epidemiology of viruses. Serologic testing, in addition to virus detection tests, may be needed to support or refute the association of a virus isolate with a disease.

Viruses in tissue and body fluids. In general, the detection of any virus in host tissues, CSF, blood, or vesicular fluid is significant.[14] Recovery of adenovirus or mumps virus in urine is usually diagnostic of disease, as opposed to the culture of CMV, which may merely reflect asymptomatic reactivation. Occasionally enteroviruses are detected in urine as a result of fecal contamination or HSV from symptomatic or asymptomatic infection of the external urogenital tract is recovered in urine. Interpretation of these culture results requires correlation with clinical data. CMV viruria during the first 3 weeks of life establishes a diagnosis of congenital CMV infection, whereas detection at 4 weeks or later suggests intrapartum or postpartum acquisition.

Viruses in the respiratory tract. Detection of measles, mumps, influenza, parainfluenza, and respiratory syncytial viruses is very significant because asymptomatic carriage and prolonged shedding is unusual.[8] Conversely, HSV, CMV, and adenoviruses can be shed in the absence of symptoms for periods ranging from a few days to many months. Adenoviruses are isolated commonly from asymptomatic infants and young children. Simultaneous isolation of these agents from both throat and feces in febrile and respiratory syndromes increases the probability of association with illness. Isolation from throat but not feces has a lesser probability of association, and isolation from feces alone has the least diagnostic significance.

Viruses in the eye. Isolation of adenoviruses, HSV, VZV, and some enteroviruses from diseased cornea and conjunctiva usually establishes the etiology of the infection. Enterovirus

TABLE 43.19

SEROLOGY PANELS AND IMMUNE STATUS TESTS FOR COMMON VIRAL SYNDROMES

SITUATION	VIRUSES UNDER CONSIDERATION	COMMON METHODS OF CHOICE
AIDS	Human immunodeficiency virus (HIV)	EIA, Western Blot
CNS infections	Western equine encephalitis	HI, CF, EIA
	Eastern equine encephalitis	HI, CF, EIA
	California encephalitis virus	HI, CF, EIA
	St. Louis encephalitis virus	HI, CF, EIA
	Lymphocytic choriomeningitis virus	CF, IFA
	Measles	HI, CF
	Mumps	EIA, CF, HI
	Epstein-Barr virus	IFA
	Rabies	IFA, Neut, CF
Exanthems	Measles	HI, CF
	Rubella	HI, RIA, EIA
	Parvovirus	EIA
Vesicular	Herpes simplex virus	IFA, ACIF
	Varicella-zoster virus	IFA, EIA, FAMA, ACIF
Hepatitis A	Hepatitis A virus	EIA, RIA
Hepatitis B	Hepatitis B virus	EIA
Heterophile-negative	Cytomegalovirus	IFA, ACIF, EIA
infectious mononucleosis	Epstein-Barr virus	IFA, EIA
syndromes		
Myocarditis-pericarditis	Group B coxsackievirus, types 1-5	Neut, CF
	Influenza A, B	CF
	Cytomegalovirus	IFA, ACIF, EIA
Respiratory	Influenza A, B	CF
	Respiratory syncytial virus	CF
	Parainfluenza 1-3	CF
	Adenovirus	CF
Serology needed to determine immune status (single serum)		
Rubella		HI, EIA, latex agglutination
Hepatitis B		RIA, EIA
Varicella-zoster		IFA, EIA, FAMA, ACIF
CMV		EIA, latex agglutination, IFA
Measles		EIA, IFA

Neut, neutralizing antibody; *RIA,* radioimmunoassay; *HI,* hemagglutination inhibition; *CF,* complement fixation; *IFA,* indirect immunofluorescence; *EIA,* enzyme immunoassay; *ACIF,* anticomplementary immunofluorescence; *FAMA,* fluorescent antibody to membrane antigen.

type 70 is known to cause a particularly contagious form of viral conjunctivitis.

Detection of enteroviruses. Enteroviruses are most commonly found in asymptomatic infants and children, particularly during the late summer and early fall.[19,81] A knowledge of the relative frequency of virus shedding is extremely helpful in assessing significance of results of throat or stool cultures. Prevalence of enteroviruses in the stools of infants and toddlers may approach 30% during peak periods. Shedding of enteroviruses in the throat is relatively transient, usually 1 to 2 weeks, whereas fecal shedding may last 4 to 16 weeks. Thus isolation of an enterovirus from the throat supports etiology of a clinically compatible illness more than isolation from the feces alone.

Detection of VZV and HSV. The detection of VZV is always significant. Asymptomatic shedding does not appear to occur with this virus as it does with other herpesviruses. The detection of HSV from cutaneous or mucocutaneous vesicles is also significant, implying primary or reactivation disease. HSV may be detected in respiratory secretions during asymptomatic "stress" reactivation, unless typical vesicles or ulcers are present. HSV in stool usually represents either severe disseminated infection or infection of the anus or perianal areas. Isolation from any newborn infant specimen suggests potentially severe infection.

Detection of CMV. Interpretation of results of specimens containing CMV is most difficult. Primary CMV infection is usually asymptomatic and is commonly followed by silent reactivation of the latent virus throughout the patient's remaining life. CMV disease in immunocompromised patients can be life threatening, and ganciclovir antiviral therapy may be warranted. Detection of CMV in urine or respiratory secretions, however, is not diagnostic of significant disease. Detection of CMV in uncontaminated tissue, such as lung, or blood collected by venipuncture, suggests an active role in disease. The presence of virus-specific IgM or a fourfold increase in IgG antibodies may indicate disease. However, positive serology results must be interpreted with caution. False-positive IgM results have been attributed to infections by other viruses, such as EBV, and rises of both IgM and IgG may result from transfusions or the use of immunoglobulin therapy.[16,23]

ACKNOWLEDGMENT

The author wishes to acknowledge the extensive contributions made by those with whom he worked in the clinical virology laboratory at Children's Hospital Medical Center of Akron. In particular, Figures 43.25, 43.26, 43.30, and 43.31, and Procedures 43.1, 43.2, 43.3, 43.5, and 43.7 are a result of their help and expertise.

REFERENCES

1. Arens, M.Q., Swierkosz, E.M., Schmidt, R.R., et al. 1986. Strategy for efficient detection of respiratory viruses in pediatric clinical specimens. Diagn. Microbiol. Infect. Dis. 5:307.
2. Bean, B., 1992. Antiviral therapy: Current concepts and practices. Clin. Microbiol. Rev. 5:146.
3. Blumer, S.O., Hearn, T.L., Schalla, W.O., et al. 1992. Human T-lymphotropic virus type I/II. Status of enzyme immunoassay and Western Blot testing in the United States in 1989 and 1990. Arch. Pathol. Lab. Med. 116:471.
4. Burgard, M., Mayaux, M.J., Blanche, S., et al. 1992. The use of viral culture and p24 antigens testing to diagnose human immunodeficiency virus infection in neonates. N. Engl. J. Med. 327:1192.
5. Calisher, C.H., 1988. Host responses to arthropod-borne virus infection and vaccination. Clin. Immunol. Newsl. 9:2.
6. Calisher, C.H., Berardi, V.P., Muth, D.J., et al. 1986. Specificity of immunoglobulin M and G antibody responses in humans infected with Eastern and Western equine encephalitis viruses: Application to rapid serodiagnosis. J. Clin. Microbiol. 23:369.
7. Calisher, C.H., Pretzman, C.I., Muth, D.J., et al. 1986. Serodiagnosis of LaCrosse virus infections in human by detection of immunoglobulin M class antibodies. J. Clin. Microbiol. 23:667.
8. Cates, K.L. 1986. Clinical considerations in the diagnosis of viral respiratory infections. Diagn. Microbiol. Infect. Dis. 4:23S.
9. Centers for Disease Control. 1991. Update on adult immunization. Recommendations of the immunization practices advisory committee (ACIP). U.S. Department of Health and Human Services. 40:1.
10. Centers for Disease Control. 1992. Testing for antibodies to HIV type 2 in the United States. Lab. Med. 23:819.
11. Chang, R.S., Arnold, D., Chang, Y.Y., et al. 1986. Relative sensitivity of cell culture systems in the detection of herpes simplex viruses. Diagn. Microbiol. Infect. Dis. 5:135.
12. Chernesky, M.A., Ray, C.G., and Smith, T.F. 1982. Cumitech 15: Laboratory diagnosis of viral infections. American Society for Microbiology, Washington, D.C.
13. Chokephalibulkit, K., Painter, P.C., and Patamasucon, P. 1992. Overview of hepatitis C. Lab. Med. 23:798.
14. Chonmaitree, T., Baldwin, C.D., and Lucia, H.L. 1989. Role of the virology laboratory in diagnosis and management of patients with central nervous system disease. Clin. Microbiol. Rev. 2:1.
15. Chonmaitree, T., Ford, C., and Sanders, C. 1988. Comparison of cell cultures for rapid isolation of enteroviruses. J. Clin. Microbiol. 26:2576.
16. Cockerill, F.R. 1985. Diagnosing cytomegalovirus infection. Mayo Clin. Proc. 60:636.
17. Costello, M.J., Morrow, S.L., Laney, S., et al. 1993. Guidelines for specimen collection, transportation, and test selection. Lab. Med. 24:19.
18. Cruz, J.R., Quinonez, E., Fernandez, A., et al. 1987. Isolation of viruses from nasopharyngeal secretions: Comparison of aspiration and swabbing as means of sample collection. J. Infect. Dis. 156:415.
19. Dagan, R., Hall, C.B., Powell, K.R., et al. 1989. Epidemiology and laboratory diagnosis of infection with viral and bacterial pathogens in infants hospitalized for suspected sepsis. J. Pediatr. 115:351.
20. Dennehy, P.H., Guantlett, D.R., and Tente, W.E. 1988. Comparison of nine commercial immunoassays for the detection of rotavirus in fecal specimens. J. Clin. Microbiol. 26:1630.
21. Dock, N.L. 1991. The ABC's of viral hepatitis. Clin. Microbiol. Newsl. 13:17.
22. Drew, W.L. 1986. Controversies in viral diagnosis. Rev. Infect. Dis. 8:814.
23. Drew, W.L. 1992. Nonpulmonary manifestations of cytomegalovirus infection in immunocompromised patients. Clin. Microbiol. Rev. 5:204.
24. Elder, B.L., and Smith, T.F., 1987. Comparison of Isolab Quik-sep and Quik-sep System II for separation of immunoglobulins G and M. Diagn. Microbiol. Infect. Dis. 6:125.
25. Erice, A., Holm, M.A., Gill, P.C., et al. 1992. Cytomegalovirus (CMV) antigenemia assay is more sensitive than shell vial cultures for rapid detection of CMV in polymorphonuclear blood leukocytes. J. Clin. Microbiol. 30:2822.
26. Espy, M.J., Smith, T.F., Harmon, M.W., et al. 1986. Rapid detection of influenza virus by shell vial assay with monoclonal antibodies. J. Clin. Microbiol. 24:677.
27. Fayram, S.L., Aarnaes, S.L., Peterson, E.M., et al. 1986. Evaluation of five cell types for the isolation of

herpes simplex virus. Diagn. Microbiol. Infect. Dis. 5:127.

28. Fung, J.C. 1985. Pitfalls of IgM antibody testing. Clin. Microbiol. Newsl. 7:41.

29. Gilchrist, M.J.R., Bretl, T.S., Moultney, K., et al. 1987. Comparison of seven kits for detection of rotavirus in fecal specimens with a sensitive, specific enzyme immunoassay. Diagn. Microbiol. Infect. Dis. 8:221.

30. Gleaves, C.A., Lee, C.F., Bustamante, C.I., et al. 1988. Use of murine monoclonal antibodies for laboratory diagnosis of varicella-zoster virus infection. J. Clin. Microbiol. 26:1623.

31. Gleaves, C.A., Smith, T.F., Shuster, E.A., et al. 1984. Rapid detection of cytomegalovirus in MRC-5 inoculated with urine specimens by using low-speed centrifugation and monoclonal antibody to an early antigen. J. Clin. Microbiol. 19:917.

32. Greenberg, S.B., and Krilov, L.R. 1986. Cumitech 21: Laboratory diagnosis of viral respiratory disease. American Society for Microbiology, Washington, D.C.

33. Grody, W.W., Cheng, L., and Lewin, K.J. 1987. In situ viral DNA hybridization in diagnostic surgical pathology. Human Pathol. 18:535.

34. Hall, C.B., and Douglas, R.G. 1975. Clinically useful method for the isolation of respiratory syncytial virus. J. Infect. Dis. 131:1.

35. Halstead, D.C., Todd, S., and Fritch, G. 1990. Evaluation of five methods for respiratory syncytial virus detection. J. Clin. Microbiol. 28:1021.

36. Hojvat, S.A. 1989. Diagnostic test for viral hepatitis. Clin. Microbiol. Newsl. 11:33.

37. Howell, C.L., and Miller, M.J. 1983. Effect of sucrose phosphate and sorbitol on infectivity of enveloped viruses during storage. J. Clin. Microbiol. 18:658.

38. Howell, C.L., Miller, M.J., and Martin, W.J. 1979. Comparison of rates of virus isolation from leukocyte populations separated from blood by conventional and Ficoll-paque/Macrodex methods. J. Clin. Microbiol. 10:533.

39. Huntoon, C.J., House, R.F., and Smith, T.F. 1981. Recovery of viruses from three transport media incorporated into culturettes. Arch. Pathol. Lab. Med. 105:436.

40. Johnson, F.B. 1990. Transport of viral specimens. Clin. Microbiol. Rev. 3:120.

41. Johnson, R.B., and Libby, R. 1980. Separation of immunoglobulin M (IgM) essentially free of IgG from serum for use in systems requiring assay of IgM-type antibodies without interference from rheumatoid factor. J. Clin. Microbiol. 12:451.

42. Johnston, S.L.G., and Siegel, C.S. 1990. Presumptive identification of enteroviruses with RD, HEp-2, and RMK cell lines. J. Clin. Microbiol. 28:1049.

43. Johnston, S.L.G., and Siegel, C.S. 1990. Evaluation of direct immunofluorescence, enzyme immunoassay, centrifugation culture, and conventional culture for the detection of respiratory syncytial virus. J. Clin. Microbiol. 28:2394.

44. Kaplan, J.E., and Khabbaz, R.F. 1989. HTLV-I: Newest addition to blood donor screening. Am. Fam. Phy. 40:189.

45. Kaplan, M.H. 1990. Diagnosis of some common viral infections: Role of polymerase chain reaction. In Mandell, G.L., Douglas, R.G., and Bennett, J.E., editors. Principles and practice of infectious disease, ed 3. update 6. Churchill Livingstone, New York.

46. Kotloff, K.L., Losonsky, G.A., Morris, J.G., et al. 1988. Enteric adenovirus infection and childhood diarrhea: An epidemiologic study in three clinical settings. Pediatr. 84:219.

47. Lipson, S.M., and Zelinsky-Papez, K.A. 1989. Comparison of four latex agglutination (LA) and three enzyme-linked immunosorbent assays (ELISA) for the detection of rotavirus in fecal specimens. Am. J. Clin. Pathol. 92:637.

48. Mathewson, J.J., Winsor, D.K., DuPont, H.L., et al. 1989. Evaluation of assay systems for the detection of rotavirus in stool specimens. Diagn. Microbiol. Infect. Dis. 12:139.

49. Merlin, T.L. 1986. Chronic mononucleosis: Pitfalls in the laboratory diagnosis. Human Pathol. 17:2.

50. Miller, S.E. 1988. Diagnostic virology by electron microscopy. ASM News. 54:475.

51. Minnich, L.L., and Ray, C.G. 1987. Early testing of cell cultures for detection of hemadsorbing viruses. J. Clin, Microbiol. 25:421.

52. Minnich, L.L., Smith, T.F., and Ray, C.G. 1988. Cumitech 24: Rapid detection of viruses by immunofluorescence. American Society for Microbiology, Washington, D.C.

53. Monath, T.P., Nystrom, R.R., Bailey, R.E., et al. 1984. Immunoglobulin M antibody capture enzyme-linked immunosorbent assay for diagnosis of St. Louis encephalitis. J. Clin. Microbiol. 20:784.

54. Muul, L.M. 1990. Current status of polymerase chain reaction assay in clinical research of human immunodeficiency virus infection. AIDS Updates 3:1.

55. Okano, M., Thiele, G.M., Davis, J.R., et al. 1988. Epstein-Barr virus and human diseases: Recent advances in diagnosis. Clin. Microbiol. Rev. 1:300.

56. Paya, C.V., Wold, A.D., and Smith T.S. 1988. Detection of cytomegalovirus from blood leukocytes separated by Sepracell-MN and Ficoll-paque/Macrodex methods. J. Clin. Microbiol. 26:2031.

57. Payne, C.M., Ray, C.G., Borduin, V., et al. 1986. An eight-year study of the viral agents of acute gastroenteritis in humans: Ultrastructural observations and seasonal distribution with a major emphasis on coronavirus-like particles. Diagn. Microbiol. Infect. Dis. 5:39.

58. Persing, D.H. 1991. Polymerase chain reaction: Trenches to benches. J. Clin. Microbiol. 29:1281.

59. Peter, G. 1992. Childhood immunizations. N. Engl. J. Med. 327:1794.

60. Ray, C.G., and Minnich, L.L. 1987. Efficiency of immunofluorescence for rapid detection of common respiratory viruses. J. Clin. Microbiol. 25:355.

61. Rotbart, H.A. 1991. Nucleic acid detection systems for enteroviruses. Clin. Microbiol. Rev. 4:156.

62. Ryan-Poirier, K.A., Katz, J.M., Webster, R.G., et al. 1992. Application of directigen FLU-A for the detection of influenza A virus in human and nonhuman specimens. J. Clin. Microbiol. 30:1072.

63. Salmon, V.C., Stanberry, L.R., and Overall, J.C. 1984. More rapid isolation of herpes simplex virus in a continuous line of mink lung cells than in vero or human fibroblast cells. Diagn. Microbiol. Infect. Dis. 2:317.

64. Schmidt, N.J. 1984. Update on class-specific viral antibody assays. Clin. Immunol. Newsl. 5:81.

65. Scott-Taylor, T., Ahluwalia, G., Klisko, B., et al. 1990. Prevalent enteric adenovirus variant not detected by

commercial monoclonal antibody enzyme immunoassay. J. Clin. Microbiol. 28:2797.

66. Sever, J.L. 1985. TORCH tests and what they mean. Am. J. Obst. Gynecol. 152:495.

67. Shalit, T., McKee, P.A., Beauchamp, H., et al. 1985. Comparison of polyclonal antiserum versus monoclonal antibodies for the rapid diagnosis of influenza A virus infections by immunofluorescence in clinical specimens. J. Clin. Microbiol. 22:877.

68. Shamberger, R.J. 1984. ELISA on the trail of viral diseases. Diagn. Med. October:52.

69. Sherlock, C.H., Brandt, C.J. Middleton, P.J., et al. 1989. Cumitech 26: Laboratory diagnosis of viral infections producing enteritis. American Society for Microbiology, Washington, D.C.

70. Shuster, E.A., Beneke, J.S., Tegtmeier, G.E., et al. 1985. Monoclonal antibody for rapid laboratory detection of cytomegalovirus infections: Characterization and diagnostic application. Mayo Clin. Proc. 60:577.

71. Slifkin, M., and Cumbie, R. 1992. Comparison of the Histopaque-1119 method with the Plasmagel method for separation of blood leukocytes for cytomegalovirus isolation. J. Clin. Microbiol. 30:2722.

72. Smith, T.F. 1987. Rapid methods for the diagnosis of viral infections. Lab. Med. 18:16.

73. Solomon, A.R., Rasmussen, J.E., Varani, J., et al. 1984. The Tzanck smear in the diagnosis of cutaneous herpes simplex. JAMA 251:633.

74. Steckelberg, J.M. and Cockerill, F.R. 1988. Serologic testing for human immunodeficiency virus antibodies. Mayo Clin. Proc. 63:373.

75. Stout, C., Murphy, M.D., Lawrence, S., et al. 1989. Evaluation of a monoclonal antibody pool for rapid diagnosis of respiratory viral infections. J. Clin. Microbiol. 27:448.

76. Subbarao, E.K., Griffis, J., and Waner, J.L. 1989. Detection of multiple viral agents in nasopharyngeal specimens yielding respiratory syncytial virus (RSV). An assessment of diagnostic strategy and clinical significance. Diagn. Microbiol. Infect. Dis. 12:327.

77. Tenover, F.C. 1988. Diagnostic deoxyribonucleic acid probes for infectious diseases. Clin. Microbiol. Rev. 1:82.

78. Thomson, R.B., Bostick, C.C., and McLachlan, S.S. 1991. Clinical virology for community hospitals. Clin. Microbiol. Newsl. 13:45.

79. Waner, J.L., Whitehurst, N.J., Todd, S.J., et al. 1990. Comparison of directigen RSV with viral isolation and direct immunofluorescence for the identification of respiratory syncytial virus. J. Clin. Microbiol. 28:480.

80. Wiedbrauk, D.L., and Chuang, Y.K. 1988. Managing IgM serologies: Getting the IgG out. Lab. Management 26:24.

81. Wildin, S., and Chonmaitree, T. 1987. The importance of the virology laboratory in the diagnosis and management of viral meningitis. Am. J. Dis. Child. 141:454.

82. Wolcott, M.J. 1992. Advances in nucleic acid–based detection methods. Clin. Microbiol. Rev. 5:370.

83. Woods, G.L., and Proffitt, M.R. 1987. Comparison of Plasmagel with LeucoPREP Macrodex methods for separation of leukocytes for virus isolation. Diagn. Microbiol. Infect. Dis. 8:123.

BIBLIOGRAPHY

Balows, A., Hausler, W.J., Herrmann, K.L., et al. editors. 1991. Manual of clinical microbiology, ed 5. American Society for Microbiology, Washington, D.C.

Gardner, P.S. and McQuillin, J. 1974. Rapid virus diagnosis application of immunofluorescence. Butterworth & Co. London.

Isenberg, H.D. editor. 1992. Clinical microbiology procedures handbook. American Society for Microbiology, Washington, D.C.

Lennette, E.H. and Schmidt, N.J., editors. 1979. Diagnostic procedures for viral, rickettsial, and chlamydial infections, ed 5. American Public Health Association, Washington, D.C.

Levine, A.J. 1991. Viruses. Scientific American Library, New York.

Specter, S., and Lancz, G.J. 1986. Clinical virology manual. Elsevier, New York.

44

LABORATORY METHODS IN BASIC MYCOLOGY

Historically, the fungi have been regarded as relatively insignificant causes of infection, and until recently clinical laboratories offered little more than interest in providing mycological services. During the past few years, however, literature has shown a sharp increase in the number of case reports of fungal infections. The fungi are now well-recognized causes of infection, and physician awareness of their importance is much more common. Some clinical microbiology laboratories have kept pace with the changing times, but unfortunately many still offer only minimal mycological services. The reasons for this are not easily defined, but the lack of emphasis in the area of mycology by medical technology educational programs has played a major role. In addition, the myth that mycology is too hard to learn has probably been partly responsible for the lack of interest.

The emergence of continuing education programs presenting the practical aspects of clinical mycology and a number of recent textbooks in the same area, along with the increased number of scientific publications, now makes mycology available to all clinical microbiology laboratories where it should be included as an important component.[43]

IMPORTANCE OF FUNGI

There are more than 50,000 valid species of fungi, but only 50 to 75 are generally recognized as being pathogenic for humans. These organisms normally live a saprophytic existence in nature, enriched by decaying nitrogenous matter, where they are capable of maintaining a separate existence, with a parasitic cycle in humans or animals. The mycoses are not communicable in the usual sense of person-to-person or animal-to-person transfer; humans become an accidental host by the inhalation of spores or by their introduction into tissue through trauma. With the exception of the dimorphic fungi, humans are relatively resistant to infections caused by these organisms. However, an alteration in the immune system of the host (Chapter 24) may lead to infection by fungi that are normally considered to be nonpathogenic. Such infections may occur in patients with debilitating diseases such as HIV infection, diabetes, or impaired immunological function resulting from corticosteroid or antimetabolite therapy. Other predisposing factors include long-term intravenous cannulation, gastrointestinal surgical procedures, and long-term antimicrobial chemotherapy. During recent years, the number of infections caused by saprobic fungi in immunocompromised patients has increased markedly. This has resulted in an obligation for the laboratory to identify and report all organisms recovered from clinical specimens submitted for fungal culture.

LEVELS OF LABORATORY SERVICE

To ensure adequate patient care, mycological services should be offered by all clinical microbiology laboratories. The extent to which these services are offered depends on the individual laboratory setting. The following are proposed levels of service:

I. Performance of direct examination of clinical specimens for the presence of fungi, and collection, culturing, and transport of appropriate clinical specimens to a reference laboratory.

II. Performance of all functions of level I laboratories as well as the identification of yeasts using commercially available systems, performance of the latex test for cryptococcal antigen, and performance of commercially available fungal immunodiffusion tests.

III. Performance of all functions of level II laboratories as well as the identification of commonly encountered filamentous fungi.

IV. Performance of all functions of laboratories at lower levels, the identification of all fungi, performance of all fungal serologic tests, and antifungal susceptibility tests.

Laboratories should choose the extent to which they are proficient and forward all additional procedures to a qualified reference laboratory. With the current emphasis on cost containment (Chapter 4), smaller clinical laboratories should carefully consider the extent to which they should provide mycological services. If a reference laboratory is selected, constant close communication is essential.

FUNGI COMMONLY ENCOUNTERED IN A CLINICAL LABORATORY

Of the more than 50,000 valid species of fungi, fewer than 100 are likely to be encountered in clinical specimens. Laboratorians just beginning to work in clinical mycology can feel intimidated by this large number; however, it is common for a laboratory to encounter many of the same fungi

TABLE 44.1

FUNGI MOST COMMONLY RECOVERED FROM CLINICAL SPECIMENS AT MAYO CLINIC

BLOOD	CEREBROSPINAL FLUID	GENITOURINARY TRACT	RESPIRATORY TRACT	SKIN
Candida albicans	Cryptococcus neoformans	Candida albicans	Yeast, not Cryptococcus	Trichophyton rubrum
Candida tropicalis	Candida albicans	Candida glabrata	Penicillium species	Trichophyton mentagrophytes
Candida parapsilosis	Candida parapsilosis	Candida tropicalis	Aspergillus species	Alternaria species
Cryptococcus neoformans	Candida tropicalis	Candida parapsilosis	Aspergillus fumigatus	Candida albicans
Histoplasma capsulatum	Coccidioides immitis	Penicillium species	Cladosporium species	Penicillium species
Candida lusitaniae	Histoplasma capsulatum	Candida krusei	Alternaria species	Scopulariopsis species
Candida krusei		Cryptococcus neoformans	Aspergillus niger	Epidermophyton floccosum
Saccharomyces species		Saccharomyces species	Geotrichum candidum	Candida parapsilosis
Candida pseudotropicalis		Histoplasma capsulatum	Fusarium species	Aspergillus species
Candida zeylanoides		Cladosporium species	Aspergillus versicolor	Acremonium species
Trichosporon beigelii		Aspergillus species	Aspergillus flavus	Aspergillus versicolor
Coccidioides immitis		Trichosporon beigelii	Acremonium species	Cladosporium species
Candida guilliermondii		Alternaria species	Scopulariopsis species	Fusarium species
Malasezzia furfur			Beauveria species	Trichosporon beigelii
			Trichosporon beigelii	Phialophora species

frequently. During a recent 9-year period, 267,861 clinical specimens were submitted for fungal culture at the Mayo Clinic. Of these, 75,258 (28.1%) recovered one or more fungi. Table 44.1 presents the fungi most commonly recovered from clinical specimens at the Mayo Clinic. This listing is derived from the 94,358 fungi recovered from various clinical sources, and organisms are listed in order of their frequency of occurrence. It is well known that there are differences in the geographical distribution of certain fungi; however, perhaps these data will be of value to most laboratories in the interpretation of fungal culture results.

COLLECTION, TRANSPORT, AND CULTURING OF CLINICAL SPECIMENS

The diagnosis of fungal infections depends entirely on the selection and collection of an appropriate clinical specimen for culture. Many fungal infections are similar clinically to mycobacterial infections, and often the same specimen is cultured for both fungi and mycobacteria. Most infections have a primary focus in the lungs; respiratory tract secretions are almost always included among the specimens selected for culture. It should be emphasized that dissemination to distant body sites often occurs, and fungi may be commonly recovered from nonrespiratory sites. The proper collection of specimens and their transport to the clinical laboratory are of major importance for the recovery of fungi. In many in-

stances, specimens not only contain the etiological agents but also contaminating bacteria or fungi that will rapidly overgrow some of the slower growing pathogenic fungi. Since overgrowth with contaminating organisms is common, it is important to transport the specimens to the clinical laboratory as soon as possible. Although collection methods are discussed in Chapter 7, a few specific comments concerning specimen collection and culturing are included in this chapter.

RESPIRATORY TRACT SECRETIONS
Respiratory tract secretions (sputum, induced sputum, bronchial washings, bronchoalveolar lavage, and tracheal aspirations) are perhaps the most commonly submitted specimens for fungal culture.[82] To ensure the optimal recovery of fungi and prevent overgrowth by contaminants, antibacterial antibiotics should be included in the battery of media to be used. Cycloheximide, an antifungal agent that prevents rapid overgrowth by rapidly growing molds, should be included in at least one of the culture media used. As much specimen as possible (0.5 ml) should be used to inoculate each medium.

CEREBROSPINAL FLUID
Cerebrospinal fluid (CSF) collected for culture should be filtered through a 0.45-μm pore size membrane filter attached to a sterile syringe. After filtration, the filter is removed and is placed onto the surface of an appropriate culture medium with the "organism" side down. Cultures

should be examined daily and the filter moved to another location on an every-other-day basis. If less than 1.0 ml of specimen is submitted for culture, it should be centrifuged and 1-drop aliquots of the sediment should be placed onto several areas on the agar surface. Media used for the recovery of fungi from CSF should contain no antibacterial or antifungal agents. Once submitted to the laboratory, CSF specimens should be processed promptly. If prompt processing is not possible, samples should be kept at room temperature or placed in a 30° C incubator, since most organisms will continue to replicate in this environment.

BLOOD

Disseminated fungal infections are more prevalent than previously recognized, and blood cultures provide an accurate method for determining their cause. Only a few fungal blood culture systems have been available over past years, and most are not used routinely by clinical microbiology laboratories. Currently, several automated blood culture systems including the BACTEC (Becton-Dickinson Diagnostic Instrument Systems, Towson, Md.), Bact T/Alert (Organon-Teknika Corporation, Durham, N.C.), and ESP (Difco Laboratories, Detroit, Mich.) all offer promise of being adequate systems for fungal recovery. However, until evaluations are completed, larger laboratories are encouraged to use a lysis-centrifugation system, the Isolator (Wampole Laboratories, Cranbury, N.J.), since it has been shown to be the optimal fungal blood culture system[7,8,27] (Chapter 15). Using this system, red blood cells and white blood cells that may contain the microorganisms are lysed, and centrifugation serves to concentrate fungi and bacteria before culturing. The concentrate is inoculated onto the surface of appropriate culture media, and most fungi are detected within the first 4 days of incubation. However, an occasional isolate of *Histoplasma capsulatum* may require approximately 10 to 14 days for recovery. The optimal temperature for fungal blood cultures is 30° C, and the suggested incubation time is 21 days.

HAIR, SKIN, AND NAIL SCRAPINGS

These specimens are usually submitted for dermatophyte culture and are contaminated with bacteria or rapidly growing fungi or both. Samples collected from lesions may be obtained by scraping the skin or nails with a scalpel blade or microscope slide, and infected hairs are removed by plucking them with forceps. These specimens should be placed in a sterile Petri dish or paper envelope before culturing; they should not be refrigerated. Either Mycobiotic or Mycosel agars (Chapter 9), which contain chloramphenicol and cycloheximide, are satisfactory for the recovery of dermatophytes from hair, skin, or nails. Cultures should be incubated for a minimum of 30 days at 30° C before being reported as negative.

URINE

Urine samples collected for fungal culture should be processed as soon as possible after collection. Twenty-four-hour urine samples are unacceptable for culture. After considering the arguments concerning the value of quantitating the growth of fungi recovered from urine cultures, it is concluded that quantitation should not be performed except as part of the routine urine culture procedure. All urine samples should be centrifuged and the sediment cultured using a loop to provide adequate isolation of colonies. Since urine is often contaminated with gram-negative bacteria, it is necessary to use media containing antibacterial agents to ensure the recovery of fungi.

TISSUE, BONE MARROW, AND STERILE BODY FLUIDS

All tissues should be processed before culturing by mincing or grinding or placing in a Stomacher (Tekmar; Chapter 22). Currently many laboratories use the Stomacher, which expresses the cytoplasmic contents of cells by pressure exerted from the action of rapidly moving metal paddles against the tissue in a broth suspension. After processing, at least 1 ml of specimen should be spread onto the surface of appropriate culture media, and incubation should be at 30° C for 30 days.

Bone marrow may be placed directly onto the surface of appropriate culture media and incubated in the manner previously mentioned. Sterile body fluids should be concentrated by centrifugation before culturing, and at least 1 ml of specimen should be placed onto the surface of appropriate culture media. An alternative is to place body fluids in an Isolator tube and process it as a blood culture. All specimens should be cultured as soon as they are received by the laboratory to ensure the recovery of fungi from these important sources. Chapter 22 discusses laboratory handling of these specimens.

CULTURE MEDIA AND INCUBATION REQUIREMENTS

A number of fungal culture media are satisfactory for use in the clinical microbiology laboratory.

TABLE 44.2

FUNGAL CULTURE MEDIA: INDICATIONS FOR USE

MEDIA	INDICATIONS FOR USE
Primary recovery media	
Brain-heart infusion agar	Primary recovery of saprobic and pathogenic fungi
Brain-heart infusion agar with antibiotics	Primary recovery of pathogenic fungi exclusive of dermatophytes
Brain-heart infusion biphasic blood culture bottles	Recovery of fungi from blood
Buffered-charcoal yeast extract agar	Primary recovery of *Nocardia* species
Dermatophyte test medium	Primary recovery of dermatophytes, recommended as screening medium only
Inhibitory mold agar	Primary recovery of pathogenic fungi exclusive of dermatophytes
Potato flake agar	Primary recovery of saprobic and pathogenic fungi
Mycosel or mycobiotic agar	Primary recovery of dermatophytes
SABHI agar	Primary recovery of saprobic and pathogenic fungi
Yeast-extract phosphate agar	Primary recovery of pathogenic fungi exclusive of dermatophytes
Differential test media	
Ascospore agar	Detection of ascospores in ascosporogenous yeasts such as *Saccharomyces* sp.
Cornmeal agar with Tween 80 and trypan blue	Identification of *C. albicans* by chlamydospore production; identification of *Candida* by microscopic morphology
Cottonseed conversion agar	Conversion of dimorphic fungus *B. dermatitidis* from mold to yeast form
Czapek's agar	Recovery and differential identification of *Aspergillus* sp.
Niger seed agar	Identification of *C. neoformans*
Nitrate reduction medium	Detection of nitrate reduction in confirmation of *Cryptococcus* sp.
Potato dextrose agar	Demonstration of pigment production by *T. rubrum;* preparation of microslide cultures
Rice medium	Identification of *M. audouinii*
Trichophyton agars 1-7	Identification of members of *Trichophyton* genus
Urea agar	Detection of *Cryptococcus* sp.; differentiate *T. mentagrophytes* from *T. rubrum;* detection of *Trichosporon* sp.
Yeast fermentation broth	Identification of yeasts by determining fermentation
Yeast nitrogen base agar	Identification of yeasts by determining carbohydrate assimilation

From Koneman, E.W., and Roberts, G.D. 1985. Practical laboratory mycology, ed 3. Williams & Wilkins, Baltimore.

Most are adequate for the recovery of fungi, and selection is usually left up to each laboratory director. Table 44.2 lists various fungal culture media and the indications for their use. Ideally, a battery of media should be used, and the following recommendations are suggested:

1. Media with and without blood enrichment should be used.
2. Media with and without cycloheximide should be used.
3. All media should contain antibacterial agents; if *Nocardia* is suspected, mycobacterial culture media or media used for the recovery of *Legionella* should be used for this purpose.

Culture dishes or screw-capped culture tubes are satisfactory for the recovery of fungi. Culture dishes are optimal, since they provide better aeration of cultures, a large surface area for better isolation of colonies, and greater ease of handling by technologists making microscopic preparations for examination. Agar in culture dishes has a marked tendency to dehydrate during the extended incubation period required for fungi; however, this problem can be minimized by using culture dishes containing at least 40 ml of agar and placing them in an incubator that contains a pan of water to humidify the environment. Dishes should be opened and examined only within a certified biological safety cabinet (Chapter 2). Many laboratories discourage the use of culture dishes; however, the advantages of using them outweigh the disadvantages.

Culture tubes are more easily stored, require less space for incubation, and are easily handled. In addition, they have a lower dehydration rate, and laboratory workers feel that cultures are less hazardous to handle when in tubes. The disadvantages, which include relatively poor isolation of colonies, a reduced surface area for growth, and a tendency to promote anaerobiosis, discourage their routine use in a clinical microbiology laboratory. Cotton-plugged tubes are unsatisfactory for fungal cultures.

Cultures should be incubated at room temperature or preferably at 30° C for 30 days before reporting as negative. A relative humidity of 40% to 50% can be achieved by placing an open pan of water in the incubator. Cultures should be examined at least three times weekly during incubation.

As previously mentioned, most clinical specimens are contaminated with bacteria and/or rapidly growing fungi. The need for antibacterial and antifungal agents is obvious. The addition of 0.5 µg/ml of cycloheximide and 16 µg/ml of chloramphenicol to media has been traditionally advocated to inhibit the growth of contaminating molds and bacteria.[72] Better results have been achieved using a combination of 5 µg/ml of gentamicin and 16 µg/ml of chloramphenicol as antibacterial agents. Also, ciprofloxacin, 5 µg/ml, may be used and appears to be superior.

Certain fungi seem to have a requirement for blood, and enrichment with 5% to 10% sheep blood is recommended for at least one of the culture media used. Cycloheximide may be added to any of the media that contain or lack antibacterial antibiotics. However, if cycloheximide is included as a part of the battery of culture media used, a medium lacking this ingredient should also be included. Certain pathogenic fungi are partially or completely inhibited by this compound, including *Cryptococcus neoformans, Candida krusei* and other species of *Candida, Trichosporon beigelii, Pseudallescheria boydii,* and species of *Aspergillus.*

The use of antibiotics in fungal culture media is necessary for the optimal recovery of organisms; however, the use of decontamination and concentration methods advocated for the recovery of mycobacteria is not appropriate for fungal cultures.[66]

DIRECT MICROSCOPIC EXAMINATION OF CLINICAL SPECIMENS

Direct microscopic examination of clinical specimens has been used for many years; however, its usefulness should be reemphasized.[56] Since the mission of a clinical microbiology laboratory is to provide a rapid and accurate diagnosis, the mycology laboratory can provide this service in many instances by direct examination of the clinical specimen (particularly the Gram stain) submitted

for culture. Microbiologists are encouraged to become familiar with the diagnostic features of fungi commonly encountered in clinical specimens. This very important procedure can often provide the first microbiological proof of etiology in patients with fungal infection. Currently, this is the most rapid method available.

Tables 44.3 and 44.4 present the methods available for the direct microscopic detection of fungi in clinical specimens and a summary of the characteristic microscopic features of each method. More information regarding specific procedures is given in Chapter 8. Figures 44.1 to 44.18 present photomicrographs of some of the fungi commonly seen in clinical specimens.

Traditionally, the potassium hydroxide preparation has been the recommended method for the direct microscopic examination of specimens.[62] However, it is currently felt that the calcofluor white stain[29] (Chapter 8) is preferred. Slides prepared by this method may be observed using fluorescent microscopy or brightfield microscopy as used for the potassium hydroxide preparation; the former is optimal since fungal cells will fluoresce.

CHARACTERISTICS OF FUNGI (INFORMAL GLOSSARY)

The fungi, a term that includes both yeasts and molds, differ significantly from the bacteria. The fungi are eukaryotic and contain a definite nu-

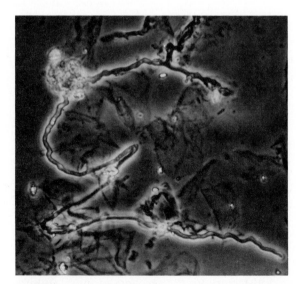

FIGURE 44.1

Potassium hydroxide preparation of skin, phase-contrast, dermatophyte, showing septate hyphae intertwined among epithelial cells (500×).

Text continued on p. 698.

TABLE 44.3

SUMMARY OF METHODS AVAILABLE FOR THE DIRECT MICROSCOPIC DETECTION OF FUNGI IN CLINICAL SPECIMENS

METHOD	USE	TIME REQUIRED	ADVANTAGES	DISADVANTAGES
Acid-fast stain	Detection of mycobacteria and *Nocardia*	12 min	Detects *Nocardia** and *B. dermatitidis*	Tissue homogenates are difficult to observe because of background staining
Calcofluor white	Detection of fungi	1 min	Can be mixed with KOH: detects fungi rapidly because of bright fluorescence	Requires use of a fluorescence microscope; background fluorescence prominent, but fungi exhibit more intense fluorescence; vaginal secretions are difficult to interpret
Gram stain	Detection of bacteria	3 min	Is commonly performed on most clinical specimens submitted for bacteriology and will detect most fungi, if present	Some fungi stain well; however, others (e.g., *Cryptococcus* sp.) show only stippling and stain weakly in some instances; some isolates of *Nocardia* fail to stain or stain weakly
India ink	Detection of *C. neoformans* in CSF	1 min	When positive in CSF, is diagnostic of meningitis	Positive in less than 50% of cases of meningitis; not reliable in non–HIV-infected patients
Potassium hydroxide (KOH)	Clearing of specimen to make fungi more readily visible	5 min; if clearing is not complete, an additional 5-10 min is necessary	Rapid detection of fungal elements	Experience required since background artifacts are often confusing, clearing of some specimens may require an extended time
Methenamine silver stain	Detection of fungi in histologic section	1 h	Best stain to detect fungal elements	Requires a specialized staining method that is not usually readily available to microbiology laboratories
Papanicolaou stain	Examination of secretions for presence of malignant cells	30 min	Cytotechnologist can detect fungal elements	
Periodic acid-Schiff (PAS) stain	Detection of fungi	20 min; 5 min additional if counterstain is employed	Stains fungal elements well; hyphae of molds and yeasts can be readily distinguished	*Nocardia* sp. do not stain well; *B. dermatitidis* appears pleomorphic
Wright stain	Examination of bone marrow or peripheral blood smears	7 min	Detects *H. capsulatum* and *C. neoformans*	Detection is limited to *H. capsulatum* and *C. neoformans*

*Acid-fast bacterium.
From Merz, W.G., and Roberts, G.D. 1991. Detection and recovery of fungi in clinical specimens. In Balows, A., Hausler, W.J., Jr., Herrman, K., Isenberg, H.D., and Shadomy, H.J., editors. Manual of clinical microbiology, ed 5. American Society for Microbiology, Washington, D.C.

TABLE 44.4

SUMMARY OF CHARACTERISTIC FEATURES OF FUNGI SEEN IN DIRECT EXAMINATION OF CLINICAL SPECIMENS

MORPHOLOGICAL FORM FOUND IN SPECIMENS	ORGANISM(S)	SIZE RANGE (DIAMETER, μm)	CHARACTERISTIC FEATURES
Yeastlike	*Histoplasma capsulatum*	2-5	Small; oval to round budding cells; often found clustered within histiocytes; difficult to detect when present in small numbers
	Sporothrix schenckii	2-6	Small; oval to round to cigar-shaped; single or multiple buds present; uncommonly seen in clinical specimens
	Cryptococcus neoformans	2-15	Cells exhibit great variation in size; usually spherical but may be football-shaped; buds single or multiple and "pinched off"; capsule may or may not be evident; occasionally pseudohyphal forms with or without a capsule may be seen in exudates of CSF
	Malasezzia furfur (in fungemia)	1.5-4.5	Small; bottle-shaped cells, buds separated from parent cell by a septum; emerge from a small collar
	Blastomyces dermatitidis	8-15	Cells are usually large, double refractile when present; buds usually single; however, several may remain attached to parent cells; buds connected by a broad base
	Paracoccidiodes brasiliensis	5-60	Cells are usually large and are surrounded by smaller buds around the periphery ("mariner's wheel appearance"); smaller cells may be present (2-5 μm) and resemble *H. capsulatum;* buds have "pinched-off" appearance
Spherules	*Coccidioides immitis*	10-200	Spherules vary in size; some may contain endospores, others may be empty; adjacent spherules may resemble *B. dermatitidis;* endospores may resemble *H. capsulatum* but show no evidence of budding; spherules may produce multiple germ tubes if a direct preparation is kept in a moist chamber for ≥24 h
	Rhinosporidium seeberi	6-300	Large, thick-walled sporangia containing sporangiospores are present; mature sporangia are larger than spherules of *C. immitis;* hyphae may be found in cavitary lesions
	Pseudallescheria boydii (cases other than mycetoma)		Hyphae are septate and are impossible to distinguish from those of other hyaline molds
Yeast and pseudohyphae or hyphae	*Candida* sp.	3-4 (yeast) 5-10 (pseudohyphae)	Cells usually exhibit single budding; pseudohyphae, when present, are constricted at the ends and remain attached like links of sausage; hyphae, when present, are septate
	M. furfur (in tinea versicolor)	3-8 (yeast) 2.5-4 (hyphae)	Short, curved hyphal elements are usually present along with round yeast cells that retain their spherical shape in compacted clusters
Nonseptate hyphae	Zygomycetes; *Mucor, Rhizopus,* and other genera	10-30	Hyphae are large, ribbonlike, often fractured or twisted; occasional septa may be present; smaller hyphae are confused with those of *Aspergillus* sp., particularly *A. flavus*
Hyaline septate hyphae	Dermatophytes		
	Skin and nails	3-15	Hyaline, septate hyphae are commonly seen; chains of arthroconidia may be present
	Hair	3-15	Arthroconidia on periphery of hair shaft producing a sheath are indicative of ectothrix infection; arthroconidia formed by fragmentation of hyphae within the hair shaft are indicative of endothrix infection
		3-15	Long hyphal filaments or channels within the hair shaft are indicative of favus hair infection
	Aspergillus sp.	3-12	Hyphae are septate and exhibit dichotomous, 45-degree angle branching; larger hyphae, often disturbed, may resemble those of Zygomycetes

TABLE 44.4

SUMMARY OF CHARACTERISTIC FEATURES OF FUNGI SEEN IN DIRECT EXAMINATION OF CLINICAL SPECIMENS—cont'd

MORPHOLOGICAL FORM FOUND IN SPECIMENS	ORGANISM(S)	SIZE RANGE (DIAMETER, μm)	CHARACTERISTIC FEATURES
	Geotrichum sp.	4-12	Hyphae and rectangular arthroconidia are present and are sometimes rounded; irregular forms may be present
	Trichosporon sp.	2-4 by 8	Hyphae and rectangular arthroconidia are present and sometimes rounded; occasionally blastoconidia may be present
Dematiaceous septate hyphae	*Bipolaris* sp. *Cladosporium* sp. *Curvularia* sp. *Drechslera* sp. *Exophiala* sp. *Exserohilum* sp. *Phialophora* sp. *Wangiella dermatitidis*	2-6	Dematiaceous polymorphous hyphae are seen; budding cells with single septa and chains of swollen rounded cells are often present; occasionally aggregates may be present in infection caused by *Phialophora* and *Exophiala* sp.
	Phaeoannellomyces (Exophiala) werneckii	1.5-5	Usually large numbers of frequently branched hyphae are present along with budding cells
Sclerotic bodies	*Cladosporium carrionii* *Fonsecaea compacta* *Fonsecaea pedrosoi* *Phialophora verrucosa* *Rhinocladiella aquaspersa*	5-20	Brown, round to pleomorphic, thick-walled cells with transverse septations; commonly, cells contain two fission planes that form a tetrad of cells; occasionally, branched septate hyphae may be found along with sclerotic bodies
Granules	*Acremonium* A. *falciforme* A. *kiliense* A. *recifei*	200-300	White, soft granules without a cementlike matrix
	Curvularia C. *geniculata* C. *lunata*	500-1,000	Black, hard grains with a cementlike matrix at periphery
	Aspergillus A. *nidulans*	65-160	White, soft granule without a cementlike matrix
	Exophiala E. *jeanselmei*	200-300	Black, soft granules, vacuolated, without a cementlike matrix, made of dark hyphae and swollen cells
	Fusarium		White, soft granules without a cementlike matrix
	F. *moniliforme*	200-500	
	F. *solani*	300-600	
	Leptosphaeria		
	L. *senegalensis*	400-600	Black, hard granules with cementlike matrix present
	L. *tompkinsii*	500-1,000	Periphery composed of polygonal swollen cells and center of a hyphal network
	Madurella		
	M. *grisea*	350-500	Black, soft granules without a cementlike matrix, periphery composed of polygonal swollen cells and center of a hyphal network
	M. *mycetomatis*	200-900	Black to brown, hard granules of two types: (1) Rust brown, compact, and filled with cementlike matrix (2) Deep brown, filled with numerous vesicles, 6-14 μm in diameter, cementlike matrix in periphery, and central area of light-colored hyphae.

Continued.

TABLE 44.4

SUMMARY OF CHARACTERISTIC FEATURES OF FUNGI SEEN IN DIRECT EXAMINATION OF CLINICAL SPECIMENS—cont'd

MORPHOLOGICAL FORM FOUND IN SPECIMENS	ORGANISM(S)	SIZE RANGE (DIAMETER, μm)	CHARACTERISTIC FEATURES
	Neotestudina		White, soft granules with cementlike matrix present at periphery
	N. rosatti	300-600	
	Pseudallescheria		White, soft granules composed of hyphae and swollen cells at periphery in cementlike matrix
	P. boydii	200-300	
	Pyrenochaeta		Black, soft granules composed of polygonal swollen cells at periphery, center is network of hyphae, no cementlike matrix present
	P. romeri	300-600	

From Merz, W.G., and Roberts, G.D. 1991. Detection and recovery of fungi in clinical specimens, In Balows, A., Hausler, W.J., Jr., Herrmann, K., et al., editors. Manual of clinical microbiology, ed 5. American Society for Microbiology, Washington, D.C.

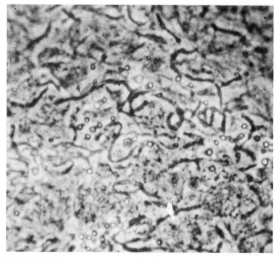

FIGURE 44.2

Potassium hydroxide preparation of skin, phase-contrast, *Malasezzia furfur,* showing spherical yeast cells and short hyphal fragments (500×).

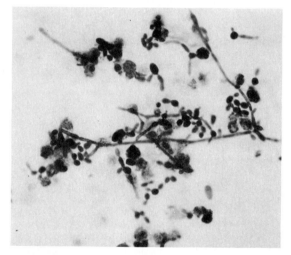

FIGURE 44.3

Gram stain of urine, *Candida albicans,* showing blastoconidia and pseudohyphae (500×).

cleus with a surrounding nuclear membrane; their cell wall is primarily composed of glucose and mannose polymers (chitin); their cell membrane contains sterols, and the organisms are resistant to antibacterial antibiotics. They show marked susceptibility to polyene antifungal agents, and all reproduce asexually; however, some reproduce sexually as well.

The fungi seen in the clinical laboratory can easily be separated into two groups based on the macroscopic appearance of the colonies formed. The yeasts produce moist, creamy, opaque, or pasty colonies on culture media, whereas the

molds (filamentous fungi) produce fluffy, cottony, woolly, or powdery colonies. Certain of the slow-growing pathogenic fungi are dimorphic and produce a mold form at 25° to 30° C and a yeast or spherule form at 35° to 37° C under certain circumstances.

The identification of the yeasts requires not only recognition of certain microscopic features, but also the use of biochemical tests to provide a definitive species identification. The definitive identification of the molds requires recognition of the characteristic microscopic features of the organism. The following paragraphs describe the basic microscopic morphological features of fungi,

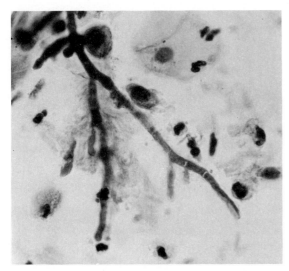

FIGURE 44.4

Papanicolaou stain of sputum, *Aspergillus fumigatus,* showing dichotomously branching septate hyphae (500×).

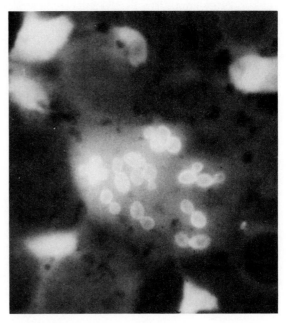

FIGURE 44.6

Calcofluor white stain of sputum with *Histoplasma capsulatum,* showing intracellular yeast cells 2 to 5 µm in diameter (500×).

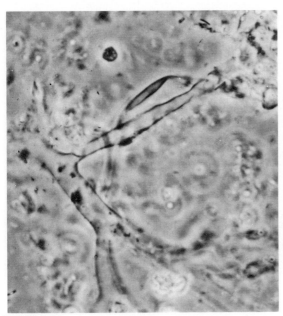

FIGURE 44.5

Potassium hydroxide preparation of sputum, phase contrast, *Rhizopus* species, showing fragmented portions of nonseptate hyphae of varying size (500×).

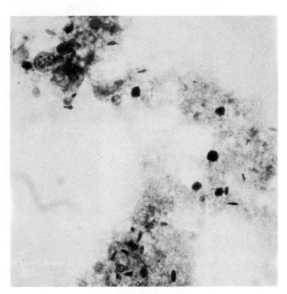

FIGURE 44.7

Periodic acid-Schiff stain of exudate with *Sporothrix schenckii,* showing cigar- to oval-shaped yeast cells (500×).

and representative photomicrographs are used to illustrate them.

The yeasts are unicellular organisms that reproduce by budding; their microscopic morphological features usually appear similar for different genera and are not particularly helpful in their separation. The basic structural units of the molds are tubelike projections known as hyphae (Figures 44.19 to 44.21). As the **hyphae** grow, they become intertwined to form a loose network called the **mycelium,** which penetrates the substrate from which it obtains the necessary nutrients for growth. The nutrient-absorbing and water-exchanging portion of the fungi is called the vegetative mycelium. The portion projecting above the substrate surface is known as the aerial mycelium; aerial mycelia often give rise to

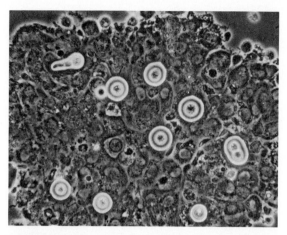

FIGURE 44.8

Potassium hydroxide preparation of pleural fluid, phase contrast micrograph with *Cryptococcus neoformans,* showing encapsulated, spherical yeast cells of varying sizes (500×).

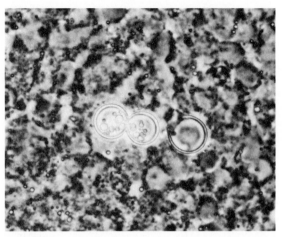

FIGURE 44.9

Potassium hydroxide preparation of exudate, phase-contrast micrograph with *Blastomyces dermatitidis,* showing large budding yeast cells with a distinct broad base between the cells (500×).

FIGURE 44.10

Auramine-rhodamine preparation of bone lesion. *Blastomyces dermatitis* showing characteristic broad-based budding yeast cell (500×).

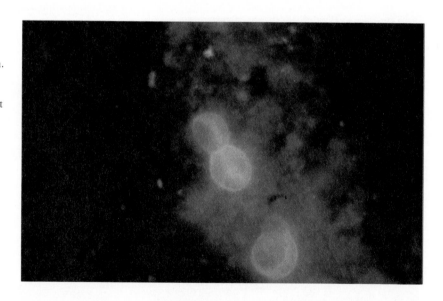

fruiting bodies from which the asexual spores are borne. Recognition of certain types of vegetative hyphae is helpful in placing the organism into a certain group. For example, dermatophytes often produce several types of hyphae, including antler hyphae (Figure 44.22) that are curved, freely branching, and antlerlike in appearance. Racquet hyphae are often found that are enlarged, club-shaped hyphae with the smaller end attached to the large end of an adjacent club-shaped hyphal strand (Figure 44.23). In addition, certain dermatophytes produce spiral hyphae that are coiled or corkscrewlike turns seen within the hyphal strand (Figure 44.24). These structures are not characteristic for any

certain group of organisms; however, they are usually found only in the dermatophytes.

The features by which most fungi are identified are related to the mode of sporulation and the morphology and arrangement of the spores produced. Fungi may reproduce sexually, asexually, or by both means. Sexual reproduction is associated with the production of specialized structures that result in nuclear fusion and the production of specialized spores. Some species of fungi produce sexual spores in a large saclike structure called an **ascocarp** (Figure 44.25). This, in turn, contains smaller sacs called asci, each of which contains four or eight ascospores. This type of sexual reproduction is sometimes seen in the fungi recovered

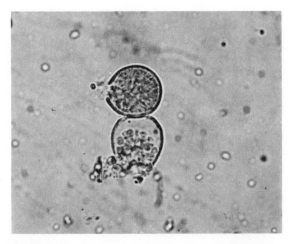

FIGURE 44.11

Potassium hydroxide preparation of sputum, bright field micrograph with *Coccidiodes immitis,* showing two spherules, filled with endospores, lying adjacent to each other that resemble *Blastomyces dermatitidis* (500×).

in the clinical microbiology laboratory. Most of the fungi of clinical importance reproduce asexually. There is a consensus that perhaps all fungi possess a sexual form; however, this form has not been observed on artificial culture media for most of the clinically important species. Asexual sporulation is the type of sporulation seen in most fungi encountered in a clinical laboratory. The type of spores, their morphology, and arrangement are important criteria for establishing the definitive identification of an organism.

The simplest type of sporulation is the development of the spore directly from the vegetative hyphae. Three types of spores are recognized. **Arthroconidia** are formed directly from the hyphae by fragmentation through the points of septation. When mature, they appear as square, rectangular, or barrel-shaped, thick-walled cells (Figure 44.26). These result from the simple fragmentation of the hyphae into spores, which are easily dislodged and disseminated into the environment. **Chlamydospores** (chlamydoconidia) are round, thick-walled resistant spores formed directly from the differentiation of the hyphae in which there is a concentration of protoplasm and nutrient material (Figure 44.27). These spores appear to be resistant resting spores produced by the rounding up and enlargement of the terminal cells of the hyphae. **Blastoconidia,** most often found in the yeasts, are cells produced by budding with the daughter cells being pinched off from portions of the mother cell through a constricted area (Figure 44.28). In some species of yeasts, most commonly *Candida,* blastoconidia may elongate and

remain attached and form structures called **pseudohyphae** (Figure 44.29).

A variety of other types of spores occur with many species of fungi. Conidia are asexual spores produced singly or in groups by specialized vegetative hyphal strands called **conidiophores.** In some instances, the conidia are freed from their point of attachment by pinching off, or abstriction. Some conidiophores terminate in a swollen structure called a vesicle. From the surface of the vesicle are formed secondary small flask-shaped phialides, which in turn give rise to long chains of conidia. This type of fruiting structure is characteristic of the aspergilli (Figure 44.30). A single slender, tubular conidiophore (phialide) that supports a cluster of conidia, held together as a gelatinous mass, is characteristic of certain fungi, including the genus *Acremonium* (Figure 44.31). In other instances, conidiophores branch into a structure known as a penicillus in which each branch terminates in secondary branches (metulae) and phialides from which chains of conidia are borne (Figure 44.32). Species of *Penicillium* and *Paecilomyces* are representative of this type of sporulation. In other instances, fungi may produce conidia of two sizes: **microconidia** are small, unicellular, round, elliptical, or pyriform in shape (Figure 44.33); **macroconidia** are large, usually multiseptate, and club- or spindle-shaped (Figure 44.34). Microconidia may be borne directly on the side of the hyphal strand or at the end of a long or short conidiophore. Macroconidia are usually borne on a short to long conidiophore and may be smooth-walled or rough-walled. Microconidia and macroconidia are seen in many fungal species and are not specific, except when they are used to differentiate the different genera of dermatophytes.

One group of fungi, the **Zygomycetes,** produces structures different from the ones previously described. The hyphae are sparsely septate, and asexual spores are produced in a large saclike structure called a **sporangium,** which is borne on the tip of a supporting structure (sporangiophore) (Figure 44.35). Sporulation takes place by progressive cleavage during maturation within the sporangium; the spores are produced and released by the rupture of the sporangial wall. This type of sporulation is characteristic of many of the Zygomycetes.

CLASSIFICATION OF FUNGI

The botanical taxonomical scheme for grouping the fungi has very little value in a clinical microbi-

FIGURE 44.12

Sporothrix schenkii in mouse testis. Methenamine silver stain (440×).

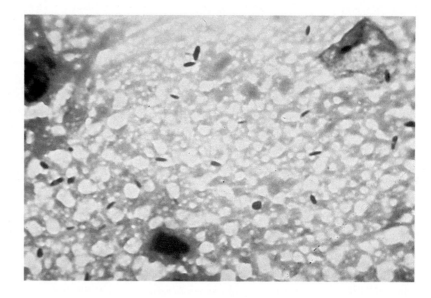

FIGURE 44.13

Sclerotic bodies of chromoblastomycosis in tissue (400×). (Courtesy Upjohn Co.)

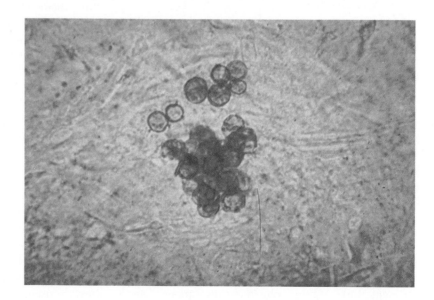

ology laboratory. Table 44.5 is a simplified taxonomical scheme illustrating the major groups to which the clinically important fungi belong. Using this classification scheme, the fungi can be included within four major divisions: Zygomycota, Ascomycota, Basidiomycota, and Deuteromycota. The division Zygomycota includes organisms that have aseptate hyphae, reproduce asexually by spores produced inside of sporangia, and reproduce sexually by the production of zygospores. *Mucor, Rhizopus,* and *Absidia* are the most common members of this group. The Ascomycota includes organisms that have septate hyphae, produce conidia, or reproduce sexually by means of ascospores produced within an ascus. Common members of this group include *Pseudallescheria boydii* and species of *Aspergillus.* The Basidiomy-

cota includes a group of organisms that have septate hyphae and reproduce asexually by conidia or sexually by means of basidiospores. *Filobasidiella (Cryptococcus)* is the major genus within this division that is encountered in a clinical laboratory. The major group to which most of the clinically important fungi belong is the Deuteromycota. This group of organisms has septate hyphae and reproduces asexually with the production of conidia. Whether they reproduce sexually is not known. As previously mentioned, the botanical taxonomical schema has little role in the clinical laboratory. Clinicians find more value in categorizing the fungi into three or four categories of mycoses: (1) the superficial or cutaneous mycoses; (2) subcutaneous mycoses; (3) the systemic mycoses; and (4) the opportunistic mycoses.

FIGURE 44.14

Candida albicans in urine, calcofluor white stain (500×).

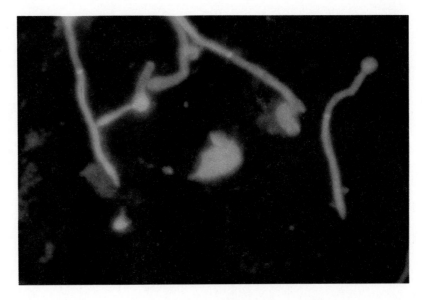

FIGURE 44.15

Coccidioides immitis. Ruptured spherule with endospores (500×).

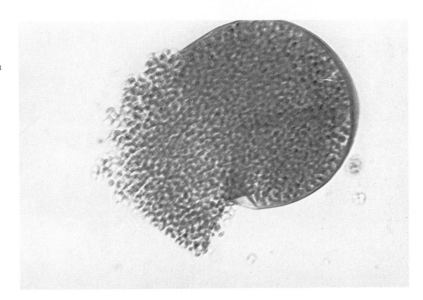

FIGURE 44.16

Histoplasma capsulatum in lung. Methenamine silver stain (440×).

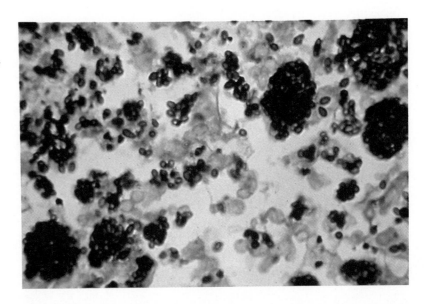

FIGURE 44.17

Histoplasma capsulatum in neutrophil on peripheral blood smear (1000×).

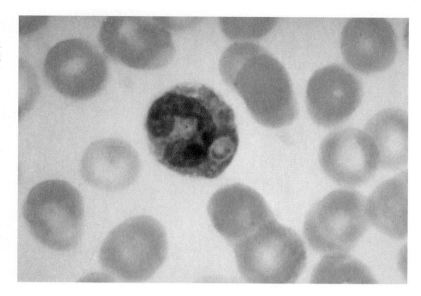

FIGURE 44.18

Blastomyces dermatitidis in tissue. Methenamine silver stain (440×).

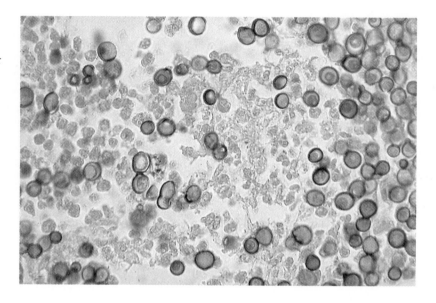

The superficial or cutaneous mycoses are fungal infections that involve the hair, skin, or nails without invasion of the tissue. The group of fungi in this category that are most commonly recovered are the dermatophytes; agents of infections such as tinea versicolor, tinea nigra, and piedra, which involve the outermost keratin layer of the skin, are also included.

Subcutaneous mycoses are infections confined to the subcutaneous tissue without dissemination to distant sites. Examples include sporotrichosis, chromoblastomycosis, mycetoma, and phaeohyphomycosis.

The traditionally defined systemic mycoses are caused primarily by fungi belonging to the genera *Blastomyces, Coccidioides, Histoplasma,* and *Paracoccidioides.* Infections caused by these organisms involve the lungs primarily, but also may become widely disseminated and involve any organ of the body.

The group of fungi included in the opportunistic mycoses includes an ever-expanding list of organisms. These infections are commonly found in patients who are immunocompromised, usually by an underlying disease process or by immunosuppressive agents. Fungi previously thought to be nonpathogenic can cause infection in the compromised host; commonly encountered infections include aspergillosis, zygomycosis, candidosis, cryptococcosis, and others, all of which may cause disseminated (systemic) disease.

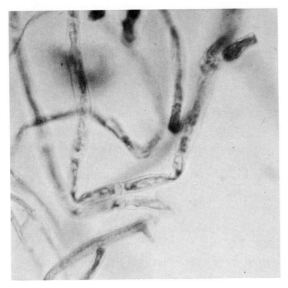

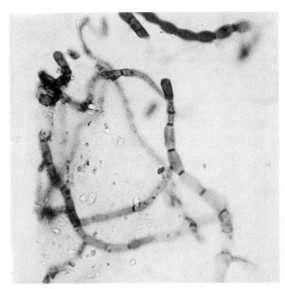

FIGURE 44.19
Hyaline hyphae showing septations (440×).

FIGURE 44.20
Dematiaceous hyphae showing pigmentation and septations (440×).

FIGURE 44.21
Hyaline hyphae lacking septations (aseptate) (400×).

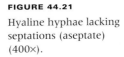

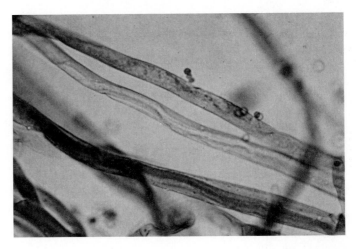

This type of classification allows the clinician to attempt to categorize organisms logically into groups having clinical relevance. Table 44.6 presents an example of the clinical classification of fungi that will be useful to the clinician. This type of classification also plays little role in the identification of fungi in the clinical microbiology laboratory.

CLINICAL LABORATORY WORKING SCHEMA

To assist persons working in clinical microbiology laboratories with the identification of clinically important fungi, Koneman and Roberts[44] have suggested the use of a practical working

schema designed to (1) assist in the recognition of fungi most commonly encountered in clinical specimens, (2) assist with the recognition of fungi recovered on culture media that belong to the strictly pathogenic fungi, and (3) provide a pathway that is easy to follow and that allows a positive identification of an organism to be made on the basis of a few colonial and microscopic features of the organism. Table 44.7 presents these features; however, it must be recognized that the table includes only those organisms commonly seen in the clinical laboratory. With practice, most laboratorians should be able to recognize the fungi commonly encountered on a day-to-day basis. The identification of others will require the use of various texts con-

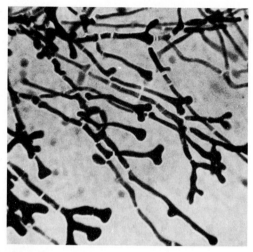

FIGURE 44.22

Antler hyphae showing swollen hyphal tips resembling antlers, with lateral and terminal branching (favic chandeliers) (500×).

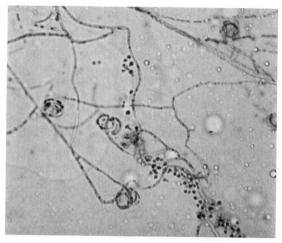

FIGURE 44.24

Spiral hyphae exhibiting corkscrewlike turns (440×).

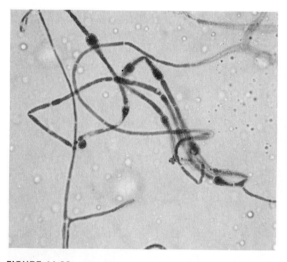

FIGURE 44.23

Racquet hyphae showing a swollen area resembling a tennis racquet (400×).

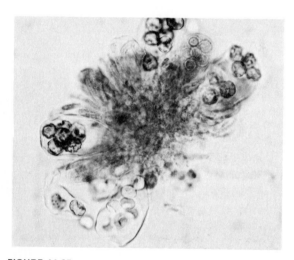

FIGURE 44.25

Ascocarp showing dark-appearing ascospores (440×).

taining photomicrographs useful for their identification.

The use of this schema requires that one first examine the culture for the presence or absence of septae. If the hyphae appear to be predominantly nonseptate, zygomycetes should be considered. If the hyphae are septate, they must be examined further for the presence or absence of pigment. If a dark pigment is present in the conidia or the hyphae, the organism is considered to be dematiaceous and the conidia are then examined for their morphological features and arrangement on the hyphae. If the hyphae are nonpigmented, they are considered to be hyaline. They are then

examined for the type and the arrangement of the conidia produced. The molds are identified by recognition of their characteristic microscopic features.

The yeasts may be presumptively identified by the presence or absence of hyphae, pseudohyphae, blastoconidia, and arthroconidia on cornmeal agar. The identification of yeasts is discussed later in this chapter.

EXTENT OF IDENTIFICATION OF FUNGI RECOVERED FROM CLINICAL SPECIMENS

The question of when and how far to go with identification of fungi recovered from clinical specimens presents an interesting challenge. The

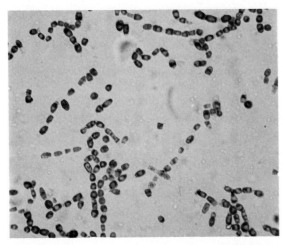

FIGURE 44.26

Arthroconidia formation produced by the breaking down of a hyphal strand into individual rectangular units (440×).

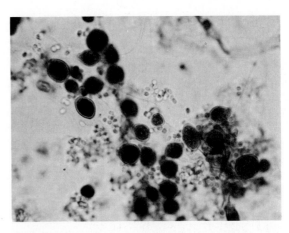

FIGURE 44.27

Chlamydospores composed of thick-walled spherical cells (440×).

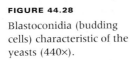

FIGURE 44.28

Blastoconidia (budding cells) characteristic of the yeasts (440×).

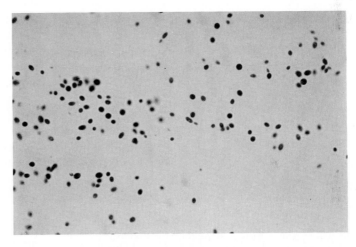

current emphasis on cost containment (Chapter 4) and the ever-increasing number of opportunistic fungi causing infection in compromised patients cause one to consider whether all fungi recovered from clinical specimens should be thoroughly identified and reported. Murray et al.[57] in 1977 were concerned with the time and expense associated with the identification of yeasts from respiratory tract specimens. Since these are the specimens most commonly submitted for fungal culture, they questioned whether it was important to provide an identification for every organism recovered. After evaluating the clinical usefulness of information provided by the identification of yeast recovered from respiratory tract specimens, they suggested that (1) the routine identification of yeasts recovered in culture from respiratory secretions is not warranted, but that all yeasts should be screened for the presence of

Cryptococcus neoformans; (2) all respiratory secretions submitted for fungal culture, regardless of the presence or absence of oropharyngeal contamination, should be cultured since common pathogens such as *Histoplasma capsulatum, Blastomyces dermatitidis, Coccidioides immitis,* and *Sporothrix schenckii* may be recovered; and (3) the routine identification of yeast in respiratory secretions is of little or no value to the clinician and probably represents "normal flora" with the exception of *Cryptococcus neoformans.*

When and how far to proceed with an identification of the molds is a much more difficult question to answer. A consensus of a number of clinical mycologists is that all commonly encountered molds should be identified and reported regardless of the clinical source. Those organisms that fail to sporulate after a reasonable time should be reported as being present but the iden-

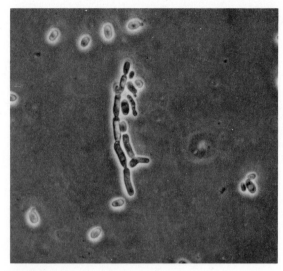

FIGURE 44.29

Pseudohyphae consisting of elongated cells with constrictions where attached (440×).

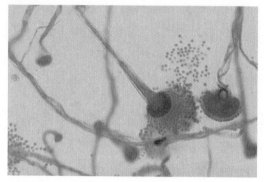

FIGURE 44.30

Conidia (asexual spores) produced on specialized structures (conidiophores) of *Aspergillus* (440×).

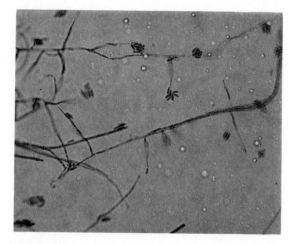

FIGURE 44.31

Simple tubular phialide with a cluster of conidia at its tip, characteristic of *Acremonium* (440×).

tification need not be attempted if the dimorphic fungi have been ruled out or if the clinician feels that the organism is not clinically significant. Ideally, all laboratories should identify all fungi recovered from clinical specimens; however, the limits of practicality and economic considerations play a definite role in making a decision as to how far to go with their identification. Each individual laboratory director, in consultation with the clinicians being served, will have to make this decision after considering the patient population, laboratory practice, and economic impact.

LABORATORY SAFETY CONSIDERATIONS

Although there are risks associated with the handling of fungi recovered from clinical specimens, a common-sense approach concerning their handling will protect the laboratory from contamination and workers from infection.

It is necessary that all mold cultures and clinical specimens be handled in a Class II, B3 biological safety cabinet (Chapter 2), **with no exceptions.** Some laboratory directors feel that it is necessary to handle mold cultures within an enclosed biological safety cabinet equipped with gloves; however, this is not necessary if a laminar flow biological safety cabinet is used. It is permissible, however, to handle yeast cultures on the bench top; they must be treated as infectious agents. The use of an electric incinerator or a gas flame is suitable for the decontamination of a loop used for transfer of yeast cultures. Cultures of organisms suspected of being pathogens should be sealed with tape to prevent laboratory contamination and should be autoclaved as soon as a definitive identification of the organism is made.

Few problems concerning laboratory contamination or infection of laboratory personnel will be associated with the handling of fungi if common safety precautions and careful handling of cultures are observed.

GENERAL CONSIDERATIONS FOR THE IDENTIFICATION OF MOLDS

The identification of molds is made using a combination of (1) the growth rate, (2) colonial morphological features, and (3) microscopic morphological features. In most instances the latter provides the most definitive means for identification. The determination of the growth rate of a culture can be one of the most helpful observations made when examining a mold culture. However, this

FIGURE 44.32

Complex method of sporulation in which conidia are borne on phialides produced on secondary branches (metulae), characteristic of *Penicillium* (440×).

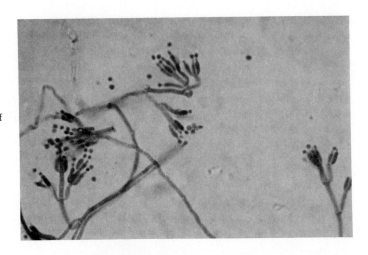

FIGURE 44.33

Numerous small, spherical microconidia contrasted with a large, elongated macroconidium (440×).

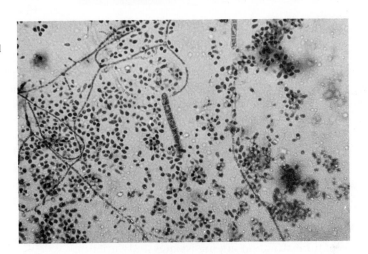

FIGURE 44.34

Large, rough-walled macronidia of *Microsporum canis* (440×).

information may be of limited value since the growth rate of certain fungi is variable, depending on the amount of inoculum present in a clinical specimen. In general, the growth rate for the dimorphic fungi, including *Blastomyces dermatitidis, Coccidioides immitis, Histoplasma capsulatum,* and *Paracoccidioides brasiliensis,* is slow; 1 to 4 weeks are usually required before colonies become visible. In some instances, however, cultures of *B. dermatitidis, H. capsulatum* and *C. immitis* may be detected within 3 to 5 days. This is a somewhat uncommon circumstance and is encountered only when large numbers of the organism are present on the culture medium. In contrast, colonies of

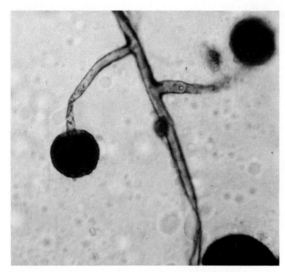

FIGURE 44.35

Large, saclike sporangium that contains sporangiospores, characteristic of the zygomycetes (250×).

the zygomycetes may appear within 24 hours, whereas the other hyaline and dematiaceous fungi often exhibit growth within 1 to 5 days. The growth rate of an organism is an important observation, but it must be used in combination with other features before the definitive identification of an organism can be made.

The colonial morphological features are also of limited value in identifying the molds because of natural variation among isolates and variation of colonies on different culture media. Although it may be possible to recognize certain organisms recovered repeatedly in the laboratory, based on their colonial morphological features, this is an unreliable criterion and should be used only to supplement the information obtained by microscopic examination of the culture. When examining the colonial morphological features, the type of culture media used and the incubation conditions must be considered. For example, *Histoplasma capsulatum,* which appears as a white to tan fluffy mold on brain-heart infusion agar, may appear yeast-like in appearance when grown on the same medium containing blood enrichment. It is recommended that the colonial morphological features be used to supplement the information obtained from the growth rate and microscopic morphological features.

In general, the microscopic morphological features of the molds are stable and exhibit minimal variation. The definitive identification is based on the characteristic shape, method of production, and arrangement of spores; however, the size of the hyphae also provides helpful informa-

tion. The large ribbon-like hyphae of the zygomycetes are easily recognized, whereas small hyphae, 1 to 2 μm in size, may suggest the presence of one of the dimorphic fungi.

The fungi may be prepared for microscopic observation using several techniques. The procedure traditionally used by most laboratories is the wet mount examination (Procedure 44.1). The wet mount can be prepared easily and quickly and often is sufficient to identify many of the fungi commonly encountered in the clinical laboratory.

The easiest, most economical, and most suitable method for the microscopic identification of fungi is the adhesive tape preparation (Procedure 44.2). This method is recommended for routine use in clinical microbiology laboratories of all sizes.

Some laboratories prefer to use the microslide culture (Procedure 44.3) for making the microscopic identification of an organism. This method might appear to be the most suitable since it allows one to observe microscopically the fungus growing directly underneath the coverslip. Microscopic features should be easily discerned, structures should be intact, and a large number of representative areas of growth are available for observation.

SUPERFICIAL AND CUTANEOUS MYCOSES

The superficial fungal infections are produced by fungi of very low virulence. They exhibit little tissue invasion and evoke minimal host response in immunocompetent patients; the infections they produce are asymptomatic, and aesthetically displeasing lesions are usually produced. Lesions usually exhibit hyperpigmentation, hypopigmentation, or nodular mass production on the hair shaft distal to the skin. Most are uncommonly seen except for tinea versicolor.

TINEA VERSICOLOR (PITYRIASIS VERSICOLOR)

Tinea versicolor is a skin infection characterized by superficial brownish scaly areas on light-skinned persons and lighter areas on dark-skinned persons. The lesions occur on the smooth surfaces of the body (i.e., the trunk, arms, shoulders, and face). It has a worldwide distribution. It is caused by *Malassezia furfur,* an organism not usually cultured in the clinical laboratory. Recovery of the organism is not required to establish a diagnosis, and it is seldom attempted; if one wishes to culture the organism, an agar medium

TABLE 44.5

PHYLOGENETIC POSITION OF MEDICALLY SIGNIFICANT FUNGI

CLASS	ORDER	GENUS/SPECIES
Phylum Zygomycota		*Basidiobolus*
Zygomycetes	Entomophthorales	*Absidia*
	Mucorales	*Cunninghamella*
		Mucor
		Rhizopus
		Syncephalastrum
Phylum Ascomycota		
Hemiascomycetes	Endomycetales	*Endomyces (Geotrichum* sp.)*
		*Kluyveromyces (Candida pseudotropicalis)**
Loculoascomycetes	Myriangiales	*Piedraia hortae*
	Microascales	*Pseudallescheria boydii*
		Scedosporium inflatum
Plectomycetes	Eurotiales	*Emericella (Aspergillus nidulans)**
	Onygerales	*Ajellomyces (Histoplasma capsulatum, Blastomyces dermatitidis)**
		Arthroderma (Trichophyton sp. and *Microsporum* sp.)*
Phylum Basidiomycota		
Teliomycetes	Filobasidiales	*Filobasidiella (Cryptococcus neoformans)**
Phylum Deuteromycota		
Blastomycetes		*Candida*
Hyphomycetes	Moniliales	*Acremonium*
		Aspergillus
		Blastomyces
		Chrysosporium
		Coccidioides
		Epidermophyton
		Geotrichum
		Gliocladium
		Histoplasma
		Microsporum
		Paecilomyces
		Paracoccidioides
		Penicillium
		Sepedonium
		Scopulariopsis
		Sporothrix
		Trichoderma
		Trichophyton

Continued.

overlaid with a fatty acid (olive oil) is used. Most often, the detection of *M. furfur* is made by the direct microscopic examination of skin scales. Here the organism will easily be recognized as oval- or bottle-shaped cells that exhibit monopolar budding in the presence of a cell wall with a septum at the site of the bud scar (see Figure 44.2). In addition, small hyphal fragments are observed. *Malassezia pachydermatis*, another species, may be recovered from skin lesions of patients; it requires no lipid additives.

It should be mentioned that *M. furfur* is currently being seen as the cause of disseminated infection in infants and young children and even adults given lipid replacement therapy.[52] The organism is readily cultured from blood or skin lesions without the use of a fatty acid additive. *M. pachydermatis* is now seen as the cause of fungemia in immunocompromised patients. The morphological form seen in the direct microscopic examination of specimens from these patients is the yeast form without the presence of pseudohyphae.

TINEA NIGRA

Tinea nigra is an infection manifested by blackish-brown macular patches on the palm of the hand or sole of the foot. Lesions have been compared

TABLE 44.5

PHYLOGENETIC POSITION OF MEDICALLY SIGNIFICANT FUNGI—cont'd

CLASS	ORDER	GENUS/SPECIES
Coelomycetes	Sphaeropsidales	*Alternaria*
		Aureobasidium
		Bipolaris
		Cladosporium
		Curvularia
		Dreschslera
		Exophiala
		Exserohilum
		Fonsecaea
		Helminthosporium
		Madurella
		Nigrospora
		Phaeoannellomyces
		Phialophora
		Rhinocladiella
		Stemphylium
		Ulocladium
		Wangiella
		Xylohypha
		Epicoccum
		Fusarium
		Phoma

*When the sexual form is known.
Modified from Kwon-Chung, K.J., and Bennett, J.E. 1992. Medical mycology. Lea & Febiger, Philadelphia.

TABLE 44.6

GENERAL CLINICAL CLASSIFICATION OF PATHOGENIC FUNGI

CUTANEOUS	SUBCUTANEOUS	OPPORTUNISTIC	SYSTEMIC
Superficial mycoses	Chromoblastomycosis	Aspergillus	Aspergillosis
Tinea	Sporotrichosis	Candidosis	Blastomycosis
Piedra	Mycetoma (eumycotic)	Cryptococcosis	Candidosis
Candidosis	Phaeohyphomycosis	Geotrichosis	Coccidioidomycosis
Dermatophytosis		Zygomycosis	Histoplasmosis
		Fusarium infection	Cryptococcosis
		Trichosporonosis	Geotrichosis
			Paracoccidioidomycosis
			Zygomycosis
			Fusarium infection
			Trichosporonosis

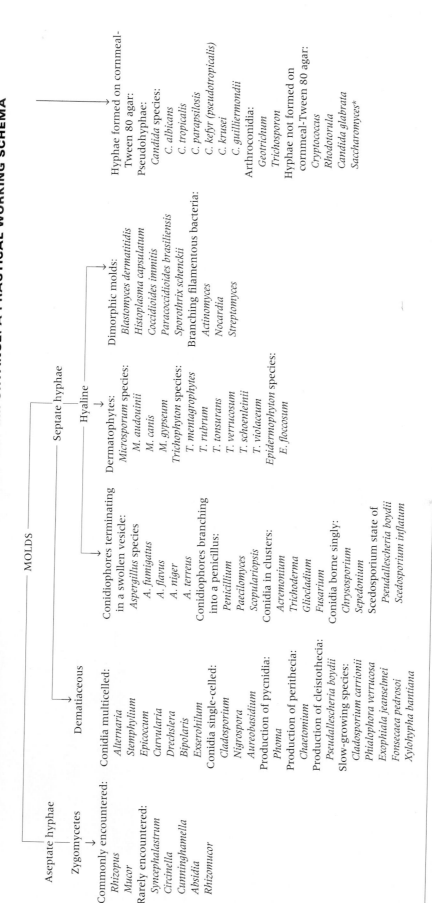

TABLE 44.7

MOST COMMONLY ENCOUNTERED FUNGI OF CLINICAL LABORATORY IMPORTANCE: A PRACTICAL WORKING SCHEMA

MOLDS

Aseptate hyphae

Zygomycetes

Commonly encountered:
Rhizopus
Mucor
Rarely encountered:
Syncephalastrum
Circinella
Cunninghamella
Absidia
Rhizomucor

Septate hyphae

Dematiaceous

Conidia multicelled:
Alternaria
Stemphylium
Epicoccum
Curvularia
Drechslera
Bipolaris
Exserohilum
Conidia single-celled:
Cladosporium
Nigrospora
Aureobasidium
Production of pycnidia:
Phoma
Production of perithecia:
Chaetomium
Production of cleistothecia:
Pseudallescheria boydii
Slow-growing species:
Cladosporium carrionii
Phialophora verrucosa
Exophiala jeanselmei
Fonsecaea pedrosoi
Xylohypha bantiana

Hyaline

Conidiophores terminating in a swollen vesicle:
Aspergillus species
A. fumigatus
A. flavus
A. niger
A. terreus
Conidiophores branching into a penicillus:
Penicillium
Paecilomyces
Scopulariopsis
Conidia in clusters:
Acremonium
Trichoderma
Gliocladium
Fusarium
Conidia borne singly:
Chrysosporium
Sepedonium
Scedosporium state of
Pseudallescheria boydii
Scedosporium inflatum

Dermatophytes:
Microsporum species:
M. audouinii
M. canis
M. gypseum
Trichophyton species:
T. mentagrophytes
T. rubrum
T. tonsurans
T. verrucosum
T. schoenleinii
T. violaceum
Epidermophyton species:
E. floccosum

Dimorphic molds:
Blastomyces dermatitidis
Histoplasma capsulatum
Coccidioides immitis
Paracoccidioides brasiliensis
Sporothrix schenckii
Branching filamentous bacteria:
Actinomyces
Nocardia
Streptomyces

Hyphae formed on cornmeal-Tween 80 agar:
Pseudohyphae:
Candida species:
C. albicans
C. tropicalis
C. parapsilosis
C. kefyr (pseudotropicalis)
C. krusei
C. guilliermondii
Arthroconidia:
Geotrichum
Trichosporon
Hyphae not formed on cornmeal-Tween 80 agar:
Cryptococcus
Rhodotorula
Candida glabrata
Saccharomyces*

From Koneman, E.W., and Roberts, G.D. 1985. Practical laboratory mycology, ed 3. Williams & Wilkins, Baltimore.
*Rudimentary hyphae may be present.

PROCEDURE 44.1

WET MOUNT

1. With a wire bent at a 90-degree angle, cut out a small portion of an isolated colony. The portion should be removed from a point intermediate between the center and the periphery. The portion removed should contain a small amount of the supporting agar.

2. Place the portion of the culture onto a slide to which has been added a drop of lactophenol cotton or aniline blue (Figure 44.36).

3. Place a coverslip into position and apply gentle pressure with a pencil eraser or other suitable object to disperse the growth and agar. Examine microscopically.

The major disadvantage of the wet mount is that the characteristic arrangement of spores is disrupted when pressure is applied to the coverslip. This method is suitable in many instances since characteristic spores are often seen but, when their arrangement cannot be determined, it is not adequate to make a definitive identification.

---

with silver nitrate staining of the skin. The etiological agent is *Phaeoannellomyces werneckii,* a dematiaceous fungus. Initial colonies of *P. werneckii* may be black, shiny, and yeastlike in appearance. With age, colonies become filamentous with velvety-gray aerial hyphae. Microscopically, the yeastlike growth consists of olive-colored budding cells that are one- or two-celled. Older colonies exhibit one- or two-celled conidia that are produced by annelides, which bear successive rings (annelations) that are difficult to see microscopically.

PIEDRA

Black piedra is a fungal infection of the hair of the scalp and rarely of axillary and pubic hair. The etiological agent is a dematiaceous fungus called *Piedraia hortae.* The disease occurs primarily in tropical areas of the world; cases have been reported in Africa, Asia, and Latin America. Portions of hair are examined (in wet mounts using potassium hydroxide that is gently heated) for the presence of nodules composed of cemented mycelium. When mature nodules are crushed, oval asci, containing two to eight aseptate ascospores, 19 to 55 μm long by 4 to 8 μm in diameter are seen. The asci are spindle shaped and have a filament at each pole. The organism is easily cultured on any fungal culture medium lacking cycloheximide.

White piedra is an uncommon fungal infection found in both tropical and temperate regions of the world. It is characterized by the development of soft, yellow or pale brown aggregations around hair shafts in the axillary, facial, genital, and scalp regions of the body. The etiological agent is *Trichosporon beigelii.* It frequently invades the cortex of the hair and causes damage. White nodules are removed and observed using the potassium hydroxide preparation after applying light pressure to the coverslip so that crushing of the nodule occurs. Hyaline hyphae, 2 to 4 μm in width, and arthroconidia are found in the preparation of the cementlike material binding the hyphae together. The organism may be identified in culture by blastoconidia and arthroconidia formation. *T. beigelii* may be distinguished from the other species in the genus by its inability to ferment carbohydrates and ability to aerobically utilize certain substrates. *T. beigelii* may produce sys-

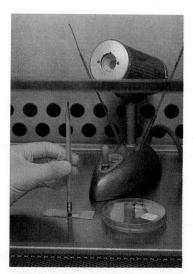

FIGURE 44.36

Performance of a wet mount, showing dispersion of growth under coverslip using pressure.

PROCEDURE 44.2

ADHESIVE (SCOTCH) TAPE PREPARATION

1. Touch the adhesive side of a small length of transparent tape to the surface of the colony.

2. Adhere the length of tape to the surface of a microscope slide to which has been added a drop of lactophenol cotton or aniline blue (Figure 43.37).

3. Observe microscopically for the characteristic shape and arrangement of the spores.

The transparent adhesive tape preparation allows one to observe the organism microscopically approximately the way it sporulates in culture. The spores are usually intact, and the microscopic identification of an organism can be made with ease. If the tape is not pressed firmly enough to the colonies' surface, the sample may not be adequate for an identification. In instances when spores are not observed, a wet mount should be made as a backup step. There have been situations in which the macroconidia of *Histoplasma* *capsulatum* were seen in wet mount preparations when the adhesive tape preparation revealed only hyphal fragments. On the contrary, there have been instances in which cultures have sporulated heavily and revealed only the presence of conidia when the adhesive tape preparation was observed. In this type of situation, a second adhesive tape preparation should be made from the periphery of the colony where sporulation is not as heavy.

---

temic infection in immunocompromised hosts, particularly patients with leukemia.

CUTANEOUS INFECTIONS (DERMATOMYCOSES)

Dermatomycoses are fungal infections that involve the superficial areas of the body, including the hair, skin, and nails. The genera *Trichophyton, Microsporum,* and *Epidermophyton* are the only fungi associated with the dermatomycoses. Such cutaneous mycoses are perhaps the most common fungal infections of humans and are usually referred to as **tinea** (Latin for "worm" or "ringworm"). The gross appearance of the lesion is that of an outer ring of an active, progressing infection with central healing within the ring. These infections may be characterized by another Latin noun to designate the area of the body involved. For example, tinea corporis (body), tinea cruris (groin), tinea capitis (scalp and hair), tinea barbae (beard), and tinea unguium (nail) are used to designate the type of infection observed. The group of fungi commonly called dermatophytes break down and utilize keratin as a source of nitrogen but are usually incapable of penetrating the subcutaneous tissue. The genus *Trichophyton* is capable of invading the hair, skin, and nails, whereas the genus *Microsporum* involves only the hair and skin; the genus *Epidermophyton* involves the skin and nails.

COMMON SPECIES

Common species of dermatophytes recovered from clinical specimens, in order of frequency, include *Trichophyton rubrum, Trichophyton mentagrophytes, Epidermophyton floccosum, Trichophyton tonsurans, Microsporum canis,* and *Trichophyton verrucosum.*[4]

Since the dermatophytes generally present a similar microscopic appearance within infected hair, skin, or nails, the final identification can be made only by culture. A summary of the colonial and microscopic morphological features of these fungi is presented in Table 44.8. Figure 44.39 presents an identification schema that will be useful to the clinical microbiologist for the identification of commonly encountered dermatophytes.

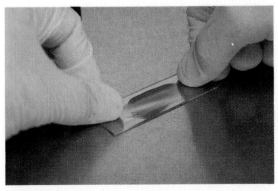

FIGURE 44.37

Scotch tape preparation, showing placement of tape onto slide containing lactophenol cotton blue.

PROCEDURE 44.3

MICROSLIDE CULTURE

1. Cut a small block of a suitable agar medium that has been previously poured into a culture dish to a depth of approximately 2 mm. The block may be cut using a sterile scalpel blade or with a sterile test tube that has no lip (which produces a round block).

2. Place a sterile microscope slide onto the surface of a culture dish containing sterile 2% agar. As an alternative, place a round piece of filter paper or paper towel into a sterile culture dish, add two applicator sticks, and position the microscope slide on top.

3. Add the agar block to the surface of the sterile microscope slide.

4. With a right-angle wire, inoculate the four quadrants of the agar plug with the organisms (Figure 44.38).

5. Apply a sterile coverslip onto the surface of the agar plug.

6. If the filter paper applicator stick method is used, add a small amount of sterile water to the bottom of the culture dish. Replace the lid of the culture and allow it to incubate at 30° C.

7. After a suitable incubation period, remove the coverslip (working inside of a biological safety cabinet) and place it on a microscope slide containing a drop of lactophenol cotton or aniline blue. It is often helpful to place the coverslip near the opening of an incinerator-burner to allow rapid drying of the organism on the coverslip before adding it to the stain.

8. Observe microscopically for the characteristic shape and arrangement of spores.

9. The remaining agar block may be used later (if the microslide culture is unsatisfactory for the microscopic identification) if it is allowed to incubate further. The agar plug is then removed and discarded, and a drop of lactophenol cotton or aniline blue is placed on the area of growth and a coverslip is positioned into place. Many laboratorians like to make two cultures on the same slide so that if characteristic microscopic features are not observed on examination of the first culture, the second will be available after an additional incubation period.

Although this method is ideal for making a definitive identification of an organism, it is the least practical of all the methods described. It should be reserved for those instances in which an identification cannot be made on the basis of an adhesive tape preparation or wet mount. **Caution: do not make slide cultures of slowly growing organisms suspected to be dimorphic pathogens such as *H. capsulatum, B. dermatitidis, C. immitis, P. brasiliensis*, or *S. schenckii*.** Microslide cultures must be observed only after a coverslip has been removed from the agar plug and not while it is in position on top of the agar plug. The latter method of observation is highly dangerous and is not to be used in the clinical laboratory.

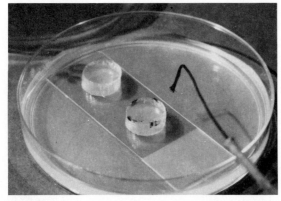

FIGURE 44.38

Microslide culture, showing inoculation of agar plug.

Descriptions of the common species of fungi causing dermatomycoses in the United States follow; other geographically limited species that are uncommonly encountered are described in the references cited.[4,79]

Figure 44.39 begins with the microscopic features of the dermatophytes as they might be observed in an initial examination of the culture. In many instances the primary recovery medium fails to function well as a sporulation medium. It is commonly necessary to subculture the initial growth onto cornmeal agar or potato dextrose agar so that sporulation will occur.

The genus *Trichophyton* is recognized by the presence of microconidia borne either laterally

TABLE 44.8

CHARACTERISTICS OF DERMATOPHYTES COMMONLY RECOVERED IN THE CLINICAL LABORATORY

DERMATOPHYTE	COLONIAL MORPHOLOGY	GROWTH RATE	MICROSCOPIC IDENTIFICATION
Microsporum audouinii *	Downy white to salmon pink colony: reverse tan to salmon pink	2 weeks	Sterile hyphae: terminal chlamydospores, favic chandeliers, and pectinate bodies; macroconidia rarely seen—bizarre-shaped if seen; microconidia rare or absent
Microsporum canis	Colony usually membranous with feathery periphery; center of colony white to buff over orange-yellow; lemon-yellow or yellow-orange apron and reverse	1 week	Thick-walled, spindle-shaped, multiseptate, rough-walled macroconidia, some with a curved tip; microconidia rarely seen
Microsporum gypseum	Cinnamon-colored, powdery colony; reverse light tan	1 week	Thick-walled rough, elliptical multiseptate macroconidia; microconidia few or absent
Epidermophyton floccosum	Center of colony tends to be folded and is khaki green; periphery is yellow; reverse yellowish brown with observable folds	1 week	Macroconidia large, smooth-walled, multiseptate, clavate, and borne singly or in clusters of two or three; microconidia not formed by this species
Trichophyton mentagrophytes	Different colonial types; white to pinkish, granular and fluffy varieties; occasional light yellow periphery in younger cultures; reverse buff to reddish brown	7-10 days	Many round to globose microconidia most commonly borne in grapelike clusters or laterally along the hyphae; spiral hyphae in 30% of isolates; macroconidia are thin-walled, smooth, club-shaped, and multiseptate; numerous or rare depending upon strain
Trichophyton rubrum	Colonial types vary from white downy to pink granular; rugal folds are common; reverse yellow when colony is young; however, wine red color commonly develops with age	2 weeks	Microconidia usually teardrop, most commonly borne along sides of the hyphae; macroconidia usually absent, but when present are smooth, thin-walled, and pencil-shaped
Trichophyton tonsurans	White, tan to yellow or rust, suedelike to powdery; wrinkled with heaped or sunken center; reverse yellow to tan to rust red	7-14 days	Microconidia are teardrop or club-shaped with flat bottoms; vary in size but usually larger than other dermatophytes; macroconidia rare and balloon forms found when present
Trichophyton schoenleinii *	Irregularly heaped, smooth white to cream colony with radiating grooves; reverse white	2-3 weeks	Hyphae usually sterile; many antler-type hyphae seen (favic chandeliers)
Trichophyton violaceum *	Port wine to deep violet colony, may be heaped or flat with waxy-glabrous surface; pigment may be lost on subculture	2-3 weeks	Branched, tortuous hyphae that are sterile; chlamydospores commonly aligned in chains
Trichophyton verrucosum	Glabrous to velvety white colonies; rare strains produce yellow-brown color; rugal folds with tendency to sink into agar surface	2-3 weeks	Microconidia rare; large and tear-drop when seen; macroconidia extremely rare, but form characteristic "rat-tail" types when seen; many chlamydospores seen in chains, particularly when colony is incubated at 37° C

From Koneman, E.W., and Roberts, G.D. 1985. Practical laboratory mycology, ed 3. Williams & Wilkins, Baltimore.
*These organisms are rarely recovered in the United States.

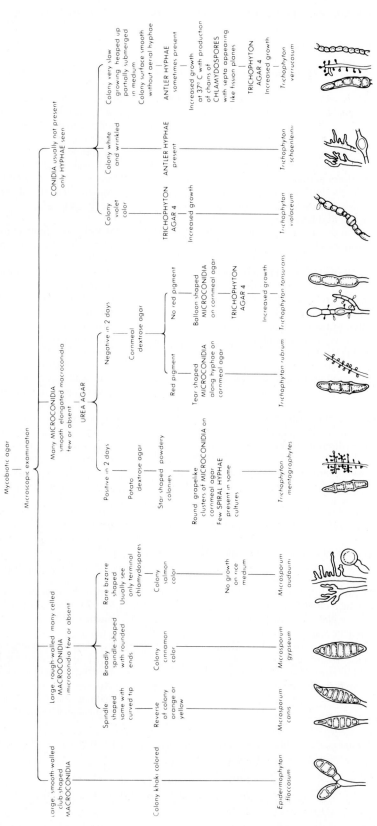

Schema used for dermatophytes commonly recovered by Mayo Clinic Mycology Laboratory

FIGURE 44.39

Dermatophyte identification schema. (From Koneman, E.W., and Roberts, G.D. 1985. Practical laboratory mycology, ed 3. Williams & Wilkins, Baltimore.)

along the side of the hyphae or in clusters. Micro-conidia are usually numerous, except for some of the slower growing species such as *T. violaceum, T. schoenleinii,* and *T. verrucosum,* which rarely sporulate. Macroconidia are uncommonly produced by most species of *Trichophyton.* When present, they are smooth walled, elongated, pencil- to cigar-shaped, and are borne on short, delicate conidiophores.

The genus *Epidermophyton* is characterized by the presence of large, club-shaped, multiseg-mented, smooth-walled macroconidia that are commonly produced either singly or in clusters of two to three from the tips of short conidiophores; microconidia are not produced. This organism is easily identified.

The genus *Microsporum* is characterized by the production of large, rough-walled, multiseg-mented macroconidia that are produced singly from short conidiophores or are produced directly from the hyphae. Macroconidia may be numerous and are usually spindle-shaped to elongated. Microconidia are uncommonly produced except for the occasional isolate in which their production may predominate.

GENUS *TRICHOPHYTON*

Species of this genus are widely distributed and are the most important and common causes of infections of the feet and nails; they may be responsible for tinea corporis (Figure 44.40), tinea capitis, and tinea barbae. They are most commonly seen in adult infections, which vary considerably in their clinical manifestations. Most cosmopolitan species are anthropophilic, or "man-loving"; few are zoophilic, primarily infecting animals.

Generally, hairs infected with members of the genus *Trichophyton* do not fluoresce under a Wood's lamp; the demonstration of fungal elements inside, surrounding, and penetrating the hair shaft, or within a skin scraping is necessary to make a diagnosis of the dermatophyte infection. The recovery and identification of the causative organism are necessary for confirmation.

Microscopically, the genus *Trichophyton* is characterized by smooth, club-shaped, thin-walled macroconidia with eight to ten septa ranging in size from 4 × 8 μm to 8 × 15 μm. The macroconidia are borne singly at the terminal ends of hyphae or on short conidiophores; the microconidia predominate and are usually spherical, pyriform (teardrop-shaped), or clavate (club-shaped), and 2 to 4 μm in size. Only the common species of *Trichophyton* are described in this chapter.

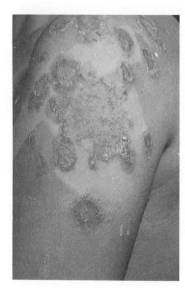

FIGURE 44.40

T. verrucosum, ringworm of skin. (Courtesy Upjohn Co.)

T. rubrum and *T. mentagrophytes* are the most common species recovered in the clinical laboratory. *T. rubrum* is a slow-growing organism that produces a flat or heaped-up colony that is generally white to reddish with a cottony or velvety surface. The characteristic cherry-red color is best observed on the reverse side of the colony; however, this is produced only after 3 or 4 weeks of incubation. Occasional strains may lack the deep red pigmentation on first isolation. Colonies may be of two types, fluffy and granular. Microconidia are uncommon in most of the fluffy strains but are more common in the granular strains and occur as small, teardrop-shaped conidia often borne laterally along the sides of the hyphae (Figure 44.41). Macroconidia are seen uncommonly, although they are sometimes found in the granular strains, where they appear as thin-walled, smooth-walled, multicelled, pencil-shaped conidia with three to eight septa. *T. rubrum* has no specific nutritional requirements. It does not perforate hair in vitro (Procedure 44.4) or produce urease.[5]

T. mentagrophytes produces two distinct colonial forms: the downy variety recovered from cases of tinea pedis and the granular variety recovered from lesions acquired by contact with animals.

T. mentagrophytes produces rapidly growing colonies that appear as white, cottony or downy colonies to cream-colored or yellow colonies that are coarsely granular to powdery. Granular colonies may show evidence of red pigmentation. The reverse side of the colony is usually rose-

PROCEDURE 44.4

HAIR PERFORATION TEST

1. Place a filter paper disk into the bottom of a sterile culture dish.

2. Cover surface of paper disk with sterile distilled water.

3. Add a small portion of sterilized hair into the water.

4. Inoculate a portion of the colony to be studied directly onto the hair.

5. Incubate at 25° C for 10 to 14 days.

6. Observe hairs at regular intervals by placing them into a drop of water on a microscope slide. Position a coverslip and examine microscopically for the presence of conical perforations of the hair shaft (Figure 44.43).

Modified from Ajello, L., and Georg, L.K. 1957. Mycopathologia 8:3.

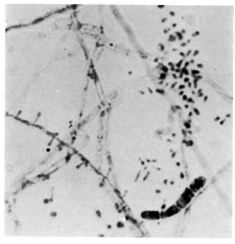

FIGURE 44.41

T. rubrum, showing an elongated macroconidium and numerous pyriform microconidia borne singly on hyphae (750×).

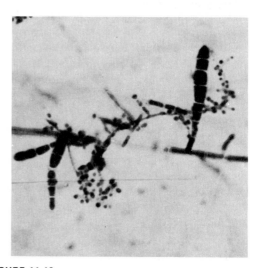

FIGURE 44.42

T. mentagrophytes showing numerous microconidia in grapelike clusters. Also shown are several thin-walled macroconidia (500×).

brown, occasionally orange to deep red, and may be confused with *T. rubrum*. The white, downy colonies produce only a few spherical microconidia; the granular colonies sporulate freely, with numerous small, spherical microconidia produced in grapelike clusters and thin-walled, smooth-walled, cigar-shaped macroconidia measuring 6 × 20 μm to 8 × 50 μm in size, with two to five septae (Figure 44.42). Macroconidia characteristically exhibit a definite narrow attachment to their base. Spiral hyphae may be found in one third of the isolates recovered.

T. mentagrophytes produces urease within 2 to 3 days after inoculation onto Christensen's urea agar (Chapter 10). In addition, *T. mentagrophytes*

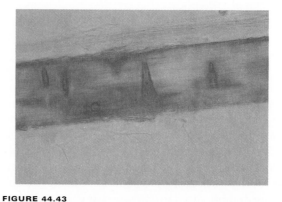

FIGURE 44.43

Hair perforation by *T. mentagrophytes;* wedge-shaped areas illustrate hair perforation (100×).

perforates hair, in contrast to *T. rubrum,* which does not.[5] This latter criterion may be used when there is difficulty in distinguishing between the two species.

Trichophyton tonsurans is responsible for an epidemic form of tinea capitis occurring most commonly in children, but occurring occasionally in adults. It has displaced *Microsporum audouinii* as a primary cause of tinea capitis in most of the United States. The fungus causes a low-grade superficial lesion of varying severity and produces circular, scaly patches of alopecia (loss of hair). The stubs of hair remain in the epidermis of the scalp after the brittle hairs have broken off and may give the typical "black dot" ringworm appearance. Since the infected hairs do not fluoresce under a Wood's lamp, a careful search for the embedded stub should be carried out by the physician with the use of a bright light.

The direct microscopic examination of infected hairs in a potassium hydroxide preparation (Chapter 8) reveals the hair shaft to be filled with masses of large (4 to 7 μm) arthroconidia in chains, characteristic of an **endothrix** type of invasion. Cultures of *T. tonsurans* develop slowly and are typically buff to brown, wrinkled and suedelike in appearance. The colony surface shows radial folds, often developing a craterlike depression in the center with deep fissures. The reverse side of the colony is yellowish to reddish brown. Microscopically, numerous microconidia with flat bases borne on the sides of hyphae are observed. With age, the microconidia tend to become pleomorphic, are swollen to elongated, and are referred to as "balloon forms" (Figure 44.44). Chlamydospores are abundant in old cultures; swollen and fragmented hyphal cells resembling arthroconidia may be seen. *T. tonsurans* grows poorly on media lacking vitamins; however, growth is greatly enhanced by the presence of thiamine.

T. verrucosum causes a variety of lesions in cattle and humans; it is most often seen in farmers who acquire their infection from cattle. The lesions are found chiefly on the beard, neck, wrist, and back of the hands; they are deep, pustular, and inflammatory. With pressure, short stubs of hair may be recovered from the purulent lesion. Direct examination of the outside of the hair shaft reveals sheaths of isolated chains of large (5 to 10 μm) spores **(ectothrix)** and hyphae within the hair **(endothrix).** Masses of these conidia may also be seen in exudate from the lesions.

T. verrucosum grows slowly (14 to 30 days), and growth is enhanced at 35° to 37° C and also on media enriched with thiamine and inositol. *T. verrucosum* may be suspected when colonies appear to embed themselves into the agar surface. Kane and Smitka[38] described a medium for the early detection and identification of *T. verrucosum.* The ingredients for this medium are 4% casein and 0.5% yeast extract. The organism is recognized by its early hydrolysis of casein and very slow growth rate. Chains of chlamydospores are formed regularly at 37° C. The early detection of hydrolysis, formation of characteristic chains of chlamydospores (not seen in Figure 44.45), and the restrictive slow growth rate of *T. verrucosum* differentiate it from *T. schoenleinii,* another slowly growing organism. Colonies are small, heaped, and folded, occasionally flat and disk shaped. At first they are glabrous and waxy, with a short aerial mycelium. Colonies range from gray and waxlike to a bright ochre. The reverse of the colony is most often nonpigmented but may be yellow.

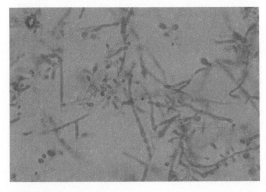

FIGURE 44.44

T. tonsurans, showing numerous pleomorphic microconidia borne singly or in clusters.

FIGURE 44.45

T. verrucosum, showing numerous microconidia, which are rarely seen (500×).

Chlamydospores produced in chains and antler hyphae may be the only structures observed microscopically. Chlamydospores may be abundant at 35° to 37° C. Microconidia may be produced by some cultures if the medium is enriched with yeast extract or a vitamin. Conidia, when present, are borne laterally from the hyphae and are large and clavate. Macroconidia are rarely formed, vary considerably in size and shape, and are referred to as "rat tail."

T. schoenleinii causes a severe type of infection called "favus." It is characterized by the formation of yellowish cup-shaped crusts or scutulae, resulting in considerable scarring of the scalp and sometimes permanent alopecia. Infections are common among members of the same family. A distinctive invasion of the infected hair, the favic type, is demonstrated by the presence of large inverted cones of hyphae and arthroconidia at the base of the hair follicle and branching hyphae throughout the length of the hair shaft. Longitudinal tunnels or empty spaces appear in the hair shaft where the hyphae have disintegrated. In potassium hydroxide preparations, these tunnels are readily filled with fluid; air bubbles may also be seen in these tunnels. *T. schoenleinii* is a slowly growing organism (30 days or longer) and produces a white to light gray colony that has a waxy surface. Colonies have an irregular border that consists mostly of submerged hyphae and that tends to crack the agar. The surface of the colony is usually nonpigmented or tan, furrowed, and irregularly folded. The reverse side of the colony is usually tan or nonpigmented. Microscopically, conidia are not formed commonly. The hyphae tend to become knobby and club-shaped at the terminal ends, with the production of many short lateral and terminal branches (antler hyphae; Figure 44.46). Chlamydospores are generally numerous. All strains of *T. schoenleinii* may be grown in a vitamin-free medium and grow equally well at room temperature or at 35° to 37° C.

T. violaceum produces an infection of the scalp and body and is seen primarily in persons living in the Mediterranean region, the Middle and Far East, and Africa. Hair invasion is of the endothrix type; the typical "black dot" type of tinea capitis is observed clinically. Direct microscopic examination of the potassium hydroxide preparation of the nonfluorescing hairs shows dark, thick hairs filled with masses of arthroconidia arranged in chains, similar to those seen in *T. tonsurans* infections.

Colonies of *T. violaceum* are very slow growing, beginning as cream-colored, glabrous, cone-shaped, and later becoming heaped up, verrucous (warty), and violet to purple in color and waxy in consistency. Colonies may often be described as being "port wine" in color. The reverse side of the colony is purple or nonpigmented. Older cultures may develop a velvety area of mycelium and sometimes lose their pigmentation. Microscopically, microconidia and macroconidia are generally not present; only sterile, distorted hyphae and chlamydospores are found. In some instances, however, swollen hyphae containing cytoplasmic granules may be seen. The growth of *T. violaceum* is enhanced on media containing thiamine.

GENUS *MICROSPORUM*

The genus *Microsporum* is immediately recognized by the presence of large (8 × 8 to 15 μm × 35 to 150 μm), spindle-shaped, rough-walled macroconidia with thick (up to 4 μm) walls that contain 4 to 15 septa. The exception is *M. nanum*, which characteristically produces macroconidia having two cells. The microconidia, when present, are small (3 to 7 μm) and club-shaped and are borne on the hyphae, either laterally or on short conidiophores. Species of *Microsporum* develop either slowly or rapidly and produce aerial hyphae that may be velvety, powdery, glabrous, or cottony, varying in color from whitish, buff, to a cinnamon brown, with varying shades on the reverse side of the colony.

M. audouinii was, in past years, the most important cause of epidemic tinea capitis among schoolchildren in the United States. This organism

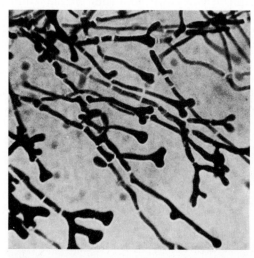

FIGURE 44.46

T. schoenleinii, showing swollen hyphal tips with lateral and terminal branching (favic chandeliers). Microconidia and macroconidia are absent (500×).

is anthropophilic and is spread directly by means of infected hairs on hats, caps, upholstery, combs, or barber clippers. The majority of infections are chronic; some heal spontaneously, whereas others may persist for several years. Infected hair shafts fluoresce yellow-green using a Wood's lamp. Colonies of *M. audouinii* generally grow more slowly than other members of the genus *Microsporum* (10 to 21 days), and they produce a velvety aerial mycelium that is colorless to light gray to tan. The reverse side often appears salmon-pink to reddish-brown. Colonies of *M. audouinii* do not usually sporulate in culture. The addition of yeast extract may stimulate growth and the production of macroconidia in some instances. Most commonly, atypical vegetative forms such as terminal chlamydospores and antler and racquet hyphae are the only clues to the identification of this organism. It is common to identify *M. audouinii* by exclusion of all the other dermatophytes as a cause of infection.

M. canis is primarily a pathogen of animals; it is the most common cause of ringworm infection in dogs and cats in the United States. Children and adults acquire the disease through contact with infected animals, particularly puppies and kittens, although human-to-human transfer has been reported. Hairs infected with *M. canis* fluoresce a bright yellow-green using a Wood's lamp, which is a useful tool for screening pets as possible sources of human infection. Direct examination of a potassium hydroxide preparation of infected hairs reveals small spores (2 to 3 μm) outside the hair, although culture procedures must be performed to provide specific identification.

Colonies of *M. canis* grow rapidly, are granular or fluffy with a feathery border, are white to buff, and characteristically have a lemon-yellow or yellow-orange fringe at the periphery. On aging, the colony becomes dense and cottony and a deeper brownish-yellow or orange and frequently shows an area of heavy growth in the center. The reverse side of the colony is bright yellow, becoming orange or reddish-brown with age. Rarely, strains are recovered that show no reverse side pigment. Microscopically, *M. canis* shows an abundance of large (15 μm × 60 to 125 μm), spindle-shaped, multisegmented (four to eight) macroconidia with curved ends (Figure 44.47). These are thick-walled with warty (echinulate) projections on their surfaces. Microconidia are usually few in number; however, large numbers may occasionally be seen.

Microsporum gypseum, a free-living organism of the soil (geophilic) that only rarely causes hu-

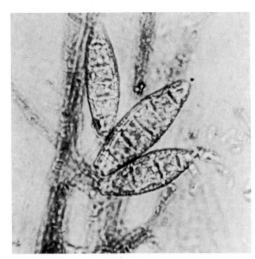

FIGURE 44.47

M. canis, showing several spindle-shaped, thick-walled, multicelled macroconidia (500×).

man or animal infection, may be seen occasionally in the clinical laboratory. Infected hairs generally do not fluoresce using a Wood's lamp. However, microscopic examination of the infected hairs shows them to be irregularly covered with clusters of spores (5 to 8 μm), some in chains. These arthroconidia of the ectothrix type are considerably larger than those of other *Microsporum* species.

M. gypseum grows rapidly as a flat, irregularly fringed colony with a coarse powdery surface that appears to be buff or cinnamon color. The underside of the colony is conspicuously orange to brownish. Microscopically, macroconidia are seen in large numbers and are characteristically large, ellipsoidal, have rounded ends, and are multisegmented (three to nine) with echinulated surfaces (Figure 44.48). Although they are spindle-shaped, these macroconidia are not as pointed at the distal ends as those of *M. canis.* The appearance of the colonial and microscopic morphological features is sufficient to make the distinction between *M. gypseum* and *M. canis.*

GENUS *EPIDERMOPHYTON*

E. floccosum, the only member of the genus *Epidermophyton,* is a common cause of tinea cruris and tinea pedis. In direct examination of skin scrapings using the potassium hydroxide preparation, the fungus is seen as fine branching hyphae. *E. floccosum* grows slowly, and growth appears as an olive-green to khaki color, with the periphery surrounded by a dull orange-brown color. After several weeks, colonies develop a cottony white aerial mycelium that completely overgrows the

FIGURE 44.48

M. gypseum, showing ellipsoidal, multicelled macroconidia (750×).

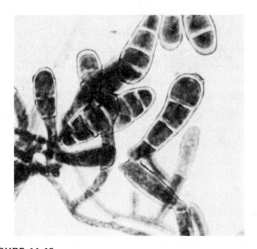

FIGURE 44.49

E. floccosum, showing numerous smooth, multiseptate, thin-walled macroconidia that appear club-shaped (1000×).

colony and is sterile, and remains so, even after subculture.

Microscopically, numerous smooth, thin-walled, club-shaped, multiseptate (2 to 4 μm) macroconidia are seen (Figure 43.49). They are rounded at the tip and are borne singly on a conidiophore or in groups of two or three. Microconidia are absent, spiral hyphae are rare, and chlamydospores are usually numerous.

This organism is susceptible to cold temperatures and for this reason it is recommended that specimens submitted for dermatophyte culture not be refrigerated before culture and that cultures not be stored at 4° C.

SUBCUTANEOUS MYCOSES

Subcutaneous mycoses are infections that involve the skin and subcutaneous tissue, generally without dissemination to other organs of the body. This classification is artificial; for example, sporotrichosis occasionally involves lungs, other viscera, meninges, or joints and may disseminate. The other agents causing subcutaneous infections may also produce disseminated infection occasionally. The etiological agents are found in several unrelated fungal genera, all of which exist in nature. Humans and animals serve as accidental hosts after traumatic inoculation of the organism into cutaneous and subcutaneous tissues. Several subcutaneous mycoses are considered here: sporotrichosis, chromoblastomycosis, mycetoma, and phaeohyphomycosis.

SPOROTRICHOSIS

Sporotrichosis is a chronic infection of worldwide distribution caused by the dimorphic fungus,

Sporothrix schenckii, whose natural habitat is living or dead vegetation. Humans acquire the infection through trauma (thorns, splinters), usually to the hand, arm, or leg. The primary lesion begins as a small, nonhealing ulcer, commonly of the index finger or the back of the hand. With time, the infection is characterized by the development of nodular lesions of the skin or subcutaneous tissues at the point of contact and later involves the lymphatic channels and lymph nodes draining the region. The subcutaneous nodules break down and ulcerate to form an infection that becomes chronic. Only rarely is the disease disseminated; pulmonary infection may be seen. The infection is an occupational hazard for farmers, nursery workers, gardeners, florists, and miners. It is commonly known as "rose gardener's" disease.

Exudate from unopened subcutaneous nodules or from open draining lesions is often submitted for culture and direct microscopic examination. Direct examination of this material is usually of little diagnostic value because it is difficult to demonstrate the characteristic yeast forms, even with special stains; reasons for this are unclear.

Colonies of *S. schenckii* grow rapidly (3 to 5 days) and are usually small, moist, and white to cream-colored in appearance. On further incubation, these become membranous, wrinkled, and coarsely matted, with the color becoming irregularly dark brown or black and the colony becoming leathery in consistency. It is not uncommon for the clinical microbiology laboratory to mistake a young culture of *S. schenckii* for that of a yeast until the microscopic features are observed. One of the commercially available yeast identification systems, API20C, has a characteristic biochemical profile for the identification of *S. schenckii*. This method is not advocated for the identification of this organism.

Microscopically, hyphae are delicate (1 to 2 μm thick), septate, exhibit branching, and bear one-celled conidia, 2 to 5 μm in diameter. These are borne in clusters from the tips of single conidiophores. Each conidium is attached to the conidiophore by an individual, delicate, threadlike structure (denticle) that may require examination under oil immersion to be visible. As the culture ages, single-celled, thick-walled, black pigmented conidia may also be borne along the sides of the hyphae, simulating the arrangement of microconidia produced by *Trichophyton rubrum*. The common designations for these types of sporulation are the "flowerette" and "sleeve" arrangements, respectively (Figure 44.50).

Because of similar morphological features, saprophytic species of the genus *Sporotrichum* may be confused with *S. schenckii,* and it is necessary to distinguish between them. During incubation of a culture at 37° C, the colony of *S. schenckii* transforms to a soft, cream-colored to white, yeast-like colony. Microscopically, singly or multiply budding, spherical, oval, or cigar-shaped yeast cells are observed without difficulty (Figure 44.51). Conversion from the mold form to the yeast form is easily accomplished and usually occurs within 1 to 5 days after transfer of the culture to a medium containing blood enrichment; most isolates of *S. schenckii* are converted to the yeast form within 12 to 48 hours at 37° C.

CHROMOBLASTOMYCOSIS

Chromoblastomycosis is a chronic fungal infection acquired via traumatic inoculation of an organism, primarily into the extremities. The infection is characterized by the development of a

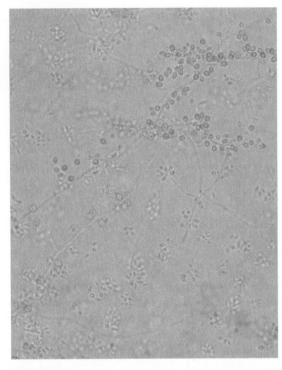

FIGURE 44.50

S. schenckii, mycelial form, showing pyriform to ovoid microconidia borne in a flowerette at the tip of the conidiophore (1500×).

FIGURE 44.51

S. schenckii, yeast form, showing cigar-shaped and oval budding cells (1000×).

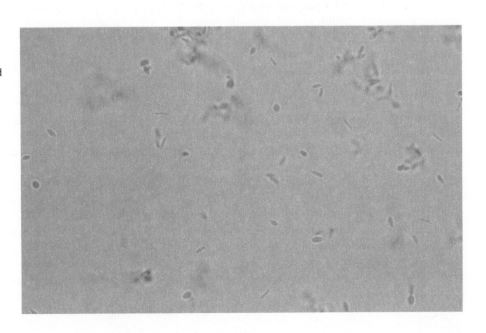

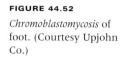

FIGURE 44.52

Chromoblastomycosis of foot. (Courtesy Upjohn Co.)

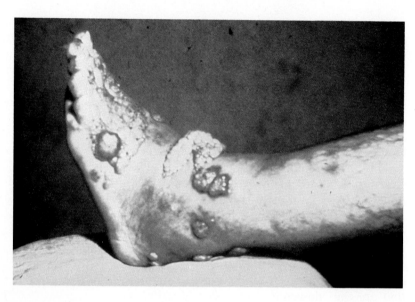

papule at the site of the traumatic insult that slowly spreads to form warty or tumorlike lesions characterized as "cauliflower-like" (Figure 44.52). There may be secondary infection and ulceration. The lesions are usually confined to the feet and legs but may involve the head, face, neck, and other body surfaces. Histologic examination of lesion tissue reveals characteristic **sclerotic** bodies, copper-colored, septate cells that appear to be dividing (see Figure 44.13). Brain abscess (cerebral chromoblastomycosis) caused by the dematiaceous fungi has been reported with some frequency and will be discussed with phaeohyphomycosis.[9]

The disease is widely distributed, but most cases occur in tropical and subtropical areas of the world. Occasional cases are reported from temperate zones, including the United States. The infection is seen most often in areas where agricultural workers fail to wear protective clothing and suffer thorn or splinter puncture wounds through which an organism enters from the soil.

The group of fungi known to cause chromoblastomycosis are dematiaceous. All are slow-growing and produce heaped-up and slightly folded, darkly pigmented colonies with a gray to olive to black velvety appearance. The reverse side of the colonies is jet black.

The taxonomy of the organisms that cause chromoblastomycosis is complex.[53] Their identification is based on distinct microscopic morphological features. Three genera, *Cladosporium, Phialophora,* and *Fonsecaea,* are known to cause chromoblastomycosis.

The genus *Cladosporium* includes those species that produce long chains of conidia (blastoconidia) that have a dark septal scar present.

The genus *Phialophora* includes those species that produce short, flask-shaped to tubular phialides, each having a well-developed collarette. Clusters of conidia are produced by the phialides through an apical pore and often remain aggregated when microscopic mounts are observed.

The genus *Fonsecaea* includes those organisms that exhibit a mixed type of sporulation, which uniquely includes one-celled primary conidia that are produced on either side of conidiophores resembling a series of bent knees (sympodially). Conidia are produced sympodially. The primary conidia give rise to secondary conidia that appear to occur in loose heads. This is known as the rhinocladiella type of sporulation and predominates, depending on the strain isolated. A mixture of the rhinocladiella and cladosporium types may occur; moreover, phialides with collarettes may also be present. Two species, *Fonsecaea pedrosoi* and *Fonsecaea compacta,* are etiological agents of chromoblastomycosis. Both are morphologically distinct: *F. pedrosoi* is differentiated from *F. compacta* by the production of loose heads, in contrast to the more compact heads produced by *F. compacta.*

The diagnostic features of the three genera are summarized as follows:

1. *Cladosporium (Cladosporium carrionii):* Cladosporium type of sporulation with long chains of elliptical conidia (2 to 3 μm × 4 to 5

FIGURE 44.53

Cladosporium species, showing *Cladosporium* type of sporulation with chains of elliptical conidia (440×).

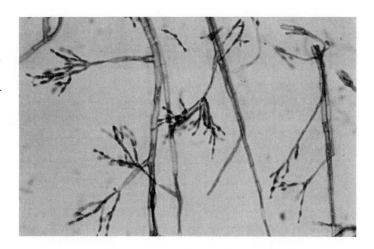

FIGURE 44.54

P. verruscosa, showing flask-shaped phialide with a distinct collarette and conidia near its tip (750×).

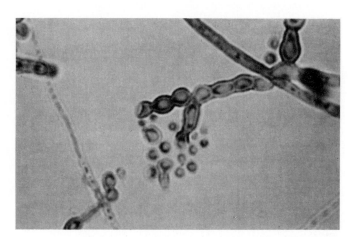

μm) borne from erect, tall, branching conid-iophores (Figure 44.53).

2. *Phialophora (Phialophora verrucosa):* Tubelike or flask-shaped phialides, each with a distinct collarette. Conidia are produced endogenously and occur in clusters at the tip of the phialide (Figure 44.54).

3. *Fonsecaea (F. pedrosoi* and *F. compacta):* Conidial heads with sympodial arrangement of conidia, with primary conidia giving rise to secondary conidia (Figure 44.55). Cladosporium type of sporulation may occur and phialides with collarettes may also be present.

The laboratory diagnosis of chromoblastomycosis is made easily. Scrapings from crusted areas added to 10% potassium hydroxide show the presence of the sclerotic bodies, which appear rounded, brown, and 4 to 10 μm in diameter and resemble "copper pennies" (see Figure 44.13).

MYCETOMA

Mycetoma is a chronic granulomatous infection that usually involves the lower extremities but may occur in any part of the body. The infection is characterized by swelling, purplish discoloration, tumorlike deformities of the subcutaneous tissue, and multiple sinus tracts that drain pus containing yellow, white, red, or black granules. The infection gradually progresses to involve the bone, muscle, or other contiguous tissue and ultimately requires amputation in most cases. Occasionally there may be dissemination to other organs, including the brain; however, this type of infection is relatively uncommon.

Mycetoma is common among persons who live in tropical and subtropical regions of the world, whose outdoor occupations and failure to wear protective clothing predispose them to trauma.

Two types of mycetoma are described: actinomycotic mycetoma, which is caused by species of the aerobic actinomycetes, including *Nocardia,*

FIGURE 44.55

F. pedrosoi, showing conidial heads with sympodial arrangement of conidia (440×).

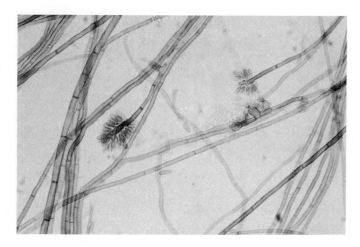

Actinomadura, and *Streptomyces* (Chapter 34); and eumycotic mycetoma, caused by a heterogeneous group of species having true septate hyphae.[51] Other causes of mycetoma include *Exophiala jeanselmei, Acremonium, Curvularia, Madurella mycetomatis* and numerous other genera and species. The most common etiological agent of mycetoma in the United States is *Pseudallescheria boydii,* a member of the Ascomycota (because it produces ascospores). The organism is commonly found in soil, standing water, and sewage; humans acquire the infection by traumatic implantation of the organism into the skin and subcutaneous tissues.

Macroscopic examination of granules from lesions of mycetoma caused by *P. boydii* reveal them to be white to yellow, and 0.2 to 2 mm in diameter. Microscopically, the granules of *P. boydii* consist of loosely arranged, intertwined hyphae, cemented together.

P. boydii is an organism that grows rapidly (5 to 10 days) on common laboratory media. Initial growth begins as a white, fluffy colony that changes in several weeks to a brownish-gray (mousey) mycelium. The reverse of the colony is black. *P. boydii* has undergone several name changes in the past and is an example of an organism that reproduces both asexually and sexually. The asexual form (anamorph) is called *Scedosporium apiospermum* and microscopically produces golden brown elliptical (sperm-shaped), single-celled conidia borne singly from the tips of long or short conidiophores (annellophores) (Figure 44.56). Clusters of conidiophores with conidia produced at the ends sometimes occur and are referred to as "coremia." This form (anamorph) is often referred to as the "Graphium" stage of *P. boydii.*

P. boydii is the sexual form (telomorph) of the organism. The sexual form exhibits cleistothecia,

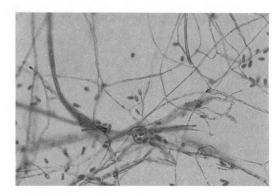

FIGURE 44.56

Scedosporium apiospermum, showing asexually produced conidia borne singly on long or short conidiophores (annellophores) (440×).

which are saclike structures that contain asci and ascospores. When the latter are fully developed, the large (50 to 200 μm), thick-walled cleistothecia rupture and liberate the asci and ascospores (Figure 44.57). The ascospores are oval, delicately pointed at each end, and resemble the conidia of the asexual form. Isolates of *P. boydii* may be induced to form cleistothecia by culturing on plain water agar.

P. boydii is also involved in causing a variety of infections elsewhere in the body. Included are infections of the nasal sinuses and septum, meningitis, arthritis, endocarditis, mycotic keratitis, external otomycosis, and brain abscess. Most of these more serious infections occur primarily in immunocompromised patients. Recently, another species of *Scedosporium, S. inflatum* has been associated with infections other than mycetoma, such as arthritis and sinusitis.[69]

It should be mentioned that it is impossible to predict the specific etiological agent of mycetoma. Culture media containing antibiotics should not

FIGURE 44.57

Pseudallescheria boydii, showing cleistothecia and numerous ascospores (150×).

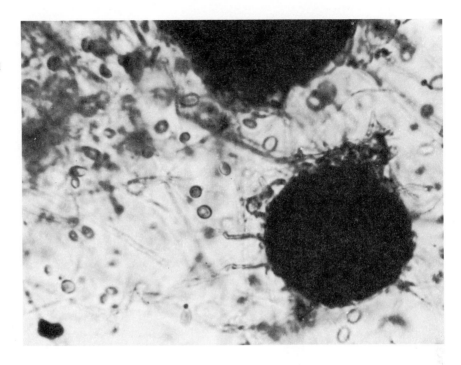

be used alone for culturing clinical specimens, since species of the aerobic actinomycetes are susceptible to antibacterial antibiotics and may be inhibited by these agents incorporated into routine culture media. For further information on other fungi involved in mycetoma, one may refer to references listed at the end of this chapter.

PHAEOHYPHOMYCOSIS

Phaeohyphomycosis is a general term used to describe any infection caused by a dematiaceous organism. The strict definition does not include chromoblastomycosis or mycetoma. However, the term used within this chapter refers to subcutaneous, localized, or systemic infection caused by any number of dematiaceous fungi. Infections include subcutaneous abscess, brain abscess, sinusitis, endocarditis, mycotic keratitis, pulmonary infection, and systemic infection.[1,18,54,89]

In general, dematiaceous fungal hyphae are seen in clinical specimens by direct microscopic examination or by histopathologic examination of tissue obtained at surgery or autopsy. In some instances, yeast-like elements may also be observed.

Common etiological agents of subcutaneous infection include *Exophiala jeanselmei, Phialophora richardsiae, Wangiella dermatitidis, Alternaria, Bipolaris,* and *Exserohilum.* Common agents associated with sinusitis include *Bipolaris, Exserohilum,* and *Curvularia.* Common etiological agents of brain abscess include *Xylohypha bantiana, Wangiella dermatitidis,* and *Bipolaris.* Mycotic keratitis and en-

dophthalmitis have been caused by *Wangiella dermatitis* and *Curvularia.*

The laboratory diagnosis is made by observing yellowish-brown septate, moniliform hyphae with or without budding cells present. The methenamine silver stain of tissue often overstains fungal elements and the observation of pigmented hyphae in hematoxylin-eosin or unstained histopathologic sections will provide the diagnosis of this clinical entity. Culture of the specific etiological agent is necessary for final confirmation.

Exophiala jeanselmei is a dematiaceous organism that grows slowly (7 to 21 days) and produces shiny black, yeastlike colonies initially. With age, colonies become filamentous, velvety, and gray to olive to black. The microscopic features of young colonies exhibit dematiaceous budding yeast cells with annellations (difficult to observe). Filamentous colonies produce dematiaceous hyphae and conidiogenous cells that are cylindrical and have a tapered tip. Annellations may be visible at the tip and clusters of oval to round conidia are apparent (Figure 44.58). Cultures can grow at 37° C and not at 40° C and can utilize inorganic KNO_3.

Wangiella dermatitidis is a dematiaceous organism that grows slowly (7 to 21 days) and produces shiny, black, yeastlike colonies initially. With age, colonies become filamentous, velvety, and gray to olive to black in color. The microscopic features of young colonies exhibit dematiaceous budding yeast cells. Filamentous colonies produce dematiaceous hyphae and long tubular phialides lacking

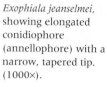

FIGURE 44.58

Exophiala jeanselmei, showing elongated conidiophore (annellophore) with a narrow, tapered tip. (1000×).

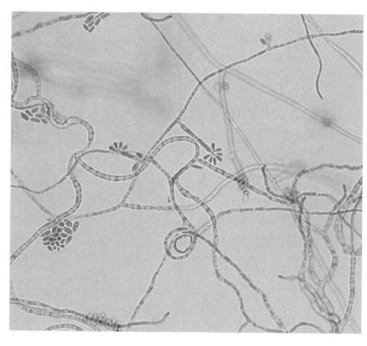

FIGURE 44.59

Wangiella dermatitidis, showing elongated tubular phialides without a collarette. (500×).

collarettes. Phialides produce dematiaceous single-celled conidia in clusters (Figure 44.59). Balls of conidia may appear to slide down the hyphae. Cultures can grow at 40° C and cannot utilize KNO_3 in contrast to *E. jeanselmei.*

Phialophora richardsiae is a dematiaceous organism that grows rapidly and produces colonies that are wooly, olive-brown to brownish-gray; some may appear to have concentric zones of color. Microscopically, hyphae are dematiaceous, and sporulation is common. Phialides with distinct "saucer-like" collarettes are readily observed (Figure 44.60). Pleomorphic phialides may also be seen; however, all produce either or both hyaline elliptical conidia or brown elliptical conidia produced within the phialides.

SYSTEMIC MYCOSES

Systemic fungal infections may involve any of the internal organs of the body as well as lymph nodes, bone, subcutaneous tissue, and skin. Asymptomatic infection is common, may go unrecognized clinically, and may be detected only by serologic testing; in some cases roentgenographic examination may reveal healed lesions. Symptomatic infections may present signs of only a mild or more severe but self-limited disease, with positive supportive evidence from cultural or immunological findings. Disseminated or progressive infection may reveal severe symptoms, with spread of the initial disease to several organs as well as to the bone, skin, and subcutaneous tissues. Some cases of disseminated infection may exhibit little in the way of signs or symptoms of disease for long periods, only to exacerbate later. Immunocompromised patients, particularly those having infection with HIV or those receiving long-term corticosteroid therapy, most often show disseminated infection.[85]

Traditionally, the systemic mycoses have included only blastomycosis, coccidioidomycosis, histoplasmosis, and paracoccidioidomycosis. The fungi responsible for these infections, although unrelated generically and dissimilar morphologically, have one characteristic in common—dimorphism. The dimorphic fungi involved exist in nature as a mold form, which is quite distinct from the parasitic, or invasive, form, sometimes called the tissue

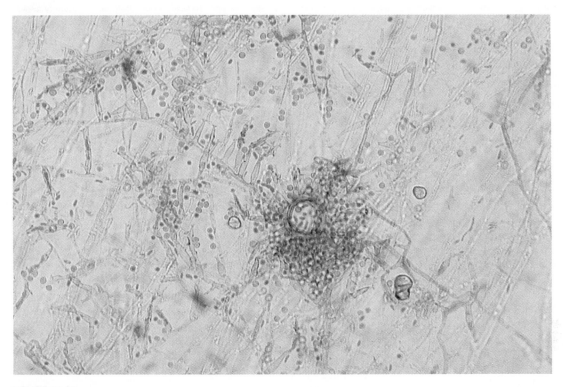

FIGURE 44.60

Phialophora richardsiae, showing phialides having prominent saucer-like collarettes. (1000×).

form. Distinct morphological differences may be observed with the dimorphic fungi both in vivo and in vitro. Temperature (35° to 37° C), certain nutritional factors, and stimulation of growth in tissue independent of temperature are among the factors necessary to initiate the transformation of the mold form to the parasitic form.[26,80]

GENERAL FEATURES OF DIMORPHIC FUNGI

Commonly, the dimorphic fungi are regarded as slow-growing organisms that require 7 to 21 days for visible growth to appear. However, exceptions to this rule occur with some frequency. Cultures of *Blastomyces dermatitidis* and *Histoplasma capsulatum* are recovered in as short a time as 2 to 5 days when large numbers of organisms are present in the clinical specimen. In contrast, when small numbers of colonies of *B. dermatitidis* and *H. capsulatum* are present, sometimes 21 to 30 days of incubation are required before they are detected. *Coccidioides immitis* is consistently recovered within 3 to 5 days of incubation, but when large numbers of organisms are present, colonies may be detected within 48 hours. Single colonies may require 14 to 21 days before visible growth can be detected. Cultures of *Paracoccidioides brasiliensis* are commonly recovered within 5 to 25 days, with a

usual incubation period of 10 to 15 days. As one can see, the growth rate, if slow, might lead one to suspect the presence of a dimorphic fungus; however, a great deal of variation in the time for recovery exists.

Textbooks present descriptions for the dimorphic fungi that the reader assumes are typical for each particular organism. As is true in other areas of microbiology, a great deal of variation in the colonial morphological features also occurs, commonly because of the type of medium used. One must be aware of this variation and must not rely heavily on colonial morphological features for the identification of members of this group of fungi.

The pigmentation of colonies is sometimes helpful but also varies widely; colonies of *B. dermatitidis* and *H. capsulatum* are described as being fluffy white with a change in color to tan or buff with age. Some isolates initially appear darkly pigmented with colors ranging from gray or dark brown to red. On media containing blood enrichment, these organisms appear heaped, wrinkled, glabrous, and neutral in color, and yeastlike in appearance; often tufts of aerial hyphae project from the top of the colony. Some colonies may appear pink to red because of the adsorption of hemoglobin from the blood in the medium. *C. im-*

PROCEDURE 44.5

IN VITRO CONVERSION OF DIMORPHIC MOLDS

PRINCIPLE

Dimorphic molds exist in the yeast or spherule form in infected tissue. Proof that a mold is actually one of the systemic dimorphic fungi can be achieved by simulating the environment of the host and converting the mold to the yeast form. This method is recommended only for the identification of *B. dermatitidis* or *S. schenckii*. Conversion of *C. immitis* to its spherule form requires special media and is not recommended for clinical laboratories.

METHOD

1. Transfer a large inoculum of the mold form of the culture onto the surface of a fresh, moist slant of brain-heart infusion agar containing 5% to 10% sheep blood. If *B. dermatitidis* is suspected, a tube of cottonseed conversion medium[87] should be inoculated.

2. Add a few drops of sterile distilled water to provide moisture if the surface of the culture medium appears dry.

3. Leave the cap of the screw-capped tube slightly loose to allow the culture to have adequate oxygen exchange.

4. Incubate cultures at 35° to 37° C for several days; observe for the appearance of yeastlike portions of the colony. It may be necessary to make several subcultures of any growth that appears, since several transfers are often required to accomplish the conversion of many isolates. Cultures of *B. dermatitidis*, however, are usually easily converted and require 24 to 48 hours on cottonseed agar medium. *C. immitis* may be converted in vitro to the spherule form using a chemically defined medium[80]; however, this method is of little use to the clinical laboratory and it should not be attempted.

QUALITY CONTROL

Because of their hazardous nature, it is not recommended that stock cultures be tested routinely. The exoantigen test (Procedure 44.6) can serve as one method to identify the dimorphic fungi.

---

mitis is described as being fluffy white with scattered areas of hyphae that are adherent to the agar surface, giving an overall "cobweb" appearance to the colony. However, numerous morphological forms, including textures ranging from woolly to powdery and pigmentation ranging from pink-lavender or yellow to brown or buff, have been reported. Although the growth rate and colonial morphological features may help one to recognize the presence of a dimorphic fungus, they should be used in combination with the microscopic morphological features to make a tentative identification.

The definitive identification of a dimorphic fungus has traditionally been made by observing both the mold and parasitic forms of the organism. Cultures of the mold form are easily recovered at 25° to 30° C; however, the tissue form is usually not recovered from the clinical specimen. Previously the definitive identification was made by the in vitro conversion of a mold form to the corresponding yeast or spherule form by animal inoculation or by in vitro conversion on a blood-enriched medium incubated at 35° to 37° C. The conversion of dimorphic molds to the yeast form (except for *C. immitis*) can be accom-

plished with some difficulty, as outlined in Procedure 44.5.

Since the conversion of the dimorphic molds to the corresponding yeast or spherule forms is technically cumbersome to perform and long delays are often experienced, the effort to convert the dimorphic fungi is not recommended and organisms can be identified by **exoantigen testing** (Procedure 44.6). This technique is used in many laboratories to make a definitive identification of *B. dermatitidis*, *C. immitis*, *H. capsulatum*, and *P. brasiliensis*. The exoantigen test relies on the principle that soluble antigens are produced and can be extracted from fungi; they are concentrated and subsequently reacted with serum known to contain antibodies directed against the specific antigenic components of the organism being tested. Reagents and materials for the exoantigen test are currently available commercially.

The exoantigen test is rapidly being replaced by specific nucleic acid probes that rapidly and specifically identify *H. capsulatum*, *B. dermatitidis* and *C. immitis*.

Perhaps the most significant advance in clinical mycology is the development of specific nucleic acid probes for the identification of the di-

---

PROCEDURE 44.6

EXOANTIGEN TEST[41,42,74,75]

PRINCIPLE

Antibodies developed against particular mycelial antigens will react specifically in a gel immunodiffusion precipitin test. The mold forms of the dimorphic fungi can be identified definitively by an antigen-antibody reaction, negating the need for conversion to the yeast form.

METHOD

1. A mature fungus culture on a Sabouraud's dextrose agar slant is covered with an aqueous solution of merthiolate (1:5000 final concentration), which is allowed to remain in contact with the culture for 24 hours at 25° C. The entire surface of the colony must be covered so that effective killing of the organism is ensured.

2. Filter the aqueous solution through a 0.45 μm size membrane filter. This should be performed inside a biological safety cabinet.

3. Five milliliters of this solution is concentrated using a Minicon Macrosolute B-15 Concentrator (Amicon Corporation). The solution is concentrated 50× when testing with *H. capsulatum* and *B. dermatitidis* antiserum and 5× and 25× for reaction with *C. immitis* antiserum.

4. The concentrated supernatant is used in the microdiffusion test. The supernatant is placed into wells punched into a plate of buffered, phenolized agar adjacent to the control antigen well and is tested against positive control antiserum obtained from commercial sources (Immunomycologics, Meridian Diagnostics, and Gibson Laboratories).

5. The immunodiffusion test is allowed to react for 24 hours at 25° C, and the plate is observed for the presence of precipitin bands of identity with the reference reagents. The sensitivity of the exoantigen test for the identification of *B. dermatitidis* may be increased by incubating the immunodiffusion plates at 37° C for 48 hours; however, bands appear sharper at 25° C after 24 hours. It is recommended that exoantigen plates from any culture suspected of being *B. dermatitidis* be incubated at both temperatures.

6. *C. immitis* may be identified by the presence of the CF, TP, or HL antigens, whereas *H. capsulatum* and *B. dermatitidis* may be identified by the presence of H or M bands, or both, and the A band, respectively. Detailed instructions for the performance and interpretation of the tests are included with the manufacturers' package insert.

QUALITY CONTROL

Extracts from known fungi are tested each time the test is performed. Lines of identity with the unknown strain are necessary for identification.

---

morphic fungi and *C. neoformans* (Procedure 44.7). Probes commercially available from Gen-Probe Inc., San Diego, Calif. are DNA probes that are complementary to species-specific ribosomal RNA. Fungal cells are disrupted by a lysing agent and sonication, heat-killed, and the nucleic acid is exposed to DNA that has been labeled with a chemiluminescent tag (acridinium ester). The labeled DNA probe combines with an organism's ribosomal RNA to form a stable DNA:RNA hybrid. The unbound DNA probe is "quenched" and DNA:RNA hybrids are measured in a luminometer. Total time for testing is less than one hour, and young colonies 2 mm or larger in size may be tested.[76]

Nucleic acid probe identification is sensitive, specific, and rapid. Contaminated colonies may be tested; however, results from colonies recovered on blood-enriched media must be interpreted with caution since hemin may cause false-positive chemiluminescence. The major disadvantage of nucleic acid probe identification relates to the cost per test. Based on extensive experience, this author recommends that nucleic acid probes be used whenever an organism is suspected of being *H. capsulatum*, *B. dermatitidis*, or *C. immitis*. The decision of whether to use nucleic acid probe testing for the identification of *C. neoformans* should be the choice of an individual laboratory.

The inclusion of only the dimorphic fungi in the group of systemic mycoses is a bit artificial, since other fungi including *Cryptococcus neoformans* and species of *Candida* and *Aspergillus* also cause disseminated infection. They are not discussed in

PROCEDURE 44.7

NUCLEIC ACID PROBE TESTING

PRINCIPLE

The Gen-Probe Accuprobe System for culture identification of fungi is based on DNA probes that are complementary to species-specific ribosomal RNA. The fungal cells are lysed by sonication, heat killed, and exposed to DNA that has been labeled with a chemiluminescent tag. The labeled DNA probe combines with the organism's ribosomal RNA to form a stable DNA:RNA hybrid. A selection reagent "kills" the signal on all the unbound DNA probe. The labeled DNA:RNA hybrids are measured in a luminometer. Probes are available for the culture identification of *B. dermatitidis, C. immitis, H. capsulatum* and *C. neoformans.*

SPECIMEN

Any actively growing culture less than one month old present on any solid medium such as Sabouraud's, Inhibitory Mold, Brain Heart Infusion, Cornmeal or Potato Dextrose agar. A medium containing blood is not recommended for direct probe testing.

PROCEDURE

1. Add one capful of ultrasonic enhancer to the water in the sonicator. De-gas the water for 5 minutes. Turn on 95° C water bath (or heat block) and 60° C water bath (or heat block).

2. Add 100 µl of Reagent 1 (lysis reagent) and 100 µl Reagent 2 (hybridization buffer) to each lysing tube.

3. Working in a biological safety cabinet, transfer a full 1 µl loop of organism to be tested into the lysing reagent tube. Twirl the loop against the side of the tube to remove the inoculum from the loop.

4. Place the lysing reagent tubes in the sonicator. Be sure the water level is high enough so the contents of the tube are below the water level. Do not allow tubes to touch the sides of the sonicator. Sonicate at room temperature for 15 minutes.

5. Place the lysing tubes in the 95° C water bath (or heat block) for 15 minutes.

6. Allow tubes to cool at room temperature for 5 minutes.

7. Pipet 100 µl of the cell lysate into the Probe Reagent Tube.

8. Incubate the tubes for 15 minutes at 60° C in the water bath (or heat block).

9. Pipet 300 µl of Reagent 3 (Selection Reagent) into each tube. Vortex, recap, and place the tube immediately back into the 60° C water bath (or heat block). Incubate for 5 minutes.

10. Prepare the luminometer for operation by completing two wash cycles. Using a damp tissue, wipe each tube before inserting into the luminometer (to prevent static build-up). Read each tube and controls. The luminometer records relative light units (RLU).

FREQUENCY AND TOLERANCE OF CONTROLS

Run a positive and negative control with each batch of organisms tested.

EXPECTED VALUES

A positive result is more than 50,000 relative light units (RLU). Signals below 50,000 RLU are considered negative.

LIMITATIONS

This method cannot be used on fresh clinical specimens. It is for culture identification only. The use of nucleic acid probes with cultures grown on a blood-containing medium is unsatisfactory for testing, since hemin will chemiluminesce and yield a false-positive result.

SAFETY PRECAUTIONS

1. If culture dishes are used, they should be handled only in a biological safety cabinet. Cultures should be sealed with tape if the specimen is suspected of containing *C. immitis.*

2. The use of cotton plug test tubes is discouraged, and screw-capped tubes should be used if culture tubes are preferred. All handling of cultures of *C. immitis* in screw-capped tubes should be performed inside a biological safety cabinet.

3. All microscopic preparations for examination should be prepared in a biological safety cabinet.

4. Cultures should be autoclaved as soon as the final identification of *C. immitis* is made.

Compiled from Hall, G.S., Pratt-Rippin, K., and Washington, J.A. 1992. J. Clin. Microbiol. 30:3003 and Stockman, L., Clark, K.M., Hunt, J.M., and Roberts, G.D. 1993. J. Clin. Microbiol. 31:850.

this portion of the chapter, since they are not dimorphic, as previously defined, and since they usually cause infection only in immunocompromised patients. They are discussed in a subsequent section on the opportunistic mycoses.

BLASTOMYCOSIS

Blastomycosis is a chronic suppurative and granulomatous infection caused by the dimorphic fungus *Blastomyces dermatitidis.* The disease is most commonly found on the continent of North America and extends southward from Canada to the Mississippi, Ohio, and Missouri River valleys, Mexico, and Central America. Some isolated cases have also been reported in Africa. The largest number of cases occur in the Mississippi, Ohio, and Missouri River valley regions.

Blastomycosis begins as a respiratory infection and is probably acquired by inhalation of the conidia or hyphal fragments of the organism. The exact ecological niche for this organism in nature has not been determined; however, patients with a history of exposure to soil or wood have the highest incidence of infection. The infection may spread and involve the lungs, long bones, soft tissue, and skin. It is not transmitted from person to person and generally occurs as sporadic cases. Several outbreaks have been reported, however, and have been related to a common exposure.

Blastomycosis is more common in men and seems to be associated with outdoor occupations.

The diagnosis of blastomycosis may easily be made when a clinical specimen is observed by direct microscopy. *B. dermatitidis* appears as large, spherical, thick-walled cells 8 to 20 μm in diameter, usually with a single bud connected to the parent cell by a broad base.

B. dermatitidis commonly requires 5 days to 4 weeks or longer for growth to be detected but may be detected in as short a time as 2 to 3 days. On enriched culture media, the mold form develops initially as a glabrous- or waxy-appearing colony that may become off-white to white in color. With age, the aerial hyphae often turn gray to brown; however, on media enriched with blood, colonies appear to have a more waxy, yeastlike appearance. Tufts of hyphae often project upward from the colonies and are referred to as the "prickly state" of the organism.

Microscopically, hyphae of the mold form are septate and delicate and measure 1 to 2 μm in diameter. Commonly, ropelike strands of hyphae are seen; however, these are found with most of the dimorphic fungi. The characteristic microscopic morphological features are single, pyriform conidia produced on long to short conidiophores that resemble lollipops (Figure 44.61). The production of conidia in some isolates is minimal or

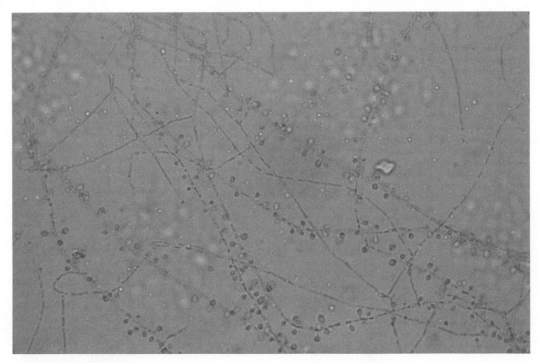

FIGURE 44.61

B. dermatitidis, mycelial form, showing oval conidia borne laterally on branching hyphae (2000×).

absent, particularly on a medium containing blood enrichment.

Colonies of the yeast form develop within 7 days and appear waxy and wrinkled, and cream to tan in color.[87] Microscopically, large, thick-walled yeast cells (8 to 13 μm) with buds attached by a broad base are seen (Figure 44.62). During the conversion process, swollen hyphal forms and immature cells with rudimentary buds may be present. Since the conversion of *B. dermatitidis* is easily accomplished, this is feasible in the clinical laboratory; however, this is the only instance in which mold-to-yeast conversion should be attempted. Some isolates may produce small yeast cells (2 to 5 μm) in size.

B. dermatitidis may also be identified by the presence of a specific A band in the exoantigen test or by nucleic acid probe testing. In some instances, *H. capsulatum, P. boydii,* and *T. rubrum* might be confused microscopically with *B. dermatitidis.* The relatively slow growth rate of *B. dermatitidis* and careful examination of the microscopic morphological features will usually differentiate these fungi from *B. dermatitidis.*

COCCIDIOIDOMYCOSIS

Coccidioidomycosis is a fungal infection primarily limited to the desert southwestern portion of the United States as well as semiarid regions of Mexico and Central and South America. Although the geographical distribution of the organism is well defined, cases of coccidioidomycosis may be seen in any part of the world because of the ease of travel. The infection is acquired by inhalation of the infective arthroconidia of *Coccidioides immitis.* Approximately 60% of the cases are asymptomatic and self-limited respiratory tract infec-

tions. The infection, however, may become disseminated, with extension to other organs, meninges, bone, skin, lymph nodes, and subcutaneous tissue. Fewer than 1% of persons who acquire coccidioidomycosis ever become seriously ill; dissemination does, however, occur most frequently in persons of dark-skinned races. This infection has been known to occur in epidemic proportions and persons who visit endemic areas and return to a distant location will present to their local physician. All laboratories should be prepared to deal with the laboratory diagnosis of coccidioidomycosis. In 1992 an epidemic occurred in northern California with more than 4000 cases seen in one single county near Bakersfield.

In direct microscopic examinations of sputum or other body fluids, *C. immitis* appears as a non-budding, thick-walled spherule, 20 to 200 μm in diameter, containing either granular material or numerous small (2 to 5 μm in diameter) endospores (Figure 44.63). The endospores are freed by rupture of the cell wall of spherules; therefore empty and collapsed "ghost" spherules may be present. Small, immature spherules measuring 10 to 20 μm may be confused with *B. dermatitidis* when two are lying adjacent. In instances when the identification of *C. immitis* is questionable, a wet preparation of the clinical specimen may be made using sterile saline, and the edges of the coverglass may be sealed with petrolatum and incubated overnight. When spherules are present, multiple hyphal strands will be produced from the endospores.

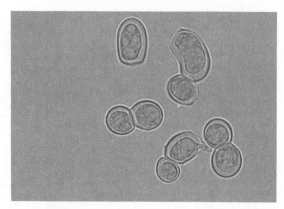

FIGURE 44.62

B. dermatitidis, yeast form, showing thick-walled, oval to round, single-budding yeastlike cells (1000×).

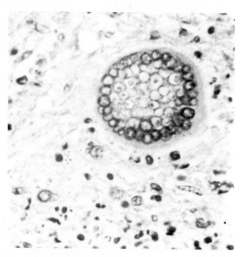

FIGURE 44.63

Methenamine silver stain of *C. immitis,* tissue form, showing spherule containing numerous spherical endospores (1000×).

Cultures of *C. immitis* represent a biohazard to laboratory workers, and strict safety precautions must be followed when examining cultures. Mature colonies may appear within 3 to 5 days of incubation and may be present on most media, including those used for bacteriology. **Laboratory workers are cautioned not to open cultures of fluffy white molds unless they are placed inside of a biological safety cabinet.** Colonies of *C. immitis* often appear as delicate, cobweblike growth after 3 to 21 days of incubation. Some portions of the colony will exhibit aerial hyphae, whereas others will have the hyphae adherent to the agar surface. Most isolates appear fluffy white; however, colonies of varying colors have been observed. On blood agar, some colonies exhibit a greenish discoloration, whereas others appear yeastlike, smooth, wrinkled, and tan in color. Isolates of *C. immitis* have been reported to range in color from pink to yellow to purple and black.[35]

Microscopically, some cultures show small septate hyphae that often exhibit right-angle branches and racquet forms. With age, the hyphae form arthroconidia that are characteristically rectangular or barrel-shaped in appearance. The arthroconidia are larger than the hyphae from which they are produced and stain darkly with lactophenol cotton or aniline blue. The arthroconidia are separated from one another by clear or lighter staining nonviable cells and are referred to as *alternate arthroconidia* (Figure 44.64). Arthroconidia have been reported to range in size from 1.5 to 7.5 μm in width and 1.5 to 30 μm in length, whereas most are 3 to 4.5 μm in width and 3.1 μm in length. Variation has been reported in the shape of arthroconidia and ranges from rounded to square or rectangular to curved; however, most are barrel-shaped. Even if alternate arthroconidia are observed microscopically, the definitive identification should be made using nucleic acid probe testing.

If a culture is suspected of being *C. immitis,* **it should be sealed with tape to prevent chances of laboratory-acquired infection.** Since *C. immitis* is the most infectious of all the fungi, extreme caution should be used when handling cultures of this organism. Safety precautions should be observed (see box on p. 738).

Some members of the Gymnoascaceae are found in the environment and may resemble *C. immitis* microscopically. Some species produce alternate arthroconidia that tend to be more rectangular, and it is necessary to consider them when making an identification. *Geotrichum candidum* and species of *Trichosporon* produce hyphae that disassociate into arthroconidia (Figures 44.65 and 44.66). The colonial morphological features of older cultures may resemble *C. immitis.* Microscopically, the arthroconidia do not appear to be alternate and are even-staining. Confusion rarely occurs between members of the Gymnoascaceae and *C. immitis.* It is also important to remember that occasional strains of *C. immitis* fail to sporulate and may be identified only by the exoantigen test or nucleic acid probe testing.

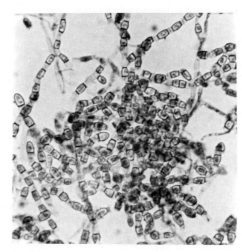

FIGURE 44.64

C. immitis, mycelial form, showing numerous thick-walled, rectangular or barrel-shaped alternate arthroconidia (500×).

HISTOPLASMOSIS

Histoplasmosis is a chronic, granulomatous infection that is primary in the lung and eventually invades the reticuloendothelial system. Approximately 95% of cases are asymptomatic and self-limited. Chronic pulmonary infections occur, and dissemination to the lymphatic tissue, liver, spleen, kidneys, meninges, and heart is becoming more common. Ulcerative lesions of the upper respiratory tract may occur. Histoplasmosis is another infection that is prevalent in the Ohio, Mississippi River, and Missouri River valleys, where conditions are optimal for growth of the organism in soil enriched with bird manure or bat guano.

Outbreaks of histoplasmosis have been associated with activities that disperse the microconidia or hyphal fragments into the air. Infection is acquired via inhalation of these infective units from the environment. The severity of the disease is generally related directly to the inoculum size. Numerous cases of histoplasmosis have been re-

Safety Precautions

1. If culture dishes are used, they should be handled only in a biological safety cabinet. Cultures should by sealed with tape if the specimen is suspected of containing *C. immitis.*
2. The use of cotton plug test tubes is discouraged, and screw-capped tubes should be used if this mode of culture is preferred. If *C. immitis* is suspected, the culture should be overlayed with sterile water or saline before using an inoculating wire to remove a portion for microscopic examination. All handling of cultures of *C. immitis* in screw-capped tubes should be performed inside a biological safety cabinet.
3. All micoscopic preparations for examination should be prepared in a biological safety cabinet.
4. Cultures should be autoclaved as soon as the final identification of *C. immitis* is made.

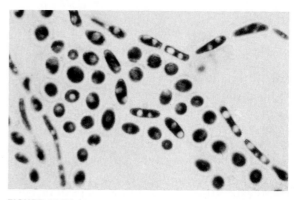

FIGURE 44.65

Trichosporon species, showing arthroconidia and an occasional blastoconidium.

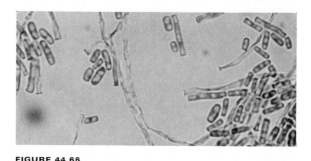

FIGURE 44.66

Geotrichum candidum, showing numerous arthroconidia. Note that arthroconidia do not alternate with a clear cell as in the case of *Coccidiodes immitis* (440×).

ported in persons cleaning a chicken house or barn that has been undisturbed for long periods or from working in or cleaning those areas that have served as roosting places for starlings and similar birds.

An estimated 500,000 persons are infected annually by *Histoplasma capsulatum,* the etiological agent of histoplasmosis. It is perhaps one of the most common fungal infections seen in the midwestern and southern parts of the United States. However, a history of exposure is often impossible to document.

The direct microscopic examination of respiratory tract specimens and other similar specimens is usually unsatisfactory for the detection of *H. capsulatum.* The organism, however, may be detected by an astute laboratorian when examining Wright- or Giemsa-stained specimens of bone marrow or peripheral blood. *H. capsulatum* is found intracellularly within mononuclear cells as small, round to oval yeast cells 2 to 5 µm in diameter.

H. capsulatum is easily cultured from clinical specimens; however, it may be overgrown by bacteria or rapidly growing molds. A procedure useful for the recovery of *H. capsulatum, B. dermatitidis,* and *C. immitis* from contaminated specimens utilizes a yeast extract phosphate medium and a drop of concentrated ammonium hydroxide (NH$_4$OH) placed on one side of the inoculated plate of medium.[72] In the past, it has been recommended that specimens not be kept at room temperature before culture, since *H. capsulatum* would

not survive. It has been shown that *H. capsulatum* will survive transit in the mail for as long as 16 days.[31] It is, however, recommended that specimens be cultured as soon as possible to ensure the optimal recovery of *H. capsulatum* and other dimorphic fungi.

H. capsulatum is usually considered to be a slow-growing mold at 25° to 30° C and commonly requires 2 to 4 weeks or more for colonies to appear. It is not uncommon to recover the organism in 5 days or less if large numbers of cells are present in the clinical specimen. Isolates of *H. capsulatum* have been reported to be recovered from blood cultures with the Isolator in a mean time of 8 days.[7,8] Textbooks describe the colonial morphology of *H. capsulatum* as being a white, fluffy mold that turns brown to buff with age. Some isolates ranging from gray to red have also been reported. The organism commonly produces wrinkled, moist, heaped, yeastlike colonies that are soft and cream, tan, or pink in color. Tufts of hyphae often project upwards from the colonies as described with *B. dermatitidis*.

Microscopically, the hyphae of *H. capsulatum* are small (1 to 2 μm in diameter) and are often intertwined to form ropelike strands. Commonly, large (8 to 14 μm in diameter) spherical or pyriform, smooth-walled macroconidia are seen in young cultures. With age, the macroconidia become roughened or tuberculate and provide enough evidence to make a tentative identification (Figure 44.67). The macroconidia are produced either on short or long lateral branches of the hyphae. Some isolates produce round to pyriform, smooth microconidia (2 to 4 μm in diameter) in addition to the characteristic tuberculate macroconidia. Some isolates of *H. capsulatum* fail to sporulate despite numerous attempts to induce sporulation.

Conversion of the mold form to the yeast form is usually difficult and is not recommended. Microscopically, a mixture of swollen hyphae and small budding yeast cells 2 to 5 μm in size should be observed. These are similar to the intracellular yeast cells seen in mononuclear cells in infected tissue (Figure 44.68). The yeast form of *H. capsulatum* cannot be recognized unless the corresponding mold form is present on another culture or unless the yeast form is converted directly to the mold form by incubation at 25° to 30° C after yeast cells have been observed. The exoantigen test can be used for identification but nucleic acid probe testing is recommended as the most definitive means of providing a rapid identification of this organism (Table 44.7).[30]

Sepedonium, an environmental organism found growing on mushrooms, is always mentioned as being confused with *H. capsulatum*, since it produces similar tuberculate macroconidia. This organism is almost never recovered from clinical specimens, does not have a yeast form, fails to produce characteristic bands seen in the exoantigen test with *H. capsulatum*, and does not react in nucleic acid probe tests.

PARACOCCIDIOIDOMYCOSIS

Paracoccidioidomycosis is a chronic granulomatous infection that begins as a primary pulmonary infection. It is often asymptomatic and then disseminates to produce ulcerative lesions of the mucous membranes. Ulcerative lesions are commonly present in the nasal and oral mucosa, gingivae, and less commonly in the conjunctivae. Lesions occur most commonly on the face in as-

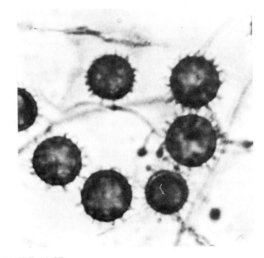

FIGURE 44.67

H. capsulatum, mycelial form, showing characteristic tuberculate macroconidia (1000×).

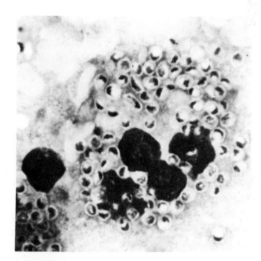

FIGURE 44.68

H. capsulatum, yeast form, showing intracellular, oval yeast cells, deeply stained (2000×).

sociation with oral mucous membrane infection. The lesions are characteristically ulcerative, with a serpiginous active border and a crusted surface. Lymph node involvement in the cervical area is common. Pulmonary infection is seen most often, and progressive chronic pulmonary infection is found in approximately 50% of cases. Dissemination to other anatomic sites, including the lymphatic system, spleen, intestines, liver, brain, meninges, and adrenal glands, occurs in some patients.

The infection is most commonly found in South America, with the highest prevalence in Brazil, Venezuela, and Colombia. It also has been seen in many other areas, including Mexico, Central America, and Africa. Occasional imported cases are seen in the United States and Europe.

The exact mechanism by which paracoccidioidomycosis is acquired is unclear; however, it is speculated that its origin is pulmonary and that it is acquired by inhalation of the organism from the environment. Since mucosal lesions are an integral part of the disease process, it is also speculated that the infection may be acquired through trauma to the oropharynx caused by vegetation that is commonly chewed by residents of the endemic areas. The specific ecological niche of the organism in nature is undetermined.

Specimens submitted for direct microscopic examinations are important for the diagnosis of paracoccidioidomycosis. Large, round or oval, multiply budding yeast cells (8 to 40 μm in diameter) are usually recognized in sputum, mucosal biopsy, and other exudates. Characteristic multiply budding yeast forms resemble a "mariner's wheel." The yeast cells surrounding the periphery of the parent cell range from 8 to 15 μm in diameter.

Colonies of *Paracoccidioides brasiliensis* grow very slowly (21 to 28 days) and are heaped, wrinkled, moist, and yeastlike. With age, colonies may become covered with a short aerial mycelium and turn tan to brown in color. The surface of colonies is often heaped with crater formations.

Microscopically, the mold form is similar to that seen with *B. dermatitidis*. Small hyphae (1 to 2 μm in diameter) are seen, along with numerous chlamydospores. Small, delicate (3 to 4 μm globose or pyriform conidia may be seen rising from the sides of the hyphae or on very short conidiophores (Figure 44.69). Most often, cultures reveal only fine septate hyphae and numerous chlamydospores.

After conversion on a blood-enriched medium, the colonial morphology of the yeast form is characterized by smooth, soft-wrinkled, yeastlike colonies that are cream to tan in color. Microscopically, the colonies are composed of yeast cells 10 to 40 μm in diameter that are surrounded by narrow-necked yeast cells around the periphery, as previously described (Figure 44.70).

If in vitro conversion to the yeast form is unsuccessful, the exoantigen test[74,75] (Procedure 44.6) should be used to make the definitive identification of *P. brasiliensis*. Nucleic acid probe testing is not available for this organism.

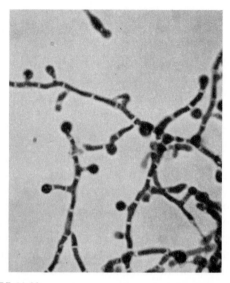

FIGURE 44.69

Paracoccidioides brasiliensis, mycelial form, showing septate hyphae and pyriform conidia singly borne (440×).

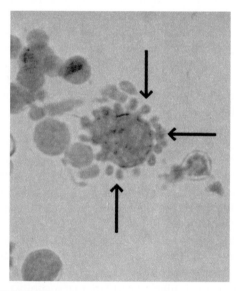

FIGURE 44.70

P. brasiliensis. Numerous small buds (440×).

Table 44.9 presents a summary of the colonial and microscopic morphological features of the dimorphic fungi.

OPPORTUNISTIC MYCOSES

The opportunistic mycoses are a group of fungal infections that occur almost exclusively in immunocompromised patients. The type of patient who acquires an opportunistic fungal infection is one who is compromised by some underlying disease process such as HIV infection, lymphoma, leukemia, diabetes mellitus, or another underlying defect of the immune system. Many of these patients have been placed on treatment with corticosteroids, cytotoxic drugs, or other immunosuppressive agents. Many fungi previously thought to be nonpathogenic are now recognized as etiological agents of the opportunistic fungal infections. Since most of the organisms known to cause infection in this group of patients are commonly encountered in the clinical laboratory as saprobes, it is impossible for the laboratory to determine the clinical significance of an isolate recovered from the specimens of immunocompromised patients. The laboratory must identify and report completely the presence of all fungi recovered from immunocompromised patients, since every organism is a potential pathogen. Many of the organisms associated with opportunistic infections are frequently acquired during construction, demolition, or remodeling of buildings. Some are hospital-acquired.

Although there is an ongoing list of opportunistic mycoses, some are seen more often and include aspergillosis, zygomycosis, candidosis (candidiasis), cryptococcosis, and trichosporonosis. The etiological agents of each will be discussed; however, other fungi also associated with opportunistic mycoses are presented later in this chapter.

ASPERGILLOSIS

Several species of aspergilli are among the most frequently encountered organisms in the clinical laboratory; some are pathogenic, whereas others are infrequently associated with infection or do not cause infection at all. The aspergilli are widespread in the environment, where they colonize grain, leaves, soil, and living plants. Conidia of the aspergilli are easily dispersed into the environment, and humans become infected by inhaling them. These organisms are capable not only of causing disseminated infection, as is seen in immunocompromised patients, but also of causing a host of other types of infections, including invasive lung infection, pulmonary fungus ball, allergic bronchopulmonary aspergillosis, external otomycosis, mycotic keratitis, onychomycosis, sinusitis, endocarditis, and central nervous system infection. Most often, immunocompromised patients acquire a primary pulmonary infection that rapidly disseminates and causes infection in virtually every organ.

Assessing the significance of members of the genus *Aspergillus* in a clinical specimen is difficult. They are found frequently in cultures of respiratory secretions, skin scrapings, and other specimens. Table 44.10 presents the Mayo Clinic experience with the recovery of species of *Aspergillus* from clinical specimens. It has been reported that *Aspergillus* is significant in only 10% of cases[78]; however, this depends on the hospital setting and type of patient population seen. Table 44.10 illustrates the diversity of aspergilli seen in clinical specimens.

Since aspergilli are recovered frequently, it is imperative that the organism be demonstrated in the direct microscopic examination of fresh clinical specimens and that it be recovered repeatedly from patients having a compatible clinical picture to ensure that the organism is clinically significant. Most species of *Aspergillus* are susceptible to cycloheximide; therefore specimens submitted for recovery of *Aspergillus* or subcultures should not be inoculated onto media containing this compound.

Aspergillus fumigatus is most commonly recovered from immunocompromised patients; moreover, it is the species most often seen in the clinical laboratory. In addition, *Aspergillus flavus* is sometimes recovered from immunocompromised patients and represents a frequent isolate in the clinical microbiology laboratory. The recovery of *A. fumigatus* or *A. flavus* from surveillance nasal cultures is strongly correlated with subsequent invasive aspergillosis.[83] The absence of a positive nasal culture does not preclude infection, however. *Aspergillus niger* is seen commonly in the clinical laboratory, but its association with clinical disease is somewhat limited. *Aspergillus terreus* is now recognized as a significant cause of infection in immunocompromised patients, but its frequency of recovery is much lower than the previously mentioned species.

Specimens submitted for direct microscopic examination may contain septate hyphae that usually show evidence of dichotomous branching. In addition, some hyphae may show the presence of rounded, thick-walled cells.

A. fumigatus is a rapidly growing mold (2 to 6 days) that produces a fluffy to granular, white to blue-green colony. Mature sporulating colonies most often exhibit the blue-green powdery appearance. Microscopically, *A. fumigatus* is characterized by the presence of septate hyphae and short or long conidiophores having a characteristic cell at their base. The tip of the conidiophore expands into a large, dome-shaped vesicle that has bottle-shaped phialides covering the upper half or two thirds of its surface. Long chains of small (2 to 3 μm in diameter), spherical, rough-

TABLE 44.9

SUMMARY OF THE CHARACTERISTIC FEATURES OF FUNGI KNOWN TO BE COMMON CAUSES OF FUNGAL INFECTION IN HUMANS

INFECTION	ETIOLOGICAL AGENT	GROWTH RATE (DAYS)	CULTURAL CHARACTERISTICS AT 30° C	
			BLOOD-ENRICHED MEDIUM	MEDIUM LACKING BLOOD ENRICHMENT
Blastomycosis	*Blastomyces dermatitidis*	2-30	Colonies are cream to tan, soft, moist, wrinkled, waxy, flat to heaped, and yeastlike; "tufts" of hyphae often project upward from colonies	Colonies are white to cream to tan, some with drops of exudate present, fluffy to glabrous, and adherent to the agar surface
Cryptococcosis	*Cryptococcus neoformans*	3-10	Colonies are usually dome-shaped, dry, cream to tan, and smaller than those on media lacking blood enrichment	Colonies are dry to mucoid and shiny, dome-shaped, smooth and cream to tan in color; however, some isolates appear golden to orange on certain media (i.e., inhibitory mold agar)
Histoplasmosis	*Histoplasma capsulatum*	3-45	Colonies are heaped, moist, wrinkled, yeastlike, soft, and cream, tan or pink in color; "tufts" of hyphae often project upward from colonies	Colonies are white, cream, tan or gray, fluffy to glabrous; some colonies appear yeastlike and adherent to the agar surface; many variations in colonial morphology occur
Paracoccidioidomycosis	*Paracoccidioides brasiliensis*	21-28	Colonies are heaped, wrinkled, moist, and yeastlike; with age, colonies may become covered with short aerial mycelium and may turn brown	
Candidosis	*Candida albicans* and other *Candida* species	2-4	Colonies vary in their morphology but are usually white to tan, shiny to dull, flat to heaped, smooth to wrinkled, and moist to dry; some colonies produce pseudohyphal fringes at the periphery; colonies growing on blood-enriched media are somewhat smaller and drier than on non–blood-containing media	

From Thomson, R.B., and Roberts, G.D. 1982. A practical approach to the diagnosis of fungal infections of the respiratory tract. Clin. Lab. Med. 2:321.

MICROSCOPIC MORPHOLOGICAL FEATURES

BLOOD-ENRICHED MEDIUM	NON–BLOOD ENRICHED MEDIUM	RECOMMENDED SCREENING TESTS	MICROSCOPIC MORPHOLOGICAL FEATURES OF TISSUE FORM	RECOMMENDED CONFIRMATORY TESTS FOR IDENTIFICATION
Hyphae 1-2 μm in diameter are present; some are aggregated in ropelike clusters; sporulation is rare	Hyphae 1-2 μm in diameter are present; single pyriform conidia are produced on short to long conidiophores; some cultures produce few conidia	Not available	8-15 μm, broad-based budding cells with double-contoured walls are seen; cytoplasmic granulation is often obvious	1. Specific nucleic acid probe 2. Broad-based budding cells may be seen after in vitro conversion on cottonseed agar 3. Exoantigen test
Cells are usually spherical, vary in size, and may be encapsulated; cells may have more than one "pinched-off" bud present on parent cell		1. Urease production 2. Phenol oxidase production 3. Nitrate reductase production	2-15 μm, single or multiply budding spherical cells that vary in size are seen; evidence of encapsulation may be present. Some pseudohyphal forms may be seen.	1. Specific nucleic acid probe 2. Substrate utilization detected by commercial identification systems 3. Pigment on niger seed agar
Hyphae 1-2 μm in diameter are present; some are aggregated in ropelike clusters; sporulation is rare	Young cultures usually have a predominance of smooth-walled macroconidia that become tuberculate with age; macroconidia may be pyriform or spherical; some isolates produce small pyriform microconidia in the presence or absence of macroconidia	Not available	2-5 μm, small, oval to spherical budding cells often seen inside of mononuclear cells	1. Specific nucleic acid probe 2. Exoantigen test
Hyphae 1-2 μm in diameter are present; some isolates produce conidia similar to those of *B. dermatitidis;* chlamydospores may be numerous, and multiple budding yeast cells 10-25 μm in diameter may be present		Not available	10-25 μm, multiply budding cells (buds 1-2 μm) resembling a "mariner's wheel" may be present; buds are attached to the parent cell by a narrow neck	Exoantigen test
Most species produce either blastoconidia, pseudohyphae, or true hyphae; chlamydospores are produced by *C. albicans* and certain isolates of *C. tropicalis*		Germ tube production by *C. albicans*	Blastoconidia 2-5 μm in diameter and pseudohyphae are present	1. Germ tube production for *C. albicans* 2. Substrate utilization as detected by commercial identification systems

Continued.

TABLE 44.9

SUMMARY OF THE CHARACTERISTIC FEATURES OF FUNGI KNOWN TO BE COMMON CAUSES OF FUNGAL INFECTION IN HUMANS—cont'd

INFECTION	ETIOLOGICAL AGENT	GROWTH RATE (DAYS)	CULTURAL CHARACTERISTICS AT 30° C	
			BLOOD-ENRICHED MEDIUM	MEDIUM LACKING BLOOD ENRICHMENT
Zygomycosis	*Rhizopus* species, *Mucor* species, and other *Zygomycetes*	1-3	Colonies are extremely fast growing, woolly, and gray to brown to gray-black in color	
Aspergillosis	*Aspergillus fumigatus* *Aspergillus flavus* *Aspergillus niger* *Aspergillus terreus* *Aspergillus* species	3-5	Colonies of *A. fumigatus* are usually blue-green to gray-green whereas those of *A. flavus* and *A. niger* are yellow-green and black, respectively; colonies of *A. terreus* resemble powdered cinnamon; other species of *Aspergillus* exhibit a wide range of colors; blood-enriched media usually have little effect on the colonial morphological features	
Coccidioidomycosis	*Coccidioides immitis*	2-21	Colonies may be white and fluffy to greenish on blood-enriched media; some isolates are yeastlike, heaped, wrinkled, and membranous	Colonies usually are fluffy white but may be pigmented gray, orange, brown, or yellow; mycelium is adherent to the agar surface in some portions of the colony

walled, green conidia form a columnar mass on the vesicle (Figure 44.71). Cultures of *A. fumigatus* are thermotolerant and are able to withstand temperatures up to 45° C.

A. flavus, somewhat more rapidly growing (1 to 5 days), produces a yellow-green colony. Microscopically, vesicles are globose and phialides are produced directly from the vesicle surface (uniserate) or from a primary row of branches (metulae; biserate). The phialides give rise to short chains of yellow-orange elliptical or spherical conidia that become roughened on the surface with age (Figure 44.72).

A. niger, sometimes referred to as being dematiaceous, produces mature colonies within 2 to 6 days. Growth begins initially as a yellow colony that soon develops a black, dotted surface as conidia are produced. With age, the colony becomes jet black and powdery, whereas the reverse remains

buff or cream color; this occurs on any culture medium.

Microscopically, *A. niger* exhibits septate hyphae, long conidiophores that support spherical vesicles that give rise to large metulae and smaller phialides from which long chains of brown, rough-walled conidia are produced (Figure 44.73). The entire surface of the vesicle is involved in sporulation.

A. terreus, less commonly seen in the clinical laboratory, produces colonies that are tan and resemble cinnamon. Vesicles are hemispherical, as seen microscopically, and phialides cover the entire surface and are produced from a primary row of branches (metulae; biserate). Phialides produce globose to elliptical conidia arranged in chains. This species produces the unique structures of larger aleuriospores, which are found on submerged hyphae (Figure 44.74). The reader is

MICROSCOPIC MORPHOLOGICAL FEATURES		RECOMMENDED SCREENING TESTS	MICROSCOPIC MORPHOLOGICAL FEATURES OF TISSUE FORM	RECOMMENDED CONFIRMATORY TESTS FOR IDENTIFICATION
BLOOD-ENRICHED MEDIUM	NON-BLOOD ENRICHED MEDIUM			
1. *Rhizopus* species—rhizoids are produced at the base of sporangiophore 2. *Mucor* species—no rhizoids are produced		Not available	Large ribbonlike (10-30 μm), twisted, often distorted pieces of aseptate hyphae may be present; septa may occasionally be seen	Identification is based on characteristic morphological features
1. *A. fumigatus*—uniserate heads with phialides covering the upper one half to two thirds of the vesicle 2. *A. flavus*—uniserate or biserate or both with phialides covering the entire surface of a spherical vesicle 3. *A. niger*—biserate with phialides covering the entire surface of a spherical vesicle; conidia are black 4. *A. terreus*—biserate with phialides covering the entire surface of a hemispherical vesicle; aleuriospores are formed on submerged hyphae		Not available	Septate hyphae 5-10 μm in diameter that exhibit dichotomous branching	Identification is based on microscopic morphological features and colonial morphology; *A. fumigatus* can tolerate elevated temperatures ≥ 45° C
Chains of alternate, barrel-shaped arthroconidia are characteristic; some arthroconidia may be elongated; hyphae are small and often arranged in ropelike strands and racquet forms are seen in young cultures		Not available	Round spherules 30-60 μm in diameter containing 2-5 μm endospores are characteristic Empty spherules are commonly seen	1. Specific nucleic acid probe 2. Exoantigen test

referred to the chapter by Rogers et al.[68] for further information regarding other species.

ZYGOMYCOSIS

Zygomycosis is a somewhat less common infection when compared with aspergillosis; however, it is a significant cause of morbidity and mortality in immunocompromised patients. The organisms involved have a worldwide distribution and are commonly found on decaying vegetable matter or in soil. The infection is generally acquired by inhalation of spores followed by subsequent development of infection. Immunocompromised patients, particularly those having uncontrolled diabetes and patients receiving prolonged corticosteroid, antibiotic, or cytotoxic therapy, are at greatest risk. The organisms involved in causing zygomycosis have a marked propensity for vascular invasion and rapidly produce thrombosis and necrosis of tissue. One of the most common forms

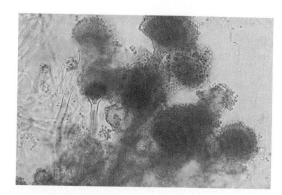

FIGURE 44.71

A. fumigatus, conidiophores and conidia (400×).

observed is the rhinocerebral form in which the nasal mucosa, palate, sinuses, orbit, face, and brain are involved; each shows massive necrosis with vascular invasion and infarction. Other types of infection involve the lungs and gastrointestinal

TABLE 44.10

SPECIES OF *ASPERGILLUS* RECOVERED FROM CLINICAL SPECIMENS DURING A 10-YEAR PERIOD AT MAYO CLINIC

ORGANISM	CLINICAL SPECIMEN SOURCE				
	RESPIRATORY SECRETIONS	GASTRO-INTESTINAL	GENITO-URINARY	SKIN, SUBCUTANEOUS TISSUE	BLOOD, BONE, CNS, ETC.
A. clavatus	97/93*	1/1	—	1/1	—
A. flavus	1298/740	10/10	11/11	177/131	2/2
A. fumigatus	3247/2656	11/9	14/14	175/137	8/8
A. glaucus	503/307	1/1	—	8/8	1/1
A. nidulans	52/48	—	—	5/3	—
A. niger	1484/1376	18/18	17/17	151/124	11/11
A. terreus	164/146	—	—	23/21	3/3
A. versicolor	1237/1202	6/6	24/22	226/224	16/16
Other species of *Aspergillus*	3463/3418	18/14	32/32	319/314	16/16

*Numerator, number of cultures; denominator, number of patients.

tract; some patients develop disseminated infection. Organs including the liver, spleen, pancreas, and kidney may be involved. The Zygomycetes have also been reported to be the etiological agent of infection in the skin of burn patients and in subcutaneous tissue of patients undergoing surgery.

The rapid diagnosis of zygomycosis may be made by examination of tissue specimens or exudate from infected lesions using the potassium hydroxide preparation. Branching, predominantly nonseptate hyphae will be observed. It is very important that the laboratory notify the clinician of these findings since Zygomycetes grow very rapidly and vascular invasion occurs at about the same rate.

The colonial morphological features of the Zygomycetes allow one to immediately suspect organisms belonging to this group. Colonies characteristically produce a fluffy, white to gray or brown hyphal growth that diffusely covers the surface of the agar within 24 to 96 hours (Figure 44.75). The hyphae appear to be coarse and fill the entire culture dish or tube rapidly with loose grayish hyphae dotted with brown or black sporangia. It is impossible to distinguish between the different genera and species of Zygomycetes based on their colonial morphological features since most are identical in appearance.

Microscopically, the Zygomycetes characteristically produce large ribbonlike hyphae that are irregular in diameter and nonseptate; however, occasional septa may exist in older cultures. The specific identification of these organisms is confirmed by observing the characteristic saclike

fruiting structures *(sporangia)*, which produce internally spherical, yellow or brown spores *(sporangiospores)* (Figure 44.76). Each sporangium is formed at the tip of a supporting structure *(sporangiophore)*. During maturation, the sporangium becomes fractured and sporangiospores are released into the environment. Sporangiophores are usually connected to each other by occasionally septate hyphae called *stolons*, which attach at contact points where rootlike structures *(rhizoids)* anchor the organism to the agar surface. The identification of the three most common Zygomycetes, *Mucor, Rhizopus,* and *Absidia,* is based on the presence or absence of rhizoids and the position of the rhizoids in relation to the sporangiophores. *Mucor* is characterized by sporangiophores that are singularly produced or branched and have at their tip a round sporangium filled with sporangiospores. *Mucor* does not have rhizoids or stolons and this distinguishes it from the other genera of Zygomycetes (see Figure 44.76). *Rhizopus* has unbranched sporangiophores with rhizoids that appear at the point where the stolon arises (Figure 44.77).

Absidia, an uncommon isolate in the clinical laboratory, is characterized by the presence of rhizoids that originate between sporangiophores (Figure 44.78). The sporangia are pyriform and have a swollen portion of the sporangiophore at the junction of the sporangium. Usually a septum is formed in the sporangiophore just below the sporangium.

The other genera of Zygomycetes that may be encountered in the clinical laboratory include *Saksenaea, Cunninghamella, Conidiobolus,* and *Basid-*

FIGURE 44.72

A. flavus, showing spherical vesicles that give rise to metulae and phialides that produce chains of conidia (440×).

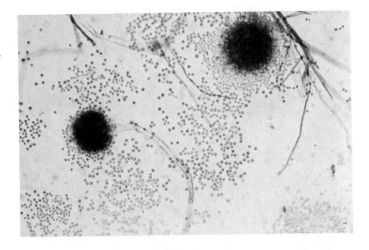

FIGURE 44.73

A. niger, showing larger spherical vesicle that gives rise to metulae, phialides, and chains of dark brown to black conidia (750×).

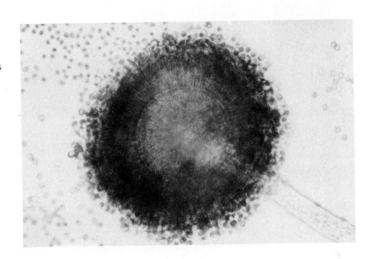

iobolus. For additional information on these and other species, the reader is referred to Goodman and Rinaldi.[24]

CANDIDOSIS (CANDIDIASIS)

Candidosis is the most frequently encountered opportunistic fungal infection. It is caused by a variety of species of *Candida,* with *Candida albicans* being the most frequent etiological agent, followed by *Candida tropicalis* and *Candida (Torulopsis) glabrata.* A number of other species have been involved in infection in immunocompromised patients; however, their incidence is not nearly as high, as previously mentioned. *C. albicans* and others are a part of the normal endogenous microbial flora, and infections are believed to be endogenous in origin. The organisms may be recovered from the oropharynx, gastrointestinal tract, genitourinary tract, and skin. *Candida* species are responsible for several different types of infections in normal and immunocompromised patients. Included are intertriginous candidosis in which skin

folds are involved, paronychia, onychomycosis, perlèche, vulvovaginitis, thrush, pulmonary infection, eye infection, endocarditis, meningitis, fungemia, and disseminated infection. The latter two types of infection are most commonly seen in immunocompromised patients, whereas the others mentioned may occur in normal hosts. Onychomycosis and esophagitis produced by *C. albicans* are very common in patients having acquired immunodeficiency syndrome (AIDS).

The clinical significance of *Candida* recovered from respiratory tract secretions is difficult to determine, since it is considered to be part of the normal flora of humans. A study at the Mayo Clinic evaluated the clinical significance of yeasts recovered from respiratory secretions, except for *Cryptococcus neoformans,* and concluded that they are part of the normal flora and that their routine identification is unnecessary.[57] The repeated recovery of different species of yeasts from multiple specimens from the same patient usually indicates colonization. The simultaneous recovery of the

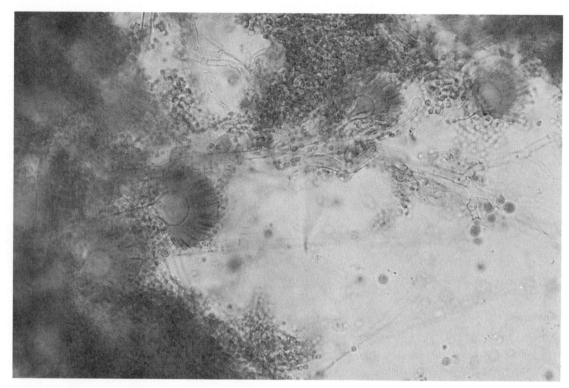

FIGURE 44.74

A. terreus, showing typical head of aspergillus and aleuriospores found on submerged hyphae of this species. (500×).

FIGURE 44.75

Rhizopus colony.

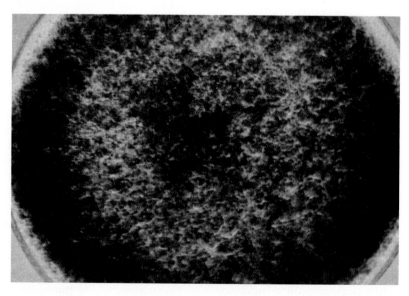

same species of yeast from several body sites, including urine, is a good indicator of disseminated infection and the subsequent development of fungemia.

The direct microscopic examination of clinical specimens containing *Candida* will reveal budding yeast cells (blastoconidia) 2 to 4 μm in diameter or pseudohyphae showing regular points of constriction, resembling lengths of sausages; or true sep-

tate hyphae. The blastoconidia, hyphae, and pseudohyphae are strongly gram positive. It is advisable to report the approximate number of such forms since the presence of large numbers in a fresh clinical specimen may be of diagnostic significance.

The colonial and microscopic morphological features of the species of *Candida* are of little value in making a definitive identification. *C. albicans,*

FIGURE 44.76

Mucor species, showing numerous sporangia in the absence of rhizoids (440×).

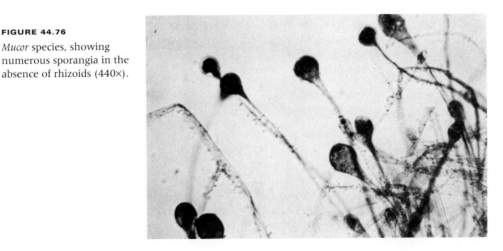

FIGURE 44.77

Rhizopus species, showing sporangium on long sporangiophore arising from nonseptate hyphae. Note presence of characteristic rhizoids at the base of the sporangiophore (250×).

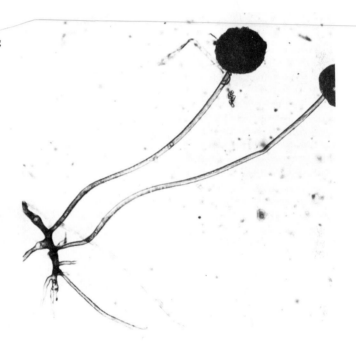

however, may be identified by the production of germ tubes or the presence of chlamydospores (Figure 44.79). Other species of *Candida* must be identified by the utilization of, and sometimes fermentation of, specific substrates.

CRYPTOCOCCOSIS

Cryptococcosis, specifically caused by *C. neoformans,* is a subacute or chronic fungal infection that has several manifestations. In the immunocompromised patient, it is not uncommon to see disseminated disease with or without meningitis; meningitis occurs in approximately two thirds of patients with disseminated infection. Disseminated cryptococcosis is becoming a very common clinical entity in patients with AIDS. Occasionally, patients with disseminated infection will exhibit

painless skin lesions that may ulcerate. Other uncommon manifestations of cryptococcosis include endocarditis, hepatitis, renal infection, and pleural effusion. It is of interest to note that a review of patient records at the Mayo Clinic revealed more than 100 cases of colonization of the respiratory tract with *C. neoformans* without subsequent development of infection. Follow-up on these patients was as long as 6 years; none in this group was considered to be immunocompromised. This makes the clinical significance of *C. neoformans* somewhat difficult to assess; however, its presence in clinical specimens from immunocompromised patients should be considered to be significant. In many instances the clinical symptoms are suppressed by corticosteroid therapy, and culture or serologic evidence provides the earliest

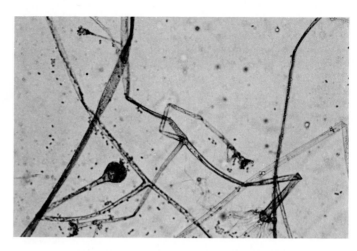

FIGURE 44.78

Absidia species, showing sporangia on long sporangiophores arising from nonseptate hyphae. Note that rhizoids are produced between sporangiophores and not at their bases (250×).

FIGURE 44.79

Candida albicans, Blastoconidia and chlamydospores (400×).

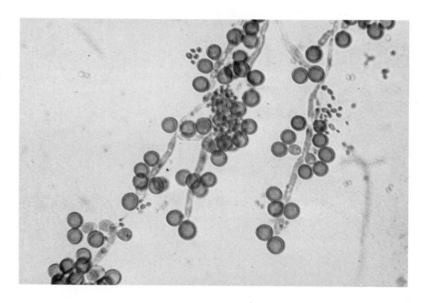

proof of infection. There is a strong association of cryptococcal infection with such debilitating diseases as leukemia and lymphoma and the immunosuppressive therapy that may be required for these and other underlying diseases. The presence of *C. neoformans* in clinical specimens in some instances precedes the symptoms of an underlying disease. The infection is probably more frequent than is commonly recognized; it is estimated that there may be 300 new cases each year in the United States.[32]

Four serotypes of *C. neoformans* have been described (A, B, C, and D) with somewhat different geographical distribution.[47] The organism has been proved to be a basidiomycete by the discovery of its sexual form. The name of the sexual form is *Filobasidiella,* with two varieties: *F. neoformans var. neoformans* and *F. neoformans var. gattii;*

the latter includes serotypes B and C.[46] The term *Cryptococcus* will continue to be used since it is well established by tradition. Despite the discovery of two varieties of *Cryptococcus,* there is no difference in disease produced or in the response to chemotherapy between the two species.

C. neoformans exists as a saprobe in nature. It is most often found associated with the excreta of pigeons. The hypothesis that pigeon habitats serve as reservoirs for human infection is substantiated by numerous reports; the pigeon manure apparently serves as an enrichment for *C. neoformans* because of its chemical makeup. It is believed that *C. neoformans* is widely distributed in nature, becomes aerosolized, and is inhaled before infection.

Traditionally, the India ink preparation has been the most widely used method for the rapid detection of *C. neoformans* in clinical specimens.

An evaluation of 39 consecutive non-AIDS patients with cryptococcal meningitis seen at the Mayo Clinic showed that only 40% gave positive India ink preparations of the cerebrospinal fluid. The purpose of the method is to delineate the large capsule of *C. neoformans,* since the ink particles cannot penetrate the capsular polysaccharide material. **Because of the low positivity rate of the India ink preparation, it is not recommended as a routine tool in the clinical microbiology laboratory. It should be replaced with the cryptococcal latex test for antigen that will be described in a subsequent section.** However, the India ink preparation is commonly positive in specimens from patients having AIDS. Laboratories examining large numbers of specimens from these patients may wish to retain the use of this procedure in combination with the cryptococcal latex test for antigen.

The microscopic examination of other clinical specimens, including respiratory secretions, can also be of value in making a diagnosis of cryptococcosis. *C. neoformans* appears as a spherical, single or multiple budding, thick-walled, yeastlike organism 2 to 15 μm in diameter, usually surrounded by a wide, refractile polysaccharide capsule. Perhaps the most important characteristic of *C. neoformans* is the extreme variation in the size of the yeast cells; this is unrelated to the amount of polysaccharide capsule present. It is important to remember that not all isolates of *C. neoformans* exhibit a capsule.

C. neoformans is easily cultured on routine fungal culture media without cycloheximide. The organism is inhibited by the presence of cycloheximide at 25° to 30° C. For the optimal recovery of *C. neoformans* from cerebrospinal fluid, it is recommended that a 0.45-μm pore size membrane filter be used with a sterile syringe. The filter is placed on the surface of the culture medium and is removed at daily intervals so that growth under the filter can be visualized. An alternative to the membrane filter technique is the use of centrifugation.

Colonies of *C. neoformans* usually appear on culture media within 1 to 5 days and begin as a smooth white to tan colony that may become mucoid and cream to brown in color. It is important to recognize the colonial morphology on different culture media since variation does occur; for example, on inhibitory mold agar, *C. neoformans* appears as a golden yellow, nonmucoid colony. Textbooks typically characterize the colonial morphology as being *Klebsiella*-like because of the large amount of polysaccharide capsule material

present (Figure 44.80). In reality, most isolates of *C. neoformans* do not have large capsules and may not have the typical mucoid appearance.

The microscopic examination of colonies of *C. neoformans* may be of help in providing a tentative identification of *C. neoformans* since the cells will be spherical and exhibit a wide variation in their size. A presumptive identification of *C. neoformans* may be based on urease production and failure to utilize an inorganic nitrate substrate. The final identification of *C. neoformans* is usually based on typical substrate utilization patterns and pigment production on niger seed agar. A commercially available nucleic acid probe method is available for the definitive identification of *C. neoformans.* This method is extremely reliable; however, costs may prevent some laboratories from using it on a routine basis. A discussion of tests useful for the identification of *C. neoformans* and other species of cryptococci will be presented in a subsequent section.

TRICHOSPORONOSIS

Trichosporonosis, caused by *Trichosporon beigelii,* a yeastlike organism, occurs almost exclusively in immunocompromised patients, particularly those having leukemia.[84] Disseminated trichosporonosis is the most common clinical manifestation. Skin lesions accompanied by fungemia are frequently seen. Endocarditis, endophthalmitis, and brain abscess have been reported. *Trichosporon beigelii* is commonly recovered from respiratory tract secretions, skin, the oropharynx and stool of patients who have no evidence of infection

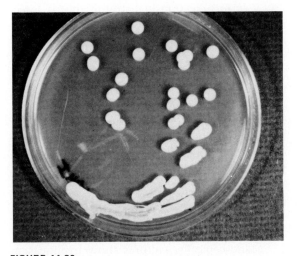

FIGURE 44.80

C. neoformans, showing colonies that appear shiny and mucoid because of the presence of a polysaccharide capsule.

and may represent colonization of those individuals.

The microscopic examination of clinical specimens reveals hyaline hyphae, numerous round to rectangular arthroconidia and, occasionally, a few blastoconidia. Usually hyphae and arthroconidia predominate.

Colonies of *T. beigelii* vary in their morphology; however, most are cream-colored, heaped, dry to moist and wrinkled. Some may appear white, dry, powdery, and wrinkled.

Arthroconidia, which are rectangular, often with rounded ends, predominate along with septate hyaline hyphae. Blastoconidia are sometimes present but are not seen in all cultures. Urease is produced and is often helpful in making a tentative identification. The final identification is based on characteristic substrate utilization profiles.

Of interest is the sharing of an antigen that is similar to that produced by *C. neoformans*. Sera from patients having trichosporonosis often yield false positive cryptococcal antigen tests when latex agglutination methods are used.

FUSARIUM INFECTION

Infection caused by any number of species of *Fusarium* is becoming more common, particularly in immunocompromised patients.[6] The organisms are common environmental flora and have long been known to cause mycotic keratitis after traumatic implantation into the cornea. However, cases of disseminated disease accompanied by fungemia are becoming more common. This entity is limited almost exclusively to severely immunocompromised patients. Necrotic skin lesions are common. Other types of infection caused by *Fusarium* include sinusitis, wound (burn) infection, and endophthalmitis. *Fusarium* species are commonly recovered from respiratory tract secretions, skin, and other specimens of patients having no evidence of infection. Interpretation of culture results rests with the clinician.

The direct examination of clinical specimens, including tissues, reveals the presence of septate hyaline hyphae that exhibit branching. These cannot be distinguished from those of *Aspergillus*, *Pseudallescheria boydii* or other fungi.

Cultures of *Fusarium* grow rapidly, within 2 to 5 days, and may produce colonies that show variation in color from white, pink, violet, to yellow, tan, or gray, depending on the species.

Microscopically, cultures show hyaline septate hyphae and either hyaline, sickle-shaped macroconidia (single to multicelled) or oval to cylindrical microconidia produced in phialides (see Figure 44.93). The latter type of sporulation may resemble that of acremonium. The most common medium used to induce sporulation of *Fusarium* is cornmeal agar. Keys for identification of species by *Fusarium* are based on growth on potato dextrose agar.

GENERAL CONSIDERATIONS FOR THE IDENTIFICATION OF YEASTS

During recent years, a significant increase in the number of fungal infections caused by yeasts and yeast-like fungi has been reported in the literature. Numerous species of *Candida* and other yeasts have been implicated. These infections are primarily seen in immunocompromised patients (including those with AIDS), and some of the yeasts have proved to be resistant to antifungal therapy.* Most infections are seen in patients having surgery and patients on long-term intravenous therapy without adequate catheter care. Microbiologists should be aware that any of the genera and species of yeast are potential pathogens in this group of patients.

Just how far the laboratory should go with a complete identification of all yeast species is of question. It is suggested that:

1. All yeasts recovered from sterile body fluids, including cerebrospinal fluid, blood, urine, paracentesis, and other fluids, should be identified to the species level.
2. Yeasts from all seriously ill or immunocompromised patients or in whom a mycotic infection is suspected should be identified to the species level.
3. Yeasts from respiratory secretions should not be identified on a routine basis; however, they should be screened for the presence of *Cryptococcus neoformans*.
4. Yeasts recovered in large amounts from any clinical source should be identified to the species level.
5. Yeasts recovered from several successive specimens, except respiratory secretions, should be identified to the species level. Each laboratory director will have to decide how much time, effort, and expense is to be spent on the identification of yeasts in the laboratory.

The development of commercially available yeast identification systems provides laboratories

---

* References 6, 16,17, 28, 50, 58, 81.

of all sizes with the availability of accurate and standardized methods. Some have extensive computer data bases that include biochemical profiles of thousands of isolates of yeasts. Variations in the reactions of carbohydrates and other substrates utilized are considered in the identification of yeasts provided by these systems. Commercially available systems are recommended for all laboratories; however, they may be used in conjunction with some less expensive and rapid screening tests that will provide the presumptive identification of *Cryptococcus neoformans* and a definitive identification of *Candida albicans*. In addition, some laboratories might prefer the use of conventional systems[17]; therefore the information presented within this section discusses rapid screening methods for the presumptive identification of yeasts, commercially available systems, and a conventional schema that will provide for the identification of commonly encountered species of yeast seen in the clinical laboratory. Nucleic acid probe identification of *C. neoformans* is certainly a viable and reasonable option for all laboratories.

RAPID SCREENING TESTS FOR THE IDENTIFICATION OF YEASTS

RAPID UREASE TEST

The rapid urease test (Procedure 44.8) is a most useful tool for screening for urease-producing yeasts recovered from respiratory secretions and other clinical specimens.[65] Alternatives to this method include the heavy inoculum of the tip of a slant of Christensen's urea agar and subsequent incubation at 35° to 37° C. In many instances, a positive reaction will occur within several hours; however, 1 to 2 days of incubation may be required. It is of interest to note that strains of *Rhodotorula, Candida,* and *Trichosporon* occasionally hydrolyze urea. Therefore the microscopic morphological features will be of help in interpreting the usefulness of the urease test for the detection of *Cryptococcus*. Another alternative method is the rapid selective urease test (Procedure 44.9). This method appears to be useful in detecting *C. neoformans* from specimens cultured onto Sabouraud's dextrose agar. It has not been evaluated using isolates tested from other media.

All of the tests mentioned provide a tentative identification of *C. neoformans;* however, they must be supplemented with new additional tests before a preliminary identification can be reported. These include the rapid nitrate reduction test and the detection of phenol oxidase production.

RAPID NITRATE REDUCTASE TEST

A characteristic feature of *C. neoformans* is that it does not utilize an inorganic nitrate substrate. A rapid method (Procedure 44.10) is recommended for the detection of nitrate reductase by species of cryptococci; however, it is not useful for the pink yeasts, including *Rhodotorula*. This test may be used to supplement other conventional methods, but its use is not essential.

It is helpful to use these screening tests to make a presumptive identification of *C. neoformans*. In instances where inoculum is limited, the laboratorian must use tests that he or she feels can be performed and then prepare a subculture so that additional tests can be performed at a later time. The Mycology Laboratory at Mayo Clinic has found that it is often just as fast to inoculate the organism onto the surface of a plate of niger seed agar; results may be obtained during the same day of incubation at 25° C. This will be discussed further within this section. This laboratory now uses nucleic acid probe testing for identification of all isolates suspected to be *C. neoformans*.

GERM-TUBE TEST (SPECIES OF *CANDIDA*)

The germ-tube test (Procedure 44.11) is the most generally accepted and economical method used in the clinical laboratory for the identification of yeasts.[3] Approximately 75% of the yeasts recovered from clinical specimens are *Candida albicans*, and the germ-tube test usually provides a definitive identification of this organism within 3 hours.

Germ tubes appear as hyphal-like extensions of yeast cells, produced usually without a constriction at the point of origin from the cell (Figure 44.81). In the past, it has been emphasized that *Candida stellatoidea* also produces germ tubes, and the distinction between *C. albicans* should be made. *The Yeasts: A Taxonomic Study,* written by Kreger-Van Rij,[45] indicates that *C. stellatoidea* is no longer a valid species and has been combined with *C. albicans*. The germ-tube test is specific for the identification of *C. albicans,* with the exception of an occasional isolate of *C. tropicalis* that may rarely produce germ tubes.

Another method of identification of *C. albicans* is based on the presence of chlamydospores (see Figure 44.79) on cornmeal agar containing 1% Tween 80 and trypan blue incubated at room temperature for 24 to 48 hours. In addition, the appearance of spiderlike colonies on eosin methylene blue agar is characteristic of *C. albicans* and may be used to make a final identification by persons having experience with this method.

PROCEDURE 44.8

RAPID UREASE TEST

PRINCIPLE

The hydrolysis of urea by the enzyme urease produces ammonia and carbon dioxide. The ammonia produces alkaline conditions in the medium and changes the indicator (phenol red) from yellow to pink.

METHOD

1. Reconstitute a vial of dehydrated urea R broth with 3 ml of sterile distilled water on the day of use.

2. Dispense three to four drops into each well of a microdilution plate. Determine the exact number of wells to be used for the day, and use only the number necessary.

3. Transfer a heavy inoculum of a yeast colony (not including pink yeasts) to a well containing the urea broth. Colonies tested should be no older than 7 days and should be free of contamination with bacteria. In some instances it might be necessary to make a subculture to obtain enough growth to provide the inoculum for the test.

4. Include positive and negative controls using *Cryptococcus neoformans* and *Candida albicans*, respectively.

5. Seal the microdilution wells with plastic tape and incubate 4 hours at 37° C.

6. Observe for the production of a pink to purple color, which is indicative of urease production.

QUALITY CONTROL

C. neoformans and *C. albicans* are used as the positive and negative controls, respectively.

EXPECTED RESULTS

C. neoformans should produce urease, whereas *C. albicans* will not.

PERFORMANCE SCHEDULE

Controls should be performed with each test run.

_____

Compiled from Roberts, G.D., Horstmeier, C.D., Land, G.A., and Foxworth, J.A. 1978. J. Clin. Microbiol. 7:584.

PROCEDURE 44.9

RAPID SELECTIVE UREASE TEST

PRINCIPLE

Benzalkonium chloride (1%) is added to the test medium to disassociate the cell wall of yeasts to allow the endogenous urease to be released into the test medium. The hydrolysis of urea is detected by the presence of ammonia, which changes the indicator to pink because of the alkaline conditions produced.

METHOD

1. Sweep a cotton-tipped applicator impregnated with the dehydrated urea substrate over the surface of two or three colonies so that the tip is well covered with the organism.

2. Place the applicator containing the yeast into a tube containing three drops of 1% benzalkonium chloride (pH 4.86 ± 0.01) and whirl firmly against the bottom of the tube to place the organisms into contact with the cotton fibers.

3. Add a cotton plug to the tube and incubate at 45° C for up to 30 minutes.

4. Examine after 10, 15, 20, and 30 minutes for the presence of a color change from yellow to purple. A red or purple color indicates urease production by *Cryptococcus neoformans*.

QUALITY CONTROL

Same as for Procedure 44.8.

_____

From Zimmer, B.L., and Roberts, G.D. 1979. J. Clin. Microbiol. 10:380.

PROCEDURE 44.10

RAPID NITRATE REDUCTION TEST

PRINCIPLE

Benzalkonium chloride (1%) is added to the test medium to disassociate the cell wall of yeasts to allow the endogenous nitrate reductase to be released into the test medium. The test reagents turn pink in the presence of nitrite. Addition of zinc dust to detect unreacted nitrate prevents a false-negative reaction resulting from complete reduction of nitrate to ammonia (and thus a lack of nitrite).

METHOD

1. Sweep the tip of an applicator impregnated with the nitrate reduction test reagents across two or three colonies of yeast.

2. Swirl the applicator containing the yeast against the bottom of an empty test tube to get the yeast cells in contact with the cotton fibers.

3. Incubate the tube and swab at 45° C for 10 minutes.

4. Remove the swab and add two drops each of *N*-naphthylethylenediamine and sulfanilic acid reagents to the tube and replace the applicator. A change in color to red indicates a positive test. If no color appears after 10 minutes, add a pinch of zinc dust. A change in color to red indicates the presence of previously unreacted nitrate and a negative test.

QUALITY CONTROL

Cryptococcus albidus and *C. neoformans* are used as positive and negative controls, respectively.

EXPECTED RESULTS

C. albidus will reduce inorganic nitrate, whereas *C. neoformans* will not.

PERFORMANCE SCHEDULE

Controls should be performed with each test run.

From Hopkins, J.M., and Land, G.A. 1977. J. Clin. Microbiol. 5:497.

PROCEDURE 44.11

GERM-TUBE TEST

PRINCIPLE

Strains of *Candida albicans* produce germ tubes from their yeast cells when placed in a liquid nutrient environment and incubated at 35° C for 3 hours (similar to the in vivo state).

METHOD

1. Suspend a very small inoculum of yeast cells obtained from an isolated colony in 0.5 ml of sheep serum (or rabbit plasma).

2. Incubate the tubes at 35° to 37° C for no longer than 3 hours.

3. After incubation, remove a drop of the suspension and place on a microscope slide. Examine under low-power magnification for the presence of germ tubes. A germ tube is defined as an appendage that is one half the width and three to four times the length of the yeast cell from which it arises (see Figure 44.81). In most instances, there is no point of constriction at the origin of the germ tube from the cell.

QUALITY CONTROL

C. albicans and *C. tropicalis* are used as positive and negative controls, respectively.

EXPECTED RESULTS

C. albicans will produce germ tubes, usually within 2 hours, whereas *C. tropicalis* will not.

PERFORMANCE SCHEDULE

Controls should be performed with each test run.

COMMERCIALLY AVAILABLE YEAST IDENTIFICATION SYSTEMS

As previously mentioned, commercially available yeast identification systems have provided laboratories of all sizes with standardized identification methods. The methods for the most part are rapid, and results are available within 72 hours. The major advantage is that the systems provide an identification based on a data base of thousands of yeast biotypes that considers a number of variations and substrate utilization patterns. Another advantage is that manufacturers of these products provide computer consultation services to help the laboratorian with the identification of isolates that give an atypical result.

API-20C YEAST SYSTEM

The API-20C yeast identification system (bio-Mérieux Vitek, Hazelwood, Mo.) has perhaps the most extensive computer-based data set of all commercial systems available. The system consists of a strip that contains 20 microcupules, 19 of which contain dehydrated substrates for determining utilization profiles of yeasts (Figure 44.82). Reactions are compared to growth in the first cupule, which lacks a carbohydrate substrate. Reactions are read and results are converted to a seven-digit biotype profile number, and the yeast identification is made from a profile register. Chapter 10 discusses the principle of such systems. Most of the yeasts are identified within 48 hours; however, species of *Cryptococcus* and *Trichosporon* may require up to 72 hours. The API-20C yeast identification system, as well as all the other commercially available products, requires that the microscopic morphological features of yeast grown on cornmeal agar containing 1% Tween 80 and trypan blue be used in conjunction with the substrate utilization patterns. This is particularly helpful when more than one possibility for an identification is provided and the micro-

scopic morphological features can be used to distinguish between the possibilities given by the profile register. Several evaluations of the API-20C yeast identification system have been made and results have all been favorable.[12,48,67] This system has the limitation of not being able to identify unusual species; however, most of those seen in the clinical laboratory are accurately identified to the species level.

UNI-YEAST TEK SYSTEM

The Uni-Yeast Tek yeast identification system (Remel Laboratories, Lexena, Kan.) consists of a sealed, multicompartment plate containing media used to indicate carbohydrate utilization, nitrate utilization, urease production, and cornmeal agar morphology (Figure 44.83). Past evaluations of this product have shown that it is generally satisfactory for the identification of commonly en-

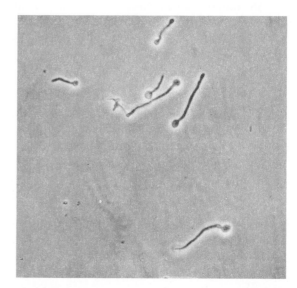

FIGURE 44.81

C. albicans, germ-tube test, showing yeast cells with germ tubes present (440×).

FIGURE 44.82

API-20C yeast identification system.

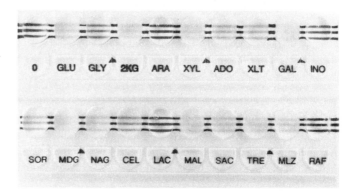

FIGURE 44.83
Uni-Yeast Tek (Remel; Lenexa, Kan.) system.

countered yeasts.[10,13,70] A major disadvantage of this system is that it requires up to 7 days for the complete identification of some yeasts. It is currently in the process of redevelopment.

BAXTER YEAST IDENTIFICATION PANEL

The Baxter Yeast Identification Panel is a 96-well microtiter plate containing 27 dehydrated substrates and was introduced by Baxter-MicroScan, W. Sacramento, Calif., as an alternative to the API-20C yeast identification system. It utilizes chromogenic substrates to assess specific enzyme activity detected within 4 hours. Specific enzyme profiles have been generated for many of the yeasts commonly encountered in the clinical microbiology laboratory. The most recent evaluation of the method showed an overall accuracy of 78% within 4 hours using no supplementary tests. When supplementary tests were used, the system identified 96.6% compared with 98.5% for API-20C. The accuracy for the identification of common and uncommon yeasts was 99.5% and 92.1%, respectively.[73] It is recommended that Cornmeal Tween-80 agar be used in conjunction with the Yeast Identification Panel.

VITEK BIOCHEMICAL CARD

The Yeast Biochemical Card (bioMérieux Vitek, Hazelwood, Mo.) is a 30-well disposable plastic card that contains 26 conventional biochemical tests and 4 negative controls. The Biochemical Card is used with the automated AutoMicrobic System, which is utilized for bacterial identification and susceptibility testing in many laboratories. The most recent evaluation of this system showed an overall accuracy of identification of 97.2% when compared with API-20C. Only 21.3% of the yeast required supplemental biochemical or morphological features to confirm their identification. Of all correctly identified yeasts, 70.8% were reported after 24 hours of incubation.[19,49] The accuracy of identification of common and uncommon species was 99.2% and 94.1%, respectively. It is not necessary to identify germ tube positive yeasts using this system. For laboratories already utilizing this system, accurate and reliable identification of most commonly encountered yeasts can be accomplished.

It appears that interest in commercially available yeast identification systems has taken precedence over the more cumbersome and labor intensive conventional yeast identification methods. Currently the rapid identification methods are financially feasible and provide the capability of identifying yeasts to laboratories of all sizes. Commercially available systems are recommended and provide accurate and rapid identification of yeasts and yeastlike organisms. In general, the systems are easy to use, easy to interpret, and relatively inexpensive when compared with testing using conventional methods. In most instances, they are faster than conventional systems, provide more standardized results and require less technical skills.

CONVENTIONAL YEAST IDENTIFICATION METHODS[2,86]

A few laboratories still prefer to use conventional methods for the identification of yeasts. Regardless of the type of identification system used, the germ-tube test is a beginning step in screening a large number of isolates. As previously mentioned, approximately 75% of yeasts recovered in the clinical laboratory can be identified using the germ-tube test.

CORNMEAL AGAR MORPHOLOGY

The second major step using this practical identification schema is to use cornmeal agar morphology as a means to determine if the yeast produces blastoconidia, arthroconidia, pseudohyphae, true hyphae, or chlamydospores (Procedure 44.12). In the past, cornmeal agar morphology was used successfully for the detection of characteristic chlamydospores produced by *C. albicans*. This method is currently satisfactory for the definitive identification of *C. albicans* when the germ-tube test is negative. In other instances, microscopic morphological features on cornmeal agar differentiate the genera *Cryptococcus, Saccharomyces, Candida, Geotrichum,* and *Trichosporon.* Previously, it was believed that the morphological features of the common species of *Candida* were distinct enough to provide a presumptive identification. This can be accomplished for *C. albicans, C. (Torulopsis) glabrata, C. krusei, C. parapsilosis, C. tropicalis,* and *C. pseudotropicalis* if one keeps in mind that there are numerous other species, uncommonly recovered in the clinical laboratory, that might resemble microscopically any of the previously mentioned species. In general, this method performs well since the previously mentioned genera and species are more commonly seen in clinical laboratories. For the uncommonly encountered isolates, cornmeal agar morphology will have less value. It is, however, recommended for use with most commercially available yeast identification systems and plays a major role in the differentiation between genera that yield similar biochemical profiles.

CARBOHYDRATE UTILIZATION

Carbohydrate utilization patterns are the most commonly used conventional methods for the definitive identification of yeast recovered in a clinical laboratory. A number of different methods have been advocated for use in determining carbohydrate utilization patterns by clinically important yeast, and all work equally well.[2,48,67] Procedure 44.13 outlines the method previously found to be most useful by the Mayo Clinic Mycology Laboratory.

Once the carbohydrate utilization profile is obtained, reactions may be compared to those listed in tables in most mycology laboratory manuals.[44] In most instances, carbohydrate utilization tests provide the definitive identification of an organism, and additional tests are unnecessary. Carbohydrate fermentation tests are preferred by some laboratories and are simply performed using purple broth containing different carbohydrate substrates. In general, carbohydrate fermentation tests are unnecessary and are not recommended for routine use.

PHENOL OXIDASE DETECTION USING NIGER SEED AGAR (PROCEDURE 44.14)

A simplified *Guizotia abyssinica* medium (niger seed medium) is a definitive method for detection of phenol oxidase production on yeasts.[40,59] Most isolates of *C. neoformans* readily produce phenol oxidase; however, some have been described that are deficient in their ability to produce this enzyme. In addition, there have been instances in which cultures of *C. neoformans* have been shown to contain both phenol oxidase–producing and phenol oxidase–deficient colonies within the same culture. If conventional methods are used, it is necessary to use all the criteria, including the nitrate reductase test, urease production, carbohydrate utilization, and the phenol oxidase test, before making a final identification of *C. neoformans.*

ENVIRONMENTAL FUNGI COMMONLY ENCOUNTERED IN THE CLINICAL LABORATORY

As shown in Table 44.1, a growing number of fungi are seen in the clinical microbiology laboratory that have not been discussed thus far. They are considered to be environmental flora, but in reality they must be regarded as potential pathogens since infections with a number of these organisms have been reported, including *Fusarium,*[6] *P. boydii,*[14,20] *Bipolaris,*[1,18,54] *Exserohilum,*[1,18,54] *Trichosporon,*[34] *Aureobasidium,*[37] *Geotrichum,*[39] and others.[58] The laboratory must identify and report all organisms recovered from clinical specimens so that their clinical significance can be determined. In many instances, the presence of environmental fungi is unimportant; however, that is not always the case. The following section will present the molds most commonly recovered from clinical specimens and a brief description of their colonial and morphological features. In addition, Tables 44.12 and 44.13 present the common molds and yeasts implicated in causing human infection, the time required for their identification,

PROCEDURE 44.12

CORNMEAL AGAR MORPHOLOGY

PRINCIPLE

Polysorbate (Tween) 80 is added to cornmeal agar to reduce the surface tension to allow for development of pseudohyphal, hyphal, and blastoconidial growth of yeasts. Certain species of yeasts develop characteristic morphological features on this medium.

METHOD

1. Obtain an isolated colony from the primary culture medium.

2. Inoculate a plate of cornmeal agar containing 1% Tween 80 and trypan blue by making three parallel cuts about ½ inch apart at a 45 degree angle to the culture medium. A sterile coverslip may be added to one area.

3. Incubate the cornmeal agar plate at 30° C for 48 hours.

4. After 48 hours, remove and examine the areas where the cuts into the agar were made for the presence of blastoconidia, arthroconidia, pseudohyphae, hyphae, or chlamydospores.[17] Table 44.11 presents the microscopic morphological features of the commonly encountered yeasts on cornmeal Tween 80 agar. This method is required for use with commercial systems for yeast identification.

QUALITY CONTROL

C. albicans is tested for production of characteristic features.

EXPECTED RESULTS

C. albicans will produce chlamydospores and clusters of blastoconidia arranged at regular intervals along the pseudohyphae.

PERFORMANCE SCHEDULE

Test a strain of *C. albicans* each time new media are received in the laboratory or produced, and monthly thereafter.

TABLE 44.11

CHARACTERISTIC MICROSCOPIC FEATURES OF COMMONLY ENCOUNTERED YEASTS ON CORNMEAL TWEEN 80 AGAR

ORGANISM	ARTHROCONIDIA	BLASTOCONIDIA	PSEUDOHYPHAE OR HYPHAE
Candida albicans	—	Spherical clusters at regular intervals on pseudohyphae	Chlamydospores present on pseudohyphae
C. glabrata	—	Small, spherical, and tightly compacted	—
C. krusei	—	Elongated; clusters occur at septa of pseudohyphae	Branched pseudohyphae
C. parapsilosis	—	Present but not characteristic	Sagebrush appearance; large (giant) hyphae present
C. kefyr (pseudotropicalis)	—	Elongated, lie parallel to pseudohyphae	Pseudohyphae present but not characteristic
C. tropicalis	—	Produced randomly along hyphae or pseudohyphae	Pseudohyphae present but not characteristic
Cryptococcus	—	Round to oval, vary in size, separated by a capsule	Rare
Saccharomyces	—	Large and spherical	Rudimentary hyphae sometimes present
Trichosporon	Numerous; resemble *Geotrichum*	May be present but difficult to find	Septate hyphae present

---

PROCEDURE 44.13

CARBOHYDRATE UTILIZATION TESTS

PRINCIPLE

Yeasts and yeastlike fungi utilize specific carbohydrate substrates. Organisms are inoculated onto a carbohydrate-free medium. Carbohydrate-containing filter paper disks are added, and utilization is determined by the presence of growth around the disk. Characteristic carbohydrate utilization profiles are used to identify species of yeasts.

METHOD

1. Prepare a suspension of the yeast in saline or distilled water to a density equivalent to a McFarland No. 4 standard.

2. Cover the surface of a yeast nitrogen base agar plate containing bromcresol purple with the suspension of the yeast cells.

3. Remove the excess inoculum and allow the surface of the agar medium to dry.

4. With sterile forceps, place selected carbohydrate disks (Difco Laboratories and BBL Laboratories) onto the surface of the agar. These should be spaced approximately 30 mm apart from each other.

5. Incubate the carbohydrate utilization plate at 30° C for 24 to 48 hours.

6. Remove the plate and observe for the presence of a color change around the carbohydrate-containing disks or the presence of growth surrounding them.

QUALITY CONTROL

Choose a combination of control yeasts that utilize each of the carbohydrates being tested. Often two or three organisms may be necessary to control each carbohydrate (use published tables[44]). Test control yeasts with each new lot number of carbohydrate disks and monthly thereafter.

---

most likely site for their recovery, and the clinical implications of each.

DEMATIACEOUS (DARKLY PIGMENTED) MOLDS

ALTERNARIA SPECIES

Colonies of *Alternaria* are rapidly growing and appear to be fluffy and gray to gray-brown or dark to gray-green in color. Microscopically, hyphae are septate and golden-brown–pigmented, and conidiophores are simple but sometimes branched. Conidiophores bear a chain of large brown conidia resembling a drumstick and contain both horizontal and longitudinal septa (Figure 44.85). It is difficult to observe chains of conidia since they are easily dislodged as the culture mount is prepared.

BIPOLARIS SPECIES

Colonies of *Bipolaris* are gray-green to dark brown and slightly powdery. Hyphae are dematiaceous and septate, as observed microscopically. Conidiophores are twisted at the ends where conidia are

attached; conidia are arranged sympodially, are oblong to fusoid, and the hilum protrudes only slightly (Figure 44.86). Germ tubes are formed at one or both ends of the conidium when the fungus is incubated in water at 25° C for up to 24 hours.

CLADOSPORIUM SPECIES

Colonies of *Cladosporium* most commonly appear as velvety or suede-like, heaped, and folded. Microscopically, hyphae are septate and brown in color. Conidiophores are long and branched and give rise to chains of darkly pigmented conidia. Conidia are usually single-celled and exhibit prominent attachment scars (disjunctors) that may resemble "shield" cells (Figure 44.87). This organism also fails to reveal chains of conidia on wet mounts because of the ease of dislodging of conidia.

CURVULARIA SPECIES

Colonies of *Curvularia* are rapidly growing and appear to resemble those of *Alternaria*. Most are fluffy and gray to gray-brown or black to gray-

Text continued on p. 766.

PROCEDURE 44.14

PHENOL OXIDASE DETECTION USING NIGER SEED AGAR

PRINCIPLE

Cryptococcus neoformans is the only species that produces 3,4-dihydroxyphenylalanine-phenol oxidase. When reacted with *L*-beta-3,4-dihydroxyphenylalanine and an iron compound (ferric citrate), *C. neoformans* oxidizes *O*-diphenol to melanin, which produces a brown to black color.

METHOD

1. Place a heavy inoculum onto the surface of a plate of niger seed agar medium.

2. Incubate at 25° C for up to 7 days.

3. Observe daily for the presence of a dark brown or black pigment, which is indicative of phenol oxidase specifically produced by *C. neoformans* (Figure 44.84). Many isolates of *C. neoformans* will produce phenol oxidase on this medium within 2 to 24 hours.

QUALITY CONTROL

C. neoformans and *C. albidus* are used as positive and negative controls, respectively.

EXPECTED RESULTS

C. neoformans should exhibit a positive test (brown color development on the disk surface), whereas *C. albidus* will not.

PERFORMANCE SCHEDULE

Controls should be performed with each test run.

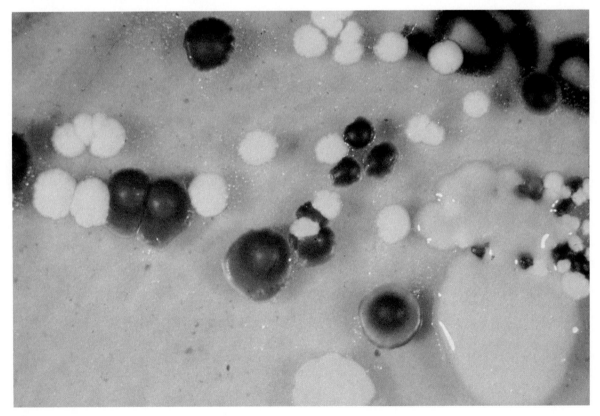

FIGURE 44.84

Niger seed agar. *C. neoformans* (dark colonies) and *C. albicans.*

TABLE 44.12

COMMON FILAMENTOUS FUNGI IMPLICATED IN HUMAN MYCOTIC INFECTIONS

ETIOLOGICAL AGENT	TIME REQUIRED FOR IDENTIFICATION	PROBABLE RECOVERY SITES	CLINICAL IMPLICATION
Acremonium	2-6 days	Skin, nails, respiratory secretions, cornea, vagina, gastric washings	Skin and nail infections, mycotic keratitis, mycetoma
Alternaria species	2-6 days	Skin, nails, conjunctiva, respiratory secretions	Skin and nail infections, sinusitis, conjunctivitis, hypersensitivity pneumonitis, skin abscess
Aspergillus flavus	1-4 days	Skin, respiratory secretions, gastric washings, nasal sinuses	Skin infections, allergic bronchopulmonary infection, sinusitis, myocarditis, disseminated infection, renal infection, subcutaneous mycetoma
Aspergillus fumigatus	2-6 days	Respiratory secretions, skin, ear, cornea, gastric washings, nasal sinuses	Allergic bronchopulmonary infection, fungus ball, invasive pulmonary infection, skin and nail infections, external otomycosis, mycotic keratitis, sinusitis, myocarditis, renal infection
Aspergillus niger	1-4 days	Respiratory secretions, gastric washings, ear, skin	Fungus ball, pulmonary infection, external otomycosis, mycotic keratitis
Aspergillus terreus	2-6 days	Respiratory secretions, skin, gastric washings, nails	Pulmonary infection, disseminated infection, endocarditis, onychomycosis, allergic bronchopulmonary infection
Bipolaris species	2-6 days	Respiratory secretions, skin, nose, bone	Sinusitis, brain abscess, peritonitis, subcutaneous abscess, pulmonary infection, osteomyelitis, encephalitis
Blastomyces dermatitidis	6-21 days (recovery time) (additional 3-14 days required for confirmatory identification)	Respiratory secretions, skin, oropharyngeal ulcer, bone, prostate	Pulmonary infection, skin infection, oropharyngeal ulceration, osteomyelitis, prostatitis, arthritis, CNS infection, disseminated infection
Cladosporium species	6-10 days	Respiratory secretions, skin, nails, nose, cornea	Skin and nail infections, mycotic keratitis, chromoblastomycosis caused by *Cladosporium carrionii*
Coccidioides immitis	3-21 days	Respiratory secretions, skin, bone, cerebrospinal fluid, synovial fluid, urine, gastric washings	Pulmonary infection, skin infection, osteomyelitis, meningitis, arthritis, disseminated infection
Curvularia species	2-6 days	Respiratory secretions, cornea, brain, skin, nasal sinuses	Pulmonary infection, disseminated infection, mycotic keratitis, brain abscess, mycetoma, endocarditis
Drechslera species	2-6 days	Respiratory secretions, skin, peritoneal fluid (following dialysis)	Pulmonary infection (rare)
Epidermophyton floccosum	7-10 days	Skin, nails	Tinea cruris, tinea pedis, tinea corporis, onychomycosis
Exserohilum species	2-6 days	Eye, skin, nose, bone	Keratitis, subcutaneous abscess, sinusitis, endocarditis, osteomyelitis
Fusarium species	2-6 days	Skin, respiratory secretions, cornea	Mycotic keratitis, skin infection (in burn patients), disseminated infection, endophthalmitis
Geotrichum species	2-6 days	Respiratory secretions, urine, skin, stool, vagina, conjunctiva, gastric washings, throat	Bronchitis, skin infection, colitis, conjunctivitis, thrush, wound infection

ETIOLOGICAL AGENT	TIME REQUIRED FOR IDENTIFICATION	PROBABLE RECOVERY SITES	CLINICAL IMPLICATION
Histoplasma capsulatum	10-45 days (recovery time) (additional 7-21 days required for confirmatory identification)	Respiratory secretions, bone marrow, blood, urine, adrenals, skin, cerebrospinal fluid, eye, pleural fluid, liver, spleen, oropharyngeal lesions, vagina, gastric washings, larynx	Pulmonary infection, oropharyngeal lesions, CNS infection, skin infection (rare), uveitis, peritonitis, endocarditis, brain abscess, disseminated infection
Microsporum audouinii	10-14 days (recovery time) (additional 14-21 days required for confirmatory identification)	Hair	Tinea capitis
Microsporum canis	5-7 days	Hair, skin	Tinea corporis, tinea capitis, tinea barbae, tinea manuum
Microsporum gypseum	3-6 days	Hair, skin	Tinea capitis, tinea corporis
Mucor species	1-5 days	Respiratory secretions, skin, nose, brain, stool, orbit, cornea, vitreous humor, gastric washings, wounds, ear	Rhinocerebral infection, pulmonary infection, gastrointestinal infection, mycotic keratitis, intraocular infection, external otomycosis, orbital cellulitis, disseminated infection
Penicillium species	2-6 days	Respiratory secretions, gastric washings, skin, urine, ear, cornea	Pulmonary infection, skin infection, external otomycosis, mycotic keratitis, endocarditis
Pseudallescheria boydii	2-6 days	Respiratory secretions, gastric washings, skin, cornea	Pulmonary fungus ball, mycetoma, mycotic keratitis, endocarditis, disseminated infection, brain abscess
Phialophora species	6-21 days	Respiratory secretions, gastric washings, skin, cornea, conjunctiva	Some species produce chromoblastomycosis or mycetoma; mycotic keratitis, conjunctivitis, intraocular infection
Rhizopus species	1-5 days	Respiratory secretions, skin, nose, brain, stool, orbit, cornea, vitreous humor, gastric washings, wounds, ear	Rhinocerebral infection, pulmonary infection, mycotic keratitis, intraocular infection, orbital cellulitis, external otomycosis, disseminated infection
Scedosporium inflatum	2-6 days	Respiratory secretions, skin, nasal sinuses, bone	Arthritis, osteomyelitis, sinusitis, endocarditis
Wangiella dermatitidis	5-21 days	Respiratory secretions, skin, eye	Phaeohyphomycosis, endophthalmitis, pneumonia
Scopulariopsis species	2-6 days	Respiratory secretions, gastric washings, nails, skin, vitreous humor, ear	Pulmonary infection, nail infection, skin infection, intraocular infection, external otomycosis
Sporothrix schenckii	3-12 days (recovery time) (additional 2-10 days required for confirmatory identification)	Respiratory secretions, skin, subcutaneous tissue, maxillary sinuses, synovial fluid, bone marrow, bone, cerebrospinal fluid, ear, conjunctiva	Pulmonary infection, lymphocutaneous infection, sinusitis, arthritis, osteomyelitis, meningitis, external otomycosis, conjunctivitis, disseminated infection
Trichophyton mentagrophytes	7-10 days	Hair, skin, nails	Tinea barbae, tinea capitis, tinea corporis, tinea cruris, tinea pedis, onychomycosis
Trichophyton rubrum	10-14 days	Hair, skin, nails	Tinea pedis, onychomycosis, tinea corporis, tinea cruris
Trichophyton tonsurans	10-14 days	Hair, skin, nails	Tinea capitis, tinea corporis, onychomycosis, tinea pedis
Trichophyton verrucosum	10-18 days	Hair, skin, nails	Tinea capitis, tinea corporis, tinea barbae
Trichophyton violaceum	14-18 days	Hair, skin, nails	Tinea capitis, tinea corporis, onychomycosis

From Koneman, E.W. and Roberts, G.D. 1991. Mycotic disease, ed 18. In Henry, J.B., editor. Clinical diagnosis and management by laboratory methods. W.B. Saunders Co., Philadelphia.

TABLE 44.13

COMMON YEASTLIKE ORGANISMS IMPLICATED IN HUMAN INFECTION*

ETIOLOGICAL AGENT	PROBABLE RECOVERY SITES	CLINICAL IMPLICATION
Candida albicans	Respiratory secretions, vagina, urine, skin, oropharynx, gastric washings, blood, stool, transtracheal aspiration, cornea, nails, cerebrospinal fluid, bone, peritoneal fluid	Pulmonary infection, vaginitis, urinary tract infection, dermatitis, fungemia, mycotic keratitis, onychomycosis, meningitis, osteomyelitis, peritonitis, myocarditis, endocarditis, endophthalmitis, disseminated infection, thrush, arthritis
Candida glabrata	Respiratory secretions, urine, vagina, gastric washings, blood, skin, oropharynx, transtracheal aspiration, stool, bone marrow, skin (rare)	Pulmonary infection, urinary tract infection, vaginitis, fungemia, disseminated infection, endocarditis
Candida tropicalis	Respiratory secretions, urine, gastric washings, vagina, blood, skin, oropharynx, transtracheal aspiration, stool, pleural fluid, peritoneal fluid, cornea	Pulmonary infection, vaginitis, thrush, endophthalmitis, endocarditis, arthritis, peritonitis, mycotic keratitis, fungemia
Candida parapsilosis	Respiratory secretions, urine, gastric washings, blood, vagina, oropharynx, skin, transtracheal aspiration, stool, pleural fluid, ear, nails	Endophthalmitis, endocarditis, vaginitis, mycotic keratitis, external otomycosis, paronychia, fungemia
Saccharomyces species	Respiratory secretions, urine, gastric washings, vagina, skin, oropharynx, transtracheal aspiration, stool	Pulmonary infection (rare), endocarditis
Candida krusei	Respiratory secretions, urine, gastric washings, vagina, skin, oropharynx, blood, transtracheal aspiration, stool, cornea	Endocarditis, vaginitis, urinary tract infection, mycotic keratitis
Candida guilliermondii	Respiratory secretions, gastric washings, vagina, skin, nails, oropharynx, blood, cornea, bone, urine	Endocarditis, fungemia, dermatitis, onychomycosis, mycotic keratitis, osteomyelitis, urinary tract infection
Rhodotorula species	Respiratory secretions, urine, gastric washings, blood, vagina, skin, oropharynx, stool, cerebrospinal fluid, cornea	Fungemia, endocarditis, mycotic keratitis
Trichosporon species	Respiratory secretions, blood, skin, oropharynx, stool	Pulmonary infection, brain abscess, disseminated infection, piedra
Cryptococcus neoformans	Respiratory secretions, cerebrospinal fluid, bone, blood, bone marrow, urine, skin, pleural fluid, gastric washings, transtracheal aspiration, cornea, orbit, vitreous humor	Pulmonary infection, meningitis, osteomyelitis, fungemia, disseminated infection, endocarditis, skin infection, mycotic keratitis, orbital cellulitis, endophthalmic infection
Cryptococcus albidus subsp. albidus	Respiratory secretions, skin, gastric washings, urine, cornea	Meningitis, pulmonary infection
Candida kefyr (pseudotropicalis)	Respiratory secretions, vagina, urine, gastric washings, oropharynx	Vaginitis, urinary tract infection
Cryptococcus luteolus	Respiratory secretions, skin, nose	Not commonly implicated in human infection
Cryptococcus laurentii	Respiratory secretions, cerebrospinal fluid, skin, oropharynx, stool	Not commonly implicated in human infection
Cryptococcus albidus subsp. diffluens	Respiratory secretions, urine, cerebrospinal fluid, gastric washings, skin	Not commonly implicated in human infection
Cryptococcus terreus	Respiratory secretions, skin, nose	Not commonly implicated in human infection

From Koneman, E.W., and Roberts, G.D. 1991. Mycotic disease, ed 18. In Henry J.B., editor. Clinical diagnosis and management by laboratory methods. W.B. Saunders Co., Philadelphia.

* Arranged in order of occurrence in the clinical laboratory.

FIGURE 44.85

Alternaria species, showing muriform, dematiaceous conidia with horizontal and longitudinal septa.

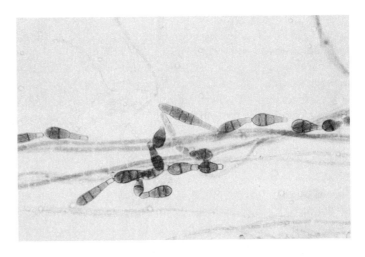

FIGURE 44.86

Bipolaris species, showing dematiaceous multicelled conidia produced sympodially (440×).

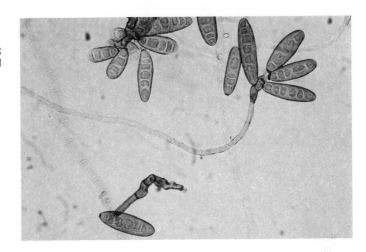

FIGURE 44.87

Cladosporium species, showing branching chains of dematiaceous blastoconidia that are easily dislodged during the preparation of a microscopic mount (440×).

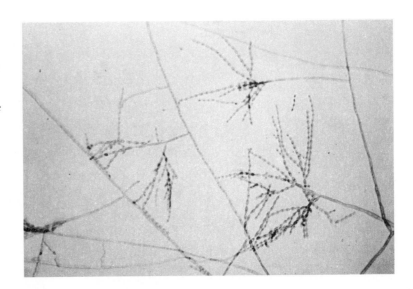

green in color. Hyphae are dematiaceous and septate, as observed microscopically. Conidiophores are twisted at the ends where conidia are attached; conidia are arranged sympodially, are golden-brown, multicelled, curved with a central swollen cell (Figure 44.88). End cells are lighter in color.

DRECHSLERA SPECIES

Drechslera is a dematiaceous organism that has been described erroneously as *Helminthosporium* or *Bipolaris* in many textbooks. Colonies of *Drechslera* are fluffy to velvety and gray to brown or black in color. Microscopically, the hyphae are septate and darkly pigmented, and conidiophores are twisted where the large multiseptate elongate conidia are attached. A zigzag appearance is pro-

duced by the production of conidia on alternate sides of the conidiophore where growth continued until a conidium was formed on the alternate side (Figure 44.89). However, sporulation is generally sparse with this organism, and it is not commonly seen.

EXSEROHILUM SPECIES

Colonies of *Exserohilum* resemble those of *Bipolaris*. Hyphae are dematiaceous and septate when viewed under the microscope. Conidiophores are twisted at the ends where conidia are attached sympodially. Conidia are ellipsoid to fusoid and exhibit a prominent hilum that is truncated and protruding (Figure 44.90).

WANGIELLA DERMATITIDIS

Colonies of *Wangiella dermatitidis* initially may be moist, black, shiny, and yeastlike. Older colonies appear gray, olive green or black and velvety to feltlike.

Young colonies produce brown to black, one-celled, budding yeasts (Figure 44.91). Older filamentous colonies produce tube-like phialides without distinct collarettes. Conidia are produced in clusters and slip down phialides (see Figure 44.59).

HYALINE MOLDS

ACREMONIUM

Colonies of *Acremonium* are rapidly growing and often appear yeastlike when initial growth is observed. Mature colonies become white to gray to

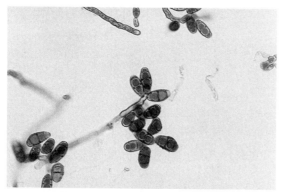

FIGURE 44.88
Curvularia species, showing twisted conidiophore and curved conidia having a swollen central cell (500×).

FIGURE 44.89
Drechslera species, showing dematiaceous multicelled conidia. Most isolates produce only a few conidia (500×).

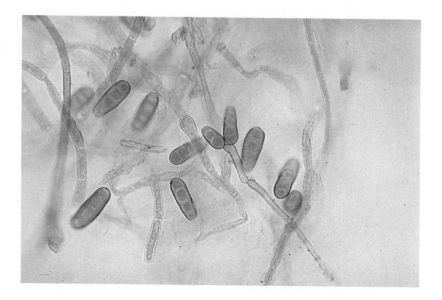

FIGURE 44.90

Exserohilum species, showing elongated multicelled conidia with prominent hila (500×).

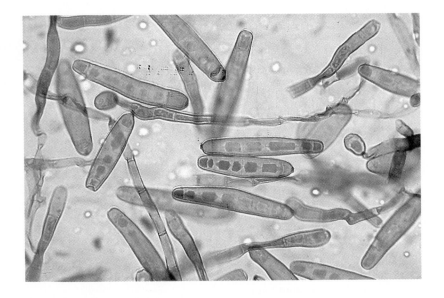

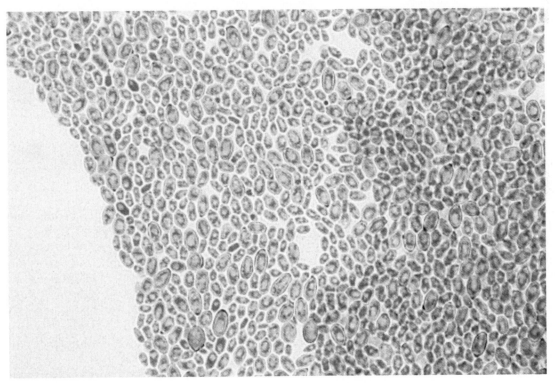

FIGURE 44.91

Wangiella dermatitidis, showing dematiaceous budding yeast cells from a young culture (1000×).

rose or reddish-orange in color. Microscopically, small septate hyphae that produce single unbranched tube-like phialides are observed. Phialides give rise to clusters of elliptical, single-celled conidia contained in a cluster at the tip of the phialide (Figure 44.92).

FUSARIUM

Colonies of *Fusarium* are fluffy to cottony and may appear in colors of pink, purple, yellow, green, and other shades. Microscopically, the hyphae are small and septate and give rise to phialides that produce either single-celled micro-

FIGURE 44.92

Acremonium species, showing single unbranched phialides that give rise to clusters of elliptical conidia at their tips (500×).

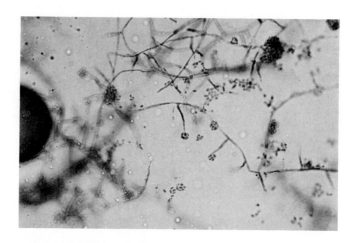

FIGURE 44.93

Fusarium species, showing characteristic multicelled, sickle-shaped macroconidia (500×).

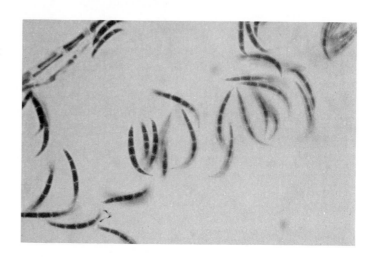

conidia, usually borne in gelatinous heads similar to those seen in *Acremonium*, or large macroconidia that are sickle- or boat-shaped and contain numerous septa (Figure 44.93). It is common to find numerous chlamydospores produced by some cultures of *Fusarium*.

PENICILLIUM

Colonies of *Penicillium* may be shades of green, blue-green, pink, white, or other colors. The surface of the colonies may be velvety to powdery because of the presence of conidia. Microscopically, hyphae are hyaline and septate and produce brushlike conidiophores. Conidiophores exhibit branching metulae from which phialides producing chains of conidia arise (Figure 44.94).

SCOPULARIOPSIS

Colonies of *Scopulariopsis* initially appear white but later become light brown and powdery in appearance. Colonies often resemble those of *Microsporum gypseum*. Microscopically, *Scopulariopsis* resembles a large *Penicillium* at first glance. The

hyaline and septate conidiophores are branched and produce a penicillium-like structure. Conidia are produced from annellides. Annellides may be produced singly directly from the hyphae. Conidia are large, have a flat base, and are rough-walled (Figure 44.95). *Scopulariopsis brumptii* is a dematiaceous species and is occasionally recovered in the clinical laboratory.

A number of other saprobic fungi may be encountered in the clinical laboratory but are seen less commonly than those previously mentioned. The chapters by McGinnis[53] and Rogers et al.[68] provide more detailed information on their identification. In many instances, it will be necessary to use reference texts to make an identification.

FUNGAL SEROLOGIC TESTS

A confirmed diagnosis of mycotic infection is based on the recovery and identification of a specific etiological agent from a clinical specimen. However, the presence and identification of characteristic etiological agents in clinical specimens

FIGURE 44.94

Penicillium species, showing typical brushlike conidiophores (penicillus) (440×).

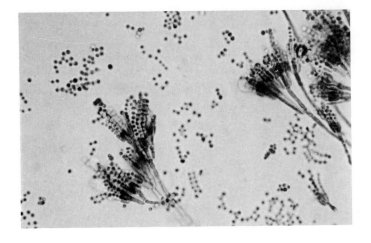

FIGURE 44.95

Scopulariopsis species, showing a large penicillus with echinulate conidia having a flat base (440×).

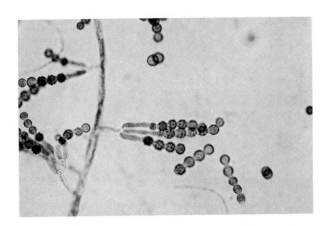

observed microscopically also allow for the definitive diagnosis. There are instances, however, when either cultural proof or pathological proof of infection cannot be obtained, and other laboratory information must be used. Fungal serologic tests often provide this supplemental information for the diagnosis of mycotic infections and in some instances provide the only means of diagnosis.[64]

Most clinical microbiology laboratories do not offer fungal serologic testing since the methods require precise technical expertise, difficult reagent preparation, and costly reagents. In the past few years, commercial sources for reagents have become available, and now fungal serologic tests can be offered by laboratories if they wish. Most often, these tests are performed by reference laboratories.

The usefulness of fungal serologic tests is hampered by false-negative results when blood is drawn at an inappropriate time or if blood is drawn from immunocompromised patients before antibodies are produced to the infection. Antigens used in fungal serologic testing are not purified and contain many cross-reacting components in common with other fungi. Antigens of *H. capsulatum, B. dermatitidis*, and *C. immitis* often cross-react and provide for difficulty in interpretation of results. A battery of antigens should be used for testing to simplify the interpretation of these cross-reactions. As with other tests, false-negative and false-positive results can occur, but in most instances fungal serologic tests provide helpful and reliable results, particularly when used in combination with cultural and histopathological proof of the etiological agent.[61,63]

The most simple and useful fungal serologic test available to the clinical laboratory is the cryptococcal latex test for antigen. Determination of the presence of cryptococcal polysaccharide by this method has proved to be extremely reliable and sensitive.[88] All laboratories are encouraged to provide this test on a routine basis because of its simplicity, sensitivity, and reliability. As previously mentioned, when compared with the India ink preparation, positive results are obtained in approximately 90% of the cases of cryptococcal meningitis when cerebrospinal fluid from the

TABLE 44.14

COMMONLY AVAILABLE FUNGAL SEROLOGIC TESTS

INFECTION	ANTIGEN(S)	TEST(S)	INTERPRETATION
Aspergillosis	*Aspergillus fumigatus* *Aspergillus niger* *Aspergillus flavus*	Immunodiffusion	One or more precipitin bands suggestive of active infection; precipitin bands shown to correlate with complement-fixation titers—the greater the number of bands, the higher the titer; when cultural proof is present in the presence of a positive test, it is diagnostic of active infection Precipitins can be found in 95% of the fungus ball cases and 50% of the allergic bronchopulmonary cases; sometimes positive in invasive infection, depending on the immunological status of the patient
Blastomycosis	*Blastomyces dermatitidis* Yeast form	Complement-fixation	Titers of 1:8 to 1:16 are highly suggestive of active infection; titers of 1:32 or greater are indicative; cross-reactions occur in patents having coccidioidomycosis or histoplasmosis; however, titers are usually lower; a decreasing titer is indicative of regression; most patients (40-60%) with blastomycosis have negative tests
	Yeast culture filtrate	Immunodiffusion	Preliminary results show that it is more sensitive than complement-fixation—52%-79% detection rate
Candidiosis	*Candida albicans*	Immunodiffusion, counterimmuno-electrophoresis, latex agglutination	Test difficult to interpret because precipitins are found in 20%-30% of the normal population, and reports in the literature are conflicting; clinical correlation must exist for the test to be useful
Coccidioidomycosis	Coccidioidin	Complement-fixation	Titers of 1:2 to 1:4 have been seen in active infection; low titers should be followed by repeat testing at 2- to 3-week intervals; titers of greater than 1:16 are usually indicative of active infection; cross-reactions occur in patients with histoplasmosis, and false-negative results occurs in patients with solitary pulmonary lesions; titer parallels severity of infection; antibodies may persist for some months
	Coccidioidin	Immunodiffusion	Results correlate with complement-fixation test and can be used as a screening test—should be confirmed by performing complement-fixation test; antibodies may persist for ≤6 months
	Coccidioidin	Latex agglutination	Precipitins occur during first 3 weeks of infection and are diagnostic, but not prognostic; useful as a screening test for precipitins in early infection; false-positive tests frequent when diluted serum or cerebrospinal fluid specimens are used
Cryptococcosis	No antigen—latex particles coated with hyperimmune anticryptococcal globulin	Latex agglutination for cryptococcal antigen	Presence of cryptococcal polysaccharide in body fluids is indicative of cryptococcosis; rheumatoid factor presents false-positive reactions; decrease in antigen titer indicates regression; positive tests (in CSF) have been seen in 95% of cryptococcal meningitis cases and 30% of non-meningitis cases; serum is less frequently positive than CSF; disseminated infections usually present positive results in serum; test may be performed using serum, CSF, and urine; test more sensitive than India ink preparation; false-positive tests occur in patients with *Trichosporon* infection

TABLE 44.14

COMMONLY AVAILABLE FUNGAL SEROLOGIC TESTS—cont'd

INFECTION	ANTIGEN(S)	TEST(S)	INTERPRETATION
Histoplasmosis	Histoplasmin and yeast form of *Histoplasma capsulatum*	Complement-fixation	Titers of 1:8 to 1:16 are highly suspicious of infection; however titer of 1:32 or greater are usually indicative of active infection; cross-reactions occur in patients with aspergillosis, blastomycosis, and coccidioidomycosis, but titers are usually lower; several follow-up serum samples should be tested—drawn at 2- to 3-week intervals
			Rising titers indicate progressive infection, and decreasing titers indicate regression; some disseminated infections are nonreactive to the complement-fixation test
			Recent skin tests in persons who have had prior exposure to *H. capsulatum* will cause an elevation in the complement-fixation titer; this occurs in 17%-20% of persons tested
			The yeast antigen gives positive reactions in 75%-80% of cases, and the histoplasmin gives positive reactions in 10%-15% of cases; in 10% of cases both are positive simultaneously
	Histoplasmin	Immunodiffusion	H and M bands appearing simultaneously are indicative of active infection; occur in ≤20% of patients
			M band may appear alone and can indicate early infection or chronic infection; also the M band may appear after a recent skin test
			The H band appears later than the M band and disappears earlier, and its disappearance may indicate regression of the infection; H band appearance alone is of indeterminate significance
	Histoplasmin	Latex agglutination	Test is unreliable; many false-positive and false-negative tests may be observed; any positive test should be confirmed by the complement-fixation test
Sporotrichosis	Yeast of *Sporothrix schenckii*	Agglutination	Titers of 1:80 or greater usually indicative of active infection; some cutaneous infections present negative tests; however, extracutaneous infections present positive tests

From Koneman, E.W. and Roberts, G.D. 1991. Mycotic disease. In Henry, J.B., editor. Clinical diagnosis and management by laboratory methods. W.B. Saunders Co., Philadelphia.

first lumbar puncture is tested. Most false-positive reactions caused by interference factors have been eliminated using an enzymatic method[25,77] that makes the test highly specific. However, the latex test is positive in serum from patients having disseminated *Trichosporon* infection.[55] The decision of whether to offer additional tests other than the cryptococcal latex test for antigen depends on the individual situation for each laboratory. Table 44.14 presents a summary of the common fungal serologic tests available and information regarding their interpretation.[43] More

detailed methods can be found in the text by Palmer et al.[60] and the chapter written by Kaufman and Reiss.[41] At present, enzyme immunoassays for antibody detection during histoplasmosis, blastomycosis and coccidioidomycosis are currently available and their clinical usefulness has not yet been determined; results look promising.[22,36] In addition, an enzyme immunoassay is available to detect cryptococcal antigen; it offers the possibility of being more sensitive than the latex test.[21] Other methods of detecting antigen and/or antibody are numerous;

however, most are not available to the routine clinical laboratory.[11,15]

ANTIFUNGAL SUSCEPTIBILITY TESTING

Antifungal susceptibility tests are designed to provide information that will allow the physician to select the appropriate antifungal agent useful for treating a specific infection. Unfortunately, antifungal susceptibility testing has not progressed as far as tests used for determining susceptibility of bacteria to antimicrobial agents. There is no standard method used by all laboratories, and there is disagreement concerning specific conditions of incubation and other variables necessary for performing the test.[23] The National Committee for Clinical Laboratory Standards is examining the issue, and standards for testing have been drafted.

Despite the problems associated with antifungal susceptibility testing, many physicians believe that these tests are important for the selection of an appropriate antifungal agent and as a method to detect the development of resistance of certain organisms during chemotherapy.[16] The method described by Shadomy et al.[71] is perhaps a commonly used procedure in the United States. A laboratory not equipped for such studies should be prepared to send an important isolate to a reference laboratory for testing. Amphotericin B, 5-fluorocytosine, miconazole, ketoconazole, itraconazole and fluconazole are common antifungal agents available at the present time. Demand for susceptibility testing of itraconazole and fluconazole is rapidly increasing.

REFERENCES

1. Adam, R.D., Paquin, M.L., Petersen, E.A., et al. 1986. Phaeohyphomycosis caused by the fungal genera *Bipolaris* and *Exserohilum.* A report of 9 cases and review of the literature. Medicine 65:203.
2. Adams, E.D., Jr., and Cooper, B.H. 1984. Evaluation of modified Wickerham medium for identification of medically important yeasts. Am. J. Med. Technol. 40:377.
3. Ahearn, D.G. 1973. Identification and ecology of yeasts of medical importance. In Prior, J.E., and Friedman, H., editors. Opportunistic pathogens. University Park Press, Baltimore.
4. Ajello, L. 1968. A taxonomic review of the dermatophytes and related species. Sabouraudia 6:147.
5. Ajello, L., and Georg, L.K. 1957. In vitro hair cultures for differentiating between atypical isolates of *Trichophyton mentagrophytes* and *Trichophyton rubrum.* Mycopathologia 8:3.
6. Anaissie, E., Kantarjian, H., and Jones, P. 1987. Fungal infections caused by *Fusarium* in cancer patients. Am. J. Clin. Oncol. 10:86.
7. Bille, J., Edson, R.S., and Roberts, G.D. 1984. Clinical evaluation of the lysis-centrifugation blood culture system for detection of fungemia and comparison with a conventional biphasic broth blood culture system. J. Clin. Microbiol. 19:126.
8. Bille, J., Stockman, L., Roberts, G.D., et al. 1983. Evaluation of a lysis-centrifugation system for recovery of yeasts and filamentous fungi from blood. J. Clin. Microbiol. 18:469.
9. Binford, C.H., Thompson, R.K., Gorham, M.E., and Emmons, C.W. 1952. Mycotic brain abscess due to *Cladosporium trichoides,* a new species. Am. J. Clin. Pathol. 22:535.
10. Bowman, P.I., and Ahearn, D.G. 1975. Evaluation of the Uni-Yeast-Tek kit for the identification of medically important yeasts, J. Clin. Microbiol. 2:354.
11. Brock, E.G., Reiss, E., Pine, L., and Kaufman, L. 1984. Effect of periodate-oxidation on the detection of antibodies against the M-antigen of histoplasmosis by enzyme immunoassay (EIA) inhibition. Curr. Microbiol. 10:177.
12. Buesching, W.J., Kurek, K., and Roberts, G.D. 1979. Evaluation of the modified API 20C system for identification of clinically important yeasts. J. Clin. Microbiol. 9:565.
13. Cooper, B.H., Johnson, J.B., and Thaxton, E.S. 1978. Clinical evaluation of the Uni-Yeast-Tek system for rapid presumptive identification of medically important yeasts. J. Clin. Microbiol. 7:349.
14. Davis, W.A., Isner, J.M., Bracey, A.W., et al. 1980. Disseminated *Petriellidium boydii* and pacemaker endocarditis. Am. J. Med. 69:929.
15. deRepentigny, L., Marr, L.D., Keller J., et al. 1985. Comparison of enzyme immunoassay and gas-liquid chromatography for the rapid diagnosis of invasive candidiasis in cancer patients. J. Clin. Microbiol. 21:972.
16. Dick, J.D., Merz, W.G., and Saral, R. 1980. Incidence of polyene-resistant yeasts recovered from clinical specimens. Antimicrob. Agents Chemother. 18:158.
17. Dolan, C.T. 1971. A practical approach to identification of yeast-like organisms. Am. J. Clin. Pathol. 55:580.
18. Douer, D., Goldschmied-Reouven A., Segea, S., and Benbassat, I. 1987. Human *Exserohilum* and *Bipolaris* infections: report of *Exserohilum* nasal infection in a neutropenic patient with acute leukemia and review of the literature. J. Med. Vet. Mycol. 25:235.
19. El-Zaatari, M., Pasarell, L., McGinnis, M.R., Buckner, J., Land, G.A., and Salkin, I.F. 1990. Evaluation of the updated Vitek Yeast Identification data base. J. Clin. Microbiol. 28:1938.
20. Enggano, I.L., Hughes, W.T., Kalasinsky, D.K., et al. 1984. *Pseudallescheria boydii* in a patient with acute lymphoblastic leukemia. Arch. Pathol. Lab. Med. 108:619.
21. Gade, W., Hinnefeld, S.W., Babcock, L.S., Gilligan, P., Kelly, W., Wait, K., Greer, D., Pinilla, M., and Kaplan, R.L. 1991. Comparison of the Premier cryptococcal antigen enzyme immunoassay and the latex

agglutination assay for the detection of cryptococcal antigens. J. Clin. Microbiol. 29:1616.

22. Gade, W., Ledman, D.W., Wethington, R., and Yi, A. 1992. Serological responses to various coccidioides antigen preparations in a new enzyme immunoassay. J. Clin. Microbiol. 30:1907.

23. Galgiani, J.N., Rinaldi, M.G., Polak, A.M., and Pfaller, M.A. 1992. Standardization of antifungal susceptibility testing. J. Med. Vet. Mycol. 30:213.

24. Goodman, N.L., and Rinaldi, M.G. 1991. Agents of zygomycosis. In Balows, A., Hausler, W.J., Jr., Herrmann, K. L., Isenberg, H.D., and Shadomy, H.J., editors. Manual of clinical microbiology, ed 5. American Society for Microbiology, Washington, D.C.

25. Gray, L.D., and Roberts, G.D. 1988. Experience with the use of pronase to eliminate interference factors in the latex agglutination test for cryptococcal antigen. J. Clin. Microbiol. 26:2450.

26. Gray, L.D., and Roberts, G.D. 1988. Laboratory diagnosis of systemic fungal diseases. Inf. Dis. Clin. N. Amer. 2:779.

27. Guerra-Romero, L., Edson, R.S., Cockerill, F.R. III, et al. 1987. Comparison of duPont Isolator and Roche Septi-Chek for detection of fungemia. J. Clin. Microbiol. 25:1623.

28. Guinet, R.J., Chanas, J., Goullier, A., et al. 1983. Fatal septicemia due to amphotericin B–resistant *Candida lusitaniae*. J. Clin. Microbiol. 18:443.

29. Hageage, G.J., and Harrington, B.J. 1984. Use of calcofluor white in clinical mycology. Lab. Med. 15:109.

30. Hall, G.S., Pratt-Rippin, K., and Washington, J.A. 1992. Evaluation of a chemiluminescent probe assay for identification of *Histoplasma capsulatum* isolates. J. Clin. Microbiol. 30:3003.

31. Harari, A.R., Hempel, H.O., Kimberlin, C.L., and Goodman, N.L. 1982. Effects of time lapse between sputum collection and culturing on isolation of clinically significant fungi. J. Clin. Microbiol. 15:425.

32. Hoeprich, P.D. 1977. Cryptococcosis. In Hoeprich, P.D., editor. Infectious diseases. Harper & Row, Hagerstown, Md.

33. Hopkins, J.M., and Land, G.A. 1977. Rapid method for determining nitrate utilization by yeasts. J. Clin. Microbiol. 5:497.

34. Hoy, J., Hsu, K.C., Rolston, K., et al. 1986. *Trichosporon beigelii* infection: a review. Rev. Infect. Dis. 8:959.

35. Huppert, M., Sun, S.H., and Bailey, J.W. 1967. Natural variability in *Coccidioides immitis*. In Ajello, L., editor. Coccidioidomycosis. University of Arizona Press, Tucson.

36. Johnson, J.E. 1992. Serologic diagnosis of histoplasmosis and coccidioidomycosis using the Premier microwell enzyme immunoassay. Abstracts of the Annual Meeting of the American Society for Microbiology, No. C-133.

37. Kaczmarski, E.B., Liu Yin, J.A., Tooth, J.A., et al. 1986. Systemic infection with *Aureobasidium pullulans* in a leukemic patient. J. Infect. 13:209.

38. Kane, J., and Smitka, C. 1978. Early detection and identification of *Trichophyton verrucosum*. J. Clin. Microbiol. 8:740.

39. Kassamali, H., Anaissie, E., Ro, J., et al. 1987. Disseminated *Geotrichum candidum* infection. J. Clin. Microbiol. 25:1782.

40. Kaufman, C.S., and Merz, W.G. 1982. Two rapid pigmentation tests for identification of *Cryptococcus neoformans*. J. Clin. Microbiol. 15:339.

41. Kaufman, L., and Reiss, E. 1992. Serodiagnosis of fungal diseases. In Rose, N.R., Conway de Macario, E., Fahey, J.L., Friedman, H., and Penn, G.M., editors. Manual of clinical immunology, ed 4. American Society for Microbiology, Washington, D.C.

42. Kaufman, L., and Standard, P. 1978. Improved version of the exoantigen test for identification of *Coccidioides immitis* and *Histoplasma capsulatum* cultures. J. Clin. Microbiol. 8:42.

43. Koneman, E.W., and Roberts, G.D. 1991. Mycotic disease. In Henry, J.B., editor. Clinical diagnosis and management by laboratory methods. W.B. Saunders Co., Philadelphia.

44. Koneman, E.W., and Roberts, G.D. 1985. Practical laboratory mycology, ed 3. Williams & Wilkins, Baltimore.

45. Kreger-Van Rij, N.J.N. 1984. The yeasts: a taxonomic study. Elsevier Science Publishing Co., New York.

46. Kwon-Chung, K.J., Bennett, J.E., and Theodore, T.S. 1978. *Cryptococcus neoformans* sp. nov.: serotype B-C of *Cryptococcus neoformans*. Int. J. Sys. Bacteriol. 28:616.

47. Kwon-Chung, K.S. 1976. A new species of *Filobasidiella*, the sexual state of *Cryptococcus neoformans* B and C serotypes. Mycologia 68:942.

48. Land, G.A., Harrison, B.A., Hulme, K.L., et al. 1979. Evaluation of the new API 20C strip for yeast identification against a conventional method. J. Clin. Microbiol. 10:357.

49. Land, G.A., Stotler, R., Land, K.J., and Staneck, J.L. 1984. Updating and evaluation of the Vitek AMS yeast identification system. J. Clin. Microbiol. 4:649.

50. Libertin, C.R., Wilson, W.R., and Roberts, G.D. 1985. *Candida lusitaniae*—an opportunistic pathogen. Diag. Microbiol. Infect. Dis. 3:69.

51. Mahgoub, E.S., and Murray, I.G. 1973. Mycetoma. William Heineman Medical Books, London.

52. Marcon, M.J., and Powell, D.A. 1987. Epidemiology, diagnosis and management of *Malassezia furfur* systemic infection. Diagn. Microbiol. Infect. Dis. 7:161.

53. McGinnis, M.R., Salkin, I.F., Schell, W.A., and Pasarell, L. 1991. Dematiaceous fungi. In Balows, A., Hausler, W.J., Jr., Herrmann, K.L., Isenberg, H.D., and Shadomy, H.J., editors. Manual of clinical microbiology, ed 5. American Society for Microbiology, Washington, D.C.

54. McGinnis, M.R., Rinaldi, M.G., and Winn, R.E. 1986. Emerging agents of phaeohyphomycosis: pathogenic species of *Bipolaris* and *Exserohilum*. J. Clin. Microbiol. 24:250.

55. McManus, E.J., and Jones, J.M. 1985. Detection of *Trichosporon beigelii* antigen cross-reactive with *Cryptococcus neoformans* capsular polysaccharide in serum from a patient with disseminated *Trichosporon* infection. J. Clin. Microbiol. 21:681.

56. Merz, W.G., and Roberts, G.D. 1991. Detection and recovery of fungi from clinical specimens. In Balows, A., Hausler, W.J., Jr., Herrmann, K.L., Isenberg, H.D., and Shadomy, H.J., editors. Manual of clinical microbiology, ed 5. American Society for Microbiology, Washington, D.C.

57. Murray, P.R., Van Scoy, R.E., and Roberts, G.D. 1977. Should yeasts in respiratory secretions be identified? Mayo Clin. Proc. 52:42.

58. Musial, C.E., Cockerill, F.R. III, and Roberts, G.D. 1988. Fungal infections of the immunocompromised host: clinical and laboratory aspects. Rev. Infect. Dis. 1:349.

59. Paliwal, D.K., and Randhawa, H.S. 1978. Evaluation of a simplified *Guizotia abyssinica* seed medium for differentiation of *Cryptococcus neoformans*. J. Clin. Microbiol. 7:346.

60. Palmer, D.F., Kaufman, L., Kaplan, W., and Cavallano, J.J. 1977. Serodiagnosis of mycotic diseases. Charles C Thomas, Publisher, Springfield, Ill.

61. Pappagianis, D., and Zimmer, B.L. 1990. Serology of coccidioidomycosis. Clin. Microbiol. Rev. 3:247.

62. Roberts, G.D. 1975. Detection of fungi in clinical specimens by phase-contrast microscopy. J. Clin. Microbiol. 2:261.

63. Roberts, G.D. 1976. Laboratory diagnosis of fungal infections. Hum. Pathol. 7:161.

64. Roberts, G.D. 1979. Serodiagnosis of mycotic infections. Ill. Med. J. 156:406.

65. Roberts, G.D., Horstmeier, C.D., Land, G.A., and Foxworth, J.H. 1978. Rapid urea broth test for yeasts. J. Clin. Microbiol. 7:584.

66. Roberts, G.D., Karlson, A.G., and DeYoung, D.R. 1976. Recovery of pathogenic fungi from clinical specimens submitted for mycobacterial culture. J. Clin. Microbiol. 3:47.

67. Roberts, G.D., Wang, H.S., and Hollick, G.E. 1976. Evaluation of the API 20C microtube system for the identification of clinically important yeasts. J. Clin. Microbiol. 3:302.

68. Rogers, A.L., and Kennedy, M.J. 1991. Opportunistic hyaline hyphomycetes. In Balows, A., Hausler, W.J., Jr., Herrmann, K.L., Isenberg, H.D., and Shadomy, H.J., editors. Manual of clinical microbiology, ed 5. American Society for Microbiology, Washington, D.C.

69. Salkin, I.F., McGinnis, M.R., Dykstra, M.J., and Rinaldi, M.G. 1988. *Scedosporium inflatum*, an emerging pathogen. J. Clin. Microbiol. 26:498.

70. Salkin, I.F., Land, G.A., Hurd, N.J., et al. 1987. Evaluation of Yeast Ident and Uni-Yeast-Tek Yeast Identification Systems. J. Clin. Microbiol. 25:624.

71. Shadomy, S., and Pfaller, M.G. 1991. Laboratory studies with antifungal agents: susceptibility testing and quantitation in body fluids. In Balows, A., Hausler, W.J., Jr., Herrmann, K.L., Isenberg, H.D., and Shadomy, H.J., editors. Manual of clinical microbiology, ed 5. American Society for Microbiology, Washington, D.C.

72. Smith, C.D., and Goodman, N.L. 1975. Improved culture method for the isolation of *Histoplasma capsulatum* and *Blastomyces dermatitidis* from contaminated specimens. Am. J. Clin. Pathol. 68:276.

73. St. Germaine, G., and Beauchesne, D. 1991. Evaluation of the MicroScan Rapid Yeast Identification Panel. J. Clin. Microbiol. 29:2296.

74. Standard, P.G., and Kaufman, L. 1980. A rapid and specific method for the immunological identification of mycelial form cultures of *Paracoccidioides brasiliensis*. Curr. Microbiol. 4:297.

75. Standard, P.G., and Kaufman, L. 1982. Safety considerations in handling exoantigen extracts from pathogenic fungi. J. Clin. Microbiol. 15:663.

76. Stockman, L., Clark, K.A., Hunt, J.M., and Roberts, G.D. 1993. Evaluation of commercially available acridinium-ester-labeled chemiluminescent DNA probes for culture identification of *Blastomyces dermatitidis, Coccidioides immitis, Cryptococcus neoformans* and *Histoplasma capsulatum*. J. Clin. Microbiol., 31:850.

77. Stockman, L., and Roberts, G.D. 1983. Corrected version. Specificity of the latex test for cryptococcal antigen: a rapid, simple method for eliminating interference factors. J. Clin. Microbiol. 17:945.

78. Strimlan, C.V., Dines, D.E., Rodgers-Sullivan, R.F., et al. 1980. Respiratory tract *Aspergillus*—clinical significance. Minn. Med. 63:25.

79. Summerbell, R.C., Rosenthal, S.A., and Kane, J. 1988. Rapid method for differentiation of *Trichophyton rubrum, Trichophyton mentagrophytes,* and related dermatophyte species. J. Clin. Microbiol. 26.2279.

80. Sun, S.H., Huppert, M., and Vukovich, K.R. 1976. Rapid in vitro conversion and identification of *Coccidioides immitis*. J. Clin. Microbiol. 3:186.

81. Terreni, A.A., Strohecker, J.S., and Dowa, H. Jr. 1987. *Candida lusitaniae* septicemia in a patient on extended home intravenous hyperalimentation. J. Med. Vet. Mycol. 25:63.

82. Thomson, R.B., and Roberts, G.D. 1982. A practical approach to the diagnosis of fungal infections of the respiratory tract. Clin. Lab. Med. 2:321.

83. Treger, T.R., Visscher, D.W., Bartlett, M.S., and Smith, J.W. 1985. Diagnosis of pulmonary infection caused by *Aspergillus*: usefulness of respiratory cultures. J. Infect. Dis. 152:572.

84. Walsh, T.J. 1989. Trichosporonosis. Infect. Dis. Clin. North Amer. 3:43.

85. Utz, J.P. 1967. Recognition and current management of the systemic mycoses. Med. Clin. North Am. 51:519.

86. Warren, N.G., and Shadomy, H.J. 1991. Yeasts of medical importance. In Balows, A., Hausler, W.J., Jr., Herrmann, K.L., Isenberg, H.D., and Shadomy, H.J., editors. Manual of clinical microbiology, ed 5. American Society for Microbiology, Washington, D.C.

87. Weeks, R.J. 1964. A rapid simplified medium for converting the mycelial phase of *Blastomyces dermatitidis* to the yeast phase. Mycopathologia 22:153.

88. Wu, T.C., and Koo, S.Y. 1983. Comparison of three commercial cryptococcal latex kits for detection of cryptococcal antigen. J. Clin. Microbiol. 18:1127.

89. Zackheim, H.S., Halde, C., Goodman, R.S., et al. 1985. Phaeohyphomycotic cyst of the skin caused by *Exophiala jeanselmei*. J. Am. Acad. Dermatol. 12:207.

90. Zimmer, B.L., and Roberts, G.D. 1979. Rapid selective urease test for presumptive identification of *Cryptococcus neoformans*. J. Clin. Microbiol. 10:380.

BIBLIOGRAPHY

Balows, A., Hausler, W.J., Jr., Herrmann, K.L., Isenberg, H.D., and Shadomy, H.J. 1991. Manual of clinical microbiology, ed 5. American Society for Microbiology, Washington, D.C.

Beneke, E.S., and Roger, A.L. 1980. Medical mycology manual, ed 4. Burgess Publishing Co., Minneapolis.

Koneman, E.W., and Roberts, G.D. 1985. Practical laboratory mycology, ed 3. Williams & Wilkins, Baltimore.

Kwon-Chung, K.J., and Bennet, J.E. 1992. Medical mycology, Lea & Febiger, Philadelphia.

Larone, D.H. 1987. Medically important fungi: a guide to identification, ed 2. ASM Press, Washington, D.C.

McGinnis, M.R. 1980. Laboratory handbook of medical mycology. Academic Press, New York.

Rippon, J.W. 1988. Medical mycology—the pathogenic fungi and pathogenic actinomycetes, ed 3. W.B. Saunders Co., Philadelphia.

45

LABORATORY METHODS FOR DIAGNOSIS OF PARASITIC INFECTIONS

DIAGNOSTIC TECHNIQUES

The field of parasitology is often associated with tropical areas; however, many parasitic organisms that infect humans are worldwide in distribution and occur with some frequency in the temperate zones. Many organisms endemic elsewhere are seen in the United States in persons who have lived or traveled in those areas. The influx of refugee populations into the United States has also served as a source of both infected patients and possible transmission of certain parasites. Another factor to be considered is the increase in numbers of compromised patients, particularly those who are immunodeficient or immunosuppressed. Those persons are greatly at risk for certain parasitic problems. Consequently, clinicians and laboratory personnel should be aware of the possibility that these organisms may be present

and should be trained in the ordering and the performance of appropriate procedures for their recovery and identification. The clinician must be able to recognize and interpret the relevance of the laboratory data, and the clinical laboratory scientist must be able to review the pros and cons of each procedure, selecting those that will provide the most accurate diagnostic test results.

Although common names are frequently used to describe parasitic organisms, these names may represent different parasites in different parts of the world. To eliminate these problems, a binomial system of nomenclature is used in which the scientific name consists of the genus and species. On the basis of life histories and morphology, systems of classification have been developed to indicate the relationship among the various parasite species. With increased emphasis on organism differences at the molecular level, classification may dramatically change during the coming years.

Parasites of humans are classified into five major subdivisions: (1) the Protozoa (amebae, flagellates, ciliates, sporozoa, coccidia, microsporidians), (2) the Platyhelminthes or flatworms (cestodes, trematodes), (3) the Acanthocephala or thorny-headed worms, (4) the Nematoda or roundworms, and (5) the Arthropoda (insects, spiders, mites, ticks). The main groups presented here include Protozoa, Nematoda, Platyhelminthes, and Arthropoda. This information is presented in the box on pp. 778 and 779. This classification scheme is designed to provide some order and meaning to a widely divergent group of organisms. No attempt has been made to include every possible organism, but only those that are considered to be clinically relevant in the context of human parasitology. It is hoped this list provides some insight into the parasite groupings, thus leading to a better understanding of organism morphology, parasitic infections, and the appropriate clinical diagnostic approach.

The identification of parasitic organisms depends on morphological criteria; these criteria in turn depend on correct specimen collection and adequate fixation. Improperly submitted specimens may result in failure to find the organisms or in their misidentification. Tables 45.1 and 45.2 contain information on the body sites and possible parasites recovered and the most frequently used specimen collection approaches.

The information presented in this chapter should provide the reader with appropriate laboratory techniques and examples of morphological criteria to permit the correct identification of the more common human parasites. Several excellent resource texts are also available.*

FECAL SPECIMENS

COLLECTION
The ability to detect and identify intestinal parasites, particularly protozoa, is directly related to the quality of the specimen submitted to the laboratory. Certain guidelines (see following discussion) are recommended to ensure proper collection and accurate examination of specimens.

Collection of fecal specimens for intestinal parasites should always be performed *before* radiological studies involving barium sulfate. Because of the excess crystalline material in the stool specimen, the intestinal protozoa may be impossible

*References 7, 10, 18, 29, 32, 57, 58, 60, 69.

to detect for at least 1 week after the use of barium. Certain medications may also prevent the detection of intestinal protozoa; these include mineral oil, bismuth, nonabsorbable antidiarrheal preparations, antimalarials, and some antibiotics (e.g., tetracyclines). The organisms may be difficult to detect for several weeks after the medication is discontinued. Fecal specimens should be collected in clean, wide-mouthed containers; most laboratories use a waxed, cardboard half-pint container with a tight-fitting lid. The specimen should not be contaminated with water that may contain free-living organisms. Contamination with urine should also be avoided to prevent destruction of motile organisms in the specimen. All specimens should be identified with the patient's name, physician's name, hospital number if applicable, and the time and date collected. Every fecal specimen represents a potential source of infectious material (e.g., bacteria, viruses, parasites) and should be handled accordingly.

The number of specimens required to demonstrate intestinal parasites will vary depending on the quality of the specimen submitted, the accuracy of the examination performed, and the severity of the infection. For a routine examination for parasites *before treatment, a minimum of three fecal specimens* is recommended: two specimens collected from normal movements and one specimen collected after a cathartic, such as magnesium sulfate or Fleet Phospho-Soda. A cathartic with an oil base should not be used, and all laxatives are contraindicated if the patient has diarrhea or significant abdominal pain. Stool softeners are inadequate for producing a purged specimen. The examination of at least six specimens ensures detection of 90% of infections; six are usually recommended when amebiasis is suspected.

Many organisms do not appear in fecal specimens in consistent numbers on a daily basis; thus collection of specimens on *alternate days* tends to yield a higher percentage of positive findings. The series of three specimens should be collected within no more than 10 days and a series of six within no more than 14 days.

The number of specimens to be examined after therapy will vary depending on the diagnosis; however, a series of three specimens collected as previously outlined is usually recommended. A patient who has received treatment for a protozoan infection should be checked 3 to 4 weeks after therapy. Patients treated for helminth infections may be checked 1 to 2 weeks after therapy and those treated for *Taenia* infections 5 to 6 weeks after therapy. Since the age of the speci-

Classification of Human Parasites

A. Protozoa
  1. Amebae: intestinal
     Entamoeba histolytica
     Entamoeba hartmanni
     Entamoeba coli
     Entamoeba polecki
     Endolimax nana
     Iodamoeba bütschlii
     Blastocystis hominis
  2. Flagellates: intestinal
     Giardia lamblia
     Chilomastix mesnili
     Dientamoeba fragilis
     Trichomonas hominis
     Enteromonas hominis
     Retortamonas intestinalis
  3. Ciliates: intestinal
     Balantidium coli
  4. Coccidia: intestinal
     Cryptosporidium sp.
     CLBS (Cyanobacteria-like or coccidia-like
       organisms *(Cyclospora)*
     Isospora belli
     Sarcocystis hominis
     Sarcocystis suihominis
     Sarcocystis "lindemanni"
     Microsporidians *(Enterocytozoon, Septata)*
  5. Sporozoa, flagellates: from blood and tissues
     Sporozoa (malaria, *Babesia*)
       Plasmodium vivax
       Plasmodium ovale
       Plasmodium malariae
       Plasmodium falciparum
       Babesia sp.
     Flagellates (leishmaniae, trypanosomes)
       Leishmania tropica complex
       Leishmania mexicana complex
       Leishmania braziliensis complex
       Leishmania donovani complex
       Leishmania peruviana
       Trypanosoma gambiense
       Trypanosoma rhodesiense
       Trypanosoma cruzi
       Trypanosoma rangeli
  6. Amebae, Flagellates: from other body sites
     Amebae
       Naegleria fowleri
       Acanthamoeba/Hartmanella group
       Entamoeba gingivalis
     Flagellates
       Trichomonas vaginalis
       Trichomonas tenax
  7. Coccidia, Sporozoa, Microsporidia: from other
     body sites
     Coccidia
       Toxoplasma gondii
     Sporozoa
       *Pneumocystis carinii**
     Microsporidia
       Nosema
       Pleistophora
       Encephalitozoon
       Septata
       "Microsporidium"

B. Nematodes
  1. Intestinal
     Ascaris lumbricoides
     Enterobius vermicularis
     Ancylostoma duodenale
     Necator americanus
     Strongyloides stercoralis
     Trichuris trichiura
  2. Tissue
     Trichinella spiralis
     Visceral larva migrans *(Toxocara canis* or *T. cati)*
     Ocular larva migrans *(Toxocara canis* or *T. cati)*
     Cutaneous larva migrans *(Ancylostoma braziliense*
       or *A. caninum)*
     Angiostrongylus cantonensis
     Angiostrongylus costaricensis
     Gnathostoma spinigerum
     Dracunculus medinensis
     Anisakis sp. (larvae from saltwater fish)
     Phocanema sp. (larvae from saltwater fish)
     Contracaecum sp. (larvae from saltwater fish)
  3. Filarial worms: from blood and tissues
     Wuchereria bancrofti
     Brugia malayi
     Loa loa
     Onchocerca volvulus
     Mansonella ozzardi
     Mansonella streptocerca
     Mansonella perstans
     Dirofilaria immitis (usually lung lesion/dog
       heartworm)
     Dirofilaria sp. (may be found in subcutaneous
       nodules)
C. Cestodes
  1. Intestinal
     Diphyllobothrium latum
     Dipylidium caninum
     Hymenolepis nana
     Hymenolepis diminuta
     Taenia solium
     Taenia saginata
  2. Larval Forms: tissue
     Taenia solium
     Echinococcus granulosus
     Echinococcus multilocularis
     Multiceps
     Diphyllobothrium spp.
D. Trematodes
  1. Intestinal
     Fasciolopsis buski
     Echinostoma ilocanum
     Heterophyes heterophyes
     Metagonimus yokogawai
  2. Liver/lung
     Clonorchis sinensis
     Opisthorchis viverrini
     Fasciola hepatica
     Paragonimus westermani
  3. Blood
     Schistosoma mansoni
     Schistosoma haematobium
     Schistosoma japonicum
     Schistosoma intercalatum
     Schistosoma mekongi

Classification of Human Parasites—cont'd

E. Arthropoda
  1. Arachnida
     Scorpions
     Spiders (black widow, brown recluse)
     Ticks *(Dermacentor, Ixodes, Argas, Ornithodoros)*
     Mites *(Sarcoptes)*
  2. Crustacea
     Copepods *(Cyclops)*
     Crayfish, lobsters, crabs
  3. Pentastomida
     Tongue worms
  4. Diplopoda
     Millipedes

  5. Chilopoda
     Centipedes
  6. Insecta
     Anoplura: sucking lice *(Pediculus, Phthirus)*
     Dictyoptera: cockroaches
     Hemiptera: true bugs *(Triatoma)*
     Coleoptera: beetles
     Hymenoptera: bees, wasps, etc.
     Lepidoptera: butterflies, moths, etc.
     Diptera: Flies, mosquitos, gnats, midges
       *(Phlebotomus, Aedes, Anopheles, Glossina,
       Simulium,* etc.)
     Siphonaptera: Fleas *(Pulex, Xenopsylla,* etc.)

From Garcia, L.S., and Bruckner, D.A. 1993. Diagnostic medical parasitology, ed 2. American Society for Microbiology, Washington, D.C.
*May be reclassified with the fungi.

TABLE 45.1

BODY SITES AND POSSIBLE PARASITES RECOVERED*

SITE	PARASITES	SITE	PARASITES
Blood		Intestinal tract—cont'd	*Blastocystis hominis*
Red cells	*Plasmodium* sp.		*Isospora belli*
	Babesia sp.		*Ascaris lumbricoides*
White cells	*Leishmania donovani*		*Enterobius vermicularis*
	Toxoplasma gondii		Hookworm
Whole blood/plasma	*Trypanosoma* sp.		*Strongyloides stercoralis*
	Microfilariae		*Trichuris trichiura*
Bone marrow	*Leishmania donovani*		*Hymenolepis nana*
Central nervous	*Taenia solium* (cysticercosis)		*Taenia saginata*
system	*Echinococcus* sp.		*Taenia solium*
	Naegleria fowleri		*Diphyllobothrium latum*
	Acanthamoeba/Hartmanella sp.		*Opisthorchis sinensis (Clonorchis)*
	Toxoplasma gondii		*Paragonimus westermani*
	Trypanosoma sp.		*Schistosoma* sp.
Cutaneous ulcers	*Leishmania* sp.	Liver, spleen	*Echinococcus* sp.
Eye	*Acanthamoeba* spp.		*Entamoeba histolytica*
Intestinal tract	*Entamoeba histolytica*		*Leishmania donovani*
	Entamoeba coli	Lung	*Pneumocystis carinii*
	Entamoeba hartmanni		*Paragonimus westermani*
	Endolimax nana	Muscle	*Taenia solium* (cysticerci)
	Iodamoeba bütschlii		*Trichinella spiralis*
	Giardia lamblia		*Onchocerca volvulus*
	Chilomastix mesnili	Skin	*Leishmania* sp.
	Dientamoeba fragilis		Microfilariae
	Trichomonas hominis	Urogenital system	*Trichomonas vaginalis*
	Cryptosporidium sp.		*Schistosoma* sp.

* This table does not include every possible parasite that could be found in a particular body site. However, the most likely organisms have been listed.

TABLE 45.2

BODY SITES AND SPECIMEN COLLECTION

SITE	SPECIMEN OPTIONS	COLLECTION METHODS
Blood	Smears of whole blood	Thick and thin films (first choice)
	Anticoagulated blood	Anticoagulated blood (second choice)
		EDTA (first choice)
		Heparin (second choice)
Bone marrow	Aspirate	Sterile
Central nervous system	Spinal fluid	Sterile
Cutaneous ulcers	Aspirates from below surface	Sterile plus air-dried smears
	Biopsy	Sterile; nonsterile to histology
Eye	Corneal scraping	Sterile saline, air-dried smear
	Corneal biopsy	Sterile saline
Intestinal tract	Fresh stool	Half-pint waxed container
	Preserved stool	5% or 10% formalin, MIF, SAF, Schaudinn's, PVA*
	Sigmoidoscopy material	Fresh, PVA, or Schaudinn's smears
	Duodenal contents	Entero-Test or aspirates
	Anal impression smear	Cellophane tape (pinworm examination)
	Adult worm/worm segments	Saline, 70% alcohol
Liver, spleen	Aspirates	Sterile, collected in four separate aliquots (liver)
	Biopsy	Sterile; nonsterile to histology
Lung	Sputum	True sputum (not saliva); no preservative (10% formalin if time delay)
	Lavage (centrifuged sediment)	No preservative
	Transbronchial aspirate	Air-dried smears
	Tracheobronchial aspirate	Same as above
	Brush biopsy	Same as above
	Open lung biopsy	Same as above
Muscle	Biopsy	Fresh-squash preparation, nonsterile to histology
Skin	Scrapings	Aseptic, smear or vial
	Skin snip	No preservative
	Biopsy	Nonsterile to histology
Urogenital system	Vaginal discharge	Saline swab, Culturette (no charcoal), culture medium
	Urethral discharge	Same as above
	Prostatic secretions	Same as above
	Urine	Single unpreserved specimen (24 h unpreserved specimen)

* *MIF,* Merthiolate-iodine-formalin; *SAF,* sodium acetate–acetic acid–formalin; *PVA,* polyvinyl alcohol.

men directly influences the recovery of protozoan organisms, the *time the specimen was collected* should be recorded on the laboratory request form. Freshly passed specimens are mandatory for the detection of trophic amebae or flagellates. *Liquid specimens should be examined within 30 minutes of passage* (not 30 minutes from the time they reach the laboratory), or the specimen should be placed in polyvinyl alcohol (PVA) fixative or another suitable preservative (see following section on preservation of specimens). *Semiformed or soft specimens should be examined within 1 hour of passage;* if this is not possible, the stool material should be preserved. Although the time limits are not as critical for the examination of a formed specimen, it is recommended that the material be examined on the day of passage. If these time limits cannot be met, portions of the sample should be preserved. Stool specimens should not be held at

room temperature but should be refrigerated at 3° to 5° C and stored in closed containers to prevent desiccation. At this temperature, eggs, larvae, and protozoan cysts remain viable for several days. Fecal specimens should never be incubated or frozen before examination. When the proper criteria for collection of fecal specimens are not met, the laboratory should request additional samples.

COLLECTION KIT FOR OUTPATIENT USE Various commercial collection systems are presently available, and one can select which vials are appropriate for a particular institution. Most manufacturers will package the kit according to individual specifications and will include complete multilanguage instructions (see following section on PVA fixative for representative suppliers). PVA solution is a combination of Schaudinn's fixative plus plastic PVA resin, which serves as an adhesive used to glue the stool onto the slide. Because PVA may contain the mercuric chloride base in the Schaudinn's fixative, most manufacturers have available collection vials with child-proof caps for institutions that select that option. All commercially available vials have appropriate labels related to toxicity.

A PVA-preserved portion of the specimen may be used for the complete examination, although many laboratories prefer to use the sample in 5% or 10% formalin for the concentration procedure. Other laboratories may prefer to use Schaudinn's fixative for collection. The standard two-vial collection kit contains one vial of 5% or 10% formalin and one vial of PVA. The benefit of the specimen vial collection system is a reduction in the lag time between when the specimen is passed and when it is fixed, thus providing better organism morphology used for identification. The unpreserved portion of the specimen (many laboratories do not request this sample unless an occult blood procedure is requested) may be examined to determine the specimen type (e.g., liquid, soft, or formed).

PRESERVATION

Depending on specimen-to-laboratory transportation time, the laboratory workload, and the availability of trained personnel, it may often be impossible to examine the specimen within specified time limits. To maintain protozoan morphology and prevent further development of certain helminth eggs and larvae, the fecal specimen should be placed in an appropriate preservative for examination at a later time. Several preservatives are available; four of these methods—PVA, formalin, Merthiolate (thimerosal)-iodine-formalin (MIF), and sodium acetate–acetic acid–formalin (SAF)—are discussed. When selecting an appropriate fixative, it is important to realize the limitations of each. PVA and SAF are the only fixatives included here from which a permanent stained smear can be easily prepared. The stained smear is extremely important in providing a complete and accurate examination for intestinal protozoa. The other fixatives mentioned permit the examination of the specimen as a wet mount only, a technique much less accurate than the stained smear for the identification of protozoa.

With the introduction of new regulatory requirements related to waste disposal, great interest has arisen in developing a preservative without the use of mercury compounds. However, studies indicate that substitute compounds may not provide the same quality of preservation necessary for good protozoan morphology on the trichrome or iron hematoxylin permanent stained smear. Copper sulfate is the compound that has been tried most frequently as a substitute in Schaudinn's fixative, but it does not provide results equal to those seen with mercuric chloride. The use of a zinc base compound as a substitute has been evaluated and provides better results than organism fixation seen with copper sulfate.[36] If you must eliminate mercury compounds from the laboratory, use SAF coupled with iron hematoxylin stain or PVA and trichrome stain, although none of the substitutes provides the overall consistent quality seen with mercuric chloride. Permanent stain results, however, are much better than those seen when a combination of trichrome and copper sulfate–based Schaudinn's fixative is used. Work will probably continue in this area to provide additional solutions related to the problem of mercury disposal, a problem many laboratories are trying to solve.

PVA FIXATIVE PVA fixative solution is highly recommended as a means of preserving protozoan cysts and trophozoites for examination at a later time.[9] The use of PVA also permits specimens to be shipped by regular mail service to a laboratory for subsequent examination. PVA, which is a combination of modified Schaudinn's fixative and a water-soluble resin, should be used in the ratio of three parts PVA to one part fecal material. Perhaps the greatest advantage in the use of PVA is that permanent stained slides can be prepared from PVA-preserved material.[9] This is not the case with many other preservatives; some permit the specimen to be examined as a wet preparation only, a technique that may not be adequate for the correct identification of protozoan organisms. PVA can be prepared in the laboratory or purchased commercially.[32]* This fixative remains stable for long periods (months to years) when kept in *sealed containers* at room temperature. Commercial packaging and distribution of these vials have essentially eliminated the problem of moisture loss and shelf life.

* Elvanol, grades 71-24, 71-30, and 90-25, can be obtained from E.I. duPont de Nemours & Co., Inc. (or their local representatives) in a minimum of 50-pound bags. You can check with a local chemical supply house to see if small quantities might be available. When ordering PVA powder, be sure to specify the pretested powder for use in PVA fixative. The PVA powder should be water soluble and of medium viscosity. Prepared liquid PVA (ready for use) can be obtained from Hardy Media, Medical Chemical Corporation, Meridian Diagnostics, Remel, Trend Scientific, Medical Media Lab, and Bio Spec. Most laboratories will purchase collection vial kits ready for distribution to patients.

Formalin Preservation Solution for Parasites (5% or 10% Formalin)

10% formalin:

Formaldehyde (USP)	100 ml
0.85% NaCl solution	900 ml

5% formalin:

Formaldehyde (USP)	50 ml
0.85% NaCl solution	950 ml

Dilute 100 ml (or 50 ml for 5%) of formaldehyde with 900 ml of 0.85% NaCl solution (distilled water may be used instead of saline).

FORMALIN Protozoan cysts, helminth eggs, and larvae are well preserved for long periods in *5% or 10% formalin* (see the box above). Although it is impractical for most laboratories, hot formalin (60° C) can be used to prevent further development of helminth eggs to the infective stage; this is important when bulk specimens are being saved for teaching purposes, particularly when using 5% formalin. Formalin should be used in the ratio of at least three parts formalin to one part fecal material; thorough mixing of the fresh specimen and fixative is necessary to ensure good preservation.

MIF SOLUTION The MIF solution of Sapèro and Lawless[70] can be used as a stain preservative for most types and stages of intestinal parasites and may be helpful in field surveys. Helminth eggs, larvae, and certain protozoa can be identified without further staining in wet mounts, which can be prepared immediately after fixation or several weeks later. This type of wet preparation may not be adequate for the diagnosis of all intestinal protozoa. Although some workers prepare a permanent stained smear from MIF-preserved material, another fixative preparation may be preferred for the precise morphology necessary for protozoan identification on the permanent stained smear. The MIF method has certain disadvantages, which may include the instability of the iodine component of the fixative.

SAF FIXATIVE SAF fixative contains formalin combined with sodium acetate, which acts as a buffer. It is a liquid fixative very similar to 10% aqueous formalin. When the sediment is used to prepare permanent stained smears, one may have some difficulty in getting material to adhere to the slide. Mayer's albumin has been recommended as an adhesive.[71,72] This fixative tends to give better results when used with iron hematoxylin rather than trichrome stain.

MACROSCOPIC EXAMINATION

The consistency of the stool (formed, semiformed, soft, or liquid) may give an indication of the protozoan stages present. When the moisture content of the fecal material is decreased during *normal passage* through the intestinal tract, the trophozoite stages of the protozoa encyst to survive. Trophozoites (motile forms) of the intestinal protozoa are usually found in soft or liquid specimens and occasionally in a semiformed specimen; the cyst stages are normally found in formed or semiformed specimens, rarely in liquid stools.

Helminth eggs or larvae may be found in any type of specimen, although the chances of finding any parasitic organism in a liquid specimen are reduced because of the dilution factor.

Occasionally, adult helminths, such as *Ascaris lumbricoides* or *Enterobius vermicularis* (pinworm), may be seen in or on the surface of the stool. Tapeworm proglottids may also be seen on the surface, or they may actually crawl under the specimen and be found on the bottom of the container. Other adult helminths, such as *Trichuris trichiura* (whipworm), hookworms, or perhaps *Hymenolepis nana* (dwarf tapeworm), may be found in the stool, but usually this occurs only after medication. The presence of blood in the specimen may indicate several factors and should always be reported. Dark stools may indicate bleeding high in the gastrointestinal tract, whereas fresh (bright-red) blood most often is the result of bleeding at a lower level. In certain parasitic infections, blood and mucus may be present; a soft or liquid stool may be highly suggestive of an amebic infection. These areas of blood and mucus should be carefully examined for the presence of trophic amebae. Occult blood in the stool may or may not be related to a parasitic infection and can result from several different conditions. Ingestion of various compounds may give a distinctive color to the stool (iron, black; barium, light tan to white).

MICROSCOPIC EXAMINATION

The identification of intestinal protozoa and helminth eggs is based on recognition of specific morphological characteristics. These studies require a good binocular microscope, good light source, and the use of a calibrated ocular micrometer.

The microscope should have 5× and 10× oculars (wide-field oculars are often recommended) and three objectives: low power (10×), high dry (40×), and oil immersion (100×). Some laboratories are using a 50× or 60× oil immersion lens in combination with the standard 100× oil immersion lens to screen all permanent stained fecal and blood smears. This approach prevents the contamination of the high-dry lens (40×) with oil when switching back and forth between high-dry (400×) and oil immersion (1000×) magnifications.

CALIBRATION OF MICROSCOPE Parasite identification depends on several parameters, including size; any laboratory doing diagnostic work in parasitology should have a calibrated microscope available for precise measurements.

Measurements are made by means of a micrometer disk placed in the ocular of the microscope; the disk is usually calibrated as a line divided into 50 units. Since the divisions in the disk represent different measurements, depending on the objective magnification used, the ocular disk divisions must be compared with a known calibrated scale, usually a stage micrometer with a scale of 0.1 and 0.01 mm divisions. Specific directions may be found in the work of Garcia and Bruckner.[32]

NOTE: After each objective power has been calibrated on the microscope, *the oculars containing the disk or these objectives cannot be interchanged with corresponding objectives or oculars on another microscope.* Each microscope that will be used to measure organisms must be calibrated as a unit; *the original oculars and objectives that were used to calibrate the microscope must also be used when an organism is measured.*

To comply with quality control requirements, each calibrated microscope should be recalibrated once a year. The calibrations do change, and a slight error can be magnified, particularly using the oil immersion objective, when measurements can be critical to accurate speciation.

DIAGNOSTIC PROCEDURES
A combination of techniques yields a greater number of positive specimens than does any one technique alone. Procedures recommended for a complete ova and parasite examination are discussed in this section.

DIRECT SMEARS Normal mixing in the intestinal tract usually ensures even distribution of helminth eggs or larvae and protozoa. However, examination of the fecal material as a direct smear may or may not reveal organisms, depending on the parasite density. The direct smear is prepared by mixing a small amount of fecal material (approximately 2 mg) with a drop of physiological saline; this mixture provides a uniform suspension under a 22 × 22 mm coverslip. Some workers prefer a 1½- × 3-inch (4 × 7 cm) glass

slide for the wet preparations, rather than the standard 1- × 3-inch (2.5 × 7 cm) slide most often used for the permanent stained smear. A 2 mg sample of fecal material forms a low cone on the end of a wooden applicator stick. If more material is used for the direct mount, the suspension is usually too thick for an accurate examination; less than 2 mg results in the examination of too thin a suspension, thus decreasing the chances of finding any organisms. If present, blood and mucus should always be examined as a direct mount. The entire 22 × 22 mm coverslip should be systematically examined using the low-power objective (10×) and low light intensity; any suspect objects may then be examined under high-dry power (40×). The use of the oil immersion objective on mounts of this type is not recommended unless the coverslip (no. 1 thickness coverslip is recommended when the oil immersion objective is used) is sealed to the slide with a cotton-tipped applicator stick dipped in equal parts of heated paraffin and petroleum jelly. Many workers believe the use of oil immersion on this type of preparation is impractical, especially since morphological detail is most easily seen and the diagnosis confirmed with oil immersion examination of the permanent stained smear.

The direct wet mount is used primarily to detect motile trophozoite stages of the protozoa. These organisms are very pale and transparent, two characteristics that require the use of low light intensity. Protozoan organisms in a saline preparation usually appear as refractile objects. If suspect objects are seen on high-dry power, one should allow at least 15 seconds to detect motility of slow-moving protozoa. Heat applied by placing a hot penny on the edge of a slide may enhance the motility of trophic protozoa.

NOTE: *With few exceptions, protozoan organisms should not be identified on the basis of a wet mount alone. Permanent stained smears should be examined to confirm the specific identification of suspected organisms.*

Helminth eggs or larvae and protozoan cysts may also be seen on the wet film, although these forms are more often detected after fecal concentration procedures.

After the wet preparation has been thoroughly checked for trophic amebae, a drop of iodine may be placed at the edge of the coverslip, or a new wet mount can be prepared with iodine alone. A weak iodine solution is recommended; too strong a solution may obscure the organisms. Several types of iodine solutions are available: Dobell and O'Connor's, Lugol's, and D'Antoni's.[32] *Gram's iodine used in bacteriological work is not recommended for staining parasitic organisms.*

NOTE: Many laboratories believe that the benefits gained in organism recovery and identification from receipt of preserved stool specimens far outweigh the disadvantages of not receiving a fresh specimen that can be examined for motile organisms. This is particularly true with the intestinal protozoa. No need exists to perform the direct wet smear if the specimen has been submitted to the laboratory in preservatives; any organisms present would be killed, and thus no motility would be seen. One should proceed directly to the concentration and permanent stained smear.

Protozoan cysts correctly stained with iodine contain yellow-gold cytoplasm, brown glycogen material, and paler refractile nuclei. The chromatoidal bodies may not be as clearly visible as they were in the saline mount.

Several staining solutions are available that may be used to reveal nuclear detail in the trophozoite stages. Nair's buffered methylene blue stain is effective in showing nuclear detail when used at a low pH.[63] Methylene blue (0.06%) in an acetate buffer at pH 3.6 usually gives satisfactory results; the mount should be examined within 30 minutes.[32]

CONCENTRATION PROCEDURES Often a direct mount of fecal material fails to reveal the presence of parasitic organisms in the gastrointestinal tract. Fecal concentration procedures should be included for a complete examination for parasites; these procedures allow the detection of small numbers of organisms that may be missed using only the direct mount. Helminth eggs can usually be identified when recovered from a concentration procedure. Generally the identification of intestinal protozoa should be considered tentative and confirmed with the permanent stained smear. Those protozoa that might be routinely identified from a concentration procedure include cysts of *Giardia lamblia, Entamoeba histolytica, Entamoeba coli,* and *Iodamoeba bütschlii* (trophozoites are rarely identified from a concentrate). Commercially available concentrating devices may assist a laboratory in standardizing the methodology.*

Various concentration procedures are available; they are either *flotation* or *sedimentation techniques* designed to separate the parasitic components from excess fecal debris through differences in specific gravity. A flotation procedure (Proce-

dure 45.1) permits the separation of protozoan cysts and certain helminth eggs through the use of a liquid with a high specific gravity. The parasitic elements are recovered in the surface film, whereas the debris is found in the bottom of the tube. This technique yields a cleaner preparation than does the sedimentation procedure; however, some helminth eggs (operculated eggs or very dense eggs, such as unfertilized *Ascaris* eggs) and some protozoa do not concentrate well with the flotation method. The specific gravity may be increased, although this may produce more distortion in the eggs and protozoa. Any laboratory that uses only a flotation procedure may fail to recover all the parasites present; to ensure detection of all organisms in the sample, both the surface film and the sediment should be carefully examined.

NOTE: Directions for any flotation technique must be followed exactly to produce reliable results.

Sedimentation procedures (using gravity or centrifugation) allow the recovery of all protozoa, eggs, and larvae present; however, the sediment preparation contains more fecal debris. If a single technique is selected for routine use, the *sedimentation procedure* (Procedure 45.2) is recommended as the easiest to perform and least subject to technical error.

PERMANENT STAINED SMEARS The detection and correct identification of intestinal protozoa frequently depend on the examination of the permanent stained smear. These slides not only provide the microscopist with a permanent record of the protozoan organisms identified, but also may be used for consultations with specialists when unusual morphological characteristics are found. In view of the number of morphological variations possible, organisms may be found that are very difficult to identify and do not fit the pattern for any one species.

Most identifications should be considered tentative until confirmed by the permanent stained slide. For these reasons, *the permanent stain is recommended for every stool sample submitted for a routine examination for parasites.*

Many staining techniques are available; those included here generally tend to give the best and most reliable results with both fresh and PVA-preserved specimens.

PREPARATION OF FRESH MATERIAL. When the specimen arrives, use an applicator stick or brush to smear a small amount of stool on two clean slides, and immediately immerse them in Schaudinn's fixative. If the slides are prepared

---

* ALPHA-TEC, Evergreen Scientific (FPC, fecal parasite concentrator), Hardy Diagnostics, Medical Chemical Corp., Meridian Diagnostics, Remel, Scientific Device Lab, Trend Scientific (FeKal, CON-Trate System).

PROCEDURE 45.1

ZINC SULFATE FLOTATION PROCEDURE (MODIFIED)

PRINCIPLE

Parasitic cysts and some helminth eggs will rise to the surface of a liquid with a high specific gravity, such as zinc sulfate, because of their buoyant properties in that solution.

METHOD

1. Prepare a fecal suspension of ¼ to ½ teaspoon of feces (more if the specimen is diarrheal) in 10 to 15 ml of tap water.

2. Filter this material through two layers of gauze into a small tube (Wassermann tube; round bottom tube is recommended rather than centrifuge tube). The gauze should never be more than two layers thick; more gauze may trap mucus (containing *Cryptosporidium* sp. oocysts and/or microsporidial spores). When using "pressed" (cannot see the actual threads) rather than "woven" gauze, use a single layer. Fill the tube with tap water to within 2 to 3 mm of the top, and centrifuge for 10 min at 500 × *g*.

3. Decant the supernatant fluid, fill the tube with water, and resuspend the sediment by stirring with an applicator stick. Centrifuge for 10 min at 500 × *g*.

4. Decant the water, add 2 to 3 ml zinc sulfate solution, resuspend the sediment, and fill the tube with zinc sulfate solution to within 0.5 cm of the top.

5. Centrifuge for 1 to 2 min at 500 × *g*. Do not "brake" the centrifuge; allow the tubes to come to a stop without interference or vibration.

6. Without removing the tubes from the centrifuge, touch the surface film of the suspension with a wire loop (diameter, 5 to 7 mm; loop should be parallel with the surface of the fluid). *Do not go below the surface of the film with the loop*. Add the material in the loop to a slide containing a drop of dilute iodine or saline.

NOTE: *Material recovered from the zinc sulfate flotation procedure must be examined within several minutes; prolonged contact with the high–specific gravity solution leads to organism distortion.* Protozoan cysts may be more distorted using this method, and the technique is unsuitable for stool specimens containing large amounts of fatty material. The specific gravity of the zinc sulfate should be 1.18 and should be checked frequently with a hydrometer. If this technique is used to

concentrate formalin-preserved material, the specific gravity should be increased to 1.20. Zinc sulfate solution ($ZnSO_4 \cdot 7H_2O$, specific gravity 1.18) consists of approximately 330 g of dry crystals in 670 ml of distilled water. A 33% solution usually approximates the correct specific gravity but may be adjusted to 1.18 by the addition of zinc sulfate or distilled water.

QUALITY CONTROL

The specific gravity of the solution should be checked. Formalin-fixed stools containing known parasites can be purchased and should be concentrated along with patient specimens to check for the detection of the parasites.

EXPECTED RESULTS

Known parasites should be detected readily.

PERFORMANCE SCHEDULE

The specific gravity of the zinc sulfate should be checked when prepared and at least monthly thereafter. Quality control samples should be concentrated at least four times yearly. Participation in a proficiency survey program that includes formalin-preserved stools would fulfill this need.

---

correctly, you should be able to read newsprint through the fecal smear. The smears should fix for a minimum of 30 minutes; fixation time may be decreased to 5 minutes if the Schaudinn's solution is heated to 60° C. However, this approach is not that practical for most clinical laboratories and is rarely used.

If a liquid specimen is received, mix 3 or 4 drops of PVA[32] with 1 or 2 drops of fecal material directly on a slide, spread the mixture, and allow

the slides to dry for several hours at 35° C or overnight at room temperature.

PREPARATION OF PVA-PRESERVED MATERIAL. Stool specimens preserved in PVA should be allowed to fix at least 30 minutes. *After fixation, the sample should be thoroughly mixed and a small amount of the material poured onto a paper towel to absorb excess PVA.* This is an important step in the procedure; allow the PVA to soak into the paper towel for 2 to 3 minutes before preparing the

PROCEDURE 45.2

FORMALIN-ETHER (FORMALIN-ETHER SUBSTITUTE) SEDIMENTATION TECHNIQUES

PRINCIPLE

Formalin fixes the eggs, larvae, and cysts, so that they are no longer infectious, as well as preserving their morphology. Fecal debris is extracted into the ether phase of the solution, freeing the sedimented parasitic elements from at least some of the artifactual material in stool.

METHOD

1. Transfer ¼ to ½ teaspoon of fresh stool into 10 ml of 5% or 10% formalin in a 15 ml shell vial, unwaxed paper cup, or 16 × 125 mm tube (container may vary depending on individual preferences) and comminute thoroughly. Let stand 30 min for adequate fixation.

2. Filter this material (funnel or pointed paper cup with end cut off) through two layers of gauze into a 15 ml centrifuge tube.

3. Add physiological saline or 5% or 10% formalin to within ½ inch (1.5 cm) of the top, and centrifuge for 10 min at 500 × g.

4. Decant, resuspend the sediment (should have 0.5 to 1 ml sediment) in saline to within ½ inch (1.5 cm) of the top, and centrifuge again for 10 min at 500 × g. This second wash may be eliminated if the supernatant fluid after the first wash is light tan or clear.

5. Decant and resuspend the sediment in 5% or 10% formalin (fill the tube only half full). If the amount of sediment left in the bottom of the tube is very small, do not add ether in step 6; merely add the formalin, then spin, decant, and examine the remaining sediment.

6. Add approximately 3 ml of ether substitute, insert stopper, and shake vigorously for 30 s. Hold the tube so that the stopper is directed away from your face; remove stopper carefully to prevent spraying of material caused by pressure within the tube.

7. Centrifuge for 10 min at 500 × g. Four layers should result: a small amount of sediment in the bottom of the tube, containing the parasites; a layer of formalin; a plug of fecal debris on top of the formalin layer; and a layer of ether substitute at the top.

8. Free the plug of debris by ringing with an applicator stick, and decant all the fluid. After proper decanting, a drop or two of fluid remaining on the side of the tube will drain down to the sediment. Mix the fluid with the sediment, and prepare a wet mount for examination.

The formalin-ether sedimentation procedure may be used on PVA-preserved material. Steps 1 and 2 differ as follows:

1. Fixation time with PVA should be at least 30 min. Mix contents of PVA bottle (stool-PVA mixture: 1 part stool to 2 or 3 parts PVA) with applicator sticks. Immediately after mixing, pour approximately 2 to 5 ml (amount will vary depending on the viscosity and density of the mixture) of the stool-PVA mixture into a 15 ml shell vial, 16 × 125 mm tube, or something similar, and add approximately 10 ml physiological saline.

2. Filter this material (funnel or paper cup with pointed end cut off) through two layers of gauze into a 15 ml centrifuge tube.

Steps 3 through 8 are the same for both fresh and PVA-preserved material.

NOTE: Tap water may be substituted for physiological saline throughout this procedure; however, saline is recommended. Some workers prefer to use 5% or 10% formalin for all the rinses (steps 3 and 4).

NOTE: The introduction of ethyl acetate (or Hemo-De, other substitute) as a substitute for diethyl ether (ether) in the formalin-ether sedimentation concentration procedure provides a much safer chemical for the clinical laboratory. Tests comparing the use of these two compounds on formalin-preserved and PVA-preserved material indicate that the differences in organism recovery and identification are minimal and probably do not reflect clinically relevant differences. When examining the sediment in the bottom of the tube:

1. Prepare a saline mount (1 drop of sediment and 1 drop of saline solution mixed together), and scan the whole 22 × 22 mm coverslip under low power for helminth eggs or larvae.

2. Iodine may then be added to aid in the detection of protozoan cysts and should be examined under high-dry power. If iodine is added before low-power scanning, be certain that the iodine is not too strong; otherwise, some of the helminth eggs will stain so darkly that they will be mistaken for debris.

PROCEDURE 45.2

FORMALIN-ETHER (FORMALIN-ETHER SUBSTITUTE) SEDIMENTATION TECHNIQUES— cont'd

3. Occasionally a precipitate is formed when iodine is added to the sediment obtained from a concentration procedure with PVA-preserved material. The precipitate is formed from the reaction between the iodine and excess mercuric chloride

that has not been thoroughly rinsed from the PVA-preserved material. The sediment can be rinsed again to remove any remaining mercuric chloride, or the sediment can be examined as a saline mount without the addition of iodine.[32]

QUALITY CONTROL
The same recommendations for testing prepared known samples as mentioned for the zinc flotation procedure are applicable.

---

slides. With an applicator stick, apply some of the stool material from the paper towel to two slides, and let them dry for several hours at 37° C or overnight at room temperature. The PVA-stool mixture should be spread to the edges of the glass slide; this causes the film to adhere to the slide during staining. It is also important to dry the slides thoroughly to prevent the material from washing off during staining.

TRICHROME STAIN. This stain was originally developed by Gomori[40] for tissue differentiation and was adapted by Wheatley[87] for intestinal protozoa (Procedure 45.3). It is an uncomplicated procedure that produces well-stained smears from both fresh and PVA-preserved material.

The trichrome stain can be used repeatedly, and stock solution may be added to the jar when the volume is decreased. Periodically, the staining strength can be restored by removing the lid and allowing the 70% alcohol carried over from the preceding jar to evaporate. Each lot number or batch of stain (either purchased commercially or prepared in the laboratory) should be checked to determine the optimal staining time, which is usually a few minutes longer for PVA-preserved material.

The 90% acidified alcohol is used as a destaining agent that will provide good differentiation; however, prolonged destaining (more than 3 seconds) may result in a poor stain. To prevent continued destaining, the slides should be quickly rinsed in 100% alcohol and then dehydrated through two additional changes of 100% alcohol. For any alcohol solution in the staining protocol other than absolute alcohol, isopropyl alcohol can be used.

INTERPRETATION OF STAINED SMEARS. Many problems in interpretation may arise when

poorly stained smears are examined; these smears are usually the result of inadequate fixation or incorrect specimen collection and submission. An old specimen or inadequate fixation may result in organisms that fail to stain or that appear as pale-pink or red objects with very little internal definition. This type of staining reaction may occur with *Entamoeba coli* cysts, which require a longer fixation time; mature cysts in general need additional fixation time and therefore may not be as well stained as immature cysts. Degenerate forms or those that have been understained or destained too much may stain pale green.

When the smear is well fixed and correctly stained, the background debris is green, and the protozoa have a blue-green to purple cytoplasm with red or purple-red nuclei and inclusions. The differences in colors between the background and organisms provide more contrast than in hematoxylin-stained smears.

Helminth eggs and larvae usually stain dark red or purple; they are usually distorted and difficult to identify. White blood cells, macrophages, tissue cells, yeast cells, and other artifacts still present diagnostic problems, since their color range on the stained smear approximates that of the parasitic organisms (Figures 45.1 to 45.3).

IRON HEMATOXYLIN STAIN. Although the original method produces excellent results, most laboratories that use an iron hematoxylin stain select one of the shorter methods. Several procedures are available; both those presented here can be used with either fresh or PVA-preserved material. Both background debris and the organisms stain gray-blue to black, with the cellular inclusions and nuclei appearing darker than the cytoplasm.

The method described by Spencer and Monroe[76] is slightly longer than the trichrome proce-

PROCEDURE 45.3

TRICHROME STAIN

PRINCIPLE

The internal elements that distinguish among cysts and trophozoites can be best visualized with a stain that enhances the morphological features. In addition, such a stained smear provides a permanent record of the results.

METHOD

1. Formula

Chromotrope 2R	0.6 g
Light green SF	0.3 g
Phosphotungstic acid	0.7 g
Acetic acid (glacial)	1 ml
Distilled water	100 ml

2. Preparation. The stain is prepared by adding 1 ml of glacial acetic acid to the dry components. Allow the mixture to stand for 15 to 30 min to "ripen"; then add 100 ml of distilled water. This preparation gives a highly uniform and reproducible stain; the stain should be purple. Store in Coplin jars.

3. Procedure
   a. Prepare fresh fecal smears or PVA smears as described.
   b. Place in 70% ethanol for 5 min.* (This step may be eliminated for PVA smears.)
   c. Place in 70% ethanol plus D'Antoni's iodine (dark reddish brown) for 2 to 5 min.
   d. Place in two changes of 70% ethanol, one for 5 min* and one for 2 to 5 min.
   e. Place in trichrome stain solution for 10 min.
   f. Place in 90% ethanol, acidified (1% acetic acid) for up to 3 s (do not leave the slides in this solution any longer).
   g. Dip once in 100% ethanol.
   h. Place in two changes of 100% ethanol for 2 to 5 min each.*
   i. Place in two changes of xylene or toluene for 2 to 5 min each.*
   j. Mount in Permount or some other mounting medium; use a no. 1 thickness coverglass.

   NOTE: If using PVA with the zinc base, begin the staining process at step e.

QUALITY CONTROL

Using the same timing and reagents as for patient samples, stain slides prepared from fixed stool samples with known parasites, available commercially or as part of a proficiency testing survey program. Negative preserved stool (PVA) seeded with buffy coat cells from sedimented whole blood can also be used.[32]

EXPECTED RESULTS

Background debris is green and protozoa shows blue-green to purple cytoplasm. The nuclei and inclusions are red or purple-red and sharply delineated from background.

PERFORMANCE SCHEDULE

Stain quality control samples when new stain solution is prepared and at least four times yearly.

*At this stage, slides can be held several hours or overnight.

---

dure. Although the slides do not require destaining, decolorizing in 0.5% hydrochloric acid after a longer initial staining time may provide better differentiation.

Another iron hematoxylin method, described by Tompkins and Miller,[84] includes the use of phosphotungstic acid as a destaining agent. This procedure also gives good, reproducible results. Another modification of the iron hematoxylin staining method incorporates a carbolfuchsin step into the protocol; in addition to the intestinal protozoa, acid-fast organisms could also be seen on the permanent stained smear.[68] This iron hematoxylin stain is being used with SAF-fixed stool specimens as an alternative to using PVA-preserved specimens (Procedure 45.4). Any fecal specimen submitted in SAF fixative can be used.

Fresh fecal specimens after fixation in SAF for 30 minutes can also be used. This combination stain approach is not recommended for specimens preserved in Schaudinn's fixative or PVA.

GENERAL INFORMATION The most important step in preparing a well-stained fecal smear is adequate fixation of a specimen that has been submitted within specified time limits. To ensure best results, the acetic acid component of Schaudinn's fixative should be added just before use; fixation time (room temperature) may be extended overnight with no adverse effects on the smears.

After fixation, it is very important to remove completely the mercuric chloride residue from the smears. The 70% alcohol-iodine mixture removes the mercury complex; the iodine solution should

FIGURE 45.1
Charcot-Leyden crystals.

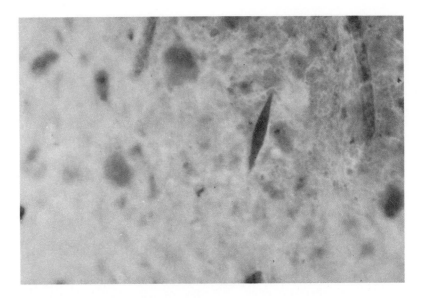

FIGURE 45.2
Polymorphonuclear
leukocytes.

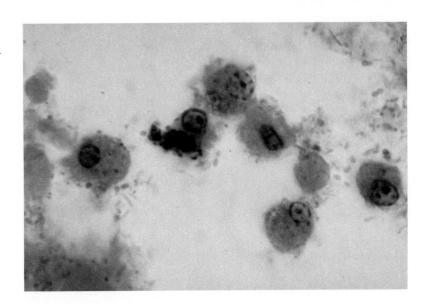

FIGURE 45.3

Blastocystis hominis (larger
objects) and yeast cells
(smaller, more
homogeneous objects).

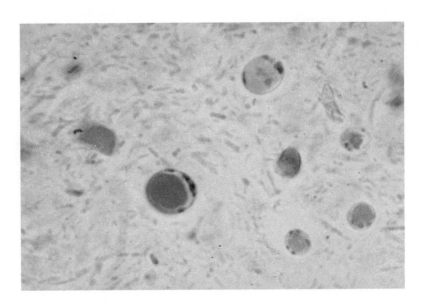

PROCEDURE 45.4

MODIFIED IRON HEMATOXYLIN STAIN

PRINCIPLE

The internal elements that distinguish among cysts and trophozoites can be best visualized with a stain that enhances the morphological features. Also, the stained smear provides a permanent record of the method.

Mayer's albumin

Add an equal quantity of glycerine to a fresh egg white. Mix gently and thoroughly. Store at 4° C, and indicate an expiration date of 3 months. Mayer's albumin from commercial suppliers can normally be stored at 25° C for 1 year (example: Product #756, E.M. Diagnostic Systems, Inc., 480 Democrat Rd., Gibbstown, N.J. 08027; 800-443-3637).

Stock solution of hematoxylin stain

Hematoxylin powder   10 g
Ethanol (95% or
    100%)   1000 ml

1. Mix well until dissolved.

2. Store in a clear glass bottle in a light area. Allow to ripen for 14 days before use.

3. Store at room temperature with an expiration date of 1 year.

Mordant

Ferrous ammonium sulfate
[Fe(NH$_4$)$_2$(SO$_4$)$_2$•6H$_2$O]   10 g
Ferric ammonium sulfate
    [FeNH$_4$(SO$_4$)$_2$•12H$_2$O] 10 g
Hydrochloric acid (HCl)
    (concentrated)   10 ml
Add distilled water to   1000 ml

Working solution of hematoxylin stain

1. Mix equal quantities of stock solution of stain and mordant.

2. Allow mixture to cool thoroughly before use (prepare at least 2 hours before use). The working solution should be made fresh every week.

Picric acid

Mix equal quantities of distilled water and an aqueous saturated solution of picric acid to make a 50% saturated solution.

Acid-alcohol decolorizer

Hydrochloric acid (HCl)
(concentrated)   30 ml
Alcohol to   1000 ml

70% alcohol and ammonia

70% alcohol   50 ml
Ammonia   0.5-1.0 ml
Add enough ammonia to bring the pH to approximately 8.0.

Carbolfuchsin

Basic fuchsin (solution A): dissolve 0.3 g basic fuchsin in 10 ml of 95% ethanol. Phenol (solution B): dissolve 5 g of phenol crystals in 100 ml distilled water. (Gentle heat may be needed.)

1. Mix solution A with solution B.

2. Store at room temperature; stable for 1 year.

PROCEDURE FOR MODIFIED IRON HEMATOXYLIN STAIN (CARBOLFUCHSIN STEP)

1. Slide preparation
   a. Place 1 drop of Mayer's albumin on a labeled slide.
   b. Mix the sediment from the SAF concentration well with an applicator stick.
   c. Add approximately 1 drop of the fecal concentrate to the albumin, and spread the mixture over the slide.

2. Allow slide to air dry at room temperature (smear will appear opaque when dry).

3. Place slide in 70% alcohol for 5 min.

4. Wash in container (not running water) of tap water for 2 min.

5. Place slide in Kinyoun stain for 5 min.

6. Wash slide in running tap water (constant stream of water into container) for 1 min.

7. Place slide in acid/alcohol decolorizer for 4 min.*

8. Wash slide in running tap water (constant stream of water into container) for 1 min.

9. Place slide in iron hematoxylin working solution for 8 min.

10. Wash slide in distilled water (in container) for 1 min.

11. Place slide in picric acid solution for 3-5 min.

12. Wash slide in running tap water (constant stream of water into container) for 10 min.

13. Place slide in 70% alcohol plus ammonia for 3 min.

14. Place slide in 95% alcohol for 5 min.

15. Place slide in 100% alcohol for 5 min.

16. Place slide in two changes of xylene for 5 min.

*This step can also be performed as follows:
   a. Place slide in acid/alcohol decolorizer for 2 min.
   b. Wash slide in running tap water (constant stream of water into container) for 1 min.
   c. Place slide in acid/alcohol decolorizer for 2 min.

PROCEDURE 45.4

MODIFIED IRON HEMATOXYLIN STAIN—cont'd

d. Wash slide in running tap water (constant stream of water into container) for 1 min.

e. Continue staining sequence with step 9 (iron hematoxylin working solution).

PROCEDURE NOTES FOR MODIFIED IRON HEMATOXYLIN STAIN (CARBOLFUCHSIN STEP)

1. The first 70% alcohol step acts with the Mayer's albumin to "glue" the specimen to the glass slide.

The specimen may wash off if insufficient albumin is used or if the slides are not completely dry before staining.

2. The working hematoxylin stain should be checked each day of use by adding a drop of stain to alkaline tap water. If a blue color does not develop, prepare fresh working stain solution.

3. The picric acid differentiates the hematoxylin stain by removing more stain from

fecal debris than from the protozoa and removing more stain from the organism cytoplasm than the nucleus. When properly stained, the background should be various shades of gray-blue, and protozoa should be easily seen with medium-blue cytoplasm and dark blue-black nuclei.

---

be changed often enough (at least once a week) to maintain a dark reddish brown color. If the mercuric chloride is not completely removed, the stained smear may contain varying amounts of highly refractive granules, which may prevent finding or identifying any organisms present.

Good results in the final stages of dehydration (100% alcohol) and clearing (xylene or xylene substitute) depend on the use of *fresh reagents*. It is recommended that solutions be changed at least weekly and more often if large numbers of slides (10 to 50 per day) are being stained. Stock containers and staining dishes should have well-fitting lids to prevent evaporation and absorption of moisture from the air. If the clearing agent turns cloudy on addition of the slides from 100% alcohol, there is water in the solution. When clouding occurs, immediately return the slides to 100% alcohol, replace all dehydrating and clearing agents with fresh stock, and continue with the dehydration process.

MODIFIED ACID-FAST STAIN FOR *CRYPTOSPORIDIUM* SP., *ISOSPORA* SP. AND THE CLBs (CYANOBACTERIA-LIKE OR COCCIDIA-LIKE ORGANISMS; NOW IDENTIFIED AS COCCIDIA IN THE GENUS *CYCLOSPORA*) Although different procedures have been tried for the recovery and identification of *Cryptosporidium* sp. in humans, the best approach seems to be hot or cold modified acid-fast stains (Procedure 45.5). The hot modified acid-fast method presented here consistently provides excellent results.[34] In patients with few oocysts in the stool, the highly

specific and sensitive fluorescent method using a monoclonal antibody reagent is more likely to reveal the organisms.[30,38,78]

ADDITIONAL TECHNIQUES FOR GASTROINTESTINAL TRACT SPECIMENS

SIGMOIDOSCOPY MATERIAL

When repeated fecal examinations fail to reveal the presence of *Entamoeba histolytica,* material obtained during sigmoidoscopy may be valuable in the diagnosis of amebiasis. However, this procedure does not take the place of routine fecal examinations; a series of at least three (six is preferable) fecal specimens should be submitted for each patient having a sigmoidoscopy examination. Material from the mucosal surface should be obtained by aspiration or scraping, not with a cotton-tipped swab. If swabs must be used, most of the cotton should be removed (leave just enough to safely cover the end) and should be tightly wound to prevent absorption of the material to be examined.

The specimen should be processed and examined immediately; the number of techniques used will depend on the amount of material obtained. If the specimen is sufficient for both wet preparations and permanent stained smears, proceed as follows. The direct mount should be examined immediately for the presence of moving trophozoites; it may take time for the organisms to become acclimated to this type of preparation, so

PROCEDURE 45.5

MODIFIED ACID-FAST STAINS FOR *CRYPTOSPORIDIUM* SP.

KINYOUN COLD STAIN

Carbolfuchsin

Basic fuchsin	4 g
Phenol	8 ml
Alcohol (95%)	20 ml
Distilled water	100 ml

Dissolve the basic fuchsin in the alcohol, and add the water slowly while shaking. Melt the phenol in a 56° C water bath, and add 8 ml to the stain, using a pipette with a rubber bulb.

Decolorizer

Ethanol (95%)	97 ml
Concentrated, hydrochloric acid (HCl)	3 ml

Add the hydrochloric acid to the alcohol slowly, working under a chemical fume hood.

Counterstain

Methylene blue	0.3 g
Distilled water	100 ml

1. Spin an aliquot of 10% formalinized stool for 10 min at 500 × g.

2. Remove upper layer of sediment with pipette, and place a thin layer onto a microscope slide.
   NOTE: If the stool specimen contains a large amount of mucus, 10 drops of 10% potassium hydroxide (KOH) can be added to the sediment (step 2), vortexed, rinsed with 10% formalin, and respun before smear preparation. Some laboratories use this approach routinely before smear preparation.

3. Heat fix the smear at 70° C for 10 min.

4. Stain the fixed smear for 3 to 5 min (no heat necessary).

5. Wash in distilled, filtered water, and shake off excess water.

6. Flood with decolorizer for approximately 1 min. Check to see that no more red color runs when the slide is tipped. Add slightly more decolorizer for very thick slides or those that continue to bleed red dye.

7. Wash thoroughly with filtered water as above, and shake off excess.

8. Flood with counterstain for approximately 1 min.

9. Wash with distilled water, and drain by standing slides upright. Do not blot dry.

By the addition of a detergent or wetting agent, the staining of acid-fast organisms may be accelerated. Tergitol No. 7 (Sigma Chemical Co.) may be used. Add 1 drop of Tergitol No. 7 to every 30 to 40 ml of the Kinyoun carbolfuchsin stain.

Acid-fast bacteria stain red with carbolfuchsin stains. The background color depends on the counterstain; methylene blue imparts a blue color to non–acid-fast material, whereas brilliant green results in green background and picric acid results in yellow.

HOT STAIN

1. Spin an aliquot of 10% formalinized stool for 10 min at 500 × g.

2. Remove upper layer of sediment with pipette, and place a thin layer onto a microscope slide.
   NOTE: If the stool specimen contains a large amount of mucus, 10 drops of 10% KOH can be added to the sediment (step 2), vortexed, rinsed with 10% formalin, and respun before smear preparation. Some laboratories use this approach routinely before smear preparation.

3. Heat fix the smear at 70° C for 10 min.

4. Place slide on staining rack and flood with carbolfuchsin.

5. Heat to steaming and allow to stain for 5 min. If the slide begins to dry, more stain is added without additional heating.

6. Rinse the smear with tap or distilled water.

7. Decolorize with 5% aqueous sulfuric acid for 30 s (thicker smears may require longer).

8. Rinse smear with tap or distilled water, drain, and flood smear with methylene blue counterstain for 1 min.

9. Rinse with tap or distilled water, drain, and air dry.

---

motility may not be obvious for several minutes. Care should be taken not to confuse protozoan organisms with macrophages or other tissue cells; any suspect cells should be confirmed with the use of the permanent stained slide. The smears for permanent staining should be prepared at the same time the direct mount is made by gently smearing some of the material onto several slides and immediately placing them into Schaudinn's fixative. The slides can then be stained by any of the techniques mentioned for routine fecal smears. If the material is bloody, contains a large amount of mucus, or is a "wet" specimen, 1 or 2 drops of the sample can be mixed with 3 or 4 drops of PVA directly on the slide. Allow the smears to dry (overnight if possible) before staining.

DUODENAL CONTENTS

In some instances, repeated fecal examinations may fail to confirm a diagnosis of *Giardia lamblia* and *Strongyloides stercoralis* infections. Since these two parasites are normally found in the duodenum, the physician may submit duodenal drainage fluid to the laboratory for examination. The specimen should be submitted without preservatives and should be received and examined within 1 hour after being taken. The amount of fluid may vary. It should be centrifuged and the sediment examined as wet mounts for the detection of motile organisms. Several mounts should be prepared and examined; because of the dilution factor, the organisms may be difficult to recover with this technique.

Another convenient method of sampling duodenal contents, which eliminates the necessity for intubation, is the use of the Entero-Test. This device consists of a gelatin capsule containing a weighted, coiled length of nylon yarn. The end of the line protrudes through the top of the capsule and is taped to the side of the patient's face. The capsule is then swallowed, the gelatin dissolves, and the weighted string is carried by peristalsis into the duodenum. After approximately 4 hours the string is recovered and the bile-stained mucus attached to the string is examined as a wet mount for the presence of organisms. This type of specimen should also be examined immediately after the string is recovered. When the mucus is examined from either duodenal drainage or the Entero-Test capsule, typical "falling leaf" motility of *G. lamblia* trophozoites is usually not visible. The flagella usually are visible as a rapid "flutter," with the organism remaining trapped in the mucus.

ESTIMATION OF WORM BURDENS

Although these procedures are not routinely performed, circumstances may arise when it is helpful to know the degree of infection in a patient or perhaps to follow the effectiveness of therapy.[5,6,79] In certain helminth infections that have little clinical significance, the patient may not be given treatment if the numbers of parasites are small. The parasite burden may be estimated by counting the number of eggs passed in the stool.[32]

RECOVERY OF LARVAL-STAGE NEMATODES

Nematode infections that give rise to larval stages, which hatch either in the soil or in tissues, may be diagnosed by using culture techniques designed to concentrate the larvae.[32,41] These procedures are used in hookworm, *Strongyloides*, and *Trichostrongylus* infections. Some of these techniques, which permit the recovery of infective-stage larvae, may be helpful because the eggs of many species are identical, and specific identifications are based on larval morphology.

BAERMANN TECHNIQUE   When the stools from a patient suspected of having strongyloidiasis are repeatedly negative, the Baermann technique may be helpful in recovering larvae. The apparatus is designed to allow the larvae to migrate from the fecal material through several layers of damp gauze into water, which is centrifuged, thus concentrating the larvae in the bottom of the tube.[32] Specimens for this technique should be collected after a mild saline cathartic, not a stool softener.

AGAR PLATE CULTURE FOR *STRONGYLOIDES*   Agar plate cultures for the detection of *Strongyloides stercoralis* larvae are available as another method for the diagnosis of strongyloidiasis. Stool or duodenal contents are placed onto agar plates, and the plates are sealed to prevent accidental infections and held for 2 days at room temperature. As the larvae crawl over the agar, they carry bacteria with them, thus creating visible "tracks" over the agar. The plates are examined under the microscope for confirmation of larvae, the surface of the agar is then washed with 10% formalin, and final confirmation of larval identification made via wet examination of the sediment from the formalin washings. The agar consists of 1.5% agar, 0.5% meat extract, 1.0% peptone, and 0.5% NaCl. This approach is thought to be more sensitive and may become more widely used.[2,48]

HATCHING PROCEDURE FOR SCHISTOSOME EGGS

When schistosome eggs are recovered from either urine or stool, they should be carefully examined to determine viability. The presence of living miracidia within the eggs indicates an active infection, which may require therapy. The viability of the miracidium larvae can be determined in two ways:

1. The cilia on the flame cells (primitive excretory cells) may be seen on high-dry power and are usually actively moving.

2. The larvae may be released from the eggs with the use of a hatching procedure.[32]

The eggs usually hatch within several hours when placed in 10 volumes of dechlorinated or spring water. The eggs, which are recovered in the urine, are easily obtained from the sediment and can be examined under the microscope to determine viability.

CELLOPHANE TAPE PREPARATIONS

Enterobius vermicularis is a roundworm that is worldwide in distribution. It is very common in children and is known as the "pinworm" or the "seatworm." The adult female migrates from the anus during the night and deposits her eggs on the perianal area. Since the eggs are deposited outside the gastrointestinal tract, examination of a stool specimen may produce negative results. Although some laboratories use the anal swab technique, pinworm infections are most frequently diagnosed by using the cellophane tape method (Procedure 45.6) for egg recovery. Another collection procedure is illustrated in Figure 45.4. Occasionally the adult female may be found on the surface of a formed stool or on the cellophane tape. Specimens should be taken in the morning *before bathing* or going to the bathroom. For small children, it is best to collect the specimen early in the morning, just before the child awakens.

A series of at least *four to six consecutive negative tapes* should be obtained before ruling out infection with pinworms. However, most laboratories rarely receive more than one or two tapes, and the clinician may decide to treat the patient on the basis of symptoms or recurrent symptoms after the initial diagnosis.

IDENTIFICATION OF ADULT WORMS

Most adult worms or portions of worms that are submitted to the laboratory for identification are *Ascaris lumbricoides, Enterobius vermicularis,* or segments of tapeworms. The adult worms present no particular problems in identification (see later discussion on intestinal helminths); however, identification of the *Taenia* sp. tapeworms depends on the gravid proglottids, which contain the fully developed uterine branches. Identification as to species is based on the number of lateral uterine branches that arise from the main uterine stem in the gravid proglottids. Often the uterine branches are not clearly visible; one technique that can be used is injecting the branches with India ink (Procedure 45.7), allowing them to be easily seen and counted.

UROGENITAL SPECIMENS (TRICHOMONIASIS)

The identification of *Trichomonas vaginalis* is usually based on the examination of wet preparations of vaginal and urethral discharges and prostatic secretions. These specimens are diluted with a drop of saline and examined under low power with reduced illumination for the presence of actively motile organisms; urine sediment can be examined in the same way. As the jerky motility of the organisms begins to diminish, it may be possible to observe the undulating membrane, particularly under high-dry power (Figure 45.5). Stained smears are usually not necessary for the identification of this organism; often the number of false-positive and false-negative results reported on the basis of stained smears would strongly suggest the value of confirmation (i.e., observation of the motile organisms).

Some studies indicate that the most sensitive method of detecting *T. vaginalis* is culture. This technique may not be the most practical, and expense may limit its availability. To improve the approach to culture, a plastic envelope method (PEM) has been developed.[4] The medium consists of dry ingredients that are reconstituted with water before use, with a shelf life of at least 1 year. The PEM pouch is composed of clear, transparent plastic that minimizes oxygen and water vapor transmission. The oxygen content within the medium is reduced further through the reducing action of the components, ascorbic acid and cysteine. Distilled water is added to the upper chamber of the pouch to reconstitute the medium; the medium is then pushed into the lower chamber. The patient specimen is inoculated into the small amount of fluid left in the upper chamber, then pushed into the lower chamber. The pouch comes with a small rigid plastic frame that holds the pouch in a horizontal position for viewing directly under the microscope. Some advantages of this system follow: (1) the pouch containing dry medium can be held at room temperature for at least a year before use; (2) holding medium is not necessary for specimen transport to the laboratory; the pouch can be inoculated with the specimen at the time of collection; and (3) the system is inexpensive to manufacture and purchase. Another excellent system, In Pouch TV test, containing liquid medium, is also available.[8a] Positive control strains should be used each time patient material is cultured.

More sensitive and specific methods using monoclonal antibodies are also available. The specimen is smeared onto a glass slide, allowed to air dry, and then transported to the laboratory be-

PROCEDURE 45.6

EXAMINATION OF CELLOPHANE* TAPE PREPARATIONS

PRINCIPLE
The female adult pinworm deposits her eggs during the night on the surface of the skin surrounding the anus. The eggs adhere to the sticky surface of clear cellophane tape, on which they can be visualized microscopically.

METHOD
1. Place a strip of cellophane tape on a microscope slide, starting ½ inch (1.5 cm) from one end and, running toward the same end, continuing around this end across the slide; tear off the strip even with the other end. Place a strip of paper, ½ × 1 inch (1.5 × 2.5 cm), between the slide and the tape at the end where the tape is torn flush.

2. To obtain the sample from the perianal area, peel back the tape by gripping the label, and, with the tape looped adhesive side outward over a wooden tongue depressor held against the slide and extended about 1 inch (2.5 cm) beyond it, press the tape firmly against the right and left perianal folds.

3. Spread the tape back on the slide, adhesive side down.

4. Place name and date on the label.
NOTE: *Do not use Magic transparent tape; use regular clear cellophane tape.*

5. Lift one side of the tape and apply 1 small drop of toluene or xylene; press the tape down onto the glass slide.

6. The tape is now cleared; examine under low power and low illumination. The eggs should be visible if present; they are described as football shaped with one slightly flattened side.

QUALITY CONTROL
The laboratory should retain control slides with known pinworm eggs for comparison and should participate in proficiency testing programs that include unknown parasites.

*Refers to cellulose (Scotch) tape.

PROCEDURE 45.7

IDENTIFICATION OF ADULT TAPEWORMS

PRINCIPLE
The uterus of gravid tapeworm proglottids (*Taenia* sp.) can be injected with India ink for visualization of the branches; the number of branches determines the species of tapeworm.

METHOD
1. Using a 1 ml syringe and a 25- to 26-gauge needle,

inject India ink into the central stem or into the uterine pore, filling the uterine branches with ink.

2. Press the proglottid between two slides, hold it up to the light, and count the branches (Figure 45.71). (Besides Permount, Euparol is another mounting medium that can be used for

tapeworm proglottids.)
NOTE: *Caution should be used in handling proglottids of Taenia sp., since the eggs of T. solium are infective for humans.*

fore testing by fluorescent methodology (Figure 45.6).[14]

SPUTUM (PARAGONIMIASIS)

When sputum is submitted for examination, it should be "deep sputum" from the lower respiratory passages, not a specimen that is mainly saliva. The specimen should be collected early in the morning (before eating or brushing teeth) and immediately delivered to the laboratory. Sputum is usually examined as a saline or iodine wet mount under low and high-dry microscope power. If the quantity is sufficient, the formalin-ether sedimentation technique can be used. A very mucoid or thick sputum can be centrifuged

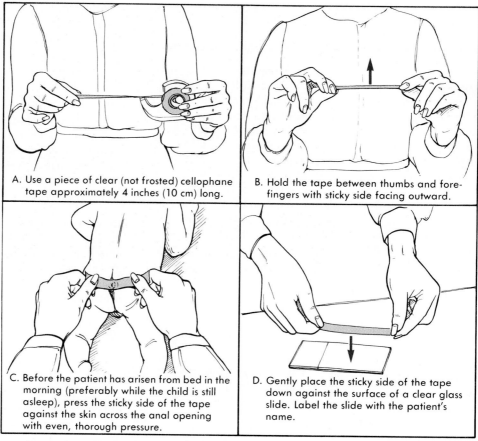

A. Use a piece of clear (not frosted) cellophane tape approximately 4 inches (10 cm) long.

B. Hold the tape between thumbs and forefingers with sticky side facing outward.

C. Before the patient has arisen from bed in the morning (preferably while the child is still asleep), press the sticky side of the tape against the skin across the anal opening with even, thorough pressure.

D. Gently place the sticky side of the tape down against the surface of a clear glass slide. Label the slide with the patient's name.

FIGURE 45.4

Method for collection of a cellulose (Scotch) tape preparation for pinworm diagnosis. This method dispenses with the tongue depressor (Procedure 45.6), requiring only tape and a glass microscope slide. The tape must be pressed deep into the anal crack.

after the addition of an equal volume of 3% sodium hydroxide. With any technique, the sediment should be carefully examined for the presence of brownish spots or "iron filings," which may be *Paragonimus* eggs.

NOTE: Care should be taken not to confuse *Entamoeba gingivalis*, which may be found in the mouth and might be seen in the sputum, with *E. histolytica* from a pulmonary abscess. *E. gingivalis* contains ingested polymorphonuclear neutrophils (PMNs); *E. histolytica* does not.

ASPIRATES

The diagnosis of certain parasitic infections may be based on procedures using aspirated material. These techniques include microscopic examination, animal inoculation, and culture.

Examination of aspirated material from lung or liver abscesses may reveal the presence of *Entamoeba histolytica*; however, the demonstration of these parasites is often extremely difficult for several reasons. Hepatic abscess material taken from

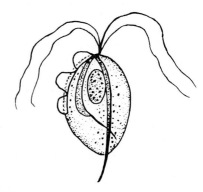

FIGURE 45.5

Trichomonas vaginalis trophozoite. (Illustration by Nobuko Kitamura.)

the peripheral area, rather than the necrotic center, may reveal organisms, although they may be trapped in the thick pus and not exhibit any motility. The Amoebiasis Research Unit, Durban, South Africa, has recommended using proteolytic enzymes, such as 10 units of streptodornase per milliliter of thick pus, to free the organisms from

FIGURE 45.6

Fluorescein-conjugated, monoclonal antibody–stained *Trichomonas vaginalis* in vaginal discharge. (Courtesy of Meridian Diagnostics, Cincinnati.)

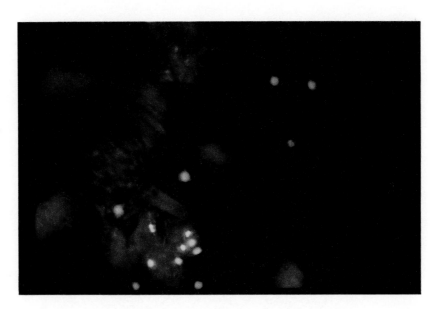

the aspirate material.[32] The enzyme is incubated with the specimen for 30 minutes at 37° C with repeated shaking, the suspension is centrifuged, and the sediment is examined as for other direct examination methods.

Aspiration of cyst material, usually liver or lung, for the diagnosis of hydatid disease is usually performed when open surgical techniques are used for cyst removal. The aspirated fluid is submitted to the laboratory and examined for the presence of hydatid sand (scolices) or hooklets; the absence of this material does not rule out the possibility of hydatid disease, since some cysts are sterile (Figure 45.7).

Material from lymph nodes, spleen, liver, bone marrow, or spinal fluid may be examined for the presence of trypanosomes or leishmanial forms. Part of the specimen should be examined as a wet preparation to demonstrate motile organisms. Impression smears can also be prepared and stained with Giemsa stain (see later section on detection of blood parasites). This type of material can also be cultured (see later section on culture techniques).

Specimens obtained from cutaneous ulcers should be aspirated from below the ulcer bed rather than the surface; this type of sample is more likely to contain the intracellular leishmanial organisms and is free of bacterial contamination. A few drops of sterile saline may be introduced under the ulcer bed (through uninvolved tissue) by needle (25 gauge) and syringe (1 or 2 ml). The aspirated fluid should be examined as stained smears and should be inoculated into appropriate media (see culture techniques).

SPINAL FLUID

Cases of primary meningoencephalitis are infrequently seen, but the examination of spinal fluid may reveal the causative agent, *Naegleria fowleri* (Figure 45.8), if present. The spinal fluid may range from cloudy to purulent, with or without red blood cells. The cell count ranges from a few hundred to more than 20,000 white blood cells per milliliter, primarily neutrophils; the failure to find bacteria in this type of spinal fluid should alert one to the possibility of primary meningoencephalitis. Motile amebae may be found in unstained spinal fluid; however, one should be very careful not to confuse organisms with various blood and tissue cells that may also be motile, particularly if the spinal fluid is in a counting chamber.

The classification, transmission, virulence, and disease pathogenesis of the free-living amebic genera *Naegleria* and *Acanthamoeba* are receiving considerable attention at this time.[1,27,45,54,77] The question of increasing infections has not been answered, although studies of thermally polluted or enriched waters indicate the presence of such amebae in these situations. Studies have also shown that these organisms have been recovered from asymptomatic human carriers (nasal passages, nasopharyngeal secretions) and have been implicated in chronic or subacute meningoencephalitis and acute primary amebic meningoencephalitis.

BIOPSY MATERIAL

In some patients, biopsy material may be used to confirm the diagnosis of certain parasitic infec-

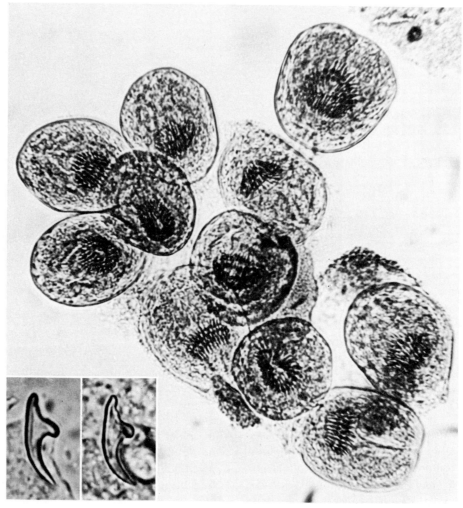

FIGURE 45.7

Echinococcus granulosus, hydatid sand (300×). *Inset,* Two individual hooklets (1000×).

FIGURE 45.8

Naegleria in brain tissue. Hematoxylin and eosin stain.

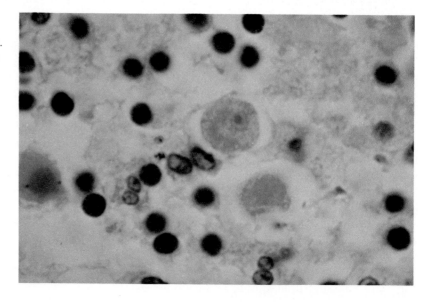

tions. Most of these specimens are sent for routine tissue processing (fixation, embedding, sectioning, staining). However, fresh material may be sent directly to the laboratory for examination; it is imperative that these specimens be received immediately to prevent deterioration of any organisms present. *Pneumocystis carinii* is recognized as an important cause of pulmonary infection in patients who are immunosuppressed as a result of therapy or in patients with congenital or acquired immunological disorders, including acquired immunodeficiency syndrome (AIDS).[59] It is also becoming more widely recognized that *Pneumocystis* can present as disseminated disease found in various tissues, particularly in severely compromised patients.[32] The organisms can be demonstrated in stained impression smears of lung material obtained by open, transbronchial, or brush biopsy. *Pneumocystis* can be seen in stained smears of tracheobronchial aspirates, although preparations of lung tissue are more likely to reveal the organisms. Bronchoalveolar lavage has a lower yield than transbronchial biopsy, but it is helpful when biopsy is contraindicated. The usefulness of induced sputum has been evaluated, and it is becoming more widely used, particularly when obtained from patients with AIDS and when coupled with the monoclonal antibody diagnostic procedure (Figure 45.9).[8,39,49,80] Sputum specimens are generally considered unacceptable for the detection of *Pneumocystis*.

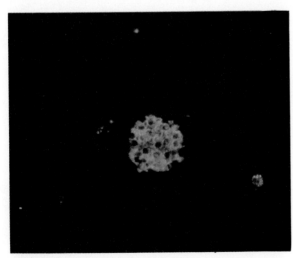

FIGURE 45.9

Cluster of *Pneumocystis carinii* cysts and trophozoites from a bronchoalveolar lavage specimen, stained with a monoclonal antibody–fluorescent stain (Merifluor *Pneumocystis,* Meridian Diagnostics). Note the "honeycomb" appearance caused by peripheral staining of cysts. (Courtesy of Meridian Diagnostics, Cincinnati.)

Regardless of the technique used, multiple specimens may have to be examined to confirm the presence of *P. carinii* organisms. Although several different stains are available, the stain most often recommended is Gomori's methenamine silver (GMS), which clearly stains *Pneumocystis* organisms in dark brown or black (Figure 45.10, *A*). Preparation of the silver stain is relatively complex (Procedure 45.8), and positive control slides should be included with each specimen to ensure accurate identification of the smears. Although the silver stain yields the most definitive results, toluidine blue O stains are much easier and faster to perform and yield reliable results in the hands of skilled technologists (Figure 45.10, *B*). One modified toluidine blue O method developed at the National Institutes of Health is described in Procedure 45.9. Although not as sensitive as the monoclonal antibody procedures, a rapid method for the detection of *Pneumocystis* cysts is the use of calcofluor white (CFW). This is a chemofluorescent dye with an affinity for the polysaccharide polymers of amebic cysts.[81a]

Cryptosporidium sp. are coccidial protozoa usually found in the intestinal tract. Respiratory cryptosporidiosis has also been reported with infection of the gallbladder and biliary tree. Human cryptosporidiosis has been confirmed from gastrointestinal biopsies and stool from patients whose immune responses were diminished, including AIDS patients, as well as those with normal immune systems. Cryptosporidia differ from other coccidia found in the intestine in that they are limited to the microvillus layer of gastrointestinal cells. Tissue diagnosis has been based on electron microscopy (EM) studies and Giemsa-stained smears from jejunal biopsy material and routine tissue processing (paraffin embedding and sectioning). Currently, more than 1000 published papers appear in the literature related to cryptosporidiosis, including information on nosocomial transmission, day-care center outbreaks, and waterborne outbreaks.[19] This is remarkable, since before 1980 less than 30 publications were available. This organism has definitely become more widely recognized as a potential pathogen in both immunosuppressed and immunocompetent hosts, and other diagnostic procedures have been developed to facilitate organism recovery and identification, that is, modified acid-fast staining methods of stool (modified Kinyoun's stain; see Procedure 45.5 and later section on coccidia). The organisms may also be visible as refractile bodies in smears made directly or from concentrated feces.

Biopsy specimens from the intestinal tract, particularly in AIDS patients, may reveal mi-

FIGURE 45.10

A, Cysts of *Pneumocystis carinii* in bronchial lavage, stained with modified Gomori methenamine silver stain (1000×). **B,** Toluidine blue O–stained cysts of *P. carinii* in smear made from centrifuged sediment from bronchial lavage specimen (1000×).

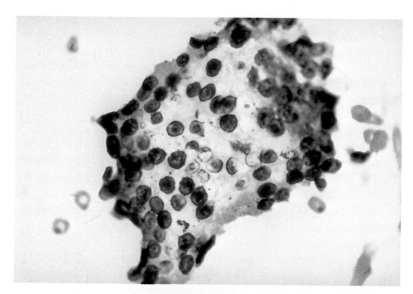

A

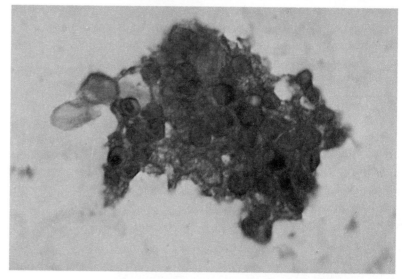

B

crosporidian organisms, although electron microscopy (EM) may be necessary for confirmation. Techniques that do not require tissue embedding or EM are now becoming more popular. Touch preparations of fresh biopsy material, which are air dried, methanol fixed, and Giemsa stained, have been used; however, screening must be performed at 1000× magnification. Cytocentrifugation followed by Giemsa staining has also been used; however, all results were confirmed using EM. Another more recent approach recommends a modified trichrome stain in which the chromotrope 2R component being added to the stain is 10 times the concentration normally used in the routine trichrome stain for stool.[86] It is important to remember that the stool preparations must be *very* thin, the staining time is 90 minutes, and the slide must be examined at 1000× magnifi-

cation (or higher). Unfortunately, many objects within stool material are oval, stain pinkish with trichrome, and measure approximately 1.5 to 3µm. If this stain is used for the identification of microsporidia in stool, positive control material should be available for comparison.

The newest approach for the identification of spores in clinical specimens uses antiserum in an indirect fluorescent antibody (IFA) procedure.[93] Fluorescing microsporidial spores were distinguished by a darker cell wall and internal visualization of the polar tubule as diagonal lines or cross-lines within the cell. In another study using this same antiserum, 9/27 (30%) patients who had already been diagnosed as having cryptosporidiosis (eight with AIDS, one without AIDS but immunodeficient), were found to have *Enterocytozoon bieneusi* or *Septata intestinalis* in the

PROCEDURE 45.8

GOMORI METHENAMINE SILVER (GMS) STAIN (80° C METHENAMINE–SILVER NITRATE STAIN)

PRINCIPLE

Hot, acidified metal ions will precipitate on the cell walls of parasites and fungi, rendering them visible in specimen material.

METHOD

1. Prepare reagents as follows:
   a. Chromic acid (5% solution)
      Chromic acid (Fisher Scientific)   50 g
      Distilled deionized water   1000 ml
      Weigh out 50 g chromic acid and dissolve in 1 L distilled water. Solution usable for up to 1 year.
   b. Methenamine (3% solution)
      Hexamethylenetetramine (Fisher Scientific)   12 g
      Distilled deionized water   400 ml
      Weigh out 12 g hexamethylenetetramine and dissolve in 400 ml distilled water. Solution usable for up to 6 months.
   c. Silver nitrate (5% solution)
      Silver nitrate (Fisher Scientific)   5 g
      Distilled deionized water   100 ml
      Weigh out 5 g silver nitrate and dissolve in 100 ml distilled water. Solution usable for up to 6 months.
   d. Stock methenamine–silver nitrate
      Methenamine (3% solution)   400 ml
      Silver nitrate (5% solution)   20 ml
      A white precipitate forms when reagents are added together but immediately clears with mixing. Store

at 4° C. Solution usable for up to 1 month.
   e. Borax (5% solution)
      Borax (sodium borate) (Fisher Scientific)   5 g
      Distilled deionized water   100 ml
      Dissolve sodium borate in distilled water. Solution usable for up to 1 year.
   f. Sodium bisulfite (1% solution)
      Sodium bisulfite (Fisher Scientific)   10 g
      Distilled deionized water   1000 ml
      Weigh out sodium bisulfite and add to distilled water. Solution usable for up to 1 year.
   g. Gold chloride (0.2% solution)
      Gold chloride (chloroauric acid; Fisher Scientific)   0.5 g
      Distilled deionized water   250 ml
      Weigh out gold chloride and add to distilled water. Solution usable for up to 1 year.
   h. Sodium thiosulfate (2% solution)
      Sodium thiosulfate (Fisher Scientific)   20 g
      Distilled deionized water   1000 ml
      Solution usable for up to 1 year.
   i. Stock light-green solution
      Light-green SF-Yellowish (Harleco)   0.2 g
      Distilled deionized water   100 ml

Glacial acetic acid   0.2 ml
      Solution usable for up to 1 year.
   j. Working light-green solution
      Stock light-green solution   10 ml
      Distilled deionized water   40 ml
      Solution usable for up to 1 month.
   k. Fill two Coplin jars each with absolute ethanol and 95% ethanol. These jars may be sealed with screw caps, and the reagents can be reused until they become cloudy.
   l. Fill two Coplin jars with xylene. These jars may be sealed, and the xylene can be reused.
2. Immediately before each staining procedure, prepare working methenamine–silver nitrate as follows:
   Stock methenamine–silver nitrate   25 ml
   Distilled deionized water   25 ml
   Borax (5% solution)   2 ml
   This solution must be prepared fresh in a Coplin jar for each set of slides stained. It is not reusable, even for a second staining procedure performed immediately afterward.
3. Staining procedure
   a. Turn on water bath, allowing 45 to 60 min to reach 80° C.
   b. Prepare smears and air dry. Heat-fix all smears at 70° C for 10 min on a heating block.
   c. Fix smears in absolute *methanol* for a minimum of 3 min.

PROCEDURE 45.8

GOMORI METHENAMINE SILVER (GMS) STAIN (80° C METHENAMINE–SILVER NITRATE STAIN)—cont'd

d. Place two Coplin jars, one containing 5% chromic acid (approximately 50 ml) and the other containing working methenamine solution into 80° C water bath for 5 min.

e. Place smears into heated 5% chromic acid. Incubate in water bath at 80° C for 2 min.

f. Remove slides from chromic acid and place in Coplin jar with distilled water. Rinse slides in three changes of distilled water.

g. Place slides in 1% sodium bisulfite for 30 s.

h. Rinse in three changes of distilled water.

i. Place slides in heated methenamine working solution. Incubate in water bath at 80° C for 4 ½ min.

j. Place slides in Coplin jar with distilled water. Rinse in three changes of distilled water.

k. Wipe the back of slides with a paper towel to remove excess methenamine.

l. Tone in 0.2% gold chloride for 30 s; background lightens in this step.

m. Rinse in three changes distilled water.

n. Place slides in 2% sodium thiosulfate for 30 s to remove unreduced silver.

o. Rinse in three changes of distilled water.

p. Counterstain in light-green working solution for 30 s.

q. Rinse in three changes of distilled water.

r. Dehydrate and clear for 30 s intervals in two changes each of 95% ethanol, 100% ethanol, and xylene, respectively.

s. Place coverslips on slides with Permount while the slides are still wet with xylene. Slides are ready to be read.

4. Interpretation: Background is green, and *Pneumocystis* cysts appear gray to black. They are spherical, may be punched in and cup shaped, and may show black parentheses-like structures in the center (Figure 45.10, *A*). The cysts tend to occur in groups.

QUALITY CONTROL

Specimens from positive patients should be smeared onto numerous slides, which are fixed and stored. One positive control slide should be stained along with each batch of patient slides. The proper appearance of cysts on the control slide is a requirement for interpretation and reporting of the patient slides.

Modified by the UCLA parasitology laboratory from Pintozzi, R. 1978. J. Clin. Pathol. 31:803.

stool.[11a,37] Although some cross-reactivity occurs with bacteria, this technique offers a more sensitive approach than routine staining methods currently available for the examination of stool specimens. As clinicians begin to suspect these infections and laboratorians become more familiar with the diagnostic methods, the number of positive patients may increase dramatically, particularly those who are immunocompromised.

Corneal scrapings or biopsy specimens can be examined by routine histology, CFW stain, or culture for the presence of *Acanthamoeba*.* Although these infections are relatively rare, the medical community has become much more aware of *Acanthamoeba* keratitis, especially among contact lens wearers, during the last few years.†

Skin biopsy specimens for the diagnosis of cutaneous amebiasis or cutaneous leishmaniasis should be submitted for tissue processing. A portion of the tissue for the diagnosis of leishmaniasis can be teased apart with sterile needles and inoculated into appropriate culture media (see later section on culture techniques).

The diagnosis of onchocerciasis (*Onchocerca volvulus*) may be confirmed by the examination of "skin snips," very thin slices of skin, which are teased apart in saline to release the microfilariae.

*References 1, 32, 45, 54, 77, 90.

†References 1, 3, 15, 16, 24, 27, 45, 51, 54, 62, 77.

PROCEDURE 45.9

MODIFIED TOLUIDINE BLUE O STAIN FOR *PNEUMOCYSTIS CARINII*

PRINCIPLE
The thick cyst wall of parasites and the cell wall of some fungi, after sulfation treatment, will absorb and retain toluidine blue O stain.

METHOD
1. Prepare the specimen as follows:
   a. For tissue, make touch preparations on clean slides by touching the cut surface of tissue to the slide and squeezing the tissue down to express infectious agents. Allow the smears to air dry.
   b. For sputum, if thin and watery, simply make thin smears. If thick and mucuslike, combine with equal amount of Sputolysin (diluted 1:10 in water), vortex, and allow to stand 30 min before revortexing and preparing thin smears.
   c. For bronchoalveolar lavage or washing, centrifuge for at least 10 min at 3000 rpm. Pour off supernatant. Prepare slides from sediment; if thick, treat as sputum. If possible, choose mucoid-appearing flecks.

2. Prepare solution A, sulfation reagent:
   a. Add 15 ml 1 N NaOH to 35 ml ether (USP) in a separating funnel and shake vigorously for 1 min. Release the stopcock occasionally to relieve pressure. Drain the NaOH phase out the bottom of the funnel. Repeat this procedure once again to remove impurities and preservatives from the ether.
   b. Add 10 ml distilled water to the 35 ml ether (USP) in the separating funnel and shake vigorously for 1 min. Release the stopcock occasionally to relieve pressure. Repeat this procedure once again. The ether should be fully saturated with water at this point. Allow the two solutions to separate and discard the water layer (the lower phase) by allowing it to run out the bottom of the funnel. Allow a few milliliters of the ether phase also to run out, to ensure that no water is left.
   c. Transfer the remaining 25 ml of water-saturated ether to a 100 ml Erlenmeyer flask immersed in an ice water bath. Allow to cool for 5 min. Place the ice bath under a chemical fume hood. Slowly add 25 ml concentrated sulfuric acid while continuously shaking the flask. Mix contents well and pour the reagent into a Coplin jar.
   d. Store the solution at room temperature and prepare fresh weekly. The jar is sealed with stopcock grease to prevent evaporation. After 10 slides have been stained, discard the reagent and use fresh reagent.

3. Prepare solution B, toluidine blue O:
   a. In a Coplin jar under a fume hood, place 0.06 g toluidine blue O (Fisher Scientific Co., Aldrich, and others) and add 15 ml distilled water. Swirl vigorously to dissolve the dye in the water. Add, in order:
      Hydrochloric acid (concentrated) 0.5 ml
      Absolute ethanol 35 ml
   b. Screw the cap of the Coplin jar tightly closed. The stain solution can be kept at room temperature and reused for 1 mo. Add fresh stain when the level drops.

4. Fix slides for 1 min in absolute *ethanol.* Allow to dry.

5. Place slides in solution A for 10 min, dipping them twice when they are initially immersed. Stir with a glass stirring rod after each 5 min period.

6. Place them into a Coplin jar filled with tap water. Allow the jar to sit under running cold tap water for 5 min.

7. Place slides into solution B for 3 min.

8. Dehydrate slides by dipping them for approximately 10 s into a jar of 95% alcohol or until they are clean.

9. Dip slides into a jar of absolute ethanol for another 10 s for further decolorizing.

Continued

PROCEDURE 45.9

MODIFIED TOLUIDINE BLUE O STAIN FOR *PNEUMOCYSTIS CARINII*—cont'd

10. Clear slides by dipping them into a jar of xylene (10 s).

11. Mount coverslips with Permount or similar mounting medium. Slides can be examined immediately.

12. Observe for cysts of *Pneumocystis,* which appear reddish blue or purple

against a blue background (Figure 45.10, *B*). The organisms are often seen in clumps, particularly in areas of frothy looking material. If there is no difference in color between cysts and background, try using toluidine blue from another source. Cysts are occasionally punched in, appearing crescent shaped.

QUALITY CONTROL

Stain a known positive control slide with each patient specimen run.

Modified by Beverly Tenenbaum, North Shore University, Manhasset, N.Y., from Witebsky, F.G., et al. 1988. J. Clin. Microbiol. 26:774.

Biopsy specimens taken from lymph nodes are submitted for routine tissue processing; impression smears can also be prepared and stained with Giemsa stain (see later section on detection of blood parasites). The diagnosis of trichinosis is usually based on clinical findings; however, confirmation may be obtained by the examination of a muscle biopsy (Figure 45.11). The encapsulated larvae can be seen in small pieces of fresh tissue, which are pressed between two slides and examined under low power of the microscope. At necropsy the larvae are most abundant in the diaphragm, masseter muscle, or tongue. Larvae can also be recovered from tissue that has undergone digestion in artificial digestive fluid at 37° C. Figure 45.12 shows the life cycle of this organism.

Tapeworm larvae may occasionally be recovered from a muscle specimen and should be carefully dissected from the capsules. They should then be pressed between two slides and examined under low power for the presence of a scolex with four suckers and a circle of hooks. If no hooks are present, it may be a species other than *Taenia solium.*

In some patients with schistosomiasis, the eggs may not be recovered in the stool or urine; however, examination of the rectal or bladder mucosa may reveal eggs of the appropriate species. The mucosal tissue should be compressed between two slides and examined under low power and decreased illumination. The eggs should be carefully examined to determine viability (see earlier discussion of schistosome eggs). Small pieces of tissue may be digested with 4% sodium hydroxide for 2 to 3 hours at 60° to 80° C.

The eggs, which are recovered by sedimentation or centrifugation, can be examined under the microscope.

CULTURE TECHNIQUES

Most clinical laboratories do not provide culture techniques for the diagnosis of parasitic organisms; however, the lack of culture procedures should not prevent the correct identification of most parasites. Isolation of intestinal amebae yields a higher number of positive results, provided fresh specimens are received by the laboratory within specified time limits. Most of the intestinal amebae do not culture as well as *Entamoeba histolytica;* however, once the organisms are established in culture, they must be speciated on the basis of morphology. Accurate identification of the organisms can be determined by the examination of permanent stained smears of culture sediment material, although the morphology may not appear typical from stained culture sediment.

Many different media have been developed for the culture of protozoan organisms (some of which are available commercially), and specific directions for their preparation are available in the literature.*

ANIMAL INOCULATION

Most laboratories have neither the time nor the facilities for animal care to provide animal inoculation procedures for the diagnosis of parasitic infections. Host specificity for many parasites also

---

*References 4, 26, 32, 43, 57, 58.

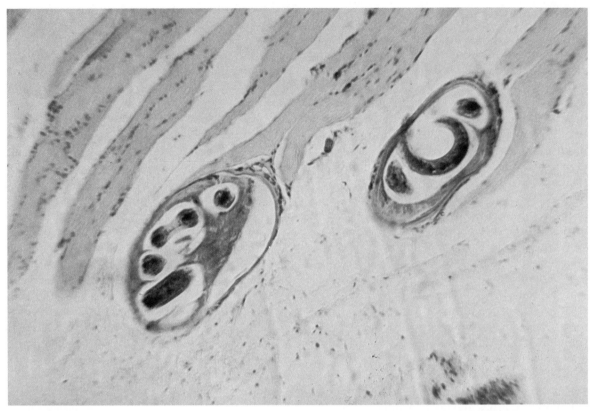

FIGURE 45.11

Trichinella spiralis larvae encysted in muscle.

FIGURE 45.12

Life cycle of *Trichinella spiralis.*

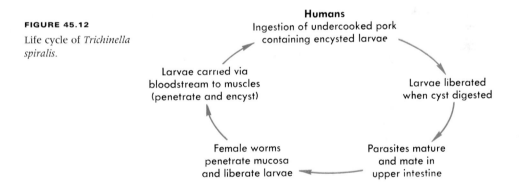

limits the types of laboratory animals available for these procedures. Occasionally, animal studies may be requested; included here are several procedures that can be used.

The hamster is the animal of choice for inoculation procedures designed to recover leishmanial organisms. After intraperitoneal or intratesticular inoculation, the infection may develop very slowly over several months; in some cases a generalized infection develops more quickly, and the animal may die in several days. Splenic and testicular aspirates should be examined for the presence of intracellular organisms; stained smears

should be prepared and carefully examined with the oil immersion lens.

Mice are generally used for the isolation of *Toxoplasma gondii,* although most cases are diagnosed on clinical and serologic findings. Mice inoculated through the peritoneum develop a fulminating infection that leads to death within a few days. Organisms can be easily recovered from the ascitic fluid and should be examined as stained smears. Giemsa stain is recommended for both intraperitoneal and intratesticular inoculation studies (specific staining techniques are discussed later).

SERODIAGNOSIS

Although serologic procedures for the diagnosis of parasitic diseases have been available for many years, they are generally not performed by most clinical laboratories. The procedures vary in both sensitivity and specificity and at times may be difficult to interpret. The Centers for Disease Control (CDC) offers several serologic procedures for diagnostic purposes, some of which are still in the experimental stages and not available elsewhere. (For a discussion of specific procedures, see Chapter 13). Although commercial antigens and diagnostic kits are available, results from different reagents have been variable.[44] Developmental work using monoclonal reagents is in progress, and some commercially prepared kits and components are now available.[32]

The enzyme-linked immunosorbent assay (ELISA) procedure is presently being evaluated for various parasitic infections. A significant contribution of the ELISA method will be detection of antigen in the body fluids of patients with parasitic diseases. With purified antigens, the usefulness of techniques such as radioimmunoassay (RIA), ELISA, FIAX, and defined antigen substrate spheres (DASS) will be greatly expanded. At present, immunodiagnostic procedures are most widely used for amebiasis, toxoplasmosis, leishmaniasis, Chagas' disease, trichinosis, schistosomiasis, cysticercosis, and hydatid disease.

DETECTION OF BLOOD PARASITES

LABORATORY DIAGNOSIS OF MALARIA

Malaria is one of the few acute parasitic infections that can be life-threatening. For this reason, any laboratory that offers this type of diagnostic service must be willing to provide technical expertise on a 24-hour basis, 7 days per week.

The definitive diagnosis of malaria is based on the demonstration of the parasites in the blood. Two types of blood films are used. The "thick smear" (Figure 45.13) allows the examination of a larger amount of blood and is used as a screening procedure; the "thin film" allows species identification of the parasite.

Blood films are usually prepared when the patient is admitted; samples should be taken at intervals of 6 to 18 hours for at least 3 successive days. From 200 to 300 microscopic fields should be examined before a film is signed out as negative. If possible, the smears should be prepared from blood obtained from the finger or earlobe; the blood should flow freely. If patient contact is not possible and the quality of the submitted slides may be poor, a tube of fresh blood should be requested (ethylenediaminetetraacetic acid [EDTA] anticoagulant is recommended), and smears should be prepared immediately after the blood is received. It is even better to use the blood remaining in the needle from a venipuncture for smear preparation because this blood has not been in contact with any anticoagulant.

To prepare the thick film, place 2 or 3 small drops of fresh blood on an alcohol-cleaned slide. With the corner of another slide, and using a circular motion, mix the drops and spread the blood over an area about 2 cm in diameter. Continue stirring for about 30 seconds to prevent formation of fibrin strands, which may obscure the parasites after staining. If the blood is too thick or any grease remains on the slide, the blood will flake off during staining. Allow the film to air dry (room temperature) in a dust-free area. Never apply heat to a thick film because heat will fix the blood, causing the red blood cells to remain intact during staining; the result is stain retention and subsequent inability to identify any parasites present.

The thin blood film is used primarily for specific parasite identification, although the number of organisms per field is much reduced compared with the thick film. The thin film is prepared in exactly the same manner used for the differential blood count. After the film has air dried (do not apply heat), it may be stained. The necessity for fixation before staining depends on the stain selected.

STAINING BLOOD FILMS

For accurate identification of blood parasites, it is very important that a laboratory develop proficiency in the use of at least one good staining method. As a general rule, blood films should be stained as soon as possible, since prolonged storage results in stain retention. The stains used are generally of two types. One has the fixative in combination with the staining solution so that both fixation and staining occur at the same time. Wright stain is an example of this type of staining solution. Giemsa stain represents the other type of staining solution, in which the fixative and stain are separate; thus the thin film must be fixed before staining (Procedures 45.10 and 45.11). Methods are also available for the identification of malarial parasites with the use of acridine orange and other deoxyribonucleic acid (DNA)–binding dyes, with smears examined using fluorescence microscopy. The QBC tube is available from Becton-Dickinson; this is a qualitative screening

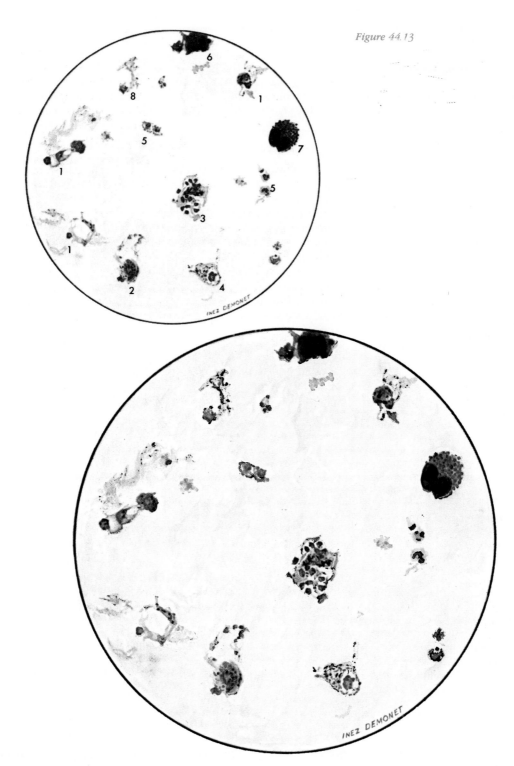

Figure 44.13

FIGURE 45.13

Plasmodium vivax in thick smear. **1,** Ameboid trophozoites. **2,** Schizont, two divisions of chromatin. **3,** Mature schizont. **4,** Microgametocyte. **5,** Blood platelets. **6,** Nucleus of neutrophil. **7,** Eosinophil. **8,** Blood platelet associated with cellular remains of young erythrocytes. (From Wilcox, A. 1960. Manual for the microscopical diagnosis of malaria in man. Public Health Service, Washington, D.C.)

PROCEDURE 45.10

STAINING THIN FILMS

PRINCIPLE
By spreading the blood cells in a thin layer, the size of red cells, inclusions, and extracellular forms can be more easily visualized.

METHOD
1. Fix blood films in absolute methanol (acetone-free) for 30 s.

2. Allow slides to air dry.

3. Immerse slides in a solution of 1 part Giemsa stock (commercial liquid stain or stock prepared from powder) to 10 to 50 parts of Triton-buffered water (pH 7.0 to 7.2). Stain 10 to 60 min (see note). Fresh working stain should be prepared from stock solution each day.

4. Dip slides briefly in Triton X-100–buffered water.

5. Drain thoroughly in vertical position and allow to air dry.

NOTE: *A good general rule for stain dilution versus staining time is that if dilution is 1:20, stain for 20 min; if 1:30, stain for 30 min; and so forth.* However, a series of stain dilutions and staining times should be tried to determine the best dilution/time for each batch of stock stain.

EXPECTED RESULTS
Giemsa stain colors the components of blood as follows: erythrocytes, pale gray-blue; nuclei of white blood cells, purple and pale purple cytoplasm; eosinophilic granules, bright purple-red; neutrophilic granules, deep pink-purple. Parasitic forms are blue to purple, with reddish nuclei. Their characteristic morphologies are used for differentiation. Inexperienced workers may confuse platelets with parasites.[89]

QUALITY CONTROL
Known positive blood films should be stained periodically along with patient specimens. The laboratory should maintain control slides for comparison and participate in a proficiency testing program for blood parasites.

PROCEDURE 45.11

STAINING THICK FILMS

PRINCIPLE
A large amount of blood can be examined for parasitic forms by lysing the red cells and staining for parasites. The lack of methanol fixation allows lysis of red cells by the aqueous stain solution. Although parasites can be found in the larger volume of blood, definitive morphological criteria necessary for specific organism identification may be more difficult to see.

METHOD
The procedure to be followed for thick films is the same as for thin films, except that the first two steps are omitted. If the slide has a thick film at one end and a thin film at the other, fix only the thin portion, and then stain both parts of the film simultaneously.

QUALITY CONTROL
Follow the same program as outlined for Procedure 45.10.

method for rapidly detecting the presence of malaria and other blood parasites in centrifuged capillary or venous blood. Although parasites are detected by observing fluorescence of acridine orange within the tube, speciation requires examination of routine stained blood films.[32]

When slides are removed from either type of staining solution, they should be dried in a vertical position. After being air dried, they may be examined under oil immersion by placing the oil directly on the uncovered blood film.

Specimens may be submitted from patients with *Plasmodium falciparum* infections who do not yet have gametocytes in the blood. Consequently, a low-level parasitemia with delicate ring forms might be missed without extensive oil immersion examination of the blood films (at least 200 to 300 oil immersion fields).

NOTE: Remember that automated differential instruments used in hematology laboratories are not designed to recognize intracellular (red blood cell) parasites. Any suspected parasitic infection or presumptive diagnosis of fever of unknown origin mandates a manual blood smear examination.[32]

Many studies have used immunodiagnostic procedures for the diagnosis of malaria; however, these procedures are not routinely performed in most laboratories. Sulzer and Wilson[82] have reported the use of thick-smear antigens prepared from washed parasitized blood cells. This type of antigen is used in the IFA procedure, which has a 95% sensitivity and a false-positive rate of 1% at a titer of 1:16.[83] The indirect hemagglutination assay (IHA) has also been used and evaluated by various workers. The newly developed polymerase chain reaction has even been used for amplification and subsequent detection of malarial DNA in blood.

IDENTIFICATION OF ANIMAL PARASITES

INTESTINAL PROTOZOA

The protozoa are unicellular organisms, most of which are microscopic. They possess a number of specialized organelles, which are responsible for life functions and which allow further division of the group into classes.

The class Sarcodina contains the organisms that move by means of cytoplasmic protrusions called pseudopodia. Included in this group are free-living organisms, as well as nonpathogenic and pathogenic organisms found in the intestinal tract and other areas of the body.

The Mastigophora, or flagellates, contain specialized locomotor organelles called flagella: long, thin cytoplasmic extensions that may vary in number and position depending on the species. Different genera may live in the intestinal tract, the bloodstream, or various tissues. Detection of the blood- and tissue-dwelling flagellates is discussed in the previous section.

The class Ciliata contains species that move by means of cilia, short extensions of cytoplasm that cover the surface of the organism. This group contains only one organism that infects humans: *Balantidium coli* infects the intestinal tract and may produce severe symptoms.

Members of the class Sporozoa are found in the blood and other tissues and have a complex life cycle that involves both sexual and asexual generations. The four species of *Plasmodium*, the cause of malaria, are found in this group; their diagnosis is discussed in the previous section. Members of the genera *Isospora* and *Cryptosporidium* and

the microsporidia can be found in the intestinal mucosa. These organisms have been seen with increasing frequency in specimens from immunosuppressed patients, particularly those with AIDS.[17,64,65]

Isospora, *Cryptosporidium*, and *Cyclospora* sp. (formerly called CLBs)[67] are passed in the stool as oocysts; the other members of the protozoa exist in the intestinal tract in the trophozoite or cyst stages. With *Enterocytozoon* and *Septata*, two of the genera of microsporidia that can infect humans, spore stages in the life cycle are passed in the stool.[11a,22] Within the last few years, diagnostic methods have become available for the identification of microsporidial spores in stool specimens.[37,86,93] However, considering the spore size (approximately 1 to 4 μm), the clinical laboratory may have to wait until monoclonal antibodies are available commercially for detection of spores.[37,93] The coccidian parasite *Toxoplasma gondii* is acquired by humans via ingestion, although its life cycle includes stages in several animal hosts, particularly cats (Figure 45.14). The trophozoites may be seen in squash preparations from lymph nodes or brain tissue, usually examined by pathologists.

The important characteristics of the intestinal protozoa are found in Tables 45.3 to 45.8. The clinically important intestinal protozoa are generally considered to be *Entamoeba histolytica*, *Dientamoeba fragilis*, *Giardia lamblia*, *Isospora belli*, *Cryptosporidium*, *Cyclospora*, and *Balantidium coli*. *E. histolytica* is the most important species and may invade other tissues of the body, resulting in severe symptoms and possible death. *D. fragilis* has been associated with diarrhea, nausea, vomiting, and other nonspecific abdominal complaints. *G. lamblia* is probably the most common protozoan organism found in persons in the United States and is known to cause symptoms ranging from mild diarrhea, flatulence, and vague abdominal pains to acute, severe diarrhea to steatorrhea and a typical malabsorption syndrome. Various documented waterborne and foodborne outbreaks have occurred during the past several years, and the beaver has been implicated as an animal reservoir host for *G. lamblia*. It is speculated that other animals may be involved as well. With present improved culture techniques and the ability to harvest material for antigen, serologic tests for giardiasis have been developed for both antibody and antigen. An ELISA method has also been marketed.

Another organism, *Blastocystis hominis* (Figures 45.3 and 45.15), is also now considered to be

FIGURE 45.14

Life cycle of *Toxoplasma gondii.*

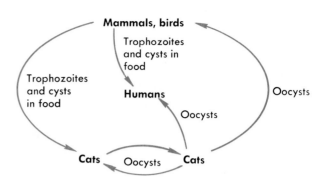

potentially pathogenic for humans. It has been re-classified as a protozoan, and there are reports of *B. hominis* as an agent of human enteric disease.[75,91,92] However, another report indicated that *B. hominis* is an incidental finding where other proven pathogens may be present in low numbers.[56] We recommend that it be reported and quantitated; this information may be valuable in helping assess the pathogenicity of the organism.[32]

Sarcocystis sp. appear in Table 45.8 but are not discussed in detail. According to the literature, ex-traintestinal human sarcocystosis is rare, with a much lower incidence than that seen with the in-testinal infection.

The identification of intestinal protozoan par-asites is difficult at best, and the importance of the permanent stained slide should be reemphasized (Procedure 45.3). It is important to remember that many artifacts (vegetable material, debris, cells of human origin) may mimic protozoan or-ganisms on a wet mount (Figure 45.15). The im-portant diagnostic characteristics are visible on the stained smear, and the *final identification of pro-tozoan parasites should be confirmed with the perma-nent stain.*

AMEBAE

Occasionally when fresh stool material is exam-ined as a direct wet mount, motile trophozoites may be observed. *Entamoeba histolytica* is described as having directional and progressive motility, whereas the other amebae tend to move more slowly and at random. The cytoplasm usually ap-pears finely granular, less frequently coarsely granular, or vacuolated. Bacteria, yeast cells, or debris may be present in the cytoplasm. The pres-ence of red blood cells in the cytoplasm is usually considered to be diagnostic for *E. histolytica* (Fig-ures 45.16 and 45.17).

Nuclear morphology is one of the most im-portant criteria used for identification; nuclei of the genus *Entamoeba* contain a relatively small

karyosome and have chromatin material arranged on the nuclear membrane (Figures 45.16 to 45.25). The nuclei of the other two genera, *En-dolimax* and *Iodamoeba,* tend to have very large karyosomes with no peripheral chromatin on the nuclear membrane (Figures 45.26 to 45.30).

The trophozoite stages may often be pleomor-phic and asymmetrical, whereas the cysts are usu-ally less variable in shape, with more rigid cyst walls. The number of nuclei in the cysts may vary, but their general morphology is similar to that found in the trophozoite stage. There are various inclusions in the cysts, such as chromatoidal bars or glycogen material, which may be helpful in identification.

Members of the *Acanthamoeba* and *Naegleria* genera may be identified from stained smears of culture material when specific organisms are sus-pected based on the original clinical specimen. Key characteristics are the typical pseudopods, *Naegleria* having lobed pseudopods and *Acan-thamoeba* having the spiky acanthapodia. Also, when organisms from the culture plates are placed in distilled water, *Naegleria fowleri* under-goes transformation within a few hours to a pear-shaped flagellate, usually with two flagella, occa-sionally with three or four flagella. The flagellate stage is a temporary nonfeeding stage and usually reverts back to the trophozoite stage.[85] Both gen-era of free-living amebae are also characterized by having the typical hexagonal double-walled cyst that can be seen using calcofluor white stain.[90]

FLAGELLATES

Four common species of flagellates are found in the intestinal tract: *Giardia lamblia, Chilomastix mesnili, Trichomonas hominis,* and *Dientamoeba frag-ilis* (Figures 45.31 to 45.38). Several other smaller flagellates, such as *Enteromonas hominis* and *Retor-tamonas intestinalis* (Figure 45.31), are rarely seen, and none of the flagellates in the intestinal tract, except for *G. lamblia* and *D. fragilis,* is considered pathogenic. *Trichomonas vaginalis* is pathogenic

Text continued on p. 816.

TABLE 45.3

MORPHOLOGICAL CRITERIA USED TO IDENTIFY INTESTINAL AMEBIC TROPHOZOITES

	ENTAMOEBA HISTOLYTICA	*ENTAMOEBA HARTMANNI*	*ENTAMOEBA COLI*	*ENDOLIMAX NANA*	*IODAMOEBA BÜTSCHLII*
Size (diameter or length)*	12-60 μm; usual range, 15-20 μm; invasive forms may be >20 μm	5-12 μm; usual range, 8-10 μm	15-50 μm; usual range, 20-25 μm	6-12 μm; usual range, 8-10 μm	8-20 μm usual range, 12-15 μm
Motility	Progressive with hyaline, fingerlike pseudopods; motility may be rapid	Usually nonprogressive	Sluggish, nondirectional, with blunt pseudopods	Sluggish, usually nonprogressive	Sluggish, usually nonprogressive
Number of nuclei	One; difficult to see in unstained preparations, usually not seen	One; usually not seen in unstained preparations	One; often visible in unstained preparations	One; occasionally visible in unstained preparations	One; usually not visible in unstained preparations
Nucleus: peripheral chromatin (stained)	Fine granules, uniform in size, usually evenly distributed; may have beaded appearance	May appear as solid ring rather than beaded; nucleus may stain more darkly than *E. histolytica*, although morphology is similar	May be clumped and unevenly arranged on membrane; may also appear as solid, dark ring with no beads or clumps	No peripheral chromatin	No peripheral chromatin
Karyosome (stained)	Small, usually compact; centrally located but may also be eccentric	Usually small and compact; may be centrally located or eccentric	Large, not compact; may or may not be eccentric; may be diffuse and darkly stained	Large, irregularly shaped; may appear "blotlike"; many nuclear variations common	Large, may be surrounded by refractile granules that are difficult to see
Cytoplasm: appearance (stained)	Finely granular, "ground-glass" appearance; clear differentiation of ectoplasm and endoplasm; if present, vacuoles usually small	Finely granular	Granular with little differentiation into ectoplasm and endoplasm; usually vacuolated	Granular, vacuolated	Coarsely granular; may be highly vacuolated
Inclusions (stained)	Noninvasive organism may contain bacteria; presence of red blood cells diagnostic	May contain bacteria; no red blood cells	Bacteria, yeast, other debris	Bacteria	Bacteria, yeast, other debris

* These sizes refer to wet preparation measurements. Organisms on a permanent stained smear may be 1 to 1.5 μm smaller because of artificial shrinkage.

TABLE 45.4

MORPHOLOGICAL CRITERIA USED TO IDENTIFY INTESTINAL FLAGELLATE TROPHOZOITES

	DIENTAMOEBA FRAGILIS	*TRICHOMONAS HOMINIS*	*GIARDIA LAMBLIA*	*CHILOMASTIX MESNILI*	*ENTEROMONAS HOMINIS*	*RETORTAMONAS INTESTINALIS*
Shape and size*	Shaped like amebae; 5-15 µm; usual range, 9-12 µm	Pear shaped; 8-20 µm; usual range, 11-12 µm	Pear shaped; 10-20 µm; usual range, 12-15 µm	Pear shaped; 6-24 µm; usual range, 10-15 µm	Oval; 4-10 µm; usual range, 8-9 µm	Pear shaped or oval; 4-9 µm; usual range, 6-7 µm
Motility	Usually nonprogressive; pseudopodia angular, serrated, or broad lobed and almost transparent	Jerky and rapid	"Falling leaf," organisms may be trapped in mucous; only flagella flutter seen	Stiff, rotary	Jerky	Jerky
Number of nuclei	Percentage may vary, but approximately 40% of organisms have 1 nucleus and 60% 2 nuclei; not visible in unstained preparations; no peripheral chromatin, karyosome composed of cluster of 4-8 granules	One; not visible in unstained mounts	Two; not visible in unstained mounts	One; not visible in unstained mounts	One; not visible in unstained mounts	One; not visible in unstained mounts
Number of flagella (usually difficult to see)	No visible flagella	3-5 anterior, 1 posterior	4 lateral, 2 ventral, 2 caudal	3 anterior, 1 in cytostome	3 anterior, 1 posterior	1 anterior, 1 posterior
Other features	Cytoplasm finely granular and may be vacuolated with ingested bacteria, yeasts, and other debris	Axostyle (slender rod) protrudes beyond posterior end and may be visible; undulating membrane extends length of body	Sucking disk occupying 1/3 to 1/2 of ventral surface; pear-shaped front view, spoon-shaped side view	Prominent cytostome extending 1/3 to 1/2 length of body; spiral groove across ventral surface	One side of body flattened; posterior flagellum extends free posteriorly or laterally	Prominent cytostome extending approximately 1/2 length of body

* These sizes refer to wet preparation measurements. Organisms on a permanent stained smear may be 1 to 1.5 µm smaller because of artificial shrinkage.

TABLE 45.5

MORPHOLOGICAL CRITERIA USED TO IDENTIFY INTESTINAL AMEBIC CYSTS

	ENTAMOEBA HISTOLYTICA	ENTAMOEBA HARTMANNI	ENTAMOEBA COLI	ENDOLIMAX NANA	IODAMOEBA BÜTSCHLII
Size*	10-20 μm; usual range, 12-15 μm	5-10 μm; usual range, 6-8 μm	10-35 μm; usual range, 15-25 μm	5-10 μm; usual range, 6-8 μm	5-20 μm; usual range, 10-12 μm
Shape	Usually spherical	Usually spherical	Usually spherical; occasionally oval, triangular, or other shapes; may be distorted on stained slide if fixation poor	Spherical, ovoidal, or ellipsoidal	Ovoidal, ellipsoidal, or other shapes
Number of nuclei	Mature cyst, 4; immature, 1 or 2 nuclei may be seen; nuclear characteristics difficult to see on wet preparation	Mature cyst, 4; immature, 1 or 2 nuclei may be seen; 2 nucleated cysts very common	Mature cyst, 8; occasionally 16 or more nuclei may be seen; immature cysts with 2 or more nuclei occasionally seen	Mature cyst, 4; immature cysts, 2 (very rarely seen and may resemble cysts of Enteromonas hominis)	Mature cyst, 1
Nucleus: peripheral chromatin (stained)	Peripheral chromatin present; fine, uniform granules, evenly distributed; nuclear characteristics may not be as clearly visible as in trophozoite	Fine granules evenly distributed on membrane; nuclear characteristics may be difficult to see	Coarsely granular and may be clumped and unevenly arranged on membrane; nuclear characteristics not as clearly defined as in trophozoite; may resemble E. histolytica	No peripheral chromatin	No peripheral chromatin
Karyosome (stained)	Small, compact, usually centrally located	Small, compact, usually centrally located	Large, may or may not be compact or eccentric; occasionally appears to be centrally located	Smaller than karyosome seen in trophozoite, but generally larger than those of genus Entamoeba	Large, usually eccentric refractile granules may be on one side of karyosome ("basket nucleus")
Cytoplasm: chromatoidal bodies (stained)	May be present; bodies usually elongate with blunt, rounded, smooth edges	Often present; bodies elongate with blunt, rounded, smooth edges	May be present (less frequently than E. histolytica); splinter shaped with rough, pointed ends	No chromatoidal bodies present; occasionally small granules or inclusions seen; also, fine linear structures may be faintly visible on well-stained smears	No chromatoidal bodies present; occasionally small granules may be present
Glycogen (stained)	May be diffuse or absent in mature cyst; clumped chromatin mass may be present in early cysts (stains reddish brown in iodine)	May or may not be present, as in E. histolytica	May be diffuse or absent in mature cysts; clumped mass occasionally seen in immature cysts (stains reddish brown in iodine)	Usually diffuse if present (stains reddish-brown in iodine)	Large, compact, well-defined mass (stains reddish brown in iodine)

* These sizes refer to wet preparation measurements. Organisms on a permanent stained smear may be 1 to 1.5 μm smaller because of artificial shrinkage.

TABLE 45.6

MORPHOLOGICAL CRITERIA USED TO IDENTIFY INTESTINAL FLAGELLATE CYSTS

	DIENTAMOEBA FRAGILIS	TRICHOMONAS HOMINIS	GIARDIA LAMBLIA	CHILOMASTIX MESNILI	ENTEROMONAS HOMINIS	RETORTAMONAS INTESTINALIS
Shape	No cyst	No cyst	Oval, ellipsoidal, or may appear round	Lemon shaped with anterior hyaline knob	Elongate or oval	Pear shaped or slightly lemon shaped
Size*			8-19 μm; usual range, 11-12 μm	6-10 μm; usual range, 8-9 μm	4-10 μm; usual range, 6-8 μm	4-9 μm; usual range, 4-7 μm
Number of nuclei			Four; not distinct in unstained preparations; usually located at one end	One, not visible in unstained preparations	One to four; usually 2 lying at opposite ends of cyst; not visible in unstained mounts	One; not visible in unstained mounts
Other features			Longitudinal fibers in cyst may be visible in unstained preparations; deep-staining fibers usually lie across longitudinal fibers; shrinkage often occurs, and cytoplasm pulls away from cyst wall; may also be "halo" effect around outside of cyst wall	Cytostome with supporting fibrils, usually visible in stained preparation; curved fibril along side of cytostome usually referred to as "shepherd's crook"	Resembles *Endolimax nana* cyst; fibrils or flagella usually not seen	Resembles *Chilomastix* cyst; shadow outline of cytostome with supporting fibrils extending above nucleus

* These sizes refer to wet preparation measurements. Organisms on a permanent stained smear may be 1 to 1.5 μm smaller because of artificial shrinkage.

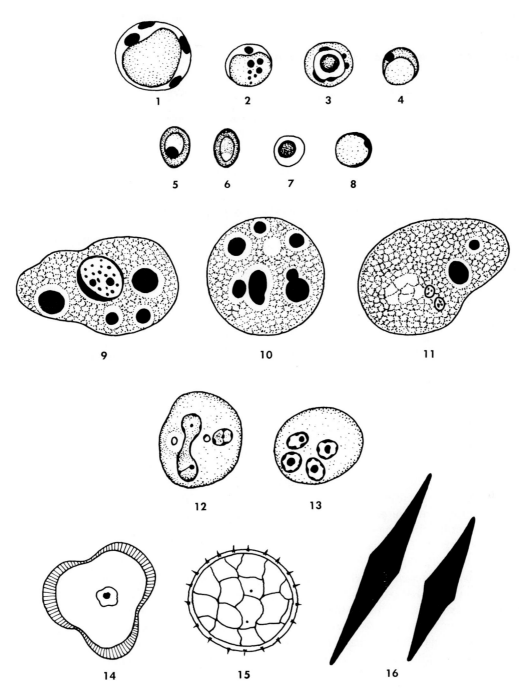

FIGURE 45.15

Various structures that may be seen in stool preparations. **1, 2,** and **4,** *Blastocystis hominis.* **3,** and **5** to **8,** Various yeast cells. **9,** Macrophage with nucleus. **10** and **11,** Deteriorated macrophage without nucleus. **12** and **13,** Polymorphonuclear leukocytes. **14** and **15,** Pollen grains. **16,** Charcot-Leyden crystals. (Modified from Markell, E.K., and Voge, M. 1981. Medical parasitology, ed 5. W.B. Saunders, Philadelphia. Illustration by Nobuko Kitamura.)

TABLE 45.7

MORPHOLOGICAL CRITERIA OF *BALANTIDIUM COLI*

	TROPHOZOITE	CYST
Shape and size*	Ovoid with tapering anterior end; 50-100 μm in length, 40-70 μm wide, usual range, 40-50 μm	Spherical or oval; 50-70 μm; usual range, 50-55 μm
Motility	Rotary or boring or both; may be rapid	
Number of nuclei	One large kidney-shaped macronucleus; 1 small round micronucleus, which is difficult to see even in stained smear; macronucleus may be visible in unstained preparation	One large macronucleus visible in unstained preparation
Other features	Body covered with cilia, which tend to be longer near cytostome; cytoplasm may be vacuolated	Macronucleus and contractile vacuole visible in young cysts; in older cysts, internal structure appears granular

* These sizes refer to wet preparation measurements. Organisms on a permanent stained smear may be 1 to 1.5 μm smaller because of artificial shrinkage.

but occurs in the urogenital tract. *Trichomonas tenax* is occasionally found in the mouth and may be associated with poor oral hygiene.

With the exception of *Dientamoeba*, the flagellates can be recognized by their characteristic rapid motility, which has been described as a "falling leaf" motion for *Giardia* and a jerky motion for the other species. Most flagellates have a characteristic pear shape and possess different numbers and arrangements of flagella, depending on the species. The sucking disk and axonemes of *Giardia,* the cytostome and spiral groove of *Chilomastix,* and the undulating membrane of *Trichomonas* are all distinctive criteria for identification (Figures 45.31 to 45.35).

Until recently, *Dientamoeba* was grouped with the amebae; however, EM studies have confirmed its correct classification with the flagellates, specifically the trichomonads. *Dientamoeba* has no known cyst stage and is characterized by having one or two nuclei, which have no peripheral chromatin and which have four to eight chromatin granules in a central mass. This organism is quite variable in size and shape and may contain large numbers of ingested bacteria and other debris. *Dientamoeba* is inconspicuous in the wet mount and is consistently overlooked without the use of the stained smear (Figures 45.36 to 45.38).

Organisms can be recovered in fecal specimens from asymptomatic persons, but reports in the literature describe a wide range of symptoms, which include intermittent diarrhea, abdominal pain, nausea, anorexia, malaise, fatigue, poor weight gain, and unexplained eosinophilia.

CILIATES

Balantidium coli is the largest protozoan and the only ciliate that infects humans (Figure 45.39). The living trophozoites have a rotatory, boring motion, which is usually rapid. The organism's surface is covered by cilia, and the cytoplasm contains both a kidney-shaped macronucleus and a smaller, round micronucleus that is often difficult to see. The number of nuclei in the cyst remains the same as that in the trophozoite. *B. coli* infections are rarely seen in the United States; however, the organisms are quite easily recognized (Figure 45.40).

COCCIDIA

Isospora belli is now considered to be the only valid species of the genus *Isospora* that infects humans, since *Isospora hominis* has been placed in the genus *Sarcocystis*. *I. belli* is released from the intestinal wall as immature oocysts, so all stages from the immature oocyst containing a mass of undifferentiated protoplasm to those oocysts containing fully developed sporocysts and sporozoites are found in the stool (Figure 45.41). If passed in the immature condition, they mature within 4 or 5 days to form sporozoites. *I. belli* infections are becoming increasingly important as a cause of diarrhea in immunodeficient and immunosuppressed patients.[65] Many recently reported infections have occurred in patients with AIDS. The diagnosis of *Isospora* infection is usually accomplished by finding oocysts in stool concentrates or rarely by direct wet-mount examination during routine ova and parasite examinations. Multiple stool examinations are recommended, since the oocysts

TABLE 45.8

MORPHOLOGICAL CRITERIA USED TO IDENTIFY INTESTINAL PROTOZOA (COCCIDIA, MICROSPORIDIA, *BLASTOCYSTIS HOMINIS*)

SPECIES	SHAPE AND SIZE	OTHER FEATURES
Cryptosporidium sp.	Oocyst generally round, 4-6 µm, each mature oocyst containing 4 sporozoites	Oocyst, usual diagnostic stage in stool, sporozoites occasionally visible within oocyst wall; acid-fast positive using modified acid-fast stains; various other stages in life cycle can be seen in biopsy specimens taken from gastrointestinal tract (brush border of epithelial cells) (intestinal tract) and other tissues; disseminated infection well documented in compromised host
Isospora belli	Ellipsoidal oocyst; range, 20-30 µm in length, 10-19 µm in width; sporocysts rarely seen broken out of oocysts but measure 9-11 µm	Mature oocyst contains 2 sporocysts with 4 sporozoites each; usual diagnostic stage in feces is immature oocyst containing spherical mass of protoplasm (intestinal tract)
Sarcocystis hominis S. *suihominis* S. *bovihominis*	Thin-walled oocyst containing 2 mature sporocysts, each containing 4 sporozoites; thin oocyst wall frequently ruptures; ovoidal sporocysts each measure 10-16 µm in length and 7.5-12 µm in width	Thin-walled oocyst or ovoid sporocysts occur in stool (intestinal tract)
S. *"lindemanni"*	Shapes and sizes of skeletal and cardiac muscle sarcocysts vary considerably	Sarcocysts contain from several hundred to several thousand trophozoites, each of which measures 4-9 µm in width and 12-16 µm in length; sarcocysts may also be divided into compartments by septa, not seen in *Toxoplasma* cysts (tissue, muscle)
CLBs (Cyanobacteria-like or coccidia-like bodies); now identified as coccidia in genus *Cyclospora*	Oocyst generally round, 8-10 µm, internal morphology difficult to see	Oocyst, usual diagnostic stage in stool; acid-fast variable using modified acid-fast stains; oocysts may appear wrinkled; mimics *Cryptosporidium* oocysts but twice as large
Microsporidia	Small, oval spores (1-4 µm) found in routine histological sections; however, electron microscopy used most succesfully	Spores shed from enterocytes now identified in stool specimens; in additon to the modified trichrome procedure, preliminary results indicate monoclonal antibodies may provide more sensitive detection method
Blastocystis hominis	Organisms generally round, measure approximately 6-40 µm, and usually characterized by a large, central body (resembles a large vacuole)	More amebic form seen in diarrheal fluid but difficult to identify

are not continually shed in the stool during infections. Oocysts can be stained with rhodamine-auramine, modified acid-fast procedures, or Giemsa. The oocyst wall and sporocyst fluoresce bright yellow when stained with rhodamine-auramine. With the modified acid-fast stain (see Procedure 45.5), the oocyst appears pink with bright-red sporocysts.[65] Giemsa (Procedure 45.10) stains both the oocyst and the sporocyst pale blue; however, trichrome stain is not recommended because it stains the oocyst poorly or not at all. Multiple biopsies of the jejunal-duodenal brush border can also lead to diagnosis, since in many

patients apparently only the asexual stages are seen.

Cryptosporidium sp. is another coccidian parasite that has been implicated in intestinal disease, primarily in immunosuppressed patients and particularly those with AIDS.[31,33-35] These organisms may not be host specific, are probably transmitted by the fecal-oral route, and may also infect persons with a competent immune system. The developmental stages occur within a vacuole of host cell origin, which is located at the microvillous surface of the host epithelial cell (Figure 45.42). Although the infection is generally self-limiting in

FIGURE 45.16

Entamoeba histolytica trophozoite containing ingested red blood cells.

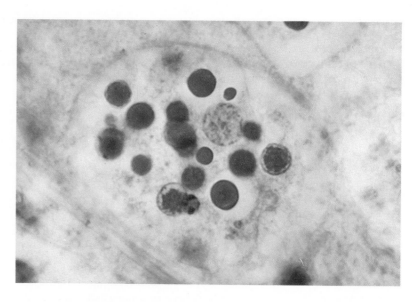

FIGURE 45.17

Entamoeba histolytica trophozoite.

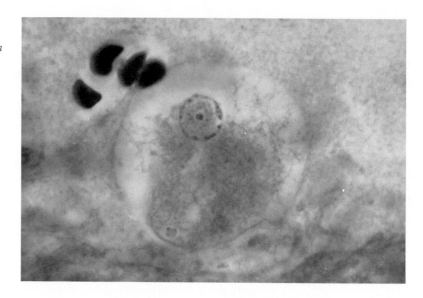

the immunocompetent person, the presence of autoinfective oocysts may explain why a small inoculum can lead to an overwhelming infection in compromised patients and why they may have persistent, life-threatening infections in the absence of documentation of repeated exposure to oocysts. In the immunodeficient individual, *Cryptosporidium* is not always confined to the gastrointestinal tract and has been associated with cholecystitis (biliary tree and gallbladder epithelium) and respiratory tract infections.

Previously, most patients had been diagnosed by light or electron microscopic examination of small or large bowel biopsy material. However, recent cases have been diagnosed from stool specimens and suggest that the use of specific stains and concentration techniques may be more sensi-

tive than intestinal biopsies in diagnosing *Cryptosporidium*. The oocysts range from 4 to 6 μm and can be confused with yeast or overlooked without special techniques; various modified acid-fast stains are one of the best approaches.[34] In wet preparations when iodine is added, yeast cells stain with the iodine, but *Cryptosporidium* does not take up the stain. However, in light infections with few oocysts and many artifacts in the stool, this difference is very difficult to visualize. Sheather's sugar flotation is no longer routinely used by most laboratories and has been replaced by the routine ova and parasite concentration (Figure 45.43).[32]

NOTE: The number of oocysts per stool specimen varies considerably, both from patient to patient and from specimen to specimen during an

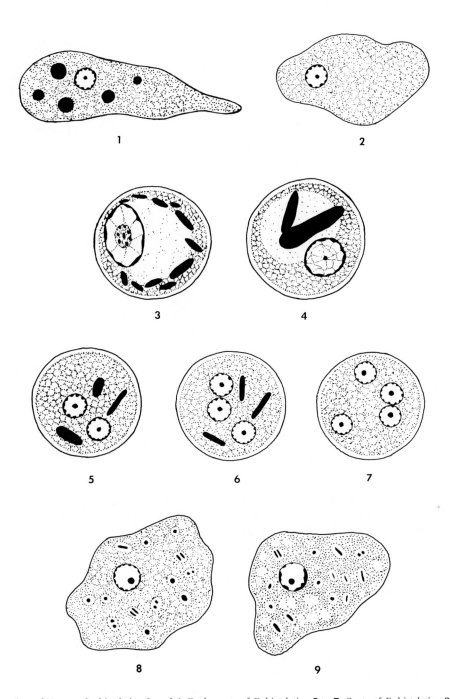

FIGURE 45.18

1 and **2,** Trophozoites of *Entamoeba histolytica.* **3** and **4,** Early cysts of *E. histolytica.* **5** to **7,** Cysts of *E. histolytica.* **8** and **9,** Trophozoites of *Entamoeba coli.* **10** and **11,** Early cysts of *E. coli.* **12** to **14,** Cysts of *E. coli.* **15** and **16,** Trophozoites of *Entamoeba hartmanni.* **17** and **18,** Cysts of *E. hartmanni.* (From Garcia, L.S., and Bruckner, D.A. 1993. Diagnostic medical parasitology. ed 2. ASM Press, Washington, D.C. Illustrations **4** and **11** by Nobuko Kitamura.)

Continued on p. 820.

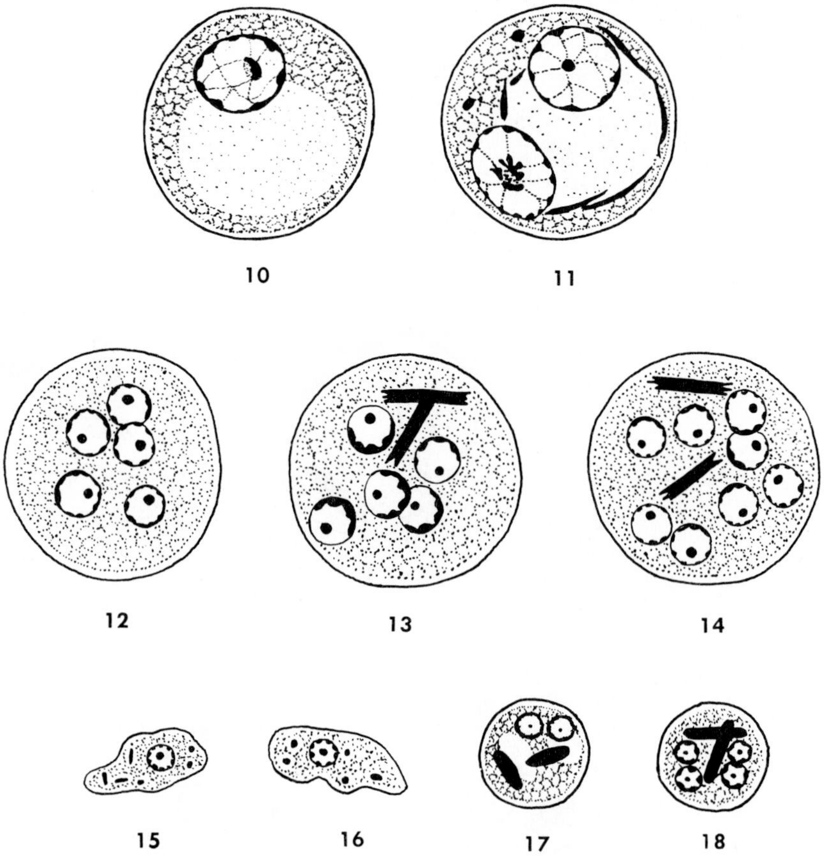

FIGURE 45.18

For legend see page 819.

infection in a single patient. Since therapy for this infection is less than optimal, even with the use of spiramycin, one should check multiple specimens before assuming the patient is cured of the infection. With the development of very specific monoclonal reagents directed against components of the oocyst wall (Figure 45.44), the ability to screen large numbers of specimens is now available.[30,78]

During the past few years, several outbreaks of diarrhea have been associated with a spherical organism measuring 8 to 10 μm in diameter; the distribution of case reports is worldwide. The organisms were subsequently thought to be a new pathogen, possibly an oocyst, a flagellate, an unsporulated coccidian, a large *Cryptosporidium*, a blue-green alga (cyanobacterium-like body), or a coccidian-like body.[47,52,53] An epidemiological report of one outbreak implicated exposure to a contaminated water source.[47] These organisms are now thought to be coccidia in the genus *Cyclospora*. The name *Cyclospora cayetanensis* has been proposed for this newly described, disease-producing coccidian organism from humans.[67] In clean wet mounts the organisms are seen as non-refractile spheres, stain orange with safranin, and are acid-fast variable with the modified acid-fast

stain; those that are unstained appear as glassy, wrinkled spheres. Modified acid-fast methods stain the oocysts from light pink to deep red, some of which contain granules or a bubbly appearance (Figure 45.45). It is very important to be aware of these organisms when using the modified acid-fast stain for *Cryptosporidium parvum*. If other similar but larger structures (approximately twice the size of *Cryptosporidium* oocysts) are seen in the stained smear, laboratories should measure them. Also, the *Cyclospora* oocysts autofluoresce (strong green or intense blue) under ultraviolet (UV) epifluorescence (Figure 45.46).

MICROSPORIDIA

The microsporidia are obligate intracellular parasites that can infect both animals and humans, probably through the gastrointestinal tract, and have become associated with infections in the compromised host.* To date, six genera have been recognized in humans: *Encephalitozoon, Nosema, Pleistophora, Enterocytozoon, Septata,* and *"Microsporidium"*, a "catchall" genus for those organisms not yet classified. Classification criteria include spore size, configuration of the nuclei

---

*References 11, 11a, 12, 23, 50, 66, 74, 81.

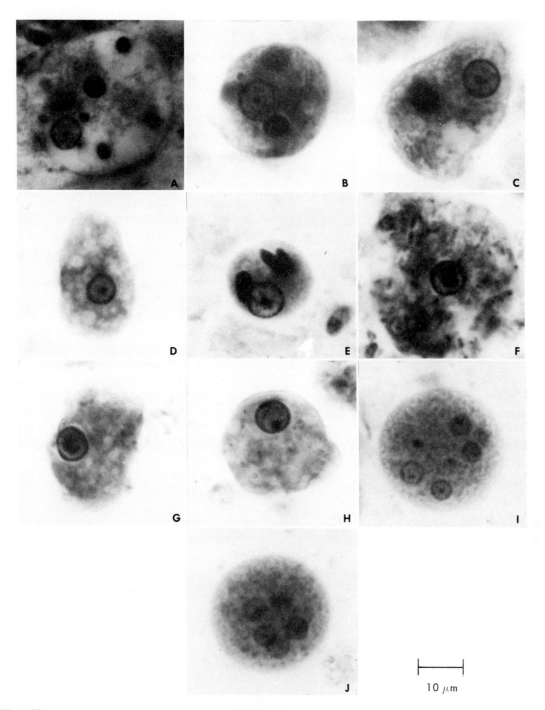

FIGURE 45.19

A to **D,** Trophozoites of *Entamoeba histolytica.* **E,** Early cysts of *E. histolytica.* **F** to **H,** Trophozoites of *Entamoeba coli.* **I** and **J,** Cysts of *E. coli.*

within the spores and developing forms, the number of polar tubule coils within the spore, and the relationship between the organism and host cell. Infection occurs with the introduction of infective sporoplasm through the polar tubule into the host cell.[28] The microsporidia multiply extensively

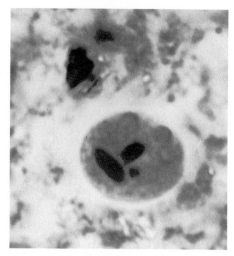

FIGURE 45.20
Entamoeba histolytica cyst.

within the host cell cytoplasm; the life cycle includes repeated divisions by binary fission (merogony) or multiple fission (schizogony) and spore production (sporogony) (Figure 45.47). Both merogony and sporogony can occur in the same cell at the same time. During sporogony, a thick spore wall is formed, thus providing environmental protection for this infectious stage of the parasite. An example of infection potential is illustrated by *Enterocytozoon bieneusi,* an intestinal pathogen.[13,25] The spores are released into the intestinal lumen and are passed in the stool. These spores are environmentally resistant and can then be ingested by other hosts. There is also evidence for inhalation of spores[73] and evidence in animals that suggests human microsporidiosis may also be transmitted via the rectal route.[88]

The number of documented cases is dramatically increasing, with more than 100 cases reported in AIDS patients; infection with *E. bieneusi* and *Septata intestinalis* have been reported in AIDS patients.[13] Chronic intractable diarrhea accompanied by fever, malaise, and weight loss are symptoms of *E. bieneusi* and *S. intestinalis* infections, similar to those seen with cryptosporidiosis or isosporiasis. Patients diagnosed with AIDS tend to

FIGURE 45.21
Entamoeba coli trophozoite.

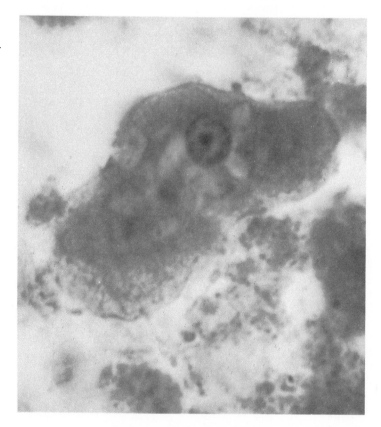

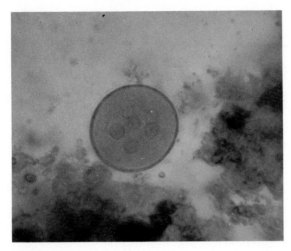

FIGURE 45.22

Entamoeba coli cyst, iodine stain.

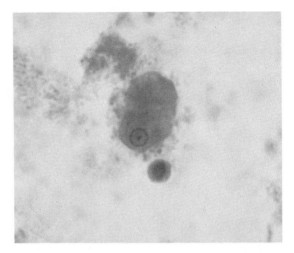

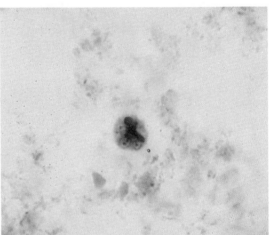

FIGURE 45.24

Entamoeba hartmanni trophozoite. *Bottom, E. hartmanni* cyst.

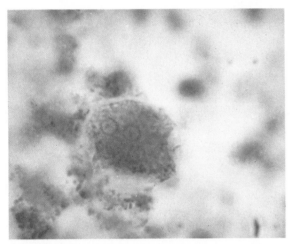

FIGURE 45.23

Entamoeba coli cyst, trichrome stain (poor preservation).

have four to eight watery, nonbloody stools daily, which can be accompanied by nausea and anorexia.

The diagnosis can be made by identifying the spores in biopsy or autopsy specimens; all body organs have been involved, including the eye (Figure 45.48). The resistant spores measure approximately 1 to 4 µm and do not stain well with hematoxylin-eosin. They are occasionally acid fast and have a periodic acid-Schiff (PAS)–positive polar granule at the anterior end.[81] Definitive diagnosis can also be made using electron microscopy (Figure 45.49). (See earlier section on biopsy material for information on the concentrated trichrome stain and IFA procedures.) Two modified trichrome stains have been used with some success for visualizing microsporidian spores in unconcentrated, formalin-fixed stool.[69a,86]

BLOOD PROTOZOA

MALARIA

Malaria is caused by four species of the protozoan genus *Plasmodium: P. vivax, P. falciparum* (Figure 45.50), *P. malariae*, and *P. ovale* (Table 45.9). Humans become infected when the sporozoites are introduced into the blood from the salivary secretion of the infected mosquito when the mosquito vector takes a blood meal. These sporozoites then leave the blood and enter the parenchymal cells of the liver, where they undergo asexual multiplication. This development in the liver before red cell invasion is called the *preerythrocytic cycle;* if further liver development takes place after red cell invasion, it is called the **exoerythrocytic cycle.** The length of time for the preerythrocytic cycle

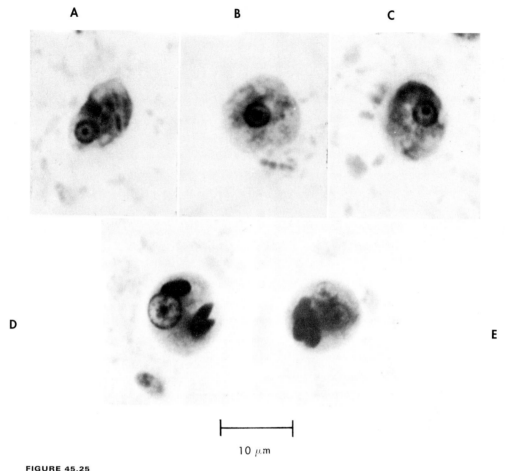

FIGURE 45.25

A to **C,** Trophozoites of *Entamoeba hartmanni.* **D** and **E,** Cysts of *E. hartmanni.*

and the number of asexual generations vary depending on the species; however, the schizonts eventually rupture, releasing thousands of merozoites into the bloodstream, where they invade the erythrocytes. The early forms in the red cells are called ring forms or young trophozoites (Figure 45.51). As the parasites continue to grow and feed, they become actively ameboid within the red cell. They feed on hemoglobin, which is incompletely metabolized; the residue that is left is called malarial pigment and is a compound of hematin and protein.

During the next phase of the cycle, the chromatin (nuclear material) becomes fragmented throughout the organism, and the cytoplasm begins to divide, each portion being arranged with a fragment of nuclear material. These forms are called mature schizonts and are composed of individual merozoites. The infected red cell then ruptures, releasing the merozoites and also metabolic products into the bloodstream. If large numbers of red cells rupture simultaneously, a malarial paroxysm may result from the amount of toxic materials released into the bloodstream. In the early stages of infection or in a mixed infection

with two species, rupture of the red cells is usually not synchronous; consequently, the fever may be continuous or daily rather than intermittent. After several days, a 48- or 72-hour periodicity is usually established.

After several generations of erythrocytic schizogony, the production of gametocytes begins. These forms are derived from merozoites, which do not undergo schizogony but continue to grow and form the male and female gametocytes that circulate in the bloodstream. When the mature gametocytes are ingested by the appropriate mosquito vector, the sexual cycle is initiated within the mosquito with the eventual production of the sporozoites, the infective stage for humans (Figure 45.52).

The asexual and sexual forms just described circulate in the human bloodstream in three species of *Plasmodium.* However, in *P. falciparum* infections, as the parasite continues to grow, the red cell membrane becomes sticky, and the cells tend to adhere to the endothelial lining of the capillaries of the internal organs. Thus, only the ring forms and crescent-shaped gametocytes occur in the peripheral blood.[32,58,89] Interference

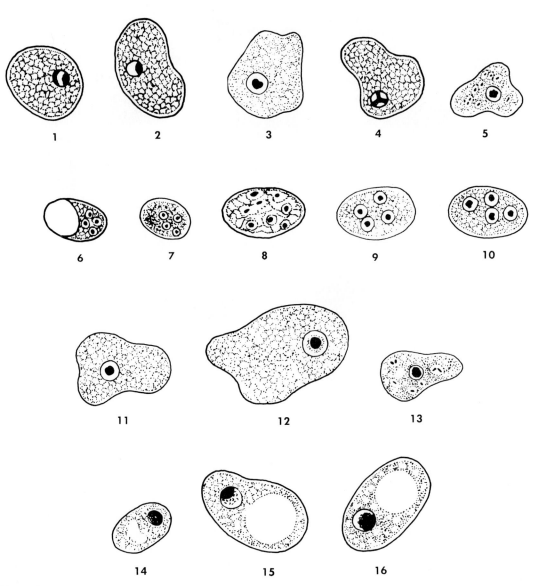

FIGURE 45.26

1 to **5,** Trophozoites of *Endolimax nana.* **6** to **10,** Cysts of *E. nana.* **11** to **13,** Trophozoites of *Iodamoeba bütschlii.* **14** to **16,** Cysts of *I. bütschlii.* (From Garcia, L.S., and Bruckner, D.A. 1993. Diagnostic medical parasitology. 2nd ed. ASM Press, Washington, D.C.)

with normal blood flow in these vessels gives rise to additional problems, which are responsible for the different clinical manifestations of this type of malaria.

In some areas of the world where *P. falciparum* is endemic, many people are carriers of hemoglobin S (HbS), thalassemia, and glucose-6-phosphate dehydrogenase (G6PD) deficiency. These factors are associated with increased resistance to falciparum malaria.

Miller et al.[61] suggest that the resistance of many west Africans and approximately 70% of American blacks to *P. vivax* may be related to the high incidence of Duffy-negative erythrocytes in these groups.

BABESIOSIS

Babesia are tickborne (also can be transmitted via a blood transfusion) sporozoan parasites that have generally been considered parasites of animals (Texas cattle fever) rather than humans. However, there are now documented human cases, with some infections occurring in splenectomized patients and others in patients with intact spleens.[7] *Babesia* organisms infect the red blood cells and appear as pleomorphic ringlike structures when stained with any of the recommended stains used for blood films (Figure 45.53). They may be confused with the ring forms in *Plasmodium* infections; however, in a *Babesia* infection, many rings (four or five) often are seen per red

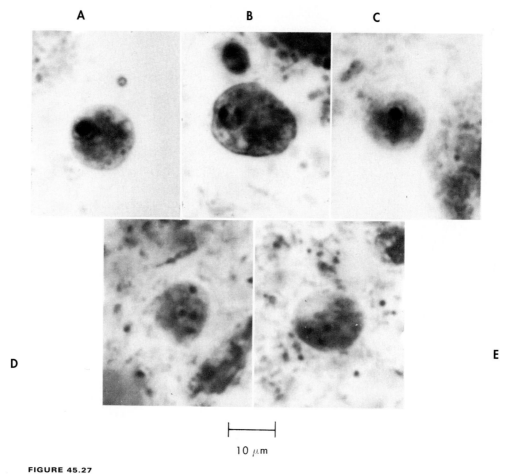

A   B   C

D   E

|———————|

10 μm

FIGURE 45.27

A to **C,** Trophozoites of *Endolimax nana,* **D** and **E,** Cysts of *E. nana.*

cell, and the individual rings are quite small compared with those found in malaria infections.[42]

HEMOFLAGELLATES

Hemoflagellates are blood and tissue flagellates, two genera of which are medically important for humans: *Leishmania* and *Trypanosoma.* Some species may circulate in the bloodstream or at times may be present in lymph nodes or muscle. Other species tend to parasitize the reticuloendothelial cells of the hematopoietic organs. The hemoflagellates of human beings have four morphological types (Figure 45.54): amastigote (leishmanial form, or Leishman-Donovan [L-D] body), promastigote (leptomonal form), epimastigote (crithidial form), and trypomastigote (trypanosomal form).

The amastigote form is an intracellular parasite in the cells of the reticuloendothelial system and is oval, measuring approximately 1.5 to 5 μm, and contains a nucleus and kinetoplast. *Leishmania* sp. usually exist as the amastigote form in humans and in the promastigote form in the insect host. The life cycle is essentially the same for all

three species, and the clinical manifestations vary depending on the species involved. As the vector takes a blood meal, the promastigote form is introduced into a human, thus initiating the infection. Depending on the species, the parasites then move from the site of the bite to the organs of the reticuloendothelial system (liver, spleen, bone marrow) or to the macrophages of the skin (Figure 45.55).

Species based on clinical grounds are morphologically the same; however, differences exist in serologies and in growth requirements for culture. Great biological variation occurs among the many strains that make up these groups. *Leishmania tropica* causes oriental sore or cutaneous leishmaniasis of the Old World; *L. braziliensis* causes mucocutaneous leishmaniasis of the New World; and *L. donovani* causes visceral leishmaniasis (Dumdum fever, or kala-azar) (Figure 45.56). Additional species have been delineated based on the buoyant density of kinetoplast DNA, the isoenzyme patterns, and serologic testing.

In tissue impression smears or sections, *Histoplasma capsulatum* must be differentiated from the

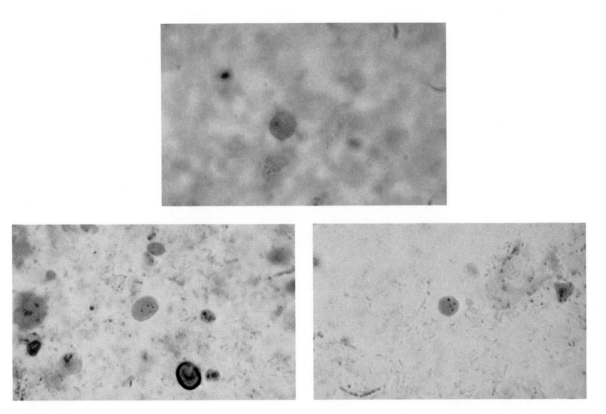

FIGURE 45.28

Top, Endolimax nana trophozoite. *Bottom left, E. nana* cyst. *Bottom right, E. nana* cyst.

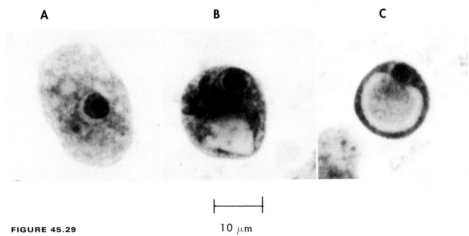

FIGURE 45.29   10 μm

A, Trophozoites of *Iodamoeba bütschlii.* **B** and **C,** Cysts of *I. bütschlii.*

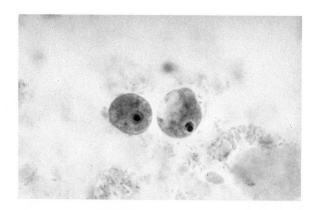

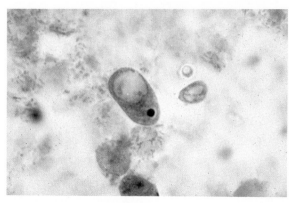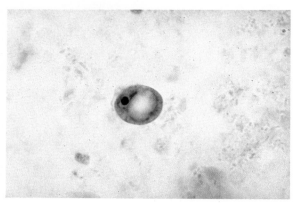

FIGURE 45.30

Top, Iodamoeba bütschlii trophozoites. *Bottom left, I. bütschlii* cyst. *Bottom right, I. bütschlii* cyst.

L-D bodies. *H. capsulatum* does not have a kineto-plast and stains with both PAS stain and GMS stain, neither of which stains L-D bodies. Diagnosis of leishmanial organisms is based on the demonstration of the L-D bodies or the recovery of the promastigote in culture.

Three species of trypanosomes are pathogenic for humans: *Trypanosoma gambiense* (Figure 45.57) causes West African sleeping sickness; *Trypanosoma rhodesiense* causes East African sleeping sickness; and *Trypanosoma cruzi* causes South American trypanosomiasis, or Chagas' disease (Figure 45.58). The first two species are morphologically similar and produce African sleeping sickness, an illness characterized by both acute and chronic stages. In the acute stage of the disease, the organisms can usually be found in the peripheral blood or lymph node aspirates. As the disease progresses to the chronic stage, the organisms can be found in the cerebrospinal fluid (CSF) (comatose stage: "sleeping sickness"). *T. rhodesiense* produces a more severe infection, usually resulting in death within 1 year.

In the early stages of infection with *T. cruzi,* the trypomastigote forms appear in the blood but do not multiply (Figure 45.59, *A*). They then in-

vade the endothelial or tissue cells and begin to divide, producing many L-D bodies, which are most often found in cardiac muscle (Figure 45.59, *B*). When these forms are liberated into the blood, they transform into the trypomastigote forms, which are then carried to other sites, where tissue invasion again occurs.

Diagnosis of the infection is based on demonstration of the parasites, most often on wet unstained or stained blood films. Both thick and thin films should be examined; these can be prepared from peripheral blood or buffy coat. The sediment recovered from CSF can also be examined for the presence of trypomastigotes. Specific techniques for culture, animal inoculation, handling of aspirate and biopsy material, and serologic procedures are presented in earlier sections of this chapter.

Another technique often used in endemic areas for the diagnosis of Chagas' disease is xenodiagnosis. Triatomids, the insect vector, are raised in the laboratory and are determined to be free from infection with *T. cruzi*. These insects are allowed to feed on the blood of an individual suspected of having Chagas' disease, and after 2 weeks the intestinal contents are checked for the presence of the epimastigote forms.[55]

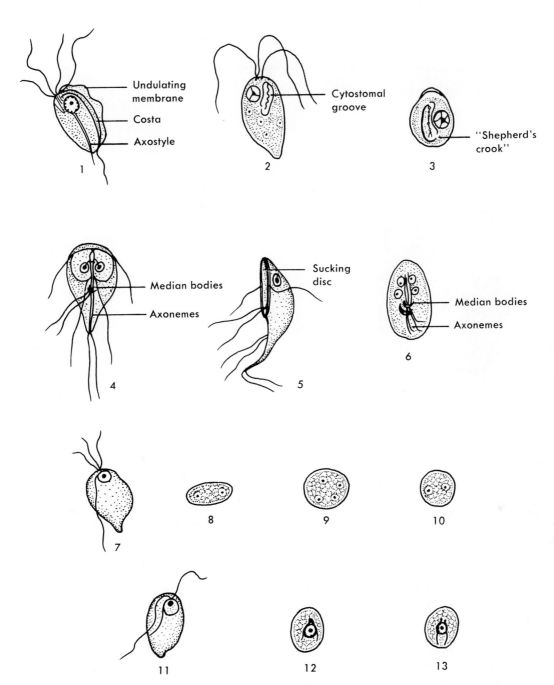

FIGURE 45.31

1, Trophozoite of *Trichomonas hominis.* **2,** Trophozoite of *Chilomastix mesnili.* **3,** Cyst of *C. mesnili.* **4,** Trophozoite of *Giardia lamblia* (front view). **5,** Trophozoite of *G. lamblia* (side view). **6,** Cyst of *G. lamblia.* **7,** Trophozoite of *Enteromonas hominis.* **8** to **10,** Cysts of *E. hominis.* **11,** Trophozoite of *Retortamonas intestinalis.* **12** and **13,** Cysts of *R. intestinalis.* (From Garcia, L.S., and Bruckner, D.A. 1993. Diagnostic medical parasitology. ASM Press, Washington, D.C. Illustration **5** by Nobuko Kitamura. Illustrations **7** to **13** modified from Markell, E.K., and Voge, M. 1981. Medical parasitology, ed 5. W.B. Saunders, Philadelphia.)

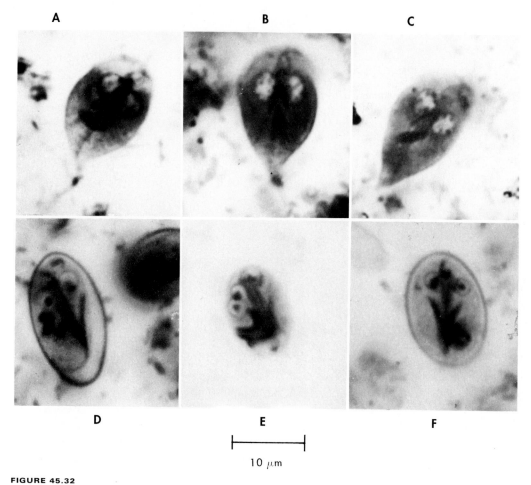

FIGURE 45.32

A to **C,** Trophozoites of *Giardia lamblia.* **D** to **F,** Cysts of *G. lamblia.*

INTESTINAL HELMINTHS

The intestinal helminths that infect humans belong to two phyla: the Nematoda, or roundworms, and the platyhelminths, or flatworms. The platyhelminths, most of which are hermaphroditic, have a flat, bilaterally symmetrical body. The two classes, Trematoda and Cestoda, contain organisms that are parasitic for human beings.

The trematodes (flukes) are leaf-shaped or elongate, slender organisms (blood flukes: *Schistosoma* sp.) that possess hooks or suckers for attachment. Members of this group, which parasitize humans, are found in the intestinal tract, liver, blood vessels, and lungs.

The cestodes (tapeworms) typically have a long, segmented, ribbonlike body, which has a special attachment portion, or scolex, at the anterior end. Adult forms inhabit the small intestine; however, humans may be host to either the adult or the larval forms, depending on the species. The cestodes, as well as the trematodes, require (with few exceptions) one or more intermediate hosts for the completion of the life cycle (Figures 45.60 and 45.61).

The nematodes, or roundworms, are elongate, cylindrical worms with a well-developed digestive tract. The sexes are separate, with the male usually smaller than the female. Intermediate hosts are required for larval development in certain species; many species parasitize the intestinal tract and certain tissues of humans (Figures 45.62 to 45.64).

Diagnosis of most intestinal helminth infections is based on the detection of the characteristic eggs and larvae in the stool; occasionally, adult worms or portions of worms may also be found. No permanent stains are required, and most diagnostic features can easily be seen on direct wet mounts or in mounts of the concentrated stool material.

NEMATODES

Most nematodes are diagnosed by finding the characteristic eggs in the stool (Figures 45.65 and 45.66). The eggs of *Ancylostoma duodenale* and *Necator americanus* are essentially identical, so an infection with either species is reported as "hookworm eggs present." *Trichostrongylus* eggs may easily be mistaken for those of hookworms; however, the eggs of *Trichostrongylus* are somewhat larger, and one end tends to be more pointed.

Strongyloides stercoralis is passed in the feces as the noninfective rhabditiform larva (Figure

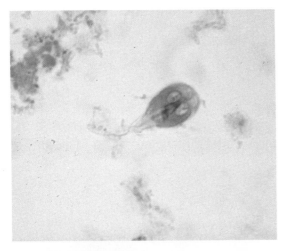

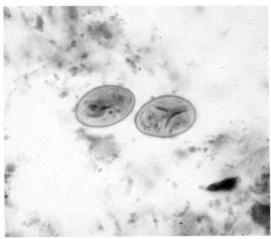

FIGURE 45.33

Top, Giardia lamblia trophozoite. *Bottom, G. lamblia* cysts.

45.67). Although hookworm eggs are normally passed in the stool, these eggs may continue to develop and hatch if the stool is left at room temperature for several days. These larvae may be mistaken for those of *Strongyloides*. Figure 45.68 shows the morphological differences between the rhabditiform larvae of hookworm and *Strongyloides*. Recovery of *Strongyloides* larvae in duodenal contents is mentioned earlier in this chapter.

The appropriate techniques for recovery of *Enterobius vermicularis* (pinworm) eggs are given earlier in the section on cellophane tape preparations (Procedure 45.6). Other worms do not deposit eggs in the same site; the characteristic morphology of pinworm eggs is shown in Figure 45.69. Eggs of the other nematodes are relatively easy to find and differentiate from one another.

CESTODES

With the exception of *Diphyllobothrium latum*, tapeworm eggs are embryonated and contain a six-hooked oncosphere (Figure 45.70 and Table

45.10). *Taenia saginata* and *T. solium* cannot be speciated on the basis of egg morphology; gravid proglottids (Figure 45.71) or the scolices must be examined. *T. saginata* (beef tapeworm) proglottids have approximately 15 to 30 main lateral branches, and the scolex has no hooks. The proglottids of *T. solium* have seven to 12 main lateral branches, and the scolex has a circle of hooks.

Hymenolepis nana has an unusual life cycle in that the ingestion of the egg can lead to the adult worm in humans (Figure 45.72). The eggs of *H. nana* (more common) and *Hymenolepis diminuta* are very similar; however, *H. nana* eggs are smaller and have polar filaments, which are present in the space between the oncosphere and the eggshell (Figure 45.70).

Eggs of *Dipylidium caninum* are occasionally found in humans, particularly children, and are passed in the feces in packets of five to 15 eggs each (Figures 45.70 and 45.73). The proglottids may also be found; they may resemble cucumber seeds or, when dry, rice grains (white).

The fish tapeworm, *Diphyllobothrium latum*, does not have embryonated eggs. These eggs have a somewhat thicker shell and are operculated, similar to the trematode eggs (Figure 45.70). Humans acquire the parasite by eating undercooked or raw freshwater fish (Figure 45.74).

TREMATODES

Humans acquire most fluke infestations by ingesting the encysted metacercariae (Figure 45.75). Most trematodes have operculated eggs, which are best recovered by the sedimentation concentration technique rather than the flotation method. Many of these eggs are very similar, both in size and morphology, and often careful measurements must be taken to speciate the eggs (Figure 45.76). The eggs of *Clonorchis (Opisthorchis)*, *Heterophyes*, and *Metagonimus* are very similar and quite small; they are easily missed if the concentration sediment is examined with the 10× objective only. The eggs of *Fasciola hepatica* and *Fasciolopsis buski* are also very similar but much larger than those just mentioned.

Paragonimus westermani eggs are not only found in the stool but also may be found in sputum (Figure 45.76). These eggs are very similar in size and shape to the egg of the fish tapeworm, *Diphyllobothrium latum*. Schistosomiasis, a great source of morbidity in much of the developing world, is acquired during activities (e.g., bathing, swimming, doing laundry, planting rice, fishing) that involve contact with water infested with the intermediate snail host of the parasite. The free-swimming cercarial form attaches to a

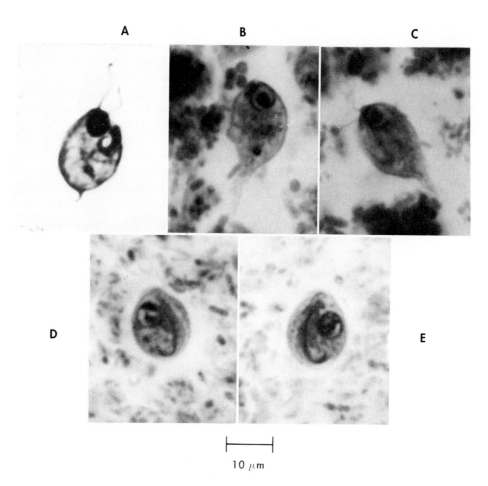

10 μm

FIGURE 45.34

A to **C,** Trophozoites of *Chilomastix mesnili* (**A,** silver stain). **D** and **E,** Cysts of *C. mesnili.*

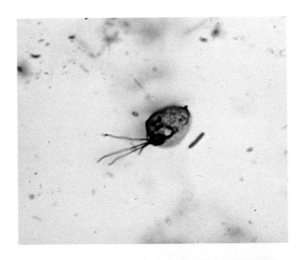

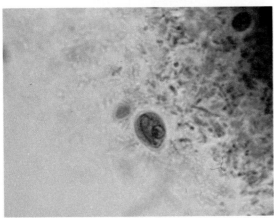

FIGURE 45.35

Top, Chilomastix mesnili trophozoite, silver stain. *Bottom, C. mesnili* cyst.

FIGURE 45.37

Top, Dientamoeba fragilis, two nuclei. *Bottom, D. fragilis,* one nucleus.

FIGURE 45.36

Trophozoites of *Dientamoeba fragilis.* (From Garcia, L.S., and Bruckner, D.A. 1993. Diagnostic medical parasitology. 2nd ed. ASM Press, Washington, D.C.)

vertebrate host and penetrates through intact skin (Figure 45.77). In temperate zones, such as in the areas near the Great Lakes, cercariae of other trematodes that normally infect birds or other animals can penetrate human skin and cause a strong local reaction (swimmer's itch), although they cannot go on to develop into adult worms in the unnatural human host. Probably the easiest trematode eggs to identify are those of

the schistosomes: *Schistosoma mansoni* eggs are characterized by having a very prominent lateral spine, *S. haematobium* a terminal spine, and *S. japonicum* a small lateral spine that may be difficult to see (Figure 45.76). These eggs are nonoperculated. Schistosome eggs stain acid fast. Specific procedures for their recovery and identification are found in the earlier section on the hatching procedure.

BLOOD HELMINTHS

FILARIAE

The filarial worms are long, thin nematodes that inhabit parts of the lymphatic system and the subcutaneous and deep connective tissues. A life cycle has been included here to show the various stages present in the human host (Figure 45.78). Most species produce microfilariae, which can be found in the peripheral blood; two species, *On-*

Text continued on p. 858.

FIGURE 45.38

A and **B,** Trophozoites of *Dientamoeba fragilis.*

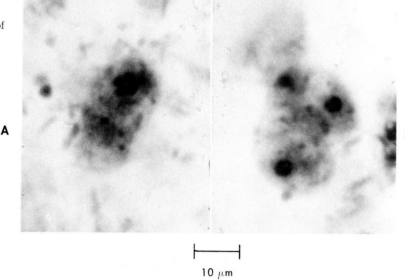

10 μm

FIGURE 45.39

Balantidium coli trophozoite, iodine stain.

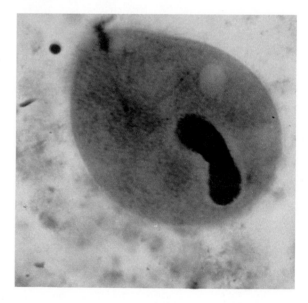

FIGURE 45.40

1, Trophozoite of *Balantidium coli.* **2,** Cyst of *B. coli.* (From Garcia, L.S., and Bruckner, D.A. 1993. Diagnostic medical parasitology. 2nd ed. ASM Press. Washington, D.C.)

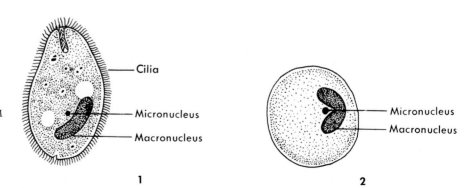

FIGURE 45.41

1, Immature oocyst of *Isospora belli.* **2,** Mature oocyst of *I. belli.* (Illustration by Nobuko Kitamura.).

1

2

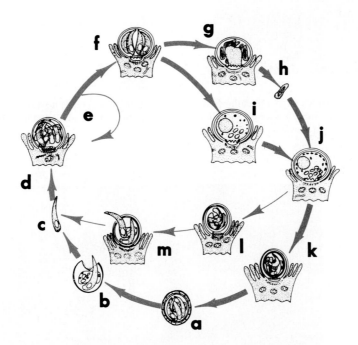

FIGURE 45.42

Proposed life cycle of *Cryptosporidium. (a)* Sporulated oocyst in feces. *(b)* Excystation in intestine. *(c)* Free sporozoite in intestine. *(d)* Type I meront (six or eight merozoites). *(e)* Recycling of Type I merozoite. *(f)* Type II meront (four merozoites). *(g)* Microgametocyte, with approximately 16 microgametes. *(h)* Microgamete fertilizes macrogamete *(i)* to form zygote *(j).* Approximately 80% of the zygotes form thick-walled oocysts *(k),* which sporulate within the host cell. About 20% of the zygotes do not form an oocyst wall; their sporozoites are surrounded only by a unit membrane *(l).* Sporozoites within "autoinfective," thin-walled oocysts *(l)* are released into the intestinal lumen *(m)* and reinitiate the endogenous cycle (at *c*). (Life cycle from William L. Current, Lilly Research Laboratories, Greenfield, Ind.)

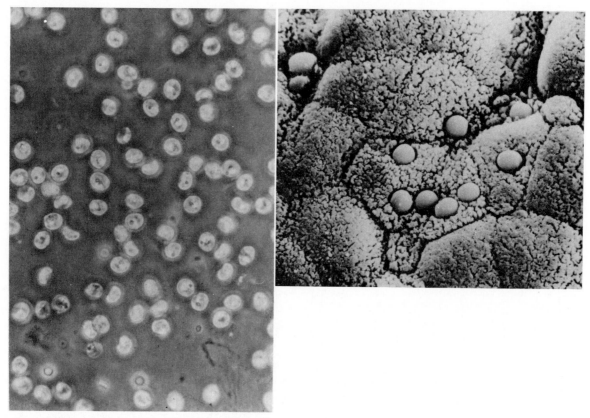

FIGURE 45.43

Cryptosporidium. **A,** Oocysts recovered from a Sheather's sugar flotation; organisms measure 4 to 6 μm. **B,** Scanning electron microscopy view of organisms at brush border of epithelial cells. (From Garcia, L.S., and Bruckner, D.A. 1993. Diagnostic medical parasitology. 2nd ed. ASM Press. Washington D.C.)

FIGURE 45.44

Cryptosporidium oocysts and *Giardia* cysts stained with monoclonal antibody–conjugated fluorescent reagent. (Courtesy Merifluor, Meridian Diagnostics, Cincinnati.)

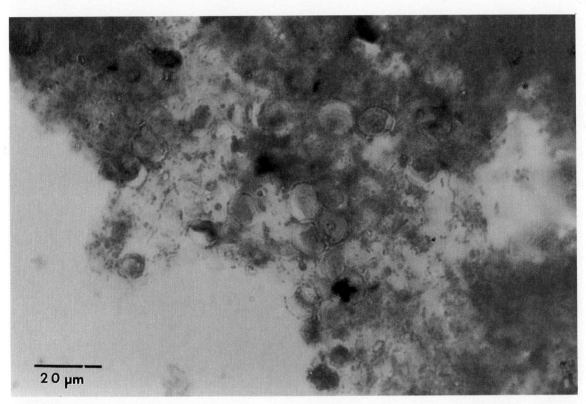

FIGURE 45.45

Cyclospora sp. (formerly called CLBs) oocysts after modified acid-fast staining. Note the variability in the intensity of stain. These oocysts are approximately twice the size of *Cryptosporidium* sp. (Photo courtesy of Charles R. Sterling, University of Arizona.)

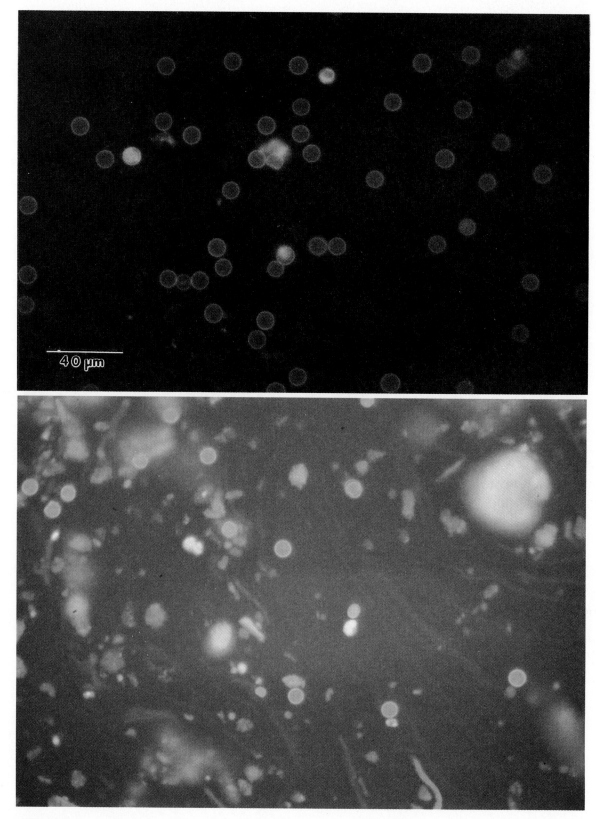

FIGURE 45.46

Cyclospora sp. (formerly called CLBs) oocysts exhibiting autofluorescence. (Top photograph courtesy of Charles R. Sterling, University of Arizona; bottom photograph courtesy of Earl G. Long, Centers for Disease Control.)

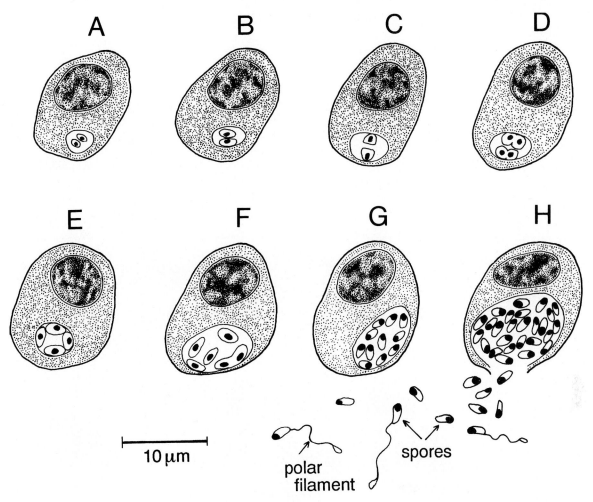

FIGURE 45.47

Life cycle diagram of the microsporidia. **A** to **G,** Asexual development of sporoblasts. **H,** Release of spores. (Modified from Gardiner, C.H., Fayer, R., and Dubey, J.P. 1988. An atlas of protozoan parasites in animal tissues. U.S. Dept of Agriculture, Agriculture Handbook No. 651. Illustration by Sharon Belkin.)

FIGURE 45.48

Diagram illustrating the polar tubule within a microsporidian spore.(Modified from Bryan, R.T., Cali, A., Owen, R.L., and Spencer, H.C. In Sun, T., editor. 1991. Progress in clinical parasitology, vol II, Field and Wood Medical Publishers, distributed by W.W. Norton, New York. Illustration by Sharon Belkin.)

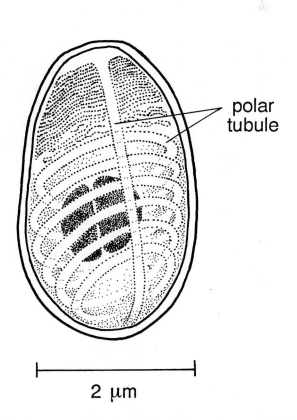

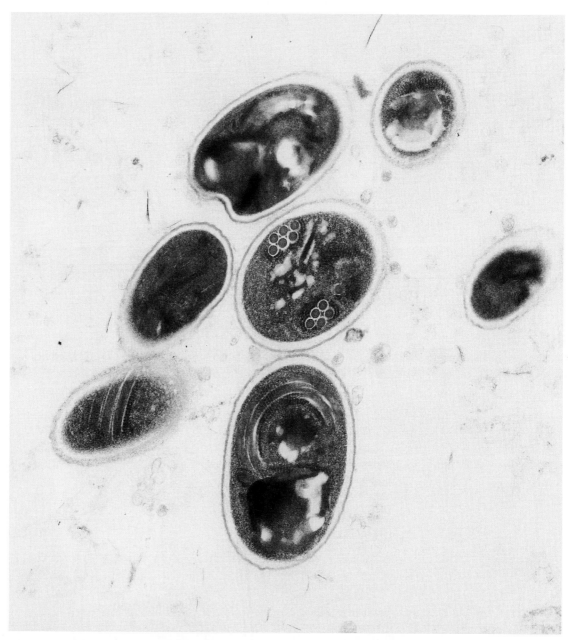

FIGURE 45.49

Transmission electron micrograph of six *Enterocytozoon bieneusi* spores in a stool specimen. Note the cross sections of the polar tubule. (From Orenstein, J.M. 1991. J. Parasitol. 77:843.)

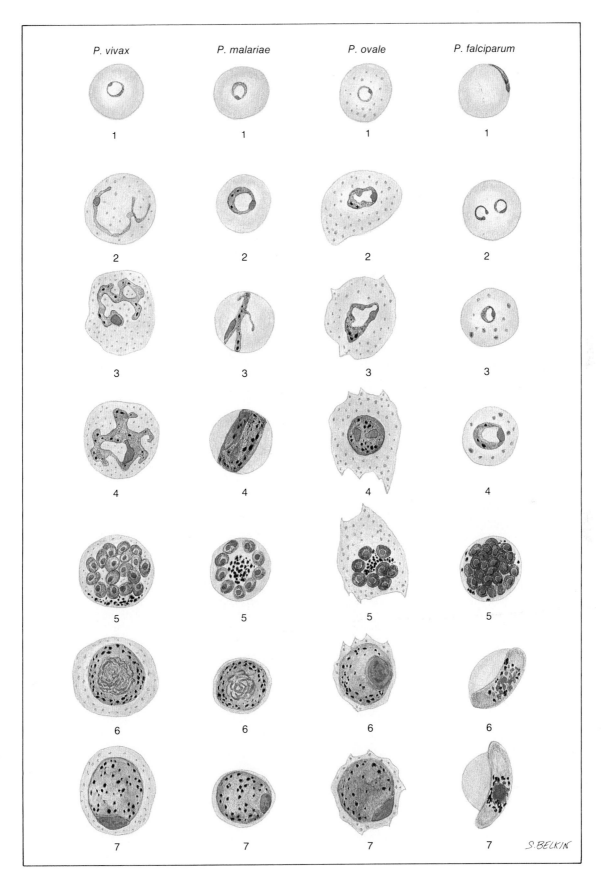

FIGURE 45.50

For legend see p. 842.

FIGURE 45.50

The morphology of malaria parasites.

Plasmodium vivax

1, Early trophozoite (ring form)

2, Late trophozoite with Schüffner's dots (note enlarged red blood cell)

3, Late trophozoite with ameboid cytoplasm (very typical of *P. vivax*)

4, Late trophozoite with ameboid cytoplasm

5, Mature schizont with merozoites (18) and clumped pigment

6, Microgametocyte with dispersed chromatin

7, Macrogametocyte with compact chromatin

Plasmodium malariae

1, Early trophozoite (ring form)

2, Early trophozoite with thick cytoplasm

3, Early trophozoite (band form)

4, Late trophozoite (band form) with heavy pigment

5, Mature schizont with merozoites (9) arranged in rosette

6, Microgametocyte with dispersed chromatin

7, Macrogametocyte with compact chromatin

Plasmodium ovale

1, Early trophozoite (ring form) with Schüffner's dots

2, Early trophozoite (note enlarged red blood cell)

3, Late trophozoite in red blood cell with fimbriated edges

4, Developing schizont with irregular shaped red blood cell

5, Mature schizont with merozoites (8) arranged irregularly

6, Microgametocyte with dispersed chromatin

7, Macrogametocyte with compact chromatin

Plasmodium falciparum

1, Early trophozoite (accolé or appliqué form)

2, Early trophozoite (one ring is in headphone configuration/double chromatin dots)

3, Early trophozoite with Maurer's dots

4, Late trophozoite with larger ring and Maurer's dots

5, Mature schizont with merozoites (24)

6, Microgametocyte with dispersed chromatin

7, Macrogametocyte with compact chromatin

NOTE: Without the appliqué form, Schüffner's dots, multiple rings/cell, and other developing stages, differentiation among the species can be very difficult. It is obvious that the early rings of all four species can mimic one another very easily. *Remember: one set of negative blood films cannot rule out a malaria infection.* (Reprinted by permission of the publisher from Garcia, L.S., and Bruckner, D.A. 1993. Diagnostic medical parasitology, 2nd ed. p. 126. Copyright by American Society for Microbiology, Washington, D.C.)

TABLE 45.9

MICROSCOPIC IDENTIFICATION OF PLASMODIA OF HUMANS IN GIEMSA-STAINED THIN BLOOD SMEARS

	PLASMODIUM VIVAX	PLASMODIUM MALARIAE	PLASMODIUM FALCIPARUM	PLASMODIUM OVALE
Appearance of parasitized red blood cells: size and shape	1½ to 2 times larger than normal; oval to round	Normal shape; size may be normal or slightly smaller	Both normal	60% of cells larger than normal and oval; 20% have irregular, frayed edges (fimbriated)
Schüffner's dots (eosinophilic stippling)	Usually present in all infected cells except early ring forms	None	None; occasionally commalike red dots are present (Maurer's dots)	Present in all stages, including early ring forms; dots may be larger and darker than in *P. vivax*
Color cytoplasm	Decolorized, pale	Normal	Normal, bluish tinge at times	Decolorized, pale
Multiple infections	Occasional	Rare	Common	Occasional
All developmental stages present in peripheral blood	All stages present	Ring forms few because ring stage brief; mostly growing and mature trophozoites and schizonts	Young ring forms and very rarely older stages in moribund patients; few gametocytes	All stages present
Appearance of parasite: young trophozoite (early ring form)	Ring is ⅓ diameter of cell; cytoplasmic circle around vacuole; heavy chromatin dot	Ring often smaller than in *P. vivax*, occupying ⅙ of cell; heavy chromatin dot; vacuole at times "filled in"; pigment forms early	Delicate, small ring with small chromatin dot (frequently 2); scanty cytoplasm around small vacuoles; sometimes at edge of red cell (appliqué form) or filamentous slender form; may have multiple rings per cell	Ring is larger and more ameboid than in *P. vivax*, otherwise similar to *P. vivax*
Growing trophozoite	Multishaped irregular ameboid parasite; streamers of cytoplasm close to large chromatin dot; vacuole retained until close to maturity; increasing amounts of brown pigment	Nonameboid rounded or band-shaped solid forms; chromatin may be hidden by coarse dark-brown pigment	Heavy ring forms fine pigment grains	Ring shape maintained until late in development
Mature trophozoite	Irregular ameboid mass; 1 or more small vacuoles retained until schizont stage; fills almost entire cell; fine brown pigment	Vacuoles disappear early; cytoplasm compact, oval, band shaped, or nearly round, almost filling cell; chromatin may be hidden by peripheral, coarse, dark-brown pigment	Not seen in peripheral blood (except in severe infections); development of all phases after ring form occurs in capillaries of viscera	Compact vacuoles disappear; pigment dark brown, less than in *P. malariae*

Continued.

TABLE 45.9

MICROSCOPIC IDENTIFICATION OF PLASMODIA OF HUMANS IN GIEMSA-STAINED THIN BLOOD SMEARS—cont'd

	PLASMODIUM VIVAX	PLASMODIUM MALARIAE	PLASMODIUM FALCIPARUM	PLASMODIUM OVALE
Schizont (presegmenter)	Progressive chromatin division; cytoplasmic bands containing clumps of brown pigment	Similar to *P. vivax* except smaller, darker, larger pigment granules peripheral or central	Not seen in peripheral blood (see above)	Smaller and more compact than *P. vivax*
Mature schizont	Sixteen (12-24) merozoites, each with chromatin and cytoplasm, filling entire red cell, which can hardly be seen	Eight (6-12) merozoites in rosettes or irregular clusters filling normal-sized cells, which can hardly be seen; central arrangement of brown-green pigment	Not seen in peripheral blood (rare exceptions)	Three-quarters of cells occupied by 8 (8-12) merozoites in rosettes or irregular clusters
Macrogametocyte	Rounded or oval homogeneous cytoplasm; diffuse, delicate, light-brown pigment throughout parasite; eccentric compact chromatin	Similar to *P. vivax*, but fewer in number, pigment darker and more coarse	Sex differentiation difficult; "cresent" or "sausage" shapes characteristic; may appear in "showers"; black pigment near chromatin dot, which is often central	Smaller than *P. vivax*
Microgametocyte	Large pink to purple chromatin mass surrounded by pale or colorless halo; evenly distributed pigment	Similar to *P. vivax*, but fewer in number, pigment darker and more coarse	See above	Smaller than *P. vivax*
Main criteria	Large, pale red cell; trophozoite irregular; pigment usually present; Schüffner's dots not always present; several phases of growth seen in one smear; gametocytes appear early	Red cell normal in size and color; trophozoites compact, stain usually intense, band forms not always seen; coarse pigment; no stippling of red cells; gametocytes appear late	Development after ring stage occurs in blood vessels in internal organs; delicate ring forms and crescent-shaped gametocytes are only forms normally seen in peripheral blood	Red cell enlarged, oval, with fimbriated edges; Schüffner's dots seen in all stages

FIGURE 45.51

Plasmodium falciparum,
early ring forms.

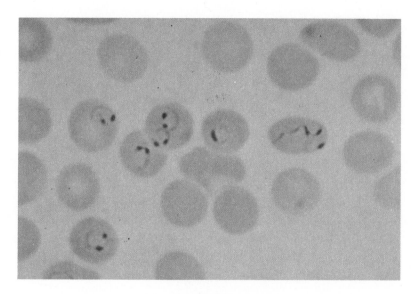

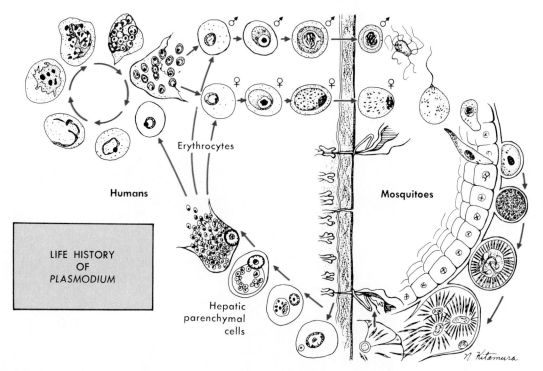

FIGURE 45.52

Life cycle of *Plasmodium.* (Modified from Wilcox, A. 1960. Manual for the microscopical diagnosis of malaria in man. U.S. Public Health Service, Washington, D.C. Illustration by Nobuko Kitamura.)

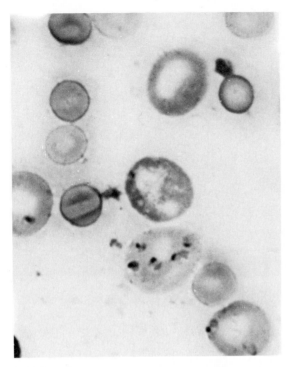

FIGURE 45.53

Babesia in red blood cells. (Photomicrograph by Zane Price. From Markell, E.K., and Voge, M. 1981. Medical parasitology, ed 5. W.B. Saunders, Philadelphia.)

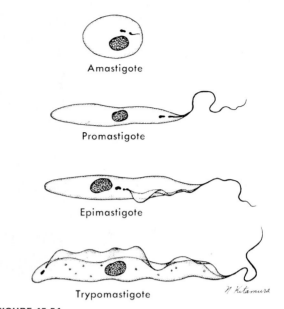

Amastigote

Promastigote

Epimastigote

Trypomastigote

N. Kitamura

FIGURE 45.54

Characteristic stages of species of *Leishmania* and *Trypanosoma* in human and insect hosts. (Illustration by Nobuko Kitamura.)

FIGURE 45.55

Leishmania donovani parasites in Küpffer cells of liver (2000×).

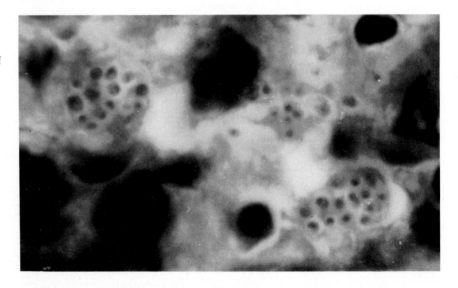

FIGURE 45.56
Leishmania donovani amastigotes.

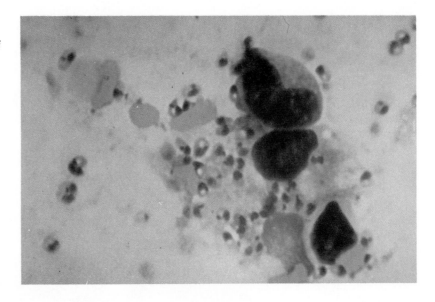

FIGURE 45.57
Trypanosoma gambiense in blood film (1600×).

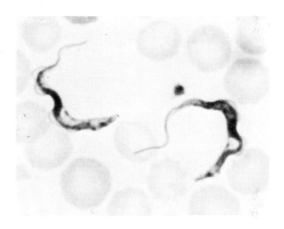

FIGURE 45.58
Trypanosoma cruzi trypomastigote.

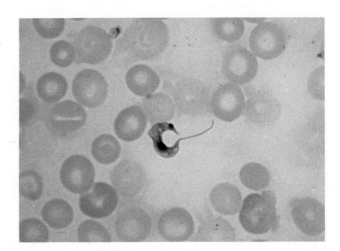

FIGURE 45.59

A, *Trypanosoma cruzi* in blood film (1600×). **B,** *Trypanosoma cruzi* parasites in cardiac muscle (2500×). (From Markell, E.K., and Voge, M. 1981. Medical parasitology, ed 5. W.B. Saunders, Philadelphia.)

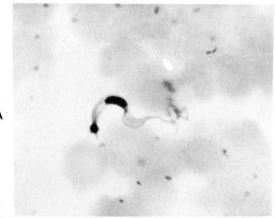

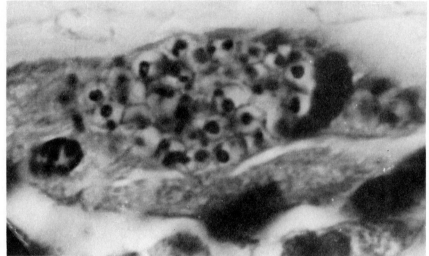

FIGURE 45.60

Life cycle of *Taenia saginata* and *Taenia solium*.

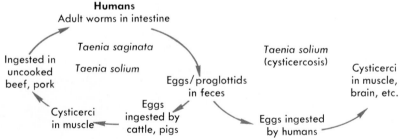

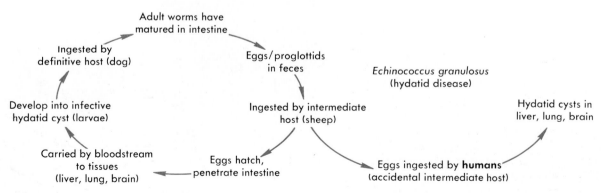

FIGURE 45.61

Life cycle of *Echinococcus granulosus* (hydatid disease).

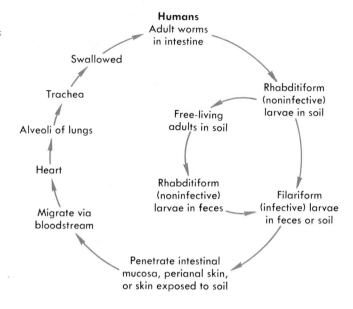

FIGURE 45.62

Life cycle of *Strongyloides stercoralis*.

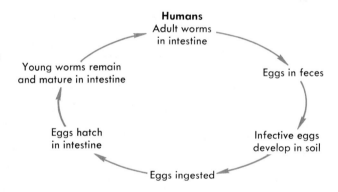

FIGURE 45.63

Life cycle of *Enterobius vermicularis* and *Trichuris trichiura* (direct type of cycle).

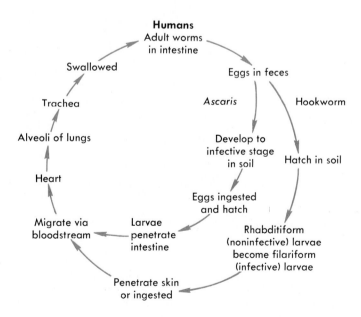

FIGURE 45.64

Life cycle of *Ascaris lumbricoides* and hookworms (indirect type of cycle).

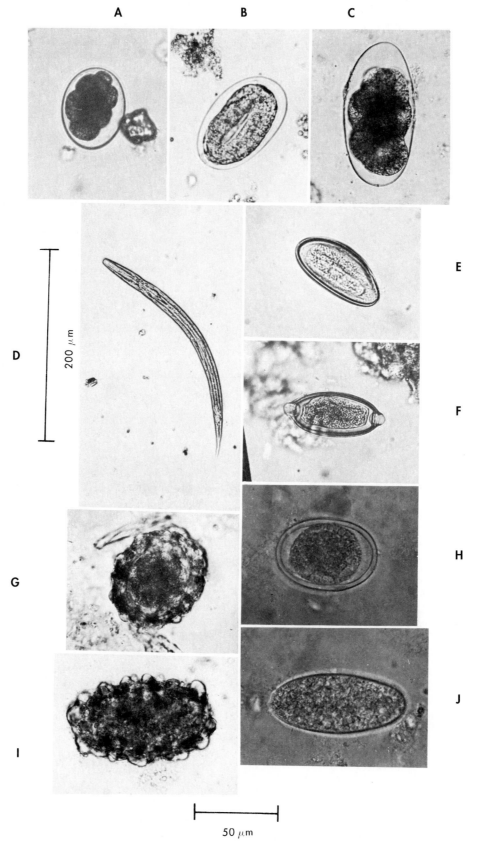

FIGURE 45.65

A, Immature hookworm egg. **B,** Embryonated hookworm egg. **C,** *Trichostrongylus orientalis,* immature egg. **D,** *Strongyloides stercoralis,* rhabditiform larva (200 μm). **E,** *Enterobius vermicularis* egg. **F,** *Trichuris trichiura* egg. **G,** *Ascaris lumbricoides,* fertilized egg. **H,** *A. lumbricoides,* fertilized egg, decorticate. **I,** *A. lumbricoides,* unfertilized egg. **J,** *A. lumbricoides,* unfertilized egg, decorticate.

FIGURE 45.66
Hookworm egg, iodine stain.

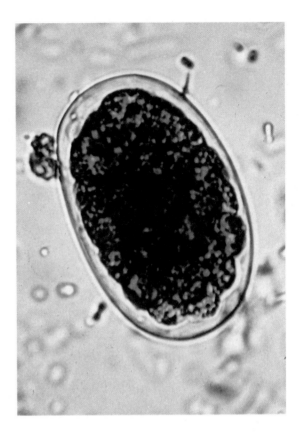

FIGURE 45.67
Strongyloides stercoralis larva, iodine stain.

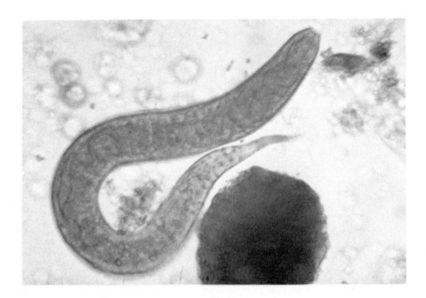

FIGURE 45.68

Rhabditiform larvae. **A,** *Strongyloides.* **B,** Hookworm. **C,** *Trichostrongylus. bc,* Buccal cavity; *es,* esophagus; *gp,* genital primordia; *cb,* beadlike swelling of caudal tip. (Illustration by Nobuko Kitamura.)

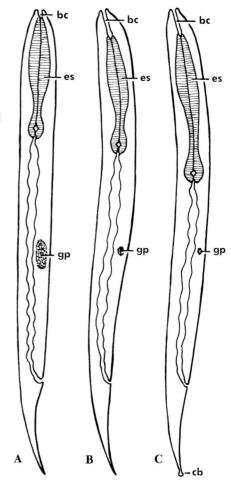

FIGURE 45.69

Enterobius vermicularis eggs (cellophane [Scotch] tape preparation).

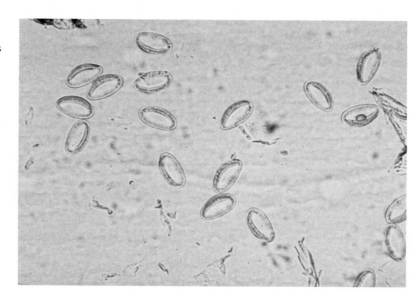

FIGURE 45.70

A, *Taenia* spp. egg. **B,** *Diphyllobothrium latum* egg. **C,** *Hymenolepis diminuta* egg. **D,** *Hymenolepis nana* egg. **E,** *Dipylidium caninum* egg packet.

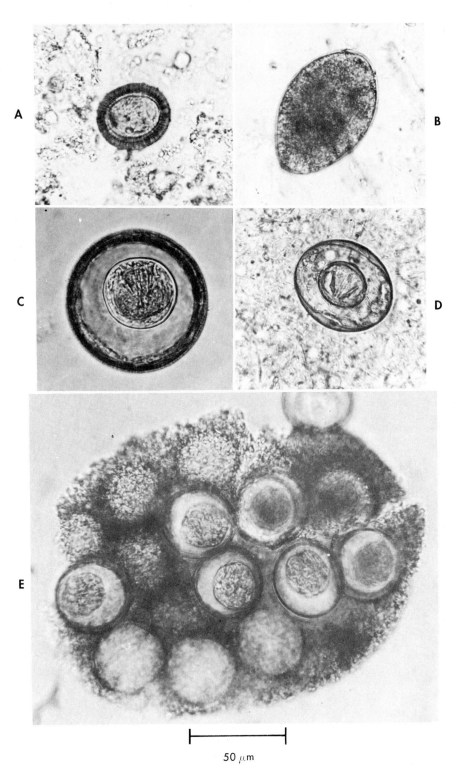

50 μm

TABLE 45.10

DIFFERENTIAL CHARACTERISTICS OF SOME IMPORTANT TAPEWORMS OF HUMANS

	TAENIA SAGINATA	*TAENIA SOLIUM*	*HYMENOLEPIS NANA*	*DIPHYLLOBOTHRIUM LATUM*
Length	4-8 m	3-5 m	2.5-4 cm	4-10 m
Scolex				
Shape	Quadrilateral	Globular	Usually not seen	Almondlike
Size	1 × 1.5 mm	1 × 1 mm		3 × 1 mm
Rostellum and hooklets	No	Yes		No
Suckers	Four	Four		Two (grooves)
Terminal proglottids (gravid)				
Size	19 × 7 mm, longer than wide	11 × 5 mm	Usually not seen	3 × 11 mm, wider than long
Primary lateral uterine branches	Fifteen to 30 on each side	Six to 12 on each side		Rosette shaped
Color	Milky white	Milky white		Ivory
Appearance in feces	Usually appear singly	Five or six segments		Varies from a few inches to a few feet in length
Ova				
Shape	Spheroid	Spheroid	Broadly oval	Oval
Size	35 μm	35 μm	30 × 47 μm	45 × 70 μm
Color	Rusty brown	Rusty brown	Pale	Yellow-brown
Embryo with hooklets	Yes	Yes	Yes	No
Operculum	No	No	No	Yes (difficult to see)

FIGURE 45.71

Gravid proglottids. **1,** *Taenia saginata.* **2,** *Taenia solium.* **3,** *Diphyllobothrium latum.* **4,** *Dipylidium caninum.* (From Garcia, L.S., and Bruckner, D.A. 1993. Diagnostic medical parasitology. 2nd ed. ASM Press, Washington D.C.)

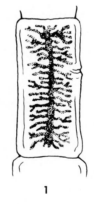
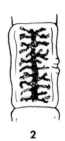

1   2   3   4

FIGURE 45.72

Life cycle of *Hymenolepis nana.*

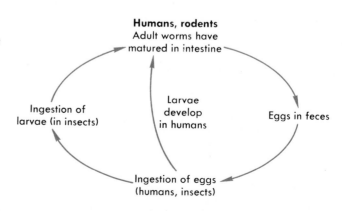

FIGURE 45.73

Dipylidium caninum egg packet. (Illustration by Nobuko Kitamura.)

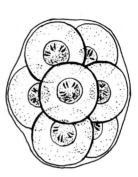

FIGURE 45.74

Life cycle of *Diphyllobothrium latum.*

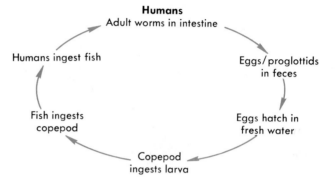

FIGURE 45.75

Life cycle of trematodes acquired by humans through ingestion of raw fish, crabs, or crayfish and vegetation.

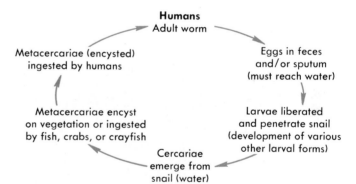

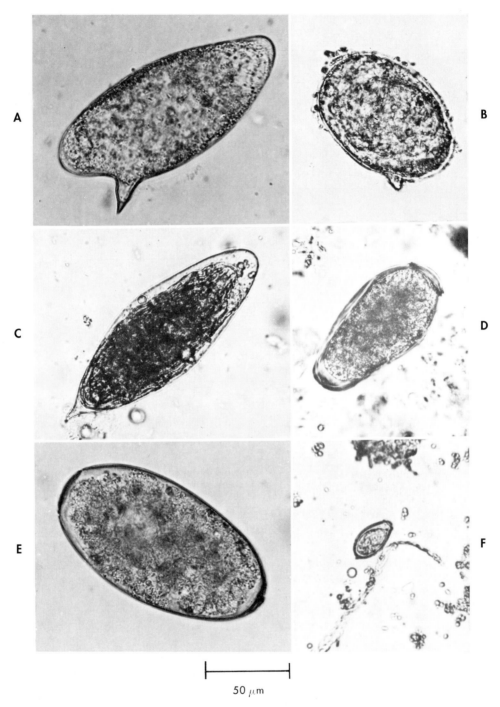

50 μm

FIGURE 45.76

A, *Schistosoma mansoni* egg. **B,** *Schistosoma japonicum* egg. **C,** *Schistosoma haematobium* egg. **D,** *Paragonimus westermani* egg. **E,** *Fasciola hepatica* egg. **F,** *Clonorchis (Opisthorchis) sinensis* egg.

FIGURE 45.77

Life cycle of human schistosomes.

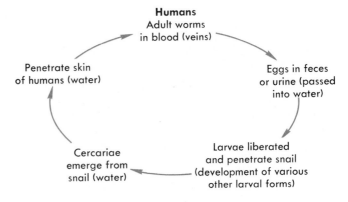

Humans
Adult worms
in blood (veins)

Penetrate skin
of humans (water)

Eggs in feces
or urine (passed
into water)

Cercariae
emerge from
snail (water)

Larvae liberated
and penetrate snail
(development of various
other larval forms)

FIGURE 45.78

Life cycle of human filarial worms.

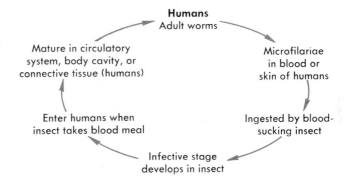

Humans
Adult worms

Mature in circulatory
system, body cavity, or
connective tissue (humans)

Microfilariae
in blood or
skin of humans

Enter humans when
insect takes blood meal

Ingested by blood-
sucking insect

Infective stage
develops in insect

FIGURE 45.79

Anterior and posterior ends of microfilariae found in humans. **A,** *Wuchereria bancrofti.* **B,** *Brugia malayi.* **C,** *Loa loa.* **D,** *Onchocerca volvulus.* **E,** *Mansonella perstans,* **F,** *Mansonella streptocerca.* **G,** *Mansonella ozzardi.*

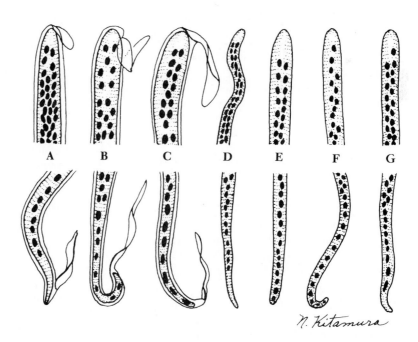

A   B   C   D   E   F   G

N. Kitamura

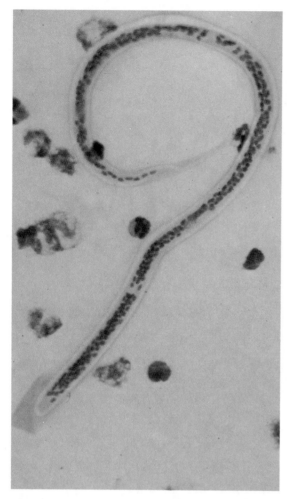

FIGURE 45.80

Microfilaria of *Wuchereria bancrofti* in thick blood film. (From Markell, E.K., and Voge, M. 1981. Medical parasitology, ed 5. W.B. Saunders, Philadelphia.)

chocerca volvulus and *Dipetalonema streptocerca,* produce microfilariae found in the subcutaneous tissues and dermis.

Diagnosis of filarial infections is often based on clinical grounds, but demonstration of the parasite is the only accurate means of confirming the diagnosis (Figure 45.79). Fresh blood films may be prepared; actively moving microfilariae can be observed in a preparation of this type. If the patient has a light infection, thick blood films can be prepared and stained. The Knott concentration procedure[46] and the membrane filtration technique[20,21] may also be helpful in recovering the organisms. Microfilariae of some strains tend to exhibit nocturnal periodicity; thus the time the blood is drawn may be critical in demonstrating the parasite. The microfilariae of *O. volvulus* and *D. streptocerca* are found in "skin snips," very thin slices of skin, which are teased apart in normal saline to release the organisms. Differentiation of

the species depends on (1) the presence or absence of the sheath and (2) the distribution of nuclei in the tail region of the microfilaria (Figure 45.80).

REFERENCES

1. Anderson, D., Soo, S.S., and Towler, H. 1991. *Acanthamoeba* keratitis: experience in a non-specialist microbiology laboratory. J. Clin. Pathol. 44:699.
2. Arakaki, T., Iwanaga, M., Kinjo, F., et al. 1990. Efficacy of agar-plate culture in detection of *Strongyloides stercoralis* infection. J. Parasitol. 76:425.
3. Auran, J.D., Starr, M.B., and Jakobiec, F.A. 1987. *Acanthamoeba* keratitis. Cornea 6:2.
4. Beal, C., Goldsmith, R., Kotby, M., et al. 1992. The plastic envelope method, a simplified technique for culture diagnosis of trichomoniasis. J. Clin. Microbiol. 30:2265.
5. Beaver, P.C. 1949. A nephelometric method of calibrating the photoelectric meter for making egg-counts by direct fecal smear. J. Parasitol. 35(sect. 2):13.
6. Beaver, P.C. 1950. The standardization of fecal smears for estimating egg production and worm burden. J. Parasitol. 36:451.
7. Beaver, P.C., Jung, R.C., and Cupp, E.W. 1984. Clinical parasitology, ed 9. Lea & Febiger, Philadelphia.
8. Bigby, T.V., Margolskee, D., Curtis, J.L., et al. 1986. The usefulness of induced sputum in the diagnosis of *Pneumocystis carinii* pneumonia in patients with the acquired immunodeficiency syndrome. Am. Rev. Respir. Dis. 133:515.
8a. Borchardt, K.A., and Smith, R.F. 1991. An evaluation of an InPouch TV culture method for diagnosing *Trichomonas vaginalis* infection. Genitourin. Med. 67:149.
9. Brooke, M.M., and Goldman, M. 1949. Polyvinyl alcohol-fixative as a preservative and adhesive for protozoa in dysenteric stools and other liquid material. J. Lab. Clin. Med. 34:1554.
10. Brooke, M.M., and Melvin, D. 1969. Morphology of diagnostic stages of intestinal parasites of man. DHEW Pub. No. (HSM)72-8116. U.S. Government Printing Office, Washington, D.C.
11. Bryan, R.T., Cali, A., Owen, R.L., and Spencer, H.C. 1991. Microsporidia: opportunistic pathogens in patients with AIDS. In Sun, T., editor. Progress in clinical parasitology, vol II, Field and Wood Medical Publishers, distributed by W.W. Norton, New York.
11a. Cali, A., Kotler D.P., and Orenstein, J.M. 1993. *Septata intestinalis* N.G., N.Sp., an intestinal microsporidian associated with chronic diarrhea and dissemination in AIDS patients. J. Euk. Microbiol. 40:101.
12. Cali, A., Miesler, D.M., Rutherford, I., et al. 1991. Corneal microsporidiosis in a patient with AIDS. Am. J. Trop. Med. Hyg. 44:463.
13. Cali, A., and Owen, R.L. 1990. Intracellular development of *Enterocytozoon*, a unique microsporidian in the intestine of AIDS patients. J. Protozool. 37:145.

14. Chang, T.H., Tsing, S.Y., and Tzeng, S. 1986. Monoclonal antibodies against *Trichomonas vaginalis.* Hybridoma 5:43.

15. Cohen, E.J., Buchanan, H.W., Laughrea, P.A., et al. 1985. Diagnosis and management of *Acanthamoeba* keratitis. Am. J. Ophthalmol. 100:389.

16. Cohen, E.J., Parlato, C.J., Arentsen, J.J., et al. 1987. Medical and surgical treatment of *Acanthamoeba* keratitis. Am. J. Ophthalmol. 103:615.

17. Cohen, J.D., Ruhlig L., Jayich, S.A., et al. 1984. *Cryptosporidium* in acquired immunodeficiency syndrome. Dig. Dis. Sci. 29:773.

18. Committee on Education, American Society of Parasitologists. 1977. Procedures suggested for use in examination of clinical specimens for parasitic infection. J. Parasitol. 63:959.

19. Current, W.L., and Garcia, L.S. 1991. Cryptosporidiosis. Clin. Microbiol. Rev. 3:325.

20. Dennis, D.T., and Kean, B.H. 1971. Isolation of microfilariae: report of a new method. J. Parasitol. 57:1146.

21. Desowitz, R.S., and Hitchcock, J.C. 1974. Hyperendemic bancroftian filariasis in the Kingdom of Tonga: the application of the membrane filter concentration technique to an age-stratified blood survey. Am. J. Trop. Med. Hyg. 23:877.

22. Desportes, I., Le Charpentier, Y., Galian, A., et al. 1985. Occurrence of a new microsporidian: *Enterocytozoon bieneusi* n.g., n.sp., in the enterocytes of a human patient with AIDS. J. Protozool. 32:250.

23. Didier, E.S., Didier, P.J., Friedberg, D.N., et al. 1991. Isolation and characterization of a new human microsporidian, *Encephalitozoon hellem* (n.sp), from three AIDS patients with keratoconjunctivitis. J. Infect. Dis. 163:617.

24. Donzis, P.B., Mondino, B.J., Weissman, B.A., and Bruckner, D.A. 1987. Microbial contamination of contact lens care systems. Am. J. Ophthalmol. 104:325.

25. Eeftinck Schattenkerk, J.K.M., van Gool, T., van Ketel, R.J., et al. 1991. Clinical significance of small intestinal microsporidiosis in HIV-1 infected individuals. Lancet 337:895.

26. Feinberg, J.G., and Whittington, M.J. 1957. A culture medium for *Trichomonas vaginalis* Donne and species of *Candida.* J. Clin. Pathol. 10:327.

27. Fritsche, T.R., and Bergeron, D.L. 1987. *Acanthamoeba* keratitis. Clin. Microbiol. Newsletter 9:109.

28. Frixione, E., Ruiz, L., Santillan, M., et al. 1992. Dynamics of polar filament discharge and sporoplasm expulsion by microsporidian spores. Cell Motil. Cytoskeleton 22:38.

29. Garcia, L.S., and Ash, L.R. 1979. Diagnostic parasitology: clinical laboratory manual. ed 2. Mosby, St. Louis.

30. Garcia, L.S., Brewer, T.C., and Bruckner, D.A. 1987. Fluorescence detection of *Cryptosporidium* oocysts in human fecal specimens by using monoclonal antibodies. J. Clin. Microbiol. 25:119.

31. Garcia, L.S., Brewer, T.C., and Bruckner, D.A. 1989. Incidence of *Cryptosporidium* in all patients submitting stool specimens for ova and parasite examination: monoclonal antibody IFA method. Diagn. Microbiol. Infect. Dis. 11:25.

32. Garcia, L.S., and Bruckner, D.A. 1993. Diagnostic medical parasitology, ed 2. American Society for Microbiology, Washington, D.C.

33. Garcia, L.S., Bruckner, D.A., and Brewer, T.C. 1988. Cryptosporidiosis in patients with AIDS. Am. Clin. Prod. Rev. 7:38.

34. Garcia, L.S., Bruckner, D.A., Brewer, T.C., and Shimizu, R.Y. 1983. Techniques for the recovery and identification of *Cryptosporidium* oocysts from stool specimens. J. Clin. Microbiol. 18:185.

35. Garcia, L.S., and Current, W.L. 1989. Cryptosporidiosis: clinical features and diagnosis. Clin. Rev. Clin. Lab. Sci. 27:439.

36. Garcia, L.S., Shimizu, R.Y., Shum, A.C. and Bruckner, D.A. 1993. Evaluation of intestinal protozoan morphology in polyvinyl alcohol preservative: comparison of zinc-based and mercuric chloride-based compounds for use in Schaudinn's fixative. J. Clin. Microbiol. 31:307.

37. Garcia, L.S., Shimizu, R.Y., Shum, A.C. and Zierdt, C.H. 1992. Diagnosis of *Enterocytozoon bieneusi* microsporidiosis by recovery of spores from human fecal specimens. Abstr. No. 801, ICAAC Meeting, Anaheim, Calif.

38. Garcia, L.S., Shum, A.C., and Bruckner, D.A. 1992. Evaluation of a new monoclonal antibody combination reagent for the direct fluorescent detection of *Giardia* cysts and *Cryptosporidium* oocysts in human fecal specimens. J. Clin. Microbiol. 30:3255.

39. Gill, V.J., Evans, G., Stock, F., et al. 1987. Detection of *Pneumocystis carinii* by a fluorescent antibody stain using a combination of three monoclonal antibodies. J. Clin. Microbiol. 25:1837.

40. Gomori, G. 1950. A rapid one-step trichrome stain. Am. J. Clin. Pathol. 20:661.

41. Harada, Y., and Mori, O. 1955. A new method for culturing hookworm. Yonago Acta Med. 1:177.

42. Healy, G.R., and Ruebush, T.K., II. 1980. Morphology of *Babesia microti* in human blood smears. Am. J. Clin. Pathol. 73:107.

43. Hendricks, L., and Wright, N. 1979. Diagnosis of cutaneous leishmaniasis by in vitro cultivation of saline aspirates in Schneider's Drosophila medium. Am. J. Trop. Med. Hyg. 28(6):962.

44. Kagan, I.G. 1980. Serodiagnosis of parasitic diseases. In Lennette, E.H., Balows, A., Hausler, W.J., Jr., and Truant, J.P., editors. Manual of clinical microbiology, ed 3. American Society for Microbiology, Washington, D.C.

45. Kilvington, S., Larkin, D.F.P., White, D.G., and Beeching, J.R. 1990. Laboratory investigation of *Acanthamoeba* keratitis. J. Clin. Microbiol. 28:2722.

46. Knott, J.I. 1939. A method for making microfilarial surveys on day blood. Trans. R. Soc. Trop. Med. Hyg. 33:191.

47. Kocka, F., Peters, C., Dacumos, E., et al. 1991. Outbreaks of diarrheal illness associated with Cyanobacteria (Blue-Green Algae)-like bodies— Chicago and Nepal, 1989 and 1990. M.M.W.R. 40:325.

48. Koga, K., Kasuya, S., Khamboonruang, C., et al. 1991. A modified agar plate method for detection of *Strongyloides stercoralis.* Am. J. Trop. Med. Hyg. 45:518.

49. Kovacs, J.A., Ng, V.L., Masur, H., et al. 1988. Diagnosis of *Pneumocystis carinii:* improved detection in sputum using monoclonal antibodies. N. Engl. J. Med. 318:589.

50. Ledford, D.K., Overman, M.D., Gonzalvo, A., et al. 1985. Microsporidiosis myositis in a patient with acquired immunodeficiency syndrome. Ann. Intern. Med. 102:628.

51. Lindquist, T.D., Sher, N.A., and Doughman, D.J. 1988. Clinical signs and medical therapy of early *Acanthamoeba* keratitis. Arch. Ophthalmol. 106:73.

52. Long, E.G., Ebrahimzadeh, A., White, E.H., et al. 1990. Alga associated with diarrhea in patients with acquired immunodeficiency syndrome and in travelers. J. Clin. Microbiol. 28:1101.

53. Long, E.G., White, E.H., Carmichael, W.W., et al. 1991. Morphologic and staining chracteristics of a *Cyanobacterium*-like organism associated with diarrhea. J. Infect Dis. 164:199.

54. Ma, P., Visvesvara, G.S., Martinez, A.J., et al. 1990. *Naegleria* and *Acanthamoeba* infections: review. Rev. Infect. Dis. 12:490.

55. Maekelt, G.A. 1964. A modified procedure of xenodiagnosis for Chagas' disease. Am. J. Trop. Med. Hyg. 13:11.

56. Markell, E.K., and Udkow, M.P. 1986. *Blastocystis hominis:* pathogen or fellow traveler? Am. J. Trop. Med. Hyg. 35:1023.

57. Markell, E.K., and Voge, M. 1981. Medical parasitology, ed 5. W.B. Saunders, Philadelphia.

58. Markell, E.K., Voge, M., and John, D.T. 1992. Medical parasitology, ed 7. W.B. Saunders, Philadelphia.

59. McCauley, D.I., Naidick, D.P., Leitman, B.S., et al. 1982. Radiographic pattern of opportunistic lung infections and Kaposi sarcoma in homosexual men. Am. J. Radiol. 139:653.

60. Melvin, D.M., and Brooke, M.M. 1974. Laboratory procedures for the diagnosis of intestinal parasites. DHEW Pub. No. (CDC) 75-8282, U.S. Government Printing Office, Washington, D.C.

61. Miller, L.H., Mason, S.J., Dvorak, J.A., et al. 1975. Erythrocyte receptors for *(Plasmodium knowlesi)* malaria: Duffy blood group determinants. Science 189:561.

62. Moore, M.B., McCulley, J.P., Luckenbach, M., et al. 1985. *Acanthamoeba* keratitis associated with soft contact lenses. Am. J. Ophthalmol. 100:396.

63. Nair, C.P. 1953. Rapid staining of intestinal amoebae on wet mounts. Nature 172:1051.

64. Navin, T.R., and Juranek, D.D. 1984. Cryptosporidiosis: clinical, epidemiologic and parasitologic review. Rev. Infect. Dis. 6:313.

65. Ng, E., Markell, E.K., Fleming, R.L., and Fried, M. 1984. Demonstration of *Isospora belli* by acid-fast stain in a patient with acquired immune deficiency syndrome. J. Clin. Microbiol. 20:384.

66. Orenstein, J.M. 1991. Microsporidiosis in the acquired immunodeficiency syndrome. J. Parasitol. 77:843.

67. Ortega, Y., Sterling, C.R., Gilman, R.H., et al. 1993. *Cyclospora cayetanensis:* a new protozoan pathogen of humans. N. Eng. J. Med. 328:1308.

68. Palmer, J. 1991. Modified iron hematoxylin/kinyoun stain. Clin. Microbiol. Newsletter 13:39 (letter).

69. Parasitology Subcommittee, Microbiology Section of Scientific Assembly, American Society for Medical Technology. 1978. Recommended procedures for the examination of clinical specimens submitted for the diagnosis of parasitic infections. Am. J. Med. Technol. 44:1101.

69a. Ryan, N.J. et al. 1993. A new trichome-blue stain for detection of microsporidial species in urine, stool, and nasopharyngeal specimens. J. Clin. Microbiol. 31: 3264-3269.

70. Sapèro, J.J., and Lawless, D.K. 1953. The MIF stain-preservation technique for the identification of intestinal protozoa. Am. J. Trop. Med. Hyg. 2:613.

71. Scholten, T.H. 1972. An improved technique for the recovery of intestinal protozoa. J. Parasitol. 58:633.

72. Scholten, T.H., and Yang, J. 1974. Evaluation of unpreserved and preserved stools for the detection and identification of intestinal parasites. Am. J. Clin. Pathol. 62:563.

73. Schwartz, D.A., Bryan, R.T., Hewanlowe, K.O., et al. 1992. Disseminated microsporidiosis *(Encephalitozoon hellem)* and acquired immunodeficiency syndrome—autopsy evidence for respiratory acquisition. Arch. Pathol. Lab. Med. 116:660.

74. Shadduck, J.A., and Greeley, E. 1989. Microsporidia and human infections. Clin. Microbiol. Rev. 2:158.

75. Sheehan, D.J., Raucher, B.G., and McKitrick, J.C. 1986. Association of *Blastocystis hominis* with signs and symptoms of human disease. J. Clin. Microbiol. 24:548.

76. Spencer, F.M., and Monroe, L.S. 1976. The color atlas of intestinal parasites, ed 2. Charles C Thomas, Springfield, Ill.

77. Stehr-Green, J.K., Bailey, T.M., and Visvesvara, G.S. 1989. The epidemiology of *Acanthamoeba* keratitis in the United States. Am. J. Ophthalmol. 107:331.

78. Sterling, C.R., and Arrowood, M.J. 1986. Detection of *Cryptosporidium* sp. infections using a direct immunofluorescent assay. Pediatr. Infect. Dis. 5:S139.

79. Stoll, N.R., and Hausheer, W.C. 1926. Concerning two options in dilution egg counting: small drop and displacement. Am. J. Hyg. 6:134.

80. Stover, D.E., Zaman, M.B., Hadju, S.I., et al. 1984. Bronchoalveolar lavage in the diagnosis of diffuse pulmonary infiltrates in the immunosuppressed host. Ann. Intern. Med. 101:1.

81. Strano, A.J., Cali, A., and Neafie, R.C. 1976. Microsporidiosis. p. 336. In Binford, C.H., and Connor, D.H., editors. Pathology of tropical and extraordinary diseases. Armed Forces Institute of Pathology, Washington, D.C.

81a. Stratton, N., Hryniewicki, J., Aarnaes, S.L., et al. 1991. Comparison of monoclonal antibody and calcofluor white stains for the detection of *Pneumocystis carinii* from respiratory specimens. J. Clin. Microbiol. 29:645.

82. Sulzer, A.J., and Wilson, M. 1967. The use of thick-smear antigen slides in the malaria indirect fluorescent antibody test. J. Parasitol. 53:1110.

83. Sulzer, A.J., Wilson, M., and Hall, E.C. 1969. Indirect fluorescent antibody tests for parasitic diseases. V. An evaluation of a thick-smear antigen in the IFA for malaria antibodies. Am. J. Trop. Med. Hyg. 18:199.

84. Tompkins, V.N., and Miller, J.K. 1947. Staining intestinal protozoa with iron-hematoxylin-phosphotungstic acid. Am. J. Clin. Pathol. 17:755.

85. Visvesvara, G.S. In Isenberg, H.D., editor. 1992. Clinical microbiology procedures handbook (parasitology section). American Society for Microbiology, Washington, D.C.
86. Weber, R., Bryan, R.T., Owen, R.L., et al. 1992. Improved light-microscopical detection of microsporidia spores in stool and duodenal aspirates. N. Engl. J. Med. 326:161.
87. Wheatley, W.B. 1951. A rapid staining procedure for intestinal amoebae and flagellates. Am. J. Clin. Pathol. 21:990.
88. Wicher, V., Baughn, R.E., Fuentealba, C., et al. 1991. Enteric infection with an obligate intracellular parasite, *Encephalitozoon cuniculi,* in an experimental model. Infect. Immun. 59:2225.
89. Wilcox, A. 1960. Manual for the microscopical diagnosis of malaria in man. Public Health Service, Washington, D.C.
90. Wilhelmus, K.R., Osato, M.S., Font, R.L., et al. 1986. Rapid diagnosis of *Acanthamoeba* keratitis using calcofluor white. Arch. Ophthalmol. 104:1309.
91. Zierdt, C.H. 1978. *Blastocystis hominis,* an intestinal protozoan parasite of man. Public Health Lab. 36:146.
92. Zierdt, C.H. 1988. *Blastocystis hominis,* a long-misunderstood intestinal parasite. Parasitol. Today 4:15.
93. Zierdt, C.H., and Zierdt, W.S. 1992. IFA identification of the microsporidian *Enterocytozoon bieneusi* in clinical samples using antisera to *Encephalitozoon cuniculi* and *Encephalitozoon hellem.* Abstr. No. C-454, ASM General Meeting, New Orleans.

GLOSSARY

A Acid.

abscess Localized collection of pus.

accessioning Receipt and recording of specimens delivered to laboratory

accolé Early ring form of *Plasmodium falciparum* found at margin of red cell.

acid-fast Characteristic of certain bacteria, such as mycobacteria, that involves resistance to decolorization by acids when stained by an aniline dye, such as carbolfuchsin.

acquired immunodeficiency syndrome (AIDS) Severe immune deficiency disease caused by human immunodeficiency virus (HIV-1) infection of the T cells, characterized by opportunistic infections and other complications.

acute serum Serum collected for antibody determination early in course of an illness when there would have been little or no antibody produced.

acute urethral syndrome Lower urinary tract infection that may be difficult to differentiate from cystitis; seen most commonly in younger, sexually active females and caused by *Escherichia coli* (counts as low as 100 per milliliter may be significant in this situation), *Chlamydia,* and other organisms.

aerobe, obligate Microorganism that lives and grows freely in air and cannot grow anaerobically.

aerogenic Producing gas (in contrast to anaerogenic: non–gas-producing).

aerosol Atomized particles suspended in air; in context of this book, microorganisms suspended in air.

aerotolerant Ability of an anaerobic microorganism to grow in air, usually poorly, especially after initial anaerobic isolation.

AFB Acid-fast bacilli.

affinity Inherent attraction and relationship.

agarose gel electrophoresis Separation of proteins based on molecular weight by electrical-current-stimulated movement through a semisolid gel matrix.

agglutination Aggregation or clumping of particles, such as bacteria when exposed to specific antibody.

AIDS Acquired immunodeficiency syndrome.

AIDS-related complex (ARC) Prodromal symptoms in patients infected with HIV virus, including lymphadenopathy, fever, weight loss, and malaise.

Alk (or K) Alkaline.

alopecia Baldness.

AMI Antibody-mediated immune response.

aminoglycosides Group of related antibiotics including streptomycin, kanamycin, neomycin, tobramycin, gentamicin, and amikacin.

amniotic Pertaining to the innermost fetal membrane forming a fluid-filled sac.

anaerobe, obligate Microorganism that grows only in complete or nearly complete absence of air or molecular oxygen.

anaerogenic Non–gas-producing.

analytic reagent (AR) Grade of chemical.

anamnestic response More rapid production of antibodies in response to exposure to an antigen previously encountered.

anergy Absence of reaction to antigens or allergens.

antagonism Diminution of activity of one drug by a second one.

antibiogram Distinctive pattern of susceptibility of an organism to a battery of antimicrobial agents.

antibiotic Substance, produced by a microorganism, that inhibits or kills other microorganisms; a broad-spectrum antibiotic is therapeutically effective against a wide range of bacteria.

antibody Substance (immunoglobulin), formed in blood or tissues that interacts only with antigen that induced its synthesis (e.g., agglutinin).

antigen Molecular structure that is capable of stimulating production of antibody.

antigenic determinant Antigen.

antimicrobial Chemical substance, either produced by a microorganism or by synthetic means, that is capable of killing or suppressing growth of microorganisms.

antiseptic Compound that stops or inhibits growth of bacteria without necessarily killing them.

arboviruses Arthropod-borne viruses.

ART Automated reagin test for syphilis.

arthritis, septic Infection of synovial tissue and joint fluid of one or more joints; characterized by joint pain, stiffness, swelling, and fever.

arthroconidium Spore formed by septation of a hypha and subsequent separation of septa.

ascitic fluid Serous fluid in peritoneal cavity.

assimilation Utilization of nutrients. Assimilation tests are used to determine whether yeasts are able to grow with only a single carbohydrate or nitrate; these tests are useful for classification of yeasts.

ATCC American Type Culture Collection.

autoclave Instrument consisting of a double-walled, sealable enclosure in which steam heat at greater than atmospheric pressure is used to sterilize biologically contaminated material.

autotroph Organism that can utilize inorganic carbon sources (CO_2).

auxotroph Differing from the wild strain (prototroph) by an additional nutritional requirement.

avid The property of binding strongly, such as an antibody that strongly binds to an antigen.

avidity Firmness of union of two substances; used commonly to describe union of antibody to antigen.

B cells Lymphocytes involved in antibody production.

bacteremia Presence of viable organisms in blood.

bacterial vaginosis Noninflammatory condition in vagina characterized by foul-smelling vaginal discharge and presence of mixed bacteria.

bactericidal Term used to describe a drug that kills microorganisms.

bacteriocins Antibiotic-like substances, produced by bacteria, that exert a lethal effect on other bacteria.

bacteriophage Virus that infects a bacterial cell, sometimes bringing about its lysis.

bacteriostatic Term used to describe a drug that inhibits growth of an organism without killing it.

bacteriuria Presence of bacteria in urine.

BAP Blood agar plate.

BCG Bacille Calmette-Guérin, an attenuated strain of *Mycobacterium tuberculosis* used for immunization.

benign tertian malaria Malaria caused by *Plasmodium vivax*.

Beta-lactamases Enzymes that destroy penicillins and/or cephalosporins and are produced by a variety of bacteria.

BFP Biological false-positive.

bifurcated Divided into two branches.

biological safety cabinet Enclosure in which one can work with relatively dangerous organisms without risk of acquiring or spreading infection caused by them. These cabinets, also called biosafety hoods, vary in design according to nature of agents to be worked with. The simpler ones maintain a negative pressure within the work area and a laminar air curtain, both of which operate to prevent escape of organisms from interior of hood. Air that is exhausted may be passed through a high-efficiency bacterial filter that will trap all microorganisms that are antici-

pated or may be passed through a furnace that will incinerate any organisms. (See Chapter 2 for more information.)

bioluminescence Light generation by living organisms.

biopsy Removal of tissue from a living body for diagnostic purposes (e.g., lymph node biopsy).

biotin Small vitamin with two binding sites, one of which can bind covalently with nucleic acid leaving the other free to form a strong bond with the protein avidin, which in turn can be bound to enzymes. The system is used as a label for nucleic acid probe detection.

biotype Biological or biochemical type of an organism. Organisms of the same biotype will display identical biological or biochemical characteristics. Certain key markers are used to define and recognize biotypes in tracing the spread of organisms in the environment and in epidemics or outbreaks.

blackwater fever Condition in which the diagnostic symptom is passage of reddish or red-brown urine, which indicates massive intravascular hemolysis *(Plasmodium falciparum)*.

blastoconidium A spore formed by budding, as in yeasts.

blepharitis Inflammation of eyelids.

blind loop A condition in which continuity of the bowel has been interrupted in such a way that a segment of bowel receives material from higher in the bowel in the usual way but cannot empty normally because it ends in a blind sac. This leads to proliferation of microorganisms, especially anaerobes, in the blind loop of bowel and may produce profound pathophysiologic effects. There are other situations in which there may be bacterial overgrowth in a segment of the bowel by other mechanisms.

B-lymphocytes (or B cells) Bursa-derived lymphocytes important in humoral immunity.

breakpoint Level of an antibacterial drug achievable in serum; organisms inhibited by this level of drug are considered susceptible. In certain situations, clinicians strive to achieve serum or body fluid levels several times that of the breakpoint.

brightfield microscopy Conventional microscopy in which the object to be viewed is illuminated from below.

bronchial lavage Similar to bronchial washings but this term implies instillation of a larger volume of fluid before aspiration. Alveolar organisms may be present in the lavage.

bronchial washings Fluid that may be aspirated from bronchial tree during bronchoscopy.

bronchitis Inflammation of mucous membranes of bronchi; often caused by infectious agents, viruses in particular.

bronchoscopy Examination of bronchi through a bronchoscope, a tubular illuminated instrument introduced through the trachea (windpipe).

BSC Biological safety cabinet.

bubo Inflammatory enlargement of lymph node, usually in the groin or axilla.

buffy coat Layer of white blood cells and platelets above red blood cell mass when blood is sedimented.

bullae Large blebs or blisters, filled with fluid, in or just beneath the epidermal layer of skin.

bursitis Inflammation of a bursa, which is a small sac lined with synovial membrane and filled with fluid interposed between parts that move on each other.

butt Lower portion or medium in a tube that has the medium dispensed in such a way that the lower portion fills the tube entirely while the upper portion is distributed in the form of a slanted surface leaving an air space between the slant and the opposite wall of the tube.

butyrous Butterlike consistency.

C & S Culture and sensitivity.

calibrated loop Bacteriological loop that is carefully calibrated to deliver a specified volume of fluid providing that directions are followed carefully and the loop has not been damaged; used as a simple means of quantitating the number of organisms present, especially for urine culture.

CAMP Lytic factor named after Christie, Atkins, and Munch-Peterson.

candle jar A jar with a lid providing a gas-tight seal in which a small white candle is placed and lit after the culture plates have been placed inside. Candle will burn only until the oxygen concentration has been lowered to the point where it will no longer support the flame. Atmosphere of such a jar has a lower oxygen content than room air and a carbon dioxide content of about 3%.

cannula An artificial tube for insertion into a tube or cavity of the body.

CAP Chocolate agar plate.

CAPD Chronic ambulatory peritoneal dialysis.

capsid Protein layer or coat surrounding viral nucleic acid core.

capneic incubation Incubation under increased CO_2 tension, as in a candle extinction jar (approximate 3% CO_2).

capnophilic Term used to describe microorganisms that prefer an incubation atmosphere with increased carbon dioxide concentration.

capsomere Protein subunits that serve as components of the viral capsid.

capsule Gelatinous material surrounding bacterial cell wall. Usually of polysaccharide nature.

carrier One who harbors a pathogenic organism but is not affected by it.

ceseation necrosis Tissue death with loss of cell outlines and a cheeselike, amorphous appearance.

catalase Bacterial enzyme that breaks down peroxides with liberation of free oxygen.

catheter Flexible tubular (rubber or plastic) instrument used for withdrawing fluids from (or intro-ducing fluids into) a body cavity or vessel (e.g., urinary bladder catheter).

cation A positive ion.

CDC Centers for Disease Control.

cell line A cell culture that has been passed (subcultured) in vitro.

cell line, continuous Line of tissue cells that is maintained by serial culture of an established cell line.

cell line, primary Line of tissue cells established by cutting up fresh tissue, often kidney, into tiny pieces, trypsinizing, and putting in a flask with appropriate medium.

cellulitis Inflammation of subcutaneous tissue.

cerebriform With brainlike folds.

cervical Pertaining either to the neck or to the cervix of the uterus.

CF Complement fixation.

CFU Colony-forming unit (i.e., colony count).

Charcot-Leyden crystals Slender crystals shaped like a double pyramid with pointed ends, formed from the breakdown products of eosinophils and found in feces, sputum, and tissues: indicative of an immune response that may have parasitic or nonparasitic causes.

chemotherapeutic Chemical agent used in the treatment of infections (e.g., sulfonamides).

chlamydospore Thick-walled spore formed from a vegetative cell.

chromatography Method of chemical analysis by which a mixture of substances is separated by fractional extraction or adsorption or ion exchange on a porous solid.

chromogen Bacterial species whose colonial growth is pigmented (e.g., *Flavobacterium* species, yellow).

chromogenic Giving rise to color, as chromogenic substrates for colored products of biochemical reactions or chromogenic bacteria that produce pigmented colonies.

CIE Counterimmunoelectrophoresis.

clavate Club-shaped.

clone Group of microorganisms of identical genetic makeup derived from a single common ancestor.

CMI Cell-mediated immunity.

CNS Central nervous system.

coagglutination Agglutination of protein A–containing cells of *Staphylococcus aureus* coated with antibody molecules when exposed to corresponding antigen.

colitis Inflammation of mucosa of colon.

colon Macroscopically visible growth of a microorganism on a solid culture medium.

colorimetry Chemical analysis by color determination.

commensal Microorganism living on or in a host but causing the host no harm.

complement fixation test Antigen-antibody test based on fixation of complement in the presence of both elements and use of an indicator system to determine whether or not complement has been fixed.

congenital Existing before or at birth.

conjugation Passing genetic information between bacteria by transferring chromosomal material, often via pili.

conjunctivitis Inflammation of the conjunctivae or membranes of the eye and eyelid.

convalescent serum Serum collected later in the course of an illness than the acute serum, usually at least 2 weeks after initial collection.

counterimmunoelectrophoresis (CIE) Detection of antibody or antigen by precipitin reaction that occurs in solid-phase (gel or paper) system that uses electric current to carry reactants toward each other.

CPC Clinical pathologic conference.

CPE Cytopathogenic (cytopathic) effect; visual effect of virus infection on cell culture.

Creutzfeldt-Jakob disease (CIE) Debilitating prion-caused disease characterized by dementia, ataxia, delirium, stupor, coma, and death: has been transmitted by organ transplant.

croup Inflammation of upper airways (larynx, trachea) with respiratory obstruction, often caused by virus infections in children.

CRP C-reactive protein.

CSF Cerebrospinal fluid.

culdocentesis Aspiration of fluid from the cul-de-sac by puncture of the vaginal vault.

CYE Charcoal yeast extract (agar plate).

cystitis Inflammation of urinary bladder, most often caused by bacterial infection

cytopathic effect (CPE) Alteration in cell morphology resulting from viral infection of a cell culture monolayer.

cytotoxin Toxin that produces cytopathic effects in vivo or in a tissue culture system.

darkfield microscopy Technique used to visualize very small microorganisms or their characteristics by a system that permits light to be reflected or refracted from surface of objects being viewed.

debridement Surgical or other removal of nonviable tissue.

decontamination Process of rendering an object or area safe for unprotected people by removing or making harmless biological or chemical agents.

decubitus ulcer A craterlike defect in skin and subcutaneous tissue caused by prolonged pressure on the area. This occurs primarily over bony prominences of the lower back and hips in individuals who are unable to care for themselves well and unable to roll or move periodically; also known as pressure sore or bedsore.

definitive host Host in which the sexual reproduction of a parasite occurs.

dermatophyte A fungus parasitic on skin, hair, or nails.

desquamation Shedding or scaling of skin or mucous membrane.

DFA Direct fluorescent antibody test.

DIC Disseminated intravascular coagulation.

dichotomous Branching in two directions.

diluent Fluid used to dilute a substance.

dimorphic fungi Fungi with both a mold phase and a yeast phase.

direct wet mount A preparation from clinical material suspended in sterile saline or other liquid medium on a glass slide and covered with a coverslip; used for microscopic examination to detect microorganisms in clinical material and, in particular, to detect motility directly.

disinfectant Agent that destroys or inhibits microorganisms that cause disease.

DNA Deoxyribonucleic acid, the lipoprotein molecule that contains the genetic code for most living things.

DNase Deoxyribonuclease, an enzyme that depolymerizes DNA.

DRGs Diagnosis related groups; system introduced to control rising medical costs under which hospitals are reimbursed for costs of caring for patients based on the primary diagnoses of the patients and what is considered to be a fair dollar value for treatment of usual patients with those diagnoses.

droplet nucleus A tiny aerosolized particle that, because of its lack of mass, may stay suspended in air for extended periods of time.

Durham tube Small tube placed in an inverted position below the surface of a growth-supportive broth and filled completely with the broth solution. Gas produced by the organism displaces the broth in the Durham tube, and the resulting gas bubble is visual evidence of gas production.

Dx Diagnosis.

dysentery Inflammation of the intestinal tract, particularly the colon, with frequent bloody stools (e.g., bacillary dysentery).

dysgonic Growing poorly (bacterial cultures).

ectoparasite Organism that lives on or within skin.

ectothrix Outside of hair shafts.

edema Excessive accumulation of fluid in tissue spaces.

effusion Fluid escaping into a body space or tissue (e.g., pleural effusion).

Eh Oxidation-reduction potential.

elementary body The infectious stage of *Chlamydia* or a cellular inclusion body of a viral disease.

elephantiasis Condition caused by inflammation and obstruction of the lymphatic system, resulting in hypertrophy and thickening of the surrounding tissues, usually involving the extremities and external genitalia (often as a result of filariasis).

ELISA Enzyme-linked immunosorbent assay.

elution Process of extraction by means of a solvent.

EMB Eosin-methylene blue (agar plate).

emetic Inducing vomiting.

empyema Accumulation of pus in a body cavity, particularly empyema of the thorax or chest.

encephalitis Inflammation of the brain.

endocarditis A serious infection of the endothelium of the heart, usually involving leaflets of the heart valves where destruction of valves or distortion of them by formation of vegetations may lead to serious physiological disturbances and death: also, an inflammation of the endocardial surface (much less common).

endocervix Mucous membrane of the cervical canal.

endogenous Developing from within the body.

endoparasite Parasite that lives within the body.

endophthalmitis Inflammation of internal tissues of eye; may rapidly destroy the eye.

endothelium Squamous epithelium lining blood vessels.

endothrix Within the hair shaft.

endotoxin Substance containing lipopolysaccharide complexes found in the cell wall of bacteria, principally gram-negative bacteria; felt to play an important role in many of the complications of sepsis such as shock, DIC, and thrombocytopenia.

engineering controls Physical devices, such as needle resheathing holders and sharps containers, that help to remove biohazardous material from the workplace.

enteric fever Typhoid fever; paratyphoid fever.

enteroinvasive Capable of invading the mucosal surface and sometimes the deeper tissues of the bowel.

enteropathogenic *Escherichia coli* **(EPEC)** Specific serotypes of *E. coli* isolated from feces of patients with diarrhea, usually associated with nursery epidemics.

enterotoxigenic Producing an enterotoxin (e.g., enterotoxigenic *Escherichia coli*).

enterotoxin Toxin affecting the cells of the intestinal mucosa.

enzyme-linked immunosorbent assay (ELISA) An immunological assay similar to RIA that uses an enzyme conjugated to antibodies to produce a visible endpoint.

EPEC Enteropathogenic *Escherichia coli*.

epidemiology The study of the occurrence and distribution of disease and factors that control presence or absence of disease.

epididymitis Inflammation of the epididymis characterized by fever and pain on one side of the scrotum; seen as a complication of prostatitis and cystitis.

epiglottitis Inflammation of the epiglottis, a structure that prevents aspirating swallowed food and fluids into the tracheobronchial tree; a serious infection because the swollen epiglottis may block the airway.

epithelium Tissue composed of contiguous cells that forms the epidermis and lines hollow organs and all passages of the respiratory, digestive, and genitourinary systems.

erysipelas An acute cellulitis caused by group A streptococci.

erythema Redness of the skin from various causes.

erythema migrans Characteristic radial skin lesion of Lyme disease.

erythrasma A minor, superficial skin infection caused by *Corynebacterium minutissimium.*

erythrocytic cycle Developmental cycle of malarial parasites within red blood cells.

eschar A dry scar, particularly one related to a burn.

ETEC Enterotoxigenic *Escherichia coli.* Strains of *E. coli* that produce a choleralike toxin; have been implicated as causes of diarrhea.

etiology Cause or causative agent.

eugonic Growing luxuriantly (bacterial cultures).

eukaryotic Organisms with a true nucleus, in contrast to bacteria and viruses.

exanthem Skin eruption as a symptom of an acute disease, usually viral.

exoantigen test In vitro immunodiffusion test method for identifying fungal hyphae as *Histoplasma, Blastomyces,* or *Coccidioides.*

exoerythrocytic cycle Portion of the malarial life cycle occurring in the vertebrate host in which sporozoites, introduced by infected mosquitoes, penetrate the parenchymal liver cells and undergo schizogony, producing merozoites, which then initiate the erythrocytic cycle.

exogenous From outside the body.

exotoxin A toxin produced by a microorganism that is released into the surrounding environment.

exudate Fluid that has passed out of blood vessels into adjacent tissues or spaces; high protein content.

facultative anaerobe Microorganism that grows under either anaerobic or aerobic conditions.

fascia Membranous covering of muscle.

favus Dermatophyte infection of the scalp produced by *Trichophyton schoenleinii.*

fermentation Anaerobic decomposition of carbohydrate.

filamentous Threadlike.

fimbriae Proteinaceous fingerlike surface structures of bacteria that provide for adherence to host surfaces.

fistula Abnormal communication between two surfaces or between a viscus or other hollow structure and the exterior.

Fitz-Hugh-Curtis syndrome Inflammation of the capsule of the liver that may be seen in course of gonococcal or chlamydial infection in the female.

floccose Cottony, in tufts.

flocculation test Antigen-antibody test in which a precipitin end product forms macroscopically or microscopically visible clumps.

fluorescent Emission of light by a substance (or a microscopic preparation) while acted on by radiant energy, such as ultraviolet rays, as in the immunofluorescent procedure.

fluorescent immunoassays (FIAs) Similar to RIA but employ fluorescent molecules as labels.

fluorochrome A dye that becomes fluorescent or self-luminous after exposure to ultraviolet light.

fomite Any inanimate object that may be contaminated with disease-causing microorganisms and thus serves to transmit disease.

FTA Fluorescent treponemal antibody.

FTA-ABS Fluorescent Treponemal Antigen-Antibody absorption test; indirect fluorescent antibody stain used to detect antibodies directed against whole cell antigens of *Treponema pallidum* (syphilis bacillus).

fungemia Presence of viable fungi in blood.

FUO Fever of unknown origin.

fusiform Spindle-shaped, as in the anaerobe *Fusobacterium nucleatum*.

gangrene Death of a part of tissue resulting from disease, injury, or failure of blood supply.

gas-liquid chromatography (GLC) A method for separating substances by allowing their volatile phase to flow through a heated column with a carrier gas and measuring the time required to detect their presence at the distal end of the column.

gastric aspirate Fluid that may be aspirated from the stomach via a tube placed in the stomach by way of the nose or mouth.

gastroenteritis Inflammation of the mucosa of the stomach and intestines.

GC Gonococcus.

germicide An agent that destroys germs; disinfectant.

germ tube Tubelike process, produced by a germinating spore, that develops into mycelium.

glabrous Smooth.

GLC Gas-liquid chromatography.

glove box A device with flexible, semirigid, or rigid plastic walls used for cultivation of anaerobic bacteria. Bacteriologist must work by way of glove ports, and materials are brought into and out of the anaerobic work area by way of an interchange (airlock) that can be evacuated and refilled with an anaerobic atmosphere repeatedly until it itself is anaerobic. The interchange has one door leading into the glove box per se and another door to the outside. Some glove boxes have incubators within them. Also called anaerobic chamber or anaerobic cabinet.

GN Gram-negative (broth).

granulocytopenia Reduced number of granulocytic white blood cells in the blood.

granuloma Aggregation and proliferation of macrophages to form small (usually microscopic) nodules.

HAA Hepatitis-associated antigen.

HAI Hemagglutination inhibition.

halophilic Preferring high halide (salt) content.

HATTS The hemagglutination treponemal test for syphilis.

headspace gas Gas generated during growth of an organism in a closed tube that accumulates between the upper surface of the broth in the tube and the top of the tube.

hemadsorption Ability of certain virally infected cells to bind erythrocytes; mediated by glycopeptide adherence molecules (induced by viral activities within the cell) on the cell's surface.

hemagglutination Agglutination of red blood cells caused by certain antibodies, virus particles, or high molecular weight polysaccharides.

hematogenous Disseminated by the bloodstream.

hemolysis, alpha Partial destruction of, or enzymatic damage to, red blood cells in a blood agar plate, leading to greenish discoloration about the colony of the organism producing the alpha hemolysin.

hemolysis, beta Total lysis of red blood cells about a colony on a blood agar plate, leading to a completely clear zone surrounding the colony.

hemolysis, gamma No hemolysis is seen with organisms classed as gamma hemolytic; nonhemolytic would be a better designation.

HEPA High-efficiency particulate air filter; used in biological safety cabinets to trap pathogenic microorganisms.

herpes Inflammation of the skin characterized by clusters of small vesicles (e.g., caused by herpes simplex); disease caused by herpes simplex virus.

heterotroph Organism that requires an organic carbon source.

hidradenitis suppurativa Inflammation of certain sweat glands leading to formation of abscesses and draining sinuses.

high-pressure liquid chromatography (HPLC) Similar to GLC but capable of higher resolution because of increased pressure of liquid carrier that runs through the column.

hilum of lung Central area where major vessels and bronchi begin to branch.

HPLC High-pressure (or performance) liquid chromatography.

humidophilic Organisms preferring or requiring an increased moisture content; can be achieved by placing cultures in sealed jars or bags. For some organisms that do better in a candle jar than in the ambient atmosphere, it is the increased moisture content rather than the change in the gas content that is important.

humoral immunity Immunity effected by antibody.

hyaline Colorless, transparent.

hybridoma The product of fusion of an antibody-producing cell and an immortal malignant antibody-producing cell.

hydrolysis Breakdown of a substrate by an enzyme that adds the components of water to key bonds within the substrate molecule.

hyperalimentation Process by which nutrition (literally, extra nutrition) is provided; typically done by the intravenous route in subjects who are not able to absorb foods well from the gut because of disease of the bowel, in subjects in whom it is desirable to put the bowel "at rest" to promote healing, and in malnourished individuals to improve the nutritional status (e.g., before surgery), usually done over an extended period of time and requires use of a special access IV catheter such as a Hickman catheter.

hyperemia Increased blood in a part, resulting in distention of blood vessels.

hypertonic Hyperosmotic.

hypertrophy Increased size of an organ resulting from enlargement of individual cells.

hypha Tubular cell making up the vegetative portion of mycelium of fungi.

hypoxia Decreased oxygen content of tissues.

ID Infectious disease; identification; immunodiffusion.

IE Infectious endocarditis.

IFA Indirect fluorescent antibody; test that detects antibody by allowing an antibody to react with its substrate and adding a second fluorescein dye–labelled antibody that will bind to the first.

Ig, IgG, etc. Immunoglobulin, immunoglobulin G, etc.

immunodiffusion Detection of antigen or antibody by observing the precipitin line formed in a semisolid gel matrix when homologous antigens and antibodies are allowed to diffuse toward each other and react.

immunofluorescence Microscopic method of determining the presence or location of an antigen (or antibody) by demonstrating fluorescence when the preparation is exposed to a fluorescein-tagged antibody (or antigen) using ultraviolet radiation.

immunoglobulin Synonymous with antibody; five distinct classes have been isolated: IgG, IgM, IgA, IgE, and IgD.

immunoperoxidase stain Combination of an enzyme that catalyzes production of a colored product with an antibody to facilitate detection of certain antigens, particularly viral antigens.

immunosuppression Depression of the immune response caused by disease, irradiation, or administration of antimetabolites, antilymphocyte serum, or corticosteroids.

impedance Apparent resistance of a circuit to the flow of an alternating electric current.

impetigo Acute inflammatory skin disease, caused by streptococci or staphylococci, characterized by vesicles and bullae that rupture and form yellow crusts.

inclusion bodies Microscopic bodies, usually within body cells; thought to be virus particles in morphogenesis.

indigenous flora Normal or resident flora.

induced malaria Malaria infection acquired by parenteral inoculation (e.g., blood transfusion or sharing of needles by drug addicts).

induration Abnormal hardness of a tissue or part resulting from hyperemia or inflammation, as in a reactive tuberculin skin test.

infection Invasion by and multiplication of microorganisms in body tissue resulting in disease.

inhibitory quotient Ratio of the average peak achievable level of antibiotic in a body fluid from which an organism was isolated to the MIC of that organism.

in situ hybridization Detection of nucleic acid of a pathogenic organism in tissue sections by separating the DNA into single-stranded molecules and allowing a labeled strand of homologous DNA to bind to the target. The target is visualized by developing the label (either enzymatic precipitate, fluorescence, or radiolabel).

inspissation Process of making a liquid or semisolid medium thick by evaporation or absorption of fluid.

intermediate host Required host in the life cycle in which essential larval development must occur before a parasite is infective to its definitive host or to additional intermediate hosts.

intertrigo Erythematous skin eruption of adjacent skin parts.

intramuscular (intraperitoneal, intravenous) Within the muscle (peritoneum, vein), as in intramuscular injection.

in vitro Literally, within glass (i.e., in a test tube, culture plate, or other nonliving material).

in vivo Within the living body.

involution forms Abnormally shaped bacterial cells occurring in an aging culture population.

ion-exchange chromatography Separation of components of a solution by chromatography based on the reversible exchange of ions in the solution with ions present in or on an external matrix.

isotonic Of the same osmolality of body tissues, red blood cells, bacteria, etc.

K Alkaline.

Kawasaki disease Mucocutaneous lymph node syndrome; disease of children that may involve more serious phenomena such as cardiac disease.

keratitis Inflammation of the cornea.

KIA Kligler's iron agar (tube).

KOH Potassium hydroxide.

Köhler illumination Modification of brightfield microscopy in which a substage condenser is used to avoid glare from illuminating source.

lag phase Period of slow microbial growth that occurs following inoculation of the culture medium.

laked blood Hemolyzed blood; hemolysis may be effected in various ways, but alternate freezing and thawing is a simple method.

laminar flow Nonturbulent flow of air in layers (flowing in a vertical direction in the case of a biosafety hood).

latent Not manifest; potential.

latex agglutination Agglutination of latex particles coated with antibody molecules when exposed to the corresponding antigen.

lectin Naturally produced proteins or glycoproteins that can bind with carbohydrates or sugars to form stable complexes.

Leishman-Donovan (L-D) body Small, round intracellular form (called amastigote or leishmanial stage) of *Leishmania* species and *Trypanosoma cruzi.*

leukocytosis Elevated white blood cell count.

leukopenia Low white blood cell count.

L form Cell wall–deficient form of a bacterium.

LGV Lymphogranuloma venereum; the name for certain strains of *Chlamydia trachomatis* that cause a systemically expressed sexually transmitted disease.

Limulus **amebocyte lysate** Product of white blood cells of the horseshoe crab used in an assay for endotoxin.

lipopolysaccharide Carbohydrate-lipid complex; integral substance in gram-negative cell walls. Also known as "endotoxin."

liposome Small closed vesicle consisting of a single lipid bilayer.

logarithmic phase Period of maximal growth rate of a microorganism in a culture medium.

low passage cell line Cell culture used for virus isolation that is sensitive to the virus only after the cells have been subcultured (passaged) between 20 and 50 times; further passages result in cell with lower affinity for the virus.

LPA Latex particle agglutination.

LPS Lipopolysaccharide; see **endotoxin.**

lysis Disintegration or dissolution of bacteria or cells.

lysogeny Process by which a viral genome is integrated into that of its host bacterium.

MAC MacConkey (agar plate).

macroconidia Large, usually multiseptate, club- or spindle-shaped fungal spores.

malignant tertian malaria Malaria caused by *Plasmodium falciparum.*

mass spectrometry Method for determining composition of a substance by observing its volatile products during disintegration and comparing them with known standards.

MBC Minimum bactericidal concentration.

meconium Pasty greenish mass in intestine of fetus; made up of mucus, desquamated cells, bile, and such.

media, differential Media that permit ready recognition of a particular organism or group of organisms by virtue of facilitating recognition of a natural product of the organism being sought or by incorporating an appropriate substrate and indicator system so that organisms possessing certain enzymes will be readily recognized.

media, enrichment Media, usually liquid, that favor the growth of one or more organisms while suppressing most of the competing flora in a specimen with a mixture of organisms.

media, selective Culture media that contain inhibitory substances or unique growth factors such that one particular organism or group of organisms is conferred a real advantage over other organisms that may be found in a mixture. Efficient selective media will select out only the organism or organisms being sought, with little or no growth of other types of organisms.

media, supportive Culture media that provide adequate nutrition for most nonfastidious microorganisms and that permit each organism to grow normally without conferring any special advantage to one organism over another.

mediastinum Space in the middle of the chest between the medial surfaces of the two pleurae.

Meleney's ulcer A chronic undermining ulcerative lesion of the subcutaneous tissue, usually caused by streptococcal infection.

meningitis Inflammation of the meninges, the membranes that cover the brain and spinal cord (e.g., bacterial meningitis).

merozoite Product of schizogonic cycle in malaria that will invade red blood cells.

mesenteric adenitis Inflammation of mesenteric lymph nodes.

mesentery A fold of the peritoneum that connects the intestine with the posterior abdominal wall.

metastatic Spread of an infectious (or other) process from a primary focus to a distant one via the bloodstream or lymphatic system.

MHA-TP Microhemagglutination test for antibody to *Treponema pallidum.*

MIC Minimum inhibitory concentration.

microaerobic Requiring a partial pressure of oxygen less than that of atmospheric oxygen for growth. New term for "microaerophilic."

microaerophile, obligate Microorganism that grows only under reduced oxygen tension and cannot grow aerobically or anaerobically.

microaerophilic See "microaerobic."

microcalorimetry Study of microcalories, a measure of heat production.

microconidia Small, single-celled fungal spores.

microfilaria Embryos produced by filarial worms and found in the blood or tissues of individuals with filariasis.

miliary Of the size of a millet seed (0.5 to 1.0 mm); characterized by the formation of numerous lesions of the above size distributed rather uniformly throughout one or more organs.

minimum bactericidal concentration (MBC) The minimum concentration of antimicrobial agent needed to yield a 99.9% reduction in viable colony forming units of a bacterial or fungal suspension.

minimum inhibitory concentration (MIC) The minimum concentration of antimicrobial agent needed to prevent visually discernible growth of a bacterial or fungal suspension.

mixed culture (pure culture) More than one organism growing in or on the same culture medium, as opposed to a single organism in pure culture.

monoarticular Occurring in only one joint.

monoclonal antibody Antibody that is derived from a single cell producing one antibody molecule type that reacts with a single epitope.

monolayer A confluent layer of tissue culture cells one cell thick.

MOTT Mycobacteria other than *Mycobacterium tuberculosis.*

mucopurulent Term used to describe material containing both mucus and pus (e.g., mucopurulent sputum).

mucosa A mucous membrane.

multiple myeloma Malignancy involving antibody-producing plasma cells.

mycelium Mass of hyphae making up a colony of a fungus.

mycetoma Chronic infection, usually of feet, caused by various fungi or by *Nocardia* or *Streptomyces,* resulting in swelling and sinus tracts; pulmonary mycetoma is a mass of fungal hyphae ("fungus ball") growing in a cavity formed during previous tuberculosis infection or other pathological condition.

mycoses Diseases caused by fungi (e.g., dermatomycosis, fungal infection of the superficial skin).

myocarditis Inflammation of the heart muscle.

myositis Inflammation of a muscle, sometimes caused by infection as in pyomyositis, an infection caused by *Staphylococcus aureus* that leads to small abscesses within the muscle substance.

nares External openings of nose (i.e., nostrils).

nasopharyngeal Pertaining to the part of the pharynx above the level of the soft palate.

necrosis Pathological death of a cell or group of cells.

necrotizing fasciitis A very serious, painful infection involving the fascia (membranous covering) of one or more muscles; may spread widely in short periods of time since there is no anatomic barrier to spread in this type of infection.

Negri bodies Characteristic virally-induced inclusions present in rabies-infected brain cells.

neonatal First 4 weeks after birth.

nephelometry Determination of the degree of turbidity of a fluid by light scatter.

neuritis Inflammation of a nerve; most often not caused by an infectious agent.

neurotrophic Having a selective affinity for nerve tissue. Rabies is caused by a neurotrophic virus.

NFB Glucose nonfermenting gram-negative bacteria.

NGU Nongonococcal urethritis.

nick translation Use of enzymes to break DNA and repolymerize small sections of the molecule, usually for purposes of labeling the DNA with a radioactive nucleotide.

nonphotochromogens Slow-growing, nonpigmented mycobacteria.

nonsporulating Does not produce spores.

nosocomial Pertaining to or originating in a hospital, as nosocomial infection.

nucleic acid hybridization Process by which the single-stranded probe unites with complementary DNA in an unknown sample.

nucleic acid probe Piece of labeled single-stranded DNA used to detect complementary DNA in clinical material or a culture and thus to specifically identify the presence in these materials of an organism identical to that used to make the probe.

nucleocapsid Name of viral particle that includes virus nucleic acid core enclosed in the protein capsid coat.

O & P Ova and parasites.

O-F Oxidation-fermentation medium.

octal numbers Numbers employed in computer data bases to identify biochemical profiles of organisms and thus their identification.

oil immersion microscopy Use of immersion oil to fill the space between the slide being studied and the special objective of the microscope; this keeps the light rays from dispersing and provides good resolution at high magnification (total magnification of $1000\times$).

oncogenic Possessing the potential to cause normal cells to become malignant; causing cancer.

ONPG o-nitrophenol-β-galactopyranoside (β-galactosidase test).

operculated ova Ova possessing a cap or lid.

opsonize To facilitate destruction of pathogens by phagocytic ingestion or lysis by complement through the action of adherent antibodies.

osteomyelitis Inflammation of the bone and the marrow.

otitis Inflammation of the ear from a variety of causes, including bacterial infection; otitis media: inflammation of the middle ear.

oxidation A metabolic pathway of the microorganism that involves use of oxygen as a terminal electron acceptor. This type of reaction occurs in air.

oxidation-reduction potential Electromotive force exerted by a nonreacting electrode in a solution containing the oxidized and reduced forms of a chemical, relative to a standard hydrogen electrode; the more negative the value, the more anaerobic conditions are.

pandemic Epidemic over a wide geographical area, or even worldwide.

paracentesis Surgical transcutaneous puncture of the abdominal cavity to aspirate peritoneal fluid.

parasite Organism that lives on or within and at the expense of another organism.

parenteral Route of administration of a drug other than by mouth; includes intramuscular and intravenous administration.

paronychia Purulent inflammation about margin of a nail.

parotitis Inflammation of the parotid gland, the largest of the salivary glands; mumps is the most common cause of this.

paroxysm Rapid onset (or return) of symptoms; term usually applies to cyclic recurrence of malaria symptoms, which are chills, fever, and sweating.

pathogenic Producing disease.

pathological Caused by or involving a morbid condition, as a pathological state.

PCR Polymerase chain reaction.

penicillinase (β-lactamase I) Enzyme produced by some bacterial species that inactivates the antimicrobial activity of certain penicillins (e.g., penicillin G).

percutaneous Performed through the skin (e.g., percutaneous bladder aspiration).

pericarditis Inflammation of the covering of the heart (pericardium).

perineum The portion of the body bound by the pubic bone anteriorly, the coccyx posteriorly, and the bony prominences (tuberosities) of the ileum on both sides.

peritoneal cavity Space between the visceral and parietal layers of the peritoneum; the serous membrane lining the abdominal cavity and surrounding the contained viscera.

peritonitis Inflammation of the peritoneal cavity, most often caused by bacterial infection.

petechiae Tiny hemorrhagic spots in the skin or mucous membranes.

phase-contrast microscopy Technique for direct observation of unstained material in which light beams pass through the object to be visualized and are partially deflected by the different densities of the object. These light beams are deflected again when they impinge on a special objective lens, increasing in brightness when aligned in phase.

photochromogens Mycobacteria that produce pigment after exposure to light but whose colonies remain buff-colored in the dark.

phycomycosis Serious infection involving fungi of the zygomycete group, often beginning with necrotic lesions in the nasal mucous or palate but rapidly spreading to involve other tissues. Seen in immunocompromised patients.

PID Pelvic inflammatory disease.

pili Structures in bacteria similar to fimbriae that participate in bacterial conjugation and transfer of genetic material.

pilonidal cyst Hair-containing cyst in the skin or subcutaneous tissue, often with a sinus tract, commonly in the sacrococcygeal area.

plasma Fluid portion of blood; obtained by centrifuging anticoagulated blood.

plasmids Extrachromosomal DNA elements of bacteria carrying a variety of determinants that may permit survival in an adverse environment or successful competition with other microorganisms of the same or different species.

pleomorphic Having more than one form, usually widely different forms, as in pleomorphic bacteria.

pleura The serous membrane enveloping the lung and lining the internal surface of the thoracic cavity.

pleuropulmonary Pertaining to the lungs and pleura.

PMC Pseudomembranous colitis.

pneumonia Inflammation of the lungs, primarily caused by infectious agents.

pneumonia, aspiration Pneumonia caused by aspiration of oropharyngeal or gastric contents.

pneumothorax Introduction of air (usually inadvertently) into the pleural space, leading to collapse of the lung on that side.

polyarticular Occurring in more than one joint.

polymerase chain reaction (PCR) A method for expanding small discrete sections of DNA by binding DNA primers to sections at the ends of the DNA to be expanded and using cycles of heat (to create single-stranded DNA) and cooler temperatures (to allow a DNA polymerase enzyme to create new sections of DNA between the primer ends).

PPD Purified protein derivative (skin test antigen for tuberculosis).

PPNG Penicillinase-producing (i.e., penicillin-resistant) *Neisseria gonorrhoeae*.

PRAS Prereduced, anaerobically sterilized.

precipitin test Detection of antigen by allowing specific antibody to diffuse through liquid or gel until an antigen-antibody complex forms, which can be viewed as a line of precipitated material.

prion Proteinaceous infectious agent associated with Creutzfeldt-Jakob disease and perhaps other chronic, debilitating central nervous system diseases.

proctitis Inflammation of the rectum.

prodromal Early manifestations of a disease before specific symptoms become evident.

proglottid Segments of the tapeworm containing male and female reproductive systems; may be immature, mature, or gravid.

prognosis Forecast as to the possible outcome of a disease.

progressive bacterial synergistic gangrene Serious polymicrobial infection of skin and soft tissue, often a postoperative complication, involving a mixture of microaerobic streptococci and *Staphylococcus aureus* classically. At times the streptococci are anaerobic or gram-negative bacilli are seen in lieu of the staphylococci.

prokaryotic Organisms without a true nucleus.

prophylaxis Preventive treatment (e.g., the use of drugs to prevent infection).

prostatitis Inflammation of the prostate gland, usually caused by infection, characterized by fever, low back or perineal pain, and at times urinary frequency and urgency; a common background factor for recurrent cystitis in males.

prosthesis An artificial part such as a hip joint or eye.

protein A A protein on the cell wall of strains of *Staphylococcus aurus* (Cowan strain) that binds the Fc portion of antibodies.

prototroph Naturally occurring or wild strain.

pseudomembrane Necrosis of mucosal surface simulating a membrane.

pseudomembranous colitis (PMC) Syndrome in the large bowel characterized by a layer of necrotic tissue and dead inflammatory cells often caused by the toxin of *Clostridium difficile.*

psychrophilic Cold-loving (e.g., microorganisms that grow best at low [4° C] temperatures).

purulent Consisting of pus.

pus Product of inflammation, consisting of fluid and many white blood cells; often bacteria and cellular debris are also present.

pyelonephritis Infection of the kidney and renal pelvis and the late effects of such infection.

pyocin Pigment produced by a bacterium that has antibacterial properties against other strains or species of bacteria.

pyoderma Any of the pus-producing lesions of the skin such as boils or impetigo.

pyogenic Pus-producing.

QC Quality control.

QNS Quantity not sufficient.

quartan malaria Malaria caused by *Plasmodium malariae.*

radioimmunoassay (RIA) A very sensitive competitive binding assay that uses radioactively labeled antigen and specific homologous antibody to determine the amount of antigen in a sample. The antigen could also be an antibody.

radioisotope Unstable molecule that emits detectable radiation (gamma rays, X-rays, etc.) for a known period of time (half-life). Can be incorporated into other compounds as a label for later detection by X-ray film exposure or by measurement in a scintillation counting instrument.

radiometric analysis Determination of an element that is not itself radioactive by means of an interaction with a radioactive element.

reagin An antibody that reacts in various serologic tests for syphilis.

reservoir Source from which an infectious agent may be disseminated; for example, humans are the only reservoir for *Mycobacterium tuberculosis.*

resin Plant product composed largely of esters and ethers of organic acids and acid anhydrides.

restriction endonuclease Enzyme that breaks nucleic acid (usually DNA) at only one specific sequence of nucleotides.

reticulate body The metabolically more active form of elementary bodies of *Chlamydia* sp.

reticuloendothelial system Macrophage system, which includes all the phagocytic cells of the body except for the granulocytic leukocytes.

R factor Plasmid that carries gene coding for resistance to one or more antibacterial agents.

rheumatoid factor IgM antibodies produced by some patients against their own IgG.

rhizoid Rootlike absorbing organ.

RIA Radiometric immunoassay (radioimmunoassay).

RNA Ribonucleic acid.

RPR Rapid plasma reagin, nontreponemal test for antibodies developed in response to syphilis infection.

saccharolytic Capable of breaking down sugars.

saprophytic Nonpathogenic.

SBA Suprapubic bladder aspiration.

SBE Subacute bacterial endocarditis.

schizogony Stage in the asexual cycle of the malaria parasite that takes place in the red blood cells of humans.

Schlichter test Synonym for the serum bactericidal level test.

sclerotic Hard, indurated.

scolex (pl., scolices) Head portion of a tapeworm; may attach to the intestinal wall by suckers or hooklets.

scotochromogens Mycobacteria that are pigmented even in the absence of exposure to light.

SEM Scanning electron micrograph.

sensitivity Ability of a test to detect all true cases of the condition being tested for; absence of false-negative results. (Also see "specificity.")

septate Having cross walls.

septic shock Acute circulatory failure caused by toxins of microorganisms; often leads to multiple organ failure and is associated with a relatively high mortality.

septicemia (sepsis) Systemic disease associated with presence of pathogenic microorganisms or their toxins in the blood.

sequestrum A detached or dead piece of bone within a cavity, abscess, wound, or area of osteomyelitis.

Sereny test Test for bacterial invasiveness; involves applying a supension of the organism to the conjunctiva of a small mammal and observing for development of conjunctivitis.

serosanguineous Like serous but with some blood present grossly.

serous Like serum.

serum Cell- and fibrinogen-free fluid remaining after whole blood clots.

serum bactericidal level Lowest dilution of a patient's serum that kills a standard inoculum of an organism isolated from that patient; this, of course, is related to antibiotic level achieved in the patient's serum and the bactericidal activity of the drug being employed.

sinus Suppurating tract; paranasal sinus, hollows, or cavities near the nose (e.g., frontal and maxillary sinuses).

slant See definition of "butt." The slant is the upper surface of the medium in the tube described. It is exposed to air in the tube.

solid-phase immunosorbent assay (SPIA) ELISA test in which the capture antigen or antibody is attached to the inside of a plastic tube, microwell, or to the outside of a plastic bead, in a filter matrix, or some other solid support. Allows faster interaction between reactants and more concentrated visual end products than ELISA tests performed in liquid.

somatic Pertaining to the body (of a cell) (e.g., the somatic antigens of *Salmonella* species).

Southern blot Identification of specific genetic sequences by separating DNA fragments by gel electrophoresis and transferring them to membrane filters in situ. Labeled complementary DNA applied to the filter will bind to homologous fragments, which can then be identified by detecting the presence of the labeled DNA in association with bands of certain molecular size. Named after its discoverer, E.M. Southern.

specificity Ability of a test to correctly yield a negative result when the condition being detected is absent; absence of false-positive results. (Also see "sensitivity.")

SPIA Solid-phase immunosorbent assay.

spore Reproductive cell of bacteria, fungi, or protozoa; in bacteria, may be inactive, resistant forms within the cell.

sporogony Stage in the sexual cycle in the malarial parasite that takes place in the mosquito.

sporozoite Slender, spindle-shaped organism that is the infective stage of the malarial parasite; it is inoculated into humans by an infected mosquito and is the result of the sexual cycle of the malarial parasite in the mosquito.

sputum Material discharged from the surface of the lower respiratory tract air passages and expectorated (or swallowed).

stab culture Culture in which the inoculation of a tube of solid medium is made by stabbing with a needle to encourage anaerobic growth in the bottom.

stat Statim (Latin); immediately.

stationary phase Stage in the growth cycle of a bacterial culture in which the vegetative cell population equals the dying population.

STD Sexually transmitted disease.

sterile (sterility) Free of living microorganisms (the state of being sterile).

strobila Entire chain of tapeworm proglottids, excluding the scolex and neck.

substrate A substance on which an enzyme acts.

sulfur granule Small colony of organisms with surrounding clublike material; yellow-brown; resembles grain of sulfur.

superantigen Molecules produced by microbes (viruses, bacteria, and perhaps parasites) that act independently to stimulate T-cell activities, including cytokine release. Among the most potent T-cell mitogens, superantigen stimulation can result in anergy, or alternatively, systemic immune system activation.

superinfection Strictly speaking, this refers to a new infection superimposed on another being treated with an antimicrobial agent. The new infecting agent is resistant to the therapy initially employed and thus survives and causes persistence of the infection (now resistant to the treatment) or a new infection at a different site. Term is also used to indicate persistence or colonization with a new organism without any evidence of resulting infection.

suppuration Formation of pus.

suprapubic bladder aspiration (SBA) Obtaining urine by direct needle puncture of the full bladder through the abdominal wall above the pubic bone.

syncytia Structure resulting from fusion of cell membranes of several cells to form a multinucleated cellular structure; usually the result of viral infection of the cells.

syndrome Set of symptoms occurring together (e.g., nephrotic syndrome).

synergism Combined effect of two or more agents that is greater than the sum of their individual effects.

synovial fluid Viscid fluid secreted by the synovial membrane; formed in joint cavities, bursae, and so forth.

T cells Lymphocytes involved in cellular immunity.

TB Tuberculosis.

TDM Therapeutic drug monitoring, testing serum for levels of antibiotics or other therapeutic agents.

TEM Transmission electron micrograph.

T-M Thayer-Martin (agar plate).

tenesmus Painful, unsuccessful straining in an attempt to empty the bowels.

therapy, antimicrobial Treatment of a patient for the purpose of combating an infectious disease.

thermolabile Adversely affected by heat (as opposed to thermostable, not affected by heat).

thoracentesis Drainage of fluid from the pleural space.

thoracic Pertaining to the chest cavity.

thrush A form of *Candida* infection that typically produces white plaquelike lesions in the oral cavity.

tinea Dermatophyte infection (tinea capitis, tinea of scalp; tinea corporis, tinea of the smooth skin of the body; tinea cruris, tinea of the groin; tinea pedis, tinea of the foot).

titer Level of substance such as antibody or toxin present in material such as serum; reciprocal of the highest dilution at which the substance can still be detected.

T-lymphocytes (or T cells) Thymus-derived lymphocytes important in cell-mediated immunity.

tolerance A form of resistance to antimicrobial drugs; of uncertain clinical importance. *See* tolerant.

tolerant Characteristic of an organism that requires a great deal more antimicrobial agent to kill it than to inhibit its growth.

TORCH Toxoplasmosis, rubella, cytomegalovirus infection, and herpes infection; acronym used for illnesses or tests.

TPI *Treponema pallidum* immobilization test, a test for antibodies against the agent of syphilis that uses live treponemes.

trachoma Serious eye infection caused by *Chlamydia trachomatis;* often leads to blindness.

transduction Moving genetic material from one prokaryote to another via a bacteriophage or viral vector.

transposon Genetic material from a plasmid that can move between plasmids or from a plasmid to a chromosome; so-called "jumping genes."

transtracheal aspiration Passage of needle and plastic catheter into the trachea for obtaining lower respiratory tract secretions free of oral contamination.

transudate Similar to exudate but with low protein content.

trophozoite Feeding, motile stage of protozoa.

tropism Preferred environment or destination. In viral infection, preference for a particular tissue site (rabies viruses have a tropism for neural tissue).

TSI Triple sugar iron (agar tube).

TSS Toxic shock syndrome.

TTA Transtracheal aspiration.

typing Methods of grouping organisms, primarily for epidemiological purposes (e.g., biotyping, serotyping, bacteriophage typing, and the antibiogram).

Tzanck test Stained smear of cells from the base of a vesicle examined for inclusions produced by HSV or VZV.

urethritis Inflammation of urethra, the canal through which urine is discharged (e.g., gonococcal urethritis).

URI Upper respiratory tract infection.

UTI Urinary tract infection.

VD Venereal disease.

VDRL Veneral Disease Research Laboratory; classic nontreponemal serologic test for syphilis antibodies. Uses cardiolipin, lecithin, and cholesterol as cross-reactive antigen that flocculates in the presence of "reaginic" antibodies produced by patients with syphilis. Best test for cerebrospinal fluid in cases of neurosyphilis.

vector An arthropod or other agent that carries microorganisms from one infected individual to another.

vegetation In endocarditis, the aggregates of fibrin and microorganisms on the heart valves or other endocardium.

vesicle A small bulla or blister containing clear fluid.

villi Minute, elongated projections from the surface of intestinal mucosa that are important in absorption.

Vincent's angina An old term, seldom used presently, referring to anaerobic tonsillitis.

viremia Presence of viruses in the bloodstream.

virion The whole viral particle, including nucleocapsid, outer membrane or envelope, and all adherence structures.

virulence Degree of pathogenicity or disease-producing ability of a microorganism.

viscus (pl., viscera) Any of the organs within one of the four great body cavities (cranium, thorax, abdomen, and pelvis).

V-P Voges-Proskauer.

Weil-Felix test Whole cell agglutination assay that uses *Proteus* cell surface antigens to detect cross-reactive rickettsial antibodies.

Western blot Similar to Southern blot, except that antigenic proteins of an organism are separated by gel electrophoresis and transferred to membrane filters. Antiserum is allowed to react with the filters, and specific antibody bound to its homologous antigen is detected using labeled antiantibody detectors.

Whipple's disease A disease caused by an uncultured bacterium, *Trophyrema whippeli,* and characterized by diarrhea, arthritis, and neurologic manifestations.

xenodiagnosis Procedure involving the feeding of laboratory-reared triatomid bugs on patients suspected of having Chagas' disease; after several weeks, the feces of the bugs are checked for intermediate stages of *Trypanosoma cruzi.*

zoogleal mass Jellylike matrix in which microorganisms may be embedded.

zoonoses Diseases of lower animals transmissible to humans (e.g., tularemia).

Zygomycetes Group of fungi with nonseptate hyphae and spores produced within a sporangium.

INDEX

Autopsy
  cultures and, 293, 294
  plating media for, 60
AutoReader, 139-140
AutoSceptor system, 140
Avid, definition of, 154
Avidin in biotin-avidin-enzyme-conjugated stain; *see*
    Biotin-avidin-enzyme-conjugated stain
Avidin-labeled probe, preparation and use of, 131, 132
Azidothymidine, 637
AZT; *see* Azidothymidine

B
B lymphocytes, 154, 156
Babesia, 825-826, 846
  in blood, 195
  body sites of, 779
  classification of, 778
  in red blood cells, 846
  sporozoa of, 778
Babesiosis, 825-826, 846
Bacillary peliosis, 581, 582, 583
Bacille Calmette-Guérin, 592
Bacilli, 451-456
  acid-fast; *see* Acid-fast bacilli
  asaccharolytic or weakly saccharolytic anaerobic, 532-
    536
  Bacillus anthracis in, 452-454
  catalase test for, 100
  characteristics of, 451-452
  differentiation of, 452
  Gram stain for, 99
  gram-negative; *see* Gram-negative bacilli
  gram-positive; *see* Gram-positive bacilli
  identification of, 453
  morphology of, 451-452
  in muscle pathology, 279
  processing clinical specimens for non–spore-forming,
    496
Bacillus alvei, 453
Bacillus anthracis
  biological safety cabinet and, 14
  differentiation of, 453
  hemagglutination test for, 159
  identification methods for, 454
  in respiratory tract, 222, 227
  in skin and soft tissue, 274
  soybean agglutinin binding to, 126
  structure and extracellular products of, 454
  susceptibility of, to antimicrobial agents, 454
Bacillus brevis, 455
  differentiation of, 453
Bacillus cereus, 454-455
  differentiation of, 453
  in endophthalmitis, 299
  enterotoxins from, 239
  in food poisoning, 237-238
  laboratory identification of, 455
  susceptibility of, to antimicrobial agents, 455
Bacillus circulans, 455
  differentiation of, 453
Bacillus coagulans, 455
  differentiation of, 453
Bacillus firmus, 453
Bacillus fragilis, 528-529
  antimicrobial susceptibility of, 528-529

Bacillus fragilis—cont'd
  identification of, 528
  morphology and characteristics of, 528
Bacillus laterosporus, 453
Bacillus licheniformis, 455, 456
  differentiation of, 453
Bacillus macerans, 455
Bacillus megaterium, 455
  differentiation of, 453
Bacillus mycoides, 455, 456
  differentiation of, 453
Bacillus polymyxa, 453
Bacillus pumilus, 455, 456
  differentiation of, 453
Bacillus sphaericus, 455, 456
  differentiation of, 453
Bacillus stearothermophilus, 453
Bacillus subtilis, 455
  classification of, by hazard level, 13
  differentiation of, 453
Bacillus thuringiensis, 455
  differentiation of, 453
Bacitracin
  clostridial susceptibility to, 515
  in differentiation of coagulase-negative *Staphylococcus*,
    328
  susceptibility test for, 329
BacT/Alert, 206, 692
BACTEC; *see* Becton-Dickinson Diagnostic Instrument
    Systems
Bacteremia, 197; *see also* Bacterial infections in blood
  in acquired immunodeficiency syndrome, 315
  Clostridium tertium in, 507
  in elderly, 308
  in extremity infections, 276
  identification of pathogens in, 5
  muscle pathology and, 279
  nonconventional methods for detecting, 206-208
  nosocomial, 42-43
  in pyoderma or cellulitis, 277
  in septic arthritis, 288
Bacteria
  adenosine triphosphate of, in urine, 252
  aerobic; *see* Aerobes
  agglutination test for, 158
  anaerobic; *see* Anaerobes
  antimicrobial agents for systemic infection from,
    169
  darkfield microscopy for, 67
  disk diffusion methods for, 175-177, 178
  in endophthalmitis, 298
  extracellular toxins in, 221
  gastrointestinal tract cultures of, 244-247
  in joint fluid, 288
  latex particle agglutination tests for, 126
  in meningitis; *see* Meningitis, bacterial
  microbiological methods for, 97-122; *see also*
    Microbiology
  in middle ear infection, 301
  in pericardial fluid, 287
  in pneumonia, 226-228
  replica plating system for, 143
  in respiratory tract, 220; *see also* Respiratory tract
  thermophilic; *see* Thermophilic bacteria
  unnamed, 578

Germicides
  for equipment decontamination, 12
  for skin decontamination before specimen collection, 54
Germ-tube test, 753, 755
Gerstmann-Straussler disease, 585-586
Giardia
  cysts of, 830, 831, 836
  motility of, 816
  in neoplastic disease and bone marrow transplants, 309
  stool specimen of
    direct wet mount for, 242
    monoclonal antibody fluorescent stain for, 243
Giardia lamblia, 810-816, 829-834
  body sites for, 779
  classification of, 778
  cysts of, 830, 831
  direct wet mount and, 66
  in duodenal contents, 793
  fecal specimens of, 242, 243
    concentration procedures for, 784
  in gastrointestinal disease, 240
  in homosexual male, 312
  importance of, 809
  morphology of, 812, 814
  in sexually transmitted disease, 260
  trophozoites of, 829, 830, 831
Giemsa stain
  for *Calymmatobacterium granulomatis*, 583
  for eye specimens, 302, 303
    from newborn, 307
  in intestinal parasitic infection, 71-72, 799, 800, 817
  for *Plasmodium*, 843-844
Giménez stain, 72
Gingivitis, 223-224, 300
Glanders, 387
GLC; *see* Gas-liquid chromatography
Gliocladium, 711, 713
Glomerulonephritis, 335
Glomerulosclerosis, 223
Glossina, 779
Glove box, 86-87
Gloves for protection, 10, 12
Glucan production, 348
Glucose
  fermentation of, Enterobacteriaceae in, 379, 380
  triple sugar iron agar and Kligler's iron agar and, 110-111
  utilization of
    by Enterobacteriaceae, 373
    gram-negative bacilli in, 391-392
Glutamic acid decarboxylase, 491
Glutamic acid decarboxylase test, 491
Glycerol
  for stool specimen, 241
  in urine preservation containers, 252
Glycoprotein spikes, 634
GMS; *see* Gomori methenamine silver
GN; *see* Gram-negative broth
Gnathostoma spinigerum, 778
Gnats, 779
Goggles for protection, 12
Gomori methenamine silver
  in *Nocardia*, 467, 468
  in parasitic lung infection, 799, 801-802
  in *Pneumocystic carinii* cysts, 800, 801-802
Gonococcal agar base for *Helicobacter pylori*, 245

Gonococci; *see also Neisseria*
  cost control and, 31
  prompt delivery of specimen and, 55
Gonorrhea in homosexual male, 312; *see also Neisseria*
Gonozyme test, 271
Gowns for protection, 12
Gradient strip method in antimicrobial susceptibility testing, 182-183
Grafts
  graft-versus-host disease and, 310
  host vulnerability and, 311-312
Gram stain, 69-70
  for anaerobic bacteria, 476-477, 479
    in isolates subculture, 481
  automatic, 144
  of cerebrospinal fluid, 214
  cost control and, 28
  for *Eubacterium*, 518-519
  for eye specimens, 302, 303
  for fungi, 694, 695, 698
  for genital tract specimens, 263, 264, 267, 268
  for joint fluid, 289
  in mycobacteriology, 605
  for newborn conjunctival smears, 307
  for peritoneal dialysis fluid, 287
  quantitation of organisms and, 6
  respiratory tract specimens and, 224-225
    lower, 230, 231
  for stool specimens, 243, 244
  as traditional microbiological method, 98-100
  for urethral discharge, 261, 262
  for urine specimen, 252
  for uterine specimens, 265
  for vaginal discharge, 263, 264
  for wound culture, 27
Gram-negative bacilli
  anaerobic, 493-495, 524-542
    asaccharolytic or weakly saccharolytic, 532-536; *see also* Bacilli, asaccharolytic or weakly saccharolytic anaerobic
    Bacteroidaceae in, 524-532, 540-541; *see also* Bacteroidaceae
    in blood, 194
    differentiation of genera of, 525
    facultative, 406-428; *see also* Gram-negative bacilli, facultatively anaerobic
    flow chart for, 535
    Fusobacterium in, 536-540; *see also Fusobacterium*
    nonpigmented bile-sensitive, 537
    pigmented, 534
    taxonomic changes in, 528
  bone biopsy in, 290
  in burn infections, 280
  in cellulitis, 280
  classification of, 99
  coccobacilli morphologically similar to, 421
  in draining sinuses or fistulas, 277
  in drug abusers, 312
  in elderly, 307, 308
  in endocarditis, 199
  in external ear infections, 301
  facultatively anaerobic, 406-428
    Actinobacillus in, 419-420, 423; *see also Actinobacillus*
    Bordetella in, 409-413
    Brucella in, 408-409, 410
    Capnocytophaga in, 424-426

Mobiluncus—cont'd
  dot-blot assay and, 135
  in genital tract, 263
  identification of, 521
  in infection, 521
  morphology and characteristics of, 521
  as non–spore-forming bacilli, 515
  in normal flora, 521
Mobiluncus curtisii, 521
Mobiluncus mulieris, 521
Modified acid-fast stain; *see* Acid-fast stain, modified
Modified oxidase test, 330
Modified Thayer-Martin agar; *see* Thayer-Martin agar
Modified zephiran-trisodium phosphate method, 598, 601, 602
Moellerella wisconsensis, 364
Moeller's decarboxylase medium, 375
Molds
  dematiaceous, 760-766, 767
  extent of identification of, 706-708
  hyaline, 766-768, 769
  identification of, 708-710
  macroscopic appearance of, 698
Molluscum contagiosum, 260
Monobactams as bactericides, 174
Monoclonal antibodies, 129-131
  fluorescein-conjugated stains and, 75-76
    for *Giardia* and *Cryptosporidium* cysts, 836
    lower respiratory tract organisms and, 231, 799
    in parasitic lung infection, 799
    for stool specimens, 243
    in urinary tract infection, 262-263
Monolayer of growth, 93, 94
Monosporium apiospermum, 299
Moraxella
  differentiation of, 383, 401
  drug susceptibility of, 403
  growth characteristics of, 353
  identification of, 400-401
  in keratitis, 298
  morphology of, 353
  in periocular infection, 299
  vancomycin and, 100
Moraxella catarrhalis, 353, 358-360
  characteristics of, 358
  commercial biochemical systems for, 360
  growth characteristics of, 353
  in head and neck flora, 297
  in infectious arthritis, 289
  in middle ear infection, 301
  in pneumonia, 227
  in respiratory tract, 220, 221, 227
  in sinusitis, 300
Moraxella lacunata
  in conjunctivitis, 298, 303
  differentiation of, 401
Moraxella osloensis
  differentiation of, 401
  in infectious arthritis, 289
Moraxella phenylpyruvica, 401
Morbidity and Mortality Weekly Report, 212
Morganella
  triple sugar iron agar and, 111
  in wound infections and abscesses, 275
Morganella morganii, 364
  identification of, 371

Mortality from nosocomial infections, 42
Mosquitos, 779
Moths, 779
Motility
  of anaerobic bacteria, 490
  of Enterobacteriaceae, 378-379
  in pure culture isolate, 490
Mouth infections, 300
Mouth pipetting prohibition, 12
Moxalactam
  flavobacteria susceptible to, 404
  solvent and diluent for, 172
MS-2 system for urine specimen, 252
Mucicarmine stain, 72
Mucocutaneous leishmaniasis, 826
Mucocutaneous lymph node syndrome, 587
Mucor, 763
  as causes of fungal infection, 744
  characteristic features of, 696-697
  definitive identification of, 746
  in infection, 744-745
  in malignancy, 309
  in neoplastic disease and bone marrow transplants, 310
  phylogenetic position of, 711
  schema of, 713
  in skin and soft tissue infection, 279
  sporangia of, 749
Mucormycosis, 279; *see also Mucor*
  in periocular infection, 299
Mucous membrane
  of genital tract, 266-267, 268
  of intestinal tract, 235
  pathogenic flora of, 266-267, 268
  virology and, 647, 648
Mueller-Hinton medium
  agar dilution susceptibility testing and, 182
  broth dilution methods and, 170
  Haemophilus ducreyi in, 415
  in septicemia diagnosis, 204
MUG test; *see* 4-Methylum-belliferyl-—D-glucuronide test
Multiceps, 778
Multiple sclerosis, 587-588
Multitest systems, 113-115
Mumps virus
  diseases of, 642
  encephalitis after, 212-213
  fluorescent antibody tests for, 656, 659
  hemadsorption of primary monkey kidney monolayers in detection of, 672-673
  hemagglutination inhibition tests for, 160
  in infectious arthritis, 289
  isolation and identification of, 666
  in salivary gland infection, 300
  serology panels and immune status tests for, 685
  specimens for, 647
  in viral syndromes, 639
Muscle biopsy for *Trichinella spiralis*, 805
Muscle pathology, 279
Mycelium, 699-700
Mycetoma, 469
  subcutaneous, 727-729
Mycobacteria, 590-633
  antimicrobial susceptibility testing for, 179-182, 616-618, 626
    indirect, 182
    Middlebrook 7H11 agar, 181

Nitrate reduction test—cont'd
  for mycobacteria, 609, 614-615, 623
    in differential diagnosis, 621
  with paper strips, 617
  for yeasts, 753, 755
Nitrocefin
  cost control and, 31
  rapid antimicrobial susceptibility testing and, 185
Nitrofurantoin
  solvent and diluent for, 172
  swarming *Proteus* resistant to, 384
Nitroimidazole, 442
o-Nitrophenyl-—ᴅ-galactopyranoside test
  for Enterobacteriaceae, 379-380, 381
  for *Nocardia*, 469
NNIS; *see* National Nosocomial Infections Study
Nocardia, 467-469, 471
  acid-fast stains for, 70, 72, 605, 608
  in chronic meningitis, 212
  culture media for, 693
  draining sinuses or fistulas and, 278
  in endophthalmitis, 299
  identification of, 467-469
  infection from
    epidemiology and pathogenesis of, 467
    therapy for, 469
  in keratitis, 298
  in malignancy, 309
  in mycetoma, 727-729
  in neoplastic disease and bone marrow transplants, 309
  as non–spore-forming bacilli, 458
  in pleural fluid, 285
  in pneumonia, 227in respiratory tract, 222
    in chronic infections, 228
  schema of, 713
  in skin and soft tissue infection, 281
  versus *Streptomyces*, 469, 470
Nocardia asteroides
  colony of, 469
  differentiation of, 471
  epidemology and pathogenesis of, 467
Nocardia brasiliensis
  differentiation of, 471
  epidemology and pathogenesis of, 467
Nocardia caviae, epidemology and pathogenesis of, 467
Nocardia dassonvillei, differentiation of, 471
Nocardia otitidis-caviarum
  differentiation of, 471
  epidemology and pathogenesis of, 467
Nocardiopsis, 467, 469
  identification of, 467, 469
  in infections, 469
  lysozyme in, 469-470
  as non–spore-forming bacilli, 458
Nonchromogens, 592, 628-629
  Mycobacterium ulcerans as, 621
  pigment production of, 610
Noncultivatable disease agents, 580-589
Nonfermentative gram-negative bacilli and coccobacilli, 383, 386-405
  identification of, 388-403
    Acinobacter in, 400, 401
    Agrobacterium in, 402-403
    Alcaligenes in, 399-400
    Chryseomonas luteoloa in, 398
    Eikenella in, 401-402

Nonfermentative gram-negative bacilli and coccobacilli—cont'd
  identification of—cont'd
    Flavimonas oryzihabitans in, 398
    Flavobacterium in, 402
    Gram stain, 99
    Moraxella in, 400-401
    nonconventional methods for, 403
    Ochrobactrum in, 402-403
    Oligella in, 400-401
    pseudomallei group in, 398-399
    Pseudomonas aeruginosa in, 389-396, 397
    Pseudomonas fluorescens in, 399
    Pseudomonas putida in, 399
    Pseudomonas stutzeri in, 399
    Psychrobacter in, 402-403
    Shewanella putrefaciens in, 399
    Sphingobacterium in, 402
    Weeksella in, 402
    Xanthomonas maltophilia in, 396-398
  infections from
    epidemiology and pathogenesis of, 387-388
    treatment of, 403-404
Nonfermenters; *see* Nonfermentative gram-negative bacilli and coccobacilli
Nongonococcal urethritis, 554
Nonhemolytic streptococci; *see* Streptococci, nonhemolytic
Non-*Legionella pneumophila*, 570
Nonoxidizers among gram-negative bacilli, 389
Non–spore-forming aerobic, gram-positive bacilli, 457-473; *see also* Gram-positive bacilli, aerobic, non–spore-forming
Non–spore-forming anaerobic bacilli
  in genital tract, 258
  in wound infections and abscesses, 275
Non–spore-forming bacilli; *see* Gram-positive anaerobes, non–spore-forming
Normal flora
  anaerobic, 55, 474, 475
    gram-positive, 505, 516, 517-518, 520-521
  Bacteroidaceae as, 527-528, 530-531
  gram-negative, 475, 527-528, 530-531
  intestinal, 235
  in oral cavity, 531
  reporting of, 37
  in respiratory tract, 220
  in urinary tract, 249, 250
Norwalk agent virus
  in diarrhea, 240
  electron microscopy of, 653, 655
  specimens for, 647
    immunoelectron microscopy of, 242
  in viral syndromes, 639
Nose
  infections of, 299-301
  specimen from, 62
Nosema, 820
  classification of, 778
Nosocomial infection, 41-49
  antimicrobial therapy in, 235
  control programs and, 43-44
  Enterobacteriaceae in, 384
  epidemic strains and, 44
  epidemiological typing systems and, 47
  host vulnerability in, 310-311
  incidence of, 41-42

P

Paecilomyces
  characteristics of, 701, 709
  phylogenetic position of, 711
  schema of, 713
Paecilomyces lilacinus, 299
Pantoea agglomerans, 365
Pantoea dispersa, 365
Papanicolaou stain for fungi detection, 695
  in sputum, 699
Papillomavirus, human; *see* Human papillomavirus
Papovavirus, 637
Para-aminosalicylic acid, 180, 181
Paracentesis, 287
Paracoccidioides
  immunodiffusion assay for, 160
  phylogenetic position of, 711
Paracoccidioides brasiliensis, 740
  characteristic features of, 696-697
  definitive identification of, 732, 733
  general features of, 731
  growth rate for, 709
  in infection, 742-743
  Loboa loboi versus, 585
  in pneumonia, 227
  schema of, 713
Paracoccidioidomycosis, 739-740, 742-744
Paragonimiasis, 795-796, 798
Paragonimus, 795-796, 798
Paragonimus westermani
  body sites of, 779
  in chronic meningitis, 212
  classification of, 778
  eggs of, 831, 856
  in pneumonia, 227
Parainfluenza virus
  diseases of, 642
  isolation and identification of, 666
  in pharyngo-tonsillitis, 223
  in pneumonia, 227
  in respiratory tract, 222, 227
  in sinusitis, 300
  specimens for, 647
  tests for, 680-681
    fluorescent antibody, 225, 656, 659
    hemadsorption of primary monkey kidney
      monolayers in, 672-673
    hemagglutination inhibition, 160
    serology panels and immune status, 685
  in viral syndromes, 639
Paramyxovirus, 637
Parasites, 776-861
  adult worm identification in, 794, 795
  agglutination tests for, 158
  animal inoculation for, 804-805
  in aspirates, 796-797, 798
  biopsy material for, 797-804, 805
  blood, 194-195, 806-809
    babesiosis from, 825-826, 846
    helminths as, 833, 858
    hemoflagellates in, 826-828, 846-848
    malaria from, 823-825, 841-845
    protozoal, 823-829
  body sites of, 779
  cellophane tape preparations for, 794, 795, 796
  classification of, 777, 778

Parasites—cont'd
  culture techniques for, 804
  detection of, 195
  in diarrhea, 240
  differential stains for, 71-72
  duodenal specimens for, 793
  fecal specimens for, 241, 777-791
    collection of, 777-781
    concentration procedures for, 784, 785, 786-787
    diagnostic procedures for, 783-791
    direct smears of, 783-784
    general information for, 788-791
    macroscopic examination of, 782
    microscopic examination of, 782-783
    modified acid-fast stain for, 791, 792
    permanent stained smears for, 784-788, 789
    preservation of, 781-782
  hatching procedure for schistosome eggs and, 793-794
  intestinal helminths as, 830-833, 848, 849
  intestinal protozoa as, 809-823; *see also* Protozoa,
      intestinal
  laboratory safety and, 14
  in larval-stage nematode recovery, 793
  latex particle agglutination tests for, 126
  in meningitis, 213in myositis, 274
  pathogenic mechanisms of, 194-195
  in respiratory tract, 222
  serodiagnosis of, 806
  sigmoidoscopy specimens for, 791-793
  in spinal fluid, 797, 798
  sputum for paragonimiasis in, 795-796, 798
  stains for, 67, 70-71
  in tissue cultures, 284
  trichomoniasis from, 794-795, 796, 797
  urogenital specimens for, 794-795, 796, 797
  worm burden estimations and, 793
Parenteral solutions, 48
Paronychia, 277
Parotitis, 300
Partial acid fast-stain for *Nocardia*, 467, 468
Particle, viral, 635
Particle agglutination, 124-125, 126
  coagglutination, 124, 125
  in cost-limiting, 28-29
  in immunology, 158-159
  latex; *see* Latex agglutination
  lectin assays in, 125
Parvovirus
  in birth defects, 587
  in congenital and neonatal infections, 269, 306
  human importance of, 637
  serology panels and immune status tests for, 685
  specimens for, 647
  in viral syndromes, 639
Parvovirus B19, 586-587
  diseases of, 643
  in sickle cell disease, 308-329
PAS stain; *see* Periodic acid-Schiff stain
Pasco panel, 140
Passive hemagglutination test, 159
Pasteurella, 420-422, 423
  in bite infections, 280
  characteristics of, 423
  differentiation of, 383
  drug susceptibility of, 420-422
  identification of, 420-422

Quality control; *see also* Quality assessment
    for cultures, 92-93, 95
      for media requiring in-house testing, 95
      mycobacteria and, 594, 597
Qualture method for urine filtration, 253
Quantitation of results; *see* Results, quantitation of
Quantitative susceptibilty testing in cost control, 31-32
Quantity of specimen, 55
Quantum II kit for Enterobacteriaceae, 116, 120
Quellung reaction, 67
    cerebrospinal fluid and, 214
    for β-hemolytic streptococci, 246-347
    for lower respiratory tract organisms, 230
Quink black ink, 67
Quinolones
    Acinetobacter and, 403
    Aeromonas and, 437
    in bacillary dysentery, 384
    as bactericides, 174
    Enterobacteriaceae and, 384
    in nosocomial infection, 384
    Salmonella and, 384

R

Rabbit blood agar
    in anaerobic isolate subculture, 481
    control organism for, 95
    kanamycin-vancomycin laked, 89
    in pigment demonstration, 529, 536
Rabbit serum in leptospirosis diagnosis, 205
Rabies virus
    diseases of, 639, 644
    serology panels and immune status tests for, 685
Radioimmunoassay, 164
    automation in, 146
    for lower respiratory tract organisms, 232
    for parasites, 806
    for viruses, 655-658
Radioimmunoprecipitation assay, 314
Radiometric detection, 138
    cost control and, 32
Radish *Bacillus*, 629
Rahnella aquatilis, 365
Ranitidine, 275
RapID ANA panel for anaerobic bacteria, 498, 499-500
    Mobiluncus and, 521
Rapid antigen extraction for cerebrospinal fluid, 216
    217
Rapid detection, 28-32
    advantages and disadvantages of, 500
    of anaerobic bacteria, 499-500
    of Enterobacteriaceae, 121, 382
    of single colonies growing on primary media,
      100-107
    of yeasts, 753, 754, 755
Rapid growers in mycobacteria, 592, 629-630
Rapid hippurate hydrolysis test, 106, 108
Rapid identification of specimen, 29-31
RapID N/H, 115
Rapid nitrate reductase test, 753, 755
Rapid plasma reagin
    sensitivity and, 270
    in syphilis, 159, 160
RapID SS/U kit, 121
    for urinary tract bacteria, 115
RapID STR for streptococci, 115

Rapid susceptibility testing, 186-187
    cost control and, 31-32
Rapid thermonuclease test, 106, 107
Rapid urease test, 106, 753, 754
Rapitex ASO kit, 350
Rash specimen, plating media for, 63
Rat-bite fever, 576-578, 581
    diagnosis of, 205
R/B Enteric kit for Enterobacteriaceae, 118
Reagin
    flocculation tests and, 159
    rapid plasma; *see* Rapid plasma reagin
Reaginic antibodies, 269, 270
REAP; *see* Rapid antigen extraction for cerebrospinal fluid
Rectal specimen
    plating media for, 62
    storage of, 59
    for virology, 647, 648, 650
Reference laboratories
    for *Francisella tularensis*, 407
    sending specimens to, 59
      cost control and, 27
Refrigeration
    prompt delivery of specimen and, 55
    for specimen storage and shipping, 59
Regan-Lowe agar, 411
Reimbursement, cost control and, 24
Rejection
    of requests for tests, 19
    of specimen, 18, 20
      criteria for, 36-37
      procedure for, 55
Relapsing fever, 447
Relatedness determinations for epidemiologic studies,
      44-45, 46, 47
Renibacterium, 458
Reovirus, 637
Replica plating system for bacteria, 143
Reporting of results
    characteristics of, 25
    to clinician, 7
    laboratory organization and, 18-19
    in nosocomial infections, 44
Requisition form
    design of, 36
    for tests, 18
Resheathing of needles, 11
Resin-containing media, 203
Respiratory epithelium, 298
Respiratory syncytial virus
    antiviral chemotherapy for, 638
    in disease, 639, 642
      in cystic fibrosis, 308
      of middle ear, 301
      in nosocomial infection, 43, 311
      in pharyngo-tonsillitis, 223
      in pneumonia, 227
    isolation and identification of, 666
    specimens for, 647
    tests for, 680-681
      enzyme-linked immunosorbent assays as, 126
      fluorescent antibody, 225, 656, 657-658, 659
      serology panels and immune status, 685
Respiratory tract
    anatomy of, 219-220
    infection of; *see* Respiratory tract infection

Streptococci—cont'd
  in head and neck flora, 297
  α-hemolytic; *see also* Streptococci, viridans
    in bite infections, 280
    in dental and oral infection, 300
    epidemiology and pathogenic mechanisms of, 334-336
    in head and neck flora, 297
    laboratory identification of, 342-344, 345, 347
    screening for, 347-349
  β-hemolytic
    in blood, 194
    epidemiology and pathogenic mechanisms of, 336
    in genital tract, 258
    in head and neck flora, 297
    in infectious arthritis, 289
    laboratory identification of, 341-342, 343
    in respiratory tract, 220, 221, 223, 225, 226, 227
    screening for, 349
    in systemic infection with cutaneous lesions, 282
    in urinary tract infection, 250
  in hidradenitis suppurativa, 281
  identification of
    initial characterization in, 338-341
    laboratory, 338-344, 345, 347
    rapid methods and commercial systems for, 344-347
  immunoglobulin M separation systems and, 166
  in infectious arthritis, 289
  in malignancy, 309
  membrane-bound solid-phase immunoassay and, 127
  microaerophilic
    in dental and oral infection, 300
    in drug abusers, 312
    in pneumonia, 227
    in progressive bacterial synergistic gangrene, 280
  in myositis, 279
  in neck space infection, 302
  nonhemolytic
    in cellulitis, 280
    epidemiology and pathogenic mechanisms of, 336-337
    laboratory identification of, 342-344, 345, 347
    in respiratory tract, 220
    in urethra, 249
  nucleic acid hybridization and, 134
  nutritionally variant
    in blood, 199, 205epidemiology and pathogenic mechanisms of, 337
  particle agglutination tests for, 158
  phenylethyl alcohol agar for, 90
  in pleural fluid, 285
  in progressive bacterial synergistic gangrene, 280
  related genera of
    epidemiology and pathogenic mechanisms of, 338
    laboratory identification of, 342-344, 345, 347
  satelliting, 205, 341
  screening methods for, 347-349
  serologic diagnosis of, 349, 350
  synergism and antagonism and, 185
  treatment of infections from, 360
  viridans
    in blood, 194, 198, 199
    in bronchial specimens, 229
    differentiation of, 347
    epidemiology and pathogenic mechanisms of, 336-337

Streptococci—cont'd
  viridans—cont'd
    in grafts, shunts, catheters, or prostheses, 312
    identification of, 347
    in pharyngitis and tonsillitis, 299
    in respiratory tract, 220
    in systemic infection with cutaneous lesions, 282
    in urethra, 249
  in wound infections and abscesses, 275
Streptococcus adjacens
  in blood, 205
  identification of, 341
Streptococcus agalactiae; see Streptoccus group B
  in cerebrospinal fluid, 216
  cost control and rapid testing for, 28
  detection of, 349
  differentiation of, 461
  epidemiology and pathogenic mechanisms of, 336
  in genital tract, 258
  hydrolysis test for, 108
  identification of, 341
  versus *Listeria monocytogenes*, 460
  in nosocomial infection, 310
  snail agglutinin lectin binding to, 126
  streptococcal selective agar for, 91
Streptococcus anginosus
  epidemiology and pathogenesis of, 336
  in metastatic abscess, 198
  in wound infections and abscesses, 275
Streptococcus bovis
  in blood, 199
  in endocarditis, 199
  epidemiology and pathogenesis of, 337
  identification of, 6, 343
  in neoplastic disease and bone marrow transplants, 310
Streptococcus bovis I, 347
Streptococcus bovis II, 347
Streptococcus canis, 336
Streptococcus constellatus
  epidemiology and pathogenesis of, 336
  in wound infections and abscesses, 275
Streptococcus defectivus
  in blood, 205
  identification of, 341
Streptococcus dysgalactiae, 336
Streptococcus equi, 337
  epidemiology and pathogenesis of, 336
  identification of, 343
Streptococcus equinus; see *Streptococcus equi*
Streptococcus equisimilis
  epidemiology and pathogenesis of, 336
  identification of, 343
Streptococcus faecalis; see *Enterococcus faecalis*
Streptococcus faecium; see *Enterococcus faecium*
Streptococcus gordonii, 347
Streptococcus intermedius
  epidemiology and pathogenesis of, 336
  in wound infections and abscesses, 275
Streptococcus "milleri"; see also Streptococcus *anginosus, intermedius, constellatus,* 349
  epidemiology and pathogenesis of, 336
  identification of, 342
  rapid tests for, 346
Streptococcus mitis
  differentiation of, 347

Weeksella
  differentiation of, 383
  identification of, 402
Weeksella zoohelcum
  in bite infections, 280
  identification of, 402
  in wounds, 388
Weil-Felix test, 158
West African sleeping sickness, 828
Western blot assays, 134-135, 164-166
  in acquired immunodeficiency syndrome, 314
  for genital pathogens, 271
  for syphilis antibodies, 270
Western equine encephalitis, 685
Wet mounts
  direct, 66
    for fecal specimen, 242
  for fungi, 710, 714
Wheat germ agglutinin binding, 126
Wheatley-trichrome stain, 71
Whipple's disease, 241, 586
Whipworm; *see Trichuris trichiura*
White piedra, 714
Whole blood, specimen shipping and, 59
Whole pathogen agglutination for antibody detection, 157-158
Whooping cough, 411
Wilkins-Chalgren agar, 179
Wire applicators for specimen collection, 57
Wolinella
  characteristics of, 534, 537
  differentiation of, 525
  in family Bacteroidaceae, 526
  in head and neck flora, 297
  level II identification of, 488, 496
Wolinella curva, 429
Wolinella recta, 429
Wolinella succinogenes
  as saccharolytic gram-negative bacilli, 532
  taxonomic status of, 534
Wood pigeon, 631
Word processors, 150
Workbench decontamination, 11
Workload calculations, computers in, 150
Worms; *see also* Parasites
  adult, identification of fecal, 794, 795
  estimation of burdens of, 793
Wound, 274-275
  infections and abscesses of
    Clostridium in, 507
    organisms generally encountered in, 275
  specimen from
    cultures of, 27
    plating media for, 62
    storage of, 59
Wound botulism, 237
  Clostridium in, 506
Wright stain
  Calymmatobacterium granulomatis and, 583
  for fungi detection, 695
  for parasites, 71-72
Written reports to clinician, 7
Wuchereria, 195
Wuchereria bancrofti
  classification of, 778
  microfilaria of, 857, 858

X

X factor
  in chocolate agar, 89
  Haemophilus and, 415
  porphyrin test for, 415-416, 417
Xanthomonas
  in blood, 275
  identification of, 396-398
Xanthomonas maltophilia, 388
  identification of, 396-398
  typing systems and, 44
Xenopsylla, 779
Xenorhabdus, 365
XLD agar; *see* Xylose-lysine desoxycholate agar
Xylene
  in fecal specimen, 791
  in indole production, 392
Xylohypha, 712
Xylohypha bantiana
  in mycetoma, 729
  schema of, 713
Xylose-lysine desoxycholate agar, 91-92
  components and purpose for, 83
  for stool specimens, 244

Y

Y-1 adrenal cell assay, 239
Yaws, 446
  in external ear infections, 301
Yeast
  Bacteroides bile esculin agar, growth on, 88
  in blood culture, 203, 207
  in cerebrospinal fluid, 214
  Columbia colistin-nalidixic acid agar with blood for, 89
  commercially available systems for, 756-757
  on cornmeal Tween 80 agar, 759
  in endocarditis, 199
  in genital tract, 258, 260
  identification of, 698-700, 706-708, 752-753
    conventional methods for, 757-758
  in vitro susceptibility testing and, 183
  macroscopic appearance of, 698
  microscopic appearance of, 759
  in neoplastic disease and bone marrow transplants, 309
  in osteomyelitis, 290
  rapid screening for, 753, 754, 755
  stains for, 67
  structural units of, 699
  in urinary tract infection, 249, 250
  vaginal, cultures of, 261, 264
Yeast fermentation broth
  for fungi, 693
  indications for, 693
Yeast nitrogen base agar
  for fungi, 693
  indications for, 693
Yeast-extract phosphate agar
  for fungi, 693
  indications for, 693
Yeastlike organisms, 764
Yersinia
  in bacteremia, 205
  DNA probes for, 243
  drug susceptibility of, 384